IMPORTANT:

HERE IS YOUR REGISTRATION CODE TO ACCESS YOUR PREMIUM McGRAW-HILL ONLINE RESOURCES.

For key premium online resources you need THIS CODE to gain access. Once the code is entered, you will be able to use the Web resources for the length of your course.

If your course is using **WebCT** or **Blackboard**, you'll be able to use this code to access the McGraw-Hill content within your instructor's online course.

Access is provided if you have purchased a new book. If the registration code is missing from this book, the registration screen on our Website, and within your WebCT or Blackboard course, will tell you how to obtain your new code.

Registering for McGraw-Hill Online Resources

To gain access to your McGraw-Hill web resources simply follow the steps below:

1. USE YOUR WEB BROWSER TO GO TO: **http://www.mhhe.com/botany/**
2. CLICK ON **FIRST TIME USER**.
3. ENTER THE REGISTRATION CODE* PRINTED ON THE TEAR-OFF BOOKMARK ON THE RIGHT.
4. AFTER YOU HAVE ENTERED YOUR REGISTRATION CODE, CLICK **REGISTER**.
5. FOLLOW THE INSTRUCTIONS TO SET-UP YOUR PERSONAL UserID AND PASSWORD
6. WRITE YOUR UserID AND PASSWORD DOWN FOR FUTURE REFERENCE. KEEP IT IN A SAFE PLACE.

TO GAIN ACCESS to the McGraw-Hill content in your instructor's **WebCT** or **Blackboard** course simply log in to the course with the UserID and Password provided by your instructor. Enter the registration code exactly as it appears in the box to the right when prompted by the system. You will only need to use the code the first time you click on McGraw-Hill content.

Thank you, and welcome to your McGraw-Hill online resources!

Mc Graw Hill **Higher Education**

* YOUR REGISTRATION CODE CAN BE USED ONLY ONCE TO ESTABLISH ACCESS. IT IS NOT TRANSFERABLE.

0-07-290941-2 STERN: INTRODUCTORY PLANT BIOLOGY, 9E

McGRAW-HILL

ONLINE RESOURCES

REGISTRATION CODE

conexiones-28468017

Mc Graw Hill **Higher Education**

Introductory

Plant Biology

Edition Nine

Introductory Plant Biology

Kingsley R. Stern
California State University, Chico

Shelley Jansky
University of Wisconsin—Stevens Point

James E. Bidlack
University of Central Oklahoma

McGraw Hill

Boston Burr Ridge, IL Dubuque, IA Madison, WI New York San Francisco St. Louis
Bangkok Bogotá Caracas Kuala Lumpur Lisbon London Madrid Mexico City
Milan Montreal New Delhi Santiago Seoul Singapore Sydney Taipei Toronto

McGraw-Hill Higher Education
A Division of The ***McGraw-Hill*** *Companies*

INTRODUCTORY PLANT BIOLOGY, NINTH EDITION

Published by McGraw-Hill, a business unit of The McGraw-Hill Companies, Inc., 1221 Avenue of the Americas, New York, NY 10020.

This book is printed on recycled, acid-free paper containing 10% postconsumer waste.

International 2 3 4 5 6 7 8 9 0 QPV/QPV 0 9 8 7 6 5 4 3 2
Domestic 2 3 4 5 6 7 8 9 0 QPV/QPV 0 9 8 7 6 5 4 3 2

ISBN 0–07–290941–2
ISBN 0–07–119900–4 (ISE)

Publisher: *Margaret J. Kemp*
Senior developmental editor: *Kathleen R. Loewenberg*
Marketing manager: *Heather K. Wagner*
Senior project manager: *Kay J. Brimeyer*
Senior production supervisor: *Sandy Ludovissy*
Media project manager: *Sandra M. Schnee*
Media technology producer: *Renee Russian*
Coordinator of freelance design: *Michelle D. Whitaker*
Cover/interior designer: *Jamie E. O'Neal*
Cover image: © *Paul Martin Brown*
Interior design graphics: © *Visual Language Library*
Senior photo research coordinator: *Lori Hancock*
Supplement producer: *Brenda A. Ernzen*
Compositor: *Precision Graphics*
Typeface: *10/12 Times Roman*
Printer: *Quebecor World Versailles Inc.*

Library of Congress Cataloging-in-Publication Data

Stern, Kingsley Rowland.
Introductory plant biology / Kingsley R. Stern, Shelley Jansky, James E. Bidlack — 9th ed.
p. cm.
Includes index.
ISBN 0–07–290941–2 — ISBN 0–07–119900–4 (ISE)
1. Botany. I. Jansky, Shelley. II. Bidlack, James E. III. Title.

QK47 .S836 2003
580—dc21 2002070050
CIP

INTERNATIONAL EDITION ISBN 0–07–119900–4

www.mhhe.com

Contents in Brief

Contents

Preface

This book is designed as an introductory text in botany. It assumes little knowledge of the sciences on the part of the student. It includes sufficient information for some shorter introductory botany courses open to both majors and nonmajors, but it is arranged so that certain sections—for example, "Soils," "Molecular Genetics," "Division Psilotophyta"—can be omitted without disrupting the overall continuity of the course.

Botany instructors vary greatly in their opinions concerning the depth of coverage needed for the topics of photosynthesis and respiration in a text of this type. Some feel that nonmajors, in particular, should have a brief introduction only, while others consider a more detailed discussion essential. In this text, photosynthesis and respiration are discussed at three levels. Some may find one or two levels sufficient, and others may wish their students to become familiar with the processes at all three levels.

Despite eye-catching chapter titles and headings, many texts for majors and nonmajors give relatively minor coverage of the current interests of a significant number of students. This text emphasizes current interests without giving short shrift to botanical principles. Present interests of students include topics such as global warming, ozone layer depletion, acid rain (acid deposition), genetic engineering, organic gardening, Native American and pioneer uses of plants, pollution and recycling, house plants, backyard vegetable gardens, natural dye plants, poisonous and hallucinogenic plants, and the nutritional values of edible plants. The rather perfunctory coverage or absence of such topics in many botany texts has occurred partly because botanists previously have tended to believe that some of the topics not covered in their departmental economic botany courses are more appropriately addressed in anthropology and horticulture courses. I have found, however, that both majors and nonmajors in botany, who may be initially disinterested in the subject matter of a required course, frequently become engrossed if the material is repeatedly related to such topics. Accordingly, a considerable amount of ecological and ethnobotanical material has been included with traditional botany throughout the book—without, however, resorting to excessive use of technical terms.

ORGANIZATION OF THE TEXT

A relatively conventional sequence of botanical subjects is included. Chapters 1 and 2 cover introductory and background information; Chapters 3 through 11 deal with structure and function; Chapters 12 and 13 introduce meiosis and genetics. Chapter 14 discusses plant propagation and biotechnology; Chapter 15 introduces evolution. Chapter 16 presents a six-kingdom system of classification; Chapters 17 through 23 stress, in phylogenetic sequence, the diversity of organisms traditionally regarded as plants, and Chapter 24 deals with ethnobotanical aspects and information of general interest pertaining to sixteen major families of flowering plants. Chapters 25 and 26 constitute an overview of the vast topic of ecology, although ecological topics and applied botany are included in most of the preceding chapters as well. Some of these subjects are broached in anecdotes that introduce the chapters, while others are mentioned in the ecological review summaries and in the human and ecological relevance sections (with which most of the chapters in the latter half of the book conclude.)

AIDS TO THE READER

A chapter outline, review questions, discussion questions, learning online topics, and additional reading lists are provided for each chapter. New terms are defined as they are introduced, and those used more than once are boldfaced and included in a pronunciation glossary. The use of the scientific names throughout the body of the text has been held to a minimum, but a list of the scientific names of all organisms mentioned is given in Appendix 1. Appendix 2 deals with the biological controls and companion planting; Appendix 3 lists wild edible plants, poisonous plants, medicinal plants, hallucinogenic plants, spices, tropical fruits, and natural dye plants. Appendix 4 gives horticultural information on house plants; information on the cultivation and nutritional value of vegetables is included. Appendix 5 gives some metric equivalents.

NEW TO THIS EDITION

In addition to a complete updating of all chapters, based on over 100 reviewers' comments, the following major changes can be found in this edition:

- Chapter 3 (Cells) has been modified to include several rewritten sections, the incorporation of changes suggested by reviewers, more logical organization of topics, and many new photographs and illustrations.
- Chapter 10 (Plant Metabolism) revisions include new sections on energy transfer in the introduction and secondary metabolism, as well as several sections rewritten to simplify complex topics. Additional changes suggested by reviewers were made, a more consistent

use of terminology was introduced, several new/revised figures were added, and a summary table for ATP production in cellular respiration has been included.

- Chapter 13 (Genetics) has been completely reorganized to present molecular genetics first. The chapter now has new sections on cytogenetics, extranuclear DNA and quantitative genetics, a completely revised section on gene expression, revised and expanded content on mutation, new material on Barbara McClintock's discovery of transposable elements, a new illustration on the linkage map of pea, and a new Awareness Box on the Polymerase Chain Reaction.
- Chapter 14 (Plant Propagation and Biotechnology) has also been completely rewritten. This chapter includes a new section on crop plant evolution, including discussions of the origins of agriculture and steps in domestication; a new section on plant breeding, including breeding strategies, gene banks, protoplast fusion and transgenic plants; an updated discussion of the pros and cons of transgenic plants; over 20 new photos, and a new illustration on the origins of crop plants.
- A new feature, Learning Online, can now be found at the end of every chapter. This list of chapter-related topics mimics an identical list on the Online Learning Center, which is linked to crucial information on each of these topics. The links are updated periodically through our database to help students stay on top of their research and study responsibilities.
- Much of the appendix information has received extensive revision, particularly in the area of medicinal plants.
- Glossary definitions have been updated and over 40 new words have been added.

ACKNOWLEDGMENTS

The valuable contributions of Drs. James Bidlack and Shelley Jansky in rewriting Chapters 3 and 10, and 13 and 14 respectively, are gratefully acknowledged. Dr. Martha Phillips' input on Chapter 25 was also much appreciated. All but two of the boxed readings are written by Dr. Daniel Scheirer, whose contributions continue to add value to the text. We also are very grateful to Dr. Manuel Molles for his highlighted summaries of the ecological aspects of each chapter.

Additional persons who read parts of the manuscripts of various editions and made many helpful criticisms and suggestions include Richard S. Demaree, Jr., Robert I. Ediger, Larry Hanne, Donald T. Kowalski, Robert B. McNairn, Patricia Parker, and Robert A. Schlising. Others whose encouragement and contributions are deeply appreciated include Isabella A. Abbott, Donald E. Brink, Jr., Gerald Carr, William F. Derr, Timothy Devine, Beverly Marcum, Robert McNulty, Paul C. Silva, Lorraine Wiley, the faculty and staff of the Department of Biological Sciences, California State University, Chico, my many inspiring students, the Lyon Arboretum of the University of Hawaii, the editorial, production, and design staffs of McGraw-Hill Publishers, and most of all, my family. Special thanks are due the artists, Denise Robertson Devine, Janet Monelo, and Sharon Stern.

Finally, the ninth edition of *Introductory Plant Biology* would not have been possible without the considerable feedback from the following reviewers, to whom I am indebted:

Ligia Arango, *Stone Child College*

Mark H. Armitage, *Azusa Pacific University*

Janice Asel, *Mitchell Community College*

Randy G. Balice, *New Mexico Highlands University*

Paul W. Barnes, *Southwest Texas State University*

Robert W. Bauman, Jr., *Amarillo College*

Dorothea Bedigian, *Washington University*

Cynthia A. Bottrell, *Scott Community College*

Richard R. Bounds, *Mount Olive College*

Richard G. Bowmer, *Idaho State University*

Rebecca D. Bray, *Old Dominion University*

James A. Brenneman, *University of Evansville*

George Murchie Briggs, *State University of New York*

Brad S. Chandler, *Palo Alto College*

Stephen S. Daggett, *Avila College*

Bill D. Davis, *Rutgers University*

John W. Davis, *Benedictine College*

Semma Dhir, *Fort Valley State University*

Rebecca M. DiLiddo, *Mount Ida College*

Susan C. Dixon, *Walla Walla College*

Ben L. Dolbeare, *Lincoln Land Community College*

Jan Federic Dudt, *Bartlesville Wesleyan College*

Diane Dudzinski, *Washington State Community College*

Kerry B. Dunbar, *Dalton State College*

H. Herbert Edwards, *Western Illinois University*

Inge Eley, *Hudson Valley Community College*

Frederick B. Essig, *University of South Florida*

G. F. Estabrook, *The University of Michigan*

James Ethridge, *Joliet Junior College*

Rosemary H. Ford, *Washington College*

Stephen W. Fuller, *Mary Washington College*

Sibdas Ghosh, *University of Wisconsin, Whitewater*

Mike Gipson, *Oklahoma Christian University*

Charles Good, *Ohio State University*

Sharon Gusky, *Northwestern Connecticut Community Technical College*

Mark Hammer, *Wayne State College*

Laszlo Hansely, *Northern Illinois University*

Nancy E. Harris, *Elon College*

Jill F. Haukos, *South Plains College*

Peter Heywood, *Brown University*

H. H. Ho, *State University of New York, New Patlz*

Susan Houseman, *Southeastern Community College*

Lauren D. Howard, *Norwich University*

Patricia L. Ireland, *San Jacinto College, South*

William A. Jensen, *Ohio State University*

Cindy Johnson-Groh, *Gustavus Adolphus College*

Sekender A. Khan, *Elizabeth City State University*

Joanne M. Kilpatrick, *Auburn University, Montgomery*

Kaoru Kitajima, *University of Florida*

Roger C. Klockziem, *Martin Luther College*

Robert N. Kruger, *Mayville State University*

Brenda Price Latham, *Merced College*

Peter J. Lemay, *College of the Holy Cross*

Barbara E. Liedl, *Central College*

John F. Logue, *University of South Carolina, Sumter*

Elizabeth L. Lucyszyn, *Medaille College*

Karen Lustig, *Harper College*

Paul Mangum, *Midland College*

Steve Manning, *Arkansas State University, Beebe*

Bernard A. Marcus, *Gensee Community College*

David Martin, *Centralia College*

William J. Mathena, *Kaskaskia College*

Alicia Mazari-Andersen, *Kwantlen University College*

Joseph H. McCulloch, *Normandale Community College*

Julie A. Medlin, *Northwestern Michigan College*

Richard G. Merritt, *Houston Community College*

David W. Miller, *Clark State Community College*

Lillian W. Miller, *Florida Community College, Jacksonville*

Beth Morgan, *University of Illinois, Urbana-Champaign*

Lytton John Musselman, *Old Dominion University*

Chuks A. Ogbonnaya, *Mountain Empire College*

Sebastine O. Onwuka, *Lesley College*

Martha M. Phillips, *The College of St. Catherine*

Indiren Pillay, *Southwestern Tennessee Community College*

Mary Ann Polasek, *Cardinal Stritch University*

Dr. Robert J. Porra, *CSIRO*

Kumkum Prabhakar, *Nassau Community College*

Tyre J. Proffer, *Kent State University*

Mohammad A. Rana, *St. Joseph College*

W. T. Rankin, *University of Montevallo*

Tom Reynolds, *University of North Carolina, Charlotte*

Maralyn A. Renner, *College of the Redwoods*

Stanley A. Rice, *Southeastern Oklahoma State University*

Dennis F. Ringling, *Pennsylvania College of Technology*

Darryl Ritter, *Okaloosa-Walton Community College*

Suzanne M. D. Rogers, *Salem International University*

Wayne C. Rosing, *Middle Tennessee State University*

Robert M. Rupp, *Ohio State University, Agricultural Technical Institute*

Thomas H. Russ, *Charles County Community College*

Michael A. Savka, *University of West Florida*

Neil Schanker, *College Of The Siskiyous*

Bruce S. Serlin, *DePauw University*

Brian Shmaefsky, *Kingwood College*

Shaukat M. Siddiqi, *Virginia State University*

Dilbagh Singh, *Blackburn College*

Marshall D. Sundberg, *Emporia State University*

Donald D. Sutton, *California State University, Fullerton*

Mesfin Tadesse, *Ohio State University*

Max R. Terman, *Tabor College*

R. Dale Thomas, *Northeast Louisiana University*

Stephen L. Timme, *Pittsburg State University*

Rani Vajravelu, *University of Central Florida*

John Vanderploeg, *Ferris State University*

C. Gerald Van Dyke, *North Carolina State University*

Christopher R. Wenzel, *Eastern Wyoming College*

Ingelia White, *Windward Community College*

Garrison Wilkes, *University of Massachusetts, Boston*

Donald L. Williams, *Sterling College*

Dwina W. Willis, *Freed-Hardeman University*

Richard J. Wright, *Valencia Community College*

Rebecca R. Zamora, *South Plains College*

Teaching and Learning Supplements

Digital Content Manager

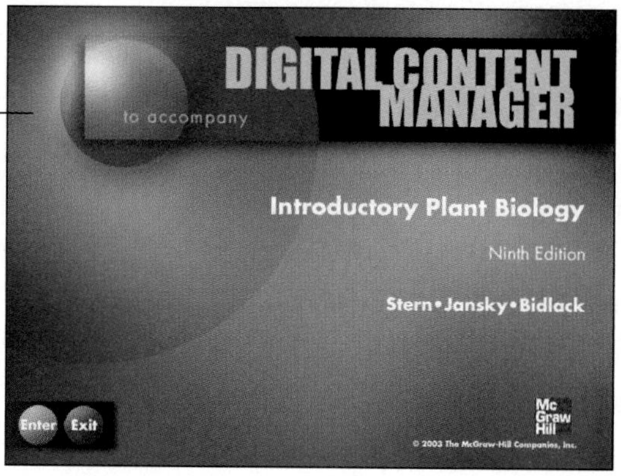

This multimedia collection of visual resources allows instructors to utilize artwork from the text in multiple formats to create customized classroom presentations, visually based tests and/or quizzes, dynamic course web site content, or attractive printed support materials. The digital assets on this cross-platform CD-ROM are grouped within the following easy-to-use folders:

- **Art Library.** Full-color digital files of all illustrations in the book.
- **Photo Library.** Hundreds of discipline-appropriate photos in digital files
- **PowerPoint Lecture Outline.** Ready-made presentations combine art from the text with customized lecture notes covering all 26 chapters.
- **Table Library.** Every table that appears in the text is provided in electronic form.
- **Active Art.** These special art pieces consist of key images that are converted to a format that allows instructors to break the art down into core elements and then group the various pieces and create customized images. This is especially helpful with difficult concepts; they can be presented step by step.
- **Animations Library.** Numerous full-color animations illustrating many different concepts covered in the study of botany are provided. The visual impact of motion will enhance classroom presentations and increase comprehension.

Call the McGraw-Hill Customer Service Department at 800-338-3987, or contact your local sales representative to obtain this valuable teaching aid. (ISBN 0-07-256589-6)

The Amazing Lives of Plants: The Reproductive Lives of Mosses, Pines, Ferns, and Flowers

Available upon adoption, *The Amazing Lives of Plants* includes four independent segments: "Mosses," "Ferns," "Pines," and "Flowers." Their reproductive lives are presented in a vivid, full-color combination of live video footage and sharp animation. Subtitled text makes it easy to cue up for use in lecture, and the pace of the program is suitable for students taking notes.
Videotape (ISBN 0-07-256393-1)
CD-ROM (ISBN 0-07-266394-X)

Transparencies

A set of 100 Transparencies is available to users of the text. These acetates include key figures from the text, including new art from this edition. (ISBN 0-07-290943-9)

Classroom Testing Software

Available on a cross-platform CD-ROM, this test bank utilizes Brownstone Diploma© testing software to quickly create customized exams. This user-friendly program allows instructors to search for questions by topic, or format; edit existing questions or add new ones; and scramble questions and answer keys for multiple versions of the same test. (ISBN 0-07-290945-5)

Introductory Plant Biology Lab Manual

The laboratory manual that accompanies *Introductory Plant Biology* has been revised according to reviewer feedback. It is written for the student entering the study of botany for the first time. The exercises utilize plants to introduce biological principles and the scientific method. They are written to allow for maximum flexibility in sequencing. (ISBN 0-07-290946-3)

Online Learning Center

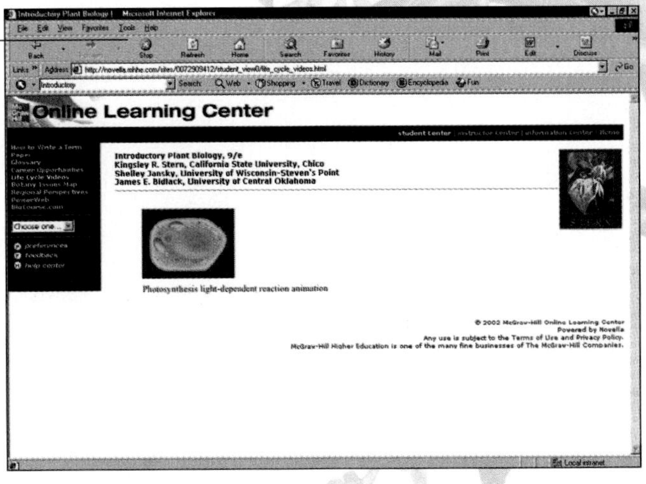

The Online Learning Center that accompanies this text provides abundant resources for both instructor and student.

For each chapter of the textbook, students will have access to:

- Chapter Summaries
- Chapter Web links Correlated to each chapter of the ninth edition, these regularly updated URLs provide additional learning opportunities for research, writing papers, or tutorial purposes.
- Practice Quizzing Designed as a study aid, these chapter-by-chapter quizzes help students review text material and prepare for upcoming exams.
- Key Term Flash Cards Yet another great study tool that tests students' knowledge of important botany terms.

Additional resources available to students include:

- Global Botany Issues World Map Updated every six months, a world map is tagged with hot points that link to case studies and articles complete with photos, references back to the text, and additional links to appropriate Internet sites. Past issues are indexed in a convenient chart for easy access.
- Guide to Electronic Research
- Metric Equivalents and Conversions Tables
- Career Opportunities
- Animations
- How to Write a Paper
- Scientific Names of Organisms
- Biological Controls
- Useful Plants and Poisonous Plants
- House Plants and Home Gardening

As an instructor you'll receive complete access to all of the above plus:

- PowerPoint Lecture Notes/Art
- Answers to Discussion Questions

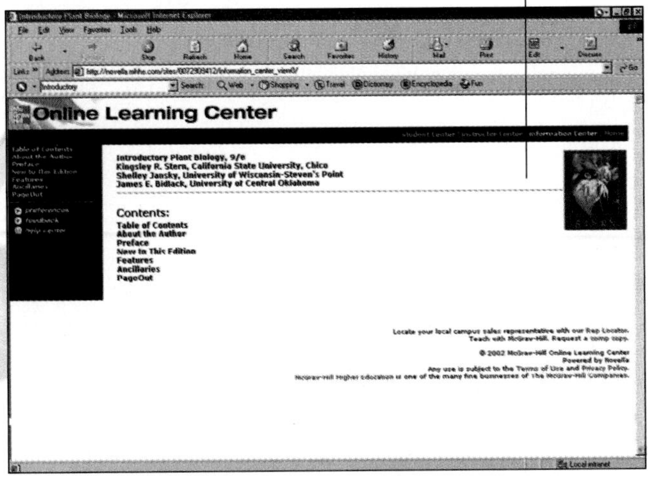

PowerWeb

What a great way to get the information you need quickly and easily! For a nominal fee, students can access PowerWeb for Botany. Here is what you will find:

- Botany articles from current magazines, newspapers, and journals
- Weekly updates of current issues
- Interactive exercises
- Web research tips
- An online library of updated research links to help you find the right information
- Up-to-the-minute headlines from around the world including course-specific and general news
- Online quizzing and assessment to measure your understanding of course material, and more!

For more information, visit http://www.dushkin.com/powerweb/

Botany PowerWeb Password Code Card (ISBN 0-07-254905-X)

PageOut

Need a course web site? PageOut offers instant access to fully loaded course web sites with no work required on your part. Courses can now be password protected, and you can now upload, store, and manage up to 10MB of data. Copy your course and share it with colleagues or use it as a foundation for next semester. Short on time? Let us do the work. Our McGraw-Hill service team is ready to build your PageOut Web site—and provide content—and provide any necessary training. Learn more about PageOut and other McGraw-Hill digital solutions at www.mhhe.com/solutions.

Steershead (*Dicentra uniflora*), a diminutive relative of bleeding hearts, native to the mountains of the western United States and Canada.

What Is Plant Biology?

Chapter Outline

1

Overview

This chapter introduces you to botany: what it is, how it developed, how it relates to our everyday lives, and what its potential is for the future. The discussion includes a brief introduction to some common questions about plants and their functions, an examination of the scientific method, and a brief look at botany after the invention of the microscope. It concludes with a brief survey of the major disciplines within the field of botany.

Some Learning Goals

1. Understand how humans have impacted their environment, particularly during the past century.
2. Explain briefly what the scientific method is and what hypotheses are.
3. Explain how and why all life is dependent on green organisms.
4. Be able to indicate briefly the particular aspects of botany with which each of the major botanical disciplines is concerned.

Upas trees (*Antiaris toxicaria*), which are relatives of the common fig tree, flourish in the jungles of Java and some of the neighboring islands. There are legendary tales of people having died while sleeping beneath these tall trees or even from merely being downwind from them. Whether or not there is truth to the stories, we do know that upas trees produce a deadly sap, which for centuries has been used to tip poison arrows used in hunting. On the other hand, the cow trees of Venezuela and Brazil (e.g., *Brosimum utile*; *Mimusops huberi*) produce a sweet, nutritive latex that is relished by the natives of the region. Still other plants, such as opium poppies, produce latex that contains narcotic and medicinal drugs (Fig. 1.1). Why do plants such as upas trees produce poisons, while parts of so many other plants are perfectly edible, and some produce spices, medicines, and a myriad of products useful to humans?

In late 1997, a fast-food chain began airing a television commercial that showed a flower of a large potted plant gulping down a steak sandwich. Most of us have seen at least pictures of Venus's flytraps and other small plants that do, indeed, trap insects and other small animals, but are there larger carnivorous plants capable of devouring big sandwiches or animals somewhere in remote tropical jungles?

Occasionally we hear or read of experiments—often associated with school science fairs—that suggest plants respond in some positive way to good music or soothing talk; conversely, some plants are said to grow poorly when exposed to loud rock music or to being harshly yelled at. Do plants really respond to their surroundings, and, if so, how and to what extent?

When a botanist friend of mine invited me to his office to see a 20-gallon glass fish tank he had on his desk, I expected to find a collection of house plants or tropical fish. Instead, I saw what at first appeared to be several small, erect sticks that had been suspended in midair with large rubber bands; there were also beakers of water in the corners. When I got closer, I could see that the "sticks" were cuttings (segments) of poplar twigs that were producing roots at one end and new shoots at the other end. The roots, however, were growing *down* from the tops of the cuttings, and the shoots were growing *upward* from the bottoms (Fig. 1.2). My friend had originally suspended the cuttings upside down, and new roots and shoots were being produced in the humid, lighted surroundings of the fish tank—regardless of the orientation of the cuttings. If I'd seen such bizarre plants in a movie, I might have assumed that the fiction writers had imagined something that didn't exist. There right in front of me, however, were such plants, and they were real! When

Figure 1.1 Immature opium poppy capsules that were gashed with a razor blade. Note the opium-containing latex oozing from the gashes.

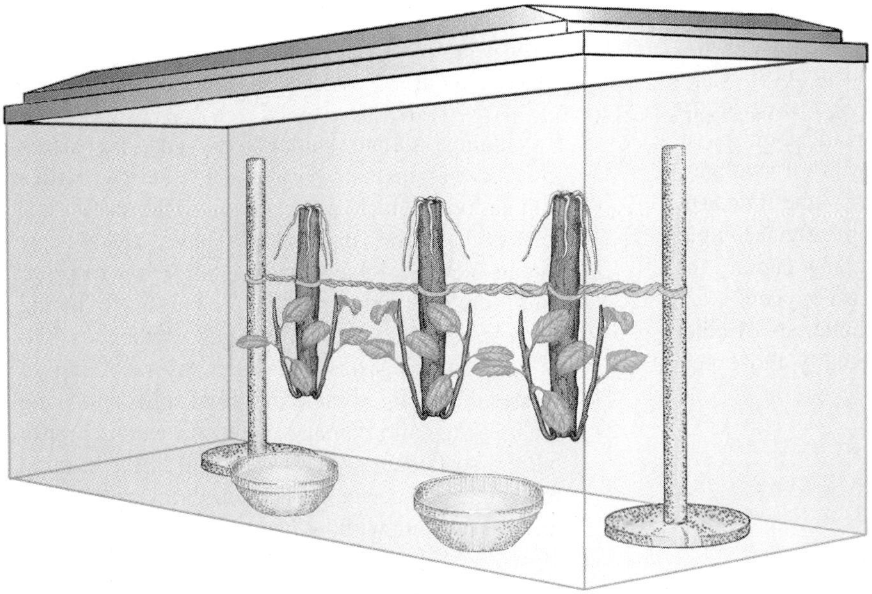

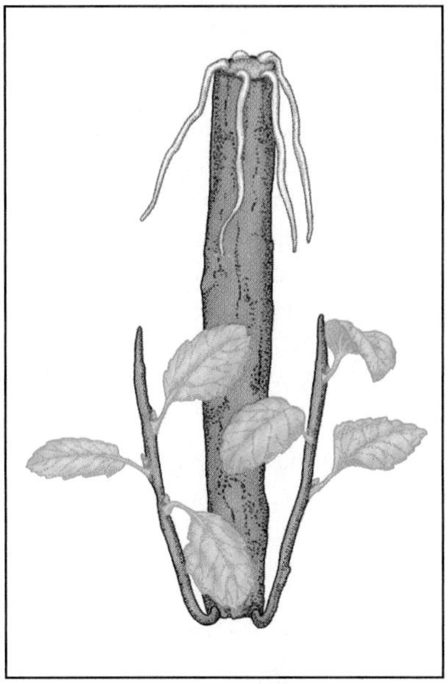

Figure 1.2 Cuttings (segments) of twigs that were suspended upside down in a humid, lighted glass tank. New roots are growing down from the top ends, and new shoots are growing up from the bottom.

cuttings are separated from the parent plant how do they "know" which end is up, and why would the roots and shoots grow the way they did?

California's huge coastal redwoods and Tasmania's giant gum trees can grow to heights of 60 to 90 or more meters (200 to 300 or more feet). When these giant trees are cut down, there is no evidence of pumps of any kind within them. How then does water get from the roots below ground to the tops of these and other trees? How does food manufactured in the leaves get down to the roots (Fig. 1.3)?

Our tropical rain forests, which occupy about 5% of the earth's surface, are disappearing at the rate of several acres a minute as the plant life is cleared for agriculture, wood supplies (primarily for fuel), cattle ranching, and for other human activities such as mining for gold. Is the dwindling extent of our rain forests, which are home to 50% of all the species of living organisms, cause for alarm? Or will the same plant and animal life simply return if the human activities cease?

There is currently much debate about global warming and the potential effects on life as we know it. Are those who proclaim that global warming will eventually have disastrous effects on modern civilization and living organisms simply exaggerating, or is there a scientific basis for the claims? What about the many forms of pollution that exist? Will we be able to either ignore or overcome the effects?

Plant life constitutes more than 98% of the total biomass (collective dry weight of living organisms) of the earth. Plants and other green organisms have the exclusive capacity to produce oxygen while converting the sun's energy into forms vital to the existence of both plant and animal life. At the same time, plants remove the large amounts of carbon dioxide given off by all living organisms

Figure 1.3 California coastal redwoods (*Sequoia sempervirens*). Coastal redwoods may grow for thousands of years and some may reach heights of nearly 100 meters (330 feet).

as they respire. In other words, virtually all living organisms are totally dependent on green organisms for their existence. If some major disease were to kill off all or most of the green organisms on land and in the oceans and lakes, all the animals on land, in the sea, and in the air would soon starve. Even if some alternative source of energy were available, animal life would suffocate within 11 years—the time estimated for all the earth's oxygen to be completely used up if it were not replaced. Just how do green plants capture the sun's energy, use carbon dioxide, and give off oxygen?

This book tries to answer these and hundreds of other questions about living organisms—particularly those pertaining to plants, fungi, and bacteria.

THE RELATIONSHIP OF HUMANS TO THEIR ENVIRONMENT

It has been estimated that the total human population of the world was less than 20 million in 6000 B.C. During the next 7,750 years, it rose to 500 million; by 1850, it had doubled to 1 billion; and 70 years later, it had doubled again to 2 billion. The 4.48-billion mark was reached in 1980, and within 5 years, it had grown to 4.89 billion. It is presently increasing by nearly 100 million annually, and estimates for the year 2000 are 6.25 billion. The earth remains constant in size, but humans obviously have occupied a great deal more of it over the past few centuries or at least have greatly increased in density of population.

In feeding, clothing, and housing ourselves, we have had a major impact on our environment. We have drained wetlands and cleared natural vegetation from vast areas of land. California, for example, now has less than 6% of the wetland it had 100 years ago. We have dumped wastes and other pollutants into rivers, oceans, lakes; we have added pollutants to the atmosphere; and we have killed pests and plant disease organisms with poisons. These poisons have also killed natural predators and other useful organisms and, in general, have thoroughly disrupted the delicate balance of nature that existed before humans began degrading their natural surroundings.

If we are to survive on this planet beyond the 21st century, there is little question that humans have to stop increasing in numbers, and the many unwise agricultural and industrial practices that have accompanied the mushrooming of human populations must be replaced with practices more in tune with restoring some ecological balance. Agricultural practices of the future will have to include the return of organic material to the soil after each harvest, instead of adding only inorganic fertilizers. Harvesting of timber and other crops will have to be done in a manner that prevents topsoil erosion, and the practice of clearing brush with chemicals will have to be abolished. Industrial pollutants will have to be rendered harmless and recycled whenever possible.

Ecological Review

Expanding human populations and increasing intensity of human activity now threaten the earth's populations, which are critical to the ecological integrity of the biosphere. These global-scale threats include global warming, numerous forms of pollution, and widespread land clearing. Reducing or reversing these environmental challenges will require applying measures such as recycling of wastes, returning organic matter to soils, and using plants to reclaim damaged land. As we attempt to build a sustainable future, we should bear in mind that while plants can live without humans, we cannot live for long without plants.

Many products that now are still largely discarded (e.g., garbage, paper products, glass, metal cans) will also have to be recycled on a much larger scale. Biological controls (discussed in Appendix 2) will have to replace the use of poisonous controls whenever possible. Water and energy conservation will have to be universally practiced, and rare plant species, with their largely unknown gene potential for future crop plants, will need to be saved from extinction by preservation of their habitats and by other means. The general public will have to be made even more aware of the urgency for wise land management and conservation—which will be especially needed when pressures are exerted by influential forces promoting unwise measures in the name of "progress"—before additional large segments of our natural resources are irreparably damaged or lost forever. Alternatives appear to be nothing less than death from starvation, respiratory diseases, poisoning of our food and drink, and other catastrophic events that could ensure the premature demise of large segments of the world's population.

In recent years, scientists, and increasingly the general public, have become alarmed about the effects of human carelessness on our environment. It wasn't until the 1980s, however, that damage to forests and lakes caused by acid rain, the "greenhouse effect," contamination of ground water by nitrates and pesticides, reduction of ozone shield, major global climatic changes, loss of biodiversity in general, and loss of tropical rain forests in particular gained widespread publicity.

Human and Animal Dependence on Plants

Our dependence on green organisms to produce the oxygen in the air we breathe and to remove the carbon dioxide we give off doesn't stop there. Plants are also the sources of products that are so much a part of human society that we largely take them for granted. We know, of course, that rice, corn, potatoes, and other vegetables are plants (Fig. 1.4); but all foods, including meat, fish, poultry, eggs, cheese, and milk, to mention just a few, owe their existence to plants. Condiments such as spices (Fig. 1.5) and luxuries such as perfumes are produced by plants, as are some dyes, adhesives, digestible surgical stitching fiber, food stabilizers, beverages (Fig. 1.6), and emulsifiers.

Our houses are constructed with lumber from trees, which also furnish the cellulose for paper, cardboard, and synthetic fibers. Some of our clothing, camping equipment, bedding, draperies, and other textile goods are made from fibers of many different plant families (Fig. 1.7). Coal is fossilized plant material, and oil probably came from microscopic green organisms or animals that either directly or indirectly were plant consumers. All medicines and drugs at one time came from plants, fungi, or bacteria, and many important ones, including most of the antibiotics, still do (Fig. 1.8).

Figure 1.4A Rice cakes being manufactured. Unprocessed rice is poured into small ovens where the kernels are expanded. The kernels are then compressed into cakes, which are conveyed by belt to a packaging area.

Figure 1.4B Part of a produce section in a supermarket.

Figure 1.5 Some of the spices derived from plants.

Microscopic organisms play a vital role in recycling both plant and animal wastes and aid in the building of healthy soils. Others are responsible for human diseases and allergies.

Although gluts or shortages of oil and other fossil fuels may at present be politically or economically manipulated, there is no question that these fuels are finite and eventually will disappear. Accordingly, the development of alternative energy sources is receiving increased attention.

Methane gas, which can be used as a substitute for natural gas, has been produced from animal manures and decomposed plants in numerous villages in India and elsewhere for many years, and after several years of trial on a small scale in the United States, the production of methane on a larger scale from human sewage is being investigated. Potatoes, grains, and other sources of carbohydrates are currently used in the manufacture of alcohols, some of which are being blended with gasoline ("gasohol"), and such uses may increase in the future. In fact, electric cars, as well as buses and automobiles that can run on propane and either methanol (wood alcohol) or gasoline—or a mixture of both—are now in use in many communities in the United States and other parts of the world.

Figure 1.6A Ripening coffee berries. They are picked by hand when they are red. The seeds are extracted for roasting after the berries have been fermented.

Figure 1.6B Coffee beans cooling after being roasted.

Figure 1.7 Cotton plants. The white fibers, in which seeds are embedded, are the source of textiles and fabrics. The seeds are the source of vegetable oils used in margarine and shortening. After the oils have been extracted, the remaining "cotton cake" is used for cattle feed.

Figure 1.8 A *Penicillium* colony. The tiny beads of fluid on the surface contain penicillin, widely used as an antibiotic.

What of plants and the future? As you read this, the population of the earth already has exceeded 6 billion persons, every one of whom needs food, clothing, and shelter in order to survive. To ensure survival, a majority of us eventually may need to learn not only how to cultivate food plants but also how to use plants in removing pollutants from water (Fig. 1.9), in making land productive again, and in renewing urban areas. In addition, many more of us may need to be involved in helping halt the destruction of plant habitats caused primarily by the huge increase in the number of earth's inhabitants. This subject and related matters are further discussed in Chapter 25.

Although at present the idea that humanity may not be able to save itself from itself may seem radical, there are a few who have suggested that it might become necessary in the future to emigrate to another planet. If so, microscopic algae could play a vital role in space exploration. Experiments with portable oxygen generators have been in progress for many years. Tanks of water teeming with tiny green algae are taken aboard a spacecraft and installed so that they are exposed to light for at least part of the time. The algae not only produce oxygen, which the spacecraft inhabitants can breathe, but they also utilize the waste carbon dioxide produced by respiration. As the algae multiply, they can be fed to a special kind of shrimp, which, in turn, multiply and become food for the space travelers. Other wastes are recycled by different microscopic organisms. When this self-supporting arrangement, called a *closed system,* is perfected, the range of spacecraft should greatly increase because heavy oxygen tanks will not be necessary, and the amount of food reserves needed will be reduced.

Inhabitants of undeveloped areas still use plants for food, shelter, clothing, and medicine and also in hunting and fishing. Today, small teams of botanists, anthropologists, and medical doctors are interviewing medical practitioners and herbal healers in remote tropical regions and taking notes on various uses of plants by primitive peoples. These scientists are doing so in the hope of preserving at least some plants with potential for modern civilization before disruption of their habitats results in their extinction.

Figure 1.9 A polluted waterway in an urban area.

BOTANY AS A SCIENCE

The study of plants, called **botany**—from three Greek words *botanikos* (botanical), *botane* (plant or herb), and *boskein* (to feed) and the French word *botanique* (botanical)—appears to have had its origins with Stone Age peoples who tried to modify their surroundings and feed themselves. At first, the interest in plants was mostly practical and centered around how plants might provide food, fibers, fuel, and medicine. Eventually, however, an intellectual interest arose. Individuals became curious about how plants reproduced and how they were put together. This inquisitiveness led to plant study becoming a science, which broadly defined is simply "a search for knowledge of the natural world."

A science is distinguished from other fields of study by several features. It involves the observation, recording, organization, and classification of facts, and more importantly, it involves what is done with the facts. Scientific procedure involves the process of experimentation, observation, and the verifying or discarding of information, chiefly through inductive reasoning from known samples. There is no universal agreement on the precise details of the process. A few decades ago, scientific procedure was considered to involve a routine series of steps that involved first asking a *question,* then formulating a *hypothesis,* followed by experiments, and finally developing a *theory.* This series of steps came to be known as the *scientific method,* and there are still instances where such a structured approach works well. In general, however, the scientific method now describes the procedures of assuming and testing hypotheses.

Hypotheses

A **hypothesis** is simply a tentative, unproven explanation for something that has been observed. It may not be the correct explanation—testing will determine whether it is correct or incorrect. To be accepted by scientists, the results of any experiments designed to test the hypothesis must be repeatable and capable of being duplicated by others.

The nature of the testing will vary according to the circumstances and materials, but good experiments are run in two forms, the second form being called a *control.* In the first form, a specific aspect, or *variable,* is changed. The control is run in precisely the same way but *without* changing the specific aspect, or variable. The scientist then can be sure that any differences in the results of the parallel experiments are due to the change in the variable.

For example, we may *observe* that a ripe orange we have eaten tastes sweet. We may then make the hypothesis that all ripe citrus fruits taste sweet. We may test the hypothesis by tasting oranges and other citrus fruits such as tangerines and lemons. As a result of our testing (since lemons taste sour), we may *modify the hypothesis* to state that only some ripe citrus fruits are sweet. In such an experiment, the variable involves more than one kind of ripe citrus fruit; the control, on the other hand, involves only ripe oranges.

When a hypothesis is tested, *data* (bits of information) are accumulated and may lead to the formulation of a useful generalization called a *principle.* Several related principles may lend themselves to grouping into a *theory,* which is not simply a guess. A theory is a group of generalizations (principles) that help us understand something. We reject or modify theories only when new principles increase our understanding of a phenomenon.

Microscopes

The microscope is an indispensable tool of most botanists, and biologists in general. This instrument traces its origin to 1590 when a family of Dutch spectacle makers found they could magnify tiny objects more than 30 times when they combined two convex lenses in a tube; they also found they could make minute objects visible with the magnification their instrument achieved. A few decades later, a Dutch draper—Anton van Leeuwenhoek (1632–1723)—ground lenses and eventually made 400 microscopes by hand, some of which could magnify up to 200 times. Modern microscopes, discussed in Chapter 3, can produce magnifications of more than 200,000 times and are leading almost daily to new discoveries in biology.

DIVERSIFICATION OF PLANT STUDY

Plant anatomy, which is concerned chiefly with the internal structure of plants, was established through the efforts of several scientific pioneers. Early plant anatomists of note included Marcello Malpighi (1628–1694) of Italy, who discovered various tissues in stems and roots, and Nehemiah Grew (1628–1711) of England, who described the structure of wood more precisely than any of his predecessors (Fig. 1.10).

Today, a knowledge of plant anatomy is used to help us find clues to the past, as well as for many practical purposes. For example, the related discipline of *dendrochronology* deals with determining past climates by examining the width and other features of tree rings. We can also learn much from archaeological sites by matching tree rings found in the wood of ancient buildings to the rings of wood of known age. Plant anatomy is also used to solve crimes. Forensic laboratories may use fragments of plant tissues found on clothing or under fingernails to determine where a crime took place or if certain persons could have been present where the crime was committed. The anatomy of leaves,

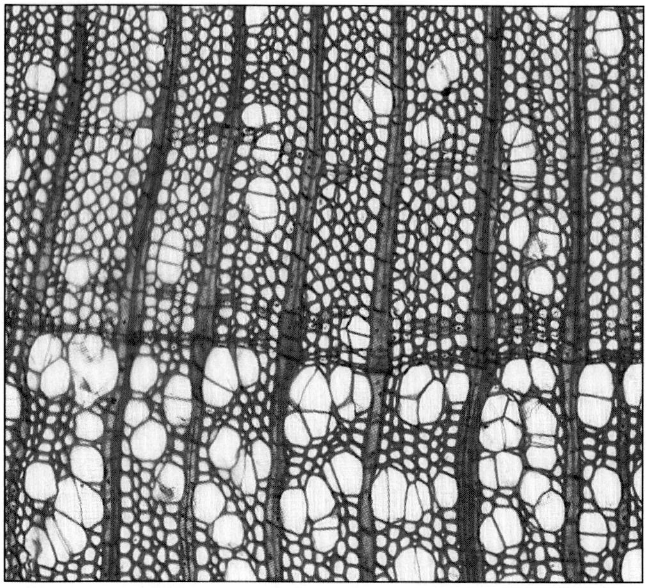

Figure 1.10 A thin section of *Magnolia* wood as seen through a microscope, ×40.

stems, and other plant parts is currently being used to unravel and sort out relationships among plants. A form of plant anatomy, known as *paleobotany,* involves the study of plant fossils.

Plant physiology, which is concerned with plant function, was established by J. B. van Helmont (1577–1644), a Flemish physician and chemist, who was the first to demonstrate that plants do not have the same nutritional needs as animals. In a classic experiment, van Helmont planted a willow branch weighing 5 pounds in an earthenware tub filled with 74.4 kilograms (200 pounds) of dry soil. He covered the soil to prevent dust settling on it from the air, and after 5 years, he reweighed the willow and the soil. He found that the soil weighed only 56.7 grams (2 ounces) less than it had at the beginning of the experiment but that the willow had gained 164 pounds. He concluded that the tree had added to its bulk and size from the water it had absorbed. We know now that most of the weight came as a result of photosynthetic activity (discussed in Chapter 10), but van Helmont deserves credit for landmark experimentation in plant physiology.

Modern plant physiologists have learned how to isolate and clone *genes* (units of heredity that are found within the nuclei of cells) and are using the knowledge gained to learn in precise detail much more about plant functions, including how plants conduct materials internally; how temperature, light, and water are involved in growth; why plants flower; and how plant growth regulatory substances are produced, to mention just a few.

European explorers of other continents during past centuries took large numbers of plants back home with them, and it soon became clear to those working with the plants that some sort of formalized system was necessary just to keep the collections straight. Several *plant taxonomists* (botanists who specialize in the identifying, naming, and classifying of plants) proposed ways of accomplishing this, but we owe our present system of naming and classifying plants to the Swedish botanist Carolus Linnaeus (1707–1778) (see Fig. 16.2).

Plant taxonomy (also called *plant systematics*), which is the oldest branch of plant study, began in antiquity, but Linnaeus did more for the field than any other person in history. Thousands of plant names in use today are those originally recorded in Linnaeus's book *Species Plantarum,* published in 1753. An expanded account of Linnaeus and his system of classification is given in Chapter 16.

There are still thousands of plants, fungi, and other organisms that have not yet been described or even discovered. Although it obviously is already too late to identify species that were not described before they became extinct, plant taxonomists around the world have united to try to identify and describe as many new organisms—many with food, medicinal, and other useful potential—before much more of their natural habitat disappears. Other plant taxonomists, through the use of *cladistics* (analysis of shared features) and molecular techniques, are refining our knowledge of plant relationships and are contributing to the improvement of many of our food crops.

Plant taxonomists often specialize in certain groups of plants. For example, *pteridologists* specialize in the study of ferns, while *bryologists* study mosses and plants with similar life cycles.

The discipline of **plant geography,** the study of how and why plants are distributed where they are, did not develop until the 19th century (Fig. 1.11). The allied field of **plant ecology,** which is the study of the interaction of plants with one another and with their environment, also developed in the 19th century.

After the publication in 1962 of a best-seller entitled *Silent Spring* (authored by Rachel Carson), public awareness of the field of ecology as a whole increased considerably. In this book, based on more than 4 years of literature research, Ms. Carson called attention to the fact that more than 500 new toxic chemicals annually are put to use as pesticides in the United States alone, and she detailed how these chemicals and other pollutants were having an insidious impact on all facets of human life and the environment.

The study of the form and structure of plants, **plant morphology,** was developed during the 19th century, and during the 20th century, much of our basic knowledge about the form and life cycles of plants was incorporated in the plant sciences as we know them today. During this time, the number of scientists engaged in investigating plants also greatly increased.

Genetics, the science of heredity, was founded by the Austrian monk Gregor Mendel (1822–1884), who performed classic experiments with pea plants. Today, various branches of genetics include *plant breeding,* which has greatly improved yields and quality of crop plants, and *genetic engineering.* Genetic engineering includes the introduction of

Figure 1.11 Ecologists, plant geographers, and other biologists recognize large communities of plants and animals that occur in areas with distinctive combinations of environmental features. These areas, called *biomes,* are represented here by the tropical rain forest, which, although occupying less than 5% of the earth's surface, is home to more than half of the world's species of organisms.

genes from one organism to another and has already improved the pest, frost, and disease resistance, as well as yields, of some crop plants. It holds enormous potential for continued development of better agricultural, medicinal, and other useful plants. Future control of human, animal, and plant diseases is also anticipated.

Cell biology (previously called **cytology**), the science of cell structure and function, received a boost from the discovery of how cells multiply and how their various components perform and integrate a variety of functions, including that of sexual reproduction. The mid-20th-century development of *electron microscopes* (see Chapter 3) further spurred cell research and led to a vast new insight into cells and new forms of cell research that continues to the present.

Economic Botany and **ethnobotany,** which involve practical uses of plants and plant products, had their origin in antiquity as humans discovered, used, and eventually cultivated plants for food, fiber, medicines, and other purposes. Today, research is being conducted on the uses of plants by indigenous peoples with an eye to discovering medicines and other useful plant products previously unknown in developed countries.

There is still a vast amount of botanical information to be discovered. For example, 11,000 papers on botanical subjects were published in 1938 alone, and the number per year in recent times is much greater. It also appears probable that at the start of the 21st century, at least one-third of all the organisms regarded in the past as plants (particularly algae and fungi) have yet to be named, let alone thoroughly investigated and understood.

Summary

1. Why do some plants produce poisons while others are edible and useful? Are there large carnivorous plants? How and why do plants respond to their environment? What is the future of tropical rain forests? What can be done about pollution and other environmental problems?
2. Human populations have increased dramatically in the past few centuries, and the disruption of the balance of nature by the activities directly or indirectly associated with the feeding, clothing, and housing of billions of people threatens the survival of not only humans but many other living organisms.
3. We are totally dependent on green organisms because they alone can convert the sun's energy into forms that are usable by, and vital to the very existence of, animal life.
4. We largely take plants and plant products for granted. Animals, animal products, many luxuries and condiments, and other useful substances, such as fibers, lumber, coal, medicines, and drugs, either depend on plants or are produced by them.
5. To ensure human survival, all persons soon may need to acquire some knowledge of plants and how to use them. Plants will undoubtedly play a vital role in space exploration as portable oxygen generators.
6. Teams of scientists are interviewing medical practitioners and herbal healers in the tropics to locate little-known plants used by primitive peoples before the plants become extinct.
7. Botany, the study of plants, apparently began with Stone Age peoples' practical uses of plants. Eventually, botany became a science as intellectual curiosity about plants arose.
8. A science involves observation, recording, organization, and classification of facts. The verifying or discarding of facts is done chiefly from known samples through inductive reasoning. The scientific method involves specifically following a routine series of steps and generally assuming and testing hypotheses.
9. The microscope has had a profound effect on studies in the biological sciences and led to the discovery of cells.
10. Plant anatomy and plant physiology developed during the 17th century. J. B. van Helmont was the first to demonstrate that plants have nutritional needs different from those of animals. During the 17th century, Europeans engaged in botanical exploration on other continents and took plants back to Europe.
11. During the 18th century, Linnaeus produced the elements of our present system of naming and classifying plants.
12. During the 19th century, plant ecology, plant geography, and plant morphology developed, and by the beginning of the 20th century, genetics and cell biology became established. Much remains yet to be discovered and investigated.

Review Questions

1. How and to what extent have humans affected their natural environment?
2. What is meant by the scientific method?
3. To what extent is animal life dependent on green organisms for its existence?
4. In terms of biological experiments, what are hypotheses and controls?
5. What is the oldest branch of botany, and why did it precede other branches?
6. What are the basic features of each of the other branches of botany?

Discussion Questions

1. Since humans survived on wild plants for thousands of years, might it be desirable to return to that practice?
2. What factors are involved in possibly determining if and when humans might not be able to sustain themselves on this planet?
3. How would you guess that Stone Age peoples discovered medicinal uses for plants?
4. Many of the early botanists were also medical doctors. Why do you suppose this is no longer so?
5. Consider the following hypothesis: "The majority of mushrooms that grow in grassy areas are not poisonous." How could you go about testing this hypothesis scientifically?

Learning Online

Visit our webpage at http://www.mhhe.com/botany for interesting case studies, practice quizzes, current articles, and animations within the Online Learning Center to help you understand the material in this chapter. You'll also find active links to these topics:

Introductory Materials and Governmental Sites
General References in Botany
Writing Papers and Study Tips
Glossaries and Dictionaries
Careers in Science
Utility and Organizational Sites
Scientific Method
General References in Botany
Microscopy
Economic and Ecological Importance of Angiosperms
Origins of Agriculture

Additional Reading

Carson, R. L. 1999. *Silent spring.* Boston: Houghton Mifflin.

Davis, E. B., and D. Schmidt. 1995. *Guide to information sources in the botanical sciences,* 2d ed. Englewood, CO: Libraries Unlimited.

Ewan, J. (Ed.). 1969. *A short history of botany in the United States.* Forestburgh, NY: Lubrecht & Cramer.

Johnson, T. 1998. *CRC ethnobotany desk reference.* Boca Raton, FL: CRC Press.

McCarthy, S. 1993. *Ethnobotany and medicinal plants.* Upland, PA: Diane Publishing.

Morton, A. G. 1981. *History of botanical science: An account of the development of botany from the ancient time to the present.* San Diego, CA: Academic Press.

Nordenskiold, E. 1988 reprint of 1935 ed. *The history of biology.* Irvine, CA: Reprint Services Corp.

Swift, L. H. 1974 reprint of 1970 ed. *Botanical bibliographies: A guide to bibliographic materials applicable to botany.* Ann Arbor, MI: University Microfilms International.

A flowering head of burdock (*Arctium lappa*), a weed whose burs catch in clothing and the fur and hide of animals. Burdock is cultivated in Japan for its edible roots.

The Nature of Life

2

Chapter Outline

Overview

This chapter begins with a discussion of the attributes of living organisms. These include growth, reproduction, response to stimuli, metabolism, movement, complexity of organization, and adaptation to the environment. Then it examines the chemical and physical bases of life. A brief look at the elements and their atoms is followed by a discussion of compounds, molecules, valence, bonds, ions, acids, bases, and salts. Forms of energy and the chemical components of cells are examined next. The chapter concludes with an introduction to macromolecules: carbohydrates, lipids, proteins, and nucleic acids.

Some Learning Goals

1. Learn the attributes of living organisms.
2. Define matter; describe its basic state.
3. Understand the nature of compounds and describe acids, bases, and salts.
4. Know the various forms of energy.
5. Learn the elements found in cells.
6. Understand the nature of carbohydrates, lipids, and proteins.

Have you ever dropped a pellet of dry ice (frozen carbon dioxide) into a pan of water and watched what happens? The solid pellet darts randomly about the surface, looking like a highly energetic bug waterskiing, as the warmer water rapidly converts it to a gas. Does all that motion make the dry ice alive? Hardly; yet one of the attributes of living things is the capacity to move. But if living things move, what about plants? If a tree remains fixed in one place and doesn't crawl down the sidewalk, does that mean it isn't alive? Again the answer is no, but these questions do serve to point out some of the difficulties encountered in defining *life.* In fact, some argue that there is no such thing as life—only living organisms—and that life is a concept based on the collective attributes of living organisms.

ATTRIBUTES OF LIVING ORGANISMS

Composition and Structure

The activities of living organisms originate in tiny structural units called *cells*, which consist of *cytoplasm* (a souplike fluid) bounded by a very thin membrane. All living cells contain genetic material that controls their development and activities. In the cells of many organisms, this genetic material, known as *DNA* (an abbreviation for deoxyribonucleic acid), is housed in a somewhat spherical structure called the *nucleus*, which is suspended in the cytoplasm. In bacteria and other simple cells, however, the DNA is distributed directly in the cytoplasm. The cells of plants, algae, fungi, and many simpler organisms have a *cell wall* outside of the membrane that bounds the cytoplasm. The cell wall provides support and rigidity. Cells are discussed in more detail in Chapter 3.

Growth

Some have described **growth** as simply an increase in *mass* (a body of *matter*—the basic "stuff" of the universe), usually accompanied by an increase in *volume*. Most growth results from the production of new cells and includes variations in *form*—some the result of inheritance, some the result of response to the environment. What happens, for example, if you plant two apple seeds of the same variety in poor soil but don't give them the same care? If you water one just enough to allow it to germinate and grow, while you water the other one freely and work fertilizers and conditioners into the soil around it, you might expect the second one to grow larger and produce more fruit than the first. In other words, although the two apple trees grew from the same variety of apple seed, they differ in form, following patterns of growth dictated by the DNA and the environment. Various aspects of growth are discussed in Chapter 11.

Reproduction

Dinosaurs were abundant 160 million years ago, but none exist today. Hundreds of mammals, birds, reptiles, plants, and other organisms are now listed as endangered or threatened species, and many of them will become extinct within the next decade or two. All these once-living or living things have one feature in common: it became impossible or it has become difficult for them to reproduce. **Reproduction** is such an obvious feature of living organisms that we take it for granted—until it no longer takes place.

When organisms reproduce, the offspring always resemble the parents: guppies never have puppies—just more guppies—and a petunia seed, when planted, will not develop into a pineapple plant. Also, offspring of one kind tend to resemble their parents more than they do other individuals of the same kind. The laws governing these aspects of inheritance are discussed in Chapter 13.

Response to Stimuli

If you stick a pin into a pillow, you certainly don't expect any reaction from the pillow, but if you stick the same pin into a friend, you know your friend will react immediately (assuming he or she is conscious) because responding to stimuli is a major characteristic of all living things. You might argue, however, that when you stuck a pin into your house plant, nothing happened, even though you were fairly certain the plant was alive. You might not have been aware that the house plant did indeed respond but in a manner very different from that of a human. Plant responses to stimuli are generally much slower than those of animals and usually are of a different nature. If the house plant's food-conducting tissue was pierced, it probably responded by producing a plugging substance called **callose** in the affected cells. Some studies have shown that callose may form within as little as 5 seconds after wounding. Also, an unorganized tissue called **callus,** which forms much more slowly, may be produced at the site of the wound. Responses of plants to injury and to other stimuli, such as light, temperature, and gravity, are discussed in Chapters 9 through 11.

Metabolism

Definitions of metabolism vary somewhat but are mostly based on the observation that metabolism is *the collective product of all the biochemical reactions taking place within an organism.* All living organisms undergo various metabolic activities, which include the production of new cytoplasm, the repair of damage, and normal cell maintenance. The most important activities include **respiration,** an energy-releasing process that takes place in all living things; **photosynthesis,** an energy-harnessing process in green cells that is, in turn, associated with energy storage; **digestion,** the conversion of large or insoluble food molecules to smaller soluble ones; and **assimilation,** the conversion of raw materials into cytoplasm and other cell substances. These topics are discussed in Chapters 9 through 11.

Movement

At the beginning of this chapter, we mentioned that plants generally don't move from one place to another (although their reproductive cells may do so). This does not mean, however, that plants do not exhibit movement, a universal characteristic of living things. The leaves of sensitive plants (*Mimosa pudica*) fold within a few seconds after being disturbed or subjected to sudden environmental changes, and the tiny underwater traps of bladderworts (*Utricularia*) snap shut in less than one-hundredth of a second. But most plant movements, when compared with those of animals, are slow and imperceptible and are mostly related to growth phenomena. They become obvious only when demonstrated experimentally or when shown by time-lapse photography. Time-lapse photography often reveals many types and directions of motion, particularly in young organs. Movement is not confined to the organism as a whole but occurs down to the cellular level. For example, the cytoplasm of living cells constantly flows like a river within cells; this streaming motion is called *cyclosis*, or *cytoplasmic streaming.* Cyclosis usually appears to run clockwise or counterclockwise within the boundaries of each cell, but movement may actually be in various directions.

Complexity of Organization

The cells of living organisms are composed of large numbers of **molecules** (the smallest unit of an element or compound retaining its own identity). Typically there are more than 1 trillion molecules in a single cell. The molecules are not simply mixed, like the ingredients of a cake or the concrete in a sidewalk, but are organized into compartments, membranes, and other structures within cells and tissues. Even the most complex nonliving object has only a tiny fraction of the types of molecules of the simplest living organism. Furthermore, the arrangements of these molecules in living organisms are highly structured and complex. Bacteria, for example, are considered to have the simplest cells known, yet each cell contains a minimum of 600 different kinds of protein as well as hundreds of other substances, with each component having a specific place or being a part of a specific structure within the cell. When flowering plants and other larger living objects are examined, the complexity of organization is overwhelming, and the number of molecule types can run into the millions.

Adaptation to the Environment

If you skip a flat stone across a body of water and it lands on the opposite shore, the stone is not affected by the change from air to water to land during its brief journey; it does not respond to its environment. Living organisms, however, do respond to the air, light, water, and soil of their environment, as will be explained in later chapters. They are also, after countless generations of natural selection (as discussed in Chapter 15), genetically adapted to their environment in many subtle ways. Some weeds (e.g., dandelions) can thrive in a wide variety of soils and climates, while many species now threatened with extinction have adaptations to their environment that are so specific they cannot tolerate even relatively minor changes.

CHEMICAL AND PHYSICAL BASES OF LIFE

The Elements: Units of Matter

The basic "stuff of the universe," called matter, occurs in three states—*solid, liquid,* and *gas.* In simple terms, matter's characteristics are as follows:

1. It occupies space.
2. It has *mass,* which we commonly associate with weight.
3. It is composed of **elements.** There are 93 elements that occur naturally on our planet. At least 19 more elements have been produced artificially. Only a few of the natural elements (e.g., nitrogen, oxygen, gold, silver, copper) occur in pure form; the others are found combined together chemically in various ways. Each element has a designated symbol, often derived from its Latin name: The symbol for copper, for example, is **Cu** (from the Latin *cuprum*); and for sodium, **Na** (from the Latin *natrium*). The symbols for carbon, hydrogen, and oxygen are **C, H,** and **O,** respectively.

The smallest stable subdivision of an element that can exist is called an **atom.** Atoms are so minute that until the mid-1980s, individual atoms were not directly visible to us with even the most powerful electron microscopes. We have known for over 100 years, however, that atoms consist of several kinds of subatomic particles. Each atom has a tiny **nucleus** consisting of **protons,** which are particles with positive electrical charges, and other particles called **neutrons,** which have no electrical charges. Both protons and neutrons have a small amount of mass. If the nucleus, which contains nearly all of the atom's mass, were enlarged so that it was as big as a beach ball, the atom, which is mostly space, would be larger than a professional football stadium (Fig. 2.1).

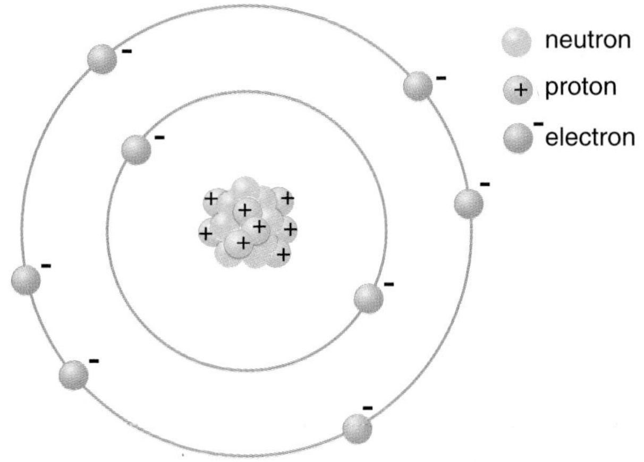

Figure 2.1 Model of an oxygen atom. The nucleus in the center consists of eight electrically neutral neutrons and eight positively charged protons. Eight negatively charged electrons whirl around the nucleus. In a real atom, the electrons would not be spaced or confined as shown in this simple diagram.

Each atom of an element has a specific number of protons in its nucleus, ranging from 1 in hydrogen, the lightest element, to 92 in uranium, the heaviest natural element. This number is referred to as the *atomic number.* The atomic number is often shown as a subscript to the left of the chemical symbol. For example, nitrogen, which has seven protons in its nucleus, has its atomic number of 7 shown as ${}_{7}N$. The combined number of protons and neutrons in a single atom is referred to as its *atomic mass* (Table 2.1). The atomic mass number is commonly shown as a superscript to the left of the chemical symbol. For example, the atomic mass of nitrogen, which has seven protons and seven neutrons in its nucleus, is shown as ${}^{14}N$, and when both the atomic number and the atomic mass are shown, the chemical symbol appears as ${}^{14}_{7}N$.

Electrons, which are little more than negative electric charges, whirl around an atom's nucleus. Electron masses are about 1,840 times lighter than those of both protons and neutrons and are so minute that their masses are generally disregarded. Since opposite electric charges attract each other, the positive electric charges of protons attract the negative electric charges of electrons and determine the paths of the electrons whirling around the nucleus.

The region in which electrons whirl around the nucleus is called an **orbital.** Each orbital has an imaginary axis and is somewhat cloudlike, but it doesn't have a precise boundary, and so we can't be certain of an electron's position within an orbital at any time. This has led to an orbital being defined as *a volume of space in which a given electron occurs 90% of the time.*

Electrons may be located in one or more energy levels of an atom, and their distance from the nucleus depends on their energy level. Each energy level is usually referred to as an *electron shell.* The outermost electron shell determines how or if an atom reacts with another atom. Only two electrons can occupy the first and lowest energy level associated with the innermost orbital; this orbital is more or less spherical and is so close to the nucleus that it is often not shown on diagrams of atoms. One to several additional orbitals, which are mostly spindle shaped (like the tips of cotton swabs), generally occupy much more space. Up to eight electrons can be held by the second energy level, and although the third and fourth energy levels can hold more than eight electrons each, they can become unstable if more than eight electrons are present. If an electron in one orbital is provided with more energy, it can jump to an orbital farther away from the nucleus. Conversely, if an electron releases energy, it drops to an energy level closer to the nucleus. The electrons of each orbital tend to repel those of other orbitals, so that the axes of all the orbitals of an atom are oriented as far apart from each other as possible; the outer parts of the orbitals, however, actually overlap more than shown in diagrams of them. Orbitals usually have diameters thousands of times more extensive than that of an atomic nucleus (Fig. 2.2).

Because each atom usually has as many electrons as it does protons, the negative electric charges of the electrons balance the positive charges of the protons, making the

TABLE 2.1

Atomic Numbers, Masses, and Functions of Some Elements Found in Plants

ELEMENT	ATOMIC NUMBER	USUAL ATOMIC MASS	SOME FUNCTIONS
Hydrogen (H)	1	1	Part of nearly all organic molecules
Carbon (C)	6	12	Forms skeleton of organic molecules
Nirogen (N)	7	14	Part of amino acids, nucleic acids, and chlorophyll
Oxygen (O)	8	16	Essential for most respiration; part of most organic molecules
Magnesium (Mg)	12	24	Basic element of chlorophyll
Phosphorus (P)	15	31	Part of ATP (a molecule involved in energy storage and exchange); part of nucleic acid molecules
Sulfur (S)	16	32	Stabilizes a protein's three-dimensional structure
Potassium (K)	19	39	Helps stabilize balance between ions in cells
Calcium (Ca)	20	40	Important in the structure of cell walls
Iron (Fe)	26	56	Involved in electron transport during respiration

atom electrically neutral. The number of neutrons in the atoms of an element can vary slightly, so the element may occur in forms having different weights but with all forms behaving alike chemically. Such variations of an element are called **isotopes.** The element oxygen (Fig. 2.3), for example, has seven known isotopes. The nucleus of one of these isotopes contains eight protons and eight neutrons; the nucleus of another isotope holds eight protons and ten neutrons, and the nucleus of a third isotope consists of eight protons and nine neutrons. If the number of neutrons in an isotope of a particular element varies too greatly from the average number of neutrons for its atoms, the isotope may be unstable and split into smaller parts, with the release of a great deal of energy. Such an isotope is said to be *radioactive*.

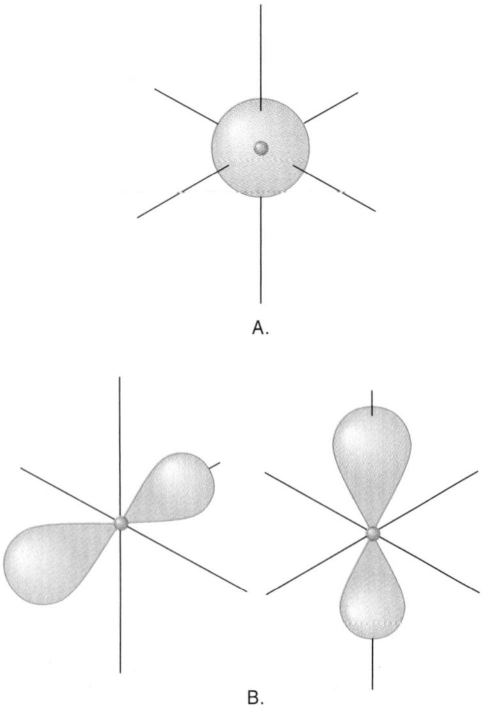

Figure 2.2 Models of orbitals. *A*. The two electrons closest to the atom's nucleus occupy a single spherical orbit. *B*. Additional orbitals are dumbbell-shaped, with axes that are perpendicular to one another. The atom's nucleus is at the intersection of the axes.

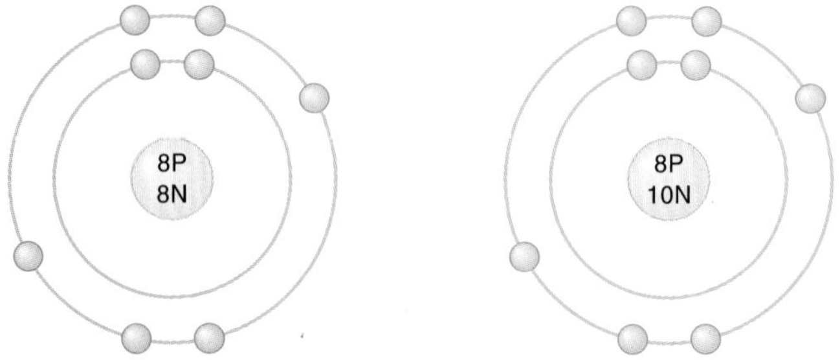

Figure 2.3 Isotopes of oxygen portrayed two dimensionally.

Molecules: Combinations of Elements

The atoms of most elements can combine with other atoms of the same or different elements; in fact, most elements do not exist independently as single atoms. When two or more elements are united in a definite ratio by chemical bonds, the substance is called a **compound.** Table salt (sodium chloride, NaCl), for example, is a compound consisting of sodium and chlorine atoms combined in a 1:1 ratio.

A **molecule** consists of two or more atoms bound together and is the smallest independently existing particle of a compound or element. The molecules of the gases oxygen and hydrogen, for example, exist in nature as combinations of two atoms of oxygen (O_2) or two atoms of hydrogen (H_2), respectively. Water molecules (H_2O) consist of two atoms of hydrogen and one atom of oxygen (Fig. 2.4). Molecules are in constant motion, with an increase or decrease in temperature speeding up or slowing down the motion. The more molecular movement there is, the greater the chances are that some molecules will collide with each other. Also, the chances of random collisions increase in proportion to the density of the molecules (i.e., the number of molecules present in a given space).

Random collisions between molecules capable of sharing electrons are the basis for all chemical reactions. The reactions often result in new molecules being formed. Each chemical reaction in a cell usually takes place in a watery fluid and is controlled by a specific *enzyme.* Enzymes are organic *catalysts* (a catalyst speeds up a chemical reaction without being used up in the reaction. Enzymes are discussed on page 24).

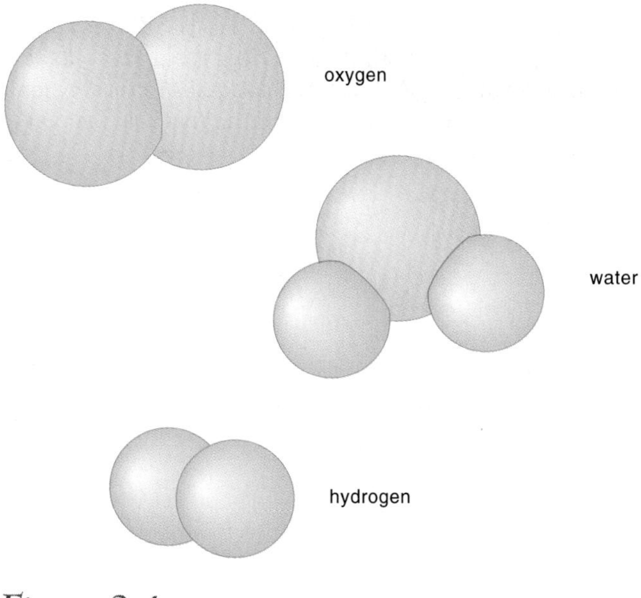

Figure 2.4 Models of oxygen, water, and hydrogen molecules. A water molecule is 0.6 nanometer in diameter.

When a water molecule is formed, two hydrogen atoms become attached to an oxygen atom at an angle averaging 105° in liquid water (for ice, the angle is precisely 105°). The electrons of the three atoms are shared and form an electron cloud around the core, giving the molecule an asymmetrical shape. Although the electron and proton charges balance each other, the asymmetrical shape and unequal sharing of the electrons in the bond between oxygen and hydrogen cause one side of the water molecule to have a slight positive charge and the other a slight negative charge. Such molecules are said to be *polar.* Since negative charges attract positive charges, polarity affects the way in which molecules become aligned toward each other; polarity also causes molecules other than water to be water soluble.

Water molecules form a cohesive network as their slightly positive hydrogen atoms are attracted to the slightly negative oxygen atoms of other water molecules (Fig. 2.5). The cohesion between water molecules is partly responsible for their movement through fine (capillary) tubes, such as those present in the wood and other parts of plants. The attraction between the hydrogen atoms of water and other negatively charged molecules, such as those of fibers, also causes *adhesion* (attraction of dissimilar molecules to each other) and is the basis for water wetting substances. When

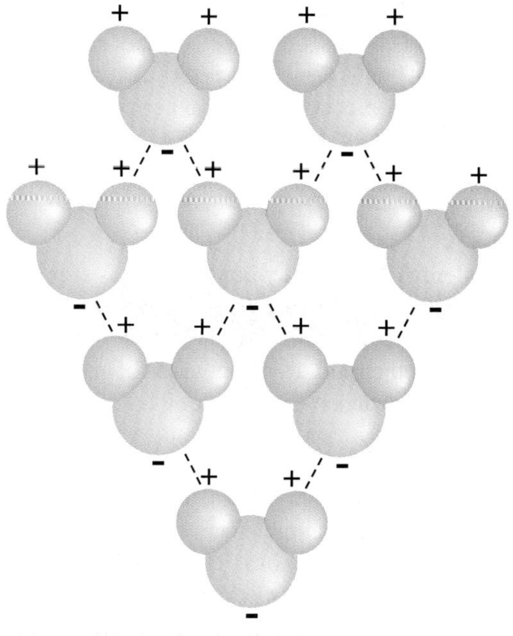

Figure 2.5 The asymmetrical shape of water molecules and the resulting unequal sharing of electrons in the bond between the oxygen and hydrogen atoms cause one side of a water molecule to have a slight positive charge and the other side a slight negative charge. Such molecules are said to be *polar.* The polarity of water molecules cause them to be attracted to one another in a cohesive network. The cohesion of water molecules is partly responsible for their movement through fine (capillary) tubes such as those of living wood.

there is no attraction between water and other substances (e.g., between water and the waxy surface of a cabbage leaf), the cohesion between the water molecules results in droplets beading in the same way that raindrops bead on a freshly waxed automobile.

Valence

The combining capacity of an atom or ion is called **valence.** For example, atoms of the element calcium, an important element in cell walls and in transmitting chemical "messages" in plant cells, have a valence of two, while those of the element chlorine have a valence of one. In order for the atoms of these two elements to combine, there must be a balance between electrons lost or gained (i.e., the valences must balance); it takes *two* chlorine atoms, for example, to combine with *one* calcium atom. The compound formed by the union of calcium and chlorine is called *calcium chloride.* It is customary to use standard abbreviations taken from the Latin names of the elements when giving chemical formulas or equations. Calcium chloride is shown as $CaCl_2$, indicating that one atom of calcium (Ca^{++}) required two atoms of chlorine (Cl^-) to form a calcium chloride molecule.

Bonds and Ions

Bonds are forces that form molecules by attracting and holding atoms together in several different ways. The number of electrons in an atom's outermost orbital determines how many chemical bonds can be formed by that particular atom. If the number of electrons in the outermost orbital is less than eight, the atom may lose, gain, or share electrons, resulting in an outermost energy level of eight. Three types of chemical bonds are of particular significance for living organisms:

1. **Covalent bonds** form when two atoms complete their outermost energy level by sharing a pair of electrons in the outermost orbital; they hold two or more atomic nuclei together and travel between them, keeping them at a stable distance from each other. For example, the single orbital of a hydrogen atom, which has just one electron, is usually filled by attracting an electron from another hydrogen atom. As a result, two hydrogen atoms share their single electrons, making a combined orbital with two electrons. The combined orbital, with its two hydrogen atoms, forms a molecule of hydrogen gas. The covalent bond is shown as a single line, so that hydrogen gas (H_2) is depicted as H–H.

   Except for hydrogen and helium, which have only one orbital, elements can have up to four more orbitals. Carbon atoms, for example, have six electrons—two in the innermost orbital and one in each of the four outer orbitals; by covalent bonding, carbon can share four electrons. When four hydrogen atoms bond to one carbon atom, a molecule of methane gas (CH_4) is formed. To illustrate the bonds, the structural formula for CH_4 is shown as follows:

   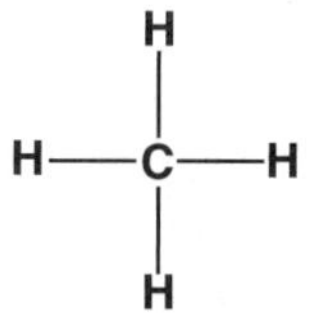

   When one pair of electrons is shared, the bond is said to be *single.* When two pairs of electrons are shared, the bond is referred to as *double,* and *triple* bonds are formed when three pairs of electrons are shared. Double bonds are shown in structural formulas with double lines (e.g., C=C), and triple bonds are shown with three lines (e.g., C≡N). In covalent bonds involving molecules such as those of hydrogen (H_2), where electrons are shared equally, the bonds are said to be *nonpolar.* However, *polar covalent bonds* (e.g., those of a water molecule) are formed when electrons are closer to one atom than to another and therefore are unequally shared. Because the electrons are shared unequally, parts of the molecule are not electrically neutral and are slightly charged. Covalent bonds are the strongest of the three types of bonds discussed here and are the principal force binding together atoms that make up some important biological molecules discussed later in this chapter (Fig. 2.6).

2. **Ionic bonds.** In nature, some electrons in the outermost orbital are not really shared but instead are completely removed from one atom and transferred to another, particularly between elements that can strongly attract or easily give up an electron. Molecules that lose or gain electrons become positively or negatively charged particles called **ions.** Ionic bonds form whenever one or more electrons are donated to another atom and result whenever two oppositely charged ions come in contact. Ions are shown with their charges as superscripts. For example, table salt (sodium chloride) is formed by ionic bonding between an ion of sodium (Na^+) and an ion of

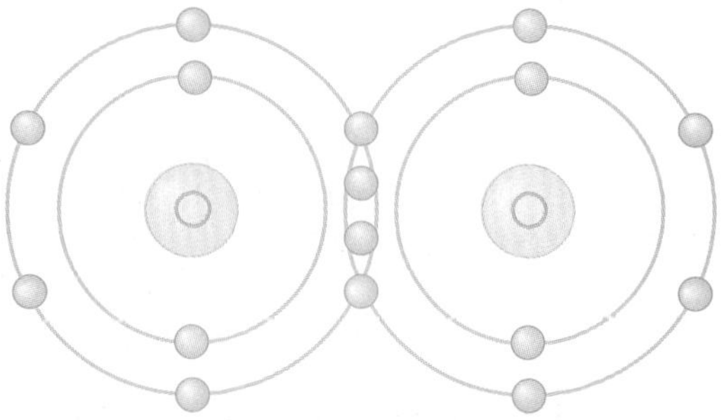

Figure 2.6 A covalent bond between two oxygen atoms. In a covalent bond, electrons are shared as outer shells of atoms overlap. In this instance, two pairs of electrons are shared between the two atoms, and the shared electrons are counted as belonging to each atom.

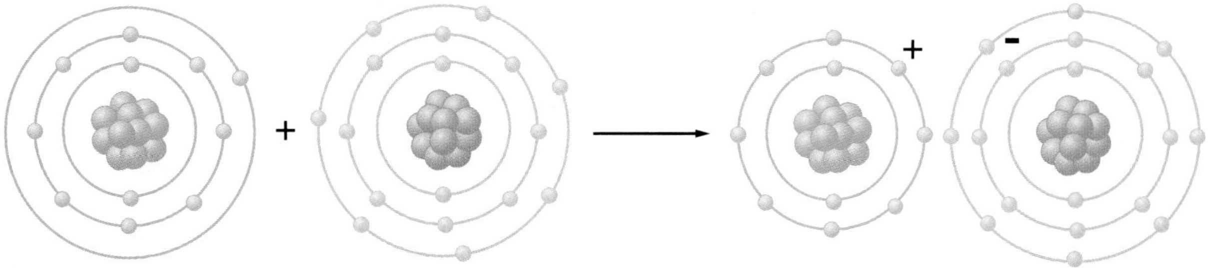

Figure 2.7 Ionic bonding between a sodium atom and a chlorine atom. The sodium becomes positively charged when it loses one of its electrons, which is gained by an atom of chlorine. The gained electron makes the chlorine ion negatively charged, and the two ions become bonded together by the attraction of opposite charges.

chlorine (Cl^-). The sodium becomes a positively charged ion when it loses one of its electrons, which is gained by an atom of chlorine. This extra electron makes the chlorine ion negatively charged, and the sodium ion and chlorine ion become bonded together by the force of the opposite charge (Fig. 2.7).

Some ions, such as those of magnesium (Mg^{++}), give up two electrons and therefore have two positive charges. Such ions can form ionic bonds with two single negatively charged ions such as those of chlorine (Cl^-), forming magnesium chloride ($MgCl_2$). Many biologically important molecules exist as ions in living matter.

3. **Hydrogen bonds** form as a result of attraction between positively charged hydrogen atoms in polar molecules and negatively charged atoms in other polar molecules. Negatively charged oxygen and/or nitrogen atoms of one molecule may attract positively but weakly charged hydrogen atoms of other molecules, forming a weak bond. Hydrogen bonds are very important in nature because of their abundance in many biologically significant molecules. They have, however, only about 7% to 10% of the strength of covalent bonds. Hydrogen bonds help cellular processes by maintaining the shapes of proteins such as enzymes, which make different compounds fit together precisely to complete a chemical reaction.

Acids, Bases, and Salts

Water molecules are held together by weak hydrogen bonds. In pure water, however, a few molecules sometimes dissociate into hydrogen (H^+) and hydroxyl (OH^-) ions, with the number of H^+ ions precisely equaling the number of OH^- ions.

Acids, which include things that taste sour, like cranberry or lemon juice, are chemicals that release hydrogen ions (H^+) when dissolved in water, resulting in proportionately more hydrogen than hydroxyl ions being present. Some acids, such as the acetic acid of vinegar, release relatively few hydrogen ions and are said to be weak. Strong acids such as sulfuric acid dissociate almost completely into hydrogen and sulfate ions.

Bases (also referred to as *alkaline compounds*) usually feel slippery or soapy. They are defined as *compounds that release negatively charged hydroxyl ions (OH^-) when dissolved in water.* Caustic soda, which is sodium hydroxide (NaOH), is a base that dissociates in water to positively charged sodium ions (Na^+) and negatively charged hydroxyl ions (OH^-). Bases can also be defined as *compounds that accept H^+ ions.*

The acidity or alkalinity of the soil or water in which a plant occurs affects how it lives and grows or even if it can exist in a particular environment. Similarly, the acidity or alkalinity of the fluids inside cells has to be stable or various chemical reactions vital to life itself can't take place.

The pH Scale

The concentration of H^+ ions present is used to define degrees of acidity or alkalinity on a specific scale, called the **pH scale.** The scale ranges from 0 to 14, with each unit representing a tenfold change in H^+ concentration. Pure water has a pH of 7—the point on the scale where the number of H^+ and OH^- ions is exactly the same, or the neutral point.[1] The lower a number is below 7, the higher the degree of acidity; conversely, the higher a number is above 7, the higher the degree of alkalinity. Vinegar, for example, has a pH of 3, tomato juice has a pH of 4.3, and egg white has a pH of 8. Precipitation with a pH of less than 4.5 is now commonly referred to as *acid rain* (acid deposition). Acid rain (discussed in Chapter 25) is associated with industrial emissions and appears to be causing damage to vegetation, soil organisms, and buildings in some parts of the world, including North America.

When an acid and a base are mixed, the H^+ ions of the acid bond with the OH^- ions of the base, forming water (H_2O). The remaining ions bond together, forming a salt. If hydrochloric acid (HCl) is mixed with a base—for example,

1. Note that although distilled water is theoretically "pure," its pH is always less than 7 because carbon dioxide from the air dissolves in it, forming carbonic acid (H_2CO_3); the actual pH of distilled water is usually approximately 5.7.

sodium hydroxide (NaOH)—water (H_2O) and sodium chloride (NaCl), a salt, are formed. The reaction is represented by symbols in an equation that shows what occurs:

$$\mathbf{HCl + NaOH \longrightarrow H_2O + NaCl}$$

Energy

Energy is the ability or capacity to do work or to produce a change in motion or matter. Energy exists in several forms and is required for growth, reproduction, movement, cell or tissue damage repair, and other activities of whole organisms, cells, or molecules. On earth, the sun is the ultimate source of life energy.

Thermodynamics is the study of energy and its conversions from one form to another. Scientists apply two laws of thermodynamics to energy. The *first law of thermodynamics* states that energy is constant—it cannot be increased or diminished—but it can be converted from one form to another. Among its forms are *chemical, electrical, heat,* and *light* energy.

The *second law of thermodynamics* states that when energy doesn't enter or leave a given system and is converted from one form to another, it (energy) flows in one direction. Furthermore, there will always be less energy remaining after the conversion than existed before the conversion. The total amount of energy in the universe, however, remains constant. For example, heat will always flow from a hot iron to cold clothing but never from the cold clothing to the hot iron. Such energy-yielding reactions are vital to the normal functions of cells and provide the energy needed for other cell reactions that require energy. Both types of reactions are discussed in Chapter 10.

Forms of energy include *kinetic* (motion) and *potential* energy. Potential energy is defined as the "capacity to do work owing to the position or state of a particle." For example, if a cart resting at the top of a hill rolls down the hill, the cart's potential energy is converted to kinetic energy. Some chemical reactions release energy, and others require an input of energy (Fig. 2.8).

Although all electrons have the same weight and electrical charge, their amount of potential energy varies. Electrons with the least potential energy are located within the single spherical orbital closest to the atom's nucleus, and electrons with the most potential energy are in the outermost orbital (Fig. 2.9). Some of the numerous energy exchanges and carriers that occur in living cells are discussed in later chapters.

Chemical Components of Cells

The living substance of cells consists of cytoplasm and the structures within it. The numerous internal structures, which vary considerably in size, are discussed in Chapter 3. About 96% of cytoplasm and its included structures are composed of the elements carbon, hydrogen, oxygen, and nitrogen; 3% consists of phosphorus, potassium, and sulfur. The remaining 1% includes calcium, iron, magnesium, sodium, chlorine, copper, manganese, cobalt, zinc, and minute quantities of other elements. When a plant first absorbs these elements from the soil or atmosphere or when it uses breakdown products within the cell, the elements are in the form of simple molecules or ions. These simple forms may be converted to very large, complex molecules through the metabolism of the cells.

The large molecules invariably have "backbones" of carbon atoms within them and are said to be organic. The structure and the myriad of chemical reactions of living organisms are

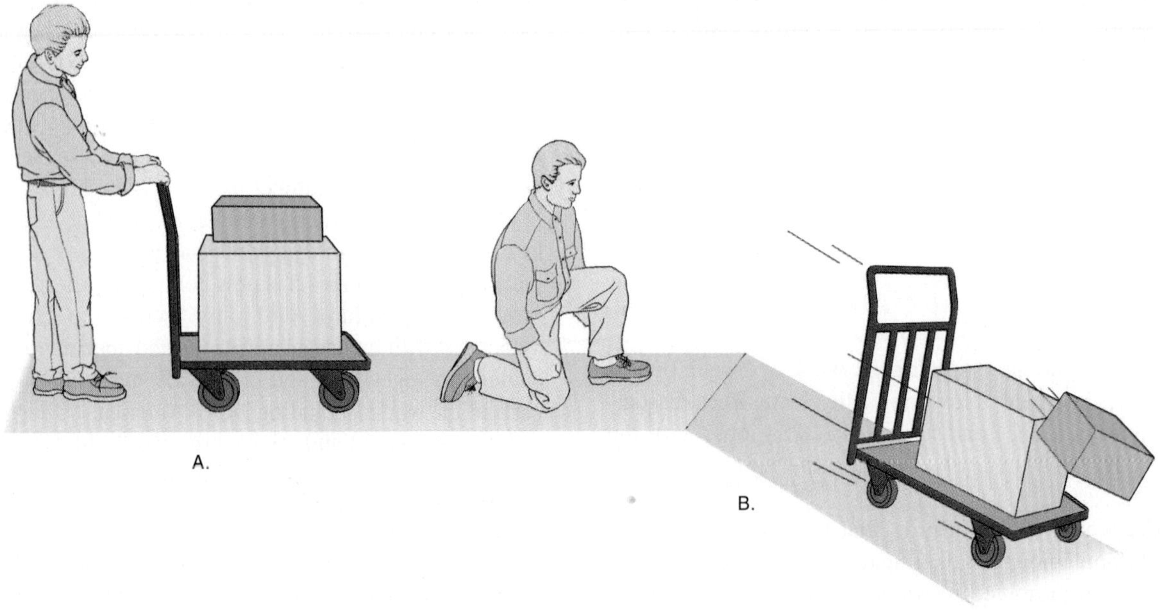

Figure 2.8 A. The cart resting on top of the hill has *potential energy* (capacity to do work owing to its position). *B.* The potential energy is converted to *kinetic energy* when the cart rolls down the hill.

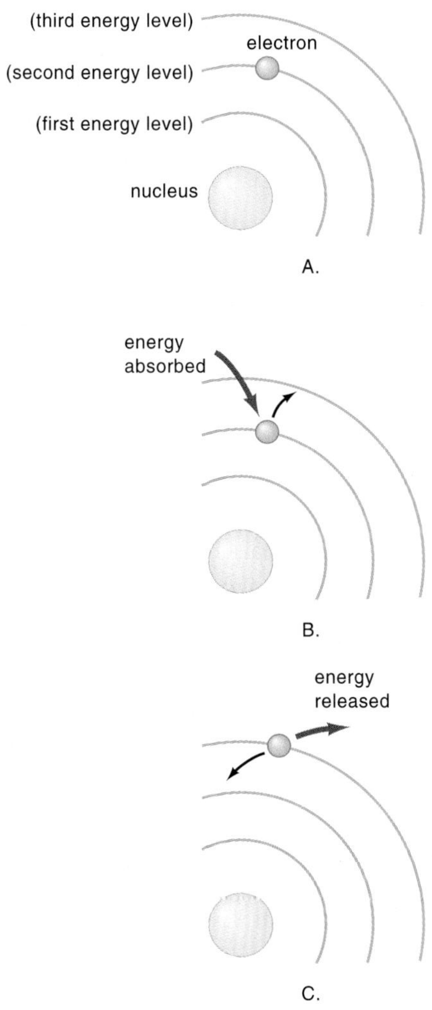

Figure 2.9 Energy levels of electrons. The closer electrons are to the nucleus, the less energy they possess, and vice versa. The energy levels are referred to as *electron shells. A.* An electron at a second energy level. *B.* An electron can absorb energy from sunlight or some other source and be boosted to a higher energy level. *C.* The absorbed energy can be released, with the electron dropping back to its original level (see Fig. 10.8).

based on organic compounds. Most other molecules that contain no carbon atoms are called inorganic. Exceptions include carbon dioxide (CO_2) and sodium bicarbonate ($NaHCO_3$). The name organic was given to most of the chemicals of living things when it was believed that only living organisms could produce molecules containing carbon. Today, many organic compounds can be produced artificially in the laboratory, and scientists sometimes hesitate to classify as either organic or inorganic some of the 4 million carbon-containing compounds thus far identified. Most scientists, nevertheless, agree that inorganic compounds usually do not contain carbon.

Monomers and Polymers

The large molecules comprising the majority of cell components are called *macromolecules,* or **polymers.** Polymers are formed when two or more small units called **monomers** bond together. The bonding between monomers occurs when a hydrogen (H^+) is removed from one monomer and a hydroxyl (OH^-) is removed from another, creating an electrical attraction between them. Since the components of water (H^+ and OH^-) are removed (*dehydration*) in the formation (*synthesis*) of a bond, the process is referred to as *dehydration synthesis.* Dehydration synthesis is controlled by an enzyme (see page 24).

Hydrolysis, which is essentially the opposite of dehydration synthesis, occurs when a hydrogen from water becomes attached to one monomer and a hydroxyl group to the other. Energy is released when a bond is broken by hydrolysis. This energy may be stored temporarily or used in the manufacture or renewal of cell components.

Four of the most important classes of polymers found in cells are *carbohydrates, lipids, proteins,* and *nucleic acids.*

Carbohydrates *Carbohydrates* are the most abundant organic compounds in nature. They include sugars and starches and contain C, H, and O in a ratio of, or close to a ratio of, 1C:2H:1O (CH_2O). The number of CH_2O units in a carbohydrate can vary from as few as three to as many as several thousand. There are three basic kinds of carbohydrates:

1. *Monosaccharides* are simple sugars with backbones consisting of three to seven carbon atoms. Among the most common monosaccharides are *glucose* ($C_6H_{12}O_6$) and *fructose,* which is an *isomer* of glucose. Isomers are molecules with identical numbers and kinds of atoms but with different structures and shapes. Accordingly, fructose, which is found in fruits, has the same $C_6H_{12}O_6$ formula as glucose, but the different arrangement of its atoms gives it different properties, such as a slightly sweeter taste. Glucose, which is produced by photosynthesis in green plant cells, is a primary source of energy in the cells of all living organisms (Fig. 2.10).
2. *Disaccharides* are formed when two monosaccharides become bonded together by dehydration synthesis. The common table sugar **sucrose** ($C_{12}H_{22}O_{11}$) is a disaccharide formed from a molecule of glucose and a molecule of fructose; a molecule of water is removed during synthesis. The removal of a molecule of water during the formation of a larger molecule from smaller molecules is referred to as a *condensation reaction.* Sucrose is the form in which sugar is usually transported throughout plants and is also the form of sugar stored in the roots of sugar beets and the culms (stems) of sugar cane.
3. *Polysaccharides* are formed when several to many monosaccharides bond together. Polysaccharide polymers sometimes consist of thousands of simple sugars attached to one another in long branched or unbranched chains or in coils. For example, *starches,* which are the main carbohydrate reserve of plants, are polysaccharides that usually consist of several hundred to several thousand coiled glucose units. When many glucose molecules become a starch molecule, each glucose gives up a molecule of water. The formula for starch is $(C_6H_{10}O_5)n$, the *n* representing many units. In order for

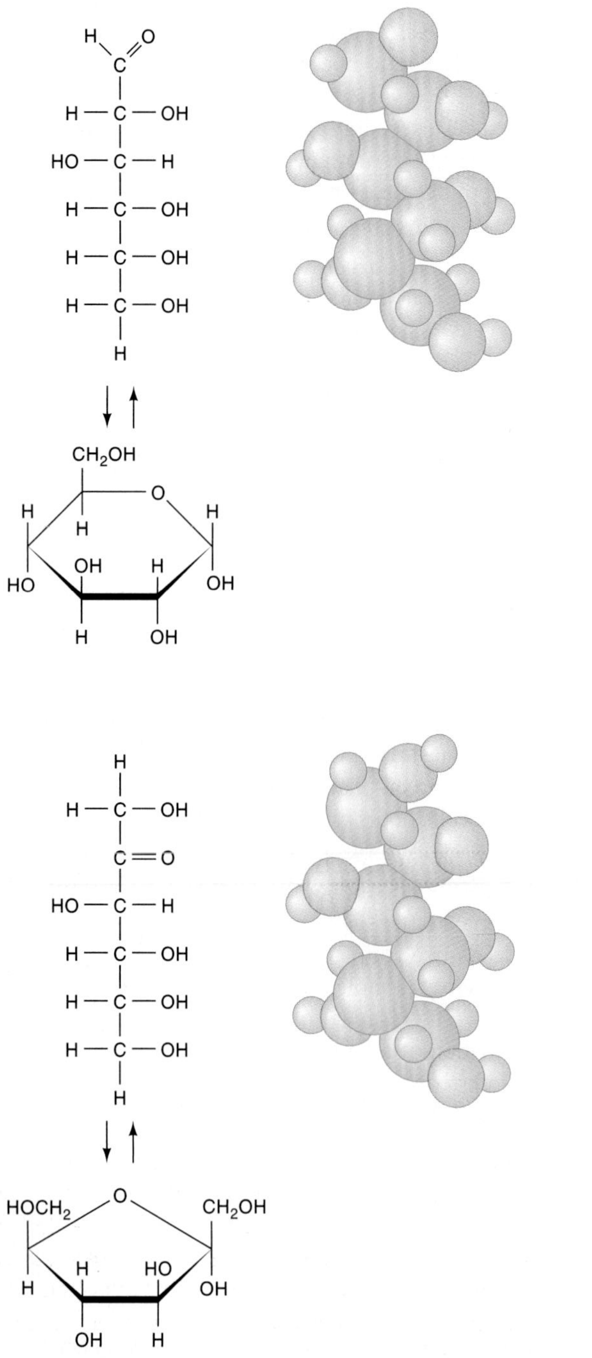

Figure 2.10 Structures of glucose (*top*) and fructose (*bottom*) molecules. The numbers of atoms and locations of bonds are easy to see in the upper linear diagrams, but when these molecules are in solution, they are in the form of rings, as shown in the lower diagrams.

a starch molecule to become available as an energy source in cells, it has to be hydrolyzed; that is, it has to be broken up into individual glucose molecules through the restoration of a water molecule for each unit.

Throughout the world, starches are major sources of carbohydrates for human consumption—the principal starch crops being potatoes, wheat, rice, and corn in temperate areas and cassava and taro in tropical areas.

Cellulose, the chief structural polymer in plant cell walls, is a polysaccharide consisting of 3,000 to 10,000 unbranched chains of glucose molecules. Although cellulose is very widespread in nature, its glucose units are bonded together differently from those of starch, and most animals digest it much less readily than they do starch. Organisms that do digest cellulose, such as the protozoans living in termite guts, caterpillars, and some fungi, produce special enzymes capable of facilitating the breakdown of bonds between the carbons and the glucose units of the cellulose; the organisms then can digest the released glucose.

Lipids *Lipids* are fatty or oily substances that are mostly insoluble in water because they have no polarized components. They typically store about twice as much energy as similar amounts of carbohydrate and play an important role in the longer-term energy reserves and structural components of cells. Like carbohydrates, lipid molecules contain carbon, hydrogen, and oxygen, but there is proportionately much less oxygen present. Examples of lipids include **fats**, which are solid at room temperature (Fig. 2.11), and **oils,** which are liquid. Oil molecules are formed from sugars and are composed of a unit of *glycerol*—a three-carbon compound that has three hydroxyl (–OH) groups—with three *fatty acids* attached to the glycerol unit. A fatty acid has a carboxyl (–COOH) group at one end and typically has an even number of carbon atoms to which hydrogen atoms can become attached.

Most fatty acid molecules consist of a chain with 16 to 18 carbon atoms. If hydrogen atoms are attached to every available attachment point of these fatty acid carbon atoms, as in most animal fats such as butter and those found in meats, the fat is said to be *saturated.* If there is at least one double bond between two carbons and there are fewer hydrogen atoms attached, the fat is said to be *unsaturated.* If there are three or more double bonds between the carbons of a fatty acid, as in some vegetable oils such as those of canola, olive, or safflower, the fat is said to be *polyunsaturated.* Unsaturated vegetable oils can become saturated by bubbling hydrogen gas through them, as is done in the manufacture of margarine. Human diets high in saturated fats often ultimately lead to clogging of arteries and other heart diseases, while diets low in saturated fats promote better health. However, some fat in the diet appears to be essential to normal animal and human absorption of nutrients, and there is concern that consumption of "fake" fat introduced to the public in the late 1990s could lead to health problems.

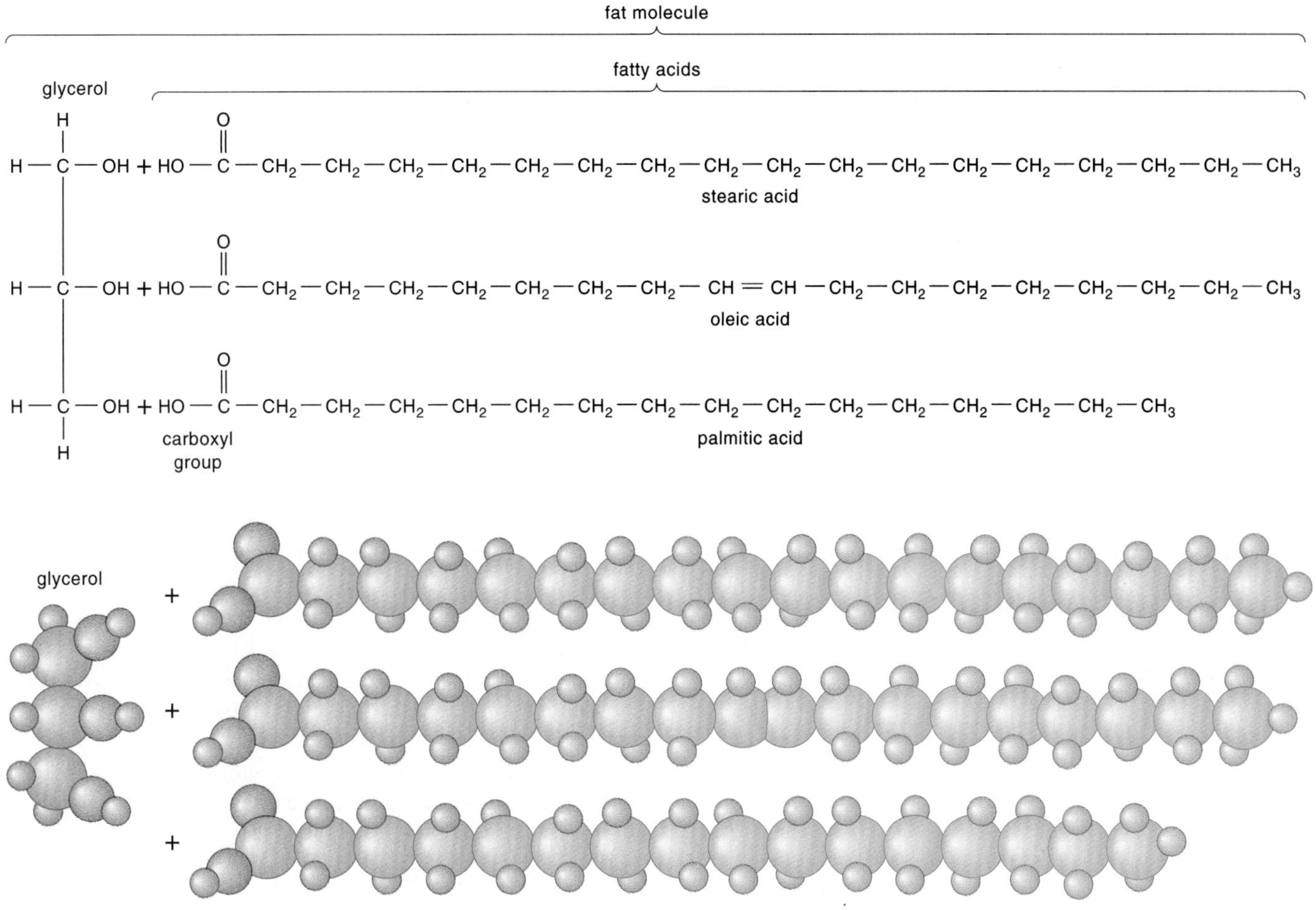

Figure 2.11 Structural formula and model of a fat molecule. H = hydrogen, C = carbon, O = oxygen. A typical fatty acid is 4 nanometers long.

Like polysaccharides and proteins (discussed in the next section, Proteins, Polypeptides, and Amino Acids), lipids are broken down by hydrolysis.

Waxes are lipids consisting of very long-chain fatty acids bonded to a very long-chain alcohol other than glycerol. Waxes, which are solid at room temperature, are found on the surfaces of plant leaves and stems. They are usually embedded in a matrix of *cutin* or *suberin*, which are also lipid polymers that are insoluble in water. The combinations of wax and cutin or wax and suberin function in waterproofing, reduction of water loss, and protection against microorganisms and small insects.

Phospholipids are constructed like fats, but one of the three fatty acids is usually replaced by a phosphate group; this can cause the molecule to become a polarized ion. When phospholipids are placed in water, they form a double-layered sheet resembling a membrane. Indeed, phospholipids are important components of all membranes found in living organisms.

Proteins, Polypeptides, and Amino Acids The cells of living organisms contain from several hundred to many thousands of different kinds of **proteins,** which are second only to cellulose in making up the dry weight of plant cells. Each kind of organism has a unique combination of proteins that give it distinctive characteristics. There are, for example, hundreds of kinds of grasses, all of which have certain proteins in common and other proteins that make one grass different from another. The hundreds of kinds of daisies are distinguished from each other and from grasses by their particular combinations of proteins. Analysis of proteins helps evolutionary botanists sort out relationships and heredity among plants and is a popular, current area of research.

Proteins consist of carbon, hydrogen, oxygen, and nitrogen atoms, and sometimes also sulfur atoms. Proteins regulate chemical reactions in cells and comprise the bulk of protoplasm, apart from water. Protein molecules are usually very large and consist of one or more *polypeptide* chains with, in some instances, simple sugars or other smaller molecules attached.

Polypeptides are chains of **amino acids** (Fig. 2.12). There are 20 different kinds of amino acids, and from 50 to 50,000 or more of them are present in various combinations in each protein molecule. Each amino acid has two special function groups of atoms plus a remainder called the *R group*. One

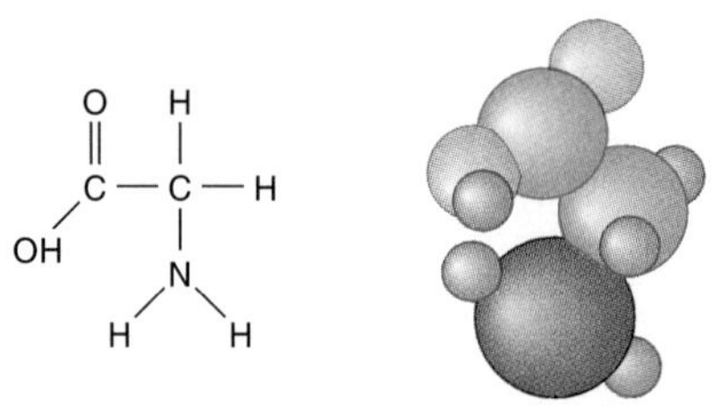

Figure 2.12 Structural formula and model of the amino acid glycine.

functional group is called the *amino group* ($-NH_2$); the other, which is acid, is called the *carboxyl group* ($-COOH$). Amino acids are linked to other amino acids by *peptide bonds*, which are covalent bonds formed between the carboxyl carbon of one amino acid and the nitrogen of the amino group of another, a molecule of water being removed in the process. The makeup of the R group is distinctive for each of the 20 amino acids. Some R groups (which can consist of just a single hydrogen atom or, in others, a complex ring structure) are polar while some are not. The general structural formula for an amino acid is

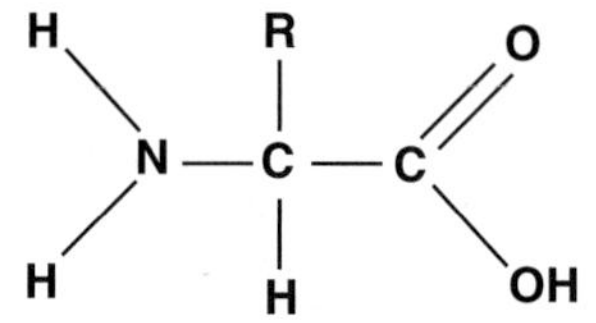

Plants can synthesize amino acids they need from raw materials in their cells, but animals have to supplement from plant sources some amino acids they need, since they can manufacture only a few amino acids themselves.

Each polypeptide usually coils, bends, and folds in a specific fashion within a protein, which characteristically has three levels of structure and sometimes four:

1. A sequence of amino acids fastened together by peptide bonds forms the *primary structure* of a protein.
2. As hydrogen bonds form between oxygen and hydrogen atoms or nitrogen and hydrogen atoms of different amino acids, the polypeptide chain coils like a spiral staircase, and the *secondary structure* develops. Some secondary structures include polypeptide chains that double back and form hydrogen bonds between the two lengths in what is referred to as a *beta sheet,* or *pleated sheet.*
3. *Tertiary structure* develops as the polypeptide further coils and folds. The tertiary structure is maintained by bonds between R groups.
4. If a protein happens to have more than one kind of polypeptide, a fourth, or *quaternary structure*, may form (Fig. 2.13).

The three-dimensional structure of a protein may be somewhat flexible in solution, but chemicals or anything that disturbs the normal pattern of bonds between parts of the protein molecule can *denature* the protein. Denaturing alters the characteristic coiling and folding and adversely affects the protein's function or properties. Denaturing may be reversible, but if it is brought about by high temperatures or harsh chemicals, it may kill the cell of which the protein is a part. For example, boiling an egg, which is mostly protein, brings about an irreversible denaturing; the solid egg proteins simply can't be restored to their original semiliquid condition.

Storage Proteins Some plant food-storage organs, such as potato tubers and onion bulbs, store small amounts of proteins in addition to large amounts of carbohydrates. Seeds, in particular, however, usually contain proportionately larger amounts of proteins in addition to their complement of carbohydrates and are very important sources of nutrition for humans and animals. One example of an important protein source in human and animal diets is wheat gluten (to which, incidentally, some humans become allergic). The gluten consists of a complex of more than a dozen different proteins.

A seed's proteins get used during germination and its subsequent development into a seedling. Some legume seeds may contain more than 40% protein, but legumes are deficient in certain amino acids (e.g., methionine), and a human diet based on beans needs to be balanced with other storage proteins (e.g., those found in unpolished rice) to furnish a complete complement of essential amino acids. Some seed proteins, such as those of jequirity beans (*Abrus precatorius*—used in India to induce abortions and as a contraceptive), are highly poisonous.

Enzymes *Enzymes* are mostly large, complex proteins (a few are ribonucleic acids, which are discussed on p. 26) that function as organic catalysts under specific conditions of pH and temperature. By breaking down bonds and allowing new bonds to form, they facilitate cellular chemical reactions, even at very low concentrations, and are absolutely essential to life. None of the 2,000 or more chemical reactions in cells can take place unless the enzyme specific for each one is present and functional in the cell in which it is produced. They can increase the rate of reaction as much as a billion times, and without them, the chemical reactions in cells would take place much too slowly for living organisms to exist. Enzymes are often used repeatedly and usually do not break down during the reactions they accelerate.

Enzyme names normally end in *-ase* (e.g., maltase, sucrase, amylase). The material whose breakdown is catalyzed by an enzyme is known as the *substrate.* Maltose is a very common disaccharide composed of two glucose monomers. The enzyme maltase catalyzes the hydrolysis of maltose (its substrate) to glucose. Enzymes work by lowering the *energy of activation,* which is the minimal amount of energy needed to cause molecules to react with one another. An enzyme brings about its effect by temporarily bonding with potentially reactive molecules at a surface site. The reactive molecules temporarily fit into the active site, where a short-lived complex is formed. The reaction occurs rapidly, often at rates of more than 500,000 times per second. The complex then breaks down as the products of the reaction are released, with the enzyme remaining unchanged and capable of once more catalyzing the reaction (Fig. 2.14).

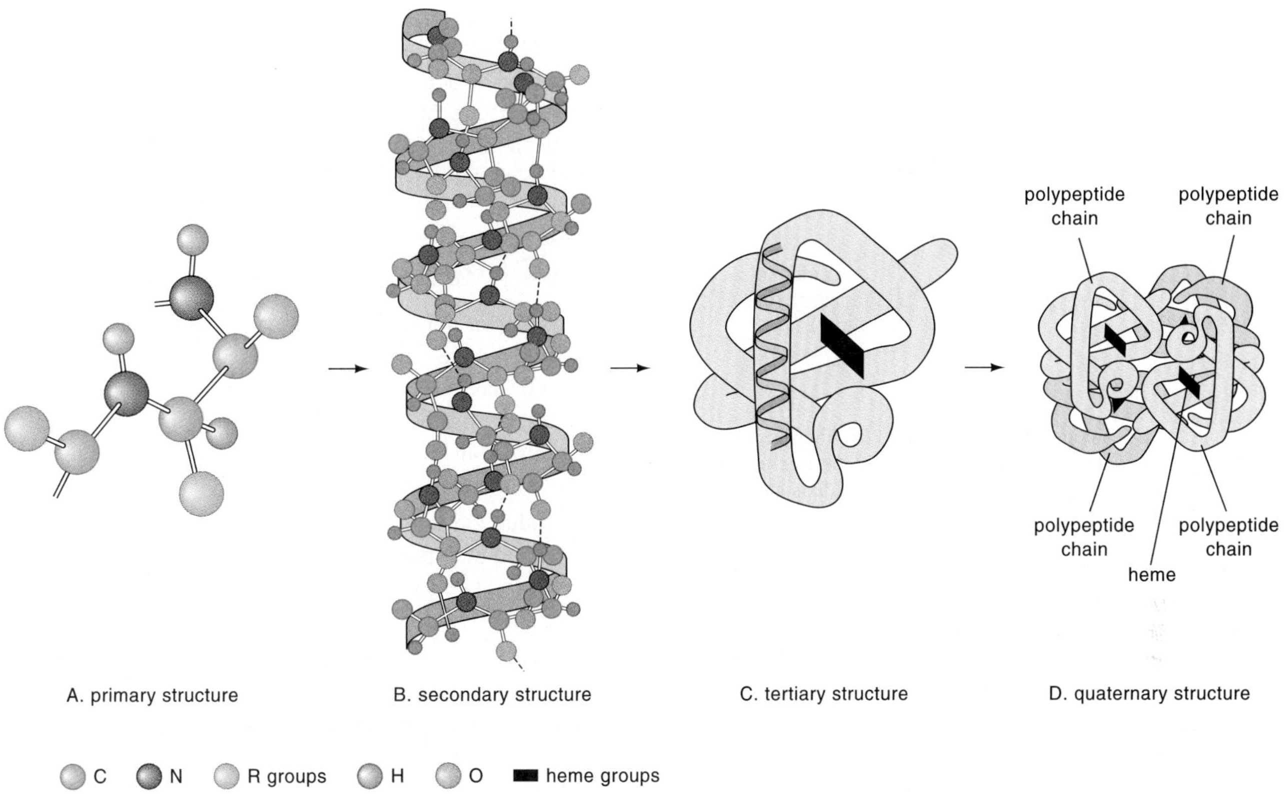

Figure 2.13 The four levels of protein structure. The example shown is for hemoglobin. *A.* The primary structure consists of a chain of amino acids bonded together. *B.* As the amino acid chain grows, rotation of the chain occurs to form an alpha helix, which is stabilized by hydrogen bonds. *C.* The coil or helix forms further, forming a somewhat globular structure. *D.* Several chains combine into a single functional protein molecule. *After Caret, R. L., K. J. Denniston, and J. J. Topping, 1993.* Organic and Biological Chemistry. *Copyright 1993 by Wm C. Brown Publishers, Dubuque, IA.*

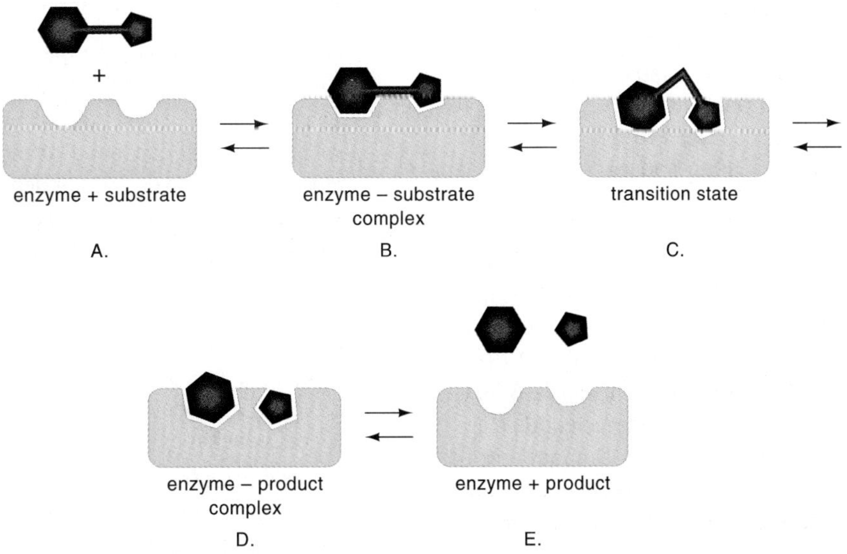

Figure 2.14 How an enzyme facilitates a reaction. *A.* An enzyme and the raw material (substrate) for which it is specific. *B.* The substrate fits into the active site on the enzyme. *C.* The enzyme then changes shape, putting stress on the linkage between parts of the substrate. *D.* The bonds (linkage) are broken. *E.* The enzyme returns to its original shape, and the products are released. When an enzyme is combining substrates, the events shown proceed in reverse.

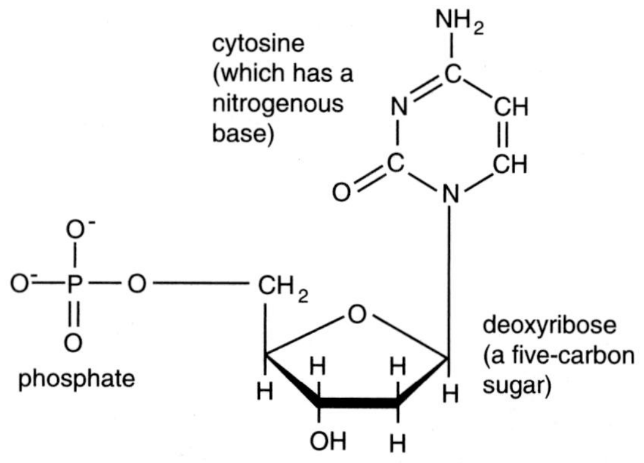

Figure 2.15 DNA consists of double coils of subunits called *nucleotides*, which consist of a nitrogenous base, a five-carbon sugar, and a phosphate. The phosphate of one nucleotide is attached to the sugar of the next nucleotide.

Many enzymes, derived mostly from bacteria and fungi, have very important industrial uses. For example, waste treatment plants, the dairy industry, and manufacturers of detergents all use enzymes that have been mass-produced by microorganisms in large vats. One such commercially marketed enzyme, known as Beano, is produced by the activities of *Aspergillis,* a mold. Beano breaks down complex sugars found in beans, broccoli, and many other vegetables consumed by humans. A few drops of the enzyme placed on these foods while they are being consumed effectively reduces the gas produced when enzymes in human digestive tracts are otherwise unable to accomplish the breakdown.

Nucleic Acids *Nucleic acids* are exceptionally large, complex polymers originally thought to be confined to the nuclei of cells but now known also to be associated with other cell parts. They are vital to the normal internal communication and functioning of all living cells. The two types of nuclcic acids—deoxyribonucleic acid (DNA) and ribonucleic acid (RNA)—are briefly introduced here and discussed in more detail in Chapter 13.

Deoxyribonucleic acid (DNA) molecules consist of double helical (spiral) coils of repeating subunits called **nucleotides.** Each nucleotide is composed of three parts: (1) a base containing nitrogen, (2) a five-carbon sugar, and (3) a phosphate (phosphoric acid) molecule. The phosphate of one nucleotide is attached to the sugar of the next nucleotide (Fig. 2.15). Four kinds of nucleotides, each with a unique nitrogenous base, occur in DNA. DNA molecules contain, in units known as **genes,** the coded information that precisely determines the nature and proportions of the myriad substances found in cells and also the ultimate form and structure of the organism itself. If this coded information were written out, it would fill over 1,000 books of 300 pages each—at least for the more complex organisms. DNA molecules can replicate (duplicate themselves) in precise fashion. When a cell divides, the hereditary information contained in the DNA of the new cells is an exact copy of the original and can be passed on from generation to generation without change, except in the event of a *mutation* (discussed in Chapter 13).

Ribonucleic acid (RNA) is similar to DNA but differs in its sugar and one of its nucleotide components. It usually occurs as a single strand. Different forms of RNA are involved in protein synthesis.

Summary

1. Activities of living organisms originate in cells. Structure and growth are among the attributes of living organisms. Growth has been described as an increase in volume; it results primarily from the production of new cells. Variations in form may be inherited or result from response to the environment.
2. Reproduction involves offspring that are always similar in form to their parents; if reproduction ceases, the organism becomes extinct. Plants generally respond to stimuli more slowly and in a different fashion from animals.
3. All living organisms exhibit metabolic activities, including respiration, digestion, assimilation, production of new cytoplasm, and, in green organisms, photosynthesis; they also all exhibit movement. Cyclosis is the streaming motion of cytoplasm within living cells. Living organisms have a much more complex structure than nonliving objects and are adapted to their individual environments.
4. The basic "stuff of the universe" is called matter, which occurs in solid, liquid, or gaseous form. It is composed of elements, of which the smallest stable subdivision is an atom. Atomic nuclei contain positively charged protons and uncharged neutrons; the nuclei are surrounded by much larger orbitals of whirling, negatively charged electrons. Isotopes are forms of elements that have slight variations in the number of neutrons in their atoms.
5. The combining capacities of atoms or ions are called valence. Atoms can bond to other atoms, and those of most elements do not exist independently; compounds are substances composed of two or more elements combined in a definite ratio by chemical bonds; molecules are the smallest independently existing particles. In a covalent bond, pairs of electrons link two or more atomic nuclei; nitrogen and/or oxygen atoms of one molecule may form weak hydrogen bonds with hydrogen atoms of other molecules. If a molecule loses or gains electrons, it becomes an ion, which may form an ionic bond with another ion.
6. Water molecules are polar because they are asymmetrical in shape. Water molecules cohere to each other and adhere to other molecules.
7. Acids release positively charged hydrogen ions when dissolved in water. Bases release negatively charged hydroxyl ions when dissolved in water. The pH scale is used to measure degrees of acidity or alkalinity. Salts and water are formed when acids and bases are mixed.
8. Energy can be defined as "ability to produce a change in motion or matter" or as "ability to do work." Its forms

include chemical, electrical, heat, light, kinetic, and potential. The farther away from the nucleus an electron is, the greater the amount of energy required to keep it there.

9. Cells are composed of carbon, hydrogen, oxygen, and nitrogen, with a little phosphorus and potassium, plus small amounts of other elements. A plant may convert the simple molecules or ions it recycles or absorbs from the soil to very large, complex molecules. Organic molecules are usually large polymers that have a "backbone" of carbon atoms.
10. Carbohydrates contain carbon, hydrogen, and oxygen in a ratio of 1C:2H:1O. Carbohydrates occur as monosaccharides (simple sugars) and disaccharides (two simple sugars joined together). Polysaccharides may consist of many simple sugars condensed together; others are more complex. Simple sugars, when they are attached to one another, each give up a molecule of water, forming starch. Hydrolysis involves restoring a water molecule to each simple sugar when starch is broken down during digestion.
11. Lipids (e.g., fats, oils, and waxes), which are insoluble in water, consist of a unit of glycerol or other alcohol with three fatty acids attached. They contain carbon, hydrogen, and oxygen, with proportionately much less oxygen than is found in carbohydrates. Saturated fats have hydrogen atoms attached to every available attachment point of their carbon atoms; if there are very few places for hydrogen atoms to attach, the fat is said to be polyunsaturated. Phospholipids have a phosphate group replacing one fatty acid.
12. Proteins are usually large molecules composed of subunits called amino acids. Each amino acid has an amino group ($-NH_2$) and a carboxyl group ($-COOH$); these groups bond amino acids together, forming polypeptide chains; the bonds are called peptide bonds. Enzymes are large protein molecules that function as organic catalysts. Their names end in -ase. Some have important industrial uses.
13. There are two nucleic acids (DNA and RNA) associated primarily with cell nuclei. DNA and RNA molecules consist of chains of nucleotides. Four kinds of nucleotides, each with a unique nitrogenous base, occur in DNA. Helical coils of DNA contain coded information determining the nature and proportions of substances in cells and the ultimate form and structure of the organism. RNA has a different sugar and nucleotide.

Review Questions

1. What distinguishes a living organism from a nonliving object, such as a rock or a tin can?
2. What is meant by the term organic?
3. How are acids, bases, and salts distinguished from one another?
4. Distinguish among carbohydrates, lipids, and proteins.
5. What is energy, and what forms does it take?
6. How are polymers formed?
7. How is a protein molecule different from a nucleic acid molecule?

Discussion Questions

1. Can part of an organism be alive while another part is dead? Explain.
2. What is the difference between inherited form and form resulting from response to the environment?
3. What might happen if all enzymes were to work at half their usual speed?

Learning Online

Visit our web page at http://www.mhhe.com/botany for interesting case studies, practice quizzes, current articles, and animations within the Online Learning Center to help you understand the material in this chapter. You'll also find active links to these topics:

Chemistry of Biology
Inorganic Chemistry
Atoms
Molecules
Bonds
Carbon
Water
Organic Chemistry
Lipids
Nucleic Acids
Proteins

Additional Reading

Boyer, P. D. 1998. *Introductory biochemistry.* Belmont, CA: Wadsworth.

Day, W. 1996. *Bridge from nowhere,* vol. II, *The photonic origin of matter.* Cambridge, MA: Rhombics.

Lehninger, A. L., D. L. Nelson, and M. M. Cox. 2000. *Principles of biochemistry,* 3d ed. New York: St. Martin's Press.

Lewis, R. 1994. *The beginnings of life.* Dubuque, IA: McGraw-Hill.

Margulis, L., et al. (Eds.). 2000. *Environmental evolution: Effects of the origin and evolution of life on planet earth,* 2d ed. Cambridge, MA. MIT Press.

Raven, P. H., R. F. Evert, and S. E. Eichhorn. 1999. *Biology of plants,* 6th ed. New York: Worth.

Sackheim, G. 1998. *Introduction to chemistry for biology students,* 6th ed. Redwood City, CA: Benjamin/Cummings.

Smith, C. A., and E. J. Wood (Eds.). 1991. *Biological molecules.* New York: Chapman and Hall.

Timberlake, K. C. 1999. *An introduction to general organic and biological chemistry,* 7th ed. Old Tappan, NJ: Addison Wesley.

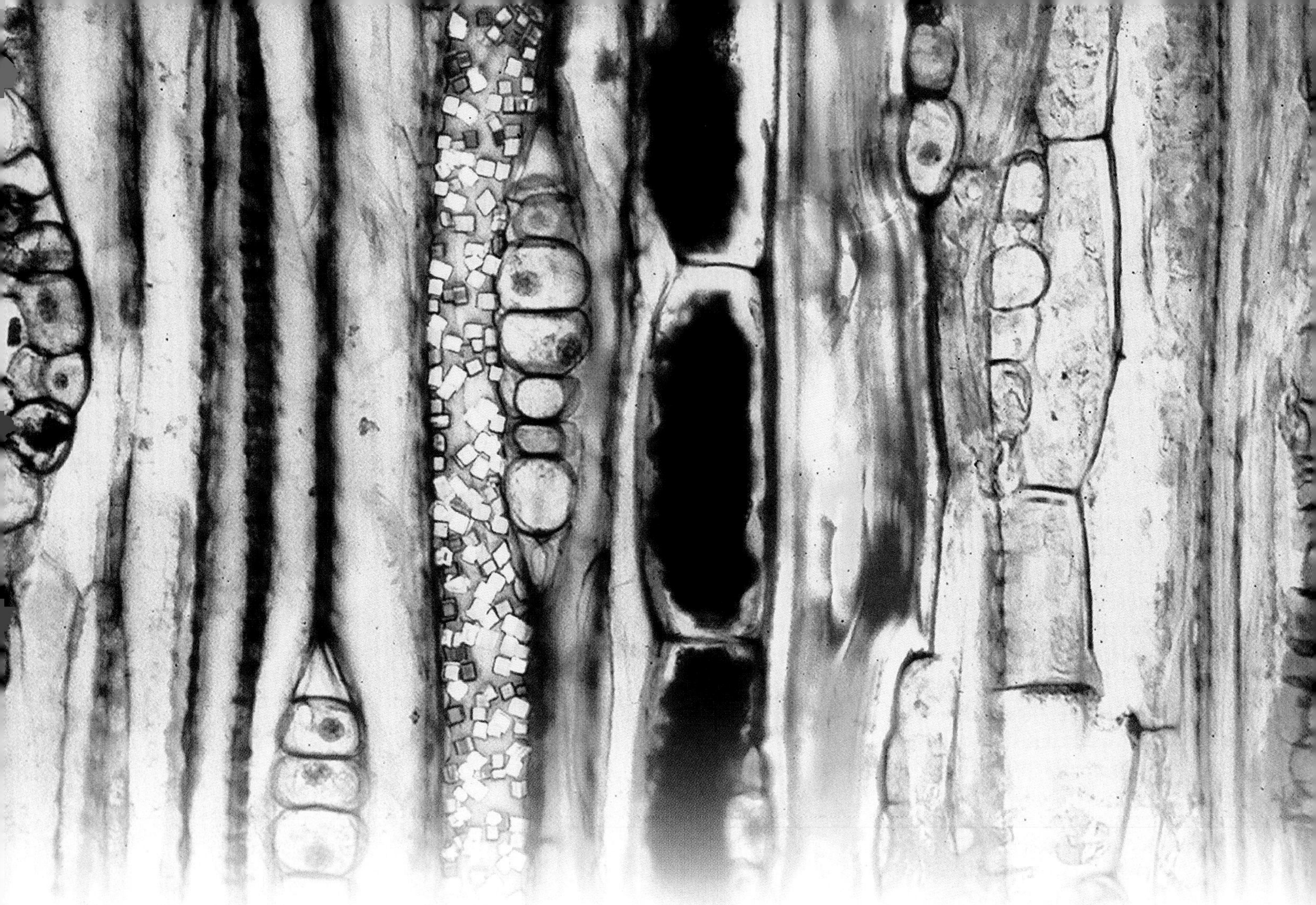

Phloem of Douglas fir (*Pseudotsuga menziesii*), ×500.
(Polarized light photomicrograph by G. S. Ellmore.)

Cells

3

Chapter Outline

Overview

This chapter provides a brief review of the history of cell discovery and the development of the cell theory. Differences between prokaryotic and eukaryotic cells are mentioned, and observations on cell structure and communication follow. Descriptions are provided for the plasma membrane, nucleus, endoplasmic reticulum, ribosomes, dictyosomes, plastids, mitochondria, microbodies, vacuoles, and the cytoskeleton. The chapter next discusses the cell cycle, including interphase as well as mitosis and cytokinesis, and concludes with a brief comparison of plant and animal cells.

Some Learning Goals

1. Trace the development of modern cell theory and show how early researchers have led us to our current understanding.
2. Explain the unique structure of plant cells and how communication between and within cells occurs.
3. Know the following cell structures and organelles and indicate the function of each: plasma membrane, nucleus, endoplasmic reticulum, ribosomes, dictyosomes, plastids, mitochondria, microbodies, vacuoles, and cytoskeleton.
4. Describe how information contained in the nucleus relates to other parts of the cell.
5. Understand the cell cycle and the events that take place in each phase of mitosis.

All living organisms, from aardvarks to zinnias, are composed of cells, and all living organisms, including each of us, also begin life as a single cell. This single cell divides repeatedly until it develops into an organism often consisting of billions of cells. During the first few hours of an organism's development, the cells all look alike, but changes soon take place, not only in the appearance of the cells but also in their function. Some modifications, for example, equip cells to transport food and water, while other cells become modified for secretion of various fluids such as resin or nectar, and still others give strength to tissues such as wood. Some cells may live and function for many years; others mature and degenerate in just a few days. Even as you read this, millions of new cells are being produced in your body. Some cells add to your total body mass (if you have not yet stopped growing), but most replace the millions of older cells that are destroyed every second you remain alive. The variety and form of cells seem almost infinite, but certain features are shared by most of them. A discussion of these features forms the body of this chapter.

Figure 3.1 Robert Hooke's microscope, as illustrated in one of his works. (*Courtesy National Library of Medicine*)

CELLS

History

Imagine the excitement of the first scientist who observed cells! This discovery was made in 1665 by the English physicist Robert Hooke, who used a primitive microscope (Fig. 3.1) to examine thin slices of cork found in stoppered wine bottles:

> I took a good clear piece of Cork, and with a Penknife sharpen'd as keen as Razor, I cut a piece of it off, and thereby left the surface of it exceedingly smooth, then examining it very diligently with a Microscope, methought I could perceive it to appear a little porous . . . these pores, or cells . . . were indeed the first microscopical pores I ever saw, and perhaps that were ever seen, for I had not met with any Writer or Person, that had made mention of them before this . . . I had with the discovery of them, presently hinted to me the true and intelligible reason of all the Phaenomena of Cork.

Hooke compared the boxlike compartments he saw to the surface of a honeycomb and is credited with applying the term *cell* to those compartments. He also estimated that a cubic inch of cork would contain approximately 1,259 million such cells. What Hooke saw in the cork were really only the walls of dead cells, but he also saw "juices" in living cells of elderberry plants and thought he had found something similar to the veins and arteries of animals.

Two physicians, Marcello Malpighi in Italy and Hooke's compatriot Nehemiah Grew in England, along with Anton van Leeuwenhoek, reported for 50 years on the organization of cells in a variety of plant tissues. In the 1670s, they also reported on the form and structure of single-celled organisms, which they referred to as "animalcules."

After this period, little more was reported on cells until the early 1800s. This lack of progress was mainly due to imperfections in the primitive microscopes and the crude way in which tissues were prepared for microscopic examination. Microscopes and tissue preparations both slowly improved, however, and by 1809, the famous French biologist Jean Baptiste de Lamarck had seen a wide enough variety of cells and tissues to conclude that "no body can have life if its constituent parts are not cellular tissue or are not formed by cellular tissue." In 1824, René J. H. Dutrochet, also of France, reinforced Lamarck's conclusions that all animal and plant tissues are composed of cells of various kinds. Neither of them, however, realized that each cell could, in many cases, reproduce itself and exist independently.

In 1831, the English botanist Robert Brown discovered that all cells contain a relatively large body that he called the *nucleus.* Soon after the discovery of the nucleus, the German botanist Matthias Schleiden observed a smaller body within the nucleus that he called the *nucleolus.* Schleiden and German zoologist Theodor Schwann were not the first to understand the significance of cells, but they explained them more clearly and with greater insight than others before them had done. They are generally credited with developing the **cell theory,** beginning with their publications of 1838 to 1839. This theory, in essence, states that all living organisms are composed of cells and that cells form a unifying structural basis of organization.

In 1858, another German scientist, Rudolf Virchow, argued persuasively in a classic textbook that every cell comes from a preexisting cell (*"omnis cellula e cellula"*) and that there is no spontaneous generation of cells. Virchow's publication stirred up much controversy because there had previously been a widespread belief among scientists and nonscientists alike that animals could originate spontaneously from dust. Many who had microscopes were thoroughly convinced they could see "animalcules" appearing in decomposing substances.

The controversy became so heated that in 1860, the Paris Academy of Sciences offered a prize to anyone who could experimentally prove or disprove spontaneous generation. Just two years later, the brilliant scientist Louis Pasteur of France was awarded the prize. Pasteur, using swan-necked flasks, demonstrated convincingly that boiled media remained sterile indefinitely if microorganisms from the air were excluded from the media.

In 1871, Pasteur proved that natural alcoholic fermentation always involves the activity of yeast cells. In 1897, the German scientist Eduard Buchner accidentally discovered that the yeast cells did not need to be alive for fermentation to occur. He found that extracts from the yeast cells would convert sugar to alcohol. This discovery was a big surprise to the biologists of the time and quickly led to the identification and description of enzymes (discussed in Chapter 2), the organic *catalysts* (substances that aid chemical reactions without themselves being changed) found in all living cells. This also led to the belief that cells were little more than miniature packets of enzymes. During the first half of the 20th century, however, further advances were made in the refinement of microscopes and in tissue preparation techniques. Many structures and bodies, besides the nucleus, were observed in cells, and the relationship between structure and function came to be realized and understood on a much broader scale than previously had been possible.

Modern Microscopes

Our investigation of life is greatly assisted by various types of microscopes that can magnify our images of cells and tissues up to hundreds or even thousands of times their actual size. A better understanding of cell structure and function is also provided by preparing in different ways the tissues that are to be examined. While **light microscopes,** similar to the one used by Hooke in 1665, provide basic information about cell structure and some of the bodies within cells, the development of **electron microscopes** has revealed detailed images of tiny structures within cells.

Light microscopes increase magnification as light passes through a series of transparent lenses, presently made of various types of glass or calcium fluoride crystals. The curvatures of the lens materials and their composition are designed to minimize distortion of image shapes and colors.

Light microscopes are of two basic types: *compound microscopes,* which require most material being examined to be sliced thinly enough for light to pass through, and *dissecting microscopes* (*stereomicroscopes*), which allow three-dimensional viewing of opaque objects. Microscopic organisms such as bacteria and protozoans often consist of a single cell and are thin enough to be viewed without being sliced. The best compound microscopes in use today can produce magnifications of up to 1,500 times under ideal conditions. Many dissecting microscopes used in teaching laboratories magnify up to 30 times, but higher magnifications are possible with both types of microscopes. Light microscope magnifications of more than 1,500 times, however, are considered "empty" because *resolution* (the capacity of lenses to aid in separating closely adjacent tiny objects) does not improve with magnification beyond a certain point. In general, when using a compound light microscope, one can distinguish *organelles* (bodies within cells) only if they are 2 micrometers or larger in diameter. In this chapter, the structures discussed that are most commonly observed with a light microscope include cell walls, nuclei, nucleoli, cytoplasm, chloroplasts, and vacuoles. Light microscopes (Fig. 3.2) will continue in the foreseeable future to be useful, particularly for observing living cells.

Since the 1950s, the development of high-resolution electron microscopes has resulted in observation of much greater detail than is possible with light microscopes.

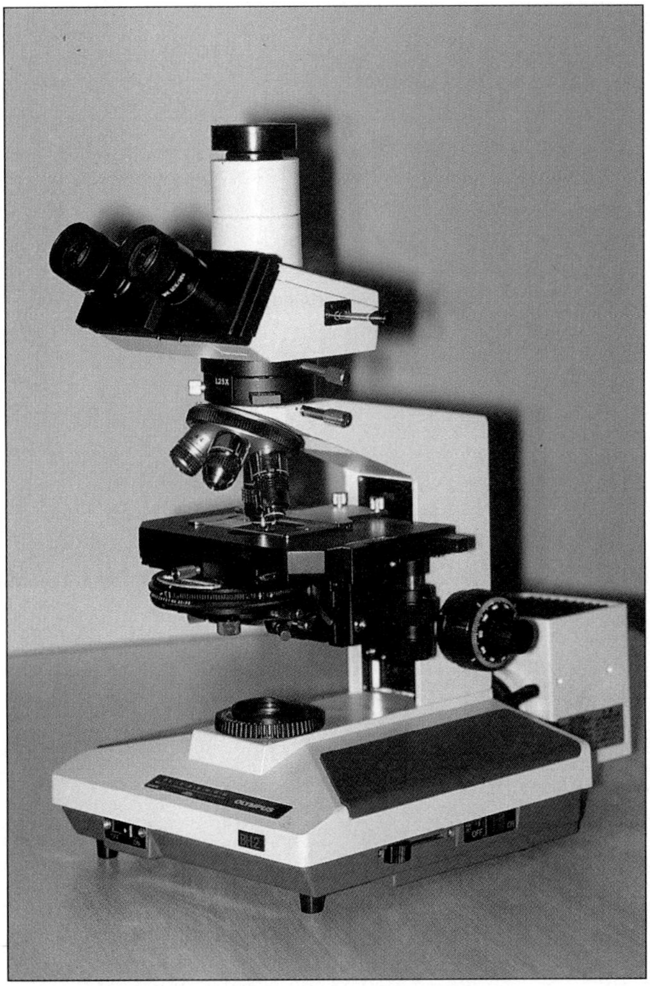

Figure 3.2A A compound light microscope.

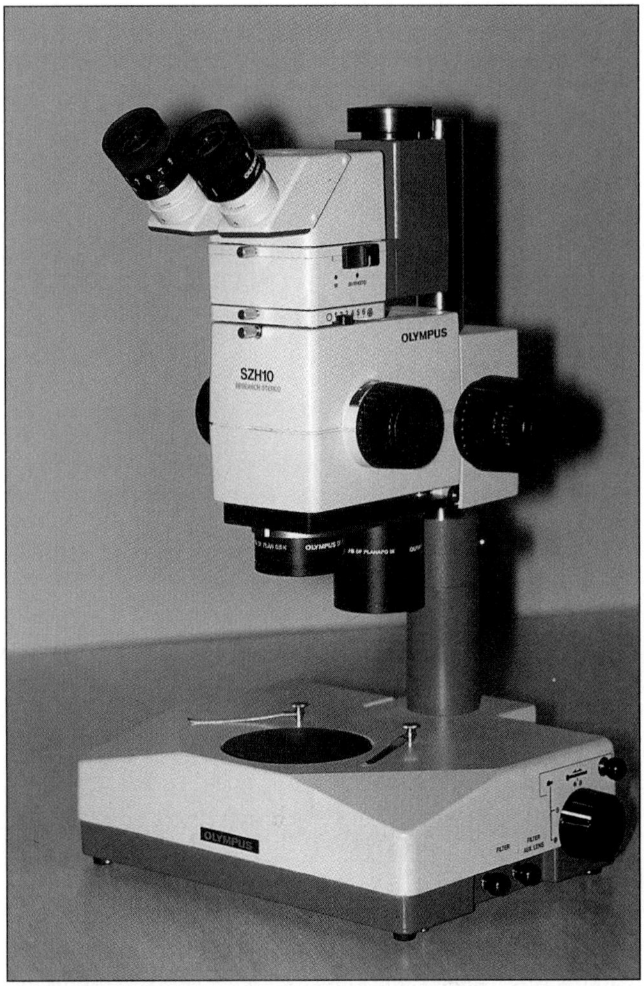

Figure 3.2B A stereomicroscope (dissecting microscope).

Instead of light, electron microscopes use a beam of electrons produced when high-voltage electricity is passed through a wire. This electron beam is directed through a vacuum in a large tube or column. When the beam passes through a specimen, an image is formed on a plate toward the base of the column. Magnification is controlled by powerful electromagnetic lenses located on the column.

Like light microscopes, electron microscopes are of two basic types. *Transmission electron microscopes* (Fig. 3.3A) can produce magnifications of 200,000 or more times, but the material to be viewed must be sliced extremely thin and introduced into the column's vacuum so that living objects can't be observed.

Scanning electron microscopes (Fig. 3.3B) usually do not attain such high magnifications (3,000 to 10,000 times is the usual range), but surface detail of thick objects can be observed when a scanner makes the object visible on a cathode tube like a television screen. The techniques for observation with electron microscopes have become so refined that even preserved material can appear exceptionally lifelike, and high-resolution, three-dimensional images can be obtained.

In 1986, the Nobel Prize in physics was awarded to two IBM scientists, Gerd Binnig and Heinrich Rohrer, for their invention in 1982 of a *scanning tunneling microscope*. This microscope uses a minute probe that "tunnels" electrons upon a sample. As the probe is moved across the surface, its height is continually adjusted to keep the flow of electrons constant, and the fluctuations in height are recorded to produce a map of the sample surface. Without doing any damage to the probed area, this microscope reproduces an image of such high magnification that even atoms can become discernible. The probe can scan areas barely twice the width of an atom and theoretically could be used to print on the head of an ordinary pin the words contained in more than 50,000 single-spaced pages of books. Early in 1989, the first picture of a segment of DNA showing its helical structure was taken with a scanning tunneling microscope by an undergraduate student associated with the Lawrence Laboratories in northern California. Several variations of this microscope, each using a slightly different type of probe, have now been produced. Significant new discoveries by cell biologists using one or more of all three types of microscopes in their research have become frequent events.

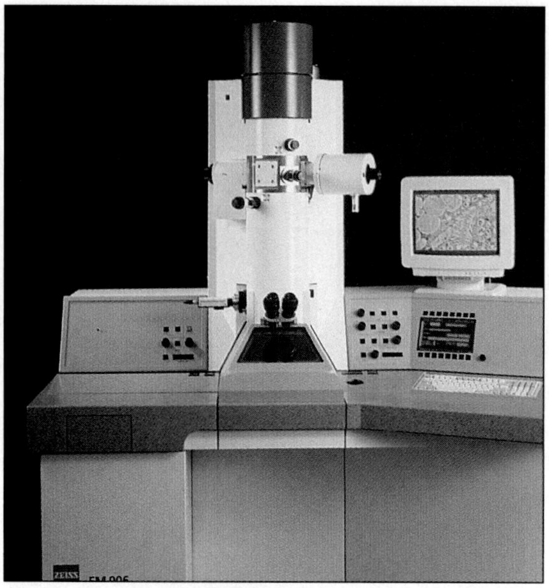

Figure 3.3A A transmission electron microscope. (*Courtesy LEO Electronenmikroskopie GmbH, Germany*)

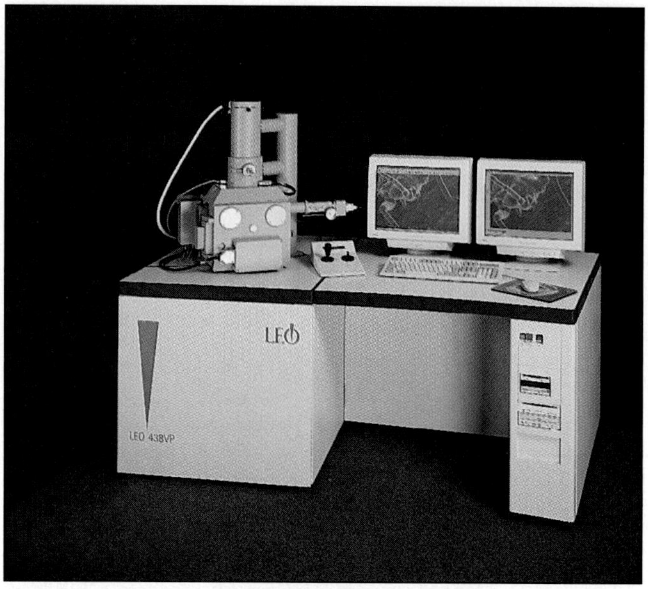

Figure 3.3B A scanning electron microscope. (*Courtesy LEO Electron Microscopy Ltd, England*)

EUKARYOTIC VERSUS PROKARYOTIC CELLS

Nearly all higher plant and animal cells share most of the various features discussed in this chapter. Some of these features (e.g., nuclei, plastids) are, however, lacking in the cells of some very simple organisms such as bacteria. Cells without nuclei and some other features are called **prokaryotic** (*pro* = before; *karyon* = nucleus) to distinguish them from typical **eukaryotic** (*eu* = well or good; *karyon* = nucleus) cells discussed here. Prokaryotic cells are covered in more detail in Chapter 17. **Organelles** (membrane-bound bodies found within eukaryotic cells), **cell walls,** and other cellular components are discussed in the sections that follow.

CELL STRUCTURE AND COMMUNICATION

Plant cells typically have a *cell wall* surrounding the **protoplasm,** which consists of all the living components of a cell. These living components are bounded by a membrane called the *plasma membrane.* All cellular components between the plasma membrane and a relatively large body called the *nucleus* are known as **cytoplasm.** Within the cytoplasm is a souplike fluid called *cytosol,* in which various bodies called *organelles* are dispersed. Organelles are persistent structures of various shapes and sizes with specialized functions in the cell; most, but not all, are bounded by membranes (Figs. 3.4 and 3.5).

Cell Size

Most plant and animal cells are so tiny they are invisible to the unaided eye. Cells of higher plants generally vary in length between 10 and 100 micrometers.[1] Since there are roughly 25,000 micrometers to the inch, it would take about 500 average-sized cells to extend across 2.54 centimeters (1 inch) of space; 30 of them could easily be placed across the head of a pin. Some prokaryotic (bacterial) cells are less than one-half micrometer wide, while cells of the green alga, mermaid's wineglass (*Acetabularia*), are mostly between 2 and 5 centimeters long, and fiber cells of some nettles are about 20 centimeters long.

Why are cells so small, and why aren't they larger? Consider that as a cell increases in size, its volume grows much more than its surface area. The increase in surface area of a spherical cell, for example, is equal to the square of its increase in diameter, but its increase in volume is equal to the cube of its increase in diameter. This means that a cell whose diameter increases 10 times would increase in surface area 100 times (10 squared) but in volume 1,000 times (10 cubed). Since all substances enter or leave cells through their surfaces, which are the only contact areas with their surroundings, larger cells are at a disadvantage. Furthermore, the nucleus regulates all aspects of a cell's activities, and the greater the volume of the cell, the longer it can take for instructions from the nucleus to reach the surface. On the other hand, smaller cells have a clear advantage because they have relatively larger surface area to volume ratios, thereby enabling faster and more efficient communication between the nucleus and other parts of the cell.

1. See Appendix 6 for metric conversion tables.

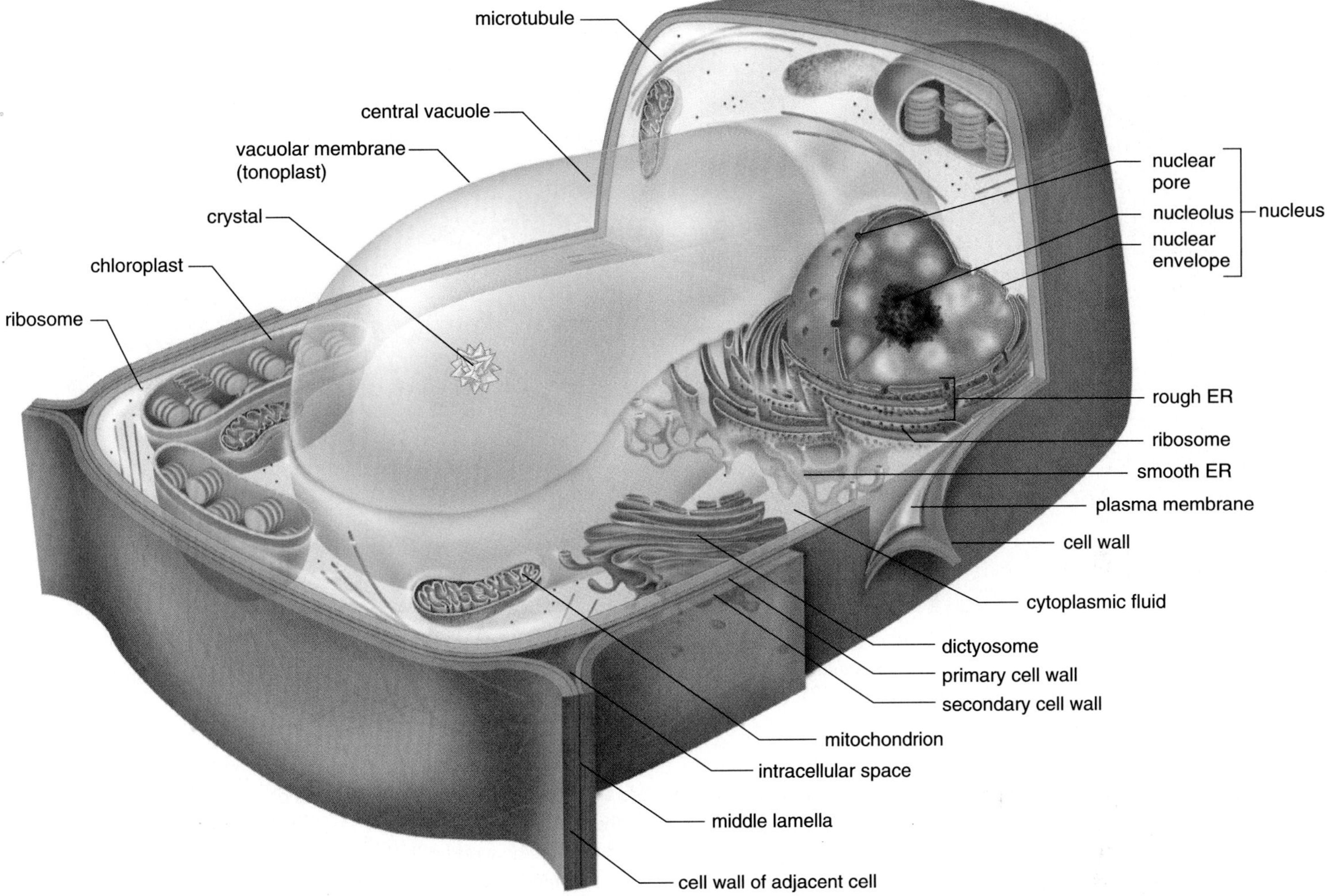

A.

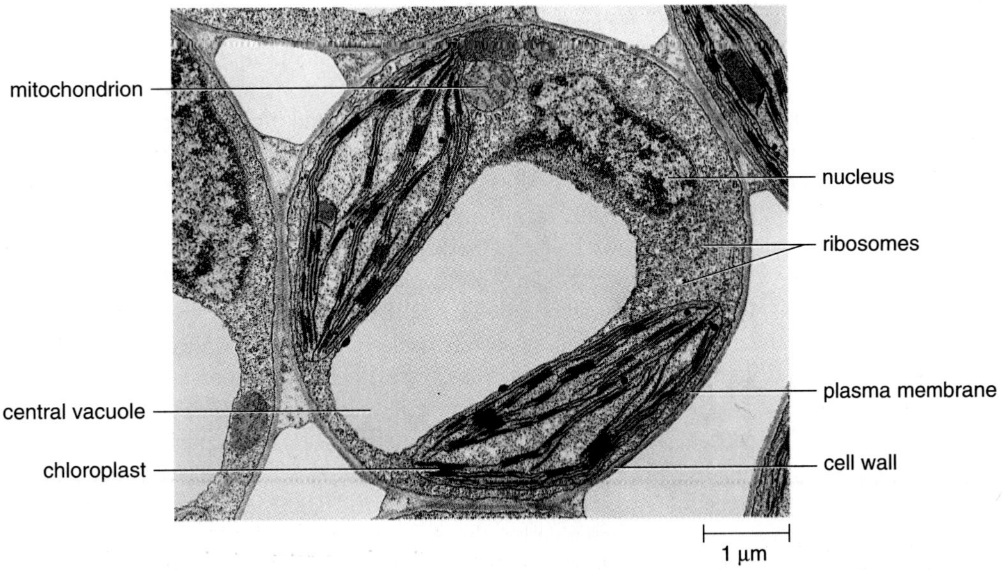

B.

Figure 3.4 Anatomy of a young leaf cell. *A.* Generalized drawing. *B.* Transmission electron micrograph of a small leaf cell with cross sections of two chloroplasts visible, ×20,000. (*A. From Sylvia S. Mader, Biology, 7th edition. © 2001 The McGraw-Hill Companies. All rights reserved. B. © Newcomb/Wergin/BPS.*)

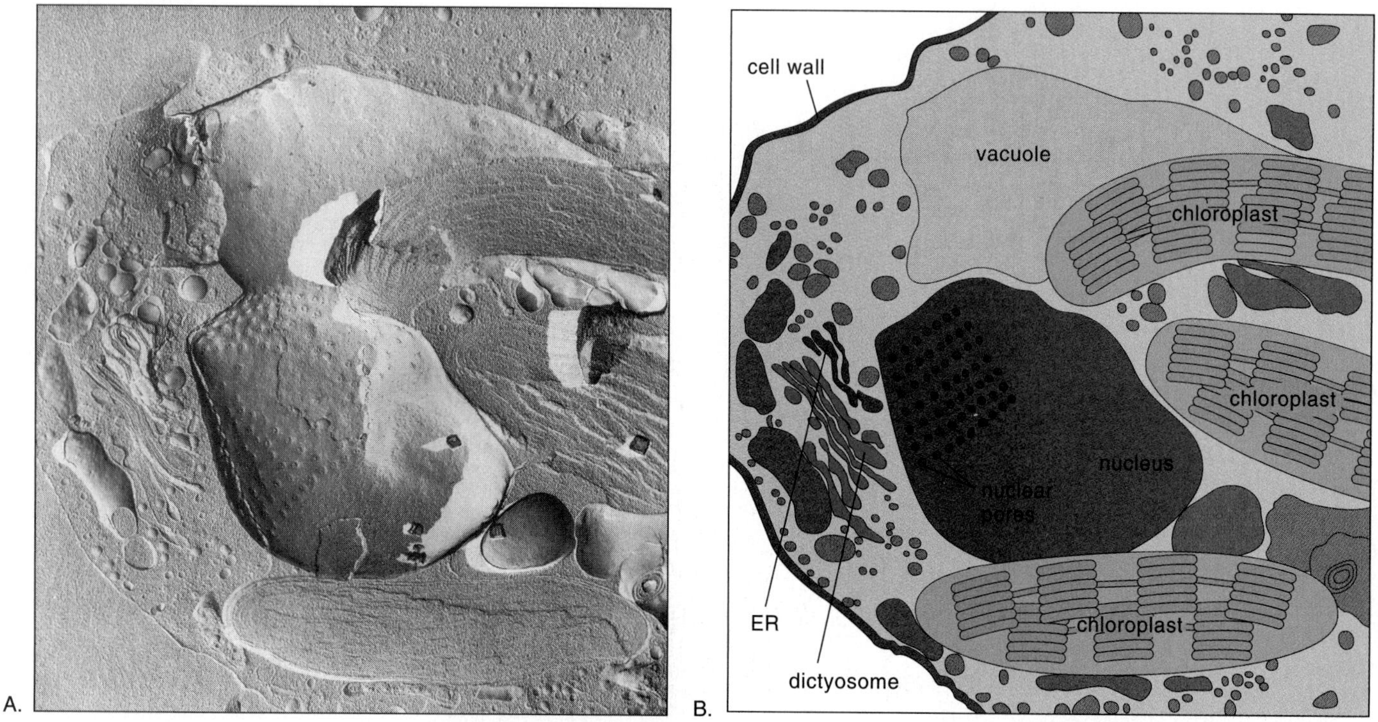

Figure 3.5 Anatomy of a plant cell. *A.* Scanning electron micrograph, ×20,200. *B.* Diagram showing interpretation of structures in the micrograph. (*Electron micrograph © Gary T. Cole/BPS*) (*From Moore, Clark, and Vodopich, Botany, 2nd edition. © 1998 The McGraw-Hill Companies. All rights reserved.*)

Full-grown organisms have astronomical numbers of cells. For example, it has been calculated that a single mature leaf of a pear tree contains 50 million cells and that the total number of cells in the roots, stem, branches, leaves, and fruit of a full-grown pear tree exceeds 15 trillion. Can you imagine how many cells there are in a 3,000-year-old redwood tree of California, which may reach heights of 90 meters (300 feet) and measure up to 4.5 meters (15 feet) in diameter near the base?

Some cells are boxlike with six walls, but others assume a wide variety of shapes, depending on their location and function. The most abundant cells in the younger parts of plants and fruits may be more or less spherical, like bubbles, when they are first formed, but as they press against each other, they commonly end up with an average of 14 sides by the time they are mature. These cell types are discussed in Chapter 4.

The Cell Wall

A novelty song of more than 50 years ago listed food items the writer said he disliked. Each verse ended with the line, "But I like bananas because they have no bones!" Indeed, bananas and all plants differ from larger animals in having no bones or similar internal skeletal structures. Yet large trees support branches and leaves weighing many tons. They can do this because most plant cells have either rigid walls that provide the support afforded to animals by bones or semirigid walls that provide flexibility. At the same time, the walls protect delicate cell contents within. When millions of these cells function together as a tissue, their collective strength is enormous. The redwoods and Tasmanian *Eucalyptus* trees, which are the largest trees alive today, exceed the mass and volume of the largest land animals, the elephants, by more than a hundred times. The wood of one giant redwood tree could support the combined weight of a thousand elephants.

The first cell structure discovered by Robert Hooke in 1665 was the *cell wall,* and among plant cell structures observed with a microscope, the cell wall is the most obvious because it defines the shape of the cell. Many of the prepared specimens observed with a microscope in plant biology are merely stained remnants of once-living cells. But the vast diversity of cell walls within and among species tells a story about the structure and function of each cell. For instance, *epidermal* cells, which form a thin layer on the surfaces of all plant organs, often have unusual shapes and sizes. Some such cells form hairs that may secrete substances that discourage animals from grazing on the plants producing them. Thin-walled cells found beneath the epidermis of leaves are specialized for their function of photosynthesis (discussed in Chapter 10); and thick-walled cells of wood help to transport water without collapsing. The structure and function of different plant cells, and the tissues they form, are addressed in Chapter 4.

The main structural component of cell walls is *cellulose,* which is composed of 100 to 15,000 glucose monomers in long chains, and is the most abundant polymer on earth. As a primary food source for grazing animals and at least indirectly for nearly all other living organisms, it could be said that most life on earth relies directly or indirectly on the cell wall. Humans also depend on cell walls because they provide clothing, shelter, furniture, paper, and fuel.

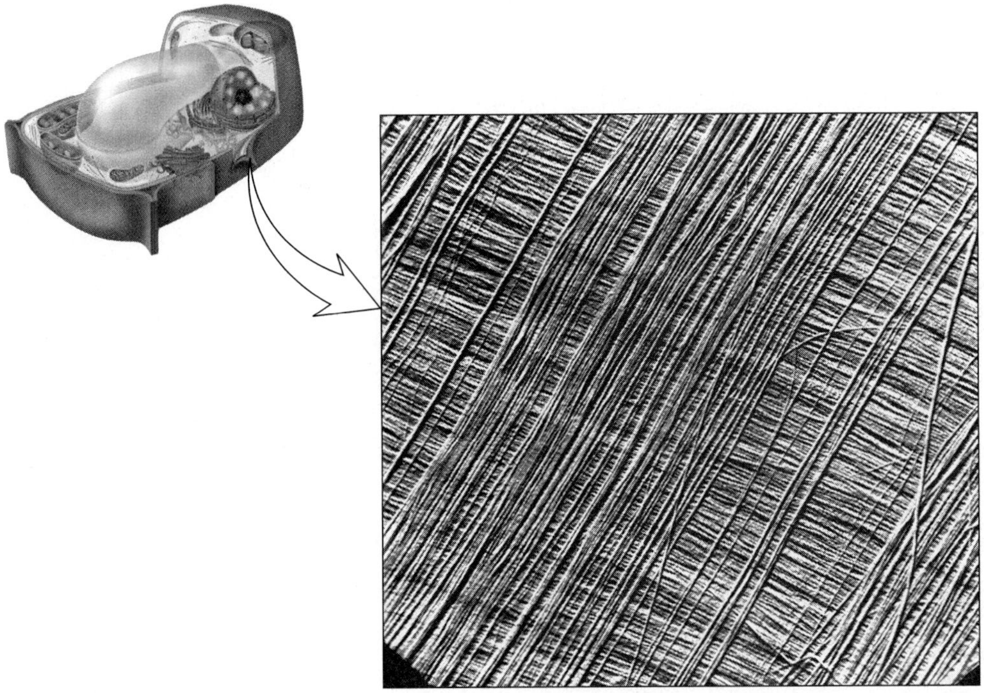

Figure 3.6A A small portion of a cell wall of the green alga Chaetomorpha melagonium, showing how cellulose microfibrils are laid down. Each microfibril is composed of numerous molecules of cellulose, ×24,000. (*Electron micrograph courtesy Eva Frei and R. D. Preston*)

In addition to the cellulose, cell walls typically contain a matrix of *hemicellulose* (a gluelike substance that holds cellulose fibrils together), *pectin* (the organic material that gives stiffness to fruit jellies), and *glycoproteins* (proteins that have sugars associated with their molecules).

A **middle lamella,** which consists of a layer of pectin, is first produced when new cell walls are formed. This middle lamella is normally shared by two adjacent cells and is so thin that it may not be visible with an ordinary light microscope unless it is specially stained. A flexible primary wall, consisting of a fine network of cellulose, hemicellulose, pectin, and glycoproteins, is laid down on either side of the middle lamella (Fig. 3.6A). Reorganization, synthesis of new molecules, and insertion of new wall polymers lead to rearrangement of the cell wall during growth. *Secondary walls,* which are produced inside the primary walls, are derived from primary walls by thickening and inclusion of *lignin,* a complex polymer.

Secondary cell walls of plants generally contain more cellulose (40% to 80%) than primary walls. As the cell ages, wall thickness can vary, occupying as little as 5% to more than 95% of the volume of the cells. During secondary wall formation, cellulose microfibrils become embedded in lignin, much like steel rods are embedded in concrete to form prestressed concrete (Fig. 3.6B).

Communication Between Cells

Cells that store, manufacture, or process food have thin walls, while those involved in support usually have relatively thick walls. Although each living cell is capable of independently carrying on complex activities, it is essential that these activities be coordinated through some means of communication among all the living cells of an organism. Fluids and dissolved substances can pass through primary walls of adjacent cells via **plasmodesmata** (singular: **plasmodesma**), which are tiny strands of cytoplasm that extend between the cells through minute openings (see Fig. 3.20). The translocation of sugars, amino acids, ions, and other substances occurs through the plasmodesmata. The middle lamellae and most cell walls are, however, permeable and permit slower movement of water and dissolved substances between cells.

CELLULAR COMPONENTS

Most chemical reactions that take place in cells occur in the protoplasm, as part of a dynamic series of events that make the plant a living entity. Each organelle within the protoplast has a primary function, and the flow of *metabolites* (products of chemical synthesis or breakdown) from one organelle to another is necessary for a balance of events that take place.

Envision a journey through the plant cell as an exciting voyage in which information is stored primarily in the nucleus, processed in the cytoplasm, and sent on to different parts of the cell. This information can bring about the synthesis of proteins in the cytoplasm where they become involved in *metabolic reactions* (see Chapter 10), or they may be destined for use in other cellular locations. The

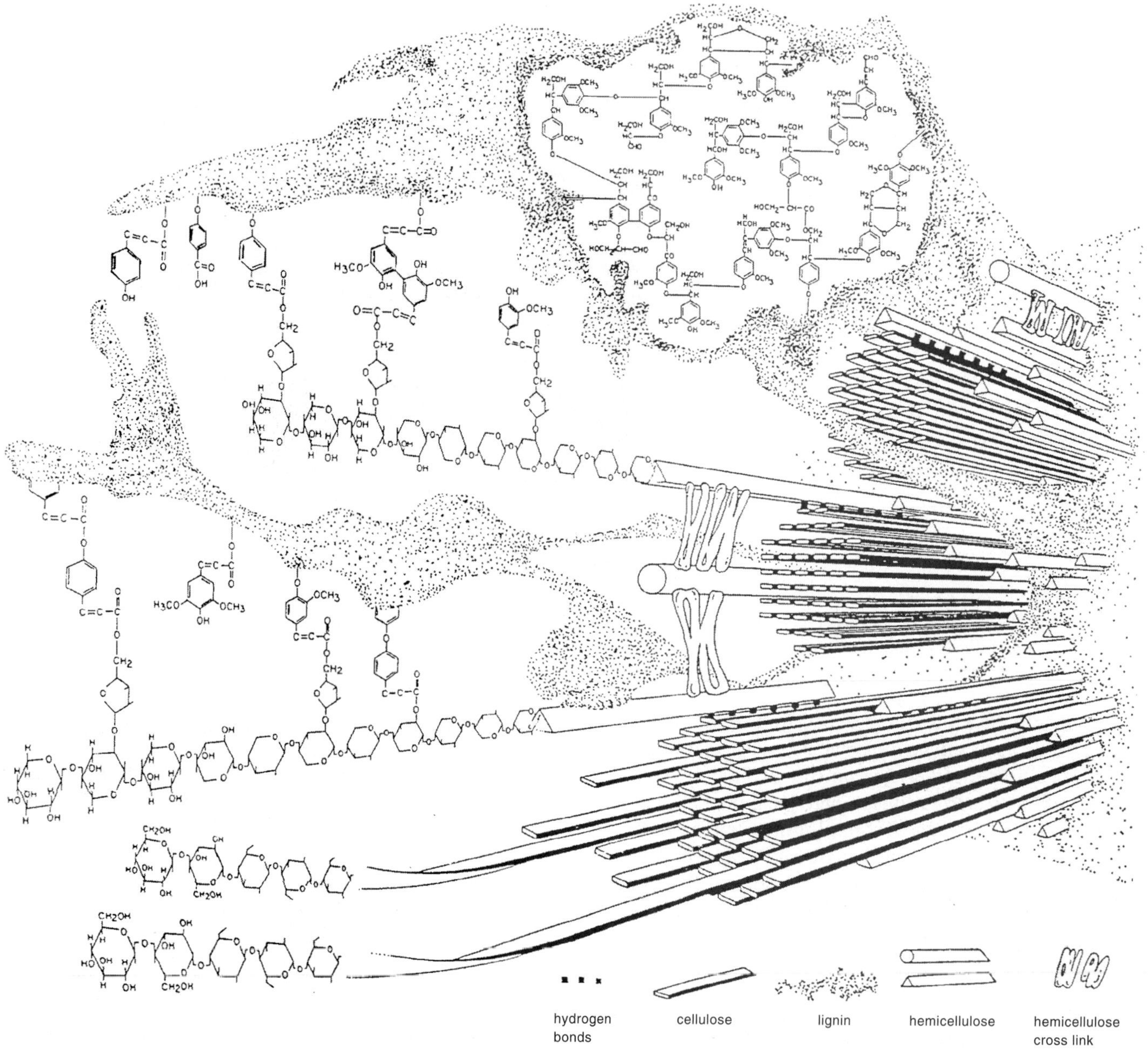

Figure 3.6B Secondary cell wall structure. Components are arranged so that the cellulose microfibrils and hemicellulose chains are embedded in lignin. (*Reproduced by permission of the Oklahoma Academy of Science; Figure provided by Dr. James E. Bidlack.*)

packaged proteins may be incorporated in membranes or organelles, and other compounds may be manufactured in specific organelles or enter from an adjacent cell.

The Plasma Membrane

The outer boundary of the living part of the cell, the **plasma membrane,** is roughly eight-millionths of a millimeter thick. To get an idea of how incredibly thin that is, consider that it would take 12,500 such membranes neatly stacked in a pile to achieve the thickness of an ordinary piece of writing paper. Yet this delicate, semipermeable structure is of vital importance in regulating the movement of substances into and out of the cell. While the plasma membrane may inhibit movement of some substances, it can otherwise allow free movement and can even control movement of other substances into and out of the cell. As a result, the proportions and makeup of chemicals within a cell become quite different from those outside the cell. The plasma membrane is also involved in the production and assembly of cellulose for cell walls.

Evidence obtained since the early 1970s indicates that the plasma membrane and other cell membranes are composed of *phospholipids* arranged in two layers, with proteins interspersed throughout (Fig. 3.7). Covalent bonds link carbohydrates to both lipids and proteins on the outer surfaces

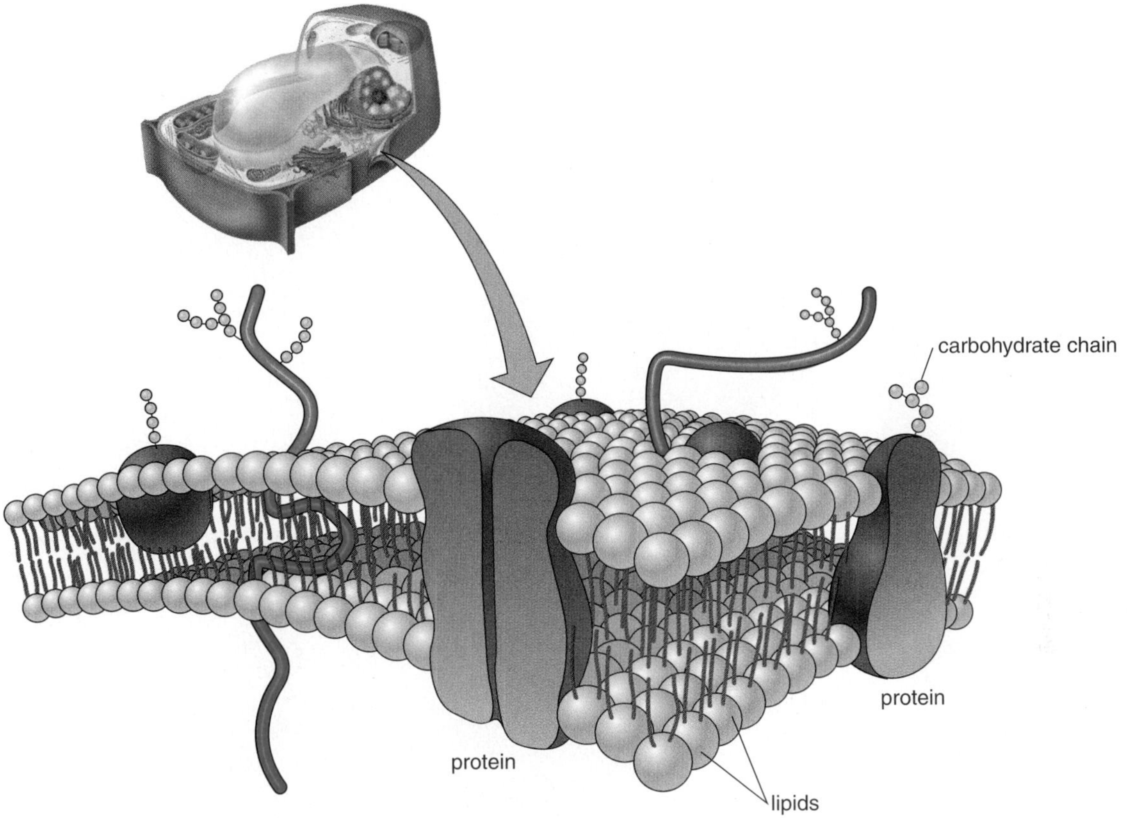

Figure 3.7 A model of a small portion of a plasma membrane, showing its fluid-mosaic nature. The proteins, which are coiled chains of polypeptides, are either embedded or on the surfaces. Some of the embedded proteins extend all the way through and may serve as conduits for diffusion of certain ions. In cells, and other places where there are watery fluids, a double layer of *phospholipids* forms. The heads point outward toward the water. The tails, which are long-chain fatty acids, are hydrophobic (i.e., they "dislike" water) and point inward away from the water. The membrane is about 8 nanometers thick. *(Inset from Sylvia S. Mader, Biology, 7th edition. © 2001 The McGraw-Hill Companies. All rights reserved.)*

of membranes. Some proteins extend across the entire width of the membrane, while others are embedded or apparently are loosely bound to the outer surface.

The remainder of cell contents usually push the plasma membrane up against the cell wall because of pressures developed by *osmosis* (see Chapter 9), but the membrane is quite flexible and often forms folds, which may, in turn, become little hollow spheres or vesicles that float off into the cell. In fact, experiments have shown that by adding detergents to a continuous membrane, it can be broken up and dispersed, yet it can partially reform when the detergents are removed. The membrane may even shrink away from the wall temporarily, but if it ever ruptures, the cell soon dies.

The Nucleus

The **nucleus** is the control center of the cell. In some ways, it functions like a combination of a computer program and a dispatcher that sends coded messages or "blueprints" originating from DNA in the nucleus with information that will ultimately be used in other parts of the cell. In other words, the DNA in the nucleus provides the original information needed to fulfill the cell's needs. This nuclear information contributes toward growth, differentiation, and the myriad activities of the complex cell "factory." The nucleus also stores hereditary information, which is passed from cell to cell as new cells are formed.

The nucleus often is the most conspicuous object in a living cell, although in green cells, chloroplasts may obscure it. In living cells without chloroplasts, the nucleus may appear as a grayish, spherical to ellipsoidal lump, sometimes lying against the plasma membrane to one side of the cell or toward a corner. Some nuclei are irregular in form, and they can vary greatly in size. They are, however, generally from 2 to 15 micrometers or larger in diameter. Certain fungi and algae have numerous nuclei within a single extensively branched cell, but the cells of more complex plants usually have a single nucleus.

Each nucleus is bounded by two membranes, which together constitute the **nuclear envelope.** Structurally complex pores, about 50 to 75 nanometers apart, occupy up to one-third of the total surface area of the nuclear envelope (Fig. 3.8). Proteins that act as channels for molecules are embedded within the pores. The pores apparently permit only certain kinds of molecules (for example, proteins being carried into the nucleus and RNA being carried out) to pass between the nucleus and the cytoplasm.

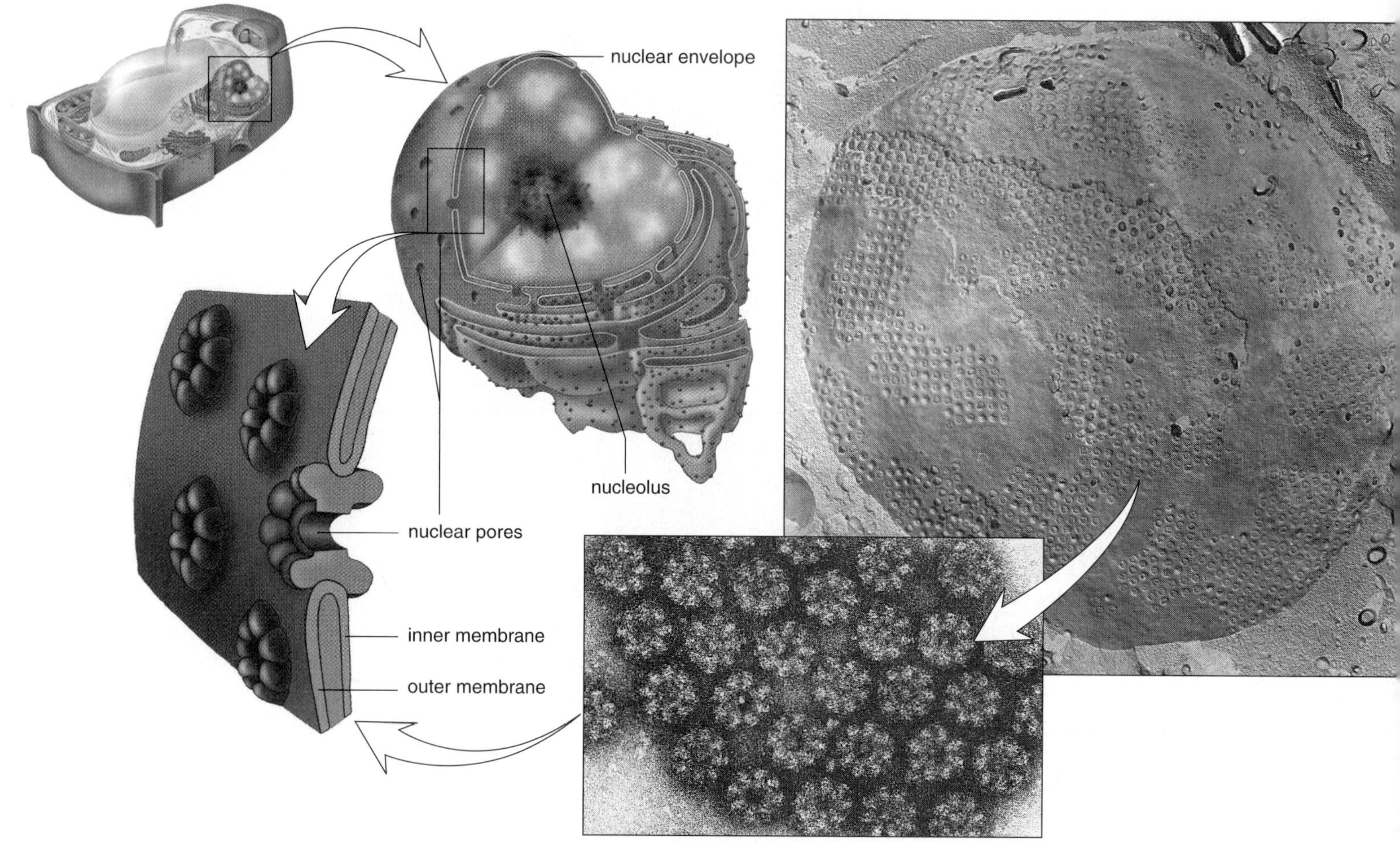

Figure 3.8 Drawing of the nucleus showing the inner and outer membranes, nuclear pores, and nucleolus. The electron micrographs show detail of the nuclear pores, which are about 60 nanometers in diameter. *Larger photo courtesy E. G. Pollock; smaller photo courtesy Ron Milligan; Inset from Sylvia S. Mader, Biology, 7th edition. © 2001 The McGraw-Hill Companies. All rights reserved.*

The nucleus contains a granular-appearing fluid called *nucleoplasm,* which is packed with short fibers that are about 10 nanometers in diameter; several different larger bodies are suspended within it. Of the larger nuclear bodies, the most noticeable are one or more **nucleoli** (singular: **nucleolus**), which are composed primarily of RNA and are not bounded by membranes. Additional discussion of the activities of nucleoli is found in Chapter 13.

Other important nuclear structures, which are not apparent with light microscopy unless the cell is stained or is in the process of dividing, include thin strands of **chromatin.** When a nucleus divides, the chromatin strands coil, becoming shorter and thicker, and in their condensed condition, they are called **chromosomes.** Chromatin is composed of protein and DNA (discussed in Chapters 2 and 13). Each cell of a given plant or animal species has its own fixed number and composition of chromosomes; the cells involved in sexual reproduction have half the number found in other cells of the same organism. The number of chromosomes present in a nucleus normally bears no relation to the size and complexity of the organism. Each body cell of a radish, for example, has 18 chromosomes in its nucleus, while a cell of one species of goldenweed has 4, and a cell of a tropical adder's tongue fern has over 1,000. Humans have 46 chromosomes in each body cell.

The Endoplasmic Reticulum

The outer membrane of the nucleus is connected and continuous with the **endoplasmic reticulum.** The endoplasmic reticulum facilitates cellular communication and channeling of materials. Many important activities, such as the synthesis of membranes for other organelles and modification of proteins from components assembled from elsewhere within the cell, occur either on the surface of the endoplasmic reticulum or within its compartments.

The endoplasmic reticulum (often referred to simply as **ER**) is an enclosed space consisting of a network of flattened sacs and tubes that form channels throughout the cytoplasm, the amount and form varying considerably from cell to cell. Transmission electron micrographs of sectioned ER give it the appearance of a series of parallel membranes that resemble long, narrow tubes or sacs, creating subcompartments within the cell.

Ribosomes (discussed in the section that follows) may be distributed on the outer surface (i.e., the surface in contact

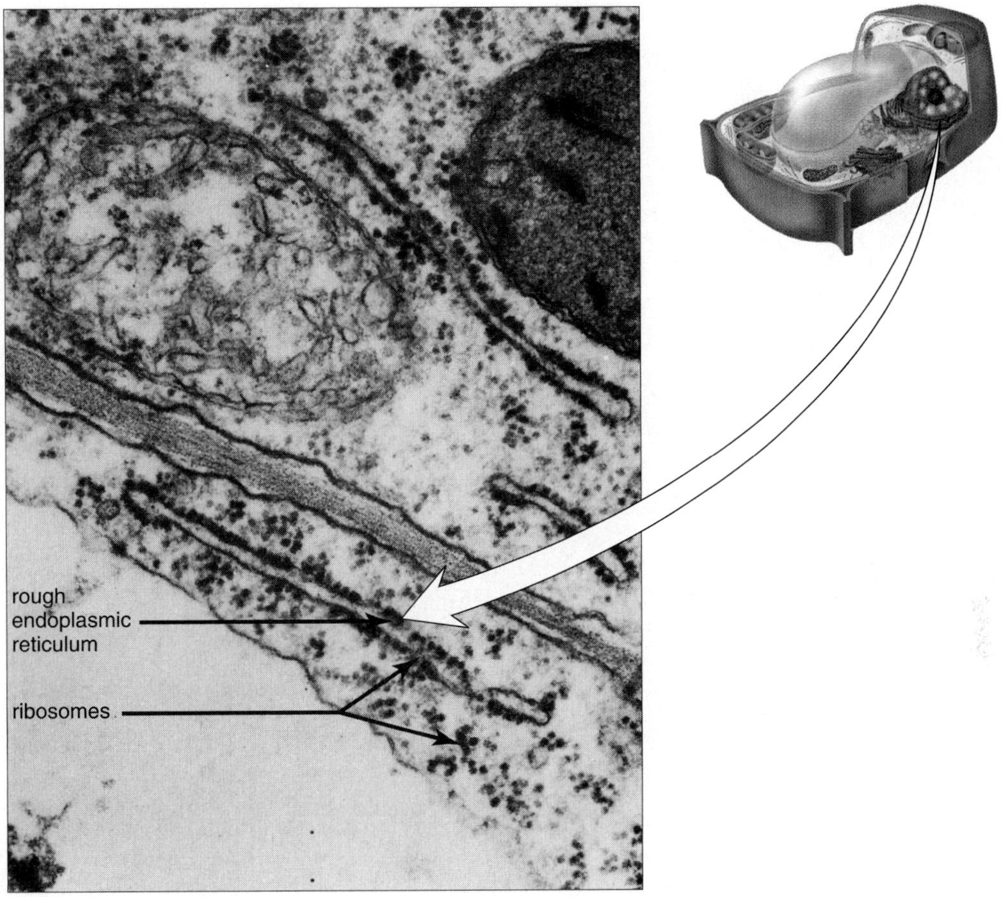

Figure 3.9 A small portion of the endoplasmic reticulum and ribosomes in a young leaf cell of corn *(Zea mays)*. ×100,000. The ribosomes are 20 nanometers in diameter. *(Electron micrograph courtesy Jean Whatley; Inset from Sylvia S. Mader, Biology, 7th edition. © 2001 The McGraw-Hill Companies. All rights reserved.)*

with the cytoplasm) of the endoplasmic reticulum. Such endoplasmic reticulum is said to be *rough* and is primarily associated with the synthesis, secretion, or storage of proteins (Fig. 3.9; see also Chapter 13). This contrasts with *smooth* endoplasmic reticulum, which has few, if any, ribosomes lining the surface, and is associated with lipid secretion. Both types of endoplasmic reticulum can occur in the same cell and can be interconverted, depending on the demands of the cell. Many enzymes involved in the process of cellular respiration are synthesized on the surface of the endoplasmic reticulum. The enzymes, however, enter other organelles (primarily mitochondria, which are discussed later in this chapter) without passing through the endoplasmic reticulum. The endoplasmic reticulum also appears to be the primary site of membrane synthesis within the cell.

Ribosomes

Ribosomes are tiny bodies that are visible with the aid of an electron microscope. They are typically roughly ellipsoidal in shape with apparently varied and complex surfaces. Each ribosome is composed of two subunits that are composed of RNA and proteins; the subunits, upon close inspection, can be differentiated by a line or cleft toward the center. Ribosomes average only about 20 nanometers in diameter in most plant cells. Unattached ribosomes often occur in clusters of 5 to 100, particularly when they are involved in linking amino acids together in the construction of the large, complex protein molecules that are a basic part of all living organisms.

Ribosomal subunits are assembled within the nucleolus, released, and in association with special RNA molecules, they initiate protein synthesis. Once assembled, complete ribosomes may line the outside of the endoplasmic reticulum but can also occur unattached in the cytoplasm, chloroplasts, or other organelles. About 55 kinds of protein are found in each ribosome of prokaryotic cells and a slightly higher number in those of eukaryotic cells (see the discussion of various types of RNA in Chapter 13). Unlike other organelles, ribosomes have no bounding membranes, and because of this, some scientists prefer not to call them organelles.

Dictyosomes

Cells may contain from several to hundreds of groups of roundish, flattened-appearing sacs scattered throughout the cytoplasm. These sacs, which in plant cells are known as **dictyosomes** (*Golgi bodies* in animal cells), are often bounded by branching tubules that originate from the endoplasmic

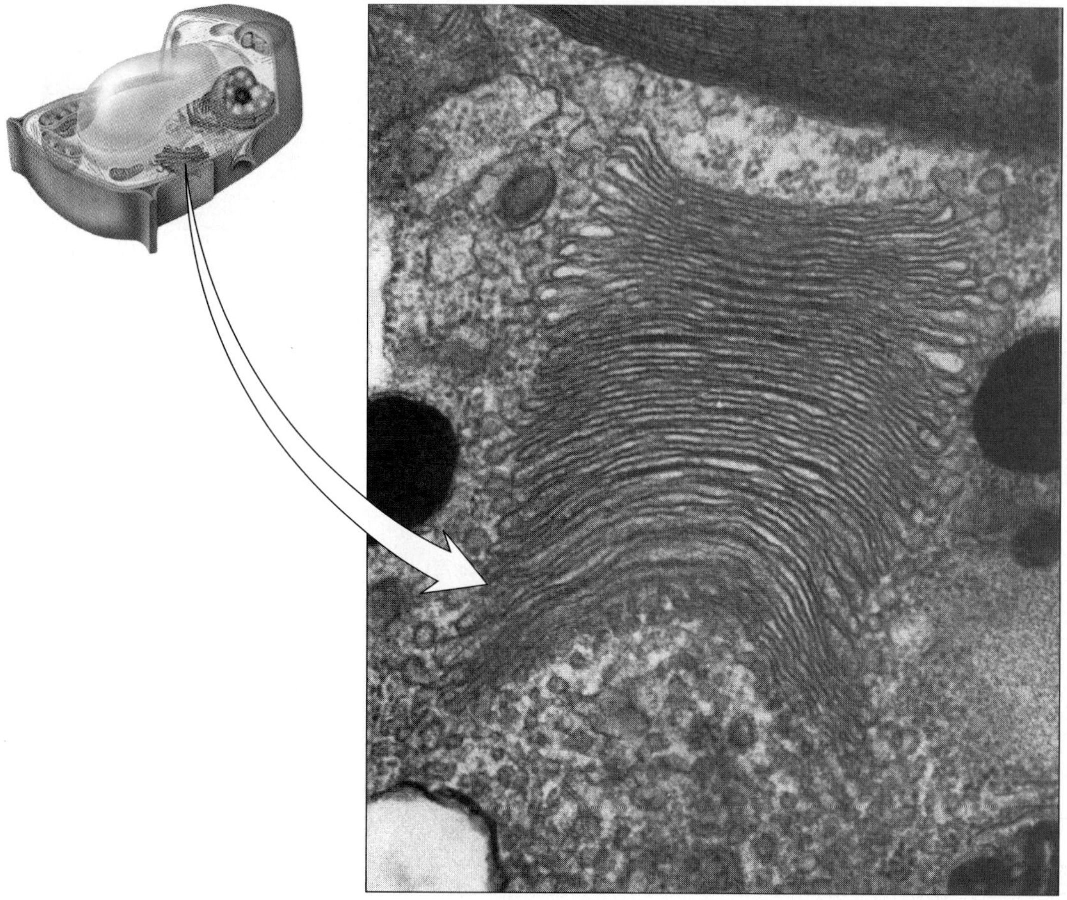

Figure 3.10 A dictyosome from the alga Euglena, ×40,000. *(Electron micrograph courtesy John Z. Kiss; Inset from Sylvia S. Mader, Biology, 7th edition. © 2001 The McGraw-Hill Companies. All rights reserved.)*

reticulum but are not directly connected to it (Fig. 3.10). Frequently, 5 to 8 dictyosomes are organized in stacks, but stacks of up to 30 or more are common in simpler organisms.

Dictyosomes are involved in the modification of carbohydrates attached to proteins that are synthesized and packaged in the endoplasmic reticulum. Complex polysaccharides are also assembled within the dictyosomes and collect in small *vesicles* (tiny blisterlike bodies) that are pinched off from the margins. These vesicles migrate to the plasma membrane, fuse with it, and secrete their contents outside of the cell. Substances secreted by vesicles may include cell-wall polysaccharides, floral nectars, and essential oils found in herbs.

The enzymes needed for the packaging of proteins are produced by the endoplasmic reticulum and further modified within the dictyosomes. One might describe dictyosomes as collecting, packaging, and delivery centers or, perhaps, as "post offices" of the cell.

Plastids

Most living plant cells have several kinds of **plastids,** with the **chloroplasts** (Fig. 3.11A) being the most conspicuous. They occur in a variety of shapes and sizes, such as the beautiful corkscrew-like ribbons found in cells of the green alga *Spirogyra* (see Fig. 18.6) and the bracelet-shaped chloroplasts of other green algae, such as *Ulothrix* (see Figs. 18.2D and 18.5). The chloroplasts of higher plants, however, tend to be shaped somewhat like two Frisbees glued together along their edges, and when they are sliced in median section, they resemble the outline of a rugby football.

Although several algae and a few other plants have only one or two chloroplasts per cell, the number of chloroplasts is usually much greater in a green cell of higher plants. Seventy-five to 125 is quite common, with the green cells of a few plants having up to several hundred. The chloroplasts may be from 2 to 10 micrometers in diameter, and each is bounded by an envelope consisting of two delicate membranes. The outer membrane apparently is derived from endoplasmic reticulum, while the inner membrane is believed to have originated from the cell membrane of a cyanobacterium (discussed in Chapter 17). Within is a colorless fluid matrix, the **stroma,** which contains enzymes. Genes in the nucleus dictate most of the activities of chloroplasts, but each chloroplast contains a small circular molecule of DNA that encodes instructions for production of proteins related to photosynthesis and other activities in the chloroplast and cell.

Grana (singular: **granum**), which are formed from membranes, have the appearance of independent stacks of coins with double membranes within each chloroplast.

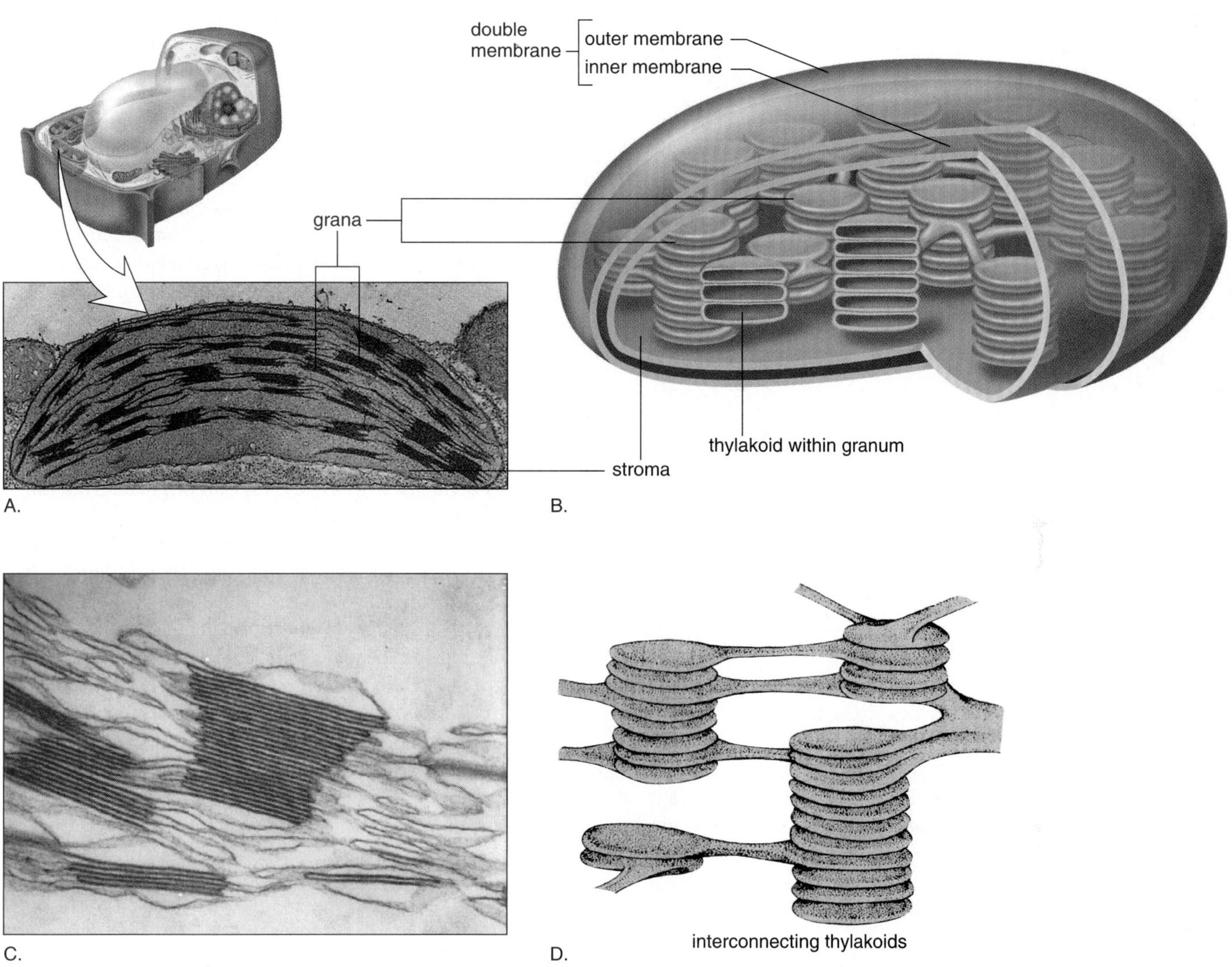

Figure 3.11 Drawings of leaf mesophyll cell chloroplasts and transmission electron micrographs showing variation in chloroplast structure. *A.* A chloroplast ×20,000. *B.* Cutaway of a chloroplast. *C.* Grana ×40,000. *D.* A few thylakoids. (*A. Courtesy Herbert W. Israel; C. Courtesy Blake Rowe; B., D., and Inset from Sylvia S. Mader, Biology, 7th edition. © 2001 The McGraw-Hill Companies. All rights reserved.*)

There are usually about 40 to 60 grana linked together by arms in each chloroplast, and each granum may contain from 2 or 3 to more than 100 stacked **thylakoids.** In reality, thylakoids are part of an overlapping and continuous membrane system suspended in the stroma (Fig. 3.11B). The thylakoid membranes contain green **chlorophyll** and other pigments. These "coin stacks" of grana are vital to life as we know it, for it is within the thylakoids that the first steps of the important process of *photosynthesis* (see Chapter 10) occur. In photosynthesis, green plants convert water and carbon dioxide (from the air) to simple food substances, harnessing energy from the sun in the process. The existence of human and all other animal life depends on the activities of the chloroplasts.

There are usually four or five *starch grains* in the stroma, as well as oil droplets and enzymes. Some plastids (e.g., those of tobacco) store proteins. The stroma contains ribosomes in addition to the DNA molecule.

Chromoplasts are another type of plastid found in some cells of more complex plants. Although chromoplasts are similar to chloroplasts in size, they vary considerably in shape, often being somewhat angular. They sometimes develop from chloroplasts through internal changes that include the disappearance of chlorophyll. Chromoplasts are yellow, orange, or red in color due to the presence of carotenoid pigments, which they synthesize and accumulate. They are most abundant in the yellow, orange, or some red parts of plants, such as ripe tomatoes, carrots, or red peppers (Fig. 3.12). These carotenoid pigments, which are lipid soluble, are not, however, the predominant pigments in most red flower petals. The anthocyanin pigments of most red flower petals are water soluble.

Leucoplasts are yet another type of plastid common to cells of higher plants. They are essentially colorless and include *amyloplasts,* which synthesize starches, and *elaioplasts,* which synthesize oils. If exposed to light, some leucoplasts will develop into chloroplasts, and vice versa.

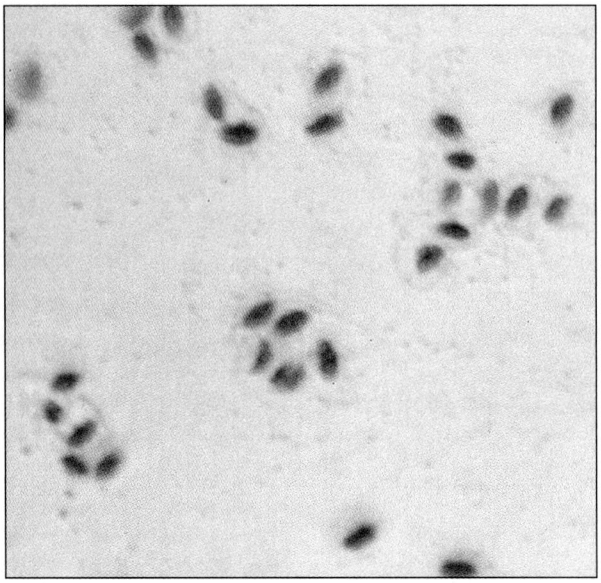

Figure 3.12 Chromoplasts in the flesh of a red pepper, ×100.

Plastids of all types develop from **proplastids,** which are small, pale green or colorless organelles having roughly the size and form of mitochondria (discussed in the next section). They are simpler in internal structure than plastids and have fewer thylakoids, the thylakoids not being arranged in grana stacks. Proplastids frequently divide and become distributed throughout the cell. After a cell itself divides, each daughter cell has a proportionate share. Plastids also arise through the division of existing mature plastids.

Mitochondria

Mitochondria (singular: **mitochondrion**) are often referred to as the powerhouses of the cell, for it is within them that energy is released from organic molecules by the process of *cellular respiration* (the role of mitochondria in respiration is further discussed in Chapter 10). This energy is needed to keep the individual cells and the plant functioning as a whole. Carbon skeletons and fatty acid chains are also rearranged within mitochondria, allowing for the building of a wide variety of organic molecules. Mitochondria are numerous and tiny, typically measuring from 1 to 3 or more micrometers in length and having a width of roughly one-half micrometer; they are barely visible with light microscopes. They appear to be in constant motion in living cells and tend to accumulate in groups where energy is needed. They often divide in two; in fact, they all originate from the division of existing mitochondria.

Mitochondria typically are shaped like cucumbers, paddles, rods, or balls. A sectioned mitochondrion resembles a scooped-out watermelon with inward extensions of the rind forming mostly incomplete partitions perpendicular to the surface (Fig. 3.13). The appearance of incomplete partitions results from the fact that each mitochondrion is bounded by two membranes, with the inner membrane forming numerous platelike folds called *cristae.* The cristae greatly increase the surface area available to the enzymes contained in a matrix fluid. The number of cristae, as well as the number of mitochondria themselves, can change over time, depending on the activities taking place within the cell. The matrix fluid also contains DNA, RNA, ribosomes, proteins, and dissolved substances.

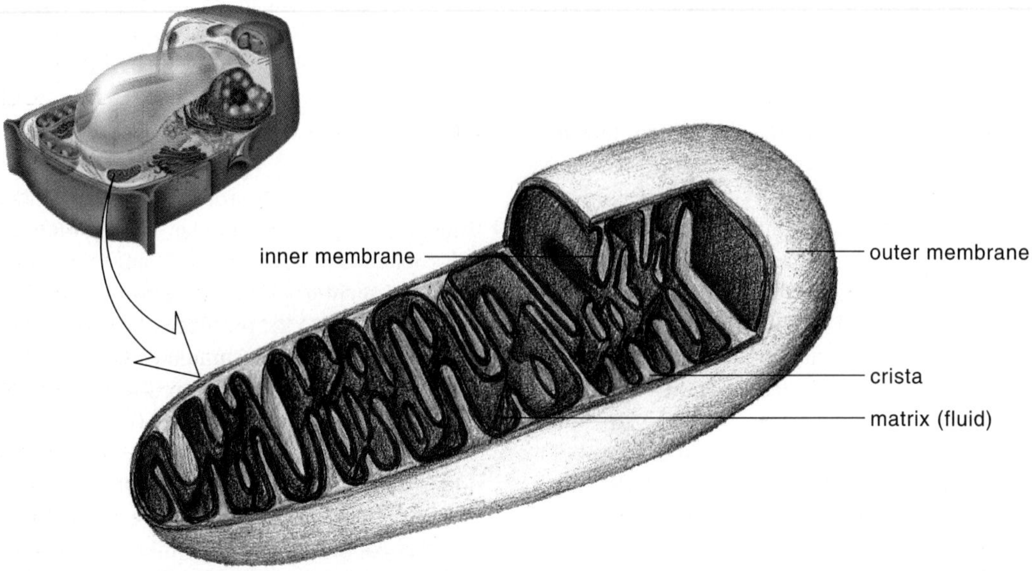

Figure 3.13 A mitochondrion greatly enlarged and cut away to show the cristae (folds of the inner membrane). A mitochondrion is about 2 micrometers long. *(Inset from Sylvia S. Mader, Biology, 7th edition. © 2001 The McGraw-Hill Companies. All rights reserved.)*

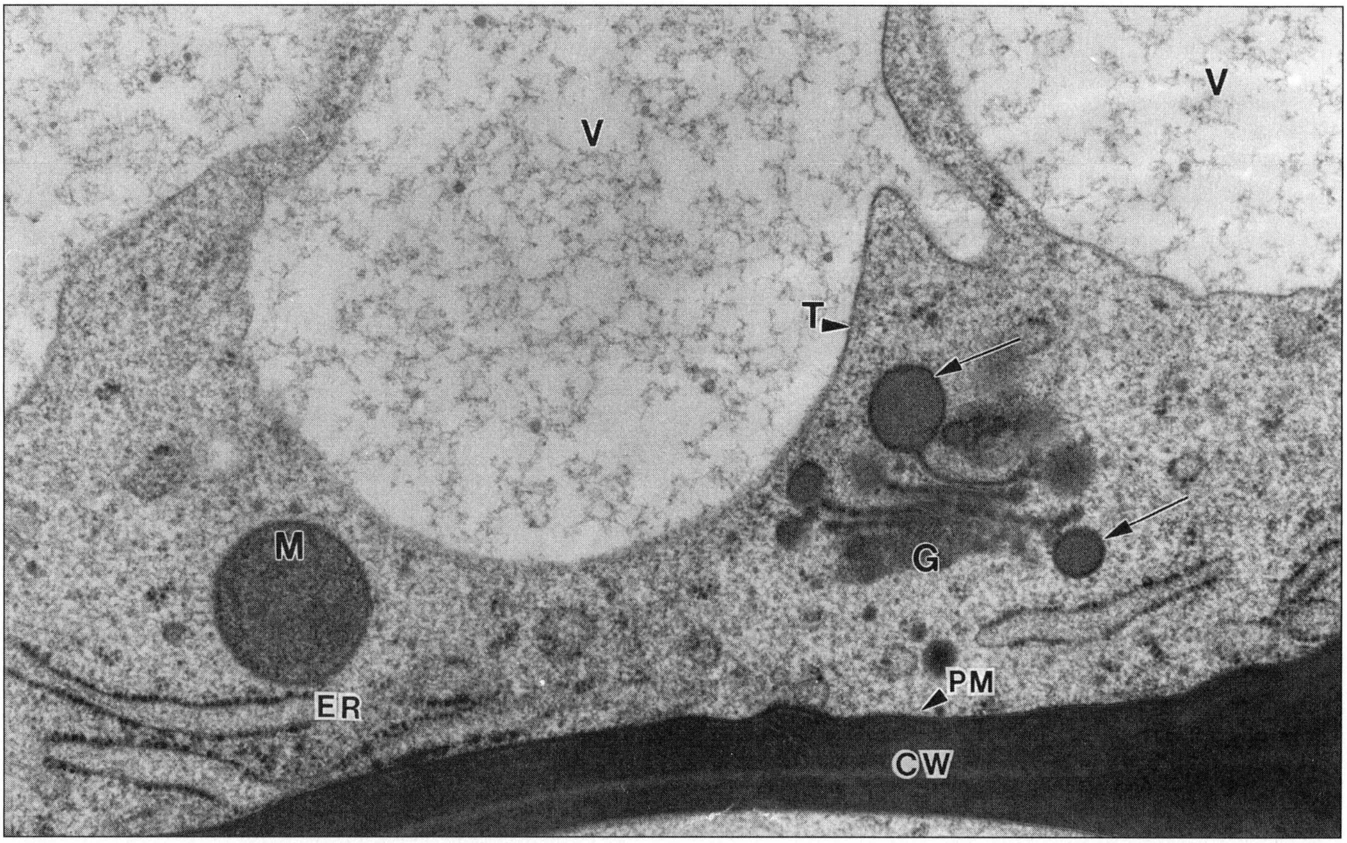

Figure 3.14 A small portion of a root cap cell of tobacco, ×100,000. V = vacuole; T = vacuolar membrane (tonoplast); G = dictyosome with vesicles (arrows); M = mitochondrion; ER = endoplasmic reticulum; PM = plasma membrane; CW = cell wall. *(Electron micrograph courtesy John Z. Kiss)*

Microbodies

Various small bodies distributed throughout the cytoplasm tend to give it a granular appearance. Examples of such components include types of small, spherical organelles called **microbodies,** which contain specialized enzymes and are bounded by a single membrane. **Peroxisomes,** for instance, contain enzymes needed by some plants to survive during hot conditions in a process called *photorespiration* (discussed in Chapter 10), whereas **glyoxisomes** contain enzymes that aid in the conversion of fats to carbohydrates during, for example, the germination of seeds containing fats. If present, peroxisomes are generally found associated with chloroplasts, and glyoxisomes usually are located near mitochondria. During a plant's life cycle, peroxisomes and glyoxisomes may increase in number at stages when the need for them is greatest.

At one time, lipid, fat, or wax droplets commonly found in cytoplasm were believed to be bounded by a membrane; recent evidence, however, suggests no membrane is present, and some, therefore, do not consider them true organelles. Another organelle, called a *lysosome,* stores enzymes that digest proteins and certain other large molecules but is apparently confined to animal cells. The digestive activities of lysosomes are similar to those of the vacuoles of plant cells (discussed next).

Vacuoles

In a mature living plant cell, as much as 90% or more of the volume may be taken up by one or two large central **vacuoles** that are bounded by **vacuolar membranes (tonoplasts)** (Fig. 3.14). The vacuolar membranes, which constitute the inner boundaries of the living part of the cell, are similar in structure and function to plasma membranes.

The vacuole evidently received its name because of a belief that it was just an empty space; hence its name has the same Latin root as the word *vacuum* (from *vacuus*—meaning "empty"). Vacuoles, however, are filled with a watery fluid called **cell sap,** which is slightly to moderately acidic. Cell sap, which helps to maintain pressures within the cell (see the discussion of *osmosis* in Chapter 9), contains dissolved substances, such as salts, sugars, organic acids, and small quantities of soluble proteins. It also frequently contains water-soluble pigments. These pigments, called *anthocyanins,* are responsible for many of the red, blue, or purple colors of flowers and some reddish leaves. In some instances, anthocyanins accumulate to a greater extent in response to cold temperatures in the fall. They should not be confused, however, with the red and orange *carotenoid* pigments confined to the chromoplasts. Yellow

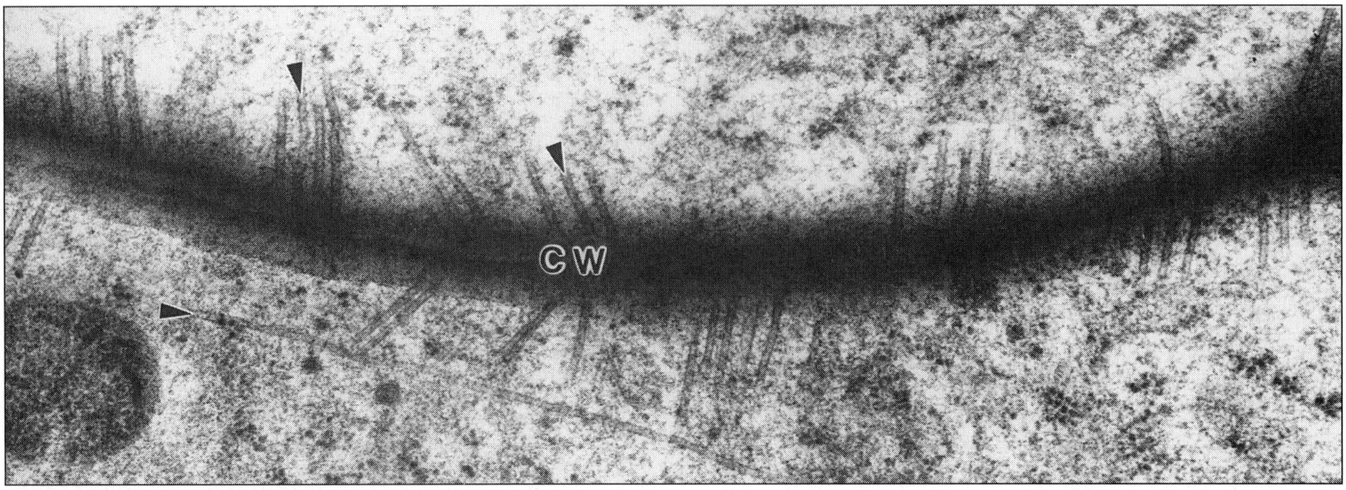

Figure 3.15 A small portion of a plant cell wall with microtubules more or less perpendicular to it, ×100,000. *(Electron micrograph courtesy John Z. Kiss)*

carotenoid pigments (*carotenes*) also play a role in fall leaf coloration (discussed in Chapter 7).

Sometimes, large crystals of waste products form within the cell sap after certain ions have become concentrated there. Vacuoles in newly formed cells are usually tiny and numerous. They increase in size and unite as the cell matures. In addition to accumulating the various substances and ions mentioned above, vacuoles are apparently also involved in the recycling of certain materials within the cell and even aid in the breakdown and digestion of organelles, such as plastids and mitochondria.

The Cytoskeleton

The **cytoskeleton** is involved in movement within a cell and in a cell's architecture. It is an intricate network constructed mainly of two kinds of fibers—*microtubules* and *microfilaments.*

Microtubules apparently control the addition of cellulose to the cell wall (Fig. 3.15). They are also involved in cell division, movement of cytoplasmic organelles, controlling the movement of vesicles containing cell-wall components assembled by dictyosomes, and movement of the tiny whiplike *flagella* and *cilia* possessed by some cells (see the section on plant movements in Chapter 11).

Microtubules are unbranched, thin, hollow, tubelike structures that resemble tiny straws. They are composed of proteins called *tubulins* and are of varying lengths, most being between 15 and 25 nanometers in diameter. They are most commonly found just inside the plasma membrane. Microtubules are also found in the special fibers that form the *spindles* and *phragmoplasts* of dividing cells discussed later in this chapter.

Microfilaments, which play a major role in the contraction and movement of cells in multicellular animals, are present in nearly all cells. They are three or four times thinner than microtubules and consist of long, fine threads of protein with an average diameter of 6 nanometers. They are often in bundles and appear to play a role in the *cytoplasmic streaming* (sometimes referred to as *cyclosis*) that occurs in all living cells. When cytoplasmic streaming is occurring, a microscope reveals the apparent movement of organelles as a current within the cytoplasm carries them around within the walls. This streaming probably facilitates exchanges of materials within the cell and plays a role in the movement of substances from cell to cell. The precise nature and origin of cytoplasmic streaming is still not known, but there is evidence that bundles of microfilaments may be responsible for it. Other evidence suggests that it may be related to the transport of cellular substances by microtubules.

CELLULAR REPRODUCTION

The Cell Cycle

When cells divide, they go through an orderly series of events known as the **cell cycle.** This cycle is usually divided into **interphase** and **mitosis,** mitosis itself being subdivided into four phases (Fig. 3.16). The length of the cell cycle varies with the kind of organism involved, the type of cell within an organism, and with temperature and other environmental factors. In most instances, however, interphase may occupy up to 90% or more of the time it takes to complete the cycle. The relatively small amount of time involved in actual division explains why most cells viewed with a microscope are in interphase, and cells in stages of mitosis can be hard to find.

Interphase

Living cells that are not dividing are said to be in *interphase,* a period during which chromosomes are not visible with light microscopes. It is such cells that have been discussed up to this point.

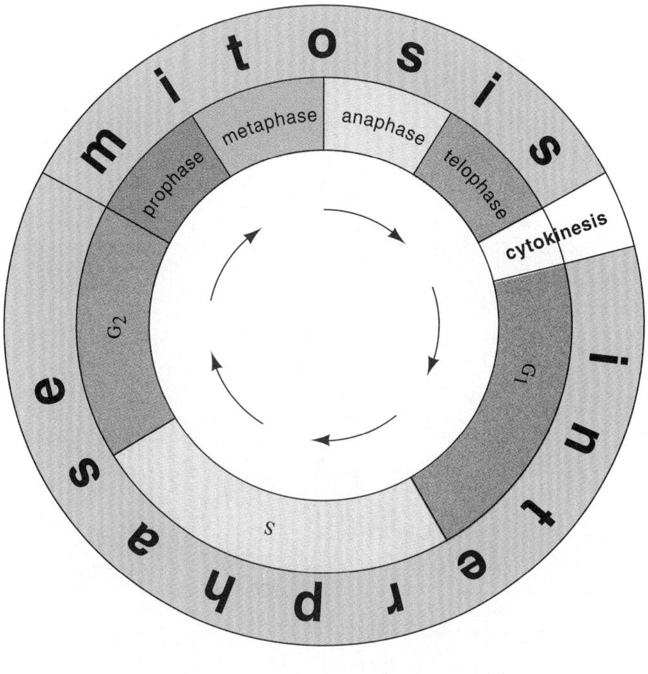

Figure 3.16 A diagram of a cell cycle showing relative amounts of time involved in interphase, mitosis, and cytokinesis.

For many years, immature cells were considered to be "resting" when they were not actually dividing, but we know now that three consecutive periods of intense activity take place during interphase. These intervals are designated as *gap* (or *growth*) *1, synthesis,* and *gap* (or *growth*) *2* periods, usually referred to as G_1, S, and G_2, respectively.

The G_1 period is relatively lengthy and begins immediately after a nucleus has divided. During this period, the cell increases in size. Also, ribosomes, RNA, and substances that either inhibit or stimulate the S period that follows are produced. During the S period, the unique process of DNA replication (duplication) takes place. Details of this process and of DNA structure are discussed in Chapter 13. In the G_2 period, mitochondria and other organelles divide, and microtubules and other substances directly involved in mitosis are produced. Coiling and condensation of chromosomes also begin during G_2.

Mitosis

All organisms begin life as a single cell. This cell usually divides almost immediately, producing two new cells. These two cells, in turn, divide, with each of them producing two more cells. This process, called *mitosis* (Fig. 3.17) occurs in an organism until it dies. It ensures that the two new cells (*daughter cells*) resulting from a cell undergoing mitosis each have precisely equal amounts of DNA and certain other substances duplicated during interphase. Strictly speaking, mitosis refers to the division of the nucleus alone, but with a few exceptions seen in algae and fungi (discussed in Chapters 18 and 19), the division of the remainder of the cell, called **cytokinesis,** normally accompanies or follows mitosis. Both processes will be considered together here.

In flowering plants, conifers, and other higher plants, mitosis occurs in specific regions, or tissues, called *meristems* (see Fig. 4.1). Meristems are found in the root and stem tips and also in a thin, perforated, and branching cylinder of tissue called the *vascular cambium* (often referred to simply as the *cambium*), located in the interior of some stems and roots a short distance from the surface. In some herbaceous and most woody plants, a second meristem similar in form to the cambium lies between the cambium and the outer bark. This second meristem is called the *cork cambium.* These specific tissues are discussed in Chapters 4, 5, and 6.

When mitosis occurs, the number of chromosomes in the nucleus, whether small or large, makes no difference in the way the process takes place. The daughter cells that result from mitosis each have exactly the same number of chromosomes and distribution of DNA as the parent cell. Mitosis is a continuous process, which may take as little as 5 minutes or as long as several hours from start to finish. Typically, however, it takes from 30 minutes to 2 or 3 hours.

Mitosis is initiated with the appearance of a ringlike *preprophase band* of microtubules just beneath the plasma membrane and is usually divided into four arbitrary phases, primarily for convenience. Descriptions of the phases follow.

Prophase

The main features of **prophase** (Fig. 3.17A) are (1) the chromosomes become shorter and thicker, and their two-stranded nature becomes apparent; (2) the nuclear envelope fragments, and the nucleolus disintegrates.

Prophase utilizes about as much time as the remaining three phases combined. Before prophase begins, a *preprophase band,* formed from microtubules and microfilaments inside the plasma membrane, develops in a narrow bundle around the nucleus. The beginning of prophase is marked by the appearance of the chromosomes as faint threads in the nucleus. These chromosomes gradually coil or fold into thicker and shorter structures, and soon, two strands, or **chromatids,** can be distinguished for each chromosome. The chromatids are themselves independently coiled and are identical to each other. The coils appear to tighten and condense until the chromosomes have become relatively short, thick, and rodlike, with areas called **centromeres** holding each pair of chromatids together.

The centromere is located at a constriction on the chromosome (Fig. 3.18). A **kinetochore,** which is a dense region composed of a protein complex, is located on the outer surface of each centromere; spindle fibers become attached to the kinetochore. When examined with a light microscope, the centromeres appear to be single structures, but they actually have become double by the G_2 stage of interphase and simply function as a single unit at this point. They may be located almost anywhere on a chromosome but tend to be toward the middle. Sometimes other constrictions may appear on individual chromosomes, usually toward one end,

Awareness

Microscapes

Scientists use the scanning electron microscope (SEM) to study the details of many different types of surfaces. Unlike the light microscope or even a transmission electron microscope that form images by passing either a beam of light or electrons through a thin slice of fixed tissue, the SEM's great advantage is its ability to allow us to look at surfaces of specimens and observe topographical detail not possible with other types of microscopy.

The basic concept of a scanning electron microscope is that a finely focused beam of electrons is scanned across the surface of the specimen. The high-velocity electrons from the beam create an energetic interaction with the surface layers. These electron-specimen interactions generate particles that are emitted from the specimen and can be collected with a detector and sent to a TV screen (cathode ray tube). Particles that form the typical scanning electron image are called secondary electrons because they come from the electrons in the specimen itself. The more electrons a particular region emits, the brighter the image will be on the TV screen. The end result, therefore, is brightness associated with surface characteristics and an image that looks very much like a normally illuminated subject. SEM images typically contain a good deal of topographical detail because the electrons that are emitted and produced on the TV screen represent a one-for-one correspondence with the contours of the specimen.

All scanning electron images have one very distinctive characteristic because of this feature of electron emission and display—the images are three dimensional rather than the flat two-dimensional images obtained from other types of microscopes. The images can be understood even by the lay person because the eye is accustomed to interpreting objects that are in three dimensions.

Take, for instance, a leaf surface, which looks smooth with an ordinary light microscope. But with a scanning electron microscope, the leaf surface is a rich composition of undulating cell walls, cells joined together like pieces of a jigsaw puzzle, squiggly ridges of waxes that look like frosting decorations on a cake, and lens-shaped stomata (Box Figure 3.1A). The stomata even provides a window into the interior of the leaf where deeper cellular layers are visible. Or look at the tentacles seen in Box Figure 3.1B that remind us of some sinister sea creature. No stinging tentacles here but rather the surface of a small flower of a common weed called mouse-ear cress *(Arabidopsis thaliana)*. The "tentacles" are actually stigmatic papillae that serve to trap pollen grains, which are released from the pollen sacs of the flower (see Chapter 8 for details of flower structure).

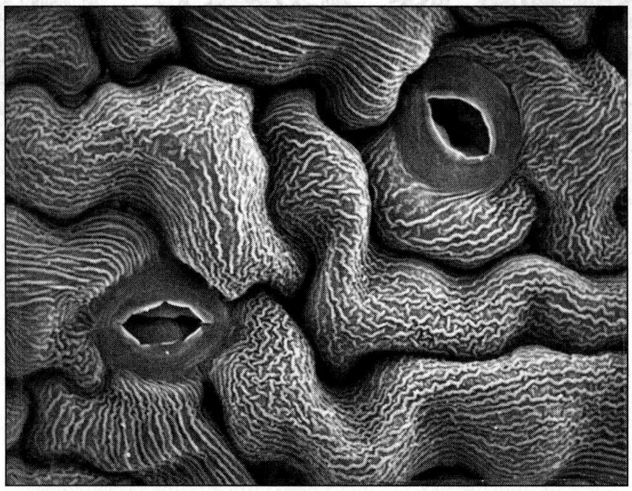

Box Figure 3.1A A scanning electron micrograph of the surface of a sepal (modified leaf) from a flower of the mouse-ear cress, *Arabidopsis thaliana,* ×2,000. *(Electron micrograph by Daniel Scheirer)*

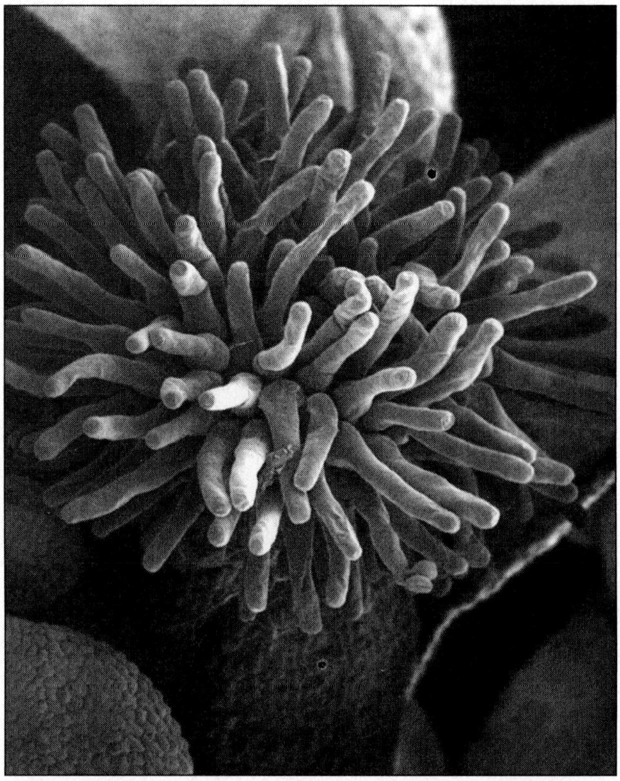

Box Figure 3.1B A scanning electron micrograph of the surface of the stigma from a flower of the mouse-ear cress, *Arabidopsis thaliana,* ×200. *(Electron micrograph by Daniel Scheirer)*

Awareness Box Continued

While biologists utilize the SEM extensively, other types of scientists put it to work in diverse ways as well, whether looking at "moon rocks" brought to Earth by the Apollo astronauts or studying the impact craters created by micrometeorite projectiles striking the space shuttle's heat-resistant tiles. Recently, a textile technologist in England examined a piece of the frayed linen tunic of King Tut, the ancient Egyptian boy-Pharaoh whose tomb was discovered in 1922. Apparently, the tunic had either been washed about 40 times in water or had been washed less frequently in a solution of sodium carbonate, a chemical used to whiten as it cleans. Additionally, unlike the clothing of ordinary people, King Tut's tunic had few mends in it—not surprising considering the wealth of the deceased. The tomb was filled with golden treasures as well as wooden chests containing his clothes and footwear.

Whether used by biologists or material scientists, the scanning electron microscope provides a stunning view of the previously unseen, but nevertheless real, world. As the beauty of nature becomes seen for the first time in startling detail, micrographs do indeed become "microscapes."

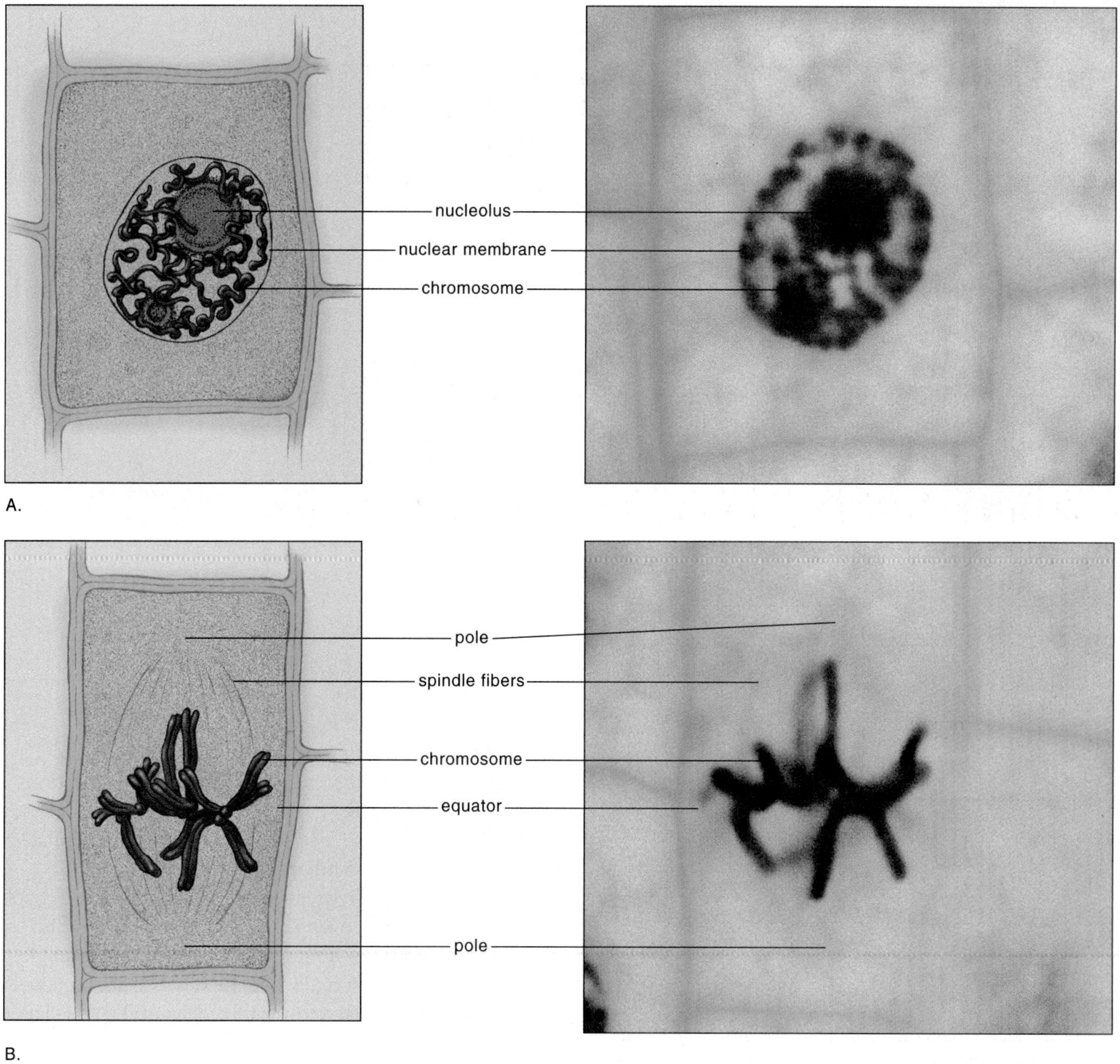

Figure 3.17 The phases of mitosis as seen in onion root-tip cells. These chromosomes are about 4 micrometers long. *A.* Cell *(center)* in prophase. *B.* Cell *(center)* in metaphase.

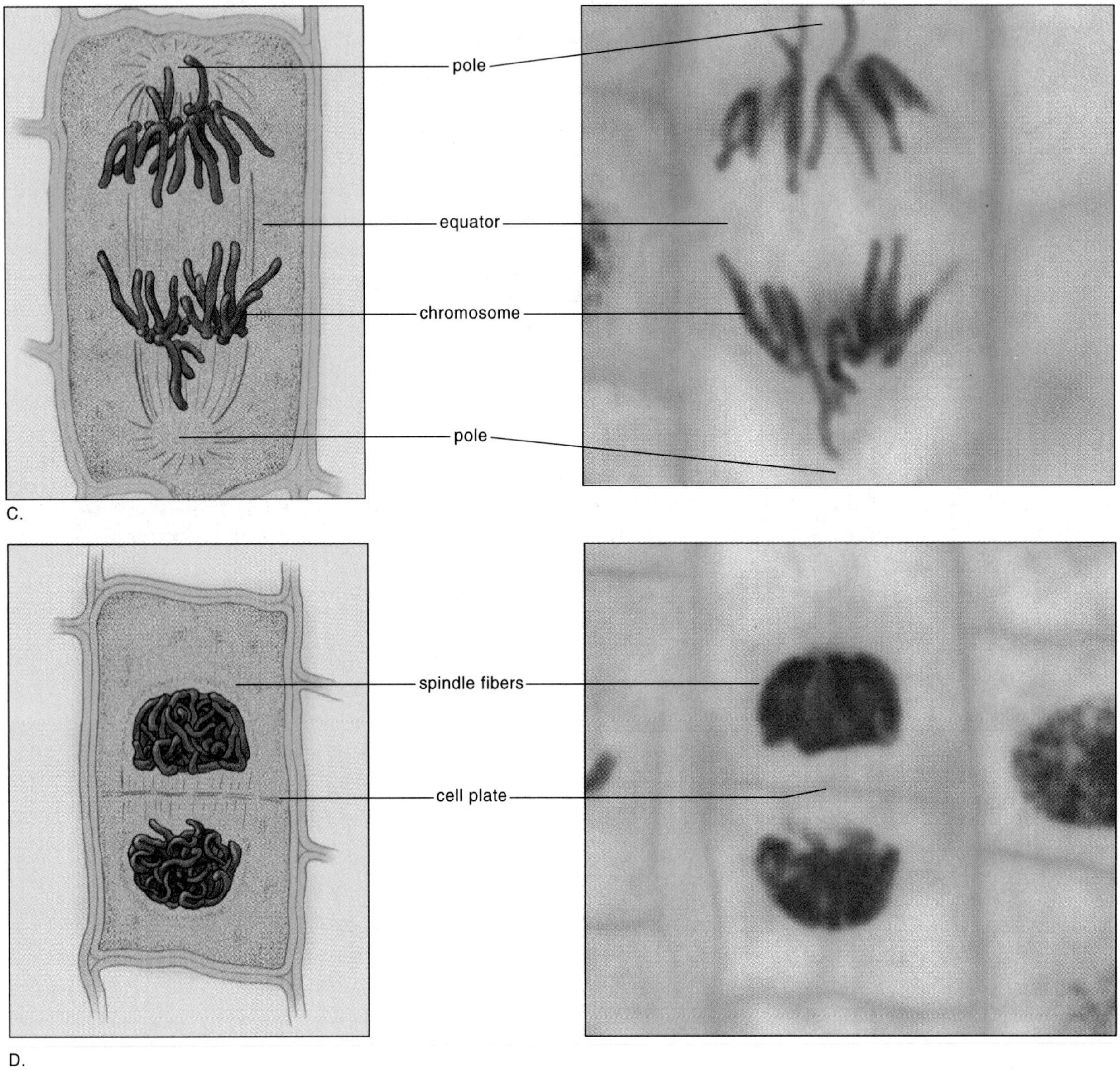

Figure 3.17 Continued. *C.* Cell *(center)* in anaphase. *D.* Cell *(center)* in telophase.

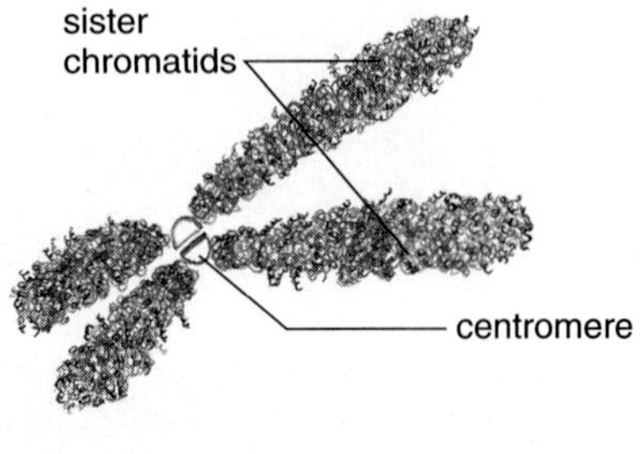

Figure 3.18 Diagram of a chromosome at metaphase. Spindle fibers from opposite ends of the cell become attached at the centromere.

giving them the appearance of having extra knobs; these knobs are referred to as *satellites.* The constrictions at the base of the satellites have no known function, but the satellites themselves are useful in helping to distinguish certain chromosomes from others in a nucleus.

As prophase progresses, the nucleolus *gradually* becomes less distinct and eventually disintegrates. By the end of prophase, **spindle fibers** consisting of microtubules have developed; these spindle fibers extend in arcs between two invisible *poles* located toward the ends of the cell. The tips of the spindle fibers become anchored at the poles. Other spindle fibers grow from each pole to the center of the cell where they become attached to a centromere. At the conclusion of prophase, the nuclear envelope has been reabsorbed into the endoplasmic reticulum and has totally fragmented.

In certain simpler organisms, such as fungi and algae, and in animal cells, the cytoplasm just outside the nucleus contains pairs of tiny keg-shaped structures called *centrioles.* The centrioles are surrounded by microtubules that radiate out from them and arrange cytoplasmic particles in the vicinity into starlike rays, each group of rays collectively called an *aster.* At the beginning of prophase, the aster divides into two parts; one part remains in its original location, while the other part migrates around the nuclear envelope to the opposite side. Centrioles have not been detected in the cells of most of the more complex members of the Plant Kingdom.

Metaphase

The main feature of **metaphase** (Fig. 3.17B) is the alignment of the chromosomes in a circle midway between the two poles around the circumference of the *spindle* and in the same plane as that previously occupied by the preprophase band. This invisible circular plate is perpendicular to the axis of the spindle and is something like the equator of a globe.

As indicated in our discussion of prophase, spindle fibers can be seen in the area previously occupied by the nucleus after the nuclear envelope has disassociated. They form a structure that looks like an old-fashioned spinning top made of fine threads. Collectively, the spindle fibers are referred to as the **spindle.** The chromosomes become aligned so that their centromeres are in a plane roughly in the center of the cell. This invisible circular plate, called the *equator,* is analogous to the equator of the earth. At the end of metaphase, the centromeres holding the two strands (*sister chromatids*) of each chromosome together separate lengthwise.

Anaphase

Anaphase—the briefest of the phases—involves the sister chromatids of each chromosome separating and appearing to be pulled to opposite poles (Fig. 3.17C).

Until the end of metaphase, the sister chromatids of each chromosome have been united at their centromeres. Anaphase begins with all the sister chromatids separating in unison and moving toward the poles. The chromatids, which after separation at their centromeres are called **daughter chromosomes,** appear to be pulled toward the poles as their spindle fibers gradually shorten. The shortening occurs as a result of material continuously being removed from the polar ends of the spindle fibers. The centromeres of the daughter chromosomes lead the way, with the chromosomes assuming V shapes as they appear to drag in the cytoplasm.

All of the chromosomes separate and move at the same time. Although experiments have shown that a chromosome will not migrate to a pole if the fiber attached to its centromere is severed, other experiments have shown that the chromosomes will separate from one another but not move to the poles, even if no spindle is present. The force or forces involved in this initial separation phenomenon have not yet been identified. It appears, however, that the main movement of the chromosomes to the poles results from a shortening of the spindle fibers.

Telophase

The five main features of **telophase** (Fig. 3.17D) are (1) each group of daughter chromosomes becomes surrounded by a reformed nuclear envelope; (2) the daughter chromosomes become longer and thinner and finally become indistinguishable; (3) nucleoli reappear; (4) many of the spindle fibers disintegrate; and (5) a *cell plate* forms.

The transition from anaphase to telophase is not distinct, but telophase is definitely in progress when elements of new nuclear envelopes appear around each group of daughter chromosomes at the poles. These elements gradually form intact envelopes as the daughter chromosomes return to the diffuse, indistinct threads seen at the onset of prophase. The new nucleoli appear on specific regions of certain chromosomes.

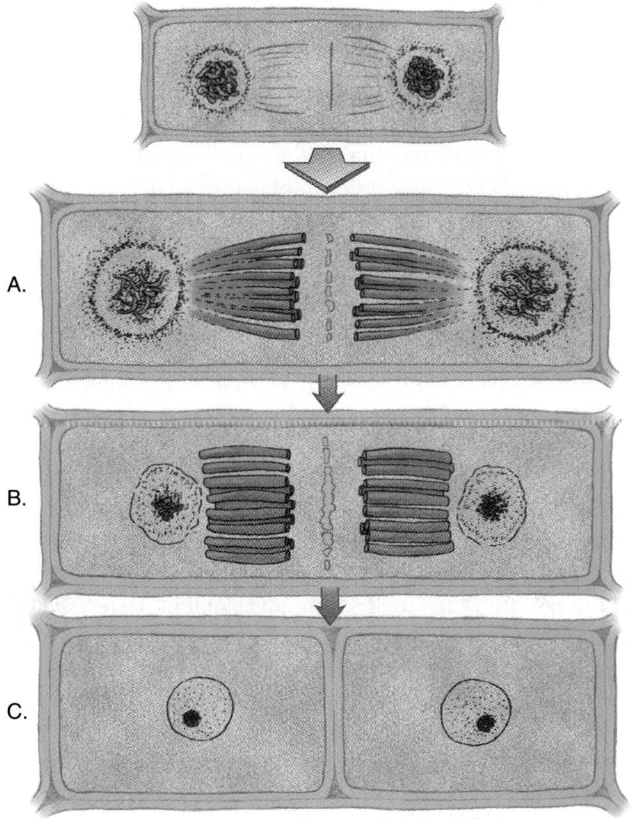

Figure 3.19 How a cell plate is formed. *A.* During telophase a *phragmoplast* (a complex of microtubles and endoplasmic reticulum) develops between the two daughter cell nuclei. The microtubules trap dictyosome vesicles along a central plane. *B.* The vesicles fuse into a flattened, hollow structure that becomes a *cell plate. C.* The cell plate grows outward; two primary cell walls and two plasma membranes form. When the cell plate reaches the mother cell walls, the plasma membranes unite with the existing plasma membrane, and the production of two daughter cells is complete.

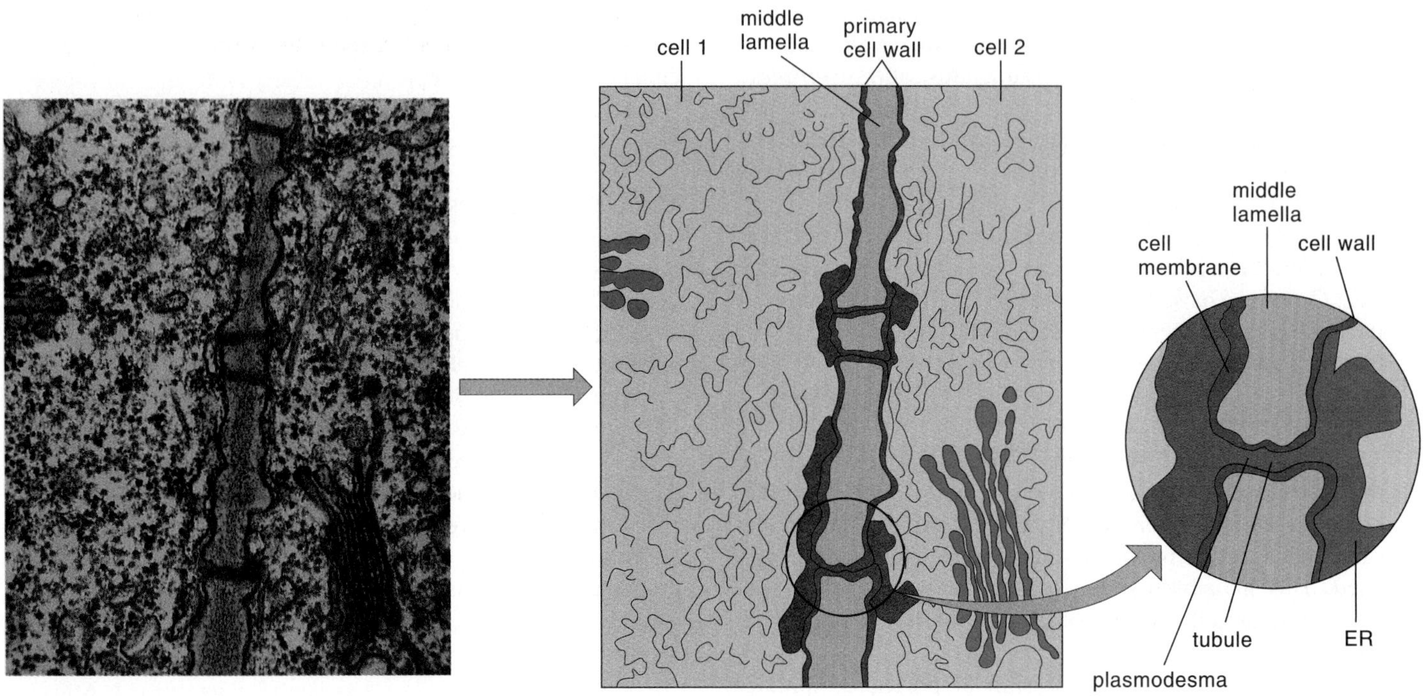

Figure 3.20 A. A diagram of two adjacent cells connected by a plasmodesma. *B.* A diagram of adjacent cells depicting the relative locations of the nucleus, endoplasmic reticulum, and a desmotubule. (*A.* © *Biophoto Assoc/Photo Researchers, Inc.*)

During telophase, the spindle microtubules gradually break down, and a set of shorter fibers (fibrils), composed of microtubules, develops in the region of the equator between the daughter nuclei. This set of fibrils, which appears somewhat keg-shaped, is called a *phragmoplast.* Dictyosomes produce small vesicles containing raw materials for the cell wall and membranes. Some of these vesicles, which resemble tiny droplets of fluid when viewed with a light microscope, are directed toward the center of the spindle (equator) by the remaining spindle fibers.

The microtubules apparently trap the dictyosome-derived vesicles, which then fuse together into one large, flattened but hollow structure called a **cell plate** (Fig. 3.19). Carbohydrates in the vesicles are synthesized into two new primary cell walls and a middle lamella. The *middle lamella* is shared by what now have become two new **daughter cells.** The cell plate grows outward until it contacts and unites with the plasma membrane of the mother cell. *Plasmodesmata* (minute strands of protoplasm that extend via tiny *desmotubules* through the walls between cells—Fig. 3.20) are formed apparently as portions of the endoplasmic reticulum are trapped between fusing vesicles of the cell plate.

New plasma membranes develop on either side of the cell plate as it forms, and new cell-wall materials are deposited between the middle lamella and the plasma membranes. These new walls are relatively flexible and remain so until the cells increase to their mature size. At that time, additional cellulose and other substances may be added, forming a secondary cell wall interior to the primary wall. In some instances, cell-plate formation does not accompany division of the nucleus.

HIGHER PLANT CELLS VERSUS ANIMAL CELLS

All animals have either internal or external skeletons or skeleton-like systems to support their tissues. Animal cells do not have cell walls; instead, the plasma membrane, called the *cell membrane* by most zoologists (animal scientists), forms the outer boundary of animal cells. Higher plant cells have walls that are thickened and rigid to varying degrees, with a framework of cellulose fibrils. Higher plant cells also have plasmodesmata connecting the protoplasts with each other through microscopic holes in the walls. Animal cells lack plasmodesmata since they have no walls. When higher plant cells divide, a cell plate is formed during the telophase of mitosis, but cell plates do not form in animal cells, which divide by pinching in two.

Other differences pertain to the presence or absence of certain organelles. Centrioles, for example, the tiny paired keg-shaped structures found just outside the nucleus, occur in all animal cells but are generally absent from higher plant cells. Plastids, common in plant cells, are not found in animal cells. Vacuoles, which are often large in plant cells, are either small or absent in animal cells.

Summary

1. All living organisms are composed of cells. Cells are modified according to the functions they perform; some live for a few days, while others live for many years.
2. The discovery of cells is associated with the development of the microscope. In 1665, Robert Hooke coined the word *cells* for boxlike compartments he saw in cork. Leeuwenhoek and Grew reported frequently during the next 50 years on the existence of cells in a variety of tissues.
3. In 1809, Lamarck concluded that all living tissue is composed of cells, and in 1824, Dutrochet reinforced Lamarck's conclusions. In 1833, Brown discovered that all cells contain a nucleus, and shortly thereafter, Schleiden saw a nucleolus within a nucleus. Schleiden and Schwann are credited with developing the cell theory in 1838 to 1839. The theory holds that all living organisms are composed of cells and that cells form a unifying structural basis of organization.
4. In 1858, Virchow contended that every cell comes from a preexisting cell and that there is no spontaneous generation of cells from dust. In 1862, Pasteur experimentally confirmed Virchow's contentions and later proved that fermentation involves activity of yeast cells. In 1897, Buchner found that yeast cells do not need to be alive for fermentation to occur. This led to the discovery of enzymes.
5. Light microscopes can magnify up to 1,500 times. Thinly sliced materials can be viewed with compound microscopes. Opaque objects can be viewed with stereomicroscopes; most magnify up to 30 times.
6. Electron microscopes have electromagnetic lenses and a beam of electrons within a vacuum that achieve magnification. Transmission electron microscopes magnify up to 200,000 or more times. Scanning electron microscopes, which can be used with opaque objects, usually magnify up to 10,000 times.
7. Scanning tunneling microscopes use a minute probe to scan surfaces at a width as narrow as that of two atoms.
8. Eukaryotic cells are the subject of this chapter. Prokaryotic cells, which lack some of the features of eukaryotic cells, are discussed in Chapter 17.
9. Cells are minute, varying in diameter between 10 and 100 micrometers. They number into the billions in larger organisms, such as trees. Plant cells are bounded by walls that surround the living protoplast. The cytoplasm contains a souplike fluid called the cytosol and all cellular components between the plasma membrane and nucleus.
10. A pectic middle lamella is sandwiched between the primary cell walls of adjacent cells. The primary wall and also the secondary cell wall, often added inside the primary wall, are composed of cellulose polymers, with hemicelluloses and glycoproteins. Secondary cell walls contain lignin that strengthens the wall.
11. Living cells are in contact with one another via fine strands of cytoplasm called plasmodesmata, which often extend through minute holes in the walls.
12. A flexible plasma membrane, which is sandwich-like and often forms folds, constitutes the outer boundary of the cytoplasm. It regulates the substances that enter and leave the cell.
13. The nucleus is bounded by a nuclear envelope consisting of two membranes that are perforated by numerous pores. Within the nucleus are a fluid called nucleoplasm, one or more spherical nucleoli, and thin strands of chromatin, which condense and become chromosomes when nuclei divide. Each species of organism has a specific number of chromosomes in each cell.
14. The endoplasmic reticulum is a system of flattened sacs and tubes associated with the storing and transporting of protein and other cell products. Granular particles called ribosomes, which function in protein synthesis, may line the outer surfaces of the endoplasmic reticulum. Ribosomes also occur independently in the cytoplasm.
15. Dictyosomes are structures that appear as stacks of sacs and function as collecting and packaging centers for the cell.
16. Plastids are larger green, orange, red, or colorless organelles. Green plastids, known as chloroplasts, contain enzymes that catalyze reactions of photosynthesis. These reactions take place in the membranes of structures that resemble stacks of coins, called thylakoids, as well as the surrounding matrix, called the stroma. Plastids develop from proplastids, which divide frequently, and also arise from the division of mature plastids.
17. Mitochondria are tiny, numerous organelles that are bounded by two membranes with inner platelike folds called cristae; they are associated with cellular respiration.
18. One or more vacuoles may occupy 90% or more of the volume of a mature cell. Vacuoles are bounded by a vacuolar membrane (tonoplast) and contain a watery fluid called cell sap. Cell sap contains dissolved substances and sometimes water-soluble red or blue anthocyanin pigments.
19. The cytoskeleton, which is involved in the architecture of cells and internal movement, is composed of microtubules and microfilaments. Microfilaments may be responsible for cytoplasmic streaming.
20. Cells that are not dividing are in interphase, which is subdivided into three periods of intense activity that precede mitosis or division of the nucleus. Mitosis is usually accompanied by division of the rest of the cell and takes place in meristems.
21. Mitosis is arbitrarily divided into four phases: (1) prophase, in which the chromosomes and their two-stranded nature become apparent and the nuclear envelope

breaks down; (2) metaphase, in which the chromosomes become aligned at the equator of the cell; a spindle composed of spindle fibers is fully developed, with some spindle fibers being attached to the chromosomes at their centromeres; (3) anaphase, in which the sister chromatids of each chromosome (now called daughter chromosomes) separate lengthwise, with each group of daughter chromosomes migrating to opposite poles of the cell; and (4) telophase, in which each group of daughter chromosomes becomes surrounded by a nuclear envelope, thus becoming new nuclei, and a wall dividing the daughter nuclei forms, creating two daughter cells.

22. Animal cells differ from those of higher plants in not having a wall, plastids, or large vacuoles. Also, they have keg-shaped centrioles in pairs just outside the nucleus and pinch in two instead of forming a cell plate when they divide.

Review Questions

1. What cellular structures can be observed with the aid of light microscopy and electron microscopy?
2. Why are cells so small, and how is this small size beneficial for transport of substances within and between cells?
3. What is the function of the plant cell wall?
4. What is the difference between protoplasm and cytoplasm?
5. What is the function of a cell nucleus? How does it perform its function?
6. What are plasmodesmata? What is their importance to living plant cells?
7. Describe the major parts and functions of a chloroplast.
8. In a typical complete cell cycle, how long, proportionately, does mitosis take?
9. What cellular structures are responsible for division of cytoplasm, and how does this occur?
10. What are the differences and similarities between plant and animal cells?

Discussion Questions

1. Would you consider any one type of cell more useful than another? Why?
2. After you have completed your introductory plant science course, do you believe you would be able to determine the function of each of a cell's organelles in a laboratory? Explain.

Learning Online

Visit our web page at www.mhhe.com/botany for interesting case studies, practice quizzes, current articles, and animations within the Online Learning Center to help you understand the material in this chapter. You'll also find active links to these topics:

Microscopy
Cell Theory
Prokaryotes and Eukaryotes
Cell Structure of Plants
Nucleus
Cellular Organelles
Endoplasmic Reticulum
Golgi Complex
Lysosomes
Mitochondria
Chloroplasts
Cell Division
Mitosis in Plant Cells
Chromosomes and Their Structure
The Cell Cycle

Additional Reading

Alberts, B., et al. 1994. *Molecular biology of the cell,* 3d ed. New York: Garland.

Argyroudi-Akoyungolo, J. H. (Ed.). 1992. *Regulation of chloroplast biogenesis.* New York: Plenum.

Becker, W. L. et al. 2000. *The world of the cell,* 4th ed. San Francisco: Benjamin Cummings.

Bidlack, J. E., et al. 1992. Molecular structure and component integration of secondary cell walls in plants. *Proceedings of the Oklahoma Academy of Science* 72: 51–56.

Buvat, R. 1989. *Ontogeny, cell differentiation and structure of vascular plants.* New York: Springer-Verlag.

Cross, P. C., and K. L. Mercer. 1995. *Cell and tissue ultrastructure: A unique perspective,* 2d ed. New York: W. H. Freeman.

Darnell, J. E., and H. Lodish. 1995. *Molecular cell biology.* New York: W. H. Freeman.

Loewy, A. G., et al. 1991. *Cell structure and function,* 3d ed. Philadelphia, PA: Saunders.

Murray, A., and T. Hunt. 1993. *The cell cycle: An introduction.* New York: Oxford University Press.

Ohki, S. (Ed.). 1992. *Cell and model membrane interactions.* New York: Plenum.

Sadava, D. E. 1993. *Cell biology: Organelle structure.* Boston, MA: Jones & Bartlett.

Sussex, I., et al. (Eds.). 1985. *Plant cell interactions.* New York: Cold Spring Harbor.

Wolfe, S. L. 1993. *Molecular and cell biology.* Belmont, CA: Wadsworth.

A heliconia (*Heliconia* sp.), native to the New World tropics. There are about 200 known species of these striking plants, many of which are becoming increasingly popular as ornamentals in tropical and subtropical areas.

Tissues

Chapter Outline

4

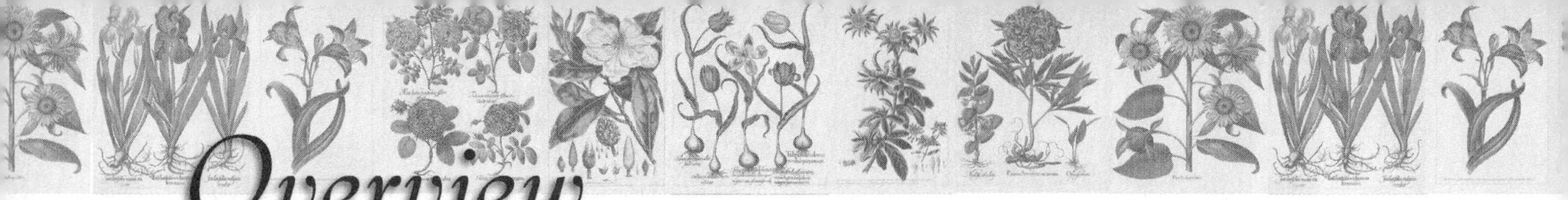

Overview

A discussion of meristems (apical meristems, vascular cambium, cork cambium, intercalary meristems) and non-meristematic tissues (parenchyma, collenchyma, sclerenchyma, secretory tissues, xylem, phloem, epidermis, periderm) forms the body of this chapter.

Some Learning Goals

1. Know the meristems present in plants and where they are found.
2. Learn the conducting tissues of plants and the function of each cell component.
3. Learn tissues of plants that are neither meristematic nor function in conduction at maturity.

I once was privileged to have a new house constructed for me on a vacant lot. I followed the stages of construction with considerable interest. First, a foundation was laid; then, trucks arrived with various building materials, and construction of a frame began. This was followed by the installation of plumbing, electrical wiring, windows, heating and air-conditioning units, vents, and various other devices. Finally, waterproof walls and a roof were added, and, upon occupation, food and other materials were stored in their appropriate niches.

In a sense, the growth of a plant from a seed is something like the construction of a house. Using raw materials from the soil and a superb manufacturing process, each plant develops a framework, "plumbing," a waterproof covering that includes "windows," vents, means of waste disposal, and food-storage areas. Even a form of air-conditioning, which enables plants to survive and thrive in the hottest summer sun, is included in each mature plant package. The building components of the framework, plumbing, and related features of plants form the body of this chapter.

There are many interesting modifications of higher plants discussed in the three chapters that follow this one, but, regardless of the outer form, most plants have three or four major groups of organs—**roots, stems, leaves,** and in some instances, **flowers.** Each of these organs is composed of **tissues,** which are defined as "groups of cells performing a similar function." Any plant organ may be composed of several different tissues; each tissue is classified according to its structure, origin, or function.

Three basic tissue patterns occur in roots and stems (see *woody dicots, herbaceous dicots* and *monocots,* discussed in Chapters 5 and 6). The following are major kinds of tissues found in higher plants. The specific types of cells associated with each tissue, as well as illustrations of them, are included in the discussions that follow the classification.

MERISTEMATIC TISSUES

Unlike animals, plants have permanent regions of growth called **meristems,** or *meristematic tissues,* where cells actively divide (Fig. 4.1). As new cells are produced, they typically are small, six-sided boxlike structures, each with a proportionately large nucleus, usually near the center, and with tiny vacuoles or no vacuoles at all. As the cells mature, however, they assume many different shapes and sizes, each related to the cell's ultimate function; the vacuoles increase in size, often occupying more than 90% of the volume of the cell.

Apical Meristems

Apical meristems are meristematic tissues found at, or near, the tips of roots and shoots, which increase in length as the apical meristems produce new cells. This type of growth is known as *primary growth.* Three *primary meristems,* as well as embryo leaves and buds, develop from apical meristems. These primary meristems are called **protoderm, ground meristem,** and **procambium.** The tissues they produce are called **primary tissues.** Note their locations in Figure 4.1; they are discussed in Chapters 5 and 6.

Lateral Meristems

The *vascular cambium* and *cork cambium,* discussed next, are **lateral meristems,** which produce tissues that increase the girth of roots and stems. Such growth is termed *secondary growth.*

Vascular Cambium

The **vascular cambium,** often referred to simply as the **cambium,** produces *secondary tissues* that function primarily in support and conduction. The cambium, which extends throughout the length of roots and stems in perennial and many annual plants, is in the form of a thin cylinder of mostly brick-shaped cells. The cambial cylinder often branches, except at the tips, and the tissues it produces are responsible for most of the increase in a plant's girth as it grows. The individual remaining cells of the cambium are referred to as *initials,* while their sister cells are called *derivatives.* The cambium and its cells and tissues are discussed in Chapters 5 and 6.

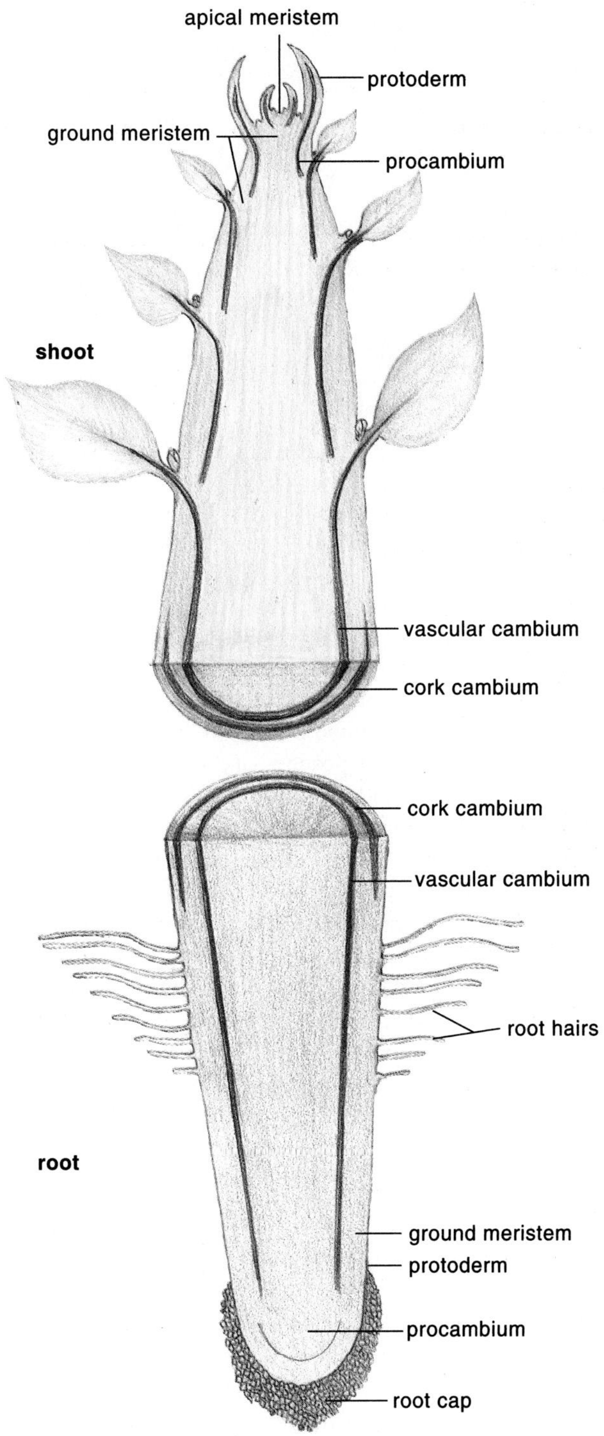

Figure 4.1 A diagram of the longitudinal axis of a plant, showing the location of meristems.

Cork Cambium

The **cork cambium,** like the vascular cambium, is in the form of a thin cylinder that runs the length of roots and stems of woody plants. It lies outside of the vascular cambium, just inside the outer bark, which it produces. The cork cambium is discussed in Chapters 5 and 6. The tissues laid down by the vascular cambium and the cork cambium are called *secondary tissues,* since they are produced *after* the primary tissues have matured.

Intercalary Meristems

Grasses and related plants have neither a vascular cambium nor a cork cambium. They do, however, have apical meristems, and, in the vicinity of **nodes** (leaf attachment areas), they have other meristematic tissues called *intercalary meristems.* The intercalary meristems develop at intervals along stems, where, like the tissues produced by apical meristems, their tissues add to stem length.

TISSUES PRODUCED BY MERISTEMS

After cells are produced by meristems, the cells assume various shapes and sizes related to their functions as they develop and mature. Some tissues consist of only one kind of cell, while others may have two to several kinds of cells. Simpler, basic types of such tissues are discussed first, followed by those that are more complex.

Simple Tissues

Parenchyma

Parenchyma tissue is composed of *parenchyma cells* (Fig. 4.2), which are the most abundant of the cell types and are found in almost all major parts of higher plants. They are more or less spherical in shape when they are first produced, but when all the parenchyma cells push up against one another, their thin, pliable walls are flattened at the points of contact. As a result, parenchyma cells assume various shapes and sizes, with the majority having 14 sides. They tend to have large vacuoles and may contain starch grains, oils, tannins (tanning or dyeing substances), crystals, and various other secretions.

More often than not, parenchyma cells have spaces between them; in fact, in water lilies and other aquatic plants, the intercellular spaces are quite extensive and form a network throughout the entire plant. This type of parenchyma tissue—with extensive connected air spaces—is referred to as *aerenchyma.*

Parenchyma cells containing numerous chloroplasts (as found in leaves) are collectively referred to as **chlorenchyma** tissue. Chlorenchyma tissues function mainly in photosynthesis, while parenchyma tissues without chloroplasts function mostly in food or water storage. For example, the soft edible parts of most fruits and vegetables consist largely of parenchyma.

Some parenchyma cells develop irregular extensions of the inner wall that greatly increase the surface area of the plasma membrane. Such cells, called *transfer cells,* are

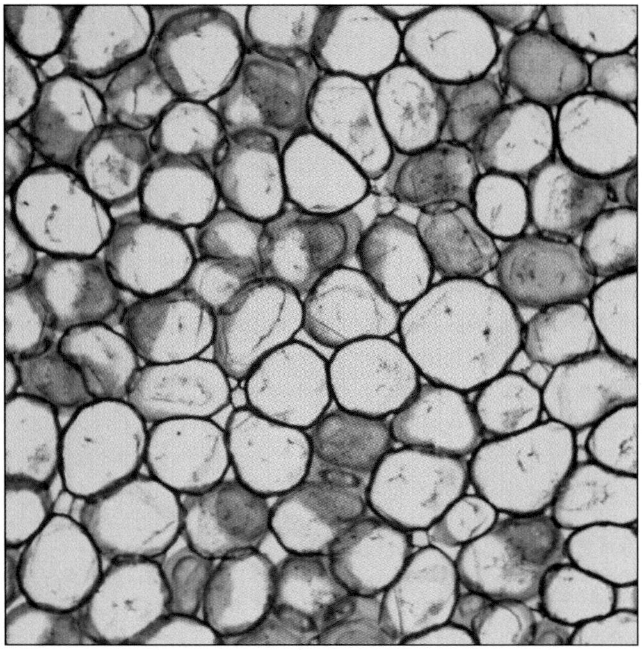

Figure 4.2 Parenchyma cells. They are more or less spherical when first formed, but as their walls touch other parenchyma cell walls, the cells end up with an average of 14 sides at maturity, ×100. (*Photomicrograph by G. S. Ellmore*)

found in nectaries of flowers and in carnivorous plants, where they apparently play a role in transferring dissolved substances between adjacent cells. Many parenchyma cells live a long time; in some cacti, for example, they may live to be over 100 years old.

Mature parenchyma cells can divide long after they were produced by a meristem. In fact, when a *cutting* (segment of stem) is induced to grow, it is parenchyma cells that start dividing and give rise to new roots. When a plant is damaged or wounded, the capacity of parenchyma cells to multiply is especially important in repair of tissues.

Collenchyma

Collenchyma cells (Fig. 4.3), like parenchyma cells, have living cytoplasm and may remain alive a long time. Their walls generally are thicker and more uneven in thickness than those of parenchyma cells. Collenchyma cells often occur just beneath the epidermis; typically, they are longer than they are wide, and their walls are pliable as well as strong. They provide flexible support for both growing organs and mature organs, such as leaves and floral parts. The "strings" of celery that get stuck in our teeth, for example, are composed of collenchyma cells.

Sclerenchyma

Sclerenchyma tissue consists of cells that have thick, tough, secondary walls, normally impregnated with **lignin.** Most sclerenchyma cells are dead at maturity and function in support. Two forms of sclerenchyma occur: **sclereids**

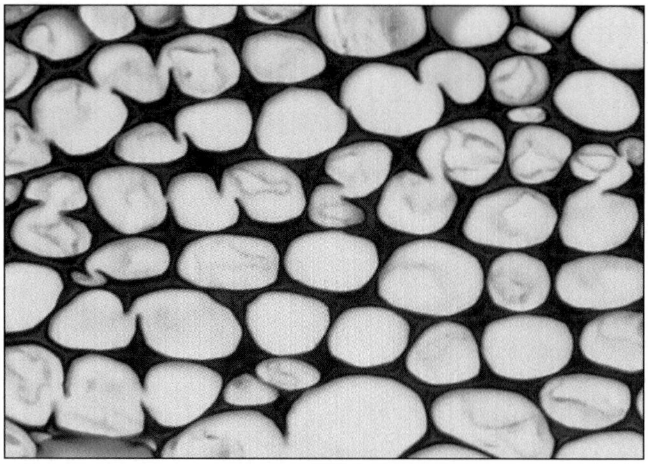

Figure 4.3 Collenchyma cells. Note the unevenly thickened walls, ×100. (*Photomicrograph by G. S. Ellmore*)

and **fibers.** Sclereids (Fig. 4.4) may be randomly distributed in other tissues. For example, the slightly gritty texture of pears is due to the presence of groups of sclereids, or *stone cells,* as they are sometimes called. The hardness of nut shells and the pits of peaches and other stone fruits is due to sclereids. Sclereids tend to be about as long as they are wide and sometimes occur in specific zones (e.g., the margins of camellia leaves) rather than being scattered within other tissues.

Fibers (Fig. 4.5) may be found in association with a number of different tissues in roots, stems, leaves, and fruits. They are usually much longer than they are wide and have a proportionately tiny cavity, or *lumen,* in the center of the cell. At present, fibers from more than 40 different families of plants are in commercial use in the manufacture of textile goods, ropes, string, canvas, and similar products. Archaeological evidence indicates that humans have been using plant fibers for at least 10,000 years.

Complex Tissues

Most of the tissues we have discussed thus far consist of one kind of cell, but a few important tissues are always composed of two or more kinds of cells and are sometimes referred to in a general sense as *complex tissues.* Two of the most important complex tissues in plants, *xylem* and *phloem,* function primarily in the transport of water, ions, and soluble food (sugars) throughout the plant. Some complex tissues are produced by apical meristems, but most complex tissues in woody plants are produced by the vascular cambium and are often referred to as *vascular tissues.*

The *epidermis,* which forms a protective layer covering all plant organs, consists primarily of parenchyma or parenchyma-like cells, but it also often includes specialized cells involved in the movement of water and gases in and out of plants, secretory glands, various hairs, cells in which crystals are isolated, and others that greatly

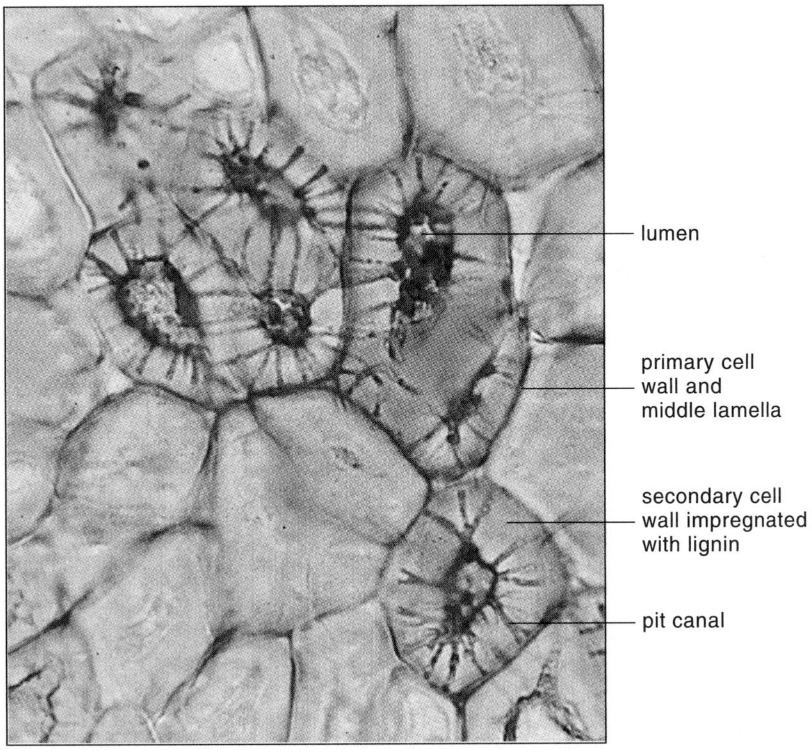

A.

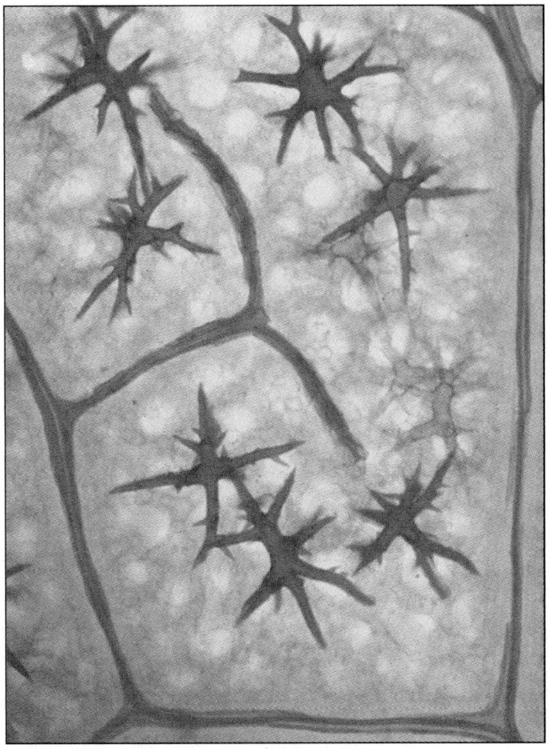

B.

Figure 4.4 A. Sclereids (stone cells) of a pear in cross section, ×1,000. *B.* Sclereids in the leaf of a wheel tree (*Trochodendron aralioides*), ×1000. (*B. Photomicrograph by G. S. Ellmore*)

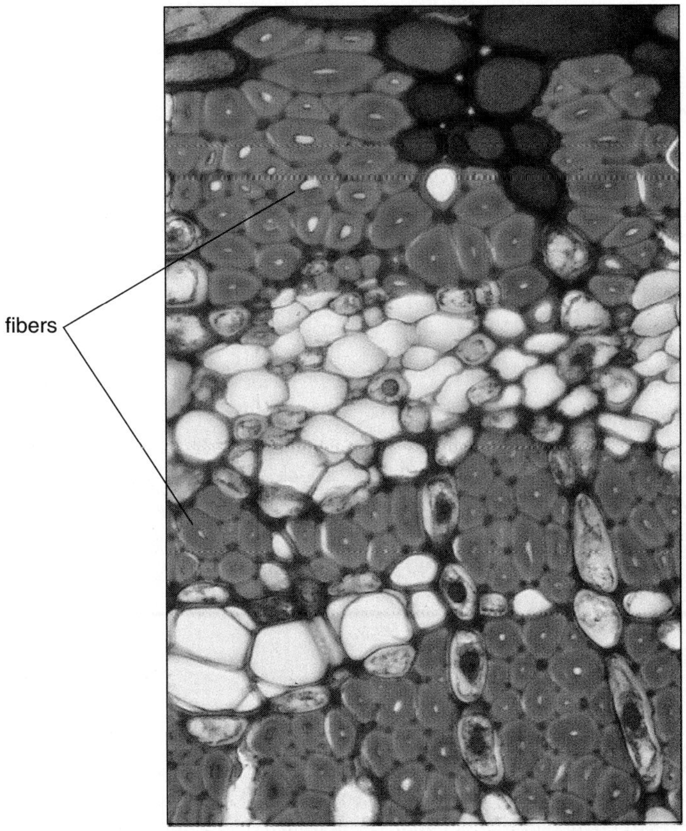

A.

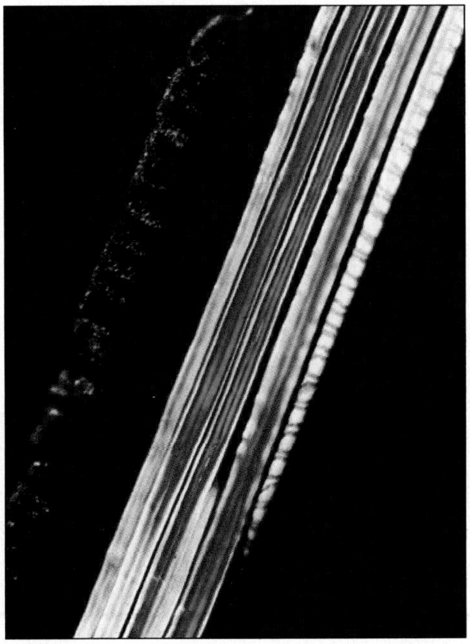

B.

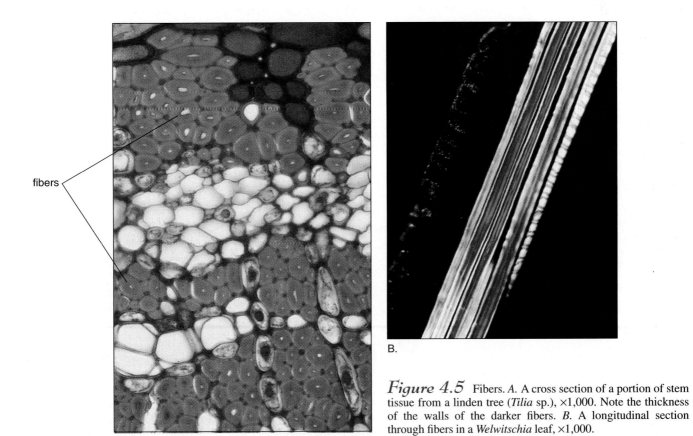

Figure 4.5 Fibers. *A.* A cross section of a portion of stem tissue from a linden tree (*Tilia* sp.), ×1,000. Note the thickness of the walls of the darker fibers. *B.* A longitudinal section through fibers in a *Welwitschia* leaf, ×1,000. (*B. Photomicrograph by G. S. Ellmore*)

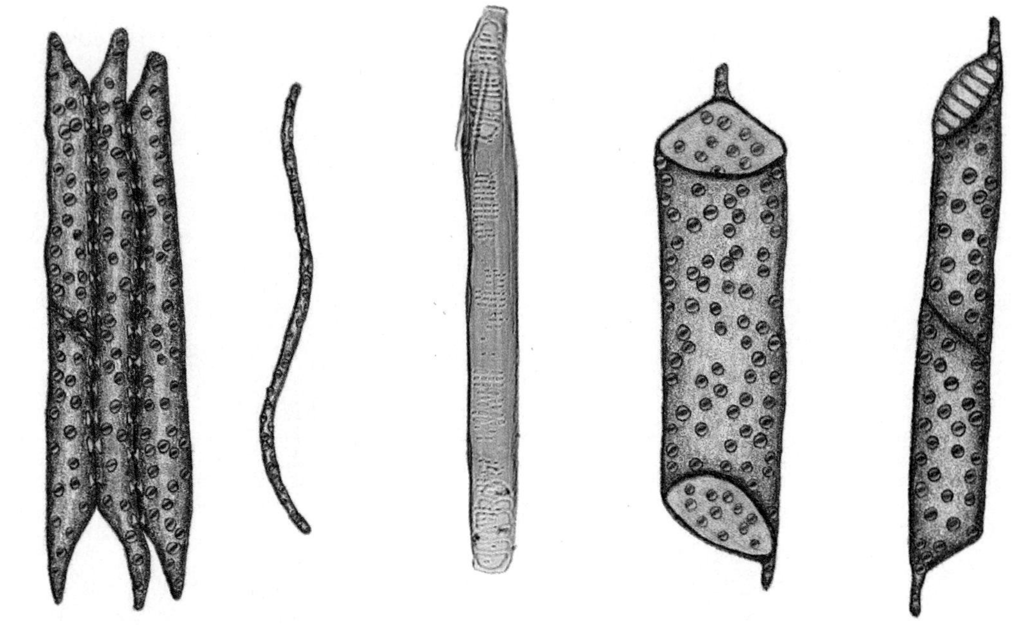

Figure 4.6 *A.* Tracheids, ×200. *B.* Vessel elements, ×200.

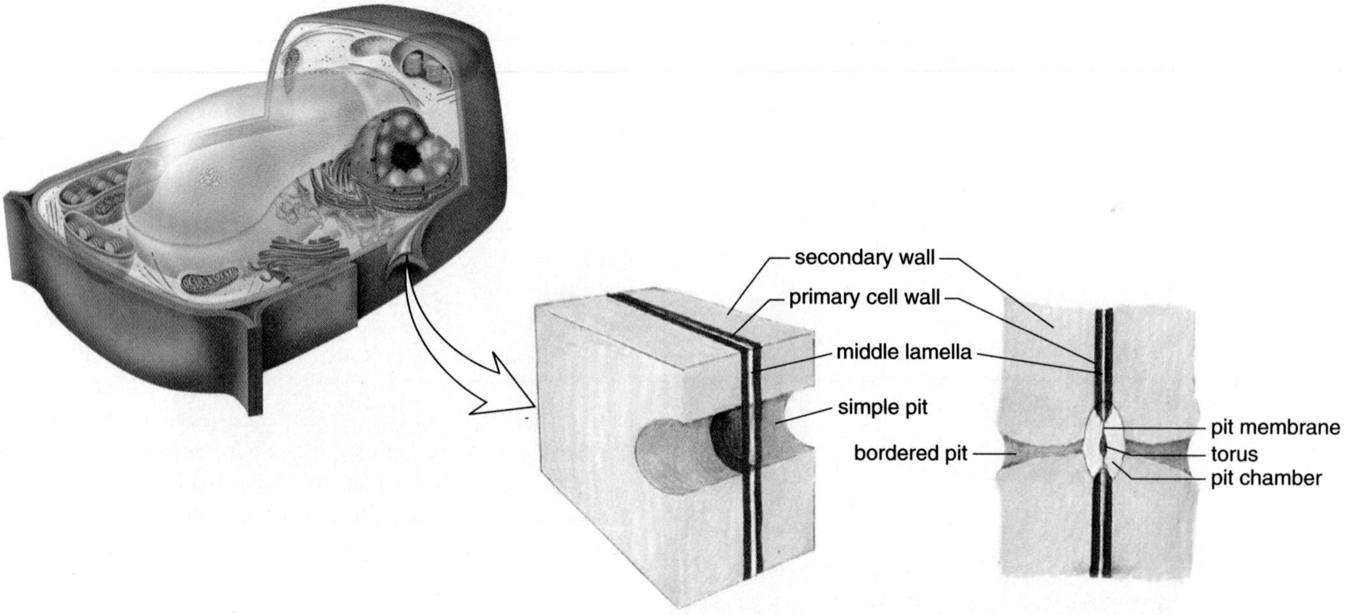

Figure 4.7 Pits. Pits are depressions or cavities in cell walls where the secondary wall does not form. There may be from one or two to several thousand in a cell. They always occur in pairs, with one on each side of the middle lamella. Some, called *bordered pits,* bulge out from the wall and resemble doughnuts in surface view, while others, called *simple pits,* do not bulge. (*Inset from Sylvia S. Mader, Biology, 7th edition. © 2001 The McGraw-Hill Companies. All rights reserved.*)

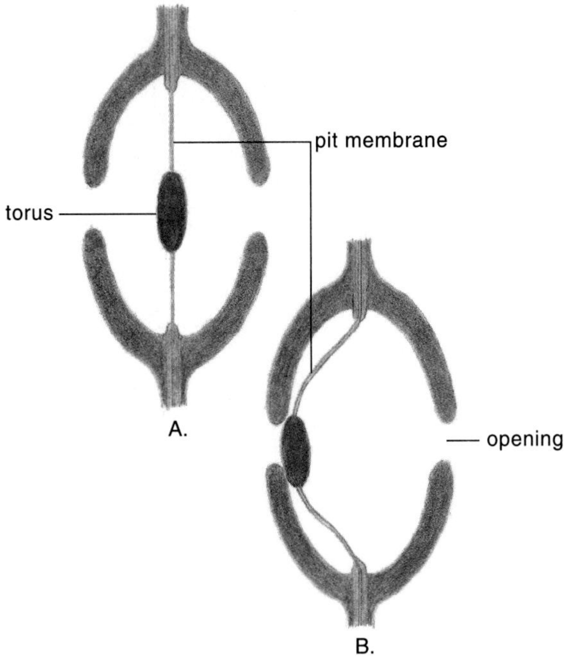

Figure 4.8 How water flow is controlled in adjacent pairs of bordered pits. The pits are separated by a *pit membrane* consisting of the middle lamella and two thin layers of primary walls. *A.* Water moves relatively freely through the pit openings and pit membrane when the *torus* (a thickened region of the pit membrane) is in the center. *B.* If the flexible pit membrane swings to one side so that the torus blocks an opening, water movement through the pit pair is restricted.

increase absorptive parts of roots. Accordingly, the epidermis and tissues with secretory cells are discussed in this section.

Periderm, which comprises the outer bark of woody plants, consists mostly of cork cells, but it is included in this discussion because it contains pockets of parenchyma-like cells.

Xylem

Xylem tissue is an important component of the "plumbing" and storage systems of a plant and is the chief conducting tissue throughout all organs for water and minerals absorbed by the roots. Xylem consists of a combination of parenchyma cells, fibers, *vessels, tracheids,* and *ray cells* (Fig. 4.6). **Vessels** are long tubes made up of individual cells called **vessel elements** that are open at each end, with barlike strips of wall material extending across the open areas in some instances. The tubes are formed as a result of the cells being joined end to end.

Tracheids, which, like vessel elements, are dead at maturity and have relatively thick secondary cell walls, are tapered at each end, the ends overlapping with those of other tracheids. Tracheids have no openings similar to those of vessels, but there are usually pairs of *pits* present wherever two tracheids are in contact with one another (Fig. 4.7). Pits are

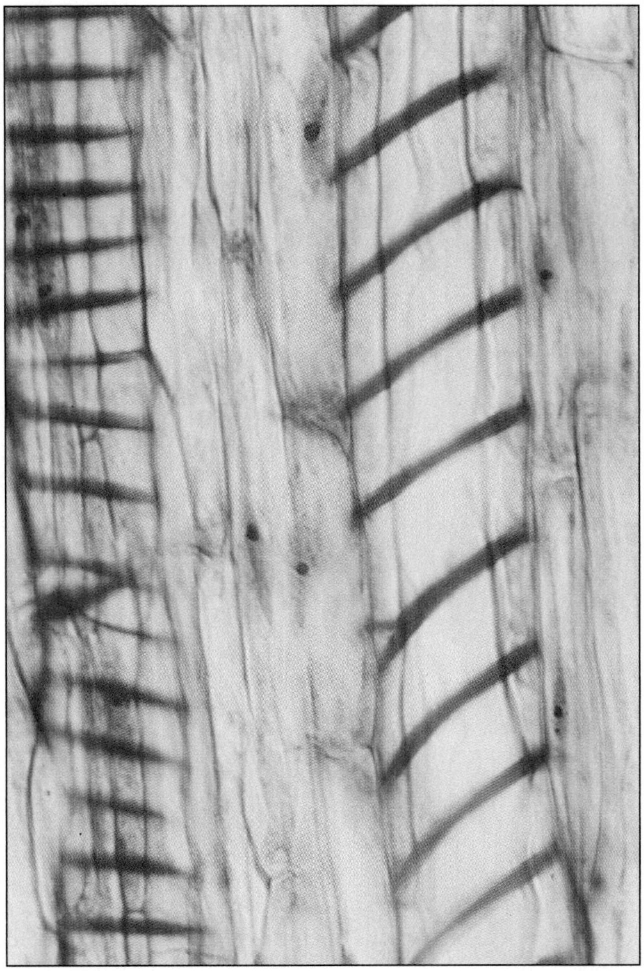

Figure 4.9 Spiral thickenings on the inside wall of a tracheid, ×400.

areas in which no secondary wall material has been deposited and, as indicated in Chapter 3, they allow water to pass from cell to cell. Figure 4.8 illustrates how pit pairs function in regulating the passage of water between adjacent cells.

In cone-bearing trees and certain other non-flowering plants, the xylem is composed almost entirely of tracheids. The walls of many tracheids have spiral thickenings on them that are easily seen with the light microscope (Fig. 4.9). Most conduction is up and down, but some is lateral (sideways). The lateral conduction takes place in the **rays.** Ray cells, which also function in food storage, are actually long-lived parenchyma cells that are produced in horizontal rows by special *ray initials* of the vascular cambium. In woody plants, the rays radiate out from the center of stems and roots like the spokes of a wheel (see Figs. 6.6 and 6.8).

Phloem

Phloem tissue (Fig. 4.10), which conducts dissolved food materials (primarily sugars) produced by photosynthesis throughout the plant, is composed mostly of two types of

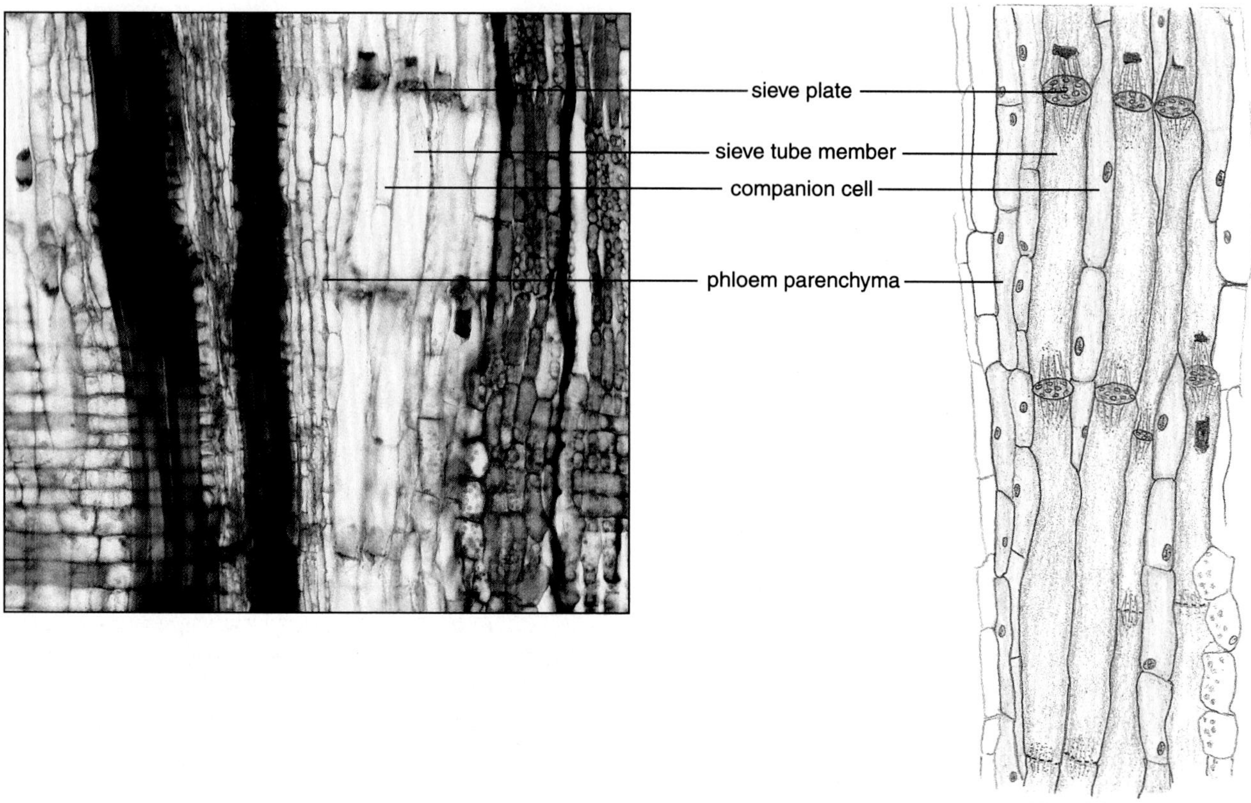

Figure 4.10 Longitudinal view of part of the phloem of a black locust tree (*Robinia pseudo-acacia*), ×1,000.

cells without secondary walls. The relatively large, more or less cylindrical **sieve tube members** have narrower, more tapered **companion cells** closely associated with them. Phloem is derived from the parent cells of the cambium, which also produce xylem cells; it often also includes fibers, parenchyma, and ray cells. Sieve tube members, like vessel elements, are laid end to end, forming **sieve tubes.** Unlike vessel elements, however, the end walls have no large openings; instead, the walls are full of small pores through which the cytoplasm extends from cell to cell. The porous regions of sieve tube members are called **sieve plates.**

Sieve tube members have no nuclei at maturity, despite the fact that their cytoplasm is very active in the conduction of food materials in solution throughout the plant. Apparently, the adjacent companion cells form a very close relationship with the sieve tubes next to them and function in some manner that brings about the conduction of the food.

Living sieve tube members contain a polymer called *callose* that stays in solution as long as the cell contents are under pressure. If an insect such as an aphid injures a cell, however, the pressure drops, and the callose precipitates. The callose and a phloem protein are then carried to the nearest sieve plate where they form a *callus* plug that prevents leaking of the sieve tube contents.

Sieve cells, which are found in ferns and cone-bearing trees, are similar to sieve tube members but tend to overlap at their ends rather than form continuous tubes. Like sieve tube members, they have no nuclei at maturity, and they also have no adjacent companion cells. They do have adjacent *albuminous cells,* which are equivalent to companion cells and apparently function in the same manner.

Epidermis

The outermost layer of cells of all young plant organs is called the **epidermis.** Since it is in direct contact with the environment, it is subject to modification by the environment and often includes several different kinds of cells. The epidermis is usually one cell thick, but a few plants produce aerial roots called **velamen roots** (e.g., orchids) in which the epidermis may be several cells thick, with the outer cells functioning something like a sponge. Such a multiple-layered epidermis also occurs in the leaves of some tropical figs and members of the Pepper Family (Piperaceae), where it protects a plant from desiccation.

Most epidermal cells secrete a fatty substance called **cutin** within and on the surface of the outer walls. Cutin forms a protective layer called the **cuticle** (Fig. 4.11). The thickness of the cuticle (or, more importantly, wax secreted on top of the cuticle by the epidermis) to a large extent determines how much water is lost through the cell walls by evaporation. The cuticle is also exceptionally resistant to bacteria and other disease organisms and has been recovered from fossil plants millions of years old. The waxes deposited on the cuticle in a number of plants (see Fig. 7.7) apparently reach the surface through microscopic

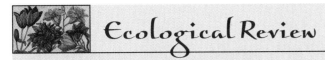

Ecological Review

The first point of contact between plants and the environment is a surface layer of cells called the epidermis. The structure of the epidermis, particularly the thickness of waxes on its surface, determines the potential rate of water exchange between a plant and the environment. The resistance of the epidermis to water loss is generally higher in the plants of arid environments. A second critical ecological function of the epidermis is a barrier to attack by pathogens. Pathogens, which may restrict the distribution of plant species and strongly influence plant population size, are an important part of a plant's biological environment. The extent of development of a plant's xylem and sclerenchyma cells also is related to the plant's environment, with, for example, aquatic plants having weakly developed xylem, large trees having well-developed xylem, and fire-resistant trees, such as redwoods, having thick bark.

channels in the cell walls. The susceptibility of a plant to herbicides may depend on the thickness of these wax layers. Some wax deposits are extensive enough to have commercial value. Carnauba wax, for example, is deposited on the leaves of the wax palm. It and other waxes are harvested for use in polishes and phonograph records. In colonial times, a wax obtained from boiling leaves and fruits of the wax myrtle was used to make bayberry candles.

In leaves, the epidermal cell walls perpendicular to the surface often assume bizarre shapes that, under the microscope, give them the appearance of pieces of a jigsaw puzzle. Epidermal cells of roots produce tubular extensions called *root hairs* (see Fig. 5.4) a short distance behind the growing tips. The root hairs greatly increase the absorptive area of the surface.

Hairs of a different nature occur on the epidermis of above-ground parts of plants. These hairs form outgrowths consisting of one to several cells (Fig. 4.12). Leaves also have numerous small pores, the **stomata,** bordered by pairs of specialized epidermal cells called **guard cells** (see Figs. 7.8 and 9.13). Guard cells differ in shape from other epidermal cells; they also differ in that chloroplasts are present within them. The stomatal apparatus is discussed in Chapters 7 and 9. Some epidermal cells may be modified as **glands**

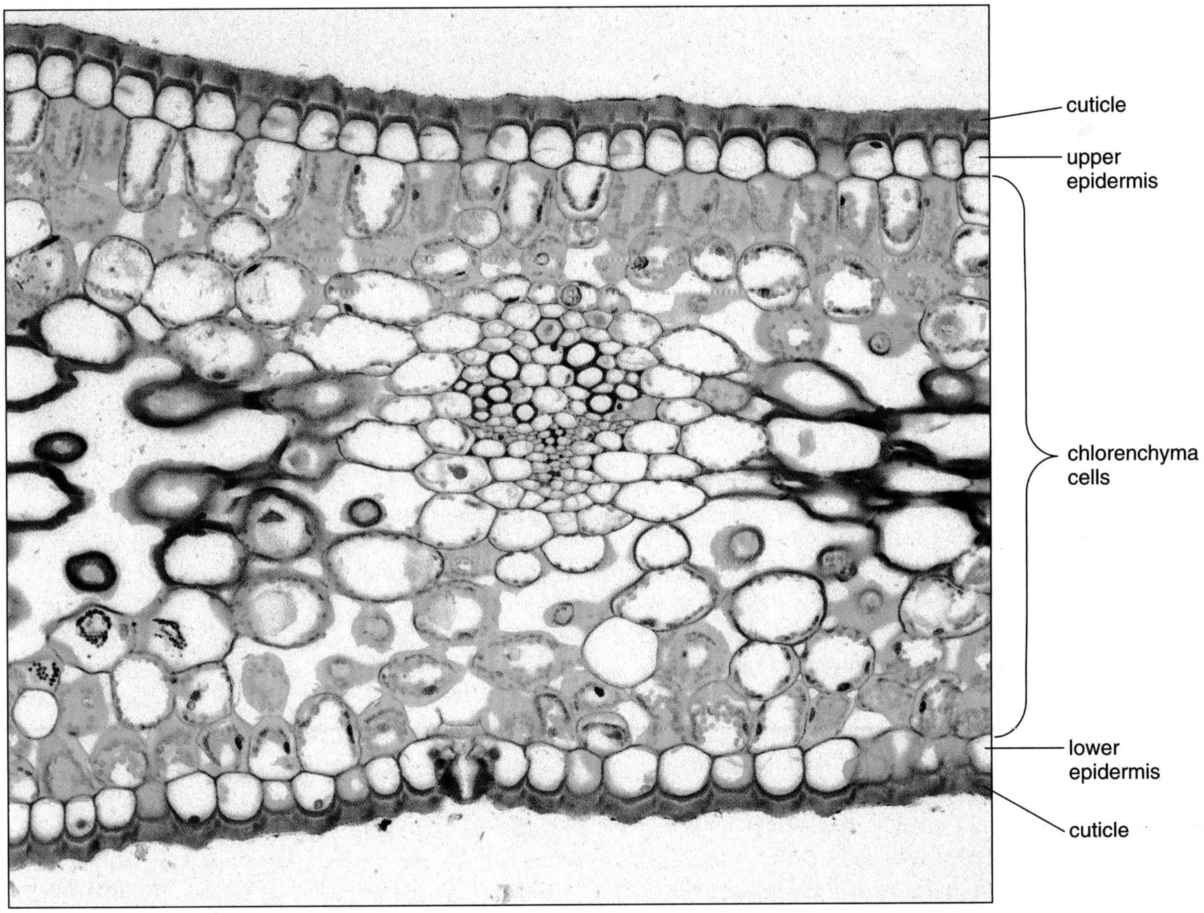

Figure 4.11 A portion of a cross section of a kaffir lily (*Clivia*) leaf, showing the thick cuticle secreted by the epidermis, ×1000.

Figure 4.12 Hairs on the surface of an ornamental mint plant, ×50.

that secrete protective or other substances, or modified as hairs that either reduce water loss or repel insects and animals that might otherwise consume them (Fig. 4.13).

Periderm

In woody plants, the epidermis is sloughed off and replaced by a **periderm** after the cork cambium begins producing new tissues that increase the girth of the stem or root. The periderm constitutes the outer bark and is primarily composed of somewhat rectangular and boxlike **cork** cells, which are dead at maturity (see Fig. 4.14). While the cytoplasm of cork cells is still functioning, it secretes a fatty substance, **suberin,** into the walls. This makes cork cells waterproof and helps them protect the phloem and other tissues beneath the bark from drying out, mechanical injury, and freezing temperatures. Some cork tissues, such as those produced by the cork oak, are harvested commercially and are used for bottle corks and in the manufacture of linoleum and gaskets.

Some parts of a cork cambium form pockets of loosely arranged parenchyma cells that are not impregnated with suberin. These pockets of tissue protrude through the surface of the periderm; they are called **lenticels** (Fig. 4.14) and function in gas exchange between the air and the interior of the stem. The fissures in the bark of trees have lenticels at their bases. The various tissues discussed are shown as they occur in a woody stem in Figure 6.6.

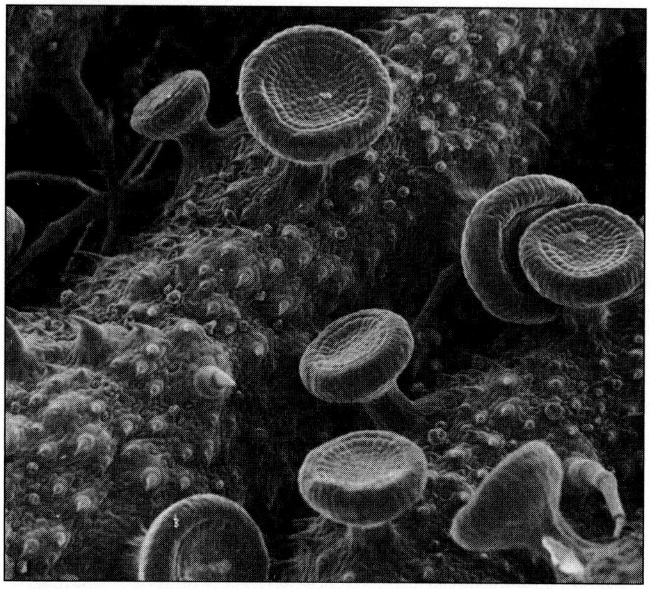

Figure 4.13A Tack-shaped glands and epidermal hairs of various sizes on the surface of flower bracts of a western tarweed. Scanning electron micrograph ca. ×200.

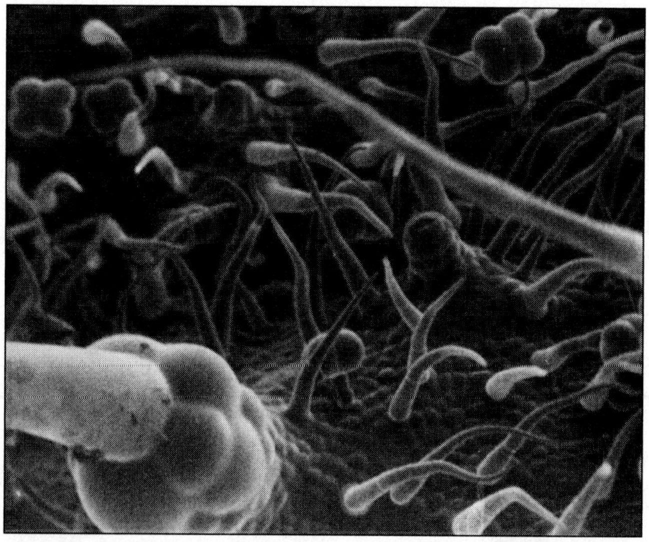

Figure 4.13B Hairs on the surface of a tomato plant stem. Note the raised stoma to the right of center. Scanning electron micrograph ca. ×300. (*A. Courtesy Dr. Charles Sternburg; B. Courtesy Dr. Tahany H. I. Sherif*)

Secretory Cells and Tissues

All cells secrete certain substances that can damage the cytoplasm, if allowed to accumulate internally. Such materials either must be isolated from the cytoplasm of the cells in which they originate or moved outside of the plant body. Often, the substances consist of waste products that are of no further use to the plant, but some substances, such as nectar, perfumes, and plant hormones (discussed in Chapter 11), are vital to normal plant functions.

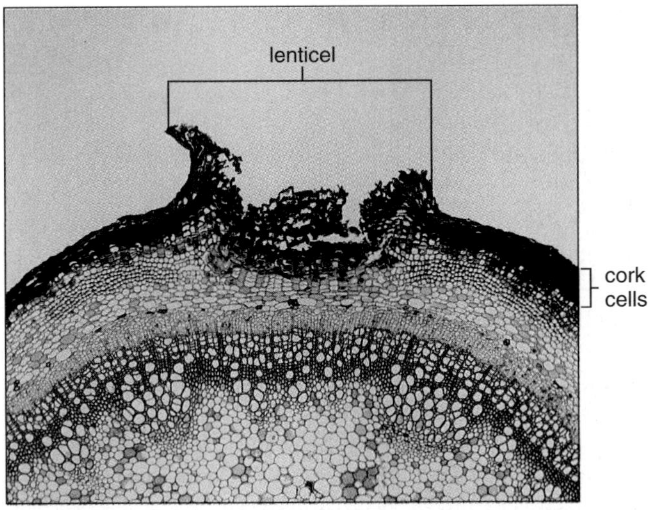

Figure 4.14 Periderm and a lenticel. A cross section through a small portion of elderberry (*Sambucus*) periderm, showing a large lenticel, ca. ×250.

Secretory cells may function individually or as part of a **secretory tissue.** Secretory cells or tissues, which often are derived from parenchyma, can occur in a wide variety of places in a plant. Among the most common secretory tissues are those that secrete nectar in flowers; oils in citrus, mint, and many other leaves; mucilage in the glandular hairs of sundews and other insect-trapping plants; latex in members of several plant families, such as the Spurge Family; and resins in coniferous plants, such as pine trees. Latex and resins are usually secreted by cells lining tubelike ducts that form networks throughout certain plant species (see Fig. 6.11). Some plant secretions, such as pine resin, rubber, mint oil, and opium, have considerable commercial value.

Summary

1. A group of cells performing a common function is called a tissue.
2. Apical meristems are found in the vicinity of the tips of roots and stems; the vascular cambium and the cork cambium occur as lengthwise cylinders within roots and stems; intercalary meristems occur in the vicinity of nodes of grasses and related plants.
3. Tissues produced by meristems consist of one to several kinds of cells. They include parenchyma, collenchyma, sclerenchyma, epidermis, xylem, phloem, periderm, and secretory tissues.
4. Parenchyma cells are thin-walled, while collenchyma cells have unevenly thickened walls that provide flexible support for various plant organs.
5. Two types of sclerenchyma occur—fibers (which are long and tapering) and sclereids (which are short in length); both types have thick walls and are usually dead at maturity.
6. Complex tissues have more than one kind of cell. The principal types are xylem, phloem, epidermis, and periderm.
7. Xylem conducts water and minerals throughout the plant. It consists of a combination of parenchyma, fibers, vessels (tubular channels), tracheids (cells with tapering end walls that overlap), and ray cells (involved in lateral conduction).
8. Phloem conducts primarily dissolved sugars throughout the plant. It is composed of sieve tubes (made up of cells called sieve tube members), companion cells (which apparently regulate adjacent sieve tube members), parenchyma, ray cells, and fibers. Callose aids in plugging injured sieve tubes. Sieve cells, which have overlapping end walls, and adjacent albuminous cells take the place of sieve tube members and companion cells in ferns and cone-bearing trees.
9. The epidermis is usually one cell thick, with fatty cutin (forming the cuticle) within and on the surface of the outer walls. The epidermis may include guard cells that border pores called stomata; root hairs, which are tubular extensions of single cells; other hairs that consist of one to several cells; and glands that secrete protective substances.
10. Periderm, which consists of cork cells and loosely arranged groups of cells comprising lenticels involved in gas exchange, constitutes the outer bark of woody plants.
11. Secretory tissues occur in various places in plants; they secrete substances such as nectar, oils, mucilage, latex, and resins.

Review Questions

1. What is the function of meristems? Where are they located?
2. How are parenchyma, collenchyma, and sclerenchyma distinguished from one another?
3. Distinguish between epidermis and periderm.
4. What are the functions of xylem and phloem? What cells are involved in their normal activities?
5. What types of substances do secretory cells secrete?

Discussion Questions

1. Most plant meristems are located at the tips of shoots and roots and in cylindrical layers within stems and roots. What could happen if they were present in leaves?
2. The cambium produces xylem toward the center of a tree and phloem toward the outside. Do you think it would make any difference if the positions of the xylem and phloem were reversed? Why?

Learning Online

Visit our web page at www.mhhe.com/botany for interesting case studies, practice quizes, current articles, and animations within the Online Learning Center to help you understand the material in this chapter. You'll also find active links to these topics:

Architectural Pattern of Plants
Meristematic Tissues
Non-meristematic Tissues
Roots
Stems and Secondary Growth
Herbaceous Stems
Leaves and the Movement of Water
Wood, Cork, and Bamboo

Additional Reading

Cutler, E. F., and K. L. Alvin. 1982. *The plant cuticle.* San Diego, CA: Academic Press.

Cutter, E. G. 1978. *Plant anatomy, Part I: Cells and tissues,* 2d ed. Reading, MA: Addison-Wesley.

Fahn, A. 1990. *Plant anatomy,* 4th ed. Elmsford, NY: Pergamon Press.

Foskett, D. E. 1994. *Plant growth and development.* San Diego, CA: Academic Press.

Lloyd, C. W. (Ed.). 1991. *The cytoskeletal basis of plant growth and form.* San Diego, CA: Academic Press.

Mauseth, J. D. 1988. *Plant anatomy.* Menlo Park, CA: Benjamin/Cummings.

Metcalfe, C. R., and L. Chalk (Eds.). 1988–1989. *Anatomy of the dicotyledons,* 2 vols. Fair Lawn, NY: Oxford University Press.

Romberger, J. A. 1993. *Plant structure: Function and development.* New York: Springer-Verlag.

Starr, C. C. 2000. *Biology concepts and applications,* 4th ed. Stamford, CT: Brooks/Cole.

Black mangrove (*Avicennia nitida*) pneumatophores. Pneumatophores are above-ground spongy outgrowths of roots in wet areas where insufficient oxygen needed for normal respiration of the roots is present. The pneumatophores allow freer exchange of carbon dioxide and oxygen. (*Courtesy Lani Stemmerman*).

Roots and Soils

5

Chapter Outline

Overview

This chapter discusses roots, beginning with the functions and continuing with the development of roots from a seed. It covers the function and structure of the root cap, region of cell division, region of elongation, and region of maturation (with its tissues). The endodermis and pericycle are also discussed.

Specialized roots (food-storage roots, water-storage roots, propagative roots, pneumatophores, aerial roots, contractile roots, buttress roots, parasitic roots) and mycorrhizae are given brief treatment. This is followed by some observations on the economic importance of roots.

After a brief examination of soil horizons, the chapter concludes with a discussion of the development of soil, its texture, composition, structure, and its water.

Some Learning Goals

1. Know the primary functions and forms of roots.
2. Learn the root regions, including the root cap, region of cell division, region of cell elongation, and region of maturation (including root hairs and all tissues), and know the function of each.
3. Discuss the specific functions of the endodermis and the pericycle.
4. Understand the differences among the various types of specialized roots.
5. Know at least 10 practical human uses of roots.
6. Understand how a good agricultural soil is developed from raw materials.
7. Contrast the various forms of soil particles and soil water with regard to specific location and availability to plants.

You have at least seen pictures of the destruction caused by a tornado as it passed through a village or a city, but have you seen what a twister can do to a forest? Large trees may be snapped off above the ground or knocked down, and branches may be stripped bare of leaves. Unless the soil in the area happens to be thin, sandy, or loose, however, you will probably see relatively few trees completely torn up by the roots and blown elsewhere. In the tropics, it is indeed rare to find healthy palm trees uprooted even after a hurricane.

Roots *anchor* trees firmly in the soil, usually through an extensive branching network that constitutes about one-third of the total dry weight of the plant. The roots of most plants do not usually extend down into the earth more than 3 to 5 meters (10 to 16 feet); those of many herbaceous species are confined to the upper 0.6 to 0.9 meter (2 to 3 feet). The roots of a few plants, such as alfalfa, however, often grow more than 6 meters (20 feet) into the earth. When the Suez Canal was being built, workers encountered roots of tamarisk at depths of nearly 30 meters (100 feet), and mesquite roots have been seen 53.4 meters (175 feet) deep in a pit mine in the southwestern United States. Some plants, such as cacti, form very shallow root systems but still effectively anchor the plants with a densely branching mass of roots radiating out in all directions as far as 15 meters (50 feet) from the stem.

Besides anchoring plants, roots *absorb water and minerals in solution,* mostly through "feeder" roots found in the upper meter (3.3 feet) of soil. Some plants have roots that, as well as anchoring and absorbing, store water or food, or perform other specialized functions.

Some aquatic plants (e.g., duckweeds and water hyacinths) normally produce roots in water, and many *epiphytes* (nonparasitic plants that grow suspended without direct contact with the ground—e.g., orchids) produce aerial roots. The great majority of vascular plants, however, develop their root systems in soils. The soils, which vary considerably in composition, texture, and other characteristics, are discussed toward the end of this chapter.

HOW ROOTS DEVELOP

When a seed germinates, the tiny, rootlike **radicle,** a part of the **embryo** (immature plantlet) within it, grows out and develops into the first root. The radicle may develop into a thick, tapered *taproot,* from which thinner *branch roots* arise, or many *adventitious roots* may arise from the stem, which is attached to the radicle and continuous with it. Adventitious roots are those that do not develop from another root but develop instead from a stem, leaf, or other plant part. A *fibrous root system,* which may have large numbers of fine roots of similar diameter, then develops from the adventitious roots (Fig. 5.1). Many mature plants have a combination of taproot and fibrous root systems.

The number of roots produced by a single plant may be prodigious. For example, a single mature ryegrass plant may have as many as 15 million individual roots and branch roots, with a combined length of 644 kilometers (400 miles) and a total surface area larger than a volleyball court, all contained within 0.57 cubic meter (20 cubic feet) of soil. *Root hairs* (discussed in the section "The region of maturation") greatly increase the total surface area of the root.

Most *dicotyledonous plants* (e.g., peas and carrots, whose seeds have two "seed leaves"—commonly referred to as *dicots*) have taproot systems with one, or occasionally more, *primary roots* from which *secondary roots* develop (see the discussion of primary and secondary tissues in

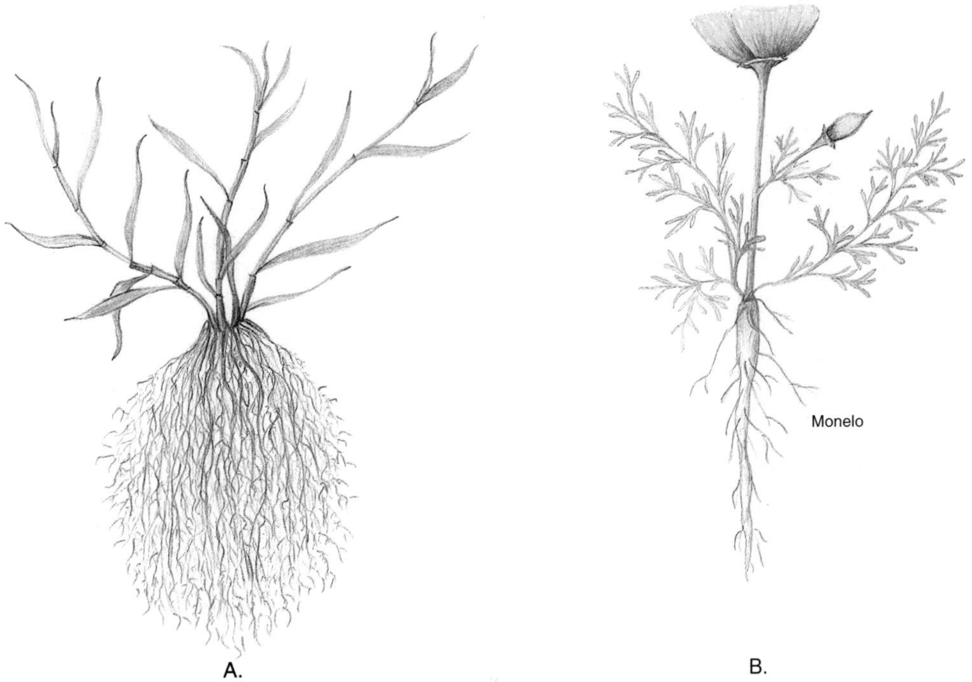

Figure 5.1 Root systems. *A.* A fibrous root system of a grass. *B.* A taproot system of a poppy.

Chapter 4). *Monocotyledonous plants* (e.g., corn and rice, whose seeds have one "seed leaf"—commonly referred to as monocots), on the other hand, have fibrous root systems. Adventitious and other types of roots may develop in both dicots and monocots. In English and other ivies, lateral adventitious roots that aid in climbing appear along the aerial stems, and in certain plants with specialized stems (e.g., rhizomes, corms, and bulbs; see Fig. 6.14), adventitious roots are the only kind produced.

ROOT STRUCTURE

Close examination of developing young roots usually reveals four regions or zones. Three of the regions are not sharply defined at their boundaries. The cells of each region gradually develop the form of those of the next region, and the extent of each region varies considerably, depending on the species involved. These regions are called (1) the *root cap,* (2) the *region of cell division,* (3) the *region of elongation,* and (4) the *region of maturation* (Fig. 5.2).

The Root Cap

The **root cap** is composed of a thimble-shaped mass of parenchyma cells covering the tip of each root. It is quite large and obvious in some plants, while in others, it is nearly invisible. One of its functions is to *protect* from damage the delicate tissues behind it as the young root tip pushes through often angular and abrasive soil particles. The root cap has no equivalent in stems. The dictyosomes of the root cap's outer cells secrete and release a slimy substance that lodges in the walls and eventually passes to the outside. The cells, which are replaced from the inside, constantly slough off, forming a slimy lubricant that facilitates the root tip's movement through the soil. This mucilaginous lubricant also provides a medium favorable to the growth of beneficial bacteria that add to the nitrogen supplies available to the plant (see the discussion of the nitrogen cycle in Chapter 25).

The root cap, whose cells have an average life of less than a week, can be slipped off or cut from a living root, and when this is done, a new root cap is produced. Until the root cap has been renewed, however, the root seems to grow randomly instead of more or less downward, suggesting that the root cap also functions in the *perception of gravity* (see *gravitropism* in Chapter 11). It is known that *amyloplasts* (plastids containing starch grains) act as gravity sensors, collecting on the sides of root-cap cells facing the direction of gravitational force. When a root that has been growing vertically is artificially tipped horizontally, the amyloplasts tumble or float down to the "bottom" of the cells in which they occur. The root begins growing downward again within 30 minutes to a few hours. The exact nature of this gravitational response is not known, but there is some evidence that calcium ions known to be present in the amyloplasts influence the distribution of growth hormones in the cells.

The Region of Cell Division

Cells in the **region of cell division,** which is composed of an *apical meristem* (a tissue of actively dividing cells) in the center of the root tip, produce the surrounding root cap. Most of the cell divisions take place next to the root cap at

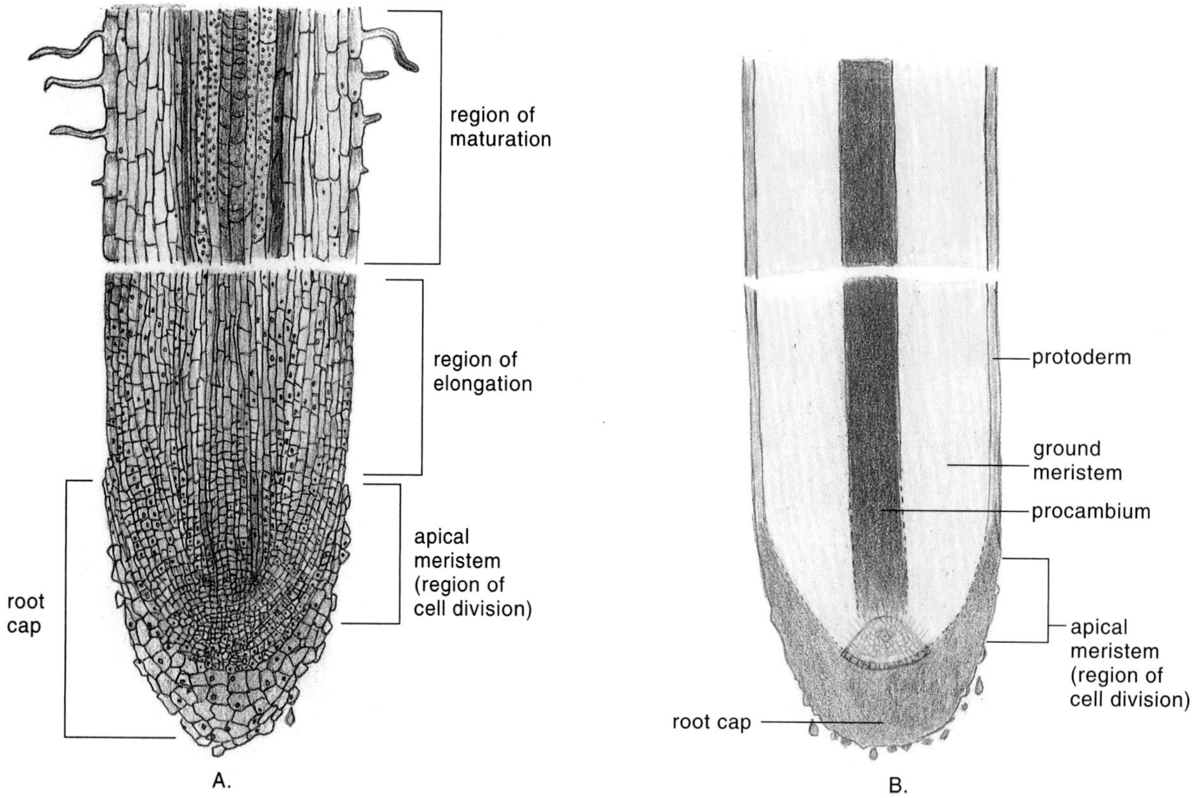

Figure 5.2 A longitudinal section through a dicot root tip. *A.* Regions of the root. *B.* Locations of the primary meristems of the root.

the edges of this inverted cup-shaped zone, located a short distance behind the actual base of the meristem. Here the cells divide every 12 to 36 hours, while at the base of the meristem, they may divide only once in every 200 to 500 hours. The divisions are often rhythmic, reaching a peak once or twice each day, usually toward noon and midnight, with relatively quiescent intermediate periods. Cells in this region are mostly cubical, with relatively large, more or less centrally located nuclei and a few very small vacuoles.

In both roots and stems, the apical meristem soon subdivides into three meristematic areas: (1) the **protoderm** gives rise to an outer layer of cells, the *epidermis*; (2) the **ground meristem,** to the inside of the protoderm, produces parenchyma cells of the *cortex*; (3) the **procambium,** which appears as a solid cylinder in the center of the root, produces *primary xylem* and *primary phloem* (Fig. 5.3). *Pith* (parenchyma) tissue, which originates from the ground meristem, is generally present in stems but is absent in most dicot roots. Grass roots and those of most other monocots, however, do have pith tissue.

The Region of Elongation

The **region of elongation,** which merges with the apical meristem, usually extends about 1 centimeter (0.4 inch) or less from the tip of the root. Here the cells become several times their original length and also somewhat wider. At the same time, the tiny vacuoles merge and grow until one or two large vacuoles, occupying up to 90% or more of the volume of each cell, have been formed. Only the root cap and apical meristem are actually pushing through the soil, since no further increase in cell size takes place above the region of elongation. The usually extensive remainder of each root remains stationary for the life of the plant. If a cambium is present, however, there normally is a gradual increase in girth through the addition of **secondary tissues** produced by the cambium.

The Region of Maturation

Most of the cells mature, or *differentiate,* into the various distinctive cell types of the primary tissues in this region, which is sometimes called the **region of differentiation,** or **root-hair zone.** The large numbers of hairlike, delicate protuberances that develop from many of the epidermal cells give the root-hair zone its name. The protuberances, called **root hairs,** which absorb water and minerals, adhere tightly to soil particles (Fig. 5.4) with the aid of microscopic fibers they produce and greatly increase the total absorptive surface of the root. Differentiation is discussed further in Chapter 11.

The root hairs are not separate cells; rather, they are tubular extensions of specialized epidermal cells. In fact, the nucleus of the epidermal cell to which each is attached often moves out into the protuberance. They are so numerous that they appear as fine down to the naked eye, typically numbering more than 38,000 per square centimeter (250,000 per square inch) of surface area in roots of plants such as corn;

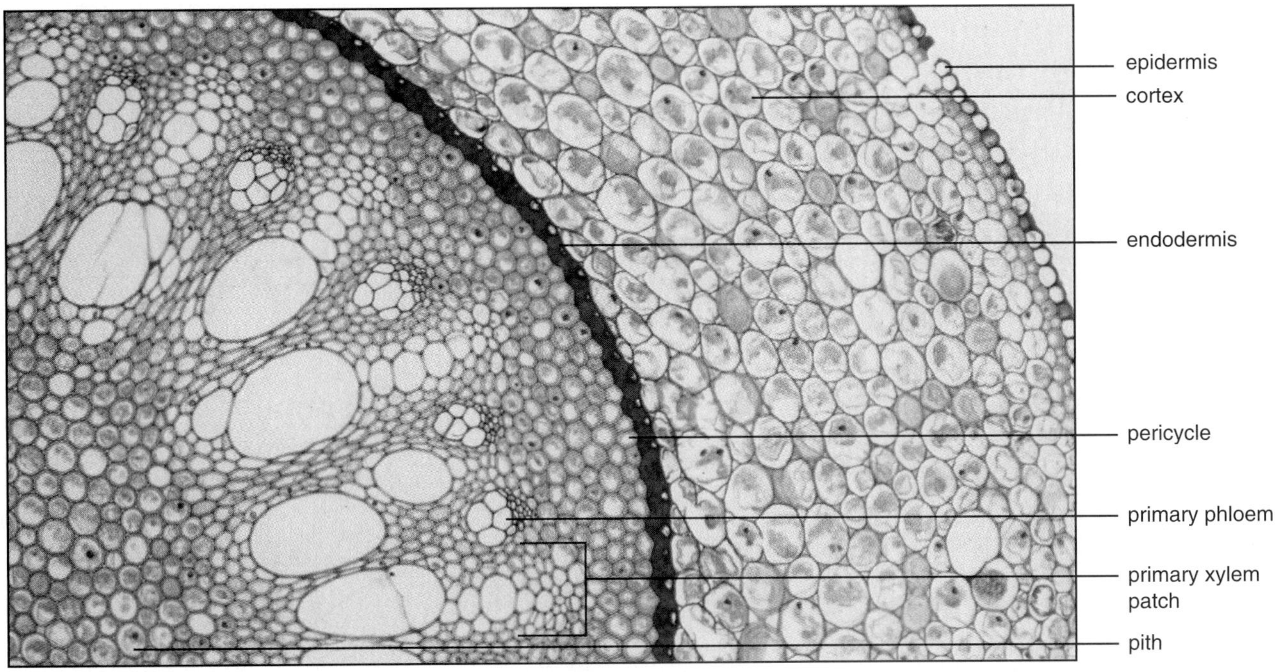

Figure 5.3 A cross section of a portion of a root of greenbrier (*Smilax*), a monocot, ×500. (*Photomicrograph by G. S. Ellmore*)

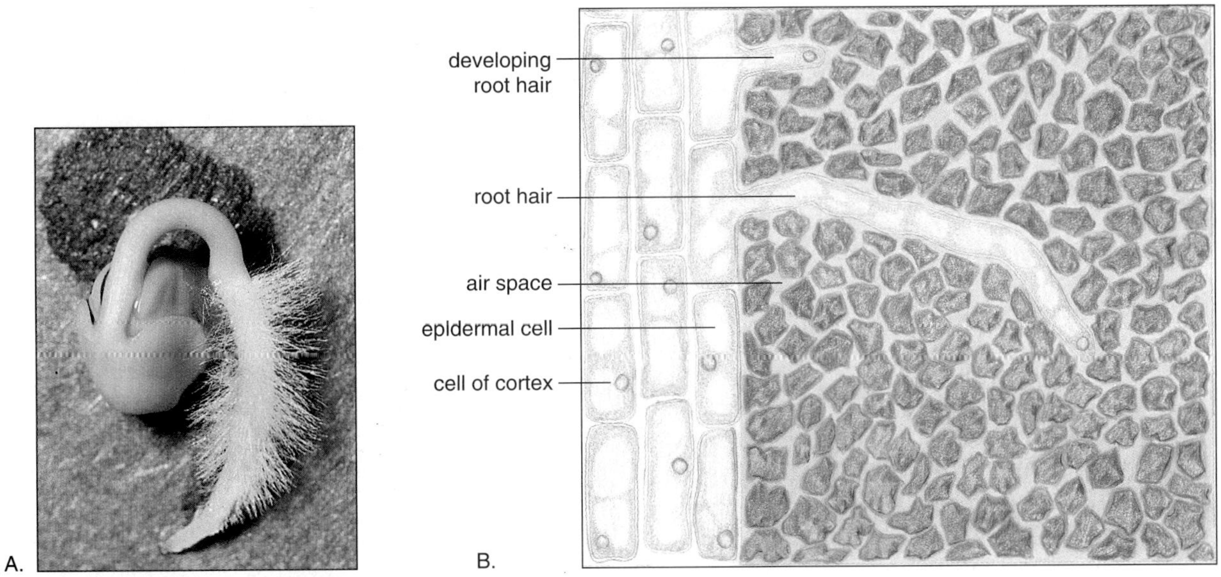

Figure 5.4 A. A radish (*Raphanus*) seedling shortly after germination, showing the root hair zone. *B.* A diagram of an enlargement of a longitudinal section through a small portion of a root hair zone, showing root hairs in contact with soil particles.

they seldom exceed 1 centimeter (0.4 inch) in length. A single ryegrass plant occupying less than 0.6 cubic meter (20 cubic feet) of soil was found to have more than 14 billion root hairs, with a total surface area almost the size of a football field.

When a seedling or plant is moved, many of the delicate root hairs are torn off or die within seconds if exposed to the sun, thereby greatly reducing the plant's capacity to absorb water and minerals in solution. This is why plants should be watered, shaded, and pruned after transplanting until new root hairs have formed. In any growing root, the extent of the root-hair zone remains fairly constant, with new root-hairs being formed toward the root cap and older root-hairs dying back in the more mature regions. The life of the average root-hair is usually not more than a few days, although a few live for a maximum of perhaps 3 weeks.

The *cuticle* (see Fig. 4.11), which may be relatively thick on the epidermal cells of stems and leaves, is thin enough on the root hairs and epidermal cells of roots in the region of maturation (Fig. 5.5) to allow water to be absorbed but still sufficient to protect against invasion by bacteria and fungi.

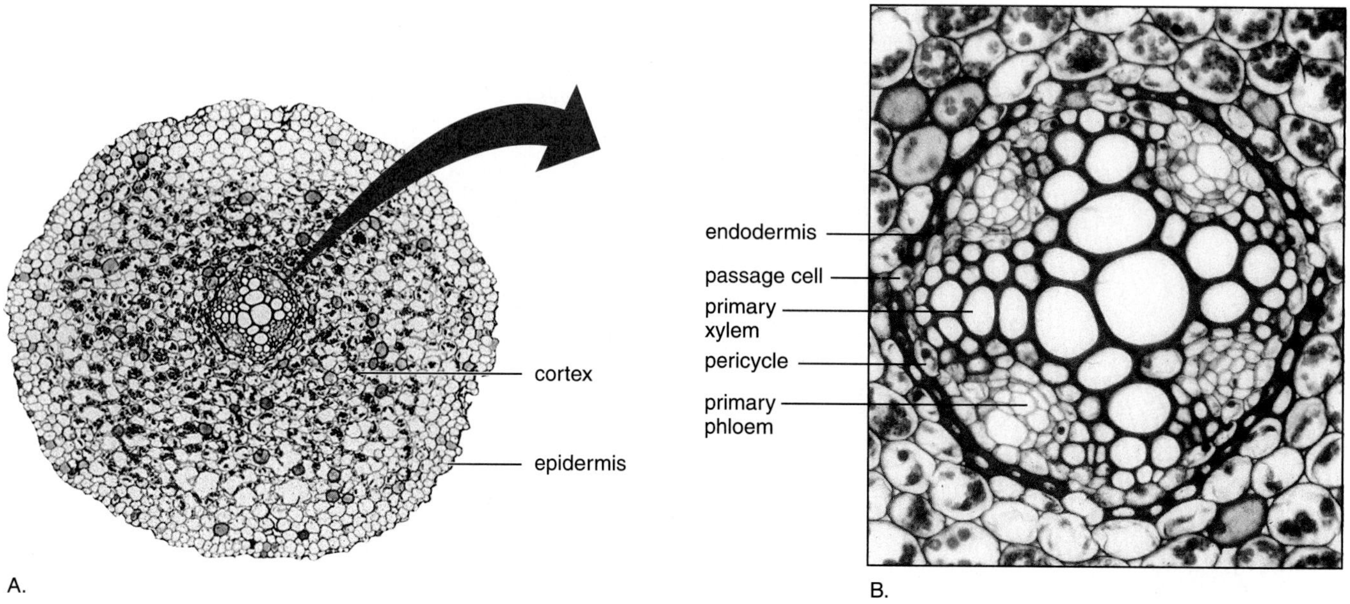

Figure 5.5 A cross section through the region of maturation of a dicot (buttercup—*Ranunculus*) root. *A*. Complete view, ×100. *B*. Enlargement of the root center, ×500.

The cells of the **cortex,** a tissue composed of parenchyma cells lying between the epidermis and inner tissues, mostly store food. This tissue, which may be many cells thick, is similar to the cortex of stems except for the presence of an **endodermis** at its inner boundary (Fig. 5.6). The endodermis consists of a single-layered cylinder of compactly arranged cells whose primary walls are impregnated with suberin. The suberin bands, called **Casparian strips,** are on the radial and transverse walls. The plasma membranes of the endodermal cells are fused to the Casparian strips, which are perpendicular to the root's surface; they prevent water from passing through the otherwise permeable (porous) cell walls. The Casparian strip barrier forces all water and dissolved substances entering and leaving the central core of tissues to pass through the plasma membranes of the endodermal cells or their plasmodesmata. This regulates the types of minerals absorbed and transported by the root to the stems and leaves. The plasma membranes tend to exclude harmful minerals while generally retaining useful ones.

An endodermis is rare in stems but so universal in roots that only three species of plants are known to lack a root endodermis. In some roots, the epidermis, cortex, and endodermis are sloughed off as their girth increases, but in those roots where the endodermis is retained, the inner walls of the endodermal cells eventually become thickened by the addition of alternating layers of suberin and wax. Later, cellulose and sometimes lignin are also deposited. Some endodermal cells, called **passage cells,** may remain thin-walled and retain their Casparian strips for a while, but they, too, eventually tend to become suberized.

A core of tissues, referred to collectively as the **vascular cylinder,** lies to the inside of the endodermis. Most of the cells of the vascular cylinder conduct water or food in solution, but lying directly against the inner boundary of the

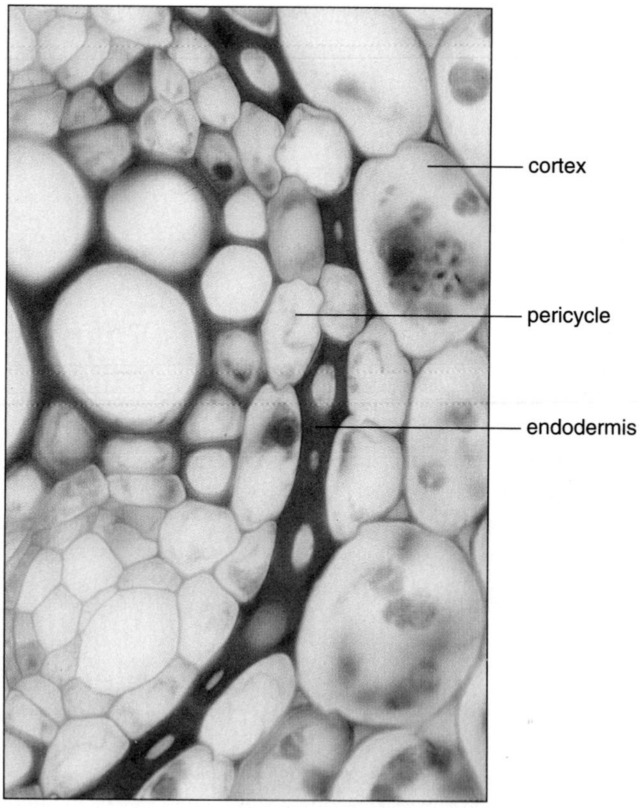

Figure 5.6 A portion of the endodermis of a buttercup (*Ranunculus*) root, ×1000.

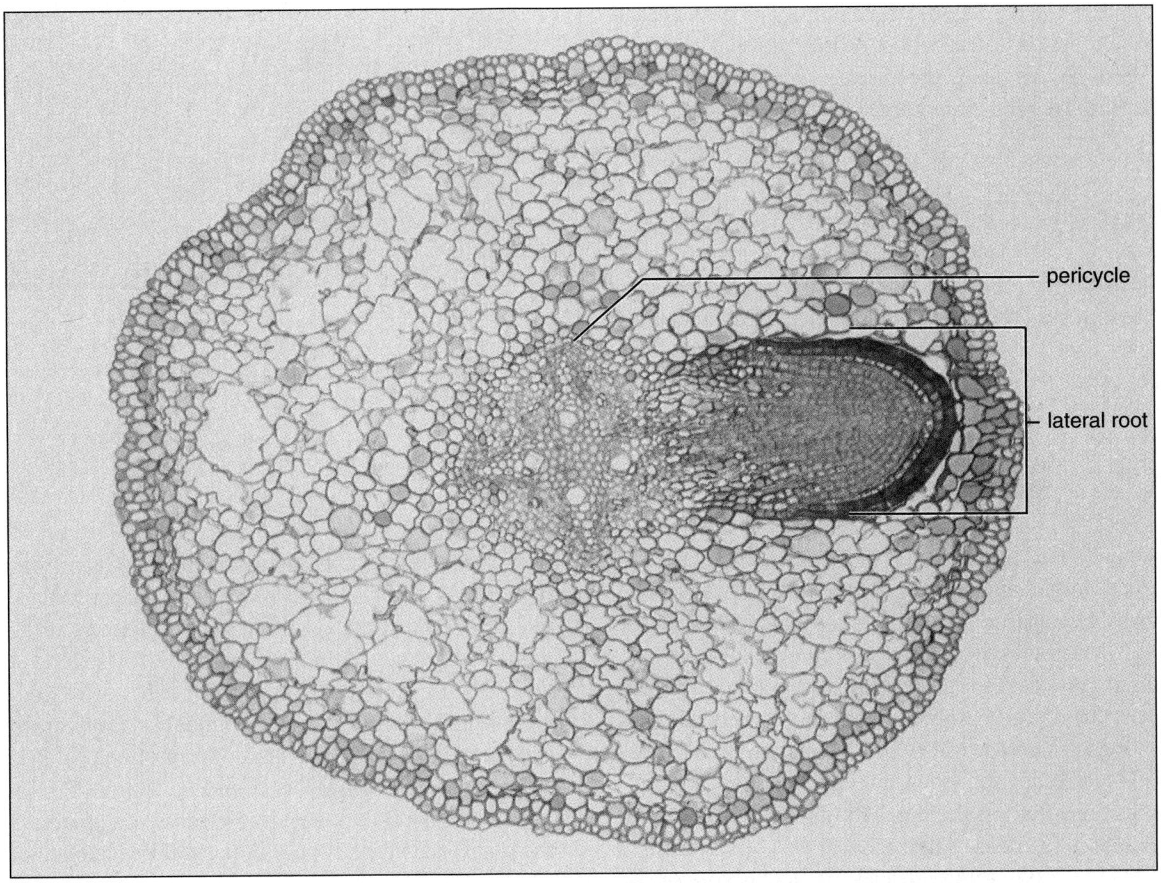

Figure 5.7 A cross section through a dicot (willow—*Salix*) root showing the origin of a lateral (branch) root. (*Photomicrograph by G. S. Ellmore*)

endodermis is an important layer of parenchyma tissue known as the **pericycle.** This tissue, which is usually one cell wide, may in some plants be a little wider. The cells of the pericycle may continue to divide even after they have matured. *Lateral (branch) roots* and part of the *vascular cambium* of dicots arise within the pericycle (Fig 5.7).

In most dicot and conifer roots, the primary xylem consists of a solid central core of water-conducting cells (e.g., tracheids; vessels). In cross section, this first root xylem usually loosely resembles the rear view of a rocket with fins. The fins, generally referred to as *arms,* tend to taper toward their tips and terminate just inside of the thin cylindrical pericycle layer. There are usually four of these arms, with some plants having two, three, or several. Branch roots begin to grow and develop in the pericycle opposite the xylem arms and push their way out to the surface through the endodermis, cortex, and epidermis.

The primary xylem surrounds pith parenchyma cells in monocot roots and those of a few dicots; in such plants, the arms may not be well defined. Primary phloem, which conducts food, forms in discrete patches between the xylem arms of both dicot and monocot roots.

A vascular cambium develops from parts of the pericycle and other parenchyma cells between the xylem arms and phloem patches in most dicots and conifers. This cambium at first follows the starlike outline of the primary xylem as it starts producing secondary phloem to the outside and secondary xylem to the inside. Eventually, however, the position of the cambium gradually shifts so that instead of appearing as patches and arms, the secondary conducting tissues appear as concentric cylinders. The primary phloem, in particular, may be sloughed off and lost as secondary tissues are added.

In woody plants, a second cambium, the *cork cambium,* normally arises in the pericycle outside of the vascular cambium and gives rise to *cork* tissue (*periderm*). Cork cells, which are dead at maturity, are impregnated with suberin and are impervious to moisture; similar tissues are produced in stems. Although there are exceptions, monocot roots generally have no secondary meristems and therefore no secondary growth.

In both roots and stems (see Chapter 6), growth may be *determinate* or *indeterminate.* Determinate growth is growth that stops after an organ such as a flower or a leaf is fully expanded or after a plant has reached a certain size. Indeterminate growth occurs in trees and other perennials where new tissues are added indefinitely, season after season.

Natural grafting can take place between roots of different trees of the same species, especially in the tropics. The roots unite through secondary growth when they come in contact with one another, but the details of the uniting process are not yet known. One unfortunate aspect of this grafting is that if one tree becomes diseased, the disease can be transmitted through the grafts to all the other trees connected to it.

SPECIALIZED ROOTS

As mentioned earlier, most plants produce either a fibrous root system, a taproot system, or, more commonly, combinations of the two types. Some plants, however, have roots with modifications that adapt them for performing specific functions as well as the absorption of water and minerals in solution.

Food-Storage Roots

Most roots and stems store some food, but in certain plants, the roots are enlarged and store large quantities of starch and other carbohydrates (Fig. 5.8), which may later be used for extensive growth. In sweet potatoes and yams, for example, extra cambial cells develop in parts of the xylem of branch roots and produce large numbers of parenchyma cells. As a result, the organs swell and provide storage areas for large amounts of starch and other carbohydrates. Similar food-storage roots are found in the deadly poisonous water hemlocks, in dandelions, and in salsify. In carrots, beets, turnips, and radishes, the food-storage tissues are actually a combination of root and stem. Although the external differences are not obvious, approximately 2 centimeters (0.8 inch) at the top of an average carrot is derived from stem tissue that merges with the root tissue below.

Figure 5.8 A sweet potato (*Ipomoea*) plant. Note the food-storage roots.

Figure 5.9 A manroot (*Marah*) water-storage root that weighed over 25.3 kilograms (60 pounds). (*Courtesy Robert A. Schlising*)

Water-Storage Roots

Some members of the Pumpkin Family (Cucurbitaceae) produce huge water-storage roots. This is particularly characteristic of those that grow in arid regions or in those areas where there may be no precipitation for several months of the year. In certain manroots (*Marah*), for example, roots weighing 30 kilograms (66 pounds) or more are frequently produced (Fig. 5.9), and a major root of one calabazilla plant (*Cucurbita perennis*) was found to weigh 72.12 kilograms (159 pounds). The water in the roots is apparently used by the plants when the supply in the soil is inadequate.

Propagative Roots

Many plants produce **adventitious buds** (buds appearing in places other than stems) along the roots that grow near the surface of the ground. The buds develop into aerial stems called *suckers,* which have additional rootlets at their bases. The rooted suckers can be separated from the original root and grown individually. Cherries, apples, pears, and other fruit trees often produce suckers. The adventitious roots of rice-paper plants (*Tetrapanax papyrifera*) and tree-of-heaven (*Ailanthus altissima*) can become a nuisance in gardens, often producing propagative roots 10 meters (33 feet) or more from the parent plant. Horseradish (*Rorippa armoracia*), Canada thistle (*Cirsium arvense*), and some other weeds have a remarkable facility to reproduce in this fashion as well as by means of seeds. In the past, this capacity has made it difficult to control them, but some biological controls being investigated (see Appendix 2) may be an answer to the problem in the future.

Pneumatophores

Water, even after air has been bubbled through it, contains less than one-thirtieth the amount of free oxygen found in the air. Accordingly, plants growing with their roots in water may not have enough oxygen available for normal respiration in their root cells. Some swamp plants, such as the black mangrove (*Avicennia nitida*) and the yellow water weed

Figure 5.10 Pneumatophores (foreground) of tropical mangroves rising above the sand at low tide. The pneumatophores are spongy outgrowths from the roots beneath the surface. Pneumatophores facilitate the exchange of oxygen and carbon dioxide for the roots, which grow in areas where little oxygen is otherwise available to them. (*Courtesy Lani Stemmerman*)

(*Ludwigia repens*), develop special spongy roots, called **pneumatophores,** which extend above the water's surface and enhance gas exchange between the atmosphere and the subsurface roots to which they are connected (Fig. 5.10). The woody "knees" of the bald cypress (*Taxodium distichum*), which occurs in southern swamps (see Fig. 22.18), were in the past believed to be pneumatophores, but there is no conclusive evidence for this theory.

Aerial Roots

Velamen roots of orchids, *prop roots* of corn and banyan trees (Fig. 5.11), *adventitious roots* of ivies, and *photosynthetic roots* of certain orchids are among various kinds of **aerial roots** produced by plants. It was formerly assumed that the epidermis of velamen roots, which is several cells thick, aided in the absorption of rain water. It appears, however, it may function more in preventing loss of moisture from the root. Corn prop roots, produced toward the base of the stems, support the plants in a high wind. Some tropical plants, including the screw pines and various mangroves, produce sizable prop roots extending for several feet above the surface of the ground or water. Debris collects between them and helps to create additional soil.

Figure 5.11 The aerial (velamen) roots of orchids have a thick epidermis that reduces water loss from internal tissues.

Many of the tropical figs or banyan trees produce roots that grow down from the branches until they contact the soil. Once they are established, they continue secondary growth and look just like additional trunks (Fig. 5.12). Banyan trees may live for hundreds of years and can become very large. In India and southeast Asia, there are several banyan trees that have almost 1,000 root-trunks and have circumferences approaching 450 meters (1,476 feet). The oldest is estimated to be about 2,000 years old.

The vanilla orchid, from which we obtain vanilla flavoring, produces chlorophyll in its aerial roots and, through photosynthesis, can manufacture food with them. The adventitious roots of English ivy, Boston ivy, and Virginia creeper appear along the stem and aid the plants in climbing.

Contractile Roots

Some herbaceous dicots and monocots have contractile roots that pull the plant deeper into the soil. Many lily bulbs are pulled a little deeper into the soil each year as new sets of contractile roots are developed (Fig. 5.13). The bulbs continue to be pulled down until an area of relatively stable

Figure 5.12 A banyan (*Ficus*) tree with many large prop roots that have developed from the branches.

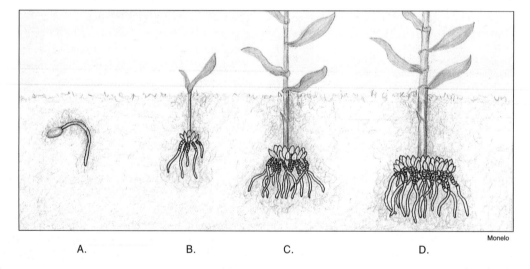

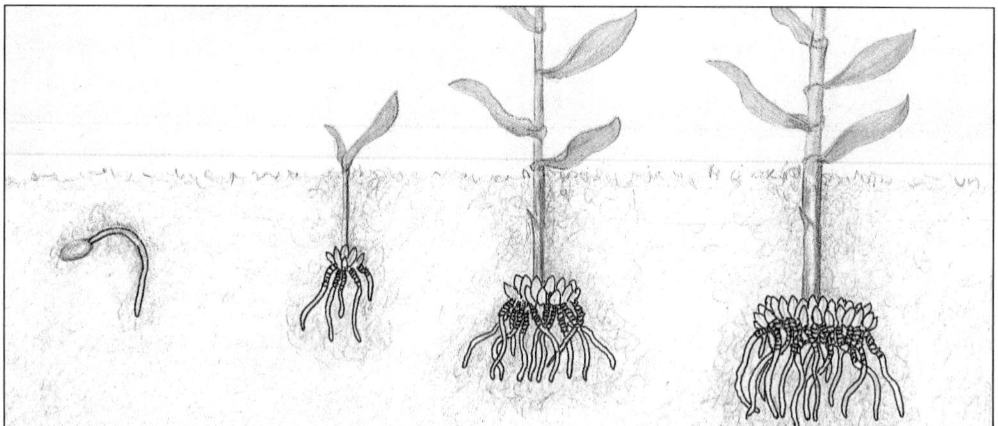

Figure 5.13 How a lily bulb over three seasons is pulled deeper into the soil by the action of contractile roots. *A.* A seed germinates. *B.* Contractile roots pull a newly formed bulb down several millimeters during the first season. *C.* The bulb is pulled down farther the second season. *D.* The bulb is pulled down even farther the third season. The bulb will continue to be pulled down in succeeding seasons until it reaches an area of relatively stable soil temperatures.

Figure 5.14 Buttress roots of a tropical fig tree.

temperatures is reached. Plants such as dandelions always seem to have the leaves coming out of the ground as the top of the stem is pulled down a small amount each year when the root contracts. The contractile part of the root may lose as much as two-thirds of its length within a few weeks as stored food is used and the cortex collapses.

Buttress Roots

Some tropical trees growing in shallow soils produce huge buttresslike roots toward the base of the trunk, giving them great stability (Fig. 5.14). Except for their angular appearance, these roots look like a part of the trunk.

Parasitic Roots

Some plants, including dodders, broomrapes, and pinedrops, have no chlorophyll (necessary for photosynthesis) and have become dependent on chlorophyll-bearing plants for their nutrition. They parasitize their host plants via peglike projections called **haustoria** (singular: **haustorium**), which develop along the stem in contact with the host. The haustoria penetrate the outer tissues and establish connections with the water-conducting and food-conducting tissues (Fig. 5.15). Some green plants, including Indian warrior and the mistletoes, also form haustoria. These haustoria, however, apparently aid primarily in obtaining water and dissolved minerals from the host plants, since the partially parasitic plants are capable of manufacturing at least some of their own food through photosynthesis. Other plants lacking chlorophyll (e.g., Indian pipes) are not parasitic at all. Instead, these plants are saprophytic, obtaining all the nutrients they require from organic materials in the soil.

MYCORRHIZAE

More than three-quarters of all seed plant species have various fungi associated with their roots. The association is *mutualistic;* that is, both the fungus and the root benefit from it but are dependent upon the association for normal development. (Mutualism is a form of *symbiosis;* see page 300.)

The fungus is able to absorb and concentrate phosphorus much better than it can be absorbed by the root hairs. In fact, if mycorrhizal fungi have been killed by fumigation or are otherwise absent, many plants appear to have considerable difficulty absorbing phosphorus, even when the element is abundant in the soil. The phosphorus is stored in granular form until it is used by the plant. The fungus also often forms a mantle of millions of threadlike strands that facilitate the absorption of water and nutrients. The plant furnishes sugars and amino acids without which the fungus cannot survive.

These "fungus-roots," or **mycorrhizae** (Fig. 5.16), are essential to the normal growth and development of forest trees and many herbaceous plants. Orchid seeds will not

A.

B.

Figure 5.15 A. Pale stems of a parasitic plant (dodder *Cuscuta*) twining about other vegetation. *B.* A close-up view of dodder, showing the peglike *haustoria* that penetrate the tissues of the host plant.

Ecological Review

Plants take up water and nutrients from soils through their roots, one of the major avenues of ecological exchange between plants and the environment. Roots also act as organs for storage of water and energy, for support, and for uptake of oxygen in some environments. Plant roots are also sites of mutualistic relationships with fungi (mycorrhizae) and with nitrogen-fixing bacteria. The roots of parasitic plants attack host plants. Soils, the primary medium with which most roots interact, are structured through complex relationships between parent mineral material, climate, organisms, and topography over the course of time.

germinate until mycorrhizal fungi invade their cells. In virtually all of the woody trees and shrubs found in forests, the fungal threads grow between the walls of the outer cells of the cortex but rarely penetrate into the cells themselves. If they should happen to penetrate, they are apparently broken down and digested by the host plants. In herbaceous plants, the fungi do penetrate the cortex cells as far as the endodermis, but they cannot grow beyond the Casparian strips. Once inside the cells, the fungi branch repeatedly but do not break down the plasma or vacuolar membranes.

Some plants do not seem to need mycorrhizae unless there are barely enough essential elements for healthy growth present in the soil. Plants with mycorrhizae develop few root hairs compared to those growing without an associated fungus. Mycorrhizae have proved to be particularly susceptible to acid rain (discussed in Chapter 25); this may signal major problems for our coniferous forests in the future if the problem of acid rain is not solved. Methyl bromide, used in the past to sterilize seed beds, kills all soil organisms, including mycorrhizae; its use in the United States has now been banned.

ROOT NODULES

Although almost 80% of our atmosphere consists of nitrogen gas, plants cannot convert the nitrogen gas to usable forms. A few species of bacteria, however, produce enzymes with which they can convert nitrogen into nitrates and other nitrogenous substances readily absorbed by roots. Members of the Legume Family (Fabaceae), which includes peas and beans, and a few other plants such as alders, form associations with certain soil bacteria that result in the production of numerous small swellings called

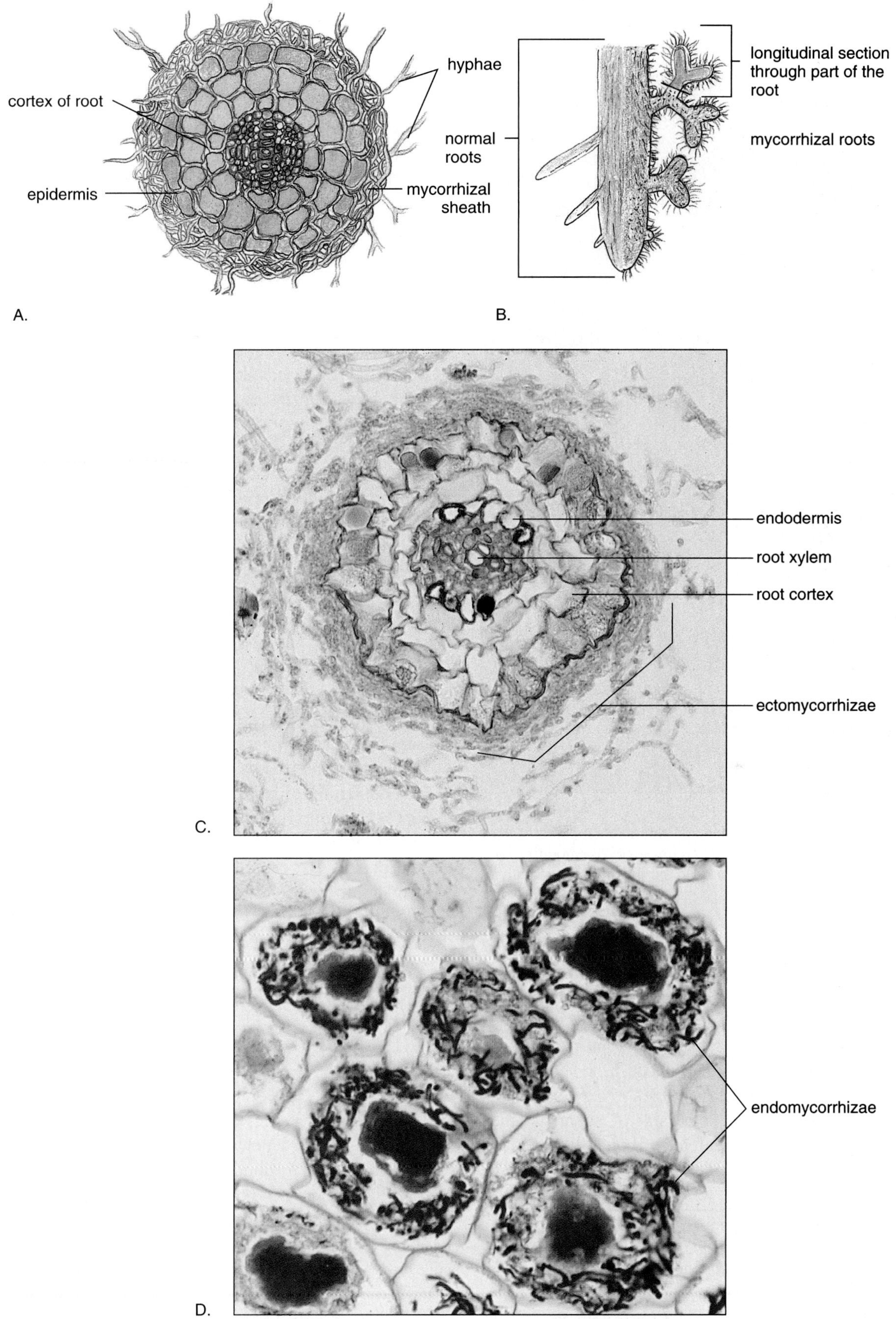

Figure 5.16 Mycorrhizae. *A.* A diagram of a cross section of a root with ectomycorrhizae (visible on the outside of the root). *B.* Comparison, in longitudinal section, of a root without ectomycorrhizae and one with ectomycorrhizae. *C.* Photomicrograph of a cross section of a root around which ectomycorrhizae have formed a mantle. The fungal cells have not penetrated deeper than the outermost layers of root cells. *D.* A cross section of a few root parenchyma cells with endomycorrhizae. The endomycorrhizae develop and flourish within the parenchyma cells.

Figure 5.17 Root nodules on the roots of bur clover (*Medicago polymorpha*). The somewhat popcornlike nodules contain bacteria that convert nitrogen from the air into forms that can be used by the plant, ×5.

root nodules that are clearly visible when such plants are uprooted (Fig. 5.17). The nodules contain large numbers of nitrogen-fixing bacteria.

A substance exuded into the soil by plant roots stimulates *Rhizobium* bacteria, which, in turn, respond with another substance that prompts root hairs to bend sharply. A bacterium may attach to the concave side of a bend and then invade the cell with a tubular infection thread that does not actually break the host cell wall and plasma membrane. The infection thread grows through to the cortex, which is stimulated to produce new cells that become a part of the root nodule; here the bacteria multiply and engage in nitrogen conversion. (See also the discussion of the nitrogen cycle in Chapter 25.)

Root nodules should not be confused with *root knots,* which are also swellings that may be seen in the roots of tomatoes and many other plants. Root knots develop in response to the invasion of tissue by small, parasitic roundworms (nematodes). Unlike bacterial nodules, root knots are not beneficial, and the activities of the parasites within them can eventually lead to the premature death of the plant.

HUMAN RELEVANCE OF ROOTS

Roots have been important sources of food for humans since prehistoric times, and some, such as the carrot, have been in cultivation in Europe for at least 2,000 years. A number of cultivated root crops involve biennials (i.e., plants that complete their life cycles from seed to flowering and back to seed in two seasons). Such plants store food in a swollen taproot during the first year of growth, and then the stored food is used in the production of flowers in the second season. Among the best-known biennial root crops are sugar beets, beets, turnips, rutabagas, parsnips, horseradishes, and carrots. Other important root crops include sweet potatoes, yams, and cassava. Cassava (Fig. 5.18), from which tapioca is made, forms a major part of the basic diet for millions of inhabitants of the tropics. With a minimum of human labor, it yields more starch per hectare (about 45 metric tons, the equivalent of 20 tons per acre) than any other cultivated crop. Minor root crops, including relatives of wild mustards, nasturtiums, and sorrel, are cultivated in South America and other parts of the world.

Several well-known spices, including sassafras, sarsaparilla, licorice, and angelica, are obtained from roots. Sweet potatoes are used in the production of alcohol in Japan. Some important red to brownish dyes are obtained from roots of members of the Madder Family (Rubiaceae), to which coffee plants belong. Drugs obtained from roots include aconite, ipecac, gentian, and reserpine, a tranquilizer. A valuable insecticide, *rotenone,* is obtained from the barbasco plant, which has been cultivated for centuries as a fish poison by primitive South American tribes. When thrown into a dammed stream, the roots containing rotenone cause the fish to float but in no way poison them for human consumption. In tobacco plants, nicotine produced in the roots is transported to the leaves. Other uses of roots are discussed in Chapter 24.

SOILS

The soil is a dynamic, complex, constantly changing part of the earth's crust, which extends from a few centimeters deep in some places to hundreds of meters deep in others. It is essential not only to our existence but also to the existence of most living organisms. It has a pronounced effect on the plants that grow in it, and they have an effect on it.

If you dig up a shovelful of soil from your yard and examine it, you will probably find a mixture of ingredients, including several grades of sand; rocks and pebbles; powdery silt; clay; humus; dead leaves and twigs; clods consisting of soil particles held together by clay and organic matter; plant roots; and small animals, such as ants, pill bugs, millipedes, and earthworms. Also present, but not visible, would be millions of microorganisms, particularly bacteria, fungi, and, of course, air and water.

Figure 5.18 Cassava (*Manihot esculenta*) plants. Note the food-storage roots on the plant that has been dug up.

Figure 5.19 A soil profile. (*Reproduced from the Marbut Memorial Slide Set, 1968, SSSA, by permission of Soil Sciences of America, Inc.*)

The soil became what it is today through the interaction of a number of factors: climate, parent material, topography of the area, vegetation, living organisms, and time. Because there are thousands of ways in which these factors may interact, there are many thousands of different soils. The solid portion of a soil consists of mineral matter and organic matter. Pore spaces, shared by variable amounts of water and air, occur between the solid particles. The smaller pores often contain water, and the larger ones usually contain air. The sizes of the pores and the connections between them largely determine how well the soil is aerated.

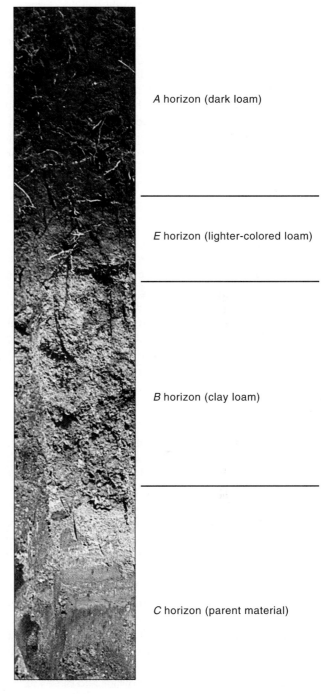

If one were to dig down 1 or 2 meters (3 to 6 feet) in an undisturbed area, a soil profile of three intergrading regions called *horizons* would probably be exposed (Fig. 5.19). The horizons show the soil in different stages of development, and the composition varies accordingly. The upper layer, usually extending down 10 to 20 centimeters (4 to 8 inches), is called the *topsoil.* It is usually subdivided into a darker upper portion called the *A horizon* and a lighter lower portion called the *E horizon.* The *A* portion contains more organic matter than the layers below.

The next 0.3 to 0.6 meter (1 to 2 feet) is called the *B horizon,* or *subsoil.* It usually contains more clay and is lighter in color than the topsoil. The *C horizon* at the bottom may vary from about 10 centimeters (4 inches) to several meters (6 to 10 feet or more) in depth; it may even be absent. It is commonly referred to as the *soil parent material* and extends down to *bedrock.*

Parent Material

The first step in the development of soil is the formation of *parent material* from existing rocks that have not yet been broken down into smaller fragments. Parent material accumulates through the weathering of three types of rock, which originate from various sources. These sources and types include volcanic activity (*igneous rocks*); fragments deposited by glaciers, water, or wind (*sedimentary rocks*); or changes in igneous or sedimentary rocks brought about by great pressures, heat, or both (*metamorphic rocks*).

Climate

Climate varies greatly throughout the globe, and its role in the weathering of rocks varies correspondingly. In desert areas, for example, there is little weathering by rain, and soils are poorly developed. In areas of moderate rainfall, however, well-developed soils are common. In some areas of high rainfall, the excessive flow of water through the soil may leach out important minerals. Similar leaching out of important minerals may occur when garden sprinklers are left on all night. Many gardeners and house-plant enthusiasts have stunted or killed the very plants they were trying to foster by "drowning" them with too much water or too frequent watering. As a general rule, plants should not be watered unless the soil surface feels dry.

In areas where there are great temperature ranges, rocks may split or crack as their outer surfaces expand or contract at different rates from the material beneath the surface. When water in rock crevices freezes, it expands and causes further cracks and splits.

Living Organisms and Organic Composition

There are many kinds of organisms in the soil, as well as roots and other parts of plants. In the upper 30 centimeters (1 foot) of a good agricultural soil, living organisms constitute about one-thousandth of the total weight of the soil. This may not sound significant, but it amounts to approximately 6.73 metric tons per hectare (3 tons per acre). Bacteria and fungi present in the soil decompose all types of organic matter, which accumulates when leaves fall and roots and animals die. (This process is further discussed under the section on composting in Chapter 17.) Roots and all other living organisms produce carbon dioxide, which combines with water and forms an acid, thereby increasing the rate at which minerals dissolve. Ants and other insects, earthworms, burrowing animals, and birds all alter the soil through their activities and add to its organic increment either through wastes that they deposit or through the decomposition of their bodies when they die. *Humus,* which consists of partially decomposed organic matter, gives some soils a dark color.

The total organic composition of a soil varies greatly. An average topsoil might consist of about 25% air, 25% water, 48% minerals, and 2% organic matter. Soils in low, wet areas, where a lack of oxygen keeps microorganisms from their normal activities, may contain as much as 90% organic matter. Except in legumes and a few other plants, almost all of the nitrogen utilized by growing plants, as well as much of the phosphorus and sulfur, comes from decomposing organic matter. In addition, as organic matter breaks down, it produces acids, which, in turn, decompose minerals. Other roles of organic matter in the soil are discussed in the section on soil structure.

Topography

If the topography (surface features) is steep, soil may wash away or erode through the action of wind, water, and ice as soon as it is weathered from the parent material. It has been estimated that more than 20 metric tons of topsoil per hectare (8.2 tons per acre) is washed away annually from some prime croplands in the central United States.

If an area is flat and poorly drained, pools and ponds may appear in slight depressions when it rains. If these bodies of water cannot drain quickly, the activities of organisms in the soil are interrupted, and the development of the soil is arrested. The ideal topography for the development of soil is one that permits drainage without erosion.

Soil Texture and Mineral Composition

Soil texture refers to the relative proportions of sand, silt, and clay in a given soil (Table 5.1).

Sands are usually composed of many small particles bound together chemically or by a cementing matrix material. Silt consists of particles that are mostly too small to be seen without a lens or a microscope.

Clay particles are so tiny that they can't be seen with even a powerful light microscope, although they can be seen with an electron microscope. Individual clay particles are called *micelles.* Micelles are somewhat sheetlike, negatively charged, and held together by chemical bonds. The negative charges attract, exchange, or retain positively charged ions. Many of the positively charged ions, such as magnesium (M^{++}) and potassium (K^{+}), which are needed for normal plant growth, are absorbed with water by the roots. Clay is

TABLE 5.1

Soil Mineral Components as Classified by the U. S. Department of Agriculture

MINERAL	DIAMETER (RANGE IN MM)	COMMENTS
Stones	>76 mm	Do not support plant growth but affect permeability and erosion of the soil
Gravel	76 mm–2.0 mm	
Very coarse sand	2.0 mm–1.0 mm	
Coarse sand	1.0 mm–0.5 mm	
Medium sand	0.5 mm–0.25 mm	Makes soil feel gritty
Fine sand	0.25 mm–0.10 mm	
Very fine sand	0.10 mm–0.05 mm	
Silt	0.05 mm–0.002 mm	A lens or microscope needed to see any but the coarsest silt particles
Clay	<0.002 mm	May absorb water, swell, and later shrink, causing the soil to crack as it dries

plastic in nature because the water that adheres tightly to the surface of the particles acts both as a binding agent and a lubricant. Physically, clay is a **colloid,** that is, a suspension of particles that are larger than molecules but that do not settle out of a fluid medium.

The best agricultural soils are usually loams, which are a mixture of sand, clay, and organic matter. The better loams have about 40% silt, 40% sand, and 20% clay. Light soils have a high sand and low clay content. Heavy soils have high clay content. Coarse soils, which have larger particles, are porous and don't retain much water, while clay soils have high water content and allow little water to pass through.

Over half the composition by weight of mineral matter is oxygen. Other elements commonly present are hydrogen, silicon, aluminum, iron, potassium, calcium, magnesium, and sodium. However, soil inherits hundreds of different mineral combinations from its parent material.

Soil Structure

Soil structure refers to the arrangement of the soil particles into groups called *aggregates.* Aggregates in sands and gravels show little cohesion, but most agricultural soils have aggregates that stick together. Structure develops when colloidal particles clump together, mostly as a result of the activities of soil organisms, freezing, and thawing. If the individual granules do not become coated with organic matter, they may continue to clump until they become clods.

Productive agricultural soils are granular soils with pore spaces that occupy between 40% and 60% of the total volume of the soil. The pores contain air and water, and their sizes are more important than their total volume. Clay soils, for example, have more pore space than sandy soils, but the pores are so small that water and air are restricted in their movement through the soil. When the pores are full of water, air is kept out, and there is not enough oxygen for root growth. Sandy soils have large pores, which drain by gravity soon after they are filled. The water is replaced by air, but too much air speeds up nitrogen release by microorganisms. Plants can't use the nitrogen that quickly, and much of it is lost.

Water itself, as we have noted, can be harmful. Under anaerobic conditions (marked by the absence of oxygen), too much water leaches mineral nutrients and slows the mineralization process. Too much water also slows the release of nitrogen, interferes with plant growth, and accelerates the breakdown of nitrates to the extent that virtually all the nitrates may be lost in as little as half an hour.

Water in the Soil

Water in the soil occurs in three forms. **Hygroscopic water** is physically bound to the soil particles and is unavailable to plants. **Gravitational water** drains out of the pore spaces after a rain. If drainage is poor, it is this water that interferes with normal plant growth. Plants are mainly dependent for their needs on the third type, **capillary water,** which is water held against the force of gravity, in pores of the soil. The structure and organic matter of the soil—which enable the soil to hold water against the force of gravity—the density and type of vegetational cover, and the location of underground water tables largely determine the amount of capillary water available to the plant.

Awareness

Metal-Munching Plants

Years ago, prospectors looking for promising sites to mine silver or gold often noticed that certain plants grew on old mine tailings. They reasoned that these plants might be indicators of the precious metals entrapped in the soil. It turns out that these prospectors had it right.

Certain plants not only grow in heavy metal-laden soils but are able to extract these metals through their root systems and accumulate them in their tissues without being damaged. Plants absorb metals because they require certain ones such as zinc and copper as components of their proteins and enzymes for normal growth and development. In certain soils, plant mineral uptake makes no distinction between heavy metals such as cadmium or selenium and these required elements. In either case, these metals are absorbed by the plant's extensive root system, which may extend a meter or more in depth. Scientists have coined a new term to describe this process—**phytoremediation**—the use of plants to facilitate the removal of toxic compounds from ground water and soil. The plants that "munch metal" so well are called "hyperaccumulating" plants.

What makes a good hyperaccumulator? Researchers are attempting to answer this question, but so far, nobody knows for sure. One clue may be metal-binding polypeptides called **phytochelatins** that sequester and detoxify heavy metals in plant tissue. A survey of numerous plants has shown that phytochelatins are produced when these plants are exposed to heavy metals in soil. Interestingly, a wide range of plants from the most advanced flowering plants (even orchids) to red, green, and brown algae produce these detoxifying polypeptides. One possible method of accumulating heavy metals is for the plant to transport them into the cell's vacuoles, a sort of waste disposal dump.

Phytoremediation is attracting the increased attention not only of scientists, but also of regulators who see it as a low-cost alternative for cleaning up contaminated sites across the country. The conventional process of soil excavation and reburial in a landfill is very expensive, typically costing about $1.5 million per acre, depending on the pollutant. The price tag opens the door for many alternative ideas. These plants can be harvested and disposed of—and in certain instances, the metals can even be recovered by sending the plants to a smelter. But can hyperaccumulators effectively get the job done?

Alpine pennycress (*Thlaspi caerulescens*) is a remarkable hyperaccumulator, able to accumulate 4% of its dry body weight in zinc. This translates into 10 metric tons of zinc per hectare. The problem, however, is that *Thlaspi* is a small and slow-growing plant. For phytoremediation to become practical, plants with metal uptake rates comparable to *Thlaspi* but with faster growth rates and larger tissue mass must be found. Screening of other plants is being done in several laboratories around the world. One plant that has been identified so far is Indian mustard, *Brassica juncea,* a relative of some highly nutritious vegetables—cabbage, cauliflower, broccoli, collard, kale, and mustard greens. Indian mustard can accumulate 3.5% of its dry body weight in lead. It also can absorb cadmium, chromium, nickel, selenium, zinc, and copper. Cattails (*Typha* spp.), which also are known to accumulate heavy metals, are already in use in some areas in the final stages of the treament of human sewage.

Despite the advantages of phytoremediation, one drawback is that multiple crops must be planted over several years to reduce contamination to acceptable levels, while removal of the soil provides an immediate resolution. Additionally, there is concern about increasing the accumulation of these metals in the food chain as wildlife and insects eat the plants, accumulating toxicity in their bodies. In this way, toxic metals could work their way up the food chain and pose a new set of problems.

Phytoremediation is being put to a big test at Chernobyl, the site of the largest environmental disaster in modern history. In 1986, the meltdown of the nuclear reactor left radioactive wastes scattered over the Ukrainian countryside. A pilot study there suggests that Indian mustard could remove radioactive strontium from the soil over a 5-year period. While scientists continue to test this promising natural process, environmental cleanups of the future could be as simple as letting the flowers grow.

D. C. Scheirer

The ancient Incas of Peru knew that water would rise just so far in some areas. Where the water table was close to the surface, they removed the upper 0.6 meter (2 feet) of soil and planted their crops down in the hollowed-out areas so that the roots would be able to reach the capillary water. They knew that in some areas having sandy soils and low annual precipitation, soils could be compacted by a heavy roller to create finer capillaries to raise water from below. This technique is effective only if the available water is within 1.5 to 3.0 meters (5 to 10 feet) of the surface and if the soils do not contain much silt or clay.

After rain or irrigation, water in the soil drains away by gravity. The water remaining after such draining is referred to as the *field capacity* of the soil. Field capacity is mainly governed by the texture of the soil, but the structure and organic content also influence it to a certain extent. Plants readily absorb water from the soil when it is at, or near, field capacity. As the soil dries, the film of water around each soil particle becomes thinner and more tightly bound to the soil particle and less likely to enter the root. If water is not added to the soil, eventually a point is reached at which the rate of absorption of water by the plant is insufficient for its needs, and the plant wilts permanently. The soil is then said to be at the *permanent wilting point.* In clay soils, the permanent wilting point is reached when the water content drops below 15%, while in sandy soils, the permanent wilting point may be as low as 4%. *Available water* is soil water between field capacity and the permanent wilting point.

Soil pH

The **pH** (acidity or alkalinity) of a soil affects both the soil and the plants growing in it in various ways. Cranberries, for example, thrive under acidic conditions, but a soil that is unusually acid or alkaline may be toxic to the roots of other plants, and mycorrhizae do not survive in soils having pH extremes. These conditions, however, do not normally directly affect plants nearly as much as they affect nutrient availability. For example, alkalinity causes minerals, such as copper, iron, and manganese, to become less available to plants, while acidity, if high enough, inhibits the growth of nitrogen-fixing bacteria. Acid soils tend to be common in areas of high precipitation where significant amounts of bases are leached from the topsoil.

It is a common agricultural practice to counteract soil acidity by adding compounds of calcium or magnesium in a process known as *liming.* Alkaline soils can be made more acid by the addition of sulfur, which is converted by bacteria to sulfuric acid. The addition of some nitrogenous fertilizers may have the same effect.

Summary

1. Roots anchor plants and absorb water and minerals in solution. A germinating seed radicle becomes the first root. Taproots with branch roots, or adventitious roots that become a fibrous root system develop from the radicle or the stem just above it. Many plants have combinations of both systems.
2. Four zones or regions of young roots are recognized: (1) A protective root cap that also aids in the perception of gravity. (2) A region of cell division. Its apical meristem subdivides into a protoderm, which produces the epidermis; a ground meristem, which produces the cortex; and a procambium, which produces primary xylem and primary phloem. (3) A region of elongation in which the cells produced by the apical meristem become considerably longer and slightly wider. (4) A region of maturation in which the cells mature into the distinctive cell types of primary tissues.
3. Some of the epidermal cells in the region of maturation develop root hairs; the root hairs greatly increase the absorptive surface of the root. The tissues that mature in this region are similar to those of stem tips, but pith is absent in most dicot roots and originates from the procambium in monocot roots.
4. The cortex has an endodermis with suberized Casparian strips at its inner boundary.
5. Next to the endodermis toward the center of the root are parenchyma cells constituting the pericycle. Branch roots and the vascular cambium arise in the pericycle.
6. In dicot roots, the primary xylem usually first forms a solid core with two to several arms in the center of the root; a pith may be present in monocot roots.
7. Primary phloem first is produced in discrete patches between the primary xylem arms, but the tissues eventually appear as concentric cylinders. In woody plants, a cork cambium usually arises in the pericycle and produces cork tissues similar to those of stems. Roots may graft together naturally. There are no leaves in roots.
8. Specialized roots include those for food- or water-storage; pneumatophores; aerial roots (velamen roots, prop roots, photosynthetic roots, and adventitious roots); contractile roots; and buttress roots. Haustoria are peglike parasitic roots. Mycorrhizae are mutualistic associations between roots and fungi. Some plants have nitrogen-fixing bacteria in nodules on their roots.
9. Root crops include sugar beets, beets, turnips, rutabagas, parsnips, carrots, sweet potatoes, yams, and cassava. Several spices are obtained from roots. Other uses of roots include the production of alcohol and the extraction of dyes, drugs, insecticides, and poisons.
10. Soils contain a mixture of ingredients, including sands, rocks, silt, clay, humus, dead organic matter, plant roots, small animals, microorganisms, plus air and water, within pore spaces of various sizes.
11. A vertical column of soil exhibits horizons; the topsoil is divided into an upper *A* horizon and a lower *E* horizon. The *B* horizon (subsoil) usually contains more clay

and is lighter in color than the topsoil. The *C* horizon (bottom portion) constitutes the weathered soil parent material.

12. Living organisms in the soil decompose organic matter, the source of most important plant nutrients. Animals also cultivate the soil. Soil erosion is affected by topography.
13. Soil texture pertains to the relative proportions of sand, silt, and clay. More than half of the composition by weight of mineral matter is oxygen.
14. Soil structure refers to the arrangement of the soil particles into aggregates. Good soils are highly granular and have pore spaces that constitute about half the total volume.
15. Water in the soil occurs as hygroscopic water, gravitational water, and capillary water.
16. The field capacity of the soil is the amount of water that remains after the rest of the water has drained away by gravity. Soil reaches the permanent wilting point when plants wilt permanently because they can no longer extract enough water from the soil for their needs. Available water is soil water between field capacity and the permanent wilting point.

Review Questions

1. Distinguish between a tiny root and a root hair. What is the function of a root hair?
2. What is the difference between parasitic roots and mycorrhizae?
3. If you were shown cross sections of a young root and a young stem from the same dicot plant, how could you tell them apart?
4. What is the function of the root cap, and from which meristem does it originate?
5. How do endodermal cells differ from other types of cells?
6. Where do branch roots originate?
7. List some spices and drugs obtained from roots.
8. What is soil parent material? Where does it come from, and how does it become soil?
9. What is the difference between soil texture and soil structure?
10. What types of soil water are recognized? What is available soil water?

Discussion Questions

1. Japanese gardeners regularly trim away parts of the root system to assist in dwarfing a plant. A plant's food is obtained through photosynthesis in the leaves; can you suggest why trimming the roots can cause dwarfing?
2. It was suggested that roots perceive gravity through the root cap. Would it really matter if roots grew randomly in the soil instead of responding to gravity?
3. From the viewpoint of the plant, can you suggest a practical reason for branch roots originating internally instead of at the surface?
4. When you eat a yam or a sweet potato, what kinds of compounds and cells are you consuming?
5. Persons associated with commercial nurseries and greenhouses often sterilize their soil by heating it to get rid of pests, but then they have to wait for a short time after it has cooled to use it. Why?

Learning Online

Visit our web page at www.mhhe.com/botany for interesting case studies, practice quizzes, current articles, and animations within the Online Learning Center to help you understand the material in this chapter. You'll also find active links to these topics:

Characteristics of Plants
Architectural Pattern of Plants
Root Systems and Plant Mineral Nutrition
Roots
Water Relations in Plants
Specialized Roots
Soil and Plant Relations
Economic Uses of Roots
Soil and Its Uses
Sustainable Agriculture
Agricultural Methods

Additional Reading

Altman, A., and Y. Waisel (Eds.). 1998. *Biology of root formation and development.* Hingham, MA: Kluwer Academic.

Brady, N. C., and R. R. Weil. 1998. *The nature and properties of soils,* 12th ed. Paramus, NJ: Prentice-Hall.

Charman, P., and B. Murphy (Eds.). 2000. *Soils: Their properties and management.* New York: Oxford University Press.

Cutter, E. G. 1971. *Plant anatomy: Experiment and interpretation. Part 2: Organs.* Reading, MA: Addison-Wesley.

Davis, T. D., and B. E. Haissig (Eds.). 1994. *Biology of adventitious root formation.* New York: Plenum.

Epstein, E. 1973. Roots. *Scientific American* 228: 48–58.

Fahn, A. 1990. *Plant anatomy,* 4th ed. Elmsford, NY: Pergamon Press.

Hatfield, J. L., and B. A. Stewart (Eds.). 1992. *Limitations to plant root growth.* New York: Springer-Verlag.

Mauseth, J. D. 1988. *Plant anatomy.* Menlo Park, CA: Benjamin/Cummings.

Miller, R., and D. T. Gardiner. 2000. *Soils in our environment,* 9th ed. Paramus, NJ: Prentice-Hall.

Mukerji, K. G., et al. 2000. *Mycorrhizal biology.* Hingham, MA: Kluwer Academic.

Raven, P. et al. 1999. *Biology of plants,* 6th ed. New York: Worth.

Singer, M. J., and D. N. Munns. 1998. *Soils: An introduction,* 4th ed. Paramus, NJ: Prentice-Hall.

Starr, C. C. 2000. *Biology concepts and applications,* 4th ed. Stamford, CT: Brooks/Cole.

Varma, A., and B. Hock (Eds.). 1998. *Mycorrhizae: Structure, function, molecular biology and biotechnology,* 2d ed. New York: Springer-Verlag.

The trunk of a Mindanao gum tree (*Eucalyptus deglupta*) of Indonesia and the Philippine Islands.

Stems

6

Chapter Outline

Overview

After a brief introduction, this chapter discusses, in general, the origin and development of stems. Items such as the apical meristem and the tissues derived from it, leaf gaps, cambia, secondary tissues, and lenticels are included. This general discussion is followed by notes on the distinctions between herbaceous and woody dicot stems and monocot stems. This section covers annual rings, rays, heartwood and sapwood, resin canals, bark, laticifers, and vascular bundles.

Next, there is a survey of specialized stems (rhizomes, stolons, tubers, bulbs, corms, cladophylls, and others). The chapter concludes with a discussion of the economic importance of wood and stems.

Some Learning Goals

1. Know the tissues that develop from shoot apices and the meristems from which each tissue is derived. Distinguish between primary tissues and secondary tissues.
2. Learn and give the function of each of the following: vascular cambium, cork cambium, stomata, lenticels.
3. Contrast the stems of herbaceous and woody dicots with the stems of monocots.
4. Understand the composition of wood and its annual rings, sapwood, heartwood, and bark. Explain how a log is sawed for commercial use.
5. Distinguish among rhizomes, stolons, tubers, bulbs, corms, cladophylls, and tendrils.
6. Learn at least 10 human uses of wood and stems in general.

A brief check of furnishings and tools around the house and garage—or even the house itself—soon reveals that, with the exception of appliances and the family car, much of what we use daily and take for granted, from pencils and pianos to newspapers and brooms, has some wood content. Most of that wood directly or indirectly involves plant stems. In fact, stems have been an integral part of human life ever since cave dwellers first used wooden clubs to kill for food.

Grafting, which usually involves artificially uniting stems or parts of stems of different but related varieties of plants, has been practiced by humans for hundreds of years. The careful matching of certain tissues is critical to its success, as is seen in the discussion of grafting in Chapter 14. To understand how and why grafts may or may not be successful and to identify which parts of stems are useful, we first need to examine the structure of stems and learn the basic functions of the various tissues.

Unlike animals, some plants have *indeterminate growth* (i.e., they can grow indefinitely), with the meristems at their tips increasing their length and other meristems increasing their girth for hundreds or even thousands of years. In stems, the cells produced by the meristems usually become the familiar, erect, aerial *shoot system* with branches and leaves. In certain plants, such as ferns or perennial grasses, this shoot system may develop horizontally beneath or at the surface of the ground; in other plants, the stem may be so short and inconspicuous as to appear nonexistent. In a number of plants, stems are modified in ways that allow specialized functions, such as climbing or the storage of food or water.

EXTERNAL FORM OF A WOODY TWIG

A woody twig consists of an axis with attached leaves (Fig. 6.1). If the leaves are attached to the twig alternately or in a spiral around the stem, they are said to be *alternate,* or *alternately arranged.* If the leaves are attached in pairs, they are said to be *opposite,* or *oppositely arranged,* or if they are in whorls (groups of three or more), their arrangement is whorled. The area, or region (not structure), of a stem where a leaf or leaves are attached is called a **node,** and a stem region between nodes is called an **internode.** A leaf usually has a flattened **blade,** and in most cases is attached to the twig by a stalk called the **petiole.**

Each angle between a petiole and the stem contains a bud. The angle is called an **axil,** and the bud located in the axil is an *axillary bud.* Axillary buds may become branches, or they may contain tissues that will develop into the next season's flowers. Most buds are protected by one to several *bud scales,* which fall off when the bud tissue starts to grow.

There often (but not always) is a *terminal bud* present at the tip of each twig. A terminal bud usually resembles an axillary bud, although it is often a little larger. Unlike axillary buds, terminal buds do not become separate branches, but, instead, the meristems within them normally produce tissues that make the twig grow longer during the growing season. The bud scales of a terminal bud leave tiny scars around the twig when they fall off in the spring. Counting the number of groups of *bud scale scars* on a twig can tell one how old the twig is.

Sometimes other scars of different origin also occur on a twig. These scars come from a leaf that has **stipules** at the

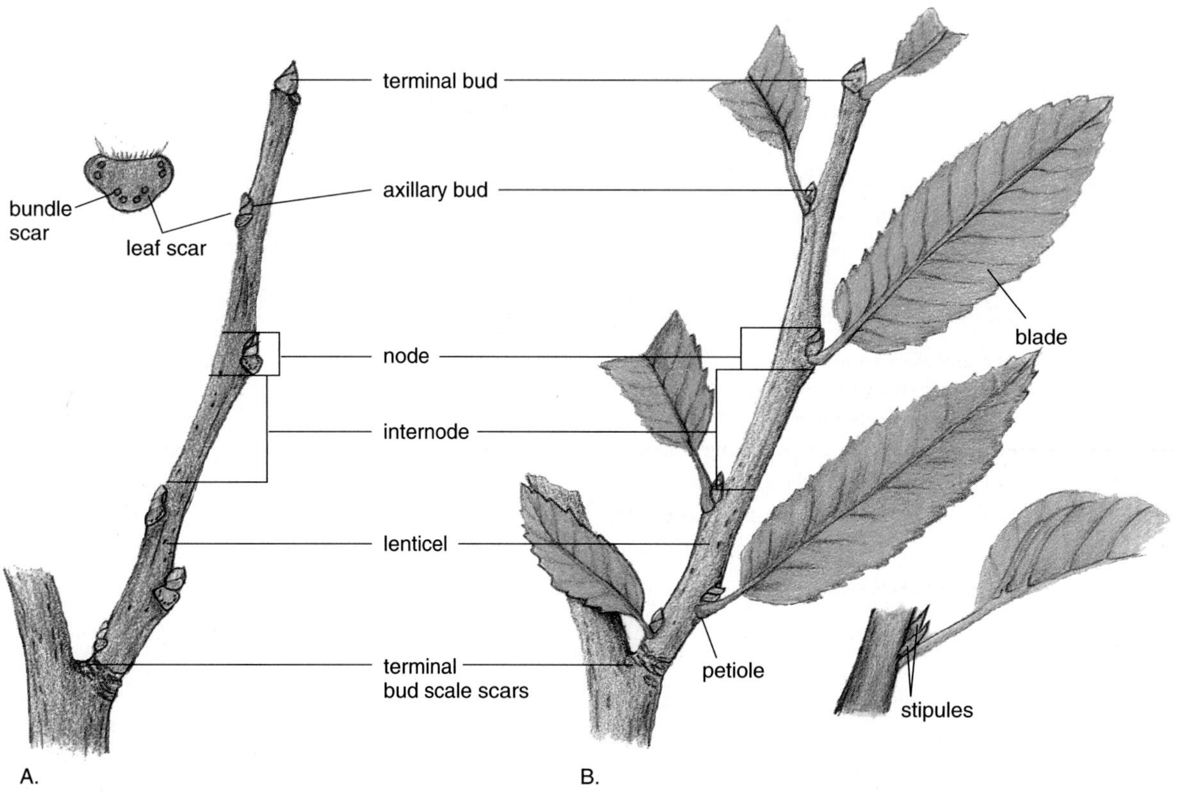

Figure 6.1 **A woody twig.** *A.* The twig in its winter condition. *B.* The twig as it appeared the summer before.

base of the petiole. Stipules are paired, often somewhat leaflike, appendages that may remain throughout the life of the leaf. In some plants, they fall off as the buds expand in the spring, leaving tiny *stipule scars.* The stipule scars may resemble a fine line encircling the twig, or they may be very inconspicuous small scars on either side of the petiole base.

Deciduous trees and shrubs (those that lose all their leaves annually) characteristically have dormant axillary buds with **leaf scars** left below them after the leaves fall. Tiny **bundle scars,** which mark the location of the water-conducting and food-conducting tissues, are usually visible within the leaf scars. There may be one to many bundle scars present, but more often than not, there are three. The shape and size of the leaf scars and the arrangement and numbers of the bundle scars are characteristic for each species. One can often identify a woody plant in its winter condition by means of scars and buds.

ORIGIN AND DEVELOPMENT OF STEMS

There is an *apical meristem* (tissue in which cells actively divide) at the tip of each stem, and it is this meristem that contributes to an increase in the length of the stem. The apical meristem is dormant before the growing season begins. It is protected by bud scales of the bud in which it is located and also to a certain extent by leaf **primordia** (singular: **primordium**), the tiny embryonic leaves that will develop into mature leaves after the bud scales drop off and growth begins. The apical meristem in the embryonic stem of a seed is also dormant until the seed begins to germinate.

When a bud begins to expand or a seed germinates, the cells of the apical meristem undergo mitosis, and soon three *primary meristems* develop from it (see Fig. 4.1). The outermost of these primary meristems, the **protoderm,** gives rise to the *epidermis.* Although there are exceptions, the epidermis is typically one cell thick and usually becomes coated with a thin, waxy, protective layer, the *cuticle.* A cylinder of strands constituting the **procambium** appears to the interior of the protoderm. (The procambium produces water-conducting *primary xylem* cells and food-conducting *primary phloem* cells.)

The remainder of the meristematic tissue, called **ground meristem,** produces two tissues composed of parenchyma cells. The parenchyma tissue in the center of the stem is the **pith.** Pith cells tend to be very large and may break down shortly after they are formed, leaving a cylindrical hollow area. Even if they do not break down early, they may eventually be crushed as new tissues produced by other meristems add to the girth of the stem, particularly in woody plants. The other tissue produced by the ground meristem is the **cortex.** The cortex may become more extensive than the pith, but in woody plants, it, too, eventually will be crushed and replaced by new tissues produced from within. The parenchyma of both the pith and the cortex function in storing food or sometimes, if chloroplasts are present, in manufacturing it.

Awareness

Standing in Fields of Stone

The combination of high altitude; dry, cool air; low rainfall; high winds; and poor soil has provided sustenance for the oldest known living species on the earth! These ancient warriors, whose great age was unknown until 1953, are the **bristlecone pines** that flourish atop the arid mountains of the Great Basin from Colorado to California. The oldest—determined to be almost 5,000 years old—is located in the Ancient Bristlecone Pine Forest high in the White Mountains of California. Some of the trees standing today were seedlings when the great pyramids were built, were middle-aged trees during the time of Christ, and today are hoary patriarchs standing in fields of stone.

There are two species of bristlecones, one living in the western-most regions (*Pinus longaeva*) and the other inhabiting the eastern regions (*Pinus aristata*). The trees do not grow very tall, with none over 60 feet and many much shorter. Typical of their girth is the bristlecone named the "Patriarch" that is just over 36.5 feet around, but it is a relative youngster at 1,500 years. With the short summer growing season high in these mountains, bristlecones typically grow 1/100th of an inch or less in diameter in any given year. The trees stand in isolation, each appearing as a sentinel, overlooking an otherwise barren rock-strewn landscape. Their needles are remarkable. While occurring five per bundle and about 1 to 1.5 inches in length, they can live for 20 to 30 years before being cast off. This extraordinary leaf longevity gives the trees a stable photosynthetic output and sustains the tree during years of unusual stress when producing new leaf tissue would be difficult. The trees must generate new leaves and cones as well as produce enough reserves for the long winter months, all on scant annual precipitation of about 10 inches.

Ecologists have long noted the peculiar distribution and growth habits of certain plants. Why do plants grow where they do? What adaptations permit them to survive in life-threatening environments? The bristlecones' age and habitat offer several insights. These trees grow in places on earth where no other plant can grow. One answer appears to be the type of soil in which they are anchored. Stands of bristlecone pine grow on outcrops of dolomite, an alkaline limestone substrate of low nutrient but higher moisture content than the surrounding sandstone. The granite and sandstone formations surrounding the dolomite outcroppings support sagebrush and Limber pine, but not bristlecones. At these altitudes, the radiant sunlight is extreme. Dolomite reflects more sunlight than other rocks and thus keeps the root zones cooler and moisture-laden during the important growing season.

Box Figure 6.1 An ancient bristlecone pine (*Pinus aristata*) from the White Mountains of California.

Another survival tactic is revealed by taking small core samples of the wood. These trees have large amounts of die-back (deadwood that is no longer functional), reducing the amount of tissue that the leaves need to supply with food. For example, one bristlecone over 4,000 years old is nearly 4 feet in diameter, but its functional conducting tissue is only 10 inches wide.

Other characteristics provide bristlecones with survival advantages. Because their dense resin-filled wood renders them inhospitable sites for colonization by pathogenic fungi or bacteria, they are relatively free from these attacks. These trees are often struck by lightning, but with the absence of ground cover and decaying leaf litter, fire rarely spreads from tree to tree.

The age of these living trees is determined, like others, by an instrument called an **increment borer,** a type of drill that is inserted into the tree trunk at its widest girth. Using a hollow drill bit, a linear core of wood is obtained that can be "read" for the number of **annual growth rings.** In this way, the age of a tree can be obtained without cutting the tree down to examine the annual rings revealed in the cross section of the stump. Each year, a tree will add a layer of wood to its trunk, and these become the annual rings that can be observed in a cross section of the trunk. During spring growth of wood, large-diametered water-conducting cells are formed; later in the summer, the water-conducting

Awareness Box Continued

cells produced have a small diameter. This difference in appearance between early (spring) wood and late (summer) wood is sufficient to make each growth increment distinctive, and thus counting of the rings possible.

In 1957, "Methuselah" was discovered and determined to be 4,723 years old. Methuselah remains today as the world's oldest living organism. But what about tomorrow? After surviving nearly five millennia, Methuselah is being protected against a more insidious enemy. Standing in its field of stone without a marker because of fear from vandalism, Methuselah serves as a reminder that the human species can be just as destructive as any microbe.

D.C. Scheirer

All five of the tissues produced by this apical meristem complex (epidermis, primary xylem, primary phloem, pith, and cortex) arise while the stem is increasing in length and are called *primary tissues.* As these primary tissues are produced, the leaf primordia and the *bud primordia* (embryonic buds in the axils of the leaf primordia) develop into mature leaves and buds (Fig. 6.2). As each leaf and bud develops, a strand of xylem and phloem, called a *trace,* branches off from the cylinder of xylem and phloem extending up and down the stem and enters the leaf or the bud. As the traces branch from the main cylinder of xylem and phloem, each trace leaves a little thumbnail-shaped gap in the cylinder of vascular tissue. These gaps, called **leaf gaps** and **bud gaps,** are filled with parenchyma tissue (Fig. 6.3).

A narrow band of cells between the primary xylem and the primary phloem may retain its meristematic nature and become the **vascular cambium,** one of the two lateral meristems. The vascular cambium is often referred to simply as the *cambium.* The cells of the cambium continue to divide indefinitely, with the divisions taking place mostly in a plane parallel to the surface of the plant. The secondary tissues produced by the vascular cambium add to the girth of the stem instead of to its length (Fig. 6.4).

Cells produced by the vascular cambium become *tracheids, vessel elements, fibers,* or other components of *secondary xylem* (*inside* of the meristem, toward the center), or they become *sieve tube members, companion cells,* or other components of *secondary phloem* (*outside* of the meristem, toward the surface). The functions of these secondary tissues are the same as those of their primary counterparts—secondary xylem conducts water and soluble nutrients, while secondary phloem conducts, in soluble form, food manufactured by photosynthesis throughout the plant.

In many plants, especially woody species, a second cambium arises within the cortex or, in some instances, develops from the epidermis or phloem. This is called the **cork cambium,** or **phellogen.** The cork cambium produces box like **cork cells,** which become impregnated with **suberin,** a waxy substance that makes the cells impervious to moisture. The cork cells, which are produced annually in cylindrical layers, die shortly after they are formed. The cork cambium may also produce parenchyma-like **phelloderm** cells to the inside. Cork tissue makes up the outer bark of woody plants; it functions in reducing water loss and in protecting the stem against mechanical injury (see the discussion of *periderm* in Chapter 4).

Cork tissue cuts off water and food supplies to the epidermis, which soon dies and is sloughed off. In fact, if the cork were to be formed as a solid cylinder covering the entire stem, vital gas exchange with the interior of the stem would not be possible. In young stems, such gas exchange takes place through the stomata, located in the epidermis (see Figs. 7.6 and 9.13). As woody stems age, **lenticels** (see Fig. 4.14) develop beneath the stomata. As cork is produced, the parenchyma cells of the lenticels remain so that exchange of gases (e.g., oxygen, carbon dioxide) can continue through spaces between the cells. Lenticels occur in the fissures of the bark of older trees and often appear as small bumps on younger bark. In birch and cherry trees, the lenticels form conspicuous horizontal lines.

Differences between the activities of the apical meristem and those of the cambium and cork cambium become apparent if one drives a nail into the side of a tree and observes it over a period of years. The nail may eventually become embedded as the stem increases in girth, but it will always remain at the same height above the ground, as the cells that increase the length of a stem are produced only at the tips.

TISSUE PATTERNS IN STEMS

Steles

Primary xylem, primary phloem, and the pith, if present, make up a central cylinder called the **stele** in most younger and a few older stems and roots. The simplest form of stele, called a *protostele,* consists of a solid core of conducting tissues in which the phloem usually surrounds the xylem. Protosteles were common in primitive seed plants that are now extinct and are also found in whisk ferns, club mosses (see Chapter 21), and other relatives of ferns. *Siphonosteles,* which are tubular with pith in the center, are common in ferns.

Most present-day flowering plants and conifers have *eusteles* in which the primary xylem and primary phloem are in discrete *vascular bundles,* as discussed in the section "Herbaceous Dicotyledonous Stems."

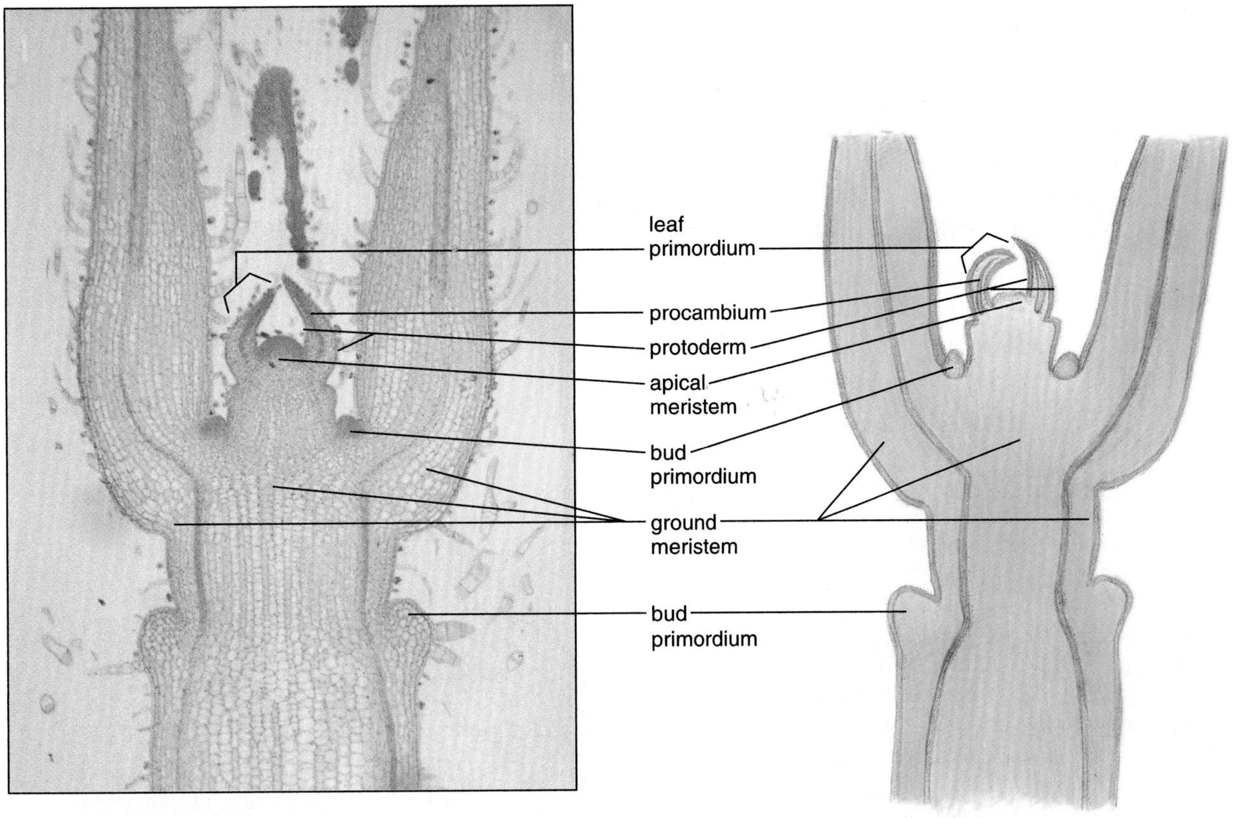

Figure 6.2 A longitudinal section through the tip of a *Coleus* stem, ca. ×800. *(Photomicrograph by G.S. Ellmore)*

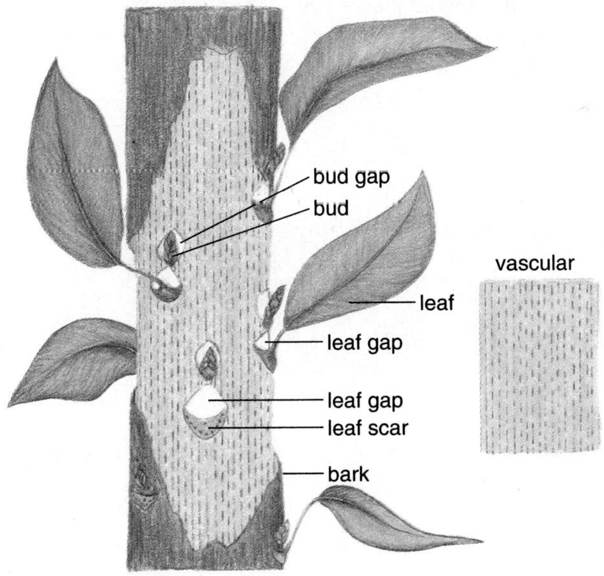

Figure 6.3 A portion of a stem showing leaf gaps and bud gaps in the cylinder of vascular tissue.

Flowering plants develop from seeds that have attached to their embryonic stems either one or two "seed leaves," called **cotyledons** (see Chapters 8 and 23). The seeds of pines and other cone-bearing trees have several (usually eight) cotyledons. The cotyledons usually store food needed by the young seedling until its first true leaves can produce food themselves.

Flowering plants that develop from seeds having two cotyledons are called **dicotyledons** (usually abbreviated to **dicots**), while those developing from seeds with a single cotyledon are called **monocotyledons** (abbreviated to **monocots**). Dicots and monocots differ from one another in several other respects; the differences in stem structure are noted in the following sections, and a summary of these and other differences in these two classes of flowering plants is given in Table 8.1.

Herbaceous Dicotyledonous Stems

In general, plants that die after going from seed to maturity within one growing season (**annuals**) have green, herbaceous (nonwoody) stems. Most monocots are annuals, but many dicots (discussed next) are also annuals.

The tissues of annual dicots are largely primary, although *cambia* (plural of cambium) may develop some secondary tissues. Herbaceous dicot stems (Fig. 6.5) have discrete **vascular bundles** composed of patches of xylem and phloem. The vascular bundles are arranged in a cylinder that separates the cortex from the pith, although in a few plants (e.g., foxgloves), the xylem and the phloem are produced as continuous rings (cylinders) instead of in separate bundles.

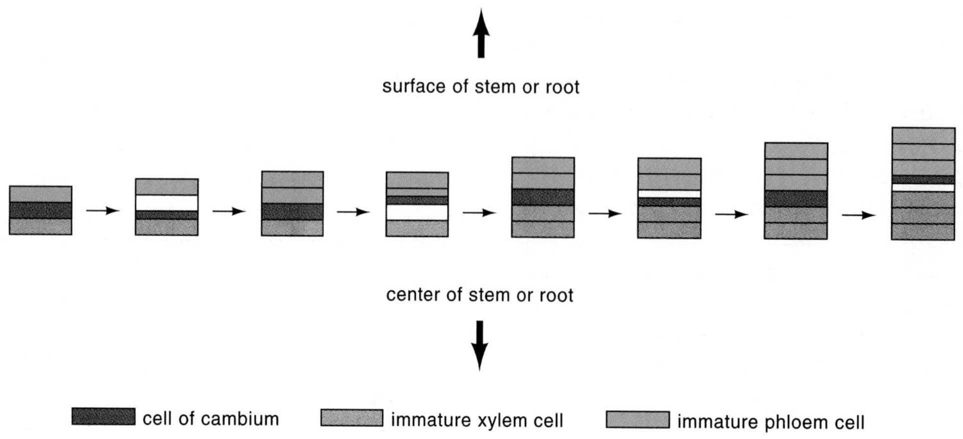

Figure 6.4 An illustration of how a cell of the vascular cambium produces new secondary phloem cells to the outside and new secondary xylem cells to the inside. Note, in cross section, that the cambium gradually becomes shifted away from the center as new cells are produced. Phloem is produced before xylem in secondary growth.

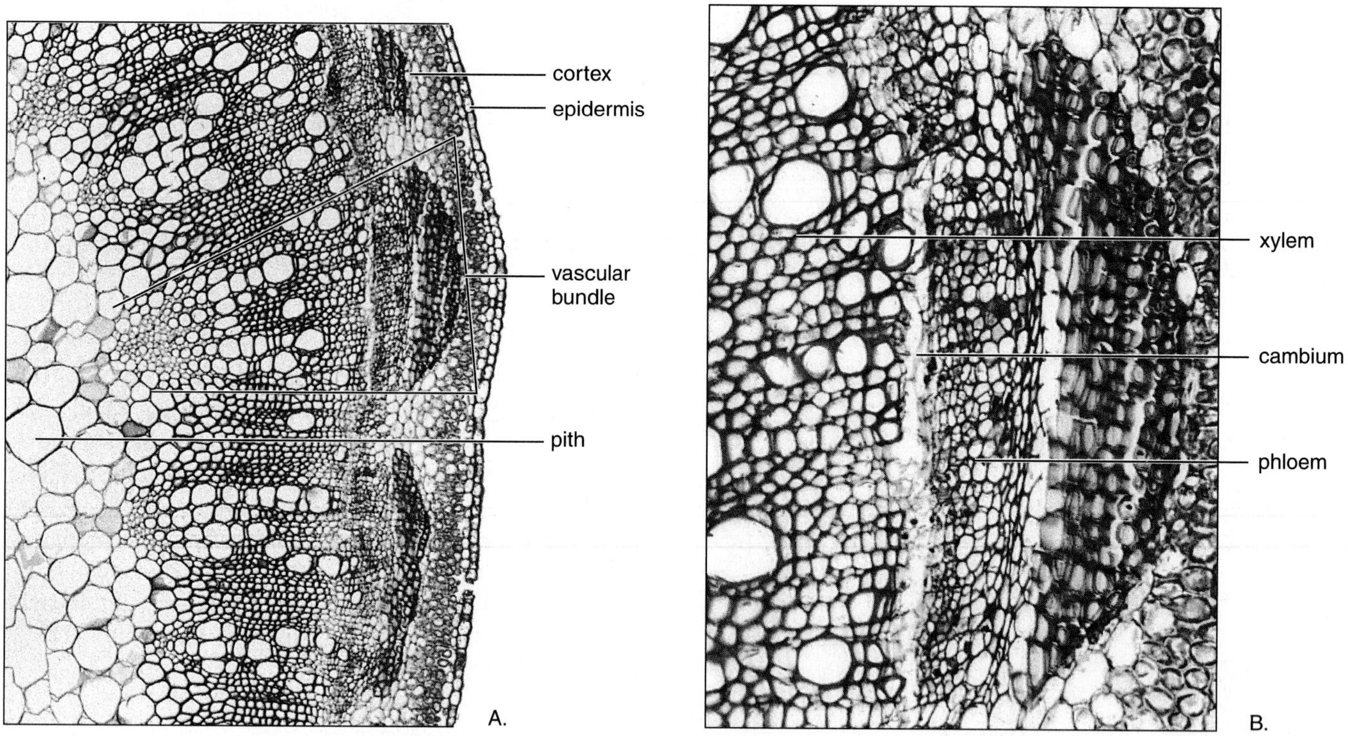

Figure 6.5 *A.* A cross section of an alfalfa (*Medicago*) stem, ×100. The tissue arrangement is typical of herbaceous dicot stems. *B.* An enlargement of a small portion of the outer part of the stem, ×500.

The procambium produces only primary xylem and phloem, but later, a vascular cambium arises between these two primary tissues and adds secondary xylem and phloem to the vascular bundles. In some plants, the cambium extends between the vascular bundles, appearing as a narrow ring, producing not only the conducting tissues within the bundles but also the parenchyma cells between them. In other plants, the cambium is not in an uninterrupted cylinder but is instead confined to the bundles, each of which has its own small band of cambium between the xylem and phloem.

Woody Dicotyledonous Stems

In the early stages of development, the primary tissues of stems of young herbaceous dicots, woody dicots, and cone-bearing trees are all arranged in similar fashion. In woody plants, however, obvious differences begin to appear as soon as the vascular cambium and the cork cambium develop. The most conspicuous differences involve the secondary xylem, or wood, as it is best known (Fig. 6.6). Some tropical trees (e.g., ebony), in which both the vascular cambium and the cork cambium are active all year, produce an ungrained, uniform wood. The wood of most trees, however, is produced seasonally. In trees of temperate climates, virtually all growth takes place during the spring and summer and then ceases until the following spring.

When the vascular cambium of a typical broadleaf tree first becomes active in the spring, it usually produces relatively large vessel elements of secondary xylem; such xylem is referred to as *spring wood.* As the season progresses, the vascular cambium may produce vessel elements whose diameters become progressively smaller in each succeeding series of cells produced, or there may be fewer vessel elements in proportion to tracheids produced until tracheids (and sometimes fibers) predominate.

The xylem that is produced after the spring wood, and which has smaller or fewer vessel elements and larger numbers of tracheids, is referred to as *summer wood.* Over a period of years, the result of this type of switch between the early spring and the summer growth is a series of alternating concentric rings of light and dark cells. One year's growth of xylem is called an **annual ring.** In conifers, the wood consists mostly of tracheids, with vessels and fibers being absent. Annual rings are still visible, however, since the first tracheids produced in the spring are considerably larger and lighter in color than those produced later in the growing season. Note that an annual ring normally may contain many layers of xylem cells and it is all the layers produced in one growing season that constitute an annual ring—not just the dark layers.

The vascular cambium produces more secondary xylem than it does phloem. Xylem cells also have stronger, more rigid walls than those of phloem cells and are less subject to collapse under tension. As a result, the bulk of a tree trunk consists of annual rings of wood. The annual rings not only indicate the age of the tree (since normally only one is produced each year), but they can also tell something of the climate and other conditions that occurred during the tree's lifetime (Fig. 6.7). For example, if the rainfall during a particular year is higher than normal, the annual ring for that year will be wider than usual. Sometimes, caterpillars or locusts will strip the leaves of a tree shortly after they have appeared. This usually results in two annual rings being very close together, since very little growth can take place under such conditions.

If there is a fire not resulting in the death of the tree, it may be possible to determine when the fire occurred, since the burn scar may appear next to a given ring. The most recent season's growth is directly next to the vascular cambium, and one need only count the rings back from the cambium to determine the actual year of the fire.

It is not necessary to cut down a tree to determine its age. Instead, botanists and foresters can employ an *increment borer* to find out how old a woody plant is. This device, which resembles a piece of pipe with a handle on one end, removes a plug of wood from the tree perpendicular to the axis, and the annual rings in the plug can then be counted. The small hole left in the tree can be treated with a disinfectant to prevent disease and covered up without harm to the tree.

Ecological Review

Plant stems, which provide support for plants, also perform many specialized functions such as food and water storage, physical protection in the form of spines, and chemical protection by tannins and oils. Stem structure and growth influence a broad range of ecological relationships, ranging from success or failure in competition with other plants for light to surviving periods of drought and protection from attack by herbivores. Since the width of stem growth rings is influenced by climate, stems also provide a temporal record of environmental conditions and can be used to date events, such as fire, that leave marks on growth rings. Much of what we know about past climates has come from studies of the growth rings on plant stems.

A count of annual rings has produced some red faces on at least one occasion. The Hooker Oak, which was named in honor of Sir Joseph Hooker, a famous British botanist who once examined it, was located in the community of Chico, California. Until its demise in 1977, thousands of visitors from all over the world visited the huge oak, which provided enough shade for 9,000 persons on a midsummer's day. A plaque indicating the tree to be over 1,000 years old was located beneath the tree. A count of rings after its death, however, revealed the Hooker Oak to have been less than 300 years old.

When a tree trunk is examined in transverse, or cross section, fairly obvious lighter streaks or lines can be seen radiating out from the center across the annual rings (see Figs. 6.6 and 6.8). These lines, called **vascular rays,** consist of parenchyma cells that may remain alive for 10 or more years. Their primary function is the lateral conduction of nutrients and water from the stele, through the xylem and phloem, to the cortex, with some cells also storing food. Any part of a ray within the xylem is called a *xylem ray,* but its extension through the phloem is called a *phloem ray.* In basswood (*Tilia*) trees, some of the phloem rays, when observed in cross section, flare out from a width of two or three cells near the cambium to many cells wide in the part next to the cortex (see Fig. 6.6).

In radial section, rays may be from 2 or 3 cells to 50 or more cells deep, but the majority of rays in both xylem and phloem are 1 or 2 cells wide. Ray cells can be seen in cross section if a woody stem is cut or split lengthwise along a ray (Fig. 6.8). Another view of rays (in tangential section) is obtained when the stem is cut at a tangent (i.e., cut lengthwise and off center) (see Fig. 6.17).

As a tree ages, the protoplasts of some of the parenchyma cells that surround the vessels and tracheids grow through the pits in the walls of these conducting cells

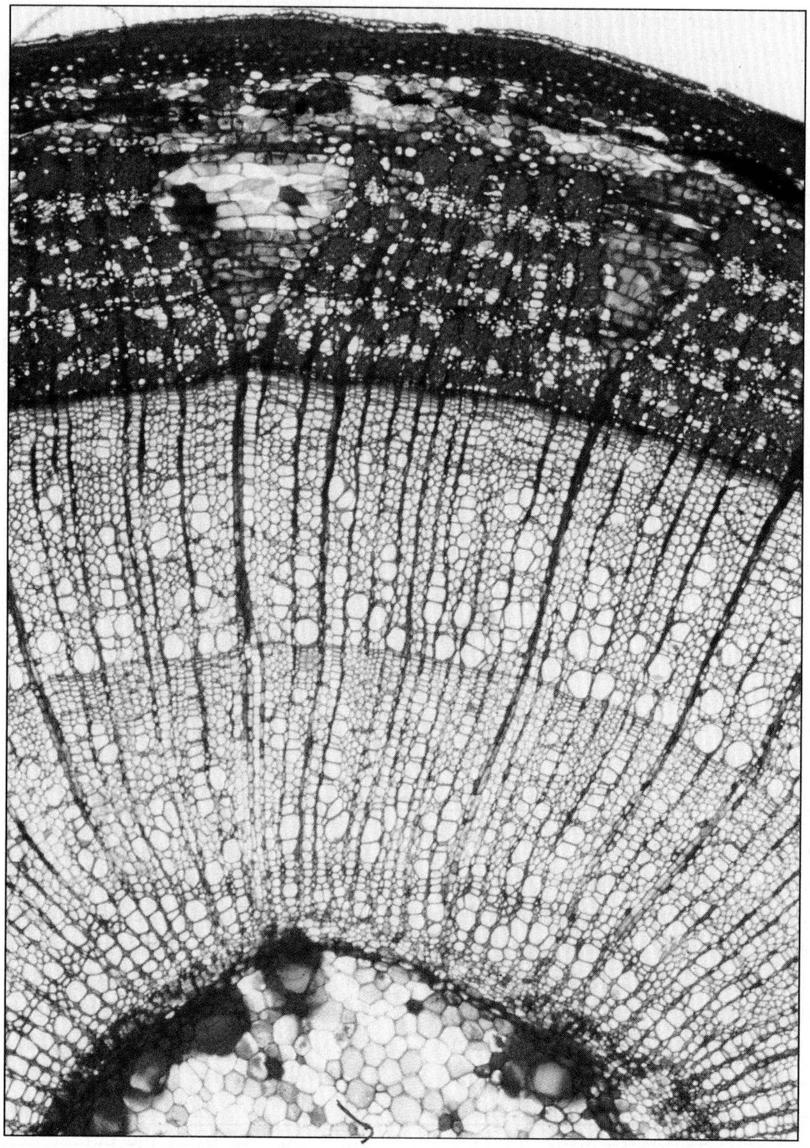

Figure 6.6 A cross section of a portion of a young linden (*Tilia*) stem, ca. ×300.

and balloon out into the cavities. As the protoplasm continues to expand, much of the cavity of the vessel or tracheid becomes filled. Such protrusions, called *tyloses* (singular: *tylosis*), prevent further conduction of water and dissolved substances. When this occurs, resins, gums, and tannins begin to accumulate, along with pigments that darken the color of the wood.

This older, darker wood at the center is called **heartwood,** while the lighter, still-functioning xylem closest to the cambium is called **sapwood** (Fig. 6.9). Except for giving strength and support, the heartwood is not of much use to the tree since it can no longer conduct materials. A tree may live and function perfectly well after the heartwood has rotted away and left the interior hollow. It is even possible to remove part of the sapwood and other tissues and apparently not affect the tree very much, as has been done with giant trees, such as the coastal redwoods of California, where holes big enough to drive a car through have been cut out without killing the trees (Fig. 6.10).

Sapwood forms at roughly the same rate as heartwood develops, so there is always enough "plumbing" for the vital conducting functions. The relative widths of the two types of wood, however, vary considerably from species to species. For example, in the golden chain tree (a native of Europe and a member of the Legume Family), the sapwood is usually only one or two rings wide, while in several North American trees (e.g., maple, ash, and beech), the sapwood may be many rings wide.

Pines and other cone-bearing trees have xylem that consists primarily of tracheids; no fibers or vessel elements are produced. Since it has no fibers, the wood tends to be softer than

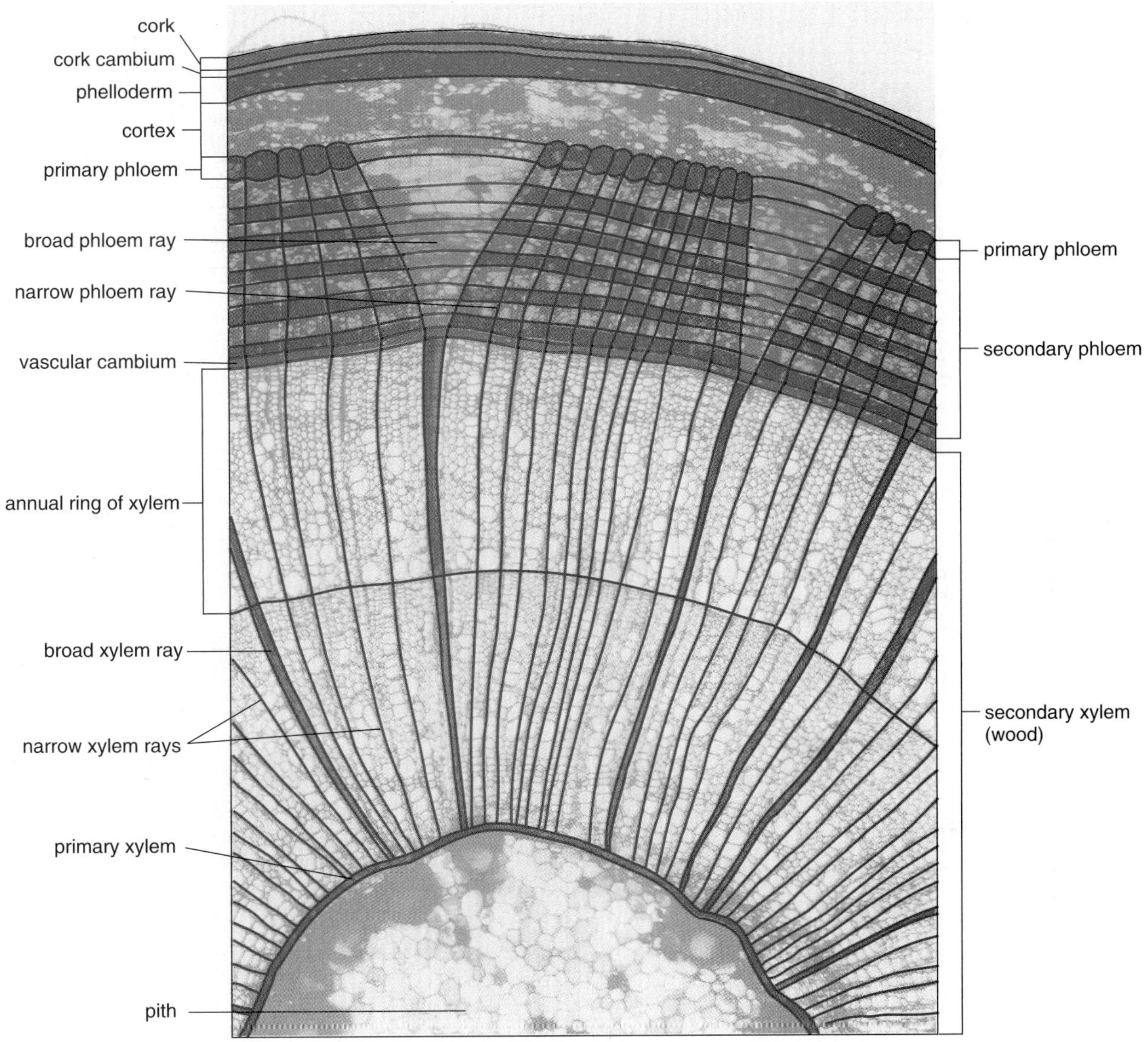

Figure 6.6 Continued

that of trees with them and is commonly referred to as *softwood,* while the wood of woody dicot trees is called *hardwood.*

In many cone-bearing trees, resin canals are scattered not only through the xylem but throughout other tissues as well. These canals are tubelike and may or may not be branched; they are lined with specialized cells that secrete resin (discussed in Chapter 22) into their cavities (Fig. 6.11). Although resin canals are commonly associated with cone-bearing trees, they are not confined to them. Tropical flowering plants, such as olibanum and myrrh trees, have resin ducts in the bark that produce the soft resins frankincense and myrrh of biblical note.

While the vascular cambium is producing secondary xylem to the inside, it is also producing secondary phloem to the outside. The term **bark** is usually applied to all the tissues outside the cambium, including the phloem. Some scientists distinguish between the *inner bark,* consisting of primary and secondary phloem, and the *outer bark* (*periderm*), consisting of cork tissue and cork cambium. Despite the presence of fibers, the thin-walled conducting cells of the phloem are not usually able to withstand for many seasons the pressure of thousands of new cells added to their interior, and the older layers become crushed and functionless.

The parenchyma cells of the cortex to the outside of the phloem also function only briefly because they too become crushed or sloughed off. Before they disappear, however, the cork cambium begins its production of cork, and since new xylem and phloem tissues produced by the vascular cambium arise to the inside of the older phloem, the mature bark may consist of alternating layers of crushed phloem and cork.

The younger layers of phloem nearest to the cambium transport, via their sieve tubes, sugars and other substances in solution from the leaves where they are made to various

This tree is 62 years old. It's been through fire and drought, plague and plenty. And all of this is recorded in its rings.
Each spring and summer a tree adds new layers of wood to its trunk. The wood formed in spring grows fast, and is lighter because it consists of large cells. In summer, growth is slower; the wood has smaller cells and is darker. So when the tree is cut, the layers appear as alternating rings of light and dark wood.
Count the dark rings, and you know the tree's age. Study the rings, and you can learn much more. Many things affect the way the tree grows, and thus alter the shape, thickness, color and evenness of the rings.
For St. Regis these rings have a special significance. They record the steady accumulation of those fibers we use to create noteworthy printing papers, kraft paper and boards, fine papers, packaging products, building materials, and products for consumers.
Essentially, then, the life of the forest is St. Regis' life. That is why we—together with the other members of the forest products industry—are vitally concerned with maintaining the beauty and utility of America's forests for the generations to come.
ST REGIS
1904
The tree—a loblolly pine—is born.
1909
The tree grows rapidly, with no disturbance. There is abundant rainfall and sunshine in spring and summer. The rings are relatively broad, and are evenly spaced.
Two-thirds typical size.

Figure 6.7 Climatic history illustrated by a cross section of a 62-year-old tree. *(Courtesy St. Regis Paper Company)*

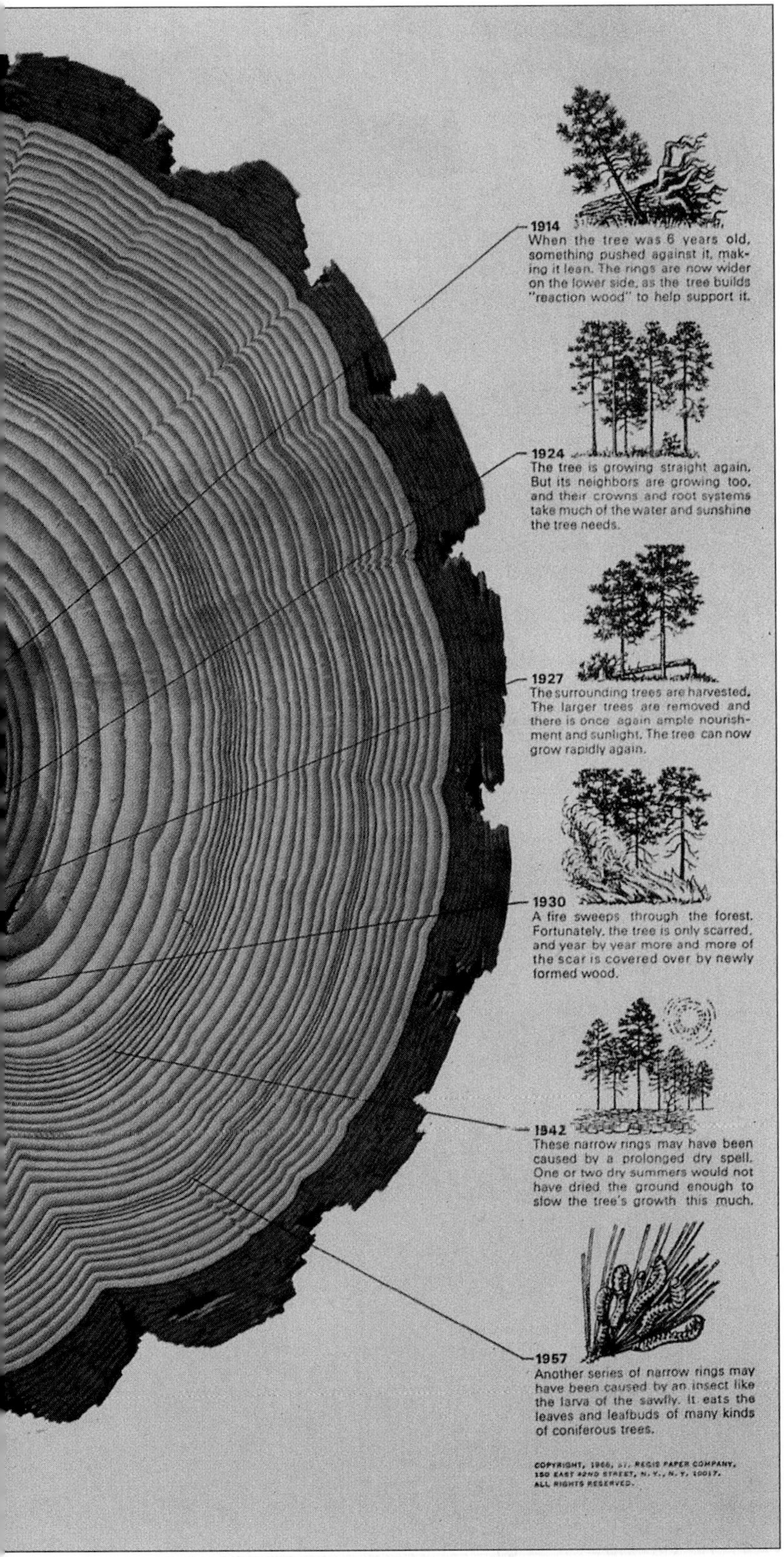
1914
When the tree was 6 years old, something pushed against it, making it lean. The rings are now wider on the lower side, as the tree builds "reaction wood" to help support it.
1924
The tree is growing straight again. But its neighbors are growing too, and their crowns and root systems take much of the water and sunshine the tree needs.
1927
The surrounding trees are harvested. The larger trees are removed and there is once again ample nourishment and sunlight. The tree can now grow rapidly again.
1930
A fire sweeps through the forest. Fortunately, the tree is only scarred, and year by year more and more of the scar is covered over by newly formed wood.
1942
These narrow rings may have been caused by a prolonged dry spell. One or two dry summers would not have dried the ground enough to slow the tree's growth this much.
1957
Another series of narrow rings may have been caused by an insect like the larva of the sawfly. It eats the leaves and leafbuds of many kinds of coniferous trees.
COPYRIGHT, 1966, ST. REGIS PAPER COMPANY, 150 EAST 42ND STREET, N.Y., N.Y. 10017. ALL RIGHTS RESERVED.

parts of the plant, where they are either stored or used in the process of *respiration* (discussed in Chapter 10). This sugar content of the phloem was in the past recognized by native Americans. Some stripped the young phloem and cambium from Douglas fir trees and used the dried strips as food for winter and in emergencies.

Specialized cells or ducts called **laticifers** are found in about 20 families of herbaceous and woody flowering plants. These cells are most common in the phloem but occur throughout all parts of the plants. The laticifers, which resemble vessels, form extensive branched networks of latex-secreting cells originating from rows of meristematic cells. Unlike vessels, however, the cells remain living and may have many nuclei.

Latex is a thick fluid that is white, yellow, orange, or red in color and consists of gums, proteins, sugars, oils, salts, alkaloidal drugs, enzymes, and other substances. Its function in the plant is not clear, although some believe it aids in closing wounds. Some forms of latex have considerable commercial value (see the discussion under "The Spurge Family" in Chapter 24). Of these, rubber is the most important. Amazon Indians utilized rubber for making balls and containers hundreds of years before Pará rubber trees were cultivated for their latex. The chicle tree produces a latex used in the making of chewing gum. Several poppies, notably the opium poppy, produce a latex containing important drugs, such as morphine and heroin. Other well-known latex producers include milkweeds, dogbanes, and dandelions.

Monocotyledonous Stems

Most monocots (e.g., grasses, lilies) are herbaceous plants that do not attain great size. The stems have neither a vascular cambium nor a cork cambium and thus produce no secondary vascular tissues or cork. As in herbaceous dicots, the surfaces of the stems are covered by an epidermis, but the xylem and phloem tissues produced by the procambium appear in cross section as discrete vascular bundles scattered throughout the stem instead of being arranged in a ring (Fig. 6.12).

Each bundle, regardless of its specific location, is oriented so that its xylem is closer to the center of the stem and its phloem is closer to the surface. In a typical monocot such as corn, a bundle's xylem usually contains two large vessels with several small vessels between them (Fig. 6.13). The first-formed xylem cells usually stretch and collapse under the stresses of early growth and leave an irregularly shaped air space toward the base of the bundle; the remnants of a vessel are often present in this air space. The phloem consists entirely of sieve tubes and companion cells, and the entire bundle is surrounded by a sheath of thicker-walled sclerenchyma cells. The parenchyma tissue between the vascular bundles is not separated into cortex and pith in monocots, although its function and appearance are the same as those of the parenchyma cells in cortex and pith.

In a corn stem, there are more bundles just beneath the surface than there are toward the center. Also, a band of sclerenchyma cells, usually two or three cells thick, develops

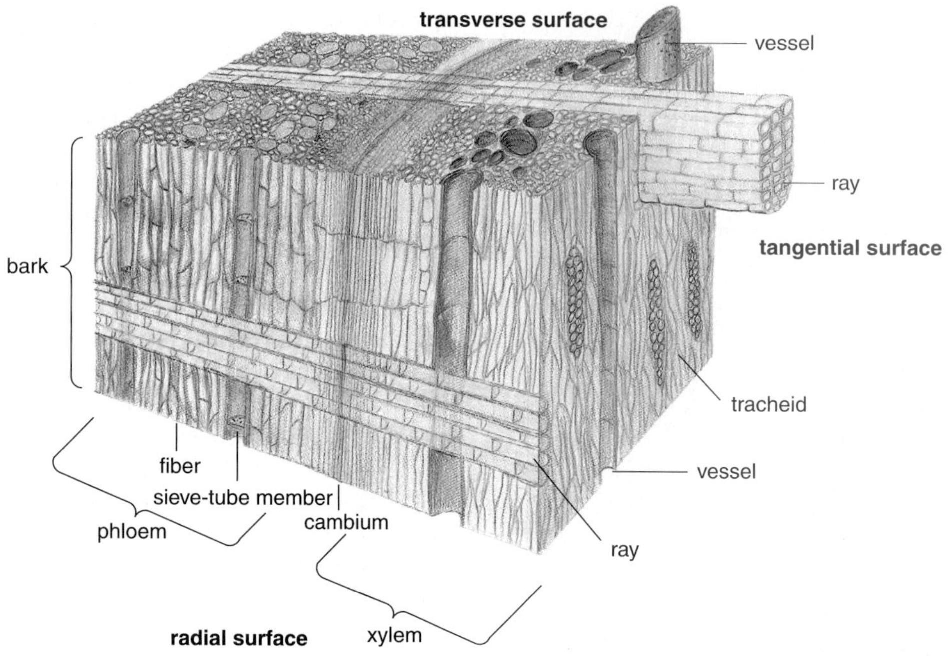

Figure 6.8 A three-dimensional, magnified view of a block of a woody dicot.

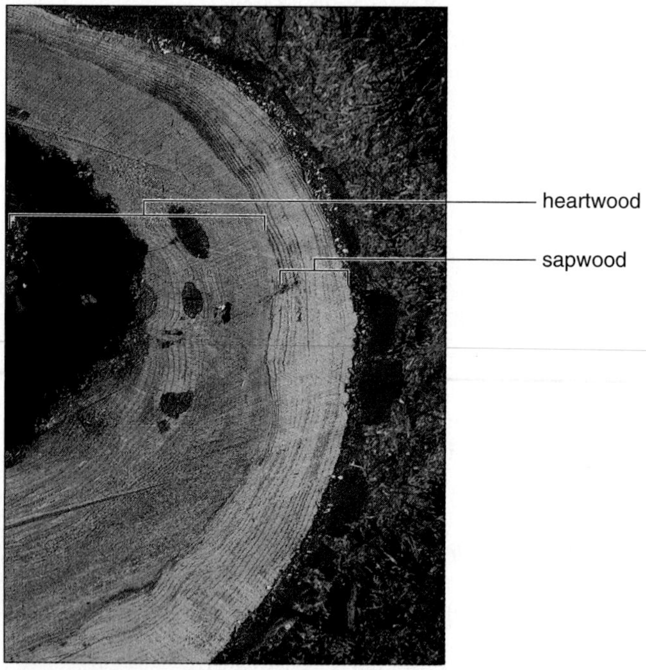

Figure 6.9 This tree was 100 years old when it was cut down. Note the proportion of *sapwood,* which consists of functional cells, to *heartwood,* in which the cells are no longer capable of conduction.

Figure 6.10 This coastal redwood is thriving despite the removal of its lower heartwood and a little of its sapwood.

Figure 6.11 Resin canals in a portion of a pine (*Pinus*) stem, ca. ×400.

immediately beneath the epidermis, and parenchyma cells in the area develop thicker walls as the stem matures. The concentration of bundles, combined with the band of sclerenchyma cells beneath the epidermis and the thicker-walled parenchyma cells, all contribute to giving the stem the capacity to withstand stresses resulting from summer storms and the weight of the leaves and the ears of corn as they mature.

In wheat, rice, barley, oats, rye, and other grasses, there is an *intercalary meristem* (discussed in Chapter 4) at the base of each internode; like the apical meristem, it contributes to increasing stem length. Although the stems of such plants elongate rapidly during the growing season, growth is columnar (i.e., there is little difference in diameter between the top and the bottom) because there is no vascular cambium producing tissues that would add to the girth of the stems.

Palm trees, which differ from most monocots in that they often grow quite large, do so primarily as a result of their parenchyma cells continuing to divide and enlarge without a true cambium developing. Several popular house plants (e.g., ti plants, *Dracaena, Sansevieria*) are monocots in which a secondary meristem develops as a cylinder that extends throughout the stem. Unlike the vascular cambium of dicots and conifers, this secondary meristem produces only parenchyma cells to the outside and secondary vascular bundles to the inside.

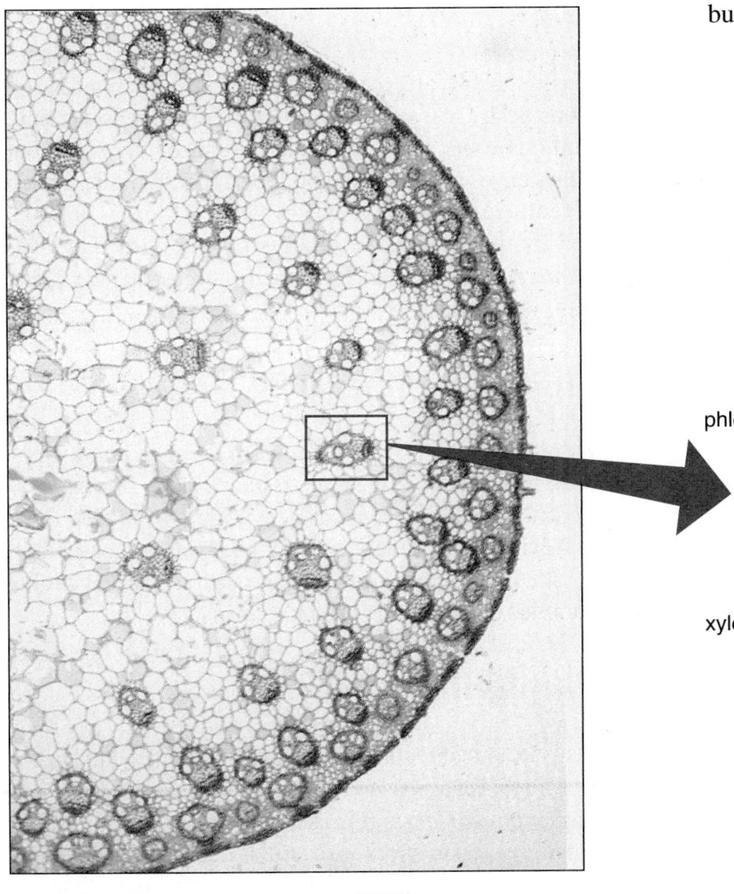

Figure 6.12 A portion of a cross section of a monocot (corn—*Zea mays*) stem, ca. ×100. *(Photomicrograph by G.S. Ellmore)*

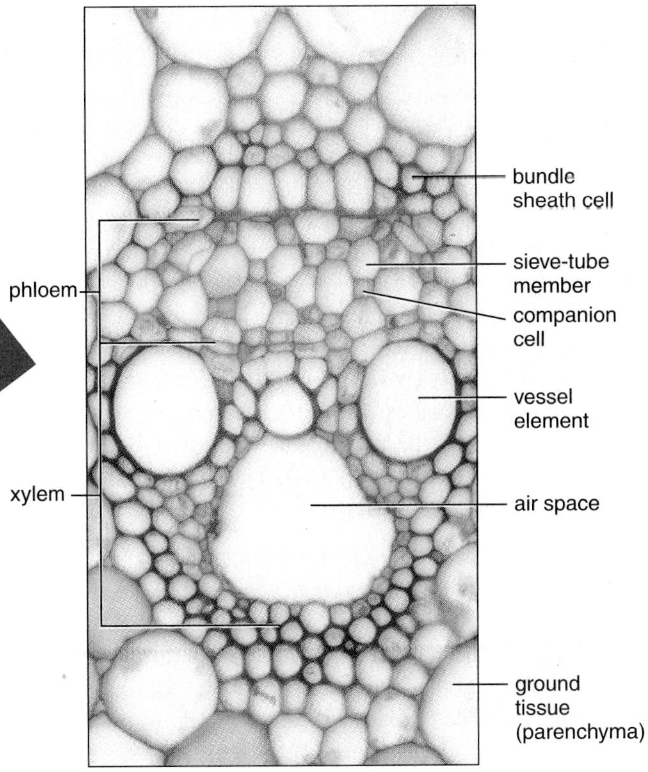

Figure 6.13 A single vascular bundle of corn (*Zea mays*) enlarged, ca. ×800.

Several commercially important cordage fibers (e.g., broomcorn, Mauritius and Manila hemps, sisal) come from the stems and leaves of monocots, but the individual cells are not separated from one another by *retting* (a process that utilizes the rotting power of microorganisms thriving under moist conditions to break down thin-walled cells) as they are when fibers from dicots are obtained. Instead, during commercial preparation, entire vascular bundles are scraped free of the surrounding parenchyma cells by hand; the individual bundles then serve as unit "fibers." If such fibers are treated with chemicals or bleached, the cementing middle lamella between the cells breaks down. Monocot fibers are not as strong or as durable as most dicot fibers.

SPECIALIZED STEMS

Although most higher plants have an erect shoot system, many species have specialized stems that are modified for various functions (Fig. 6.14). The overall appearance of specialized stems may differ markedly from that of the stems discussed so far, but all stems have *nodes, internodes,* and *axillary buds*; these features distinguish them from roots and leaves, which do not have them. The leaves at the nodes of these specialized stems are often small and scalelike. They are seldom green, but full-sized functioning leaves are also produced. Descriptions of some of the specialized stems follow.

Rhizomes

Rhizomes (Fig. 6.14) are horizontal stems that grow below ground, often near the surface of the soil. They superficially resemble roots, but close examination will reveal scalelike leaves and axillary buds at each node, at least during some stage of development, with short to long internodes in between. Adventitious roots are produced all along the rhizome, mainly on the lower surface. As indicated in Chapter 5, the word *adventitious* refers to structures arising at unusual places, such as roots growing from stems, or leaves or buds appearing at places other than leaf axils and tips of stems. A rhizome may be a relatively thick, fleshy, food-storage organ, as in irises, or it may be quite slender, as in many perennial grasses or some ferns.

Runners and Stolons

Runners are horizontal stems that differ from rhizomes in that they grow above ground, generally along the surface; they also have long internodes (Fig. 6.14). In strawberries, runners are usually produced after the first flowers of the season have appeared. Several runners may radiate out from the parent plant, and within a few weeks may grow up to 1 meter (3 feet) or more long. Adventitious buds appear at alternate nodes along the runners and develop into new strawberry plants, which can be separated and grown independently. In saxifrages and some other house plants, runners may produce new plants at intervals as they grow out and hang over the edge of the pot.

Stolons are similar to runners but are produced beneath the surface of the ground and tend to grow in different directions but usually not horizontally. In Irish potato plants, tubers are produced at the tips of stolons.

Some botanists consider stolons and runners to be variations of each other and prefer not to make a distinction between them.

Tubers

In Irish or white potato plants, several internodes at the tips of stolons become **tubers** as they swell from the accumulation of food (Fig. 6.14). The mature tuber becomes isolated after the stolon to which it was attached dies. The "eyes" of the potato are actually nodes formed in a spiral around the modified stem. Each eye consists of an axillary bud in the axil of a scalelike leaf, although this leaf is visible only in very young tubers; the small ridges seen on mature tubers are leaf scars.

Bulbs

Bulbs (Fig. 6.14) are actually large buds surrounded by numerous fleshy leaves, with a small stem at the lower end. Adventitious roots grow from the bottom of the stem, but the fleshy leaves comprise the bulk of the bulb tissue, which stores food. In onions, the fleshy leaves usually are surrounded by the scalelike leaf bases of long, green, aboveground leaves. Other plants producing bulbs include lilies, hyacinths, and tulips.

Corms

Corms resemble bulbs but differ from them in being composed almost entirely of stem tissue, except for the few papery scalelike leaves sparsely covering the outside (Fig. 6.14). Adventitious roots are produced at the base, and corms, like bulbs, store food. The crocus and the gladiolus are examples of plants that produce corms.

Cladophylls

The stems of butcher's broom plants are flattened and appear leaflike. Such flattened stems are called **cladophylls** (or *cladodes* or *phylloclades*) (Fig. 6.14). There is a node bearing very small, scalelike leaves with axillary buds in the center of each butcher's broom cladophyll. The feathery appearance of asparagus is due to numerous small cladophylls. Cladophylls also occur in greenbriers, certain orchids, prickly pear cacti (Fig. 6.15), and several other lesser-known plants.

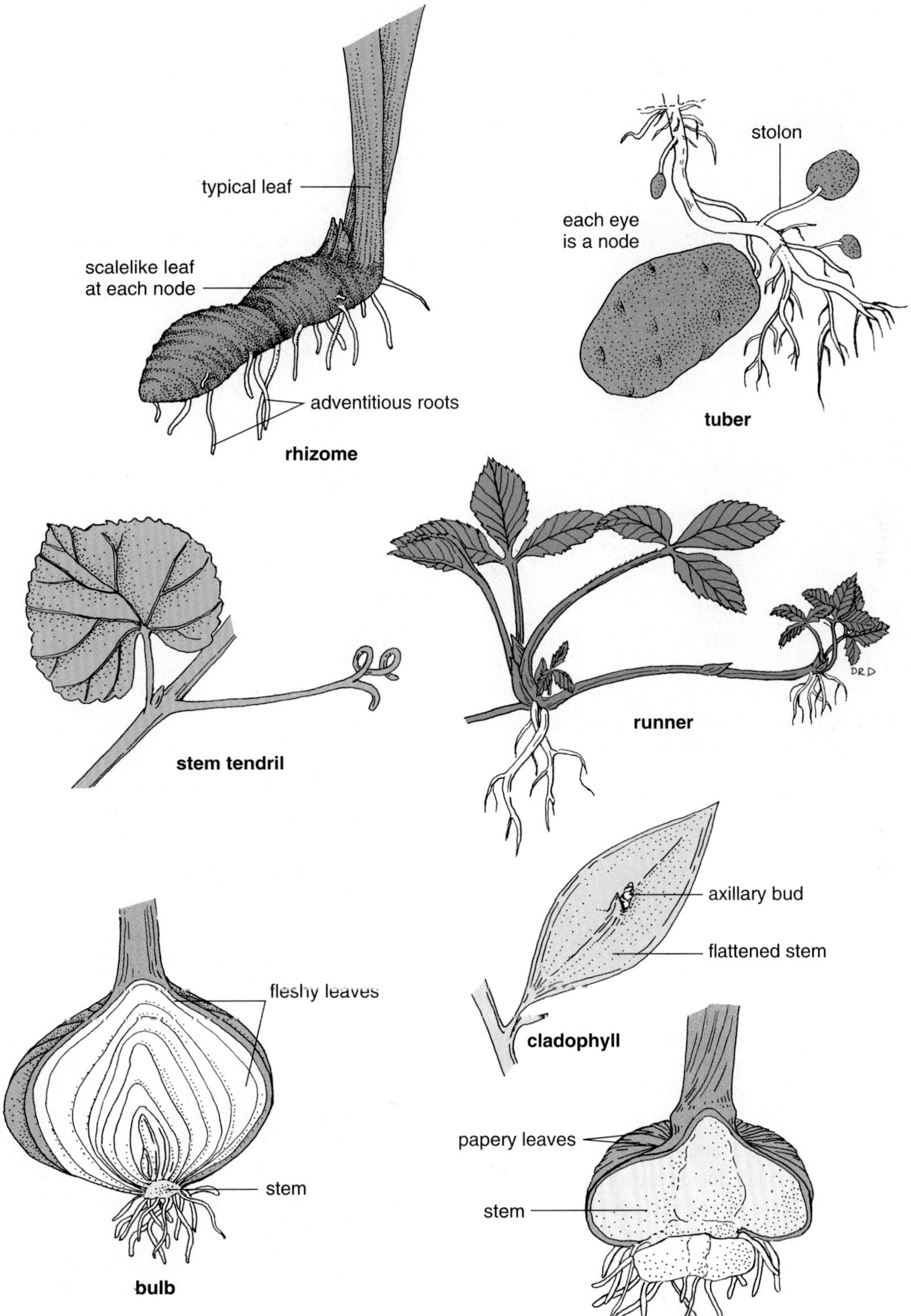

Figure 6.14 Types of specialized stems.

Figure 6.15 The flattened stems of prickly pear cacti (*Opuntia*) are cladophylls on which the leaves have been reduced to spines.

Other Specialized Stems

The stems of many cacti and some spurges are stout and fleshy. Such stems are adapted for storage of water and food. Other stems may be modified in the form of *thorns,* as in the honey locust whose branched thorns may be more than 3 decimeters (1 foot) long, but all thornlike objects are not necessarily modified stems. For example, at the base of the petiole of most leaves of the black locust are a pair of *spines* (modifed *stipules;* stipules were mentioned in the discussion of twigs and are discussed further in Chapter 7). The *prickles* of raspberries and roses, both of which originate from the epidermis, are neither thorns nor spines. Tiger lilies produce small aerial *bulblets* in the axils of their leaves.

Climbing plants have stems modified in various ways that adapt them for their manner of growth. Some stems, called *ramblers,* simply rest on the tops of other plants, but many produce **tendrils** (see Fig. 6.14). These are specialized stems in the grape and Boston ivy but are modified leaves or leaf parts in plants like peas and cucumbers. In Boston ivy, the tendrils have adhesive disks. In English ivy, the stems climb with the aid of adventitious roots that arise along the sides of the stem and become embedded in the bark or other support material over which the plant is growing.

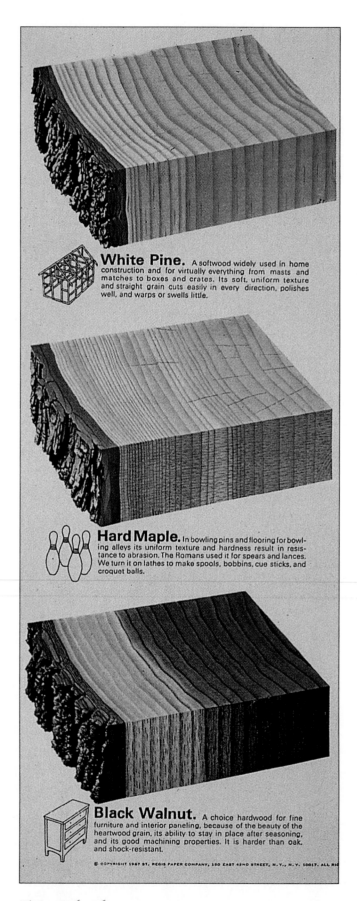

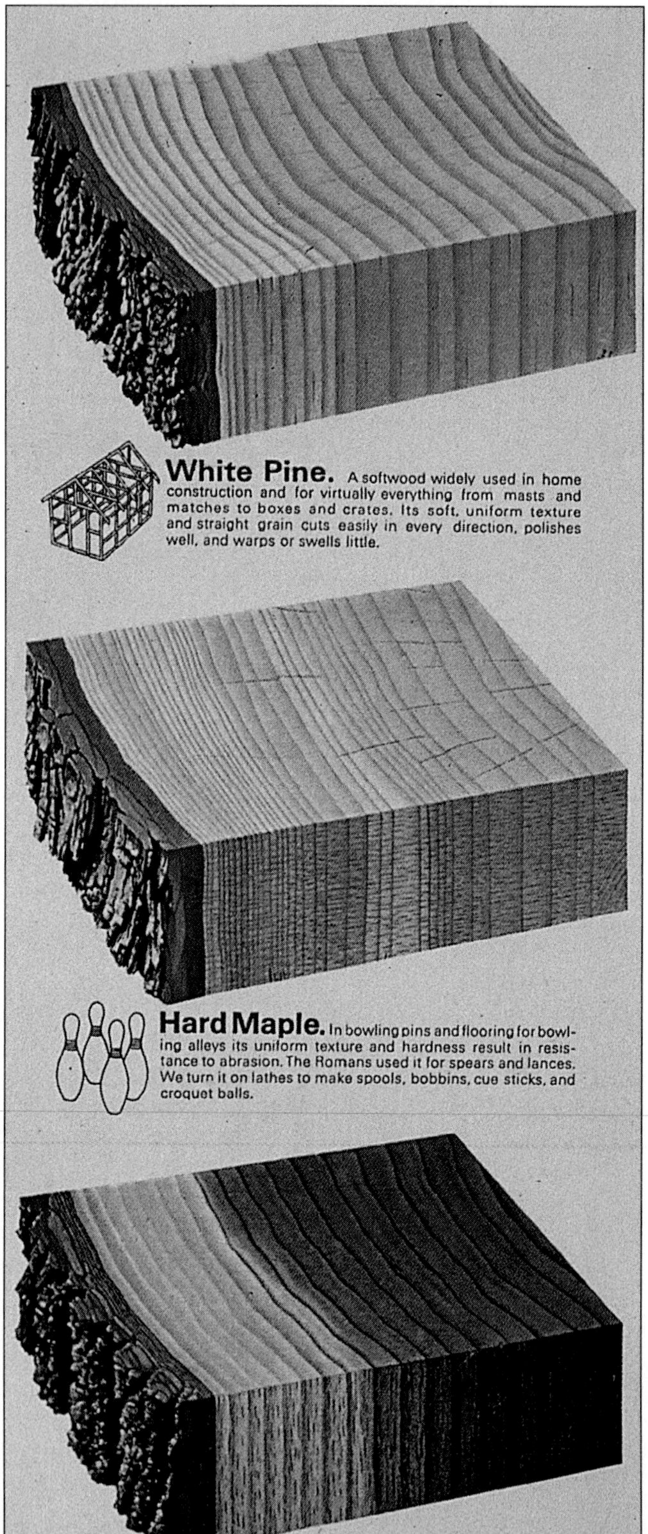

White Pine. A softwood widely used in home construction and for virtually everything from masts and matches to boxes and crates. Its soft, uniform texture and straight grain cuts easily in every direction, polishes well, and warps or swells little.
Hard Maple. In bowling pins and flooring for bowling alleys its uniform texture and hardness result in resistance to abrasion. The Romans used it for spears and lances. We turn it on lathes to make spools, bobbins, cue sticks, and croquet balls.
Black Walnut. A choice hardwood for fine furniture and interior paneling, because of the beauty of the heartwood grain, its ability to stay in place after seasoning, and its good machining properties. It is harder than oak, and shock-resistant.
© COPYRIGHT 1967 ST. REGIS PAPER COMPANY, 150 EAST 42ND STREET, N.Y., N.Y. 10017. ALL RI

Figure 6.16 Uses of some common North American woods. *(Courtesy St. Regis Paper Company)*

White Oak. Makes good barrels because the wood is resilient, durable, and impermeable to liquids. This hardwood, which is about twice as dense as white pine, has many other uses ranging from flooring to fine cabinet work.
Baldcypress. Because it is weather-resistant without treatment, this wood was widely used for cross ties in the early days of railroading. Today it is used for water tanks and other applications requiring prolonged contact with water.
White Ash. Perfect for baseball bats, tennis racquets, oars and long tool handles. This hardwood's major virtues are straight grain, stiffness, strength, moderate weight, good bending qualities, and capacity for wearing smooth.
GHTS RESERVED.
Red Spruce. A favorite for violin sounding boards because of its high resonant qualities. A softwood, it is easy to work, and is light in relation to its strength and stiffness. These qualities also make it eminently suitable for ladder rails, canoe paddles, and oars.
Hemlock. This relatively soft, light, straight-grained, resin-free wood, with its uniformly long fibers, is becoming one of the most important species for paper pulp. It is also used for structural lumber and plywood, and for boxes, barrels, and concrete forms.
J. KUNZ
Hickory. A hardwood unsurpassed for the handles of impact tools like axes and hammers, and for skis, because of its hardness, strength, toughness, and resiliency. In horse-and-buggy days it was widely used for wheel spokes and rims, singletrees, and buggy shafts.

Figure 6.16 Continued.

WOOD AND ITS USES

The use of wood by humans for fuel, clubs, and other purposes dates back into antiquity, and present uses are so numerous that it would be impossible to list in a work of this type more than the most important ones. Before discussing the economic importance of wood, let's take a brief look at its properties.

In a living tree, up to 50% of the weight of the wood comes from the water content. Before the wood can be used, seasoning reduces the moisture content to 10% or less, either by air-drying it in ventilated piles or stacks or by drying it in special ovens known as kilns. The seasoning has to be done gradually and under carefully controlled conditions, or the timber may warp and split along the rays, making it unfit for most uses. The dry part of wood is composed of 60% to 75% *cellulose* and about 15% to 25% *lignin,* an organic substance that makes the walls of xylem cells tough and hard. Other substances present in smaller amounts include resins, gums, oils, dyes, tannins, and starch. The proportions and amounts of these and other substances determine how various woods will be used (Fig. 6.16).

Density

The density of wood is among its most important physical properties. Technically, the density is the weight per unit volume. The weight is compared with that of an equal volume of water and is stated as a fraction of 1.0. Because of the considerable air space within the cells, most woods have a specific gravity, as the comparative density is called, of less than 1.0. The range of specific gravities of known woods varies from 0.04 to 1.40, the lightest commercially used wood being balsa with a specific gravity of about 0.12. Woods with specific gravities of less than 0.50 are considered light; those with specific gravities of above 0.70 are considered heavy. Among the heaviest woods are the South American ironwood and lignum vitae, with specific gravities of over 1.25. Lignum vitae, obtained from West Indian trees, is extremely hard wood and is used instead of metal in the manufacture of main bearings for drive shafts of submarines because it is self-lubricating and less noisy.

Durability

A wood's ability to withstand decay caused by organisms and insects is referred to as its *durability.* Moisture is needed for the enzymatic breakdown of cellulose and other wood substances by decay organisms, but the seasoning process usually reduces the moisture to a level below that necessary for the fungi and other decay organisms to survive. Other natural constituents of wood that repel decay organisms include tannins and oils. Wood with a tannin content of 15%

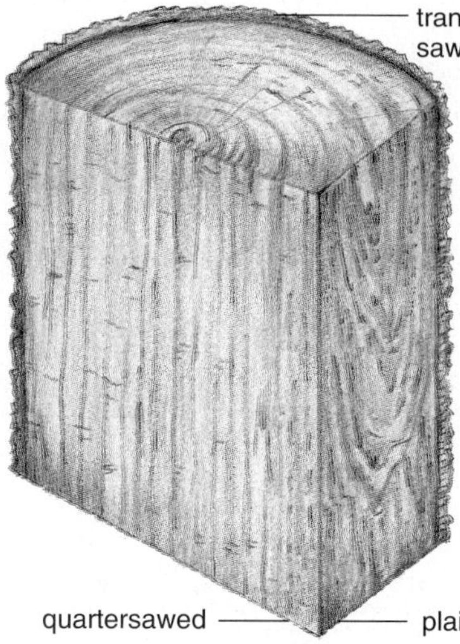

Figure 6.17 How the surfaces of plain-sawed, quartersawed, and transversely sawed wood appear. (See also Fig. 6.8.)

or more may survive on a forest floor for many years after diseases of the phloem and other causes have toppled it. Among the most durable of American woods are cedar, catalpa, black locust, red mulberry, and Osage orange. The least durable woods include cottonwood, willow, fir, and basswood.

Types of Sawing

Logs are usually cut longitudinally in one of two ways: along the radius or perpendicular to the rays (Fig. 6.17). Radially cut, or *quartersawed,* boards show the annual rings in side view; they appear as longitudinal streaks and are the most conspicuous feature of the wood. Only a few perfect quartersawed boards can be obtained from a log, making them quite expensive. Boards cut perpendicular to the rays (tangentially cut boards) are more common. In these, the annual rings appear as irregular bands of light and dark alternating streaks or patches, with the ends of the rays visible as narrower and less conspicuous vertical streaks. Lumber cut tangentially is referred to as being *plain-sawed,* or *slab cut.* Slabs are the boards with rounded sides at the outside of the log; they are usually made into chips for pulping.

Knots

Knots are the bases of lost branches that have become covered, over a period of time, by new annual rings of wood produced by the cambium of the trunk. They are found in greater concentration in the older parts of the log toward the center, because in the forest, the lowermost branches of a tree (produced while the trunk was small in girth) often die from insufficient light. When a branch dies and falls off, the cambium at its base also dies, but the cambium of the trunk remains alive and increases the girth of the tree, slowly enveloping the dead tissue of the branch base until it may be completely buried and not visible from the surface. Knots usually weaken the boards in which they occur.

Wood Products

In the United States and Canada, about half of the wood produced is used as lumber, primarily for construction; the sawdust and other waste formed in processing the boards is converted to particle board and pulp. A considerable amount of lumber goes into the making of furniture, which may be constructed of solid wood or covered with a *veneer.* A veneer is a very thin sheet of desirable wood that is glued to cheaper lumber; it is carefully cut so as to produce the best possible view of the grain (Fig. 6.18).

The next most extensive use of wood is for *pulp,* which among other things is converted by various processes to paper, synthetic fibers, plastics, and linoleum. In recent years, it has been added as a filler to commercial ice cream and bread. Some hardwoods are treated chemically or heated under controlled conditions to yield a number of chemicals, such as wood alcohol and acetic acid, but other sources of these products are now usually considered more economical. Charcoal, excelsior, cooperage (kegs, casks, and barrels), railroad ties, boxes and crates, musical instruments, bowling pins, tool handles, pilings, cellophane, photographic film, and Christmas trees are but a few of the additional wood products worth billions of dollars annually on the world market (see Fig. 6.16).

In developing countries, approximately half of the timber cut is used for fuel, but in the United States and Canada, a little less than 10% is currently used for that purpose. In colonial times, wood was the almost exclusive source of heating energy. In the 1990s, Brazil's major cities were still using scrub timber from the surrounding forests to energize their utilities, but rapidly depleting supplies and problems related to the greenhouse effect (discussed in Chapter 25) have pointed to the need for alternate sources of energy. Many types of coal are wood that has been compressed for millions of years until nearly pure carbon remains. The formation of coal and other fossils is discussed in Chapter 20. Although the world's supply of coal is still plentiful, the rate at which this fossil fuel is being consumed makes it obvious that resources will eventually be exhausted unless our energy demands find renewable or less destructive alternatives.

Some of the vast array of secondary products from stems, including dyes, medicines, spices, and foods, are discussed in later chapters and in the appendices.

Summary

1. Stem structure and function need to be examined to understand practical uses of stems.
2. The shoot system of plants is usually erect, but some stems may be horizontal or modified for climbing or food/water storage.
3. Leaves of woody twigs may be arranged alternately, oppositely, or in a whorl. Nodes are stem regions where leaves are attached; internodes occur between nodes. Most leaves have petioles and blades. Axillary buds occur in leaf axils. Most buds are protected by bud scales. Terminal buds occur at twig tips. Terminal bud scales, when they fall, leave bud scale scars that help one determine the age of the twig.
4. Stipules are paired appendages present at the base of some leaves; when they fall off, they leave small scars on the twig. When whole leaves fall, they leave leaf scars on the twig, with tiny bundle scars within the leaf-scar surfaces.
5. Each stem has an apical meristem at its tip that produces tissues resulting in increase in length. Leaf primordia develop into mature leaves when growth begins. Three primary meristems develop from an apical meristem: the protoderm gives rise to the epidermis; the procambium produces primary xylem and primary phloem; and the ground meristem produces pith and cortex.
6. As leaves and buds develop from primordia, traces of xylem and phloem branch off from the main cylinder, leaving leaf gaps or bud gaps.
7. A vascular cambium, producing secondary tissues, may arise between primary xylem and phloem. Secondary xylem cells include tracheids, vessel elements, and fibers. Secondary phloem cells include sieve tube members and companion cells.
8. In many plants, a cork cambium producing cork and phelloderm cells develops near the surface of the stem. Cork cells, which are part of the outer bark (periderm), have suberin in their walls. Suberin is impervious to moisture, and the outer bark, therefore, aids in protection. Lenticels in the bark permit gas exchange.
9. Primary vascular tissues and the pith, if present, constitute the stele. Protosteles have a solid core of xylem, usually surrounded by phloem; siphonosteles are tubular, with pith in the center; eusteles have the vascular tissues in discrete bundles.
10. Dicotyledons (dicots) are plants whose seeds have two seed leaves (cotyledons), while monocotyledons (monocots) have seeds with one seed leaf. Herbaceous dicots have vascular bundles arranged in a ring in the stem.

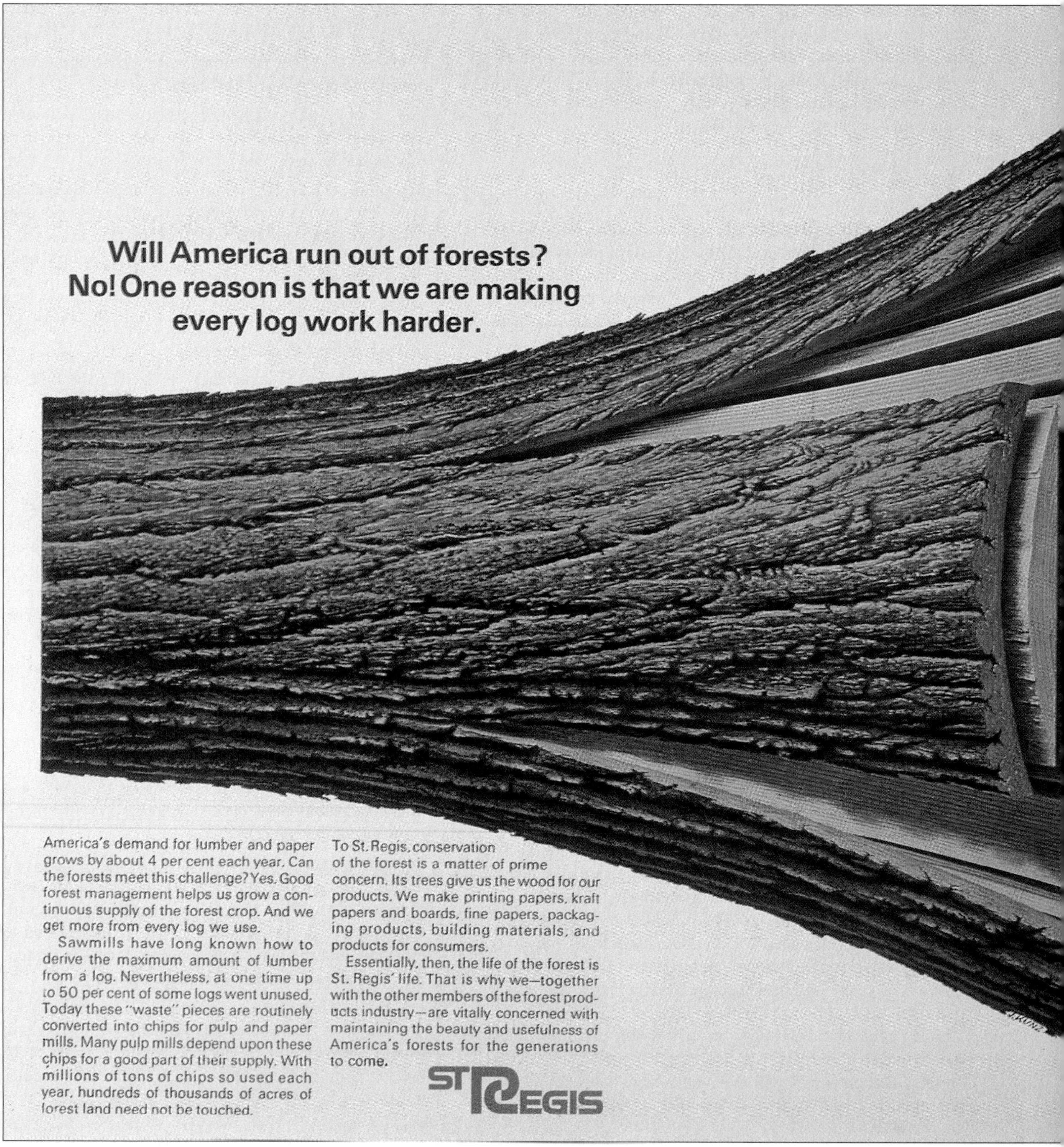

Figure 6.18 How a log is used. Note that some logging practices have become controversial due to their effects on the habitats of threatened species of living organisms, as discussed in Chapter 25. *(Courtesy St. Regis Paper Company)*

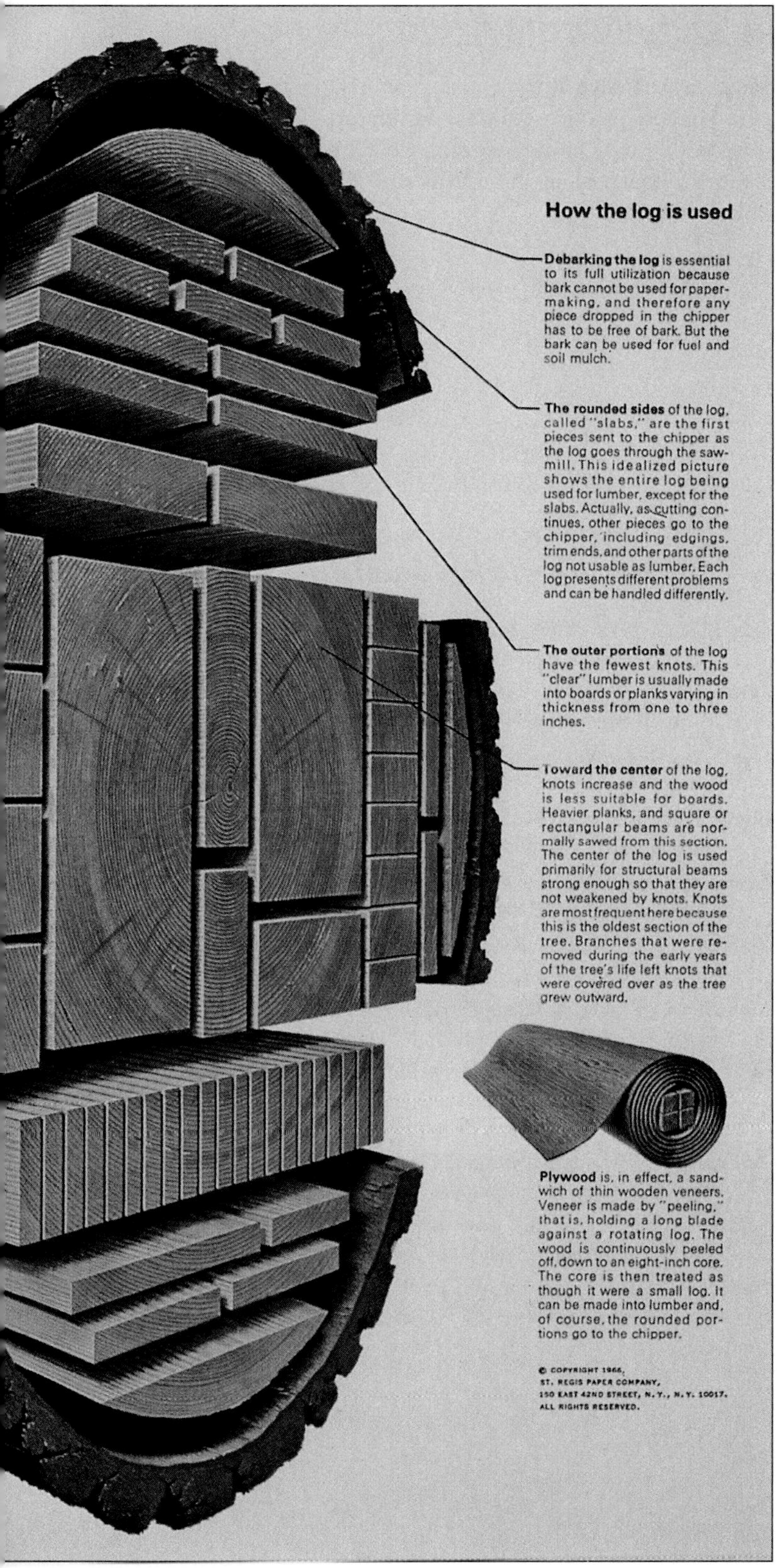
How the log is used
Debarking the log is essential to its full utilization because bark cannot be used for paper-making, and therefore any piece dropped in the chipper has to be free of bark. But the bark can be used for fuel and soil mulch.
The rounded sides of the log, called "slabs," are the first pieces sent to the chipper as the log goes through the saw-mill. This idealized picture shows the entire log being used for lumber, except for the slabs. Actually, as cutting continues, other pieces go to the chipper, including edgings, trim ends, and other parts of the log not usable as lumber. Each log presents different problems and can be handled differently.
The outer portions of the log have the fewest knots. This "clear" lumber is usually made into boards or planks varying in thickness from one to three inches.
Toward the center of the log, knots increase and the wood is less suitable for boards. Heavier planks, and square or rectangular beams are normally sawed from this section. The center of the log is used primarily for structural beams strong enough so that they are not weakened by knots. Knots are most frequent here because this is the oldest section of the tree. Branches that were removed during the early years of the tree's life left knots that were covered over as the tree grew outward.
Plywood is, in effect, a sandwich of thin wooden veneers. Veneer is made by "peeling," that is, holding a long blade against a rotating log. The wood is continuously peeled off, down to an eight-inch core. The core is then treated as though it were a small log. It can be made into lumber and, of course, the rounded portions go to the chipper.
© COPYRIGHT 1966, ST. REGIS PAPER COMPANY, 150 EAST 42ND STREET, N.Y., N.Y. 10017. ALL RIGHTS RESERVED.

11. Woody dicots have most of their secondary tissues arranged in concentric layers. The most conspicuous tissue is wood (secondary xylem). In broadleaf trees, spring wood usually has relatively large vessel members, while summer wood has smaller vessels and/or a predominance of tracheids.
12. An annual ring is 1 year's growth of xylem. A tree's age and other aspects of its history can be determined from annual rings. Rays, which function in lateral conduction, radiate out from the center of the trunk. Older wood toward the center (heartwood) ceases to function when its cells become plugged with tyloses. Younger, functioning wood (sapwood) is closer to the surface. A tree's functions are not particularly affected by the rotting of its heartwood.
13. The wood of cone-bearing trees consists primarily of tracheids, and resin canals are often present. The wood of conifers has no fibers or vessels and is called softwood, while the wood of woody dicots is called hardwood. In woody plants, older tissues composed of thin-walled cells become crushed and functionless, and some are sloughed off.
14. Laticifers are latex-secreting cells or ducts found in various flowering plants. The latex of some plants has considerable commercial value.
15. Monocot stems have scattered vascular bundles and no cambia. The parenchyma tissue is not divided into pith and cortex. Each vascular bundle is surrounded by a sheath of sclerenchyma cells. Numerous bundles and a band of sclerenchyma cells and thicker-walled parenchyma cells just beneath the surface of monocot stems aid in withstanding stresses.
16. Palm trees are monocots that become large because their parenchyma cells continue to divide. Other monocots develop a secondary meristem that produces parenchyma cells and secondary vascular bundles. Grasses have at the base of each internode intercalary meristems that contribute to rapid increases in length. Several commercially important cordage fibers are obtained from monocots.
17. Specialized stems include rhizomes, stolons, tubers, bulbs, corms, cladophylls, and tendrils. Such stems may have adventitious roots.
18. The dry part of wood consists primarily of cellulose and lignin. Resins, gums, oils, dyes, tannins, and starch are also present. Properties of wood that play a role in its use include density, specific gravity, and durability.
19. Logs are usually cut longitudinally along the radius (quartersawed) or perpendicular to the rays (tangentially, plain-sawed, or slab cut). Knots are bases of lost branches that have become covered over by new wood; they usually weaken the boards in which they occur.

20. About half the timber produced in the United States and Canada is used as lumber. Sawdust and waste are converted to particle board and pulp for paper, synthetics, and linoleum. Other timber is used for cooperage, charcoal, railroad ties, boxes, tool handles, and so forth. Developing countries use a greater proportion of their timber for fuel.

Review Questions

1. What is the function of bud scales?
2. How can you tell the age of a twig?
3. Distinguish among procambium, vascular cambium, and cork cambium.
4. How can you tell, when you look at a cross section of a young stem, whether it is a dicot or a monocot?
5. What are laticifers?
6. An Irish or white potato is a stem, but a sweet potato is a root. How can you tell?
7. Distinguish among corms, bulbs, and tubers.
8. If you were examining the top of a wooden desk, how could you tell if the wood had been radially or tangentially cut (quartersawed or plain-sawed)?
9. What differences are there between heartwood and sapwood?
10. What is meant by the specific gravity of wood?

Discussion Questions

1. If the cambium of a tropical tree were active all year long, how would its wood differ from that of a typical temperate climate tree?
2. It was mentioned that a nail driven into the side of a tree will remain at exactly the same distance from the ground for the life of the tree. Why?
3. Do climbing plants have any advantages over erect plants? Any disadvantages?
4. If two leaves are removed from a plant and one is coated with petroleum jelly while the other is not, the uncoated leaf will shrivel considerably sooner than the coated one. Would it be helpful to coat the stems of young trees with petroleum jelly? Explain.
5. Suggest some reasons for heartwood being preferred to sapwood for making furniture.

Learning Online

Visit our web page at www.mhhe.com/botany for interesting case studies, practice quizzes, current articles, and animations within the Online Learning Center to help you understand the material in this chapter. You'll also find active links to these topics:

Characteristics of Plants
Stems and Secondary Growth
Herbaceous Stems
Woody Stems
Economic Uses of Stems
Cereal Grains and Forage Grasses
Wood, Cork, and Bamboo
Economic and Ecological Importance of Angiosperms

Additional Reading

Cutter, E. G. 1978. *Plant anatomy: Experiment and interpretation. Part I: Cells and tissues,* 2d ed. Reading, MA: Addison-Wesley.

Desch, H. E. 1996. *Timber: Structure, properties, conversion and use,* 7th ed. Binghampton, NY: Haworth Press Journal Co-Editions.

Fahn, A. 1990. *Plant anatomy,* 4th ed. Elmsford, NY: Pergamon Press.

Flynn, J. H., Jr. (Ed.). 1994. *A guide to useful woods of the world.* Portland, ME: King Philip.

Lewin, M., and I. S. Goldstein (Eds.). 1991. *Wood structure and composition.* New York: Dekker, Marcel.

Mauseth, J. D. 1988. *Plant anatomy.* Menlo Park, CA: Benjamin/Cummings.

Metcalfe, C. R., and L. Chalk (Eds.). 1988–1989. *Anatomy of the dicotyledons,* 2 vols. Fair Lawn, NY: Oxford University Press.

Meylan, B. A., and B. G. Butterfield. 1972. *Three-dimensional structure of wood: A scanning electron microscope study.* Syracuse, NY: Syracuse University Press.

Perlin, J. A. 1991. *A forest journey: The role of wood in the development of civilization.* Cambridge, MA: Harvard University Press.

Schweingruber, F. H. 1993. *Trees and wood in dendrochronology.* New York: Springer-Verlag.

Steeves, T. A., and I. M. Sussex. 1989. *Patterns in plant development,* 2d ed. Englewood Cliffs, NJ: Prentice-Hall.

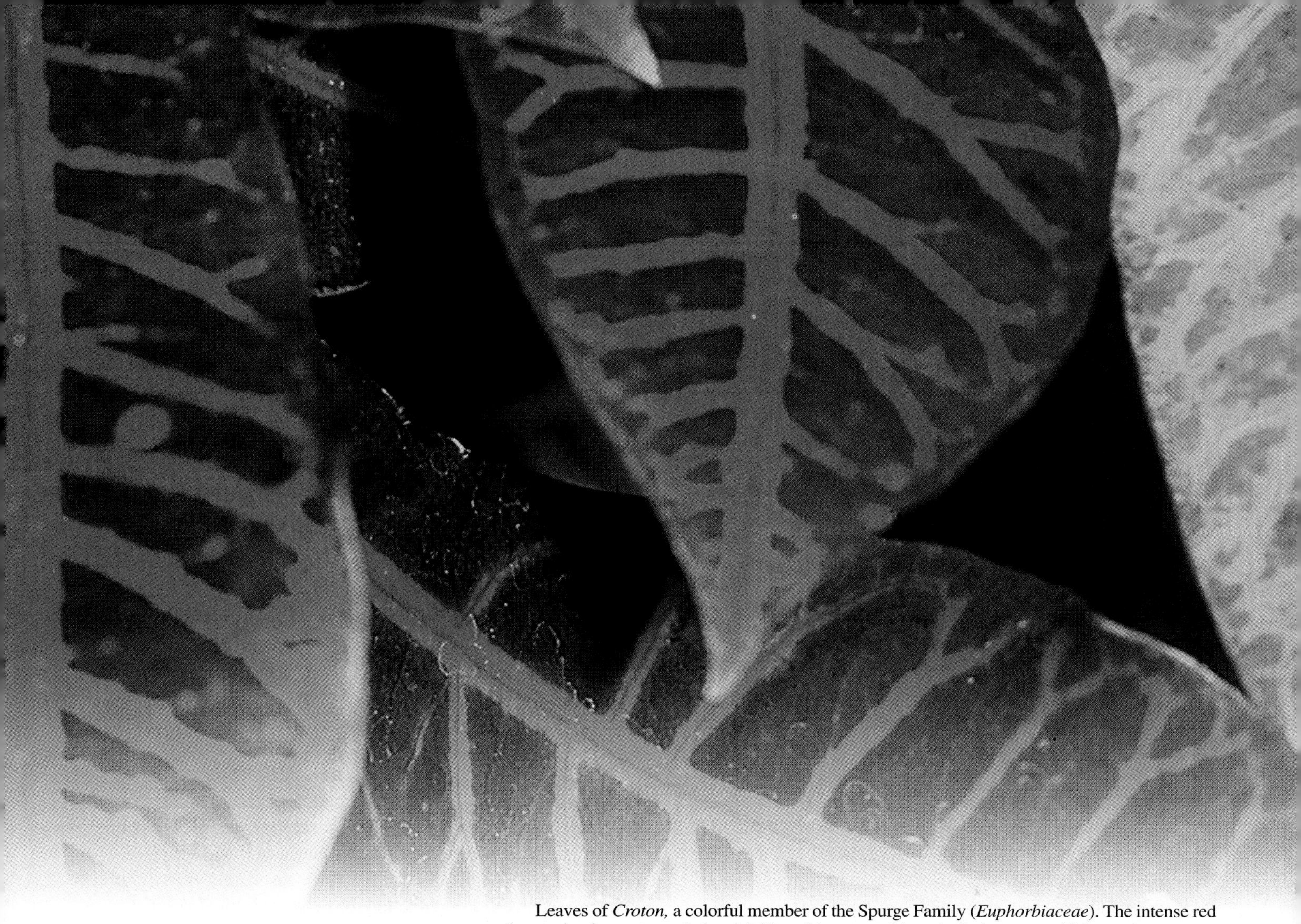

Leaves of *Croton,* a colorful member of the Spurge Family (*Euphorbiaceae*). The intense red anthocyanin pigments mask the visibility of the green chlorophyll pigments, which are also present.

Leaves

7

Chapter Outline

Overview

This chapter introduces leaves by comparing them with solar panels and by discussing their general functions, morphology, and dimensions. This is followed by descriptive information on basic leaf types and specific forms and arrangements. The chapter next discusses the internal structure of leaves, including epidermis and cuticle, stomata, glands, mesophyll, and veins.

Specialized leaves, including tendrils, spines, flower-pot leaves, window leaves, reproductive leaves, floral leaves, and insectivorous leaves, are then examined. The chapter concludes with an explanation of autumnal color changes and abscission and some observations on the human and ecological relevance of leaves.

Some Learning Goals

1. Learn the external forms and parts of leaves. Know the functions of a typical leaf and the specific tissues and cells that contribute to those functions.
2. Understand the differences among pinnate, palmate, and dichotomous venation and also the differences between simple and compound leaves.
3. Contrast tendrils, spines, storage leaves, flower-pot leaves, window leaves, reproductive leaves, floral leaves, and different types of insect-trapping leaves.
4. Explain why deciduous leaves turn various colors in the fall and how such leaves are shed.
5. Know at least 15 uses of leaves by humans.

The earliest records of glass being used by humans date back to about 2600 B.C., when the ancient Egyptians and Babylonians made beads from the material. The use of glass panes for windows, however, did not begin until the Roman Imperial period a little over 2,000 years ago. Since then, the use of glass windows for admitting light to buildings of all sizes and shapes has become almost universal.

A comparatively recent use of glass involves solar energy as an alternative to nonrenewable sources of energy such as fossil fuels. The construction industry, particularly in the southwestern United States, is building new houses with flat panels and windows inclined at angles that maximize the amount of energy captured from the sun's rays. Some buildings have mechanical devices that slowly move the solar panels so that they will follow the sun in its daily course across the sky. The use of such means of capturing solar energy is now spreading to other countries and could soon become commonplace.

Plants had a highly efficient form of solar panel that captured the sun's energy many aeons before modern civilization began to realize that fossil-fuel supplies eventually would be exhausted. These remarkably constructed solar panels are the plant organs known to us as **leaves**.

All leaves originate as **primordia** in the buds, regardless of their ultimate size or form. In early spring, a leaf primordium may consist of fewer than 200 cells, but in response to changes in temperature, day length, and availability of water, hormones are produced that stimulate these cells to begin dividing. Within a few days or weeks, the original 200 cells have multiplied, differentiated, and expanded into a structure consisting of millions of cells. In some plants (e.g., *Eucalyptus*), the first leaves produced (*juvenile leaves*) may appear quite different in form from those produced later. The juvenile form may be promoted by a class of hormones known as *gibberellins* (discussed in Chapter 11).

At maturity, most leaves have a stalk, called the **petiole**, and a flattened **blade**, or *lamina*, which has a network of *veins* (*vascular bundles*). A pair of leaflike, scalelike, or thornlike appendages, called *stipules*, are sometimes present at the base of the petiole. Occasionally, leaves may lack petioles; when they do, they are said to be *sessile*. Leaves of **deciduous** trees normally live through only one growing season, and even those of evergreen trees rarely function for more than 2 to 7 years.

Leaves of flowering plants are associated with *leaf gaps* (as illustrated in Fig. 6.3), and all have an *axillary bud* at the base. Leaves may be *simple* or *compound*. A **simple leaf** has a single blade, while the blade of a **compound leaf** is divided in various ways into *leaflets* (see Fig. 7.4). Regardless of the number of leaflets, a compound leaf still has a single axillary bud at its base, with the leaflets having no such buds. **Pinnately compound** leaves have the leaflets in pairs along an extension of the petiole called a **rachis**, while **palmately compound** leaves have all the leaflets attached at the same point at the end of the petiole. Sometimes, the leaflets of a pinnately compound leaf may be subdivided into still smaller leaflets, forming a *bipinnately compound* leaf (Fig. 7.1).

The flattened surfaces of leaves, which are completely covered with a transparent protective layer of cells, the *epidermis*, admit light to all parts of the interior. Many leaves twist daily on their petioles so that their upper surfaces are inclined at right angles to the sun's rays throughout daylight hours (Fig. 7.2).

Green leaves capture the light energy available to them by means of the most important process for life on earth, at least life as we know it today. This process, called *photosynthesis* (discussed in Chapter 10), involves the trapping and ultimate storing of energy in sugar molecules that are constructed from ordinary water and from carbon dioxide present in the atmosphere. All the energy needs of living organisms ultimately depend on photosynthesis, from the first day of their existence to the last.

Figure 7.1 A bipinnately compound leaf of *Albizia* sp.

Figure 7.2 Algerian ivy on a tree trunk. Note how each leaf blade is oriented to receives the maximum amount of light.

The lower surfaces of leaves (and in some plants, the upper surfaces as well) are dotted with tiny pores (**stomata**), which not only allow entry for the carbon dioxide gas needed for photosynthesis, but also play a role in the diffusion out of the leaf of oxygen produced during photosynthesis. Water vapor evaporating from the moist interior cell surfaces can also escape via the stomata. The evaporation of water can bring about some cooling of the leaf, but excessive water loss can result in damage to the plant. The *stomatal apparatus*, which consists of a pore bordered by a pair of sausage-shaped *guard cells*, controls the water loss when the guard cells inflate or deflate, opening or closing the pore.

Leaves also perform other functions. For example, all living cells *respire* (respiration is also discussed in Chapter 10), and in the process of this and other metabolic activities, waste products are produced. These wastes accumulate in the leaves and are disposed of when the leaves are shed, mostly in the fall. Before dropping from the plant, the leaves are sealed off at the bases of their petioles (see the discussion of leaf colors and *abscission* later in this chapter). The following season, the discarded leaves are replaced with new ones.

Leaves play a major role in the movement of water absorbed by roots and transported throughout the plant. Most of the water reaching the leaves evaporates in vapor form into the atmosphere by a process known as **transpiration** (discussed in Chapter 9). In some plants, there are special openings called *hydathodes* at the tips of leaf veins. *Root pressure* (see page 161) forces liquid water out of hydathodes, usually at night when transpiration is not occurring. The loss of water through hydathodes is called **guttation**. The expelled water may contain ions secreted by root cells. Other functions of leaves are discussed throughout this chapter.

LEAF ARRANGEMENTS AND TYPES

Many of the roughly 275,000 different species of plants that produce leaves can be distinguished from one another by their leaves alone. The variety of shapes, sizes, and textures of leaves seems to be almost infinite. The leaves of some of the smaller duckweeds are less than 1 millimeter (0.04 inch) wide. The mature leaves of the Seychelles Island palm can be 6 meters (20 feet) long, and the floating leaves of a giant water lily, which reach 2 meters (6.5 feet) in diameter (Fig. 7.3), can support, without sinking, weights of more than 45 kilograms (100 pounds) distributed over their surface. In lilies, pines, ferns, and many other plants, different forms of leaves (e.g., tiny papery scales; colored leaves called *bracts;* spines) may be produced, along with typical photosynthetic leaves, on the same plant.

In addition to flattened, variously shaped, colored, spinelike leaves and those of various textures, there are others that are tubular, feathery, cup shaped, or needlelike; in

Figure 7.3 Floating leaves of a giant water lily (*Victoria amazonica*), which sometimes attain a diameter of 2 meters (6.5 feet). The larger leaves are capable of supporting, without sinking, the weight of a child.

fact, leaves may assume virtually any form (Fig. 7.4). They may be smooth or hairy, slippery or sticky, waxy or glossy, pleasantly fragrant or foul smelling, edible or poisonous. They also may be of almost every color of the rainbow and of exquisite beauty, especially when viewed with a microscope.

Leaves are attached to stems at regions called **nodes**, with stem regions between nodes being known as **internodes**. The arrangement of leaves on a stem (*phyllotaxy*) in a given species of plant generally occurs in one of three ways. In most species, leaves are attached alternately or in a spiral along a stem, with one leaf per node, in what is called an **alternate** arrangement. In some plants, two leaves may be attached at each node, providing an **opposite** arrangement. When three or more leaves occur at a node, they are said to be **whorled**.

The arrangement of veins in a leaf or leaflet blade (*venation*) may also be either pinnate or palmate. In **pinnately veined** leaves, there is one primary vein called the **midvein,** which is included within an enlarged **midrib**; secondary veins branch from the midvein. In **palmately veined** leaves, several primary veins fan out from the base of the blade. The primary veins are more or less *parallel* to one another in monocots (Fig. 7.5) and diverge from one another in various ways in dicots (see Fig. 7.9). The branching arrangement of veins in dicots is called *netted*, or *reticulate venation*. In a few leaves (e.g., those of *Ginkgo*), no midvein or other large veins are present. Instead, the veins fork evenly and progressively from the base of the blade to the opposite margin. This is called *dichotomous venation* (Fig. 7.4K).

INTERNAL STRUCTURE OF LEAVES

If a typical leaf is cut transversely and examined with the aid of a microscope, three regions stand out: *epidermis, mesophyll*, and *veins* (referred to as *vascular bundles* in our discussion of roots and stems) (Fig. 7.6). The epidermis is a single layer of cells covering the entire surface of the leaf. The epidermis on the lower surface of the blade can sometimes be distinguished from the upper epidermis by the presence of tiny pores called *stomata*, which are discussed in the section that follows.

When seen from the top, the wavy, undulating walls of the epidermal cells often resemble pieces of a jigsaw puzzle fitted together. Except for *guard cells,* the upper epidermal cells for the most part do not contain chloroplasts, their function being primarily protection of the delicate tissues to the interior. A coating of waxy **cutin** (the **cuticle**—see Fig. 4.6) is normally present, although it may not be visible with ordinary light microscopes without being specially stained. In addition to the cuticle, many plants produce other waxy substances on their surfaces (Fig. 7.7). In studies of the effects of smog and auto exhaust fumes on plants, it was found that these waxes may be produced in abnormal fashion on beet leaves within as little as 24 hours after exposure to the pollutants. Presumably, the wax affords added protection to the leaves. Beet leaves also respond to aphid damage by producing wax around each tiny puncture.

Figure 7.4 Types of leaves and leaf arrangements. *A*. Palmately compound leaf of a buckeye. *B*. Pinnately compound leaf of a black walnut. *C*. Alternate, simple but lobed leaves of a tulip tree. *D*. Opposite, simple leaves of a dogwood. *E*. Palmately veined leaf of a maple. *F*. Globe-shaped succulent leaves of string-of-pearls. *G*. Pinnately veined, lobed leaf of an oak. *H*. Parallel-veined leaf of a grass. *I*. Whorled leaves of a bedstraw. *J*. Linear leaves of a yew. *K*. Fan-shaped leaf of a *Ginkgo* tree, showing dichotomous venation.

In some plants, waste materials occasionally accumulate and crystallize in epidermal cells. Different types of **glands** may also be present in the epidermis. Glands occur in the form of depressions, protuberances, or appendages either directly on the leaf surface or on the ends of hairs (see Fig. 4.13A). Glands often secrete sticky substances.

STOMATA

The lower epidermis of most plants generally resembles the upper epidermis, but the lower is perforated by numerous tiny pores called **stomata** (Fig. 7.8). Some plants (e.g., alfalfa, corn) have these pores in both leaf surfaces, while

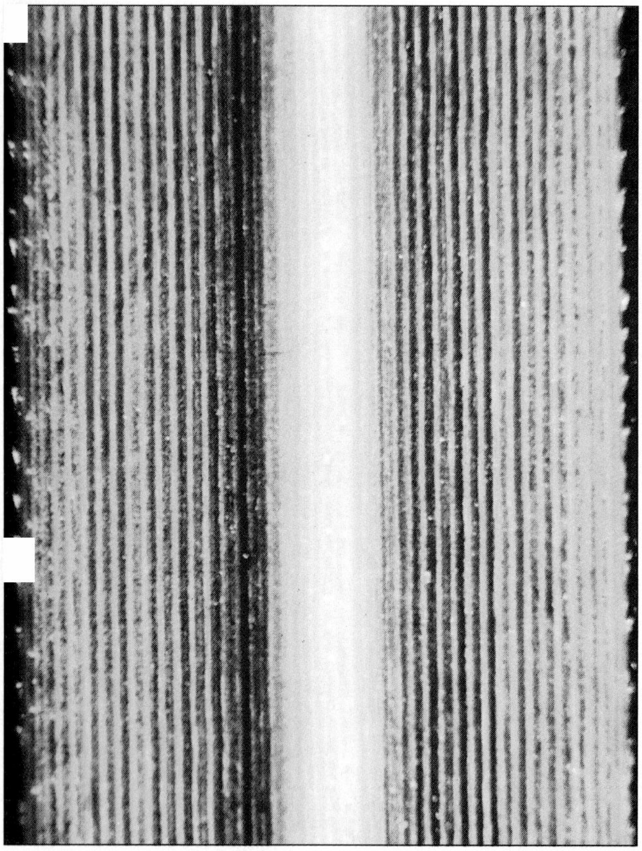

Figure 7.5 A portion of a monocot leaf showing the parallel veins.

others (e.g., water lilies) have them exclusively on the upper epidermis; they are absent altogether from the submerged leaves of aquatic plants. Stomata are very numerous, ranging from about 1,000 to more than 1.2 million per square centimeter (6,300 to 8 million per square inch) of surface. An average-sized sunflower leaf has about 2 million of these pores throughout its lower epidermis. Each pore is bordered by two sausage- or dumbbell-shaped cells that usually are smaller than most of the neighboring epidermal cells. These **guard cells,** which originate from the same parental cell, are part of the epidermis, but they, unlike most of the other cells of either epidermis, contain chloroplasts.

The functioning of guard cells is aided by the photosynthesis that takes place within them. The primary functions include regulating gas exchange between the interior of the leaf and the atmosphere and regulation of evaporation of most of the water entering the plant at the roots. Guard cell walls are distinctly thickened but quite flexible on the side adjacent to the pore. As the guard cells inflate or deflate with changes in the amount of water within the cells, their unique construction causes the stomata to open or close. When the guard cells are inflated, the stomata are open; when the water content of the guard cells decreases, the cells deflate, and the stomata close. For more detailed discussions of this stomatal mechanism, see "Regulation of Transpiration" in Chapter 9 and "Turgor Movements" in Chapter 11).

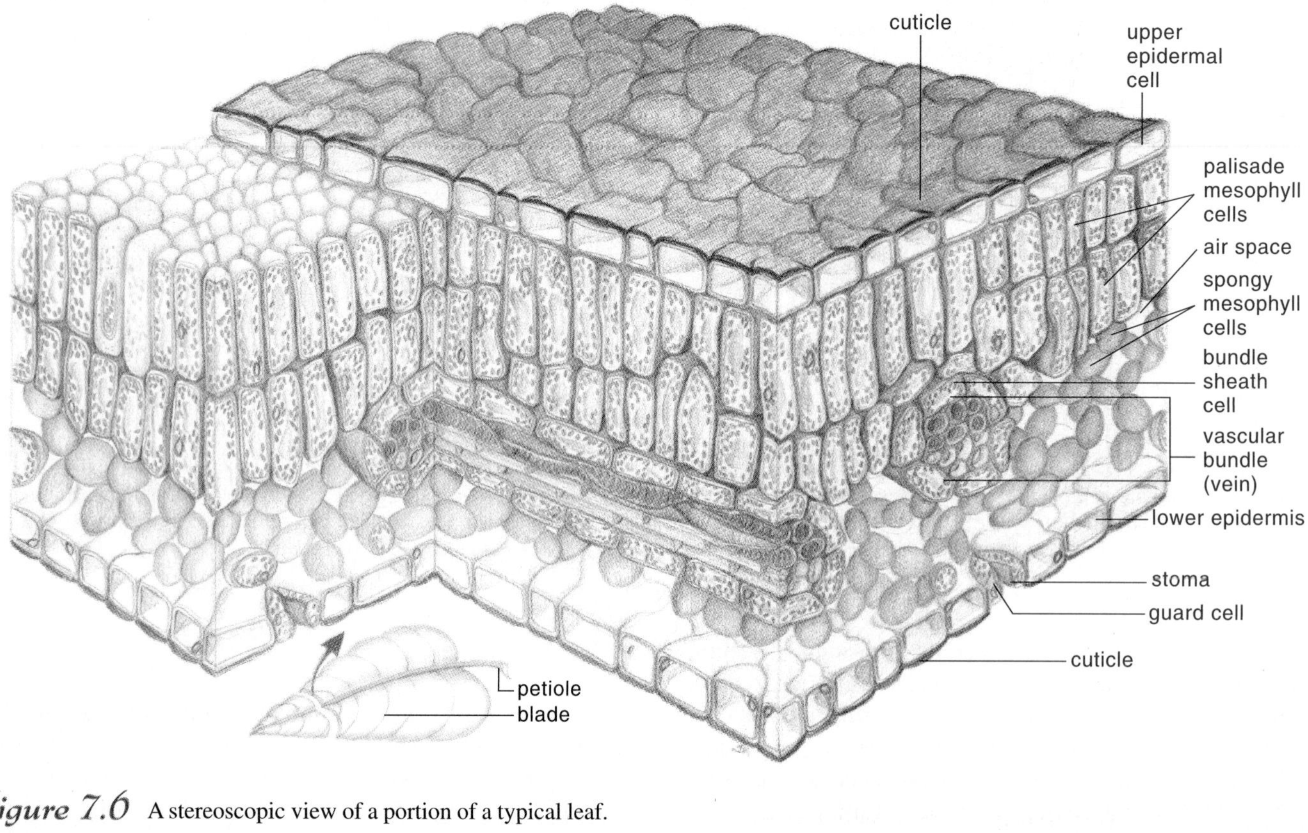

Figure 7.6 A stereoscopic view of a portion of a typical leaf.

Figure 7.7 Waxes, in addition to those of the cuticle, are sometimes produced on the surfaces of leaves, stems, and fruits, giving them a whitish appearance, as seen on this black raspberry cane.

MESOPHYLL AND VEINS

Most photosynthesis takes place in the **mesophyll** between the two epidermal layers, with two regions often being distinguishable. The uppermost mesophyll consists of compactly stacked, barrel-shaped or post-shaped parenchyma cells that are commonly in two rows. This region is called the **palisade mesophyll** and may contain more than 80% of the leaf's chloroplasts. The lower region, consisting of loosely arranged parenchyma cells with abundant air spaces between them, is called the **spongy mesophyll**. Its cells also have numerous chloroplasts.

Parenchyma tissue with chloroplasts is called *chlorenchyma*. Chlorenchyma tissue is also found in the outer parts of the cortex in the stems of herbaceous plants as well as in leaves. Inside the leaf, the surfaces of mesophyll cells in contact with the air are moist. If the moisture level decreases below a certain point, the stomata close, thereby significantly reducing further drying.

Veins (*vascular bundles*) of various sizes are scattered throughout the mesophyll (Fig. 7.9). They consist of xylem and phloem tissues surrounded by a jacket of thicker-walled parenchyma cells called the **bundle sheath**. The veins give the leaf its "skeleton." The phloem transports throughout the plant carbohydrates produced in the mesophyll cells. Water, sometimes located more than 100 meters (330 feet) away in the ground below, is brought up to the leaf by the xylem, which, like the phloem, is part of a vast network of "plumbing" throughout the plant. Since veins run in all directions in the network, particularly in dicots, it is common when examining a cross section of a leaf under the microscope, to see veins cut transversely, lengthwise, and at a tangent, all in the same section.

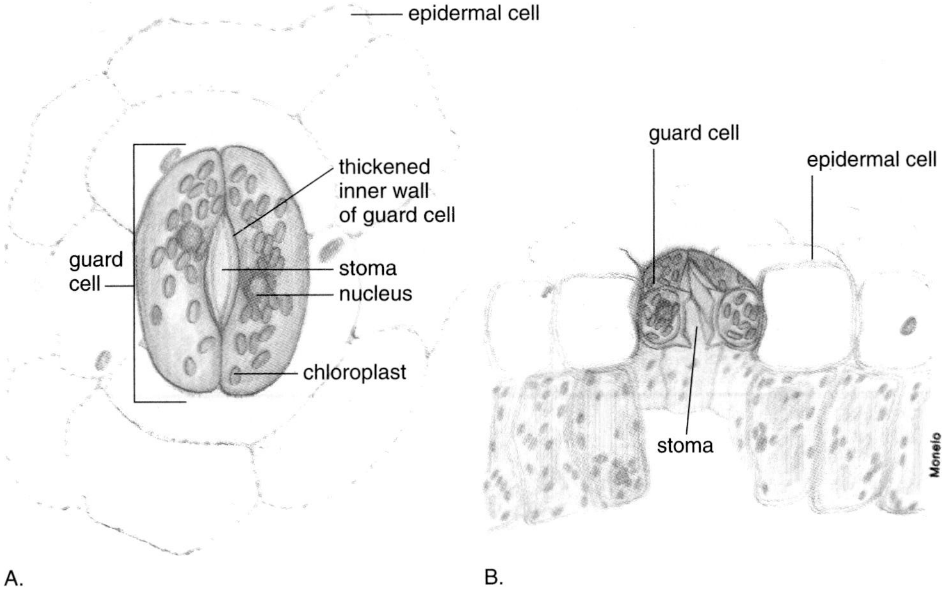

Figure 7.8. A typical dicot stoma. *A*. Surface view. *B*. View in transverse section.

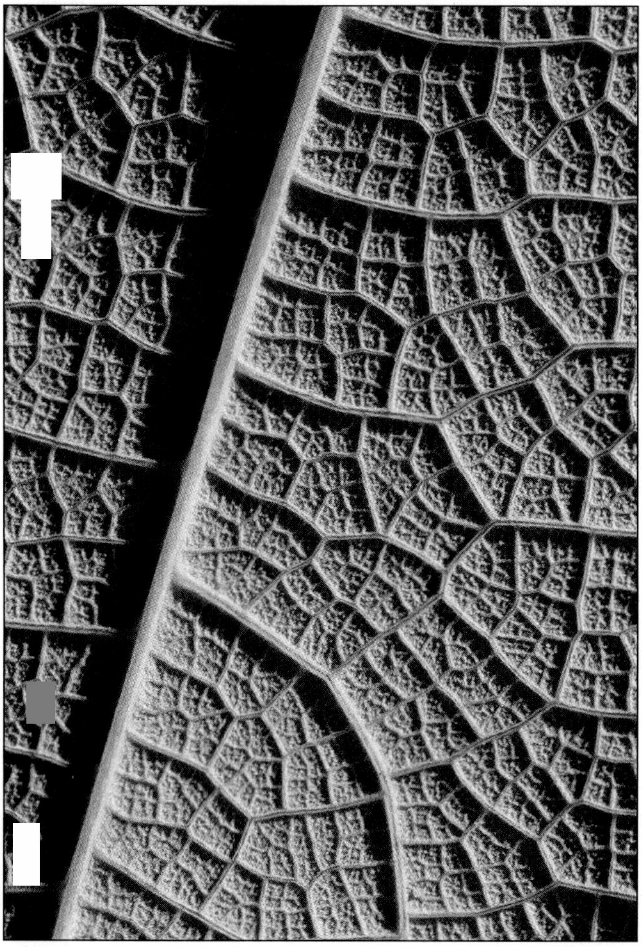

Figure 7.9 Net venation in an *Aristolochia* leaf, ×50.

Monocot leaves, besides having *parallel veins*, usually do not have the mesophyll differentiated into palisade and spongy layers. Some monocot leaves (e.g., those of grasses) have large, thin-walled *bulliform cells* on either side of the main central vein (midrib) toward the upper surface (Fig. 7.10). Under dry conditions, the bulliform cells partly collapse, causing the leaf blade to fold or roll; the folding or rolling reduces transpiration (see also Fig. 11.19).

SPECIALIZED LEAVES

If the leaves of all plants could function normally under any environmental condition, various leaf modifications would provide no special benefits to a plant. But the form and structure of tropical rain-forest plants do not adapt them to thrive in a desert, and cacti soon die if planted in a creek because their structure, form, and life cycles are attuned to specific combinations of environmental factors, such as temperature, humidity, light, water, and soil conditions. The modifications of leaves occupying any single ecological niche may be very diverse, resulting in such a rich variety of leaf forms and specializations throughout the plant kingdom that only a few may be mentioned here.

Shade Leaves

A single tree may have leaves that superficially all appear similar, but close inspection may reveal various differences. For example, because leaves in the shade receive

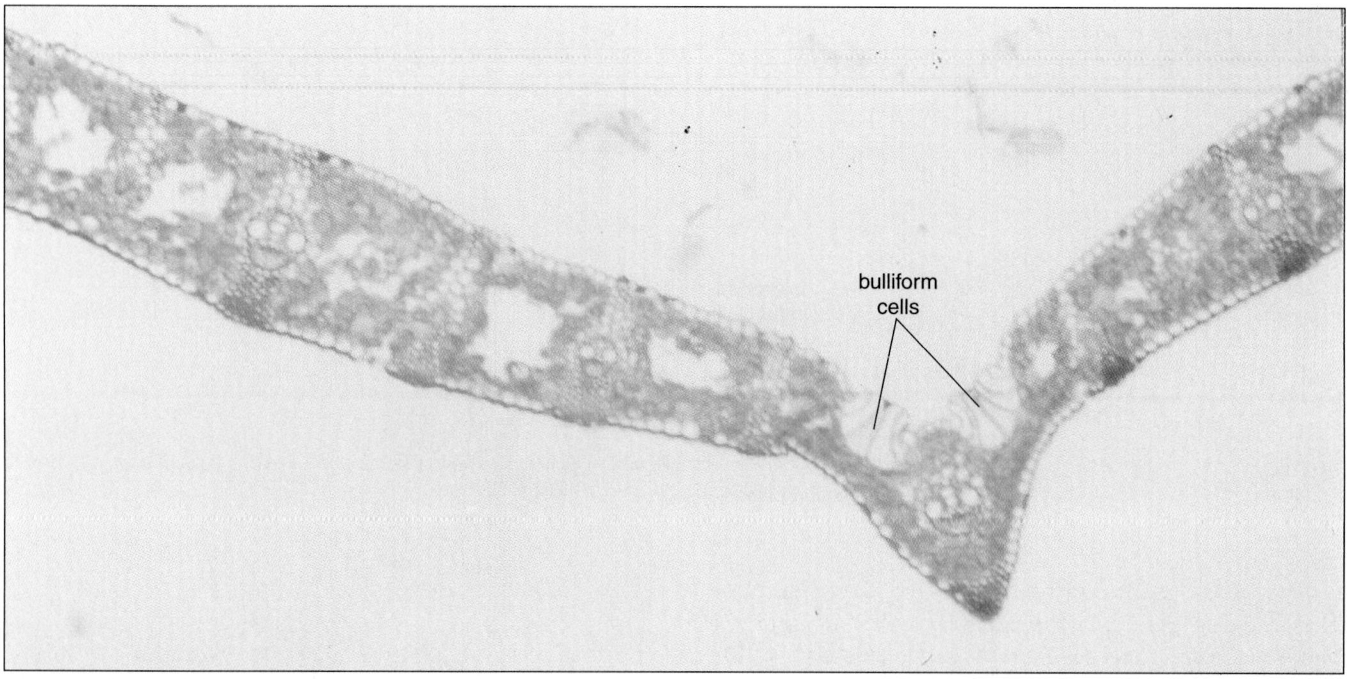

Figure 7.10 Part of a cross section of a grass leaf. When the bulliform cells are turgid, the leaf is expanded. When insufficient water is available, the bulliform cells collapse, causing the leaf blade to roll inward, thereby reducing water loss through the stomata, ×20.

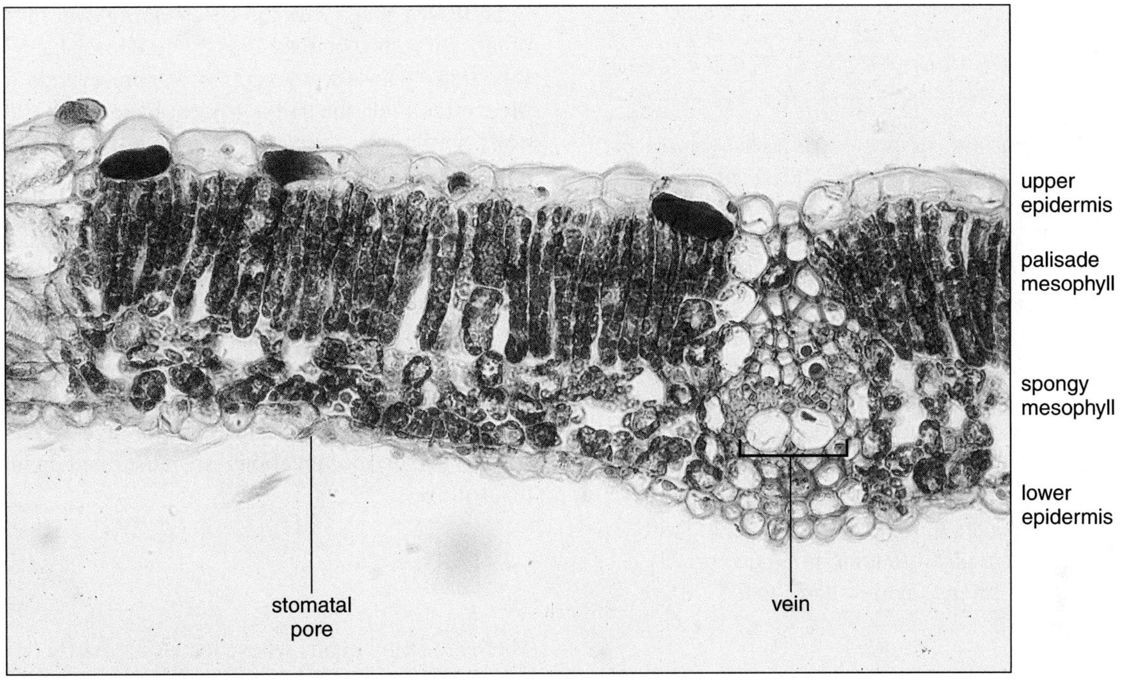

A.

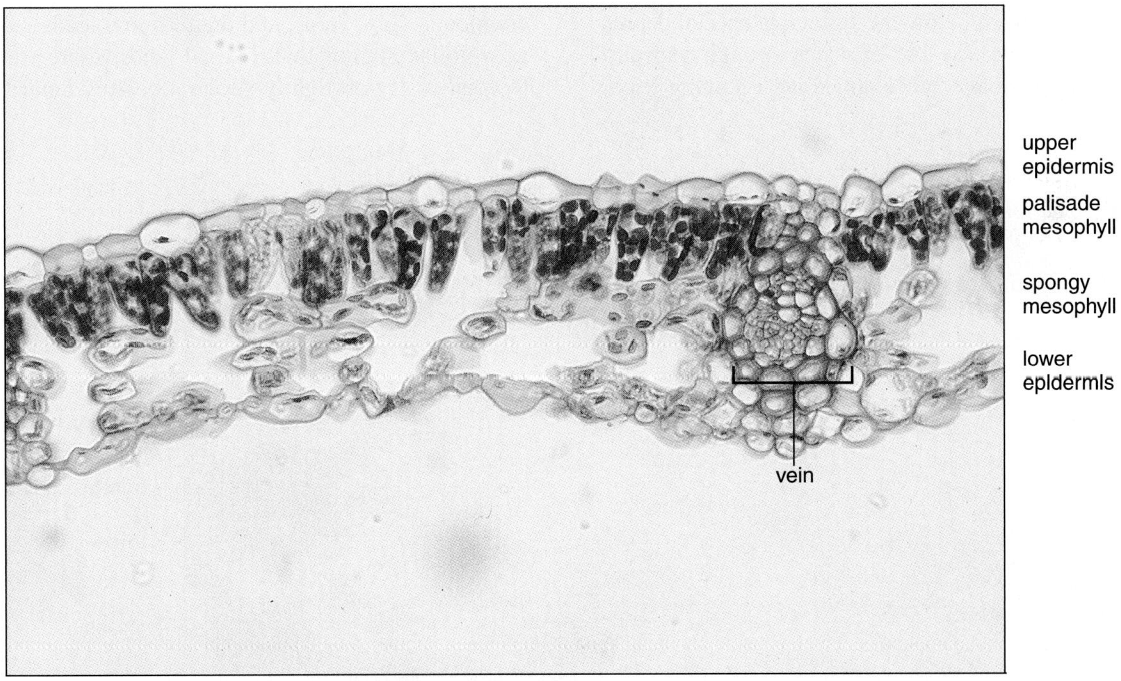

B.

Figure 7.11 Portions of cross sections of maple (*Acer*) leaves. The chloroplasts are stained red. *A*. A leaf exposed to full sun. *B*. A leaf exposed to shade. Note the reduction in mesophyll cells and chloroplasts in the shade leaf.

less total light needed for photosynthesis, they tend to be thinner and have fewer hairs than leaves on the same tree that are exposed to direct light. Shade leaves also tend to be larger and to have fewer well-defined mesophyll layers and fewer chloroplasts than their counterparts in the sun (Fig. 7.11).

Leaves of Arid Regions

Leaf modifications are generally more pronounced in different climatic zones or habitats. Because of the limited availability of water, wide temperature ranges, and high light intensities, plants growing in arid regions have developed

Ecological Review

Leaves, the main site of photosynthesis by most plants, function in light interception and gas exchange. Since the uptake of carbon dioxide through leaf stomata must be balanced against water loss, leaf structure usually shows clear connections to prevailing environmental conditions. For instance, plants in arid environments often have thick leathery leaves, with fewer stomata, and a dense covering of hairs that reflect light and may reduce water loss from the leaf surface. On an individual plant, leaves that develop in the shade (shade leaves) are generally thinner and have fewer leaf hairs than leaves that develop in the sun (sun leaves). Leaves are also often modified chemically or physically for defense against herbivores.

adaptations that allow them to thrive under such conditions. Many have thick, leathery leaves and fewer stomata, or stomata that are sunken below the surface in special depressions, all of which reduce loss of water through transpiration. They also may have succulent, water-retaining leaves or no leaves at all (with the stems taking over the function of photosynthesis), or they may have dense, hairy coverings. Pine trees, whose water supply may be severely restricted in the winter when the soil is frozen, have some leaf modifications similar to those of desert plants. The modifications include sunken stomata, a thick cuticle, and a layer of thick-walled cells (the **hypodermis**) beneath the epidermis (Fig. 7.12). The leaves of compass plants face east and west, with the blades perpendicular to the ground, so that when the sun is overhead, it strikes only the thin edge of the leaf, minimizing moisture loss (Fig. 7.13). The submerged leaves of plants that grow in water usually have considerably less xylem than phloem, and the mesophyll, which is not differentiated into palisade and spongy layers, has large air spaces. Other modifications are described in the sections that follow.

Tendrils

There are many plants whose leaves are partly or completely modified as **tendrils.** These modified leaves, when curled tightly around more rigid objects, help the plant in climbing or in supporting weak stems. The leaves of garden peas are compound (Fig. 7.14), and the terminal leaflets are reduced to whiplike strands that, like all tendrils, are very sensitive to contact. If you lightly stroke a healthy tendril, there is a

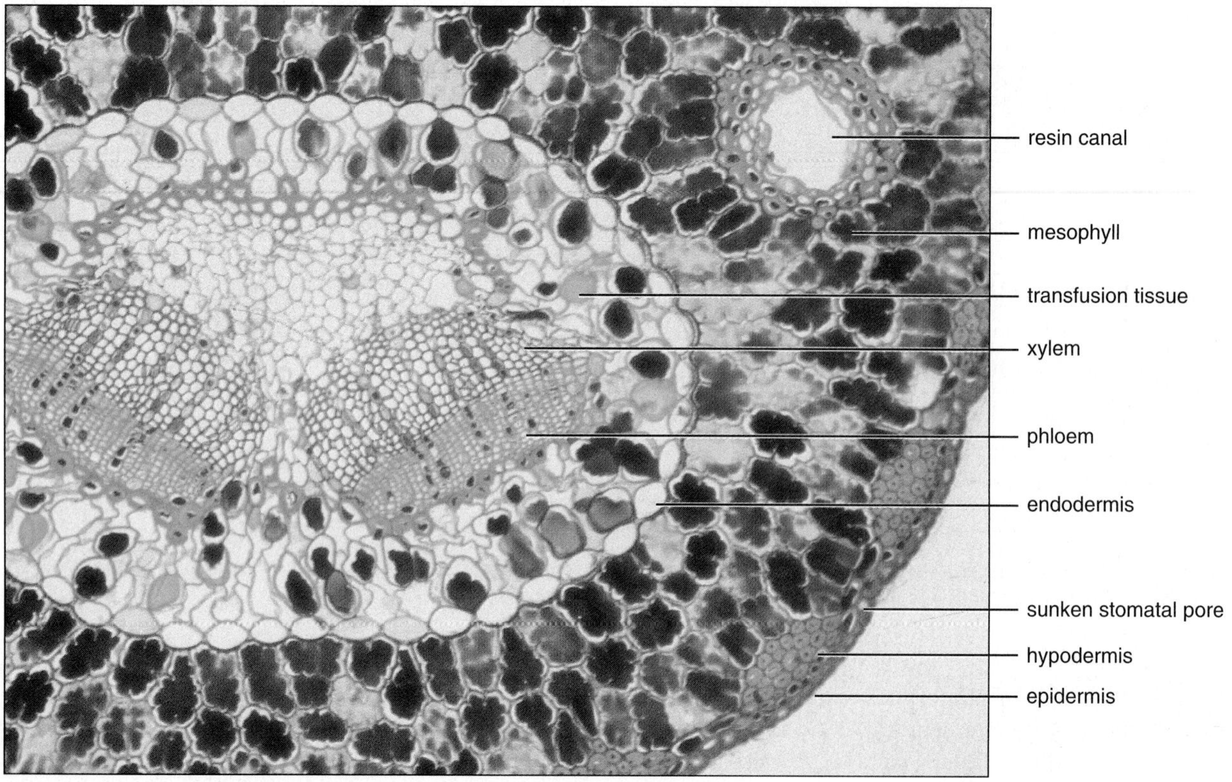

Figure 7.12 A pine needle in cross section. *(Photomicrograph by G.S. Ellmore)*

Figure 7.13 A compass plant (*Lactuca serriola*). The leaves of this and other compass plants such as *Silphium laciniatum* have their blades parallel to the sun and are oriented toward the east and west.

Figure 7.14 The terminal leaflets of this garden pea plant leaf are modified as tendrils.

sudden rapid growth of cells on the opposite side, and it starts curling in the direction of the contact within a minute or two. If the contact is very brief, the tendril reverses movement and straightens out again. If, however, the tendril encounters a suitable solid support (e.g., a twig), the stimulation is continuous, and the tendril coils tightly around the support as it grows.

Whole leaves of yellow vetchlings are modified as tendrils, and photosynthesis is carried on by the leaflike stipules at the bases. In the potato vine and the garden nasturtium, the petioles serve as tendrils, while in some greenbriers, stipules are modified as tendrils. In *Clematis*, the rachises of some of the compound leaves serve very effectively as tendrils. Members of the Pumpkin Family (Cucurbitaceae), which includes squashes, melons, and cucumbers, produce tendrils that may be up to 3 decimeters (1 foot) long.

As the tendrils develop, they become coiled like a spring. When contact with a support is made, the tip not only curls around it, but the direction of the coil reverses (see Fig. 11.8); sclerenchyma and collenchyma cells then develop in the vicinity of contact. The sclerenchyma cells provide rigid support, while the collenchyma cells impart flexibility. This makes a very strong but flexible attachment that protects the plant from damage during high winds. The tendrils of many other plants (e.g., grapes) are not modified leaves but develop instead from stems.

Spines, Thorns, and Prickles

The leaves of many cacti and other desert plants are modified as **spines**. This reduction in leaf surface correspondingly reduces water loss from the plants, and the spines also tend to protect the plants from browsing animals. In such desert plants, photosynthesis, which would otherwise take place in leaves, occurs in the green stems. Most spines are modifications of the whole leaf, in which much of the normal leaf tissue is replaced with sclerenchyma, but in a number of woody plants (e.g., mesquite, black locust), it is the stipules at the bases of the leaves that are modified as short, paired spines. Like grape and other tendrils, many spinelike objects arising in the axils of leaves of woody plants are modified stems rather than modified leaves. Such modifications should be referred to as **thorns** to distinguish them from true spines. The **prickles** of roses and raspberries, however, are neither leaves nor stems but are outgrowths from the epidermis or cortex just beneath them (Fig. 7.15).

Storage Leaves

As previously mentioned, desert plants may have *succulent leaves* (i.e., leaves that are modified for water retention). The adaptations for water storage involve large, thin-walled parenchyma cells without chloroplasts to the interior of chlorenchyma tissue just beneath the epidermis. These nonphotosynthetic cells contain large vacuoles that can store relatively

Figure 7.15A The *spines* of this barberry (*Berberis*) are modified leaves.

Figure 7.15B *Thorns* (modified stems) produced in the axils of leaves.

Figure 7.15C Prickles of a raspberry stem. The prickles are neither leaves nor stems; they are outgrowths from the epidermis or the cortex just beneath it.

substantial amounts of water. If removed from the plant and set aside, the leaves will often retain much of the water for up to several months. Many plants with succulent leaves carry on a special form of photosynthesis, discussed in Chapter 10.

The fleshy leaves of onion, lily, and other bulbs store large amounts of carbohydrates, which are used by the plant in the subsequent growing season.

Flower-Pot Leaves

Some leaves of *Dischidia* (Fig. 7.16), an *epiphyte* (a plant that grows, usually non-parasitically, on other plants) from tropical Australasia, develop into urnlike pouches that become the home of ant colonies. The ants carry in soil and add nitrogenous wastes, while moisture collects in the leaves through condensation of the water vapor coming from the mesophyll through stomata. This creates a good growing medium for roots, which

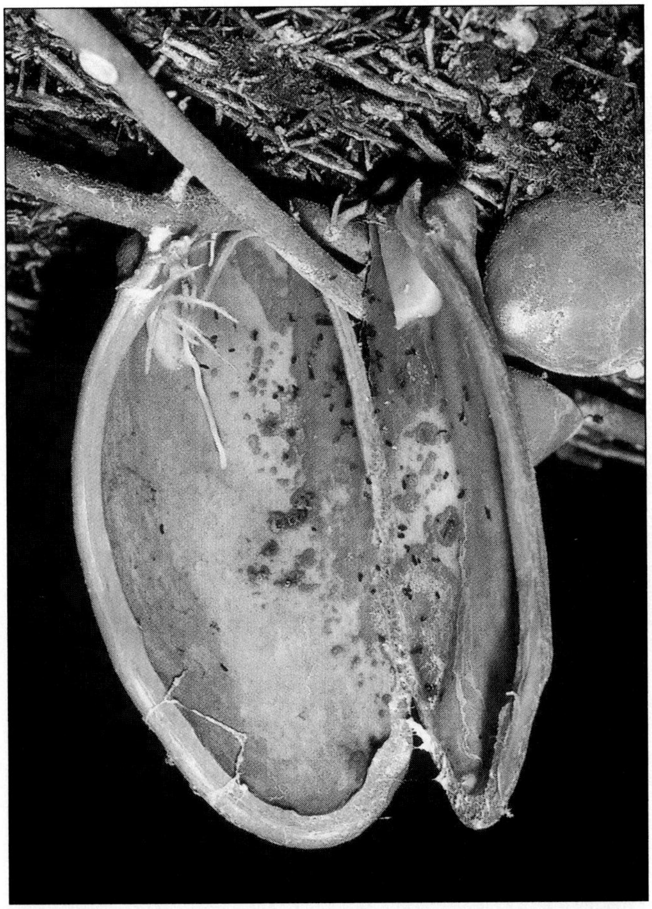

Figure 7.16 A flower-pot leaf of *Dischidia*. The leaf has been sliced lengthwise to reveal the roots inside that have started to grow down from the top.

Figure 7.17 A window plant (*Fenestraria*). Note the transparent tips of the window leaves.

Figure 7.18 A leaf of an air plant (*Kalanchoë*), showing plantlets being produced along the margins.

develop adventitiously from the same node as the leaf and grow down into the soil contained in the urnlike pouch. In other words, this extraordinary plant not only reproduces itself by conventional means but also, with the aid of ants, provides its own fertilized growing medium and flower pots and then produces special roots, which "exploit" the situation.

Window Leaves

In the Kalahari desert of Botswana and South Africa, there are at least three plants belonging to the Carpetweed Family (Aizoaceae) that have unique adaptations to living in dry, sandy areas. Their leaves, which are shaped like ice-cream cones, are about 3.75 centimeters (1.5 inches) long (Fig. 7.17) and are buried in the sand; only the dime-sized wide end of a leaf is exposed at the surface. This exposed end is covered with a relatively transparent, thick epidermis with few stomata and a waxy cuticle. There is a mass of tightly packed, transparent water-storage cells below the exposed end; these allow light coming through the "windows" to penetrate to the chloroplasts in the mesophyll, located all around the inside of the shell of the leaf. This arrangement, which keeps most of the plant buried and away from drying winds, allows the plant to thrive under circumstances that most other plants could not tolerate. Window leaves also occur in succulent plants of a few other families.

Reproductive Leaves

Some of the leaves of the walking fern are most unusual in that they produce new plants at their tips. Occasionally, three generations of plants may be found linked together. The succulent leaves of air plants (Fig. 7.18) have little notches along the leaf margins in which tiny plantlets are produced, complete with roots and leaves, even after a leaf has been removed from the parent plant. Each of the plantlets can develop into a mature plant if given the opportunity to do so.

Floral Leaves (Bracts)

Specialized leaves known as **bracts** are found at the bases of flowers or flower stalks. In the Christmas flower (poinsettia), the flowers themselves have no petals, but the brightly colored floral bracts that surround the small flowers function like petals

Figure 7.19 A poinsettia (*Euphorbia pulcherrima*) "flower." There are several flowers without petals in the center. The most conspicuous parts of the "flower" are the colored bracts (modified leaves) surrounding the true flowers.

in attracting pollinators (Fig. 7.19). In dogwoods and a few other plants, the tiny flowers in their buttonlike clusters do have inconspicuous petals. However, the large white-to-pink bracts that surround the flower clusters, which appear to the casual observer to be petals, are actually modified leaves. In Clary's annual sage (*Salvia viridis*), large colorful bracts are produced at the top of flowering stalks, well above the flowers (Fig. 7.20).

Figure 7.20 Clary's annual sage (*Salvia viridis*). Note the small flowers along the lower half of the stem and the large colorful bracts toward the top.

Insect-Trapping Leaves

Highly specialized *insect-trapping leaves* have intrigued humans for hundreds of years. Almost 200 species of flowering plants are known to have these leaves. Insectivorous plants grow mostly in swampy areas and bogs of tropical and temperate regions. In such environments, certain needed elements, particularly nitrogen, may be deficient in the soil, or they may be in a form not readily available to the plants. Some of these elements are furnished when the soft parts of insects and other small organisms trapped by the specialized leaves are broken down and digested. All the plants have chlorophyll and are able to make their own food. It has been demonstrated that they can develop normally without insects if they are given the nutrients they need. The following plants represent four types of insect-trapping mechanisms.

Pitcher Plants

The blades of leaves of many *pitcher plants* are flattened and function like those of any other leaves. Some of the leaves of these curious plants, however, are larger and cone-shaped

Figure 7.21 Insect-trapping leaves of pitcher plants (*Sarracenia*).

or vaselike. In some species, these larger pitcher leaves have umbrella-like flaps above the open ends (Fig. 7.21), but the flaps don't prevent a little rain water from accumulating at the bottom. Some Asian pitcher plants are vines with leaves whose long petioles are twisted around branches for support. Their pitchers are formed at the tips of the leaves.

Pitcher leaves have nectar-secreting glands around the rim. The distinctive odor produced by these glands attracts insects, which, while foraging, often fall into the watery fluid at the bottom. If the insects try to climb out, they find the walls highly polished and slippery. In fact, the walls of some pitcher plant leaves are coated with wax, and as the insects struggle up the surface, their feet become coated with the wax, which builds up until the victims seem to have acquired heavy clod-like boots. Most insects never make it up the walls, but even if they do, they still face a formidable barricade of stiff downward-pointing hairs near the rim. Eventually they drown, and their soft parts are digested by bacteria and by enzymes secreted by digestive glands near the bottoms of the leaves.

In North America, the pitcher plants produce their pitchers in erect clusters on the ground, but the mechanisms for trapping insects are similar to those of Asian species. Malaysian tree frogs, which have sticky pads on their feet, are undaunted by pitcher-plant leaves and lay their eggs in them. The eggs contain a chemical that neutralizes the pitcher's digestive enzymes.

Merchandisers have been shameless in their wholesale collection of these plants for sale, and pitcher plants may become extinct in the wild. The cobra plant, a pitcher plant restricted to a few swampy areas in California and Oregon, has been placed on official threatened species lists.

Sundews

The tiny plants called *sundews* (Fig. 7.22) often do not measure more than 2.5 to 5.0 centimeters (1 to 2 inches) in diameter. The roundish to oval leaves are covered with up to 200 upright glandular hairs that look like miniature clubs. There is a clear, glistening drop of sticky fluid containing digestive enzymes at the tip of each hair. As the droplets sparkle in the sun, they may attract insects, which find themselves stuck if they alight. The hairs are exceptionally sensitive to contact, responding to weights of less than one-thousandth of a milligram, and bend inward, surrounding any trapped insect within a few minutes. The digestive enzymes break down the soft parts of the insects, and after digestion has been completed (within a few days), the glandular hairs return to their original positions. If bits of nonliving debris happen to catch in the sticky fluid, the hairs barely respond, showing they can distinguish between protein and something "inedible." Some sundew owners regularly feed their plants tiny bits of hamburger and boiled egg white.

In Portugal, relatives of sundews with less specialized leaves are used in houses as flypaper. In response to contact by living insects, the edges of specialized leaves of similar plants called *butterworts* rapidly curl over and trap unwary victims.

Venus's Flytraps

The *Venus's flytrap* (Fig. 7.23), which has leaves constructed along the lines of an old-fashioned steel trap, is found in nature only in wet areas of North Carolina and South Carolina. The two halves of the blade have the appearance of being hinged along the midrib, with stiff hairlike projections located along their margins. There are three tiny trigger hairs on the inner surface of each half. If two trigger hairs are touched simultaneously or if any one of them is touched twice within a few seconds, the blade halves suddenly snap together, trapping the insect or other small animal. As the organism struggles, the trap closes even more tightly. Digestive enzymes secreted by the leaf break down the soft parts of the insect, which are then absorbed. After digestion has been completed, the trap reopens, ready to repeat the process. Venus's flytrap leaves, like those of sundews, do not normally close for bits of debris that might accidentally fall on the leaf, because nonliving material does not move about and stimulate the trigger hairs.

Bladderworts

Bladderworts (Fig. 7.24), which are found submerged and floating in the shallow water along the margins of lakes and streams, have finely dissected leaves with tiny

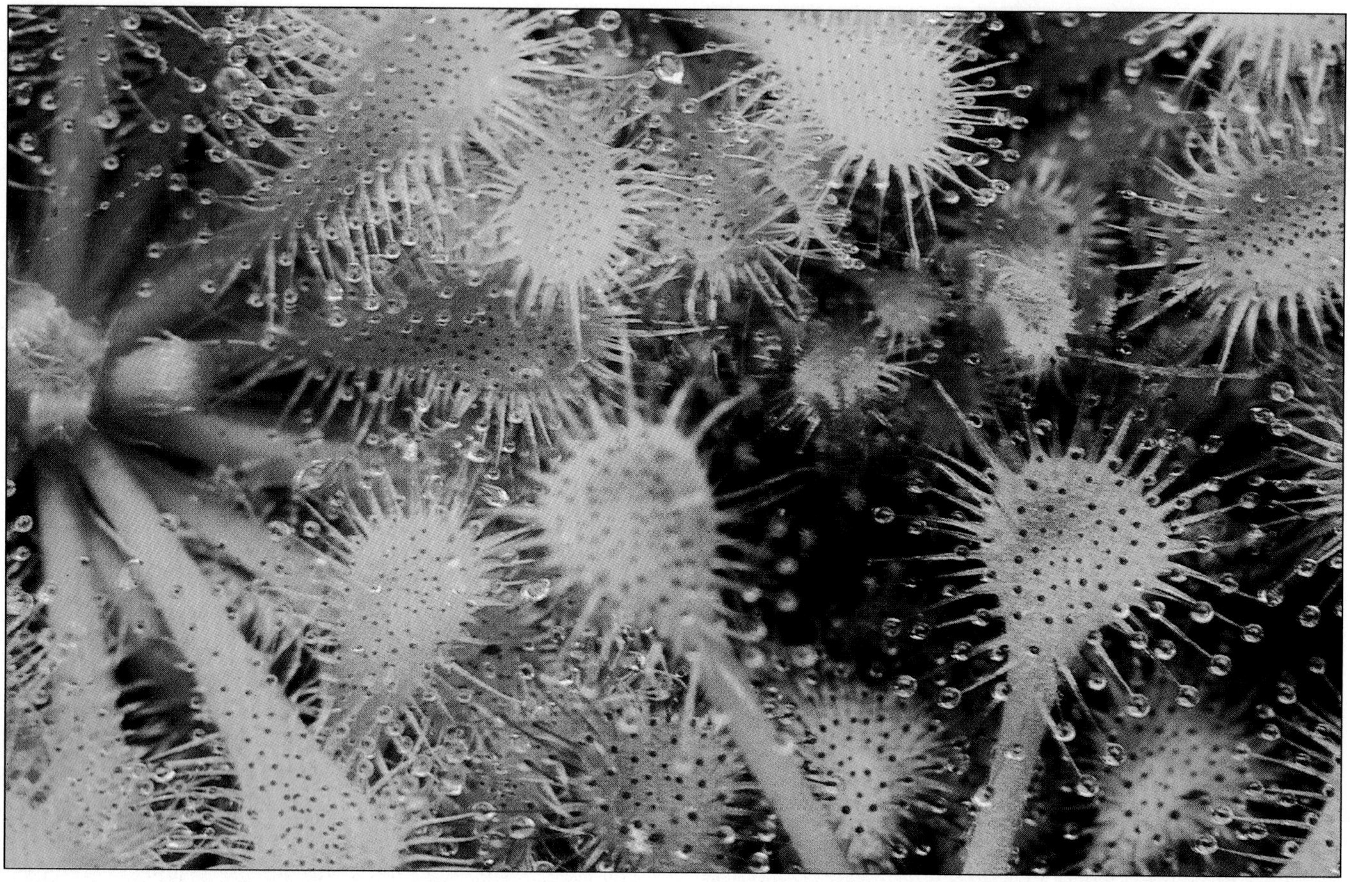

Figure 7.22 Sundew (*Drosera*) leaves.

Figure 7.23 A Venus's flytrap plant (*Dionaea muscipula*).

bladders. The stomach-shaped bladders are between 0.3 and 0.6 centimeter (0.125 to 0.25 inch) in diameter and have a trapdoor over the opening at one end. The trapping of aquatic insects and other small animals takes place through a complex mechanism. Four curled but stiff hairs at one end of the trapdoor act as triggers when an insect touches one of them. The trapdoor springs open, and water rushes into the bladder. The stream of water propels the victim into the trap, and the door snaps shut behind it. The action takes place in less than one-hundredth of a second and makes a distinct popping sound, which can be heard with the aid of a sensitive underwater microphone. The trapped insect eventually dies, is broken down by bacteria, and the breakdown products are absorbed by cells in the walls of the bladder.

Science-fiction writers have contributed to superstitions and beliefs that deep in the tropical jungles there are plants capable of trapping humans and other large animals. No such plants have been proved to exist, however. The largest pitcher plants known hold possibly 1 liter (roughly 1 quart) of fluid in their pitchers, and small frogs have been known to decompose in them, but the trapping of anything larger than a mouse or possibly a small rabbit seems very unlikely.

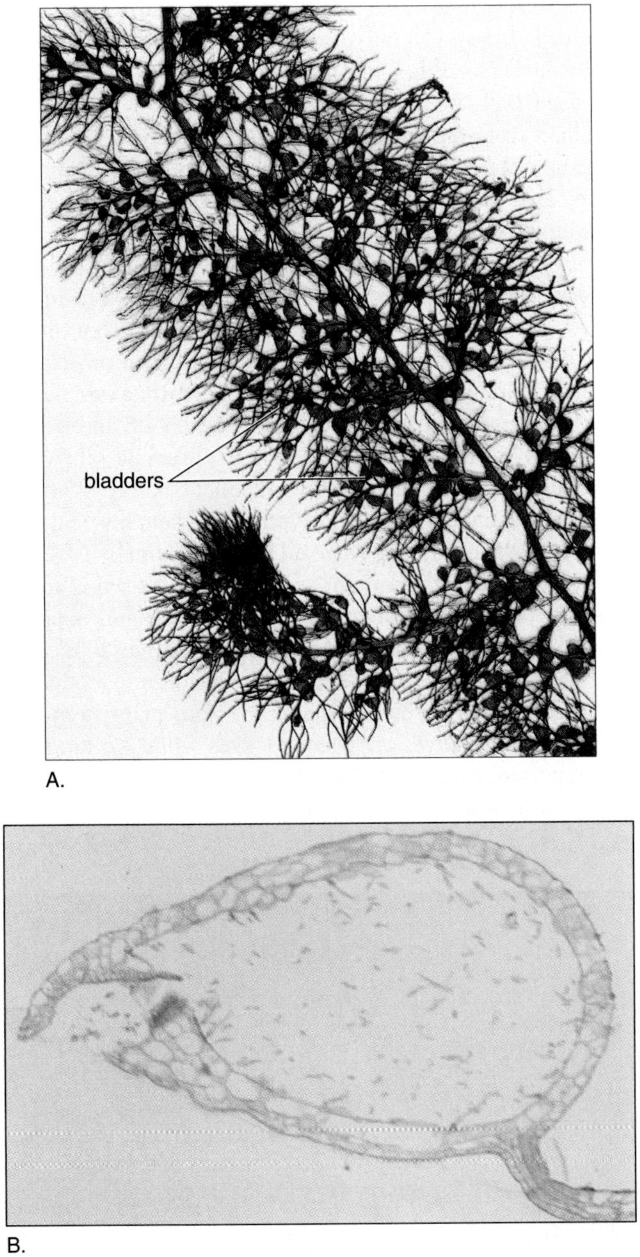

Figure 7.24 A. A bladderwort (*Utricularia*). *B.* A single bladder, ×20.

AUTUMNAL CHANGES IN LEAF COLOR

The leaves of many oaks and several other plants generally turn some shade of brown or tan when their cells break down and die, due to a reaction between leaf proteins and tannins stored in the cell vacuoles. This is similar to the formation of leather when tannins react with animal hides. Leaves of many other deciduous plants, however, exhibit a variety of colors and drop before turning brown.

The chloroplasts of mature leaves contain several groups of pigments, such as green *chlorophylls* and *carotenoids,* which include yellow *carotenes* and pale yellow *xanthophylls.* Each of these groups plays a role in photosynthesis. Usually, considerably more chlorophyll than other pigments is present, and the intense green color of the chlorophylls masks or hides the presence of the carotenes and xanthophylls. In the fall, however, the chlorophylls break down, and other colors are revealed. The exact cause of the chlorophyll breakdown is not known, but it does appear to involve, among other factors, a gradual reduction in day length.

Water-soluble *anthocyanin* and *betacyanin* pigments may also accumulate in the vacuoles of the leaf cells in the fall. Anthocyanins, the more common of the two groups, are red if the cell sap is slightly acid, blue if it is slightly alkaline, and of intermediate shades if it is neutral. Betacyanins are usually red; they apparently are restricted to several plant families, such as the cacti (Cactaceae); the Goosefoot Family (Chenopodiaceae), to which beets belong; the Four-o'clock Family (Nyctaginaceae); and the Portulaca Family (Portulacaceae).

Some plants (e.g., birch trees) consistently exhibit a single shade of color in their fall leaves, but many (e.g., maple, ash, sumac) vary considerably from one locality to another or even from one leaf to another on the same tree, depending on the combinations of carotenes, xanthophylls, and other pigments present. Some of the most spectacular fall colors in North America occur in the Eastern Deciduous Forest, particularly in New England and the upper reaches of the Mississippi Valley. In parts of Wisconsin and Minnesota, one can observe the brilliant reds, oranges, and golds of maples, the deep maroons (and also yellows) of ashes, the bright yellows of aspen, and the seemingly glowing reds of sumacs and wahoos, all in a single locality. Some fall coloration is found almost anywhere in temperate zones where deciduous trees and shrubs exist.

ABSCISSION

Plants whose leaves drop seasonally are said to be **deciduous**. In temperate climates, new leaves are produced in the spring and are shed in the fall, but in the tropics, the cycles coincide with wet and dry seasons rather than with temperature changes. Even evergreen trees shed their leaves; they do so a few at a time, however, so that they never have the bare look of deciduous trees in their winter condition. The process by which the leaves are shed is called **abscission**.

Abscission occurs as a result of changes that take place in an *abscission zone* near the base of the petiole of each leaf (Fig. 7.25). Sometimes the abscission zone can be seen externally as a thin band of slightly different color on the petiole. Hormones that apparently inhibit the formation of the specialized layers of cells that facilitate abscission are produced

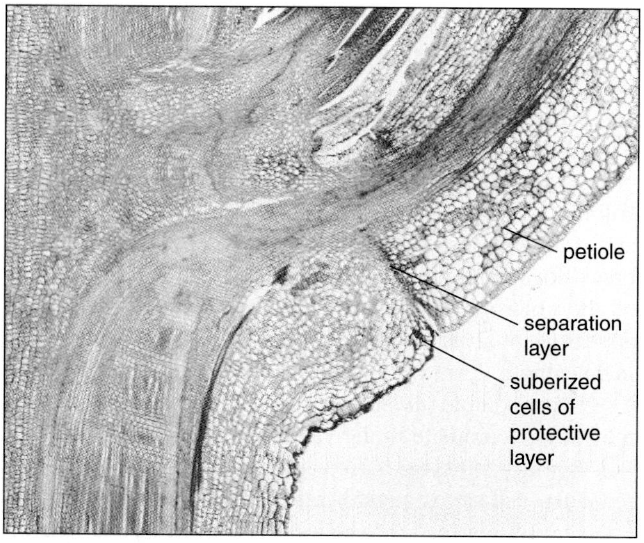

Figure 7.25 The abscission zone of a leaf.

in young leaves. As the leaf ages, hormonal changes take place, and at least two layers of cells become differentiated. Closest to the stem, the cells of the *protective layer,* which may be several cells deep, become coated and impregnated with fatty *suberin.* On the leaf side, a *separation layer* develops in which the cells swell, sometimes divide, and also become gelatinous. In response to any of several environmental changes (such as lowering temperatures, decreasing day lengths or light intensities, lack of adequate water, or damage to the leaf), the pectins in the middle lamella of the cells of the separation layer are broken down by enzymes. All that holds the leaf on to the stem at this point are some strands of xylem. Wind and rain then easily break the connecting strands, leaving tiny bundle scars within a leaf scar (see Fig. 6.1), and the leaf falls to the ground.

HUMAN AND ECOLOGICAL RELEVANCE OF LEAVES

Humans use shade trees and shrubs in landscaping for cooling as well as for aesthetic effects. The leaves of shade plants planted next to a dwelling can make a significant difference in energy costs to the homeowner. Humans also use for food the leaves of cabbage, parsley, lettuce, spinach, chard, and the petioles of celery and rhubarb, to mention a few. Many spices and flavorings are derived from leaves, including thyme, marjoram, oregano, tarragon, peppermint, spearmint, wintergreen, basil, dill, sage, cilantro, and savory.

Various dyes (e.g., a yellow dye from bearberry, a reddish dye from henna, and a pale blue dye from blue ash) can be extracted from leaves (see Appendix 3), although nearly all commercial dyes are now derived from coal tar. Many cordage fibers for ropes and twines come from leaves, with various species of *Agave* (century plants) accounting for about 80% of the world's production. Bowstring fibers are obtained from a relative of the common house plant *Sansevieria* (see Fig. 24.28), and Manila hemp fibers, which are used both in fine-quality cordage and in textiles, are obtained from the leaves of a close relative of the banana. Panama hats are made from the leaves of the panama hat palm, and palms and grasses are used in the tropics as thatching material for huts and other buildings.

In the high mountains of Chile and Peru, the leaves of the yareta plant are used for fuel. They produce a resin that causes the leaves to burn with an unusually hot flame. Leaves of many plants produce oils. Petitgrain oil, from a variety of orange tree leaves, and lavender, for example, are used for scenting soaps and colognes. Patchouli and lemongrass oils are used in perfumes, as is citronella oil, which was once the leading mosquito repellent before synthetic repellents gained favor. Eucalyptus oil, camphor, cajeput, and pennyroyal (Fig. 7.26) are all used medicinally.

Leaves are an important source of drugs used in medicine and also of narcotics and poisons. Cocaine, obtained from plants native to South America, has been used medicinally and as a local anesthetic, but its use as a narcotic has, in recent years, become a major problem in western cultures. Andean natives chew coca leaves while working and are reported capable of performing exceptional feats of labor with little or no food while under the influence of

Figure 7.26 Flowers of a pennyroyal plant (*Hedeoma*). The leaves are the source of a pungent oil.

Awareness

Glass Cuts from Grass?

We all know that if we drop a glass bottle or dish, we need to pick up the pieces so that someone won't get hurt on the sharp edges of the broken glass. But did you know that something as soft as a blade of grass can also give you a nasty cut if you don't handle it carefully?

Many leaves, including those of grasses, are produced with tiny teeth along the margins. In some instances when a person brushes against the edge of the leaf, the teeth alone have enough rigidity to act like a miniature saw that can produce a cut on human skin. One of the reasons grasses are generally not palatable to humans is that they accumulate silica—the primary ingredient of glass. In some grasses, including rice cutgrass (*Leersia oryzoides*—Box Figure 7.1), the tiny teeth along the margins are covered with a film of glass that gives the downward-pointing teeth additional rigidity. This tiny saw instantly cuts if human skin is brushed against it.

If you have the opportunity to do so, examine the leaf margins of several grasses with a microscope and see if any of them have teeth encased in glass.

Box Figure 7.1 Part of the margin of a rice cutgrass (*Leersia oryzoides*) leaf, showing the "glass"-encased teeth that easily cut human skin, ×40.

cocaine. Apparently, the drug, which is highly addictive in forms such as "crack," anesthetizes the nerves that convey hunger pangs to the brain.

Belladonna is a drug complex obtained from leaves of the deadly nightshade, a native of Europe. The plant has been used in medicine for centuries, and several drugs are now isolated from belladonna. Included among the isolates is atropine, which is used in shock treatments, to dilate eyes, to relieve pain locally, and to slow secretions during surgery. Scopolamine, also a belladonna derivative, is used in tranquilizers and sleeping aids. Another European plant, the foxglove, is the source of digitalis. This drug has been used for centuries in regulating blood circulation and heartbeat.

Tobacco is another widely used leaf. More than 940 million kilograms (2 billion pounds) of tobacco for smoking, chewing, and use as snuff are produced annually around the world. At present, about 125,000 Americans die annually from lung cancer, and almost all of them have a history of cigarette smoking. Cigarette smoking is also evidently a principal contributor to cardiovascular diseases. During recent years, the increase in the use of chewing tobacco has seen a corresponding increase in the development of mouth and throat cancers. Federal law forbids concentrations of more than five parts per billion of nitrosamines (cancer-causing chemicals) in cured meats, but the levels of nitrosamines in the five most popular brands of chewing tobacco range from 9,600 to 289,000 parts per billion. Though humans have long used tobacco, it has been in only the last two decades that its health threat has become appreciated.

Similarly, marijuana, the controversial plant widely used as an intoxicant, has been utilized in various ways for thousands of years. The active principle, tetrahydrocannabinol (THC), although found in the leaves, is concentrated in hair secretions among the female flowers. In recent decades, marijuana has found increasing acceptance in the western hemisphere; it appears, however, not to be the harmless intoxicant many thought it to be. Regular use of marijuana for a year has been shown to have the same effect on human lungs as smoking one and a half packs of cigarettes a day for 13 years. A decision to legalize its use, either smoked or in the form of prescription pills for the relief of pain and the alleviation of nausea caused by chemotherapy treatments, has been criticized by some medical and other authorities.

The drug lobeline sulfate is obtained from the leaves of a close relative of garden lobelias. It is used in compounds taken by those who are trying to stop smoking. The leaves of several species of *Aloe,* especially *Aloe vera,* yield a juice that is used to treat various types of skin burns, including those accidentally received from X-ray equipment.

Several beverages are extracted by the brewing of leaves. Numerous teas have been obtained from a wide variety of plants, but most now in use come from a close relative of the garden camellia. Maté, the popular South American tea, is brewed from the leaves and twigs of a relative of holly. The alcoholic beverages pulque and tequila find their origin in the mashed leaves of *Agave* plants, and absinthe liqueur receives its unique flavor from the leaves of wormwood, a relative of western sagebrush, and other flavorings, such as anise.

Insecticides of various types are also derived from leaves. A type of rotenone and a substance related to nicotine are obtained from tropical plants; both are effective against a variety of insects. Mexico's cockroach plant has leaves that when dried are highly effective in killing cockroaches, flies, fleas, and lice. Water extracts of the leaves (as well as other parts) of India's neem tree are reported to control more than 100 species of insects, mites, and nematodes, with little effect on useful predator insects.

Carnauba and caussu waxes are obtained from the leaves of tropical palms. Leaves are used extensively by florists in floral arrangements and bouquets, and their uses for other aesthetic purposes are legion. Leaves may find more extensive use in the future as a direct source of food. It has been shown that a curd obtained by coagulating juice squeezed from alfalfa leaves contains more than 40% protein, and juices from other leaves have yielded better than 50% protein. Experiments are under way to make the leaf curd palatable for human consumption. Additional information on past and present uses of leaves is given in Chapter 24 and in Appendices 2, 3, and 4. The scientific names of the plants that are the sources of the various substances discussed are given in Appendix 1.

Summary

1. Leaves are similar to solar panels in that they are covered with a transparent epidermis that admits light to the interior, and many twist on their petioles so that their flat surfaces are inclined to the sun throughout the day.
2. All leaves originate as primordia. Most leaves consist of a blade and a petiole that may have paired stipules at the base. Leaves may be simple or compound, and all are associated with leaf gaps.
3. The lower and often the upper surfaces of leaves have pores (stomata) that permit air circulation and facilitate photosynthesis. Leaves also respire, accumulate wastes, and eliminate excess moisture via transpiration. The shapes, sizes, and textures of leaves vary greatly.
4. The venation and the compounding of leaves may be either pinnate or palmate. The main vein is the midvein in simple leaves and the rachis in compound leaves. Monocot leaves have parallel veins. Dichotomous venation occurs in a few leaves.
5. The epidermis is a surface layer of cells that is coated with a cuticle. Waxes, glands, hairs, and an occasional cellular crystal may also be present. The lower epidermis usually contains numerous stomata, each formed by a pair of guard cells that regulate both evaporation of water vapor from the leaf and gas exchange between the interior and the atmosphere.
6. The mesophyll between the upper epidermis and the lower epidermis may be divided into an upper palisade layer, which consists of rows of parenchyma cells containing numerous chloroplasts, and a lower spongy layer in which the parenchyma cells are loosely arranged. Veins traverse the mesophyll.
7. Leaves may be specialized, usually in adaptation to specific combinations of environmental factors. Specializations include thick leaves, fewer stomata, water-storage leaves, tendrils, spines, food-storage leaves, flower-pot leaves, window leaves, reproductive leaves, bracts, and insect-trapping leaves.
8. Environmental factors and hormonal changes that occur in autumn in the abscission zone at the petiolar base of each deciduous leaf cause leaves to drop. Leaves change color as green chloroplast pigments break down, revealing pigments of other colors, and different pigments accumulate in cell vacuoles.
9. Leaves are a source of many useful products, including food, beverages, dyes, fuel, spices, flavorings, oils, medicinal and narcotic drugs, insecticides, and waxes.

Review Questions

1. Leaves have no secondary xylem and phloem. Why not?
2. What are bracts?
3. Identify or define hydathode, transpiration, guard cells, mesophyll, venation, glands, and compound leaf.
4. How can one distinguish between the upper and lower epidermis in most leaves?
5. What is the function of bundle sheaths?
6. How do leaves in shaded areas differ from leaves in sunlit areas on the same plant?
7. What leaf modifications are associated with dry areas, wet areas (e.g., lakes), climbing, and reproduction?
8. Why do leaves turn different colors in the fall?

Discussion Questions

1. Can you think of any advantages or disadvantages to a plant in having very tiny or very large leaves?
2. In Chapter 3, it was noted that living cells are connected to one another by plasmodesmata that extend through tiny holes in the walls. If this is true, does a leaf really need veins for the acquisition and distribution of materials?
3. Is there any advantage to a leaf in having palisade mesophyll on the upper half and spongy mesophyll on the lower half?

4. Pines and other plants that grow in very cold climates have sunken stomata just like plants of the desert, yet the annual precipitation where they grow may be very abundant. Is there any advantage to their having such a modification?
5. Since tendrils and spines can be either modified leaves or modified stems, how might you determine the origin of a given specimen?

Learning Online

Visit our web page at www.mhhe.com/botany for interesting case studies, practice quizzes, current articles, and animations within the Online Learning Center to help you understand the material in this chapter. You'll also find active links to these topics:

Characteristics of Plants
Architectural Pattern of Plants
Leaves and the Movement of Water
Economic Uses of Leaves
Carnivorous Plants
Photosynthesis
Economic and Ecological Importance of Angiosperms

Additional Reading

Dale, J. E. 1992. How do leaves grow? *Bioscience* 42: 323–32.

Juniper, B. E., et al. 1989. *The carnivorous plants.* San Diego, CA: Academic Press.

Lanner, R. 1990. *Autumn leaves.* Minnetonka, MN: Creative Publishing International.

Percy, K. E., et al. (Eds.). 1994. *Air pollutants and the leaf cuticle.* New York: Springer-Verlag.

Prance, G. T. 1985. *Leaves: The formation, characteristics and uses of hundreds of leaves found in all parts of the world.* New York: Crown.

Roth, I. 1984. *Stratification of tropical forests as seen in leaf structure.* Norwell, MA: Kluwer Academic Publications.

Steeves, T. A., and I. M. Sussex. 1989. *Patterns in plant development,* 2d ed. Englewood Cliffs, NJ: Prentice-Hall.

Vitole, A. T. 1997. *Leaves in magic, myth, and medicine.* New York: Stewart, Tabori, and Chang.

Zeiger, E., et al. (Eds.). 1987. *Stomatal function.* Palo Alto, CA: Stanford University Press.

A water-lily flower (*Nymphaea* sp.).

8 Flowers, Fruits, and Seeds

Chapter Outline

Overview

The structure, parts, and some modifications of flowers are described in this chapter. Dicots and monocots are briefly compared. The nature and development of fruits are discussed next, and fruit structure and parthenocarpy are described. A listing and discussion of types of dry and fleshy fruits follows, and then fruit and seed dispersal are explored. An examination of seed structure and germination and some observations on dormancy and longevity of seeds, stratification, and vivipary conclude the chapter.

Some Learning Goals

1. Know the parts of a typical flower and the function of each part.
2. Learn the features that distinguish monocots from dicots.
3. Understand the distinction between a fruit and a vegetable.
4. Know the regions of mature fruits.
5. Learn five types of fleshy and dry fruits and know how simple, aggregate, and multiple fruits are derived from the flowers.
6. Learn the adaptations of fruits and seeds to the agents by which they are dispersed.
7. Diagram and label a mature dicot seed (e.g., bean) and a monocot seed (e.g., corn) in section to show the parts and regions.
8. Understand the changes that occur when a seed germinates and note the environmental conditions essential to germination.
9. Know the types of factors that control dormancy. Learn how dormancy may be broken both naturally and artificially.

Note to the Reader:

Although the numbers of bacteria, algae, and fungi are much greater than those of flowering plants, much of this text is devoted to various aspects of flowering plants, partly because flowering plants are more abundant than other true plants and more important to humanity than all the other kinds of plants. In early editions of this book, most basic features of flowers, fruits, and seeds were discussed in a single chapter. This followed a progressive examination of members of the Plant Kingdom, beginning with the simplest forms, so that the reader could appreciate how flowering plants developed and became the most advanced and complex of all plants. Some shorter courses, however, do not cover certain reproductive and evolutionary aspects of the Plant Kingdom, and instructors differ in their assessment of the relative importance of such matters in a short course.

In recognition of this dilemma, the original single chapter is represented in this edition by Chapters 8 and 23. Chapter 8 emphasizes form and structure of flowers, fruits, and seeds, with brief reference to reproduction. Chapter 23 covers more details of the life cycle and evolution of flowering plants. Even if your course does not include Chapter 23 in the curriculum, it is recommended that you read it as an adjunct to Chapter 8.

A number of years ago, an Australian farmer, while plowing a field, glanced back and was startled to see what looked like flowers being tossed to the surface of the furrows. He climbed off his tractor to take a closer look and found that what he saw were, indeed, flowers. Furthermore, the plants to which the flowers were attached were pale and had no chlorophyll. He reported his find to a university, where botanists determined that the farmer had stumbled upon the first known underground flowering plant. The plant proved to be an orchid that lived on organic matter in the soil and was pollinated by tiny flies that gained access to the below-ground flowers via mud cracks that developed in the dry season (Fig. 8.1).

The underground-flowering orchid is only one of about 240,000 known species of flowering plants. Other unknown numbers of undescribed species are established in remote areas, particularly in the tropics. The flowering plants are vital to humanity, providing countless useful products, with just 11 species—10 of them members of the Grass Family (Poaceae)—furnishing 80% of the world's food. This subject is discussed in more detail in Chapter 24.

The flowers themselves range in size from the minute flowers of the duckweed, *Wolffia columbiana,* whose entire speck of a plant body is only 0.5 to 0.7 millimeter (0.02 to 0.03 inch) wide, with flowers little more than 0.1 millimeter long, to the enormous *Rafflesia* flowers (Fig. 8.2) of Indonesia that are up to 1 meter (3 feet 3 inches) in diameter and may weigh 9 kilograms (20 pounds) each.

Flowers may be any color or combination of colors of the rainbow, as well as black or white; they may have virtually any texture, from filmy and transparent to thick and leathery, spongy to sticky, hairy, prickly, or even dewy wet to the touch. Flowers of many trees, shrubs, and garden weeds are quite inconspicuous and lack odor (Fig. 8.3), but other flowers are strikingly beautiful, particularly when examined with a dissecting microscope. Their fragrances, which can be exhilarating to seductive or even putrid, are the basis of international perfume and pet-repellent industries.

Figure 8.1 Flowers of an Australian orchid produced by plants that complete their entire life cycle below ground. *(Drawing courtesy Karen Hamilton)*

Flowering plant habitats are as varied as their form. Besides the underground habitat mentioned in the chapter introduction, some flowering plants (*epiphytes*) go through their life cycles dangling from wires or other plants (Fig. 8.4). They also occur in both fresh and salt water, in the cracks and crevices of rocks, in deserts and jungles, in frigid arctic regions, and in areas where the temperatures regularly soar to 45°C (113°F) in the shade. In fact, they can be found almost anywhere they receive their basic needs of light, moisture, and a minimal supply of minerals. One species of chickweed survives at an altitude of 6,135 meters (over 20,000 feet) in the Himalaya Mountains, and fumitory plants in the same area flower even when the night temperatures plummet to –18°C (0°F).

Flowering plants can go from the germination of a seed to a mature plant producing new seeds in less than a month, or the process may take as long as 150 years. In **annuals,** the cycle is completed in a single season and ends with the death of the parent plant. **Biennials** take two growing seasons to complete the cycle; **perennials,** however, may take several to many growing seasons to go from a germinated seed to a plant producing new seeds, although many species that aren't annuals do produce seeds during their first growing season. Perennials may also produce flowers on new growth that dies back each winter, while other parts of the plant may persist indefinitely.

Flowering plants are currently placed in two major classes, the *Magnoliopsida* and *Liliopsida,* previously known as the *Dicotyledonae* and the *Monocotyledonae.* Despite the revised class names, the two groups are still commonly referred to as **dicots** and **monocots.** Members of the two classes are usually distinguished from one another on the basis of features listed in Table 8.1.

DIFFERENCES BETWEEN DICOTS AND MONOCOTS

Slightly less than three-fourths of all flowering plant species are dicots. Dicots include many annual plants and virtually all flowering trees and shrubs. Monocots, which are primarily herbaceous, are believed to have developed from primitive dicots. They include species that produce bulbs (e.g., lilies), grasses and related plants, orchids, irises, and palms. Palms, like other monocots, do not have a vascular cambium but become large through a *primary thickening* process of cells that occurs just below the apical meristem.

STRUCTURE OF FLOWERS

Regardless of form, all flowers share certain basic features. A typical flower develops several different parts, each with its own function (Fig. 8.5). Each flower, which begins as an embryonic *primordium* that develops into a *bud* (discussed in Chapter 6), occurs as a specialized branch at the tip of a stalk called a **peduncle,** which may in some instances have branchlets of smaller stalks called **pedicels.** A peduncle or pedicel swells at its tip into a small pad known as a **receptacle.** The other parts of the flower, some of which are in *whorls,* are attached to the receptacle (a whorl consists of three or more plant parts—such as leaves—encircling another plant part—such as a twig—at the same point on an axis).

The outermost whorl typically consists of three to five small, usually green, somewhat leaflike **sepals.** The sepals of a flower, which are collectively referred to as the **calyx,** may, in some flowers, be fused together. In many species, the calyx protects the flower while it is in the bud.

The next whorl of flower parts consists of three to many **petals;** the petals collectively are known as the **corolla.** Showy corollas attract pollinators, such as bees, moths, or birds. Bees, which can see in ultraviolet light, detect in some corollas special markings invisible to humans. The petals are distinct separate units in peach flowers, but in some flowers (e.g., petunias), the petals are fused together into a single, flared, trumpetlike sheet of tissue (Fig. 8.6). The corolla may not be showy or conspicuous in many tree and weed species and is often missing altogether or highly modified in wind-pollinated plants such as

Figure 8.2 A *Rafflesia* flower. These flowers may be as much as 1 meter (3 feet 3 inches) in diameter. *(Courtesy Charles H. Lamoureux)*

grasses. The calyx and the corolla together are referred to as the **perianth.** *Bracts* are specialized leaves that may be as colorful as petals and can attract pollinators the way petals do; they are discussed in Chapter 7.

Several to many **stamens** are attached to the receptacle around the base of the often greenish **pistil** in the center of the flower. Each stamen consists of a semirigid but otherwise usually slender **filament** with a sac called an **anther** at the top. The development and dissemination of **pollen grains** in anthers is described in Chapter 23. In most flowers, the pollen is released through lengthwise slits that develop on the anthers, but in members of the Heath Family (Ericaceae) and those of a few other groups, the pollen is released through anther pores.

The *pistil,* which often is shaped like a tiny vase that is closed at the top, consists of three regions that merge with one another. At the top is the **stigma,** which is connected by a slender, stalklike **style** to the swollen base called the **ovary.** The stigma may be little more than a point, a slight swelling, or may consist of up to several divergent arms or branches. The ovary later develops into a *fruit.*

There is evidence that ovaries first developed when the margins of leaves bearing ovules rolled inward. Such ovule-bearing leaves were called **carpels.** In some instances, two or more carpels eventually fused together, and many ovaries are now *compound,* consisting of two to several united carpels. The number of carpels present is often reflected by the number of lobes or divisions of the stigma. Each segment of a mature ovary, such as a tomato or orange, for example, represents a carpel. Carpels are discussed further in Chapter 23. Some texts refer to pistils as carpels, but since a pistil can consist of one to several carpels, such references are unfortunate and unnecessarily confusing.

The ovary is said to be **superior** if the calyx and corolla are attached to the receptacle at the base of the ovary, as in pea and primrose flowers. In other instances, the ovary becomes **inferior** when the receptacle grows up around it so that the calyx and corolla appear to be attached at the top, as in cactus and carrot flowers. A cavity containing one or more egg-shaped **ovules** lies within the ovary; ovules are attached to the wall of the cavity by means of short stalks. An ovule, the development of which takes place after *fertilization* has occurred, eventually becomes a **seed.** Details of fertilization and ovule development are given in Chapter 23.

Peach flowers are produced singly, each on its own peduncle, but many other flowers such as lilac, grape, and bridal wreath are produced in **inflorescences.** Inflorescences are groups of several to hundreds of flowers that may all

Figure 8.3 Male catkins of an alder, (*Alnus*). Each catkin consists of numerous tiny, inconspicuous, wind-pollinated flowers that have no petals.

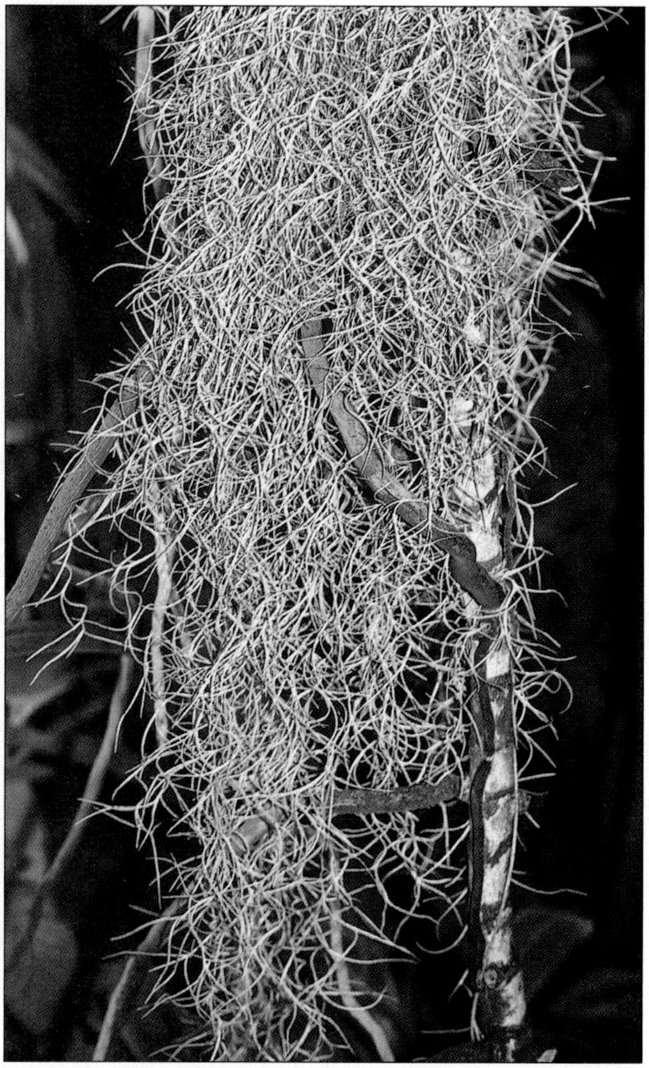

Figure 8.4 Spanish moss (*Tillandsia*). Spanish moss is a non-parasitic flowering plant that goes through its life cycle suspended from other plants or objects such as wires. This plant should not be confused with *lichens,* which consists of an alga and a fungus (see Chapter 18), and which also may hang suspended from other objects.

open at the same time, or they may follow an orderly progression to maturation (Fig. 8.7). The single peduncle of an inflorescence has many little stalks called *pedicels* attached to it—one pedicel for each flower.

A discussion of primitive and specialized flowers, along with their evolutionary development and ecological adaptations, is given in Chapter 23.

FRUITS

Introduction

In 1893, the U.S. Supreme Court, in the case of *Nix* v. *Hedden,* ruled that a tomato was legally a vegetable rather than a fruit. This was in keeping with the public's general conception of fruits as being relatively sweet and dessertlike. Vegetables, on the other hand, were considered more savory and useful as salad or main-course foods. Regardless of the court's decision, a **fruit,** botanically speaking, is any ovary and its accessory parts that has developed and matured. It also usually contains seeds. By this definition, many so-called vegetables, including tomatoes, string beans, cucumbers, and squashes, are really fruits. On the other hand, vegetables can consist of leaves (e.g., lettuce, cabbage), leaf petioles (e.g., celery), specialized leaves (e.g., onion), stems (e.g., white potato), roots (e.g., sweet potato), stems and roots (e.g., beets), flowers and their peduncles (e.g., broccoli), flower buds (e.g., globe artichoke), or other parts of the plant.

All fruits develop from flower ovaries and accordingly are found exclusively in the flowering plants. *Fertilization* (see Chapter 23) usually indirectly determines whether or not the ovary or ovaries (and sometimes the receptacle or

TABLE 8.1

Some Differences Between Dicots and Monocots

DICOTS	MONOCOTS
1. Seed with two cotyledons (seed leaves)	1. Seed with one cotyledon (seed leaf)
2. Flower parts mostly in fours or fives or multiples of four or five	2. Flower parts in threes or multiples of three
3. Leaf with a distinct network of primary veins	3. Leaf with more or less parallel primary veins
4. Vascular cambium, and frequently cork cambium, present	4. Vascular cambium and cork cambium absent
5. Vascular bundles of stem in a ring	5. Vascular bundles of stem scattered
6. Pollen grains mostly with three *apertures* (thin areas in the wall—see Figures 23.6 and 23.7)	6. Pollen grains mostly with one *aperture*

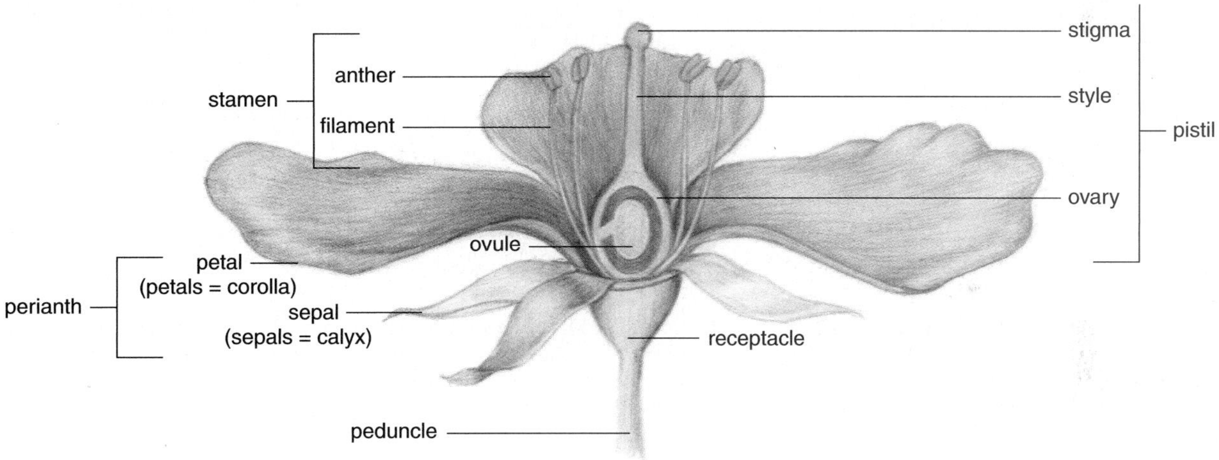

Figure 8.5 Parts of a generalized flower. The interior structure of the ovule and the sexual processes involved are discussed in Chapter 23.

Figure 8.6 A flower of jimson weed (*Datura*). The petals are united in a single, flared sheet of tissue.

other tissues) of a flower will develop into a fruit. If at least a few of the ovules are not fertilized, the flower normally withers and drops without developing further. Pollen grains contain specific stimulants called *hormones* (discussed in detail in Chapter 11) that may initiate fruit development, and sometimes a little dead pollen is all that is needed to stimulate an ovary into becoming a fruit. It is the hormones produced by the developing seeds, however, that promote the greatest fruit growth. These hormones, in turn, stimulate the production of more fruit growth hormones by the ovary wall. In a few instances (e.g., the cultivated banana), fruits develop without fertilization. Such development is called *parthenocarpic.* Parthenocarpy is discussed in Chapter 23.

Fruit Regions

Most of a mature fruit has three regions, which sometimes merge and can be difficult to distinguish from one another (Fig. 8.8). The skin forms the **exocarp,** while the inner

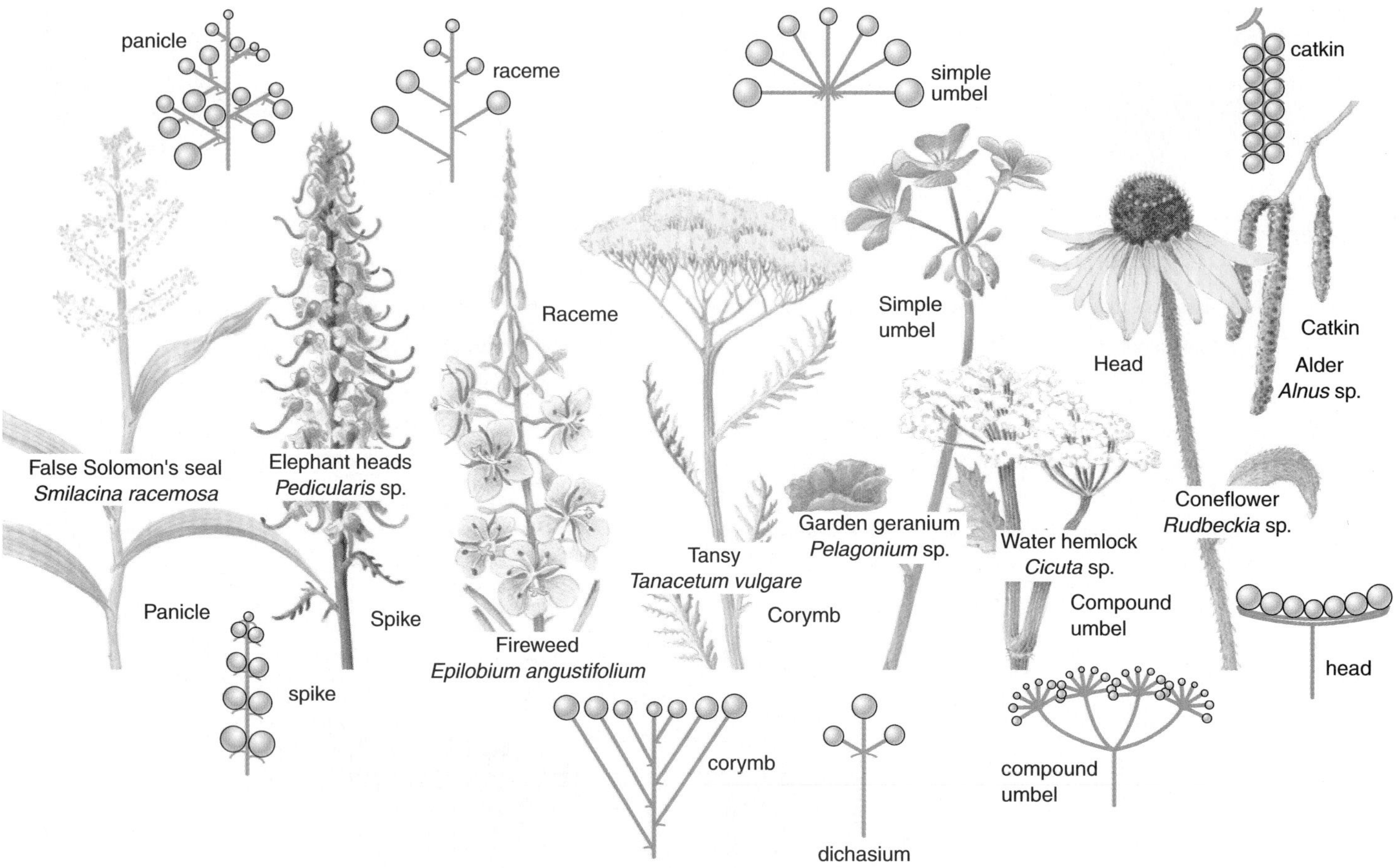

Figure 8.7 Inflorescence types. Each ball represents a flower. In all inflorescences shown, except for the dichasium and catkin, the lowermost or outermost flowers open first. The flowers then open in succession upward or inward. In a dichasium, the central flower opens first, and the side flowers open simultaneously. In a catkin, all the flowers open simultaneously. *(From Moore, Clark, and Vodopich, Botany, 2nd edition. © 1998 The McGraw-Hill Companies. All rights reserved.)*

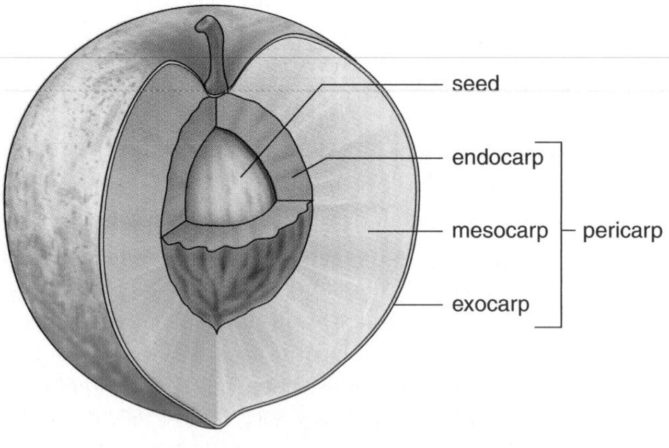

Figure 8.8 Regions of a mature peach.

boundary around the seed(s) forms the **endocarp.** The endocarp may be hard and stony (as in a peach pit around the seed). It also may be papery (as in apples), or it may not be distinct from the **mesocarp,** which is the often fleshy tissue between the exocarp and the endocarp. The three regions collectively are called the **pericarp.** In dry fruits, the pericarp is usually quite thin.

Some fruits consist of only the ovary and its seeds. Others have adjacent flower parts, such as the receptacle or calyx fused to the ovary or different parts modified in various ways. Fruits may be either fleshy or dry at maturity, and they may split, exposing the seeds, or no split may occur. They may be derived from a single ovary or from more than one. Traditionally, all these features have been used to classify fruits, but unfortunately, not all fruits lend themselves to neat pigeonholing by such characteristics. Some of these problems are pointed out in the classification that follows.

Kinds of Fruits

Fleshy Fruits

Fruits whose mesocarp is at least partly fleshy at maturity are classified as fleshy fruits.

Simple Fleshy Fruits Simple fleshy fruits develop from a flower with a single pistil. The ovary may be superior or inferior, and it may be *simple* (derived from a single modified leaf called a **carpel**), or it may consist of two or more carpels and be *compound* (for a discussion of the derivation

A.

B.

C.

Figure 8.9 Representative drupes. *A.* Peaches. *B.* Almonds. *C.* Olives.

of carpels and compound ovaries, see Chapter 23). The ovary alone may develop into the fruit, or other parts of the flower may develop with it.

Drupe A **drupe** is a simple fleshy fruit with a single seed enclosed by a hard, stony endocarp, or pit (Fig. 8.9). It usually develops from flowers with a superior ovary containing a single ovule. The mesocarp is not always obviously fleshy, however. In coconuts, for example, the *husk* (consisting of the mesocarp and the exocarp), which is usually removed before the rest of the fruit is sold in markets, is very fibrous (the fibers, incidentally, are used in making mats and brushes). The seed ("meat") of the coconut is hollow and contains a watery *endosperm* (see Chapter 23) commonly but incorrectly referred to as "milk." It is surrounded by the thick, hard endocarp typical of drupes. Other examples of drupes include the stone fruits (e.g., apricots, cherries, peaches, plums, olives, and almonds). In almonds, the husk, which dries somewhat and splits at maturity, is removed before marketing, and it is the endocarp that we crack to obtain the seed.

Berry **Berries** usually develop from a compound ovary and commonly contain more than one seed. The entire pericarp is fleshy, and it is difficult to distinguish between the mesocarp and the endocarp (Fig. 8.10). Three types of berries may be recognized.

A *true berry* is a fruit with a thin skin and a pericarp that is relatively soft at maturity. Although most contain more than one seed, notable exceptions are dates and avocados, which have only one seed. Typical examples of true berries include tomatoes, grapes, persimmons, peppers, and eggplants. Some fruits that popularly include the word *berry* in their common name (e.g., strawberry, raspberry, blackberry) botanically are not berries at all.

A.

B.

Figure 8.10 Representative berries. *A.* Grapes. *B.* Tomatoes.

Some berries are derived from flowers with inferior ovaries so that other parts of the flower also contribute to the flesh. They can usually be distinguished by the remnants of flower parts or their scars that persist at the tip. Examples of such berries include gooseberries, blueberries, cranberries, pomegranates, and bananas. Because fruit development in the cultivated banana is parthenocarpic, there are no seeds. Several other species of banana produce an abundance of seeds.

Pepos are berries with relatively thick rinds. Fruits of members of the Pumpkin Family (Cucurbitaceae), including pumpkins, cucumbers, watermelons, squashes, and cantaloupes, are pepos.

The *hesperidium* is a berry with a leathery skin containing oils. Numerous outgrowths from the inner lining of the ovary wall become saclike and swollen with juice as the fruit develops. All members of the Citrus Family (Rutaceae) produce this type of fruit. Examples include oranges, lemons, limes, grapefruit, tangerines, and kumquats.

Pome **Pomes** are simple fleshy fruits, the bulk of whose flesh comes from the enlarged floral tube or receptacle that grows up around the ovary. The endocarp around the seeds is papery or leathery. Examples include apples, pears, and quinces. In an apple, the ovary consists of the core and a little adjacent tissue. The remainder of the fruit has developed primarily from the floral tube, with a little tissue contributed by the receptacle (Fig. 8.11). Botany texts often refer to pomes, pepos, some berries, and other fruits derived from more than an ovary alone as *accessory fruits* or as fruits having *accessory tissue.*

Dry Fruits

Fruits whose mesocarp is definitely dry at maturity are classified as *dry fruits.*

Dry Fruits That Split at Maturity (Dehiscent Fruits) The fruits in this group are distinguished from one another by the way they split.

Follicle The **follicle** splits along one side or seam (*suture*) only, exposing the seeds within (Fig. 8.12). Examples include larkspur, columbine, milkweed, and peony.

Legume The **legume** splits along two sides or seams (Fig. 8.13). Literally thousands of members of the Legume Family (Fabaceae) produce this type of fruit. Examples include peas, beans, garbanzo beans, lentils, carob, kudzu, and mesquite. Peanuts are also legumes, but they are atypical in that the fruits develop and mature underground. The seeds are usually released in nature by bacterial breakdown of the pericarp instead of through an active splitting action.

Silique **Siliques** also split along two sides or seams, but the seeds are borne on a central partition, which is exposed when the two halves of the fruit separate (Fig. 8.14A). Such fruits, when they are less than three times as long as they are wide, are called *silicles* (Fig. 8.14B). Siliques and silicles are produced by members of the Mustard Family (Brassicaceae), which includes broccoli, cabbage, radish, shepherd's purse, and watercress.

Figure 8.11 Apples (representative of pomes). The bulk of the flesh is derived from the floral tube that grows up around the ovary.

A.

B.

Figure 8.12 Follicles. *A.* Milkweed. *B. Magnolia.* The fruit of the magnolia is actually an aggregate fruit consisting of approximately 40 to 80 individual one-seeded follicles on a common axis. The follicles and axis fall from the tree as a unit.

Figure 8.13 Legumes of a coral tree (*Erythrina*).

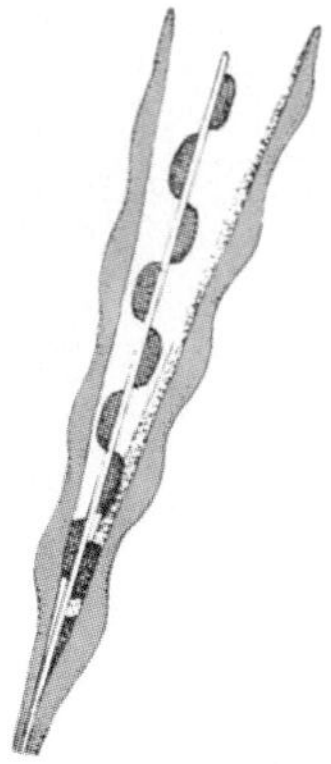

A.

B.

Figure 8.14 *A.* A silique after it has split open. The seeds are borne on a central, membranous partition. *B.* Silicles of *Lunaria* (dollar plant).

Capsule **Capsules** are the most common of the dry fruits that split (Fig. 8.15). They consist of at least two carpels and split in a variety of ways. Some split along the partitions between the carpels, while others split through the cavities (*locules*) in the carpels. Still others form a cap toward one end that pops off and releases the seeds, or they form a row of pores through which the seeds are shaken out as the capsule rattles in the wind. Examples include irises, orchids, lilies, poppies, violets, and snapdragons.

Dry Fruits That Do Not Split at Maturity (Indehiscent Fruits) In this type of dry fruit, the single seed is, to varying degrees, united with the pericarp.

Achene Only the base of the single seed of the **achene** is attached to its surrounding pericarp. Accordingly, the husk (pericarp) is relatively easily separated from the seed. Examples include sunflower "seeds" (the edible kernel plus the husk constitute the achene) (Fig. 8.16), buttercup, and buckwheat.

Nut **Nuts** are one-seeded fruits similar to achenes, but they are generally larger, and the pericarp is much harder and thicker. They develop with a cup, or cluster, of bracts at their base. Examples include acorns (Fig. 8.16), hazelnuts (filberts), and hickory nuts. Botanically speaking, many nuts in the popular sense are not nuts. We have already seen that peanuts are atypical legumes and that coconuts and almonds are drupes. Walnuts and pecans are also drupes, whose "flesh" withers and dries after the seed matures. Brazil nuts

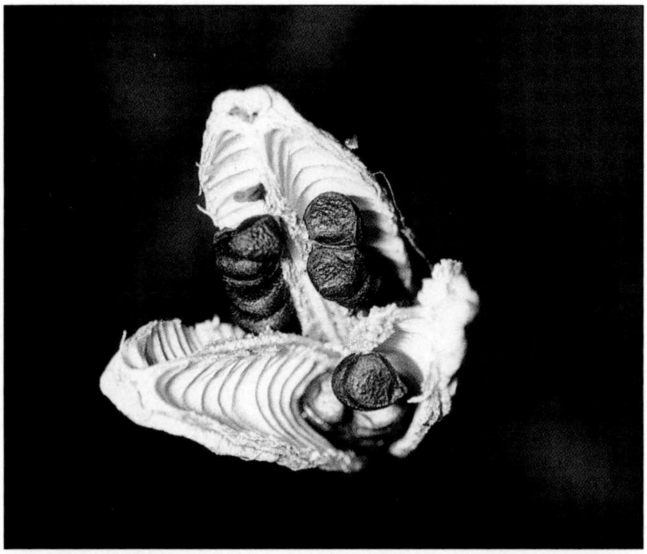

A.

C.

B.

D.

Figure 8.15 Capsules. *A.* Butterfly iris (*Moraea*). *B. Bletilla* orchid. *C.* Autograph tree (*Clusia rosea*). *D.* Unicorn plant (*Proboscidia*).

are the seeds of a large capsule, and a cashew nut is the single seed of a unique drupe. It appears as a curved appendage at the end of a swollen pedicel, which is eaten raw in the tropics or made into preserves or wine. Pistachio nuts are also the seeds of drupes.

Grain (Caryopsis) The pericarp of the **grain** is tightly united with the seed and cannot be separated from it (Fig. 8.16). All members of the Grass Family (Poaceae), including corn, wheat, rice, oats, and barley, produce grains (also called **caryopses**).

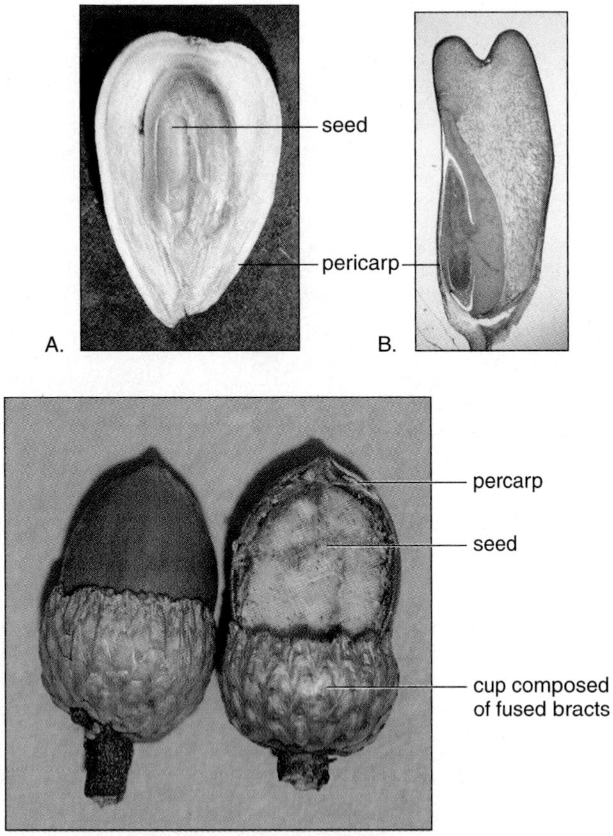

Figure 8.16 Dry fruits that do not split at maturity. *A.* Achene of a sunflower split open. *B.* Grain (caryopsis) of corn, cut lengthwise. *C.* Nuts (acorns) of an oak. The acorn on the right was cut lengthwise above the cup.

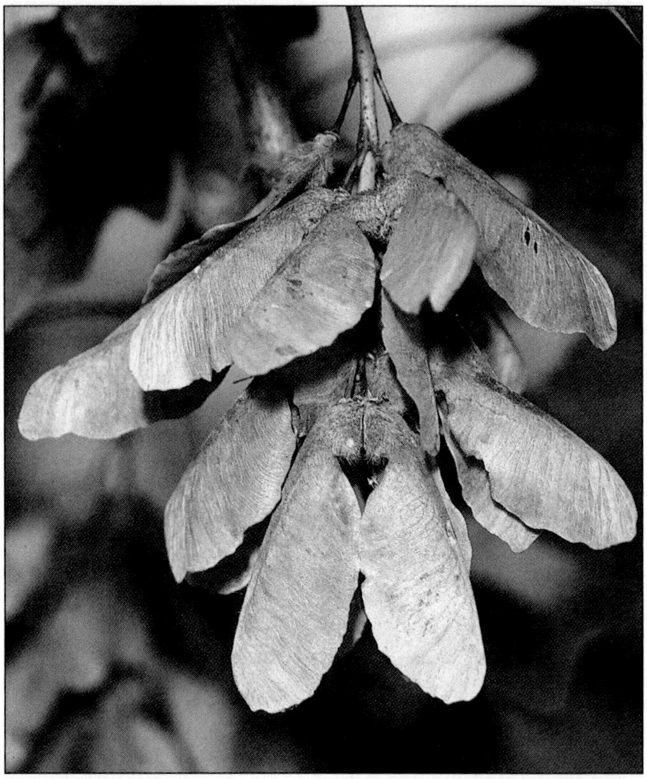

Figure 8.17 Samaras of a big-leaf maple (*Acer macrophyllum*). Maple samaras are produced in pairs that separate at maturity.

Samara In **samaras,** the pericarp surrounding the seed extends out in the form of a wing or membrane, which aids in dispersal (Fig. 8.17). In maples, samaras are produced in pairs, but in ashes, elms, and the tree of heaven, they are produced singly.

Schizocarp The twin fruit called a **schizocarp** (Fig. 8.18) is unique to the Parsley Family (Apiaceae). Members of this family include parsley, carrots, anise, caraway, and dill. Upon drying, the twin fruits break into two one-seeded segments called *mericarps.*

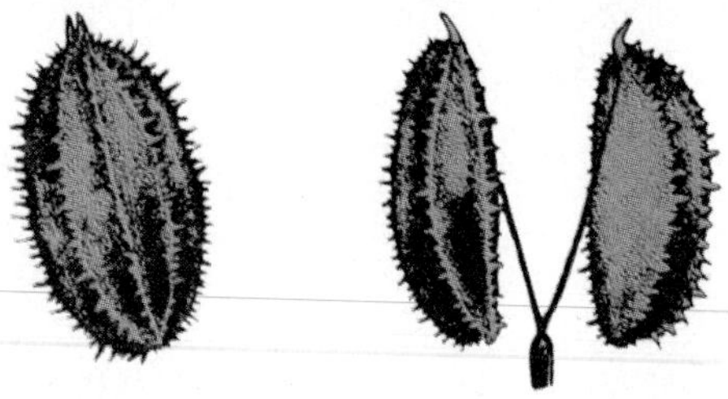

Figure 8.18 Schizocarps of carrots. A schizocarp separates at maturity into two one-seeded fruitlets.

Aggregate Fruits

An **aggregate fruit** is one that is derived from a single flower with several to many pistils. The individual pistils develop into tiny drupes or other fruitlets, but they mature as a clustered unit on a single receptacle (Fig. 8.19). Examples include raspberries, blackberries, and strawberries. In a strawberry, the cone-shaped receptacle becomes fleshy and red, while each pistil becomes a little *achene* on its surface. In other words, the strawberry, while being an aggregate fruit, is also partly composed of accessory tissue.

Multiple Fruits

Multiple fruits are derived from several to many individual flowers in a single inflorescence. Each flower has its own receptacle, but as the flowers mature separately into fruitlets, they develop together into a single larger fruit, as in aggregate fruits. Examples of multiple fruits include mulberries, Osage oranges (Fig. 8.20), pineapples, and figs. Pineapples, like bananas, usually develop parthenocarpically (see Chapter 23), and there are no seeds. The individual flowers are fused together on a fleshy axis, and the fruitlets coalesce into a single fruit.

Figs mature from a unique "outside in" inflorescence. The individual flowers of the inflorescence are enclosed by the common receptacle, which has an opening to the out-

A. B.

Figure 8.19 A. A blackberry flower. Note the numerous green pistils. *B.* Blackberries, representative of aggregate fruits. *(B. Courtesy Robert A. Schlising)*

side at the tip (Fig. 8.21). Such a multiple fruit arrangement is referred to as a *syconium.* Some fig varieties develop parthenocarpically, but others are pollinated by tiny wasps that crawl in and out through the opening. Some multiple fruits, such as those of the sweet gum, are dry at maturity.

FRUIT AND SEED DISPERSAL

Why are so many species of orchids rare, while dandelions, shepherd's purse, and other weeds occur all over the world? Why are some plants confined to single continents, mountain ranges, or small niches occupying less than a hectare (2.47 acres) of land? The answers to these questions involve many different factors, including climate, soil, the adaptability of the plant, and its means of seed dispersal. How fruits and seeds are transported from one place to another is the subject of the following sections. Other factors are discussed in Chapters 13 and 25.

Dispersal by Wind

Fruits and seeds have a variety of adaptations for wind dispersal (Fig. 8.22). The samara of a maple has a curved wing that causes the fruit to spin as it is released from the tree. In a brisk wind, samaras may be carried up to 10 kilometers (6 miles) away from their source, although usually most are relatively evenly distributed within a few meters of the tree. In hop hornbeams, the seed is enclosed in an inflated sac that gives it some buoyancy in the wind. In some members of the Buttercup and Sunflower Families (Ranunculaceae and Asteraceae), the fruits have plumes, and in the Willow Family (Salicaceae), the fruits are surrounded by cottony or woolly hairs that aid in wind dispersal. In button snakeroots and Jerusalem sage, the fruits are too large to be airborne, but they are spherical enough to be rolled along the ground by the wind.

Seeds themselves may be so tiny and light that they can be blown great distances by the wind. Orchids and heaths, for example, produce seeds with no endosperm that are as

Figure 8.20 An osage orange (*Maclura*), representative of multiple fruits. The ovaries of many flowers become united in a single unit.

Figure 8.21 Section through a developing fig.

fine as dust and equally light in weight. In catalpa and jacaranda trees, the seeds themselves are winged rather than the fruits, which remain on the branches and split, releasing their contents. Dandelion fruitlets have plumes that radiate out at the ends like tiny parachutes; these catch even a slight breeze. In tumble mustard and other tumbleweeds, the whole aboveground portion of the plant may *abscise* (separate from the root) and be blown away by the wind, releasing seeds as it bumps along.

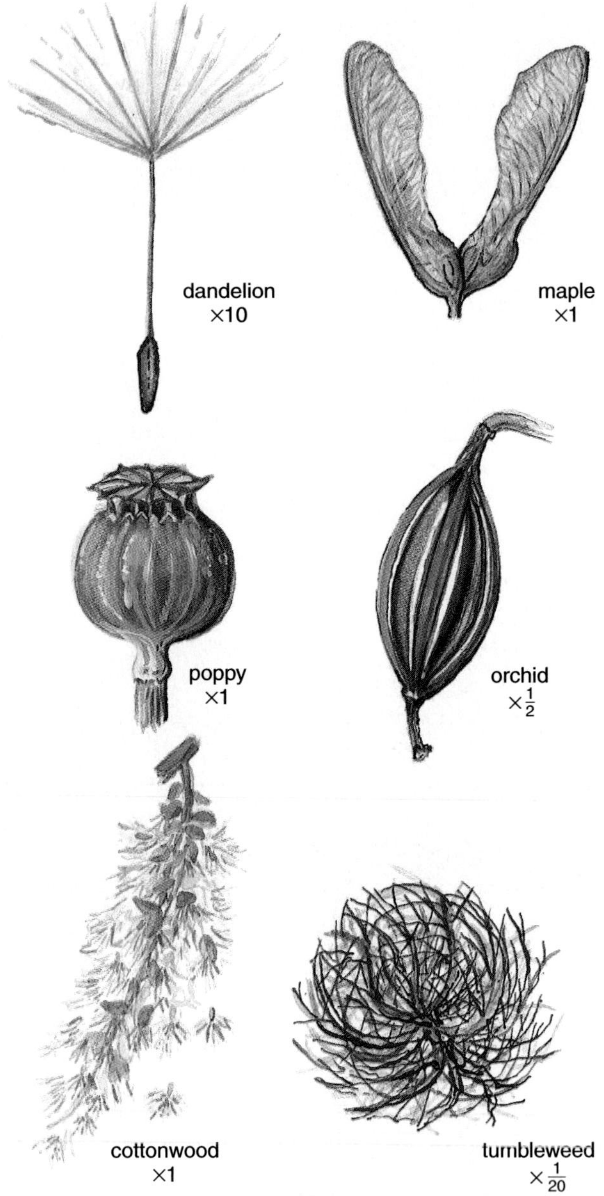

Figure 8.22 Types of seeds and fruits dispersed by wind.

Dispersal by Animals

The adaptations of fruits and seeds for animal dispersal are legion. Birds, mammals, and ants all act as disseminating agents (Fig. 8.23). Shore birds may carry seeds great distances in mud that adheres to their feet. Other birds and mammals eat fruits whose seeds pass unharmed through their digestive tracts. Some bird-disseminated fruits contain laxatives that speed their passage through the birds' digestive tracts. In blackbirds, the seeds may remain in the tract as little as 15 minutes, but in mammals, the period is more commonly about 24 hours. In the giant tortoises of the Galápagos Islands, seeds do not pass through the tract for 2 weeks or more, and the seeds usually will not germinate unless they

Figure 8.23 Types of seeds and fruits dispersed by animals and birds.

have been subjected to such treatment (see Chapter 11). Some seeds and fruits are gathered and stored by rodents, such as squirrels and mice, and then are abandoned. Blue jays, woodpeckers, and other birds carry away nuts and other fruits, which they may drop in flight and abandon.

Many fruits and seeds catch in or adhere to the fur or feathers of animals and birds. Bedstraw and bur clover fruits are covered with small hooks that catch in fur (or a hiker's socks). The large capsules of unicorn plants have two giant curved extensions about 15 centimeters (6 inches) long. These catch on the fetlock of a deer or other animal that happens to step on the fruit, and the seeds are scattered as the animal moves along. Twinflowers and flax have fruits with sticky appendages that adhere to fur on contact, and those of the puncture vine penetrate the skin and stick by means of hard little prickles (Fig. 8.24).

Figure 8.24 Fruits of a puncture vine (*Tribulus terrestris*) clinging to a bicycle tire.

Bleeding hearts, trilliums, and several dozen other plants have on their seeds appendages that contain oils attractive to ants (Fig. 8.25). The Scandinavian scientist Sernander once estimated that more than 36,000 such seeds were carried by members of a single ant colony to their nest, where the ants stripped off the appendages for food but did not harm the seeds themselves.

Dispersal by Water

Some fruits contain trapped air, adapting them to water dispersal. Many sedges, for example, have seeds surrounded by inflated sacs that enable the seeds to float (Fig. 8.26). Others have waxy material on the surface of the seeds, which temporarily prevents them from absorbing water while they are floating. Sometimes, a heavy downpour will create a torrent of water that dislodges masses of vegetation along a stream

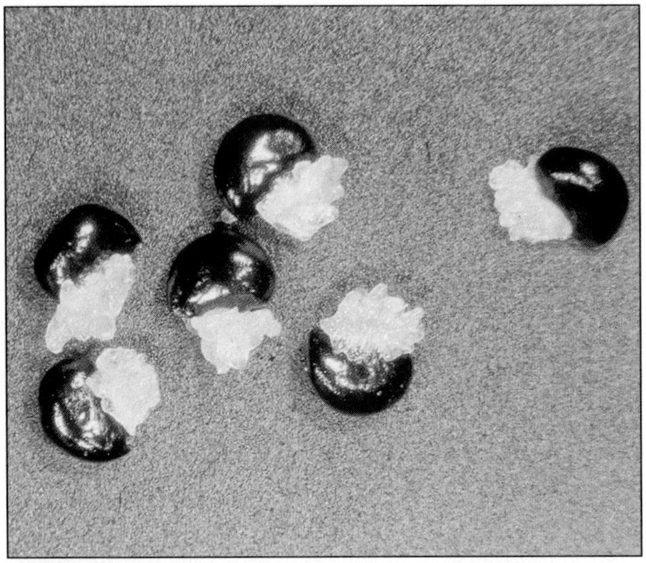

Figure 8.25 Seeds of the Pacific bleeding heart (*Dicentra formosa*). The white appendages are *elaiosomes,* which are removed from the seeds by ants and used for food.

Figure 8.26 Sedge adaptation to water dispersal. *A.* A sedge plant. *B.* A sedge fruit. The seed is enclosed within an inflated covering that enables it to float on water.

bank, carrying whole plants and their fruits to new locations. Large raindrops themselves may splash seeds out of their opened capsules.

Seeds and fruits of a few plants have thick, spongy pericarps that absorb water very slowly. Such fruits are adapted to dispersal by ocean currents, even though salt water eventually may penetrate enough to kill the delicate embryos. Enough fruits are beached before this occurs to ensure the survival of the species. The best known of ocean-dispersed plants is the coconut palm, whose large fruits apparently have been carried many hundreds of kilometers throughout the tropical seas of the world.

Other Dispersal Mechanisms and Agents

Fruits of some legumes, touch-me-nots, and members of other families mechanically eject seeds, sometimes with considerable force. For example, the splitting action of drying witch hazel capsules may fling the seeds over 12 meters (40 feet) away. Fruits of dwarf mistletoes may be violently released in response to the heat of a warm-blooded animal coming close to the plants. In fact, small welts have been raised by dwarf mistletoe fruits on the skin of humans who ventured close to the plants. In manroots and a few other members of the Pumpkin Family (Cucurbitaceae), the seed release resembles a geyser eruption as a frothy substance containing the seeds squirts out of one end of the melonlike fruits.

In filarees and other members of the Geranium Family (Geraniaceae), each carpel of the fruit splits away and curls back from a central axis. Each fruitlet consists of a single, pointed seed with a long, slender tail that is sensitive to changes in humidity (Fig. 8.27). At night when the humidity increases, the tail is relatively straight, but in the sun, it coils up like a corkscrew, literally drilling the pointed seed into the ground as it does so and effectively planting it in the process.

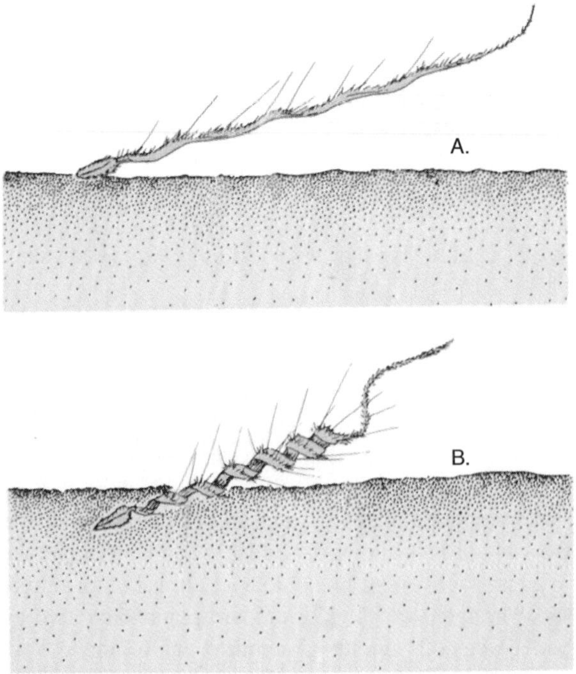

Figure 8.27 Filaree (*Erodium*) fruitlets. *A.* Under humid conditions. *B.* Under dry conditions. Alternate coiling and uncoiling causes the fruitlets to be "screwed" into the ground.

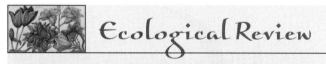

Ecological Review

Flowers, fruits, and seeds are focal points for ecological interactions between plants and biological aspects of the environment. Many types of flowers serve to attract animals that transfer pollen between flowers. Adaptations of fruits and seeds for dispersal by animals, mainly birds, mammals, and ants, are legion and include attractive colors, scents, and oils, laxatives to speed the passage of seeds through the animal gut, and adhesive structures that attach to fur or feathers. Seed germination is usually subject to very specific environmental controls.

Humans, both intentionally and unintentionally, are by far the most efficient transporters of fruits and seeds. Travelers and explorers have carried many noxious weeds and plant diseases, as well as valuable food and medicinal plants, from one continent to another. Most countries now have strict regulations barring the importation of plant materials, except by special permit, and carrying some plants across borders is not legal under any circumstances. In the United States, for example, certain fruits and seeds that might carry diseases harmful to local agriculture may be barred from entry. Arizona, California, and Hawaii have border inspections to try to prevent the importation of popcorn, citrus, and other fruits across state lines; in the past, such plants have been carriers of diseases or pests that are presently under control.

SEEDS

Structure

The concave side of an ordinary kidney bean (a dicot) has a small white scar called the *hilum.* The hilum marks the point at which the ovule was attached to the ovary wall. A tiny pore called the *micropyle* is located right next to the hilum. If this bean is placed in water for an hour or two, it may swell enough to split the seed coat. Once the seed coat is removed, the two halves, called **cotyledons,** can be distinguished (Fig. 8.28). The cotyledons, which have a tiny immature plantlet along one edge between them, are food-storage organs that also function as the first "seed leaves" of the seedling plant. The cotyledons and the tiny, rudimentary bean plant to which they are attached constitute the *embryo.* Some seeds (e.g., those of grasses and all other monocots) have only one cotyledon.

The tiny embryo plantlet has undeveloped leaves and a meristem at the upper end of the embryo axis. This embryo shoot is called a **plumule.** The cotyledons are attached just

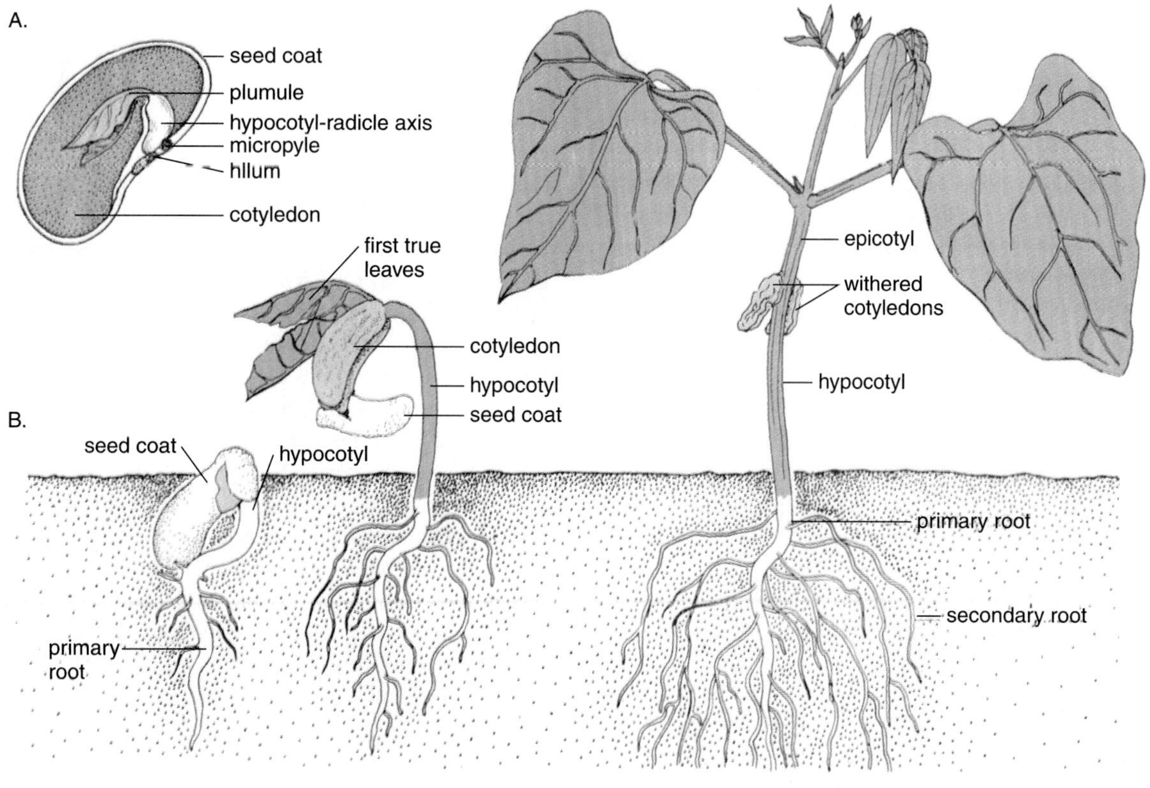

Figure 8.28 A common garden bean. *A.* Seed structure. *B.* Germination and development of the seedling.

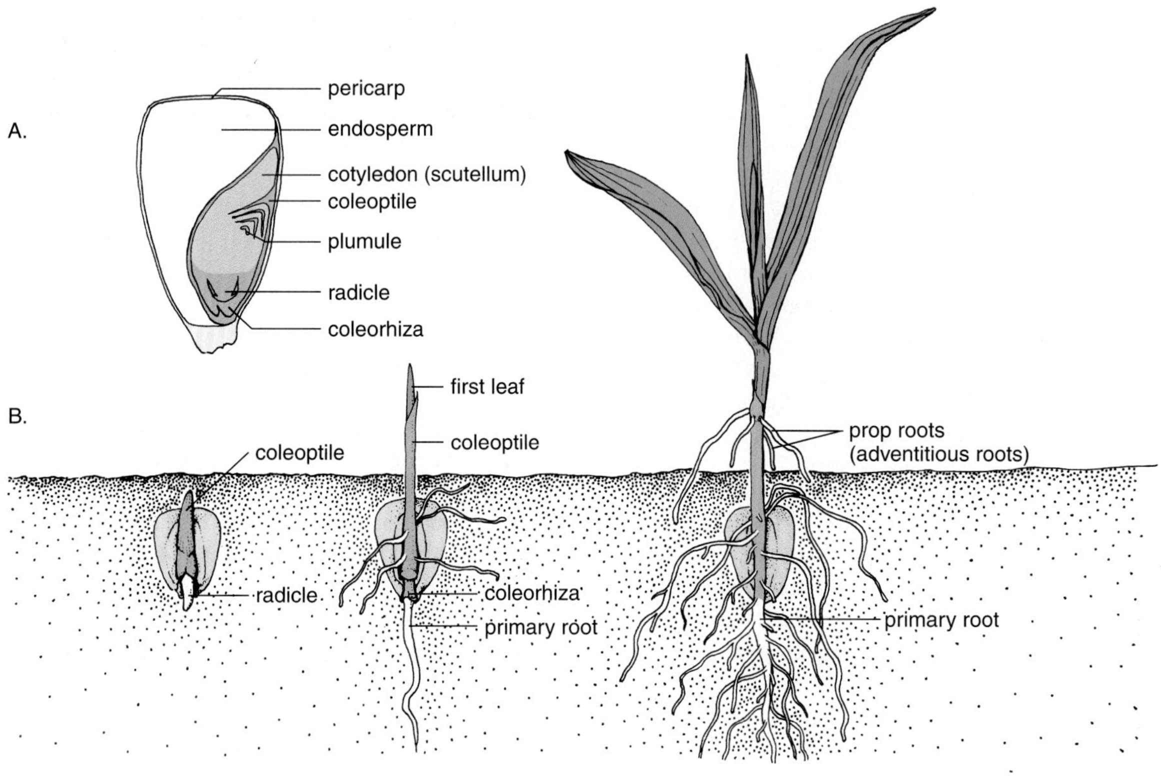

Figure 8.29 Corn. *A.* Grain structure. *B.* Germination and development of the seedling.

below the plumule. The very short part of the stem above the cotyledons is called the **epicotyl,** while the stem below the attachment point is the **hypocotyl.** In an embryo, it is often difficult to tell where the stem ends and the root begins, but the tip that will develop into a root is called a **radicle.** When a kidney bean germinates, the hypocotyl lengthens and bends, becoming hook-shaped. The top of the hook emerges from the ground, pulling the cotyledons above the ground. Once the cotyledons have emerged, the hook straightens out. In lima beans and peas, however, the hypocotyl remains short so that the cotyledons do not emerge above the surface (see Chapter 11).

In other seeds, the cotyledon(s) may not play a significant role in food storage. In corn, for example, the bulk of the food-storage tissue is *endosperm* (see Chapter 23). Corn "seeds" (Fig. 8.29) also display other features not seen in beans. The plumule and the radicle are enclosed in tubular, sheathing structures called the **coleoptile** and the **coleorhiza,** respectively. These protect the delicate tissues within as the seeds germinate. After the coleoptile and coleorhiza have become several millimeters long, their development ceases, and the plumule and radicle burst through the tips.

Germination

Germination, which is the beginning or resumption of growth of a seed, depends on the interplay of a number of factors, both internal and external. In order to germinate, a seed must first be *viable* (capable of germinating). Many seeds for various reasons (e.g., death of the embryo within) are not viable, and all lose their viability after varying periods of time. Many seeds also require a period of **dormancy** (see page 217) before they will germinate. Dormancy is brought about by either mechanical or physiological circumstances or both. In the Legume Family (Fabaceae) and others, the seeds may have seed coats so thick or tough that they prevent the absorption of water or oxygen. Some seeds even have a one-way valve that lets moisture out but prevents its uptake. Dormancy in such seeds may sometimes be broken artificially by *scarification,* which involves nicking or slightly cracking the seed coats or dipping the seeds in a concentrated acid for a few seconds to a few minutes. In nature, such seeds may remain dormant until cracks in the seed coat are brought about by the mechanical abrasion of rock particles in the soil, alternate thawing and freezing, or in some cases, bacterial action.

Dormancy may also be brought about by growth-inhibiting substances present in the seed coat, the interior of the seed, or tissues of the fruit surrounding it. Many desert plants have inhibitors in the seed coat. These have to be washed away by soaking rains before germination will occur. The inhibitors function in survival of the species by preventing germination unless there has been sufficient rainfall for a seedling to become established.

Apples, pears, citrus fruits, tomatoes, and other fleshy fruits contain inhibitors that prevent germination of the seeds within the fruits. Once the seeds are removed and washed,

Figure 8.30 Castor bean (*Ricinus*) seeds. Note the small water-absorbing appendage (caruncle) at the end of each seed.

they germinate readily. The embryos of some seeds, such as those of the American holly, consist of only a few unspecialized cells when the fruit ripens. The seeds will not germinate after the fruit has dropped until the embryo has developed fully with the aid of food materials stored in its endosperm. Such a process of development is called *after-ripening.*

In many woody plants of temperate areas, germination stimulators need to be present to initiate growth. These normally do not develop unless the seeds encounter a wet period accompanied by cold temperatures. Usually this period needs to be a minimum of 4 to 6 weeks. The dormancy of such seeds can be broken artificially by placing them in a refrigerator, preferably in damp sand, for a few weeks.

Even when mechanical and physiological barriers to germination are not present, a seed will not normally germinate unless environmental factors are favorable. Water and oxygen are essential to the completion of germination, and light or its absence also plays a role. Many seeds imbibe 10 times or more their total weight in water before the radicle emerges. Some seeds, such as those of castor beans (Fig. 8.30) and certain spurges, have appendages that function in water absorption and thereby speed up the germination process.

After water has been imbibed, enzymes begin to function in the cytoplasm, which has now been rehydrated. Some enzymes convert stored proteins to amino acids, others convert fats and oils to soluble compounds, and still other enzymes aid in the conversion of starch to sugar. The soluble substances can then be conveyed to the embryo, and respiration, which in a dormant seed is almost imperceptible, can be greatly accelerated. In a few seeds such as rice and barnyard grass, anaerobic respiration initially furnishes the energy for embryo growth, but in most seeds, the energy is released through aerobic respiration. A new plant begins to develop as mitosis and cell elongation take place. Both forms of respiration are discussed in Chapter 10.

If seeds are kept waterlogged after planting, oxygen available to them is greatly reduced and germination then may fail to be completed. Most seeds require temperatures within certain ranges to germinate. These usually need to be above freezing but below 45°C (113°F). Germination percentages tend to be low approaching either extreme, however. Most crop plants have an optimum (ideal) germination temperature of between 20°C and 30°C (68°F to 86°F).

The role of light in germination varies with the kinds of plants concerned. Seeds of some varieties of lettuce will not germinate in the dark (see the discussion of *phytochrome* in Chapter 11), while those of other seeds, such as the California poppy, germinate only in the dark. In lettuce seeds, the light apparently inactivates germination inhibitors, while in the California poppy, it stimulates inhibitor formation. (See the additional discussion of dormancy in Chapter 11.)

Longevity

From time to time, one reads or hears of seeds of wheat or other edible plants germinating after lying dormant in Egyptian pyramids or Native American tombs and caves for thousands of years, but none of these reports has been confirmed. In fact, there is evidence in a few instances that rats or rodents in recent times carried the seeds concerned to their nests. However, reports of seeds of the aquatic lotus plant germinating after a little more than 1,000 years and another documenting the germination of Arctic tundra lupine seeds that were frozen for an estimated 10,000 years have been confirmed.

Seeds remain viable (retain the capacity to germinate) for periods that vary greatly, depending on the species and the conditions of storage. Some seeds, such as those of certain willows, cottonwoods, orchids, and tea, remain viable for only a few days or weeks, regardless of how they are stored, but the period of viability of most seeds is extended by months or even years when they are stored at low temperatures and kept dry.

By law, packets of vegetable and flower seeds sold in stores are dated, giving the buyer a rough idea of how long a significant number of the seeds might be expected to remain viable. Generally, seeds of Pumpkin Family (Cucurbitaceae) members (e.g., squash, cantaloupe, cucumber) retain a relatively high percentage of viability for several years, while those of members of the Lily Family (Liliaceae) (e.g., onion, leek, chives) retain a good percentage of viability for only 2 or 3 years. Properly stored wheat seeds have been reported to retain better than 30% viability for more than 30 years, and some weed seeds stored under conditions of low oxygen, high humidity, and cool temperatures have remained viable for even longer periods.

In 1879, William J. Beal, a botanist who pioneered in the development of hybrid corn, buried 20 pint-sized bottles of weed seeds on the campus of what is now Michigan State University in East Lansing, Michigan. Each bottle contained 1,000 seeds of 20 different species of weeds.

Awareness

The Seed That Slept for 1,200 Years

Rip Van Winkle would appreciate this. An Oriental Sacred Lotus (*Nelumbo nucifera*) seed collected from the sediment of a dry lake bottom near a small village in northeastern China has germinated after being dormant for over 1,200 years. It is one of the oldest living seeds ever found. The Beijing Institute of Botany donated Sacred Lotus seeds to a team of UCLA scientists who dated the seeds with a non-destructive method called accelerator mass spectroscopy. Small amounts of tissue (less than 10 milligrams) were sampled for radiocarbon dating prior to germination studies. Before the use of this newer dating method, whole seeds had to be destroyed in the process of dating, thereby eliminating the possibility of testing the seeds for viability. After lying dormant in a bed of black clay at depths of 0.5 to 2.8 meters (1.5 to 9 feet), the germinated 1,200-year-old seed (1,288 ± 271 years) was the oldest, but not the only survivor from the subterranean tomb. Three other ancient lotus seeds found at various depths germinated, and the UCLA team determined that one was more than 600 years old and another was more than 300.

While reports have claimed seed germination of more ancient seeds recovered from dry archaeological sites in Egypt (such as King Tut's tomb in the pyramids of Giza and from the tombs of other Pharaohs), experts now agree that these reports are unreliable. Apart from these lotus seeds, the oldest documented viable seeds are lupine seeds that were frozen in Arctic tundra and from Professor Beal's seed germination study (see text section entitled "Longevity" for discussion).

These Sacred Lotus seeds have managed to ward off the ravages of time. Existing in an impenetrable seed coat and mired in an oxygen-deficient mud, the seeds have intact genetic and enzymatic systems that reactivated when split open and soaked in water. (Enzymes are proteins that speed up chemical reactions in the cell.) A key enzyme that repairs proteins was present during germination; this enzyme has been found in similar quantities in modern-day Sacred Lotus seeds. The repair enzyme functions in converting damaged amino acids back to their naturally occurring functional form. This is especially important in "repairing" the proteins of the cell membrane. Without intact membranes, cells are not able to function, and such damage will lead to the death of the cell and eventually the organism.

The architecture of the fruit no doubt plays a key role in the longevity of these seeds. The fruits of the Sacred Lotus are round to oblong, 10 to 13 centimeters (4 to 5 inches) long, 8 to 10 centimeters (3 to 4 inches) in diameter, and each contains a single seed. The fruit wall, or pericarp, which is impervious to water and is also airtight, is initially green and turns purplish brown and becomes dry and notably hard. Chinese botanists who first investigated similar ancient lotus seeds collected from the same deposits were unable to get them to germinate, even after 20 months of soaking in water. It wasn't until they scarified (filed open to permit water absorption) the seeds, as prescribed in a 1,400-year-old Chinese manuscript, *Ch'i Min Yao Su* (Important Technology for People in Harmony), that they were ultimately successful in germinating the seeds. The hard, airtight fruit walls are the most significant of the structural features that contribute to the exceptional longevity of the seeds.

Box Figure 8.1 A lotus pod with 20 fruits.

The Sacred Lotus was introduced to China following the introduction of Buddhism from India in the 1st century B.C. It is regarded as a symbol of purity and strength, emerging from the mire of lake waters and opening its crimson flowers to the heavens. The earliest dated depiction of Buddha (in A.D. 240) shows him surrounded with a halo and seated cross-legged on a lotus throne. From old Chinese manuscripts, we learn that lotus has been cultivated as a crop plant (most parts are edible) for the past 4,000 years and traditional Chinese herbal medicine considered the Sacred Lotus a mainstay of their pharmaceutical collections. The large number of lotus seeds collected from the site suggests that the plant was under cultivation in this now dried-up lotus lake. It is likely that the 1,200-year-old seed was derived from a plant cultivated by Buddhists at this site in northeastern China.

Unlike Rip Van Winkle, who, following a drink of liquor, slept for only 20 years, these Sacred Lotus seeds are remarkable in their capacity to revive after more than 1,000 years. Lessons learned from these plants can shape our thinking about our own aging and the possibilities that may exist in the future for extending life at the margins.

D.C. Scheirer

Figure 8.31 Young seedlings of red mangrove (*Rhizophora mangle*) whose seeds have no dormant period and germinate while the fruit is still on the tree. The seedlings grow to lengths of up to 25 centimeters (10 inches) before falling and becoming planted in the mud below. Ocean currents and tides also distribute mangrove seeds throughout tropical tidal zones. The dispersal has been so effective that there are 60,000 square kilometers (23,000 square miles) of mangroves in Southeast Asia alone. Mangrove wood is harvested for fuel, and in the past 40 years, the groves in some localities have been reduced in area by more than 50%. Incidentally, the fruit is sweet and edible.

Every 5 years, a bottle of seeds was dug up and the seeds were planted, until the schedule was changed in 1920 to every 10 years. When the first bottle was dug up in 1884, seeds of most of the weeds germinated; in 1960, seeds of evening primrose, curly dock, and moth mullein still germinated; and in 1980, 29 moth mullein seeds, 1 mullein seed, and 1 mallow seed germinated—101 years after they were placed in the bottles. It is of interest to note that a mallow seed previously had not germinated since 1899. Only six of the original bottles now remain; they are not scheduled to be unearthed until the year 2040. Recent evidence indicates that the timing of the digging up of the seed bottles did not take into account the fact that certain temperature patterns are critical to the germination of many weed seeds and that if Beal's experiment had been conducted in a different way, the germination results would have been quite different.

A few species of both dicots and monocots produce seeds that have no period of dormancy at all. In some instances, the embryo, which develops from the zygote, continues to grow without pause in a phenomenon known as *vivipary*. In the red mangrove, a tropical tree associated with coastal waters and estuaries, each fruit contains a single seed in which the embryo continues to grow while the fruit is still on the tree, reaching a length of 25 centimeters (10 inches) or more before the seedling becomes detached and essentially plants itself in the mud below (Fig. 8.31).

Summary

1. Flowers occur in a wide variety of sizes, colors, textures, and habitats. Annuals complete their life cycles in one growing season; biennials take two seasons; and perennials may take several to many years to complete their cycles.
2. Dicots and monocots are distinguished from one another by the differences in numbers of flower parts and cotyledons, venation, presence or absence of cambia, vascular bundle arrangement, and pollen grain apertures.
3. A typical flower consists of a peduncle and a receptacle to which are attached sepals (calyx), petals (corolla), stamens, and pistil. Flowers may have no petals, or the petals may be fused together. Many flowers may be in an inflorescence.
4. Stamens consist of a pollen-bearing anther and a stalk or filament.
5. A pistil consists of a stigma, style, and ovary. The ovary may be superior or inferior and contains one or more ovules.
6. A fruit, which is unique to flowering plants, is a mature ovary. Hormones promote the greatest fruit growth. Parthenocarpic fruits are seedless and develop without fertilization occurring, but not all seedless fruits are parthenocarpic.
7. A mature fruit has an outer exocarp, an inner endocarp around the seed(s), and a mesocarp between the exocarp and endocarp; the three regions may be fused together as a pericarp. At maturity, fruits may be fleshy or dry. A fruit may be derived from the ovary alone or from adjacent flower parts as well.
8. Fleshy fruits may develop from a flower with a single pistil (simple fruit); aggregate fruits develop from a single flower with more than one pistil; multiple fruits are derived from flowers in an inflorescence.
9. Simple fleshy fruits include drupes and berries; berries may be true berries, pepos, or hesperidiums. Pomes have flesh derived from both the receptacle and the ovary.

10. Accessory fruits consist of more than the ovary alone; some aggregate fruits (e.g., strawberries) are largely composed of accessory tissue. The individual fruitlets of a multiple fruit develop together into a single larger fruit.
11. Some dry fruits split as they mature; such fruits include follicles, legumes, siliques or silicles, and capsules.
12. Non-splitting dry fruits include achenes, nuts, grains (caryopses), samaras, and schizocarps.
13. Fruits and seeds may have wings, plumes, and other adaptations for wind dispersal. Some fruits and seeds have adaptations for animal, bird, or water dispersal. Some fruits eject seeds with force, and some have modifications that drill seeds into the ground.
14. Humans disperse many seeds, and most countries and a few states have strict regulations governing the importation of plant materials, primarily to control the spread of pests and diseases.
15. A bean seed has a hilum, a micropyle, a seed coat, two cotyledons, and an embryonic bean plant consisting of a plumule and a radicle.
16. In grains, the plumule and the radicle are protected by a coleoptile and a coleorhiza, respectively.
17. Germination of a seed depends on the cessation of dormancy. Dormancy may be sustained by mechanical circumstances; scarification may break dormancy in such seeds.
18. Dormancy may also be induced by growth or germination inhibitors, or after-ripening may need to occur. Cold temperatures may be necessary for the germination of some seeds; stratification may break the dormancy of such seeds.
19. A seed will not germinate unless environmental factors including water, oxygen, light, and certain temperature ranges are favorable.
20. Seeds remain viable for a few days to more than 100 years. The viability of most seeds is extended by storage at low temperatures under dry conditions, but some weed seeds have their viability extended by storage under humid, cool conditions that include little oxygen.
21. Some plants produce seeds that undergo no dormancy at all. The growth of the embryo while the seed and fruit are still on the plant is termed vivipary.

Review Questions

1. Define calyx, corolla, receptacle, peduncle, pedicel, pistil, filament, ovary, and carpel.
2. Indicate the features by which dicots are distinguished from monocots.
3. What is the difference between a fruit and a vegetable?
4. What causes an ovary to develop into a fruit?
5. What are the various parts of a fruit?
6. How do fleshy fruits differ from dry fruits?
7. Distinguish among simple, aggregate, and multiple fruits.
8. Distinguish among achenes, grains, samaras, and nuts.
9. What adaptations do seeds and fruits have for dispersal by water and animals?
10. Define plumule, radicle, coleoptile, coleorhiza, hypocotyl, after-ripening, stratification, and vivipary.

Discussion Questions

1. Most wind-pollinated flowering plants have inconspicuous, nonfragrant flowers. How might nature be affected if all flowers were that way?
2. Do you believe the botanical distinction between fruits and vegetables is a good one? If not, how would you change it?
3. In discussing pomes, it was observed that the bulk of the flesh in an apple comes from the floral tube. What could you do to prove or disprove this?
4. Seed and fruit dispersal is achieved with the aid of wind, water, animals, mechanical means, and humans. If you were "designing" a new plant, can you think of any new way in which it might be dispersed?
5. When volcanic activity or coral polyps cause new islands to appear in the oceans, they eventually acquire some vegetation. Would you expect the types of dispersal mechanisms for the flowering plants on these islands to be the same as they were for ancient continents?

Learning Online

Visit our web page at www.mhhe.com/botany for interesting case studies, practice quizzes, current articles, and animations within the Online Learning Center to help you understand the material in this chapter. You'll also find active links to these topics:

Characteristics of Plants
Flowers
Flowers and Fruits
Fruits and Seeds
Coevolution
Ornamental Plants
Economic and Ecological Importance of Gymnosperms
Economic and Ecological Importance of Angiosperms
Origins of Agriculture
Temperate Fruits and Nuts
Tropical Fruits and Nuts
Cereal Grains and Forage Grasses

Additional Reading

Barton, L. V. 1992. *Seed preservation and longevity.* New York: State Mutual Book and Periodical Service.

Baskin, C. C., and J. M. Baskin. 1998. *Seeds: Ecology, biogeography, and evolution of dormancy and germination.* San Diego, CA: Academic Press.

Beattie, A. J. 1990. Seed dispersal by ants. *Scientific American* 263(2): 76.

Bell, A. D. 1991. *Plant form: An illustrated guide to flowering plant morphology.* New York: Oxford University Press.

Bold, H. C., et al. 1997. *Morphology of plants and fungi,* 5th ed. Old Tappan, NJ: Addison-Wesley.

Boyston, A. 1999. *Flowers, fruits, and seeds.* Chicago, IL: Heinemann Library.

Duffus, C. M., and J. C. Slaughter. 1980. *Seeds and their uses* (reprint). Ann Arbor, MI: Books Demand.

Endress, P. K. 1996. *Diversity and evolutionary biology of tropical flowers.* New York: Cambridge University Press.

Kigel, J., and G. Galili (Eds.). 1995. *Seed development and germination.* New York: Dekker, Marcel.

Murray, D. (Ed.). 1987. *Seed dispersal.* San Diego, CA: Academic Press.

Nau, J. 1999. *Ball culture guide: The encyclopedia of seed germination,* 3d ed. Batavia, IL: Ball.

Raven, P., et al. 1999. *Biology of plants,* 6th ed. New York: Worth.

Van Rheede-Van Oudtshoorn, K., and M. W. Van Rooyen. 1999. *Dispersal biology of desert plants.* New York: Springer-Verlag.

Floating leaves of a giant water lily (*Victoria amazonica*), which sometimes attain a diameter of 2 meters (6.5 feet). The larger leaves are capable of supporting, without sinking, the weight of a child.

Water in Plants

9

Chapter Outline

Overview

This chapter begins by introducing molecular movement through a comparison between balls in motion in a room and molecular activity. This is followed by a discussion of diffusion, osmosis, turgor, plasmolysis, imbibition, and active transport.

The entry of water into the plant is then explored; this is followed by a discussion of the movement of water through the plant, the evaporation of water into leaf air spaces, and transpiration. A discussion of mineral requirements for growth concludes the chapter.

Some Learning Goals

1. In simple terms, explain diffusion, osmosis, turgor, imbibition, and active transport.
2. Discuss the pressure-flow hypothesis and the cohesion-tension theory.
3. Know the pathway, movement, and utilization of water in plants.
4. Explain how a stomatal apparatus opens and closes the pore.
5. Know and understand mineral requirements for growth.

Nearly everyone has had the experience of driving along a highway or city street when someone in the car says, "What's that smell?" Soon a bakery, or a paper mill, or perhaps a dead skunk comes into view, and the smell gets stronger. Then it fades away as the source is left behind (Fig. 9.1). We take for granted the fact that there is a correlation between our proximity to an odor source and the intensity of the odor, but how and why does the odor reach us?

By way of an answer, imagine two adjacent rooms identical in height, width, and length, with no windows, doors, fixtures, or furniture. Now suppose we lift a small flap in the ceiling of one of them and drop in 100 tennis balls. These are ordinary tennis balls except for one extraordinary feature: they have perpetual-motion motors in them that cause them to travel in any direction at 30 MPH. The tennis balls quickly make the room seem like a battlefield as they whiz around, bounce off the walls, floor, and ceiling, and also collide with one another, each time being deflected at a different angle. Shortly after they are introduced, the tennis balls will probably become randomly distributed throughout the room.

Now what would you expect to happen if we opened up a small hole in the wall between the two rooms? Eventually, a tennis ball should bounce or travel at just the right angle to go through the hole into the other room. Theoretically, it could then bounce straight back into the first room, but it seems unlikely that it would. Reason tells us that long before it might happen to strike the exact angle it needed to return, several more balls will come in from the first room. Given enough time, some balls might indeed bounce back into the

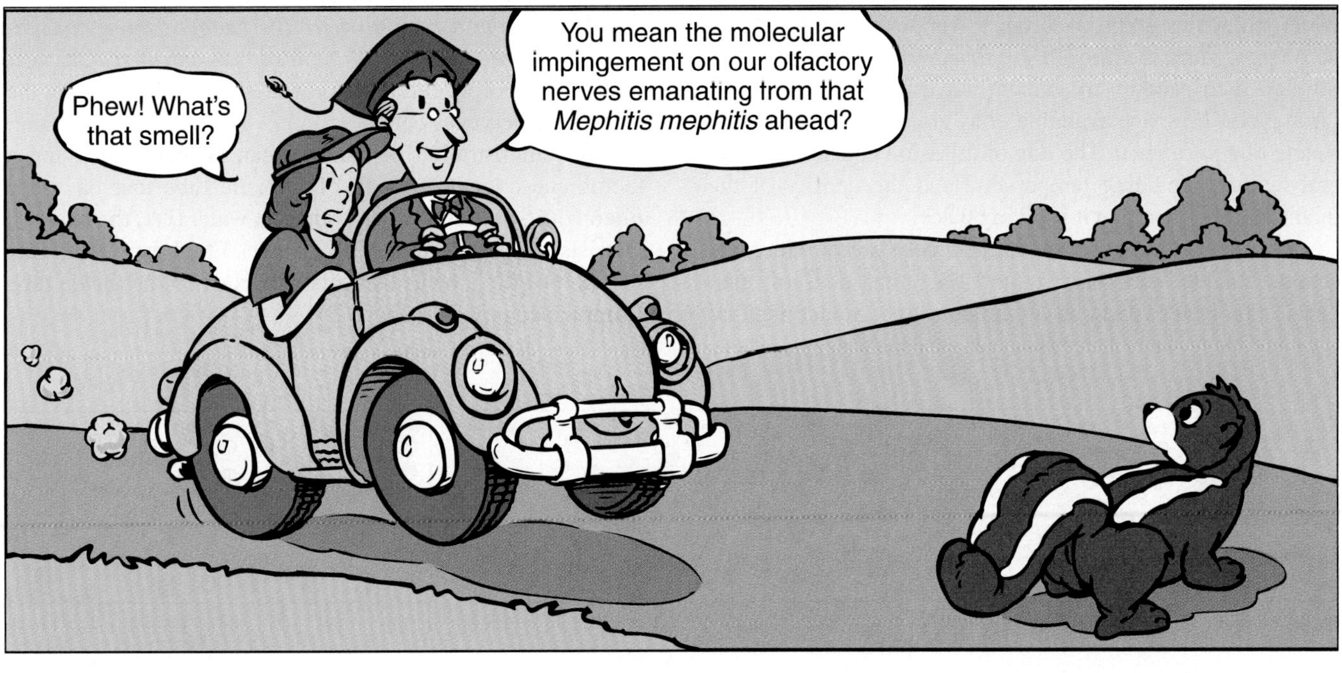

Phew! What's that smell?
You mean the molecular impingement on our olfactory nerves emanating from that *Mephitis mephitis* ahead?

Figure 9.1

first room, but in the long run, each of the two rooms would end up with roughly 50 tennis balls, with an occasional ball going between the rooms in either direction.

The situation just described is somewhat analogous to what takes place in nature on a molecular level.

MOLECULAR MOVEMENT

Molecules and ions (discussed in Chapter 2) are constantly in random motion. Visual evidence of this can be seen with an ordinary light microscope. If a drop of India ink is diluted with water and observed through a microscope under high power, the tiny carbon particles of the ink appear to be in constant motion. This motion, known as *Brownian movement,* is the result of the bombardment of the visible particles by invisible water molecules, which are in constant motion themselves.

Diffusion

The differing intensity of smells discussed earlier involves molecules behaving somewhat like the tennis balls. Through their random motion, molecules tend to become distributed throughout the space available to them. Accordingly, if perfume molecules are kept in a bottle, they will become distributed throughout the bottle, but if the stopper is removed, they will eventually become dispersed throughout the room, even if there is no fan or other device to move the air.

This movement of molecules or ions from a region of higher concentration to a region of lower concentration is called **diffusion** (Fig. 9.2). Molecules that are moving from a region of higher concentration to a region of lower concentration are said to be moving *along a diffusion gradient,* while molecules going in the opposite direction are said to be going *against a diffusion gradient.* When the molecules, through their random movement, have become distributed throughout the space available, they are considered to be in a state of *equilibrium.* The rate of diffusion depends on several factors, including temperature and the density of the medium through which it is taking place.

Except within the area immediately surrounding the source, unaided diffusion requires a great deal of time because molecules and ions are infinitesimal. Something that is less than a millionth of a millimeter in diameter is going to take a long time to move just 1 millimeter, even though the amount of movement may be great in proportion to the size of the particle concerned. In gases, there is a great deal of space between the molecules and correspondingly less chance of the molecules bumping into each other and thus being slowed down. Accordingly, gas molecules occupy a space that becomes available to them relatively rapidly, while liquids do so more slowly, and solids are slower yet.

Large molecules move much more slowly than small molecules. If you added a tiny drop of a dye (which has relatively large molecules) to one end of a bathtub of water without disturbing the water in any way, it would take years for the dye molecules to diffuse throughout the tub and reach a state of equilibrium. In nature, however, wind and water currents distribute molecules much faster than they ever could be distributed by diffusion alone.

Osmosis

Despite the fact that the cytoplasm of living cells is bounded by membranes, it is now well known that water (a **solvent**) moves freely from cell to cell. This has led scientists to believe that plasma, vacuolar, and other membranes have tiny holes or spaces in them, even though such holes or spaces are invisible to the instruments presently available. It also has led to the construction of models of such membranes (see Fig. 3.11). Membranes through which different substances diffuse at different rates are described as **differentially permeable,** or **semipermeable.** All plant cell membranes appear to be differentially permeable.

In plant cells, **osmosis** is essentially the *diffusion of water through a differentially permeable (semipermeable) membrane from a region where the water is more concentrated to a region where it is less concentrated.* Osmosis ceases if the concentration of water on both sides of the membrane becomes equal.

A demonstration of osmosis can be made by tying a membrane over the mouth of a thistle tube that has been filled with a solution of 10% sugar in water (i.e., the solution consists of 10% sugar and 90% water). Fluid rises in the narrow part of the tube as osmosis occurs when the thistle tube is immersed in water (Fig. 9.3).

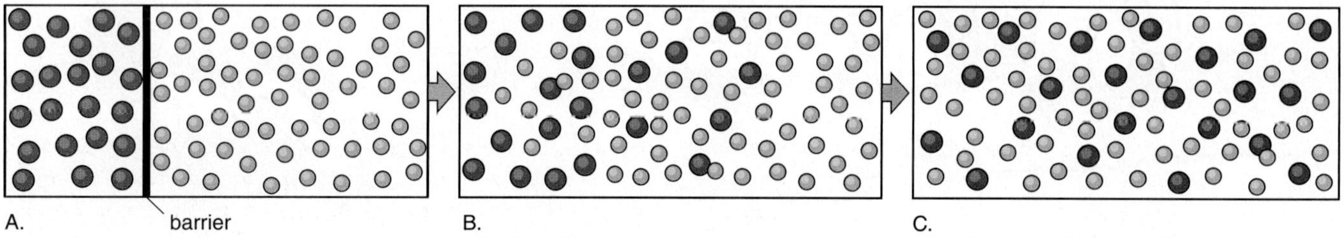

Figure 9.2 Simple diffusion. *A.* A barrier separates two kinds of molecules. *B.* When the barrier is removed, random movement of individual molecules results in both kinds moving from a region of higher concentration to a region of lower concentration. *C.* Eventually, equilibrium (even distribution) is reached. The diffusion gradually slows down as equilibrium is approached.

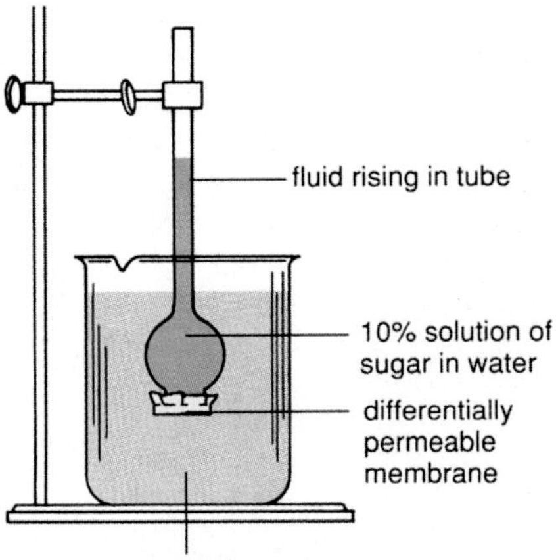

Figure 9.3 A simple osmometer, made by tying a membrane over the mouth of a thistle tube.

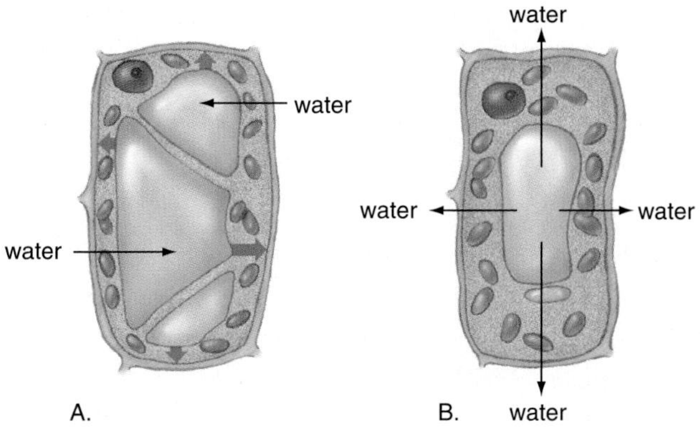

Figure 9.4 *A.* A turgid cell. Water has entered the cell by osmosis, and turgor pressure is pushing the cell contents against the cell walls. *B.* Water has left the cell, and turgor pressure has dropped, leaving the cell flaccid.

Although the previous simple definition of osmosis serves our purposes, plant physiologists prefer to define and discuss osmosis more precisely in terms of *potentials.* It is possible to prevent osmosis by applying pressure. Just enough pressure to prevent fluid from moving as a result of osmosis is referred to as the *osmotic potential* of the solution. In other words, osmotic potential is the pressure required to prevent osmosis.

Water enters a cell by osmosis until the osmotic potential is balanced by the resistance to expansion of the cell wall. Water gained by osmosis may keep a cell firm, or **turgid,** and the **turgor pressure** that develops against the walls as a result of water entering the vacuole of the cell is called **pressure potential.**

The release of turgor pressure can be heard each time you bite into a crisp celery stick or the leaf of a young head of lettuce. When we soak carrot sticks, celery, or lettuce in pure water to make them crisp, we are merely assisting the plant in bringing about an increase in the turgor of the cells (Fig. 9.4).

The *water potential* of a plant cell is essentially its osmotic potential and pressure potential combined. If we have two adjacent cells of different water potentials, water will move from the cell having the higher water potential to the cell having the lower water potential.

Osmosis is the primary means by which water enters plants from their surrounding environment. In land plants, water from the soil enters the cell walls and intercellular spaces of the epidermis and the root hairs and travels along the walls until it reaches the endodermis. Here it crosses the differentially permeable membranes and cytoplasm of the endodermal cells on its way to the xylem. Water flows from the xylem to the leaves, evaporates within the leaf air spaces, and diffuses out (*transpires*) through the stomata into the atmosphere. The movement of water takes place because there is a water potential gradient from relatively high soil water potential to successively lower water potentials in roots, stems, leaves, and the atmosphere.

Plasmolysis

If you place turgid carrot and celery sticks in a 10% solution of salt in water, they soon lose their rigidity and become limp enough to curl around your finger. The water potential inside the carrot cells is greater than the water potential outside, and so diffusion of water out of the cells into the salt solution takes place. If you were to examine such cells with a microscope, you would see that the vacuoles, which are largely water, had disappeared and that the cytoplasm was clumped in the middle of the cell, having shrunken away from the walls. Such cells are said to be *plasmolyzed.* This loss of water through osmosis, which is accompanied by the shrinkage of protoplasm away from the cell wall, is called **plasmolysis** (Fig. 9.5). If plasmolyzed cells are placed in fresh water before permanent damage is done, water reenters the cell by osmosis, and the cells become turgid once more.

Imbibition

Osmosis is not the only force involved in the absorption of water by plants. *Colloidal* materials (i.e., materials that contain a permanent suspension of fine particles) and large molecules, such as cellulose and starch, usually develop electrical charges when they are wet. The charged colloids and molecules attract water molecules, which adhere to the internal surfaces of the materials. Because water molecules are polar, they can become both highly adhesive to large organic molecules such as cellulose and cohesive with one another. As discussed in Chapter 2, polar molecules have slightly different electrical charges at each end due to their asymmetry. This process, known as **imbibition,** results in the swelling of tissues, whether they are alive or dead, often to several times their original volume. Imbibition is the initial step in the germination of seeds (Fig. 9.6).

The physical forces developed during germination can be tremendous, even to the point of causing a seed to split a rock weighing several tons (Fig. 9.7). It has been found, for

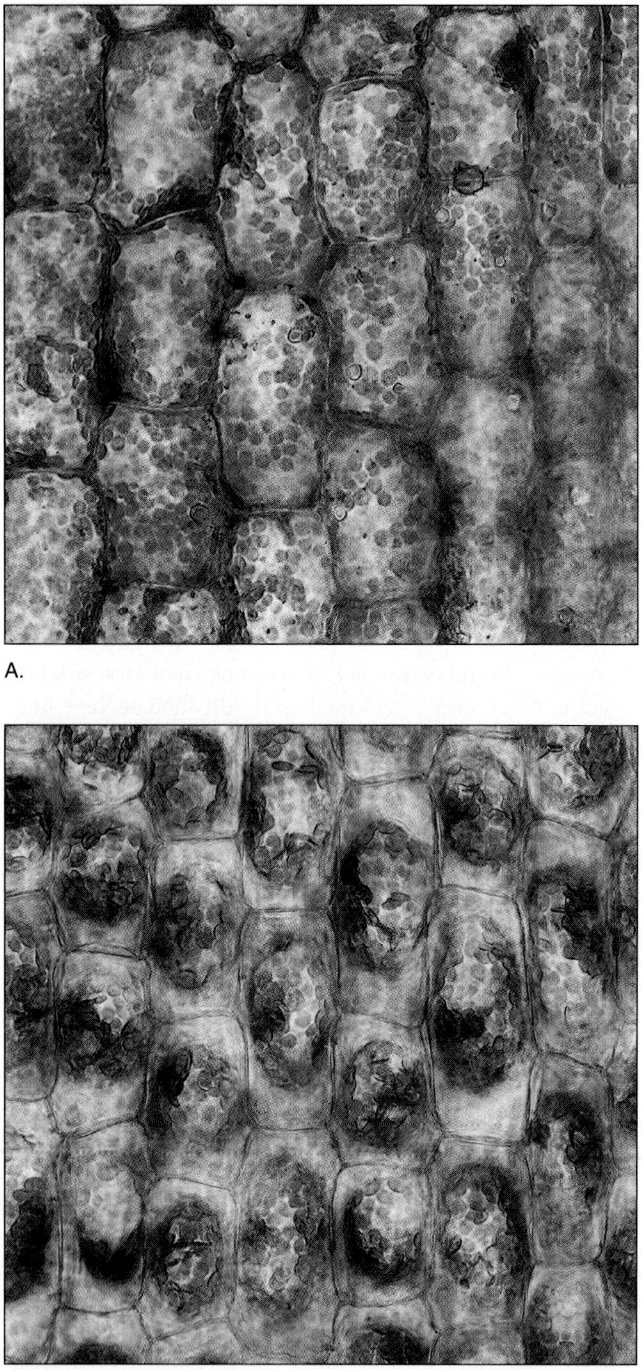

A.

B.

Figure 9.5 A portion of a leaf of the water weed *Elodea*. *A*. Normal cells. *B*. Plasmolyzed cells.

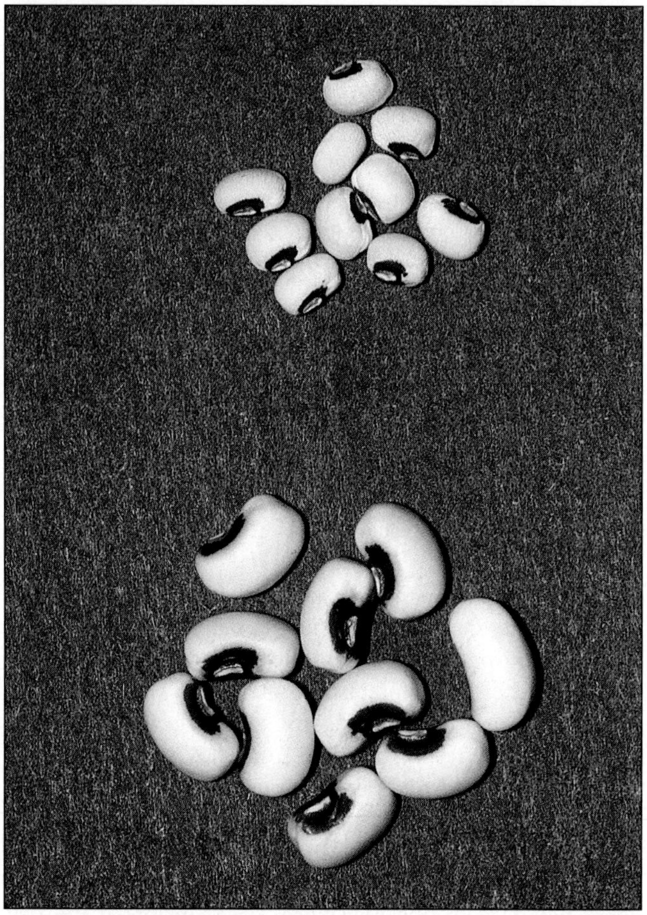

Figure 9.6 Black-eyed pea seeds before and after imbibition of water.

example, that a pressure of 42.2 kilograms per square centimeter (600 pounds per square inch) is needed to break the seed coat of a fresh walnut from within and that water being imbibed by a cocklebur seed develops a force of up to 1,000 times that of normal atmospheric pressure. Yet when water and oxygen reach walnut and cocklebur seeds, they germinate readily, as do seeds that fall into the crevices of rocks or have boulders roll over on them.

The huge stone blocks used in the construction of the pyramids of Egypt are believed to have been quarried by hammering rounded wooden stakes into holes made in the face of the stone and then soaking the stakes with water. As the stakes swelled, the force created by imbibition split the rocks.

Active Transport

Return for a moment to our two rooms with the tennis balls. Suppose that, besides the 100 tennis balls, we drop in 50 slightly underinflated basketballs; these basketballs are also extraordinary in having perpetual-motion motors that propel them in any direction at 12 MPH. They should also become randomly distributed throughout the room shortly after they are introduced. Assume, however, that the hole in the wall (which is large enough for the passage of a tennis ball) is not quite large enough to allow a basketball to pass through freely. The basketballs will then remain in the first room. However, if we were to install a mechanical arm next to the hole in the second room, and if this arm could grab basketballs that come near the hole and squeeze them through in one

Figure 9.7 A live oak that grew from an acorn lodged in a small crack in the rock. When it rained, the acorn imbibed water, and the force of the swelling split the rock. A root is now slowly widening the split.

direction, basketballs would be transported into the second room *through the expenditure of energy.* The basketballs obviously would gradually accumulate in the second room in greater numbers.

Plants expend energy, too. Plant cells generally have a larger number of mineral molecules and ions than exist in the soil immediately next to the root hairs. If it were not for the barriers imposed by the differentially permeable membranes, these molecules and ions would move from a region of higher concentration in the cells to a region of lower concentration in the soil.

Figure 9.8 A mangrove tree. Mangroves flourish in tropical tidal zones where the salt content of the water is high enough to plasmolyze the cells of most plants. The mangroves still obtain water via osmosis, which takes place because the mangrove cells accumulate an unusually high concentration of organic solutes; some are also able to excrete excess salt.

Most molecules needed by cells are polar, and those of solutes may set up an electrical gradient across a differentially permeable membrane of a living cell. To pass through the membrane, molecules require special embedded transport proteins (see Fig. 3.11). The transport proteins are believed to occur in two forms, one facilitating the transport of specific ions to the outside of the cell and the other facilitating the transport of specific ions into the cell.

The plants absorb and retain these solutes against a diffusion (or electrical) gradient through the expenditure of energy. This process is called **active transport.** The precise mechanism of active transport is not fully understood. It apparently involves an enzyme and what has been referred to as a *proton pump*. The pump involves the plasma membrane of plant and fungal cells and sodium and potassium ions in animal cells. Both pumps are energized by special energy-storing ATP molecules (discussed in Chapter 10).

Mangroves, saltbush, and certain algae thrive in areas where the water or soil contains enough salt to kill most vegetation. Such plants accumulate large amounts of organic solutes, including the carbohydrate *mannitol* and the amino acid *proline.* The organic solutes facilitate osmosis, despite the otherwise adverse environment (Fig. 9.8). The leaves of some mangroves also have salt glands through which they excrete excess salt.

WATER AND ITS MOVEMENT THROUGH THE PLANT

If you were to cover the soil at the base of a plant with foil, place the pot where it receives light, and then put the potted plant under a glass bell jar, you would notice moisture

Figure 9.9 A potted plant sealed under a bell jar. The surface of the soil has been covered with foil. Note the accumulation of moisture on the inside of the glass. The moisture came through the plant by transpiration.

accumulating on the inside of the jar within an hour or two. Because of the foil barrier, the water could not have come directly from the soil; it had to have come through the plant. More than 90% of the water entering a plant passes through and evaporates—primarily into leaf air spaces and then through the stomata into the atmosphere (see Fig. 9.10)—with usually less than 5% of the water escaping through the cuticle. This process of water vapor loss from the internal leaf atmosphere is called *transpiration* (Fig. 9.9).

The amount of water transpired by plants is greater than one might suspect. For example, mature corn plants each transpire about 15 liters (4 gallons) of water per week, while four-tenths of a hectare (1 acre) of corn may transpire more than 1,325,000 liters (350,000 gallons) in a 100-day growing season. A hardwood tree uses about 450 liters (120 gallons) of water while producing 0.45 kilogram (1 pound) of wood (or 1,800 liters while producing 0.45 kilogram of dry weight substance), and the 200,000 leaves of an average-sized birch tree will transpire from 750 to more than 3,785 liters (200 to 1,000 gallons) per day during the growing season. Humans recycle much of their water via the circulatory system, but if they were to have requirements similar to those of plants, each adult would have to drink well over 38 liters (10 gallons) of water per day.

Why do living plants require so much water? Water constitutes about 90% of the weight of young cells. The thousands of enzyme actions and other chemical activities of cells take place in water, and additional, although relatively negligible, amounts are used in the process of photosynthesis. The exposed surfaces of the mesophyll cells within the leaf have to be moist at all times, for it is through this film of water that the carbon dioxide molecules needed for the process of photosynthesis enter the cell from the air. Water is also needed for cell turgor, which gives rigidity to herbaceous plants.

Consider also what it must be like in the mesophyll of a flattened leaf that is fully exposed to the midsummer sun in areas where the air temperature soars to well over 38°C (100°F) in the shade. If it were not for the evaporation of water molecules from the moist surfaces, which brings about some cooling, and reradiation of energy by the leaf, the intense heat could damage the plant. Sometimes, the transpiration is so rapid that more water is lost than is taken in. The stomata may then close, preventing wilting. The relation and role of abscisic acid in excessive water loss is discussed in Chapter 11.

How does water travel through the roots from 3 to 6 meters (10 to 20 feet) or more beneath the surface and then up the trunk to the topmost leaves of a tree that is more than 90 meters (300 feet) tall? We know that interconnected tubes of xylem extend throughout the plant, from the young roots up through the stem and branches to the tiny veinlets of the leaves. We also know that the water, following a water potential gradient, gets to the start of this "plumbing system" by osmosis. Water is then raised through the columns apparently by a combination of factors, and the process has been the subject of much debate for the past 200 years (Fig. 9.10).

One of the earliest explanations for the rise of water in a living plant was given in 1682 by the English scientist Nehemiah Grew. He suggested that cells surrounding the xylem vessels and tracheids performed a pumping action that propelled the water along. This was questioned, however, when it was found that water will also rise in lengths of dead stems. Then, after Marcello Malpighi suggested it, the belief that capillary action moved the water became popular.

It is well known that the height water will rise in a narrow tube is inversely proportional to the diameter of the tube. It is also known that this rise occurs through the forces involved in the forming of a concave *meniscus* (curved surface) at the top of the water column (Fig. 9.11). Even though water can, indeed, rise 1 meter (3 feet) or more in a very narrow tube, air must be present above the column for the forces to work, which is not the case in a plant. In fact, any air introduced into a water column in xylem interferes with the rise of water. Also, while capillarity might produce enough force to raise water a meter or two, the diameter of the tubes is not small enough to raise it more than that.

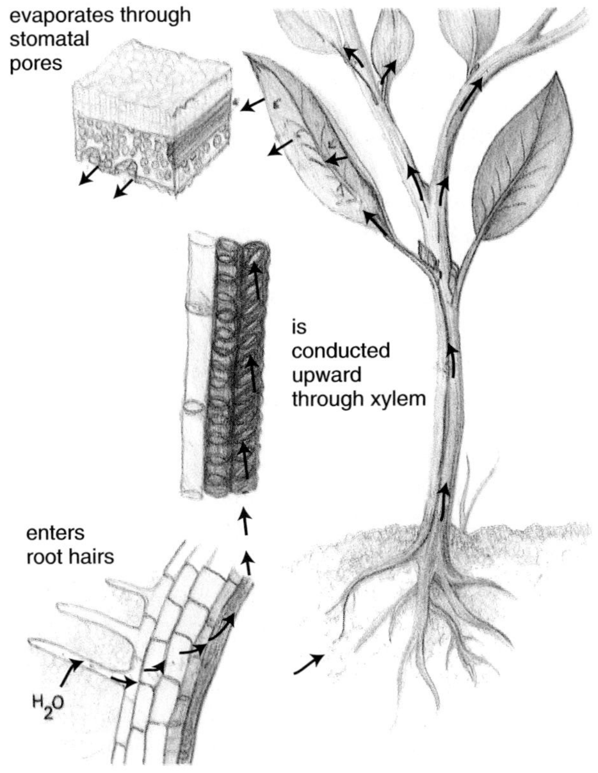

Figure 9.10 Pathway of water through a plant.

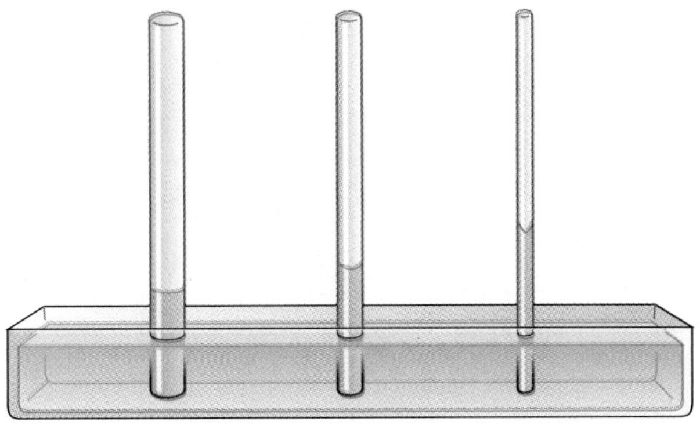

Figure 9.11 Capillarity in narrow tubes. The smaller the diameter of the tube, the greater the rise of the fluid.

The pioneer plant physiologist Stephen Hales discovered and measured *root pressure* as one means by which water moves through plants. When some plants are pruned after growth has begun in the spring, water will exude from the cut ends. This is the result of root pressure, which is discussed on page 160. Some plants do not "bleed" when they are pruned, and the force exerted by root pressure has been shown generally to be less than 30 grams per square centimeter (a few pounds per square inch). This is considerably less than what is needed to raise water to the tops of tall trees. Furthermore, root pressure seems to drop to negligible amounts in the summer, when the greatest amounts of water are moving through the plant.

The Cohesion-Tension Theory

Stephen Hales also identified a pulling force due to evaporation of water from leaves and stems. This has led to the cohesion-tension theory, the most satisfactory explanation for the rise of water in plants thus far suggested. Water molecules are electrically neutral, but they are asymmetrical in shape (see Fig. 2.4). This results in the molecules having very slight positive charges at one end and very slight negative charges at the other end. Such molecules are said to be *polar.* When the negatively charged end of one water molecule comes close to the positively charged end of another water molecule, weak hydrogen bonds hold the molecules together.

We know that water molecules adhere to capillary walls (e.g., those of xylem tracheids and vessels) and cohere to each other, creating a certain amount of tension. It is possible, for example, to fill a small glass with water, place a thin, smooth sheet of cardboard over the mouth, and invert the glass without the water spilling. This is because the adhesion of the water molecules to the cardboard and the cohesiveness of the water molecules to one another hold the cardboard against the rim of the glass.

When water evaporates from the mesophyll cells in a leaf and diffuses out of the stomata (*transpires*), the cells involved develop a lower water potential than the adjacent cells. Because the adjacent cells then have a correspondingly higher water potential, replacement water moves into the first cells by osmosis. This continues across rows of mesophyll cells until a small vein is reached. Each small vein is connected to a larger vein, and the larger veins are connected to the main xylem in the stem, and that, in turn, is connected to the xylem in the roots that receive water, via osmosis, from the soil. As transpiration takes place, it creates a "pull," or tension, on water columns, drawing water from one molecule to another all the way through an entire span of xylem cells. The cohesion required to move water to the top of a tall tree is considerable, but the cohesive strength of the water columns is usually more than adequate. Any breaking of the tension through the introduction of a gas bubble results in a temporary or permanent blocking of water transport. This seldom is a problem, however, because small bubbles may be redissolved and larger gas bubbles rarely block more than a few of the numerous capillary tubes of xylem at any time the tissue is functioning.

Water molecules move partly through cell cytoplasm and partly through spaces between cells; they also move between cellulose fibers in the walls and through spaces in the centers of dead cells. Most water and solutes can travel across the epidermis and cortex via the cell walls until they reach the endodermis. There, the water and solutes are forced by *Casparian strips* to cross the cytoplasm of the endodermal cells on their way to the vessels or tracheids of the xylem (Fig. 9.12; see also Chapter 5).

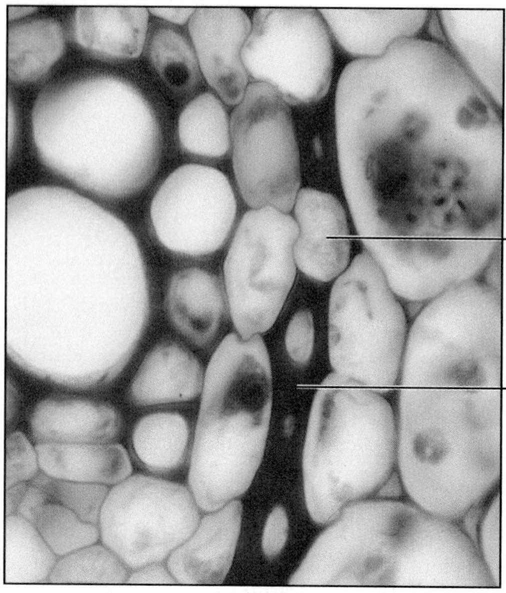

Figure 9.12 Part of the center of a buttercup root, showing endodermal cells with Casparian strips, ×600.

If rapid transpiration is occurring, the roots are likely to grow rapidly toward available water. In corn plants, for example, the main roots may grow at a rate of more than 6 centimeters (2.3 inches) a day. Solutes, as well as water, may move so rapidly during periods of rapid transpiration that there is little osmosis taking place across the endodermis. Scientists believe that at such times water may be pulled through the roots by bulk flow, which is the passive movement of a liquid from higher to lower water potential.

In summary: "columns" of water molecules move through the plant from roots to leaves, and the abundant water of a normally moist soil supplies these "columns" as the water continues to enter the root by osmosis (see Fig. 9.10); simply put, the difference between the water potentials (water "concentrations") of two areas (e.g., soil and the air around stomata) generates the force to transport water in a plant.

REGULATION OF TRANSPIRATION

The stomatal apparatuses, which often occupy 1% or more of the surface area of a leaf, *regulate transpiration and gas exchange.* Control of transpiration is, however, strongly influenced by the water-vapor concentration of the atmosphere. The guard cells bordering each stoma have relatively elastic walls with radially oriented microfibrils, making them analogous to pairs of sausage-shaped balloons joined at each end, each with a row of rubber bands around it. The part of the wall adjacent to the hole itself is considerably thicker than the remainder of the wall (Fig. 9.13). This thickness allows each stoma to be opened and closed by means of changes in the turgor of the guard cells. The stoma is closed when turgor pressure is low and open when turgor pressure is high. Changes in turgor pressures in the guard cells, which contain chloroplasts, take place when they are exposed to changes in light intensity, carbon dioxide concentration, or water concentration.

Changes in turgor pressure take place when osmosis and active transport between the guard cells and other epidermal cells bring about shifts in solute concentrations. While photosynthesis is occurring in the guard cells, they expend energy to acquire potassium ions from adjacent epidermal cells, leading to the opening of the stomata. When photosynthesis is not occurring in the guard cells, the potassium ions leave, and the stomata close. With an increase in potassium ions, the water potential in the guard cells is lowered, and the osmosis that takes place as a result brings in water that makes the cells turgid. The departure of potassium ions also results in water leaving, making the cells less turgid and causing the stomata to close (see Fig. 9.13).

Stomata will close passively whenever water stress occurs, but there is evidence that the hormone abscisic acid is produced in leaves subject to water stress and that this hormone causes membrane leakages, which induce a loss of potassium ions from the guard cells and cause them to deflate.

The stomata of most plants are open during the day and closed at night. However, the stomata of a number of desert plants are open only at night when there is less water stress on the plants. This conserves water but makes carbon dioxide needed for photosynthesis inaccessible during the day. Such plants convert the carbon dioxide available at night to organic acids, which are stored in cell vacuoles. The organic acids are then converted back to carbon dioxide during the day when photosynthesis occurs (Fig. 9.14). A specialized form of photosynthesis called *CAM photosynthesis* uses the carbon dioxide released from the organic acids. CAM photosynthesis is discussed in Chapter 10.

Other desert plants have their stomatal apparatuses recessed below the surface of the leaf or stem in small chambers. These chambers often are partially filled with epidermal hairs, which further reduce water loss. Similar recessed stomatal apparatuses are found in the leaves of pine trees, which have little water available to them in winter when the soil is frozen (see Fig. 7.12). A few tropical plants that occur in damp, humid areas (e.g., ruellias; see also Fig. 4.13B) have stomata that are raised above the surface of the leaf, while plants of wet habitats generally lack stomatal apparatuses on submerged surfaces.

Although light and carbon dioxide concentration affect transpiration rates, several other factors play at least an indirect role. For example, air currents speed up transpiration as they sweep away water molecules emerging from stomata.

Humidity plays an inverse but direct role in transpiration rates: high humidity reduces transpiration, and low humidity accelerates it. Temperature also plays a role in the movement of water molecules out of a leaf. The transpiration rate of a

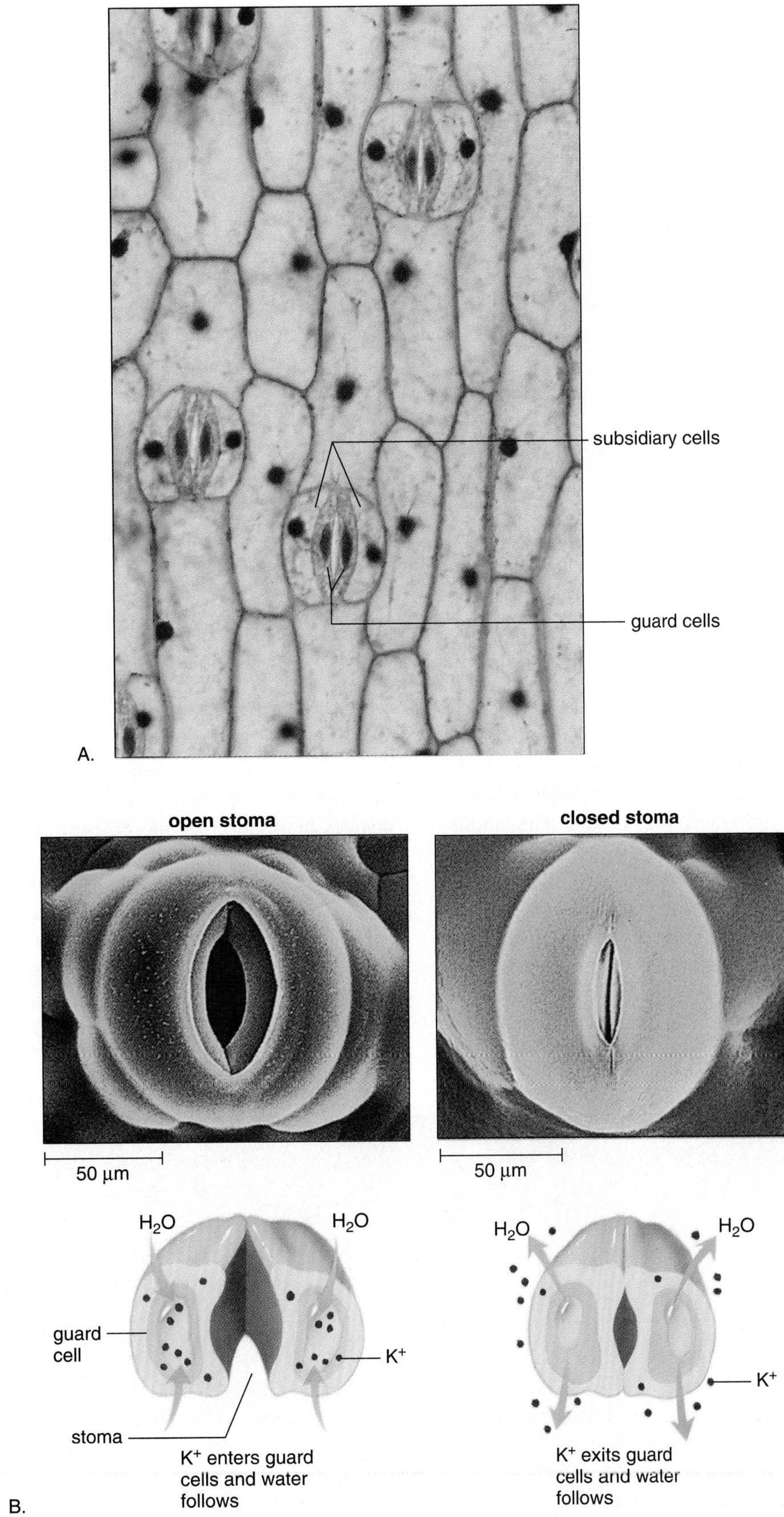

Figure 9.13 A. A small portion of the leaf epidermis of Wandering Jew (*Zebrina* sp.) with several stomata interspersed among ordinary epidermal cells. Each stoma is bordered by a pair of guard cells, which, in turn, each have a smaller epidermal cell called a *subsidiary cell* adjacent to it. *B. Left.* An open stoma. The guard cells swell when turgor pressure in them increases and the stoma opens as the thinner outer walls stretch more than the thicker inner walls. *Right.* The stoma closes when the turgor pressure in the guard cells decreases. The reasons for the changes in turgor pressure are discussed in the text. *9.13A: © BioPhot. 9.13B: © Jeremy Burgess/SPL/Photo Researchers, Inc. From Sylvia S. Mader, Inquiry into Life, 9th edition. © 2000 The McGraw-Hill Companies. All rights reserved.*

Figure 9.14 A barrel cactus (*Ferrocactus*). The stems store, in the form of organic acids, carbon dioxide taken in at night; the carbon dioxide is then released for use in photosynthesis during the day.

leaf at 30°C (86°F), for example, is about twice as great as it is for the same leaf at 20°C (68°F). The various adaptive modifications of leaves and their surfaces and the availability of water to the roots also may play important roles in influencing the amount of water transpired. Leaf modifications are discussed in Chapter 7.

If a cool night follows a warm, humid day, water droplets may be produced through structures called **hydathodes** at the tips of veins of the leaves of some plants. This loss of water in liquid form is called **guttation** (Fig. 9.15). Minerals absorbed at night are pumped into the intercellular spaces surrounding the vessels and tracheids of the xylem. As a result, the water potential of the xylem elements is lowered, and water moves into them from the surrounding cells. In the absence of transpiration at night, the pressure in the xylem elements builds to the point of forcing liquid water out of the hydathodes in the leaves. Although the droplets resemble dew, the two should not be confused. Dew is water that is condensed from the air, while guttation water is literally forced out of the plant by root pressure. As the sun strikes the droplets in the morning, they dry up, leaving a residue of salts and organic substances, one of which is used in the manufacture of commercial flavor enhancers (e.g., the monosodium glutamate in products such as Accent). In the tropics, the amount of water produced by guttation can be considerable. In taro plants, used by the Polynesians to make poi, a single leaf may overnight produce as much as a cupful (about 240 milliliters) of water through guttation.

TRANSPORT OF FOOD SUBSTANCES (ORGANIC SOLUTES) IN SOLUTION

One of thc most important functions of water in the plant involves the *translocation* (transportation) of food substances in solution by the phloem, a process that has only recently come to be better understood. Many of the studies that led to our present knowledge of the subject used aphids (small, sucking insects) and organic compounds designed as radioactive tracers.

Most aphids feed on phloem by inserting their tiny tubelike mouthparts (*stylets*) through the leaf or stem tissues until a sieve tube is reached and punctured. The turgor

Figure 9.15 Droplets of guttation water at the tips of leaves of young barley plants.

Figure 9.16 An aphid feeding on a young stem of basswood (*Tilia*). A droplet of "honeydew" is emerging from the rear of the aphid, ×10. *From Martin H. Zimmerman, "Movements of Organic Substances in Trees" Science 133: 73–79, 1961, American Association for the Advancement of Science.*

pressure of the sieve tube then forces the fluid present in the tube through the aphid's digestive tract, and it emerges at the rear as a droplet of "honeydew." In some studies, research workers anesthetized feeding aphids and cut their stylets so that much of the tiny tube remained where it had been inserted. Fluid exuded (sometimes for many hours) from the cut stylets and was then collected and analyzed (Fig. 9.16).

Carbon dioxide, a basic raw material of photosynthesis, can be synthesized with radioactive carbon. By exposing a photosynthesizing leaf to radioactive carbon dioxide, the pathway of manufactured food substances can be traced. The radioactive substances produce on photographic film an image corresponding to the food pathway. Data obtained from such studies reveal that food substances in solution are confined entirely to the sieve tubes while they are being transported. At one time, it was believed that ordinary diffusion and cyclosis (discussed in Chapter 3) were responsible for the movement of the substances from one sieve tube member to the next, but it is now known that the substances move through the phloem at approximately 100 centimeters (almost 40 inches) per hour—far too rapid a movement to be accounted for by diffusion and cyclosis alone.

The Pressure-Flow Hypothesis

At present, the most widely accepted theory for movement of substances in the phloem is called the **pressure-flow** (or **mass flow**) **hypothesis.** According to this theory, food substances in solution (organic solutes) flow from a *source,* where water enters by osmosis (e.g., a food-storage tissue, such as the cortex of a root or rhizome, or a food-producing tissue, such as the mesophyll tissue of a leaf). The water leaves at a *sink,* which is a place where food is utilized, such as the growing tip of a stem or root. Food substances in solution (organic solutes) are moved along concentration gradients between sources and sinks (Fig. 9.17).

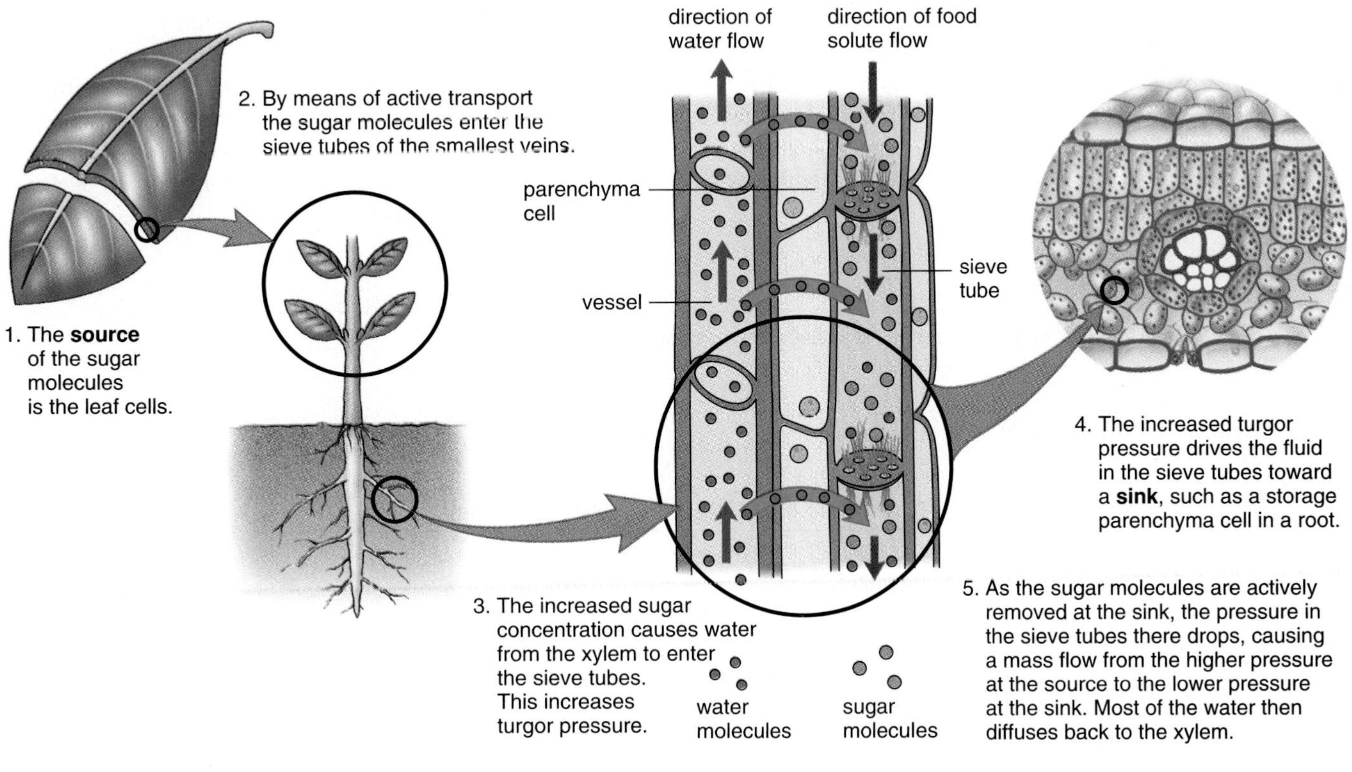

direction of water flow
direction of food solute flow
2. By means of active transport the sugar molecules enter the sieve tubes of the smallest veins.
parenchyma cell
vessel
sieve tube
1. The **source** of the sugar molecules is the leaf cells.
4. The increased turgor pressure drives the fluid in the sieve tubes toward a **sink**, such as a storage parenchyma cell in a root.
5. As the sugar molecules are actively removed at the sink, the pressure in the sieve tubes there drops, causing a mass flow from the higher pressure at the source to the lower pressure at the sink. Most of the water then diffuses back to the xylem.
3. The increased sugar concentration causes water from the xylem to enter the sieve tubes. This increases turgor pressure.
water molecules
sugar molecules

Figure 9.17 The pressure-flow hypothesis.

First, in a process called *phloem-loading,* sugar, by means of active transport, enters the sieve tubes of the smallest veinlets. This decreases the water potential in the sieve tubes, and water then enters these phloem cells by osmosis. Turgor pressure, which develops as this osmosis occurs, is responsible for driving the fluid through the sieve-tube network toward the sinks.

As the food substances (largely sucrose) in solution are actively removed at the sink, water also leaves the sink ends of sieve tubes, and the pressure in these sieve tubes is lowered, causing a mass flow from the higher pressure at the source to the lower pressure at the sink. Most of the water diffuses back to the xylem, where it then returns to the source and is transpired or recirculated. The pressure-flow hypothesis explains how nontoxic dyes applied to leaves or substances entering the sieve tubes, such as viruses introduced by aphids, are carried through the phloem.

MINERAL REQUIREMENTS FOR GROWTH

Growth phenomena are not entirely controlled by internal means. Light, temperature, soil structure, minerals, and other external factors all play a role. In fact, growth depends on a complex, interrelated combination of chemical and physical forces, both internal and external, which are in delicate balance with one another.

Plants may take up many elements from the soil, but besides the carbon and oxygen obtained from carbon dioxide and the hydrogen obtained from water, only 15 elements are essential to most plants as building blocks for the many compounds they synthesize. Sodium, a comparatively abundant element, is apparently required by few plants. The **essential elements** can be remembered by a sentence (Fig. 9.18) that includes the symbols of the elements involved.

Macronutrients and Micronutrients

The mineral elements are usually put into two categories: (1) *macronutrients,* which are used by plants in greater amounts and constitute from 0.5% to 3.0% of the dry weight of the plant; and (2) *micronutrients,* which are needed by the plant in very small amounts, often constituting only a few parts per million of the dry weight.

The macronutrients are nitrogen, potassium, calcium, phosphorus, magnesium, and sulfur, with the first four usually making up about 99% of the nutrient total. Those elements remaining, the micronutrients, are present in amounts ranging from bare traces—as in the case of sodium and cobalt, neither of which may actually be essential for some plants—to 1,500 parts per million of iron and manganese and up to 10,000 parts per million of chlorine.

In addition to these widely required elements, specific organisms may require others. For example, certain algae apparently require the elements vanadium, silicon, or iodine, while some ferns utilize aluminum. Several loco weeds absorb and accumulate selenium in amounts constituting up to 5 micrograms per gram of dry weight. Selenium, which is often fatally poisonous to livestock, appears to enhance the growth of these plants by reducing toxic effects of phosphates, but there is no direct evidence it is essential to them.

When any of these elements is deficient in the soil, the plants exhibit characteristic symptoms of the deficiency, which disappear after the problem has been corrected (Fig. 9.19). Table 9.1 shows some of the uses of essential elements in plants and describes the symptoms of deficiency for each element.

See the discussion of fertilizers and fertilizing in Appendix 4 for further information on ratios of nitrogen, phosphorus, and potassium (NPK) needed for plants and organic versus inorganic fertilizers. See also page 304 for a discussion of compost and composting.

C. H O P K'N S CaFe; Mighty good (but) Not always Clean. CuM'n CoZ'n MoB(y)?

carbon	hydrogen	oxygen	phosphorus	potassium	nitrogen	sulfur	calcium	iron	magnesium	sodium	chlorine	copper	manganese	cobalt	zinc	molybdenum	boron
C	H	O	P	K	N	S	Ca	Fe	Mg	Na	Cl	Cu	Mn	Co	Zn	Mo	B

Figure 9.18 Elements essential as building blocks for compounds synthesized by plants.

TABLE 9.1
Uses of Essential Elements in Plants

ELEMENT	SOME FUNCTIONS	DEFICIENCY SYMPTOMS
Nitrogen	Part of proteins, nucleic acids, chlorophyll	Relatively uniform loss of color in leaves, occurring first on the oldest ones
Potassium	Activates enzymes; concentrates in meristems	Yellowing of leaves, beginning at the margins and continuing toward center; lower leaves mottled and often brown at the tip
Calcium	Essential part of middle lamella; involved in movement of substances through cell membranes	Terminal bud often dead; young leaves often appearing hooked at tip; tips and margins of leaves withered; roots dead or dying
Phosphorus	Necessary for respiration and cell division; high-energy cell compounds	Plants stunted; leaves darker green than normal; lower leaves often purplish between veins
Magnesium	Part of the chlorophyll molecule; activates enzymes	Veins of leaves green but yellow between them, with dead spots appearing suddenly; leaf margins curling
Sulfur	Part of some amino acids	Leaves pale green with dead spots; veins lighter in color than the rest of the leaf area
Iron	Needed to make chlorophyll and in respiration	Larger veins remaining green while rest of leaf yellows; mainly in young leaves*
Manganese	Activates some enzymes	Dead spots scattered over leaf surface; all veins and veinlets remain green; effects confined to youngest leaves
Boron	Influences utilization of calcium ions, but functions unknown	Petioles and stems brittle; bases of young leaves break down

*NOTE: The symptoms of iron deficiency may be caused by several factors, such as overwatering, cold temperatures, and nematodes (small roundworms) in the roots. The iron may be relatively abundant in the soil, but its uptake may be prevented or sharply reduced by these environmental conditions. Iron becomes more soluble under acid conditions—so much so that it can produce toxic conditions for most plants. Acid soil plants, on the other hand, have a much higher iron requirement than plants that require more non-acid conditions; accordingly, azaleas and other plants having high iron requirements can achieve normal growth in a non-acid soil.

All the micronutrients are harmful to plants when supplied in excessive quantities. Copper will kill algae in concentrations of one part per million, and boron has been used in weed killers. Even macronutrients are harmful if present in heavy amounts, although non-essential elements sometimes can counteract their toxicity.

Summary

1. Molecules and ions are in constant random motion and tend to distribute themselves evenly in the space available to them. They move from a region of higher concentration to a region of lower concentration by simple diffusion along a diffusion gradient; they may also move against a diffusion gradient. Evenly distributed molecules are in a state of equilibrium. Diffusion rates are affected by temperature, molecule size and density, and other factors.
2. Osmosis is the diffusion of water through a differentially permeable membrane. It takes place in response to concentration differences of dissolved substances.
3. Osmotic pressure or potential is the pressure required to prevent osmosis from taking place. The pressure that develops in a cell as a result of water entering it is called turgor. Water moves from a region of higher water potential (osmotic potential and pressure potential combined) to a region of lower water potential when osmosis is occurring. Osmosis is the primary means by which plants obtain water from their environment.

Figure 9.19 Leaves of bean plants grown in media deficient in various elements to show deficiency symptoms. *A.* A normal plant that has been furnished with all the essential elements. The other plants were grown in media deficient in specific elements, as follows: *B.* Potassium. *C.* Phosphorus. *D.* Calcium. *E.* Nitrogen. *F.* Sulfur. *G.* Micronutrients. *H.* Magnesium. *I.* Iron.

4. Plasmolysis is the shrinkage of the cytoplasm away from the cell wall as a result of osmosis taking place when the water potential inside the cell is greater than outside.
5. Imbibition is the attraction and adhesion of water molecules to the internal surfaces of materials; it results in swelling and is the initial step in the germination of seeds.
6. Active transport is the expenditure of energy by a cell that results in molecules or ions entering or leaving the cell against a diffusion gradient.
7. Water that enters a plant passes through it and mostly transpires into the atmosphere via stomata. Water retained by the plant is used in photosynthesis and other metabolic activities.
8. The cohesion-tension theory postulates that water rises through plants because of the adhesion of water molecules to the walls of the capillary-conducting elements of the xylem, cohesion of the water molecules, and tension on the water columns created by the pull developed by transpiration.
9. The translocation of food substances takes place in a water solution, and according to the pressure-flow hypothesis, such substances flow along concentration gradients between their sources and sinks.
10. Transpiration is regulated by humidity and the stomata, which open and close through changes in turgor pressure of the guard cells. These changes, which involve potassium ions, result from osmosis and active transport between the guard cells and the adjacent epidermal cells.
11. Aquatic, desert, tropical, and some cold-zone plants have modifications of stomatal apparatuses or specialized forms of photosynthesis that adapt them to their particular environments.
12. Guttation is the loss of water at night in liquid form through hydathodes at the tips of leaf veins.
13. Growth phenomena are controlled by both internal and external means and by chemical and physical forces in balance with one another. Besides carbon, hydrogen, and oxygen, 15 other elements are essential to most plants. When any of the essential elements are deficient in the plant, characteristic deficiency symptoms appear.

Review Questions

1. Distinguish among diffusion, osmosis, active transport, plasmolysis, and imbibition.
2. Why do living plants need a great deal of water for their activities?
3. Explain how a tall tree gets water to its tips without the aid of mechanical pumps.
4. What is the difference between transpiration and guttation?
5. Explain the pressure-flow hypothesis.
6. What are macronutrients? List them.
7. When nutrients are deficient in the soil, how are the deficiencies manifested in plants?

Discussion Questions

1. Why is salted meat less likely to spoil than unsalted meat?
2. Why would dye molecules in a bathtub of water take a long time to diffuse completely throughout the tub, but perfume molecules released in a closed room take considerably less time to do the same thing?
3. Why does osmosis not cause submerged water plants to swell up and burst?
4. Some bodies of water, such as the Dead Sea, have considerably higher salt concentrations than those of the human body. If you were swimming in such water, would you expect your cells to become plasmolyzed? Why?

Learning Online

Visit our web page at www.mhhe.com/botany for interesting case studies, practice quizzes, current articles, and animations within the Online Learning Center to help you understand the material in this chapter. You'll also find active links to these topics:

Root Systems and Plant Mineral Nutrition
Roots
Water Relations in Plants
Herbaceous Stems
Leaves and the Movement of Water
Water
Water Use and Management
Water Pollution

Additional Reading

Borghetti, M., et al. (Eds.). 1993. *Analysis of water transport in plants.* New York: Cambridge University Press.

Kluge, M., and I. P. Ting. 1978. "Crassulacean acid metabolism." *Ecological Studies,* vol. 30. New York: Springer-Verlag.

Kramer, P. J., and J. S. Boyer. 1995. *Water relations of plants and soils.* San Diego, CA: Academic Press.

Loughman, B. C., et al. 1990. *Structural and functional aspects of transport in roots.* Norwell, MA: Kluwer Academic Publications.

Nobel, P. S. 1995. *Physiochemical and environmental plant physiology,* 2d ed. San Diego, CA: Academic Press.

Taiz, L., and E. Zeiger. 1998. *Plant physiology,* 2d ed. Redwood City, CA: Benjamin/Cummings.

Flowers of *Bougainvillea* hybrids, widely cultivated in the tropics and subtropics. Large, colorful bracts surround the three cream-colored flowers.

Plant Metabolism

10

Chapter Outline

Overview

Enzymes and energy transfer are explained first to introduce general metabolic concepts. Photosynthesis and respiration are then presented at three different levels: the essence of the process is examined, the major steps are briefly introduced, and the processes are explored in greater detail. One or two levels may be sufficient for some readers; others will want to explore all three. The chapter discusses the importance of the main features of each process and summarizes the light-dependent reactions, light-independent reactions, glycolysis, the citric acid cycle, and the electron transport system. It concludes with a tabular comparison between photosynthesis and respiration and makes a few brief observations on additional metabolic pathways.

Some Learning Goals

1. Contrast the generalized equations of photosynthesis and respiration.
2. Understand what occurs in the light-dependent and light-independent reactions of photosynthesis and know the principal products of the reactions.
3. Explain what occurs in glycolysis, the citric acid cycle, and electron transport during respiration.
4. Distinguish between aerobic respiration and fermentation.
5. Compare assimilation and digestion.

Once while I was hiking along a California forest trail, I noticed a ray of sunlight beaming down through a mist toward the tops of some coastal redwoods. The scene reminded me of what is undoubtedly the most important chemical process on earth—**photosynthesis.** In this process, which involves little more than the air we breathe, water, a green pigment, and light, parts of the water and air are converted in cells to sugar, all without the aid of any cumbersome machinery. Also, as long as any cell remains alive, stored energy is released by another process, **respiration.** Without the proper balance between photosynthesis and respiration found in nature, life on earth as we know it would not exist (Fig. 10.1). Photosynthesis and respiration form the basis for most of our discussion in this chapter.

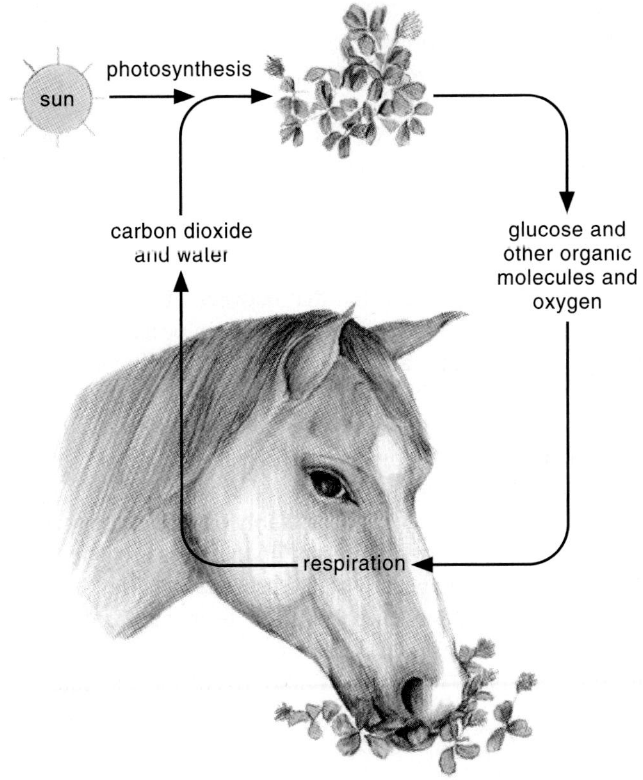

Figure 10.1 Summary of photosynthesis and aerobic respiration. Photosynthesis uses carbon dioxide and water to form carbohydrates. Aerobic respiration uses oxygen in breaking down carbohydrates to release carbon dioxide and water.

Each living cell of a plant contains genetic information that programs all the **metabolic** activities that take place in the cells and plant as a whole. All forms of **metabolism,** which may be defined as the sum of all the interrelated biochemical processes that take place in a living organism, require energy to occur. When rosebushes flower, apples are produced on an apple tree, leaves appear on a maple tree in the spring, or any other form of life activity occurs, the dynamic process of acquiring, releasing, and transferring energy from one form to another takes place. Both plants and animals release energy during their life cycles, with the energy then eventually being recycled or being used by other living organisms.

Photosynthetic cells can convert light energy to a usable form, and that usable energy may then be released during respiration, facilitating growth, development, and reproduction. Although photosynthetic organisms carry on both *photosynthesis* and *respiration,* most animals, likc humans, carry on only respiration and rely upon green plants for oxygen, food, shelter, and many other products.

ENZYMES AND ENERGY TRANSFER

Enzymes (the proteins that speed up chemical reactions in cells without being used up in the reactions) regulate just about every metabolic activity. In biochemical reactions, one or more specific enzymes are associated with the myriad forms of energy conversion that take place within cells. In some cases, these enzymes help build molecules that store energy in chemical bonds through a process called **anabolism.** Use of this energy to perform work often requires chemical bonds to be broken through **catabolism.** Most reactions of photosynthesis are anabolic because they involve construction of molecules that are stored for energy, whereas reactions in **cellular respiration** (referred to as "respiration" in the text) are generally catabolic, since in performing work, they release energy held in chemical bonds. Photosynthesis builds organic compounds by combining carbon dioxide and water, forming carbohydrates. Respiration, on the other hand, breaks down those carbohydrates, producing carbon dioxide and water, which may be used once again in photosynthesis. As we will see in the sections to follow, this photosynthesis–respiration cycle is keyed by an enzyme complex that splits water molecules and releases electrons that function, at least temporarily, in storing biochemical energy. These electrons transfer energy from one form to another through *oxidation-reduction reactions.*

Oxidation-Reduction Reactions

The processes of both photosynthesis and respiration include many **oxidation-reduction reactions. Oxidation** is the *loss of one or more electrons;* it involves removal of electrons from a compound. **Reduction** is the *gain of one or more electrons;* it involves the addition of electrons to a compound. In most oxidation-reduction reactions, oxidation of one compound is coupled with reduction of another compound catalyzed by the same enzyme or enzyme complex. When an electron is removed, a proton may follow, with the result that a hydrogen is often removed during oxidation and added during reduction. Oxygen is usually the oxidizing agent (i.e., the final acceptor of the electron), but oxidations can occur without oxygen being involved.

PHOTOSYNTHESIS

Oil and coal today provide about 90% of the energy needed to power trains, trucks, ships, airplanes, factories, computers, communication systems, and a multitude of electrically energized appliances. The energy within that oil and coal was originally captured from the sun by plants and algae growing millions of years ago and then transformed into fossil fuels by geological forces.

The energy needs of transportation, industry, and the modern household seem insignificant, however, when compared with the combined energy requirements of all living organisms. Every living cell requires energy just to remain alive, and more energy is needed for the cell to reproduce, grow, or do physical work as part of an organism. In addition, oxygen is vital to nearly all life in processes that release stored energy.

Photosynthesis, at least indirectly, is not only the principal means of keeping all forms of humanity functioning, but also the sole means of sustaining life at any level—except for a few bacteria that derive their energy from sulfur salts and other inorganic compounds. This unique manufacturing process of green plants furnishes raw material, energy, and oxygen. In photosynthesis, energy from the sun is harnessed and, with the aid of chlorophyll, is transformed from light energy to biochemical energy in the bonds between the atoms in a sugar molecule. Oxygen is given off as a by-product of the process.

It has been estimated that all of the world's green organisms (including those in the oceans) together produce between 100 billion and 200 billion metric tons (between 110 billion and 220 billion tons) of sugar each year. To visualize that much sugar, consider that it is enough to make about 300 quadrillion sugar cubes with a total volume exceeding that of two million Empire State Buildings.[1]

Much of the sugar produced by plants is converted to wood, fibers (such as cotton and linen), and other structural materials. The first products of photosynthesis may also be converted to disaccharides, polysaccharides such as starch, and other storage forms of carbohydrates. The digestive activities of living organisms break down the carbohydrates to smaller molecules.

Sugars produced by photosynthesis are also involved in the synthesis of amino acids for proteins and a host of other cell constituents. In fact, photosynthesis produces more than 94% of the dry weight substance of green organisms, with the remainder coming from the soil or dissolved matter.

The capacity of plants to meet our energy needs may well determine the ultimate size of human populations. In some heavily populated parts of the world, the food supply already is falling short of providing enough energy to sustain life, and starvation is widespread. Meanwhile, in the Western world, significant numbers of persons consume too much food and are spending large sums on weight reduction. We will, however, eventually approach a point at which human populations in general will need to stabilize, or even those in the most affluent areas could exceed the capacity of the plants to sustain them.

A great deal of photosynthesis occurs in organisms living in the oceans. It is estimated that between 40% and 50% of the oxygen in the atmosphere originates in oceans and lakes. Laboratory tests have shown, however, that pollutants—such as the PCBs (polychlorinated biphenyls) used in electrical insulators—are capable, in concentrations as low as 20 parts per billion, of stopping many delicate algae from carrying on photosynthesis. The concentration of such substances in ocean waters at present is considerably less than one part per billion, but PCB concentrations of up to five or more parts per billion have already been reported in some estuaries. The use of PCBs and related chemicals has been curtailed in the United States, but other countries are still using them, and residues are still washing off into rivers and on into the oceans. It is important to make the world community fully aware of the dangers of allowing pollutant concentrations to build up to the point of creating significant problems that could ultimately adversely affect all life.

---

1. One quadrillion is 1,000,000,000,000,000. New York City's Empire State Building has 102 stories and is 381 meters (1,250 feet) tall.

Note to the Reader *Photosynthesis is undoubtedly the most important process on earth to life as we know it. It is also a complex process that can be summarized briefly or examined in detail. What follows is a discussion of the subject at three different levels: (1) The essence of photosynthesis; (2) a brief introduction to the major steps of photosynthesis; and (3) a closer look at photosynthesis. The process of respiration, which is discussed after photosynthesis, is treated in similar fashion. The third level, in particular, contains detail that either may or may not be discussed in your course.*

1. The Essence of Photosynthesis

Cells need energy to do their work and reproduce. The energy for most of this cellular activity involves energy-storing molecules commonly known as **ATP** (*adenosine triphosphate*).[2] ATP doesn't last more than a few seconds and has to be produced constantly. Instead of ingesting food as animals do, plants can make ATP, using light as the source of energy, just like a solar-powered engine. If the light goes out, however, ATP production stops, and the cells could quickly die. In a solar-powered engine, this problem is avoided by using some of the generated electricity to charge batteries before the sun goes down. Plants also accumulate energy for later use by building sugar molecules for short-term energy storage or starch for longer-term energy storage.

The energy-storing process of photosynthesis takes place in chloroplasts and other parts of green organisms in the presence of light. The light energy stored in a simple sugar molecule is produced from carbon dioxide (CO_2) present in the air and water (H_2O) absorbed by the plant. When the carbon dioxide and water (H_2O) are combined and ultimately a sugar molecule ($C_6H_{12}O_6$, glucose) is produced in a chloroplast, oxygen gas (O_2) is released as a by-product. The oxygen diffuses out into the atmosphere. The overall process can be depicted by the equation that follows. The equation should not, however, be taken literally because there are many intermediate steps to the process, and glucose is not the immediate first product of photosynthesis.

$$\underset{\text{carbon dioxide}}{6\,CO_2} + \underset{\text{water}}{12\,H_2O} + \text{light energy} \xrightarrow[\text{enzymes}]{\text{chlorophyll}} \underset{\text{glucose}}{C_6H_{12}O_6} + \underset{\text{oxygen}}{6\,O_2} + \underset{\text{water}}{6\,H_2O}$$

2. There are millions of ATP molecules in living cells. When an ATP molecule releases its energy, it gives up the terminal phosphate group in the row of three attached to its base and becomes ADP (adenosine diphosphate). An ADP molecule becomes an ATP molecule again when it regains its third phosphate group and stores energy. ATP is an important participant in many reactions involving the transfer of energy.

Photosynthesis takes place in chloroplasts (see Figs. 3.5 and 3.8) or in cells with membranes in which chlorophyll is embedded. The principals in the process are *carbon dioxide, water, light,* and *chlorophyll;* a brief examination of each follows.

Carbon Dioxide

Our atmosphere consists of approximately 78% nitrogen and about 21% oxygen. The remaining 1% is made up of a mixture of less common gases, including 0.036% (360 parts per million) carbon dioxide and a little hydrogen, helium, argon, and neon. The carbon dioxide in the air surrounding the leaves of plants reaches the chloroplasts in the mesophyll cells by diffusing through the stomata into the leaf interior. The carbon dioxide goes into solution in a thin film of water on the outside walls of cells. It then diffuses through the cell walls, across cell membranes, and into the cytoplasm where it finally reaches the chloroplasts.

The amount of carbon dioxide constantly being taken from the atmosphere during daylight hours by all green plants is enormous. Just four-tenths of a hectare (1 acre) of corn (10,000 plants) accumulates more than 2,500 kilograms (5,512 pounds) of carbon from the atmosphere during a growing season. Over 10 metric tons (11 tons) of carbon dioxide are needed to furnish this much carbon. It also has been calculated that the total present atmospheric supply of carbon dioxide (more than 2.2 billion metric tons or about 50 metric tons over each hectare of the earth's surface) would be completely used up in about 22 years if it were not constantly being replenished.

A large reservoir of carbon dioxide in the oceans has helped to maintain atmospheric carbon dioxide levels at about 0.036% throughout most of recorded history, but in the last 50 years, the percentage has begun to climb slightly. Plant and animal respiration, decomposition, natural fires, volcanoes, and similar sources replace carbon dioxide at roughly the same rate at which it is removed during photosynthesis. The use of fossil fuels, pollution, deforestation, and other human activities, however, have disrupted this cycle by adding excess carbon dioxide to the atmosphere.

Measurements of atmospheric carbon dioxide at the Mauna Loa research station on the big island of Hawaii have demonstrated this steady increase. Increased carbon dioxide levels have the potential to cause global increases in temperatures because, in the atmosphere, this gas acts much like a glass panel in a greenhouse by trapping solar radiation from the sun. Climatic models project that the earth's average atmospheric temperature will rise from 1.0°C to 3.5°C (1.8°F to 6.3°F) by the year 2100. Since the 1980s, global warming, presumably as a result of this increase, has become a major political and scientific issue. See the "Greenhouse Effect" in Chapter 25 for a discussion of this problem.

Although increases in carbon dioxide levels may enhance photosynthesis and increase food production, insects, bacteria, and viruses that proliferate with warmer temperatures can offset these potential gains. Under carefully monitored conditions, commercial greenhouses have pumped carbon dioxide through pipes placed over plant beds to supplement their natural supply and have found that fertilizing plants with this gas has increased yields by more than 20%. Indeed, this indicates that plants have the potential temporarily to limit elevated carbon dioxide levels in the atmosphere. However, recent evidence suggests that many plants develop fewer stomata when carbon dioxide levels increase, thereby adapting them to such changes and reducing photosynthetic efficiency. The abundance of photosynthetic organisms that may or may not proliferate under these conditions is a key factor affecting global warming.

Some scientists note that increased carbon dioxide levels will cause temperatures to rise globally, which, in turn, will result in longer growing seasons in middle and high latitudes and hence increase global photosynthesis. Nonetheless, these same temperature increases may accelerate plant and animal respiration and decomposition, which would add carbon dioxide to the environment. Respiration at higher temperatures may also diminish benefits to be anticipated from an increase in numbers of photosynthetic organisms during global warming. Through careful evaluation of global trends and physiological processes, scientists should be better able to predict what is in store for our planet's future and how to improve management of the environment.

Water

Less than 1% of all the water absorbed by plants is used in photosynthesis; most of the remainder is transpired or incorporated into cytoplasm, vacuoles, and other materials. The water used is the source of electrons involved in photosynthesis, and the oxygen released is a by-product, even though carbon dioxide also contains oxygen. This has been demonstrated by conducting photosynthetic experiments using either carbon dioxide or water containing isotopes of oxygen. When the isotope is used only in the water, it appears in the oxygen gas released. If, however, it is used only in the carbon dioxide, it is confined to the sugar and water produced and never appears in the oxygen gas, demonstrating clearly that the water is the sole source of the oxygen released.

If water is in short supply, it may indirectly become a limiting factor in photosynthesis; under such circumstances, the stomata usually close and sharply reduce the carbon dioxide supply.

Light

Light exhibits properties of both waves and particles. Energy reaches the earth from the sun in waves of different lengths, the longest waves being radio waves and the shortest being gamma rays. About 40% of the radiant energy we receive is in the form of visible light. If this visible light is passed through a glass prism, it splits up into its component colors. Reds are on the longer wavelength end and violets on the shorter wave end, with yellows, greens, and blues between (Fig. 10.2). Although nearly all of the visible light colors can be used in photosynthesis, those in the violet to blue and red-orange to red wavelengths are used most extensively. Light in the green range is reflected. Leaves commonly absorb about 80% of the visible light that reaches them.

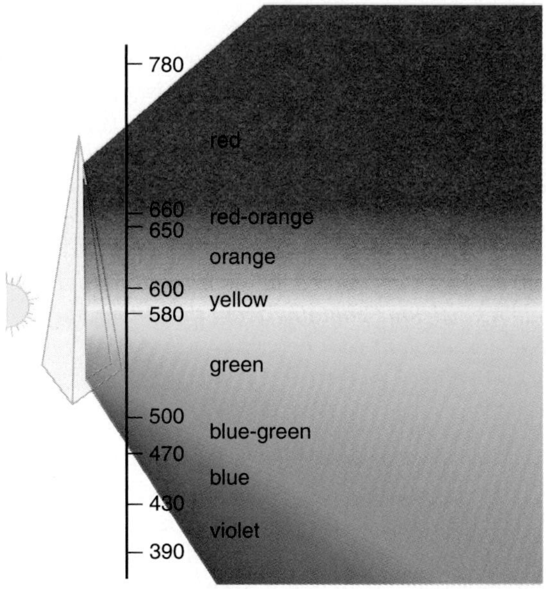

Figure 10.2 Visible light that is passed through a prism is broken up into individual colors with wavelengths ranging from 390 nanometers (violet) to 780 nanometers (red).

The intensity of light varies with the time of day, season of the year, altitude, latitude, and atmospheric conditions. On a clear summer day at noon in a temperate zone, sunlight reaches an intensity of about 2,000 µmols[3] per square meter of surface per second. In contrast, consider that a room with fluorescent lighting produces only about 40 µmols per square meter of surface per second.

Plants vary considerably in the light intensities they need for photosynthesis to occur at optimal rates. Factors such as temperature and amount of carbon dioxide available can also be limiting (Fig. 10.3). For example, an increase in photosynthesis will not occur in some plants receiving more than 670 µmols of light per square meter unless supplemental carbon dioxide is provided. With supplements, however, rates of photosynthesis will continue to increase up to about 1,000 µmols. Herbaceous plants on a forest floor can survive with less than 2% of full daylight, and some mosses are reported to thrive on intensities as low as 0.05% to 0.01%. Most land plants that grow naturally in the open need at least 30% of full daylight to thrive. The optimal amount for some species of trees approaches full daylight, while shade plants often do well in 10% of full daylight.

Light that is too intense may change the way in which some of a cell's metabolism takes place. For example, higher light intensities and temperatures may change the

3. A µmol is a measurement of the number of photons (see p. 176) striking a specified area such as a leaf surface.

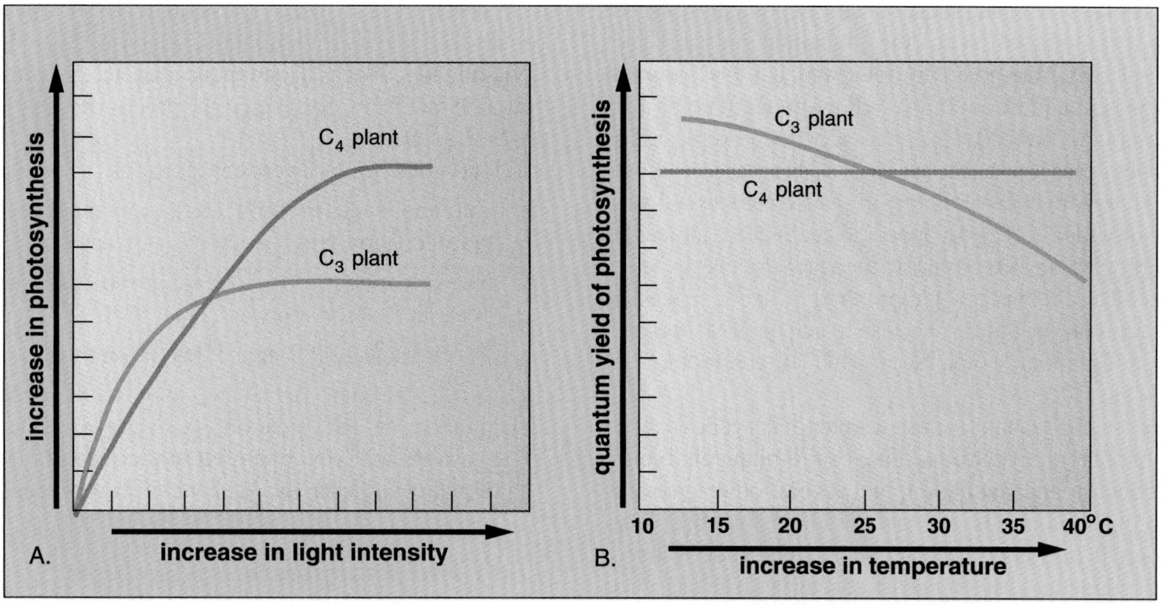

Figure 10.3 Effects of light and temperature on two forms of photosynthesis. Both forms of photosynthesis, known as C_3 and C_4, respectively, are discussed in this chapter. *A.* In some plants, the rate of photosynthesis will not increase beyond a certain intensity of light. In other plants that have additional carbon dioxide available, photosynthetic rates increase with up to a third higher light intensity. *B.* In some plants, quantum yield of photosynthesis decreases as temperatures increase, whereas in other plants, the quantum yield of photosynthesis is not significantly affected by temperature fluctuations between 10°C and 40°C.

ratio of carbon dioxide to oxygen in the interiors of leaves, which, in turn, may accelerate *photorespiration* (discussed on p. 183). Photorespiration is a wasteful process that uses oxygen and releases carbon dioxide. It differs from common aerobic respiration in its chemical pathways.

Photooxidation, which involves the destruction ("bleaching") of chlorophyll by light, may also occur. High light intensities may cause chlorophyll molecules to go to a different excited state. The energy released from the excited chlorophyll is passed to oxygen molecules, which become highly reactive and bleach the chlorophyll. In the fall, photooxidation plays a significant role in the breakdown of chlorophyll in leaves, resulting in the autumn colors discussed in Chapter 7. High light intensities may also cause an increase in transpiration, resulting in the closing of stomata. A sharp reduction in the available carbon dioxide supply inevitably follows.

Chlorophyll

There are several different types of **chlorophyll** molecules, all of which contain one atom of magnesium. They are very similar in structure to the heme of hemoglobin, the iron-containing red pigment that transports oxygen in blood. Each molecule has a long lipid tail, which anchors the chlorophyll molecule in the lipid layers of the thylakoid membranes (Fig. 10.4).

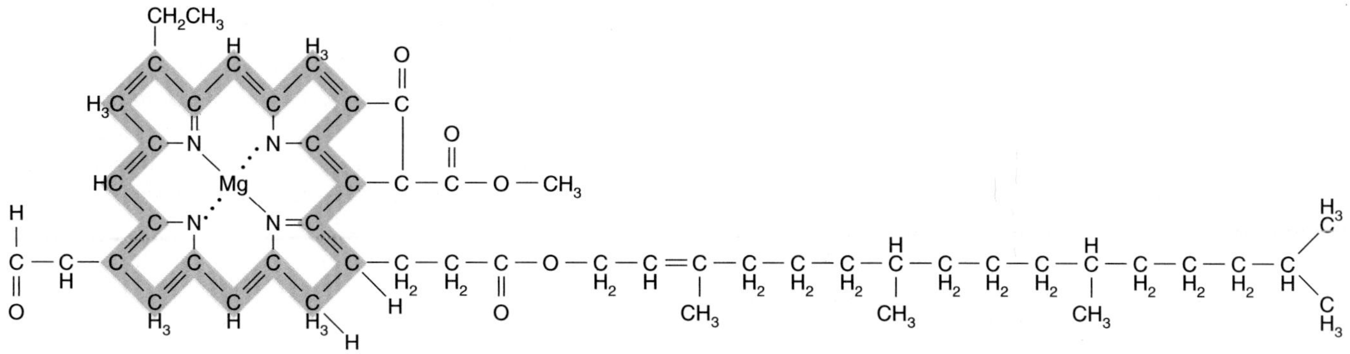

Figure 10.4 The structure of a molecule of chlorophyll *a,* the most important of the pigments involved in photosynthesis. The boxlike ring structure on the left, with magnesium and nitrogen inside, functions in capturing light energy. The tail, which extends into the interior of a thylakoid membrane, is incapable of absorbing water; all chlorophyll molecules are, however, fat soluble.

Chloroplasts of most plants contain two major kinds of chlorophyll associated with the thylakoid membranes. Chlorophyll *a* is blue-green in color and has the formula $C_{55}H_{72}MgN_4O_5$. Chlorophyll *b* is yellow-green in color and has the formula $C_{55}H_{70}MgN_4O_6$. Usually, a chloroplast has about three times more chlorophyll *a* than *b*. The more chlorophyll *a* there is in a cell, the brighter green the cell and the tissue of which it is a part appear to be. When a molecule of chlorophyll *b* absorbs light, it transfers the energy to a molecule of chlorophyll *a*. Chlorophyll *b*, then, makes it possible for photosynthesis to take place over a broader spectrum of light than would be possible with chlorophyll *a* alone (see Fig. 10.7).

Other such pigments include carotenoids (yellowish to orange pigments), phycobilins (blue or red pigments found in cyanobacteria and red algae), and several other types of chlorophyll. Chlorophylls *c, d,* and possibly *e* take the place of chlorophyll *b* in certain algae, and several other photosynthetic pigments are found in bacteria. The various chlorophylls are all closely related and differ from one another only slightly in the structure of their molecules.

In chloroplasts, about 250 to 400 pigment molecules are grouped as a light-harvesting complex called a **photosynthetic unit,** with countless numbers of these units in each granum. Two types of these photosynthetic units function together in the chloroplasts of green plants, bringing about the first phase of photosynthesis, the *light-dependent* reactions, which are discussed in the next section, "Introduction to the Major Steps of Photosynthesis."

2. Introduction to the Major Steps of Photosynthesis

The process of photosynthesis takes place in two series of steps called the **light-dependent reactions** and the **light-independent reactions.** Although the light-independent reactions use products of the light-dependent reactions, both processes occur simultaneously.

The Light-Dependent Reactions

The *light-dependent reactions* are the first major steps in the conversion of light energy to biochemical energy in the form of ATP and **NADPH** (*reduced nicotinamide adenine dinucleotide phosphate*). The ATP and NADPH run various biochemical reactions in cells. The reactions are initiated when units of light energy (*photons*) strike chlorophyll molecules embedded in the thylakoid membranes of chloroplasts.

Our knowledge of the light-dependent reactions essentially began in the 1930s in England through a discovery by Robin Hill, a biochemist. He found that a solution of fragmented and whole chloroplasts, isolated from leaves that had been ground up and centrifuged, could briefly produce oxygen if an electron acceptor was present to receive electrons from water. In 1951, it was shown that **NADP** *(nicotinamide adenine dinucleotide phosphate),* which is derived from the B vitamin *niacin,* was a natural electron acceptor in this reaction. In honor of its discoverer, the process became known as the *Hill reaction.*

During the light-dependent reactions,

1. water molecules are split apart, producing electrons and hydrogen ions, and oxygen gas is released;
2. the electrons from the split water molecules are passed along an electron transport system;
3. energy-storing **ATP** molecules are produced;
4. some hydrogen from the split water molecules is involved in the creation of NADPH, which carries hydrogen and which is used in the second major phase of photosynthesis, the *light-independent reactions.*

The Light-Independent Reactions

The *light-independent* reactions (or *carbon-fixing and reducing reactions*) complete the conversion of light energy to chemical energy in the form of ATP and NADPH. Some scientists refer to the light-independent reactions as the *dark reactions* because they don't directly require light, but darkness has nothing to do with their functioning. In fact, even though light is not directly required in the same sense as it is for the light-dependent reactions, light is nevertheless required for the activation of the enzymes involved, and the processes normally can occur only in daylight.

The light-independent reactions are a series of reactions that take place outside of the grana in the stroma of the chloroplast (see page 40), if the products of the light-dependent reactions are available. They may initially proceed in different ways, depending on the particular kind of plant involved, but they all go through the **Calvin cycle,** discovered and elucidated by Dr. Melvin Calvin of the University of California. In 1961, Calvin received a Nobel Prize for unraveling how this most widespread type of light-independent reactions takes place.

In this cycle, carbon dioxide (CO_2) from the air is combined with a 5-carbon sugar *(RuBP, ribulose bisphosphate),* and then the combined molecules are converted, through several steps, to sugars, such as glucose ($C_6H_{12}O_6$). Energy and electrons involved in these steps are furnished by the ATP molecules and NADPH produced during the light-dependent reactions. Some of the sugars that are produced during the light-independent reactions are recycled, while others are stored as starch or other polysaccharides (simple sugars strung together in chains). A summary of simplified photosynthetic reactions is shown in Figure 10.5. More detailed diagrams of the light-dependent and light-independent reactions are shown in Figures 10.8, 10.9, and 10.12.

Two molecules of a 3-carbon sugar compound *(3PGA—an abbreviation of 3-phosphoglyceric acid)* are shown as the first stable substance produced when carbon

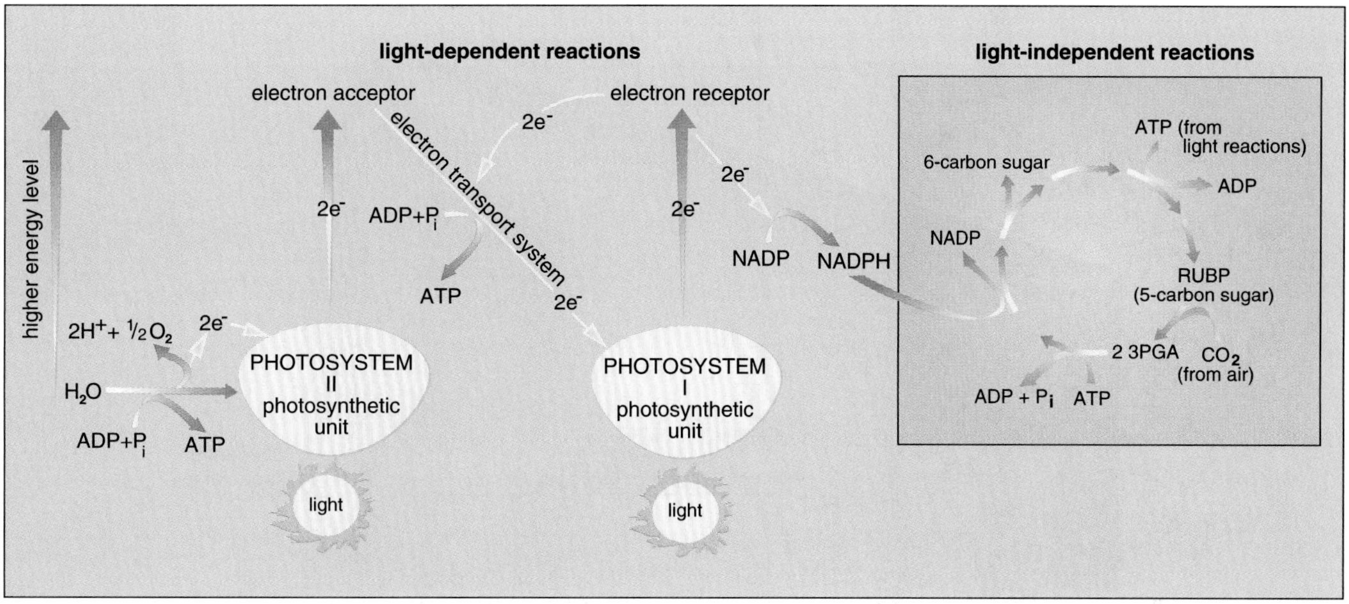

Figure 10.5 A simplified summary of photosynthetic reactions.

dioxide from the air and RuBP are combined and then converted during the light-independent reactions (Fig. 10.5). Some grasses and also many plants of arid regions fix carbon differently. They produce a 4-carbon acid as the first product, followed by the Calvin cycle. This 4-carbon pathway is discussed, along with another variation found mostly in desert plants, in the next section, "A Closer Look at Photosynthesis."

3. A Closer Look at Photosynthesis

A great deal has been learned about photosynthesis since 1772 when Joseph Priestly (1733–1804), a naturalist in England, received a medal for demonstrating that a sprig of mint "restored" oxygen so that a mouse could live in air that had been used up by a burning candle. Seven years later, Jan Ingen-Housz (1730–1799) of Holland, who visited England to treat the royal family for smallpox, carefully repeated Priestly's demonstrations. He showed that the air was restored only when green parts of plants were receiving sunlight.

In 1782, Jean Senebier (1742–1809), a Swiss pastor, discovered that the photosynthetic process required carbon dioxide, and in 1796, Ingen-Housz showed that carbon went into the nutrition of the plant. The final component of the overall photosynthetic reaction was explained in 1804 by another Swiss, Nicholas Theodore de Saussure (1767–1845), who showed that water was involved in the process.

A little current information about the details of photosynthesis is given in the following modest amplification of the preceding section, "Introduction to the Major Steps of Photosynthesis." Those who wish more information are referred to the reading list at the end of the chapter.

The Light-Dependent Reactions Reexamined

As noted earlier, light has characteristics of both waves and particles. Sir Isaac Newton, while experimenting with a prism over 300 years ago, produced a spectrum of colors from visible white light and postulated that light consisted of a series of discrete particles he called "corpuscles."

Newton's theory only partially explained light phenomena, however, and by the middle of the 19th century, James Maxwell and others showed that light and all other parts of the extensive electromagnetic spectrum travel in waves.

By the late 1800s, with the discovery that a *photoelectric effect* can be produced in all metals, the wave theory also became inadequate to explain certain attributes of light. When a metal is exposed to radiation of a critical wavelength, it becomes positively charged because the radiation forces electrons out of the metal atoms. The ability of light to force electrons from metal atoms depends on its particular wavelength—its energy content—and not on its intensity or brightness. The shorter the wavelength, the greater the energy, and vice versa.

In 1905, Albert Einstein proposed that the photoelectric effect results from discrete particles of light energy he called **photons.** In 1921, he received the Nobel Prize in physics for this work. Both waves and particles (photons) are today almost universally recognized as aspects of light. The energy (quantum) of a photon is not the same for all kinds of light; those of longer wavelengths have lower energy, and those of shorter wavelengths have proportionately higher energy.

Chlorophylls, the principal pigments of photosynthetic systems, absorb light primarily in the violet to blue and also in the red wavelengths; they reflect green wavelengths,

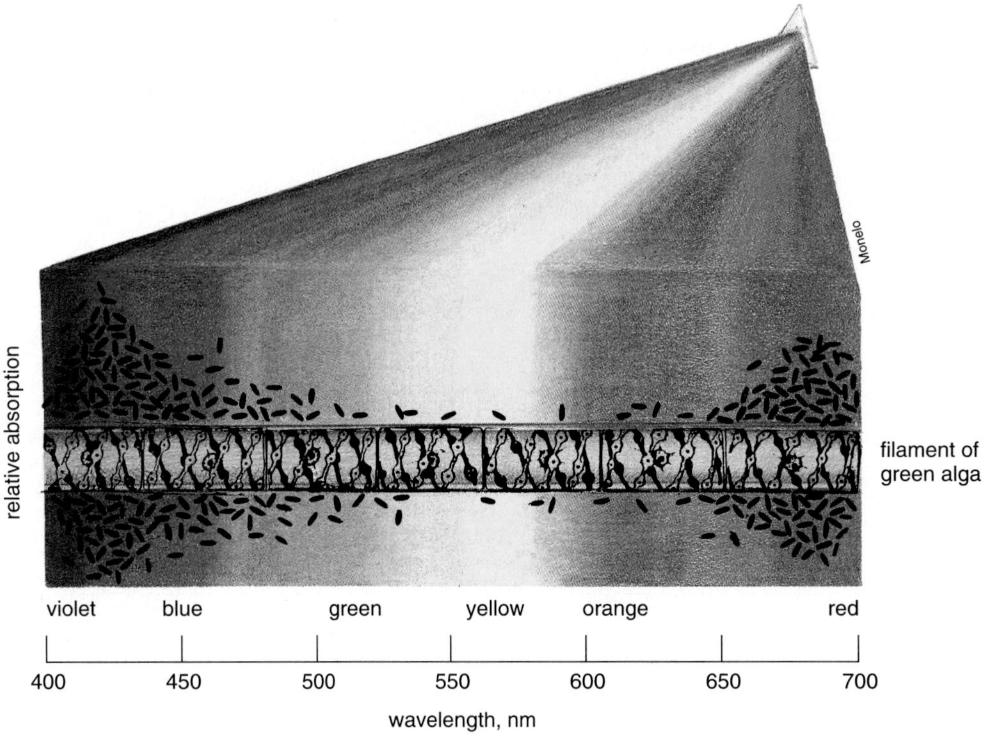

Figure 10.6 Engelmann's experiment. A tiny spectrum of light was focused on a microscope slide with a row of algal cells suspended in water containing bacteria that move toward an oxygen source. The bacteria assembled mostly in the areas exposed to the red and blue portions of the spectrum.

which is why leaves appear green. This was first ingeniously demonstrated in 1882 by T. W. Engelmann. He focused a tiny spectrum of light on a filament (single row of cells) of *Spirogyra,* a freshwater alga. The alga had been mounted in a drop of water on a microscope slide containing bacteria that move toward an oxygen source. As shown in Fig. 10.6, the bacteria assembled in greatest numbers along the algal filament in the blue and red portions of the spectrum, demonstrating that oxygen production is directly related to the light the chlorophyll absorbs. An analysis using living material is called a *bioassay.* Information gained from bioassays often is significant.

Each pigment has its own distinctive pattern of light absorption, which is referred to as the pigment's *absorption spectrum* (Fig. 10.7). When a pigment absorbs light, the energy levels of some of the pigment's electrons are raised. When this occurs, the energy may be emitted immediately as light (a phenomenon called *fluorescence*). In chlorophyll, this is characteristically in the red part of the visible light spectrum, so an extract of chlorophyll placed in light (especially ultraviolet or blue light) will appear red. The absorbed energy may also be emitted as light after a delay (a phenomenon called *phosphorescence*); it may otherwise be converted to heat. The most important use of the absorbed energy, however, is its storage in chemical bonds for photosynthesis.

Photosystems The two types of photosynthetic units present in most chloroplasts make up **photosystems** known as *photosystem I* and *photosystem II* (Fig. 10.8). Photosystem

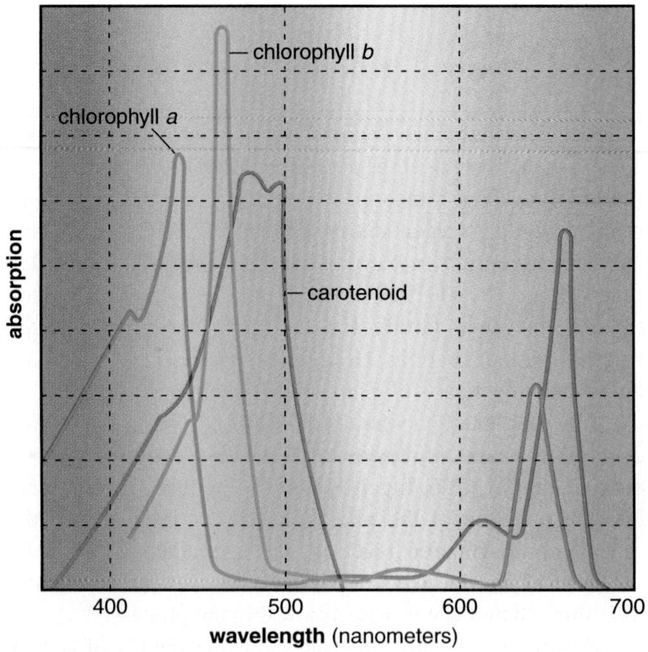

Figure 10.7 The absorption spectra of chlorophyll *a,* chlorophyll *b,* and a carotenoid. The maximum absorption of the chlorophylls is in the blue and red wavelengths. The maximum absorption of the carotenoids is in the blue-green to green parts of the visible spectrum.

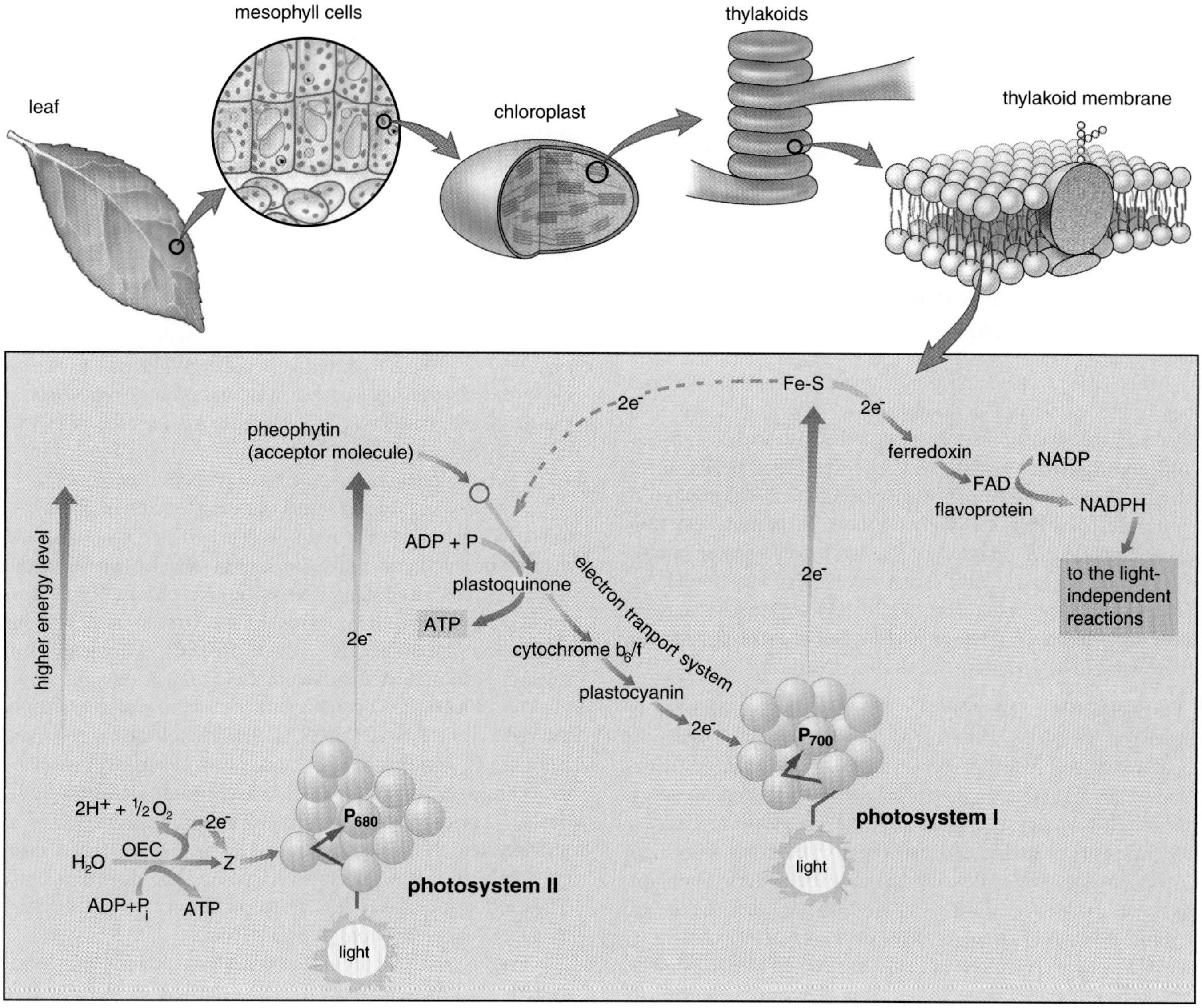

Figure 10.8 The light-dependent reactions of photosynthesis, which occur in more than one way. In noncyclic photophosphorylation, involving photosystems I and II, which convert light energy to biochemical energy in the form of ATP and NADPH, water molecules are split, releasing electrons, protons, and oxygen gas. The electrons are subsequently used to produce NADPH, whereas the protons are used, in part, to enable production of ATP. Oxygen gas is a by-product of this noncyclic photophosphorylation, although aerobic organisms rely upon this gas for respiration. The ATP and NADPH are used in the carbon-fixing reactions that convert CO_2 to sugars (see Fig. 10.10). Only photosystem I is involved in cyclic photophosphorylation. In this relatively simple system, electrons boosted from a photosystem I reaction-center molecule are shunted back into the reaction center via the electron transport system. ATP is produced from ADP, but no NADPH or oxygen is produced.

II received its "II" designation because it was discovered after photosystem I, but we know now that the events of photosynthesis that take place in photosystem II come *before* those of photosystem I.

While both photosystems produce ATP, only organisms that have photosystems I *and* II can produce NADPH and oxygen as a consequence of electron flow. It is likely that evolutionary events led to organisms that possess both photosystems. At least 2.8 billion years ago, photosynthetic forms of bacteria (cyanobacteria) are believed to have evolved from primitive bacteria with the development of chlorophyll *a* and photosystem II. Humans and other organisms dependent on oxygen are fortunate that photosynthetic organisms evolved the oxygen-generating steps of photosystem II because these reactions are what provide the oxygen we breathe today. Since photosynthetic organisms can generate oxygen from water, this very process in the future could be exploited to sustain human life during space travel and colonization of other planets.

Each photosynthetic unit of photosystem I consists of 200 or more molecules of chlorophyll *a,* small amounts of chlorophyll *b,* carotenoid pigment with protein attached, and a special reaction-center molecule of *chlorophyll* called $\mathbf{P_{700}}$. Although all pigments in a photosystem can absorb photons, the *reaction-center* molecule is the only one that can actually use the light energy. The remaining photosystem pigments are called *antenna pigments* because together they function somewhat like an antenna in gathering and passing light energy to the reaction-center molecule (see Fig. 10.9). There are also iron-sulfur proteins that are the primary electron acceptors for photosystem I (i.e., iron-sulfur proteins first receive electrons from P_{700}).

A photosynthetic unit of photosystem II consists of chlorophyll *a,* β-carotene (the precursor of vitamin A) attached to protein, a little chlorophyll *b,* and a reaction-center molecule of chlorophyll *a;* these special molecules are called $\mathbf{P_{680}}$.

The letter *P* stands for pigment and the numbers *700* and *680* of the reaction-center molecules of chlorophyll *a* refer to peaks in the absorption spectra of light with wavelengths of 700 and 680 nanometers, respectively. These peaks differ slightly from those of the otherwise identical chlorophyll *a* molecules of the photosynthetic units. A primary electron acceptor called *pheophytin* (or *Pheo)* is also present in photosystem II. One reaction-center molecule was found by Johann Deisenhofer and Hartmot Michel of Germany to have over 10,000 atoms. They received a Nobel Prize in 1988 for their work in determining the atomic structure.

Water-Splitting (Photolysis) When a photon of light is absorbed by a P_{680} molecule of a photosystem II reaction center (located near the inner surface of a thylakoid membrane), the light energy boosts an electron to a higher energy level. This is referred to as *exciting* an electron. Excited electrons are unstable and can rapidly revert back to their lower energy level, releasing the absorbed energy, perhaps as heat. However, during photosynthesis, the energy of excited electrons is transferred to the P_{680} reaction centers.

The excited electron in P_{680} is picked up by the primary acceptor molecule, pheophytin. This electron is shuttled to another acceptor, **PQ** (plastoquinone), within the thylakoid membrane.[4] PQ is mobile and moves within the lipid bilayer toward the side of the stroma, unloading the electrons to the *cytochromes* next in line. Electrons extracted from water by a manganese-containing *oxygen-evolving complex* (*OEC*) replace the electrons lost by the P_{680} molecule. It has been suggested that there is an oxidation-reduction system, usually designated **Z,** operating between water and P_{680}. Transfer of an electron from Z to P_{680} reoxidizes Z and prepares it for accepting another electron from the OEC. Recent investigations indicate that for each two water molecules that are split, one molecule of oxygen is produced, along with four protons and four electrons.

This metabolic pathway eventually evolved in photosynthetic bacteria (cyanobacteria). The abundance of water, as an electron source, probably facilitated the mechanism that generated oxygen as a by-product of photosynthesis. This process increased the supply of oxygen in earth's atmosphere and made it possible for energy-efficient aerobic respiration to evolve.

4. Except for the high speed (trillionths of a second) at which it usually carries out its function, an acceptor molecule could be compared to a pickup order telephoned to a clerk in a store in the sense that an item is picked up from a source, held temporarily, and then transferred elsewhere.

Photophosphorylation The high-energy acceptor molecule PQ releases an electron to an electron transport system, which functions something like a downhill bucket brigade that temporarily moves electrons from H_2O to a storage molecule, *nicotinamide adenine dinucelotide phosphate* (**NADP**), an electron acceptor for photosystem I. The electron transport system consists of iron-containing pigments called **cytochromes** and other electron transfer molecules, plus *plastocyanin*—a protein that contains copper. While electrons pass along the electron transport system and protons move across the thylakoid membrane by *chemiosmosis* (see the next section, "Chemiosmosis"), ATP molecules are being formed from ADP and phosphate in the process of *photophosphorylation.*

A somewhat similar series of events occurs in photosystem I. When a photon of light is absorbed by a P_{700} molecule in a photosynthetic unit, the energy excites an electron, which is transferred to an iron-sulfur acceptor molecule designated Fe-S. This then passes the electron to another iron-sulfur acceptor molecule, *ferredoxin* (**Fd**), which, in turn, releases it to a carrier molecule called *flavin adenine dinucleotide* (**FAD**). FAD contains flavoprotein, which assists in the reduction of NADP to NADPH. The electrons removed from the P_{700} molecule are replaced by electrons from photosystem II via the electron transport system outlined previously. This overall movement of electrons from water to photosystem II to photosystem I to NADP is called *noncyclic* electron flow, because it goes in one direction only. The production of ATP that correspondingly occurs is designated as *noncyclic photophosphorylation.*

Photosystem I can also work independently of photosystem II. When it does, the electrons boosted from P_{700} reaction-center molecules (of photosystem I) are passed from ferredoxin to plastoquinone (instead of NADPH) and back into the reaction center of photosystem I. This process, which does not occur in most plants, is *cyclic* electron flow. ATP generated by cyclic electron flow is called *cyclic photophosphorylation,* but water molecules are not split and neither NADPH nor oxygen are produced in this process.

Chemiosmosis The oxygen-evolving complex on the inside of a thylakoid membrane catalyzes the splitting of water molecules, producing protons and electrons, as well as oxygen gas. These electrons used to replenish those excited in chlorophyll are then transferred in bucket-brigade fashion through an electron transport system, ultimately reducing NADP to NADPH. As electrons travel through this transport system, additional protons move from the stroma to the inside of the thylakoid membrane in specific oxidation-reduction reactions when electrons pass from photosystem II to PQ. These protons join with the protons from the split water molecules and thereby contribute to an accumulation of four protons toward the inside of the thylakoid membrane (an area also known as the thylakoid *lumen*).

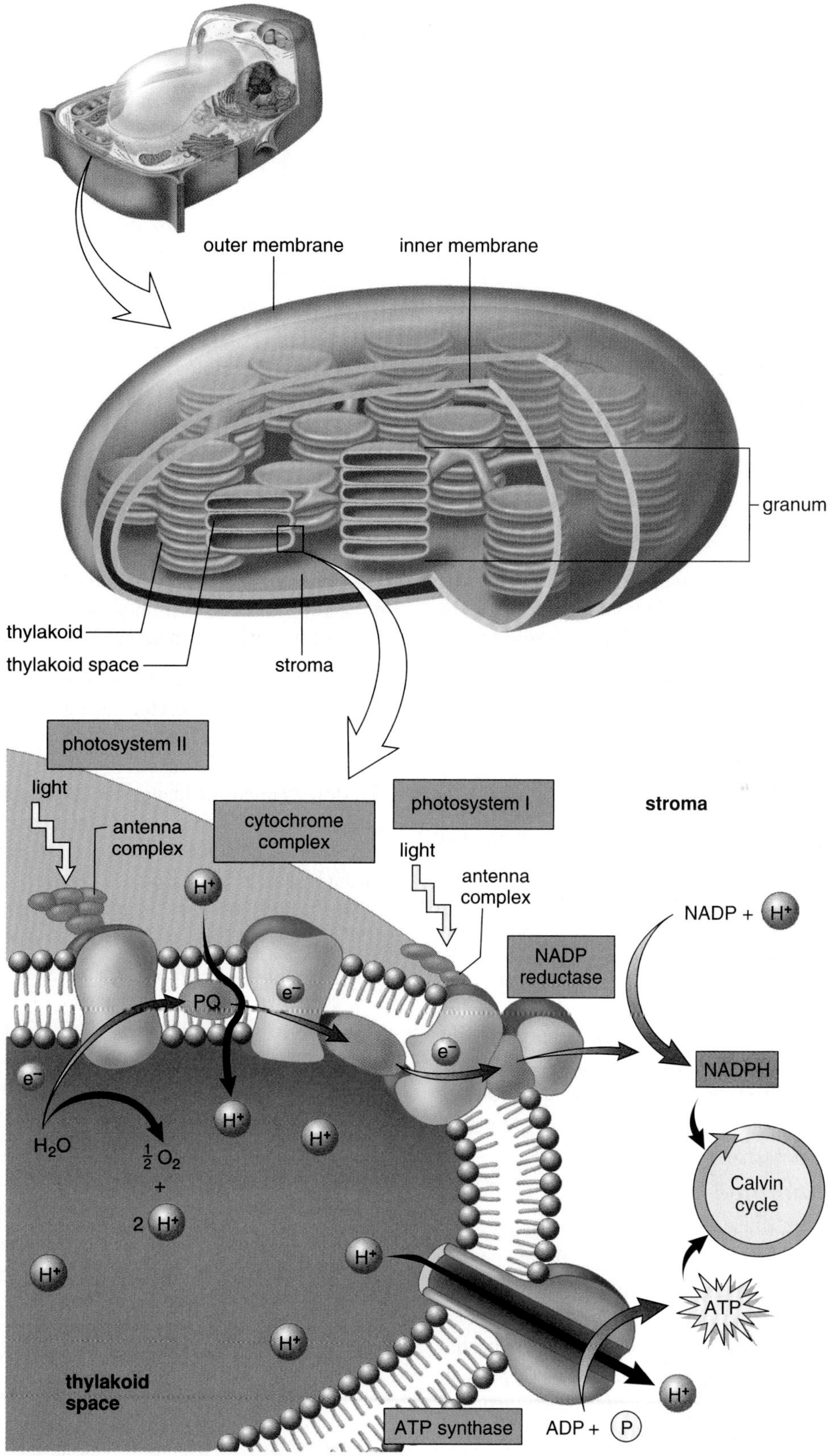

Figure 10.9 Organization of the thylakoid membrane showing relative positions of antenna complexes, photosystems, and protein complexes. Proteins of photosystems within the thylakoid membrane pump hydrogen ions from the stroma into the thylakoid space (lumen). When hydrogen ions flow back out of the thylakoid into the stroma through the ATP synthase complex, ATP is produced.
Source: From Sylvia S. Mader, Biology, 7th edition.

Awareness

Photosynthesis and Pizza

The next time you order pizza, imagine all the things that went into making that pizza. Of course, the very bread used to make the crust came from wheat seeds, produced as a result of photosynthesis. You can probably ascertain that any fruits or vegetables, including tomatoes, green peppers, onions, and spices also were produced because photosynthesis took place. The cheese, pepperoni, sausage, bacon, and, perhaps, even anchovies on the pizza came to exist as a result of animals that ate plants produced from photosynthesis. Even the pizza box itself was made as a direct result of photosynthesis, which produces the cellulose we need for paper products, clothing, and wood. As you turn the pages (also made from plants) of this chapter, try to envision the remarkable processes that enabled conversion of light energy into biological energy. It is this energy, derived from the photosynthetic light reactions, that fuels the light-independent reactions of photosynthesis and, hence, the synthesis of organic molecules needed for life on earth.

Although some protons are used in the production of NADPH on the stroma side of the thylakoid membrane, there is still a net accumulation of protons in the thylakoid lumen from the splitting of water molecules and electron transport. This establishes a proton gradient, giving special proteins in the thylakiod membrane the potential for moving protons from the thylakoid lumen back to the stroma. Movement of protons across the membrane is thought to be a source of energy for the synthesis of ATP. The action has been described as similar to the movement of molecules during osmosis and has been called *chemiosmosis,* or the *Mitchell theory,* after its author, Peter Mitchell. In this concept, protons move across a thylakoid membrane through protein channels called ATPase. With the proton movement, ADP and phosphate (P) combine, producing ATP (Fig. 10.9). This chemiosmotic mechanism for connecting electron flow with conversion of ADP to ATP is essentially the same as that of oxidative phosphorylation in mitochondria (see Fig. 10.14), except that in mitochondria, oxygen (instead of NADP) is the terminal electron acceptor.

The Light-Independent Reactions Reexamined

We have seen how synthesis of ATP and NADPH is set in motion during the *light-dependent reactions.* Both of these substances play key roles in the synthesis of carbohydrate from carbon dioxide from the atmosphere, which during the *light-independent reactions* reaches the interior of chlorenchyma tissues via stomata. As indicated earlier, the light-independent reactions are really a whole series of reactions, each mediated by an enzyme in this major phase of photosynthesis.

The light-independent reactions take place in the stroma of the chloroplasts and, as long as the products of the light-dependent reactions are present, they do not directly need light in the same sense that the light-dependent reactions do. However, they normally take place only during daylight hours because some of the enzymes involved in the light-independent reactions require light for their activation or conversion to a form in which they can actively catalyze steps of the light-independent reactions.

The Calvin Cycle The heart of the light-independent reactions is the **Calvin cycle,** during which carbon dioxide is fixed and converted to carbohydrate. The carbohydrate produced during these reactions facilitates growth, including the development of leaves, stems, roots, flowers, and other plant structures. From an ecological standpoint, this process is essential because it is the main biosynthetic pathway through which carbon enters the web of life. As discussed next, the Calvin cycle, or *photosynthetic carbon reduction (PCR)* pathway (Fig. 10.10), runs in five main steps:

1. Six molecules of carbon dioxide (CO_2) from the air combine with six molecules of ribulose 1,5-bisphosphate (RuBP, the 5-carbon sugar continually being formed while photosynthesis is occurring), with the aid of the enzyme **rubisco** (*RuBP carboxylase/oxygenase*).[5]
2. The resulting six 6-carbon unstable complexes are immediately split into twelve *3-carbon* molecules known as 3-phosphoglyceric acid (3PGA), the first stable compound formed in photosynthesis.
3. The NADPH (which has been temporarily holding the hydrogen and electrons released during the light-dependent reactions) and ATP (also from the light-dependent reactions) supply energy and electrons that chemically reduce the 3PGA to twelve molecules of glyceraldehyde 3-phosphate (GA3P, 3-carbon sugar phosphate).
4. Ten of the twelve glyceraldehyde 3-phosphate molecules are restructured, using another six ATPs, and become six 5-carbon molecules of RuBP, the sugar with which the Calvin cycle was initiated.

---

5. Rubisco is a huge, complex enzyme that may constitute up to 30% of the protein present in a living leaf. It is the most abundant protein known.

Ecological Review

Life on earth depends on photosynthesis to provide oxygen and absorb carbon dioxide. Photosynthesis is also the foundation through which carbon enters the web of life, directly or indirectly providing food and shelter for most living organisms. Photosynthetic efficiency varies with changes in carbon dioxide concentration, water availability, temperature, and light intensity. Human activities, including the use of fossil fuels, pesticides, and pollutants such as PCBs (polychlorinated biphenyls), disrupt the factors affecting photosynthesis. As a result of human impact, there have been, and continue to be, fluctuations in photosynthetic efficiency within nature, as well as changes in the way natural elements, water, and organic materials are recycled. While some of the activities mentioned may be temporarily beneficial to humans in the name of "progress," the long-term detrimental effects of industrialization, deforestation, and pollution are becoming apparent. The vulnerability of photosynthesis to changes in the environment signifies our need to do as much as possible to preserve the balance of nature on our planet.

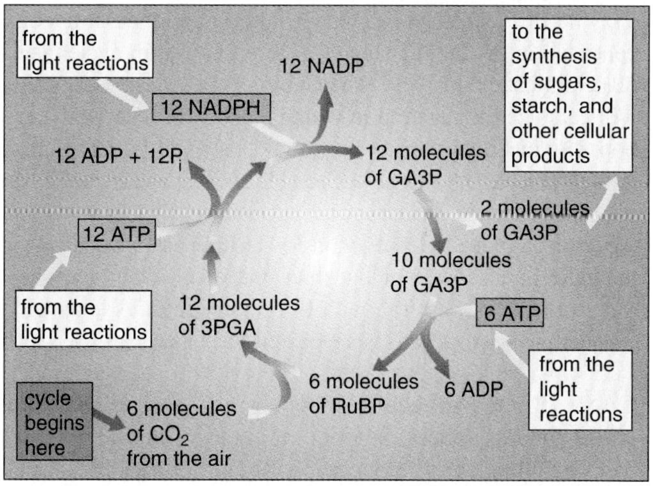

Figure 10.10 The Calvin cycle (light-independent reactions) of photosynthesis. The cycle takes place in the stroma of chloroplasts where each step is mediated by an enzyme. Carbon dioxide molecules from the air enter the cycle one at a time, making six turns of the cycle necessary to produce one molecule of a 6-carbon sugar such as glucose ($C_6H_{12}O_6$).

5. This leaves a net gain of two GA3P molecules, which can contribute either to an increase in the carbohydrate content of the plant (glucose, starch, cellulose, or related substances) or can be used in pathways that lead to the net gain of lipids and amino acids.

Since rubisco catalyzes formation of the 3-carbon compound 3PGA as the first isolated product in these light-independent reactions, plants demonstrating this process are called **C_3 plants.** However, as indicated by its name, the enzyme RuBP carboxylase/oxygenase has the potential to fix both CO_2 through its carboxylase activity described by the Calvin cycle and O_2 through its oxygenase activity. The oxygenase activity of rubisco makes C_3 plants vulnerable to a process called **photorespiration.**

Photorespiration When the weather is hot and dry, stomata are generally closed. The closed stomata prevent carbon dioxide from entering the leaf, and the concentration of carbon dioxide within drops steadily while the relative concentration of oxygen increases. When the carbon dioxide concentration drops below roughly 50 parts per million, rubisco starts fixing oxygen instead of carbon dioxide, initiating *photorespiration.* Like photosynthesis, the process of photorespiration requires light because regeneration of RuBP requires ATP and NADPH, presumably from the light-dependent reactions. When photorespiration occurs, RuBP reacts with oxygen, releases carbon dioxide, and if the Calvin cycle is functioning, is in competition with it. Accordingly, photorespiration is generally thought of as an inefficient pathway associated with C_3 plants under hot, dry conditions. Although photorespiration often increases under drought stress, the oxygen-binding properties of rubisco are not directly affected by the reduced availability of water. Rather, water stress induces stomatal closure, which then decreases the CO_2:O_2 ratio in the leaf, favoring respiration. While the stomata are closed, the concentration of carbon dioxide within the leaf drops steadily while the relative concentration of oxygen increases.

The products of photorespiration are the 2-carbon phosphoglycolic acid, which in mitochondria and peroxisomes ultimately releases carbon dioxide in mitochondria and peroxisomes, and the 3-carbon phosphoglyceric acid that can reenter the Calvin cycle. No ATP is produced by photorespiration.

The 4-Carbon Pathway Sugar cane, corn, sorghum, and at least 1,000 other species of tropical grasses or arid region plants subject to high temperature stresses have a distinctive leaf anatomy called *Kranz anatomy* (Fig. 10.11). Kranz anatomy leaves have two forms of chloroplasts. In the bundle sheath cells surrounding the veins, there are large chloroplasts, often with few to no grana. The large chloroplasts have numerous starch grains, and the grana, when present, are poorly developed. The chloroplasts of the mesophyll cells are much smaller, usually lack starch grains, and have well-developed grana.

Plants with Kranz anatomy produce a *4-carbon* compound, *oxaloacetic acid,* instead of the 3-carbon 3-phosphoglyceric acid during the initial steps of the light-independent reactions. Oxaloacetic acid is produced when a 3-carbon compound, *phosphoenolpyruvate (PEP),* and carbon dioxide are combined in mesophyll cells with the aid of a different carbon-fixing enzyme, **PEP carboxylase.** Depending on the species, the oxaloacetic acid may then be converted to

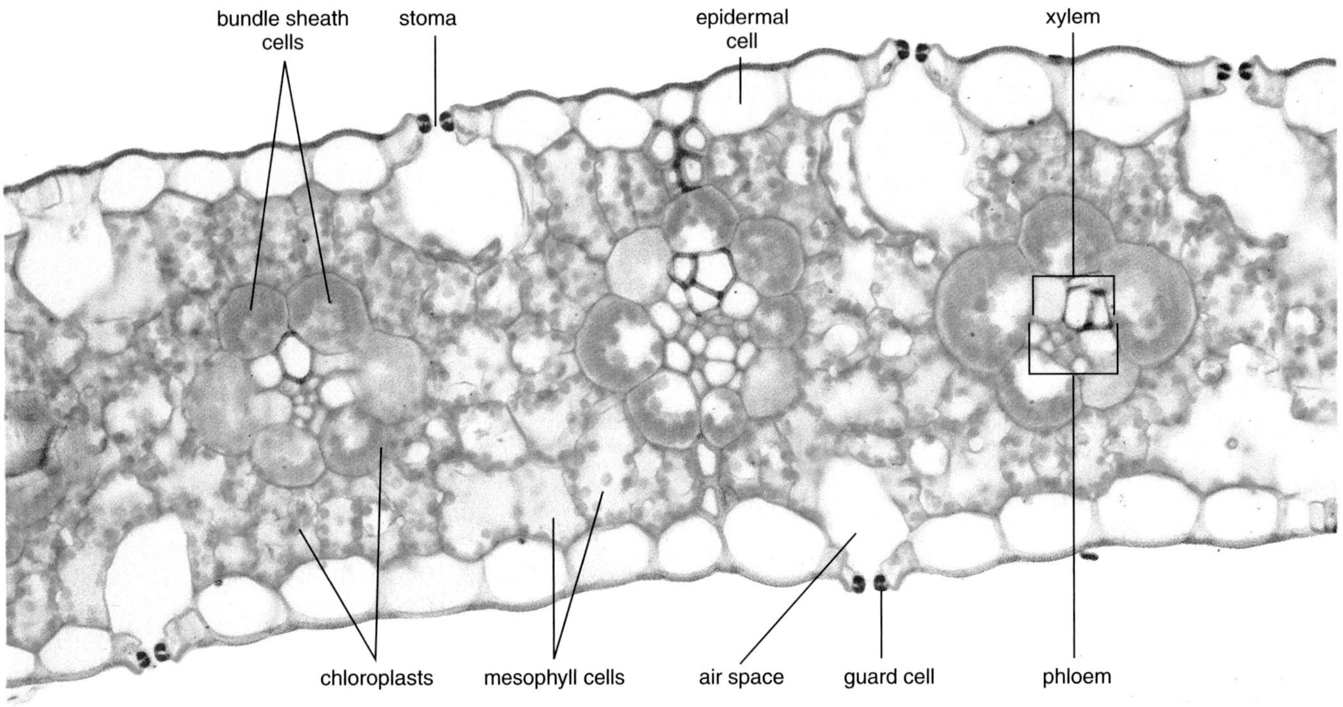

Figure 10.11 A portion of a cross section of a leaf of corn (*Zea mays*), a C_4 plant with Kranz anatomy leaves. Compare this with leaves of typical C_3 plants, as illustrated in Figures 7.6 and 7.11*A*.

aspartic, malic, or other acids. PEP carboxylase is not sensitive to oxygen and hence has a greater enzyme affinity for carbon dioxide, so there is no photorespiratory loss of carbon dioxide captured in the organic acids. The carbon dioxide is transported to the bundle sheath cells where it is released and enters the normal Calvin cycle just as in C_3 plants. The carbon dioxide concentration can be kept high in relation to the oxygen concentration in the bundle sheath cells, thereby keeping the reaction of rubisco with oxygen very low (Fig. 10.12).

Because the PEP system produces 4-carbon compounds, plants having this system are called 4-carbon, or **C_4, plants**, to distinguish them from plants that have only the 3-carbon, or C_3, system. Besides Kranz anatomy, C_4 plants have other characteristic features:

1. High concentrations of PEP carboxylase are found in the mesophyll cells. This is significant because PEP carboxylase allows the conversion of carbon dioxide to carbohydrate at much lower concentrations than does rubisco (found only in the bundle sheath cells), the corresponding enzyme in the Calvin cycle.
2. The optimum temperatures for C_4 photosynthesis are much higher than those for C_3 photosynthesis, allowing C_4 plants to thrive under conditions that would adversely affect C_3 plants.

Obviously, the C_4 pathway furnishes carbon dioxide to the Calvin cycle in a more roundabout way than does the C_3 pathway, but the advantage of this extra pathway is a major reduction of photorespiration in C_4 plants. Also, in C_4 plants, the C_4 pathway in the mesophyll cells results in carbon dioxide being picked up even at low concentrations (via the enzyme PEP carboxylase) and in carbon dioxide being concentrated in the bundle sheath, where the Calvin cycle takes place almost exclusively. The Calvin cycle enzyme, rubisco, will catalyze the reaction in which RuBP will react with carbon dioxide rather than oxygen. Consequently, C_4 plants, which typically photosynthesize at higher temperatures, have photosynthetic rates that are two to three times higher than those of C_3 plants. However, at lower temperatures, C_3 photosynthesis is more efficient than that of C_4 plants because the cost of photorespiration at those temperatures is usually less than the two extra ATPs required for the C_4 pathway.

CAM Photosynthesis *Crassulacean acid metabolism (CAM)* photosynthesis is found in plants of about 30 families, including cacti, stonecrops, orchids, bromeliads, and other succulents that are often stressed by limited availability of water. A few succulents do not have CAM photosynthesis, however, and several nonsucculent plants do. Many CAM plants are facultative C_3 plants that can switch to C_3 photosynthesis during the day after a good rain or when night temperatures are high. Plants with CAM photosynthesis typically do not have a well-defined palisade mesophyll in the leaves, and, in contrast to the chloroplasts of the bundle sheath cells of C_4 plants, those of CAM photosynthesis plants resemble the mesophyll cell chloroplasts of C_3 plants.

CAM photosynthesis is similar to C_4 photosynthesis in that 4-carbon compounds are produced during the light-independent reactions. In these plants, however, the organic acids

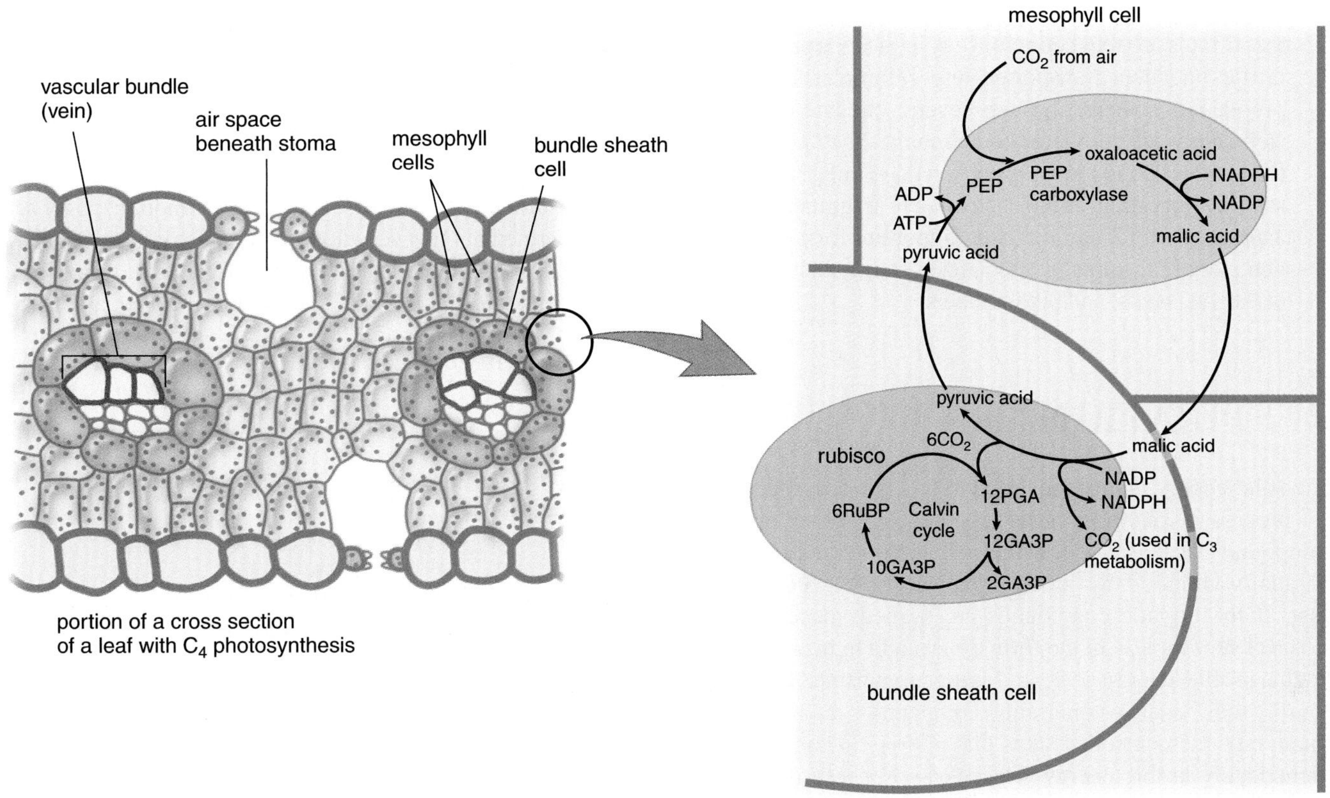

Figure 10.12 An illustration of the C_4 photosynthesis pathway. Carbon dioxide is converted to organic acids in mesophyll cells. After the acids move into bundle sheath cells, some carbon dioxide is released and enters the Calvin cycle where it becomes a 3-carbon compound that moves back to a mesophyll cell; there it is converted to PEP, which accepts carbon dioxide from the air.

(mainly malic acid) accumulate at night and break down during the day, releasing carbon dioxide. The enzyme PEP carboxylase is responsible for converting the carbon dioxide plus PEP to organic acids at night when the stomata are open. During daylight, the organic acids diffuse out of the cell vacuoles in which they are stored and are converted back to carbon dioxide for use in the Calvin cycle. A much larger amount of carbon dioxide can be converted to carbohydrate each day than would otherwise be possible, since the stomata of such plants are closed during the day to conserve water. This arrangement allows the plants to function well under conditions of both limited water supply and high light intensity (Fig. 10.13).

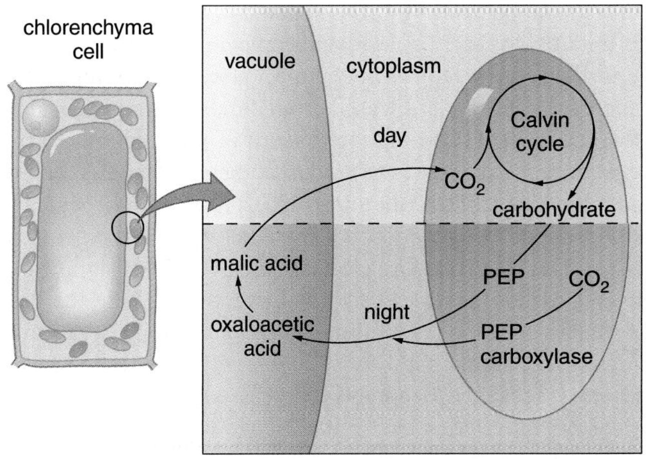

Figure 10.13 Crassulacean acid metabolism (CAM) photosynthesis found in orchids, pineapples, and many desert plants. CAM is similar to C_4 photosynthesis, but the plants have their stomata closed during daylight heat, thus conserving water. The organic acids accumulate at night and break down during the day, releasing carbon dioxide, which then enters the Calvin cycle and C_3 metabolism while the stomata are closed.

Other Significant Processes That Occur in Chloroplasts

In addition to photosynthesis, there are two very important sets of biochemical reactions that take place in chloroplasts (and also in the proplastids of roots).

1. Sulfates are reduced to sulfide via several steps involving ATP and enzymes. The sulfide is rapidly converted into important sulfur-containing amino acids, such as methionine and cysteine, which are part of the building blocks for proteins, anthocyanin pigments, chlorophylls, and several other cellular components.

2. Nitrates are reduced to organic compounds. Initially, the nitrates are reduced to nitrites in the cytoplasm. The nitrites are then transported into chloroplasts (or root proplastids) where, through several enzyme-mediated steps, they are converted to ammonia. The ammonia is then converted to amino groups that are integral parts of several important amino acids such as glutamine and aspartic acid. Glutamine is an important form of nitrogen storage in roots or specialized stems such as those of carrots, beets, and potato tubers.

RESPIRATION

The solar energy that is converted to biochemical energy by the process of photosynthesis is stored in various organic compounds such as wood, while coal and oil contain energy originally captured by green organisms in the geological past. If the organic compounds are burned, the energy is released very rapidly in the form of heat and light, and much of the usable energy is lost. Living organisms, however, "burn" their energy-containing compounds in numerous, small, enzyme-controlled steps that release tiny amounts of immediately usable energy. The released energy is usually stored in ATP molecules, which allows the available energy to be used more efficiently and the process to be controlled more precisely.

1. The Essence of Respiration

Respiration is essentially the release of energy from glucose molecules that are broken down to individual carbon dioxide molecules. The process takes place in all active cells 24 hours a day, regardless of whether or not photosynthesis happens to be occurring simultaneously in the same cells. It is initiated in the cytoplasm and completed in the mitochondria. The energy, stored in chemical bonds containing high-energy electrons, is released from simple sugar molecules that are broken down during a series of steps controlled by enzymes. No oxygen is needed to initiate the process, but in **aerobic respiration** (the most widespread form of respiration), the process cannot be completed without oxygen gas (O_2). The controlled release of energy is the significant event; carbon dioxide (CO_2) and water (H_2O) are the by-products. Aerobic respiration can be summed up in the following equation, but bear in mind that respiration, like photosynthesis, is a complex process that involves many steps not reflected in a simplified equation.

$$\underset{\text{glucose}}{C_6H_{12}O_6} + \underset{\text{oxygen}}{6\,O_2} \xrightarrow{\text{enzymes}} \underset{\text{carbon dioxide}}{6\,CO_2} + \underset{\text{water}}{6\,H_2O} + \text{energy}$$

Anaerobic respiration and **fermentation** are two forms of respiration that were probably carried on in the geological past when there was no oxygen in the atmosphere. These forms of respiration are still carried on today by certain bacteria and other organisms in the absence of oxygen gas. Anaerobic respiration and fermentation release less than 6% of the energy released from a molecule of glucose by aerobic respiration. The two forms differ from one another in the manner in which hydrogen released from the glucose is combined with other substances (see the discussion in a later section). Fermentation is very important industrially, particularly in the brewing and baking industries. Two well-known forms of fermentation are illustrated by the following equations:

$$\underset{\text{glucose}}{C_6H_{12}O_6} \xrightarrow{\text{enzymes}} \underset{\text{ethyl alcohol}}{2\,C_2H_5OH} + \underset{\text{carbon dioxide}}{2\,CO_2} + \text{energy (ATP)}$$

$$\underset{\text{glucose}}{C_6H_{12}O_6} \xrightarrow{\text{enzymes}} \underset{\text{lactic acid}}{2\,C_3H_6O_3} + \text{energy (ATP)}$$

The relatively small amount of energy released during these inefficient forms of respiration is partly stored in two ATP molecules. The actual amount of energy stored is roughly only 29% of the approximately 48 Kcals[6] of energy released in anaerobic respiration.

2. Introduction to the Major Steps of Respiration

Glycolysis

In most forms of carbohydrate respiration, the first major phase takes place in the cytoplasm and requires no oxygen gas (O_2). This phase, called **glycolysis,** involves three main steps and several smaller ones, each controlled by an enzyme. During the process, a small amount of energy is released, and some hydrogen atoms are removed from compounds derived from a glucose molecule. The essence of this complex series of steps is as follows:

1. In a series of reactions, the glucose molecule becomes a fructose molecule carrying two phosphates (P).
2. This sugar (fructose) molecule is split into two 3-carbon fragments called glyceraldehyde 3-phosphate (GA3P).
3. Some hydrogen, energy, and water are removed from these 3-carbon fragments, leaving **pyruvic acid** (Fig. 10.14).

Two ATP molecules supply the energy needed to start the process of glycolysis. By the time pyruvic acid has been formed, however, four ATP molecules have been produced from the energy released along the way, for a net gain of two ATP molecules. A great deal of the energy originally in the glucose molecule remains in the pyruvic acid. The hydrogen ions and high-energy electrons released during the process

6. A Kcal is the energy required to raise the temperature of one kilogram of water from 14.5°C to 15.5°C.

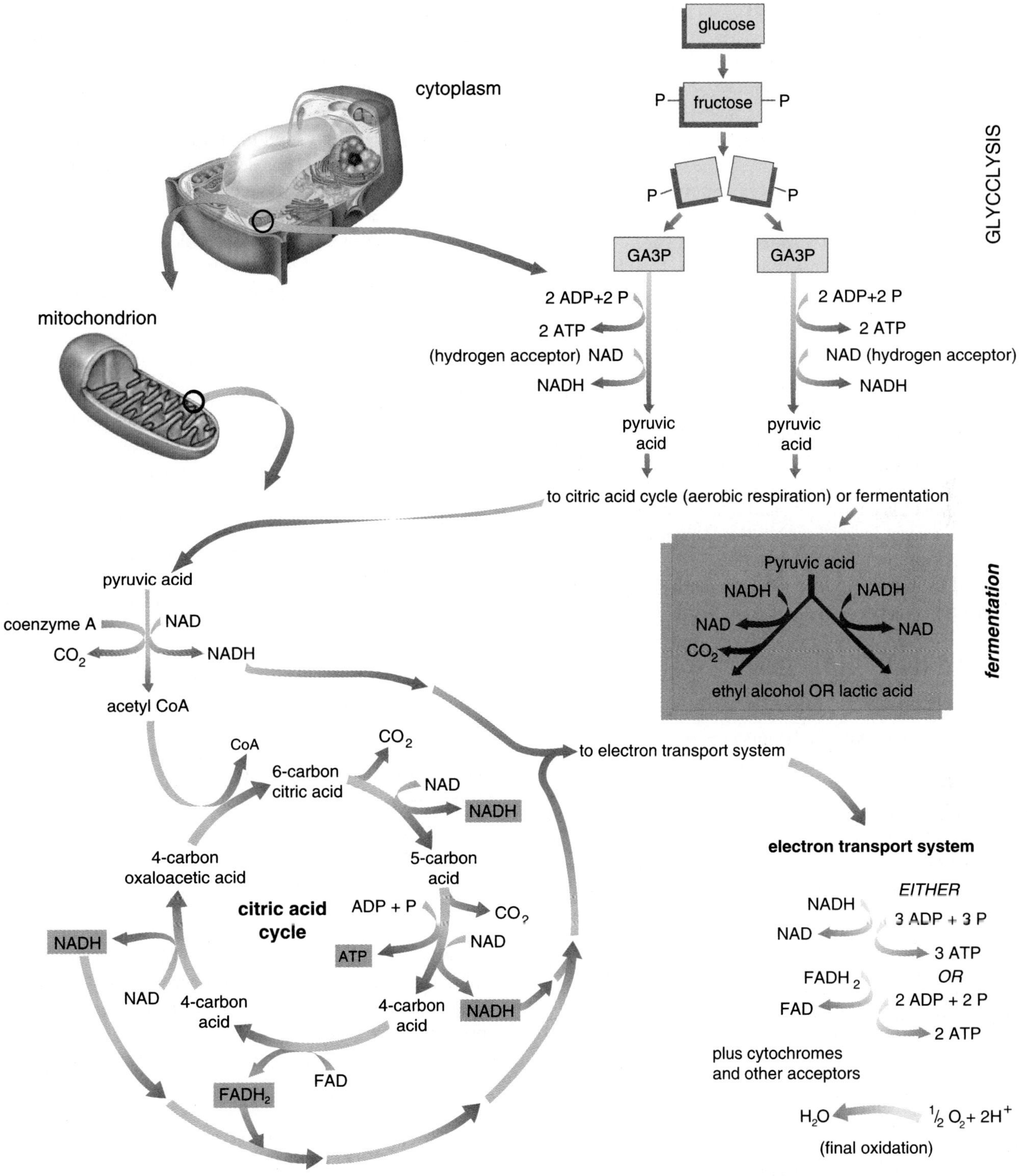

Figure 10.14 A summary of respiration. In the first phase, glycolysis, which takes place in the cytoplasm, a sugar molecule is converted to two pyruvic acid molecules. The subsequent phases take place within the matrix of a mitochondrion. In aerobic respiration, the pyruvic acid is broken down in the citric acid cycle, and energy is transferred to compounds such as NADH, ATP, and $FADH_2$. Carbon dioxide and hydrogen are also released. The released hydrogen is carried by an electron transport system and combined with oxygen, forming water. In anaerobic respiration and fermentation, pyruvic acid is converted in the absence of oxygen gas to ethyl alcohol or lactic acid with little release of energy. *Inset from Sylvia S. Mader, Biology, 7th edition.*

are picked up and temporarily held by an acceptor molecule, **NAD** (*nicotinamide adenine dinucleotide*). What happens to them next depends on the kind of respiration involved: *aerobic respiration, true anaerobic respiration,* or *fermentation.*

Aerobic Respiration

In aerobic respiration (the most common type of respiration), glycolysis is followed by two major stages: the *citric acid cycle* and *electron transport.* Both stages occur in the mitochondria and involve many smaller steps, each of which is controlled by enzymes (see Fig. 10.14).

The Citric Acid (Krebs) Cycle The **citric acid cycle** was originally named the *Krebs cycle* after Hans Krebs, a British biochemist who received a Nobel Prize in 1953 for his unraveling of many of the complex reactions that take place in respiration. The name *citric acid cycle,* or *tricarboxylic acid (TCA) cycle,* reflects the important role played by several organic acids during the process.

Before entering the citric acid cycle, which takes place in the fluid matrix located within the compartments formed by the cristae of mitochondria (see Fig. 3.14), carbon dioxide is released from pyruvic acid that was produced by glycolysis. What remains is restructured to a 2-carbon acetyl group. This acetyl group combines with an acceptor molecule called *coenzyme A (CoA).* This combination (called *acetyl CoA*) then enters the citric acid cycle, which is a series of biochemical reactions that are catalyzed by enzymes. Little of the energy originally trapped in the glucose molecule is released during glycolysis. As the citric acid cycle proceeds, however, high-energy electrons and hydrogen are successively removed. This removal takes place from a series of organic acids and, after transfer, ultimately produces compounds such as **NADH** (*reduced nicotinamide adenine dinucleotide*) and **$FADH_2$** (*reduced flavin adenine dinucleotide*), as well as a small amount of ATP. Carbon dioxide is produced as a by-product while the cycle is proceeding.

Electron Transport Much of the energy originally in the glucose molecule now has been transferred to the acceptors NAD and FAD, which became NADH and $FADH_2$, respectively. NADH and $FADH_2$ are electron donors to an *electron transport system* consisting of special acceptor molecules arranged in a precise sequence on the inner membrane of mitochondria. The electrons flow through the system down an energy gradient. Energy is extracted from the high-energy electrons to pump protons across the inner mitochondrial membrane, creating an electrochemical gradient. With the aid of an enzyme, the gradient can then be harvested to produce ATP. The production of ATP stops if there are no electron donors or electron acceptor oxygen.

The acceptor molecules include iron-containing proteins called *cytochromes.* Energy is released in small increments at each step along the system, and ATP is produced from ADP and P. As the final step in aerobic respiration (see Fig. 10.14), oxygen acts as the ultimate electron acceptor, producing water as it combines with hydrogen.

By the time the process is complete, the recoverable energy locked in a molecule of glucose has been released and is stored in ATP molecules. This stored energy is then available for use in the synthesis of other molecules and for growth, active transport, and a host of other metabolic processes. Aerobic respiration produces a net gain of 36 ATP molecules from one glucose molecule, using up six molecules of oxygen and producing six molecules of carbon dioxide and a net total of six molecules of water. For each mole (180 grams) of glucose aerobically respired, 686 Kcal of energy is released, with about 39% of it being stored in ATP molecules and the remainder being released as heat.

Anaerobic Respiration and Fermentation

In living organisms, glucose molecules often may undergo glycolysis without enough oxygen being available to complete aerobic respiration. In such cases, the hydrogen released during glycolysis is simply transferred from the hydrogen acceptor molecules back to the pyruvic acid after it has been formed, creating ethyl alcohol in some organisms, and lactic acid or similar substances in others. A little energy is released during either fermentation or true anaerobic respiration, but most of it remains locked up in the alcohol, lactic acid, or other compounds produced.

In *true anaerobic respiration,* the hydrogen removed from the glucose molecule during glycolysis is combined with an inorganic ion, as, for example, when sulfur bacteria (discussed in Chapter 17) convert sulfate (SO_4) to sulfur (S) or another sulfur compound or when certain cellulose bacteria produce methane gas (CH_4) by combining the hydrogen with carbon dioxide.

Oxygen gas is not required to make these compounds, but few organisms can live long without oxygen, and many that carry on fermentation can also respire aerobically. If oxygen becomes available, the remaining energy can be released by further breakdown of these compounds. About 7% of the total energy in a glucose molecule is removed during anaerobic respiration or fermentation. So much of that energy goes into the making of the alcohol or the lactic acid or is dissipated as heat that there is a net gain of only two ATP molecules (compared with 36 ATP molecules produced in aerobic respiration). The forms of anaerobic respiration are adaptive to the organisms that have them in that they recycle NAD and allow glycolysis to continue.

Living cells can tolerate only certain concentrations of alcohol. In media in which yeasts are fermenting sugars, for example, once the alcohol concentration builds up beyond 12%, the cells die and fermentation ceases. This is why most wines have an alcohol concentration of about 12% (24 proof).

Many bacteria carry on both fermentation and true anaerobic respiration simultaneously, making it difficult to distinguish between the two processes. Some texts use the terms *anaerobic respiration* and *fermentation* interchangeably to designate respiration occurring in the presence of little or no oxygen gas.

Factors Affecting the Rate of Respiration

Temperature

Temperature plays a major role in the rate at which the various respiratory reactions occur. For example, when air temperatures rise from 20°C (68°F) to 30°C (86°F), the respiration rates of plants double and sometimes even triple. The faster respiration occurs, the faster the energy is released from sugar molecules, with an accompanying decrease in weight. In growing plants, this weight loss is more than offset by the production of new sugar by photosynthesis. In harvested fruits, seeds, and vegetables, however, respiration continues without sugar replacement, and some water loss also occurs. Respiring cells convert energy stored as starch or sugar primarily to ATP, but much of the energy is lost in the form of heat, with only 39% being stored as ATP. Most fresh foods are kept under refrigeration, not only to lower the respiration rate and retard water loss, but also to dissipate the heat. Keeping the temperatures down is also important to prevent the growth and reproduction of food-spoiling molds and bacteria, which may thrive at warmer temperatures.

Heat inactivates most enzymes at temperatures above 40°C (104°F), but a few organisms, such as various cyanobacteria and algae in the hot springs of Yellowstone National Park and similar places, have adapted in such a way that they are able to thrive at temperatures exceeding 60°C (140°F)—heat that would kill other organisms of comparable size almost instantly.

Water

Water inside the cells and their organelles act as a medium in which the enzymatic reactions can take place. Living cells often have a water content of more than 90%, but the cells of mature seeds may have a water content of less than 10%. When water content becomes this low, respiration does not cease completely, but it continues at a drastically reduced rate, resulting in only very tiny amounts of heat being released and of carbon dioxide being given off. Seeds may remain viable (capable of germinating) for many years if stored under dry conditions. If they come in contact with water, however, they swell by imbibition. Respiration rates then increase rapidly. If the wet seeds happen to be in an unrefrigerated storage bin, the temperature may increase to the point of killing the seeds. In fact, if fungi and bacteria begin to grow on the seeds, temperatures from their respiration can become so high that spontaneous combustion can sometimes occur.

Oxygen

If flooding sharply reduces the oxygen supply available to the roots of trees and house plants, their respiration and growth rates may be decreased. They may even die if the condition persists too long. When foods are stored, however, it helps to bring about lower rates of respiration by reducing the oxygen in the storage areas. In fact, it is a common commercial practice to reduce the oxygen present in warehouses where crops are stored. The oxygen content is reduced to as little as 1% to 3% by pumping in nitrogen gas, while maintaining low temperatures and humidity. Oxygen concentration is not reduced below 1% because that can result in an undesirable increase in fermentation.

3. A Closer Look at Respiration

Respiration, like photosynthesis, is a very complex process, and, as with photosynthesis, it is beyond the scope of this book to explore the subject in great detail. The following amplification of information already discussed is modest, and those who wish further information are referred to the reading list at the end of the chapter.

Glycolysis Reexamined

As previously discussed, this initial phase of all forms of respiration brings about the conversion of each 6-carbon glucose molecule to two 3-carbon pyruvic acid molecules via three main steps, each mediated by enzymes. The three main steps are as follows:

1. *Phosphorylation,* whereby the 6-carbon sugars receive phosphates
2. *Sugar cleavage,* which involves the splitting of 6-carbon fructose into two 3-carbon sugar fragments
3. *Pyruvic acid formation,* which involves the oxidation of the sugar fragments

Energy needed to initiate the process is furnished by an ATP molecule, which also furnishes the phosphate group for the phosphorylation of the sugar glucose to yield glucose 6-phosphate. Another ATP, with the aid of the enzyme fructokinase, yields fructose bisphosphate (fructose 1,6-diphosphate). As a result of the cleavage of the fructose bisphosphate, two different 3-carbon sugars are produced, but ultimately, only two glyceraldehyde 3-phosphate (GA3P) molecules remain. These two 3-carbon sugars are oxidized to two 3-carbon acids, and, in the successive production of several of these acids, phosphate groups are removed from the acids. The phosphate groups combine with ADP, producing a net direct gain of two ATP molecules during glycolysis. In addition, hydrogen is removed as GA3P is oxidized. This hydrogen is picked up by the acceptor molecule, NAD, which becomes NADH. Glycolysis, which requires no oxygen gas, is summarized in Figure 10.14.

Transition Step to the Citric Acid (Krebs) Cycle

Before a pyruvic acid molecule enters the citric acid cycle, which takes place in the mitochondria, a molecule of carbon dioxide is removed and a molecule of NADH is produced, leaving an acetyl fragment. The 2-carbon fragment is then bonded to a large molecule called *coenzyme A.* Coenzyme A consists of a combination of the B vitamin pantothenic acid and a nucleotide. Pantothenic acid is one of several B vitamins essential to respiration in both plants and animals; others include thiamine (vitamin B_1), niacin, and riboflavin. The bonded acetyl fragment and coenzyme A molecule is referred to as *acetyl CoA.* The following equation summarizes the fate of the two pyruvic acid molecules following glycolysis and leading to the citric acid cycle:

$$\textbf{2 pyruvic acid + 2 CoA + 2 NAD} \xrightarrow{\textbf{enzymes}} \textbf{2 acetyl CoA + 2 NADH + 2 CO}_2$$

In addition to pyruvic acid, fats and amino acids can also be converted to acetyl CoA and enter the process at this point. The NADH molecules donate their hydrogen to an electron transport system (discussed in the section, "Electron Transport and Oxidative Phosphorylation"), and the acetyl CoA enters the citric acid cycle (see Fig. 10.14).

The Citric Acid (Krebs) Cycle Reexamined

In the citric acid cycle, acetyl CoA is first combined with oxaloacetic acid, a 4-carbon compound, producing citric acid, a 6-carbon compound. The citric acid cycle is kept going by oxaloacetic acid, which is produced in small amounts, but is an intermediate product rather than a starting substance or an end product of the cycle. As the cycle progresses, a carbon dioxide is removed, producing a 5-carbon compound. Then another carbon dioxide is removed, producing a 4-carbon compound. This 4-carbon compound, through additional steps, is converted back to oxaloacetic acid, the substance with which the cycle began, and the cycle is repeated.

Each full cycle uses up a 2-carbon acetyl group and releases two carbon dioxide molecules while regenerating an oxaloacetic acid molecule for the next turn of the cycle. Some hydrogen is removed during the process and is picked up by FAD and NAD. One molecule of ATP, three molecules of NADH, and one molecule of $FADH_2$ are produced for each turn of the cycle. The citric acid cycle may be summarized as follows:

$$\underset{\textbf{acid}}{\textbf{oxaloacetic}} + \underset{\textbf{CoA}}{\textbf{acetyl}} + \textbf{ADP + P + 3 NAD + FAD} \xrightarrow{\textbf{enzymes}}$$

$$\underset{\textbf{acid}}{\textbf{oxaloacetic}} + \textbf{CoA + ATP + 3 NADH + H}^+ + \textbf{FADH}_2 + \textbf{2 CO}_2$$

The hydrogen carried by NAD and FAD can mostly be traced to the acetyl groups and to water molecules added to some compounds in the citric acid cycle. The FAD and $FADH_2$ are now known to be intermediate compounds. *Ubiquinol,* a component of the electron transport system, receives electrons from either NADH or $FADH_2$.

Electron Transport and Oxidative Phosphorylation

After completion of the citric acid cycle, the glucose molecule has been totally dismantled, and some of its energy has been transferred to ATP molecules. A considerable portion of the energy was transferred to NAD and FAD when they were used to pick up hydrogen and electrons from the molecules derived from glucose as they were broken down during glycolysis and the citric acid cycle. This energy is released as the hydrogen and electrons are passed along an *electron transport system.* This system, like the electron transport system of photosynthesis, functions something like a high-speed bucket brigade in passing along electrons from their source to their destination. Several of the electron carriers in the transport system are cytochromes. They are very specific and, as electrons flow along the system, they can transfer their electrons only to other specific acceptors. When the electrons reach the end of the system, they are picked up by oxygen and combine with hydrogen ions, forming water.

Part of the energy that is released during the movement of electrons along the electron transport system can be used to make ATP in a process called *oxidative phosphorylation.* If hydrogen ion and electron transport begins with NADH, which was produced inside the mitochondria (i.e., during the conversion of pyruvic acid to acetyl CoA and during the citric acid cycle), enough energy is produced to yield three ATP molecules from each NADH molecule. Similarly, if hydrogen ion and electron transport begins either with $FADH_2$ or with NADH, produced outside the mitochondria (i.e., during glycolysis), two ATP molecules are produced.

The manner in which ATP is produced during the operation of the respiratory electron transport system involves essentially the same *chemiosmotic* concept that was applied earlier to proton movement across thylakoid membranes.

The **chemiosmosis** theory concerning electron transport and proton movement across membranes was proposed in the 1960s by Peter Mitchell, a British biochemist, and is now widely accepted as the explanation for the movements in both photosynthesis and respiration. In respiration, oxidative phosphorylation is energized by a gradient of protons (H^+) that flow by chemiosmosis across the inner membrane of a mitochondrion. Mitchell, who received a Nobel Prize for his work in 1978, surmised that protons are "pumped" from the matrix of the mitochondria to the region between the two membranes (see Fig. 3.13) as electrons flow from their source in NADH molecules along the electron transport system, which is located in the inner membrane. The protons are believed to "diffuse" back into the matrix via channels provided by an enzyme complex known as the F1 particle (an ATPase), releasing energy that is used to synthesize ATP.

If we retrace our steps through the entire process of aerobic respiration, we find that glycolysis yields four molecules of ATP and two molecules of NADH (from which more molecules of ATP are formed), for a total of eight ATP from the conversion of glucose to two pyruvic acid molecules. Two ATP are used in the process, however, leaving a net gain of six ATP.

When two pyruvic acid molecules are converted to two acetyl CoA in the mitochondria, two more NADH molecules (which will generate six molecules of ATP) are produced. The two acetyl CoA molecules metabolized in the citric acid

TABLE 10.1

Summary of ATP Synthesized from One Glucose Molecule During Respiration

METABOLIC PATHWAY (AEROBIC RESPIRATION)	CELLULAR LOCATION: CYTOPLASM	CELLULAR LOCATION: MITOCHONDRIAL MATRIX	CELLULAR LOCATION: INNER MITOCHONDRIAL MEMBRANE	ATP PRODUCED (SUBSTRATE-LEVEL AND OXIDATIVE PHOSPHORYLATION)
Glycolysis	2 ATP			2 ATP
	2 NADH		2 ATP (2)	4 ATP
Conversion of pyruvate to acetyl CoA/electron transport and oxidative phosphorylation		2 (1 NADH)	2 (3 ATP)	6 ATP
Krebs cycle/electron transport and oxidative phosphorylation		2 (1 ATP)		2 ATP
		2 (3 NADH)	2 (9 ATP)	18 ATP
		2 (1 $FADH_2$)	2 (2 ATP)	4 ATP
Total				**36 ATP**

TABLE 10.2

Summary Comparison of Photosynthesis and Respiration

PHOTOSYNTHESIS	RESPIRATION
1. Stores energy in sugar molecules	1. Releases energy from sugar molecules
2. Uses carbon dioxide and water	2. Releases carbon dioxide and water
3. Increases weight	3. Decreases weight
4. Occurs only in light	4. Occurs in either light or darkness
5. Occurs only in cells containing chlorophyll	5. Occurs in all living cells
6. Produces oxygen in green organisms	6. Utilizes oxygen (aerobic respiration)
7. Produces ATP with light energy	7. Produces ATP with energy released from sugar

cycle yield two molecules of ATP, two molecules of $FADH_2$ (from which four ATP are formed), and six molecules of NADH (which cause the formation of 18 molecules of ATP), making a citric acid cycle total of 24 ATP. A grand total of 36 ATP is produced for the aerobic respiration of one glucose molecule (Table 10.1). The 36 ATP molecules represent about 39% of the energy originally present in the glucose molecule. The remaining energy is lost as heat or is unavailable. Aerobic respiration is still about 18 times more efficient than anaerobic respiration.

A condensed comparison between photosynthesis and respiration is shown in Table 10.2.

ADDITIONAL METABOLIC PATHWAYS

While photosynthesis and respiration are the main processes through which plants grow, develop, reproduce, and survive, there are many additional processes that contribute toward these activities. Most of these use intermediate steps, but they could not function without photosynthesis and respiration. Some of the essential compounds produced from additional pathways include sugar phosphates and nucleotides, nucleic acids, amino acids,

TABLE 10.3
Examples of Plant Secondary Compounds

COMPOUND	EXAMPLE SOURCE	HUMAN USE
Alkaloids		
Codeine	Opium poppy	Narcotic pain reliever; cough suppressant
Quinine	Quinine tree	Used to treat malaria
Phenolics		
Lignin	Woody plants	Used for hardwood furniture and baseball bats
Salicin	Willow tree	Aspirin precursor
Tetrahydrocannabinol	Marijuana	Treatment for glaucoma; nausea suppressant
Terpenoids		
Camphor	Camphor tree	Component of medicinal oils and disinfectants
Menthol	Mints and eucalyptus tree	Strong aroma; used in cough medicines
Rubber	Rubber tree	Rubber tires, rubber bands, and other commercial products

proteins, chlorophylls, cytochromes, carotenoids, fatty acids, oils, and waxes.

Metabolic processes not required for normal growth and development are generally referred to as *secondary metabolism.* Although not essential, many of the products from secondary metabolism enable plants to survive and persist under special conditions. These products provide the plant with unique colors, aromas, poisons, and other compounds that may attract or deter other organisms or give them a competitive edge in nature. Humans have exploited many secondary compounds from plants for medicinal, culinary, or other purposes. It has been estimated that 50,000 to 100,000 such compounds exist in plants with only a few thousand of these thus far having been identified. Secondary metabolic products may be derived from modification of amino acids and related compounds to produce alkaloids or through specialized conversions such as the *shikimic acid pathway* (phenolics) and *mevalonic acid pathway* (terpenoids). Examples of these compounds are shown in Table 10.3. Lignin, which is a component of secondary cell walls, is, for example, synthesized through the shikimic acid pathway. Because it is hard to digest and is toxic to some predators, it protects plants from herbivorous animals.

ASSIMILATION AND DIGESTION

Sugars produced through photosynthesis may undergo many transformations. Some sugars are used directly in respiration, but others not needed for that purpose may be transformed into lipids, proteins, or other carbohydrates. Among the most important carbohydrates produced from simple sugars are sucrose, starch, and cellulose. Much of the organic matter produced through photosynthesis is eventually used in the building of protoplasm and cell walls. This conversion process is called **assimilation.**

When photosynthesis is taking place, sugar may be produced faster than it can be used or transported away to other parts of the plant. When this happens, the excess sugar may be converted to large, insoluble molecules, such as starch or oils, temporarily stored in the chloroplasts and then later changed back to a soluble form that is transported to other cells. The conversion of starch and other insoluble carbohydrates to soluble forms is called **digestion** (Fig. 10.15). The process is nearly always one of *hydrolysis,* in which water is taken up and, with the aid of enzymes, the links of the chains of simple sugars that comprise the molecules of starch and similar carbohydrates are broken by the addition of water. The disaccharide malt sugar (maltose), for example, is transformed to two molecules of glucose, with the aid of an enzyme (maltase), by the addition of one molecule of water, as follows:

$$\underset{\textbf{maltose}}{C_{12}H_{22}O_{11}} + \underset{\textbf{water}}{H_2O} \xrightarrow[\textbf{(enzyme)}]{\textbf{maltase}} \underset{\textbf{glucose}}{2\,C_6H_{12}O_6}$$

Fats are broken down to their component fatty acids and glycerol, and proteins are digested to their amino acid building blocks in similar fashion. Digestion is carried on in any cell where there may be stored food, with very little energy being released in the process. In animals, special digestive organs also play a role in digestion, but plants have no such additional "help" in the process. In both plants and animals, digestion within cells is similar and is a normal part of metabolism.

Awareness

Greenhouse Gases and Plant Growth

Like a greenhouse, the earth has an atmosphere that traps the radiant energy from the sun and warms the planet, making it suitable for life. Without this protective atmosphere, the planet would lose heat into space, with the result being a drastic cooling of the planet's surface. The gases that trap the heat, like the glass of a greenhouse, are called "greenhouse gases." Consisting of carbon dioxide, methane, nitrous oxides, and chlorofluorocarbons, these gases account for less than 1% of the atmosphere, yet are crucial to the sustenance of life on earth.

It is difficult to conceive that human activity could alter the atmosphere of the earth. But ever since the beginning of the Industrial Revolution some 200 years ago, greenhouse gas levels, especially carbon dioxide, have risen significantly and are projected to increase even more in the next century. This transformation of earth's atmosphere is taking place by the burning of fossil fuels and by destroying the vegetational cover of the earth through massive deforestation. How do scientists know that greenhouse gases are increasing? By drilling into the ice caps at the poles and taking ice-core samples that are then dated, trapped air bubbles in ice layers hundreds of years old can be analyzed and compared to recent atmospheric levels of greenhouse gases. What these studies show is that the level of carbon dioxide has dramatically increased from about 280 ppm (parts per million) in 1800 to roughly 360 ppm in 2001, with a projected doubling of CO_2 to 700 ppm by the middle to the end of the next century. This represents a yearly rate of change of approximately 1.8 ppm.

Why should we be concerned with increases in carbon dioxide and other greenhouse gases? There are several reasons. First, climate scientists tell us that with doubling of CO_2 levels, the earth's surface temperature will increase an average of between 1.5°C and 4.5°C. With warming occurring on a global scale, snow fields and glaciers will begin to melt and contribute to a rise in sea level that will inundate thousands of miles of coastal areas and force vast numbers of people to seek shelter inland. Sea-level rise will also occur by ocean water expanding as it warms. Coastal areas that now support extensive agriculture will be unfarmable because of flooding. Secondly, an increased CO_2 level will have effects on plant life that could be detrimental to world agriculture. Plant biologists studying the effects of an enriched CO_2 atmosphere tell us that some plants will grow more robust because of a "CO_2 fertilizing" effect. However, to sustain this increased plant productivity will require sufficient nutrients, light, and water—commodities that will be available only at a high price. Because farmers in most of the world cannot afford to increase fertilizer applications and buy irrigation water, plant growth in a CO_2-enriched environment will be limited to current production levels. A warmer climate may cause more frequent droughts and, therefore, reduce crop yields in today's major grain-producing regions of the world, notably the Great Plains of the United States. Extreme climate fluctuations would limit the yields of certain crops, thus negating the potential for greater productivity through the "CO_2 fertilization" effect.

How is carbon dioxide released into the atmosphere? Although all living organisms give off CO_2 during respiration, this does not affect global CO_2 levels significantly. The primary culprit is the burning of fossil fuels—oil, natural gas, and coal. Deforestation and burning of timber are other sources of CO_2 emissions. On a per capita basis, the United States is the leading nation in CO_2 emissions (19.7 tons per person), followed by Canada (17.4) and Saudi Arabia (13.1). China (1.9), Egypt (1.4), and India (0.7) have significantly lower per capita CO_2 emissions. These figures clearly reveal the United States' and Canada's dependence upon fossil fuels, while less industrialized countries such as China, Egypt, and India consume much less energy per capita.

Plants and algae created our present atmosphere by absorbing CO_2 and releasing O_2 during photosynthesis. Because of the oxygen-generating abilities of plants, the gradual buildup of an oxygen atmosphere occurred some 600 million years ago and facilitated the development of large animals on earth. Plants have maintained the delicate balance of our atmosphere for millions of years. Yet this balance is now threatened by human activity. Plant response to these perturbations of the atmosphere is one of the most important reasons for studying plants today.

D.C. Scheirer

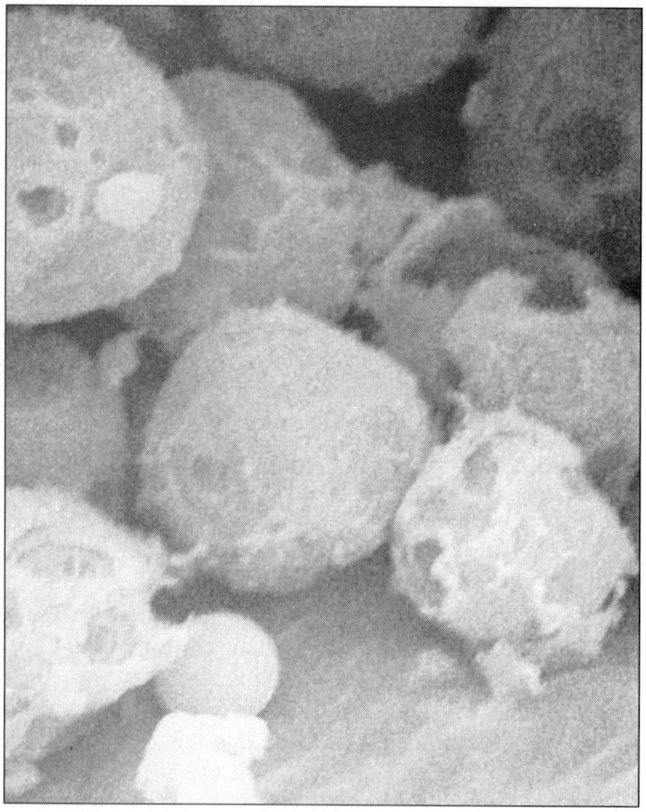

Figure 10.15 Endosperm starch grains of rice mimic grass (*Echinochloa crus-galli* var. *oryzicola*) being digested ×5,000. *Scanning electron micrograph courtesy Delmar Vander Zee.*

Except in insect-trapping plants, digestion takes place within plant cells where the carbohydrates, fats, or proteins are stored, while in animals, digestion usually occurs outside of the cells in the digestive tract. Apart from the location, the process is essentially similar in plants and animals.

Summary

1. Enzymes catalyze reactions of metabolism. Many of these include oxidation-reduction reactions. Oxidation is loss of electrons; reduction is gain of electrons.
2. Photosynthesis is an anabolic process that combines carbon dioxide and water in the presence of light with the aid of chlorophyll; oxygen is a by-product. All life depends on photosynthesis, which takes place in chloroplasts.
3. Carbon dioxide constitutes 0.036% of the atmosphere, but the percentage has been rising in recent years. Increased carbon dioxide levels have potential to elevate global temperatures through the "greenhouse effect."
4. Chlorophyll *b* and carotenoids are antenna pigments that direct light energy to chlorophyll *a*. Photosynthetic units containing chlorophylls and accessory pigments absorb units of light energy, become excited, and pass this energy to acceptors during the light-dependent reactions of photosynthesis.
5. During the light-dependent reactions of photosynthesis, which occur in thylakoid membranes of chloroplasts, water molecules are split, and oxygen gas is released. Hydrogen ions and electrons are released from water and transferred to produce NADPH and ATP.
6. The two types of photosynthetic units present in most chloroplasts are photosystems I and II. Each photosystem has a reaction-center molecule of chlorophyll *a* that boosts an electron to a higher energy level as it reacts to light energy.
7. Photosystem I boosts electron excitation energy to a level that, when it encounters photosystem II, has the potential to reduce NADP to NADPH through noncyclic electron flow. Photosystem II, by itself, can cycle electrons for generation of ATP. Electron transport while the photosystems are operating and proton movement across thylakoid membranes are both involved in ATP production.
8. The light-independent reactions occur through a series of reactions known as the Calvin cycle, which takes place in the stroma of chloroplasts. In the first step, carbon dioxide is combined with RuBP through catalytic action of the enzyme rubisco to form two molecules of the 3-carbon compound, GA3P. The ATP and NADPH from the light-dependent reactions furnish energy to eventually convert GA3P to 6-carbon carbohydrates. This cycle also regenerates RuBP to enable continued carbon fixation.
9. In the light-independent reactions of C_4 plants, 4-carbon oxaloacetic acid is initially produced instead of 3-carbon PGA. In the leaf mesophyll of C_4 plants, there are large chloroplasts, which contain rubisco in the bundle sheaths, and small chloroplasts, which contain higher concentrations of PEP carboxylase that facilitate the conversion of carbon dioxide to carbohydrate at much lower concentrations than is possible in C_3 plants.
10. CAM photosynthesis occurs in succulent plants whose stomata are closed and admit little CO_2 during the day. Regular photosynthesis occurs as the 4-carbon compounds that accumulate at night are converted back to carbon dioxide during the day.
11. Respiration is a catabolic process that takes place in the cytoplasm and mitochondria of cells. The energy is released, with the aid of enzymes, from simple sugar and organic acid molecules.
12. In aerobic respiration, stored energy release requires oxygen; CO_2 and water are by-products of the process.
13. Anaerobic respiration and fermentation do not require oxygen gas, and much less energy is released. The remaining energy is in the ethyl alcohol, lactic acid, or other such substances produced. Some released energy is stored in ATP molecules. Temperature, available water, and environmental oxygen affect respiration rates.

14. Glycolysis, which occurs in the cytoplasm, requires no molecular oxygen; two phosphates are added to a 6-carbon sugar molecule, and the prepared molecule is split into two 3-carbon sugars (GA3P). Some hydrogen, energy, and water are removed from the GA3P, producing pyruvic acid. There is a net gain of two ATP molecules. Hydrogen ions and electrons released during glycolysis are picked up by NAD, which becomes NADH.
15. In aerobic respiration, which occurs in the mitochondria, pyruvic acid loses some CO_2, is restructured, and becomes acetyl CoA. Energy, CO_2, and hydrogen are removed from the acetyl CoA in the citric acid cycle, which involves enzyme-catalyzed reactions of a series of organic acids.
16. NADH passes the hydrogen gained during glycolysis and the citric acid cycle along an electron transport system; small increments of energy are released and partially stored in ATP molecules, and the hydrogen is combined with oxygen gas, forming water in the final step of aerobic respiration.
17. Hydrogen removed from glucose during glycolysis is combined with an inorganic ion in anaerobic respiration. The hydrogen is combined with the pyruvic acid or one of its derivatives in fermentation. Both processes occur in the absence of oxygen gas, with only about 7% of the total energy in the glucose molecule being released, for a net gain of two ATP molecules.
18. Two molecules of NADH and two ATP molecules are gained during glycolysis when two 3-carbon pyruvic acid molecules are produced from a single glucose molecule. Another molecule of NADH is produced when the pyruvic acid molecule is restructured and becomes acetyl CoA prior to entry into the citric acid cycle.
19. In the citric acid cycle, acetyl CoA combines with 4-carbon oxaloacetic acid, producing first a 6-carbon compound, next a 5-carbon compound, and then several 4-carbon compounds. The last 4-carbon compound is oxaloacetic acid. Two CO_2 molecules are also released during this process.
20. Some hydrogen removed during the citric acid cycle is picked up by FAD and NAD; one molecule of ATP, three molecules of NADH, and one molecule of $FADH_2$ are produced during one complete cycle. Energy associated with electrons and/or with hydrogen picked up by NAD and FAD is gradually released as the electrons are passed along the electron transport system; some of this energy is transferred to ATP molecules during oxidative phosphorylation.
21. Energy used in ATP synthesis during oxidative phosphorylation is believed to be derived from a gradient of protons formed across the inner membrane of a mitochondrion, while electrons are moving in the electron transport system by chemiosmosis.
22. Altogether, 38 ATP molecules are produced during the complete aerobic respiration of one glucose molecule; two are used to prime the process, so there is a net gain of 36 ATP molecules.
23. In addition to photosynthesis and respiration, other metabolic pathways are required for growth, development, reproduction, and survival. Essential products of additional pathways include nucleotides, proteins, chlorophylls, and fatty acids. Secondary metabolites include alkaloids, phenolics, and terpenoids.
24. Conversion of sugar produced by photosynthesis to fats, proteins, complex carbohydrates, and other substances is termed assimilation. Digestion takes place within plant cells with the aid of enzymes. During digestion, large insoluble molecules are broken down by hydrolysis to smaller soluble forms that can be transported to other parts of the plant.

Review Questions

1. Why are enzymes important, and how do they contribute toward metabolism?
2. What happens during the light-dependent and light-independent reactions of photosynthesis?
3. What roles do water, light, carbon dioxide, and chlorophyll play in photosynthesis?
4. Which gases influence rubisco activity, and how does this activity affect photosynthesis?
5. Explain how products produced by photosynthesis relate to those used by respiration.
6. How does glycolysis contribute toward aerobic respiration, anaerobic respiration, and fermentation?
7. What do cells get out of aerobic respiration?
8. How do temperature, water, and oxygen affect respiration?
9. How is chemiosmosis in photosynthesis similar to oxidative phosphorylation in respiration?
10. What are cytochromes, and how do they function in photosynthesis and respiration?
11. What additional metabolic pathways are demonstrated by plants, and why are they necessary?

Discussion Questions

1. If space exploration revealed a planet that had water, sunlight, and climate necessary to maintain plant life, would it be possible to produce oxygen and food necessary for human life?
2. How would you genetically engineer a C_3 plant to achieve the high efficiency of C_4 photosynthesis?

3. Can growing more plants and increasing the abundance of other photosynthetic autotrophs regulate the "greenhouse effect"?
4. How does cellular respiration benefit life on earth if it is a catabolic process?
5. Explain why secondary metabolites, originally designed by plants to deter pests or attract pollinators, have commercial value to humans.

Learning Online

Visit our web page at www.mhhe.com/botany for interesting case studies, practice quizzes, current articles, and animations within the Online Learning Center to help you understand the material in this chapter. You'll also find active links to these topics:

Photosynthesis
Architectural Pattern of Plants
Cellular Respiration
Aerobic Respiration
Anaerobic Respiration and Fermentation Pathways
Enzymes

Additional Reading

Attridge, T. H. 1992. *Light and plant responses: A study of plant photophysiology and the natural environment.* New York: Cambridge University Press.

Baker, N. R. 1996. *Oxygenic photosynthesis: The light reactions.* Norwell, MA: Kluwer Academic Publications.

Deisenhofer, J., and J. R. Norris. 1993. *The photosynthetic reaction center.* Orlando, FL: Academic Press.

Dennis, D. T., and D. H. Turpin (Eds.). 1990. *Plant physiology, biochemistry, and molecular biology.* New York: John Wiley & Sons.

Douce, R., and D. A. Day (Eds.). 1985. *Higher plant cell respiration.* New York: Springer-Verlag.

Galston, A. W. 1993. *Life processes of plants: Mechanisms for survival.* New York: W. H. Freeman.

Goodwin, T. W. 1988. *Plant pigments.* San Diego, CA: Academic Press.

Goodwin, T. W., and E. I. Mercer. 1983. *Introduction to plant biochemistry.* New York: Pergamon Press.

Gutteridge, S. 1990. Limitations of the primary events of CO_2 fixation in photosynthetic organisms: The structure and mechanism of rubisco. *Biochimica et Biophysica Acta* 1015: 1–14.

Hopkins, W. G. 1999. *Introduction to plant physiology.* New York: John Wiley & Sons.

Jones, H. G. 1992. *Plants and microclimate: A quantitative approach to plant physiology.* San Diego, CA: Academic Press.

Kozlowski, T. T., and S. G. Pallady (Eds.). 1995. *Physiology of woody plants,* 2d ed. Orlando, FL: Academic Press.

Lehninger, A. L., D. L. Nelson, and M. M. Cox. 1992. *Principles of biochemistry,* 2d ed. New York: Worth.

Salisbury, F. B., and C. W. Ross. 1992. *Plant physiology,* 4th ed. Belmont, CA: Wadsworth.

Taiz, T., and E. Zeiger. 1998. *Plant physiology,* 2d ed. Sunderland, MA: Sinauer Associates.

Ting, I. P. 1985. Crassulacean acid metabolism. *Annual Review of Plant Physiology* 36: 595–622.

A single fiddleneck of a tropical tree fern (*Sadleria cyatheoides*). As the fern frond (leaf) matures, the fiddleneck uncoils and expands into a beautiful, feathery structure.

11 Growth

Chapter Outline

Overview

This chapter introduces growth phenomena with a discussion of the distinctions among growth, differentiation, and development. This is followed by a discussion of plant hormones (auxins, gibberellins, cytokinins, abscisic acid, ethylene) and their roles in plant growth and development. The chapter explores plant movements, including nutations, twining, contraction, nastic, tropic, turgor, and solar tracking movements. The discovery and functions of photoperiodism and phytochrome are briefly surveyed. The chapter concludes with a discussion of the relationship of temperature to growth and dormancy.

Some Learning Goals

1. Contrast growth, differentiation, and development and distinguish among nutrients, vitamins, and plant hormones.
2. Identify the types of plant hormones and describe the major functions of each; discuss commercial applications for each.
3. Distinguish among the various types of plant movements and identify the forces behind them.
4. Define photoperiodism and make distinctions among short-day, long-day, intermediate-day, and day-neutral plants.
5. Explain what phytochrome is and how it functions.
6. Summarize or outline the role of temperature in plant growth.
7. Describe dormancy and stratification and give examples.

From time to time, I grow a few vegetables and berries in my back yard. When Italian squash plants are producing fruit, I have occasionally seen a young squash just beginning to develop, and I have made a mental note to harvest it within a day or two. Then, other matters have distracted me, and I have forgotten to follow through, only to discover a few days later that my squash has grown into an enormous "monster" the size of a watermelon. I then have wondered, How did it grow that big that fast? The words **grow, growing,** or **growth** are used in several ways. If, for example, you see a rubber balloon being inflated with gas, you may refer to its gradual increase in diameter as its growth in size. Or if there is a leak in the roof, you might say that the puddle beneath is growing bigger. In the biological realm, however, growth is always associated with cells. It may be defined simply as an *irreversible increase in mass due to the division and enlargement of cells,* and may be applied to an organism as a whole or to any of its parts.

Many plants, such as radishes and pumpkins, go through a sequence of growth stages. They grow rapidly at first, then for a while they show little, if any, increase in volume, and eventually, they stop growing completely. Finally, tissues break down, and the plant dies. Such growth is said to be *determinate.* Parts of other plants may exhibit *indeterminate* growth and continue to be active for several to many years.

All living organisms begin as a single cell and increase in mass through cell division and enlargement until a body consisting of possibly billions of cells is formed. Once cells have enlarged, **differentiation** occurs; that is, the cells develop different forms adapted to specific functions, such as conduction, support, or secretion of special substances. If differentiation did not occur, the result would be a shapeless blob with no distinct tissues and little, if any, coordination. The coordination of growth and differentiation of a single cell into tissues, organs, and the whole organism is called **development.**

What regulates development? What determines where and when growth and differentiation occur? The answers to these questions lie in the organism's genes and its environment. In the 1950s, F. C. Steward and his colleagues at Cornell University illustrated this concept with a tissue culture experiment. They isolated parenchyma cells from a carrot root and placed them on nutrient media. The cells divided, grew, and developed into embryo-like structures. The embryos eventually grew into new carrot plants, demonstrating that a cell has all the genes necessary to form a new individual. Genes, the basic units of heredity, control the synthesis and development of *enzymes,* which catalyze every metabolic step within cells. Certain genes are programmed to synthesize specific enzymes continually in every living cell, while other genes catalyze the production of enzymes only in specific tissues or at specific times during a plant's life cycle. The environment determines which of those genes will be expressed. In an organ or an organism, the environment includes both external and internal factors. The internal environment includes nutrients, vitamins, and hormones. The external environment includes water, minerals, gases, light, and temperature, among many factors.

NUTRIENTS, VITAMINS, AND HORMONES

Nutrients are substances that furnish the elements and energy for the organic molecules that are the building blocks by which an organism develops. Lack of nutrients restricts the normal development of plants.

Vitamins play an important role in reactions catalyzed by enzymes. Most vitamins are *coenzymes* or parts of coenzymes, which are organic molecules of varied structure that participate in catalyzed reactions, mostly by functioning as an electron acceptor or donor. Vitamins are synthesized in the membranes and cytoplasm of cells. They are essential in relatively small amounts for the normal growth and development of all organisms. Most vitamins essential to all living organisms are produced by plants, although a few are also synthesized by animals. Vitamin A, for example, does not occur in plants. Carotene pigments found in chloroplasts act as *precursors* (simple molecules that produce new molecules after reacting with other molecules) for vitamin A in animals.

Genes also dictate the production of **hormones,** which influence many developmental phenomena. Hormones, which are produced mostly in actively growing regions of plants, are organic substances that differ from enzymes in structure. They are produced and are active in far smaller amounts than vitamins and enzymes. Hormones ordinarily are transported from their point of origin to another part of the plant where they have specific effects, such as causing stems to bend, initiating flowering, or even inhibiting growth.

Because some of the effects of vitamins are similar to those of hormones, they are sometimes difficult to distinguish from one another. The term *growth regulator* has been applied to both natural and synthetic compounds that have effects on plant development similar to those of naturally produced hormones and vitamins. The existence of growth regulators emphasizes the commercial importance of hormones.

Plant Hormones

A relatively small number of plant hormones regulate plant development, and each type of hormone can have multiple effects. It is believed that hormones act by chemically binding to specific receptors. The observed effect is thought to be initiated by this hormone-receptor association, which triggers a series of biochemical events, including turning genes on and off. The biochemical events are called *signal transduction* and may include changes in the complement of enzymes produced in a tissue or changes in transport across membranes.

The major known types of plant hormones are *auxins, gibberellins, cytokinins, abscisic acid,* and *ethylene.*

Auxins

In 1881, Charles Darwin and his son Francis were fascinated by the fact that *coleoptiles* (more or less tubular, closed sheaths protecting the emerging shoots of germinating seeds of the Grass Family—Poaceae) bend toward a light source. They noticed that the bending did not occur if they placed little metal foil caps over the coleoptile tips but the bending resumed when the caps were removed. They concluded that the coleoptile tips must be sensitive to light. Charles Darwin passed away in 1882, but others following up on his work demonstrated that a water-soluble "influence" was produced in the coleoptile tips and that electrical triggering of the bending, which had been suspected, did not exist.

In 1926, Frits Went, a young plant physiologist in Holland, cut off the tips of oat coleoptiles and placed the tips, cut surface down, on flat portions of a gelatinlike growth medium called **agar** (a substance obtained from marine algae). After a few hours, he removed the tips and cut the agar into little square blocks, which he then placed on decapitated coleoptiles. He found that if he placed the block squarely over the cut top of a coleoptile, it grew straight up, but if he placed the block off center, the tip bent away from the side on which the block had been placed (Fig. 11.1). He also found that the more coleoptile tips he placed on the agar to begin with, the more pronounced was the response. His

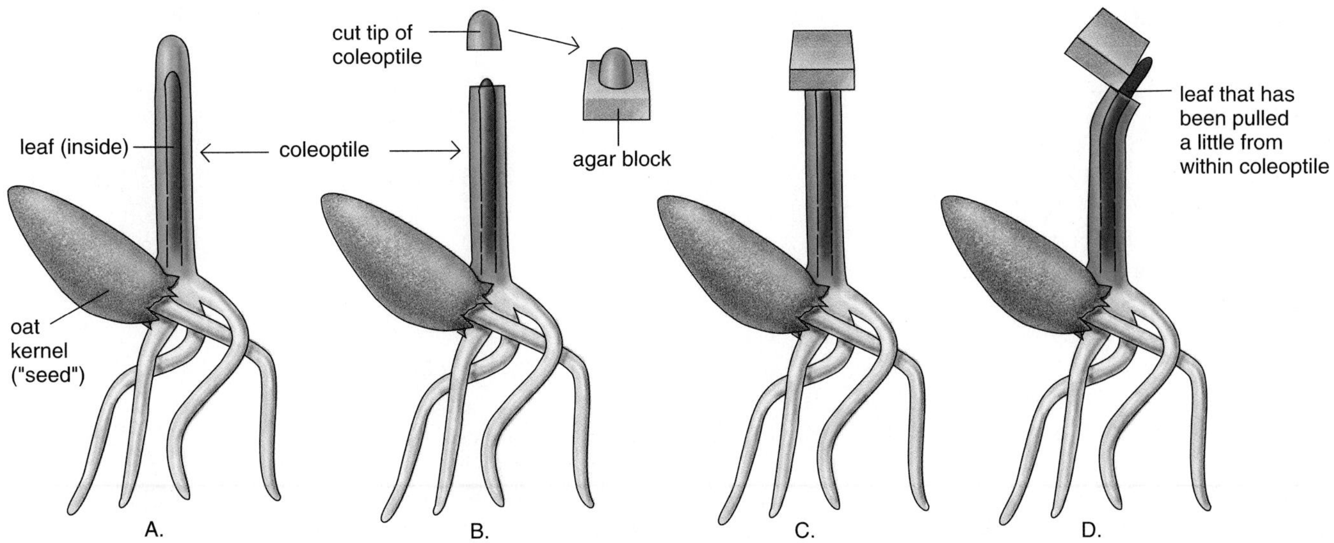

Figure 11.1 Went's experiment with oat coleoptiles. *A.* A germinated oat "seed" with an intact coleoptile. *B.* The tip of a coleoptile was cut off, placed on a small agar block, and left for an hour or two. *C.* When this agar block was placed squarely on a decapitated coleoptile, growth was vertical. *D.* When the agar block was placed off center so that only half of the decapitated coleoptile was in contact with it, the tip bent away from it. This experiment demonstrated that something affecting growth diffused from the coleoptile tip into the agar and from the agar to the decapitated coleoptile.

experiments demonstrated conclusively that something produced in the coleoptile tips moved out into the agar and that this substance was responsible for the bending observed. Went named the substance **auxin** (from the Greek word *auxein,* meaning "to increase"), a term now in widespread use.

Auxins were the first plant hormones to be discovered, with several others coming to light later. At least three major groups apparently promote (but do not control) the growth of plants. Depending on the concentration, some may also have an inhibitory effect.

Auxin molecules have a structure similar to those of the amino acid tryptophan, which is found in both plant and animal cells. For many years, auxins were thought to be synthesized from tryptophan by living cells. In 1991, however, it was demonstrated that mutant plants incapable of producing tryptophan can develop auxin without it, and the precise nature of the synthesis remains to be determined.

Production of auxins occurs mainly in apical meristems, buds, young leaves, and other active young parts of plants. It is sometimes difficult to predict how cells will respond to auxins because responses vary according to the concentration, location, and other factors. For example, if an auxin of a specific concentration promotes shoot growth in a certain plant, the same auxin of identical concentration will usually inhibit its root growth. At appropriate concentrations, auxins normally stimulate the enlargement of cells by increasing the plasticity (irreversible stretching) of cell walls.

Hormone concentrations were measured at first by *bioassays,* which relate the response of a sensitive plant part to the amount of hormone applied to the part. For example, a fixed number of coleoptile tips can be placed on a specific sample of agar for a certain time period, and a measured cut block of the agar can then be placed off center on a decapitated coleoptile. The concentration can be determined from the angle of bending of the coleoptile that occurs within a certain time frame when it is compared with the bending brought about by a similar block containing a known amount of auxin (Fig. 11.2).

Hormone concentrations today are more frequently determined by vaporizing a sample and moving it through a tube of liquid or powdered material by a technique known as *gas chromatography.* As the sample moves through the column, the hormone and other components of the sample separate, and the amount of hormone present then can be determined relatively precisely.

Auxins also may have many other effects, including triggering the production of other hormones or growth regulators (especially ethylene), causing the dictyosomes to increase rates of secretion, playing a role in controlling some phases of respiration, and influencing many developmental aspects of growth. Auxins promote cell enlargement and stem growth, cell division in the cambium, initiation of roots, and differentiation of cell types. Auxins delay developmental processes such as fruit and leaf abscission, fruit ripening, and inhibit lateral branching. Sensitivity to auxins is less in many monocots

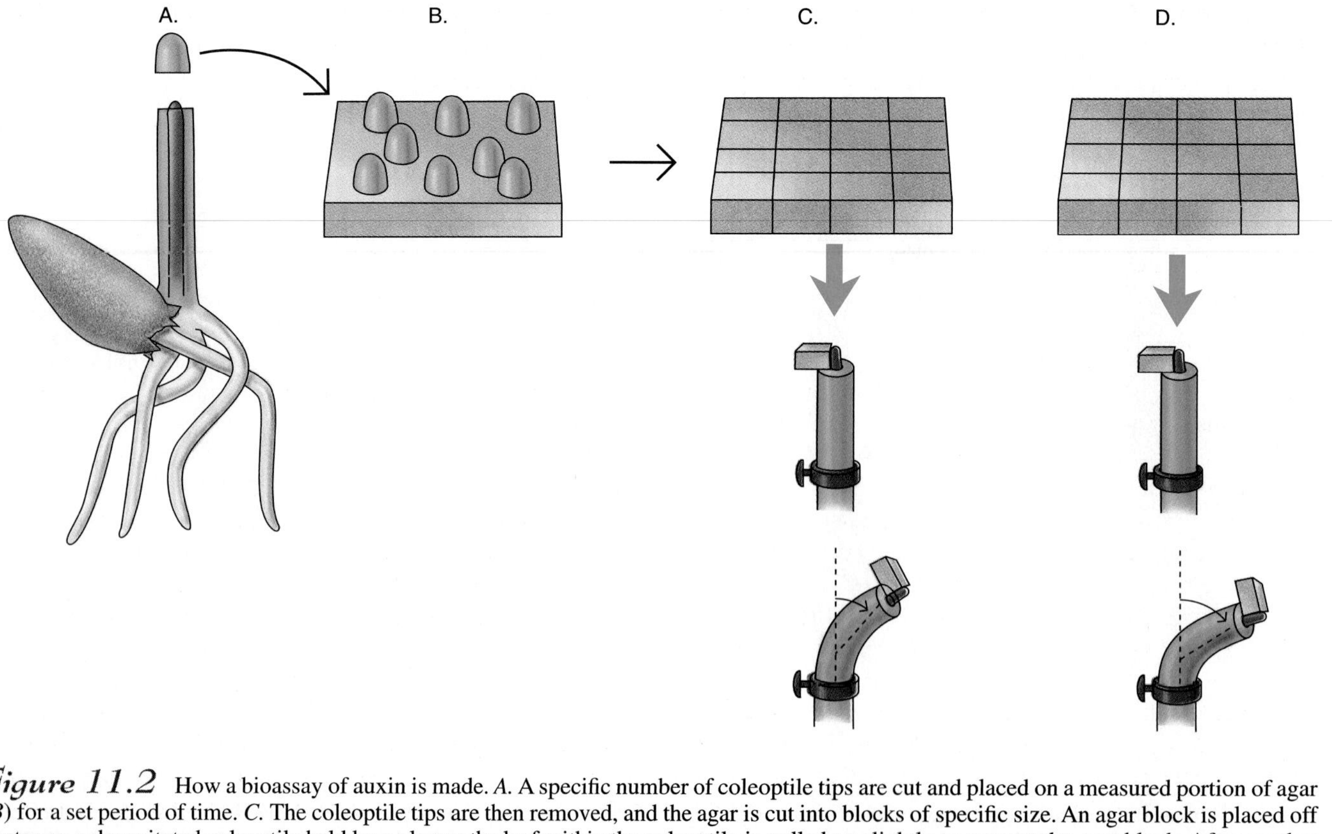

Figure 11.2 How a bioassay of auxin is made. *A.* A specific number of coleoptile tips are cut and placed on a measured portion of agar (*B*) for a set period of time. *C.* The coleoptile tips are then removed, and the agar is cut into blocks of specific size. An agar block is placed off center on a decapitated coleoptile held by a clamp; the leaf within the coleoptile is pulled up slightly to support the agar block. After another set period, the angle of curvature is measured. *D.* The angle of curvature is compared to that produced when a similar agar block containing a known amount of auxin is placed off center on another coleoptile for the same period of time.

than it is in dicots, and shoots are less sensitive than roots. Higher concentrations, however, will kill almost any living plant tissues. The effects of auxins combined with those of other regulators produce many of the growth phenomena.

Movement of auxins from the cells where they originate requires the expenditure of energy stored in ATP molecules. The migration is relatively slow (about 1 centimeter per hour), although the active transport "pumping" mechanism involved carries the hormone up to 10 times faster than it would be by simple diffusion alone. The movement is *polar,* which refers to the movement of auxins away from their source, usually from a stem tip toward the base, with movement in roots proceeding toward the tip. This polar movement occurs even when a stem is inverted. The auxins apparently are ordinarily not carried through the sieve tubes of the phloem but proceed from cell to cell, particularly through parenchyma cells surrounding vascular bundles.

In the past, it was believed that several natural auxins occurred in nature, but until relatively recently, most scientists thought that *indoleacetic acid* (IAA) was the only active auxin. Furthermore, it was believed that other organic acids that act like auxins were actually converted to IAA in the plants. Current research, however, indicates that plants do actually produce three other growth regulators that are not converted to IAA but bring about many of the same responses as IAA. One, called *phenylacetic acid* (PAA), is often more abundant but less active than IAA. Another, *4-chloroindoleacetic acid* (4-chloroIAA), is found in germinating seeds of legumes. The third, *indolebutyric acid* (IBA) occurs in the leaves of corn and various dicots. It previously was known only in synthetic form. At least three auxin precursors also have activities similar to those of auxins.

IAA and a number of organic acids that are not plant hormones regulate growth and have also been synthesized. They are used widely in agriculture and horticulture. At the proper concentrations, IAA and other growth regulators stimulate the formation of roots on almost any plant organ. Nurseries apply these substances in paste or powdered form to the bases of cut segments of stems to stimulate a more vigorous production of roots than would occur naturally (Fig. 11.3). Synthetic growth regulators presently in use include NAA (naphthalene acetic acid), 2,4-D (2,4-dichlorophenoxyacetic acid), and MCPA (2-methyl-4-chlorophenoxyacetic acid).

Orchardists spray fruit trees with auxins to promote uniform flowering and fruit set, and they later spray the fruit to prevent the formation of abscission layers and subsequent premature fruit drop. A substantial saving in labor expenses results from being able to go through an orchard or a pineapple plantation and pick all of the fruit at one time. If auxins are applied to flowers before pollination occurs, seedless fruit can be formed and developed. Some orchardists use auxins for controlling the number of fruits that will mature in order not to have too many small ones, and some even control the shapes of plants for sales appeal.

Weeds were controlled by hand labor and by the use of caustic or otherwise poisonous substances until it was discovered that synthetic auxins, including 2,4-D, 2,4,5-T, 2,4,5-TP, NAA, and MCPA, when sprayed at low concentrations, kill some weeds. Broad-leaved plants such as dandelions, plantains, and others (for reasons that are not clearly understood) are more susceptible to low concentrations of these plant growth regulators than narrow-leaved plants such as grasses. They have been used on lawns to kill weeds without noticeably affecting the grass. The precise mechanisms of action of these herbicides (plant killers) are still largely unknown.

Figure 11.3 Effect of auxin applied to the base of a *Gardenia* cutting 4 weeks after application. *Left.* A treated cutting. *Right.* An untreated cutting.

Thus far, there is little evidence that 2,4-D has directly adverse effects on humans and other animal life, but such is not the case with 2,4,5-T, which was banned in 1979 for most uses in the United States by the Environmental Protection Agency. The controversy over the use of 2,4,5-T began in the 1960s during the Vietnam War when Agent Orange, a 1-to-1 mixture of 2,4-D and 2,4,5-T, was used to defoliate jungles. Subsequent tests and experiments with 2,4,5-T have produced leukemia, miscarriages, birth defects, and liver and lung diseases in laboratory animals. The diseases and defects are apparently caused by TCDD (2,3,7,8-tetrachlorodibenzoparadioxin), a dioxin contaminant that evidently is unavoidably produced in minute amounts during the manufacture of 2,4,5-T. TCDD harms or even kills laboratory animals in doses as low as a few parts per billion.

Gibberellins

In 1926, Eiichi Kurosawa of Japan reported the discovery of a new substance that was causing what was referred to as *bakanae,* or "foolish seedling," a serious disease of rice. The stems of rice seedlings infected with a certain fungus grew twice as long as those of uninfected plants, but the stems were weakened so that they eventually collapsed and died. Kurosawa found that extracts of the fungus applied to uninfected plants brought about the same growth stimulations as the fungal disease. Nine years later, other Japanese scientists were able to crystallize the substance, which was named **gibberellin,** after the scientific name of the fungus that produced it (*Gibberella fujikuroi*). It was not until the 1950s, however, that the true chemical structure of gibberellin was elucidated by a British group headed by Brian Cross and an American group led by Frank Stodola.

There are now more than 110 known gibberellins, customarily abbreviated to **GA.** Each individual GA is identified by a subscript (e.g., GA_6). They have been isolated from immature seeds (especially those of dicots), root and shoot tips, young leaves, and fungi. Most GA produced by plants is inactive, apparently functioning as precursors to active forms. No single species has thus far been found to have more than 15 kinds of GA, which probably also occur in algae, mosses, and ferns. None are known in bacteria. GA moves through xylem and phloem and, unlike that of auxin, the movement is not polar.

Acetyl coenzyme A, which is vital to the process of respiration, functions as a precursor in the synthesis of GA. Interest in the hormonal properties of GA is high, for their ability to increase growth in plants is far more impressive than that of auxin alone, although traces of auxin apparently need to be present for GA to produce its maximum effects.

Most dicots and a few monocots grow faster with an application of GA, but coniferous trees such as pines and firs generally show little, if any, response. If a GA, at the appropriate concentration, is applied to a cabbage, the plant may become 2 meters (6 feet) tall, and bush bean plants can become pole beans with a single application. There is not usually, however, a corresponding stimulation of root growth. Many genetically dwarf plants grow as tall as their normal counterparts after an application of the appropriate GA.

Gibberellins not only dramatically increase stem growth, but they are also involved in nearly all of the same regulatory processes in plant development as auxins (Fig. 11.4). In some kinds of plants, flowering can be brought about by applications of GA, and the dormancy of buds and seeds can be broken. Some GA appears to lower the threshold of growth; that is, plants may start growing at lower temperatures than usual after an application of GA. For example, an application to a lawn could cause it to turn green 2 or 3 weeks earlier in the spring.

Figure 11.4 Effect of gibberellins on flowering. *Left.* Cabbage plants grown outside in cool temperatures. *Right.* Cabbage plants grown in a warm greenhouse. The plants grew tall but did not flower until treated with gibberellins. *(© Sylvan H. Wittwer/Visuals Unlimited)*

In some varieties of lettuce and cereals, the seeds normally germinate only after being subjected to a period of cold temperatures. An application of GA can eliminate the cold requirement for germination. It is common for plants after germination of seed to produce juvenile foliage at first and then different adult foliage as the plant matures. In such plants, GA treatment of buds that would normally develop into adult branches causes the foliage to develop in juvenile form.

Several *growth retardants* available commercially inhibit or block the synthesis of GA. Applications of growth retardants result in stunted plants because stem elongation is inhibited. When applied to certain commercially grown crops such as chrysanthemums, flowers with thicker, stronger stalks are produced.

GA has been used experimentally to increase yields of sugar cane and hops and has revolutionized the production of seedless grapes through increasing the size of the fruit and lengthening fruit internodes, which results in slightly wider spaces between grapes in the bunches. Better air circulation between grapes reduces their susceptibility to fungal diseases, and the need for hand thinning is eliminated. They have been used in navel-orange orchards to delay the aging of the fruit's skin and have been found to increase the length and crispness of celery petioles (the parts that are most in demand as a raw food). GA is now used to increase seed production in conifers and to enhance starch digestion by increasing the rate of malting in breweries.

Field experiments have shown that if GA is applied to plants at the same time selected herbicides are applied, the GA may reverse the effects of the herbicides. The relatively high cost of GA has limited the extent of its use in horticulture and agriculture.

Cytokinins

By the close of the 19th century, botanists suspected that there must be something that regulates cell division in plants. In 1913, Gottlieb Haberlandt of Germany discovered an unidentified chemical in the phloem of various plants. This chemical stimulated cell division and initiated the production of cork cambium. In 1941, Johannes van Overbeek discovered that coconut milk (a liquid *endosperm;* endosperm is discussed in Chapter 23) contained something that increased the growth rate of tissues and embryos, and within a few years, several substances that accelerated cell division were isolated from coconut milk. In 1955, a substance named *kinetin* was found, in the presence of auxin, to stimulate the proliferation of parenchyma cells in tobacco pith. Kinetin was isolated from heat-treated herring sperm and is not a naturally occurring substance. Its discovery was important, however, because it demonstrated that a simple chemical could promote cell division. In 1964, the identity of a substance very similar to kinetin was also determined in kernels of corn. These various stimulants to cell division came to be known as **cytokinins.**

The several cytokinins now known differ somewhat in their molecular structure and possibly also in origin, but they are similar in composition to adenine. You'll recall that adenine is a building block of one of the four nucleotides found in DNA, although none of the cytokinins appears to be derived from DNA. Some cytokinins do, however, occur in certain forms of RNA. Cytokinins are synthesized in root tips and germinating seeds.

If auxin is present during the cell cycle, cytokinins promote cell division by speeding up the progression from the G_2 phase to the mitosis phase, but no such effect takes place in the absence of auxin. Cytokinins also play a role in the enlarging of cells, the differentiation of tissues, the development of chloroplasts, the stimulation of cotyledon growth, the delay of aging in leaves, and in many of the growth phenomena also brought about by auxins and gibberellins. Cytokinins move throughout plants via the xylem, phloem, and parenchyma cells.

Certain bacteria stimulate the growth of galls (tumors) on plants, either by producing cytokinins and auxins that promote the unorganized growth of tumor cells or by transferring bacterial and auxin genes to the DNA of the host plant. The incorporation of the transferred genes results in host tissue that produces more auxin and cytokinin than normal, which, in turn, results in tumor growth.

Despite their role in cell division and enlargement, however, there is a total absence of evidence that cytokinins initiate or promote animal cancers or have any other effect on animal cells.

Experiments have shown that cytokinins prolong the life of vegetables in storage. Related synthetic compounds have been used extensively to regulate the height of ornamental shrubs and to keep harvested lettuce and mushrooms fresh. They also have been used to shorten the straw length in wheat, so as to minimize the chances of the plants blowing over in the wind, and to lengthen the life of cut flowers. Many have not yet been approved for general agricultural use.

Abscisic Acid

Although the promotion of growth by auxins and gibberellins had been amply demonstrated by the 1940s, plant physiologists began to suspect that something else produced by plants could have opposite (inhibitory) effects. In 1949, Torsten Hemberg, a Swedish botanist, showed that substances produced in dormant buds (i.e., buds whose development is temporarily arrested) blocked the effects of auxins. He called these growth inhibitors *dormins.*

In 1963, three groups of investigators working independently in the United States, Great Britain, and New Zealand discovered a growth-inhibiting hormone, which was in 1967 officially called **abscisic acid (ABA)**. Later, it was shown that ABA and dormins were one and the same.

ABA is synthesized in plastids, apparently from carotenoid pigments. It is found in many plant materials but is particularly common in fleshy fruits, where it evidently prevents seeds from germinating while they are still on the plant. When ABA is applied to seeds outside of the fruit, germination usually is delayed. Because the stimulatory effects of other hormones are inhibited by ABA, cell growth is usually also inhibited. Like the movement of gibberellins and cytokinins, that of ABA throughout plants is nonpolar.

ABA was originally believed to promote the formation of abscission layers in leaves and fruits. However, the evidence suggesting that ethylene (discussed in the following section, "Ethylene") is far more important than ABA in abscission now is overwhelming, and, despite its name, ABA has little, if any, influence on the process.

ABA apparently helps leaves respond to excessive water loss. When the leaves wilt, ABA is produced in amounts several times greater than usual. This interferes with the transport or retention of potassium ions in the guard cells, causing the stomata to close. When the uptake of water again becomes sufficient for the leaf's needs, the ABA breaks down, and the stomata reopen. ABA produced in times of drought is transported from the shoot to the root and causes an increase in both root growth and water uptake. Despite the growth-inhibiting effects of ABA, there is no evidence that it is in any way toxic to plants.

Ethylene

In 1901, Dimitry Neljubow, a Russian student at the St. Petersburg Botanical Institute, discovered that pea seedlings grown in the dark in a laboratory showed reduced stem elongation, increased swelling of stems, and abnormal horizontal growth. When the seedlings were placed outside in fresh air, however, they resumed normal growth. It was determined that the simple gas **ethylene,** present from gas lamps in the laboratory, had produced the abnormal growth.

In 1934, R. Gane discovered that ethylene was produced naturally by fruits, although it had been known for some time that the ripening of green fruits could be accelerated artificially by placing them in ethylene. It is now known that ethylene is produced not only by fruits but also by flowers, seeds, leaves, and even roots. Several fungi and a few bacteria are also known to produce it, and its regulatory effects make it a hormone in the broad sense of the word. Ethylene is produced from the amino acid methionine; oxygen is required for its formation.

The production of ethylene by plant tissues varies considerably under different conditions. A surge of ethylene lasting for several hours becomes evident after various tissues, including those of fruits, are bruised or cut, and applications of auxin can cause an increase in ethylene production of two to ten times. As pea seeds germinate, the seedlings produce a surge of ethylene when they meet interference with their growth through the soil. This apparently causes the stem tip to form a tighter crook, which may aid the seedling in pushing to the surface.

When a storm stirs up a green field or animals pass across it, plants respond to these and other mechanical stresses with an increase in the production of fibers and collenchyma tissue. Cell elongation is also inhibited, resulting in shorter, sturdier plants. The responses, called *thigmomorphogenesis,* are under the control of genes that are activated by touch. Several substances, including enzymes, ethylene, and a protein called *calmodulin,* are involved. Calmodulin, which may play a role in several plant growth responses, constitutes up to 2% of a plasma membrane and is activated when it binds to calcium. The calmodulin-calcium complex activates enzymes in membranes, which then activate or deactivate other enzymes, and plays a role in several kinds of plant reactions to stimuli, including thigmomorphogenesis.

Ethylene apparently can trigger its own production. If minute amounts are introduced to the tissues that produce the gas, a tremendous response by the tissues often results. These tissues may then produce so much ethylene that the part concerned can be adversely affected. Flowers, for example, may fade in the presence of excessive amounts, and leaves may abscise (Fig. 11.5).

In ancient China, growers used to ripen fruits in rooms where incense was being burned, and citrus growers used to ripen their fruits in rooms equipped with a kerosene stove. In houses where gas heating is used, occupants often experience great difficulty in growing house plants, and greenhouse heaters using such fuel create a dilemma for owners attempting to promote plant growth.

Figure 11.5 Effect of ethylene on holly twigs. Two similar twigs were placed under glass jars for a week; at the same time, a ripe apple was placed under the jar on the right. Ethylene produced by the apple caused abscission of the leaves.

In the days before electric street lights became commonplace, gas lights were used, and in some cities in Germany, leaves fell from the shade trees if they were located near a gas line that leaked. In all of these instances, minute amounts of ethylene gas resulting from the fuel combustion or leakage brought about the results, both good and bad. In fact, as little as 1 part of ethylene per 10 million parts of air may be sufficient to trigger responses.

Today, commercial use of ethylene is extensive. It has been used for many years to ripen harvested green fruits, such as bananas, mangoes, and honeydew melons, and to cause citrus fruits to color up before marketing. Since ethylene production almost ceases in the absence of oxygen, apple and other fruit growers have found that if they place unripe fruit in sealed warehouses after harvest, pump out the air, reduce temperatures to just above freezing, and replace the air with inert nitrogen gas or carbon dioxide, the fruit will remain metabolically inactive for long periods. The growers can then remove the fruit in batches throughout the year, add as little as one part per million of ethylene, and have ripe fruit at any time there is a demand. This is why apples are always available in supermarkets, even though harvesting is usually confined to a few weeks in the fall.

Fruits that respond to ethylene usually have a major increase in respiration just before ripening occurs. The increase in ethylene production at that time is often up to 100 times greater than it was a day or two earlier. The accompanying major increase in respiration is called a *climacteric,* and fruits that exhibit such phenomena are called *climacteric fruits.* Some fruits, such as grapes, are *nonclimacteric* and do not respond in this way to ethylene.

A few growers still use natural ethylene to ripen pears and peaches by wrapping each fruit individually in tissue paper. The paper retards the escape of ethylene from the fruit and hastens ripening. "Resting" potato tubers will sprout following brief applications of ethylene, and seeds may be stimulated to germinate if given a short exposure to the gas just before sprouting, although treatment after sprouting inhibits growth.

Ethylene is used in Hawaii to promote flowering in pineapples, and it causes members of the Pumpkin Family (Cucurbitaceae) to produce more female flowers and thus more fruit. Many trees in nurseries are grown in containers and are usually crowded. Consequently, they are often tall and spindly, but applications of small amounts of ethylene to the trunks while they are enclosed in plastic tubes causes a marked thickening, making the trees sturdier and less likely to break.

Other Hormones or Related Compounds

A number of compounds called *oligosaccharins,* which are released from cell walls by enzymes, influence cell differentiation, reproduction, and growth in plants and therefore must be considered hormones. However, oligosaccharins produce their effects at concentrations of up to 1,000 times less than those of auxins, and the effects (e.g., growth promotion, inhibition of flowering) are not only highly specific, but the responses to them are essentially identical in all species. *Brassinosteroids,* which have a gibberellin-like effect on plant stem elongation, are known from legumes and a few other plants. Yams, which incidentally are a source of DHEA—a hormone whose production by humans tends to decrease with age—are also the source of *batasins.* Batasins promote dormancy in *bulbils,* which are produced from axillary buds in lilies and a few other plants.

HORMONAL INTERACTIONS

Apical Dominance

For centuries, gardeners and nursery workers have often deliberately removed terminal buds of plants to promote bushier growth, knowing that the buds are involved in *apical dominance.* Apical dominance is the suppression of the growth of the lateral or axillary buds, each of which can form a branch with its own terminal bud. Apical dominance is believed to be brought about by an auxinlike inhibitor in a terminal bud. It is strong in trees with conical shapes and little branching toward the top (e.g., many pines, spruces, firs) (Fig. 11.6) and weak in trees that branch more often (e.g., elms, ashes, willows).

There is evidence that as the tissues produced by the apical meristem of the terminal bud increase in length, the source of the inhibitor moves farther away and the concentration of the inhibitor gradually drops in the vicinity of the lateral buds. We presume that the inhibitory effect eventually falls below a threshold, allowing the lateral buds to grow. Each lateral bud then itself becomes a terminal bud that produces its own inhibitor, which, in turn, suppresses the new lateral buds developed beneath it. These presumptions, however, have not yet been confirmed, and some studies suggest that a cytokinin deficiency in the lateral buds actually plays a greater role in apical dominance than auxinlike inhibitors. Experiments with pea plants have shown that the axillary buds will begin to grow as little as 4 hours after a terminal bud has been removed. If cytokinins are applied in appropriate concentration to axillary buds, however, they will begin to grow, even in the presence of a terminal bud, thereby offsetting apical dominance. The extent of the suppression depends on how close the axillary buds are to the terminal bud and on the amount of inhibitor involved. Further studies are needed before the precise causes of apical dominance are fully understood.

Senescence

The breakdown of cell components and membranes that eventually leads to the death of the cell is called **senescence.** As mentioned in Chapter 7, the leaves of deciduous trees and shrubs senesce and drop through a process of abscission every year. Even evergreen species often retain their leaves for only 2 or 3 years (with the notable exception of bristlecone pines, which hold their leaves for up to 30 years), and the above-ground parts of many herbaceous perennials senesce and die at the close of each growing season.

Figure 11.6 A Jeffrey pine tree. The trunk of this tree, which would normally be single, is forked because of the earlier removal of the terminal bud.

Why do plant parts senesce? Some studies have suggested that certain plants produce a senescence "factor" that behaves like, or is actually, a hormone, but we are not yet certain of the precise mechanisms involved. We do know, however, that both ABA and ethylene promote senescence. On the other hand, auxins, gibberellins, and cytokinins delay senescence in a number of plants that have been studied. Other factors, such as nitrogen deficiency and drought, also hasten senescence.

OTHER HORMONAL INTERACTIONS

The development of roots and shoots in both tissue culture and in girdled stems is also regulated by a combination of auxins and cytokinins. Experiments with living pith cells of tobacco plants have shown that such cells will enlarge when supplied with auxin and nutrients, but they will not divide unless small amounts of cytokinin are added. By varying the amounts of cytokinin, it is possible to stimulate the pith cells to differentiate into roots or into buds from which stems will develop.

Gibberellins and ABA are important in the regulation of germination of seeds. Gibberellins promote cell elongation, thereby enabling the developing root to break through the seed coat. After water has been imbibed by cereal seeds, gibberellins released by the embryo stimulate the secretion of enzymes that digest endosperm. ABA has been shown to inhibit the synthesis and release of these enzymes. The relative amounts of gibberellins and ABA in seeds other than cereal grains also influence whether the seeds germinate or remain dormant.

PLANT MOVEMENTS

We noted in Chapter 2 that all living organisms exhibit movement but that most plant movements are relatively slow and imperceptible unless seen in time-lapse photography or demonstrated experimentally. Now we want to examine how and why plant movements occur.

Growth Movements

Growth movements result from varying growth rates in different parts of an organ. They are mainly related to young parts of a plant and, as a rule, are quite slow, usually taking at least 2 hours to become apparent, although the plant may have begun microscopic changes within minutes of receiving a stimulus.

Movements Resulting Primarily from Internal Stimuli

Nutations Charles Darwin once attached a tiny sliver of glass to the tip of a plant growing in a pot. Then he suspended a piece of paper blackened with carbon over the tip, and as the plant grew, he raised the paper just enough to allow the tip to touch the paper without hurting the plant. He found that the growing point traced a spiral pattern in the blackened paper. We know now that such nutations (also referred to as *spiraling* movements) are common to many plants (Fig. 11.7).

Nodding Movements Members of the Legume Family (Fabaceae), such as garden beans, whose ethylene production upon germinating causes the formation of a thickened crook in the hypocotyl, exhibit a slow, oscillating movement (i.e., the bent hypocotyl "nods" from side to side like an upside-down pendulum) as the seedling pushes up through the soil. This nodding movement apparently facilitates the progress of the growing plant tip through the soil.

Twining Movements Although twining movements are mostly stimulated internally, external forces, such as gravity and contact, may also play a role. These movements occur when cells in the stems of climbing plants, such as morning

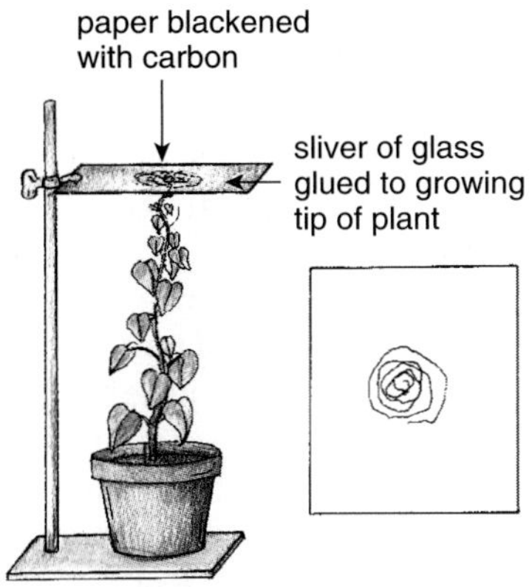

Figure 11.7 Charles Darwin's demonstration of spiraling growth.

glory, elongate to differing extents, causing visible spiraling in growth (in contrast with the spiraling movements previously mentioned, which are not visible to the eye). Tendril twining, which is initiated by contact, results from an elongation of cells on one side of the stem and a shrinkage of cells on the opposite side, followed by differences in growth rates (Fig. 11.8). Some tendrils are stimulated to coil by auxin, while others are stimulated by ethylene.

Contraction Movements In Chapter 6, we noted that the bulbs of a number of dicots and monocots have contractile roots, which pull them deeper into the ground. In lilies, for example, seeds germinating at the surface ultimately produce bulbs that end up 10 to 15 centimeters (4 to 6 inches) below ground level because of the activities of contractile roots. There is some evidence that temperature fluctuations at the surface determine how long the contracting will continue. When the bulb gets deep enough that the differences between daytime and nighttime temperatures are slight, the contractions cease. The aerial roots of some banyan trees straighten out by contraction after the roots have made contact with the ground. The "shrinking" of roots has been shown to take place at the rate of 2.2 millimeters (0.1 inch) a day in sorrel (*Oxalis*).

Nastic Movements When flattened plant organs, such as leaves or flower petals, first expand from buds, they characteristically alternate in bending down and then up as the cells in the upper and lower parts of the leaf alternate in enlarging faster than those in the opposite parts. Such non-directional movements (i.e., movements that do not result in an organ being oriented toward or away from the direction of a stimulus) are called **nastic.** Nastic movements may involve differential growth or turgor changes in special cells. *Epinasty* is the permanent downward bending of an organ, often the petiole of a leaf, in response to either an unequal flow of auxin through the petiole or to ethylene. Nastic movements that involve changes in turgor pressure include sleep movements and contact movements, which are discussed later.

Figure 11.8 Typical twining of a tendril produced by a manroot plant. Note that the direction of coiling reverses near the midpoint.

Movements Resulting from External Stimuli

Permanent movements resulting from external stimuli coming from one direction are commonly referred to as **tropisms.** Tropic movements can be divided into three phases:

1. In the initial *perception* phase, the organ receives a greater stimulus on one side.
2. Then, *transduction* occurs, during which one or more hormones become(s) unevenly distributed across the organ.
3. Finally, *asymmetric growth* occurs as a result of the uneven distribution of the hormone causing greater cell elongation on one side.

Figure 11.9 A cyclamen plant that received light from one direction for several weeks. Note how all visible plant parts are oriented on the side that received light.

Phototropism The main shoots of most plants growing in the open tend to develop vertically, although the branches often grow horizontally. If a box is placed over a plant growing vertically and a hole is cut to admit light from one side, the tip of the plant will begin to bend toward the light within a few hours. If the box is later removed, a compensating bend develops, causing the tip to grow vertically again. Such a growth movement toward light is called a *positive phototropism* (Fig. 11.9). A similar bending away from light is called a *negative phototropism.* The shoot tips of most plants are positively phototropic, while roots are either insensitive to light or negatively phototropic.

We have already noted that if the tip of a coleoptile is covered or removed, the structure will not bend toward light and that auxin is produced in the tip (see Fig. 11.1). We have also noted that auxin promotes the elongation of cells, at least in certain concentrations. For some time, it was believed that stem tips bent toward light because auxin was destroyed or inactivated on the exposed side, leaving more growth-promoting hormone on the side away from the light, causing the cells there to elongate more and produce a bend. Careful experiments have shown, however, that stem tips growing in the open have the same total amount of auxin present as stem tips from the same species receiving light from one side. Other experiments indicate that the auxin migrates away from the light, accumulating in greater amounts on the opposite side, promoting greater elongation of cells on the "dark" side. Apparently, an active transport system enables the auxin to migrate against a diffusion gradient.

Different intensities of light may bring about different phototropic responses. In Bermuda grass, for example, the stems tend to grow upright in the shade and parallel with the ground in the sun. In other plants, such as the European rock rose, which grows among rocks or on walls, the flowers are positively phototropic, but once they are fertilized, they become negatively phototropic. As the pedicels (stalks) elongate, the developing fruits are buried in cracks and crevices, where the seeds then may germinate. Phototropic responses are not confined to flowering plants. A number of mushrooms, for example, show marked positive responses to light, and certain fungi that grow on horse dung have striking phototropic movements, which are discussed in Chapter 19.

Gravitropism Growth responses to the stimulus of gravity are called **gravitropisms.** The primary roots of plants are positively gravitropic, while shoots forming the main axis of plants are negatively gravitropic (Fig. 11.10). It originally was postulated that plant organs perceive gravity through the movement of amyloplasts containing large starch grains located in special cells of the root cap (discussed in Chapter 6). The amyloplasts are also found in coleoptile tips and in the endodermis. When a potted plant is placed on its side, the starch-containing amyloplasts will, within a few minutes, begin to float or tumble down until they come to rest on the side of the cells closest to the gravity stimulus. In roots, the cells on the side opposite the stimulus begin elongating within 10 seconds to an hour or two, bringing about a downward bend, while the opposite occurs in stems (Figs. 11.11 and 11.12).

Recent experiments have cast doubt about the role of amyloplasts in bringing about gravitropic responses. It has been demonstrated that the roots of plants completely lacking root-cap amyloplasts still perceive and respond to gravity. Other recent research suggests that proteins on the outside of the plasma membrane next to the cell wall may be essential to sensing gravity and that the whole protoplast of a cell, rather than amyloplasts within, is involved in gravitropic responses.

Some cell biologists have suggested that mitochondria and dictyosomes also respond to gravity, but precisely how they bring about the response has been the subject of much conjecture. Auxin and ABA, along with calcium ions and proteins, well may all be involved in modifying cell elongation so as to produce these gravitropic bendings.

The stimulus of gravity can be negated by rotating a plant placed in a horizontal position. A simple device called a *clinostat* uses a motor and a wheel to rotate a potted plant slowly about a horizontal axis (Fig. 11.13). As the plant rotates, both the stem and roots continue to grow horizontally instead of exhibiting characteristic gravitropic responses. Obviously, neither the starch grains in the **statoliths** (gravity

Figure 11.10 This *Coleus* plant was placed on its side the day before the photograph was taken. The stems bent upward within 24 hours of the pot being tipped over.

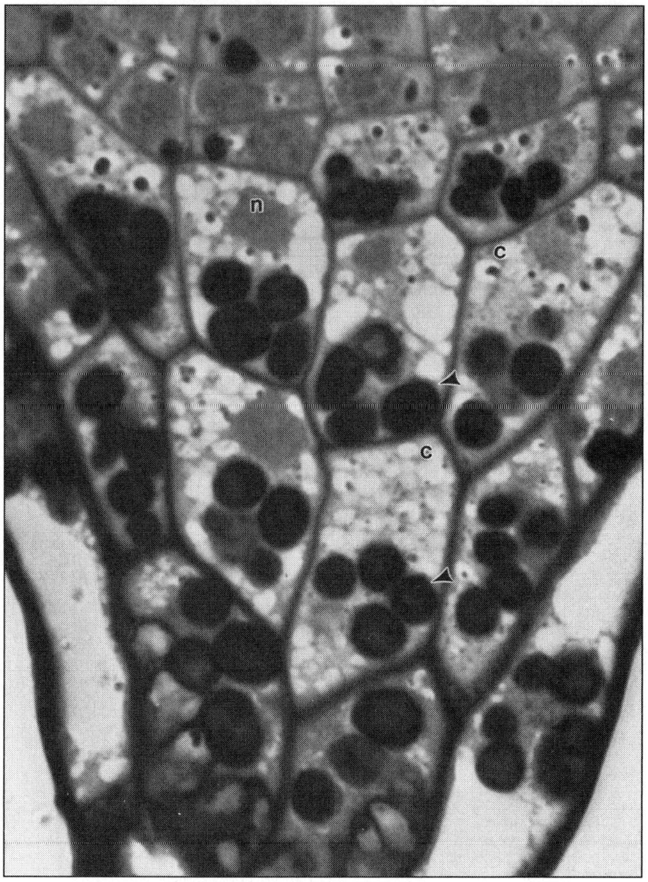

Figure 11.11 A root cap of a tobacco plant. The force of gravity is at the bottom of the picture. Note that the amyloplasts (more or less spherical dark objects) are toward the bottom of each cell. There is conflicting evidence as to whether or not the amyloplasts play a role in the perception of gravity by roots, ×2,000. *(Light micrograph courtesy John Z. Kiss)*

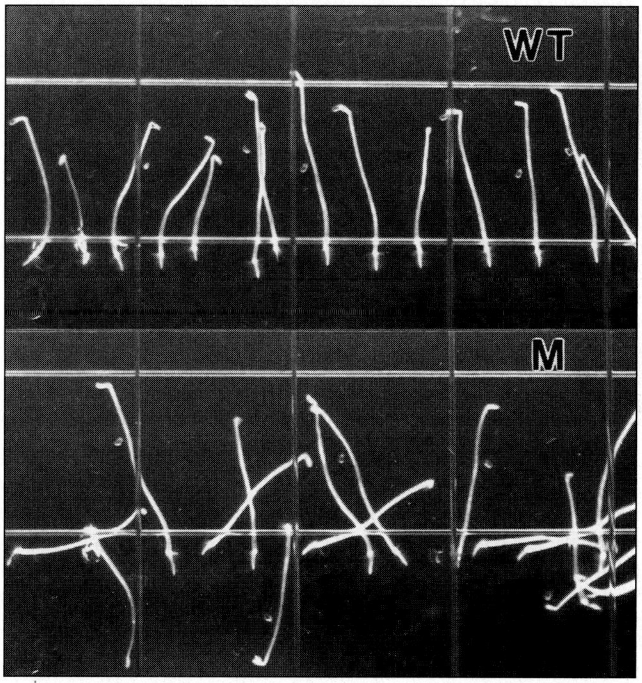

Figure 11.12 Tobacco seedlings grown in the dark. The source of gravity is at the bottom of the pictures. Normal "wild type" seedlings in the top row are all more or less perpendicular to the gravity. The seedlings in the bottom row are mutants with much less starch than normal plants. The mutant seedlings are disoriented, suggesting that any amyloplasts of typical mass function as statoliths in the perception of gravity. However, mutants of other species lacking amyloplasts do respond to gravity, suggesting that parts of cells other than amyloplasts may be involved in the perception of gravity, ×0.5. *(Courtesy John Z. Kiss)*

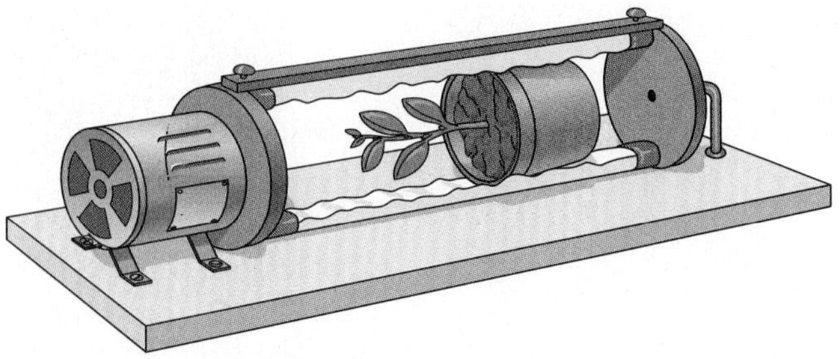

Figure 11.13 A clinostat, which is a tool used by plant biologists to negate the effects of gravity. Growing plants or seedlings are slowly rotated so that the statoliths in cells that perceive gravity do not settle to the bottom, and typical growth or bending of stems or roots away from gravity does not occur.

sensors) nor other organelles can settle while the plant is moving, and this apparently prevents transport of auxin, which would bring about cell elongations that produce curvatures of root and stem.

Other Tropisms A plant or plant part response to contact with a solid object is called a *thigmotropism.* One of the most common thigmotropic responses is seen in the coiling of tendrils and in the twining of climbing plant stems (see Figs. 7.14 and 11.8). Such responses can be relatively rapid, with some tendrils wrapping around a support two or more times within an hour. The coiling results from cells in contact becoming slightly shorter while those on the opposite side elongate.

Roots often enter cracked water pipes and sewers. In fact, roots have been known to grow upward for considerable distances in response to water leaks. Some scientists have called such growth movements *hydrotropisms,* but most plant physiologists today doubt that responses to water and several other "stimuli" are true tropisms. Other external stimuli that produce tropic responses designated as tropic by some scientists include chemicals (*chemotropism*), temperature (*thermotropism*), wounding (*traumotropism*), electricity (*electrotropism*), dark (*skototropism*), and oxygen (*aerotropism*).

Greater concentrations of roots tend to occur on the north and south sides of wheat seedlings, and it has been suggested that magnetic forces may be involved; the term *geomagnetotropism* has been proposed for this phenomenon. Some of these tropic or tropic-like responses have been artificially induced, but others take place regularly in nature. For example, germinating pollen grains produce a long tube that follows a diffusion gradient of a chemical released within a flower; this is considered a chemotropism. Thermotropic responses to cold temperatures may be seen in the shoots of many common weeds, which grow horizontally when cold temperatures prevail and return to erect growth when temperatures become warmer.

Turgor Movements

Turgor movements result from changes in internal water pressures and are often, but not always, initiated by contact with objects outside of the plant. The cells concerned may be in normal parenchyma tissue of the cortex, or they may be in special swellings called *pulvini* located at the bases of leaves or leaflets. Some turgor movements may be quite dramatic, taking place in a fraction of a second. Others may require up to 45 minutes to become visible.

The sudden movements of bladderworts (discussed in Chapter 7) involve turgor changes apparently triggered by electric charges released upon contact or as a result of variations in light or temperature. The springing of the trap of the Venus's flytrap (also discussed in Chapter 7) was thought to be brought about in similar fashion, but recent research has shown that the trap closes when its outer epidermal cells expand rapidly, and it reopens when the inner epidermal cells expand in the same way. About one-third of the ATP available in the cells is used in each movement, so that repeated stimulation of the trap by touching the trigger hairs readily fatigues the trap if sufficient time for ATP replenishment is not given between stimulations.

The sensitive plant (*Mimosa pudica*) has well-developed pulvini at the bases of its many leaflets and a large pulvinus at the base of each leaf petiole. When the leaf is stimulated by touch, heat, or wind, there is a type of chain reaction in which potassium ions migrate from one half of each pulvinus to the other half. This is followed by a rapid shuttling of water from half of the pulvinar parenchyma cells to those of the other half. The loss of turgor results in the folding of both the leaflets and the leaf as a whole (Fig. 11.14).

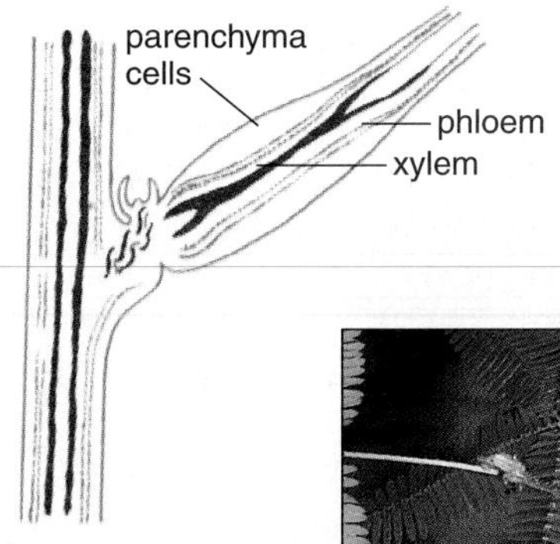

A.

B.

Figure 11.14 *A.* A longitudinal section through the pulvinus of a sensitive plant (*Mimosa pudica*). *B.* The leaflets of the leaf toward the bottom of this picture have folded upward in response to being bumped. The other leaves of this sensitive plant have remained fully expanded.

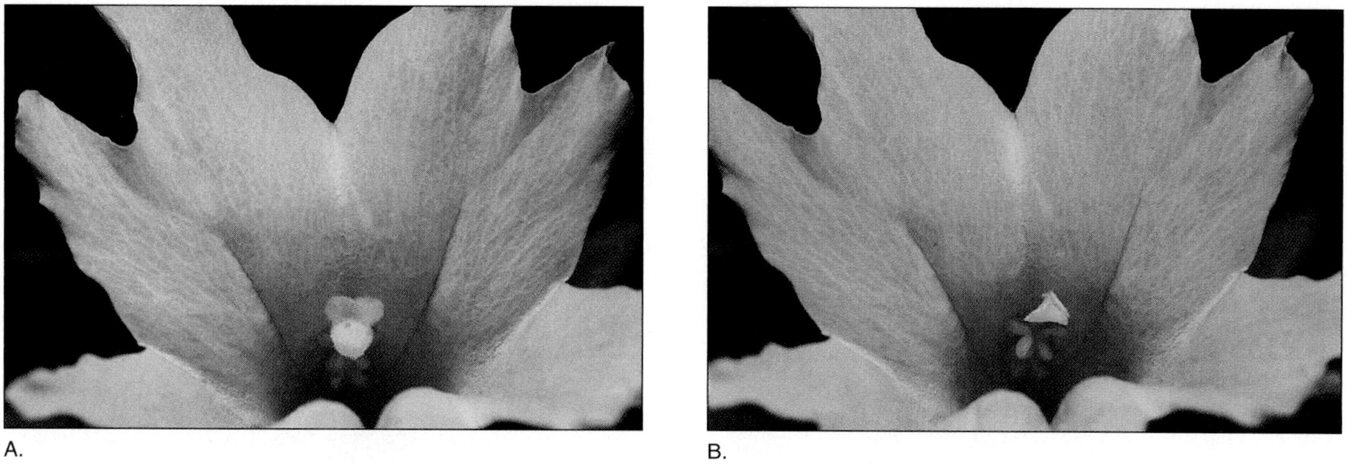

Figure 11.15 A bush monkey flower. The white, two-lobed structure in the center is the stigma. *A.* The stigma as it appears before pollination. *B.* The stigma 2 seconds after being touched by a pollinator; the lobes have rapidly folded together.

If a part of the stem of a sensitive plant (*Mimosa pudica*) is cut off and then immediately reattached with a water-filled piece of rubber tubing, it can be shown that something is transmitted through the water from above the cut. Within a few minutes after a leaf above is stimulated to fold, a leaf or two below the cut will also fold. Although potassium ions have been shown to leave cells in pulvini that are losing turgor, it is not known if the reaction is transmitted across water in the rubber tubing by these ions, by electrical charges, or by something else. A normal response to a stimulus by a sensitive plant (*Mimosa pudica*) takes from a few seconds to less than 2 minutes.

The redwood sorrel (*Oxalis oregana*), whose leaflets also have pulvini, flourishes on the floor of redwood forests where the light may be only 1/200th that of full sunlight. Occasionally, light of sufficient intensity to damage delicate leaves may temporarily penetrate the overhead canopy. When this occurs, the leaflets begin to fold downward within 10 seconds and are completely folded in about 6 minutes, unfolding once again when the brighter light is gone.

Turgor contact movements are not confined to leaves. Many flowers exhibit movements of stamens and other parts, most such movements facilitating pollination. The pollen-receptive stigmas found in flowers of the African sausage tree have two lobes. These lobes fold inward, enclosing pollen grains that have landed on them. The stigmas of monkey flowers (Fig. 11.15) and Asian cone flowers exhibit similar turgor movements upon contact, while the pollen-bearing stamens of barberry and moss rose flowers snap inward suddenly upon contact, dusting visiting insects with pollen (Fig. 11.16).

Orchid flowers exhibit some of the most spectacular of all turgor contact movements, including those by which little sacs of pollen are forcibly attached to the bodies of visiting insects. In a few instances, the benefactors even receive dunkings in small "pans" of fluid from which they escape via a trap door. Brief discussions and illustrations of a few examples are given in Figure 23.18.

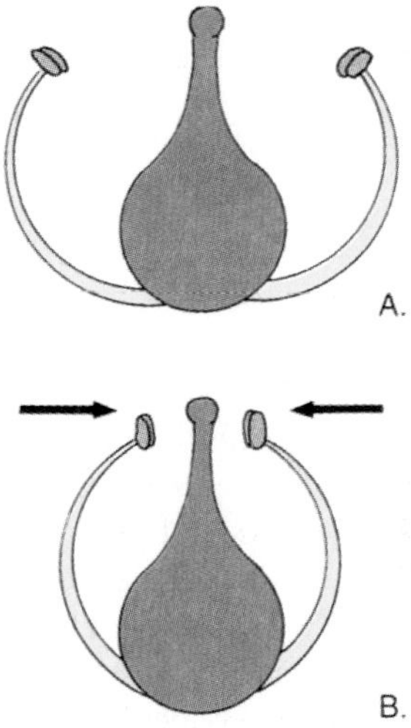

Figure 11.16 Contact movements of the stamens of a barberry flower. *A.* Position of the stamens before contact. *B.* Position of the stamens after they have jerked inward in response to contact.

"Sleep" Movements (Circadian Rhythms)

Members of several flowering plant families, particularly the Legume Family (Fabaceae), which includes peas and beans, exhibit movements in which either leaves or petals fold as though "going to sleep." The folding and unfolding usually takes place in regular daily cycles, with folding most frequently taking place at dusk and unfolding occurring in the morning. Such cycles, which have been more extensively documented in the Animal Kingdom, have come to be known as **circadian rhythms.**

In prayer plants (*Maranta*—Fig. 11.17) and others where the leaves or flowers fold up and reopen daily, turgor movements are involved, as are the external stimuli of light and temperature. These circadian rhythms are controlled by

A.

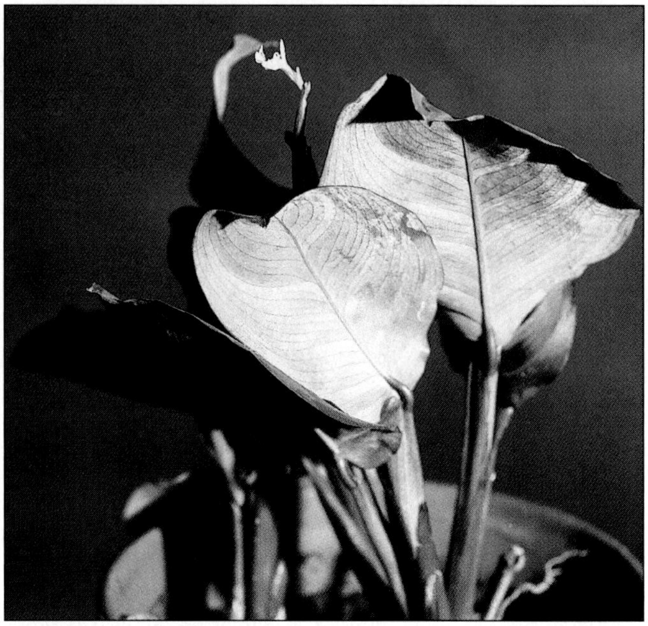

B.

Figure 11.17 A prayer plant (*Maranta*). *A*. The plant at noon. *B*. The same plant at 10 P.M., after "sleep" movements of its leaves have occurred.

a biological "clock" on cycles of approximately 24 hours. In a few instances, nastic movements apparently are initiated by external stimuli alone.

Circadian rhythms appear to be controlled internally by the plants, although they are also geared to changing day lengths and seasons. They do not generally accelerate when temperatures increase. The actual timing of circadian rhythms varies with the species, although most plants exhibiting sleep movements do so at dusk and at dawn. The flowers of several species pollinated by night-flying moths, however, open at different hours of the night.

About 200 years ago, the famous Swedish botanist Linnaeus planted wedge-shaped segments of a circular garden with plants that exhibited sleep movements. The plants were arranged in successive order of their sleep movements throughout a 24-hour day. One could tell the approximate time simply by noting which part of the garden was "asleep" and which was "awake." A few others copied the garden clock idea, but the expense and difficulty of obtaining all the plants from different parts of the world and replanting them each year proved to be too great for the practice to be continued.

The movements are produced by turgor changes caused by the passage of water in and out of cells at the bases of the leaves or leaflets. The function of these movements is not clear, and the rhythms are also not confined to sleep movements. Species of certain warmer marine-water algae called *dinoflagellates* (discussed in Chapter 18) glow in the dark through *bioluminescence,* a process by which chemical energy is reconverted to light energy. One species always glows brightly within 2 or 3 minutes of midnight, even if it is maintained in culture in continuous dim light. This particular dinoflagellate also glows when culture containers in which it is suspended are jarred, and it displays two other types of circadian rhythms. One rhythm involves cell division, with peak mitotic activity occurring just before dawn, and the other rhythm pertains to photosynthesis, which reaches a maximum around noon.

Another type of rhythm is seen in the giant bamboos of Asia, which send up huge flowering stalks every 33 or 66 years, even if the plants have been transplanted to other continents and are growing under different conditions. These flowering stalks use up all the energy reserves of the bamboo, and they die shortly after appearing. This has especially been a problem in cities where nearly all of the bamboo plantings were propagated from a single source and then all died simultaneously after flowering.

Solar Tracking

Leaves often twist on their petioles and, in response to illumination, become perpendicularly oriented to a light source. In fact, many plants have solar-tracking leaves with the blades oriented at right angles to the sun throughout the day. Some scientists have referred to solar-tracking movements as *heliotropisms,* but, unlike phototropic responses of stems and roots, growth is not involved.

It has been widely reported that sunflowers exhibit heliotropic movements and face the sun throughout the day. This is not true, except, perhaps, when the plants are very young; sunflowers face east as they develop and remain facing in that direction until the seeds are mature and the plant dies.

Strictly speaking, the twisting of petioles that facilitates heliotropic movements should be called *phototorsion,* because motor cells (in pulvini) at the junction of the blade and the petiole control the movement (see turgor movements, page 210).

Figure 11.18 Wild grape vines in the fall. Note how little overlap of leaves has occurred.

If one views a tree from directly overhead or observes a vine growing on a fence, it is perhaps surprising to note how little overlap of the leaves occurs. Each leaf is oriented so that it receives the maximum amount of light available (Fig. 11.18).

Water Conservation Movements

Leaves of many grasses have special thin-walled cells (*bulliform cells*) below parallel, lengthwise grooves in their surfaces. During periods when sufficient water is not available, these cells lose their turgor, and the leaf rolls up or folds (Fig. 11.19). Experiments with certain prairie grasses have shown that the rolling effectively reduces transpiration to as low as between 5% and 10% of normal.

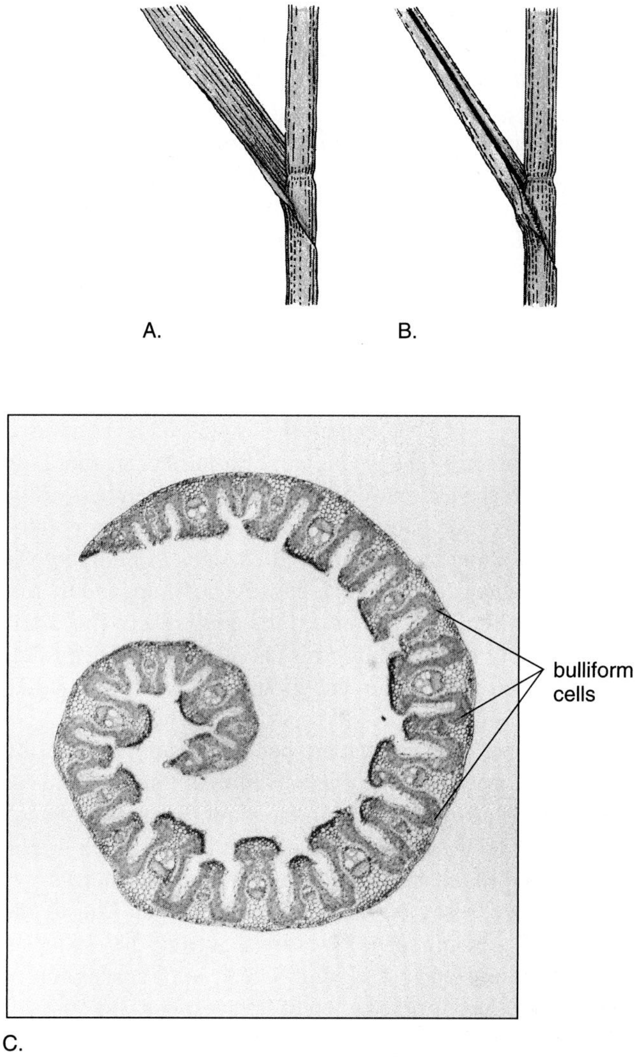

Figure 11.19 Water conservation movement in a grass leaf when insufficient water to maintain normal turgor is available. *A.* The leaf when adequate water is available. *B.* The leaf after it has rolled up. *C.* Enlargement of a cross section of a rolled leaf showing the location of the large, thin-walled *bulliform* cells, which partially collapse under dry conditions and thus bring about the rolling of the leaf blade.

Taxes (Taxic Movements)

The *taxis,* a type of movement that involves either the entire plant or its reproductive cells, occurs in several groups of plants and fungi but not among flowering plants. In response to a stimulus, the cell or organism, either propelled or pushed by **flagella** (whiplike appendages) or **cilia** (short, whiplike appendages), moves toward or away from the source of the stimulus.

Stimuli for taxic movements include chemicals, light, oxygen, and gravitational fields. In ferns, for example, the female reproductive structures produce a chemical that prompts a *chemotaxic* response in the male reproductive cells (sperms)—that is, the sperms swim toward the source of the chemical. Certain one-celled algae exhibit *phototaxic* responses, swimming either toward or away from a light source. Other similar organisms exhibit *aerotaxic* movements in response to changes in oxygen concentrations.

Miscellaneous Movements

Slime molds, discussed in Chapter 18, inhabit damp logs and debris. During their active stages, their protoplasm, which has no rigid cell walls, migrates over dead leaves and other substrates in a crawling-flowing motion, somewhat like thin, slowly moving gelatin. Certain cyanobacteria (e.g., *Oscillatoria*) wave slowly back and forth or slide up and down against each other in *gliding movements,* and diatoms (one-celled algae with thin glass shells) give the appearance of being jet-propelled. It is not certain just how these movements occur, but there is evidence that submicroscopic fibrils produce rhythmic waves that bring about the motion. In the case of diatoms, some materials are apparently forced out of the cells, and the friction set up by the process may propel the organisms through the water in which they are found.

Dehydration movements do not involve living cells or hormones, the forces being purely physical. They are caused by imbibition or by the drying out of tissues or membranes. The individual fruitlets of filarees and other members of the Geranium Family (Geraniaceae) have long, stiff, pointed extensions, which are sensitive to changes in humidity. As humidity decreases during the day, they coil up, and then at night, when humidity increases, the coils relax. This alternating coiling and uncoiling results in the pointed fruitlets planting themselves into the ground in a corkscrew-like fashion (see Fig. 8.27).

A number of fruits that are podlike and dry at maturity split in various ways with explosive force, flinging seeds as far as 12 meters (39 feet) from the plant. Examples of such fruits include those of the garbanzo bean, witch hazel, vetches, and Mexican poppies. In ice plants and stonecrops, rain or dew causes the mature fruits to open due to the swelling of membranes. Pressures inside the squirting cucumber build up to the point where, upon abscission of the fruit, the seeds are expelled from the stalk in distances of up to 10 or more meters (33 feet). Dwarf mistletoes, which are parasitic on coniferous trees, produce tiny, sticky fruits that are explosively released and adhere to the trunks and branches of trees in the vicinity.

PHOTOPERIODISM

In the early 1900s, two American plant physiologists, Wightman W. Garner and Henry A. Allard, became curious about a tobacco plant growing at a research center in Beltsville, Maryland. The plant grew 3 to 4 meters (10 to 13 feet) tall during the summer but failed to flower in August when the normal-sized tobacco plants flowered. They brought the giant plant into a greenhouse for protection during the winter and were surprised when it flowered in December.

Garner and Allard decided to conduct some experiments with the tobacco plants. If the plants were started in pots in the fall in a greenhouse, they grew only about a meter (3 feet) tall before flowering. They wondered what would happen if they kept the tobacco plants, as well as some soybean seedlings that would not bloom in Beltsville before September, in complete darkness from 4 P.M. until 9 A.M., thereby allowing them only 7 hours of daylight.

Figure 11.20 Photoperiodism. The poinsettia with red bracts received less than 8.5 hours of light per day. The green poinsettias received more than 10 hours of light per day and did not flower. *(© Sylvan Wittwer/Visuals Unlimited)*

They found that all the plants flowered, and further investigations showed that the length of day (actually the length of the night) was directly related to the onset of flowering in many plants. They published the results of their investigations in 1920 and later called the phenomenon they had discovered **photoperiodism** (Fig. 11.20).

The critical length of day (i.e., the maximum or minimum length of day) for the initiation of flowering is often about 12 to 14 hours, although it can vary considerably. Plants that will not flower unless the day length is shorter than the critical length (e.g., in the fall or spring) are called **short-day plants.** They include asters, chrysanthemums, dahlias, goldenrods, poinsettias, ragweeds, sorghums, salvias, strawberries, and violets. Plants that will not flower unless periods of light are longer than the critical length are called **long-day plants.** These include garden beets, larkspur, lettuce, potatoes, spinach, and wheat. Such plants usually flower in the summer but will also flower when left under continuous artificial illumination. Accordingly, leafy vegetables, such as lettuce and spinach, need to be harvested in the spring and grown again in the fall in temperate latitudes if *bolting* (producing flowering stalks) is to be avoided. Potato breeders in the United States grow their plants in northern states where the long summer days initiate flowering, but since the potatoes themselves are produced independently of flowering, the plants may be grown for crop purposes at any latitude where other conditions are favorable.

Indian grass and several other grasses have two critical photoperiods; they will not flower if the days are too short, and they also will not flower if the days are too long. Such species are referred to as **intermediate-day plants.** Other plants, particularly those of tropical origin, will flower under any length of day, providing, of course, they have received the minimum amount of light necessary for normal growth. Such plants are called **day-neutral plants** and include garden beans, calendulas, carnations, cyclamens, cotton, nasturtiums, roses, snapdragons, sunflowers, and tomatoes, as well as many common weeds, such as dandelions.

With some plants, small differences in day length may be critically important. Some varieties of soybeans, for example, will not flower when days are 14 hours long but will flower if the day length is increased to 14 1/2 hours. This difference could amount to less than 320 kilometers (200 miles) of latitude, with certain varieties grown in the southern states not producing fruit in the northern states, and vice versa.

Commercial florists and nursery owners have made extensive use of photoperiods, manipulating with artificial light the flowering times of poinsettias, some lilies, and other plants so as to have them flower at times of the biggest demand, such as Christmas or Mother's Day. The light intensities used to lengthen the days artificially can be very low—often less than 2 μmols per square meter per second (i.e., one-thousandth the intensity of full sunlight)—and can come from incandescent bulbs, which have more red wavelengths than fluorescent lamps and therefore have more effect on phytochromes, which are involved in photoperiodism and are discussed in the next section, "Phytochromes and Cryptochromes."

A number of vegetative activities of plants are also affected by photoperiods. In the shortening days of the fall, for example, many woody plants will begin undergoing the changes that lead to the dormancy of buds, regardless of whether or not "Indian summer" temperatures may be prevailing. In the spring, certain seeds respond to photoperiods in their germination, both long-day and short-day species having been discovered. Plants that produce tubers (e.g., Jerusalem artichoke) may develop them only under short-day conditions, even though they are long-day plants with respect to flowering. Usually, these photoperiodically controlled responses are valuable to the plants in that they prepare them for changes in the seasons and thus ensure their survival and perpetuation.

PHYTOCHROMES AND CRYPTOCHROMES

Experiments that were conducted after Garner and Allard's discoveries prompted researchers to look for a pigment that controlled photoperiodism. They had already found that the initiation of flowering could be inhibited if plants are exposed to even a brief period of red light during the night, which suggested the existence of a light-sensitive pigment. Within a few years, such a pigment was discovered. It was not visible, however, and a special pigment-analysis instrument had to be constructed to detect it. In 1959, after the pigment had been isolated, it was named **phytochrome.** Since 1959, several different phytochromes have been identified.

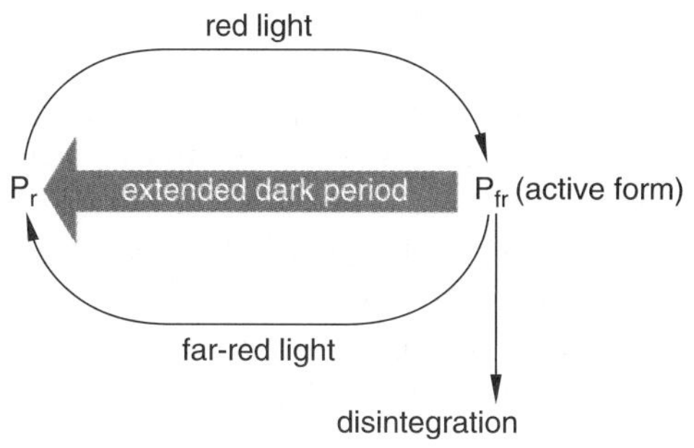

Figure 11.21 Phytochrome interconversions.

Phytochromes are extraordinary pale blue proteinaceous pigments that apparently occur in all higher plants and are associated with the absorption of light. Only minute amounts are produced, mostly in meristematic tissues. Phytochromes occur in two stable forms, either of which can be converted to the other: $\mathbf{P_{red}}$, or $\mathbf{P_r}$, is a form that absorbs red light; $\mathbf{P_{far\text{-}red}}$, or $\mathbf{P_{fr}}$, is an active related form that absorbs the far-red light found at the edge of the visible light spectrum. When either form absorbs light, it is converted to the other form, so that P_r becomes P_{fr} when it absorbs red light, and P_{fr} becomes P_r when it absorbs far-red light. P_r is stable indefinitely in the dark. The normal effect of light in nature is to cause more P_r to become P_{fr} than vice versa. P_{fr} converts back to P_r in the dark over a period of several hours or else becomes inactivated, but its conversion in the presence of appropriate light is instantaneous (Fig. 11.21).

Phytochrome pigments play a role in a great variety of plant responses: in addition to photoperiodism, they are involved with several aspects of plant development, changes in plastids, the production of anthocyanin pigments, and with the detection of shading by other plants.

One of the most studied phenomena associated with phytochrome pigments involves the germination of seeds. Some seeds, for example, do not germinate in darkness, but the red part of the spectrum in sunlight converts P_r to P_{fr}, which, in turn, somehow unblocks the germinating mechanism. If such seeds (e.g., those of "Grand Rapids" lettuce) are given suitable moisture and oxygen conditions and then exposed only to red light, they will germinate readily, but if they are given only far-red light, they will not. Furthermore, if they are given alternating flashes of red and far-red light, the type of light in the final flash determines whether or not they will germinate. If the last flash, which may be of only a few seconds duration, is red light, the seeds will germinate, but if it is far-red light, they will not.

When a seedling first emerges from the soil, light changes the P_r to P_{fr}, triggering a reduction in the production of ethylene by the cells. This, in turn, allows the crooks in the hypocotyls to relax, and the plant straightens up. Elongation of stems appears to be inhibited by P_{fr}. The mechanism is so sensitive that as little as 3 seconds of clear moonlight may produce a shortening of internodes of **etiolated** oat seedlings (those that are spindly and pale from having been grown in the dark). More far-red light reaches young growing stems of a tree that is shaded, however, causing the P_{fr} to be converted to P_r, thus lowering the inhibition so that the young stems grow longer and out from under the shade.

A second group of blue, light-sensitive pigments known as **cryptochromes** also play a role in circadian rhythms and evidently interact with phytochromes in plants in controlling reactions to light. Much of the research since cryptochromes were discovered has centered around fruit flies (*Drosophila*), cyanobacteria, and fungi. Cryptochromes have also been identified in humans.

A FLOWERING HORMONE?

After photoperiodism was discovered, many scientists, including M. H. Chailakhyan, a Russian botanist, conducted experiments to determine which part of a plant was sensitive to day length. They soon found it was mainly the leaf, although they also noted that buds of long-day plants exhibited similar responses. If they completely enclosed the leaf of a short-day plant in black paper for all but 8 hours of a 24-hour day while exposing the rest of the plant to long days, flowering was initiated, but if they treated a long-day plant the same way, flowering did not occur (Fig. 11.22). This suggested that something in the leaf was being carried to an area where flowers were initiated. In the 1930s, Chailakhyan gave support for this theory by showing that when part of a plant was exposed to the appropriate day length to initiate flowering and then grafted to a plant that had not been so exposed, something would cross the graft so that both plants would flower. Chailakhyan suggested the name *florigen* for the "something."

In the 1950s, further evidence was obtained through experiments in which the leaves of some plants were removed immediately after exposure to critical photoperiods, while the leaves of similar plants were removed several hours after exposure. The plants whose leaves were removed immediately after exposure did not flower, but those whose leaves were removed later flowered almost as well as others whose leaves were not removed. This indicated something initiating flowering moved out of the leaves, but its departure was prevented if the leaves were removed before it had a chance to do so.

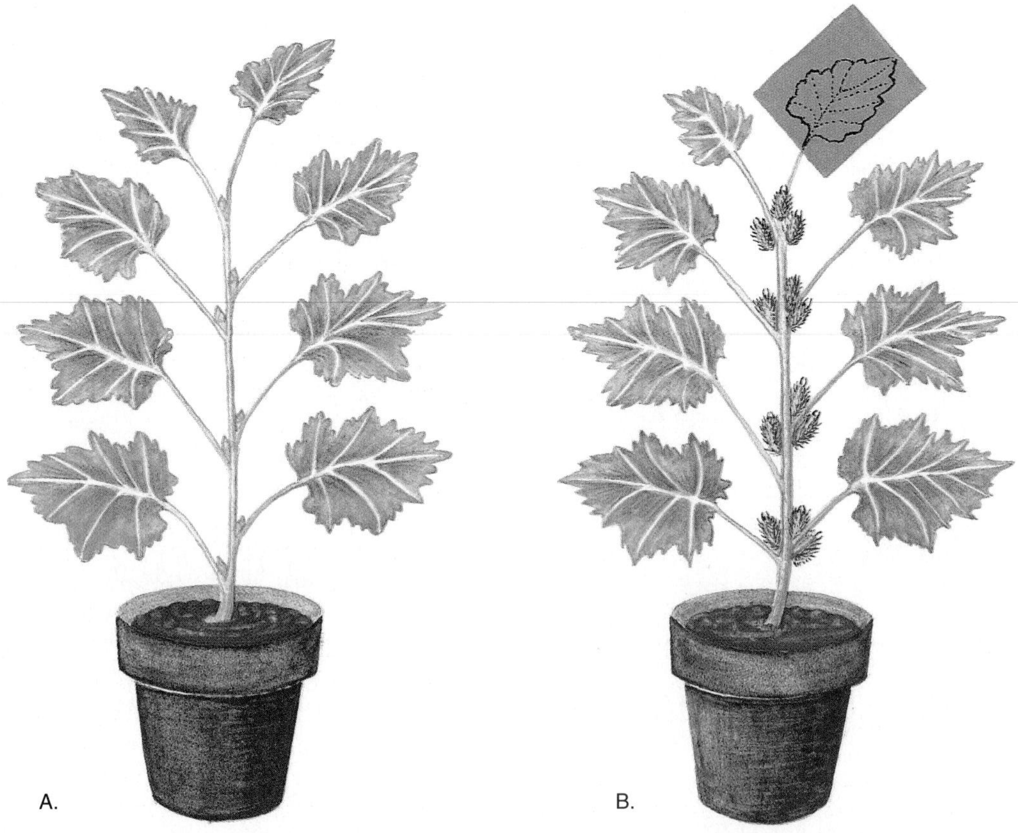

Figure 11.22 An experiment illustrating the effect of subjecting one leaf of a short-day plant (cocklebur) to short days while the rest of the plant is exposed to long days. *A.* The short-day plant exposed to long days. No flowers were produced. *B.* The same plant exposed to long days while one leaf was covered with black paper 16 hours a day for a few weeks. The plant flowered, presumably because some substance that initiates flowering was produced in the shaded leaf and then diffused or was carried to the stem tip where flower buds are produced.

After the existence of phytochrome had been demonstrated, it was theorized that it might be involved in photoperiodism, with P_{fr} inhibiting flowering in short-day plants and promoting flowering in long-day plants. Presumably the P_{fr} would accumulate in the light in short-day plants and revert back to P_r or be broken down during dark periods. After the length of the night increased sufficiently, all or most of the P_{fr} would disappear, and flowering then would cease to be inhibited, with the converse taking place for long-day plants. P_{fr} has been demonstrated to disappear in many plants in as little as 3 or 4 hours of darkness, however, indicating that phytochrome interconversions alone cannot explain photoperiodism.

On the basis of the evidence, it was theorized that plants produce one or more flowering hormones ("florigen"), which may then be transported to the apical meristems where flower buds are initiated. Despite many years of research and all the circumstantial evidence, however, a flowering hormone has never been isolated from a plant, nor has it otherwise been proved to exist. This then led to speculation that photoperiods may bring about a shunting of nutrients that initiate flowering or that flowering is triggered by changes in relative proportions of other hormones. The theory that a flowering hormone exists has now been discarded, and a simple explanation involving hormones for the phenomenon of photoperiodism has proved to be nonexistent. Instead, it is now believed that no hormones are involved in flowering and that flowering is triggered by the direct effects of phytochrome on gene expression. This leads to the accumulation of and/or disappearance of specific RNA's, which, in turn, is thought to initiate flowering.

TEMPERATURE AND GROWTH

Each species of plant has an optimum temperature for growth—although the optimum may vary throughout the life of the plant—and a minimum temperature below which growth will not occur. Each species also has a maximum temperature above which injury may result (or, at least, growth will cease).

Dr. Frits Went, the discoverer of auxin, experimented with the growth of tomato plants under conditions in which night temperatures were lowered to 17°C (63°F) while day temperatures were maintained at 25°C (77°F). These plants were found to grow better than plants kept at a steady temperature around the clock. Went applied the term **thermoperiodism** to this phenomenon. It has since been shown that the optimal thermoperiod may change with the growth stage of the plant. Young pepper plants have been shown to develop best when night temperatures are approximately 25°C (77°F), but as they grow older, the optimum night temperature for their development drops more than 11°C (20°F). Lower night temperatures often result in a higher sugar content in plants and may also produce greater root growth, although some plants, such as peas, seem to be unaffected in this regard.

The growth of many field crops is roughly proportional to prevailing temperatures—making it possible to predict harvest times—although the number of days until maturity varies considerably with the locale. In 1855, the Swiss botanist A. P. de Candolle established a basis for the harvest dates of crops. His method, based on summing the temperature means for each day, makes it possible to follow a crop's progress with some precision. This method has been refined by multiplying the temperature means by the number of hours of daylight.

In Hawaii, a high degree of precision has been attained in predicting the date of the pineapple harvest by measuring the rate of growth within the temperature range at which the plants grow and multiplying the number of hours per day at each temperature. These methods, which work best for crops that are not closely regulated by photoperiods, do not, however, take into account the requirements of some plants for different optimum temperatures at different stages of their development.

DORMANCY AND QUIESCENCE

As the days grow shorter in the fall, cells produce sugars and amino acids that lower the point at which cold damage will occur. Protective bud scales develop on buds, and leaves become senescent. As cell metabolism slows down, many plants that are now prepared for winter become *dormant.* **Dormancy** may be defined as a period of growth inactivity in seeds, buds, bulbs, and other plant organs even when environmental requirements of temperature, water, or day length are met.

Quiescence is a state in which a seed cannot germinate unless environmental conditions normally required for growth are present. We have already discussed the formation of abscission layers in the fall as plants undergo change from an active condition to either a dormant or a quiescent one. In temperate and cold climates, most plants and their seeds go through a dormant period lasting from a few days to several months, and in each case, the dormancy nearly always has some survival value for the species. Many of the stone fruits (e.g., cherries, peaches, and plums) need at least several weeks of rest at temperatures below 7°C (45°F) before the buds will develop into flowers or new growth, and in some instances, the buds will not develop before the tree has encountered a period of freezing temperatures. This dormancy adapts the plants to the conditions prevailing in the cooler temperate zones of the world. In response to prevailing conditions, seeds of desert plants germinate only after appreciable rain has fallen, and seeds of plants native to vernal pools and other wet areas may germinate only if they are immersed in water.

Depending on the species concerned, the change from dormancy to a state in which germination will occur in seeds is sometimes controlled by a variety of factors referred to as *after-ripening.* Some plants become sensitive to photoperiods

Ecological Review

Many plants survive extreme environmental conditions as a result of entering a state of dormancy or quiescence, particularly in temperate or cold climates. Dormancy may be defined as a period of growth inactivity in plant organs even when environmental requirements of temperature, water, or day length are met. Quiescence is a state in which a seed cannot germinate unless environmental conditions normally required for growth are present. Breaking dormancy or quiescence generally depends upon prolonged exposure to specific environmental conditions, such as cold temperatures, wetting, or scarification—removal of a portion of the seed coat by mechanical action or exposure to the digestive enzymes of animals. Plant dormancy, quiescence, and their breaking are under strong natural selection by the environment.

only after they have been subjected to a period of chilling, either while they are still seeds or shortly after germination. Growers often place moistened seeds of such plants in a refrigerator for several weeks. This process is known as *stratification* and is one of several practices employed to break dormancy. In some biennial plants, the cold treatment can be replaced by an application of gibberellins. Natural stratification occurs when winter grains are sown in the fall, and it has been shown that the extent of flowering and yield in such grains is profoundly reduced in the absence of a period of natural chilling.

Some plants have their dormancies broken only after they have received a series of similar stimulations, so a single stimulation, such as an unseasonal rainstorm, will not trigger growth. Other plants, such as poor man's pepper, require both certain temperatures and photoperiods for germination, and some weeds require specific thermoperiods plus some mechanical abrasion of the seed coat to break dormancy. Still other seeds may require specific enzyme action or a combination of conditions acquired only by passing through the digestive tracts of birds, bats, or other animals before they will germinate.

When a mechanical tomato picker was being developed in California during the early 1960s, tomato breeders wanted to have fruits that would stand up to the bumping they would receive in the machinery. They tried, therefore, to produce suitable varieties. Their search for new genes to use in their breeding programs led them to the wild tomatoes of the Galápagos Islands. When they brought the seeds back to California, they experienced difficulty in getting them to germinate, even after subjecting them to a wide variety of stratification treatments. Then they remembered that tomatoes are eaten by the giant tortoises of the Galápagos Islands. They tried feeding the tomatoes to giant tortoises in a zoo. After passing through the animals, the seeds germinated normally. Apparently an enzyme in the tortoises' digestive tracts broke the dormancy of the seeds.

Germination is prevented by a thick or restrictive seed coat in the seeds of plants such as cocklebur, and the seeds of some legumes are almost stonelike because of the exceptional hardness of the outer covering or seed coat. Growers often scarify (nick or file) seed coats to allow water and gases to enter. In nature, various decay organisms corrode the seed coats, or the seeds themselves may secrete enzymes that aid in this process. The seeds of certain lupines have remarkable valves that restrict the uptake of water, preventing germination before the seed coat is eroded by other means. Many plants have seeds that will not germinate until they have been exposed to fire, and one such Australian legume with a thick seed coat will not readily germinate until heat from a fire causes a plug in the seed to pop out.

Dormancy may be restricted to certain parts of the plant rather than occurring in the plant as a whole. In coffee, for example, only the flower buds may show dormancy. In other woody plants, the onset of dormancy progresses from axillary or lateral buds toward the base of a branch on to the terminal bud at the tip. Some seeds go through two cycles of dormancy before full growth occurs. In wake-robins, the radicle (embryo root) of the seed emerges and becomes established after being subjected to a period of cold, but the epicotyl does not emerge until a second season of cold has been encountered.

In many seeds, growth inhibitors, such as ABA, are present in the seed coat, often in abundance. Before germination can occur, these inhibitors may need to be leached out by rain, broken down by enzymes, or have their effects overcome by other hormones. Studies have shown that gibberellin, which has the capacity to break dormancy when applied artificially, is produced by the germinating seeds of both cone-bearing and flowering plants. Ethylene, which is also effective in breaking dormancy, is produced by various germinating seeds, including those of oats, peanuts, and clover. Cytokinins do not generally appear to be involved in breaking dormancy, but a few instances are known of their doing so artificially. Although auxins are formed during the early stages of germination, they probably are not involved in the control of dormancy.

Many aspects of growth discussed in this chapter are applied to, or are involved in, vegetative propagation, which is discussed in Chapter 14 and Appendix 4.

Summary

1. Growth is defined as an "irreversible increase in volume due to the division and enlargement of cells." As cells mature, they differentiate into forms adapted to specific functions.

2. Development is a change in form as a result of growth and differentiation combined. Cells themselves assume different shapes and forms, which adapt them to the problem involved in their total volume increasing at a greater rate than their total surface area.
3. Many growth phenomena are influenced by hormones, which are produced in minute amounts in one part of an organism and usually transported to another part, where they have specific effects on growth, flowering, and other plant activities. Vitamins are organic molecules that function in activating enzymes; they are sometimes difficult to distinguish from hormones.
4. Darwin and his son noted in 1881 that coleoptiles bend toward a light source. Frits Went, in 1926, followed up on the Darwinian observations and demonstrated that something in coleoptile tips moved out into agar when decapitated tips were placed on it. He called the substance auxin and showed that it could cause coleoptiles to bend.
5. Three groups of plant hormones that promote the growth of plants have been found; others are known only for their inhibitory effects. Auxins stimulate the enlargement of cells and are involved in many other growth phenomena. Synthetic auxinlike compounds have been used as weed killers and defoliants. Dioxin, which has caused disease and defects in laboratory animals, is a toxic contaminant of 2,4,5-T, which is now banned in the United States.
6. Gibberellins promote stem growth without corresponding root growth; cytokinins promote cell division and can be used to stimulate the growth of axillary buds. Abscisic acid causes buds to become dormant and apparently helps leaves respond to excessive loss of water. Ethylene gas hastens ripening of fruits and is used commercially to ripen green fruits.
7. Senescence is the breakdown of cell parts that leads to the death of the cell.
8. Plant movements, such as nutations, nodding, twining, contractile, and nastic, are growth movements that result primarily from internal stimuli.
9. Tropisms are permanent, directed growth movements that result from external stimuli, such as light (phototropism), gravity (gravitropism), contact (thigmotropism), and chemicals (chemotropism).
10. Turgor movements result from changes in internal water pressures; they may be very rapid or take up to 45 minutes to become visible. Turgor movements include contact movements and sleep movements in which leaves or flowers fold daily in what is known as a circadian rhythm. Other types of plant movements include water conservation movements, taxes, and dehydration movements.
11. Photoperiodism is a response of plants to the duration of day or night. Short-day plants will not flower unless the day length is shorter than a critical day length, and long-day plants will not flower unless the day length is longer. Intermediate-day plants have two critical photoperiods; the flowering of day-neutral plants is independent of day length.
12. Phytochromes are light-sensitive pigments that occur in all higher plants and play a role in many different plant responses. Phytochromes occur in two forms, each of which can be converted to the other by the absorption of light. Daylight generally results in more P_r being converted to P_{fr} (the active form) than vice versa. P_r absorbs red light and P_{fr} absorbs far-red light. P_{fr} will convert back to P_r over a period of several hours in the dark; P_r is stable indefinitely in the dark.
13. The precise mechanism of photoperiodism is not yet fully understood. Phytochrome interconversions play a role in photoperiodism but cannot be used alone to explain it. A hormone initiating flowering was once believed to be present in leaves, but it has never been isolated from plants. Cryptochromes have functions similar to phytochromes.
14. Each species of plant has an optimum temperature for growth.
15. Dormancy is a period of growth inactivity in seeds, buds, bulbs, and other plant organs even when appropriate environmental conditions are met. Quiescence is a state in which a seed is unable to germinate unless appropriate environmental conditions exist.
16. The change from dormancy to a state in which germination will occur in seeds is controlled by a variety of factors, including temperature, moisture, photoperiod, thickness of seed coat, enzymes, and growth inhibitors. Stratification is the breaking down of dormancy in seeds by keeping them refrigerated and moist for several weeks.

Review Questions

1. What are the differences among hormones, enzymes, and vitamins?
2. Auxins, gibberellins, and cytokinins all promote growth. How does one distinguish among them?
3. How do hormonal herbicides function?
4. List several commercial applications of ethylene gas.
5. What are statoliths, and what role do they play in plant growth?
6. Distinguish among internally and externally induced growth movements, turgor movements, and dehydration movements.
7. How do day-neutral and intermediate-day photoperiod plants differ in their requirements?
8. What is the difference between dormancy and quiescence? How may dormancy be artificially broken?

9. What are the requirements for a seed to germinate?
10. How does phytochrome differ from other plant pigments?

Discussion Questions

1. If green plants require light in order to produce food needed for growth, why are seedlings that are germinated in the dark taller than those of the same age grown in the light?
2. If you remove a terminal bud from a stem, will growth in length stop altogether? Explain.
3. Would it be technically correct to say that some plants go to sleep at night? Explain.
4. Many plants produce seeds that require a period of dormancy before they will germinate. Of what value is the dormancy to the plant? Where might you expect to find plants with seeds that do not undergo dormancy?
5. The rapid turgor movements of the sensitive plant (*Mimosa pudica*) are rare in the Plant Kingdom. If all plants evolved such movements, would the phenomenon have possible survival value or other value to the plants? Explain.

Learning Online

Visit our web page at www.mhhe.com/botany for interesting case studies, practice quizzes, current articles, and animations within the Online Learning Center to help you understand the material in this chapter. You'll also find active links to these topics:

Plant Growth, Development, and Responses
Plant Hormones
Meristematic Tissues

Additional Reading

Briggs, W. R., R. L. Jones, and V. Walbot (Eds.). 1992. *Annual review of plant physiology and plant molecular biology,* vol. 43. Palo Alto, CA: Annual Reviews.

Davies, P. J. 1995. *Plant hormones: Physiology, biochemistry, and molecular biology.* Norwell, MA: Kluwer Academic Publications.

Fosket, D. E. 1994. *Plant growth and development: A molecular approach.* San Diego, CA: Academic Press.

Frank Schaffer Publications Staff. 1996. *Plant growth and response.* Torrance, CA: Frank Schaffer Publications.

Hart, J. W. 1990. *Plant tropisms and other growth movements.* New York: Routledge, Chapman, and Hall.

Hopkins, W. G. 1999. *Introduction to plant physiology,* 2d ed. New York: John Wiley & Sons.

Moore, T. C. 1989. *Biochemistry and physiology of plant hormones,* 2d ed. New York: Springer-Verlag.

Satter, R. L., et al. (Eds.). 1990. *The pulvinus: Motor organ for leaf movement.* Rockville, MD: American Society of Plant Physiologists.

Steeves, T. A., and I. M. Sussex. 1989. *Patterns in plant development,* 2d ed. New York: Cambridge University Press.

Thimann, K. V. 1980. *Senescence in plants.* Elkins Park, PA: Franklin Book Co.

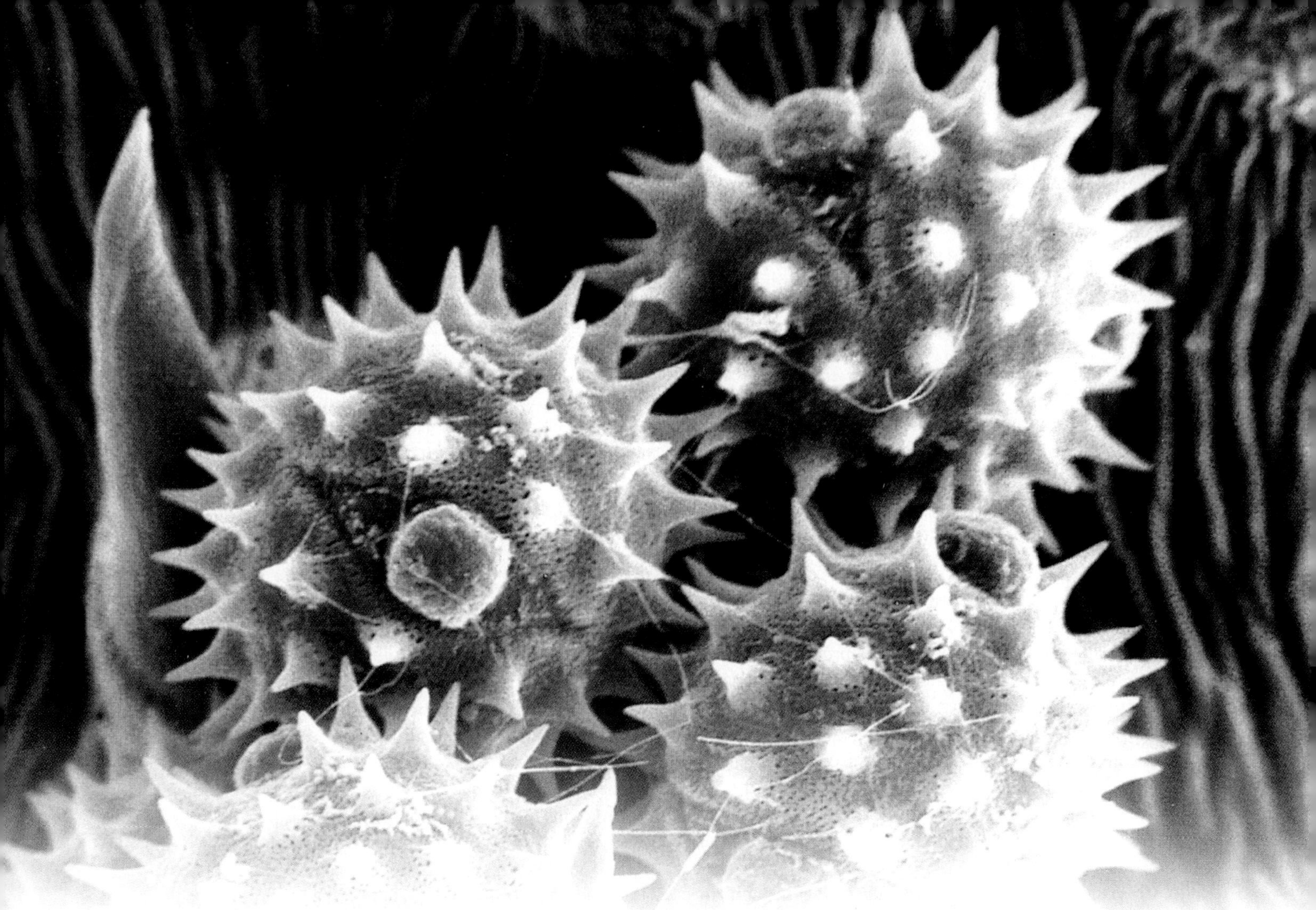

Pollen grains of a western tarweed (*Calycadenia* sp.), ×3,000.
(Scanning electron micrograph courtesy Robert L. Carr)

Meiosis and Alternation of Generations

Chapter Outline

12

Overview

After a comparison of asexual and sexual reproduction, this chapter explores the necessity for reducing the chromosome number at some stage in the life cycle of sexually reproducing organisms. The phases of meiosis are systematically covered. The chapter also presents a discussion of Alternation of Generations and general rules for understanding the process.

Some Learning Goals

1. Know the phases of meiosis and briefly describe what occurs in each of them.
2. Understand clearly what features meiosis and mitosis have in common and how they differ.
3. Explain the significance of crossing-over to offspring.
4. In Alternation of Generations, indicate at what point each of the following occurs: a change from n to $2n$; a change from $2n$ to n; initiation of the gametophyte generation.
5. Relate meiosis and Alternation of Generations to the process of DNA replication.

As a young child, I used to augment my allowance by growing a few vegetables and other plants from seed in my back yard; the produce was then sold—mostly to my mother. Occasionally, an unharvested plant would go to seed, and I would save the seeds for the next growing season. I noticed that in at least some instances, the second generation of plants did not always quite resemble the parents, either in appearance or taste. Although I was mildly curious, I did not understand until many years later that a primary basis for the variation was a process known as *meiosis.*

In Chapter 3, we noted that plants grow through an increase in the number of cells brought about by mitosis and cell division. Mitosis ensures, in very precise fashion, that the number of chromosomes and the nature of the DNA in the daughter cells will be identical with those of the parent cell. Living organisms, however, do not grow indefinitely, although a few may remain alive for a long time. In order to perpetuate the species, they must reproduce, or they will become extinct.

This reproduction may take place through natural vegetative propagation, which is discussed in Chapter 14, or by means of special cells called *vegetative spores,* which are produced by mitosis. Processes involving cells that are identical in their chromosomes with the cells from which they arise are forms of **asexual reproduction** (the prefix *a* of asexual means "without"; asexual reproduction, therefore, simply means reproduction without sex).

Nearly all plants, however, also undergo sexual reproduction, which in flowering and cone-bearing plants ultimately results in the formation of **seeds.** In sexual reproduction, sex cells called **gametes** are produced. Two gametes, called **egg** and **sperm** in higher plants and animals, form a single cell called a **zygote** when they fuse together. The zygote is the first cell of a new individual (Fig. 12.1).

If gametes had the same number of chromosomes as all the other cells of the organism, then the zygote, which results from the union of two gametes, would have double the number of its parents' chromosomes. If such zygotes were then to develop into mature organisms, each of the new organism's cells would have double the original number of chromosomes, and when they produced gametes, the zygotes resulting from the union of this next generation of gametes would have four times the number of chromosomes of the grandparents. If that sort of thing were to continue for 20 generations, a species having 16 chromosomes in each cell to begin with would end up with no fewer than 8,388,608 chromosomes per nucleus! The problem does not actually arise, however, because at a certain point in the life cycle of all sexually reproducing organisms, the unique process of **meiosis** occurs.

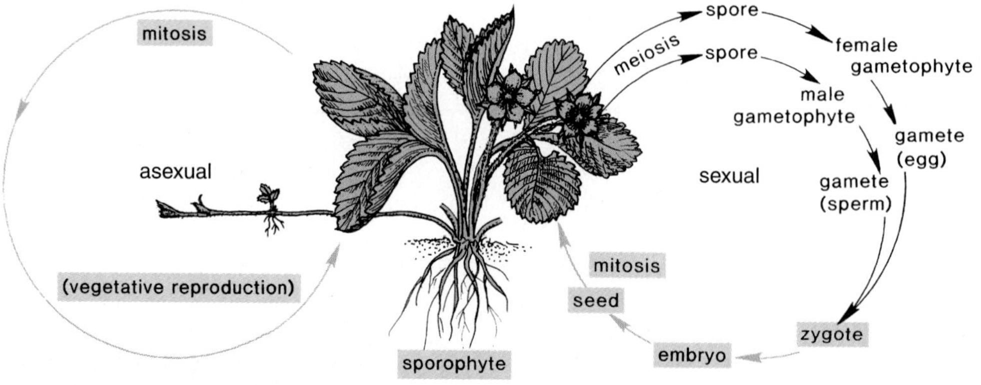

Figure 12.1 Asexual and sexual reproduction in a strawberry plant. More detailed life cycles are shown in Figure 12.6.

The process of meiosis brings about the development of gametes that have only half the number of chromosomes of any body developing from the zygote. When the gametes form zygotes as they unite in pairs, the original chromosome number is restored.

Since some aspects of meiosis are similar to those of mitosis (the two processes are compared in Fig.12.2), it is strongly recommended that you review mitosis in Chapter 3 before going through the phases of meiosis outlined in the following section.

Keep in mind that all living cells, before undergoing meiosis, have two sets of chromosomes—one from the male parent, the other from the female parent. Generally, the members of each pair of chromosomes are identical with each other in length, in the amount of DNA present, and in having a centromere at the same precise location. However, as discussed in Chapter 13, their genes may control contrasting characteristics. Such chromosomes are called *homologues,* or **homologous chromosomes.**

When a cell undergoes meiosis, four cells are produced from two successive divisions, which generally take place without pausing. Because of the remarkable events that occur during the process, the four cells, depending on the organism involved, are rarely, if ever, identical with the original cell or with each other.

THE PHASES OF MEIOSIS

As in mitosis, a doubling of the strands of DNA of each chromosome takes place before the process of meiosis actually begins so that all the chromosomes each initially have two identical parallel strands held together by a centromere. Like mitosis, meiosis is a continuous process that has been divided up into arbitrary phases for convenience. Again, bear in mind that the lines between one phase and another are indistinct.

Four main phases are recognized in each of the two divisions, and the first phase is further subdivided. Since the main phases of the first division bear the same names as those of the second division, it has become customary to designate first division phases with a Roman numeral I and those of the second division with a Roman numeral II. The number of the chromosomes is reduced to half during the first division, but no further reduction in number takes place throughout the second division. Accordingly, some have referred to division I as a *reduction division* and to division II as an *equational division.*

Meiosis generally takes much longer than mitosis to be completed. In lilies, for example, mitosis runs its full course in about 24 hours, while meiosis takes about 2 weeks. In other organisms, meiosis may take months or even years to be completed. After the process has begun in human females, for example, it sometimes takes as long as 50 years to reach its conclusion since it remains in a state of arrest for most of that time.

Division I (Meiosis I, or Reduction Division)

Prophase I

The main features of prophase I are as follows: (1) The chromosomes coil, becoming shorter and thicker as they do so, and their two-stranded nature becomes apparent; the chromosomes also become aligned in pairs. (2) The nuclear envelope and the nucleolus disassociate. (3) Parts of each closely associated pair of chromosomes may be exchanged with each other, and then the chromosomes separate (Fig. 12.2; see also Fig. 12.4A, B).

As with prophase of mitosis, the beginning of this phase is marked by the appearance of chromosomes as faint threads in the nucleus. These threads gradually coil like a spring so that they become as much as 100 times shorter in length, and correspondingly, the coil becomes obviously thicker than the thread alone. As the chromosomes become shorter and thicker, they become aligned in pairs, and eventually, two parallel threads, the **chromatids,** can be distinguished for each chromosome. Each pair of chromosomes, therefore, has four chromatids. The four chromatids of a homologous pair each have their own **centromere** (a dense, constricted area of a chromosome to which a *spindle fiber,* consisting of microtubules having the appearance of a fine thread, becomes attached; see Figs. 12.2 and 12.4B).

As prophase I progresses, parts of the chromatids of the homologous chromosomes may break and be exchanged with each other. Depending on the length of the chromosomes, this exchange may occur at one to several points throughout the length of the paired homologous chromosomes, or if the chromosomes are short, it may not occur at all. The evidence for this exchange of parts appears a little later in prophase I, when the chromosomes of each homologous pair appear to push each other apart. It can then be seen that adjacent chromatids have *crossed over* and are sticking together at one to several points.

Immediately following **crossing-over**, there is an exchange of some of the DNA contributed by the two parents, which is the basis for some of the variability seen in the offspring. An X-shaped figure called a **chiasma** (plural: **chiasmata**) results from each crossover; the relative positions of the chiasmata provide a basis for the study of where various genes are located on the chromosomes (Fig. 12.3).

After the chiasmata have appeared, the chromatids slowly separate, each remaining the same length as it was originally and with the same amount of DNA but now possessing "traded" material. The process is something like exchanging three fenders and the chrome grille of a red car with those of an otherwise identical blue car that has a black grille. When you were through with the exchange, you still would have structurally identical cars, but each would have a different combination of fender colors and grilles.

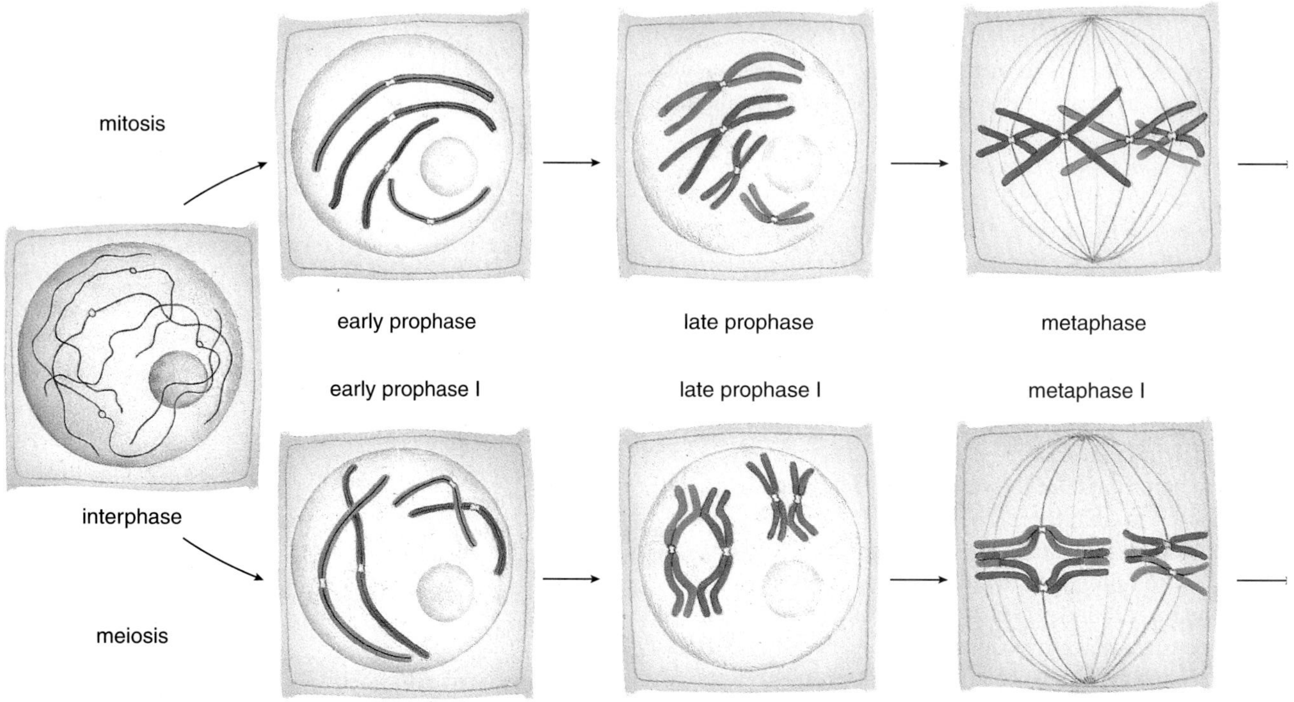

Figure 12.2 A comparison of mitosis and meiosis.

By the end of prophase I, the nuclear envelope and the nucleolus have disassociated and disappeared, and *spindle fibers* (microtubules having the appearance of fine threads) are beginning to form. As in mitosis, some spindle fibers are attached to the centromeres of the chromosomes, while others extend in arcs from pole to pole.

Metaphase I

The main features of metaphase I are as follows: (1) The chromosomes become aligned in pairs at the equator of the cell. (2) The now complete spindle becomes conspicuous (Figs. 12.2, 12.4C).

Metaphase I resembles metaphase of mitosis, except that when the chromosomes move to the invisible, circular, essentially platelike *equator,* the centromeres of the chromosomes are paired, with the homologous chromosomes lined up directly opposite one another on each side of the equator (Fig. 12.5). Also, at this stage, as in mitosis, each chromosome's two chromatids are held together at their centromeres and function as a single unit.

Anaphase I

The main feature of anaphase I is as follows: One whole chromosome (consisting of two chromatids) from each pair migrates to a **pole** (an invisible point at each end of the cell, analogous to the poles of the earth; see Figs. 12.2 and 12.4D).

In anaphase of mitosis, the two chromatids of each chromosome, which are being held together at their centromeres, separate as the centromeres become detached from each other, and a chromatid of each pair migrates to an opposite pole. Anaphase I of meiosis is fundamentally different from anaphase of mitosis in that the chromatids of each chromosome remain cohered at their centromeres and do not separate from one another. Instead, a whole, two-stranded chromosome from each pair migrates to an opposite pole. Accordingly, when the chromosomes arrive at the poles, they still consist of two chromatids, but only half the total number of chromosomes is at each pole. If a particular chromosome was involved in crossing-over in prophase I, one of the chromatids will consist of a mixture of original DNA and DNA from a homologous chromosome.

Telophase I

Depending upon the species, the chromosomes may now either partially revert back to interphase, becoming longer and thinner as they do so, or they may proceed directly to Division II (see Figs. 12.2 and 12.4E). Normally, no new nuclear envelopes are organized around the chromosomes, but the nucleoli usually do reappear. This phase has been completed when the original cell has become two cells or two nuclei.

Division II (Meiosis II, or Equational Division)

In the second division of meiosis, the events of the various phases are similar to those of mitosis except that there is no duplication of DNA during any part of the time pause that may occur between the two divisions. Also, the number of chromosomes in each cell at the end of meiosis is half the

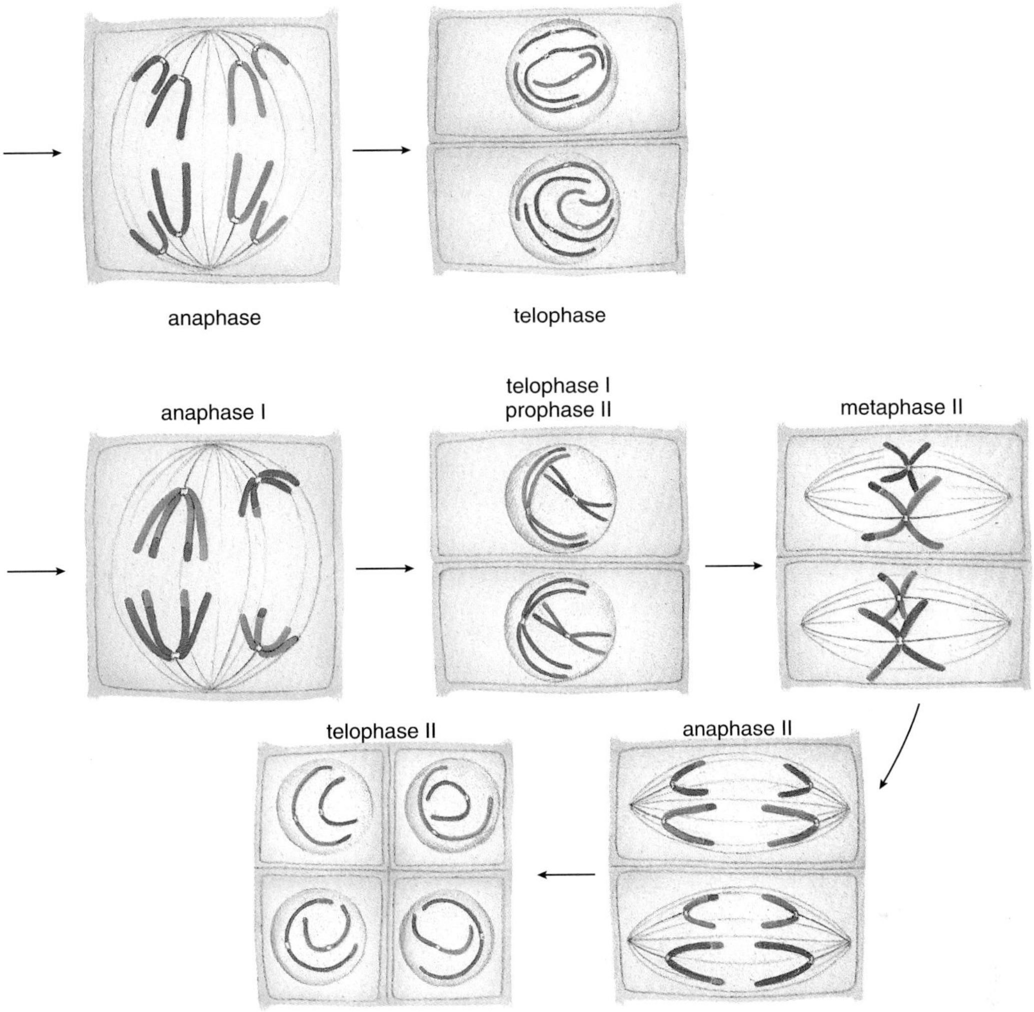

Figure 12.2 Continued

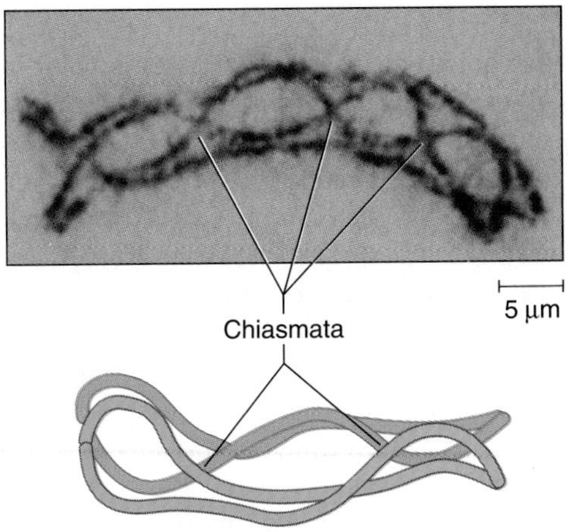

Figure 12.3 Chiasmata. Chiasmata are visible points of crossing over between homologous chromosomes.
(Micrograph source: ©Cabisco/Visuals Unlimited; Line drawing: From Lewis et al., Life, 4th edition.

number that existed at the beginning of meiosis. In some organisms where there is no interphase, the process proceeds almost directly from telophase I to prophase II or metaphase II.

Prophase II

The main feature of prophase II is as follows: The chromosomes of both groups become shorter and thicker simultaneously, and their two-stranded nature once more becomes apparent (see Figs. 12.2 and 12.4F).

Metaphase II

The main features of metaphase II are as follows: (1) The centromeres of the chromosomes become aligned along the equator. (2) New spindles become conspicuous and complete.

The two spindles may be formed either perpendicular to the ones that had formed in metaphase I or along the same plane. A spindle fiber is attached to each centromere, and other spindle fibers extend in arcs from pole to pole (see Figs. 12.2 and 12.4G).

A. Early Prophase I.
Chromosomes begin to coil and contract and appear as threads.
B. Prophase I. Homologous chromosomes become aligned in pairs, and some crossing-over is visible.
C. Metaphase I. Bivalents become aligned along the equator.
D. Anaphase I. Homologous chromosomes separate and migrate to opposite poles.
E. Late Telophase I.
Chromosomes are at the poles, and the original cell becomes two cells.

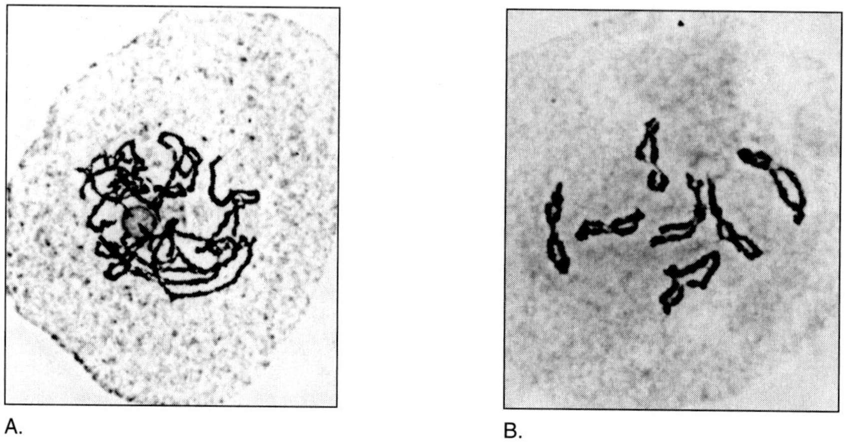

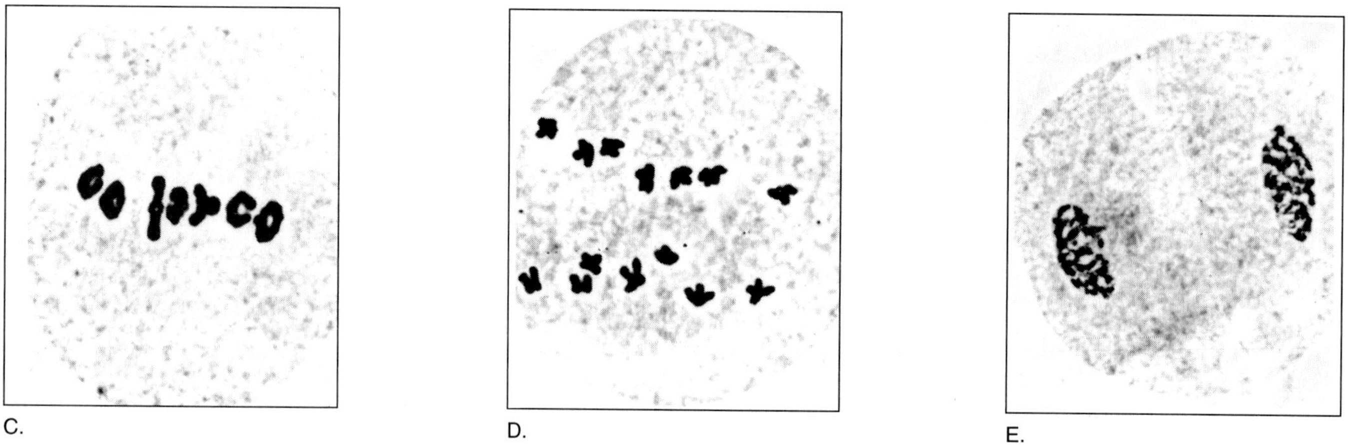

Figure 12.4 Various phases of meiosis in crested wheat grass. *(Photomicrographs by William Tai)*

Anaphase II

The main feature of anaphase II is as follows: The centromeres and chromatids of each chromosome separate and migrate to opposite poles (see Figs. 12.2 and 12.4H).

Telophase II

The main features of telophase II are as follows: (1) The coils of the chromatids (now called chromosomes again) relax so that the chromosomes become longer and thinner. (2) New nuclear envelopes and nucleoli reappear for each group of chromosomes (see Figs. 12.2 and 12.4I).

This phase is accompanied by the formation of new cell walls between each of the four groups of chromosomes. The set of chromosomes present in each of the four cells formed by the end of telophase II constitutes half the original number, and if crossing-over and exchange of material between chromatids have occurred, none of the four cells will have exactly the same combination of DNA. In many organisms, these four cells are called *meiospores;* they develop into various structures or bodies from which gametes are produced by mitosis.

ALTERNATION OF GENERATIONS

As we have noted, meiosis occurs at some point in the life cycle of all organisms that reproduce sexually. The chromosomes that result from the process constitute a complete set in each cell, since one member of every original pair ends up in each cell. The original chromosomal complement, consisting of two complete sets of chromosomes, is restored when gametes unite, forming a **zygote.**

Any cell having one set of chromosomes is said to be **haploid,** and any cell with two sets of chromosomes is said to be **diploid.** By the time meiosis is complete, four *haploid* cells have been produced from one *diploid* cell. The gametes of any organism are haploid, while a zygote of the same organism is diploid. This holds true regardless of the number of chromosomes of any given organism. We can state that an organism having *n* (a specified quantity) chromosomes in its haploid cells will have twice as many, or 2*n*, chromosomes in its diploid cells.

F. Prophase II. The chromosomes coil and contract again; because of crossing-over, the chromatids are no longer identical with each other.
G. Metaphase II. The chromosomes of each cell become aligned along their respective equators.
H. Anaphase II. The chromatids separate completely and migrate to the poles.
I. Telophase II. The four groups of chromatids (now called chromosomes again) are at the poles; new cell walls begin to form.
J. Formation of cell walls. Cell walls form; the four cells will become pollen grains.

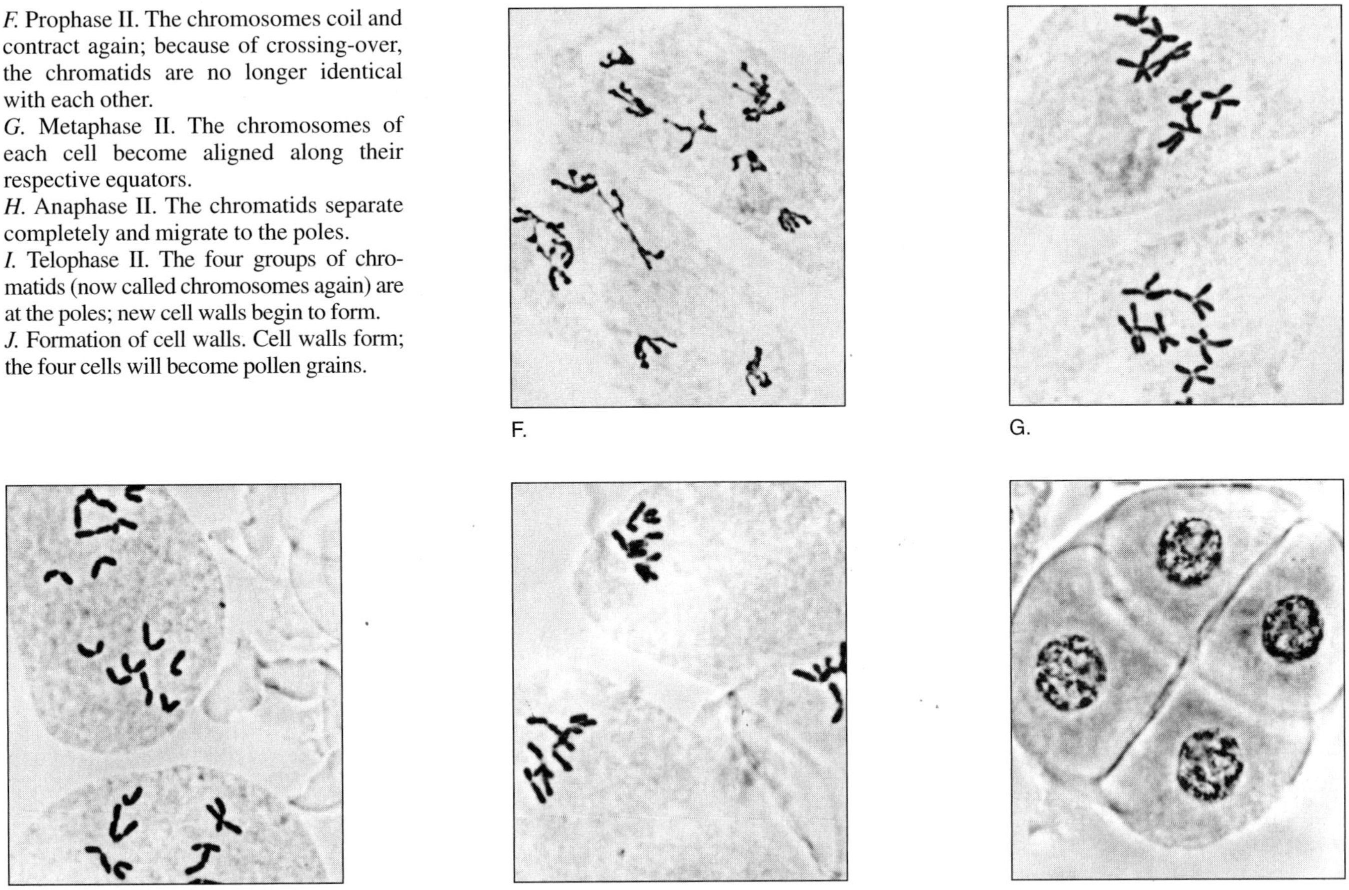

Figure 12.4 Continued

Occasionally, spindles may not form properly during meiosis or something else goes wrong, ultimately resulting in cells with more than one or two sets of chromosomes. Cells having three sets of chromosomes are said to be *triploid* ($3n$), and those with four sets are said to be *tetraploid* ($4n$). Since the homologous chromosomes of a triploid cell undergoing meiosis cannot pair properly, any resulting gametes, if they survive, invariably produce a sterile individual.

In most animals, the only haploid cells (i.e., cells with n or a single set of chromosomes) are the gametes (egg and sperm) and the cells that become the gametes. In plants and other green organisms, however, this is generally not so. In a complete life cycle involving sexual reproduction, there is an alternation between a diploid ($2n$) **sporophyte** phase and a haploid (n) **gametophyte** phase. This is commonly referred to as **Alternation of Generations.**

The diploid body itself is called a **sporophyte.** It develops from a zygote and eventually produces **sporocytes (meiocytes)**, each of which undergoes meiosis, producing four **spores.** The haploid bodies that develop from these spores are called **gametophytes.** These eventually form sex structures, or cells, in which gametes are produced by mitosis.

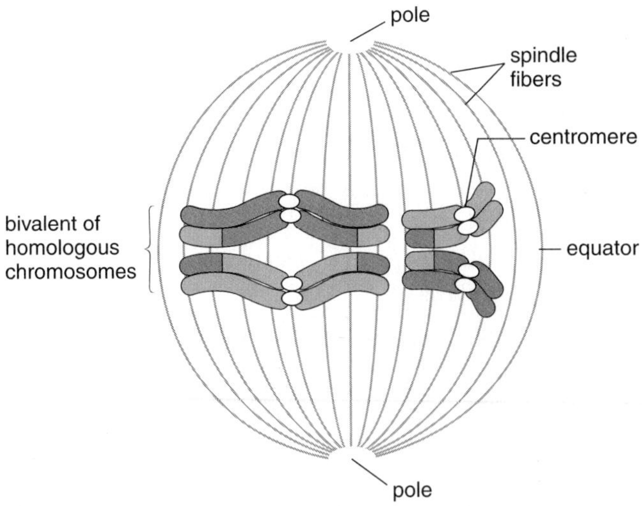

Figure 12.5 A diagram of homologous chromosomes at metaphase I of meiosis.

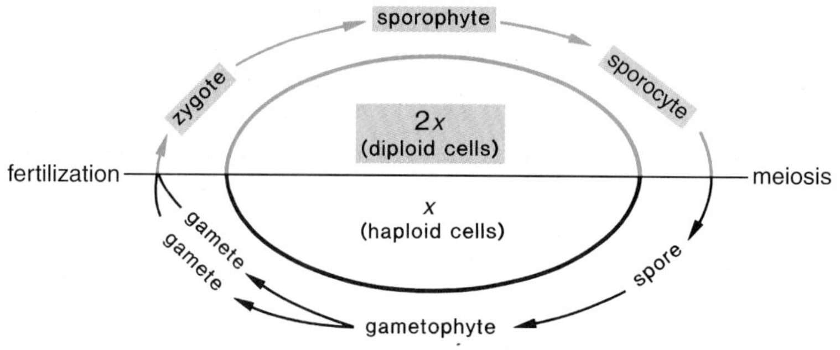

Figure 12.6 A typical life cycle of plants or related organisms that undergo sexual reproduction.

As becomes evident in chapters to follow, the gametophytes of many primitive forms constitute a large part of the visible organism, but as we progress up through the Plant Kingdom to more complex plants, they become proportionately reduced in size until they may be only microscopic. The switch from one generation to the other takes place as spores are produced when sporocytes undergo meiosis and again when a zygote is produced through fusion of gametes, or **fertilization** (also called **syngamy**).

Although the sporophyte generation of some primitive organisms may consist of a single cell (the zygote), the basic plan of Alternation of Generations can be seen in the Protistan, Fungal, and Plant Kingdoms. It becomes most conspicuous, however, in the Plant Kingdom, and it differs from one organism to the next in the forms of the various bodies and cells.

If you will note, along with the accompanying diagram (Fig. 12.6), the following six rules pertaining to Alternation of Generations in the majority of plants and other green organisms, you should have little trouble following the life cycle of any sexually reproducing organism discussed in this book.

1. The first cell of any *gametophyte generation* is normally a *spore* (*sexual spore* or *meiospore*), and the last cell is normally a *gamete.*
2. Any cell of a gametophyte generation is usually *haploid* (n).
3. The first cell of any *sporophyte generation* is normally a *zygote,* and the last cell is normally a *sporocyte* (*meiocyte*).
4. Any cell of a sporophyte generation is usually *diploid* ($2n$).
5. The change from a sporophyte to a gametophyte generation usually occurs as a result of *meiosis.*
6. The change from a gametophyte to a sporophyte generation usually occurs as a result of *fertilization* (fusion of gametes), which is also called *syngamy.*

The word *generation* as used in *Alternation of Generations* simply means phase of a life cycle and should not be confused with the more widespread use of the word pertaining to time or offspring.

Summary

1. Reproduction may take place through natural vegetative propagation or by spores (asexual reproduction) or by sexual processes (sexual reproduction). In sexual reproduction, two gametes unite, forming a zygote, which is the first cell of a new individual.
2. The process of meiosis ensures that gametes will have half the chromosome number of the zygotes and also usually ensures that offspring will not be identical with the parents in every respect.
3. Meiosis takes place by means of two successive divisions, each of which, like mitosis, is divided into arbitrary phases even though the process is continuous. In prophase I, the chromosomes become paired, often exchange parts, and then separate. The similar chromosomes of each pair are referred to as being homologous. Exchange of parts of chromatids may be affected by the parts initially crossing over one another and forming chiasmata and then tearing apart.
4. In metaphase I, the chromosomes become aligned at the equator in pairs, and in anaphase I, whole chromosomes from each pair migrate to opposite poles. In telophase I, the chromosomes either partially revert to their interphase state or initiate the second division, which is essentially similar to mitosis.
5. In prophase II, the chromosomes of each of the two groups become shorter and thicker again, with both groups becoming aligned at their respective equators in metaphase II. In anaphase II, the chromatids of each chromosome separate and migrate to opposite poles, and in telophase II, the chromatids (now called chromosomes again) lengthen, and new nuclear envelopes and nucleoli appear for all four groups. New cell walls are produced between each of the four groups.
6. If the chromatids have exchanged parts earlier, none of the four groups may have identical combinations of DNA, and each group has half the original number of chromosomes. Each of the four cells constitutes a spore (sexual spore or meiospore), which may develop into a

body or structure within which gametes may be produced by mitosis.

7. Any cell having one set of chromosomes is said to be haploid, or to have *n* chromosomes; any cell having two sets of chromosomes is said to be diploid, or to have 2*n* chromosomes.
8. In the life cycle of an organism that undergoes sexual reproduction, there is an alternation between a haploid phase and a diploid phase. The haploid body is called a gametophyte, and the diploid body is called a sporophyte.
9. The change from the haploid phase to the diploid phase occurs when two gametes (each *n*) unite, forming a zygote (2*n*) in the process of fertilization (syngamy). The change from the diploid to the haploid phase occurs as a result of a sporocyte (meiocyte) undergoing meiosis, when a 2*n* cell becomes four *n* cells. This switching of phases is referred to as Alternation of Generations.

Review Questions

1. During which phase of meiosis does each of the following events occur? (a) crossing-over (b) chromatids separate at their centromeres and migrate to opposite poles (c) chromosomes become aligned in pairs at the equator
2. Is there any difference between mitosis and the second division phases of meiosis?
3. What is the significance of meiosis with respect to sexual reproduction?
4. What is meant by saying that the cells of a sporophyte phase are diploid or 2*n*?
5. At what stage of a life cycle does the chromosome number of cells switch from *n* to 2*n*?

Discussion Questions

1. If an organism reproduces very freely by asexual means, is there any advantage to its also reproducing sexually?
2. Would it make any difference if the events of the second division of meiosis took place before the events of the first division?
3. Would it be correct to say that a bivalent has four times the amount of DNA present in the chromosomes of a cell in anaphase II?
4. Would you assume that the length of a chromosome might have anything to do with the number of crossovers it might form with its homologue? Explain.
5. Two mitotic divisions may take place in as little as a few hours, whereas meiosis may take much longer. Can you suggest any reasons for this?

Learning Online

Visit our web page at www.mhhe.com/botany for interesting case studies, practice quizzes, current articles, and animations within the Online Learning Center to help you understand the material in this chapter. You'll also find active links to these topics:

Asexual Reproduction in Plants
Cell Division
The Cell Cycle
Plant Reproduction
Meiosis

Additional Reading

Ahern, H. 1999. *Molecular and cell biology,* 2d ed. Dubuque, IA: McGraw-Hill.

Becker, W. M., and D. W. Deamer. 1991. *The world of the cell.* Redwood City, CA: Benjamin/Cummings.

Esser, K., et al. (Eds.). 1998. *Progress in botany—Cell biology & physiology—Systematics & comparative morphology—Ecology & vegetation science,* vol. 60. New York: Springer-Verlag.

Harris, H. 2000. *The birth of the cell.* New Haven, CT: Yale University Press.

John, B. 1990. *Meiosis.* New York: Cambridge University Press.

Latchman, D. 1997. *Basic molecular and cell biology,* 3d ed. Chicago, IL: Login Brothers Book Company.

Raven, P. H. 1999. *Biology of plants,* 6th ed. New York: Worth.

Wolfe, S. L. 2002. *Molecular and cellular biology,* 2d ed. Belmont, CA: Wadsworth.

Narrow-tubed monkey flowers (*Mimulus angustatus*), a vernal pool species of northern California. (Courtesy Robert A. Schlising)

Genetics

13

Chapter Outline

Overview

In this chapter, the topic of genetics is introduced with the story of Barbara McClintock's discovery of transposable elements ("jumping genes") in corn. The chapter progresses from genetics at the molecular level to the cellular level (cytogenetics) to the organismal level (Mendelian genetics and related topics) and then finishes with population genetics (the Hardy-Weinberg law).

Some Learning Goals

1. Identify components of a DNA molecule and know how they are arranged in the molecule.
2. Describe the functions of DNA.
3. Describe how a DNA molecule replicates.
4. Know the function of transcription and outline its steps.
5. Know the function of translation and outline its steps.
6. Distinguish between somatic and germ-line mutations.
7. Describe the significance of translocations and inversions.
8. Distinguish between aneuploids and polyploids.
9. Understand the significance of Mendel's experiments with peas.
10. Give the ratios of the offspring in the first two generations from a monohybrid and a dihybrid cross. Describe the genotypes involved.
11. Distinguish between genotype and phenotype; heterozygous and homozygous.
12. Be able to solve simple genetics problems involving dominance and incomplete dominance.
13. Show how genes may interact with each other to influence phenotype.
14. Explain how genotype influences phenotype.
15. Describe features of a quantitative trait.
16. Describe where extranuclear DNA is located and how it differs from nuclear DNA.
17. Explain how linkage can produce ratios that deviate from Mendelian ratios.
18. Describe the Hardy-Weinberg law.

Every so often, someone comes along who is able to push beyond our limits of understanding in science and change the way we view the natural world. Barbara McClintock was one of the most significant scientists in 20th century biology because she caused a major shift in the way we view gene organization. She was a geneticist who carried out most of her work at the Cold Spring Harbor laboratory in New York in the 1940s and 1950s. While looking at corn seedlings for one of her genetic studies, she noticed that streaks of color often appeared where they did not seem to belong. There were patches of yellow in green leaves or patches of green in white leaves. She also noticed spots of color in corn kernels that should have been colorless, based on her genetic studies (Fig. 13.1). Instead of ignoring these plants as genetic oddities, McClintock undertook years of research to identify the cause of these patterns.

In the early 1950s, McClintock published work introducing the concept of **transposition,** or the movement of a piece of a chromosome to another chromosomal location. A **transposable genetic element** (or *jumping gene* in popular literature) is a gene or a small DNA fragment that can move to a new location on the same chromosome or even to another chromosome. If it moves into an existing gene, then the function of that gene is disrupted. A transposable element can also move out of a gene and restore its original function. So, the patches of yellow in green leaves were regions where transposable elements were inserted into a gene involved in chlorophyll synthesis. Green areas on white leaves resulted when a transposable element moved out of a chlorophyll synthesis gene and allowed it to function again.

Figure 13.1 An ear of corn showing the effects of transposable elements. The speckles in the center kernel are clusters of cells in which a transposable element has removed itself from a purple pigment gene. Yellow areas in that kernel do not express purple pigment because the transposable element has interrupted that pigment gene.

Despite volumes of data to support her hypothesis, McClintock's efforts to explain her work were initially met with silence. Most of her colleagues could not understand what she was proposing, and so they ignored her work. Others ridiculed her or implied she was crazy. Gene theory at that time suggested a gene to be an indivisible unit, like a bead on a string. It was unthinkable to believe that mutations could happen within genes. And it certainly seemed absurd to propose that pieces of chromosomes could actually move to new locations. However, McClintock knew she was right, and she continued her attempts to convince colleagues to accept her ideas.

The discovery of the structure of DNA by James Watson and Francis Crick in 1953 ushered in the new field of **molecular genetics.** At that time, geneticists thought of McClintock's work as a remnant of a previous era and not relevant to modern genetics. However, studies of DNA and gene structure transformed the concept of a gene. Genes were no longer abstract entities but were actual sequences of nucleic acids that could be isolated and characterized. The concept of mutation within a gene became believable. In addition, numerous studies found transposable elements in other organisms.

McClintock's work was formally recognized in 1983 when she was awarded a Nobel Prize. We now know that transposable elements are widespread and have probably been in organisms for a long time. We do not know, however, if they represent genetic parasites causing mutations as they move about an organism's genetic material or if they perform valuable functions. One theory is that they allow nature to tinker with chromosomes much as human genetic engineers do. It may be evolutionarily beneficial to copy, move, and rearrange pieces of chromosomes, creating new and occasionally better combinations of genes within an organism.

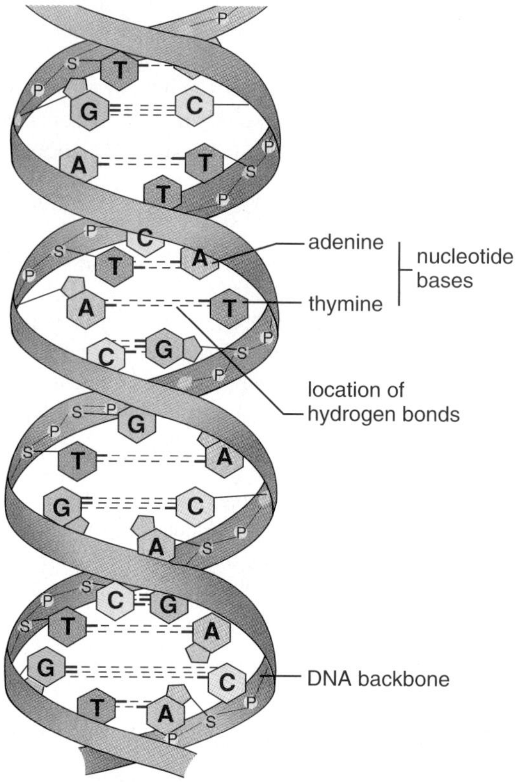

Figure 13.2 Structure of a DNA molecule. In this enlargement of a small portion of a DNA molecule, the rungs of the twisted ladder formed by the two entwined spiraling strands consist of nitrogen-containing bases supported by alternating units of sugar (S) and phosphate (P) molecules. The purines adenine or guanine (A or G) occur opposite the pyrimidines thymine or cytosine (T or C). The purines and pyrimidines opposite each other are held together by hydrogen bonds linking the nitrogenous bases of the paired molecules. The double helix is 2 nanometers wide.

MOLECULAR GENETICS

In Chapter 3, we mentioned that chromosomes in the nucleus of each cell carry genetic information that is passed on from one generation to the next. Chromosomes are composed of two types of large molecules—DNA and protein. Until the early 1950s, most scientists believed that proteins carried the genetic information because they thought that the 20 amino acids in proteins could provide more diversity than the four bases in DNA. We now know, of course, that DNA is responsible for carrying genetic information.

Structure of DNA

The millions of atoms in a DNA molecule are organized in a simple, elegant chain of building blocks called **nucleotides.** Each nucleotide consists of three parts: (1) a nitrogen-containing compound, called a **base;** (2) a 5-carbon sugar, named **deoxyribose;** and (3) a phosphate group. Both the nitrogenous base and the phosphate group are bonded to the sugar (Fig. 13.2). Four types of nucleotides occur in DNA. Each has a unique nitrogenous base, but all have the same phosphate group and deoxyribose sugar. Two of the bases—**adenine** (A) and **guanine** (G)—are called **purines.** Each has a molecular structure that resembles two linked rings. The other two bases—**cytosine** (C) and **thymine** (T)—are called **pyrimidines.** They each have a molecular structure consisting of a single ring.

The nucleotides in a DNA molecule are bonded to each other in such a way that they form a chain that looks like a ladder twisted into a spiral, or a helix. Each of the two sides of the "ladder" is composed of alternating sugar and phosphate groups (Fig. 13.3). Each sugar is also bound to a base. Hydrogen bonds hold each base on one side of the helix to another base on the other side, making the "rungs" of the ladder. Although the structure of a DNA molecule is very simple—alternating sugar and phosphate groups on the outside and pairs of bases in the middle—it provides all the genetic diversity on the planet. The variation comes in the form of base pairs. The sequence of base pairs on a DNA molecule determines whether, for example, that DNA will direct the synthesis of a chlorophyll molecule or a hemoglobin molecule.

As mentioned earlier, in the early 1950s, James Watson and Francis Crick began to unravel the mystery of DNA structure. Much of their work involved piecing together existing information to develop a model. They knew Linus Pauling had postulated that the structure of DNA might be similar to the structure of protein. Pauling had shown part of the structure of some proteins to be helical and maintained by hydrogen bonds between the amino acids. Watson and Crick also knew from X-ray work by Rosalind Franklin and Maurice Wilkins that DNA molecules are composed of regularly repeating units in a helical arrangement. Finally, studies

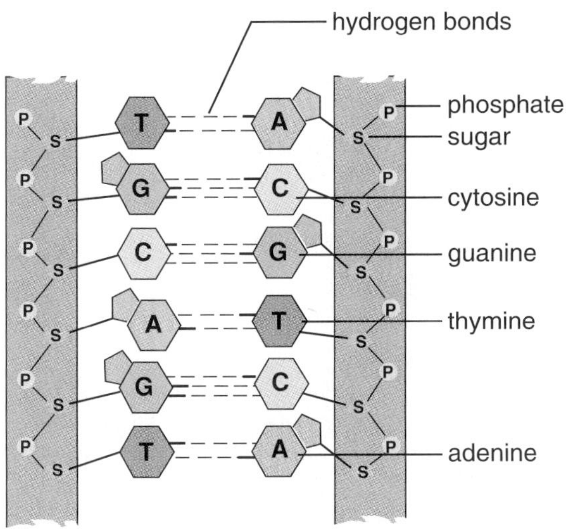

Figure 13.3 The pairing of nucleotides in a tiny portion of a molecule of DNA. The variations in sequences of pairs are virtually unlimited.

by Erwin Chargaff indicated that the amount of pyrimidine nucleotides (cytosine + thymine) in DNA equals the amount of purine nucleotides (adenine + guanine). This suggested to Watson and Crick that purines pair with pyrimidines.

When they had put together all the facts, Watson and Crick concluded that a DNA molecule consists of a double helix whose two strands appear to be wrapped around an invisible pole—one strand spiraling in one direction and the other strand spiraling at the same angle in the opposite direction. They also concluded that nucleotides are linked in ladder-like fashion between the two strands. Furthermore, in order for the nucleotides to fit precisely in the DNA molecule, two linked purines or two linked pyrimidines would either be too wide or too narrow. However, a purine and a pyrmidine linked together would fit perfectly. Accordingly, they concluded that the ladder rungs had to consist exclusively of purine-pyrimidine pairs. Guanine always pairs with cytosine, and thymine always pairs with adenine. Watson and Crick's DNA molecule is now universally accepted as an authentic representation.

DNA Functions

The DNA in a cell must perform four major tasks. It must (1) store genetic information; (2) copy that information for future generations of cells; (3) express that information; and (4) occasionally change its message, or **mutate.**

Storage of Genetic Information

Most DNA in plants resides in the nucleus. Because DNA directs all cellular activities, it is important to protect DNA and allow for its permanent storage. The nuclear envelope probably helps to protect fragile DNA molecules from mechanical shearing by components of the cytoskeleton (see Chapter 3).

The genetic information in a DNA molecule resides in the sequence of nucleotides along the rungs of the ladder. The sequence GAATCC, for example, codes for a different set of amino acids than CTTAGA. Early scientists were fooled by the simplicity of a DNA molecule and did not believe it could provide information for the survival of all organisms on earth. However, the DNA in a typical plant cell contains millions or billions of base pairs. A barley plant with 5 billion base pairs of DNA in its nucleus could produce $4^{5,000,000,000}$ different DNA sequence combinations! This provides more than enough genetic variability for the plant.

At this point, we need to introduce some important concepts, beginning with that of the gene. We are all familiar with the idea that genes control physical traits, such as eye color in people and height in plants. How do they do that? From a molecular perspective, a **gene** is a segment of DNA (several thousand base pairs long) that directs the synthesis of a protein. That protein is then used by the cell as structural or storage material, or it influences the activities of the cell by acting as an **enzyme.** As you recall from Chapter 2, enzymes are organic catalysts. Any activity in the cell, from photosynthesis to cell-wall construction, absolutely depends on enzymes to facilitate chemical reactions. Therefore, all activities of the cell are controlled by enzymes, which require genes for their synthesis. Each kind of organism has many different genes in its **genome,** which is a term for *the sum total of DNA in an organism's chromosomes.* A relatively simple organism such as a bacterium typically has several thousand genes in its genome. On the other hand, more complex eukaryotic organisms usually have between 10,000 and 100,000 genes per genome. The DNA message in a gene is read in three-nucleotide groups, called **codons.** If we compare DNA nucleotides to *letters* of the alphabet, then codons are comparable to *words* that are organized into gene *sentences*; all gene sentences in a genome make up the *book* that constitutes the whole organism. The differences between organisms lie in the differences in the composition of their genomes. Organisms that are very dissimilar—for example, bacteria and higher plants—will have many differences in their DNA sequences. The genomes of similar organisms, such as domestic plants and their wild relatives, are very similar, yet different enough to account for dramatic differences in appearance.

Replication (Duplication) of Information

DNA must duplicate itself precisely in order to pass along its information from generation to generation. DNA replication occurs during the S phase of the cell cycle (see Chapter 3). Differences between the DNA of one organism and that of another lie in the sequence of the four possible types of ladder rungs and their total number. If the four nucleotides are synthesized in living cells, then what tells the cell exactly how to put them together to form DNA molecules with proper nucleotide sequences?

Watson and Crick observed that if the two strands of DNA are "unzipped" by an enzyme breaking the hydrogen bonds between the nucleotides down the middle of the ladder,

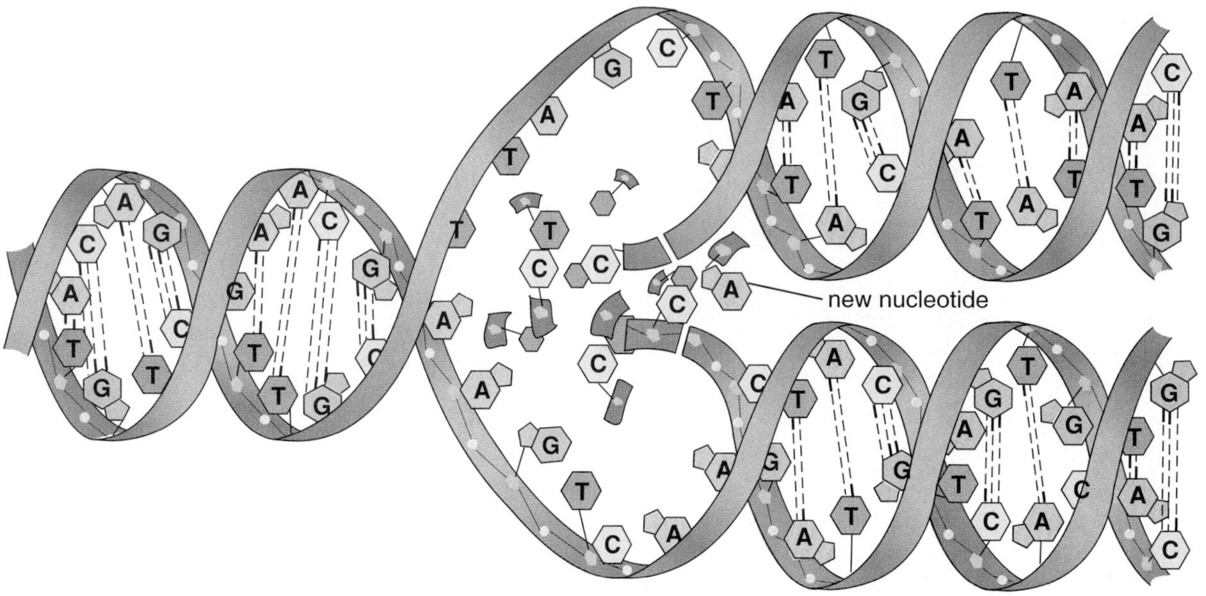

Figure 13.4 Replication (duplication) of DNA. A molecule replicates by "unzipping" along the hydrogen bonds, which link the pairs of nucleotides in the middle, and then each half serves as a template for nucleotides available in the cell. As the new nucleotides fit into the appropriate places and are linked together, a double helix is re-formed for each original half.

then each separated chain provides all the information needed to put together a new ladder (Fig. 13.4). In other words, since guanine can pair only with cytosine, and vice versa, a guanine nucleotide synthesized by the cell can lock on, or bond, only to a thymine nucleotide. If the two strands of a DNA molecule "unzip," then individual nucleotide "building blocks" can line up next to each of the single chains in precise sequence and be bonded together to form a new chain; each single chain acts as a mold or template for the creation of a new double-stranded DNA molecule. When replication is complete, two double helices have been created from a single one. Each new DNA molecule consists of one strand from the original molecule and another built using that parental strand as a template. This is called **semiconservative replication.**

Replication is a remarkably rapid and precise process. Nucleotides are added by the enzyme **DNA polymerase** at a rate of approximately 33 bases *per second* during DNA replication. However, even at this rate, a chromosome that is, for example, 200 million base pairs long, would take 70 *days* to replicate. Because a cell must replicate its DNA before it undergoes mitosis, this replication rate is far too slow for normal plant growth. How is this problem resolved for the cells? In two ways: (1) replication proceeds simultaneously in both directions on the chromosome, cutting the replication time in half; and (2) replication begins at several points on a chromosome nearly simultaneously.

The nucleus contains enzymes that are capable of "proofreading" during replication. They detect and correct mismatched base pairs, such as guanine paired with adenine or thymine paired with cytosine. Consequently, about only one in every million nucleotides added during replication is incorrectly paired. However, each nucleus of a corn plant contains 4 billion base pairs of DNA. Therefore, about 4,000 mutations (changes in DNA sequence) occur with each cell division.

Expression of Information

Since the nature and structure of DNA is of fundamental importance to its ability to control the form and function of organisms, there appears to be a paradox. If every cell in an organism contains the same genetic information, then why do cells within a plant differ from each other? That is, why is a bark cell so different from a leaf cell?

The answers to these questions stem from the fact that each cell in a multicellular plant contains the same DNA information, but different subsets of this master plan are read in each cell type. For example, a plant has epidermal cells that secrete a protective waxy cuticle, and it also has mesophyll cells that are highly specialized for photosynthesis. The epidermal cells express genes (read DNA sequences) that encode enzymes necessary to build a waxy cuticle. Mesophyll cells, on the other hand, will not express cuticle genes but do express a set of genes that encode enzymes for chlorophyll synthesis. One of the most active areas of biology research today is directed toward understanding how organisms control gene expression.

A cell's environment can influence the set of genes expressed by that cell. Many cellular changes occur as a result of the expression of new sets of genes induced by hormones (see Chapter 11). We know, for example, that an epidermal cell expresses genes for cell elongation if the hormone auxin is present; when mesophyll cells are moved from dark into light, many dramatic changes in gene expression also occur.

Awareness

The Polymerase Chain Reaction (PCR)

How many times have you said to yourself, "Why didn't I think of that?" Many scientists must have had that thought when Kary Mullis described the polymerase chain reaction (PCR) in 1984. While Mullis was driving to his cabin in northern California one evening, he was thinking about a method for determining nucleotide sequences in DNA molecules. The technique used DNA replication enzymes to synthesize short strands of DNA in a test tube. After running through several scenarios in his head, he realized that he could create a chain reaction in a test tube. Every reaction would double the amount of DNA synthesized, causing an exponential growth in the number of DNA molecules created. Twenty cycles of DNA replication would create from a single molecule a million identical DNA molecules. For years scientists had worked with replication enzymes in DNA synthesis experiments, but none of them had conceived the power of a chain reaction in a test tube.

The polymerase chain reaction allows scientists to replicate (duplicate) specific DNA sequences in small test tubes. The reaction requires DNA polymerase, nucleotides (actually nucleoside triphosphates, the high-energy precursors to nucleotides), short (approximately 20 nucleotide) single-stranded DNA molecules called *primers,* and DNA from an organism to serve as a template. The reaction follows these steps:

1. The primer is made with the aid of a DNA (gene) synthesizer. The primer DNA sequence is critical because it determines what piece of an organism's DNA will be replicated during the PCR reaction.
2. DNA polymerase, nucleotide building blocks, primers, and DNA template are mixed together in a small test tube.
3. The test tube is placed in a *thermal cycler.* This machine is capable of rapidly heating and cooling samples.
4. The tube is heated to approximately 94ºC to separate the template DNA into single-stranded molecules. In a living cell, enzymes carry out this process. In an experimental setting, however, heat provides a simple way to make the DNA single-stranded.
5. The tube is then cooled to 35°C–60°C. This allows the primers to locate and bind to complementary base sequences on the template DNA. Remember that DNA molecules are typically double helices with adenine-thymine pairs and guanine-cytosine pairs.
6. The tube is heated to 72°C, which is the optimal temperature for the DNA polymerase used in PCR reactions. Because the enzyme for PCR is derived from thermophilic (heat-loving) bacteria, it is not destroyed by heating (see Chapter 17). DNA polymerase locates the short, double-stranded regions where primers have bound to template DNA; it then uses nucleotide building blocks to make double-stranded DNA near the primer.
7. Steps 4 through 6 are repeated 20 to 30 times.

After completion of the PCR reaction, millions of copies of a double-stranded DNA molecule are present in the test tube. The real power of the reaction, though, comes from its ability to amplify only the DNA sequence near the primer binding site. This allows a researcher to make an endless supply of a specific segment of DNA. The illustration (Box Figure 13.1) shows how a target DNA sequence is amplified through cycles of PCR.

The polymerase chain reaction has become a standard tool in many research laboratories. Countless applications for this technique have been developed and more are on the way. PCR is an important tool because (1) large quantities of DNA can be synthesized from very small samples and (2) only the target DNA sequence is amplified.

How is the PCR reaction used today? Here is a short list of examples:

1. **Forensics.** Samples of hair, blood, and skin contain enough DNA to be amplified by PCR. A comparison of amplified DNA samples from a crime scene with DNA from a suspect can provide compelling evidence.
2. **Disease detection.** Primers have been developed specifically to amplify DNA from pathogens. For example, a person's blood could provide template DNA for the PCR reaction with a primer that binds only to a sequence from the human immunodeficiency virus (HIV). Then, even low levels of DNA produced by the virus would be amplified by PCR. This makes early disease detection possible. PCR is also being used to identify pathogens in plants.
3. **Food safety.** Pathogens in food can be detected quickly and accurately using PCR. Primers specific for pathogenic strains of *Escherichia coli* can detect very low levels of the pathogen in hamburger. Similar tests are available for detection of the organisms that cause *Salmonella* poisoning and botulism, among others.
4. **Fetal testing.** A small sample of fetal tissue can be amplified using primers specific to DNA regions that, in mutant form, cause human diseases. Primers are available to detect sickle cell anemia, phenylketonuria,

(Continued)

Awareness Box Continued

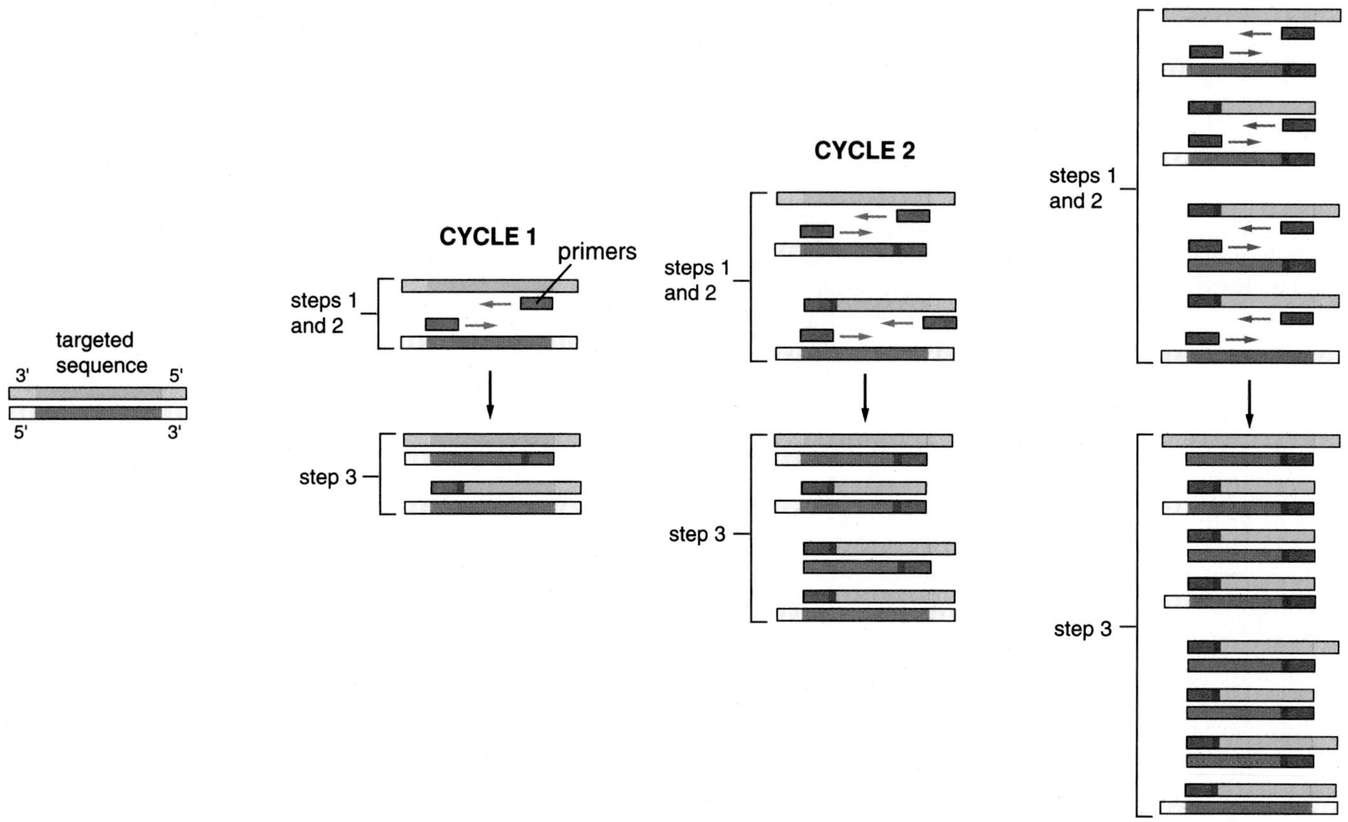

Box Figure 13.1 Polymerase chain reaction (PCR) technology.

cystic fibrosis, hemophilia, and Huntington's disease, to name a few. Of course, a moral dilemma is created if the fetus tests positive.

5. **Evolutionary relationships.** DNA from extinct animals can be amplified and compared with the DNA of modern relatives. With the use of PCR, DNA from a 120-million-year-old weevil and from a 250-million-year-old bacterium have already been extracted and amplified.
6. **Conservation efforts.** DNA from tissue samples collected at poaching sites can be amplified and compared to the DNA of pelts from suspected poachers. PCR also allows scientists to determine whether commercial products containing ivory, meat, or feathers were taken from endangered species.
7. **Genetics of dung.** Even DNA from dung can be amplified with PCR! Scientists trying to understand why the giant ground sloth became extinct have located 20,000-year-old dung in caves. PCR amplification of the dung is being used for comparison with modern relatives. In addition, careful choice of primers allows researchers to amplify DNA from the animal's food and its parasites.

Only a few examples have been presented here. Application of PCR seems to be limited only by our imagination. We are capable of performing this amazingly simple, yet powerful reaction because we understand the elements responsible for DNA replication. The next time, then, when you are asked to memorize the steps in a biological process, don't grumble about it. Instead, see if you can look at it in a unique way and make your friends wonder, "Why didn't I think of that?"

The expression of genetic information requires two processes. First, the information coded by DNA is *transcribed.* During **transcription,** a copy of the gene message is made using RNA building blocks. RNA is similar to DNA with a few important differences. First, RNA nucleotides contain **ribose** sugars instead of **deoxyribose** sugars. Ribose contains one more oxygen atom than deoxyribose. Second, thymine is replaced by **uracil,** another nitrogenous base. Finally, RNA molecules are typically single-stranded. After transcription has been completed in the nucleus, the RNA is modified and then travels to the cytoplasm, where *translation* occurs. During **translation,** the RNA message provides the information necessary to construct proteins. It might help to remember the following when you are trying to distinguish between transcription and translation. During transcription, a nucleic acid, RNA, is made using another nucleic acid, DNA, as a template. This is analogous to a stenographer transcribing shorthand to longhand. During translation, a protein is made using a nucleic acid (RNA) as a template. This is analogous to translating a document from one language to another, such as Spanish to English.

Transcription Three different types of RNA are made by transcription. Chromosomes contain genes for *messenger RNA (mRNA), transfer RNA (tRNA),* and *ribosomal RNA (rRNA).* Messenger RNA will be translated to produce proteins, while tRNA and rRNA will be used as part of the machinery for translation.

The mechanism of RNA synthesis is relatively simple. It involves the addition of RNA nucleotides to single-stranded DNA molecules, using complementary base pairing, similar to DNA replication. **RNA polymerase** is responsible for assembling the RNA strand. However, while all regions of the genome are replicated, only a very small portion of the genome is transcribed. In most eukaryotes, less than 10% of the DNA in the genome contains genes. The remainder is **noncoding DNA.** The transcription machinery must therefore search through millions of base pairs to identify small regions that contain genes. How does it do that? We are just beginning to unravel this mystery. At the beginning of every gene the transcription machinery scanning the DNA is signaled that a gene is ahead by a DNA sequence called a **promoter region**. This causes the transcription enzymes to attach to the DNA and begin to make a copy of the gene. A DNA sequence at the end of each gene tells the transcription enzymes to separate from the DNA molecule. At this point, a single-stranded RNA molecule, called a **transcript,** has been made, using one strand of a gene as a template. The transcript will undergo several modifications and move to the cytoplasm before translation begins.

As we mentioned before, not every gene is expressed in every cell. In addition to identifying genes, the transcription machinery must determine which genes to transcribe, or express. This is a very complex process and involves interactions among genes and signals from the environment.

Chromosomes contain genes for building tRNA. When transcribed, these genes produce molecules of tRNA approximately 80 nucleotides long. The single-stranded RNA produced by transcription spontaneously forms base pairs in some regions, creating a three-dimensional molecule of a characteristic shape. Transfer RNA will act as the "translator" during translation. One end of the tRNA molecule binds to the mRNA, while the other end carries a specific amino acid. There is at least one form of tRNA for each of the 20 amino acids. Each form of tRNA has a specific *anticodon* loop. The **anticodon** is a sequence of three nucleotides that recognizes a codon on mRNA and base pairs with it. An amino acid binds to the other end of the tRNA with the aid of an enzyme that specifically identifies both a single tRNA strand and its proper amino acid.

Genes for rRNA are also transcribed in the nucleus. This RNA is used to construct **ribosomes,** which act as workbenches during protein synthesis. A ribosome is composed of a large and a small subunit, each containing a specific combination of rRNA molecules and proteins. The rRNA molecules create the three-dimensional structure of a ribosome and help to orient the proteins, most of which appear to assist with the assembly of new proteins during translation. It is believed that mRNA slides between the large and small subunits during translation. All ribosomes in a cell are identical and capable of reading any mRNA.

Translation RNA transcripts that can be decoded into proteins are called *mRNAs.* It seems puzzling that a nucleic acid with only four possible nucleotides (adenine, uracil, cytosine, and guanine) can be used to make a protein polymer that may have 20 different amino acids. One nucleotide is incapable of encoding one amino acid. Even two nucleotides together could be combined in a maximum of only 16 different ways. It turns out that the genetic code is based on trios of nucleotides we earlier referred to as *codons.* Like words and sentences, codons and genes can be read from one direction only, thereby preventing the possibility of expressing the wrong amino acid. There are 64 possible combinations of all three nucleotides, which are more than enough to encode all 20 amino acids. The order of nucleotides on an mRNA molecule (determined by the sequence of nucleotides in a gene during transcription) determines the sequence of amino acids added during translation. For example, the codon AUG (adenine-uracil-guanine) on an mRNA encodes the amino acid methionine, whereas the codon AGU on an mRNA encodes the amino acid serine. The genetic code (Fig. 13.5) shows the matching of each three-nucleotide codon with a specific amino acid. With few exceptions, the genetic code is universal. In other words, the same code is used in bacteria, plants, and animals.

As mentioned before, tRNAs act as translators, or decoders, during translation. The anticodon part of the tRNA binds to the mRNA codon by base pairing with it, while the other end of the tRNA molecule has, according to the genetic code, the correct amino acid bound to it. In a sense, then, we can think of tRNAs as the dictionary used to decode an mRNA during translation.

The start of translation is usually signaled by a ribosome in the cytoplasm binding to the mRNA and scanning it for the first AUG codon. This sets the reading frame because each set of three nucleotides is translated in succession after the first AUG. A start tRNA with the amino acid methionine and an anticodon that base-pairs with the AUG on the mRNA joins the mRNA and the ribosome to form a three-part complex (Fig. 13.6). Another tRNA that base-pairs to the next codon and has its correct amino acid attached inserts in the complex next to the start tRNA. An

Codon	Amino Acid	Codon	Amino Acid	Codon	Amino Acid	Codon	Amino Acid
UUU	phe (phenylalanine)	UCU	ser (serine)	UAU	tyr (tyrosine)	UGU	cys (cysteine)
UUC	phe	UCC	ser	UAC	tyr	UGC	cys
UUA	leu (leucine)	UCA	ser	UAA	STOP	UGA	STOP
UUG	leu	UCG	ser	UAG	STOP	UGG	trp (tryptophan)
CUU	leu	CCU	pro (proline)	CAU	his (histidine)	CGU	arg (arginine)
CUC	leu	CCC	pro	CAC	his	CGC	arg
CUA	leu	CCA	pro	CAA	gln (glutamine)	CGA	arg
CUG	leu	CCG	pro	CAG	gln	CGG	arg
AUU	ile (isoleucine)	ACU	thr (threonine)	AAU	asn (asparagine)	AGU	ser
AUC	ile	ACC	thr	AAC	asn	AGC	ser
AUA	ile	ACA	thr	AAA	lys (lysine)	AGA	arg
AUG	met (methionine)	ACG	thr	AAG	lys	AGG	arg
GUU	val (valine)	GCU	ala (alanine)	GAU	asp (aspartic acid)	GGU	gly (glycine)
GUC	val	GCC	ala	GAC	asp	GGC	gly
GUA	val	GCA	ala	GAA	glu (glutamic acid)	GGA	gly
GUG	val	GCG	ala	GAG	glu	GGG	gly

Figure 13.5 The genetic code. There are 64 different codons, each of which consists of 3 bases and nucleotides, and specifies 1 of the 20 amino acids plus *stop* and *start* information.

enzyme in the ribosome then links the two adjacent amino acids together to begin to form a protein. As translation continues, new tRNAs base-pair to each mRNA codon in order, and new amino acid linkages are made. For example, an mRNA with the sequence AUG-AAG-UGU will be translated into a protein with the sequence methionine-lysine-cysteine until a stop codon (UGA, UAA, or UAG) is reached. At that point, a newly made protein is released from the complex.

To summarize, the flow of information in a cell is from the DNA genome in the nucleus, which is transcribed into mRNA, to the cytoplasm, where the mRNA is translated into protein. Because these processes occur in all living organisms, it has come to be known as the *Central Dogma of Molecular Genetics* (Fig. 13.7). Truly a cell's identity is determined by the genes it expresses, which, in turn, produce the myriad of proteins required for its survival and function.

Mutation

When my six-year-old son recently told a friend that his Mom was a scientist, the boy asked, "Can she make mutants?" While the term *mutation* may conjure up images of *Teenage Mutant Ninja Turtles,* most mutations are not so dramatic. In fact, as mentioned earlier, **mutations,** or changes in DNA sequence, happen every time DNA replicates. In addition to DNA replication errors and other spontaneous types of changes, agents called **mutagens** can alter DNA sequences. Ultraviolet light, ionizing radiation, and certain chemicals act as mutagens. DNA repair enzymes can often find and correct damaged DNA. If left uncorrected, though, a permanent change in DNA sequence, or mutation, occurs. Most mutations are *silent,* with no visible consequence of the DNA alteration. Remember that most DNA in a cell is noncoding. Even altered coding sequences may produce functional proteins.

Mutations are either *somatic* or *germ-line.* A **somatic mutation** occurs in a body cell and will exist in all cells produced by mitosis of the mutant cell. A somatic mutation may show up as a *sport,* or a branch that looks different from the others on a tree. These types of mutations are sometimes sources of new types of horticultural plants. Navel oranges and Red Delicious apples originated from sports and have been clonally propagated by grafting (see Chapter 14).

A **germ-line** mutation occurs in tissue that will produce gametes, or sex cells. Unlike somatic mutations, germ-line mutations will be passed on to future generations through seeds. A germ-line mutation is generally not apparent until it is passed on to offspring, which will carry the altered DNA in

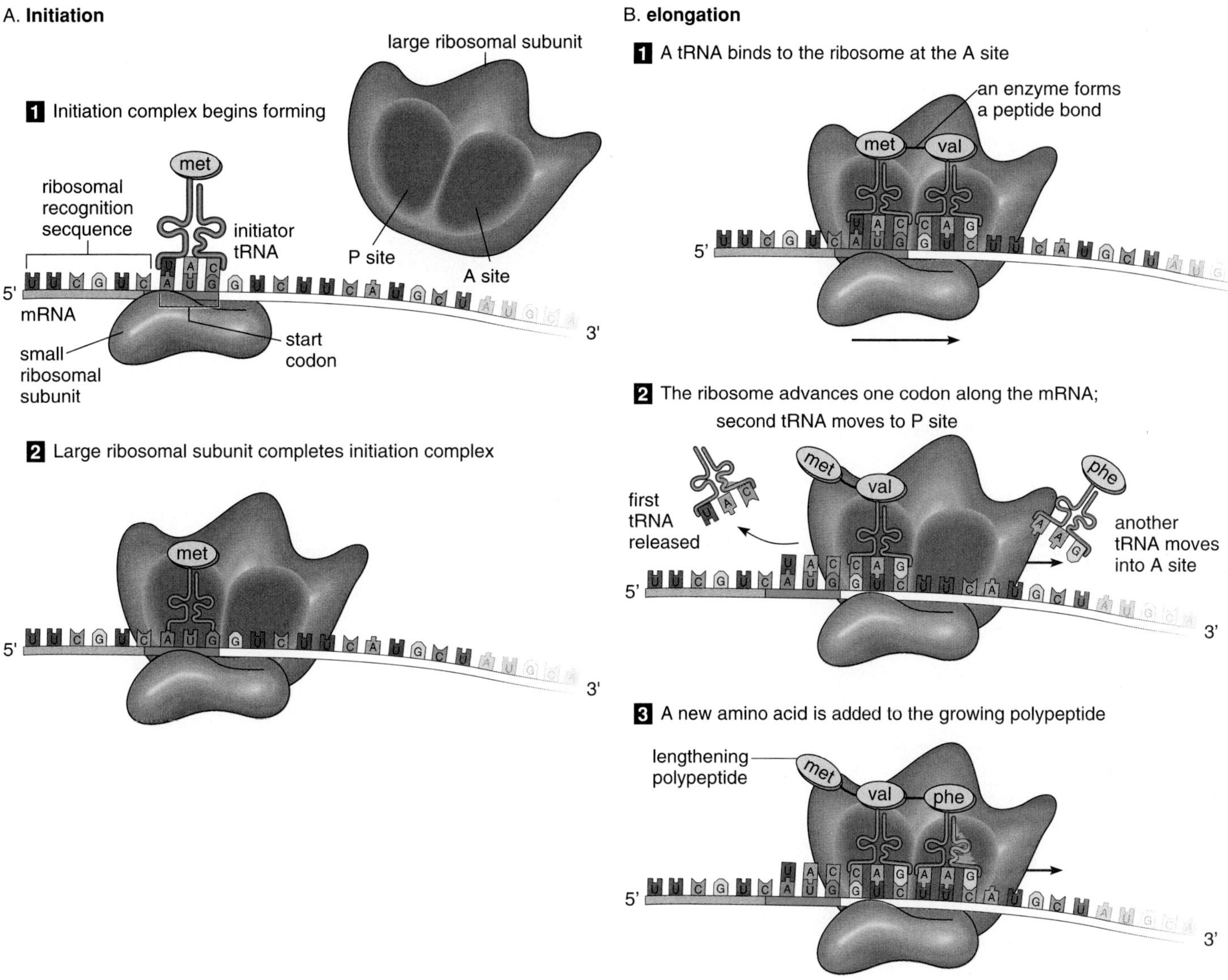
A. Initiation
1 Initiation complex begins forming
large ribosomal subunit
met
ribosomal recognition secquence
initiator tRNA
P site
A site
5'
mRNA
small ribosomal subunit
start codon
3'
2 Large ribosomal subunit completes initiation complex
met
5'
3'
B. elongation
1 A tRNA binds to the ribosome at the A site
an enzyme forms a peptide bond
met
val
5'
2 The ribosome advances one codon along the mRNA; second tRNA moves to P site
first tRNA released
met
val
phe
another tRNA moves into A site
5'
3'
3 A new amino acid is added to the growing polypeptide
lengthening polypeptide
met
val
phe
5'
3'

Figure 13.6 Translation. *A.* During initiation, the components necessary for translation bind together. These include the large and small ribosomal subunits, messenger RNA, and the initiator transfer RNA carrying the amino acid methionine. *B.* During elongation, the ribosome moves along the messenger RNA. Transfer RNA molecules bring amino acids to the ribosome, where a polypeptide chain is constructed. *C.* When the ribosome reaches a stop codon, translation terminates. The large and small ribosomal subunits, messenger RNA, transfer RNA, and polypeptide separate from each other.

all cells. This mutation has now become a permanent feature of that plant's lineage. Germ-line mutations are important for the genetic improvement of horticultural and agronomic plants. They are responsible for variability in all traits, including flower color and fragrance, fruit taste and texture, grain yield and quality, and disease and stress tolerance.

While we tend to think of mutations as "bad," they are an essential feature of DNA. All the genetic variability in nature has arisen as a result of mutation. The accumulation of mutations is the mechanism that drives evolution of species (see Chapters 15 and 16). That variability was essential for the first plants, which were in the ocean, to become established on land; it continues to be necessary for plants to survive and compete in an ever-changing environment.

CYTOGENETICS

During the middle of the 20th century, corn cytogenetics dominated the biological sciences, due in large part to Barbara McClintock's efforts. **Cytogenetics** is the study of chromosome behavior and structure from a genetic point of view. This includes chromosome movement during meiosis and mitosis, chromosome pairing during meiosis, and chromosomal variation due to changes in chromosome structure and number. Chapters 3 and 12 discuss chromosome behavior during mitosis and meiosis, respectively. Therefore, this section will emphasize changes in chromosome structure and number. These changes occur at low frequency, but since their effects are often large, they can have a dramatic influence on the evolutionary success of populations or species.

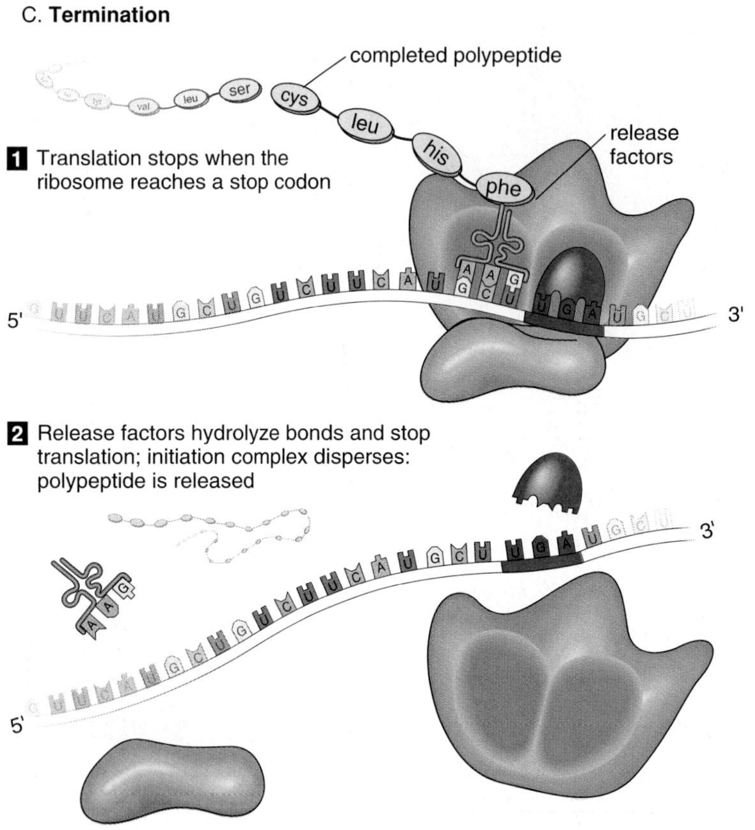
C. Termination
completed polypeptide
1 Translation stops when the ribosome reaches a stop codon
release factors
5'
3'
2 Release factors hydrolyze bonds and stop translation; initiation complex disperses: polypeptide is released
3'
5'

Figure 13.6 Continued

Changes in Chromosome Structure

Occasionally, a chromosome will break into two or more pieces. Enzymes may repair the damage, leaving the chromosome unaltered. However, sometimes a broken chromosome may become reorganized before the pieces are attached together again. The most significant events from an evolutionary perspective include *inversions* and *translocations.*

An **inversion** results when a piece of a chromosome is broken and reinserted in the opposite orientation. For example, if the chromosome contains genes ABCDE in that order, then an inversion might create a chromosome with the genes in this order: AB<u>DC</u>E. The inverted chromosome can pair up with the normal homologous one at meiosis, but crossing-over within the inverted region will create gametes that do not survive. Plants carrying an inverted chromosome therefore are less fertile than normal ones. The only gametes that are passed on from one generation to another are those in which crossovers occurred only outside the inverted region. Therefore, allele combinations within an inverted region are not rearranged by meiosis and are inherited together as blocks.

A **translocation** results when a piece of a chromosome breaks off and becomes attached to another one. A plant carrying a translocation, like one carrying an inversion, will not be as fertile as a normal one. This is because some gametes from translocation plants carry unusual combinations of chromosomes.

Translocations and inversions can be important factors in **speciation** (the formation of new species) during evolution. Plants carrying chromosomes from these events may become reproductively isolated from normal plants and will eventually develop into new species. Crosses between the new species and the old one will produce offspring with low fertility and, therefore, a poor reproductive capacity. These types of events have been important in the evolution of many plants. For example, cultivated rye (*Secale cereale*) and four wild species (*S. vavilovii, S. africanum, S. montanum, S. silvestre*) each contain 14 chromosomes. However, translocations have resulted in the wild species being reproductively isolated from the cultivated species. Genetic exchange among these species is limited by translocations.

Changes in Chromosome Number

Plants are remarkably tolerant of changes in chromosome number. Mistakes made during the pairing and separation of chromosomes during meiosis can result in gametes carrying extra or missing chromosomes. If one of these gametes is involved in fertilization, then the resulting zygote will be *aneuploid.* An **aneuploid** plant carries an extra one or more

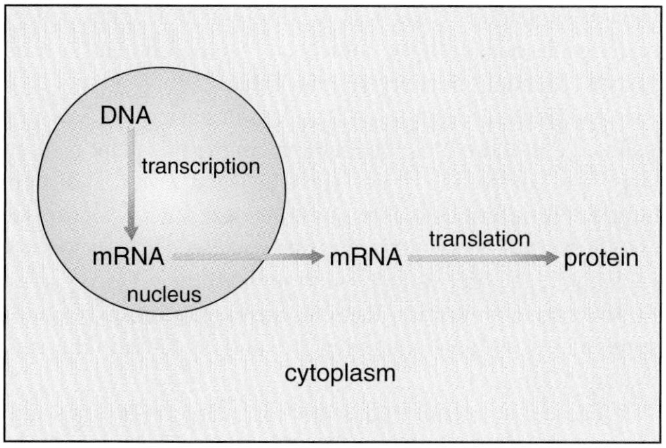

Figure 13.7 The Central Dogma of Molecular Genetics. The flow of information in a cell proceeds from the DNA in the nucleus via mRNA to the cytoplasm, where the mRNA is translated into proteins.

Figure 13.8 Leaves of potato plants. *Left:* A tetraploid. *Right:* A diploid.

chromosomes or, less commonly, is missing one or more chromosomes. For example, the cells in a carrot plant typically contain nine pairs of homologous chromosomes, for a total of 18 chromosomes. An aneuploid carrot plant would have 19 chromosomes if it had an extra copy of one chromosome or 17 chromosomes if it was missing one of a pair. As you would expect, aneuploid plants generally differ in appearance from their normal counterparts due to the effects of extra or missing doses of many genes. Geneticists use aneuploids to determine the chromosomal locations of genes.

Occasionally, meiosis may completely fail to halve the chromosome number, producing gametes with the somatic (body cell) chromosome number. These are called $2n$ gametes. If these gametes are involved in fertilization, then the resulting offspring are *polyploid.* A **polyploid** plant has at least one complete extra set of chromosomes. Using the carrot example, a polyploid carrot could have 27 chromosomes (three sets of nine chromosomes). Polyploid plants are often larger and higher-yielding than their diploid counterparts. For this reason, many of our cultivated plants are polyploid. These include potato, cotton, peanut, wheat, oats, strawberry, and sugar cane. (Fig. 13.8). The larger and longer-lasting flowers of polyploids make them attractive as ornamental plants; marigold, snapdragon, lily, and hyacinth are among the numerous examples of ornamental polyploids. In addition, triploid plants (polyploids containing three copies of each chromosome) are sterile and do not produce seeds. Seedless fruits, such as bananas and watermelons, are triploid.

Figure 13.9 Gregor Mendel. *Photo ca. 1870's, Brünn, Moravia. Courtesy Brooklyn Botanic Gardens, New York. Print at Hunt Institute, Pittsburgh, PA.*

MENDELIAN GENETICS

Genetics as a science originated about a generation before its significance became appreciated in the scientific community as a whole. An Austrian monk, Gregor Mendel (Fig. 13.9) taught between 1853 and 1868 in what became the Czechoslovakian city of Brno. In the monastery there, he carried out a wide range of studies in physics, mathematics, and natural history. He also became an authority on bees and meteorology and kept notes on experiments involving two dozen different kinds of plants. Today, he is best known for the studies he conducted with peas.

Mendel published the results of his studies on pea plants in a biological journal in 1866 and sent copies of his paper to leading European and American libraries. He also sent copies to at least two eminent botanists of the time. His work, nevertheless, was completely ignored or overlooked until 1900, when three botanists (Eric von Tschermak of Austria, Carl Correns of Germany, and Hugo de Vries of Holland), working independently in their own countries, reached the same conclusions as Mendel. Each, as a result of library research, came across Mendel's original paper.

Mendel's Studies

Peas, unlike most flowering plants, are *self-fertile.* In other words, a pollen grain of a pea flower can germinate on its own stigma, allowing a sperm cell to fertilize an egg and the ovule to develop into a viable pea seed. Although Mendel did not know about mitosis and meiosis, he had noticed that if a pea plant grew to a certain height, its offspring, if grown under the same conditions, would always reach approximately the same height, generation after generation. He wondered what would happen if he crossed a tall plant with a short one. Would the offspring be intermediate in height?

To make such a cross, Mendel needed to prevent self-pollination. He did so by reaching into a flower of one plant and removing the stamens before the pollen had matured. Then, he took pollen from another plant and applied it by hand to the stigma of the first flower. He also covered the experimental flowers with small bags to prevent insects from bringing pollen from other flowers after the cross had been made.

When Mendel made such crosses, the results were astonishing. All of the offspring were tall. There were no short or intermediate plants. He found, however, that if he allowed the offspring plants to pollinate themselves, they produced offspring in a ratio of approximately three tall plants to one short plant (Fig. 13.10).

Mendel then tried crosses between peas with smooth seeds and those with wrinkled seeds. He also crossed green-seeded plants with yellow-seeded ones and ultimately used seven different pairs of contrasting characteristics.

Mendel referred to the original plants involved in making the crosses as the **parental generation (P)** and their offspring were the **first filial (F_1)** generation. Filial means "of or relating to a son or daughter." The offspring of F_1 plants were called the **second filial (F_2)** generation.

There are several reasons why Mendel was successful in developing a model for inheritance when countless others before him failed. He performed carefully planned and executed crosses between pure-breeding parents. Each parent was genetically stable for the trait under study—a green-seeded parent produced only green-seeded offspring. Mendel also counted the number of offspring in each cross, something none of his predecessors had done. His use of his math background to apply statistical analyses to his data was unique, allowing him to develop and test inheritance models. Finally, he chose traits under simple genetic control. His predecessors usually studied traits such as weight, which is controlled by many genes and strongly influenced by environment. A summary of Mendel's data is shown in Table 13.1.

As Mendel began to collect data in his experiments, he realized that there must be elements in the pea plant responsible for controlling traits such as seed color and stem height. He referred to this unknown agent as a *factor.* He also deduced that each plant must have two such factors for each characteristic, since even though all the offspring of an F_1 generation appeared identical, the F_2 generation revealed some plants with one characteristic and others with the contrasting form.

These discoveries and deductions came to be known as the **law of unit characters.** Stated simply, this law says that *factors, which always occur in pairs, control the inheritance of various characteristics.* Today, we know that the paired factors are **alleles** of genes. Alleles are alternative forms of a gene. For example, the seed color gene has an allele for yellow color and an allele for green color. Genes are always at the same position (**locus**) on homologous chromosomes (see Chapter 12). We also now know that there are typically thousands of genes on every chromosome.

From his data and analyses, Mendel also deduced that one allele may conceal the expression of the other. For example, in the F_1 generation following a cross between green-podded and yellow-podded parents, all the plants had green pods. However, both parents obviously had to contribute something to the cross, since some of the F_2 plants had green pods while others had yellow pods. This deduction let to another principle, known as the **law of dominance.** This principle says that *for any given pair of alleles, one may mask the expression of the other.* The expressed allele is referred to as **dominant,** and the one that is not expressed (masked) is **recessive.** In the green-podded by yellow-podded cross, the green-pod allele is dominant, and the yellow-pod allele is recessive.

Mendel's crosses also made it clear that a plant's having green pods did not indicate whether or not both members of the pair of alleles controlling pod color were present, because the dominant one could mask the expression of the recessive one. To distinguish between the physical appearance of an organism and the genetic information it contains, we use the terms *phenotype* and *genotype.* **Phenotype** refers to the physical appearance of the organism, while **genotype** refers to the genetic information responsible for contributing to that phenotype. We customarily use words to describe phenotypes and letters to designate genotypes.

In Mendel's crosses, for example, seven pairs of phenotypes are shown as parents; these phenotypes include yellow seeds and green seeds, smooth seeds and wrinkled seeds, green pods and yellow pods, and so on. In designating the genotypes of each of these plants, a capital letter is used to indicate the dominant member of a pair of genes; *the same letter* in lowercase is used to indicate the corresponding recessive allele. In the green-podded × yellow-podded pea cross, for example, green is dominant over yellow. If a plant

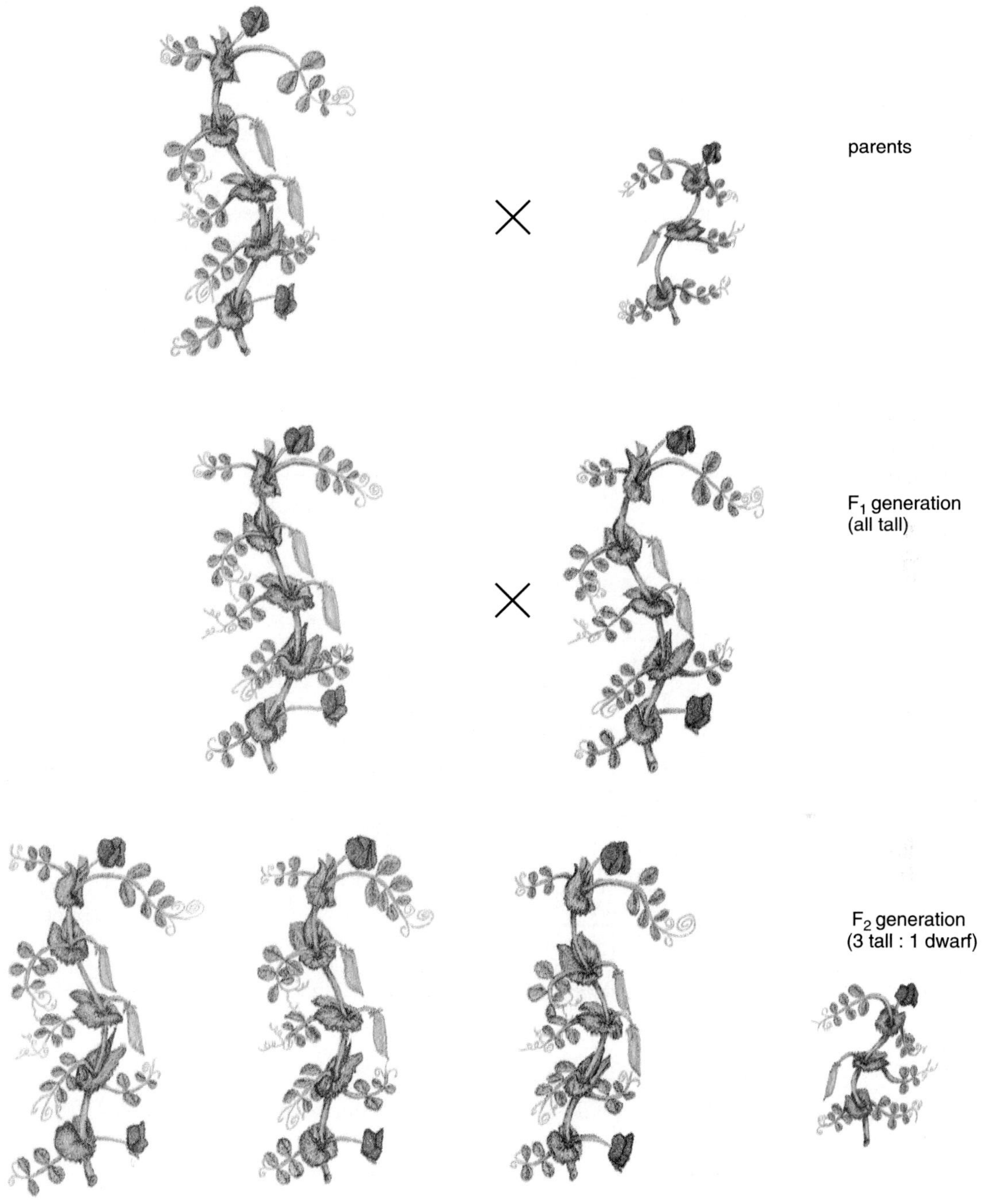

Figure 13.10 Two generations of offspring in peas. The F_1 generation is composed of tall plants and was created by crossing a true-breeding tall plant with a true-breeding short plant. The F_2 generation is composed of 75% tall plants and 25% short plants, and was created by intercrossing F_1 plants.

has a green phenotype, then its genotype could be either *GG* or *Gg*. The genotype of yellow-podded plants would always be *gg*. The plant is said to be **homozygous** if both alleles of a pair are identical (e.g., *GG* or *gg*) and **heterozygous** if the pair is composed of contrasting alleles (e.g., *Gg*). Mendel's true-breeding parents were homozygous for the traits he studied.

It is amazing that, nearly 150 years ago, Mendel was able to develop hereditary laws that are widely recognized today. Remember that he did his work before anyone knew about chromosomes, meiosis, or genes.

The Monohybrid Cross

Mendel was successful largely because he carried out well-planned crosses with true-breeding parents. Geneticists today perform similar experiments when trying to understand

TABLE 13.1
Summary of Mendel's Data

PARENTS	F_1	F_2
Yellow × green seeds	All yellow	6,022 yellow: 2,001 green
Smooth × wrinkled seeds	All smooth	5,474 smooth: 1,850 wrinkled
Green pod × yellow pod	All green	428 green: 152 yellow
Long stem × short stem	All long	787 long: 277 short
Axial flowers × terminal flowers	All axial	651 axial: 207 terminal
Inflated pods × constricted pods	All inflated	882 inflated: 299 constricted
Red flowers × white flowers	All red	705 red: 224 white

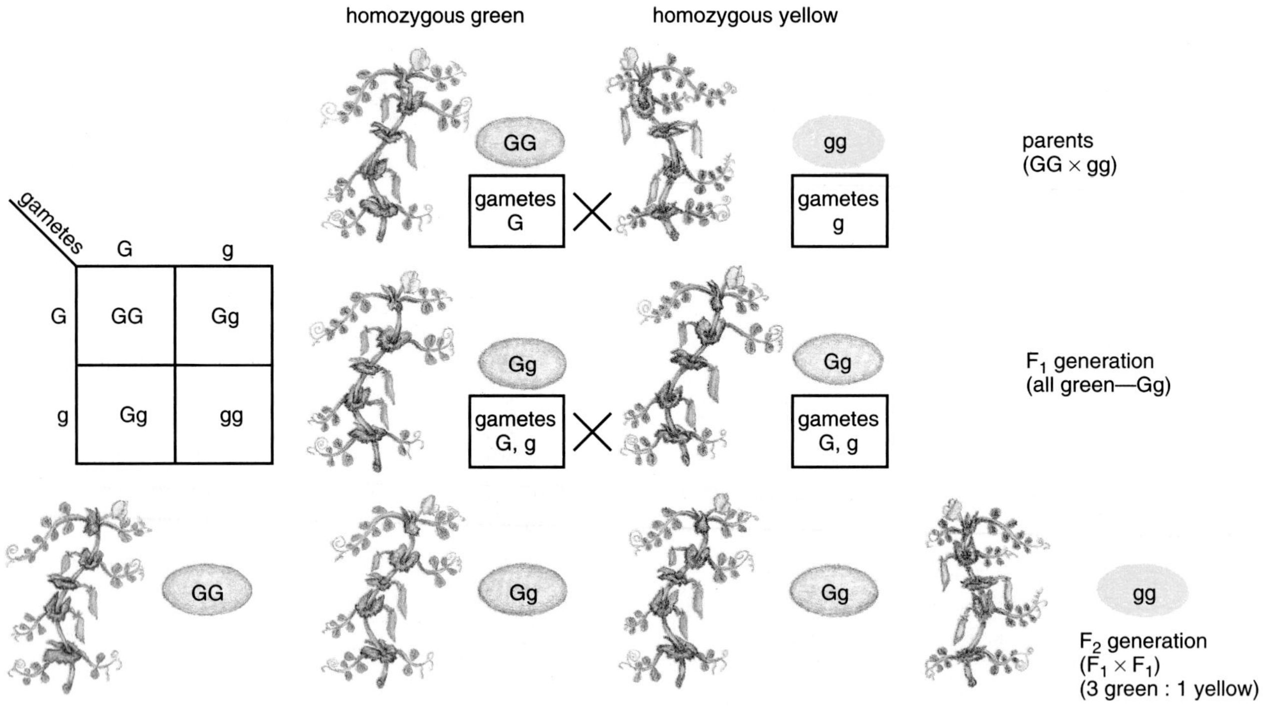

Figure 13.11 A monohybrid cross between a green-podded pea plant and a yellow-podded pea plant. Green (***G***) is dominant; yellow (***g***) is recessive.

the genetic basis of uncharacterized traits. A common strategy is to generate a **monohybrid cross.** In this scenario, a single trait is studied. A cross is made between two true-breeding parents differing for that trait, producing an F_1 generation. Then, these F_1 plants are intercrossed to produce an F_2 generation. As an example, we will examine a cross between a homozygous green-podded pea plant and a homozygous yellow-podded one (Fig. 13.11). The genotype of the homozygous dominant parent (green-podded) is *GG*; the genotype of the homozygous recessive parent (yellow-podded) is *gg*. You will recall from Chapter 12 that gametes have only one member of each pair of homologous chromosomes. The gametes of the green-podded parent (*GG*), therefore, will be *G*; the gametes of the yellow-podded parent (*gg*) will be *g*. No matter which egg unites with which sperm of the other parent, all the zygotes of this cross will be heterozygous, with the genotype *Gg*. This means that all individuals in the F_1 generation will have the green-podded phenotype, and all will be heterozygous.

The next step is the actual monohybrid cross. Members of the F_1 generation are crossed with each other (or self-pollinated) to produce an F_2 generation. All of the F_1 plants have the same phenotype (green-podded) and genotype (*Gg*). Each F_1 plant produces two kinds of gametes (*G* and *g*) in roughly equal numbers. When the gametes are produced, half will be *G*, and half will be *g*. A *G* egg may unite with a *G* sperm, producing a *GG* zygote. Alternatively, purely at random, the *G* egg may unite with a *g* sperm producing a *Gg* zygote. The same type of random combinations occur with *g* eggs so that either *Gg* or *gg* zygotes are produced. When all the offspring of a large number of such crosses are counted, a genotypic ratio of 1 *GG*: 2 *Gg*: 1 *gg* is produced. The phenotypic ratio will be approximately three green-podded plants (*GG* and *Gg*) to one yellow-podded plant (*gg*). These four equally possible F_2 offspring are shown at the bottom of Figure 13.11.

The Dihybrid Cross

So far, our examples have involved single genes. However, it is often desirable to study a pair of genes using a **dihybrid cross.** This is especially valuable in gene mapping studies. That topic is discussed later in this chapter. Mendel's **law of independent assortment** states that *the factors (genes) controlling two or more traits segregate independently of each other.* This is one law that, while true for the traits Mendel studied, often breaks down. That is because the two genes under study may be physically close to each other on a chromosome. If this is the case, then the genes are basically attached together and will follow each other through meiosis. They do not segregate independently. In this situation, the genes are said to be **linked.** In the dihybrid example that follows, however, we will assume that the pair of genes is **unlinked.** That is, they are on different chromosomes or are far apart on a single chromosome.

Let's consider the genes for plant height and pod color. Recall that tallness in peas is dominant and dwarfness is recessive, while green pod is dominant over yellow pod. The homozygous dominant (*GGTT*) parent will have a green-podded, tall phenotype, while the recessive parent (*ggtt*) will be yellow-podded and dwarf. All the gametes from the dominant parent will be *GT,* and those of the recessive parent will be *gt.* The F_1 generation will be composed of dihybrid (*GgTt*) plants that are green-podded and tall (Fig. 13.12).

When intercrossed to produce the F_2 generation, these dihybrid members of the F_1 generation will produce four types of gametes in equal proportions. The gamete genotypes are *GT, Gt, gT,* and *gt.* Remember that each gamete will carry only one allele of each gene. Since any one of the four kinds of gametes can unite randomly with any of the four kinds of gametes of the other parent, 16 combinations are possible.

In order to avoid confusion when trying to make all possible combinations of gametes, a diagram called a **Punnett square** is used to determine the genotypes of the zygotes. The Punnett square diagram looks somewhat like a checkerboard, with one set of gametes across the top and the other set down the left side. To fill in each square on the checkerboard, look at the gamete designation above the square and to the side of it. Multiply the gamete proportions (e.g. $1/4 \times 1/4$) and add together the gamete letters (e.g., *GT* + *GT*). In Figure 13.11, the first square would therefore be 1/16 *GGTT.*

Nine different zygote genotypes are possible in the 16-box Punnett square. They are produced in a ratio of 1 *GGTT*: 2 *GGTt*: 1 *GGtt*: 2 *GgTT*: 4 *GgTt*: 2 *Ggtt*: 1 *ggTT*: 2 *ggTt*: 1 *ggtt.* Four phenotypes are possible in a ratio of 9 green-podded, tall: 3 green-podded, dwarf: 3 yellow-podded, tall: 1 yellow-podded, dwarf. A geneticist observing a 9:3:3:1 ratio from a dihybrid cross can deduce that the genes controlling the two traits are unlinked and are exhibiting dominance. Genetic ratios such as the ones previously outlined are expected when large numbers of individuals are studied. Note in Mendel's data (Table 13.1) that he looked at over 8,000 F_2 offspring following a monohybrid cross involving seed color. With small numbers, chance alone may not produce the expected ratios. Human populations, for example, are divided relatively equally into males and females (a genetically controlled trait), but in smaller groups, such as families, the general population ratio of one male to one female may not be evident.

The Backcross

If a scientific theory is valid, then one should be able to test it experimentally. Mendel himself tested his predictions by means of backcrosses. A **backcross** is a cross between a hybrid and one of its parents. Mendel had found in pea flowers that red color is dominant and that white is recessive. Therefore, all F_1 offspring of a cross between a homozygous red-flowered parent (*RR*) and a homozygous white-flowered parent (*rr*) would be red (*Rr*). The gametes of these F_1 plants would be either *R* or *r,* with each type being produced in equal numbers. If the F_1 hybrid (*Rr*) is crossed to a homozygous recessive plant (*rr*), then half of the offspring would be *Rr.* These plants received the dominant allele from the heterozygous parent and a recessive allele from the homozygous parent. The remaining offspring from the backcross received a recessive allele from the heterozygous parent, so they would be *rr.* Following backcrosses of F_1 hybrids to recessive parents, Mendel observed the expected ratio of 1 red-flowered plant (*Rr*) to 1 white-flowered plant (*rr*). This provided further confirmation for his inheritance theory.

Incomplete Dominance

Although the principle of dominance was amply demonstrated by Mendel's crosses, we know now that in many other instances, neither member of a pair of genes completely dominates the other. In other words, some genes exhibit **incomplete dominance** (or absence of dominance), in

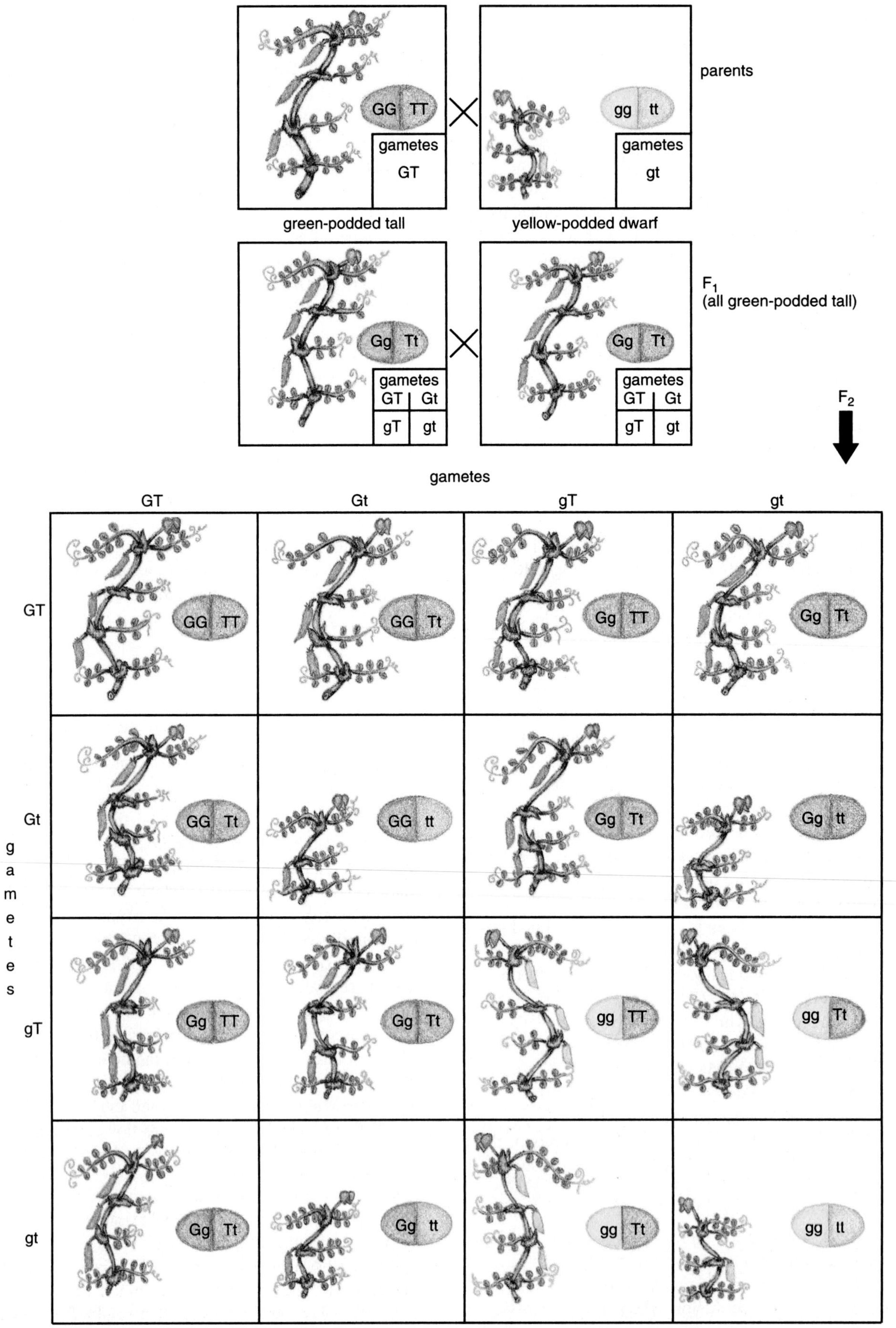

Figure 13.12 A dihybrid cross between a green-podded, tall pea plant and a yellow-podded, dwarf pea plant. Green (***G***) and tall (***T***) are dominant; yellow (***g***) and dwarf (***t***) are recessive.

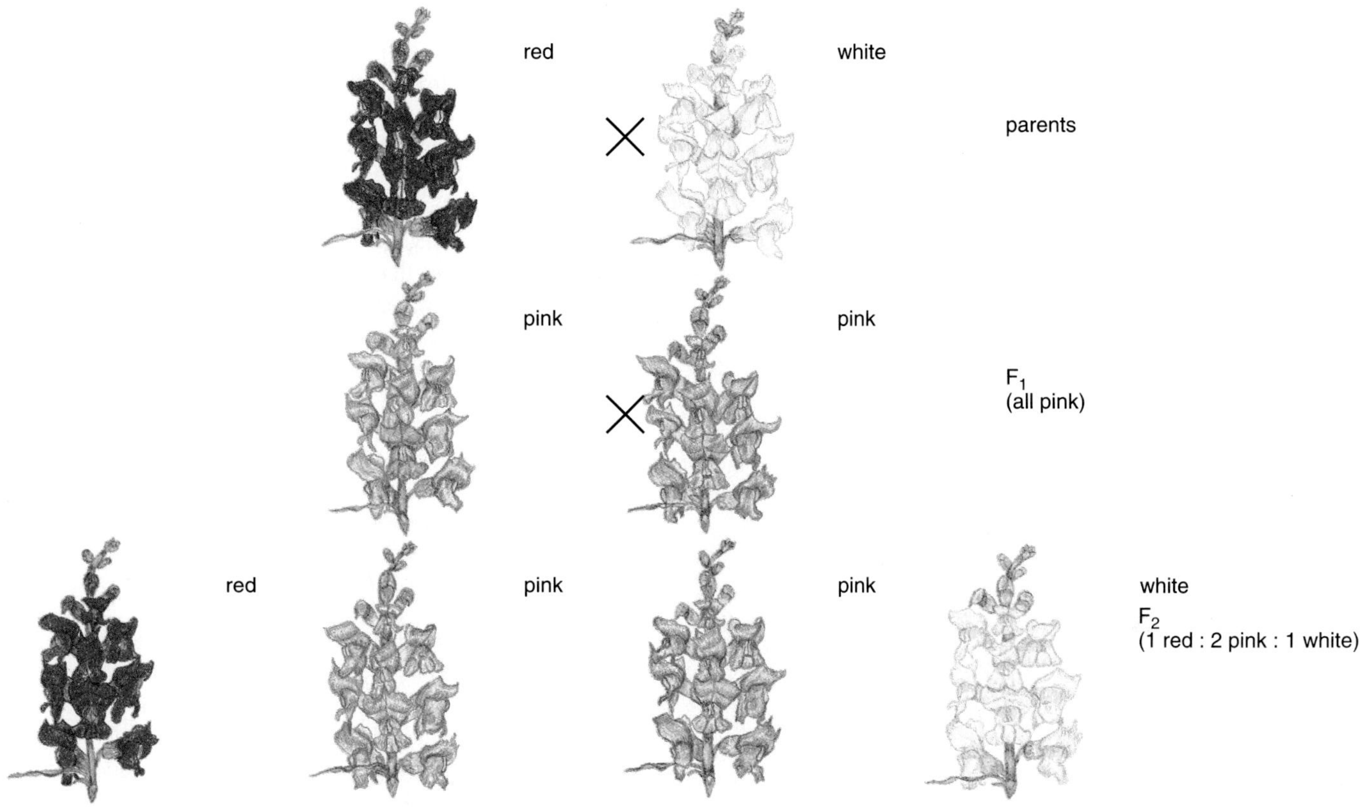

Figure 13.13 Incomplete dominance. A cross between a red-flowered and a white-flowered variety of snapdragons.

which a heterozygote is intermediate in phenotype to the two homozygotes. In snapdragons, for example, the F_1 plants of a cross between a red-flowered parent and a white-flowered parent are all pink-flowered. When an F_2 generation is produced from such a cross, the flowers of the offspring appear in a ratio of 1 red: 2 pink: 1 white (Fig. 13.13).

Interactions Among Genes

So far, our discussion of inheritance has focused on single genes or pairs of genes acting independently of each other. This type of autonomous behavior, however, is the exception rather than the rule. Genes are generally responsible for the production of proteins that are components of complex biochemical pathways. It is often necessary to consider the genotype of an individual at more than one locus before a prediction of phenotype can be made. For example, in a plant called blue-eyed Mary (*Collinsia parviflora*), two genes control flower color in a biochemical pathway as follows:

$$\textbf{colorless} \xrightarrow{\textbf{gene } W} \textbf{magenta} \xrightarrow{\textbf{gene } M} \textbf{blue}$$

If a plant contains at least one dominant *W* allele, then it can proceed through the first step of the pathway and produce either magenta or blue flowers. The flowers will be magenta if the plant is homozygous recessive for the *M* gene (*mm*) or blue if it has at least one copy of the dominant *M* allele (*MM* or *Mm*), allowing it to complete the second step in the pathway. The enzyme produced by the dominant *W* allele catalyzes a pathway that synthesizes a compound used by the gene *M* enzyme. Therefore, a homozygous recessive (*ww*) plant cannot produce colored flowers even if it contains a dominant *M* allele. Of course, a plant that is homozygous recessive for both genes (*wwmm*) will also produce colorless (white) flowers.

How Genotype Controls Phenotype

Mendel had no way of knowing how his factors (genes) could produce plants that varied in phenotype. However, as our understanding of the molecular basis of genetics increases, so does our ability to explain Mendelian inheritance patterns.

Mendel discovered that the allele for smooth seeds is dominant over the wrinkled seed allele. How does one allele mask the expression of another? Often, the dominant allele codes for a protein that can effectively catalyze a reaction in a cell and produce a phenotype we recognize. The recessive allele, on the other hand, represents a mutant form that has an altered DNA sequence. The protein product that results from transcription and translation of that recessive (mutant) allele is defective. It cannot catalyze the reaction, and, therefore, does not produce a functional product. One copy of the normal allele is usually enough to allow the cell to produce ample quantities of normal protein. Therefore, a heterozygous plant looks normal. A homozygous recessive plant, on the other hand, cannot make any normal protein from its two mutant alleles.

What is the difference between the smooth seed allele and the wrinkled seed one? The smooth pea phenotype (*RR* or *Rr*) produces an enzyme allowing the seeds to accumulate high levels of starch. Seeds with the *rr* genotype do not produce a functional form of that enzyme and so have high levels of sucrose rather than starch. The sucrose in the *rr* seeds causes them to absorb water during development. When they dry, they lose that water and shrivel. The *RR* and *Rr* seeds have lower levels of sucrose and do not absorb as much water during development; consequently, they do not shrivel when they dry. Recent molecular genetic studies have revealed that the *r* allele produces a defective enzyme because it has had an 800-nucleotide piece of DNA inserted into the gene. As you might expect, compared to the normal allele, transcription and translation of the allele carrying this insert produces a much different (and, in this case, nonfunctional) protein. How did the insert get there? The best guess is that a transposable element inserted into the pea chromosome at this location some time during the pea plant's evolutionary history.

Molecular genetics can also explain how other Mendelian traits are expressed. For example, as you recall, flower color in snapdragons is inherited as an incompletely dominant trait. In this case, one copy of the functional allele produces some red pigment, but two copies produce noticeably more pigment. Red-flowered plants therefore have two copies of an allele that is critical in the red pigment synthesis pathway and are *RR*. Pink-flowered plants have one copy of that allele, producing light red (pink) pigmentation, and are *RR′*. (We use superscripts instead of upper and lowercase letters here because we are not looking at complete dominance). Finally, white plants have two copies of the defective allele (*R′R′*) and cannot make the enzyme necessary for red pigment production.

QUANTITATIVE TRAITS

While simply inherited traits such as snapdragon flower color effectively demonstrate Mendel's principles, many traits are under much more complex genetic control. **Quantitative traits,** such as yield and days to flowering, exhibit a range of phenotypes, rather than the discrete phenotypes studied by Mendel. Typically, they are dramatically influenced by environment. Consider tomato fruit yield as an example of a quantitative trait. Under identical environments, some tomato plants produce more fruit than others due to genetic differences. However, genetically identical plants will also produce different amounts of fruit when grown under different environments. They will produce large yields under optimal growing conditions and lower yields under stressed conditions. Therefore, phenotype (yield in this example) is determined by both genotype and environment. This is the "nature versus nurture" concept. It has been estimated that the total proportion of wheat yield controlled by genotype is 30% to 60%, that controlled by environment is 10% to 30%, and that controlled by the interaction between genotype and environment is 25% to 45%. Because quantitative traits are strongly influenced by environment and are generally controlled by many genes, they do not produce typical Mendelian ratios. Instead, statistical tools such as distributions, means, and variances are used to study quantitative traits. Molecular geneticists are able to identify chromosomal fragments, called **quantitative trait loci,** or QTLs, associated with quantitative traits. Presumably, these fragments contain genes that influence the trait and behave like Mendelian genes. Using this approach, quantitative geneticists are beginning to understand the nature of complex traits such as growth rate, fruit quality, and flowering time.

EXTRANUCLEAR DNA

You may recall a statement earlier in this chapter indicating that *most* DNA in plants resides in the nucleus. Where is the remaining DNA found? In plants, **extranuclear DNA** is found in both mitochondria and chloroplasts. According to the **endosymbiont hypothesis,** mitochondria and chloroplasts were free-living bacteria at some time in their evolutionary history. They became incorporated into cells of organisms that evolved into plants and established a symbiotic association. As you might expect, then, the DNA in mitochondria and chloroplasts is similar to bacterial DNA.

Some genes involved in photosynthesis and respiration are located in the chloroplast and mitochondrial genomes, respectively. Because many copies of each organelle exist in a plant cell and multiple copies of each gene may be found in each organelle, extranuclear genes do not exhibit Mendelian segregation ratios. In addition, sperm cells rarely carry mitochondria and chloroplasts, so that extranuclear genes are typically passed on to the next generation by only the female parent. This is called **maternal inheritance.**

Leaf variegation in plants such as four-o'clocks exhibits maternal inheritance because it results from mutations in chloroplast synthesis genes. Some herbicides are effective because they inhibit enzymes encoded by chloroplast genes. Genetic engineers are currently trying to manipulate chloroplast genomes to create herbicide-resistant crop plants. See Chapter 14 for a more detailed discussion of genetically engineered plants.

LINKAGE AND MAPPING

There are typically thousands of genes on each chromosome, and the closer the genes are to one another on a chromosome, the more likely they are to be inherited together. Genes that are together on a chromosome are said to be **linked.**

In 1906, just a few years after the basic details of meiosis became known, W. Bateson and R. C. Punnett (after whom the Punnett square is named) of Cambridge University in England became the first to report on linkage in sweet peas. They knew from earlier work with sweet peas that purple flower color was dominant and red was recessive. They also knew that pollen grains with an oblong shape ("long pollen") were dominant and that spherical pollen grains ("round pollen") were recessive.

When Bateson and Punnett crossed a homozygous purple long plant (*PPLL*) with a homozygous red round plant (*ppll*), all the F_1 offspring were purple and long as expected. When they crossed the F_1 plants with one another to obtain an F_2 generation, however, the F_2 offspring phenotypes were produced in numbers that differed markedly from the expected 9:3:3:1 ratio. Puzzled by these results, they tried a backcross, crossing F_1 plants (*PpLl*) with the recessive parent (*ppll*). Again, the expected 1:1:1:1 ratio was not produced. Instead, they obtained a ratio of 7 purple long: 1 purple round: 1 red long: 7 red round. They could not adequately explain what they had observed. In 1910, however, T. H. Morgan, who had observed similarly puzzling ratios in his experiments with fruit flies, correctly theorized that linkage and crossing-over were responsible.

If the genes for purple and red and those for long and round were on separate pairs of chromosomes, then the genotypes of the F_1 would be all *PpLl,* and the homozygous recessive would be *ppll.* A backcross to the F_1 (*PpLl* × *ppll*) should have produced 1 *PpLl*: 1 *Ppll*: 1 *ppLl*: 1 *ppll* instead of the 7:1:1:7 ratio actually obtained. Apparently, the F_1 parent produced many more *PL* and *pl* gametes than *Pl* and *pL* gametes. If we assume the genes are linked, then we can use a slash (/) to separate the alleles on one chromosome from those on the homologous chromosome. The parents were *PL/PL* and *pl/pl.* The first parent would produce gametes containing the *PL* chromosome and the other would produce *pl* gametes. The F_1 plant, then, would be *PL/pl* because it received a *PL* chromosome from one parent and a homologous *pl* chromosome from the other. The F_1 plant would produce just two kinds of gametes, namely *PL* and *pl,* and the backcross (*PL/pl* × *pl/pl*) would have produced only two kinds of phenotypes, namely purple long (*PL/pl*) and red round (*pl/pl*) in equal numbers. But some purple round and red long phenotypes were also produced. The only plausible explanation for this seems to be that some crossing-over (see prophase I of meiosis in Chapter 12) occurs between genes *P* and *L.*

Crossing-over is a regular, seemingly random event during meiosis. Therefore, it may occur by chance between any pair of linked genes. Each gene has a specific location (**locus**) on a chromosome and, presumably, crossing-over is more likely to happen between two genes that are far apart on a chromosome than between two genes that are close together. We can use the frequency of crossing-over between genes to create a genetic map of each chromosome. A genetic map is similar to a map of a straight road, with genes corresponding to cities and map distances corresponding to the distances between the cities. A **map unit** equals 1% crossing-over between a pair of genes. Suppose we make a backcross (*PL/pl* × *pl/pl*), and 90% of the offspring are either *PL/pl* or *pl/pl.* These are **parental type** offspring because the gametes from the heterozygous parent (*PL* and *pl*) do not carry chromosomes for which a crossover between *P* and *L* occurred. The remaining 10% of the offspring are **recombinant types** (*Pl/pl* and *pL/pl*) because a crossover occurred between genes *P* and *L* to produce *Pl* and *pL* gametes. Remember that the heterozygous parent contains *P* and *L* on one chromosome and *p* and *l* on the homologous chromosome. So, the only way it can produce *Pl* and *pL* gametes is through crossing-over. Because 10% of the offspring (and, therefore, gametes) were recombinant, the map distance between genes *P* and *L* is 10 map units.

In Bateson and Punnett's backcross, two of every sixteen (12.5%) of the offspring were produced through recombination due to crossing-over. We can deduce from this that the genes controlling flower color and pollen shape in sweet peas are 12.5 map units apart on a chromosome. It is interesting that Mendel did not observe linkage in his studies. The seven traits he analyzed happened to be on the seven different chromosomes in peas. Imagine the difficulty he would have encountered if he had chosen to work on genes that were linked.

Linkage maps are essential tools for geneticists because they give us a physical framework in which to organize genes. Massive mapping efforts are currently underway in all major crop plants, with new genes and DNA sequences being added to maps every day. A simplified example of a genetic map of the pea is shown in Figure 13.14.

THE HARDY-WEINBERG LAW

Before Mendel's work became known, most biologists studied quantitative traits and believed that inherited characteristics resulted from a blending of those furnished by the parents. It was difficult to understand why unusual characteristics did not eventually become so diluted that they essentially disappeared. After Mendel, R.C. Punnett and other biologists asked why dominant genes did not eventually completely eliminate recessive ones in breeding populations. After all, a cross between two heterozygotes produces only 25% homozygous recessive individuals. G.H. Hardy, a mathematician, and W. Weinberg, a physician, pointed out the reason in 1908, and their observation became known as the Hardy-Weinberg law. The **Hardy-Weinberg law,** which essentially specifies the criteria for genetic equilibrium in large populations, states that the proportions of dominant alleles to recessive ones in a normally interbreeding population will remain the same from generation to generation unless there are forces that change those proportions. In small populations, for example, random losses of alleles can occur if, by chance, the individuals carrying those alleles do not mate as often as other individuals. In populations of any size, selection is the most significant cause of exceptions to the Hardy-Weinberg law and can cause dramatic changes in the proportions of dominant to recessive alleles. Discussions of artificial selection and natural selection are found in Chapters 14 and 15, respectively.

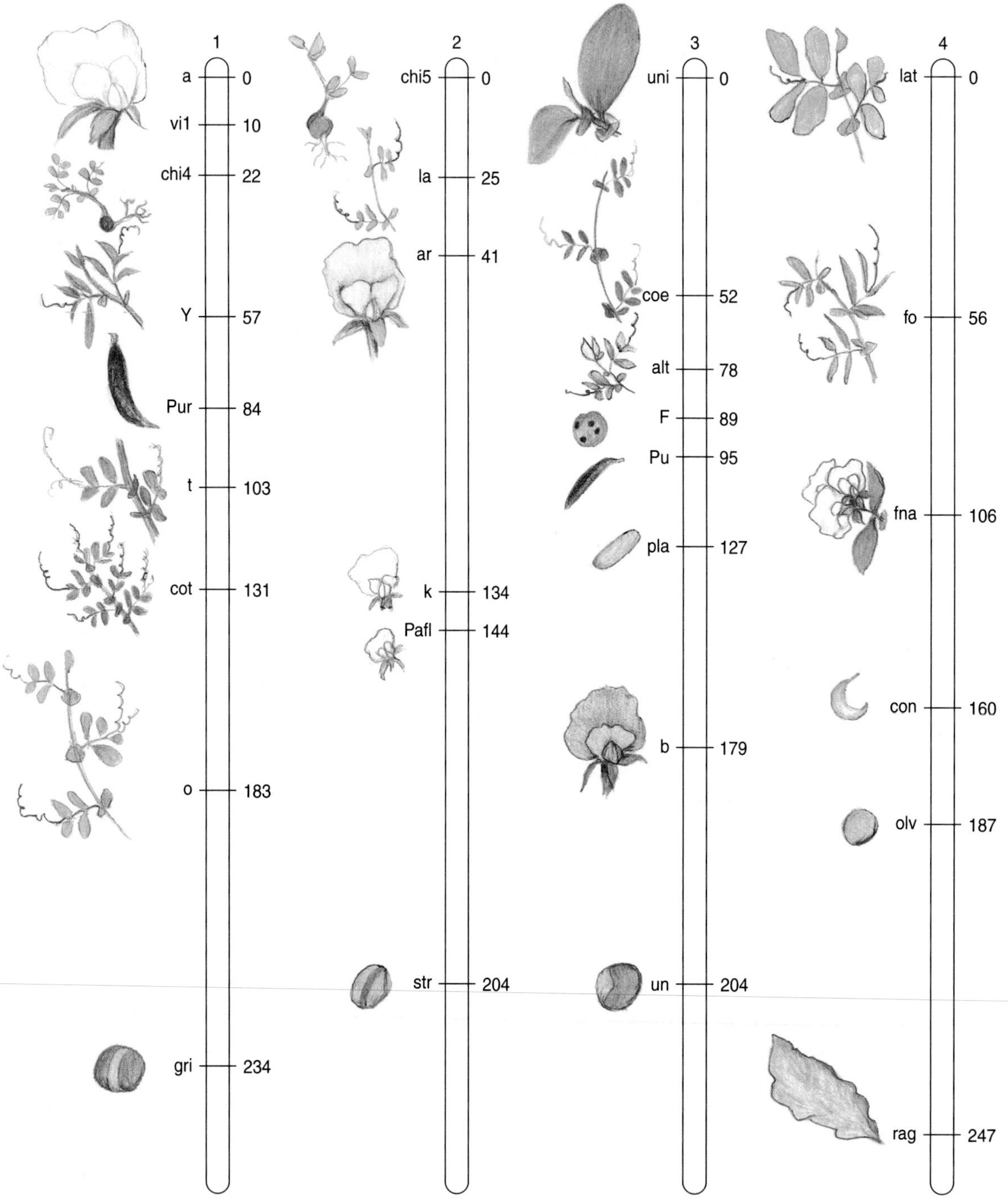

Figure 13.14 A partial genetic map of the pea. Chromosomes 1 through 7 are listed from left to right. Numbers along the right side of each chromosome are map distances. Letters along the left side are gene abbreviations. Drawings illustrate mutant phenotypes.

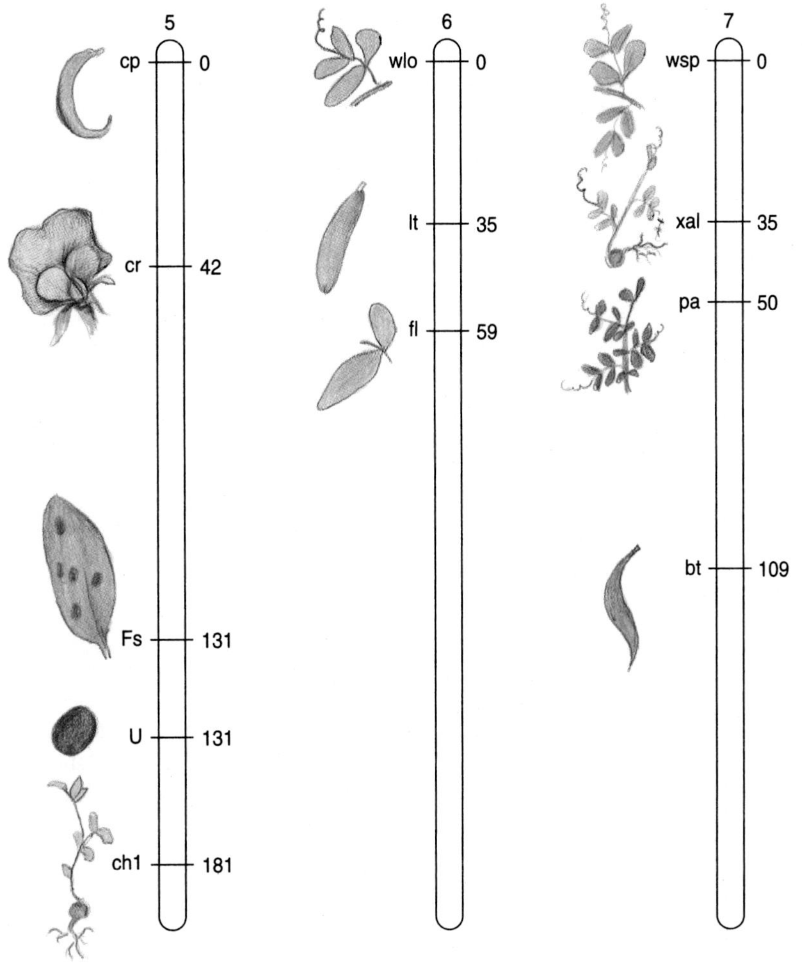

Chromosome number

1. **a** = white flowers; **vi1** and **chi4** = greenish-yellow seedlings; **Y** = narrow leaflets; **Pur** = purple pod; **t** = thick stem; **cot** = short internodes; **o** = yellowish-green foliage; **gri** = gray stripe on seed.
2. **chi5** = greenish-yellow seedlings; **la** = long internodes; **ar** = blue flowers; **k** = small flower wing petals; **Pafl** = small flowers; **str** = brown stripe on seed.
3. **uni** = leaf not divided into leaflets; **coe** = slightly longer internodes; **alt** = top of plant is white; **F** = violet spots on seeds; **Pu** = purple pods; **pla** = flattened seeds; **b** = rose-colored flowers; **un** = gray region on seed.
4. **lat** = large leaves; **fo** = narrow leaves; **fna** = large number of flowers; **con** = curved pods; **olv** = seeds olive-gray color.
5. **cp** = curved pods; **cr** = reddish-purple flowers; **Fs** = violet spots on leaves; **U** = violet-colored seeds; **ch1** = greenish-yellow seedlings.
6. **wlo** = no wax on leaf surface; **lt** = broad pods; **fl** = no air pockets under epidermis of leaves.
7. **wsp** = no wax on stems and lower surface of leaves; **xal** = reddish-yellow seedlings; **pa** = dark green foliage; **bt** = pod pointed at a pical end.

Figure 13.14 Continued

Summary

1. McClintock's work with corn changed beliefs about genes and chromosomes.
2. A DNA molecule consists of nucleotides. A typically single-stranded RNA molecule has a different sugar, and uracil instead of thymine.
3. Watson, Crick and Wilkins developed a model of a DNA molecule, now considered authentic. It is a double-stranded helix resembling a spirally twisted ladder.
4. DNA stores genetic information, replicates itself, expresses its information, and can mutate.
5. A genome is the sum total of the information in an organism's DNA.
6. A DNA molecule replicates by "unzipping;" the resulting single strands serve as templates for two double-stranded molecules.
7. With DNA as a template, mRNA, tRNA, and rRNA are made during transcription.
8. During translation mRNA from transcription is used to make proteins; tRNA decodes mRNA information. Ribosomes are made of rRNA and proteins.
9. Spontaneous or induced mutations are changes in DNA sequence.
10. Inversions and translocations may be important speciation mechanisms.
11. An aneuploid has a normal chromosome number plus or minus one to several chromosomes. A polyploid has extra complete sets of chromosomes.
12. Gregor Mendel originated the science of genetics while experimenting with peas. The pea varieties he crossed had pairs of contrasting characteristics on homologous chromosomes. Parent plant offspring were called the F_1 generation; the F_2 generation resulted from F_1 plant crosses. F_1 plants, called hybrids, all resembled their parents. F_2 offspring were produced in a 3:1 ratio of plants resembling one or the other parent.
13. The agents controlling the plants' characteristics were called "factors;" Mendel deduced that each plant had two

factors (later known as alleles) for each characteristic. His law of unit characters states that "factors, which always occur in pairs, control the inheritance of various characteristics." The suppressing factor was called dominant and its counterpart recessive.

14. Phenotypes (described with words) denote appearance; genotypes (shown with letters) designate genetic makeup. Capital letters designate dominant alleles; lower case letters designate recessive alleles.
15. Paired homozygous plant alleles are identical. Paired heterozygous plant alleles are contrasting.
16. Monohybrid cross offspring are produced in a ratio of 3 dominant phenotypes to 1 recessive phenotype. Dihybrid crosses produce a 9:3:3:1 phenotypic ratio. A backcross between an F_1 hybrid and its recessive parent produces a phenotypic ratio of 1:1 (monohybrid) or 1:1:1:1 (dihybrid).
17. Compared to homozygotes the heterozygotes are intermediate when dominance is incomplete.
18. Enzymes control biochemical reactions. Genotypes control phenotypes.
19. Quantitative trait genotypes are strongly influenced by environment.
20. Linked genes are inherited together. Bateson and Punnett observed a 7:1:1:7 ratio in a dihybrid cross, and postulated it resulted from linkage and crossing-over.
21. Chromosomal mapping requires calculating crossover percentages to determine the positions of genes on chromosomes.
22. The Hardy-Weinberg law explains why recessives in a population do not disappear. Selection is the most significant cause for 7 exceptions to the Hardy-Weinberg law.

Review Questions

1. Using S, P, and B as abbreviations for sugar, phosphate, and base, respectively, draw a DNA double helix in the arrangement you would see if you untwisted the helix so you were looking at a ladder structure.
2. If a DNA molecule had a string of cytosine-thymine base pairs, how would it differ in width from a normal molecule?
3. Explain how DNA replicates.
4. Using the genetic code, determine the amino acid sequence produced by the following gene (DNA) sequence. (Remember that during transcription, RNA forms complementary base pairs with DNA. For example, if the DNA sequence is TAG, then transcription would produce an RNA fragment with the sequence AUC.) TACACAGCAACT

Discussion Questions

1. Why do you suppose gene expression requires an intermediate RNA molecule? That is, why does the translation machinery read mRNA instead of the gene itself?
2. If proofreading enzymes becomes more accurate and efficient, then what do you suppose would happen to the speed at which organisms evolve?
3. Mendel's peas were self-pollinating. Describe advantages and disadvantages of self-pollination.

Learning Online

Visit our web page at www.mhhe.com/botany for interesting case studies, practice quizzes, current articles, and animations within the Online Learning Center to help you understand the material in this chapter. You'll also find active links to these topics.

Genetics
Mendelian Genetics
Non-Mendelian Inheritance
Molecular Basis of Genetics: DNA and RNA

Additional Reading

Frank-Kamenetskii, M.D. 1997. *Unraveling DNA: The most important molecule of life.* Reading, MA: Addison-Wesley.

Hartwell, L.H., et al. 2000. *Genetics: From genes to genomes.* Boston: McGraw-Hill.

Henig, R.M. 2000. *The monk in the garden: the lost and found genius of Gregor Mendel, the father of genetics.* Boston: Houghton Mifflin.

Keller, E.F. 1983. *A feeling for the organism: The life and work of Barbara McClintock.* New York: W.H. Freeman.

King, R.C., and W.D. Stansfield. 1997. *A dictionary of genetics,* 5th ed. New York: Oxford University Press.

Lewin, B. 2000. *Genes VII.* New York: Oxford University Press.

Watson, J.D. 1968. *The double helix.* New York: Mentor.

A flower of butterfly pea (*Clitoria ternata*), a tropical vine whose flower construction ensures that both the anthers and pistil touch the backs of visiting insects. The seeds are believed to be toxic to livestock.

Plant Breeding and Propagation

14

Chapter Outline

Overview

In this chapter, the processes that have produced and propagated domesticated plants are discussed. Proposed origins of our agricultural societies are presented along with the phenotypic changes that occurred during crop domestication. Plant breeding methods using conventional and nonconventional strategies are outlined. Then, propagation methods using seeds and asexual propagules are discussed.

Some Learning Goals

1. Provide explanations for the shift from hunter-gatherer societies to agricultural ones.
2. Describe phenotypic changes that occurred in plant populations as a result of human selection.
3. Describe breeding methods used for self-pollinating crops.
4. Describe breeding methods used for cross-pollinating crops.
5. Explain the significance of germplasm banks to crop improvement programs.
6. Describe the method used to produce protoplast fusion hybrids.
7. Outline the major steps involved in creating a transgenic plant.
8. Outline the steps involved in growing a crop from seed.
9. Describe how cutting propagation methods produce genetically identical plants.
10. Explain the benefits of grafting and outline the steps involved in making a graft.
11. Provide examples of specialized roots and stems used for asexual propagation.
12. Explain the benefits of micropropagation and outline the steps involved in micropropagation.

CROP PLANT EVOLUTION

"It is clear that the human species is currently an eater of grass seeds. We have become 'canaries.' It is also clear that the world's food supply depends on 12 or 15 plant species. It probably was not always so, although wheat, barley, rice, and maize have been the foundations of our high civilizations. The current trend is for the major crops to become even more major and for the lesser ones to dwindle."[1]

In fact, although there are approximately 200,000 species of flowering plants, only six species—wheat, rice, corn, potato, sweet potato, and cassava—provide 80% of the calories consumed by humans worldwide (Fig. 14.1). An additional eight plants—sugar cane, sugar beet, bean, soybean, barley, sorghum, coconut, and banana—complete the list of major crops grown for human consumption.

Figure 14.1 These six foods—wheat, rice, corn, cassava, potato, and sweet potato—meet most of the caloric needs of people worldwide.

In the 2 million years that we have inhabited this earth, we have cultivated plants for only the most recent 10,000 years, or less than 1% of our history. However, in that time, we have dramatically changed the plant landscape and our own lives. Hunter-gatherer societies have given way to agricultural societies and the development of cities and civilizations. Such concentrations of people are vulnerable to catastrophes such as drought and famine. Domesticated plants depend on us for their survival, but we have also become dependent on them for our survival.

We **domesticate** plants by altering them genetically to meet our needs. Strictly speaking, a domesticated plant is one whose reproductive success depends on human intervention. This is an ongoing evolutionary process, and plants are found in a continuum from purely wild to fully domesticated. Our current crop plants continue to evolve as a result of our breeding efforts. However, we are becoming increasingly dependent on fewer and fewer species of plants. It is amazing that, although humans have eaten thousands of types of plants in the past, we currently rely on only a handful to supply almost all of our nutritional needs.

Origins of Agriculture

Domestication of crop plants, and the development of industrial societies that followed, is a recent event in our history. Of the 80 billion humans that have lived on this earth, 90% were hunter-gatherers, 6% were farmers, and only 4% were industrialized.[2]

1. J.P. Harlan. 1992. *Crops and Man,* 2d ed. Madison, WI: Amer. Soc. Agron.

2. See note 1.

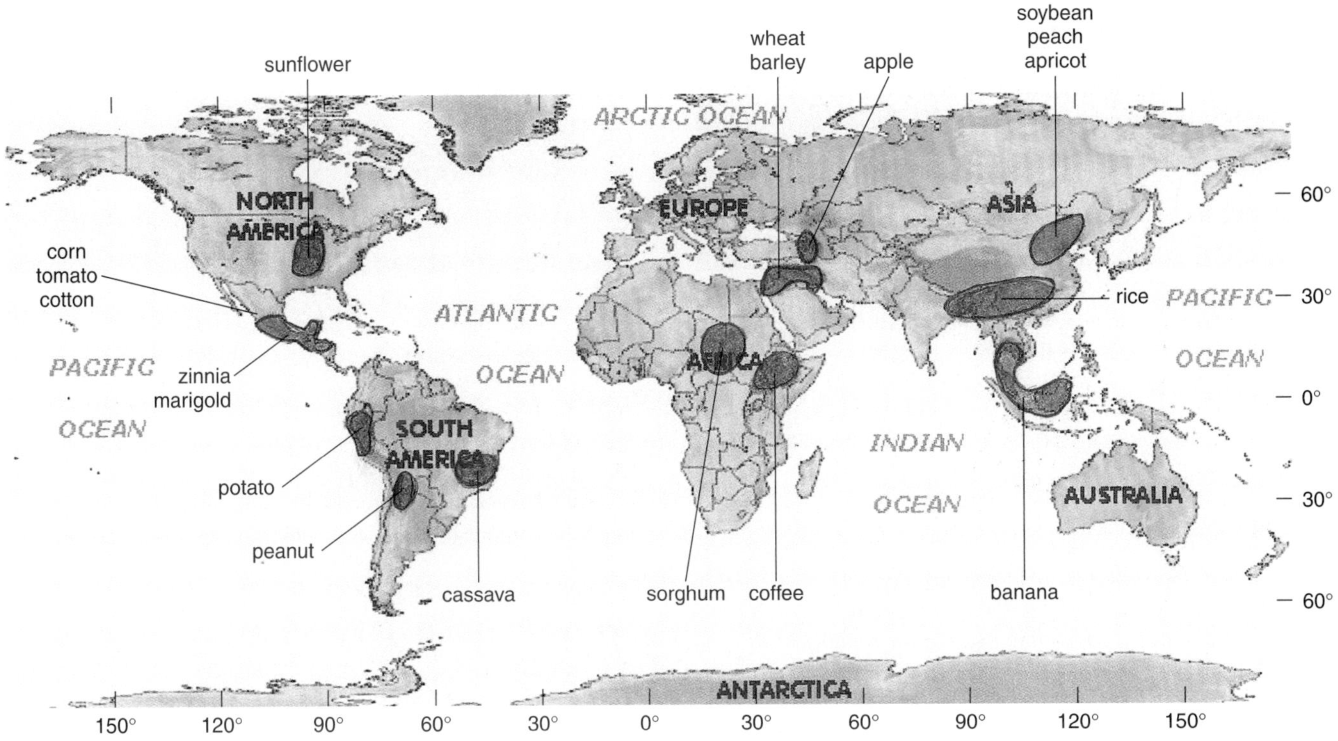

Figure 14.2 Regions of domestication of some crop plants. *Data from Jack R. Harlan. Crops and Management, 2nd edition. Am. Soc. Agronomy.*

We do not know why most humans shifted from a hunter-gatherer lifestyle to an agricultural one. Several hypotheses exist, but they are purely speculative. Hunter-gatherers seemed to live better lives than cultivators did. Their diet and health were better, their starvation rate was lower, and they worked far fewer hours. Their work week was only about 20 hours long! Because they had abundant free time, hunter-gatherers may have simply begun to grow plants in gardens as a hobby. They may have begun to cultivate a few plants to supplement lean times in the gathering schedule. Or, they may have had spiritual reasons for growing plants in a protected environment. Because agricultural practices arose independently in many parts of the world over thousands of years, each of these scenarios probably explains why humans began to domesticate plants.

People began to domesticate plants in the Near East around 10,000 years ago. Many agricultural sites from this time period have been located on the western edge of what is now Iran. Domesticated plants were developed in Asia 1,000 to 2,000 years later. Farming probably began in Africa and the New World about 4,000 years ago.

The first crops domesticated were cereal grains. Root crops and legumes, such as peas and beans, were domesticated 1,000 to 2,000 years later. These were followed by vegetables, and then oil, fiber, and fruit crops. Finally, plants used for forage, decoration, and drugs were first domesticated only about 2,000 years ago. Few crops have been brought into domestication in recent centuries. Plant breeders currently devote much more energy to improving existing crops than developing new ones. The map in Figure 14.2 shows where some of our major crops were domesticated. Sunflower is the only major crop that was domesticated in the present day United States (Fig. 14.3).

Figure 14.3 The sunflower is the only major crop that was domesticated in the United States.

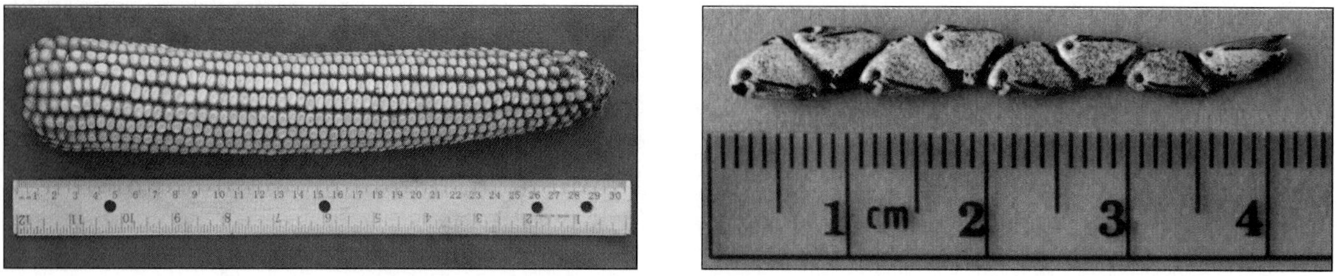

Figure 14.4 Modern corn (*left*) was probably domesticated from teosinte (*right*). *Teosinte photo courtesy John Doebley.*

Evolution Under Domestication

"It is as if man had been appointed managing director of the biggest business of all, the business of evolution . . . whether he is conscious of what he is doing or not, he is in point of fact determining the future direction of evolution on this earth. That is his inescapable destiny, and the sooner he realizes it and starts believing in it, the better for all concerned."[3]

As soon as people began sowing seeds of selected plants, instead of simply harvesting wild plants, dramatic genetic changes occurred in those populations in response to selection pressure. The seed dispersal mechanism was probably the first trait altered during domestication. The seeds of wild plants are typically carried away by wind, water, or animals. This minimizes the competition for light and nutrients that results from overcrowding. People most likely gathered seeds from plants that retained them instead of dispersing them. Early domesticators would have focused also on plants with high seed numbers and large seed size (Fig. 14.4). Competition in crowded cultivated plots would favor high seedling vigor and fast germination rate. In addition, while seed dormancy mechanisms are important in wild plants, plants that survived in cultivated gardens were those that germinated immediately upon planting. These changes accumulated over many generations and created the cultivated plants we know today. Many of these plants have been so dramatically altered from their wild forms that they cannot survive without the aid of humans. For example, if a cornfield is abandoned, it will cease to produce any corn plants within just a year or two.

The first serious attempts to understand the geographic origins of domesticated plants were made by the Russian geneticist Nikolai Ivanovich Vavilov in the 1930s. Based on extensive collecting expeditions carried out worldwide, Vavilov proposed that each crop plant was first domesticated in the region of the world where its genetic diversity is greatest. Tragically, Vavilov's career and life were cut short when he was imprisoned by the Stalin government in 1940 on charges of espionage and efforts to harm agriculture. He died in prison 3 years later. Although more recent studies indicate that patterns of crop domestication cannot be adequately explained by Vavilov's *Centers of Origin,* he can be credited with initiating studies that have led to our understanding of the relationships between cultivated plants and their wild relatives. As we will see later, wild relatives of crop plants are valuable sources of genes for modern crop improvement.

3. J. S. Huxley. 1957. *New Bottles for New Wine.* New York: Harper and Sons.

PLANT BREEDING

The systematic breeding of crop plants has been carried out only in the past 200 years. Until then, crop improvement was based on simply selecting and planting seeds from desirable plants. This method was moderately effective because plant populations were genetically diverse and could be improved gradually across generations (Fig. 14.5). However, it took an understanding of genetics and plant reproduction, and the training of scientists in these fields, to develop the sophisticated crop-breeding methods we know today.

Plant breeding is accelerated evolution guided by humans rather than nature. Breeders replace natural selection with human selection in order to modify plants genetically to meet our needs. The primary goal of plant-breeding programs is commonly improved yield, with disease resistance, pest resistance, and stress tolerance contributing to yield. With new sources of agricultural land dwindling and the human population continuing to grow, efforts must concentrate on improving the amount of food harvested per acre and developing crops that can grow on low-quality land (Fig. 14.5).

Breeding Methods Using Sexually Compatible Germplasm

Strategies

A plant breeder tends to divide crop plant species into two groups: **self-pollinating** and **cross-pollinating.** A self-pollinating plant is capable of fertilizing itself, while a cross-pollinating one cannot. Breeding methods for self-pollinating plants are radically different from those for cross-pollinating ones. A self-pollinating plant tends to be highly **homozygous** because all of its genes came from the same parent. The plant's mother was also its father! As you might expect, these plants have undergone a significant amount of **inbreeding.** As the species has evolved, it has

Figure 14.5 Potato tuber diversity. Genetic variation provides the foundation for efforts to improve plants through breeding.

adapted to and thrives on inbreeding. On the other hand, a cross-pollinating plant is likely to be highly **heterozygous,** because its mother and father were different plants. Cross-pollinating species tend to require a high level of heterozygosity in order to be productive.

Self-pollinated crop plants include wheat, rice, oats, barley, peas, beans, tomatoes, and peppers. Some fruit trees, such as apricots, nectarines, peaches, and citrus fruits are also primarily self-pollinated. The most primitive form of breeding in this group of plants is called **pure-line selection.** This method simply involves collecting seed from each of several plants, growing all the seed from each plant in a row, and selecting the most desirable row. All the plants in a row will be related to each other, since they came from the same plant. The seeds from the best row can then be propagated as a new "pure-line" variety.

A breeder with an understanding of reproductive biology can make crosses between self-pollinating plants. In fact, today, most plant breeders create hybrid populations by crossing desirable parents. Then, as the hybrid plants self-pollinate, highly diverse groups of plants will be created. These offspring of the self-pollinated hybrids provide the basis for selecting pure-line varieties. Crosses between normal wheat varieties and dwarf plants have been created to develop dwarf varieties. More of the energy in these plants is directed toward producing seeds rather than stalks, resulting in higher yields.

Norman Borlaug, who is known as the "Father of the Green Revolution," was awarded the Nobel Prize in 1970 for developing new strains of wheat in Mexico. The new high-yielding strains, which were produced from crosses with a dwarf variety from Japan, were widely planted.

Figure 14.6 Norman Borlaug, the father of the Green Revolution. © *Bill Meeks/AP Wide World Photos*

Before the project began in 1944, Mexico had imported wheat for many years, but the new varieties were so successful that, within 20 years, Mexico had quadrupled its wheat production and had enough to export to other countries. India and Pakistan experienced similar gains. It has been argued that, by increasing grain yields in developing countries, Borlaug has saved more lives than anyone in history (Fig. 14.6).

Although the Green Revolution has been a dramatic success, gains have not been unqualified. Irrigation and inorganic fertilizer is needed to support the rapid growth of these plants. Because many inorganic fertilizers are produced from fossil fuels, the cost of farming increases with fuel prices. On the other hand, the high-input system of agriculture has allowed food production to increase without a corresponding increase in the amount of land used for farming. This means that in areas where population growth is strong, high-yield farming prevents deforestation of wild areas. The key to solving the food supply problem, of course, is to control population growth on a global scale.

Cross-pollinated species include corn, rye, alfalfa, and clover, as well as most fruits, nuts, and vegetables. The simplest form of selection in cross-pollinated crops is **mass selection.** With this method, many plants from a population are selected, and seeds from these plants are then used to create the next generation. Again, the seeds from the best plants are chosen and propagated, and so on, for many generations.

This slowly molds the genetic makeup of the population to fit the breeder's preferences. For example, suppose variations in seed size are due to genetic differences and the breeder always collects from plants with the largest seeds. After several generations of selection, the breeder has genetically altered the population so that a larger proportion of plants carry the genes for large seed size.

Outcrossing (crossing between genetically different plants) in cross-pollinated crops often results in hybrid vigor, or **heterosis.** These plants are large, vigorous, fertile, and high yielding. Conversely, cross-pollinated plants exhibit **inbreeding depression** in the form of small size, poor vigor, low reproductive capacity, and a high proportion of abnormal plants. Breeders, however, do not avoid inbreeding of cross-pollinated species. In fact, a major breeding method involves forced self-pollination for several generations to create inbred lines in which deleterious alleles have been eliminated. Then, selected inbred lines are crossed to produce **hybrid seed.** Using this method, the most dramatic success story to date involves corn (maize), in which crosses between unrelated inbred lines often produce hybrids that dramatically outyield their parents. In 1908, in one of the earliest hybrid corn studies, plant breeder G.H. Shull crossed two inbred lines of corn, each of which produced 20 bushels per acre. The hybrid offspring yielded 80 bushels per acre, a quadruple increase in yield. Most of the corn in the United States is grown from hybrid seed (Fig. 14.7).

Some early American and European varieties, called **heirloom varieties,** are grown as open-pollinated populations of plants. Each variety is a mixture of genotypes, and all plants are allowed to pollinate each other, or open-pollinate, during seed production. These varieties are not as uniform as modern varieties, but their genetic variability allows them to produce a crop under many different environments. For example, a dry year and an insect pest might severely compromise all the plants in a hybrid variety if the variety is not genetically capable of surviving these conditions. An open-pollinated variety is likely to have some plants that are adapted to drought and can resist insect damage. The Seed Savers Exchange in Decorah, Iowa, has been created to preserve and distribute heirloom varieties. Heirloom varieties are important for farm market producers and growers of organic crops. They also contribute genetic diversity to breeding programs.

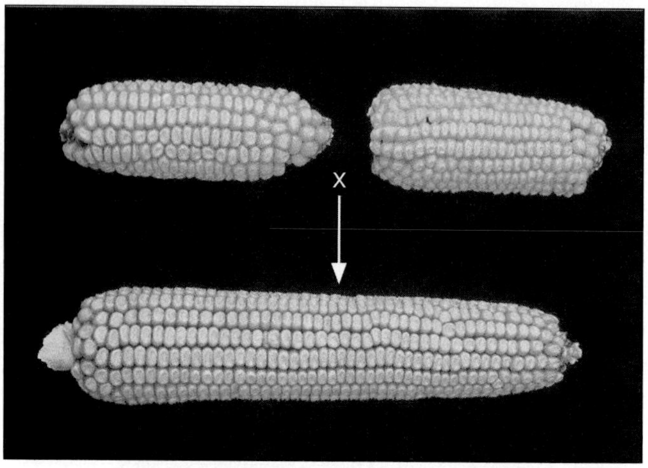

Figure 14.7A Crosses between inbred lines (*top*) often produce high-yielding hybrids (*bottom*). Hybrid vigor is called *heterosis.*

Figure 14.7B Hybrid corn plants.

Germplasm Collection and Gene Banks

Progress in plant breeding is absolutely dependent on genetic variability. It is impossible to improve a population if there is no genetic variability for the trait of interest. For example, a breeder who needs to develop rust fungus-resistant wheat must begin with a population containing at least one plant with some degree of resistance. Plant breeders are, therefore, concerned with the *germplasm* resources of crop plants. A crop plant's **germplasm** is the sum total of its genes. It is important to have access to crop plant germplasm containing genes for traits that are important for current breeding efforts. However, we must also realize that in the future, new traits may become important. For example, in a decade or two, we may need to develop plants with resistance to a new pathogen.

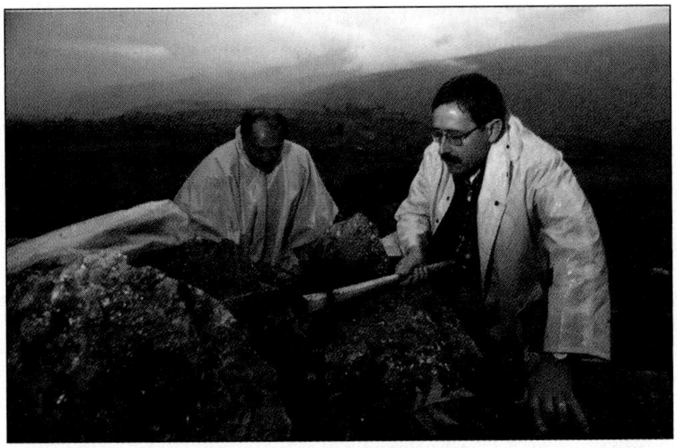

Figure 14.8 Botanists on a plant-collecting expedition in Peru. *Courtesy David Spooner.*

The plant varieties used in agriculture today have resulted from centuries of selection for specific traits. They are often genetically uniform and, therefore, may not be good sources of new genetic variability for further advances in breeding. In addition, their homogeneity makes them vulnerable to outbreaks of pests. The Irish potato famine in the mid-1800s occurred because people in that region relied on a few varieties of potatoes for most of their food supply. When a fungal pathogen spread through the fields, all the plants succumbed, and the people lost their entire potato crop. As recently as 1970, 15% of the United States corn crop was wiped out by a fungal pathogen. Genetic uniformity in the corn varieties grown at that time resulted in this significant loss.

In order to meet current and future needs for plant genetic diversity, **gene banks** have been established worldwide. The International Plant Genetic Resources Institute is responsible for coordinating international efforts to collect and conserve crop plant germplasm. The Institute also assists regional and national gene banks. The collaborative efforts of state and federal government agencies, along with private industries, comprise the National Plant Germplasm System in the United States Scientists regularly conduct collecting expeditions in regions where wild relatives of crop plants are found (Fig. 14.8). They bring plant samples back to the gene banks, where they are catalogued, propagated, and screened for desirable traits. Some seeds or other **propagules** are put into long-term storage for future needs, while others are given to plant breeders for immediate use.

Breeding Methods Using Sexually Incompatible Germplasm

In recent years, powerful new tools have been added to the toolboxes of plant breeders. Most breeding progress today is made using crosses among crop plants and their relatives, as described in the previous section, "Breeding Methods Using Sexually Compatible Germplasm." However, in many breeding programs, species boundaries no longer exist. Through techniques such as *protoplast fusion* and *gene splicing* (discussed in sections that follow), genes from virtually any living organism can be incorporated into crop plants. Everything, therefore, from an antifreeze gene in North Atlantic fish to a pesticide gene in bacteria is now accessible to breeding programs.

Protoplast Fusion

There are times when a breeder would like to tap into the genetic diversity of a plant species that is distantly related to a crop plant. However, it is impossible to make a cross between the two species and obtain viable seeds. An alternative strategy involves **protoplast fusion.** With this method, cells of each species are grown in a liquid nutrient solution. Then, their cell walls are chemically stripped off to produce protoplasts. The protoplasts of the two species are mixed together and stimulated, with the aid of an electric current or chemical solution, to fuse with each other. When screening the fusion products, a scientist must distinguish between *autofusions,* where two protoplasts of the same species have fused together, and *hybrid fusions,* in which the protoplast of one species has fused with that of the other species. One way of doing this involves dyeing the protoplasts of the each species with a different color before the fusion event. Then, the scientist can examine the fusions with a microscope and identify protoplasts that show both colors. The selected protoplasts are then carried through a series of *tissue culture* events (see "Micropropagation" on page 268) that cause them to grow into whole plants. These new plants carry genes from two distantly related species. They are called **somatic hybrids** because they resulted from combining (hybridizing) *somatic,* or body, cells from two different plants. In contrast, plants usually reproduce when gamete cells (eggs and sperm) are combined, producing sexual hybrids.

Although protoplast fusion appears to have value for the incorporation of new genes into crop plants, few success stories are currently available. Somatic hybrids have been created between potato cultivars and nontuber-bearing wild relatives. Some are highly resistant to a number of bacterial and fungal diseases, but varieties have not yet been created from the hybrids or their offspring. Tobacco is probably the best example of a commercially successful somatic hybrid. A common variety produced in Canada is the result of a fusion between cultivated tobacco and a wild relative containing a novel form of disease resistance.

Gene Splicing and Transgenic Plants

Genes from virtually any organism, from viruses to humans, can now be inserted into plants, creating **transgenic plants.** The technologies that have been developed to support this work arose from basic research to understand DNA structure and function. **Recombinant DNA** technology is based on the knowledge that DNA behaves the same no matter what organism carries it. Because the DNA code is nearly universal, a plant cell can read a DNA sequence from almost any other organism and produce a foreign gene product.

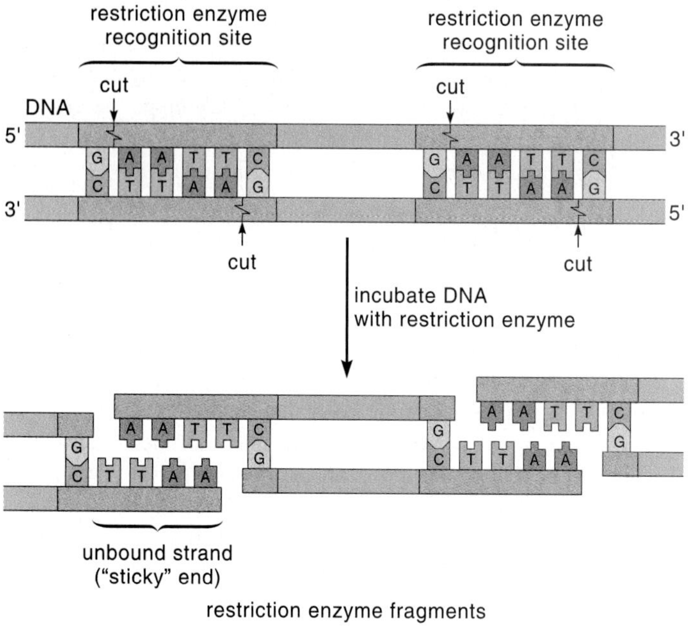

Figure 14.9 Some restriction enzymes, make a zigzag cut wherever they find a recognition site. Single-stranded portions of restriction fragments are called *sticky ends* because they can form hydrogen bonds with complementary sticky ends on other molecules of DNA. *From Moore, Clark, and Vodopich, Botany, 2nd edition.*

The serendipitous discovery of **restriction enzymes** paved the way for molecular genetic engineering. These enzymes act as molecular scissors, chopping up DNA into pieces small enough for genetic analysis. In nature, bacteria produce restriction enzymes, protecting themselves from viral pathogens by destroying pieces of foreign DNA. A key feature of these enzymes is their ability to cut DNA at specific sites and create DNA fragments with "sticky" ends. When DNA is cut by certain restriction enzymes, one strand is slightly longer than the other. Each fragment then is a double-stranded DNA molecule with a single-stranded tail at each end. This tail is called a sticky end because it will base pair with the tail of any other fragment cut with the same enzyme (Fig. 14.9).

How are transgenic plants made? As an example, let's say we would like to insert a bacterial insect resistance gene into a corn plant. First, we need to find the insect resistance gene in the bacterium. This is actually a very difficult task and beyond the scope of this book. As you might expect, finding a specific gene in the enormous genome of eukaryotes is even more difficult and is like looking for a needle in a haystack. Sometimes, if a protein gene product is known, then a machine called a **protein sequencer** can be used to determine the sequence of amino acids in the protein. Then, using information from the genetic code (see Chapter 13), a **gene synthesizer** produces an artificial gene. An alternative approach is to find the gene in an organism's genome and then use a restriction enzyme to cut the bacterial chromosome on both sides of the gene, but not within the DNA sequence of the gene. After that, the bacterial DNA and a *cloning vector* are cut with the same restriction enzyme. A *cloning vector* is a piece of DNA that can copy itself in a living cell. A common example of a cloning vector is a **plasmid,** a small circular piece of DNA naturally found in bacteria and capable of replicating independently of the bacterial chromosome. Dozens of custom-made plasmids containing restriction enzyme cutting sites have been created by genetic engineers. If the cut bacterial DNA is mixed with cut plasmids, then their sticky ends will join together and create recombinant DNA. In this case, the recombinant DNA is a plasmid with an insect resistance gene attached to it. *Escherichia coli,* the bacterial workhorse of molecular genetics, can be stimulated to take up the plasmid via a process called **transformation.** When the *E. coli* cells multiply in a flask of nutrient medium, they will replicate their recombinant plasmids. At this point, *E. coli* is cloning our insect resistance gene for us. A culture of the bacterium will contain millions of copies of the gene (Fig. 14.10).

After our insect resistance gene has been cloned, we have enough copies of it to manipulate and try to insert it into plant cells, again through transformation. We can remove the plasmids from a culture of *E. coli,* use the same restriction enzyme to cut out the resistance gene, and then purify it. Plant cells will not take up foreign DNA as readily as bacterial cells. Nevertheless, a number of techniques to transform plants with foreign DNA have been developed. We will focus on the two major ones.

The most common transformation technique relies on a natural genetic engineer, *Agrobacterium tumefaciens.* This bacterium infects plant cells and then inserts DNA (called **T-DNA,** or transfer DNA) from its plasmids into the plant's chromosomes. This causes the plant to produce cancerlike tumors that harbor and feed the bacterium (Fig. 14.11) Human genetic engineers, however, have removed the virulence (disease-causing) genes from the T-DNA region and replaced them with genes of human interest. In our example, we could insert the insect resistance gene into the T-DNA region of the *Agrobacterium* plasmid (Fig. 14.12). Then, we would allow the bacterium to infect corn cells and insert its T-DNA into the chromosomes of those cells. Transformed cells can be cultured to produce whole plants carrying and expressing the insect resistance gene. The gene has now been stably incorporated into a plant chromosome and will be passed on to subsequent generations (Fig. 14.13).

When the *Agrobacterium* transformation technique was first developed, it was not effective in monocots. As a result, an alternative strategy using a **particle gun** was developed especially for use with monocots. This technique actually first used a tiny modified shotgun, with DNA attached to the shot. Today's particle guns are much more refined (and expensive!) than that first prototype, but still rely on shooting DNA into plant tissue. With this technique, very small tungsten or gold pellets are coated with the cloned insect resistance gene. Then the DNA is shot into corn cells and, if enough plant cells are bombarded, some of them will have the gene incorporated

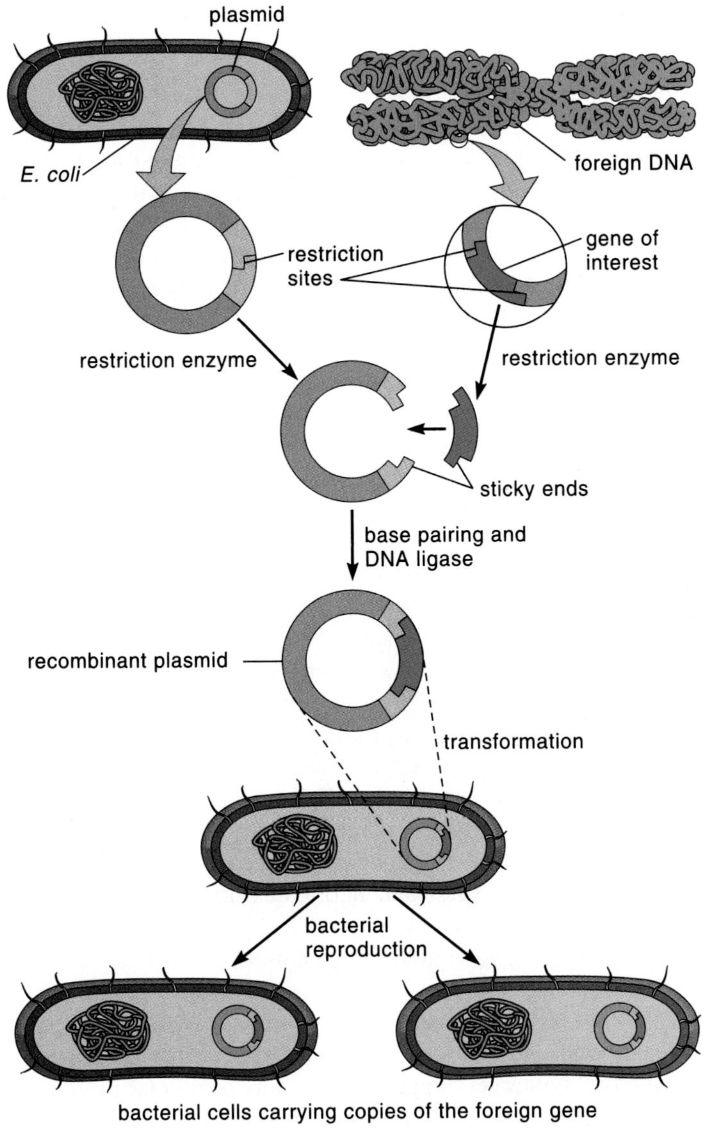

Figure 14.10 Gene cloning by bacteria. The gene of interest from another organism is cut out of its genome by a restriction enzyme and inserted into a bacterial plasmid that has been cut with the same restriction enzyme. Sticky ends of the foreign fragment bind to the complementary sticky ends of the plasmid, followed by the formation of new sugar-phosphate bonds through the action of the enzyme DNA ligase. Finally, the altered plasmid is taken up by the bacteria, which make many copies of the foreign gene by duplicating the plasmid genome during reproduction. *From Moore, Clark, and Vodopich, Botany, 2nd edition. © 1998 The McGraw-Hill Companies. All rights reserved.*

into their chromosomes. Precisely how this process works is still a mystery, but through the use of this technique, foreign genes have already been permanently inserted into crop plants. In fact, transgenic corn containing a bacterial insect resistance gene is now widely grown in the United States.

It is difficult to control the number of gene copies inserted during transformation. In addition, genes are inserted at random locations in plant chromosomes during this step. Sometimes, the foreign genes may be inserted in areas of the genome that are not expressed by the plant, or they may insert into a critical portion of a plant gene. In addition, considerable effort is often needed to get the foreign gene expressed at suitable levels in appropriate tissues. Transformation and gene expression, therefore, are the most challenging aspects of transgenic plant production.

Figure 14.11 A crown gall, caused by the bacterium *Agrobacterium tumefaciens,* on a tomato plant. *Courtesy Terese Barta.*

Transgenic corn, soybean, cotton, potato, and canola varieties containing herbicide and insect resistance are grown extensively in North America (Fig. 14.14). Traits such as disease, insect, and herbicide resistance are important for the producers of those crops, but you would not notice them as a consumer. The new generation of transgenic crops focuses on traits that will be more obvious. Transgenic plants can act as *bioreactors* to create pharmaceuticals. These plants would provide inexpensive access to vaccines and other medicines, especially in parts of the world where medical facilities are not readily available. People would simply take their medicine by eating potatoes or carrots. For example, transgenic plants are being tested for the production of vaccines against hepatitis B, rabies, cholera, tuberculosis, malaria, acute diarrhea, and even dental caries. Transgenic plants have also been made to produce a protein that can prevent or delay the onset of insulin-dependent *diabetes mellitus.* At one time, all of our drugs were derived from plants, but we have learned to synthesize many

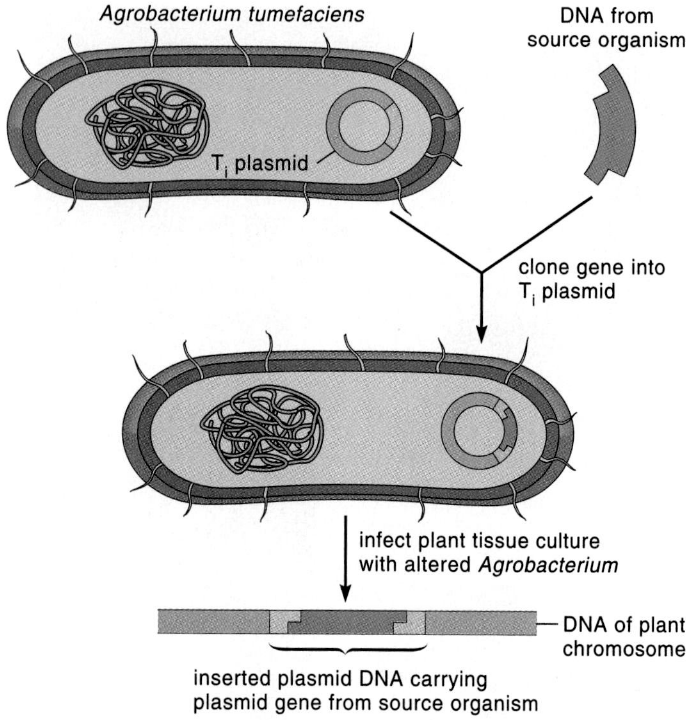

Figure 14.12 Inserting foreign genes by using the *Agrobacterium* T_i plasmid. First, the bacterium is transformed by the insertion of a foreign gene into its T_i plasmid, which carries genes for inserting itself into plant chromosomes. Then a plant cell that is infected with the plasmid receives the foreign gene into its genome. *From Moore, Clark, and Vodopich, Botany, 2nd edition.*

of them in laboratories. Now, we are completing the circle by letting the plants once again create our drugs.

Nonedible transgenic plants are also showing potential in some novel ways. Plants are being created to produce biodegradable polymers in order to reduce our dependence on plastics made from nonrenewable resources. Plants are also being engineered to sequester heavy metals. These transgenics could be grown in contaminated soils where they will pick up wastes such as copper or mercury. Then, they could be harvested and disposed of properly.

Finally, transgenic technology is providing us with ornamental plants never before seen in nature. Most popular cut flowers do not produce blue hues, and efforts are presently being directed toward the development of blue carnations, chrysanthemums, and roses. For example, a violet carnation named Moonshadow contains a petunia gene for blue color (Fig. 14.15). Genetic engineers are also trying to extend the shelf life of cut flowers, mainly by blocking genes for ethylene synthesis. In addition, "flowering-time" genes are being used to develop plants that flower under day-length conditions that would normally prevent flowering; plant architecture genes are being used to produce either compact or vine-type plants; and fragrance genes have been identified in an effort to restore fragrance quality to roses and carnations.

Pros and Cons of Transgenic Plants

In October 2000, activists called the Green Streets destroyed a test plot of genetically engineered corn at a University of California test facility. In January 2001, a group called the Nighttime Gardeners destroyed a greenhouse containing genetically engineered wheat in Albany, California. And, in February 2001, a group called the Earth Liberation Front burned a research cotton gin in Visalia, California, to protest the development of genetically engineered cotton. In less than two years, over 40 antigenetic engineering acts of vandalism have occurred in North America. Why are people so concerned about transgenic crops?

There are approximately 109 million acres of transgenic crops grown worldwide, 68% of which are in the United States. The most common transgenic crops are soybean, corn, cotton, and canola. Most often, these plants contain a gene making them resistant to the herbicide glyphosate, commercially sold as Roundup or an insect resistance gene that produces a protein called Bt toxin (because it is derived from the bacterium *Bacillus thuringiensis*).

On the positive side, proponents of transgenic crops argue that transgenic crops are environmentally friendly because they allow farmers to use fewer and less noxious chemicals for crop production. For example, a 21% reduction in the use of insecticide has been reported on Bt cotton. In addition, when glyphosate is used to control weeds, then other, more-persistent herbicides do not need to be applied.

On the negative side, opponents of transgenic crops suggest that there are many questions that need to be answered before transgenic crops are grown on a large scale. One question deals with the effects that Bt plants have on "nontarget" organisms such as beneficial insects, worms, birds, and even humans who consume the genetically engineered crop. Monarch caterpillars feeding on milkweed plants near Bt cornfields will eat some corn pollen that has fallen on the milkweed leaves. Laboratory studies indicate that the caterpillars can die from eating Bt pollen. However, field tests indicate that Bt corn is not likely to harm monarchs. Remember, too, that application of pesticides (the alternative to growing Bt plants) has been demonstrated to cause widespread harm to nontarget organisms.

Another unanswered question is whether herbicide resistance genes will move into populations of weeds. Crop plants are sometimes grown in areas where weedy relatives also live. This is especially true outside of the United States. (Remember that most of our crop plants did not originate in the United States, and they do not have wild relatives here.) If the crop plants hybridize with weedy relatives, then this herbicide-resistance gene will be perpetuated in the offspring. In this way, the resistance gene can make its way into the weed population. If this happens, a farmer can no longer use glyphosate, for example, to kill those weeds. This scenario is not likely to occur in many

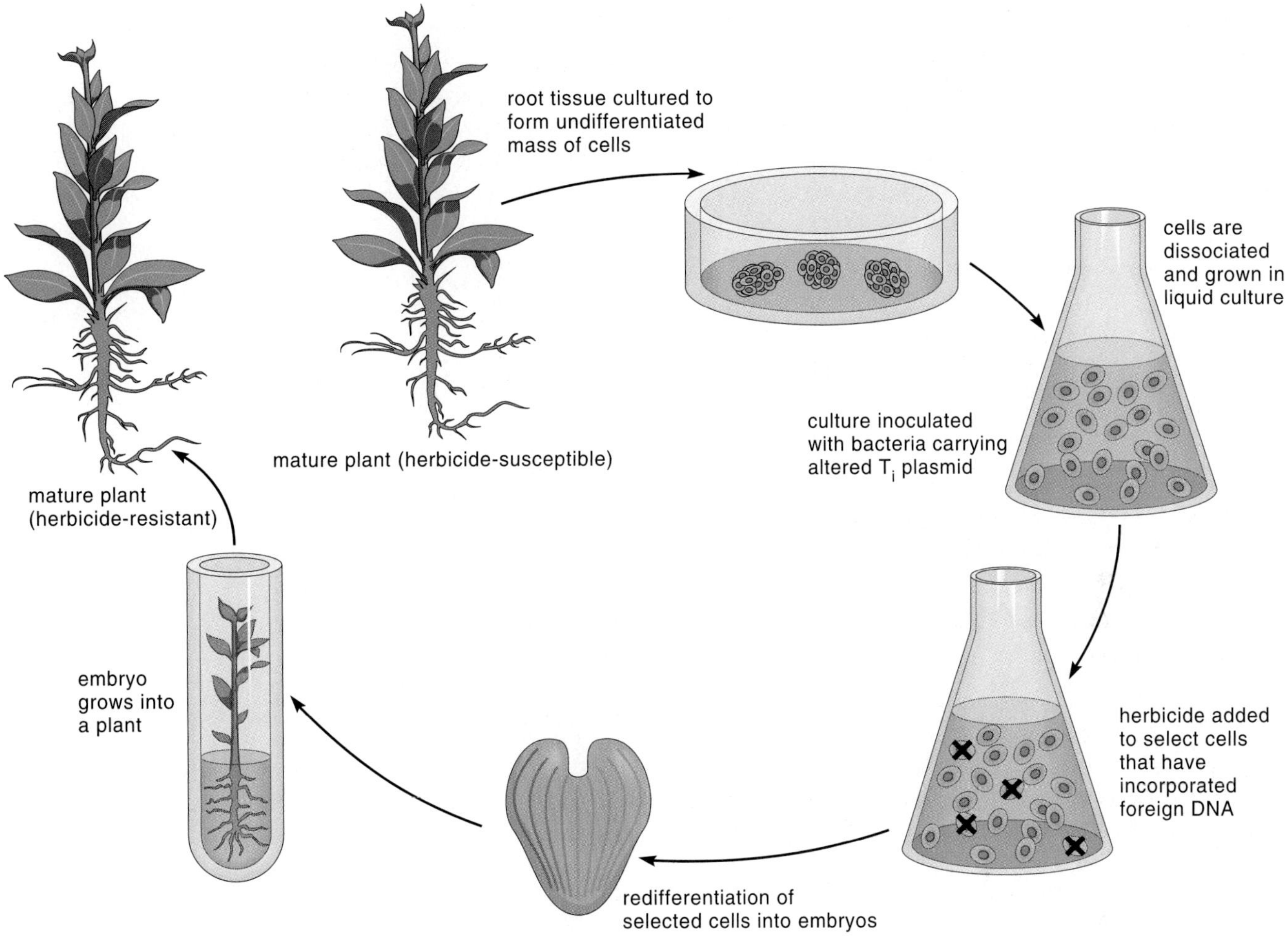
root tissue cultured to
form undifferentiated
mass of cells
cells are
dissociated
and grown in
liquid culture
culture inoculated
with bacteria carrying
altered T_i plasmid
mature plant (herbicide-susceptible)
mature plant
(herbicide-resistant)
embryo
grows into
a plant
herbicide added
to select cells
that have
incorporated
foreign DNA
redifferentiation of
selected cells into embryos

Figure 14.13 Steps in making transgenic plants. Differentiated cells from a plant tissue are put into cell culture where they grow into an undifferentiated mass of cells. The culture is transferred to a liquid medium and infected with T_i plasmids carrying a gene for herbicide resistance. Cells that survive the herbicide treatment are transgenic. Plants that are regenerated from these cells are also transgenic, making them resistant to the herbicide in the field. *From Moore, Clark, and Vodopich, Botany, 2nd edition. © 1998 The McGraw-Hill Companies. All rights reserved.*

instances because there are no weedy relatives growing near the crop plant. However, in some cases, it may become a serious problem. For example, canola readily hybridizes with mustard weed species and could transfer its herbicide-resistance genes to those weeds. In addition, Canadian farmers have reported that herbicide-resistant canola has invaded wheat fields like a weed and, of course, cannot be killed with a major herbicide.

We know that evolution will occur when transgenic plants are grown on a large scale over a period of time. Of special concern is the development of insect populations resistant to the Bt toxin. This pesticide has been applied to plants for decades without the development of insect-resistant populations. However, transgenic Bt plants express the toxin in all tissues throughout the growing season. Therefore, all insects carrying genes that make them susceptible to the toxin will die. That leaves only the genetically resistant insects to perpetuate the population. When these resistant insects mate, they will produce a high proportion of offspring capable of surviving in the presence of the Bt toxin. Farmers are attempting to slow the development of insect resistance in Bt crops by, for example, planting non-transgenic border rows to act as a refuge for susceptible insects. These insects may allow Bt susceptibility to remain in the population.

Perhaps the most serious concern about the transgenic crop plants currently in use is that they encourage farmers to head farther away from sustainable agricultural farming practices. Transgenics, at least superficially, simplify farming by reducing the choices made by the manager. The planting of a glyphosate-resistant crop commits a farmer to the use of that herbicide for the season, probably to the exclusion of all other herbicides and other weed-control practices. In the long run, though, it may be in the best interest of the farmer and the land to use more integrated, sustainable weed-control approaches. Farmers who use Bt transgenics may not feel they need to follow through with integrated pest-management practices that use beneficial insects and timely application of pesticides to control insect pests. In fact, a farmer must decide whether to plant Bt corn even before he knows whether the European corn

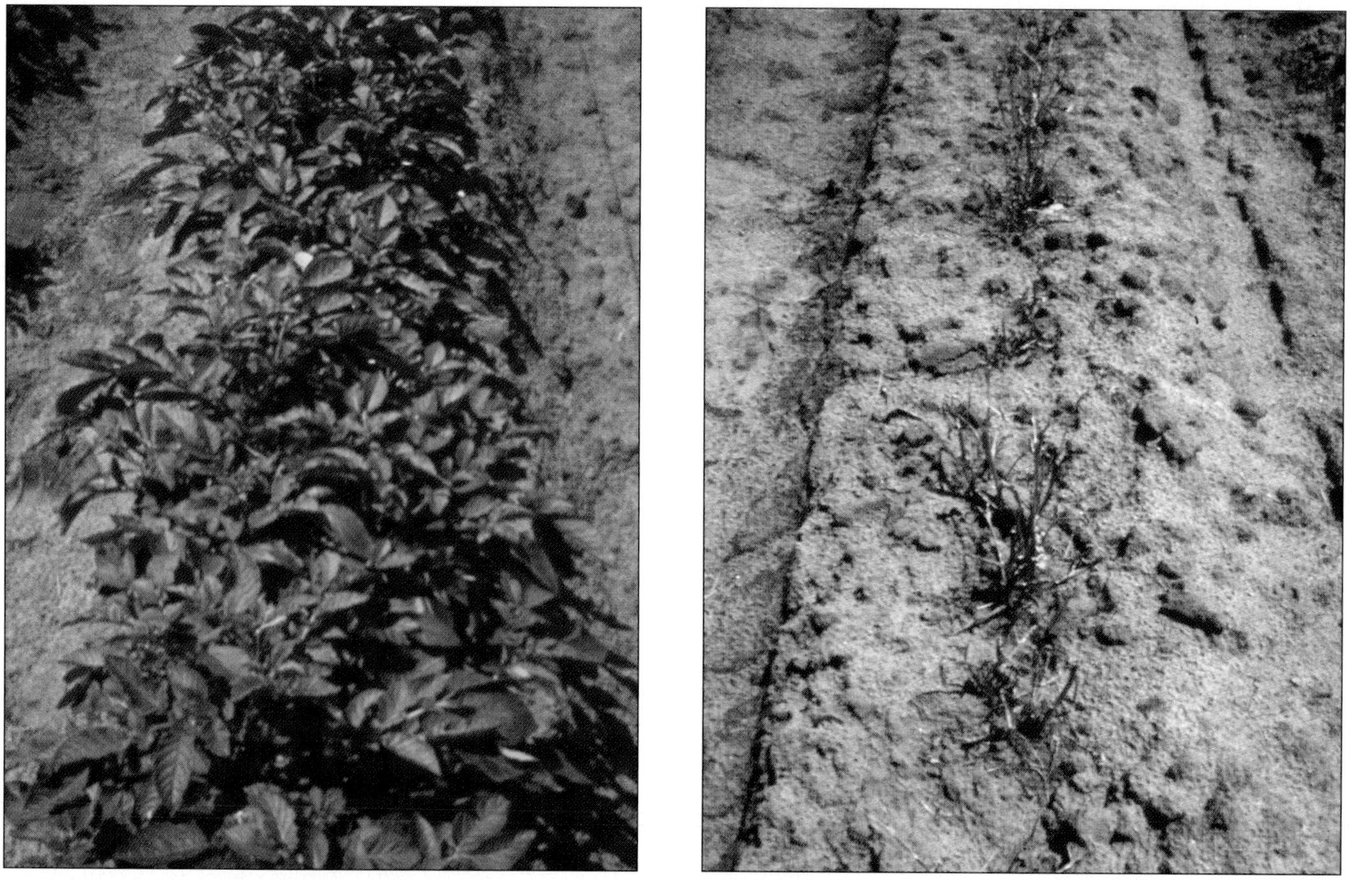

Figure 14.14 Transgenic potato plants expressing the Bt gene for insect toxin (*left*). The photo on the *right* is the same variety but has been defoliated by the Colorado potato beetle because it has not been transformed with the toxin gene. *Courtesy Jeffrey Wyman.*

borer will be a problem during the growing season. In contrast, a more sustainable approach would be to plant non-transgenic corn, monitor the fields throughout the growing season, and then apply a pesticide only if and when it is needed.

The jury is still out on the long-term effects of transgenic plants on our agricultural and natural environments. The "Catch 22" is that we will not know whether transgenic crops cause serious negative consequences until we grow them on a large scale. However, if we do grow them on a large scale and find that the concerns are valid, then we cannot reverse what has been done. Transgenic pollen has been released, and populations of everything from weeds to corn borers have been altered.

In addition to questions regarding environmental safety, human health issues must be addressed. The major concern over the consumption of foods derived from transgenic crops is the potential for the transgene protein product to cause an allergic reaction. For example, a Brazil nut gene was added to soybean to increase its methionine content. Methionine is an amino acid commonly added to animal feed. However, this gene produced a protein that caused an allergic reaction in some people. Although the transgenic soybean was being developed as an animal feed, there was concern that it might find its way into the human food chain, and consequently, it was dropped from production. How likely is it that a crop designed for animal feed will end up in human food products? Recently, Starlink corn was found in taco shells, tortilla chips, and corn dogs. Starlink is a variety of transgenic corn containing a gene for Bt toxin and was developed as an animal feed. It has not been approved for human consumption because of concerns that it might cause an allergic reaction. So far, no antibodies to the Starlink protein have been found in people who have experienced a potential allergic reaction after eating these products. The Starlink incident is just the tip of an iceberg faced by the food industry as it tries to keep track of transgenic products.

PLANT PROPAGATION

After we create genetically superior clones or populations of plants through breeding efforts, we must have a method to perpetuate, or propagate, these plants. Plant propagation techniques allow us to grow thousands of acres of a superior corn variety or to have the same day lily variety growing in thousands of yards. Some plants are most easily propagated through sexual reproduction via seed propagation. With other plants, asexual reproduction methods are used more extensively.

Figure 14.15 The Moonshadow variety of carnation contains a petunia gene for blue color.

Seed Propagation

Hybrid varieties are often grown from the seed produced by crosses between two inbred parents. A common summer job for teenagers in corn-belt regions is detassling corn (removing male flowers). Rows of one inbred parent are planted next to rows of a second inbred parent (Fig. 14.16). Then, the tassels of the first parent are removed to prevent self-pollination. Pollen from the second parent is carried by the wind to the female flowers (the ears) of the first parent and carry out fertilization. The mature ears of the detasseled parent are then collected, and the kernels are harvested as hybrid seed. That seed will be sold to farmers for planting their hybrid cornfields during the next growing season.

Inbred line varieties are typically grown from seed. It is especially easy to generate seed from these varieties. They simply are grown in a field and allowed to self-pollinate. Seeds collected from these plants will grow into a uniform population similar to the plants from which the seeds were collected. If you collect seeds from, for example, an inbred green bean variety, you could grow plants that are nearly identical to that variety in appearance and fruit quality. In contrast, seeds collected from a hybrid plant will produce a highly variable population of plants.

Figure 14.16 A field planted for hybrid corn seed production. Rows that have been detassled are adjacent to rows with tassels. Pollen from the tassels will land on ears of detassled plants. Hybrid corn will be harvested from detassled plants.

Commercial seeds usually come from fields planted solely for seed production. The fields are isolated from other fields of the same crop in order to prevent contamination by foreign pollen and mixing of seeds by harvest equipment. Seed quality depends on the health of the plants that produce it, and growing conditions are meticulously monitored. Often, the plants are grown in arid regions where irrigation can be controlled and seeds can dry out before harvest. Most grass and forage seed production is concentrated in the Pacific Northwest, while vegetable and flower seeds are generated in the coastal valleys of California. There is also a growing international seed production industry.

When seeds are mature, they are harvested and stored in a controlled environment. The **viability** of most seeds is best maintained in cool, dry storage conditions. Samples are periodically removed from a seed lot and tested for vigor and viability. You may sometimes see the results of these tests on packages of seeds you purchase for your garden. You will also see a date stamped on the seed package. Seeds of some species of plants retain their viability in storage much longer than others. Seeds of green beans, lettuce, onions, and peppers lose their viability after a year or two in storage, while those of beet and tomato may be stored for five years or more.

Before planting seeds, steps may be taken to ensure the growth of a vigorous stand of seedlings. In preparation for planting, seeds may be dusted with a **protectant,** such as a fungicide. The red coating on seed corn is such a fungicide. Sometimes, seeds are dusted with beneficial bacteria before planting. Legumes such as peas and beans establish mutualistic associations with bacteria that *fix* nitrogen (convert nitrogen from the air into forms the plants can use). Dusting seeds with these bacteria will encourage that relationship to develop.

It is important to plant seeds in a suitable bed. The soil should be moist enough to allow the seeds to imbibe water, which is essential to begin germination. However, if the soil is soggy, the new roots will not have enough oxygen to keep up with the respiration needed for active growth. Seeds of different species are adapted for growth at different soil temperatures. Celery, lettuce, and onion seeds will not germinate in hot summer soil, while tomato, bean, cucumber, and sweet-corn seeds germinate poorly in cool spring soils. Initially, seeds rely on stored food for growth. Fertilizer cannot be utilized until a root system has been established.

In their early stages of growth, seedlings are easily shaded and stunted by weed growth, so weed control is important until the plants can shade the spaces between rows. Seedlings may also compete with each other for light, water, and nutrients, making thinning necessary for production of a good crop.

Asexual Plant Propagation

In recent years, we have heard amazing stories about scientists who have cloned animals such as sheep and cows. Botanists are far ahead in that arena. They have been cloning plants for centuries! Many plants are easy to **propagate** asexually, using vegetative parts rather than seeds. For example**, crown division** is a simple technique in which a plant is separated into several pieces, each of which contains a crown and a portion of the root system (Fig. 14.17). This is commonly used for many ornamental perennial plants. A breeder who has identified a superior plant of an asexually propagated species can perpetuate that genotype indefinitely. This avoids the genetic variability and consequent unpredictability that results from seed propagation.

Cuttings

We know that nothing lives forever. Or do we? If you are particularly fond of your grandmother's African violet plant, you can use a leaf to propagate a new plant. As that plant begins to age, you can propagate a new one again from a leaf and eventually pass the plant on to your grandchildren. This plant could be propagated indefinitely and would essentially live forever.

Figure 14.17 Asexual reproduction of a daylily by crown division.

Propagation of plants from parts such as leaves is called propagation by **cuttings.** If a stem is used as a cutting, it needs to be coaxed into producing **adventitious** roots (roots produced on internodes or other parts of plant organs). Buds on the stem will grow into the shoot system. Leaf and root cuttings, though, must develop both adventitious roots and shoots (Fig. 14.18). The formation of adventitious structures requires plant cells near the wound to **dedifferentiate** (become less specialized) and create a new **meristematic** region.

After a cutting is made, it must be kept in an environment that will allow wound healing and development of new organs. The most critical step usually is prevention of water loss. Because a stem or leaf cutting has lost its root system, its tissues must be kept moist until roots can once again supply water. This is easily done by keeping the cuttings in an enclosed container to maintain high humidity and to reduce **transpiration.** In commercial production systems, frequent misting or constant fog is used to keep cuttings from drying out. Sometimes, rooting is stimulated by application of a rooting powder containing a growth regulator such as **auxin.** Rooting powders also often contain fungicides to prevent pathogens from entering wounds. The cuttings must be kept in a potting mix such as perlite or vermiculite, which holds water but also drains well. This is preferable to rooting cuttings in water, because oxygen levels in water are low and are quickly depleted as young cells in developing roots undergo respiration. Temperature is also a factor in successfully generating cuttings. Ideally, young roots should be warm enough to enhance growth, while the above-ground parts should be cool enough to minimize growth and transpiration. For this reason, cuttings are often grown on heating pads or in heated beds (Fig. 14.19). Once roots are established, fertilizer helps to stimulate growth.

Many house plants are easily propagated by cuttings. These include African violet (*Saintpaulia*), snake plant (*Sansevieria*), *Begonia, Coleus,* and *Kalanchoë.* Outdoor

Figure 14.18 Asexual reproduction by cuttings. *Left:* Stem cuttings of an ornamental fig (*Ficus*). *Right:* Leaf cuttings of *Sansevieria.* Note the adventitious roots developing from both types of cuttings.

Figure 14.19 Cuttings are often grown on a heated bed to stimulate root development.

plants propagated by cuttings include *Forsythia, Geranium,* rose (*Rosa*), *Spiraea,* raspberry (*Rubus*), and juniper (*Juniperus*). The main advantage of propagation by cuttings is that identical copies of a valuable plant can be made. A major disadvantage is that diseases carried by the mother plant, including those caused by viruses, fungi, and bacteria, are also propagated.

Layering

Layering is a modified form of cutting propagation and works well for some plants that are not easy to propagate by cuttings. This procedure allows the adventitious root system to develop before the new plant part is severed from the parent plant.

Tip layering is used with blackberries, boysenberries, and other plants with flexible stems. The canes are bent over until the tips touch the ground; the tips are then covered with a small mound of soil. Roots form on the portion of buried stem, and eventually, shoots will also appear (Fig. 14.20). These new plants can then be separated from the parent stems. Variations of tip layering include forcing a stem to lie horizontally and covering it with small mounds of soil at intervals, or heaping soil around the base of a plant so that the individual stems produce roots there. Once roots have been established, the individual plantlets or pieces of stem can be cut off from the original parent and grown independently.

Air layering is sometimes used to propagate tropical trees and shrubs. It is useful for producing a few large plants from a single plant with a rigid stem. A branch or main stem is wounded with a sterile knife and then may be dusted with a rooting powder. The wound is produced by gashing the stem or, alternatively, by *girdling.* **Girdling** is the removal of a ring of bark around the stem. Damp sphagnum moss is wrapped around the wound, and the area is covered with clear plastic film to retain moisture. Then, aluminum foil is placed over the plastic film to reflect sunlight and prevent overheating (Fig. 14.21). When roots are observed in the moss through the plastic film, the layer is removed from the parent plant and transplanted.

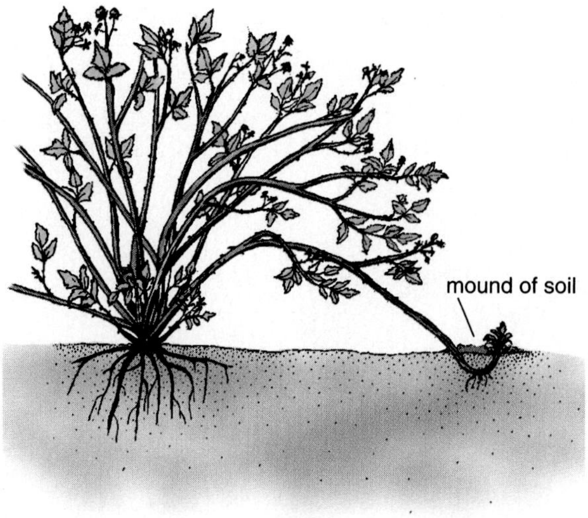

Figure 14.20 Tip layering. The tips of canes are bent to the ground and covered with a small mound of soil. When a new plant has developed at the tip, it can be cut from the parent plant and grown independently.

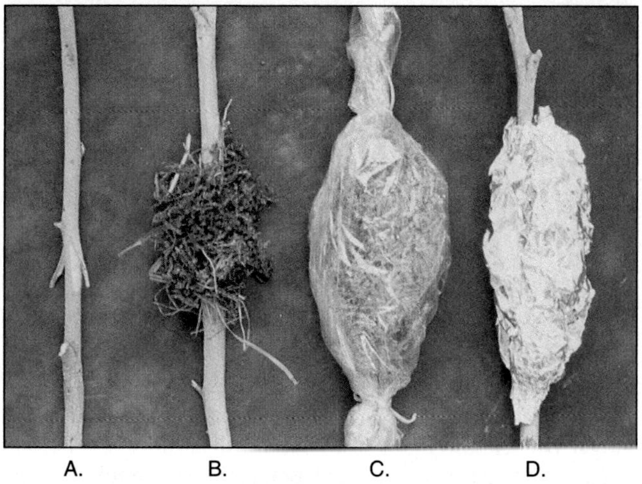

Figure 14.21 Steps in air layering. *A.* Cuts are made at an angle to the axis of a stem of a rooted plant. *B.* Damp sphagnum moss is wrapped around the cut area. *C.* Polyethylene film is wrapped around the moss, and the ends of the film are taped shut. *D.* Aluminum foil is wrapped around the film-covered moss. Within a few weeks, adventitious roots should develop and protrude through the moss. After the roots have appeared, the rooted portion can be cut off and planted.

Grafting

Imagine walking out into your yard in the morning and picking a fresh grapefruit for your breakfast. Then, you go to the same tree at lunchtime and harvest an orange. Later in the day, you pick a lemon off the same tree to squeeze on your fish dinner. Sound impossible? It is not, if you are growing a tree that was created by grafting.

Grafting is a process by which segments of different plants are connected and induced to grow together as one plant. It has been performed as a horticultural art for thousands of years, dating back to around 1560 B.C. Historically,

Figure 14.22 A brightly colored cactus stem has been grafted on to a green cactus plant.

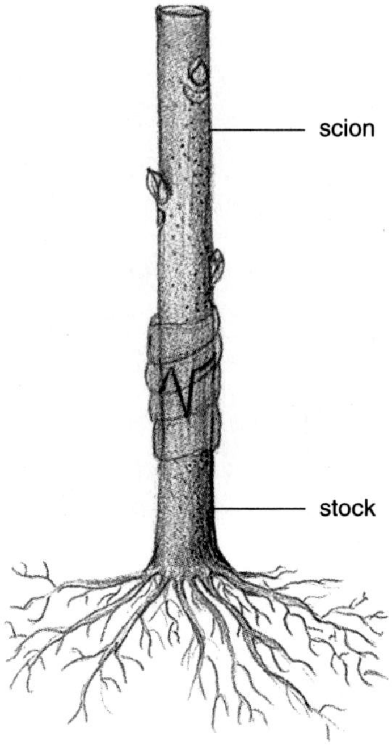

Figure 14.23 A simple graft. The root portion (*stock*) and portion to be grafted onto the stock (*scion*) are cut so that the two parts will fit together with the cambium of both portions in close contact.

grafting has been performed mainly to clone plants that are difficult to propagate as cuttings. Today, many fruit and nut trees are grafted. Trees are bred for high-quality fruits or nuts. Then, they are grafted onto root systems selected for traits such as winter hardiness, dwarfing, and disease resistance. The top part of the graft is called the **scion.** The bottom portion, that forms the root system, is the **rootstock.** To take grafting a step farther, though, it is possible to graft several different but related plants onto the same rootstock. You could, for example, have grapefruit, orange, and lemon grafted onto the same *Citrus* rootstock. Almond, plum, and apricot can be grafted onto a peach rootstock. Many nurseries sell grafted apple trees containing two or more varieties. Brightly colored novelty cactus plants, which are not capable of photosynthesis, survive because they are grafted onto normal plants (Fig. 14.22).

To describe how grafting is performed, assume we have bred a new extra-juicy variety of apple and would like to graft it onto a young native tree with a hardy root system. In late spring, we would collect some young branches from the tree with the juicy apples (the scions) and keep them in a cooler to inhibit the buds from growing. Then, later in the spring when the rootstock begins to grow, we would bring the scion branches out of the cooler. Using sterilized instruments, we would cut the top of the rootstock diagonally several inches above the soil line. At the same time we would make a diagonal cut in a scion branch that is similar in diameter to that of the rootstock. Successful grafting depends on good contact between the **vascular cambium** of the scion and that of the rootstock. Therefore, we align the scion and rootstock so their cambia are in contact and hold them together with tape or string (Fig. 14.23). The graft union is often also sealed with wax to prevent the wounded tissues from drying out. If the graft is successful, several weeks later, the vascular cambia of the scion and rootstock will have grown together. The tree now consists of a single plant with one variety of rootstock below and a different fruiting variety above.

Propagation of Specialized Stems and Roots

If you would like to create a bed of tulips in your garden, you would not plant seeds or rooted cuttings. Instead, you would rely on **bulbs,** which are natural propagules produced by tulip plants. Herbaceous perennials produce shoots that die during the winter, but in the spring, they regrow from underground storage structures. The underground structures that survive over winter may be *bulbs, corms, tuberous roots,* or *rhizomes* (discussed in Chapter 6) (Fig. 14.24). These structures are typically fleshy, with abundant food-storage tissue, and contain buds that will sprout in the spring.

Micropropagation

One of the major disadvantages of most asexual propagation techniques is that they also propagate pathogens. Once a plant is diseased, there is no immune system to activate or

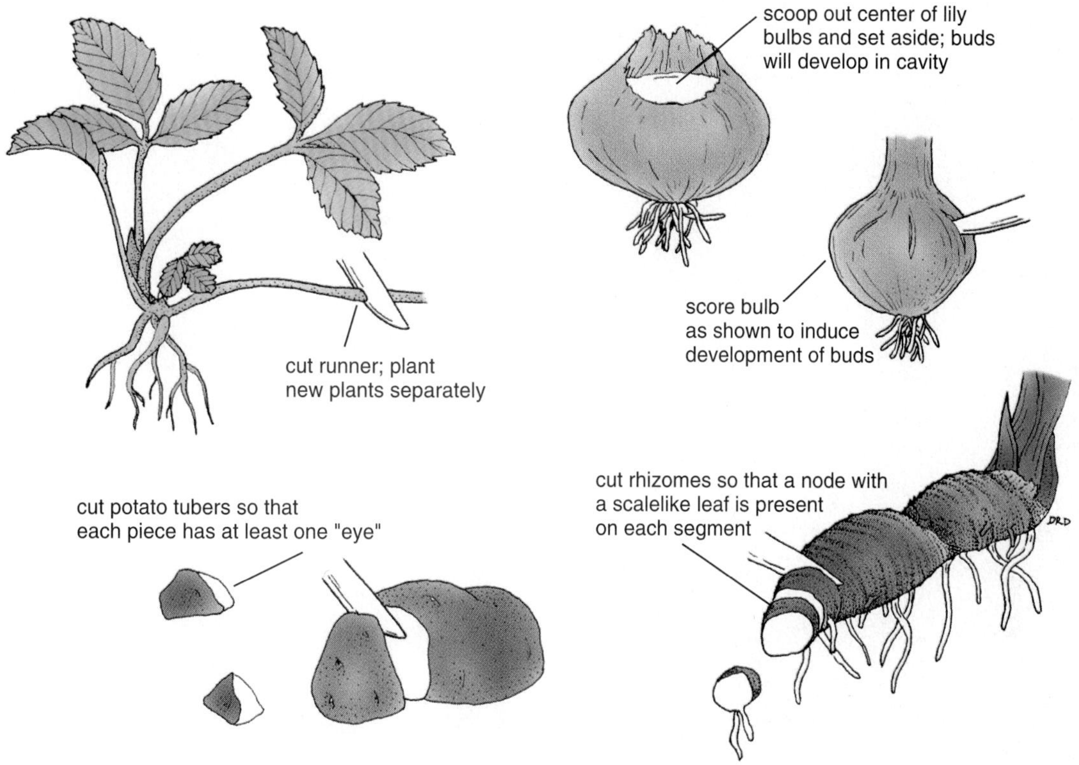

Figure 14.24A Propagation of specialized stems.

Figure 14.24B Some specific examples of specialized stems that are easily propagated asexually. *Left*: A tulip bulb. *Center*: *Crocus* corms. *Right*: A ginger rhizome.

antibiotic we can administer as a cure. The best way to eliminate the effects of plant disease is to prevent exposure to pathogens. This is nearly impossible in greenhouse and field environments. However, it is possible to maintain plants in a disease-free status if we grow them in sterile test tubes through **micropropagation.** Other advantages of micropropagation include the capacity to grow large numbers of plants in a small area, minimal maintenance required for established plants (they do not need to be watered), and rapid multiplication.

Propagating plants through micropropagation is similar to growing them as cuttings. The major difference is that the plants are grown ***in vitro*** in a sterile **medium** and maintained in special controlled environment rooms. The medium includes a support matrix composed of agar, a gelatinous material extracted from red algae. Inorganic salts are added to the medium to provide macro- and micronutrients, such as nitrogen, phosphorous, calcium, and iron. Sucrose is added to supplement the sugars produced by the plant. In addition, vitamins such as thiamine, nicotinic acid, and inositol are generally included in the growth medium. Commonly, growth regulators are also added. After the ingredients are combined and pH is adjusted, the mixture is poured into test tubes. The tubes are then capped and put in an **autoclave** (a large form of pressure cooker) to sterilize them (Fig. 14. 25). When the medium cools, it solidifies like Jell-O™. This acts as the "soil" in the micropropagation system, providing plants with support, nutrients, and water.

Micropropagation, like other forms of asexual propagation, relies on the property of **totipotency** (capacity of a cell to give rise to any structure of a mature organism) of plant cells. Each living cell has the genetic information and, therefore, the capacity to develop into any cell type. Micropropagation usually begins with an excised piece of leaf or stem tissue, or **explant,** and carries it through three steps.

The first step is establishment of explants in **tissue culture.** Micropropagation requires sterile plant material as well as growth media. Plant parts must be **disinfested** to remove surface contaminants without killing the plant tissue.

Figure 14.25 An autoclave, in which pressurized steam is used to sterilize media, glassware, and instruments.

Figure 14.26 A laminar flow hood. A sterile work area is created by blowing sterile air from the back of the hood to the opening in front.

Figure 14.27 Flasks of sterile growing medium to which meristematic tissue has been added. The flasks are slowly rotated under lights. Roots and shoots appear within a few weeks.

Common disinfestants include bleach and ethanol. This procedure does not remove internal contaminants, including pathogens, so it is important to begin with disease-free plants. After plant parts are disinfested, they are inserted into the growth medium in test tubes under sterile conditions. Often, a special reach-in chamber, called a **laminar flow hood,** is used for this step (Fig. 14.26). Filtered air is blown across the work surface in the chamber to prevent the introduction of contaminants. Test tubes containing sterile plants are then placed in a clean room with artificial lighting and temperature control. The goal of this first step is to obtain sterile, viable plant tissue cultures (Fig. 14.27).

After cultures are established, they typically will be induced to develop multiple shoots in a multiplication medium. These **microshoots** can be separated and placed in a new medium by a process called **subculturing** (Fig. 14.28).This step is similar to propagation by cuttings, except it is carried out under sterile conditions. It is not unusual to subculture plants every four weeks, making approximately four new plants from every one in a test tube. At this rate, it is theoretically possible to produce a million plants from one plant in just 10 months. Although these multiplication rates are not realized in commercial systems, many tissue culture laboratories have the capacity to produce millions of plants per year.

The third step in micropropagation is root formation. Some explants, such as those from African violets, will spontaneously produce roots in multiplication medium. In other

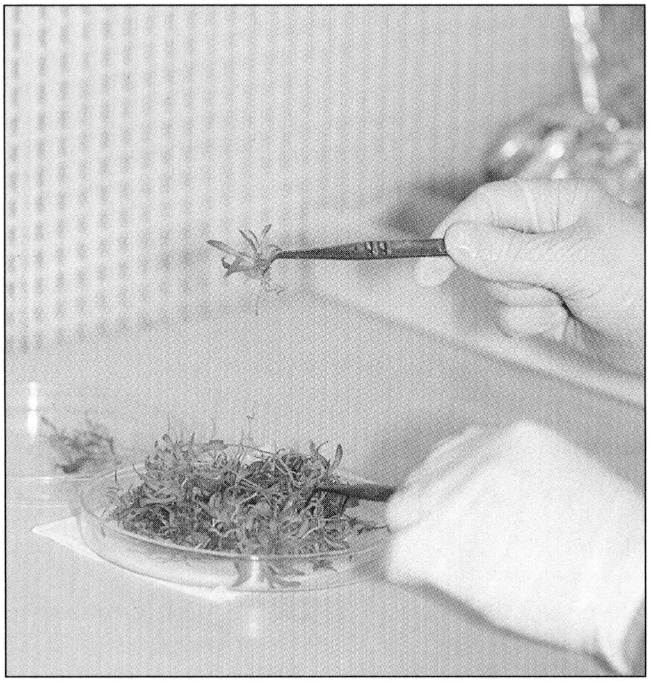

Figure 14.28 Plantlets that have developed from cultured meristematic tissue are separated and further cultured to maturity.

cases, plants are induced to form roots *in vitro* by transferring them to a rooting medium. Compared to the multiplication medium the rooting medium usually contains reduced levels of cytokinins and increased auxin levels. When possible, the most economical approach is an ***ex vitro*** one, in which microshoots are rooted in potting mix and treated as cuttings.

The last step in micropropagation is the transfer of plants back to an outdoor environment. This is often the most difficult step. Because the humidity is high in test tubes, plants grown *in vitro* do not produce as much wax on their cuticles as do those grown outdoors. In addition, their stomata do not close as readily in response to water stress. Therefore, tissue culture plantlets must be acclimatized to an outdoor environment. First, the humidity in the growth chamber is reduced for a few weeks before the plants are brought out of their test tubes. Then, when they are transferred to soil in pots, they are maintained in a high-humidity environment for several weeks.

Commercial micropropagation has become a successful venture with a number of plants. Some plants that propagate slowly by other asexual means may be rapidly increased with micropropagation. These include orchids, Boston fern, African violet, and *Hosta.* In some cases, with the use of tissue culture techniques, as many plants can be produced in a month as would be produced in a year with other techniques. New cultivars can also be rapidly multiplied to meet high market demand. For example, new apple rootstocks are often propagated *in vitro* because conventional asexual reproduction methods cannot adequately supply market needs. Tissue culture protocols, using various combinations of nutrients and growth regulators, have been developed for a number of woody plants that are otherwise difficult to root.

Tissue culture is now being used to propagate endangered plant species. Small pieces of just a few plants can be used to establish tissue cultures. Then, nearly unlimited numbers of plantlets can be produced and returned to their natural habitat.

Summary

1. Most of our food is derived from just a handful of plant species.
2. We domesticate plants by genetically altering them to meet our needs.
3. The impetus for the shift from hunter-gatherer to agricultural societies is not clear. Current hypotheses are speculative.
4. Plant domestication began in the Near East approximately 10,000 years ago and spread to Asia and then Africa and the New World.
5. During domestication, plants were selected for nonshattering seeds, high yield, seedling vigor, and absence of seed dormancy.
6. Two methods for improvement of self-pollinated crops are pure-line selection and selection within self-pollinated offspring of hybrid plants.
7. Two methods for improvement of cross-pollinated crops include mass selection and creation of hybrids from inbred lines.
8. Germplasm banks are critical repositories of genetic diversity essential for plant-breeding progress.
9. Protoplast fusion combines entire genomes of related plant species that cannot mate with each other.
10. Transgenic plants contain DNA from foreign organisms. To make a transgenic plant, a gene is spliced out of the donor and inserted into a vector. The vector is inserted into a host, such as a bacterium, which clones the foreign donor DNA. The next steps, transformation of the plant with the cloned foreign gene and expression of that gene, are the most difficult ones to perform.
11. Seed propagation requires production of high-quality seed, an adequate seed storage environment, and appropriate conditions for seedling growth.
12. Cuttings provide a method of asexual propagation for many plant species.
13. Grafting unites pieces of two different, but related plants, and can be used for cloning.
14. Another form of asexual reproduction utilizes natural propagules, such as bulbs, corms, and tubers.
15. Micropropagation is an asexual propagation technique carried out under sterile conditions.

Review Questions

1. Describe the traits that early humans selected during domestication of plants.
2 Explain why knowledge of a plant's reproductive system is important when choosing a breeding strategy.
3. Outline the steps you would use to create a petunia plant carrying a corn gene for red flower color.
4. Differentiate between a rootstock and a scion.
5. Explain how you would create an apple tree that produces both Granny Smith and Red Delicious apples.
6. Describe the four steps that are usually carried out in micropropagation.

Discussion Questions

1. Suppose you grow a hybrid tomato plant and save some of its seeds to plant in your garden next year. Would those seeds grow into plants that look like the hybrid plant from which you collected them? What if you collected seeds from an inbred line of tomato and planted them?
2. Describe advantages and disadvantages of micropropagation compared to making cuttings in a greenhouse.
3. What might be some advantages of having several varieties of apple or plum on the same tree?
4. Describe advantages and disadvantages of producing a crop of potato plants using asexual (tuber) rather than sexual (true seed) propagation.
5. Many people are opposed to the production of transgenic crop plants. Do you feel that position is justified?

Learning Online

Visit our web page at www.mhhe.com/botany for interesting case studies, practice quizzes, current articles, and animations within the Online Learning Center to help you understand the material in this chapter. You'll also find active links to these topics.

Origins of Agricultural
Asexual Reproduction in Plants
Recombinant Technology
Cloning
Plant Tissue Culture and Transgenic Crops

Additional Reading

Chrispeels, M. J., and D. E. Sadava. 1994. *Plants, genes, and agriculture.* Boston, MA: Jones & Bartlett.

Hancock, J. F. 1992. *Plant evolution and the origin of crop species.* Englewood Cliffs, NJ: Prentice-Hall.

Harlan, J. R. 1992. *Crops and man,* 2d ed. Madison, WI: American Society of Agronomy, Crop Science Society of America.

Hartmann, H. T., et al. 1997. *Plant propagation: Principles and practices.* Englewood Cliffs, NJ: Prentice-Hall.

Smith, B. D. 1995. *The emergence of agriculture.* New York: Scientific American Library.

Kahili ginger (*Hedychium gardnerianum*) and Hawaiian tree ferns (*Cibotium* sp.).

Evolution

15

Chapter Outline

Overview

An introduction to the early development of evolutionary concepts is followed by a discussion of Charles Darwin and his contributions to the theory of evolution through natural selection. Evidence for evolution is given, and mechanisms of organic evolution, including mutations, migration, and genetic drift, are discussed. The roles of reproductive isolation and hybridization in the evolution of species are explored. The chapter concludes with a discussion of some aspects of past and present controversy surrounding evolution.

Some Learning Goals

1. Be able to summarize the early development of evolutionary concepts.
2. Know the contributions of Charles Darwin to the theory of organic evolution and the principles of natural selection as he understood them.
3. Know the various lines of evidence for evolution.
4. Explain the significance of natural selection, mutation, migration, and genetic drift to evolution.
5. Outline how reproductive isolation and hybridization contribute to the evolution of species.
6. Give reasons, past and present, for the controversy over evolutionary theory.

Few historical events since 1859 have had a greater impact on society in general and the biological sciences in particular than the publication of Charles Darwin's book *On the Origin of Species by Means of Natural Selection, or the Preservation of Favoured Races in the Struggle for Life.* Darwin's theory of evolution through natural selection (the tendency of organisms with favorable adaptations to their environment to survive and produce new generations) has stimulated an enormous amount of thinking and research. It has also provided an explanation based on natural laws for the diversity of life around us.

There have been disagreements, particularly in the past, over the nature, mechanisms, and even the existence of evolution. These stemmed, in part, from a failure of people in diverse fields to define terms and to distinguish between fact and theory in matters of evolution. Evolution itself, for example, has been broadly defined by some as simply being synonymous with change. We are told that anything from cars to computers to cultures is evolving and that even our thought processes evolve. We need, therefore, to distinguish between change in inanimate objects or intangibles and *organic evolution,* which is the accumulation of genetic changes in populations of living organisms through many generations.

A BRIEF OVERVIEW OF THE EARLY DEVELOPMENT OF EVOLUTIONARY CONCEPTS

More than 2,300 years ago, Aristotle (384–322 B.C.), although not recognizing any processes of natural selection, did observe grand design in nature and arranged all organisms known to him from the simplest to the most complex in what he called the *Scala Naturae* (Scale of Nature). Aristotle's arrangement implied that all organisms were static and didn't evolve. These beliefs were widespread until at least the 17th century and were not extensively challenged before Darwin. *Fossils* (parts of previously existing organisms preserved in rocks or other substances) had been found many centuries before Darwin but were not properly identified until the 15th century when Leonardo da Vinci (1452–1519) correctly observed that they were parts of previously existing organisms that had become extinct.

Count de Buffon (1707–1788), a French naturalist, spent much of his adult life writing a natural history of 44 volumes in which he described all known plants and animals. In this work, Buffon (whose real name was Georges-Louis Leclerc) presented evidence of descent with modification in organisms and speculated on the mechanisms involved. However, he provided no theories on how evolution might take place.

Georges Cuvier (1769–1832), a French zoologist and *paleontologist* (one who studies fossils; fossils and fossilization are discussed in Chapter 21), used comparative anatomy toward the close of the 18th century to classify animals. Cuvier, however, firmly believed that organisms did not change over time. When geological finds showed apparent evolution of organisms in rock formations of certain regions, Cuvier tried to explain them away with what was called *catastrophism.* He theorized that mass extinctions or catastrophes had occurred whenever a new geological find revealed a different group of fossils, and the presence of the new fossils was due to repopulation of the region by species migrating in from surrounding areas.

By the end of the 18th century (before the principles of genetics were known), many prominent biologists had come to believe that hereditary changes in populations over long periods of time (evolution) occurred as a result of the inheritance

Figure 15.1 Lamarck and his contemporaries believed that the long neck of a giraffe developed over time as the animals stretched to reach higher leaves and that the little length increases were inherited and became cumulative. This theory of inheritance of acquired characteristics was experimentally disproved and discarded when the true mechanism of gene inheritance became known.

of acquired characteristics. One of the more prominent supporters of this widespread idea was Jean Baptiste Lamarck (1744–1829). He believed, for example, that giraffes acquired their long necks over many generations as a result of the gradual increase in neck length as shorter-necked animals stretched to reach leaves on the branches of trees. Slight stretching of the neck was supposed to have been passed on to the offspring as it occurred, and eventually, numerous tiny increases due to individual stretching added up to the present great neck length of giraffes (Fig. 15.1).

If acquired characteristics could be inherited, we should be able to demonstrate it experimentally, and indeed, many workers have attempted to do so, but all have failed. For example, one biologist surgically removed the tails of mice for many successive generations, but the average length of the tails of the last generation was exactly the same as that of the first generation. The experiment demonstrated that repeatedly removing tails in no way affects the hereditary characteristics carried in the genes within the cells. This is also the reason fruit trees that are pruned annually never produce seeds that grow into dwarfed trees, even after many generations.

Figure 15.2 Charles Darwin. *(Courtesy National Library of Medicine)*

CHARLES DARWIN

Although there are variations in some aspects of current organic evolutionary theory, that which pervades and unifies most biological thought today received its greatest impetus from the observations of Charles Darwin (1809–1882) (Fig. 15.2) and his contemporary, Alfred Wallace, who independently arrived at the same conclusions as Darwin.

Charles Darwin was born in Shrewsbury, England, in 1809. As a boy, he showed a keen interest in natural history, but in deference to his physician father's wishes, he enrolled in the University of Edinburgh medical school to study medicine; he later studied theology at Cambridge University, without, however, excelling in either subject. In 1831, when he was 22 years old, he graduated with a degree in theology but was not really sure what he wanted to do with his life.

Shortly after Darwin's graduation, King William IV of England commissioned a sailing vessel, the HMS *Beagle,* to undertake a voyage around the world to chart coastlines, particularly those of South America. Young Darwin, after a recommendation from a Cambridge biology professor, accepted an unpaid position as assistant naturalist and captain's companion on the voyage, which began December 27, 1831.

During the voyage, which lasted 5 years, Darwin read a geology book that made a profound impression on him. The book, authored by Charles Lyell, was based on a theory of John Hutton, who believed the earth was much older than the few thousand years it previously had been thought to be. Hutton suggested that the earth had been subjected to cycles of upheaval and erosion over great periods of time. The erosion resulted in rock fragments and weathered debris being washed down rivers into the oceans where the deposits gradually accumulated in thick layers. The thick layers, which slowly were converted into sedimentary rocks (many of which contained fossils), became land as they were elevated above sea level. Hutton believed that, given enough time, all geological changes could be attributed to slow, natural processes, and Lyell suggested the slow changes occurred at a uniform rate. Although some modern geologists incorporate many of Hutton's ideas about gradual change over long periods of time, they point to evidence that geological changes have not in the past always proceeded at a uniform rate. Darwin himself was not convinced geological changes occurred at a uniform rate, but he did agree with Lyell that the earth must be very old indeed and that large changes were due to gradual accumulation of small changes.

Today, we can estimate the age of the earth by determining in rocks the ratio of a radioactive element to the nonradioactive element it becomes. We know that a radioactive element gradually loses its radioactivity, although sometimes at a variable rate. The amount of time it takes for a radioactive element to lose half of its radioactivity is referred to as its **half-life.** The half-life for each radioactive element is different. For example, uranium, which becomes lead as it loses its radioactivity, has a half-life of 4.5 billion years, while radioactive carbon has a half-life of 5,730 years, and radioactive nitrogen has a half-life of only 10 minutes. By checking ratios of uranium to the lead it becomes when it loses its radioactivity, we now know that the earth is probably at least 5 billion years old.

Darwin had many opportunities to collect plants and animals on both sides of South America, as well as in the Galápagos Islands and along the coasts of Australia and New Zealand. He kept a daily journal and spent countless hours alone on horseback collecting and observing the living world around him. This gave him ample opportunity to think about the forms and distribution of the myriad new organisms he encountered. His thoughts slowly led to the development of ideas that later blossomed into his theory of evolution through natural selection.

Upon his return to England in 1836, Darwin retired to the country and began working on his collections and journal. He also carried on extensive correspondence with other biologists and made many detailed investigations into pollinating mechanisms, earthworm ecology, geographical distributions of plants and animals, and several other areas of natural history.

Throughout all of his activities, Darwin was guided by a concept he had adapted from an essay written by Malthus in 1798 on human populations and food supplies. Malthus observed that populations grow geometrically until food supplies (which, if they increase at all, do not increase geometrically), disease, wars, and other factors limit their growth. Darwin realized that although humans might artificially improve or increase their food supply through selective breeding and cultivation, wild plants and animals could not do so and were therefore vulnerable to a process of selection in nature, which would explain changes in natural populations. He was reluctant to publish his ideas, however, and did not begin putting together his book on the origin of species[1] until 1856.

Meanwhile, an English naturalist by the name of Alfred R. Wallace (1823–1913), who made major contributions to our knowledge of animal geography, independently concluded that natural selection contributed to the origin of new species and sent Darwin a brief essay on the topic in 1858. At the urging of Charles Lyell, Darwin and Wallace jointly presented a paper of their views to the Linnaean Society of London in 1858, and in 1859, Darwin's classic book *On the Origin of Species by Means of Natural Selection* was published.

EVIDENCE FOR EVOLUTION

Evidence in support of organic evolution is drawn from many areas, including similarities in the form and ecology of living organisms and the way they are related to each other today. Homologies (characteristics shared by different organisms) point to common ancestry. For example, members of the mustard family (Brassicaceae—2,500 species), which includes broccoli, cabbage, radish, and stocks, produce a pungent, watery juice that has the smell of sulfur when it breaks down. However, similarities are not always due to common ancestry. For example, a number of African spurges (species of *Euphorbia*) and American cacti have similar, succulent stems even though they are not closely related (Fig. 15.3). The plants do share similar arid habitats to which they have independently become adapted (an *adaptation* is a characteristic that makes an organism better suited to its environment). In other words, plants of very different ancestry have adapted in similar ways to common environmental conditions in different parts of the world. This type of evolution is called *convergent evolution.*

Other evidence comes from the structure and relationships of proteins, DNA, and other molecules, and the common use of ATP by living organisms. Cytochrome *c* oxidase, for example, is an enzyme that occurs universally in all living organisms, which suggests it appeared very early, probably in a single organism, and that it has been successively inherited by the myriad of organisms in existence today.

Fossils, which are remnants of previously living forms (Fig. 15.4), provide compelling evidence for descent with modification. The simplest fossils are generally found in the oldest geological strata, while more complex forms tend to be found in younger strata.

1. Our understanding of the term *species* has been modified and refined since Darwin's time, but, as discussed in Chapter 16, most biologists today think of a species in general terms of a *population of individuals of similar form and structure with a common ancestry, capable of interbreeding freely with one another in nature but not generally interbreeding with individuals of other dissimilar populations.* A shorter definition is simply to refer to a species as *a group of organisms with a common gene pool.* Note that *species* is like the word *sheep* in that it is spelled and pronounced the same in either singular or plural usage. There is no such thing as a plant or animal "specie."

Figure 15.3 Convergent evolution. Plants of different ancestry that adapt to similar habitats may evolve similar life forms. The barrel cactus on the left (*Jasminocereus howellii*) and the barrel spurge on the right (*Euphorbia obesa*) are completely unrelated but have evolved similar forms in adaptation to arid habitats.

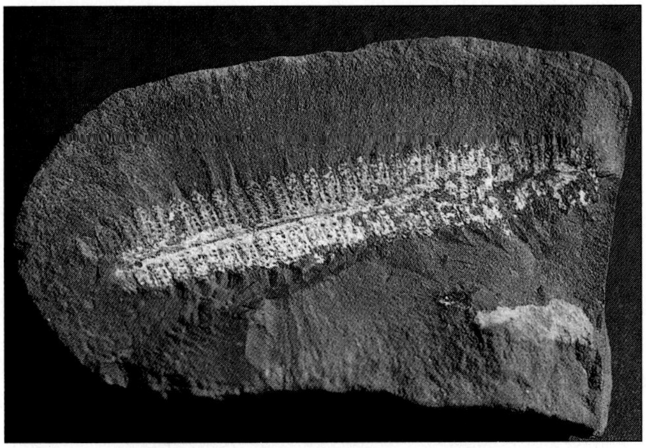

Figure 15.4 A fossil fern.

Still further evidence is drawn from the geographical distribution of organisms. Many groups are confined to a single continent or island. In some instances where similar organisms occur on more than one land mass, there is evidence that the land masses concerned were once linked, which would have permitted terrestrial migration. Other conclusions are drawn from the physiology and chemistry of the organisms.

MICROEVOLUTION—EVOLUTION WITHIN SPECIES

Natural Selection

Darwin observed that animal breeders, through a process of artificial selection, changed populations of domestic animals by retaining those with desirable traits and by destroying those with unwanted traits. Darwin also noted that in nature, some organisms do not go on to reproduce because all natural populations have a limited supply of food and other resources and that the traits of organisms vary within populations. He reasoned that those best adapted to available resources would increase in number in succeeding generations, and those less well suited would decrease. Darwin also used comparative anatomy and embryology in developing his theory of descent with modification, which he called **natural selection,** and based it on four principles:

1. *Overproduction.* Many living organisms produce enormous quantities of reproductive cells or offspring. For example, a single maple tree produces thousands of seeds each year, most being capable of becoming a new tree. A single mushroom can produce millions of spores, each with the potential to become a new mushroom-bearing fungus.
2. *Struggle for Existence.* All the germinating seeds, spores, and other reproductive cells of living organisms compete for available moisture, light, nutrients, and space. In nature, the amounts of these elements available are insufficient to support all of these organisms, and many die as a result.
3. *Inheritance and Accumulation of Favorable Variations.* All living organisms vary. Hereditary variations that have survival value or do not result in the death of the individual are inherited from generation to generation and accumulate with time, while other variations harmful to the survival and reproduction of the individual are gradually eliminated.
4. *Survival and Reproduction of the Fittest (Differential Survival and Reproduction).* Those forms of organisms best adapted to the environment ("fittest") have the best chance to survive and reproduce, while others less well adapted may die. A tree with thicker bark, for example, may have a higher probability of surviving cold temperatures to reproductive age and producing more offspring. The offspring of the thicker-barked tree may then bear its inherited features.

One of the criticisms of Darwin's theory, after it was published in 1859, was that it did not explain how hereditary variations originated and developed. It should be remembered, however, that Mendel's findings were not published until 1866, and the details of mitosis and meiosis did not become known until 1900 to 1906. Today, with our far greater knowledge of how variations occur and are inherited, we have come to understand the mechanisms of evolution in populations much better than was possible in Darwin's time.

Mutations

All wild organisms, from aardvarks to zinnias, occur in populations that are composed of a few to billions of individuals. Yet even within very large populations, it is not possible to find two individuals that are identical down to the last molecule. As humans, we are well aware of this within our own species, but variation exists in all living organisms, even though in simpler one-celled forms the differences may be much less obvious. Whether the differences are obvious or subtle, the general characteristics of a population will eventually change if the environment or other factors favor certain hereditary variations and disfavor others.

In some instances, the environment may alter the phenotype without affecting the genetic constitution of an organism. Plants that grow relatively tall at sea level may become dwarfed when they are transplanted to cooler areas in the mountains or drier areas near deserts, yet they are capable of breeding freely with plants at the original location if they are returned to that area.

If we apply fertilizers to plants, stimulate their growth with hormones, or prune them, the changes are not passed on to the offspring because no permanent change occurs in a population unless there is *inheritable variation.* The changes in transplanted, fertilized, or pruned plants are not transmitted to the offspring because the gametes of those plants will carry the same genetic information they would have carried if the transplanting, fertilizing, or pruning had not occurred. Despite this, however, dwarf fruit trees and short-tailed mice do occasionally occur, but for reasons quite different from those proposed by Lamarck and his contemporaries. They come about as a result of a sudden change in a gene or chromosome. Such a change is called a **mutation,** a term introduced in 1901 by the Dutch botanist Hugo de Vries.

Changes within chromosomes may occur in several ways. A part of a chromosome may break off and be lost (*deletion*), or a piece of a chromosome may become attached to another (*translocation*). In some instances, a part of a chromosome may break off and then become reattached in an inverted position (*inversion*) (Fig. 15.5). A mutation of a gene itself may involve a change in one or more nucleotide pairs (nucleotides are discussed in Chapter 13).

Mutation rates vary considerably from gene to gene, but mutations occur constantly in all living organisms at an average estimated to be roughly one mutant gene for every 200,000 produced. *Mutator genes,* if present, can increase the rate of mutation in other genes, but generally, the mutation rate for a specific gene remains relatively constant unless there are changes in the environment (e.g., an increase in cosmic radiation). Most mutations are harmful, many times to the point of killing the cell. However, about 1% of the mutations are either silent (have no effect on the survival of the phenotype) or produce a characteristic that may help the organism survive changes in its environment.

Migration

Gene flow between populations occurs when individuals or gametes migrate from one population to another. How much gene flow takes place depends on the size of the populations and the extent to which they may be isolated from one another. If there is a great deal of interbreeding over wide distances, as there might be in wind-pollinated plants, a single individual's genes can quickly be spread from one population to another. On the other hand, gene flow occurs at a much slower rate when interbreeding is more or less restricted to small, isolated populations. This is often the case with plants that occur only in specialized habitats, such as the vicinity of springs and seeps or on magnesium-rich serpentine soils. In general, a small amount of gene flow over longer distances can be expected to occur, even though most of the interbreeding takes place between closely associated individuals in a population.

Genetic Drift

As observed in Chapter 13, the Hardy-Weinberg law states that genes tend to remain at a constant frequency from generation to generation. However, by chance alone, genetic drift (a change in the genetic makeup of a population due to random events) may take place as a given gene fluctuates from its statistical average in any generation due to the

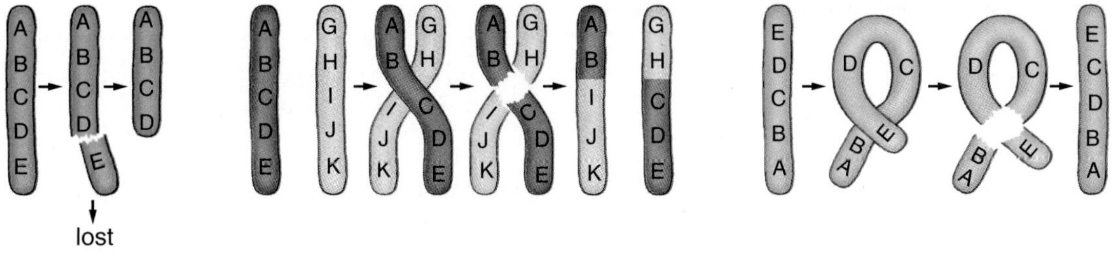

Figure 15.5 Some types of chromosomal changes that can occur. *A.* **Deletion.** Part of a chromosome breaks off and is lost. *B.* **Translocation.** Part of one chromosome becomes attached to another. *C.* **Inversion.** Parts of a chromosome breaks off, becomes inverted from its original position, and then is reattached.

events that occur during meiosis and the production of gametes. In a large population, genetic drift is unlikely to make any significant difference. In a small population, however, the chance multiplications or failures to multiply of a single genotype may cause a marked change in its frequency and, in some cases, cause it to disappear altogether or, on the other hand, have a frequency of 100%. In other words, in a small population, the random shift in gene frequencies by chance may bring about evolutionary changes.

RATES OF EVOLUTION

Darwin believed that evolution by natural selection was a slow and gradual process. A number of contemporary biologists, however, favor the *punctuated equilibrium* model, which holds that major changes have taken place in spurts of maybe 100,000 years or less, followed by periods of millions of years during which changes have been minor. They base their hypotheses on fossils, which when arrayed according to their ages reveal large gaps in the record. Although missing-link fossils are occasionally discovered, the record does little to support Darwin's concept of gradual, long-term change, even though it is believed that possibly as few as one organism in a million may have become a fossil. Others opposed to hypotheses of evolution through sudden change argue that because probably only a tiny percentage of organisms became fossilized and usually only the harder parts of organisms (e.g., bones, teeth, wood) are preserved, drawing definite conclusions from fossil evidence about evolution through either gradual or sudden change may be speculative.

The conditions necessary for an organism to become a fossil are very specialized and limited in occurrence (see Chapter 21) and probably also were in the past. This makes it quite improbable that large numbers of missing-link fossils will ever be found. Proponents of evolution through periods of rapid change argue that under conditions of changing climates or other situations exerting strong selection pressures on forms with adaptive mutations, new species of organisms could arise in less than 100 generations, making 100,000 years ample time for considerable evolution to occur. Since it is not possible to prove or disprove the various theories experimentally, the debate on whether evolution has occurred through gradual or sudden change will undoubtedly continue indefinitely until, or unless, new evidence convincingly supports one theory more than another.

MACROEVOLUTION—HOW SPECIES EVOLVE

Reproductive Isolation

If new genes are produced in a freely interbreeding population, they may gradually be spread throughout the population, and the nature of the whole population will change in time. If some barrier divides the population, however, the two new populations eventually may become distinct from each other, sometimes in a relatively short period of time.

The log of a Portuguese sailing vessel of more than 600 years ago indicates that, for unknown reasons, some rabbits native to Portugal were released on one of the Canary Islands during a visit. When 20th-century biologists examined the island descendants of those rabbits, which, in Portugal, forage during the day, they found them to be smaller than their continental ancestors, to be nocturnal in foraging habits, and to have larger eyes. In addition, attempts at breeding them back to their European ancestors failed because, in the short space of 600 years, a new species of rabbit had evolved. Although there is evidence that new plant species have evolved in as little as 50 years, it should be emphasized that the 600 years involved in the evolution of the Canary Island rabbits appears to be much less time than is typical for the evolution of many other plant or animal species.

Geographic Isolation

How do two populations of organisms that initially have the same gene pool come to have gene pools different enough to prevent their interbreeding? Geographic or other isolation of the two populations from each other prevents the flow of genes between the two populations, and random mutations, which are rarely identical, then spread only throughout the population in which they arise. Imagine, for example, that a population of white daisies occurs throughout a wide valley that is bounded on both sides by high mountains. If a mutation occurs that results in the white part of a daisy becoming red, the new color may, in time, if other environmental conditions permit, spread throughout the entire population. If, however, before the mutation occurs, there is a volcanic eruption that blocks off part of the valley, we would then have two populations of white daisies isolated from one another. If the same mutation for red color should then occur, it would spread only through one population. Meanwhile, if a mutation for hairy leaves occurs in the other population, that characteristic could spread throughout the second population but be prevented by the geographic barrier from spreading to the population of red daisies. In time, the genetic changes may become so great that even if the isolation is removed, gene flow between the two populations no longer can occur, and two distinct species of daisies would be the result.

In the United States, there are two closely related species of small trees or shrubs called redbuds (*Cercis*) that look very much alike. The eastern redbud (*Cercis canadensis*) occurs on the borders of streams, mostly east of the Mississippi River between the Canadian border and Florida, where some form of precipitation occurs throughout the year. The western redbud (*Cercis occidentalis*) is native to stream areas in California, Utah, Nevada, and Arizona, where most of the precipitation occurs in the winter and spring. The two species can be artificially hybridized, but

A.

B.

Figure 15.6 Two species of redbud, both native to North America. *A.* Eastern redbud (*Cercis canadensis*). *B.* Western redbud (*Cercis occidentalis*).

each is so adapted to its own wild habitat and associated climate that specimens of either species die when transplanted to the other's wild habitat (Fig. 15.6). Presumably, a single species of redbud once occupied riparian sites throughout much of temperate North America. When the eastern populations became geographically isolated from those of the west, however, random mutations arose independently in the east and the west, and with free gene flow between the two populations no longer possible, two species now exist where formerly there was only one.

Several other factors contribute to the development of new species from geographically isolated populations with a common ancestry. When separation first occurs, it is most unlikely that both populations will have genes that are identical in all respects, and a small population will have only a small percentage of the genetic variation present throughout the original population. In addition, geographically isolated populations normally will be subjected to selection pressures from numerous subtle to conspicuous differences in environment. The Canary Island rabbits, for example, initially probably found it difficult to compete with other animals for food during the day, and although we don't know what changes at the molecular level actually took place, it may be that mutations or recombinations for improved night vision occurred, with those acquiring new alleles or combinations being able to survive, while those without them perished.

Ecological Isolation

Isolation leading to the development of new species is not limited to physical barriers such as mountains or oceans. Ecological factors such as climate or soils may play a role, as do time and mechanical isolating factors. As a result, related species can be *sympatric* (occupy overlapping ranges of territory) without genes being exchanged. Serpentine soils, for example, have a magnesium content not tolerated by many species of flowering plants. Plants that are unable to compete with species excluded from such soils, but can, however, tolerate serpentine soils, find a niche in which they can thrive and evolve independently.

A mutant form within a population may flower at a different time, preventing exchange of genes between it and nonmutant forms. In the temperate deciduous woods of eastern North America and in the Columbia River basin in the Pacific Northwest, there are many populations of early spring-flowering herbs called *Dutchman's breeches* (*Dicentra cucullaria*), which are discussed further in Chapter 16. Dutchman's breeches and *squirrel corn* (*Dicentra canadensis*), a close relative, often grow together in the eastern deciduous forest. Both species have highly dissected leaves that are so similar in appearance many early botanists and lay persons assumed Dutchman's breeches and squirrel corn were one and the same species (Fig. 15.7).

It is believed, however, that at some point in geological history, a mutation or mutations occurred that caused some plants to begin flowering after other plants had already set seed. As a result, one group became reproductively isolated from the other group. In due course, other mutations affecting the form and fragrance of the flowers and the shape and pigmentation of food-storage bulblets beneath the surface also occurred, but the plants continued to occupy the same habitats. In short, we now have two closely related but distinct species growing together without interbreeding, simply because it is no longer possible for them to do so.

Mechanical Isolation

Other isolating mechanisms may be mechanical. In orchids, for example, the pollen is usually produced in little sacs called *pollinia* (Fig. 15.8) that stick to the heads or bodies of visiting insects. If pollination is to occur, the pollinia must be inserted within a concave stigma. Each species of orchid

A.

B.

Figure 15.7 *A.* Dutchman's breeches plant (*Dicentra cucullaria*). *B.* Squirrel corn (*Dicentra canadensis*).

is constructed so that it is highly unlikely that a pollinium of one species will be inserted within the stigma of another species. As a result, many species of related orchids can be sympatric without genes being exchanged.

Four closely related sympatric species of Peruvian *Catasetum* orchids, which can be artificially hybridized very easily, have no known natural hybrids, despite their being pollinated by a single species of bee. Microscopic examination of the pollinators has shown that the pollinium of one species is attached to the insect's head, that of another is attached to the insect's back, that of a third to the abdomen, and that of the fourth only to the left front leg. Even after visits to hundreds of flowers, none of the pollinia are misplaced! (Fig. 15.9).

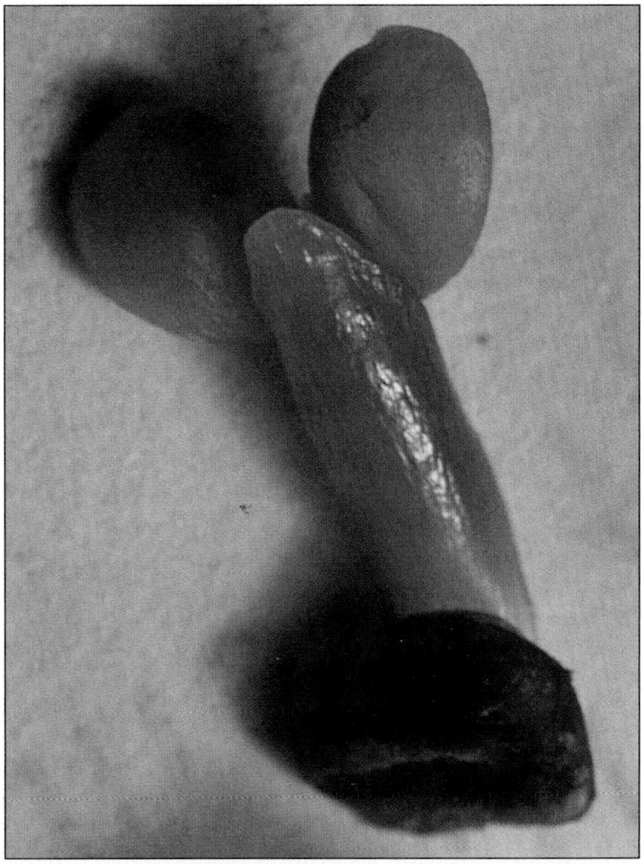

Figure 15.8 A pair of *pollinia* (sacs of pollen) produced by members of the orchid family (*Orchidaceae*). The pollinia become attached to the bodies of visiting insects, which transport them to other orchid flowers.

Other Isolating Mechanisms

Even if pollen from one species is placed on or within the stigma of another species, however, fertilization frequently does not follow because the sperm is chemically (by incompatibility) or mechanically prevented from reaching the egg. Other isolating mechanisms include the failure of hybrids to survive or breed and failure of embryos to develop.

THE ROLE OF HYBRIDIZATION IN EVOLUTION

When *hybridization* (the production of offspring from different populations by parents that differ in one or more characteristics) takes place, the hybrids may be significant or important in evolutionary change, depending on how the characteristics of the parents were combined. If, for example, the environment changes (e.g., average temperatures drop or annual precipitation increases), hybrids may have gene combinations

Figure 15.9 *Catasetum* orchids.

Figure 15.10 Fireweed (*Epilobium angustifolium*)—a polyploid found primarily in mountainous areas of North America below 3,050 meters (10,000 feet). One subspecies has four sets of chromosomes, with a diploid chromosome number of $2n = 36$. A second subspecies has 8 sets of chromosomes, with a diploid chromosome of $2n = 72$.

that are better or worse suited to the new environment than those of either parent. Two related species may hybridize occasionally, and when they do, *introgression* (backcrossing between the hybrids and the parents) may occur. If the backcrossing occurs repeatedly, some characteristics of the parents may eventually disappear from the population if the new combinations of genes in the offspring happen to be better suited to the environment than those of the parents and as natural selection favors the offspring. Both parents and hybrids may, however, also evolve in other ways.

Polyploidy occurs occasionally in nature when, during mitosis or meiosis, a new cell wall fails to develop between two daughter cells, even though the chromosomes have divided. During our discussion of plant breeding in Chapter 14, we noted that this situation can result in a cell with twice the original number of chromosomes. If mitosis were to occur normally after such a cell was formed and the cell divided repeatedly until a complete organism resulted, that organism would have double the original number of chromosomes in all its cells.

The hybrids resulting from a cross between two species are often sterile because the chromosomes do not pair up properly in meiosis. If polyploidy does occur in such a hybrid, however, the extra set of chromosomes present from each parent provides an opportunity for any one chromosome to pair with its homologue in meiosis, possibly overcoming the problem of sterility. This type of polyploidy apparently occurred frequently in the past (in terms of geological time), and it is believed that more than 40% of flowering plants that exist today originated this way (Fig. 15.10).

Apomixis

Sterile hybrids also may reproduce asexually. One form of asexual propagation, called **apomixis,** includes the development and production of seeds without fertilization. Asexual propagation also includes other forms of vegetative reproduction that are discussed in Chapters 4 through 7, Chapter 14, and Appendix 4.

When species reproduce mainly by apomixis but sometimes also hybridize so that new combinations of genes are occasionally produced, they can be highly successful in

Figure 15.11 A common dandelion. In addition to reproducing by ordinary vegetative and sexual means, dandelions reproduce apomictically. Apomixis is a form of sexual reproduction through which seeds are produced without fertilization.

Ecological Review

Because environment is a driver of natural selection, ecology is central to understanding this key evolutionary process. Darwin's theory of natural selection proposed that limiting factors in the environment produce differences in the rate of reproduction of individuals within populations. He reasoned that those individuals best adapted to the environment have the best chance for survival and reproduction and that the characteristics of these better-adapted individuals will increase in frequency within the population. Speciation, the origination of new species, is often influenced by ecological factors, and closely related species, derived from a common ancestor, often have different environmental requirements. In addition, biological diversity, a central concern of ecology, depends ultimately on a balance between rates of speciation and extinction.

nature. Dandelions and wild blackberries, for example, are among the most successful plants known. They reproduce apomictically, as well as sexually through fertilization, and also by natural vegetative propagation (Fig. 15.11).

DISCUSSION

Darwin's theory of evolution through natural selection initially caused a great deal of controversy because even though Darwin believed to his death in a Divine Creator, he also believed that the Creator had used natural laws to bring all living things into being gradually over long periods of time. Most of his contemporaries, however, were guided by a literal interpretation of the biblical account of creation and were evidently convinced that all living things had been created in 6 days and had existed unchanged since the beginning.

If we now know and understand most of the mechanisms of organic evolution, why are there still any disagreements about the broad subject itself? Obviously, lack of objective analysis of the evidence is a factor, but it is not the sole reason. Science deals with tangible facts and evidence that can be measured or experimentally tested; beliefs stemming from metaphysics or religion are outside the realm of science to prove or disprove.

When the popular *Scofield Reference Bible* was first published in 1909, it included in the margin opposite the account of creation "4004 B.C.," a date arrived at by the 17th-century Irish archbishop James Ussher, who based his calculation on faulty interpretation of biblical genealogies. Ussher's date has been deleted from the margins of the Scofield Reference Bible editions published since 1967, and the editors have observed that little evidence exists for fixing dates of biblical events prior to 2100 B.C.

A group composed mostly of non-biologists, and calling themselves *scientific creationists,* have sought since the 1970s to include a non-evolutionary interpretation of the living world in public-school biology textbooks. Scientific creationists do not necessarily believe the earth was created in 4004 B.C., and they recognize the existence of minor variations in living organisms. The majority, however, believe the earth is less than 30,000 years old and reject the foundations of evolution as incompatible with a literal interpretation of the biblical account of creation. In doing so, scientific creationists reject the evidence for the age of the earth provided by radioactive elements, and for evolution, including that which has accumulated since Darwin.

Objective scientists freely acknowledge that some problems concerning the interpretation of the geological past and the pathways of organic evolution exist, but they ask their detractors to suggest more plausible alternatives. At this point, some people apply the tenets of religious faith, which like history, is not subject to scientific experimentation. Others, including those who subscribe to the theory of *Intelligent Design,* see no conflict between science and religion. Intelligent Design proponents accept much of the evidence for organic evolution but don't believe it was possible for living cells to have arisen by chance alone. They observe that some of the complex biochemical systems of living cells, with their numerous interrelated parts, would have had to have arisen at the same time, since if only one part arose, it would not have been functional and would not have been advantageous to the organism possessing it. Their Intelligent Design theory proposes that such complex systems were designed, presumably by a supernatural power.

Others, however, are convinced that science and religion are mutually exclusive. Depending on individual points of view, an impasse may result, with persons of different persuasions—including, unfortunately, some scientists—becoming dogmatic on the topic. Virtually no objective thinkers will deny, however, the extraordinary impact theories of evolution have had on modern peoples and on their concepts of the living world around them. Nearly all scientists who have studied the evidence feel that evolutionary processes are the only plausible explanation for the unity of life at the molecular and cellular level and the extraordinary diversity of the organisms now around us. There is little unanimity of thought, however, as to the precise pathways of evolution in the past. One authority is convinced that a certain group evolved from another, while other equally eminent authorities maintain that the exact reverse occurred. Part of the reason for such paradoxes is that the historical record is quite incomplete. Although the fossil evidence for the organic evolution of a few organisms, such as certain molluscs and the horse, is fairly substantial, other fossil evidence is very fragmentary. As previously mentioned, possibly fewer than one in each million organisms that once existed ever became a fossil, and there are very few fossils of herbaceous and soft-bodied organisms. With such evidence, scientists can deal only in probabilities or possibilities, and it is inevitable that various, sometimes conflicting, interpretations result.

Summary

1. The theory of organic evolution received its greatest impetus from Charles Darwin's *On the Origin of Species*.
2. Before 1900, many biologists, notably Lamarck, believed that organic evolution occurred as a result of the inheritance of acquired characteristics. This theory was discredited experimentally.
3. Evidence in support of organic evolution is drawn from fossils and from the form, ecology, geographical distributions, and relationships of living organisms, homologies, molecular structures, and analogies such as succulent form in deserts.
4. Darwin's natural selection theory is based on four principles: (1) overproduction; (2) struggle for existence; (3) variation and inheritance; and (4) survival and reproduction of the fittest.
5. The primary mechanisms of organic evolution include natural selection, mutations, migration, and genetic drift.
6. Darwin believed that evolution through natural selection was a gradual process over great periods of time. Some contemporary biologists believe that organic evolution has taken place in spurts between long periods of little change, based on evidence from the fossil record. Interpretation of the fossil record, however, can be controversial and will be debated indefinitely.
7. If new genes are produced in a freely interbreeding population, they will gradually be spread throughout the population, and the nature of the whole population will change in time. If a population is divided by a barrier, genes occurring in the one population will not spread throughout the isolated population as before. In time, because of the isolation, each new population may develop into separate species incapable of breeding with one another.
8. Reproductive isolation and hybridization play major roles in the evolution of species, especially in plants.
9. Mechanisms of organic evolution include mutations in chromosomes or genes, hybridization, introgression, polyploidy, apomixis, and reproductive isolation.
10. Opinions and convictions on origins vary, but few can deny the major impact that theories of evolution have had on modern peoples and on their concepts of life.
11. Biologists generally feel evolution is the only plausible explanation for the unity of life at the molecular and cellular level and the great diversity of life, but there is little agreement among them as to the precise pathways of evolution in the past.

Review Questions

1. How does organic evolution differ from other forms of evolution?
2. How did Darwin's theory of evolution differ from Lamarck's?
3. What basic modifications have been made in evolutionary theory since Darwin's time?
4. What is meant by a mechanism of evolution?
5. What evidence is there to support modern concepts of evolution? Are there any problems with the evidence?

Discussion Questions

1. One of Darwin's principles was that there is a struggle for existence among living organisms. Do plants struggle with one another to survive? If so, how do they do it?
2. Some populations change noticeably in form within a hundred years. If only one gene in every 200,000 mutates and if most mutations are harmful, how is such change possible?
3. Do you think there might be scientifically supportable alternatives to organic evolutionary theory to account for the diversity of life around us? Explain.

Learning Online

Visit our web page at www.mhhe.com/botany for interesting case studies, practice quizzes, current articles, and animations within the Online Learning Center to help you understand the material in this chapter. You'll also find active links to these topics:

Evolution
Darwinian Evolutionary Theory
Neo-Darwinism
Evidences for Evolution
Natural Selection
Speciation
Macroevolution
Evolution of Plants

Additional Reading

Arnold, M. L. 1997. *Natural hybridization and evolution.* Fair Lawn, NY: Oxford University Press.

Avers, C. J. 1989. *Process and pattern in evolution.* Fair Lawn, NY: Oxford University Press.

Baltscheffsky, H., et al. (Eds.). 1996. *Origin and evolution of biological energy conversion.* Fair Lawn, NY: Oxford University Press.

Behe, M. J. 1998. *Darwin's black box: The biochemical challenge to evolution.* New York: Simon and Schuster.

Darwin, C. 1999. *The origin of species* (facsimile ed.) Westminster, MD: Bantam Books.

Eldredge, N. 2000. *The pattern of evolution.* San Francisco: W. H. Freeman.

Margulis, L. 2000. *Microcosmos, vol. 2: Evolution and diversity.* Sudbury, MA: Jones & Bartlett.

Stone, E. M., and R. J. Schwartz. 1990. *Intervening sequences in evolution and development.* Fair Lawn, NY: Oxford University Press.

Flowers of a two-toned Monkeyflower plant (*Mimulus bicolor*).

Plant Names and Classification

16

Chapter Outline

Overview

This chapter begins with a discussion of the problems involved in the use of common names for plants, as illustrated by a survey of such names for two related species of American spring-flowering perennials. It continues with a brief historical account of the events that led to the development and acceptance of Linnaeus's Binomial System of Nomenclature. The history of the development of a six-kingdom classification is presented, along with a list of the divisions and classes included in the kingdoms covered in this text. A dichotomous key to the kingdoms and divisions of organisms is provided. The chapter concludes with a brief discussion of cladistics.

Some Learning Goals

1. Understand several problems associated with the application of common names to organisms.
2. Know what the Binomial System of Nomenclature is, how it developed, and how it is currently used.
3. Learn several reasons for recognizing more than two kingdoms of living organisms.
4. Understand the bases for Whittaker's five-kingdom system.
5. Construct a dichotomous key to 10 different objects (e.g., pencils, golf balls, crayons, balloons, clocks).

Biologists sometimes are thought by the general public to be slightly pompous or weird when they refer to peanuts as *Arachis hypogaea* or an African hoopoe bird as *Upupa epops.* They aren't, however, merely showing off or being difficult. Rather, they're identifying organisms by their scientific names through a system that has evolved over the last few centuries. It has now become vital to us, regardless of the location or language, to be able to distinguish among the existing types of organisms (estimated to be at least 10 million), as well as those that have become extinct, and to do so in a way that will identify them anywhere.

At present, all living organisms are given a single two-word Latin scientific name, and many also have common names. Only one correct scientific name applies to all individuals of a specific kind of organism, no matter where they're found, but many common names may be given to the same organism, and one common name may be applied to a number of different organisms.

The scientific name *Dicentra cucullaria,* for example, was given to a pretty, spring-flowering plant native to eastern North America and the Columbia River basin in the West. Its unique flower shape, which reminds one of the baggy pants of a traditional Dutch costume, has resulted in its being given the common name *Dutchman's breeches.* It also, however, has the common names of *little-boys' breeches, kitten's breeches, breeches-flower, boys-and-girls, Indian boys-and-girls, monkshood, white eardrops, soldier's cap, colicweed, little blue stagger, white hearts, butterfly banners, rice roots,* and *meadow bells.* In addition to these English common names, the plant has Indian names, and in the Canadian province of Quebec, it has French names.

Often growing with *Dicentra cucullaria* is a related plant, with similar leaves but with slightly differently shaped flowers, having the scientific name *Dicentra canadensis.* Because of the similarities between the two plants and their close association in the woods where they occur, some people in the past assumed that they were merely two different forms of the same plant. This is reflected in the fact that some of the common names of *Dicentra cucullaria* (e.g., colicweed, Indian boys-and-girls, little blue stagger) have also been applied to *Dicentra canadensis,* in addition to the names *squirrel corn, turkey corn, turkey pea, wild hyacinth, fumitory, staggerweed,* and *trembling stagger.* However, the problem of the diversity and the overlapping of common names for these two plants doesn't stop there. The name *monkshood,* for example, is widely used for *Aconitum* species, which are in the buttercup family (Ranunculaceae)—a family only distantly related to the fumitory family (Fumariaceae) of *Dicentra.* Soldier's cap, rice roots, meadow bells, and turkey corn have also been applied to completely unrelated plants. Both species of *Dicentra* are shown in Figure 15.7.

In Europe, with its many languages, common names can become very numerous indeed. The widespread weed with the scientific name *Plantago major,* for example, is often called *broad-leaved plantain* in English, but it also has no fewer than 45 other English names, 11 French names, 75 Dutch names, 106 German names, and possibly as many as several hundred more names in other languages, with literally dozens of these common names also applying to quite different plants (Fig. 16.1). If it were not for the early recognition by biologists and others of the urgent need for worldwide uniformity in naming and classifying all organisms, utter chaos eventually might have prevailed in communications concerning them.

DEVELOPMENT OF THE BINOMIAL SYSTEM OF NOMENCLATURE

The first person known to attempt to organize and classify plants was Theophrastus, the brilliant student of Aristotle and Plato. His 4th-century B.C. classification of nearly 500 plants into trees, shrubs, and herbs, along with his distinctions between plants on the basis of leaf characteristics, was

Figure 16.1 Plantain (*Plantago major*). These plants have at least 300 different common names

used for hundreds of years. It was not until the 13th century A.D. that a distinction was made between monocots and dicots on the basis of stem structure.

After this, herbalists of the period added many plants to Theophrastus's list, and by the beginning of the 18th century, details of fruit and flower structure, in addition to form and habit, were used in classification schemes. European scholarly institutions were beginning to bulge with thousands of plants that explorers had brought back from around the world, and confusion over scientific and common names was multiplying.

The use of Latin in schools and universities had become widespread, and it was then customary to use descriptive Latin phrase names for both plants and animals. All organisms were grouped into **genera** (singular: **genus**), with the first word of the Latin phrase indicating the particular genus to which the organisms belonged. For example, all known mints were given phrase names (*polynomials*) beginning with the word *Mentha,* the name of the genus. Likewise, the phrases for lupines began with *Lupinus,* and those for poplars began with *Populus.* A complete phrase name for spearmint read *Mentha floribus spicatis, foliis oblongis serratis.* Roughly translated, it means "Mentha with flowers in a spike (an elongated but compact flower cluster); leaves oblong, saw-toothed." A phrase name for the closely related peppermint read *Mentha floribus capitatus, foliis lanceolatis serratis,* which means "leaves lance-shaped, saw-toothed, with very short petioles."

Linnaeus

At this point, the Swedish naturalist Carolus Linnaeus (1707–1778; Fig. 16.2) began improving the way organisms were named and classified. The system he established worked so well that it has persisted to the present. In fact, Linnaeus's system is now used throughout the entire world.

Linnaeus, who was nicknamed the "Little Botanist" at school, inherited his passion for plants from his father, who was a minister and an amateur gardener. He is said to have been much impressed at the age of four by his father's remarks about the uses of neighborhood plants in his home community of Råshult, located in southern Sweden. After a brief tenure as a student at the University of Lund, he spent most of his time making excursions to Lapland, Holland, France, and Germany. Eventually, he became the professor of botany and medicine at Uppsala, where he inspired large numbers of students, 23 of whom became professors themselves. He frequently led large field trips into the countryside, accompanied by a musical band.

When Linnaeus began his work, he set out to classify all known plants and animals according to their genera. In 1753, he published a two-volume work entitled *Species Plantarum,* which was later to become the most important of all works on plant names and classification. In this work, he not only included a referenced list of all the Latin phrase names previously given to the plants, but, when

Figure 16.2 Linnaeus. *(Courtesy National Library of Medicine)*

necessary, he also changed some of the phrases to reflect relationships, placing one to many specific kinds of organisms called **species**[1] in each genus. He limited each Latin phrase to a maximum of 12 words, and in the margin next to the phrase, he listed a single word, which, when combined with the generic name, formed a convenient abbreviated designation for the species. The word in the margin for spearmint was *spicata,* and the word for peppermint was *piperita.* Accordingly, the abbreviated name for spearmint was *Mentha spicata* and for peppermint, *Mentha piperita.*

Although Linnaeus originally considered the phrase names the official names of the plants, he and those who followed him eventually replaced all the phrase names with abbreviated ones.

Because of their two parts, these abbreviated names became known as *binomials*, and the method of naming became known as the *Binomial System of Nomenclature.* Today, all species of organisms are named according to this system, which also includes the authority for the name, either in abbreviated form or in full, after the Latin name. For example, the full scientific name for spearmint is written *Mentha spicata* L., the L. standing for Linnaeus; the full scientific name for the wild dandelion native to Scandinavia is written *Taraxacum officinale* Wiggers, after Fredericus Henricus Wiggers, who was the first (in his *Flora of Holstein,* published in 1780) to describe the species.

Besides establishing the Binomial System of Nomenclature, Linnaeus tried to do more than publish just a long list of plant names. After all, such lists aren't very useful if they can't be used to identify the plants concerned. Linnaeus organized all known plants into 24 classes,[2] which were based mainly on the number of *stamens* (pollen-bearing structures) in flowers. All plants with five stamens per flower, for example, were placed in one class, while those with six stamens were placed in another class. Plants and other organisms that don't produce flowers (e.g., mosses, fungi) were put in a class of their own. His arrangement was for convenience and was artificial because it did not necessarily reflect natural relationships. For the first time, however, it became possible for workers to identify plants previously unknown to them. This classification was used in *Species Plantarum* and other works by Linnaeus (Fig. 16.3), and for a short period of time was adopted by some, but not all, botanists.

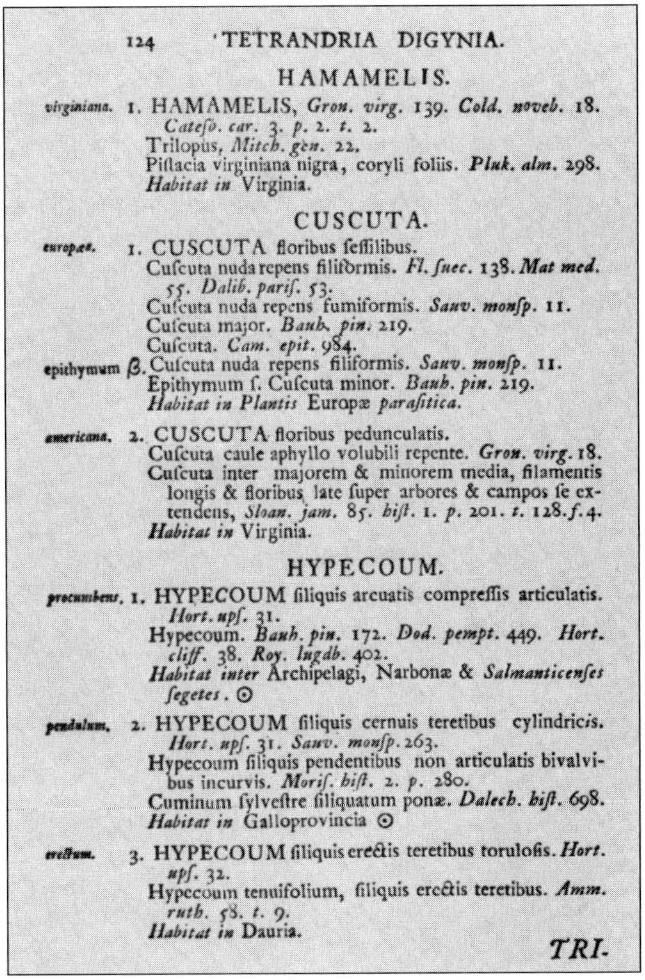

124 TETRANDRIA DIGYNIA.

HAMAMELIS.

virginiana. 1. HAMAMELIS, *Gron. virg.* 139. *Cold. noveb.* 18. *Catesb. car.* 3. *p.* 2. *t.* 2.
Trilopus, *Mitch. gen.* 22.
Pistacia virginiana nigra, coryli foliis. *Pluk. alm.* 298.
Habitat in Virginia.

CUSCUTA.

europaea. 1. CUSCUTA floribus sessilibus.
Cuscuta nuda repens filiformis. *Fl. suec.* 138. *Mat med.* 55. *Dalib. paris.* 53.
Cuscuta nuda repens fumiformis. *Sauv. monsp.* 11.
Cuscuta major. *Bauh. pin.* 219.
Cuscuta. *Cam. epit.* 984.
epithymum β. Cuscuta nuda repens filiformis. *Sauv. monsp.* 11.
Epithymum s. Cuscuta minor. *Bauh. pin.* 219.
Habitat in Plantis Europae *parasitica.*

americana. 2. CUSCUTA floribus pedunculatis.
Cuscuta caule aphyllo volubili repente. *Gron. virg.* 18.
Cuscuta inter majorem & minorem media, filamentis longis & floribus late super arbores & campos se extendens, *Sloan. jam.* 85. *hist.* 1. *p.* 201. *t.* 128. *f.* 4.
Habitat in Virginia.

HYPECOUM.

procumbens. 1. HYPECOUM siliquis arcuatis compressis articulatis. *Hort. ups.* 31.
Hypecoum. *Bauh. pin.* 172. *Dod. pempt.* 449. *Hort. cliff.* 38. *Roy. lugdb.* 402.
Habitat inter Archipelagi, Narbonae & *Salmanticenses segetes.* ⊙

pendulum. 2. HYPECOUM siliquis cernuis teretibus cylindricis. *Hort. ups.* 31. *Sauv. monsp.* 263.
Hypecoum siliquis pendentibus non articulatis bivalvibus incurvis. *Moris. hist.* 2. *p.* 280.
Cuminum sylvestre siliquatum ponae. *Dalech. hist.* 698.
Habitat in Galloprovincia ⊙

erectum. 3. HYPECOUM siliquis erectis teretibus torulosis. *Hort. ups.* 32.
Hypecoum tenuifolium, siliquis erectis teretibus. *Amm. ruth.* 58. *t.* 9.
Habitat in Dauria.

TRI-

Figure 16.3 A page from *Species Plantarum* by Linnaeus.

The International Code of Botanical Nomenclature

In 1867, more than 100 years after *Species Plantarum* was published, about 150 European and American botanists met in Paris to try to standardize rules governing the naming and the classifying of plants. They agreed to use the works of Linnaeus as the starting point for all scientific names of plants and decided that his binomials, or the earliest ones published after him, would have priority over all others.

International congresses of botanists have met at varying intervals since 1867 and have revised and expanded these rules. Today, the modified rules comprise what is known as the *International Code of Botanical Nomenclature,* which is a single book with a common index to its English, French, and German translations of the various rules and recommendations. It now specifically recognizes Linnaeus's *Species Plantarum* as the starting point for scientific names of plants and spells out details of naming and classifying, which are followed by botanists of all nationalities. A similar code for animals has been developed and established by zoologists. It, too, uses a work of Linnaeus as the starting point for scientific names.

1. As noted in Chapter 15, *species* is like the word *sheep* in that it is spelled and pronounced the same in either singular or plural usage. There is no such thing as a plant or animal "specie." Since Linnaeus's time, a species has been defined as "a population of individuals of common form, structure, and ancestry capable of freely interbreeding in nature but not generally interbreeding with individuals of other dissimilar populations."

2. Note that Linnaeus used the word *class* for a strictly artificial category of classification, resulting in unrelated organisms being grouped together simply for convenience. Modern use of the word differs from that of Linnaeus in that organisms are now assigned to a class on the basis of natural relationships.

If a botanist finds a plant that is new to science, the *International Code of Botanical Nomenclature* requires that he or she take at least two steps to have the plant officially recognized. (1) A Latin description of the plant must be published in a journal or other publication that is circulated and available to the public. Distributed mimeographed material, for example, is not an acceptable form of "publication." (2) An annotated herbarium specimen of the plant, designated by the author as a *type specimen,* must be deposited in an herbarium to which the public may have access. A discussion of herbaria and herbarium specimens is given in Chapter 23.

DEVELOPMENT OF THE KINGDOM CONCEPT

If you were to ask the average person the differences between plants and animals, you would probably be told that plants are green, don't eat each other, and don't move. Animals, on the other hand, are not green, do eat plants or each other, and do move. A distinction between plants and animals in general has probably always existed in the minds of intelligent beings, and it was natural when classification schemes were first developed that all living organisms would be placed, according to the highest category of kingdom, in either the *Plant Kingdom* or the *Animal Kingdom.*

While this distinction still works well for the more complex plants and animals, it breaks down for some of the so-called simpler organisms. For example, there are more than 300 species of single-celled organisms called *euglenoids* (see Fig. 18.23) inhabiting a variety of freshwater habitats. These microscopic creatures have small, whiplike tails called *flagella* that pull them through the water, and they also can ingest food particles through a groove called a *gullet.* Both of these features would be considered animal-like, and so euglenoids in the past have been treated as animals in some textbooks. Many of these organisms also, however, have chloroplasts, and if light is present, they can carry on photosynthesis efficiently enough to eliminate the need for ingesting food. Accordingly, they have often been treated as plants in botany books.

A similar problem has existed with slime molds, which resemble masses of protoplasm slowly creeping or flowing across dead leaves and debris, engulfing and feeding on bacteria and other substances as they go, and looking somewhat like amoebae, which botanists and zoologists alike have traditionally called animals. When slime molds reproduce, however, dramatic changes take place. They become stationary and develop reproductive bodies that are distinctly fungus-like, causing some to consider them fungi.

In an attempt to overcome this problem, the biologists J. Hogg and Ernst Haeckel proposed a third kingdom in the 1860s. All organisms that did not develop complex tissues (e.g., algae, fungi, and sponges) were placed in a third kingdom called *Protoctista.* This third kingdom included such a heterogeneous variety of organisms, however, that in 1938, another biologist by the name of Herbert F. Copeland proposed that it again be divided. He assigned the name *Monera* to all single-celled organisms with prokaryotic cells (the differences between prokaryotic and eukaryotic cells are discussed in Chapters 3 and 17), leaving the algae, fungi, and single-celled organisms with eukaryotic cells in the Kingdom Protoctista.

Although many biologists thought Copeland's four-kingdom system of classification was a definite improvement, it too had problems—particularly the fact that basic differences in the mode of nutrition existed among organisms within the Kingdom Protoctista. As a result, during the 1970s, many biologists adopted variations of a five-kingdom system proposed by R. H. Whittaker in 1969 (Table 16.1).

In Whittaker's system, three kingdoms of more complex organisms based on three basic forms of nutrition (photosynthesis, ingestion of food, and absorption of food in solution) were recognized, along with two kingdoms of protists, which were distinguished on the basis of differences in cellular structure.

The five-kingdom arrangement also has had critics and problems, particularly since the early 1980s when Carl Woese, a microbiologist and leading authority on bacteria, argued cogently that Kingdom Monera should be divided into two kingdoms. Woese based his contention on some previously unrecognized fundamental differences between two major groups of bacteria (discussed in Chapter 17). His proposal for two kingdoms of bacteria gained increasing acceptance in the 1990s and is now generally recognized.

Since acceptance of Woese's kingdom classification, James Lake and his colleagues have proposed creating yet another kingdom by dividing Kingdom Archaea into two kingdoms, making three kingdoms of prokaryotes in all. The slime molds, which have no cell walls during their active state but do develop walls when they reproduce, still do not fit well into the six-kingdom system, and other organisms traditionally regarded as fungi may be more closely related to protists. Much information still needs to be accumulated to understand natural relationships, and until this occurs, taxonomic categories will remain subject to different interpretations.

CLASSIFICATION OF MAJOR GROUPS

Since Linnaeus's time, a number of classification categories have been added between the levels of kingdom and genus. Genera are now grouped into *families,* families into *orders,* orders into *classes,* classes into *divisions* or *phyla,*[3] and divisions (phyla) into *kingdoms.*

---

3. *Division* is equivalent to the term *phylum,* which has been used universally in animal classification. Until 1993, the *International Code of Botanical Nomenclature,* which governs the naming and classification of plants, required the use of *division* for plants. In 1993, an International Botanical Congress revised the rules and permitted the use of *phylum* interchangeably with the term *division* for plants and fungi. The term *division* is now being dropped by most botanists, who are adopting *phylum* in its place.

TABLE 16.1

Five Classifications of Organisms into Kingdoms

TWO KINGDOMS (TRADITIONAL)	THREE KINGDOMS (HOGG AND HAECKEL)	FOUR KINGDOMS (COPELAND)	FIVE KINGDOMS (WHITTAKER)	SIX KINGDOMS (WOESE ET AL.)	FEATURES
		Monera	**Monera**	**Archaea**	Cells prokaryotic;
		Bacteria	Bacteria	Archaebacteria	lack muramic acid
	Protoctista	**Protoctista**	**Protista**	**Bacteria**	Cells prokaryotic;
	Bacteria	Algae	Algae	True bacteria	have muramic acid
	Algae	Slime molds	Slime molds	**Protista**	Cells eukaryotic
	Slime molds	Flagellate fungi	Flagellate fungi	Algae	
	Flagellate fungi	True fungi	Protozoa	Slime molds	
	True fungi	Protozoa	Sponges	Water molds	
	Protozoa	Sponges		Protozoa	
	Sponges			Sponges	
			Fungi	**Fungi**	Absorb food in solution
			True fungi	True fungi	
Plantae	**Plantae**	**Plantae**	**Plantae**	**Plantae**	Produce food via
Bacteria	Bryophytes	Bryophytes	Bryophytes	Bryophytes	photosynthesis
Algae	Vascular plants	Vascular plants	Vascular plants	Vascular plants	
Slime molds					
Flagellate fungi					
True fungi					
Bryophytes					
Vascular plants					
Animalia	**Animalia**	**Animalia**	**Animalia**	**Animalia**	Ingest food
Protozoa	Multicellular animals	Multicellular animals	Multicellular animals	Multicellular animals	
Sponges					
Multicellular animals					

Depending on which system of classification is used, there may be between 12 and nearly 30 divisions (phyla) of "plants" recognized.

A classification of major groups of living organisms into six kingdoms, based primarily on a modification of Whittaker's five-kingdom system, is found in Table 16.2. It is the classification followed throughout this book. Probable derivations and relationships are indicated in Figure 16.4.

Viruses, which don't have cellular structure or most of the other attributes of living organisms, are not included in this classification; they are discussed in Chapter 17 after the bacteria. Lichens, each of which is a combination of an alga and a fungus, are discussed with the fungi in Chapter 19. If we were to give a complete classification of the common onion according to this particular arrangement, it would look like this:

Kingdom: Plantae
Phylum: Magnoliophyta
Class: Liliopsida
Order: Liliales
Family: Liliaceae
Genus: *Allium*
Species: *Allium cepa* L.

TABLE 16.2

Classification of Organisms in Six Kingdoms

Kingdom Archaea
- Phylum Archaebacteria (methane, salt, and sulfolobus bacteria)

Kingdom Bacteria
- Phylum Eubacteria
  - Class Eubacteriae (unpigmented, purple, and green sulfur bacteria)
  - Class Cyanobacteriae (cyanobacteria)
  - Class Chloroxybacteriae (chloroxybacteria)

Kingdom Protista
- Phylum Chlorophyta (green algae)
- Phylum Chromophyta (yellow-green, golden-brown, and brown algae)
- Phylum Rhodophyta (red algae)
- Phylum Euglenophyta (euglenoids)
- Phylum Dinophyta (dinoflagellates)
- Phylum Cryptophyta (cryptomonads)
- Phylum Prymnesiophyta (haptophytes)
- Phylum Charophyta (stoneworts)
- Phylum Myxomycota (plasmodial slime molds)
- Phylum Dictyosteliomycota (cellular slime molds)
- Phylum Oomycota (water molds)
- [Phylum Protozoa—protozoans]
- [Phylum Porifera—sponges]

Kingdom Fungi
- Phylum Chytridiomycota (chytrids)
- Phylum Zygomycota (coenocytic fungi)
- Phylum Ascomycota (sac fungi)
  - [Lichens]
- Phylum Basidiomycota (club fungi)
- Phylum Deuteromycota (imperfect fungi)

Kingdom Plantae
- Phylum Hepaticophyta (liverworts)
- Phylum Anthocerotophyta (hornworts)
- Phylum Bryophyta (mosses)
- Phylum Psilotophyta (whisk ferns)
- Phylum Lycophyta (club mosses)
- Phylum Equisetophyta (horsetails)
- Phylum Polypodiophyta (ferns)
- Phylum Pinophyta (conifers)
- Phylum Ginkgophyta (*Ginkgo*)
- Phylum Cycadophyta (cycads)
- Phylum Gnetophyta (*Gnetum, Ephedra, Welwitschia*)
- Phylum Magnoliophyta (flowering plants)
  - Class Magnoliopsida (dicots)
  - Class Liliopsida (monocots)

(Psilotophyta through Liliopsida: Division Tracheophyta of earlier classifications)

Kingdom Animalia (multicellular animals)

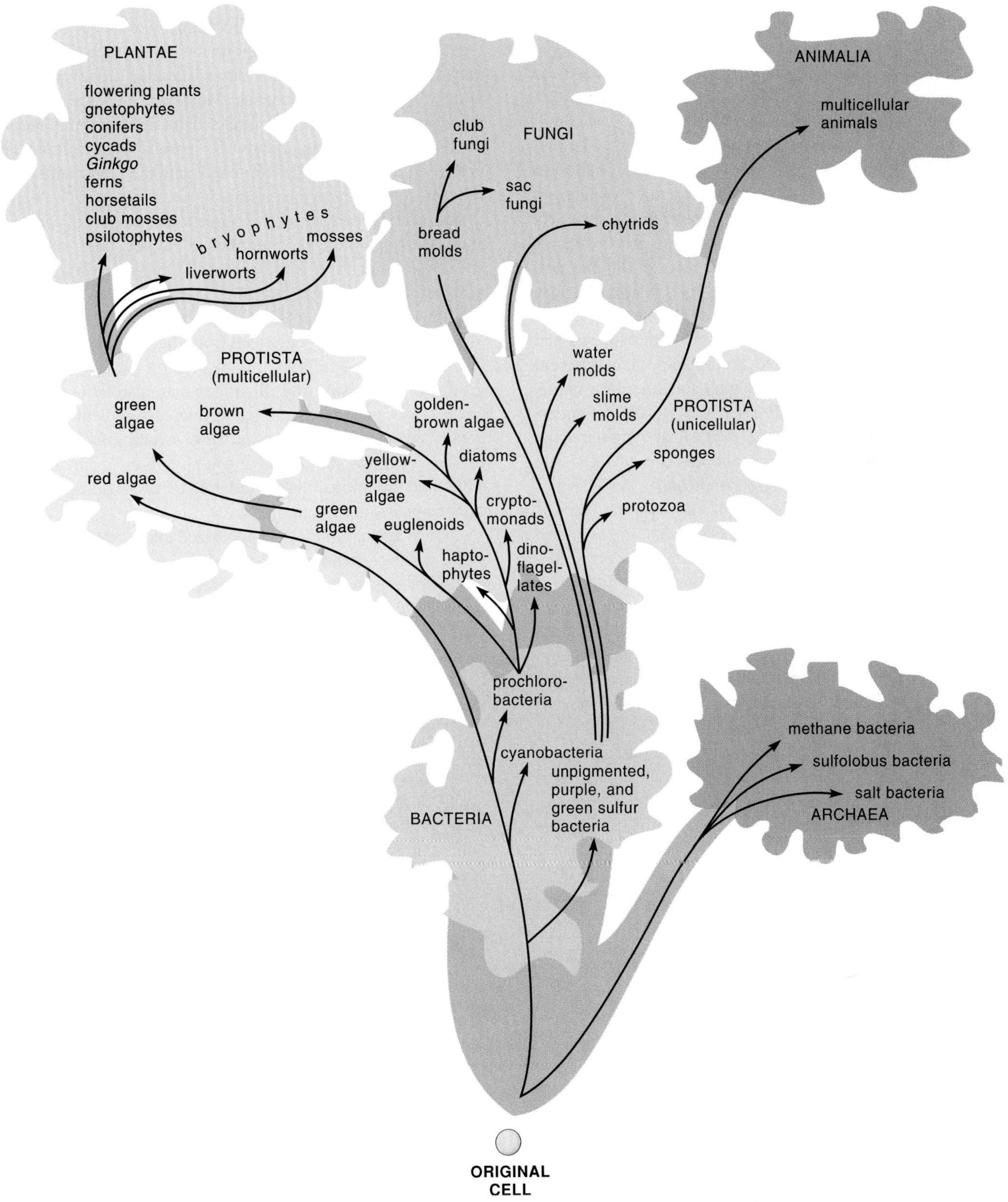

Figure 16.4 Hypothetical derivations and relationships among kingdoms and the major groups of organisms.

The second part of the scientific name of a species, called the *specific epithet,* is followed by the name of the author, mostly in abbreviated form. The author is the person (or persons) who originally named the plant or placed the species in a particular genus. Many plants have one author's name in parentheses, followed by the name of another author outside the parentheses. For example, the scientific name of mountain hemlock is *Tsuga mertensiana* (Bong.) Carr. In this instance, August Bongard (abbreviated to Bong.), a professor of botany at St. Petersburg in the early 19th century, first described the species under the genus *Pinus.* Half a century later, Êlie Carrière (abbreviated to Carr.), an authority on conifers at the Paris Natural History Museum, transferred the species to the genus *Tsuga.* Because Bongard was the first to describe the species and Carrière was the first to place the species in a different genus, both names are included with the scientific name. Binomials are printed in italics (or the words are underlined). In the case of *Allium cepa,* the L. following the name stands for Linnaeus, since this plant was one of the 7,300 species he included in his *Species Plantarum.*

In addition to these categories of classification, various "in-between" categories, such as *subphylum, subclass,* and *suborder,* have been used, and species themselves are sometimes further divided into *subspecies, varieties,* and *forms.* A few authors also recognize "super" categories, such as *superkingdom* and *superorder.* One classification widely used in the past recognized only two divisions (phyla) in Kingdom Plantae (Division [Phylum] Bryophyta for nonvascular plants and Division [Phylum] Tracheophyta, with a number of subdivisions [subphyla], for all vascular plants). These subphyla are treated as full phyla in this and a number of other texts.

Since human judgment has to enter into the compilation of classifications, there undoubtedly will never be complete agreement on them, but they are nevertheless useful, indeed necessary, for making some order of the diversity of life around us.

Those who specialize in identifying, naming, and classifying organisms are called *taxonomists.* Today, taxonomists who involve evolutionary processes in trying to sort out natural relationships refer to themselves as *systematists.* One of the activities of taxonomists and systematists is the construction of keys to help others identify organisms with which they may not be familiar. Most such keys are *dichotomous*; that is, they give the reader pairs of statements based on features of the organisms. By carefully examining an organism and choosing from each pair the statements that most closely apply to the organism, we can arrive at an identification. If, for example, we agree with the first of the two statements given for each number in the key, then we either use the identification at the end of the line or proceed to the next indented statement beneath. If we disagree with the first of the paired statements, then we look for the matching alternate statement and proceed from there.

The key that follows this discussion deals with major groups of organisms and illustrates natural relationships. It covers only general features and does not indicate various details or occasional exceptions that would appear in keys for lesser groups. All organisms in any given group are presumed to be more closely related to each other than to organisms in another group.

Once again, it must be emphasized that a key of this type, while calling attention to basic differences between major groups of organisms, may also oversimplify the differences, partly because it does not list all the characteristics of each member of a given group. Moreover, it does not necessarily identify exceptions to the rule or some of the intermediate forms that frequently are transferred back and forth as research uncovers new evidence or as new interpretations develop.

Keys at this level of classification also are not completely practical because to be able to arrive at an identification, one sometimes has to have available specific stages in the life cycles of the plants. Furthermore, chemical tests that are not easily performed or details that may be difficult to see with a light microscope may be required. Nevertheless, such a key is generally preferable to a simple listing of groups and their characteristics for purposes of distinguishing among them.

A KEY TO MAJOR GROUPS OF ORGANISMS (EXCLUSIVE OF KINGDOM ANIMALIA)

1. Unicellular, prokaryotic organisms with cell walls
    2. Cell walls with muramic acid . . . . **Kingdom Bacteria**
    2. Cell walls without muramic acid . . . . **Kingdom Archaea**

1. Unicellular, colonial, filamentous, or multicellular eukaryotic organisms, with or without cell walls
    3. Organisms whose female (and usually male) reproductive structures consist of a single cell or with sterile cells surrounding the one-celled reproductive structures; zygotes not developing into embryos
        4. Organisms unicellular, filamentous, or plasmodial (i.e., with naked protoplasm)
            5. Cell walls without chitin . . . . **Kingdom Protista**
                6. Cells with plastids
                    7. Plastids with yellow, brown, or orange pigments more conspicuous than the chlorophyll pigments

8. Food reserves oils or carbohydrates other than starch; two flagella both located at one end of the cell
   9. A haptonema (third flagellum) usually present ....Phylum Prymnesiophyta
   9. A haptonema absent
      10. The unequal flagella bearing stiff lateral hairs ....Phylum Cryptophyta
      10. Flagella not as above ....Phylum Chromophyta (in part)
8. Food reserve starch; cells with a flagellum at one end and another at right angles to it in a central groove ....Phylum Dinophyta
7. Plastids with chlorophyll pigments more conspicuous than other pigments
   11. Cells flexible; carbohydrate food reserve *paramylon* ....Phylum Euglenophyta[4]
   11. Cells not flexible; carbohydrate food reserve starch .... Phylum Chlorophyta (in part)
6. Plastids absent; vegetative bodies of organisms consisting of filaments with walls containing cellulose, multinucleate masses of protoplasm, or amoeba-like cells
   12. Cell walls with cellulose; organisms aquatic ....Phylum Oomycota
   12. Cell walls without cellulose; organisms not aquatic
      13. Vegetative bodies consisting of multinucleate masses of protoplasm (plasmodia) .... Phylum Myxomycota
      13. Vegetative bodies amoeba-like ....Phylum Dictyoliosteliomycota
5. Cell walls containing chitin .... **Kingdom Fungi**
   14. Vegetative bodies consisting primarily of a single cell, often with rhizoids; zoospores produced ....Phylum Chytridiomycota
   14. Vegetative bodies consisting primarily of branched filaments; no flagellated cells produced
      15. Filaments of the vegetative bodies containing numerous nuclei; not partitioned into individual cells .... Phylum Zygomycota
      15. Filaments of the vegetative bodies partitioned into individual cells, each with one to several nuclei
         16. Sexual reproductive cells produced within sacs ......Phylum Ascomycota
         16. Sexual reproductive cells produced externally on club-shaped structures .... Phylum Basidiomycota
4. Organisms multicellular, not filamentous or plasmodial
   17. Organisms with accessory pigments essentially similar to those of higher plants; carbohydrate food reserve starch
      18. Organisms with complex bodies differentiated into nodes and internodes ....Phylum Charophyta
      18. Organisms with bodies not differentiated into nodes and internodes ....Phylum Chlorophyta (in part)
   17. Organisms with some accessory pigments differing from those of higher plants; food reserves carbohydrates other than ordinary starch
      19. Organisms brownish in color due to presence of brown pigments; carbohydrate food reserve *laminarin* ....Phylum Chromophyta (in part)
      19. Organisms reddish in color due to presence of red pigments; carbohydrate food reserve *floridean starch* ....Phylum Rhodophyta
3. Organisms with multicellular reproductive structures ....**Kingdom Plantae**
   20. Plants without true xylem or phloem . .Phylum Bryophyta
   20. Plants with true xylem and phloem
      21. Plants with true leaves absent; *enations*[5] present; stems branching dichotomously ....Phylum Psilotophyta

---

4. Only about a third of euglenoid species develop chloroplasts

5. See page 398 for a discussion of *enations*

Key to Major Groups of Organisms *(continued)*

21. Plants with true leaves present; enations absent; stems branching in various ways or unbranched
    22. Plants with leaves having a single vein (*microphylls*)
        23. Stems not ribbed and not containing silica; leaves photosynthetic . . . . Phylum Lycophyta
        23. Stems ribbed and containing silica; leaves reduced to scales and non-photosynthetic . . . . .Phylum Equisetophyta
21. Plants with leaves usually having more than one vein (*megaphylls*)
    24. Plants reproducing by means of spores produced on the leaves . . . . Phylum Polypodiophyta
    24. Plants reproducing by means of seeds developed from ovules
        25. Plants without flowers; seeds not produced in enclosed ovaries
            26. Leaves pinnate and large, resembling those of palms . . . . . .Phylum Cycadophyta
            26. Leaves not pinnate or palmlike
                27. Leaves fan-shaped, with numerous dichotomously forking veins . . . .Phylum Ginkgophyta
                27. Leaves not fan-shaped
                    28. Wood containing no vessels . . . . . . .Phylum Pinophyta
                    28. Wood containing vessels . . . . .Phylum Gnetophyta
    24. Plants with flowers; seeds produced in enclosed ovaries . . . . Phylum Magnoliophyta
        29. Flowers with parts mostly in fours and fives; cotyledons two . . . . .Class Magnoliopsida
        29. Flowers with parts mostly in threes; cotyledon one . . . . .Class Liliopsida

CLADISTICS

Since the 1970s, taxonomists have increasingly used **cladistics** in their attempts to determine natural relationships. Cladistics is a method of examining natural relationships among organisms, based on features shared by the organisms. The relationships are portrayed in straight line diagrams (evolutionary trees) called **cladograms** (Fig. 16.5).

In cladistics, the value or form of a feature is referred to as a **character state.** In a feature such as flower color, for example, the character could be flower color, and the state could be purple or red. After the specific chemical or other nature of a character is determined, hypotheses (assumptions) are made about which states are primitive (ancestral) and which are derived (have evolved from something else). Obviously, the hypotheses need to be tested, and to test them, we often try to find additional characters that may support one hypothesis more than they support others. Those hypotheses that are not supported can be eliminated, and those that are supported can be subjected to further testing.

In trying to choose the best of several to many cladograms, taxonomists use the principle of *parsimony.* Parsimony, as it applies to cladistics, is based on a principle of logic called *Occam's razor,* which states that "one should not make more assumptions than the minimum needed to explain anything." The best cladogram is usually interpreted as that which requires the fewest evolutionary changes in the taxa involved to arrive at the present situation.

Cladistics has been used to infer evolutionary relationships among many different groups of organisms in all the kingdom survey chapters that follow in this book.

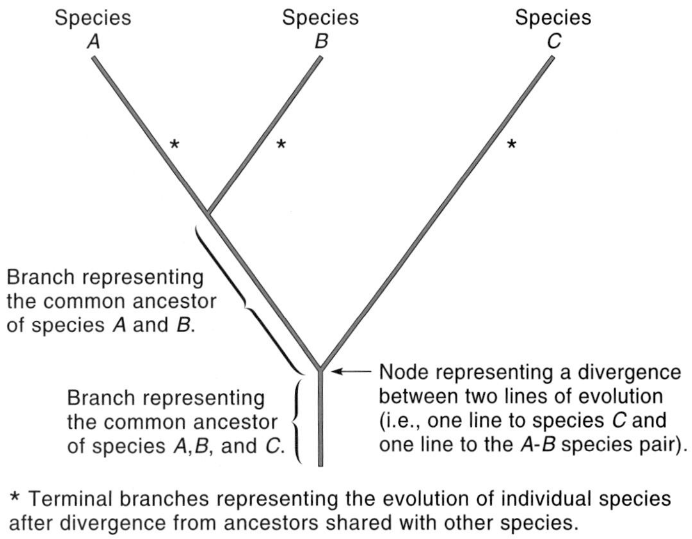

Figure 16.5 A simple cladogram showing relationships of three species derived from a common ancestor. The cladogram indicates that Species *A* and Species *B* are more closely related to each other than either is to Species *C*.

Summary

1. All organisms are identified by a two-word Latin name (binomial); many also have one to numerous common names. Biologists recognized the need for worldwide uniformity in naming and classifying organisms.
2. Theophrastus first classified plants in the 4th century B.C. Monocots and dicots were distinguished in the 13th century; herbalists later added many plants to Theophrastus's original list. Descriptive Latin phrase names were used for all organisms by the 18th century.
3. Linnaeus compiled a comprehensive list of all known Latin phrase names for plants according to their genera in *Species Plantarum,* published in 1753. Linnaeus placed specific kinds of related organisms (species) in each genus; he also listed a single word next to the Latin phrase, which, when combined with the name of the genus, formed an abbreviation for each species.
4. The abbreviated binomial, plus the authority for it, eventually replaced the phrase name; this method of naming plants became known as the Binomial System of Nomenclature. Linnaeus organized all known plants into 24 classes, based primarily on the number of stamens in flowers.
5. At a meeting in Paris in 1867, European and American botanists agreed to use Linnaeus's 1753 publication and binomials as the starting point for all scientific names of plants. The rules (*International Code of Botanical Nomenclature*) drawn up at that meeting and periodically revised are now followed for naming and classifying plants by botanists of all nationalities.
6. At first, two kingdoms (Plant and Animal) were recognized. Hogg and Haeckel proposed a third kingdom (Protoctista) in the 1860s. In 1938, H. F. Copeland, on the basis of cellular differences, divided the Protoctista into Kingdoms Monera and Protoctista.
7. In 1969, Whittaker proposed five kingdoms based both on forms of nutrition and cellular structure. In the 1980s, Carl Woese divided Whittaker's Kingdom Monera into two kingdoms, and six kingdoms are now generally recognized.
8. Since Linnaeus's time, several classifications have been added between the level of kingdom and genus.
9. The second part of a binomial, the specific epithet, is followed by the name of the author, usually in abbreviated form.
10. Subcategories (e.g., subspecies, suborder) are sometimes used in classification.
11. Taxonomists construct keys to aid in the identification of organisms. Most keys are dichotomous. Keys to major groups of organisms deal only in generalities and do not always indicate occasional exceptions or allow for intermediate forms that may be transferred back and forth between groups as new knowledge about them is gained.
12. Cladistics is a system of classifying and inferring evolutionary relationships based on an examination of shared features and differences.

Review Questions

1. What are the advantages and drawbacks of using scientific names as compared with common names?
2. What is the *International Code of Botanical Nomenclature*?
3. What is meant by the term *binomial*?

4. Which phyla include simple aquatic organisms?
5. List some characteristics by which the following may be distinguished from one another and then construct a simple dichotomous key that could aid someone from another planet in identifying them: cups, oranges, hammers, bicycles, rosebushes, automobiles, plates, cats, rats, raspberry bushes, dogs, wrenches, peaches.

Discussion Questions

1. Since phrase names are generally more descriptive of organisms than binomials, do you think they should be revived?
2. After Linnaeus organized all the genera known to him into classes, a number of other categories of classification between the level of kingdom and genus were added. Are these other categories really useful? Explain.
3. Do you think biologists are justified in dividing the traditional Plant and Animal Kingdoms into six kingdoms, and would it make any significant difference to the world at large if they had not? Explain.

Learning Online

Visit our web page at www.mhhe.com/botany for interesting case studies, practice quizzes, current articles, and animations within the Online Learning Center to help you understand the material in this chapter. You'll also find active links to these topics.

Classification and Phylogeny of Animals
Methods of Classification
Trees of Life

Additional Reading

Bisby, F. A. 1994. *Plant taxonomic database standards: Plant names in botanical databases, No. 3.* Pittsburgh, PA: Hunt Institute for Botanical Documentation.

Cronquist, A. 1992. *An integrated system of classification of flowering plants.* New York: Columbia University Press.

Index Kewensis, 1886–2001. New York: Oxford University Press.

Quattrocchi, U. 1999. *World dictionary of plant names: Common names, scientific names.* 2 vols. Boca Raton, FL: CRC Press.

Soltis, D. E., et al. 1998. *Molecular systematics of plants II: DNA sequencing.* New York: Chapman and Hall.

Stace, C. A. 1992. *Plant taxonomy and biosystematics,* 2d ed. New York: Cambridge University Press.

Stuessy, T. 1990. *Plant taxonomy: The systematic evaluation of comparative data.* New York: Columbia University Press.

Stuessy, T. 1994. *Case studies in plant taxonomy. Exercises in applied pattern recognition.* New York: Columbia University Press.

Woodland, D. W. 1997. *Contemporary plant systematics,* 2d ed. Berrien Springs, MI: Andrews University Press.

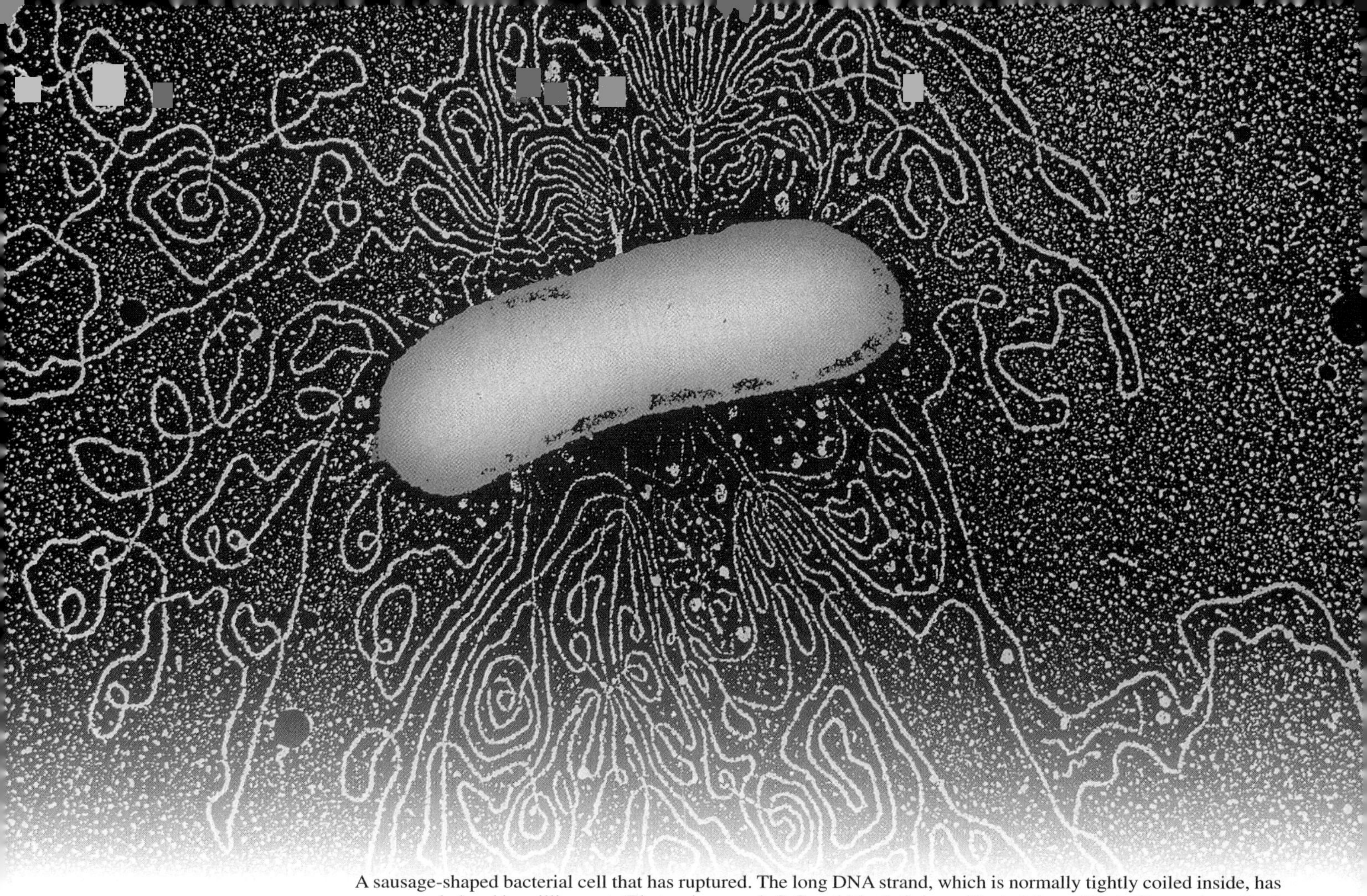

A sausage-shaped bacterial cell that has ruptured. The long DNA strand, which is normally tightly coiled inside, has unraveled and is spilling out. (Photomicrograph © Gopal Murti, Science Photo Library/Photo Researchers, Inc.)

Kingdom Bacteria, Kingdom Archaea, and Viruses

17

Chapter Outline

Overview

Following an introduction involving luminescent bacteria associated with flashlight fish, the chapter gives an overview of the features of Kingdoms Bacteria and Archaea. A brief investigation of bacteria and their reproduction is followed by a discussion of true bacteria, with representatives and human relevance of heterotrophic and autotrophic true bacteria being surveyed. Sections on composting, disease transmission, bacterial diseases, and Koch's Postulates are included. Cyanobacteria and prochlorobacteria are explored next, and human relevance of cyanobacteria is examined. Archaebacteria are then discussed. The chapter concludes with an overview of the nature, reproduction, and human relevance of viruses, viroids, and prions.

Some Learning Goals

1. Know the basic forms of bacteria. Explain how a prokaryotic cell differs from a eukaryotic cell and why prokaryotic organisms are difficult to classify.
2. Learn the forms of nutrition in bacteria.
3. Know at least 10 bacteria useful to humans and understand how they are useful.
4. Understand the various ways in which disease bacteria are transmitted; describe how each type of disease bacterium functions in causing disease.
5. Learn Koch's Postulates.
6. Know how the major groups of bacteria differ from one another in form, pigmentation, and reproduction. Also know how viruses and viroids differ from bacteria in form and reproduction.

The astonishing variety and beauty of undersea life, with representatives from all six kingdoms, have attracted the attention of snorkelers and scuba divers in ever-increasing numbers, particularly since the late Jacques Cousteau and other marine biologists have added to our knowledge of this vast domain with their sophisticated investigations and color cinematography. The shallow tropical waters of the Great Barrier Reef of Australia, the Virgin Islands National Park, Hanauma Bay of Hawaii, and countless other reefs teem with brilliantly colored marine life. Scuba divers who explore the oceans at night in areas such as Indonesia's Banda Sea, the region just north of Darwin, Australia, or Israel's Gulf of Elat are frequently rewarded with a spectacular sight not seen by those who explore by day. They find themselves in the midst of a shimmering, constantly changing light display.

Countless numbers of lights, bright enough to read by, flicker, dart, glide slowly, or even congregate in glimmering groups. The lights come from large pouches beneath the eyes of small fish, which can turn them off at will by covering them with special folds of tissue. Unlike their deep-sea relatives, these *flashlight fish,* as they are called, do not produce the light within their own bodies. Instead, the light comes from thousands of microscopic, luminescent bacteria that enjoy a symbiotic, or specifically, a mutualistic, relationship with the fish.

In this instance, the fish furnish some food and oxygen to the bacteria from numerous blood vessels in the pouches, while the bacteria produce the light used by the fish not only to search for food at night but also to attract live food and to confuse predators by flashing the lights on and off while swimming in erratic patterns. Both organisms derive benefit from the relationship.

Luminescent bacteria are minor representatives of the most numerous and geologically ancient of the earth's living organisms. Fossils of bacteria have been found in geological strata that are estimated to be 3.5 billion years old. Since fossils of the first organisms with eukaryotic cells date back "only" 1.3 billion years, bacteria are almost three times as old as any other organism known to have existed on this planet.

The approximately 5,000 species of bacteria recognized today occur in astronomical numbers in almost every conceivable natural habitat. Each gram of garden soil, for example, is estimated to contain as many as 2 billion bacteria. They are found on and in all plants and animals, in all types of soils, throughout both fresh and salt waters, in polar ice caps and bubbling hot springs, in coal and petroleum, throughout the atmosphere, in bottles of India ink, and almost anywhere else one may care to look for them.

Just how many species of bacteria there are is uncertain, since it is difficult to classify simple one-celled organisms on the basis of visible features alone. For example, a disease-causing bacterium with the scientific name of *Streptococcus pneumoniae* occurs in 84 known strains, all of which look alike but each of which causes a different form of pneumonia or related disease. Should each strain then be called a species? Microbiologists, who study microscopic organisms of all kinds, are not sure, but they tend to recognize clusters of strains based on what they do rather than how they look.

Although bacteria that cause disease and spoilage receive most of the publicity, more than 90% of bacterial species are either harmless or useful to humans. They are used in industry for a wide variety of purposes and are so important ecologically that life as we know it today would be vastly different without the activities of these tiny organisms. Both industrial uses and the role of bacteria in ecology are discussed in later sections.

FEATURES OF KINGDOMS BACTERIA AND ARCHAEA

Members of Kingdoms Bacteria and Archaea all have prokaryotic cells. Such cells have no nuclear envelopes, but each has a long strand of DNA, ribosomes, and membranes. In addition, there usually are small DNA fragments called plasmids present in bacterial cells. Membrane-bound organelles, such as plastids, mitochondria, dictyosomes, and endoplasmic reticulum are lacking (Table 17.1; see Fig. 17.13).

In a number of members of these two kingdoms, cells may occur in a common matrix of gelatinous material as variously shaped colonies or in the form of chains or **filaments** (threads), but each cell is completely independent, with protoplasmic connections between them being absent. Some species are **motile** (capable of independent movement), usually by means of simple **flagella** (whiplike tails, which often occur in pairs) that propel or pull a cell through the water. Several filamentous species exhibit a gliding motion in which the filaments glide back and forth independently or against each other, but most species are nonmotile.

Nutrition in the two kingdoms is primarily by the absorption of food in solution through the cell wall, but some bacteria are capable of obtaining their energy through chemical reactions involving various compounds or elements, while a few true bacteria, including *cyanobacteria* (blue-green bacteria) and *chloroxybacteria,* exhibit forms of photosynthesis.

Reproduction is predominantly asexual, by means of fission, a form of cell division in which one cell essentially pinches in two. Fission does not involve mitosis since there are no nuclei or other organelles. The strand of DNA duplicates itself, however, and the DNA is distributed between the two new cells originating from the parent cell. Sexual reproduction is unknown, but genetic recombination occurs in several groups. The genetic recombination is often facilitated by means of *pili* (minute tubes) that form connections between cells, but genetic recombination can still occur without pili between cells in close contact.

Cellular Detail and Reproduction of Bacteria

Although the prokaryotic cells of bacteria have no nuclear envelopes or organelles, folds of the plasma and other membranes apparently perform some of the functions of the organelles of eukaryotic cells. Ribosomes that are about half the size of those in the cytoplasm of eukaryotic cells are also present. Bacterial cells have a *nucleoid,* which is a single, long, very condensed DNA molecule in the form of a ring. The nucleoid is usually attached at one point to the plasma membrane. In addition, up to 30 or 40 small, circular DNA molecules called *plasmids* may be present. The plasmids *replicate* (produce duplicate copies of themselves) independently of the large DNA molecule or chromosome, and the entire complement of plasmids often consists of copies of one or, at most, very few different plasmids. The chromosome and sometimes all or part of the plasmids replicate before the cell divides (Fig. 17.1).

The mitosis of eukaryotic cells does not occur in bacteria. Instead, there is an internal reorganization of material during which the two DNA molecules migrate to opposite ends of the cell. Then, at approximately the middle of the cell, the plasma membrane and cell wall, as a result of forming a transverse wall, divide the cell in two. Finally, the two new cells separate and enlarge to their original size, or in some bacteria, the cells

TABLE 17.1

Some Differences Between Eukaryotic and Prokaryotic Cells

EUKARYOTIC CELLS	PROKARYOTIC CELLS
1. Have a nuclear envelope	1. Lack a nuclear envelope
2. Have two to hundreds of chromosomes per cell; DNA is double stranded	2. Have a single, closed loop of double-stranded DNA plus (usually) several to 40 plasmids
3. Have membrane-bound organelles (e.g., plastids, mitochondria)	3. Lack membrane-bound organelles
4. Have 80S ribosomes[1]	4. Have 70S ribosomes[1]
5. Have asexual reproduction by mitosis	5. Have asexual reproduction by fission
6. Have sexual reproduction by fusion	6. Sexual reproduction unknown

1. S is a Svedberg unit, which is used to measure the rate at which a particle suspended in a fluid settles to the bottom when centrifuged. An 80S ribosome is larger than a 70S ribosome.

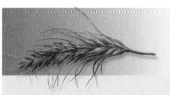

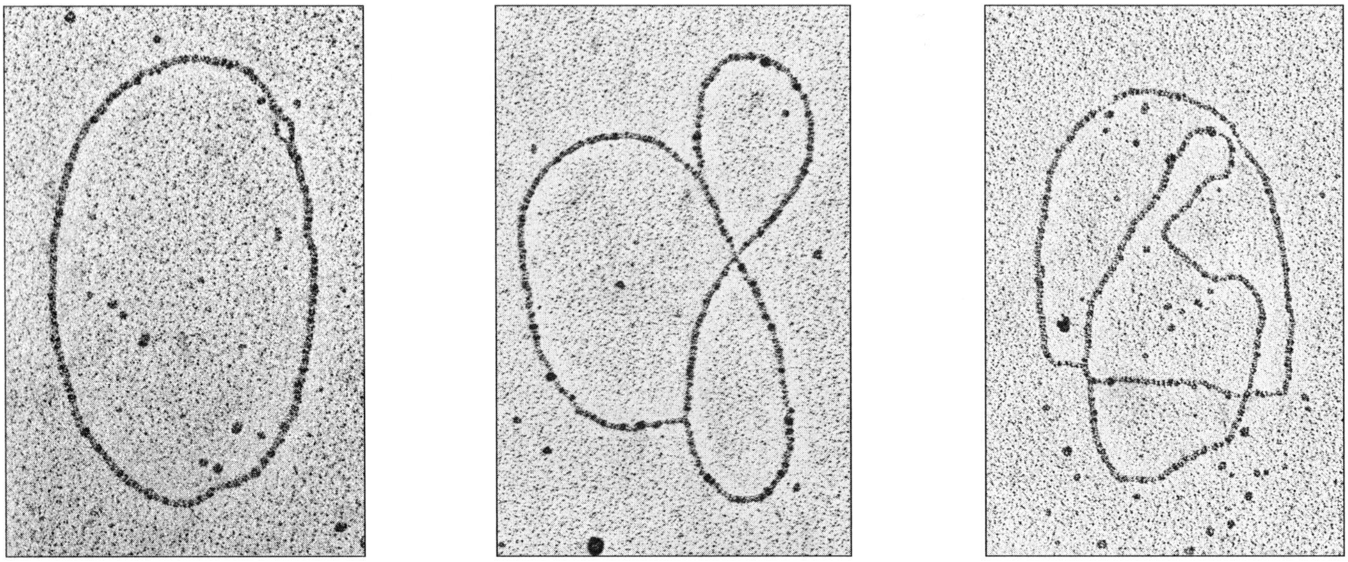

Figure 17.1 Duplication of a nucleoid in a bacterial cell. The duplication begins (*A*) at a single point or bubble (*upper right*) that expands (*B*) until the two new rings of DNA separate (*C*), ×150,000. *(Electron micrographs courtesy Tsuyoski Kakefuda)*

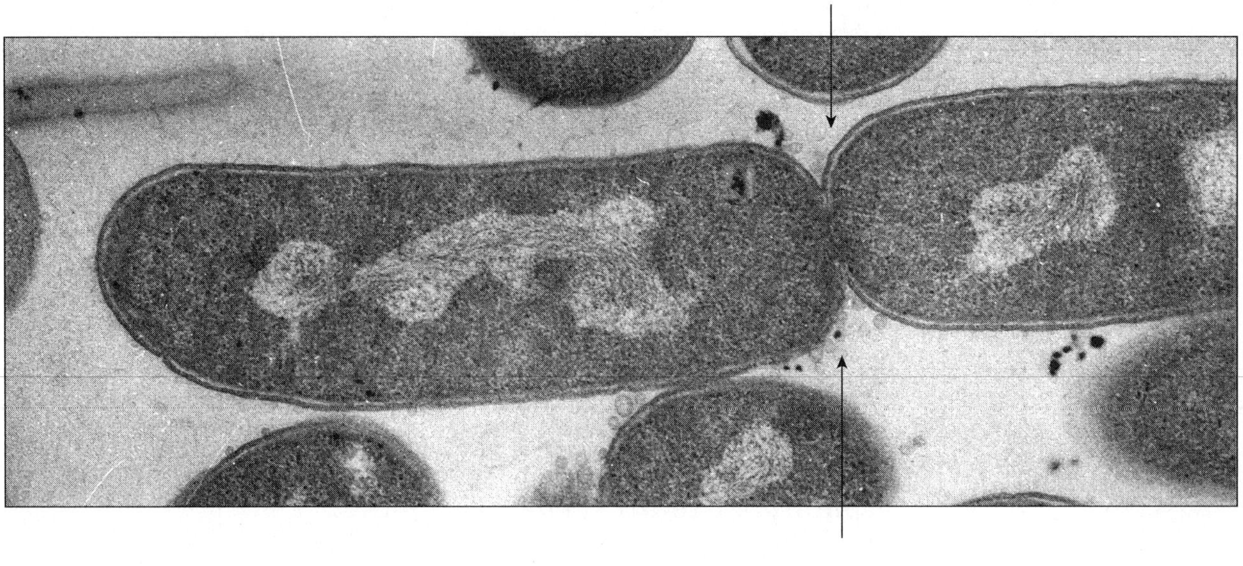

Figure 17.2 A dividing bacterial cell. The new wall is growing inward, ×48,750. *(Electron micrograph courtesy R.G.E. Murray)*

remain attached to each other in chains. This simple form of asexual reproduction, *fission,* is found so universally in bacteria that they were in the past called *schizomycetes,* which means "fission fungi" (Fig. 17.2). Bacteria, however, are not currently believed to be closely related to the fungi, as was the case when the name was originally applied.

Under ideal conditions of moisture, food supply, and temperature, a bacterium may undergo fission every 10 to 20 minutes. If a single bacterium of average size were to duplicate itself every 20 minutes for as little as 36 hours, a mass of bacteria numbering 322,981,536,679,200,000,000,000,000,000,000 and weighing 126,464,618,590 metric tons (137,438,953,472 tons) would be produced. If this mass were not compacted, its volume would exceed that of the earth. Of course, bacteria do not continue to reproduce at their maximum rate for very long because several factors prevent it. These factors include the exhaustion of food supplies and the accumulation of toxic wastes.

Bacteria do not produce gametes or zygotes, nor do they undergo meiosis. However, at least three forms of genetic recombination occur.

1. In *conjugation,* a plasmid and/or part of the DNA strand is transferred from a donor cell to a recipient cell, usually through a tiny, hollow, tubelike **pilus** while the two cells are in contact (Fig. 17.3). Once in the recipient cell, the DNA becomes part of the new cell. Bacterial

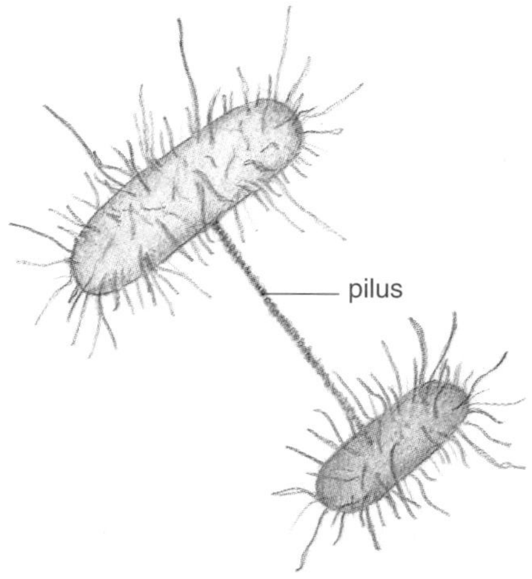

Figure 17.3 Conjugation in bacteria. Part of the DNA strand of the donor cell migrates through the hollow, tubelike *pilus* to the recipient cell, where it is incorporated into the DNA of its new cell. Shorter pili cover the surfaces of both cells.

cells that are capable of conjugating have different sexual forms of DNA, and usually only cells of different mating types undergo conjugation.

2. In *transformation,* a living cell acquires fragments of DNA released by dead cells into the medium in which it occurs and incorporates them into its own cell.
3. In *transduction,* fragments of DNA are carried from one cell to another by viruses (discussed toward the end of this chapter in the section entitled "Viruses").

Size, Form, and Classification of Bacteria

Except for the cyanobacteria, which are discussed separately, and a few giant bacteria, which attain lengths of up to 60 micrometers, most bacteria are less than 2 or 3 micrometers in diameter, and a few of the smaller species approach 0.15 micrometer. The latter are so small that 6,500 of them arranged in a row would not quite extend across the head of a pin. Because of their small size, bacteria are not visible individually to the unaided eye and are studied in the laboratory with electron microscopes or with the highest magnifications of light microscopes.

The thousands of different kinds of bacteria occur primarily in three forms: *cocci* (singular: *coccus*), which are spherical or sometimes elliptical; *bacilli* (singular: *bacillus*), which are rod shaped or cylindrical; and *spirilla* (singular: *spirillum*), which are in the form of a helix, or spiral (Fig. 17.4). Further classification is based on various visible features including, for example, development of slimy or gummy capsulelike sheaths around cells; color; presence of hairlike or budlike appendages; development of internal thicker-walled *endospores* and other cell inclusions; mechanisms of movement, including gliding movements by which threadlike groups of bacteria appear to slide back and forth

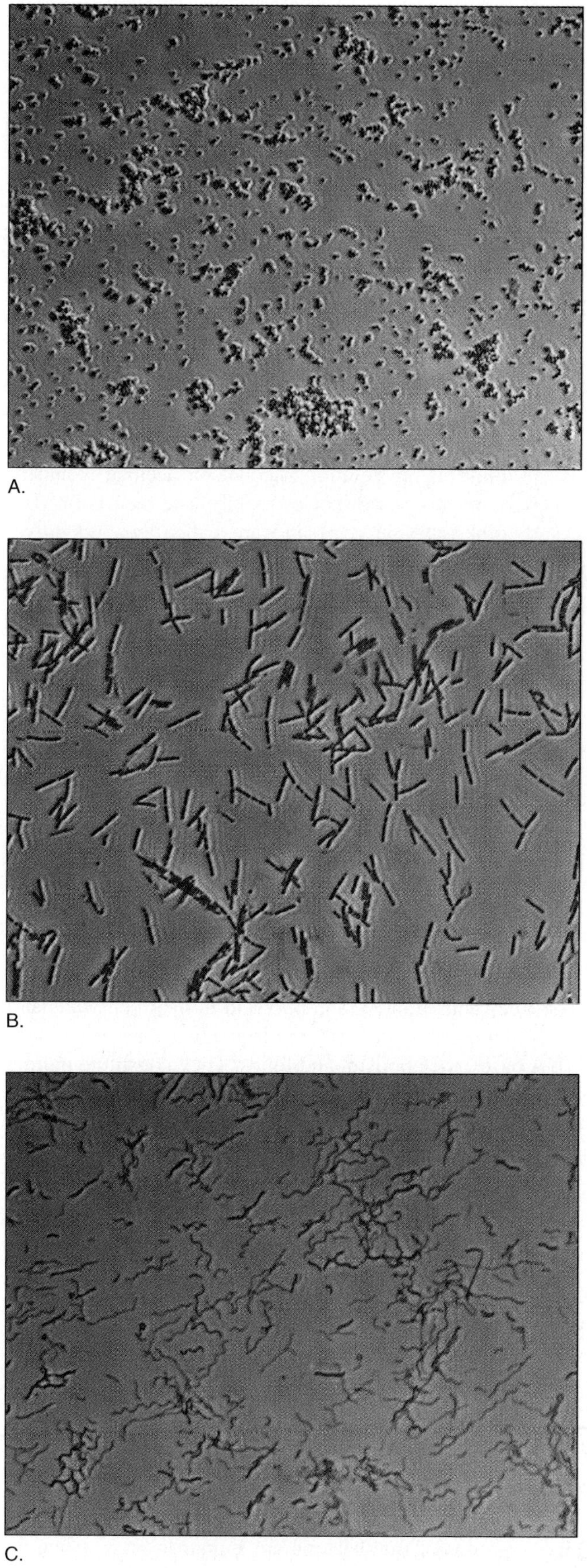

Figure 17.4 Three basic forms of bacteria. *A.* Cocci. *B.* Bacilli. *C.* Spirilli. ×1500.

lengthwise against each other; and features of colonies. Biochemical characteristics, including energy sources, cell-wall components, fermentation products, luminescence, optimum ranges of pH and temperature, oxygen relationships, amino acid sequences of proteins, and salt tolerance are also used in classification.

Some bacteria have slender flagella, usually about 5 to 10 micrometers in length, which propel them through fluid media. Others have somewhat shorter tubelike pili, which resemble flagella but do not function in locomotion. The pili apparently enable bacteria to attach themselves to surfaces or to each other, as in conjugation, which was discussed previously in the section "Cellular Detail and Reproduction of Bacteria."

True bacteria are also grouped into two large categories based on the reaction of their cell walls to a dye. After a heat-fixed smear of cells has been stained blue-black with a violet dye, a dilute iodine solution, alcohol, or acetone is added. When so treated, some species rapidly lose their color (but absorb pink safranin dye) and are called *gram-negative.* Others, called *gram-positive,* retain most of the blue-black color. Gram's stain, named after Christian Gram who discovered it in 1884, has many variations, but all the variations produce similar results. Other features used in classifying bacteria are indicated in the discussions that follow.

KINGDOM BACTERIA—THE TRUE BACTERIA

Phylum Bacteriophyta

Class Bacteriae—The Unpigmented, Purple, and Green Sulfur Bacteria

The true bacteria have muramic acid in their cell walls and are also fundamentally different from archaebacteria in their RNA bases, metabolism, and lipids. They constitute quite a heterogeneous assemblage, with most being **heterotrophic** (organisms that cannot synthesize their own food and therefore depend on other organisms for it).

The majority of heterotrophic bacteria are *saprobes* (living organisms that obtain their food from nonliving organic matter). Saprobic bacteria, along with fungi, are primarily responsible for decay and for recycling all types of organic matter in the soil. Some of their recycling activities are discussed in the section on the *nitrogen cycle* in Chapter 25. Other heterotrophic bacteria are *parasites* (living organisms that depend on other living organisms for their food).

Several parasites that cause important human diseases are discussed later in the chapter.

Autotrophic Bacteria

A few groups of true bacteria are similar to green plants in being **autotrophic;** that is, they are capable of synthesizing organic compounds from simple inorganic substances. Some carry on photosynthesis without, however, producing oxygen as a by-product. Included in the photosynthetic bacteria are the *purple sulfur bacteria,* the *purple nonsulfur bacteria,* and the *green sulfur bacteria.* The *cyanobacteria* (blue-green bacteria) and the *chloroxybacteria,* which do produce oxygen as a product of their photosynthesis, are discussed separately later in this chapter. Cells of the first two groups appear purplish, or occasionally red to brown, because of the presence of a mixture of greenish, yellow, and red pigments. Their greenish *bacteriochlorophyll* pigments (there are several closely related ones) are very similar to the chlorophyll *a* of higher plants. No plastids are present in bacteria, and their pigments are, instead, located in thylakoids or small spherical bodies. In their photosynthesis, the purple bacteria substitute hydrogen sulfide, the bad-smelling gas given off by rotten eggs, for the water used by higher plants. The generalized equation for the process is as follows:

$$\underset{\text{carbon dioxide}}{CO_2} + \underset{\text{hydrogen sulfide}}{2\,H_2S} \xrightarrow[\text{light}]{\text{bacteriochlorophyll}} \underset{\text{carbohydrate}}{(CH_2O)n} + \underset{\text{water}}{H_2O} + \underset{\text{sulfur}}{2S}$$

The equation for the purple nonsulfur bacteria is similar, but hydrogen from organic molecules is used instead of hydrogen sulfide. Green sulfur bacteria do use hydrogen sulfide, but their chlorophyll, called *chlorobium chlorophyll,* differs significantly in its chemistry from the chlorophylls of higher plants.

Other groups of bacteria are *chemoautotrophic;* that is, they are capable of obtaining the energy they require from various compounds or elements through chemical reactions involving the oxidation of reduced inorganic groups, such as NH_3, H_2S, and Fe^{++}, or the oxidation of hydrogen gas (as discussed in the section on oxidation-reduction reactions in Chapter 10).

Chemoautotrophic bacteria include *iron bacteria,* which transform soluble compounds of iron into insoluble substances that accumulate as deposits (e.g., in water pipes); *sulfur bacteria,* which can convert hydrogen sulfide gas to elemental sulfur and sulfur to sulfate; and *hydrogen bacteria,* which flourish in soils where they use molecular hydrogen derived from the activities of anaerobic and nitrogen-fixing bacteria. Nitrifying and nitrogen-fixing bacteria (some of which are heterotrophic) are a group of ecologically important bacteria that are discussed, along with their roles in the nitrogen cycle, in Chapter 25.

HUMAN RELEVANCE OF THE UNPIGMENTED, PURPLE, AND GREEN SULFUR BACTERIA

Composting and Compost

Long before the existence of bacteria was known or even suspected, primitive agriculturists made rough piles of weeds, garbage, manure, and other wastes. They watched as the piles became reduced to a fraction of their original volume and

Figure 17.5 Compost. *Left.* Garbage and leaves before composting. *Right.* After composting.

changed into **compost,** a dark, fluffy material that conditions and enriches soil when mixed with it. Today, the compost pile is at the heart of the activities of numerous organic gardeners and farmers. With many communities instituting ordinances forbidding the burning of leaves and refuse, and with space for waste disposal becoming scarce, a significant number of cities and towns have turned to the composting of street leaves and other materials. They have found that they not only save space but also are producing a useful, ecologically compatible product (Fig. 17.5).

Because bacteria are ubiquitous, a single leaf or nonliving tissues in a pile of any size will eventually be decomposed. Ideally, however, a compost pile is 2 meters wide, 1.5 meters deep, and at a minimum, 2 meters long (6 feet wide, 4.5 feet deep, and at least 6 feet long). If the pile is of lesser length or breadth, or greater depth, the conditions that favor the breakdown of materials (created by the bacteria themselves) develop at a slower pace.

Any accumulation of household garbage, leaves, weeds, grass clippings, and/or manure may be heaped together (Fig. 17.5), and it has been demonstrated that no so-called starter culture of microorganisms is needed to initiate decomposition. Once the ever-present decay organisms have material to decompose, their numbers increase rapidly, and much heat is generated. In fact, the temperature in the center of the pile often rises to 70°C (158°F). As it rises, populations of organisms adapted to higher temperatures replace those not as well-adapted. The remains of the latter are then added to the compost. The high temperatures also kill many weed seeds and most disease organisms.

The proportion of carbon to nitrogen present in a compost pile largely determines the pace at which the microbial activities proceed, a ratio of 30 carbon to 1 nitrogen being optimal. Microbial growth stops if the proportion of carbon gets much greater, and nitrogen is lost in the form of ammonia if the ratio drops lower.

If the materials are kept moist and turned occasionally to aerate the pile, composting may be completed in as little as 2 weeks. Shredding the materials exposes much more surface area for the microorganisms to work on and speeds up the process. Composters generally avoid or keep to a minimum a few materials, such as *Eucalyptus* and walnut leaves, which contain substances that inhibit the growth of other plants, when the compost is intended for use as a soil conditioner or crop fertilizer. Bamboo and some types of fern fronds are particularly resistant to decay bacteria and will decompose much more slowly than other materials. Domestic cat manure may contain stages of a parasite that can infect humans, although the organism should be killed if proper composting techniques are employed. Generally, however, almost any organic materials are suitable. Although compost adds humus to the soil and improves soil structure and moisture-holding capacity, its value as a fertilizer has certain limitations. The nitrogen content of good compost is about 2% to 3%, phosphorus 0.5% to 1%, and potash 1% to 2%, as compared with five to ten times those amounts in chemical fertilizers, which also require less labor to produce. Nevertheless, with the spiraling costs of chemical fertilizers, an increasing awareness of the problems accompanying their use, and greater public enlightenment concerning the accumulation and disposal of solid wastes, compost is undoubtedly destined to play a major role in pertinent agricultural practices of the future.

True Bacteria and Disease

It has been calculated that plant diseases alone cause American farmers losses of more than $4 billion per year. Although many plant diseases are caused by fungi, a number involve bacteria. Bacteria are involved in diseases of pears, potatoes, tomatoes, squash, melons, carrots, citrus, cabbage, and cotton. Bacteria also cause huge losses of foodstuffs after they have been harvested or processed. They are even better known, however, as the culprits in many serious diseases of animals and humans (Table 17.2).

Since they are so tiny and often incapable of independent movement, how are they transmitted, and what is it about their activities that causes them to fell creatures billions of times their size? We have learned much about their activities in the past century and know now that they can gain access to human tissues in a variety of ways.

Modes of Access of Disease Bacteria

Access from the Air Every time a person coughs, sneezes, or just speaks loudly, an invisible cloud of tiny saliva droplets is produced (Fig. 17.6). Each droplet contains bacteria or other microorganisms and a tiny amount of protein. The moisture usually evaporates almost immediately, leaving minute protein flakes to which live bacteria adhere. As normal breathing takes place, these may soon find their way into the respiratory tracts of other humans or animals, particularly if they have been confined within the air of a room. Fortunately, the natural resistance of most of those who acquire the bacteria in this way prevents the bacteria from multiplying to the point of causing a disease. When the

TABLE 17.2

Some Diseases of Humans Caused by Bacteria or Viruses

BACTERIAL DISEASES		VIRAL DISEASES	
Anthrax	Mastitis	AIDS	Meningitis (some)
Botulism	Meningitis (some)	Chicken pox/Shingles	Mononucleosis
Brucellosis	Pneumonia (some)	Common colds	Mumps
Bubonic plague	Syphilis	Diabetes (some)	Pneumonia (some)
Cholera	Tetanus	Encephalitis	Poliomyelitis
Diphtheria	Tuberculosis	Genital herpes	Rabies
Dysentery (some)	Typhoid fever	Infectious hepatitis	Rubella
Gonorrhea	Whooping cough	Influenza	Smallpox
Leprosy		Measles	Yellow fever

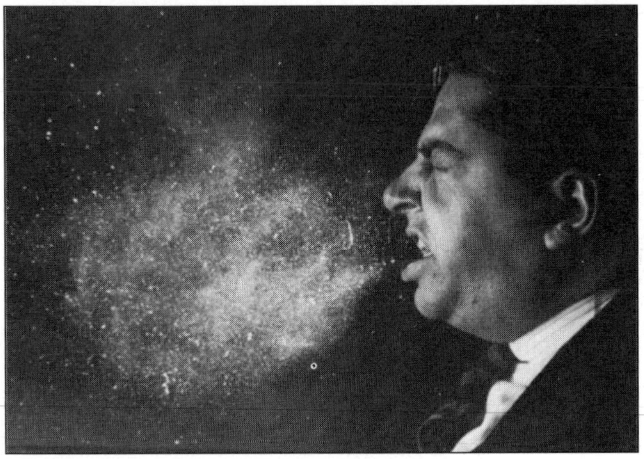

Figure 17.6 Stop-action photograph of a sneeze. *(Reprinted with the permission of Marshall W. Jennison, Syracuse University)*

resistance is not there, however, a number of diseases, including diphtheria, whooping cough, some forms of meningitis and pneumonia, and strep throat, can develop. Some bacteria such as anthrax require inhalation of a minimum of about 8,000 spores for the disease to develop internally in humans.

Psittacosis, a disease carried by birds, is caused by the inhalation of chlamydias, which are exceptionally minute organisms unable to manufacture their own ATP molecules. They are apparently energy parasites, depending on their host cells for the energy needed to carry on their own functions. Some chlamydias are transmitted sexually and in recent years have become a widespread human problem.

Access Through Contamination of Food and Drink

Food Poisoning and Diseases Associated with Natural Disasters. In the past, open sewers and unsanitary conditions for food preparation caused a number of bacterial diseases, including cholera, dysentery, *Staphylococcus* and *Salmonella* food poisoning, and typhoid and paratyphoid fevers, to reach epidemic proportions all over the world. Although *Staphylococcus* food poisoning, which is seldom fatal, is still fairly common today, developed countries now rarely see epidemics caused by other bacteria unless a natural disaster, such as a flood or a typhoon, disrupts normal sewage disposal. The diseases are more often spread by carriers who handle food, or by houseflies.

The United States and other countries with ocean shorelines ban the harvesting of shellfish at certain times because *Salmonella* bacteria may multiply enough in clams and mussels to cause illness in humans who eat them—particularly if they are not well cooked. The *Salmonella* bacteria multiply in the intestinal tract or spread from there to other parts of the body.

Legionnaire Disease. Some bacteria are very common on algae in freshwater streams, lakes, and reservoirs. One such bacterium, *Legionella pneumophila,* causes *Legionnaire disease,* which killed 34 members of the American Legion attending a convention in a Philadelphia hotel in 1976. An estimated 50,000 Americans are infected annually by Legionnaire disease bacteria, which nearly always pass through the human digestive tract harmlessly without multiplying. On rare occasions, however, something unknown triggers their reproduction, mostly in older males who are heavy smokers or alcoholics. The results are often fatal.

Botulism. The most deadly of all known biological toxins is produced by a bacterium with the scientific name of *Clostridium botulinum* (Fig. 17.7). The name comes from *botulus,* the Latin word for sausage, since the bacterium was first discovered after people had died from eating some contaminated sausage at a picnic.

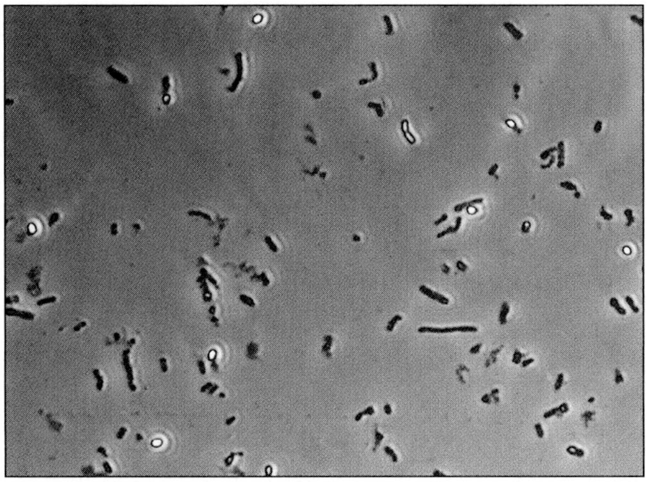

Figure 17.7 Botulism bacteria, ×1500. *(Courtesy Robert McNulty)*

Unlike *Salmonella* food poisoning, *botulism* is not an infection but is poisoning from a substance produced by bacteria that can grow and multiply anaerobically (in the absence of oxygen) in improperly processed or stored foods. Home-canned beans, beets, corn, and asparagus, in particular, have been known to permit the development of botulism bacteria. Just 1 gram (0.035 ounce) of the toxin is enough to kill 14 million adults, and a little more than half a kilogram (1.1 pound) could eliminate the entire human race.

The bacteria, which are present in most soils, produce unusually heat-resistant spores and are likely to be present on any soil-contaminated foods. They may not be destroyed during canning unless the food is heated for 30 minutes at 80°C (176°F) or boiled for 10 to 15 minutes. They ordinarily will not grow in foods that have been preserved in brines containing at least 10% salt (sodium chloride) or in fairly acid media, such as the juices produced by most stone fruits. The toxin is absorbed directly from the stomach and the intestines, affecting nerves and muscles and causing paralysis. While some antidotes are available, they are ineffective after symptoms have become advanced.

Deaths reported from botulism in the United States reached a peak of about 25 per year in the 1930s but have declined to 5 or 6 per year since then. There is evidence that some deaths of infants due to unknown causes are actually due to botulism. For reasons that are not clear, botulism bacteria, which pass harmlessly through the digestive tracts of humans over the age of 1 year, may germinate in the intestines of infants less than a year old. About 100 cases of sudden infant death reported worldwide each year appear to have been due to botulism, and the evidence now suggests that botulism is responsible for about 5% of the 8,000 cases of sudden infant death that occur each year in the United States alone.

Access Through Direct Contact

Syphilis and Gonorrhea. Bacteria responsible for diseases such as syphilis, gonorrhea, anthrax, and brucellosis enter the body through the skin or mucous membranes (i.e., those membranes lining tracts with openings to the exterior of the body). Both *syphilis* and *gonorrhea* are transmitted through sexual intercourse or other forms of direct contact, rarely through the use of public washroom towels or toilets. In recent years, gonorrhea has accounted for the largest number of reported cases of any communicable disease in the United States, and currently, despite increasing precautions taken by the public since the escalation of AIDS, more than 1 million Americans become infected every year. The symptoms of both syphilis and gonorrhea, which include persistent sores or discharges from the genitalia, sometimes disappear after a few weeks, leading victims to believe the body has healed itself. It is estimated that about 50% of females infected with gonorrhea at first exhibit no symptoms at all. More often than not, however, symptoms appear or reappear in different parts of the body at a later date. Because gonorrheal eye infections often occur in newborns as they pass through an infected birth tract, the United States, by law, requires a drop of antibiotic or silver nitrate solution be placed in the eyes of infants at birth. Both syphilis and gonorrhea are curable when treated promptly, but it is very important that such treatment be sought, since failure to do so can lead to sterility, blindness, and even death.

Anthrax. Anthrax, which is primarily a disease of cattle and other farm animals in addition to wild animals, is sometimes transmitted to humans, particularly workers in the wool and hide industries. Since September 2001, anthrax has received much publicity as an agent of bioterrorism. Like syphilis and gonorrhea, it can be effectively eliminated if treated early enough but may be fatal if allowed to progress. Brucellosis, another disease of farm animals, is occasionally transmitted to humans through direct contact or through the consumption of contaminated milk. It is sometimes called *undulant fever* because of a daily rise and fall of temperature apparently associated with the release of toxins by the bacteria.

Access Through Wounds

Tetanus and Gas Gangrene. When one steps on a dirty nail or is wounded in such a way that dirt is forced into body tissues, *tetanus* ("lockjaw") bacteria, which are common soil organisms, may gain access to dead or damaged cells. There they can multiply and produce a deadly toxin so powerful that 0.00025 gram (0.00000088175 ounce) is enough to kill an adult. In contrast, about 150 times that amount of strychnine is needed to achieve the same result. Control of tetanus through immunization is very effective and has become widespread. About 100 cases of tetanus are reported in the United States each year, most of the victims being intravenous drug users. Several related bacteria that gain access to the body in the same way are responsible for potentially fatal *gas gangrene,* which used to be feared on the battlefields in the past but is now controlled through the use of antibiotics and aseptic techniques.

Access Through Bites of Insects and Other Organisms

Bubonic Plague. Bubonic plague (the "Black Death") and tularemia are two bacterial diseases transmitted by fleas, deerflies, ticks, or lice that have been parasitizing infected animals, particularly rodents.

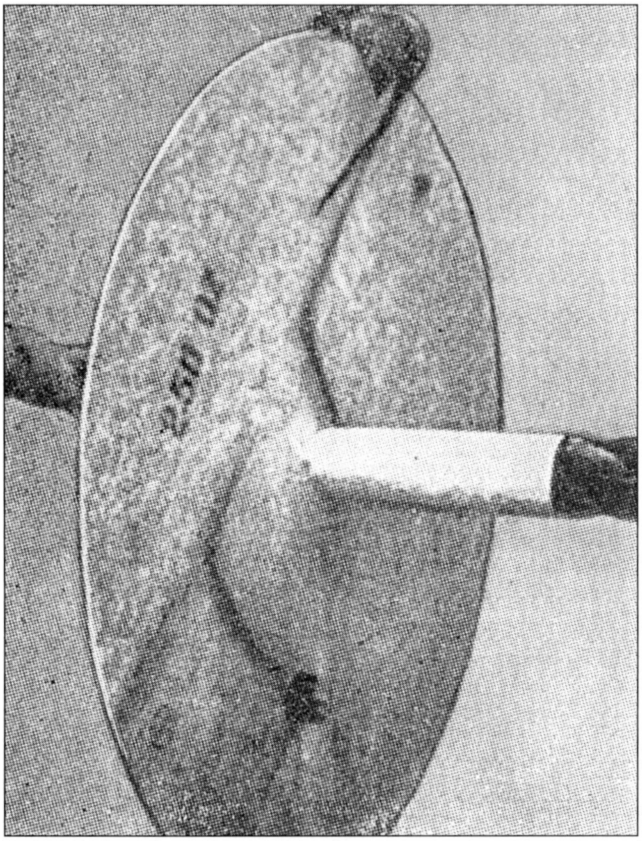

Figure 17.8 A rat climbing over a barrier on a ship's mooring line. *(Courtesy U.S. Public Health Service)*

Rat fleas, found on infected rats that inhabit dumps, sewers, barnyards, and ships (Fig. 17.8), acquire the bacteria for plague and then pass them on to humans through their bites. The disease has been found in ground squirrels and other rodents in the United States, particularly in the West, since 1900.

In the past, bubonic plague spread with great speed and reached devastating epidemic proportions. In 1665, in London, hundreds of thousands perished from the disease, and between 1347 and 1349, it is believed to have killed one-fourth of the entire population of Europe (about 25 million persons). Today, plague is rare in North America, but it still occasionally manifests itself in port cities of Asia, Europe, and South America. Control depends on control of rats and fleas, which are virtually impossible to eradicate entirely, and the use of vaccines, which produce immunity for about 6 to 12 months.

Tularemia, Rickettsias, and Mycoplasmas. Tularemia is primarily a disease of animals, but infected ticks or deerflies may transmit the disease to humans through bites. It is an occupational disease of meat handlers and is fatal in 5% to 8% of the cases. Ticks, lice, and fleas may also transmit *rickettsias,* which cause typhus and spotted fevers. Rickettsias are extremely tiny bacteria that live within eukaryotic cells.

Mycoplasmas, referred to in the past as pleuropneumonialike organisms (PPLOs), are also minute bacteria that may be transmitted in various ways. These have no cell walls and therefore are quite plastic. They are found in many plants, in hot springs, and in the moist surfaces of the respiratory and intestinal tracts of animals and humans. They are responsible for a form of human pneumonia and for numerous plant diseases. Like an increasing number of bacteria, they are resistant to penicillin.

Lyme Disease. Lyme disease, which was known in Europe before 1900, has spread rapidly throughout the United States since 1975 when an outbreak occurred at Lyme, Connecticut. Its arthritis-like effects are caused by a bacterium (*Borrelia burgdorferi*) hosted by deer and field mice. It is injected into the human bloodstream by deer ticks.

Koch's Postulates

Since there are so many bacteria present everywhere, how can we be certain that a given bacterium obtained from an infected person is actually the organism responsible for the observed disease? Robert Koch, who was a German physician, became known during the latter half of the 19th century for his investigations of anthrax and tuberculosis. As a result of his work, Koch formulated rules for proving that a particular microorganism is the cause of a particular disease. His rules, with minor modifications, are still followed today. They have come to be known as *Koch's Postulates,* and their essence is as follows:

1. The microorganism must be present in all cases of the disease.
2. The microorganism must be isolated from the victim in pure culture (i.e., in a culture containing only that single kind of organism).
3. When the microorganism from the pure culture is injected into a susceptible host organism, it must produce the disease in the host.
4. The microorganism must be isolated from the experimentally infected host and grown in pure culture for comparison with that of the original culture.

True Bacteria Useful to Humans

For many years, we controlled insect pests of food plants mostly through the use of toxic sprays. Residues of the sprays remaining on the fruits and vegetables have accumulated in human tissues, often with adverse effects, while at the same time, many organisms have become immune or resistant to the toxins. In addition, the sprays kill useful organisms, and precipitation runoff washes the toxins into streams, lakes, and oceans, harming or killing aquatic organisms.

As we have become aware of the undesirable effects of the use of toxic sprays, we have looked for alternative means of controlling crop pests. Today, many harmful pests and even weeds can be significantly limited through the use of biological controls, which are discussed in Appendix 2, and with transgenic plants, discussed in Chapter 14.

Bacillus Thuringiensis and Bacillus Popilliae

Three biological control bacteria have been registered for use by the U. S. Department of Agriculture. One, *Bacillus thuringiensis* (often referred to as Bt), has been remarkably effective against a wide range of caterpillars and worms, including peach tree borers, European corn borers, bollworms, cabbage worms and loopers, tomato and fruit hornworms (Fig. 17.9), tent caterpillars, fall webworms, leaf miners, alfalfa caterpillars, leaf rollers, gypsy moth larvae, and cankerworms.

The bacteria, which are easily mass-produced by commercial companies, are sold in the form of a stable wettable dust containing millions of spores. When the spores are sprayed on food plants, they are harmless to humans, birds, animals, earthworms, or any living creatures other than moth or butterfly larvae. When a caterpillar ingests any tissue with Bt spores on it, the bacteria quickly become active and multiply within the digestive tract, soon paralyzing the gut. The caterpillar stops feeding within 2 or 3 hours and slowly turns black, dropping off the plant in 2 to 4 days.

A.

B.

Figure 17.9 A tomato hornworm (*A*) before and (*B*) 3 days after spraying with *Bacillus thuringiensis.*

The toxin-producing gene from *Bacillus thuringiensis* was introduced into another bacterium, *Pseudomonas fluorescens,* which is used on corn to control black cutworms.

In 1983, a variety of *Bacillus thuringiensis* (var. *israelensis*) was introduced into the commercial market for the control of mosquitoes. Called BtI, the bacterium attacks only mosquito larvae ("wigglers") and one or two other pests. Six years of experiments performed on more than 70 species of fish, snails, shrimp, and insects demonstrated no adverse effects on either plants or animals other than mosquitoes (and a couple of lesser pests) even at dosages 100 times more powerful than those needed to kill mosquito larvae. Use of this bacterium to control mosquitoes in the future may constitute a significant step in lessening the ecological damage and disruption that so frequently accompanies the use of toxic chemicals for pest control.

Another bacterium, *Bacillus popilliae,* is also marketed in a powder form. When it is applied to soil where grubs of the highly destructive Japanese beetle are present, it causes what is known as "milky spore disease" in the grubs, which die in a few days. The spores are carried throughout the topsoil by rain water, foraging grubs, and by organisms that feed on the grubs.

Bioremediation

Naturally occurring bacteria have shown much potential in the developing science of *bioremediation,* which involves the study of the use of living organisms in the cleanup of toxic wastes and pollution. One bacterium produces enzymes that break down nitroglycerin and trinitrotoluene, which are contaminants in the soil around some munitions factories and explosive sites.

Preliminary tests indicate that the bacteria can decompose such waste and contaminant materials into harmless residues within 6 months. Other bacteria, such as *Pseudomonas cepacia,* can perform similar feats in oil spills and chemical dumps containing degreasers, such as trichloroethylene (TCE), creosote, and even 2,4,5-T, the Agent Orange defoliant of Vietnam fame. The pollution-fighting bacteria may, however, need a few nutrients added to the contaminants to stimulate their activities. Some scientists believe bacteria may also eventually be used to break down nuclear wastes. With the probability of the development of bacteria specifically engineered (see "Gene Splicing and Transgenic Plants" in Chapter 14) to deal with a host of pollution problems, their use for this purpose may become widespread in the future.

Other Useful Bacteria

Human eyes contain rhodopsin, a protein that is so sensitive it reacts to light in less than one millionth of a second. Certain bacteria that contain a form of rhodopsin have proved to be invaluable in research on the chemistry of vision and have led to our understanding of how eyes convert light energy into vision.

Bacteria play a major role in the dairy industry. Milk, which is composed of proteins, carbohydrates, fats, minerals, vitamins, and about 87% water, has exceptional nutritive value for animals and is also an excellent medium for the growth of many kinds of bacteria. Milk is sterile when secreted within the udder of the cow, but it picks up bacteria as it leaves the cow's body. It spoils rapidly if it is not obtained and stored under strictly sanitary conditions. Even after it has been pasteurized and refrigerated, the numbers of bacteria in it will increase the longer it stands. If milk is left in open containers in household refrigerators, for example, bacterial growth will not be kept in check for much longer than 24 hours.

Except for milk itself, either alone or in mixtures (e.g., ice cream), all dairy products are manufactured from raw or pasteurized milk by the controlled introduction of various bacteria. Such products include buttermilk, acidophilus milk, yogurt, sour cream, kefir, and cheese. Whey, the watery part of the milk separated from curd during cheese-making, is used, along with starches and molasses, for the commercial production of lactic acid. Lactic acid is used extensively in the manufacture of textile and laundry products, in the leather tanning industry, as a solvent in lacquers, and in the treatment of calcium and iron deficiencies in humans.

Beneficial bacteria, such as *Lactobacillus acidophilus* (a common organism in healthy digestive tracts), aid in digestion, reduce the risk of cancer, and may even reduce cholesterol levels. Acidophilus bacteria also have been widely used, along with antibiotics, to control or eliminate human female yeast infections. Researchers at the University of Minnesota are working on developing strains of intestinal bacteria that have an external layer of sticky polysaccharides that make them resistant to being eliminated by less desirable bacteria in the digestive tract.

In their metabolism of various sugars, proteins, and other organic substances, bacteria also produce waste products that have important industrial uses. Such products are often produced in large quantities by culturing bacteria in huge vats. These products include acetone, used in the manufacture of photographic film; explosives; solvents (e.g., nail polish remover); butyl alcohol, used in the manufacture of synthetic lacquers; dextran, used as a food stabilizer and as a blood plasma substitute; sorbose, used in the manufacture of ascorbic acid (vitamin C); and citric acid, which is the principal citrus-like tart flavoring of soft drinks, candies, and other foods. Some vitamin and medicinal preparations also involve bacterial synthesis.

Bacteria are used in the curing of vanilla pods, cocoa beans, coffee, and black tea and in the production of vinegar, sauerkraut, and dill pickles. Fibers for linen cloth are separated from flax stems by bacteria, and green plant materials are fermented in silos to produce ensilage for cattle feed. In recent years, the production of several important amino acids by bacteria has been exploited commercially. More than 6,800 metric tons (7,500 tons) of one amino acid, glutamic acid, are produced in North America each year. This is in demand as a flavor-enhancing agent in the form of monosodium glutamate.

In 1989, Patricia Mertz of the University of Miami discovered that *Brevibacterium,* the genus of bacteria responsible for the odor of Limburger and Brie cheeses, also is the source of foot odor in certain people. Bactericides are being tested to improve foot pads that merely absorb odor rather than kill the bacteria.

CLASS CYANOBACTERIAE—THE CYANOBACTERIA (BLUE-GREEN BACTERIA)

Introduction

In the past, algae as a group have been distinguished from other organisms in being photosynthetic, in having relatively simple structures, and, for those reproducing sexually, in having sex structures consisting of a single cell. As more became known about cellular details, however, it became apparent that the differences between the *cyanobacteria* (*blue-green bacteria*) (formerly known as the blue-green algae) and true algae are quite basic. Like all other members of Kingdom Bacteria, cyanobacteria have prokaryotic cells, while all algae that are assigned to Kingdom Protista have eukaryotic cells. Cyanobacteria, in fact, are so much like other true bacteria (Fig. 17.10) that biologists generally have abandoned the reference to them as algae and consider them true bacteria.

The main distinctions between organisms traditionally regarded as bacteria and cyanobacteria are (1) cyanobacteria have chlorophyll *a,* which is found in higher plants, and oxygen is produced when they undergo photosynthesis; (2) cyanobacteria also have blue *phycocyanin* and red *phycoerythrin* pigments known as *phycobilins;* (3) cyanobacteria are the only organisms that can both fix nitrogen and produce oxygen—a paradox, since nitrogen-fixation is essentially an anaerobic process. Except for the *prochlorobacteria,* other bacteria capable of carrying on photosynthesis do not produce oxygen, and they do not have chlorophyll *a.*

Distribution

Cyanobacteria are found in almost as diverse a variety of habitats as other true bacteria. They are common in temporary pools or ditches, particularly if the water is polluted. They are not found in acidic waters, but they are abundant in other fresh and marine waters around the globe, from the frozen lakes of Antarctica to warm tropical seas. In the open oceans, cyanobacteria are the principal photosynthetic organisms in plankton, the tiny cells of the cyanobacterium *Synechococcus* commonly occurring in concentrations of 10,000 cells per milliliter. *Trichodesmium* is a marine, filamentous, nitrogen-fixing cyanobacterium that forms extensive mucilage-producing colonies in some tropical waters. Other bacteria multiply in the mucilage and become food for protozoa.

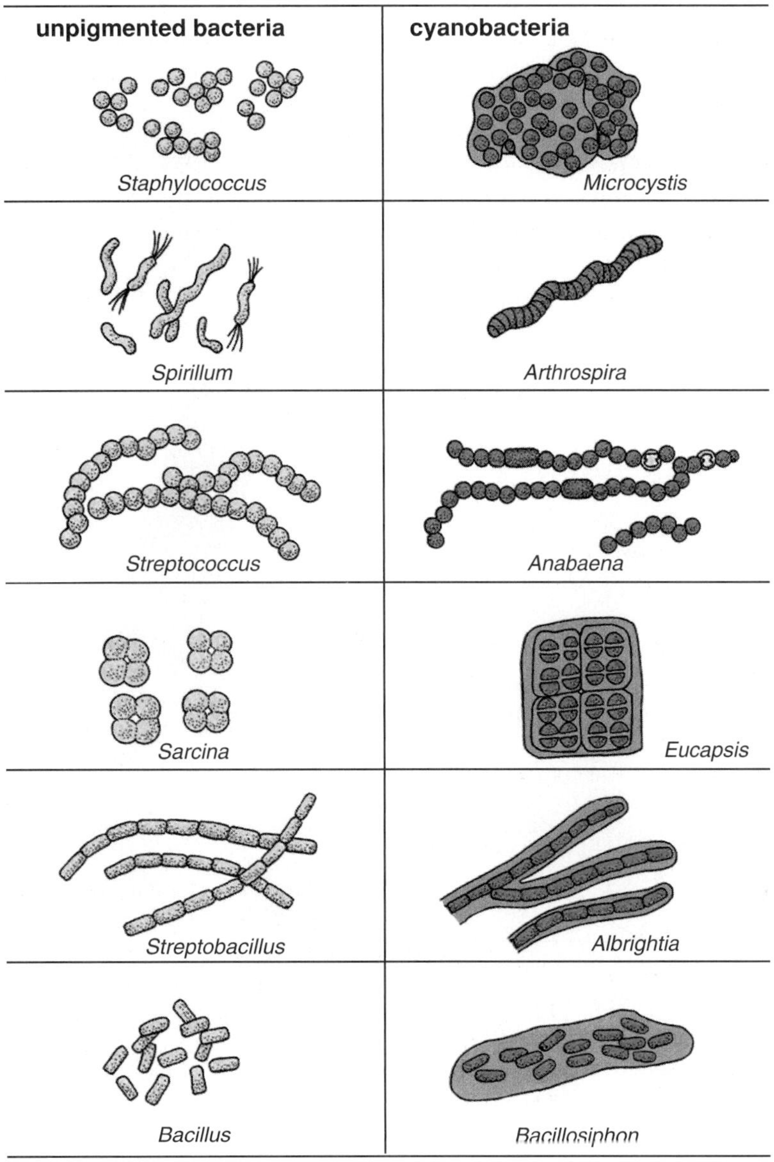

Figure 17.10 Similarity of form between various unpigmented bacteria and cyanobacteria, ×2000. *After M. J. Pelczar, Jr., and R. D. Reid. 1972.* Microbiology. *3d ed. The McGraw-Hill Companies.*

A different species of cyanobacterium is found in each temperature range of the hot springs of Yellowstone National Park, where water temperatures approach 85°C (185°F). There the bacteria precipitate chalky, insoluble carbonate deposits, which become a rocklike substance called *travertine.* The deposits accumulate at the rate of up to 2 to 4 millimeters per week, with other cyanobacteria often forming brilliantly colored streaks in the travertine.

Cyanobacteria are often the first photosynthetic organisms to appear on bare lava after a volcanic eruption, and they also thrive in the tiny fissures of desert rocks. Some are found in jungle soils or on the shells of turtles and snails, while others live symbiotically in various types of organisms, including amoebae and other protozoans, diatoms, certain sea anemones and their relatives, some fungi, and in the roots of tropical palmlike plants called *cycads.* They also flourish in tiny pools of water formed at the bases of the leaves of tropical grasses and other plants, and they are well-known components of "compound" organisms called *lichens,* which consist of a fungus and a photosynthetic partner.

Form, Metabolism, and Reproduction

The cells of cyanobacteria often occur in chains or in hairlike filaments, which are sometimes branched. Several species occur in irregular, spherical, or platelike colonies, the individual cells being held together by the gelatinous sheaths they secrete (Fig. 17.11). The sheaths may be colorless or

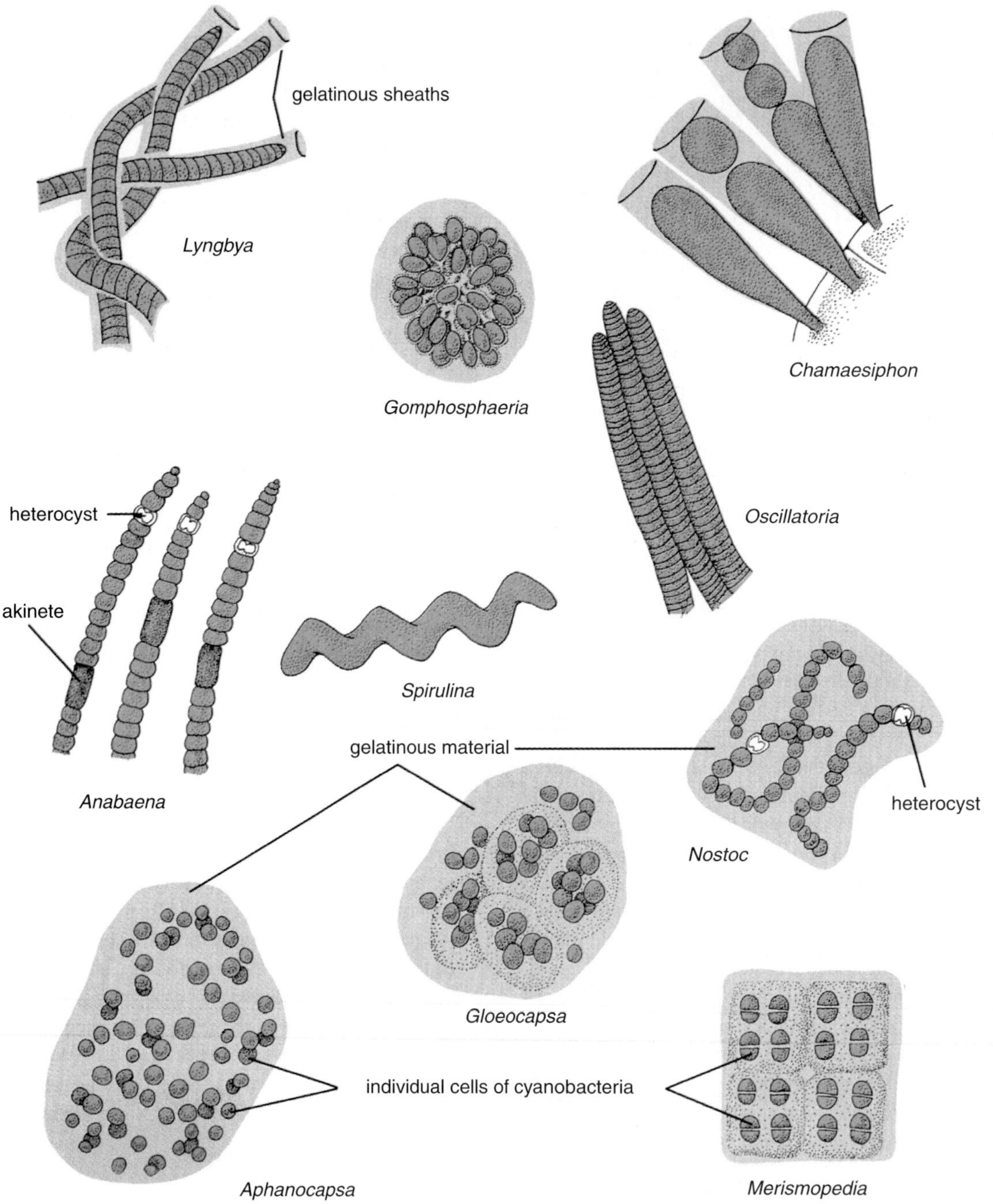

Figure 17.11 Representative cyanobacteria, ×2500.

pigmented with various shades of yellow, red, brown, green, blue, violet, or blue-black, which makes some colonies quite striking in appearance.

The cells themselves appear blue-green in color in about half of the approximately 1,500 known species. This color, which is due to the presence of green chlorophyll *a* and blue phycocyanin, is often masked by the presence of other pigments. Several yellow or orange carotenoid pigments similar to those of higher plants are usually present, and varying amounts of phycoerythrin (a red phycobilin pigment in a form unique to the cyanobacteria) may give the cells a distinct red color. The periodic appearances of large numbers of cyanobacteria with considerable amounts of phycoerythrin are believed to have given the Red Sea its name.

The organisms produce a nitrogenous food reserve called *cyanophycin.* The production of such food reserves is atypical for prokaryotic organisms. Cyanobacteria also produce and store carbohydrates and lipids. Flagella are unknown in the cyanobacteria, but some of these organisms are nevertheless capable of movement. *Oscillatoria* filaments (see Fig. 17.11), for example, seem to rotate on axes and move in a gliding fashion, apparently by the twisting of minute fibrils inside the cell walls while secreting mucilage that reduces friction. New cells are formed through fission, while new colonies or filaments may arise through fragmentation (breaking up) of older ones.

In the common genera *Nostoc* and *Anabaena* (see Fig. 17.11), which form chains of cells, fragmentation often occurs at special, larger, colorless, nitrogen-fixing cells called **heterocysts,** which are produced at intervals in the chains. Members of these two genera also may produce thick-walled cells called *akinetes,* which can resist freezing and other adverse conditions. When favorable conditions return, this survival feature enables the cells to germinate and become new chains or filaments. Some akinetes have been known to germinate after lying dormant for more than 80 years.

Cyanobacteria do not produce gametes or zygotes and do not undergo meiosis. Genetic recombination has, however, been reported—apparently taking place in similar fashion to that reported for other bacteria—but its occurrence evidently is rare.

Cyanobacteria, Chloroplasts, and Oxygen

Cyanobacteria that occur symbiotically in other organisms commonly lack a cell wall and appear to function essentially as chloroplasts. When a eukaryotic cell containing chloroplasts divides, the chloroplasts divide at the same time. The cells of cyanobacteria occurring within the cells of other organisms divide in similar fashion, leading to speculation that chloroplasts originated as cyanobacteria or *prochlorobacteria* (discussed after the cyanobacteria), living within other cells.

Fossils of photosynthetic organisms believed to be 3.5 billion years old and closely resembling present-day cyanobacteria have been found in Australia. It was not until half a billion years later, about 3 billion years ago, that these organisms began producing oxygen as a by-product of photosynthesis. The oxygen slowly began to accumulate, becoming substantial about 1 billion years ago. At the same time the oxygen was accumulating, other photosynthetic organisms appeared, and forms of aerobic respiration developed.

Within the last half billion years, enough *ozone,* which is a by-product of oxygen, accumulated to become an effective shield against most of the harmful ultraviolet radiation coming from the sun. Photosynthetic organisms, which had been protected from the radiation by their watery environments, were then able to survive on land.[2] Accordingly, cyanobacteria appear to have played a fundamental role in almost the entire history of living organisms.

2. See also the discussion in Chapter 18 of the role of certain green algae in the transition of aquatic organisms to land, and the discussion in Chapter 25 of the effects of ozone depletion on our environment.

Human Relevance of the Cyanobacteria

Cyanobacteria are included among the many aquatic and photosynthetic organisms at the bottom of various food chains. They store energy in the sugar they produce through photosynthesis. Small fish and crustaceans feed on them, only to be eaten by larger fish, which then become food for other aquatic organisms or humans.

During warmer months, various algae and cyanobacteria may become abundant in bodies of fresh water, especially if the water is polluted. A floating scum or mat called a *bloom* may extend across a square meter or two of small bodies of water to 2,000 or more square kilometers (800 square miles) of the surfaces of larger lakes in late summer (Fig. 17.12). Cyanobacteria, in particular, tend to become abundant in blooms when agricultural runoffs are high in nitrogen and phosphorus. The water in the vicinity of a bloom often acquires a "fishy" or otherwise objectionable odor or taste.

Anabaena, Microcystis, and other genera of cyanobacteria produce toxic substances that can kill both domestic and wild animals. Cattle, hogs, horses, sheep, rabbits, dogs, and

Figure 17.12 An algal bloom on a freshwater canal.

even poultry have been poisoned through drinking water that had an algal-bacterial bloom. Many poisonous fish that are immune to the toxins of certain cyanobacteria become poisonous to their predators only after feeding on the bacteria.

While the cyanobacteria are alive, they produce oxygen, and the oxygen content of the water is temporarily increased. Later, however, decay bacteria decomposing the bodies of organisms that have died may deplete the available oxygen so much that fish and other organisms are killed.

Cyanobacteria can be attacked by specific viruses called *cyanophages,* and there has been considerable interest in using the viruses to control blooms in lakes and ponds. Many of the cyanobacteria responsible for the blooms, however, have thick, mucilaginous sheaths that may prevent viruses (and, incidentally, fungi) from penetrating the cells. Blooms that develop in calm ocean waters in the tropics can be even more extensive than those of fresh waters. A marine bloom once reported between the shore and the Great Barrier Reef of Australia was 1,600 kilometers (almost 1,000 miles) long and covered 52,000 square kilometers (20,000 square miles) of ocean.

Most cyanobacteria are not palatable to humans, but there are exceptions. Species of *Spirulina* have been used for food by the natives of the Lake Chad region of central Africa and areas around Mexico City for centuries. *Spirulina,* which has a signficant vitamin content, is now cultured commercially and sold in health-food stores. The Japanese use several cyanobacteria, including two species of *Nostoc,* as side dishes, and the Chinese treat *Nostoc commune* as a delicacy. A few colorless forms are mild parasites in humans and animals.

Some strains of certain *Nostoc* species produce antibiotics that kill related strains of the same species. *Scytonema hofmannii* and other cyanobacteria produce antibiotics called *cyanobacterins* that kill many different forms of both cyanobacteria and eukaryotic algae. These particular cyanobacteria undoubtedly play a role in their own survival by inhibiting growth of competing organisms.

Swimmers in Hawaii occasionally suffer from "swimmers' itch," a severe skin inflammation that is caused by a species of *Lyngbya* that sometimes becomes abundant. Ironically, the toxin produced by these organisms has been demonstrated to suppress leukemia and several other types of cancer.

In human water supplies, cyanobacteria frequently clog filters, corrode steel and concrete, cause natural softening of water, and produce undesirable odors or coloration in the water. Many communities control cyanobacteria in reservoirs through the addition of very dilute amounts of copper sulfate.

More than 40 species of cyanobacteria are known to fix nitrogen from the air at roughly the same rates as the nitrogen-fixing bacteria of the leguminous plants discussed in Chapter 25. They may be more important than originally thought in this regard. In Southeast Asia, so much usable nitrogen is produced in the rice fields by naturally occurring cyanobacteria that rice is often grown for many years on the same land without the addition of fertilizer.

CLASS PROCHLOROBACTERIAE—THE PROCHLOROBACTERIA

In 1976, Ralph A. Lewin of the Scripps Institute of Oceanography announced the discovery of unicellular, prokaryotic organisms with bright green cells that were living on marine animals called *sea squirts* found in shallow marine waters of Baja California. These organisms, which were given the name *Prochloron,* have the chlorophylls *a* and *b* of higher plants but no trace of the phycobilin accessory pigments associated with cyanobacteria. Instead, their accessory pigments were confined to the carotenoid pigments found in higher plants. Also, unlike the single membrane thylakoids of cyanobacteria, those of the new organisms are double.

Lewin considered the pigment differences between cyanobacteria and the bright green cells of *Prochloron* to be basic enough to warrant recognition at the division level, and he proposed a new division to be known as the *Prochlorophyta.* Many microbiologists are reluctant to recognize these organisms as belonging to a separate bacterial division because the prokaryotic cell structure and chemistry is similar to that of cyanobacteria and other true bacteria (Fig. 17.13), but others

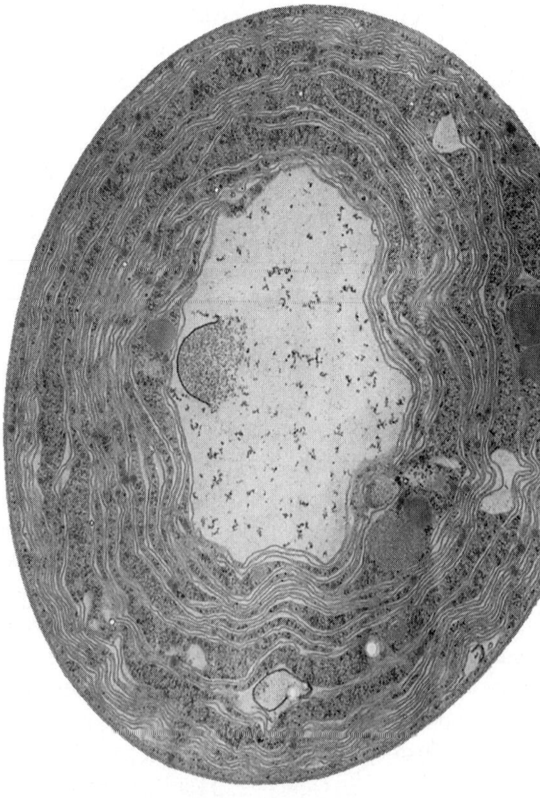

Figure 17.13 A section through a cell of the prochlorobacterium *Prochloron,* which lives on the surface of sea squirts (marine animals). Note the absence of a nucleus and other organelles, and the concentric layers of membranes that perform some of the functions of organelles, ×15,750. *(Electron micrograph courtesy Jean Whatley)*

agree with Lewin's assessment of the significance of the pigment system. While the pigment system is, indeed, significant, they are treated here as a class of true bacteria because their remaining structure and features are essentially indistinguishable from those of other true bacteria.

In 1984, Dutch biologists discovered a similar organism, which they named *Prochlorothrix,* in shallow lakes in the Netherlands. It differs from *Prochloron* in being free-living and filamentous. In the late 1980s, Sally Chisholm of the Massachusetts Institute of Technology identified yet another marine prochlorobacterium that flourishes in dim light at a depth of about 100 meters (328 feet). This organism now appears to be one of the two most numerous bacteria living in ocean waters.

The discovery of prochlorobacteria adds weight to the theory that chloroplasts may have originated from such cells living within the cells of other organisms, especially since the pigments involved are identical with those of higher plants.

KINGDOM ARCHAEA—THE ARCHAEBACTERIA

The archaebacteria represent one of two quite distinct lines of the most primitive known living organisms. They are fundamentally different in their metabolism from the other line of bacteria, the true bacteria, and also differ in the unique sequence of bases in their RNA molecules, the lack of muramic acid in their walls, and in the production of distinctive lipids.

In the early 1980s, these basic differences led University of Illinois microbiologist Carl Woese and his colleagues, who have conducted considerable research on the archaebacteria, to suggest that the organisms should be separated from the true bacteria in a kingdom of their own, a suggestion that has been widely adopted. Three distinct groups of bacteria are included in the archaebacteria.

The Methane Bacteria

The *methane bacteria,* which comprise the lion's share of the archaebacteria, are killed by oxygen and are active only under anaerobic conditions found in swamps, ocean and lake sediments, hot springs, animal intestinal tracts, sewage treatment plants, and other areas not open to the air. Their energy is derived from the generation of methane gas from carbon dioxide and hydrogen.

Methane, or "marsh gas," is a principal component of natural gas and may have been a major part of the earth's atmosphere in early geologic times. It still is present in the atmospheres of the planets Jupiter, Saturn, Uranus, and Neptune and is the main ingredient of firedamp, which causes serious explosions in mines. Methane will burn when it constitutes only 5% to 6% of the air, and a flitting, dancing light called *ignis fatuus,* or "will-o'-the-wisp," which is occasionally seen at night over swamps and marshy places, is said to be due to the spontaneous combustion of the gas.

The Salt Bacteria

Commercial salt evaporation ponds and other shallow areas in bodies of water with high salt content often have a unique appearance from above. They can be strikingly red due to the presence of a distinctive group of archaebacteria. These are the *salt bacteria,* whose metabolism enables them to thrive under conditions of extreme salinity that instantly kills other living cells. The bacteria carry on a simple form of photosynthesis with the aid of a membrane-bound red pigment called *bacterial rhodopsin.* The concentration of salt inside the cells is much lower than in their surroundings, but their metabolism is so closely tied to their environment that the bacteria die if placed in waters with lower salt concentrations (Fig. 17.14).

Figure 17.14 The north end of Utah's Lake Bonneville, as seen from the air. The water has a very high salt content. The reddish areas are due to a red pigment produced by salt bacteria. The pigment, called *bacterial rhodopsin,* is involved in a form of photosynthesis.

The Sulfolobus Bacteria

The *sulfolobus bacteria* constitute a third group of archaebacteria whose members occur in sulfur hot springs. The extraordinary metabolism of these bacteria allows them to thrive at very high temperatures—mostly in the vicinity of 80°C (176°F), with some doing very well at only 10°C below the boiling point of water. One genus (*Pyrodictium*), discovered in superheated ocean-floor areas, has a minimum temperature requirement of 82°C (179°F), an optimum growth temperature of 105°C (220°F), and can tolerate 110°C (230°F). The environment of one order of these thermophilic bacteria is also exceptionally acidic, their hot-springs habitats often having a pH of less than 2 (the neutral point on the 14-point pH scale is 7). One genus (*Thermoplasma*) in this group is bounded by only a plasma membrane and has no cell wall. It is found only in the embers of coal tailings. Another genus (*Thermoproteus*) appears to be confined to the geothermal areas of Iceland.

James Lake and his colleagues discovered that the shape of the ribosomes of sulfolobus bacteria is significantly different from those of other archaebacteria, true bacteria, and eukaryotes and that the chemistry of sulfur-dependent bacteria also distinguishes them from other archaebacteria. They have proposed a third kingdom of prokaryotes named *Eocytes* (dawn cells). Other microbiologists are hesitant to base kingdom status on ribosome shape, but the controversy over the significance of these discoveries will undoubtedly continue as further research into the matter is pursued.

Human Relevance of the Archaebacteria

Archaebacteria are significant to humans in several ways. In the future, methane bacteria may be used on a large scale to furnish energy for engine fuels and for heating, cooking, and light, since the methane gas they produce can be used to replace the methane in natural gas. Methane has an octane number of 130 and has been used as a motor fuel in Italy for over 40 years. It is clean, nonpolluting, safe, nontoxic, and prolongs the life of automobile engines, also making them easier to start. The methane is given off by the bacteria as they "digest" organic wastes in the absence of oxygen. Nine kilograms (20 pounds) of horse manure or 4.5 kilograms (10 pounds) of pig manure fed daily into a methane digester will produce all the gas needed for the average American adult's cooking needs. Considerably less green plant material is required to produce the same amount of methane.

The digester basically consists of an airtight drum connected by a pipe to a storage tank with a means of drawing off the sludge left after the gas has been produced (Fig. 17.15). The sludge itself makes an excellent fertilizer, although many sludges that originate from municipal and large-scale agricultural plants carry with them toxic levels of metals.

France had over 1,000 methane plants in operation by the mid-1950s, and in India, where cows produce over 812 million metric tons (800 million tons) of manure per year, many villages satisfy their fuel needs with manure-fed methane gas digesters. They are now being employed on a

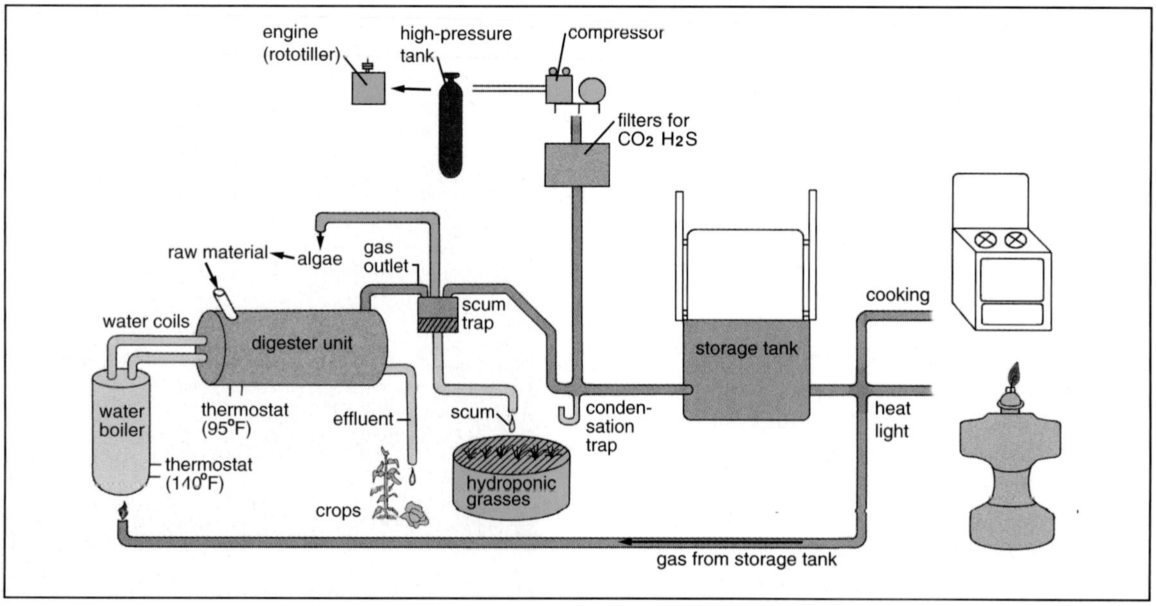

Figure 17.15 An integrated organic digester operation for the production of methane gas. *Reprinted from* Producing Your Own Power (c) *1974 by Rodale Press, Inc. Permission granted by Rodale Press, Inc., Emmaus, PA, 18049.*

small scale in rural areas in the United States. Methane is the primary source of hydrogen in the commercial production of ammonia.

VIRUSES

Introduction

Smallpox is a communicable disease that apparently was widespread for thousands of years, periodically killing countless numbers of individuals. As it developed in its victims, it appeared as blisterlike lesions on the skin, which later often became permanent pits or depressions. In 1901, an outbreak of smallpox in New York caused 720 deaths. Since that time, however, it has been eliminated in the United States through vaccination. Vaccination involves the introduction of a weakened form of disease agent to the body. The body's natural defenses, in fighting the agent, build up an immunity to the disease. The first known vaccinations were performed in England by Benjamin Jesty, a farmer, and Edward Jenner, a country physician. They both had noticed that farmhands working with cows having cowpox, a comparatively mild disease related to smallpox, did not contract smallpox itself during the devastating epidemics of the disease that occurred in Europe from time to time.

In 1796, Jenner scratched the skin of a boy with fluid he had obtained from a cowpox blister on the hand of a milkmaid. Six weeks later, he deliberately inoculated the boy with fluid from a blister of a smallpox victim, but the boy developed no symptoms of the disease. Jenner thus had performed a successful vaccination against a dread disease more than 50 years before Louis Pasteur developed the germ theory of disease.

The World Health Organization believes that as a result of vaccinations and vigilance, smallpox had essentially been eradicated throughout the world, but there has been speculation that it may have been cultured by rogue nations and could reappear as an agent of bioterrorism.

Size and Structure

We know now that smallpox was caused by something considerably smaller in size than bacteria. During Pasteur's time, virtually all infectious agents, including bacteria, protozoans, and yeast, were called **viruses.** One of Pasteur's associates, Charles Chamberland, discovered that porcelain filters would block out bacteria but would not keep an unseen agent from passing through. The agent caused rabies, another serious disease of both animals and humans. Agents of disease that could pass through filters became known as *filterable viruses,* although the word *filterable* is no longer used. Today, we know that not only smallpox and rabies are caused by these viruses but also measles, mumps, chicken pox, polio, yellow fever, influenza, fever blisters, warts, and the common cold.

Only those organisms that have certain unique features are now called viruses. These features, which include a complete lack of cellular structure, make viruses quite different from anything else in the six kingdoms of living organisms we now recognize. In fact, some question whether or not viruses are even living organisms. In 1946, Wendell Stanley, an American chemist, received a Nobel Prize for demonstrating that a virus causing tobacco mosaic, a common plant disease, could be isolated, purified, and crystalized and that the crystals could be stored indefinitely but would always produce the disease in healthy plants at any time they were placed in contact with them. We also know that viruses do not grow by increasing in size or dividing, nor do they respond to external stimuli. They cannot move on their own, and they cannot carry on independent metabolism.

Viruses are incredibly numerous. In 1989, for example, marine biologists at the University of Bergen in Norway discovered that a teaspoon of sea water typically contains more than 1 billion viruses. They are about the size of large molecules, varying in diameter from about 15 to 300 nanometers (Fig. 17.16). Thousands of the smallest ones could fit inside a single bacterium of average size.

Viruses consist of a nucleic acid core surrounded by a protein coat. The architecture of the protein coats varies considerably, but many have 20 sides and resemble tiny geodesic domes, while others have distinguishable head and tail regions. The nucleic acid core consists of either DNA or RNA—never both. Viruses have been classified in several ways. Originally, they were grouped according to their hosts and the types of tissues or organs they affected. Now they

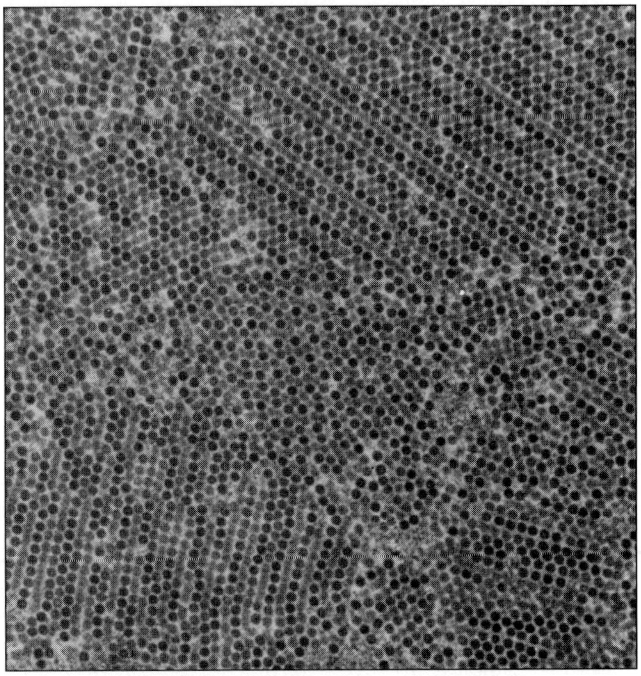

Figure 17.16 Papavoviruses in a human wart, ×52,000. *(Electron micrograph courtesy Richard S. Demaree Jr.)*

Awareness

Plant Viruses

The book *Hot Zone* and the movie *Outbreak* have created an awareness of emerging viruses and their dangers to the human population. The Ebola, Hanta, and HIV viruses are now everyday words that have become synonymous with death. There is another group of viruses that also has a significant impact—plant viruses—which cause an estimated $15 billion worth of crop loss per year worldwide. They infect plants and cause hundreds of diseases, such as tomato spotted wilt disease, tobacco mosaic disease, maize stripe disease, and apple chlorotic leaf spot disease.

In total, nearly 400 plant viruses have been identified and classified by the International Committee on Taxonomy of Viruses (ICTV). Another 320 have been identified but are awaiting final classification.

Surprisingly, the first viruses ever identified were in plants. In 1898, a Dutch professor of microbiology, Dr. Martinus Beijerinck, was working to identify the disease that caused tobacco leaves to become mottled with light green and yellow spots. He demonstrated that the condition was not caused by a bacterium as was commonly thought at the time, but rather by some other unknown pathogen in the sap of the tobacco plant. He proved this by collecting sap from a diseased plant that was then passed through a filter capable of straining out any bacteria. When the filtered solution was reinjected into the leaf veins of healthy plants and the disease was transmitted, he had made his point. He called this filtered sap a *contagium vivum fluidium* (a contagious living fluid) and introduced the term **virus** to describe its property of being able to reproduce itself within living plants. Dr. Beijerinck's virus was later named tobacco mosaic virus (TMV), consistent with the now-established practice of naming plant viruses both by the plant it infects and by describing the major disease symptom (e.g., mosaic, wilting, spotted, etc.).

Not only were the first viruses discovered in plants, but the understanding of their biochemical nature was first recognized through research on tobacco mosaic virus. Today, we know that viruses are submicroscopic, infectious particles that are composed of a protein coat and a nucleic acid center. They can be seen only with an electron microscope. As obligate parasites, they can reproduce themselves only with a living cell. The biochemical nature of viruses remained unknown until 1935 when Dr. Wendell Stanley, an organic chemist in the United States, succeeded in crystallizing the protein coat of tobacco mosaic virus (TMV). Stanley, however, did not recognize the nucleic acid content of the virus that was later shown to be RNA. The fact that RNA could exist separately from DNA was a discovery that has had great influence on the development of molecular biology thought.

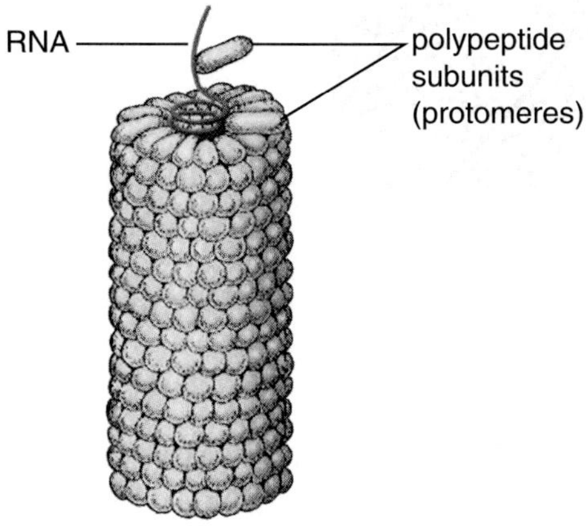

Today, we know that tobacco mosaic virus is a rigid rod, 300 nanometers by 15 nanometers, composed of a protein coat of approximately 2,100 helically arranged protein subunits surrounding an axial canal that contains a single-stranded RNA molecule consisting of 6,400 nucleotides. It, like all plant viruses, is classified according to the type of nucleic acid that it contains, either DNA or RNA but never both; whether the nucleic acid is single- or double-stranded; and the shape of the virus particle (spheres, stiff rods, flexible rods).

TMV is highly contagious, so much so that it can be transmitted to healthy plants merely from the fingers of smokers of cigarettes that were made from infected tobacco. This is unlike most other plant viruses that can survive no more than a few hours outside their living host. Plant viruses can gain entry into a plant only through an open wound or puncture and are typically transmitted by insect vectors such as aphids, leafhoppers, white flies, and mites. Aphids are the most important vectors, infecting healthy plants when they insert their mouth parts, called stylets, into phloem tubes for feeding. During feeding, they inject salivary secretions containing the virus particles into the plant's sieve tubes.

Once injected, viruses are transmitted within the phloem and move throughout the plant. However, viruses cannot move directly through cell walls. Rather, cell-to-cell movement of viral particles occurs via the *plasmodesmata* (singular: *plasmodesma*), which are membrane-lined cylindrical pores through cell walls. Plasmodesmata create cytoplasmic bridges that cross cell walls to connect adjacent cells, and this transport route explains why many viral infections are systemic, affecting the entire organism.

Awareness Box Continued

Viruses seldom kill the plant outright, but rather weaken it by causing abnormalities in leaves (such as mottling or changes in leaf color, shape, or vein patterns); changes in flower color; or irregularities in fruit size, shape, or color. Viruses can also cause fruits to ripen prematurely and to have an unpleasant taste or reduced sugar content. Crop yields of fruits and vegetables, as well as quality, can be reduced.

Few options exist for controlling plant viral diseases. The most effective control is achieved by sanitation—removing and burning diseased plants and thus killing the virus-carrying insects. Additionally, naturally resistant varieties of some plants have been developed. Chemicals remain an ineffective treatment for plant viruses because of the cost and environmental concerns.

Viral diseases affect many important agricultural crops in addition to tobacco. Crop losses worldwide are enormous each year. With the world's human population increasing at about 1.6% yearly, any disease that threatens agricultural productivity and the ability of the human population to feed itself must be taken seriously. Although not as spectacular or newsworthy as Ebola or HIV, plant viruses are silent killers because they rob humanity by directly affecting the food supply.

D.C. Scheirer

are separated first according to the DNA or RNA in their cores. Then they are grouped according to size and shape, the nature of their protein coats, and the number of identical structural units in their cores.

Bacteriophages

Viruses that attack bacteria have been studied extensively. These are called **bacteriophages,** or simply **phages** (Fig. 17.17 *A,B*). Some resemble the space exploration vehicles portrayed in the literature and films of science fiction. They consist of a head on top of a thin cylindrical core, which is surrounded by a sheathing coat. At the base of the core are six spiderlike fibril "legs," which anchor the virus in place.

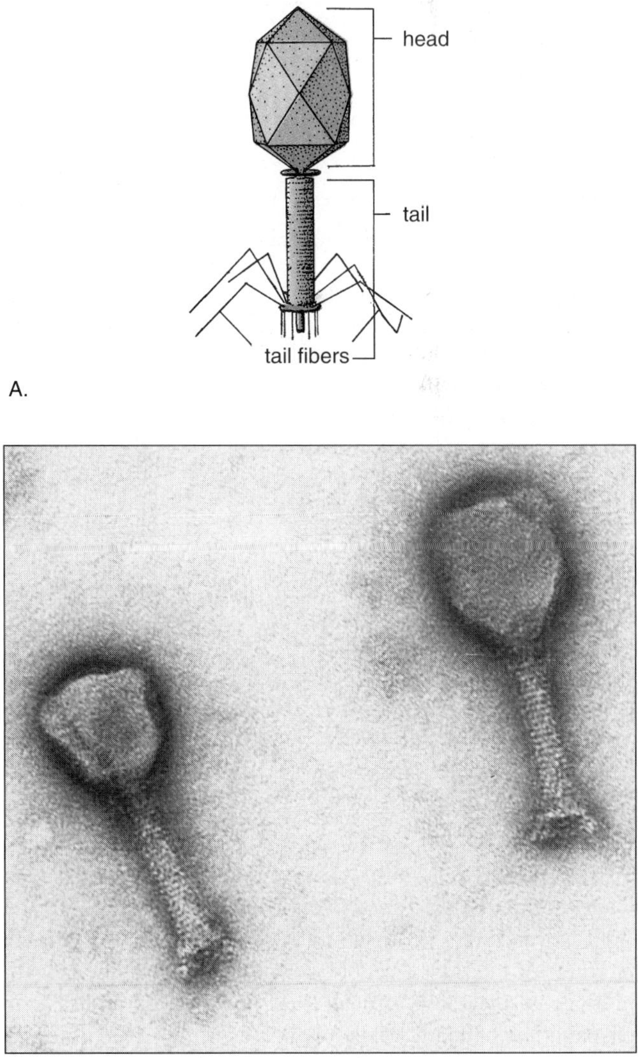

Figure 17.17 Phage viruses. Insert drawing shows some of the detail, ×20,000. *(Electron micrograph courtesy D. Kay)*

Viral Reproduction

Viruses can replicate (reproduce themselves) only at the expense of their host cells. In doing so, they first become attached to a susceptible cell. Then they penetrate to the interior, some types leaving their coats on the outside. Inside the cell, their DNA or RNA directs the synthesis of new virus molecules, which are then assembled into complete viruses. These are released from the host cell, usually as it dies (Fig. 17.18).

Some viruses (e.g., those causing influenza) can mutate (see Chapter 15) very rapidly, allowing them to attack organisms that previously had been immune to them. As a result, new vaccines constantly have to be developed to combat new strains of viruses. Some viruses greatly affect the metabolism of their host cells. For example, in botulism bacteria, the botulism toxins are produced only if specific phages are present and active. Evidence is mounting that many forms of cancer, which usually involves abnormal cell growth, are caused by viruses. Scientists also suspect that all living organisms carry viruses in an inactive form in their cells, and they are trying to discover what causes the inactive viruses suddenly to become active.

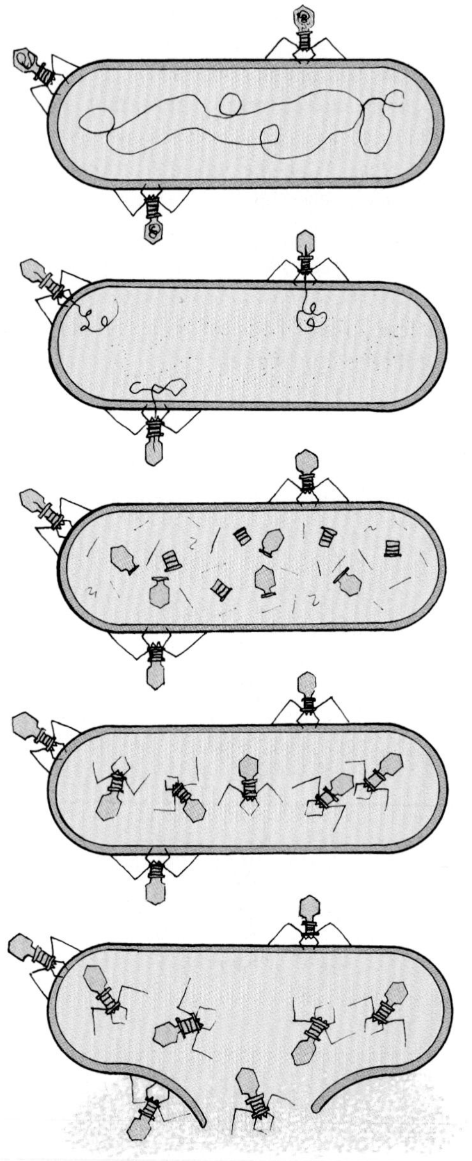

Figure 17.18 Stages in the development of a phage virus within a bacillus bacterium. *A.* The virus becomes attached to the bacterium. *B.* The DNA of the virus enters the cell. *C.* Various components of the virus are synthesized from the DNA of the bacterium. *D.* The viral components are assembled into units. *E.* The assembled viruses are released as the bacterial wall breaks down. *After M. J. Pelczar, Jr., and R. D. Reid. 1972.* Microbiology, *3d ed. The McGraw-Hill Companies. All rights reserved.*

Cells of higher animals that are invaded by viruses produce a protein called *interferon,* which is released into the fluid around the cells or into the bloodstream. Minute amounts of interferon in contact with cells cause the cells to produce a protective protein that prevents or inhibits the propagation of many types of viruses within the protected cells and also inhibits viruses from causing tumors that transform normal cells into tumor cells.

Because of these properties of interferon, it is now being produced in large quantities for use in controlling certain cancers and many other viral infections. The use of bacteria as hosts for donor DNA is especially promising for the future. This process, in effect, turns the bacteria into interferon synthesis centers (see Fig. 14.12). In 1981, scientists at the University of Washington and the Genentech Corporation of San Francisco announced that they had succeeded in producing a form of interferon by splicing interferon genes into yeast cells. Because yeast cells are larger than bacteria, the process can potentially produce much larger quantities of interferon at considerably lower cost than is possible using bacteria.

Human Relevance of Viruses

The economic impact of viruses in both third-world and industrialized countries is enormous. The annual loss in work time due to common cold and influenza viruses alone amounts to millions of hours. While discomfort, adverse effects on employment, and even deaths due to viral diseases, such as chicken pox, measles, German measles, mumps, and yellow fever, have declined dramatically since immunizations against the diseases became widespread, they still take their toll. Another viral disease, infectious hepatitis, periodically still claims victims. Guillain-Barré syndrome and Epstein-Barr are debilitating diseases caused by viruses that are apparently carried by nearly everybody, but what triggers them into action is as yet still unknown.

The virus associated with AIDS (acquired immune deficiency syndrome), a usually fatal disease, is caused by a *retrovirus* (a virus with two identical single nuclear strands) that is related to those that cause cancer. It was detected in a blood sample collected in 1959 from Zaire. AIDS was not adequately described, however, until 1981, and the virus itself was not identified until 1983. Research has shown that two forms of the AIDS virus, HIV-1 and HIV-2, diverged from a common ancestor in the early 1950s (Fig. 17.19). Thirty years later, the AIDS virus had spread all over the world. Retroviruses mutate so rapidly that they are capable of evolving about a million times faster than cellular organisms. Since the initial appearance of AIDS, increasing research is being applied to the development of a vaccine with the use of a genetically engineered virus. If successful, the vaccine will cause the human body to develop a defense against AIDS viruses without the disease itself developing.

The production of vaccines for a number of other diseases is undertaken by a flourishing worldwide multimillion dollar industry. Other mass-produced viruses are used to infect ticks, insects, and other disease organisms of both animals and plants. Some viruses cause great losses when they infest creamery vats during the manufacture of dairy products or culture vats during the production of antibiotics. One phage attacks nitrogen-fixing bacteria in the roots of leguminous plants, while other phages attack diphtheria and tuberculosis bacteria. One natural virus called *Abby* (an abbreviation of Abington, which is one of 19 strains of nuclear polyhedrosis viruses) is found only in the caterpillars of gypsy moths. Gypsy moth caterpillars have been

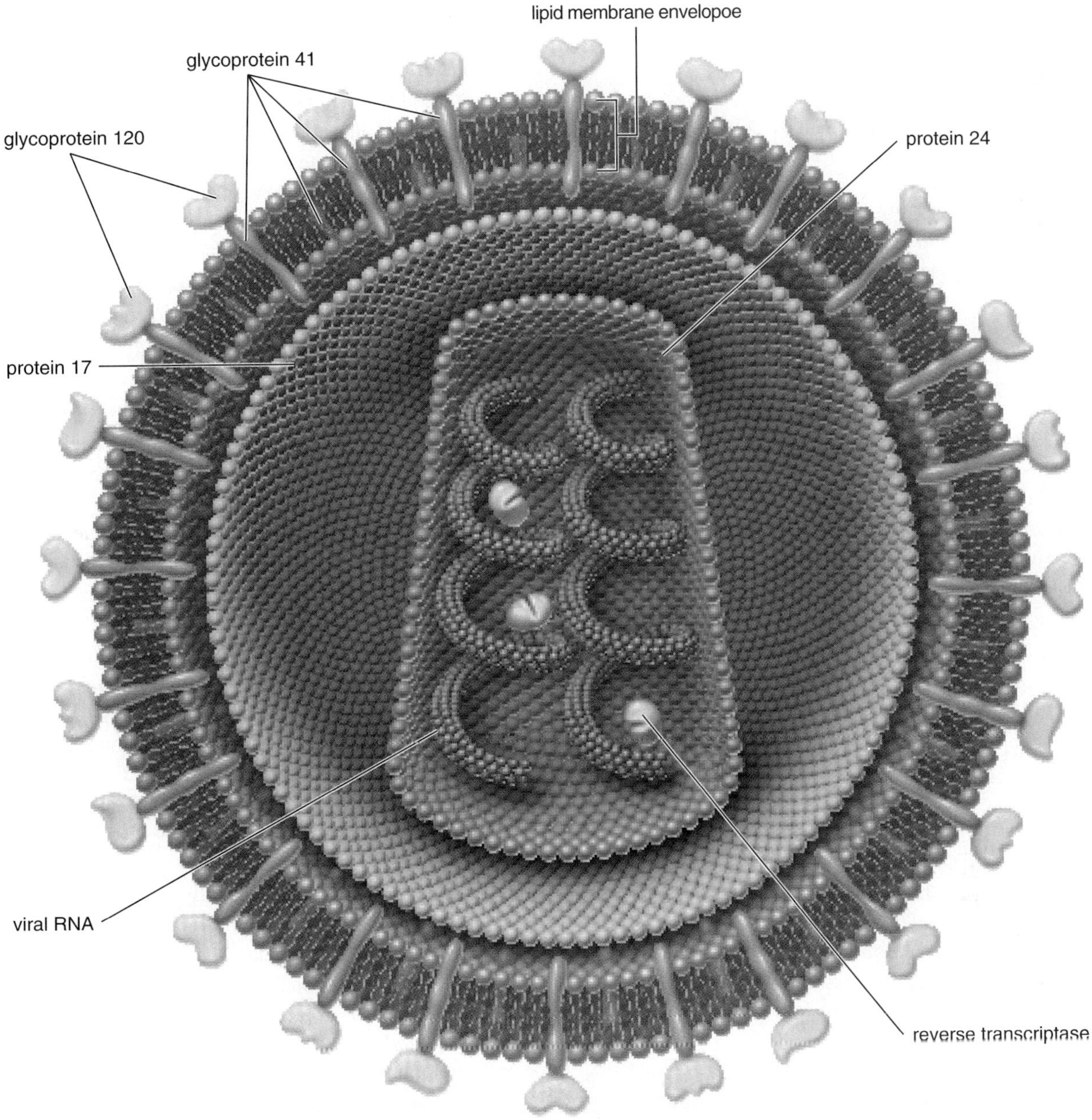

Figure 17.19 A virus is a nucleic acid coated with protein. The human immunodeficiency virus (HIV), which causes AIDS, consists of RNA surrounded by several layers of proteins. Once inside a human cell (usually a T cell, part of the immune system), the virus uses an enzyme to convert its RNA to DNA, which then inserts into the host DNA. HIV damages the human body's protection against disease by killing T cells and by using these cells to make more of itself.

particularly destructive in forests of both North America and Europe, and it is hoped that the destruction may be greatly reduced by the dissemination of this virus in the forests under attack.

VIROIDS AND PRIONS

More than a dozen plant diseases (e.g., chrysanthemum stunt disease, potato spindle-tuber disease) are caused by infectious agents called *viroids.* Viroids, which are even smaller than viruses, consist of circular strands of RNA that occur in the nuclei of infected plant cells. They may remain latent in the host nuclei without any outward manifestation of the disease in some plants but cause severe disease symptoms in others. Viroids are transmitted from plant to plant via pollen or ovules or by machinery, and it is not yet known how they produce symptoms of a disease. They do appear, however, to be synthesized from RNA templates by an enzyme.

Prions, which are also smaller than viruses, appear to be particles of protein that cause diseases of animals and humans. No nucleic acids have yet been detected in prions, and how they cause disease is still unknown.

Ecological Review

Bacteria, the most abundant and widely distributed organisms in the biosphere, are also the most metabolically diverse organisms. While most bacteria are heterotrophic, obtaining their energy from organic matter, others are photosynthetic, obtaining energy from light, or chemoautotrophic, obtaining energy from inorganic elements or compounds. Although bacteria cause many serious diseases, most do not, and the sustained ecological functioning of earth's ecosystems, especially decomposition and nutrient cycling, depends upon the activities of bacteria. The capacity of bacteria to recycle organic matter is commonly put to work in the production of compost, a valuable conditioner of soils. Viruses, which consist of a nucleic acid core surrounded by a protein coat, are among the most common disease agents and are a source of mortality in populations of all organisms from bacteria to humans.

Summary

1. Kingdoms Bacteria and Archaea consist of prokaryotic organisms (bacteria). The bacteria occur as single cells, in colonies, or in the form of chains or filaments. Some cells may be motile, or they may exhibit a gliding motion; most are nonmotile.
2. Bacterial nutrition is primarily by absorption of food in solution through the cell wall, but some are photosynthetic or chemosynthetic.
3. Reproduction is asexual by means of fission; some genetic recombination occurs by means of pili between cells.
4. Bacteria are mostly less than 2 or 3 micrometers in diameter. They occur as spheres (cocci), rods (bacilli), and in spiral forms (spirilla); they are further classified on the basis of several visible features such as sheaths, appendages, and motion. They are also classified by their chemistry and as gram-positive or gram-negative.
5. Prokaryotic cells have no nuclear envelopes or organelles. Each cell has a single, closed loop of double-stranded DNA and sometimes up to 30 or 40 small, circular DNA molecules called plasmids, which replicate independently of the large DNA molecule or chromosome.
6. Neither meiosis nor mitosis occurs, but fission takes place with the development of a transverse wall that forms near the middle of the cell; gametes and zygotes are not produced. Conjugation facilitates genetic recombination. Transformation involves the incorporation of fragments of DNA released by dead cells; transduction involves the viral transfer of fragments of DNA from one cell to another.
7. Heterotrophic bacteria are saprobes or parasites. Autotrophic bacteria are photosynthetic but do not produce oxygen; chemoautotrophic bacteria obtain their energy through oxidation of reduced inorganic groups.
8. Any nonliving organic material will eventually be decomposed to compost by bacteria and fungi. Compost is definitely good for the soil but has limited value as a fertilizer.
9. Bacteria cause huge losses through plant diseases and food spoilage and many serious diseases in animals and humans. They gain access to their hosts by various means.
10. Koch formulated postulates (rules) for proving that a particular microorganism is the cause of a particular disease.
11. Bacteria useful to humans include *Bacillus thuringiensis, Bacillus thuringiensis* var. *israelensis,* and *Bacillus popilliae.* Bacteria also play a major role in the manufacture of dairy products, such as yogurt, sour cream, kefir, and cheese.
12. Bacteria are used in the manufacture of industrial chemicals, vitamins, flavorings, food stabilizers, and a blood plasma substitute; they play a role in the curing of vanilla, cocoa beans, coffee, and tea and in the production of vinegar and sauerkraut; they aid in the extraction of linen fibers from flax stems, in the production of ensilage for cattle feed, and in the production of several important amino acids.
13. Cyanobacteria are virtually ubiquitous in their occurrence.
14. The cells of cyanobacteria occur in a variety of forms. They are distinguished from other bacteria in having chlorophyll *a,* in producing oxygen, and in having blue and red phycobilin pigments. They produce cyanophycin, a nitrogenous food reserve.
15. Cyanobacteria have no flagella, but some species have gliding movements. Fragmentation may occur at heterocysts. Akinetes may also be produced.
16. Cyanobacterial cells may have been the origin of chloroplasts, since they divide as chloroplasts do when their host cells divide.
17. Cyanobacteria may become very abundant in bodies of polluted fresh water. Toxic substances are produced when the bacteria die and are decomposed. At least 40 species of cyanobacteria are known to fix nitrogen.
18. Prochlorobacteria are similar in form to cyanobacteria but have pigmentation similar to that of higher plants and lack phycobilins.
19. Vaccination was first performed by Jesty and Jenner in connection with smallpox. Smallpox is believed to have been eradicated but may reappear as an agent of bioterrorism.

20. Viruses, which have no cellular structure, are about the size of large molecules. Some can be isolated, purified, and crystalized, yet remain virulent. They cannot grow or increase in size and cannot be replicated outside of a living cell. They depend upon DNA for their raw materials.
21. Viruses consist of a core of nucleic acid surrounded by a protein coat. They are classified on the basis of the DNA or RNA in their core, their size and shape, the number of identical structural units in their cores, and the nature of their protein coats.
22. Bacteriophages are viruses that attack bacteria. In replicating, they become attached to a susceptible cell, which they penetrate, with their DNA or RNA directing the synthesis of new virus molecules from host material; the assembled new viruses are released when the host cell dies.
23. Cells of higher animals being invaded by viruses produce interferon, a protein that causes cells to produce a protective substance that inhibits replication of viruses and also inhibits the capacity of viruses to transform normal cells into tumor cells.
24. Viral diseases, such as chicken pox, measles, mumps, and yellow fever, have declined since immunizations against the diseases have become widespread. Mass-produced viruses are used to infect insects and other pests of both plants and animals. Some viruses cause losses in creamery vats.
25. Viroids and prions are disease-causing particles that are smaller than viruses.

Review Questions

1. What is symbiosis? Give examples other than those mentioned in the text.
2. Why are bacteria not classified on the basis of visible features alone?
3. How do bacteria exchange DNA?
4. How does fission differ from mitosis?
5. Is photosynthesis the same in bacteria as it is in higher plants? Explain.
6. How do chemoautotrophic bacteria differ from photosynthetic bacteria?
7. What is the difference between nitrification and nitrogen fixation?
8. If decay bacteria use nitrogen, how does composting accumulate any nitrogen?
9. How are disease bacteria transmitted?
10. What are Koch's Postulates?
11. Why are many bacteria considered useful?
12. What do cyanobacteria and other bacteria have in common? How do they differ?
13. How do cyanobacteria survive freezing and desiccation?
14. What is an algal-bacterial bloom?
15. How do viruses differ from bacteria?
16. What is a vaccination?
17. What is a phage?
18. How do viruses multiply?

Discussion Questions

1. If a virulent phage were to eliminate all the bacteria in North America for 1 year, how would our lives be affected?
2. What would be the feasibility and the advantages or disadvantages of using only cyanobacteria and other nitrogen-fixing bacteria for our agricultural nitrogen needs?
3. Methane gas produced by bacteria is proving to be sufficient to meet all the fuel needs of villages in India. Do you think we could produce and use methane in a similar fashion in the United States?
4. If cyanobacteria are capable only of asexual reproduction, does this mean that species of these organisms can never change in form?
5. As long as viruses can multiply, why should there be any question as to whether or not they are living?

Learning Online

Visit our web page at www.mhhe.com/botany for interesting case studies, practice quizzes, current articles, and animations within the Online Learning Center to help you understand the material in this chapter. You'll also find active links to these topics:

Diversity of Eubacteria
Cyanobacteria
Diversity of Archaea
Thermophiles
Viruses

Additional Reading

Bos, L. 1992. *Introduction to plant virology.* New York: State Mutual Book and Periodical Service.

Brock, T. D. 1995. *Microorganisms: From smallpox to Lyme disease: Readings from* Scientific American. San Francisco: W. H. Freeman.

Cann, A. 1997. *Principles of molecular virology,* 2d ed. San Diego, CA: Academic Press.

Dimmock, N. J., and S. B. Primrose. 1994. *Introduction to modern virology,* 4th ed. London: Blackwell Scientific Publications.

Dixon, B. 1996. *Power unseen: How microbes rule the world.* San Francisco: W. H. Freeman.

Domingo, E. 1999. *Origin and evolution of viruses.* San Diego, CA: Academic Press.

Garrity, G. (Ed.). 2000. *Bergey's manual of systematic bacteriology,* 2d ed. New York: Springer-Verlag.

Gunter-Schlegel, H. 1994. *General microbiology,* 7th ed. New York: Cambridge University Press.

Pelczar, M. J., Jr. 1993. *Microbiology: Concepts and applications,* 5th ed. New York: McGraw-Hill.

Prescott, L., and J. P. Harley. 1996. *Microbiology,* 3d ed. Dubuque, IA: McGraw-Hill.

Tortora, G. J. 1998. *Microbiology: An introduction,* 6th ed. Old Tappan, NJ: Prentice-Hall.

Woese, C. R. 1981. Archaebacteria. *Scientific American* 244(6): 98–122.

Brown algae known as sea palms (*Postelsia*) exposed at low tide on the Pacific coast.

Kingdom Protista

18

Chapter Outline

Overview

After an introduction and summary of the features of members of Kingdom Protista, this chapter discusses the phyla of algae, slime molds, and water molds. Brief discussions of the life cycles of representatives of the principal phyla are included.

First, the Chlorophyta (green algae) are discussed, with *Chlamydomonas, Ulothrix, Spirogyra,* and *Oedogonium* being explored in some detail; mention is also made of *Chlorella,* desmids, *Acetabularia, Volvox,* and *Ulva.* The Chromophyta (yellow-green algae, golden-brown algae, diatoms, and brown algae) are taken up next. Then the Euglenophyta (euglenoids) are briefly examined, and Phylum Dinophyta (dinoflagellates) is explored. The role of members of Phylum Dinophyta in red tides and bioluminescence is noted. After a brief overview of Phyla Cryptophyta, Prymnesiophyta, and Charophyta, the chapter summarizes differences between the phyla of algae in Table 18.1, which lists food reserves, chlorophyll pigments, and flagella. This is followed with a digest of the human and ecological relevance of the algae. The chapter concludes with a brief examination of the slime molds and water molds.

Some Learning Goals

1. Know features that the members of Kingdom Protista share with one another and note the basic ways in which they differ.
2. Understand how a diatom differs in structure and form from other members of Phylum Chromophyta.
3. Diagram the life cycles of *Chlamydomonas, Ulothrix, Spirogyra,* and *Oedogonium;* indicate where meiosis and fertilization occur in each.
4. Learn at least two features that distinguish the Chlorophyta, Chromophyta, Rhodophyta, Dinophyta, and Euglenophyta from one another.
5. Know the structure and function of holdfasts, stipes, blades, bladders, and thalli.
6. Learn the function of each of the three thallus forms of a red alga, such as *Polysiphonia,* and know at least 20 economically important uses of algae, in addition to the numerous uses of algin that are given.
7. Understand why the slime molds and water molds are not considered true fungi and what they have in common with other members of Kingdom Protista.

I was once the sole passenger on a ship carrying cargo from a south Atlantic port to Boston. I had turned in early one night as we were nearing the equator, but awoke around 2:00 A.M. and went to the bathroom to get a glass of water. I had not turned on the light and was startled to notice that the water in the toilet bowl was "winking" at me. Scores of little lights in the water were flashing on and off! I remembered that ships in the open sea take in water from their surroundings for their sewage lines, and I hurried out on deck to see where we were headed. I was immediately treated to a beautiful display of *bioluminescence* (discussed in Chapter 11) caused by millions of microscopic algae called *dinoflagellates.* As the waves broke, both at the bow and in the wake, there was a sparkling, shimmering glow as the tiny organisms, through their respiration, produced light. The dinoflagellates comprise only one of several phyla of algae and other relatively simple eukaryotic organisms discussed in this chapter.

The fossil record indicates that less than 1 billion years ago, all living organisms were confined to the oceans, where they were protected from drying out, ultraviolet radiation, and large fluctuations in temperatures. They also absorbed the nutrients they needed directly from the water in which they were immersed. The fossil record also suggests that many times, beginning about 400 million years ago, green algae, which are important members of Kingdom Protista, began making the transition from water to the land, eventually giving rise to green land plants.

Coleochaete (Fig. 18.1), a tiny, freshwater green alga that grows as an **epiphyte** (an alga or plant that attaches itself in a nonparasitic manner to another living organism) on the stems and leaves of submerged plants, shares several features with higher plants and probably was an indirect ancestor of today's land plants. The features include cells that resemble parenchyma, the development of a cell plate and a phragmoplast during mitosis, a protective covering for the zygote, and the production of a ligninlike compound. Lignin adds mechanical strength to cell walls, and the discoverers of the substance suggest that it originated as a protection against microbes. The ligninlike compound was present 100 million years earlier than the presumed evolution of plant life from water to land, and its presence in algal cells could explain how such organisms adapted to land habitats.

FEATURES OF KINGDOM PROTISTA

In contrast with Kingdoms Bacteria and Archaea, whose members have prokaryotic cells, all members of Kingdom Protista have eukaryotic cells. The organisms comprising this kingdom are very diverse and heterogeneous, but none have the distinctive combinations of characteristics possessed by members of Kingdoms Plantae, Fungi, or Animalia. Many, including the euglenoids, protozoans, and

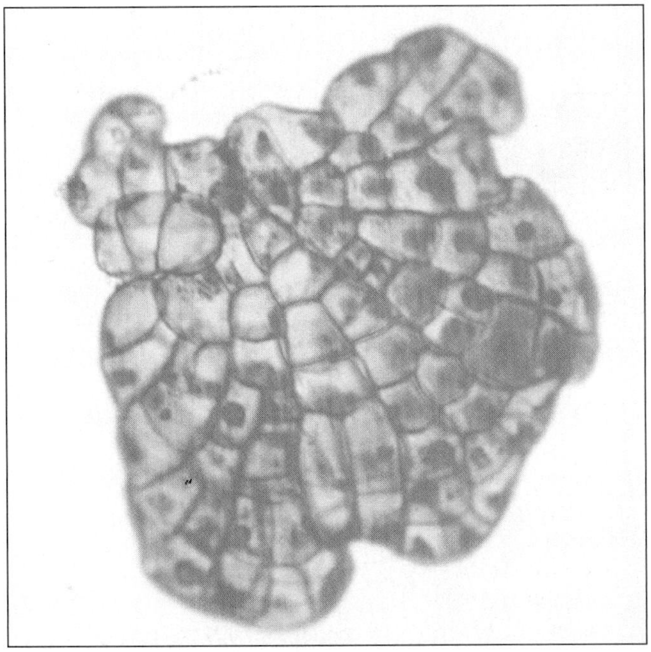

Figure 18.1 *Coleochaete,* a green alga that has several features of higher plants and is believed to be an indirect ancestor of land plants, ×500.

some algae, consist of a single cell, whereas other algae are multicellular or occur as colonies or filaments. Nutrition is equally varied, with the algae being photosynthetic, the slime molds and protozoans ingesting their food, the euglenoids either carrying on photosynthesis or ingesting their food, and the oomycetes and chytrids absorbing their food in solution.

Individual life cycles vary considerably, but reproduction is generally by cell division and sexual processes. Many protists (mostly single-celled members of Kingdom Protista) are motile, usually by means of flagella, but other members of the kingdom, especially those that are multicellular, are nonmotile, although most of the multicellular members produce some motile cells.

ALGAE

Children fortunate enough to have lived near ocean beaches where seaweeds are cast ashore by the surf have often enjoyed stamping on the bladders ("floats") of kelps to hear the distinct popping sound as they break. Some have collected and pressed beautiful, feathery red seaweeds, and others who have waded around the shores of freshwater lakes or in slow-moving streams have encountered slimy-feeling pond scums. All who have kept tropical fish in glass tanks have sooner or later had to scrape a brownish or greenish algal film from the inner surfaces of the tank, and those who have lived in homes with their own swimming pools have learned that the plaster of the pool soon acquires colored algal patches on its surface if chemicals are not regularly added to prevent them from appearing.

Although some of the seaweeds have flattened, leaflike blades, algae have no true leaves or flowers. They are involved in our everyday lives in more ways than most people realize (see the discussion in the section "Human and Ecological Relevance of the Algae," which begins on page 344). Seaweeds and some pond scums, fish-tank films, and colored patches in swimming pools include but a few of the numerous kinds of algae all assigned to Kingdom Protista. The algae are grouped into several major phyla based on the form of their reproductive cells and combinations of pigments and food reserves.

PHYLUM CHLOROPHYTA—THE GREEN ALGAE

Phylum Chlorophyta includes about 7,500 species of organisms commonly known as the **green algae.** They occur in a rich variety of forms and occur in very diverse and widespread habitats; they possess some of the most beautiful chloroplasts of all photosynthetic organisms. Some are unicellular and microscopic; in fact, the green alga *Micromonas* is only 1 μm in diameter—the smallest eukaryotic cell known. Most unicellular green algae are, however, considerably larger. Other green algae form threadlike filaments, platelike colonies, netlike tubes, or hollow spheres (Fig. 18.2).

Some green algae are seaweeds, resembling lettuce leaves or long green ropes. Several unicellular species grow in greenish patches or streaks on the bark of trees, whereas others grow in large numbers on the fur of sloths and other jungle animals. The green fur camouflages the animals in their natural habitats. Still others thrive in snowbanks, live in flatworms and sponges, on rocks, or are found on the backs of turtles. They are the most common member in lichen "partnerships" (discussed in Chapter 19). The greatest variety, however, is found in freshwater ponds, lakes, and streams. Ocean forms are also varied; there they are an important part of the **plankton** (free-floating, mostly microscopic organisms) and, therefore, of food chains.

The chlorophylls (*a* and *b*) and other pigments of green algae are similar to those of higher plants. The green algae also are believed to have been the ancestors of higher plants and, like the higher plants, they store their food in the form of starch within the chloroplasts. With the exception of *bryopsids,* which are multinucleate, most green algae have a single nucleus in each cell. Most green algae reproduce both asexually and sexually. How they do so illustrates the forms of reproduction of most of the organisms discussed in the chapters to follow, and so we'll examine several different representative green algae in some detail.

Chlamydomonas

A small, actively moving little alga, *Chlamydomonas* (Fig. 18.3) is a common inhabitant of quiet freshwater pools. It has an ancient history among eukaryotic organisms, with

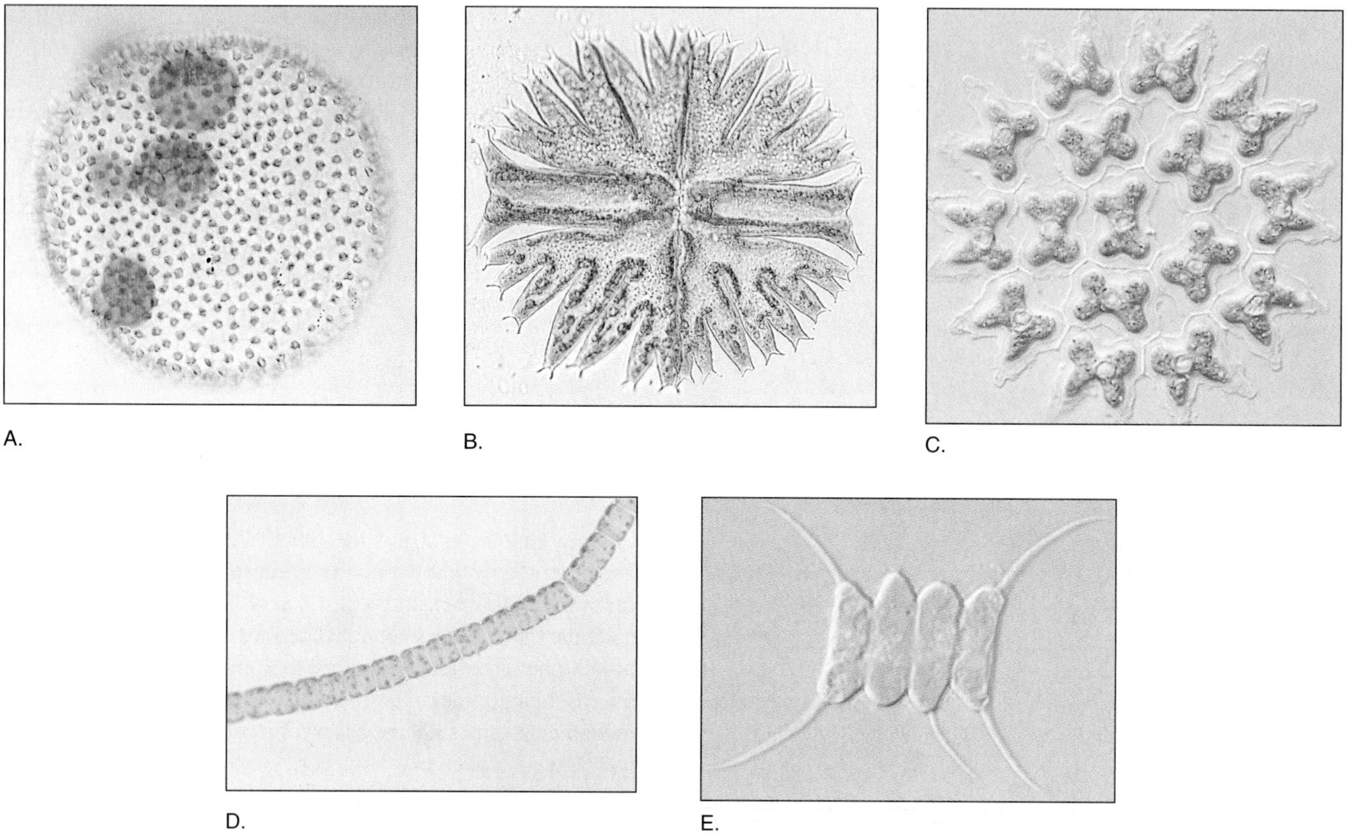

Figure 18.2 Representative green algae. *A. Volvox.* The cells form hollow, spherical colonies that spin on their axes as the flagella of each cell beat in such a way that the motion is coordinated. New colonies are produced within the older ones, ×800. *B. Micrasterias,* a desmid. These algae consist of single cells that often have a constriction in the center, ×1,200. *C. Pediastrum.* A colonial alga that forms flat plates, ×1,200. *D.* Part of a filament of *Ulothrix,* whose cells each have at their periphery a chloroplast in the form of a curved plate, ×800. *E. Scenedesmus,* a green alga that typically occurs in colonies of four cells, ×2,000.

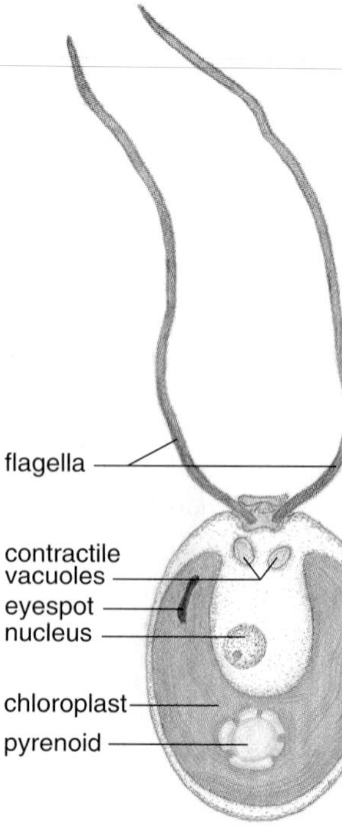

Figure 18.3 *Chlamydomonas.*

fossil relatives occurring in rock formations reported to be nearly 1 billion years old. *Chlamydomonas* is unicellular, with a slightly oval cell surrounded by a complex multilayered wall that is partially composed of glycoproteins. A pair of whiplike flagella at one end pull the cell very rapidly through the water. The flagella are, however, difficult to see with ordinary light microscopes. The cell itself is usually less than 25 micrometers (one ten-thousandth of an inch) long, which is, however, more than three times larger than a human red blood cell. Near the base of the flagella there are two or more vacuoles that contract and expand. They apparently regulate the water content of the cell.

A dominant feature of each *Chlamydomonas* is a single, usually cup-shaped chloroplast that at least partially hides the centrally located nucleus. One or two roundish **pyrenoids** are located in each chloroplast. Pyrenoids are proteinaceous structures thought to contain enzymes associated with the synthesis of starch. Most species also have a red *eyespot* on the chloroplast near the base of the flagella. The eyespot is sensitive to light; it is, however, merely part of an organelle within a single cell and is in no way complex and multicellular like an eye.

Asexual Reproduction

Before a *Chlamydomonas* reproduces asexually, the cell's flagella degenerate and drop off or are reabsorbed. Then the nucleus divides by mitosis, and the entire cell contents become two cells within the cellulose wall. The two daughter cells develop flagella, escape, and swim away as the parent cell wall breaks down. Once they have grown to their full size, they may repeat the process. Sometimes mitosis occurs more than once, so that 4, 8, or up to 32 little cells with flagella are produced inside the parent cell. Occasionally, flagella do not develop, and the cells remain together in a colony. When growth conditions change, however, each cell of the colony may develop flagella and swim away. This type of reproduction brings about no changes in the number of chromosomes present in the nucleus, and all the cells remain *haploid.*

Sexual Reproduction

Under certain combinations of light, temperature, and additional unknown environmental forces, many cells in a population of *Chlamydomonas* may congregate together. Careful study of such events has revealed that pairs of cells appear to be attracted to each other by their flagella and function as gametes that are sometimes of two types. The cell walls break down as the protoplasts slowly emerge and mate, fusing together and forming *zygotes.* A new wall, often relatively thick and ornamented with little bumps, forms around each zygote. This may remain dormant for several days, weeks, or even months, but under favorable conditions, a dramatic change occurs. The cell contents, now *diploid,* undergo meiosis, producing four haploid **zoospores** (motile cells that do not unite with other cells; many different kinds of algae produce zoospores). When the old zygote wall breaks down, the zoospores swim away and grow to full-sized *Chlamydomonas* cells (Fig. 18.4).

Ulothrix

Often, an examination of dead twigs, rocks, and other debris in cold freshwater ponds, lakes, and streams (and a few marine habitats) reveals a threadlike alga called *Ulothrix* (Fig. 18.5), whose name is derived from the Greek words *oulos* (woolly) and *thrix* (hair). Each alga consists of a single row of cylindrical cells attached end to end and forming a thread, or *filament.* The basal cell of each filament is slightly longer than the other cells and functions as an attachment cell, or **holdfast.** Around the periphery of each cell is a wide, curved, somewhat flattened chloroplast. Each chloroplast contains one to several pyrenoids. Any of the cells, except the holdfast, may divide, and as they do so, the filaments grow longer.

Asexual Reproduction

The cell contents of any cell except the holdfast appear to clump and condense inside the rigid cell wall, divide by mitosis, and become zoospores. The zoospores of *Ulothrix* are quite similar to *Chlamydomonas* cells in that they have contractile vacuoles and an eyespot, but they have four, rather than two, flagella.

Frequently, the cell contents divide one to several times before becoming zoospores, but after zoospores are formed, they usually escape from the parent cell through a pore in the wall. After swimming about for a few hours to several days, they settle on submerged objects, the flagella are shed, and the cells divide. One of the first two daughter cells becomes a holdfast, while the other continues to divide, becoming a new filament. In some instances, the protoplasts do not produce flagella after they have condensed and divided like developing zoospores, but they are otherwise capable of germinating and producing new filaments. Such cells, called *autospores,*

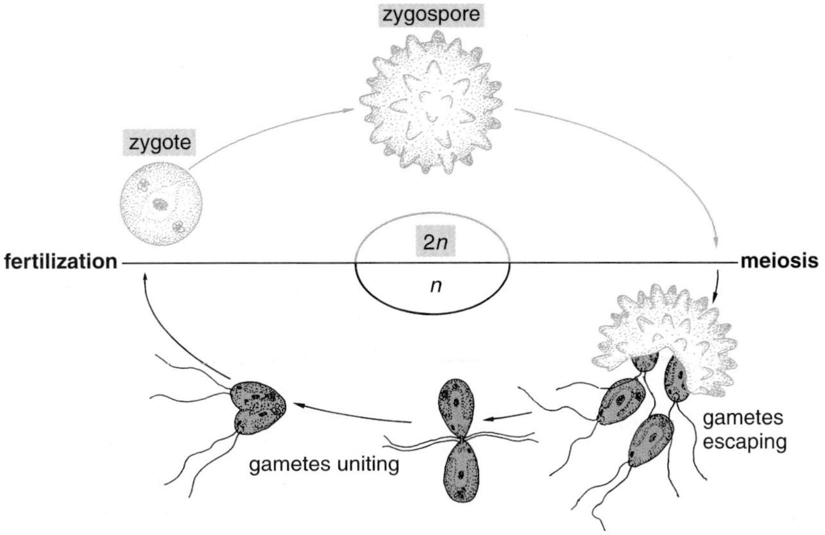

Figure 18.4 Sexual life cycle of *Chlamydomonas.*

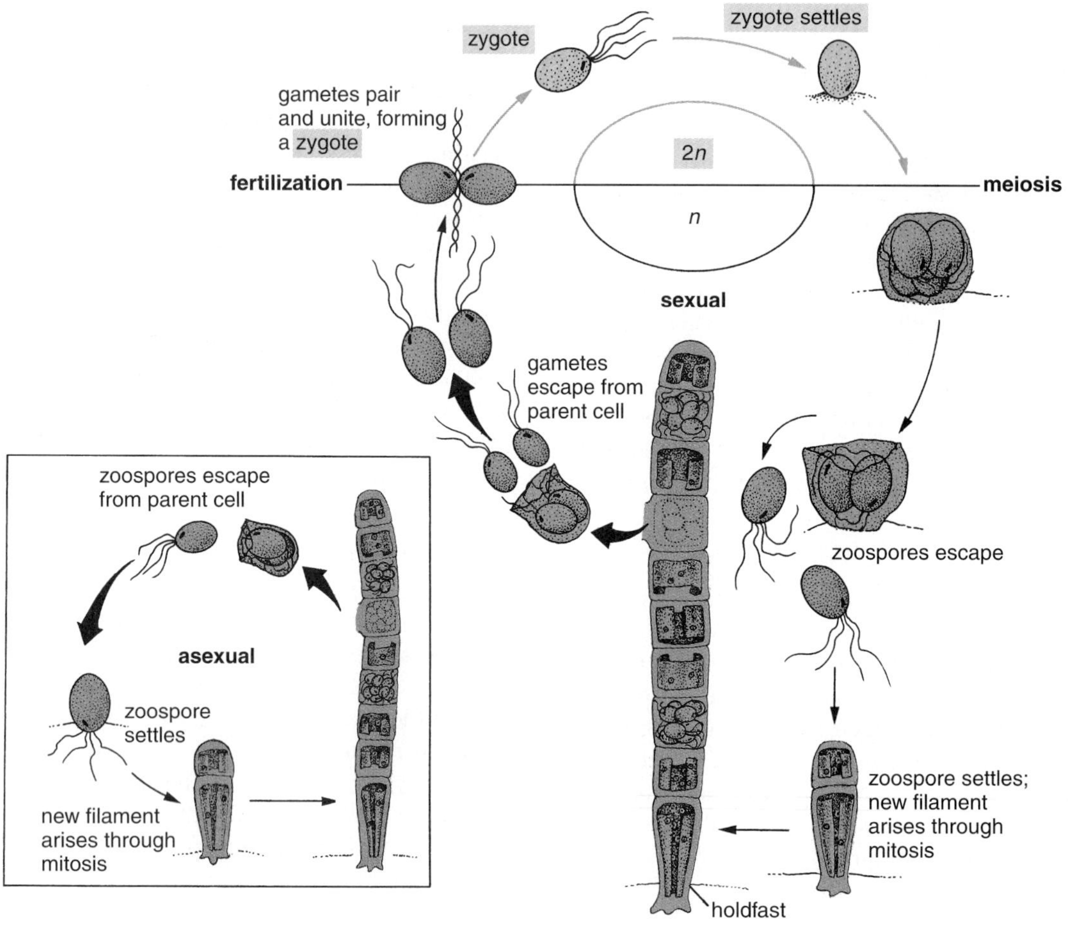

Figure 18.5 Life cycle of *Ulothrix.*

lack the capacity to become motile and are released when the parent cell wall breaks down. Similar cells that do have the capacity to become motile are called *aplanospores.*

Sexual Reproduction

Both asexual and sexual reproduction start out in the same way. The contents of any cell except the holdfast appear to condense and then divide. Up to 64 zoosporelike cells, each, however, with two flagella, may be produced. When these cells escape from the parent cell walls, they function as gametes, uniting in pairs with gametes from other filaments and forming zygotes. Thick walls form, and the zygotes become dormant. Eventually, the zygote cell contents undergo meiosis, giving rise to zoospores that then can become new filaments.

Although the gametes of *Ulothrix* come from different filaments, they are identical in size and appearance. Sexual reproduction involving such gametes is called **isogamy.** It is found only in simpler organisms. (In a few green algae, the flagellated gametes are of two slightly different sizes; sexual reproduction in such algae is referred to as *anisogamy*). As in *Chlamydomonas,* the zygotes of *Ulothrix* are the only diploid (2*n*) cells in the life cycle. All the other cells are haploid (*n*).

Spirogyra

Watersilk, as *Spirogyra* is called, has watery sheaths (which make the alga feel slimy to humans) surrounding the filaments. Examination with a microscope reveals a beautiful alga with some of the most striking chloroplasts known.

These common freshwater algae, consisting of unbranched filaments of cylindrical cells, frequently float in masses at the surface of quiet waters. Each cell contains one or more long, frilly, ribbon-shaped chloroplasts that look as though they had been spirally wrapped around an invisible pole (vacuole) occupying most of the cell's interior. Most *Spirogyra* species have one or two chloroplasts in each cell, but some have as many as 16. Every one of these elegant green ribbons has pyrenoids at regular intervals throughout its length.

Asexual Reproduction

Unlike *Chlamydomonas* and *Ulothrix, Spirogyra* produces no zoospores or other cells with flagella. Any cell is capable of dividing, but the only asexual reproduction resulting in new filaments is brought about through the breakup, or *fragmentation,* of existing filaments. Fragmentation often occurs as a result of a storm or other disturbance.

Sexual Reproduction

In colonies of *Spirogyra,* the filaments usually are produced so close to each other that they may actually be touching. When sexual reproduction begins, the individual cells of adjacent filaments form little dome-shaped bumps, or **papillae** (singular: **papilla**), opposite each other. As these papillae grow, they force the filaments apart slightly, and then the papillae fuse at their tips, forming small cylindrical **conjugation tubes** between each pair of cells. The condensed protoplasts then function as gametes. Usually, those of one filament will seem to flow or crawl like amoebae through the conjugation tubes to the adjacent cells, where each fuses with the stationary gamete, forming a zygote. Each moving protoplast is considered a male gamete, while the stationary ones function as female gametes (Fig. 18.6).

Thick walls usually develop around the zygotes, which remain dormant for some time, often over the winter. Thick-walled zygotes are characteristic of most freshwater green algae. Eventually, their cell contents undergo meiosis, producing four haploid cells. Three of these disintegrate, and a single new *Spirogyra* filament grows from the interior of the old zygote shell. The type of sexual reproduction seen in *Spirogyra* is called **conjugation.**

Oedogonium

Aquatic flowering plants and other algae often provide surfaces to which *Oedogonium* (pronounced ee-doh-góh-nee-um), a filamentous green alga, may attach itself. It is, however, strictly epiphytic and in no way parasitic. The basal cells of the unbranched filaments form holdfasts, and the terminal cell of each filament is rounded. The remaining cells are cylindrical and attached end to end. The name *Oedogonium* comes from words meaning "swollen reproductive cell" and is quite apt, as the female reproductive cells do indeed bulge noticeably. Each cell contains a large, netlike chloroplast that rolls and forms a tube something like a loosely woven basket around and toward the periphery of each protoplast. There are pyrenoids at a number of the intersections of the net.

Asexual Reproduction

Asexual reproduction in *Oedogonium* is by means of zoospores or fragmentation. Zoospores are produced singly in cells at the tips of the filaments (Fig. 18.7). Unlike the zoospores of most other algae, those of *Oedogonium* have about 120 small flagella forming a fringe around the cell toward one end. The zoospores look like tiny, balding, faceless heads. After they escape from their filaments, they eventually settle and form new filaments in the same manner as *Ulothrix.*

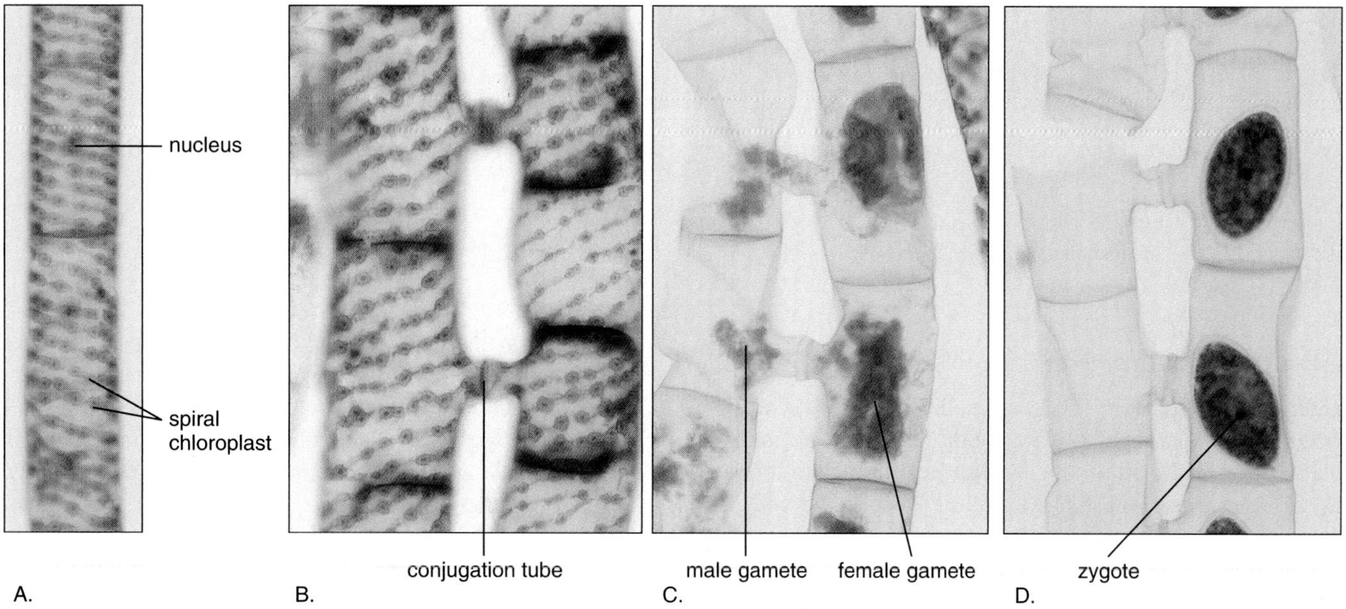

Figure 18.6 *Spirogyra* (watersilk). *A.* A portion of a vegetative filament showing the ribbonlike chloroplasts spirally arranged in each cell. The centrally located darker object in each cell is a nucleus. *B.* Papillae have grown out from opposite cells of two closely adjacent filaments and formed conjugation tubes. *C.* The condensed protoplasts in the cells on the *left* are functioning as male gametes that are migrating through the conjugation tubes to the stationary female gametes in the cells on the *right.* *D.* Zygotes have been produced in the cells on the *right* as a result of fusion of gametes.

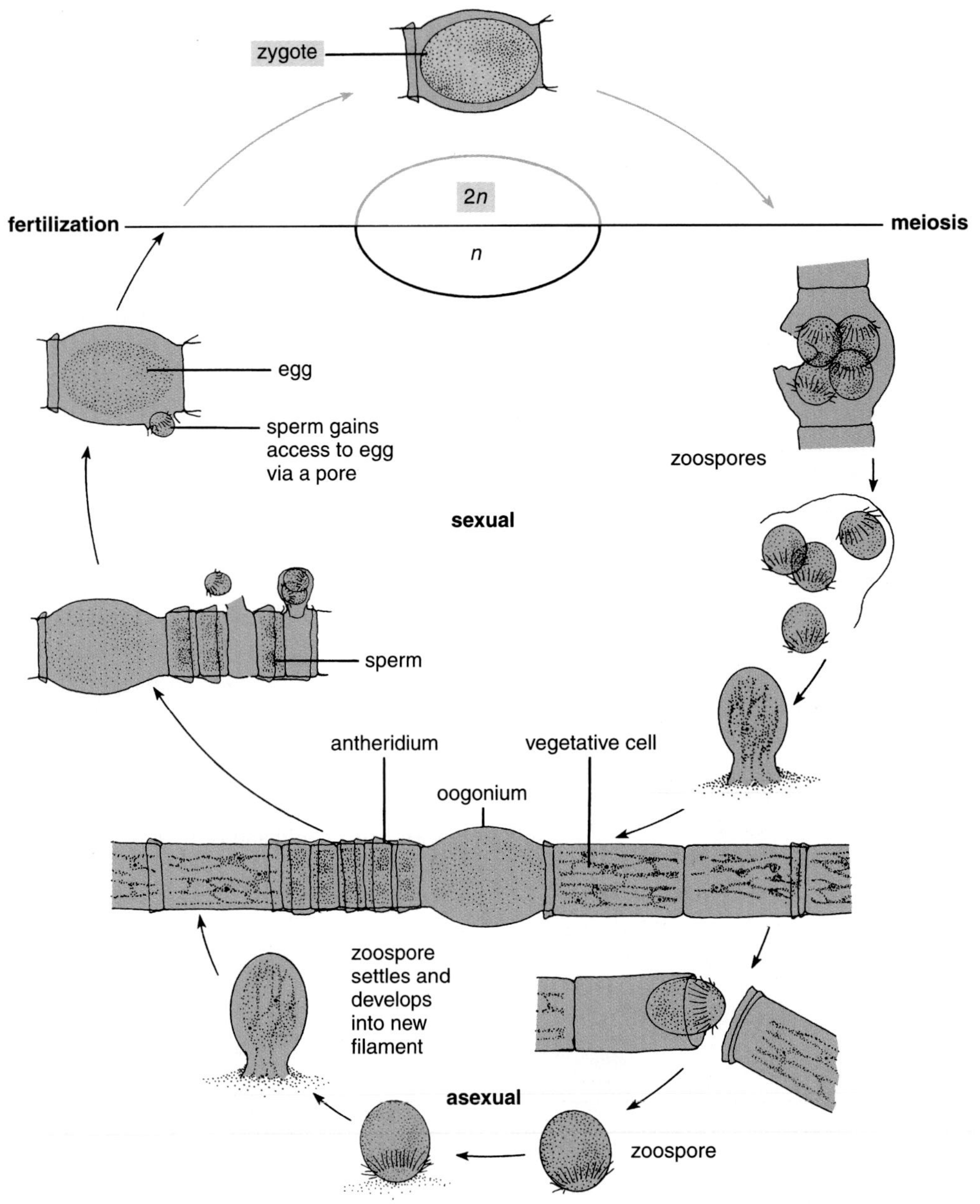

Figure 18.7 Life cycle of *Oedogonium.*

Sexual Reproduction

Oedogonium has more sexual specialization than any of the other three green algae previously discussed. Short boxlike cells called **antheridia** (singular: **antheridium**) are formed in the filaments alongside the ordinary vegetative cells. A pair of male gametes, or **sperms,** is produced in each antheridium. The sperms resemble the zoospores but are smaller. Certain cells become swollen and round to elliptical in outline. These cells, called **oogonia** (singular: **oogonium**), each contain a single female gamete, or egg, that occupies nearly all of the cell. As the egg matures, a pore develops on the side of the oogonium. When sperms escape from the antheridia, they are attracted to the oogonia by a substance released by the eggs. One sperm eventually enters the oogonium through the pore and unites with the egg, forming a zygote. After fertilization, the zygote wall thickens, and reddish oil accumulates in the protoplasm.

In some species of *Oedogonium,* oogonia are produced only on female filaments, and antheridia are produced only on male filaments. The male filaments are sometimes dwarfed. Dwarf males attach themselves to the female filaments, and then both produce hormones that influence the other's development.

Zygotes may remain dormant for a year or more, but they eventually undergo meiosis, producing four zoospores, each capable of developing into a new filament. The sexual reproduction exhibited by *Oedogonium,* in which one gamete is motile (capable of spontaneous movement) while the other gamete is larger and stationary, is called **oogamy.**

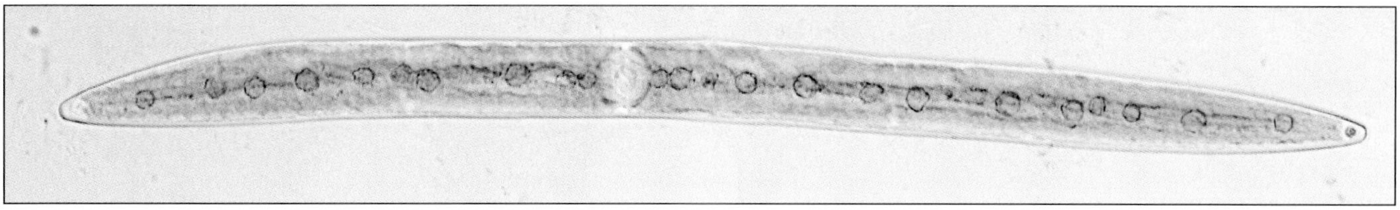

Figure 18.8 A single cell of the common desmid, *Closterium.*

Other Green Algae

As indicated earlier, the green algae constitute a very diverse group, with an extraordinary variety of forms and chloroplast shapes. Obviously, each species has to reproduce in order to perpetuate itself. Most undergo both sexual and asexual reproduction, but a few do not. For example, the globally distributed algae that make parts of some tree trunks appear as though they had received a light brushing or spattering of green paint usually are unicellular or colonial forms that reproduce only asexually.

These and *Chlorella,* another widespread green alga composed of tiny spherical cells, reproduce by forming either daughter cells or autospores through mitosis. The daughter cells often remain together in packets, while the autospores of *Chlorella* formed inside the parent cells grow to full size as the parent cell wall breaks down.

Chlorella is very easy to culture and is a favorite organism of research scientists. It has been used in many major investigations of photosynthesis and respiration, and in the future, it may become important in human nutrition. (See the section entitled "Human and Ecological Relevance of the Algae," which begins on page 344.)

Chlorella could also play a key role in long-range space exploration. Because present exploration is severely limited by the weight of oxygen tanks and food supplies needed on a spacecraft, scientists have turned to *Chlorella* and similar algae as portable oxygen generators and food sources. Future spacecraft may be equipped with tanks of such algae. These would carry on photosynthesis, using available light and carbon dioxide given off by the astronauts, while furnishing them with oxygen. As the algae multiplied, the excess could either be eaten or fed to freshwater shrimp that could, in turn, become food for the astronauts. Still other algae and bacteria could recycle other human wastes. Such a self-perpetuating *closed system,* as it is called, has already been successfully tested with mice and other animals. Many algae, however, are known to produce traces of carbon monoxide gas, which is deadly to most animal life, and until this problem is resolved, humans will not be subjected to such research.

Desmids (Fig. 18.8; see also Fig. 18.2B), whose 2,500 species of crescent-shaped, elliptical, and star-shaped cells are mostly free-floating and unicellular, reproduce by conjugation. Their chloroplasts are among the most beautiful seen in the green algae.

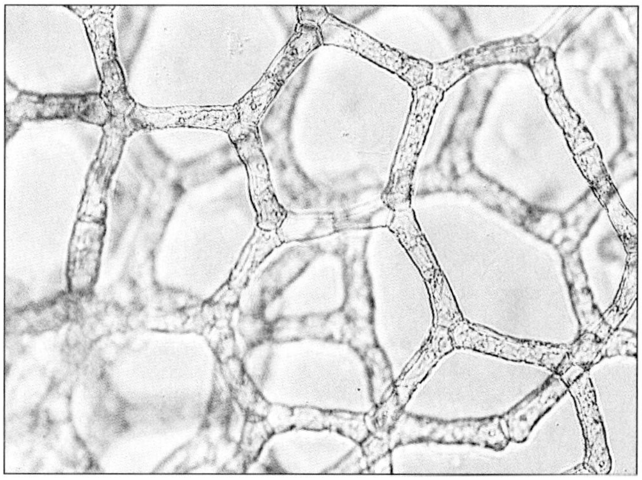

Figure 18.9 A portion of a water net (*Hydrodictyon*), a green alga whose cells form a network in the shape of a tube.

The striking *water nets* (*Hydrodictyon*) (Fig. 18.9) form netlike, tubular colonies with hexagonal or polygonal meshes. A daughter net may be asexually produced inside a parent cell and eventually released. Sexual reproduction is isogamous, with two flagellated gametes forming a zygote as they unite. The zygotes become zoospores that produce large angular cells, each new cell becoming a new netlike colony that is released from the parent cell as it breaks down.

Sexual reproduction is also isogamous in the *mermaid's wineglass* (*Acetabularia*), a marine alga consisting of a single, huge cell shaped like a delicate mushroom (Fig. 18.10). Each cell is up to 5 centimeters (2 inches) long. This alga has been used in classic experiments demonstrating the influence of the nucleus on the form of the cell. If the cap of an alga is removed and the nucleus is replaced with a nucleus taken from another species, the base regenerates a cap identical to the previous one. If this new cap is also removed, however, the next cap that develops shows form characteristics of both species. If the intermediate cap is then removed, the next cap that develops is identical to that of the species from which the nucleus originally came. Clearly, the original nucleus directs development of cytoplasmic substances regulating cap form, and when these are gone, the replacement nucleus exerts its own influence.

Volvox (see Fig. 18.2A) is representative of a line of green algae that forms colonies, apparently by means of single cells similar to those of *Chlamydomonas,* held together in a secretion of gelatinous material. In some colonies, the cells

Figure 18.10 Mermaid's wineglass (*Acetabularia*).

are actually connected to one another by cytoplasmic strands. The flagella of individual cells beat separately but pull the whole colony along. A *Volvox* colony may consist of several hundred to many thousands of cells that resemble a hollow ball spinning on its axis as it moves. Reproduction may be either asexual or sexual, with smaller daughter colonies being formed inside the parent colony. The daughter colonies are released when the parent colony breaks apart.

Sea lettuce (*Ulva*) is a multicellular seaweed with flattened, crinkly-edged green blades that may be up to 1 meter (3 feet) or more long (Fig. 18.11). A basal holdfast anchors the blades, which may be either haploid or diploid, to rocks. Diploid blades produce spores that develop into haploid blades bearing gametangia. The gametes from the haploid blades fuse in pairs, forming zygotes, that can potentially grow into new diploid blades. Except for the reproductive structures, the haploid and diploid blades of sea lettuce are indistinguishable from one another, a feature known as *isomorphism.*

Cladophora is a branched, filamentous green alga whose species are represented in both fresh and marine waters. Unlike the cells of other green algae, those of *Cladophora* and its relatives are mostly multinucleate.

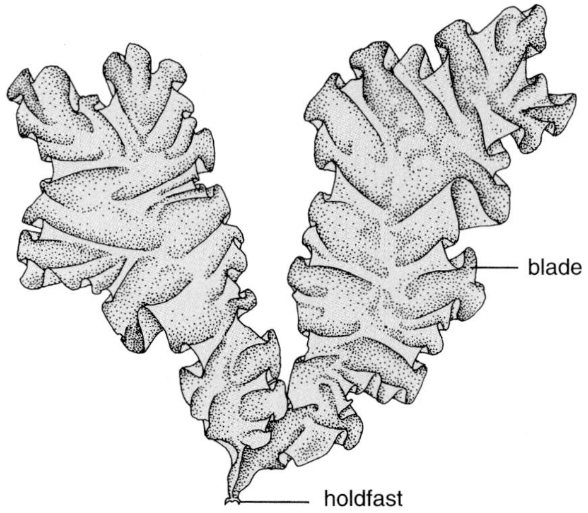

Figure 18.11 Sea lettuce (*Ulva*).

PHYLUM CHROMOPHYTA—THE YELLOW-GREEN ALGAE, GOLDEN-BROWN ALGAE, DIATOMS, AND BROWN ALGAE

About three quarters of the roughly 7,600 members of Phylum Chromophyta are primarily microscopic, but kelps and other brown seaweeds are the best-known representatives. If the microscopic forms were larger, many surely would become collectors' items in the art and antique shops of the world because of their exquisite form and ornamentation (Fig. 18.12).

The algae in this phylum can be grouped into several classes, including the *yellow-green algae* (Xanthophyceae), *golden-brown algae* (Chrysophyceae), *diatoms* (Bacillariophyceae), and *brown algae* (Phaeophyceae). Superficially, the organisms of each class may appear unrelated to each other, but they do share several features, including food reserves, specialized pigments, and other cell characteristics. Some members of the first three classes produce a unique resting cell called a *statospore* (Fig. 18.13). These cells resemble miniature glass apothecary bottles, complete with plugs that dissolve or "uncork," releasing the protoplast inside. Many statospores are striking in form, with finely sculptured ornamentations on the surface.

Yellow-Green Algae (Xanthophyceae)

Yellow-green algae are mostly freshwater organisms, with a few marine and terrestrial representatives. The two flagella of motile cells are oriented in opposite directions. The brownish pigment *fucoxanthin* (found in other members of this division) is not present except in *Vaucheria,* an oogamous, coenocytic,

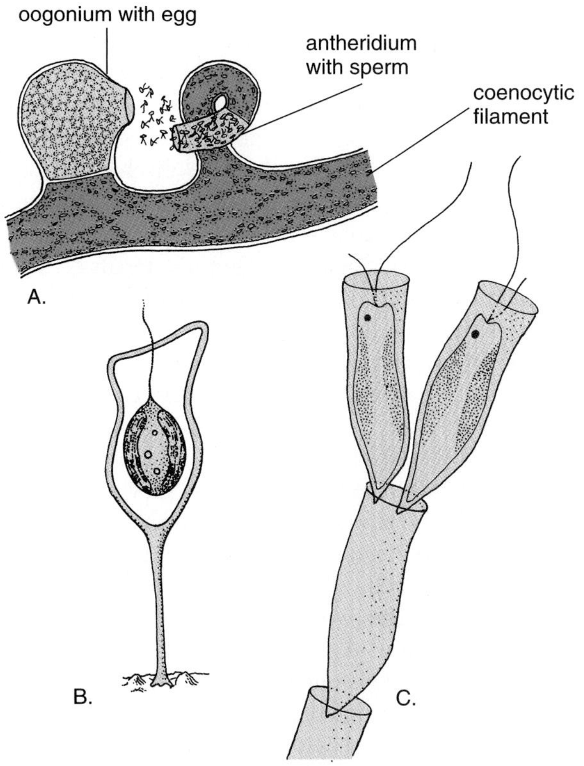

Figure 18.12 Some representatives of yellow-green algae in Phylum Chromophyta. *A. Vaucheria. B. Stipitococcus. C. Dinobryon.*

filamentous species. Aplanospores are commonly formed during asexual reproduction; less frequently, zoospores are produced. Sexual reproduction is relatively rare, but when it does occur (as in *Vaucheria*), it may exhibit specialization, with oogonia and antheridia being formed on special branches (see Fig. 18.12 A).

Golden-Brown Algae (Chrysophyceae)

Most golden-brown algae occur in the plankton of bodies of fresh water. The motile cells have two flagella of unequal length inserted at right angles to each other, with a *photoreceptor* (light-sensitive area) on the short flagellum. The photoreceptor is usually shaded by an eyespot at the flagellar end of a large chloroplast. Silica scales are present in some species.

Diatoms (Bacillariophyceae)

Diatoms (Fig. 18.14) are among the best-known and economically most important members of the phylum. These mostly unicellular algae occur in astronomical numbers in both fresh and salt water but are particularly abundant in colder marine habitats. In fact, as a rule, the colder the water, the greater the number of diatoms present, and huge populations of diatoms

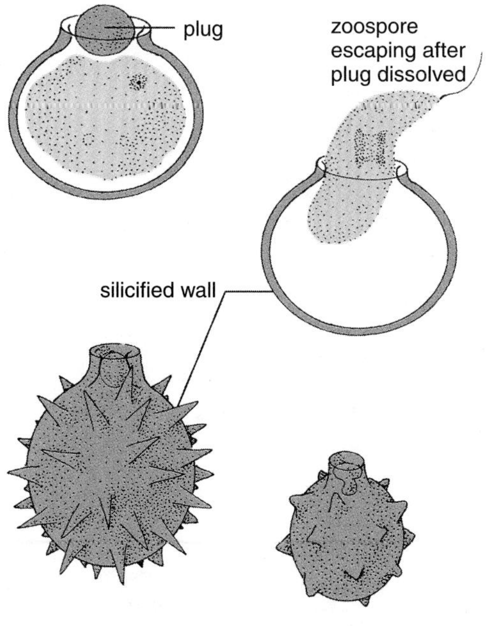

Figure 18.13 Statospores, which often resemble apothecary bottles, are formed by many of the golden-brown algae. *After G.M. Smith, 1950.* The Freshwater Algae of the United States. *2d ed. The McGraw-Hill Companies. All rights reserved.*

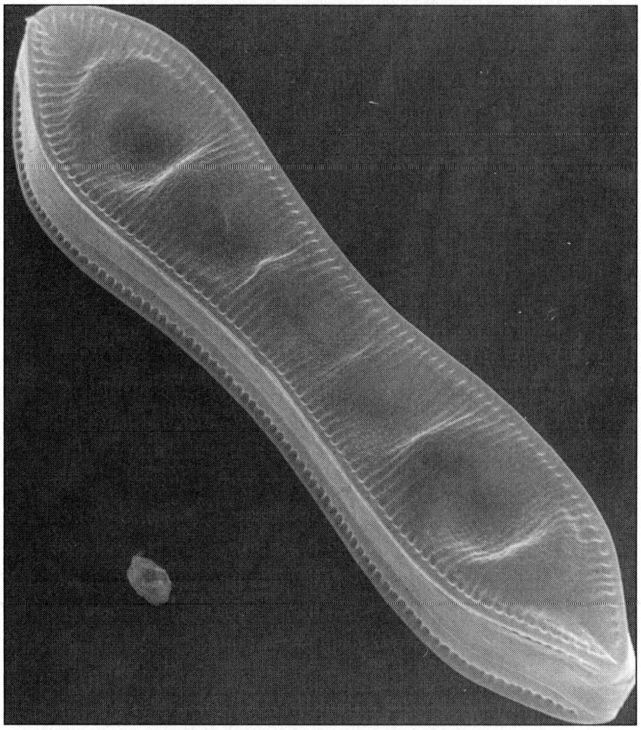

Figure 18.14 A diatom. *(Scanning electron micrograph courtesy J. D. Pickett-Heaps)*

are found on and within ice in both Antarctica and the Arctic. A major constituent of the foam that accumulates at the waveline on beaches is an oil produced by diatoms.

Diatoms usually also dominate the algal flora on damp cliffs, the bark of trees, bare soil, or the sides of buildings. More than 5,600 living species are recognized, with almost as many more known only as fossils. Some can withstand extreme drought, and one species is known to have become active after lying dormant in dry soil for 48 years.

Diatoms look like tiny, ornate, glass boxes with lids. Half of the rigid, crystal-clear wall fits inside the other overlapping half. Many marine diatoms are circular in outline when viewed from the top, while freshwater species viewed the same way tend to resemble the outline of a kayak. As much as 95% of the wall content is silica, an ingredient of glass, deposited in an organic framework of pectin or other substances.

The diatom walls usually have exquisitely fine grooves and pores that are exceptionally minute passageways connecting the protoplasm with the watery environment outside the shell. These fine grooves and pores are so uniformly spaced they have been used to test the resolution of microscope lenses. When *phycologists* (scientists who specialize in the study of algae) want to examine the markings on diatom walls, they often obtain a better view by dissolving the cell contents with hot acid.

Each diatom may have one, two, or many chloroplasts per cell. In addition to chlorophyll *a,* the accessory pigments chlorophyll c_1 and chlorophyll c_2 are typically present. The chloroplasts usually are golden-brown in color because of the dominance of *fucoxanthin,* the brownish pigment also found in the brown algae. Food reserves are oils, fats, or the carbohydrate *laminarin.*

Freshwater diatoms that have a lengthwise groove, called a *raphe,* glide backward and forward with somewhat jerky motions, at a rate of up to three times their length in 5 seconds. It is believed that the movements occur in response to external stimuli such as light. Extremely tiny fibrils apparently take up water as they are discharged into the raphe or pores through which moving cytoplasm protrudes. When they come in contact with a surface, they stick and contract, moving the cell as the caterpillar treads on a tractor do and leaving a trail something like that of a snail.

Reproduction in diatoms is unique, with half of the cells becoming progressively smaller through several generations until, through a sexual process, the original cell size is restored. Before any cell division can occur, an adequate source of silicon must be present in the surrounding medium. In culture, the number of cells produced is directly proportional to the amount of silicon added to an otherwise nutrient-balanced medium. Division in culture is also rhythmic, with all the cells dividing at the same time. After a diatom's cell contents, which are diploid, have undergone mitosis and division, the two halves of the cell separate, with a daughter nucleus remaining in each half. Then a new half wall fitting *inside* the old half is formed. This occurs for a number of generations, with the result that half of the cells become progressively smaller. Eventually, however, when the ratio of cytoplasm to nucleoplasm reaches a critical point, a cell undergoes meiosis, producing four gametes, which then escape. These fuse with other gametes, becoming zygotes called *auxospores.* Auxospores are like any other zygotes except that the original size of the diatom is restored when the auxospores balloon rapidly before rigid walls are formed (Fig. 18.15).

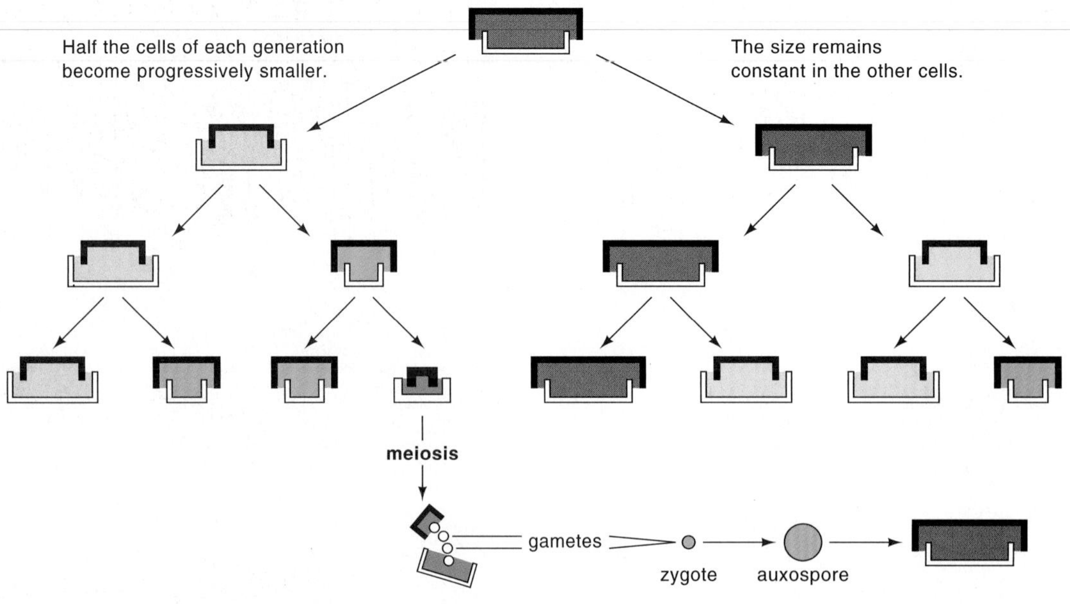

Figure 18.15 Reproduction in diatoms. The two rigid halves of each pillboxlike cell separate and a new rigid half shell forms *inside* each original half. This results in half the new cells becoming smaller with each generation. At one point, however, the diploid nucleus of a reduced-size cell undergoes meiosis, and four gametes are formed. The zygote produced when the two gametes unite becomes considerably enlarged. The enlarged cell, called an *auxospore,* develops into a diatom of the same size as the original diatom.

Brown Algae (Phaeophyceae)

More than 1,500 species of seaweeds and other algae that are brown to olive green in color are assigned to the class Phaeophyceae in this phylum. Previously, the brown algae were in a phylum of their own (*Phaeophyta*), but cladistic and structural analyses suggest relatively close relationships with other members of Phylum Chromophyta. Many brown algae are relatively large, and none are unicellular or colonial. Only 6 of the 265 known genera occur in fresh water, the vast majority growing in colder ocean waters, usually in shallower areas, although the giant kelp (*Macrocystis pyrifera*—Fig. 18.16) may be found in water up to 30 meters (100 feet) or more deep, and one brown alga (*Lobophora variegata*) was seen in clear tropical waters near the Bahamas growing at a depth of 220 meters (730 feet). One giant kelp measured 274 meters (900 feet) in length—a length believed to be a record for any single living organism.

Many of the brown algae have a **thallus** (plural: **thalli,** the term for multicellular bodies that are usually flattened and not organized into leaves, stems, and roots) that is differentiated into a **holdfast,** a **stipe,** and flattened leaflike **blades** (Fig. 18.17). The holdfast is a tough, sinewy structure resembling a mass of intertwined roots. It holds the seaweed to rocks so tenaciously that even the heaviest pounding of surf will not readily dislodge it.

The stalk that constitutes the stipe is often hollow, with a meristem either at its base or at the blade junctions. Since the meristem produces new tissue at the base, the oldest parts of the blades are at the tips.

Figure 18.16 The tip of a giant kelp (*Macrocystis pyrifera*), $\times \frac{1}{3}$.

The blades and most of the body are photosynthetic and may have gas-filled floats called *bladders* toward their bases. The bladder gases may include as much as 10% by volume of carbon monoxide. The function of the carbon monoxide, which is deadly to animal life, is not known.

The color of brown algae can vary from light yellow-brown to almost black, reflecting the presence of varying amounts of the brown pigment fucoxanthin, in addition to chlorophylls *a* and *c* and several other pigments in the chloroplasts.

The main food reserve is *laminarin,* a carbohydrate. **Algin,** or *alginic acid* (see Table 18.2 and the discussion under "Human and Ecological Relevance of the Algae," page 344) occurs on or in the cell walls and can represent as much as 40% of the dry weight of some kelps. Reproductive cells are unusual in that the two flagella are inserted laterally (i.e., on the side) instead of at the ends. The only motile cells in the brown algae are the reproductive cells.

In some localities off the coast of British Columbia and Washington, herring deposit spawn (eggs) in layers up to 2.5 centimeters (1 inch) thick on both surfaces of giant kelp blades in late spring. In the past, and to a limited extent at present, native North Americans have harvested these spawn-covered blades, sun dried them, and used them for

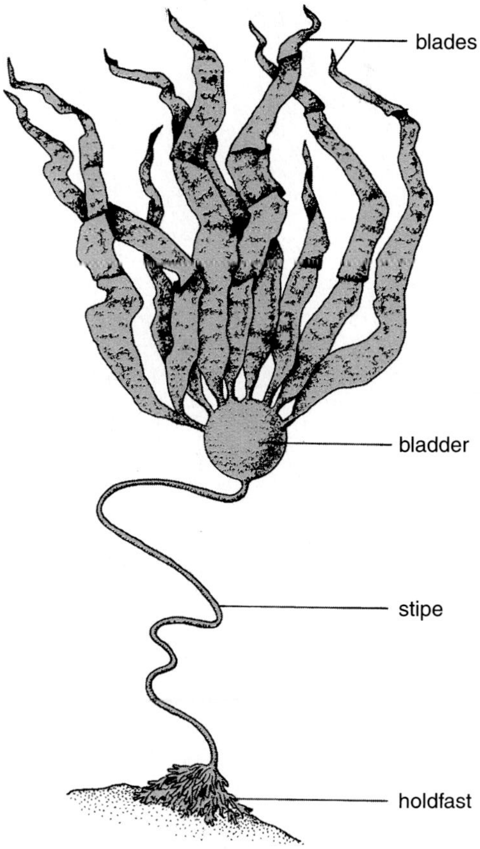

Figure 18.17 Parts of the brown alga *Nereocystis,* a kelp, $\times \frac{1}{10}$.

winter or feast food. Even today, some school children are given small pieces of the dried or preserved material for lunchbox snacks.

The Sargasso Sea gets its name from a floating brown seaweed, *Sargassum,* that is washed up in abundance on the shores along the Gulf of Mexico after tropical storms (Fig. 18.18). A species occurring in the Pacific Ocean has been used, in chopped form, as a poultice on cuts received from coral by native Hawaiians. This and several other brown algae reproduce asexually by fragmentation, while some produce autospores. Sexual reproduction takes several different forms, depending on the species. The conspicuous phases of the life cycles are usually diploid.

In the common rockweed *Fucus* (Fig. 18.19), separate male and female thalli are produced, or both sexes may develop on the same thallus. Somewhat puffy fertile areas called *receptacles* develop at the tips of the branches of the thallus. The surface of each receptacle is dotted with pores (visible to the naked eye) that open into special, spherical hollow chambers called *conceptacles.* Within the conceptacles, **gametangia** (cells or structures where gametes are produced) are formed. Eight eggs are produced in each *oogonium* (female gametangium) as a result of a single diploid nucleus undergoing meiosis followed by mitosis (Fig. 18.20). Meiosis also occurs in each *antheridium* (male gametangium), but four mitotic divisions follow meiosis so that 64 sperms are produced. Eventually, both eggs and sperms are released into the water, where fertilization takes place, and the zygotes develop into mature thalli, completing the life cycle.

Figure 18.18 *Sargassum,* a floating brown alga from which the Sargasso Sea got its name. It is also found in other marine waters, $\times\frac{1}{2}$.

PHYLUM RHODOPHYTA—THE RED ALGAE

Like the brown algae, most of the more than 5,000 species of *red algae* are seaweeds (Fig. 18.21) that tend, however, to occur in warmer and deeper waters than their brown counterparts. Some grow attached to rocks in intertidal zones, where they may be exposed at low tide. Others grow at depths of up to 200 meters (656 feet) where light barely reaches them, and in 1984, a new species of red algae was discovered at a depth of 269 meters (884 feet), where the light is only 0.0005% of peak surface sunlight. A few are unicellular, but most are filamentous. The filaments frequently are so tightly packed that the plants appear to have flattened blades or to form branching segments. Some develop as beautiful feathery structures that have the appearance of delicate works of art.[1] None match the large kelps in size, the largest species seldom exceeding a meter (3 feet) in length.

The red algae have relatively complex life cycles, often involving three different types of thallus structures. Meiosis usually occurs on a thallus called a *tetrasporophyte,* while gametes are produced on separate male and female thalli. All of the reproductive cells are nonmotile and are carried passively by water currents. Zygotes may migrate from one cell to another through special tubes, which form bizarre loops in some species.

In *Polysiphonia* (Fig. 18.22), a feathery red alga that is widespread in marine waters, the three types of thalli (male gametophyte, female gametophyte, and tetrasporophyte) all

1. Most seaweeds produce their own "glue" and are easy to mount on paper for display. Fresh specimens can be laid directly on clean, high rag-content paper, covered with a layer or two of cheesecloth, and pressed between sheets of blotting paper for a day or two until they are dry. Feathery types will make better specimens if they are placed in a shallow pan of water so that their delicate structures float out, as they do in the ocean. Paper should be slid under them and then carefully lifted so that the seaweed spreads out naturally on the surface. Once the specimens are dry, the cheesecloth (which keeps them from sticking to blotting paper) is removed, and they remain glued to the paper. They can then be displayed or stored indefinitely. Green and brown seaweeds can be treated in similar fashion.

Certain marine algae, which form crusty growths or jointed-appearing upright structures on rocks, accumulate calcium salts as they grow and often contribute to the development of coral reefs. These coralline algae need no special treatment to be displayed, although some may lose their natural pinkish or purplish color when they die.

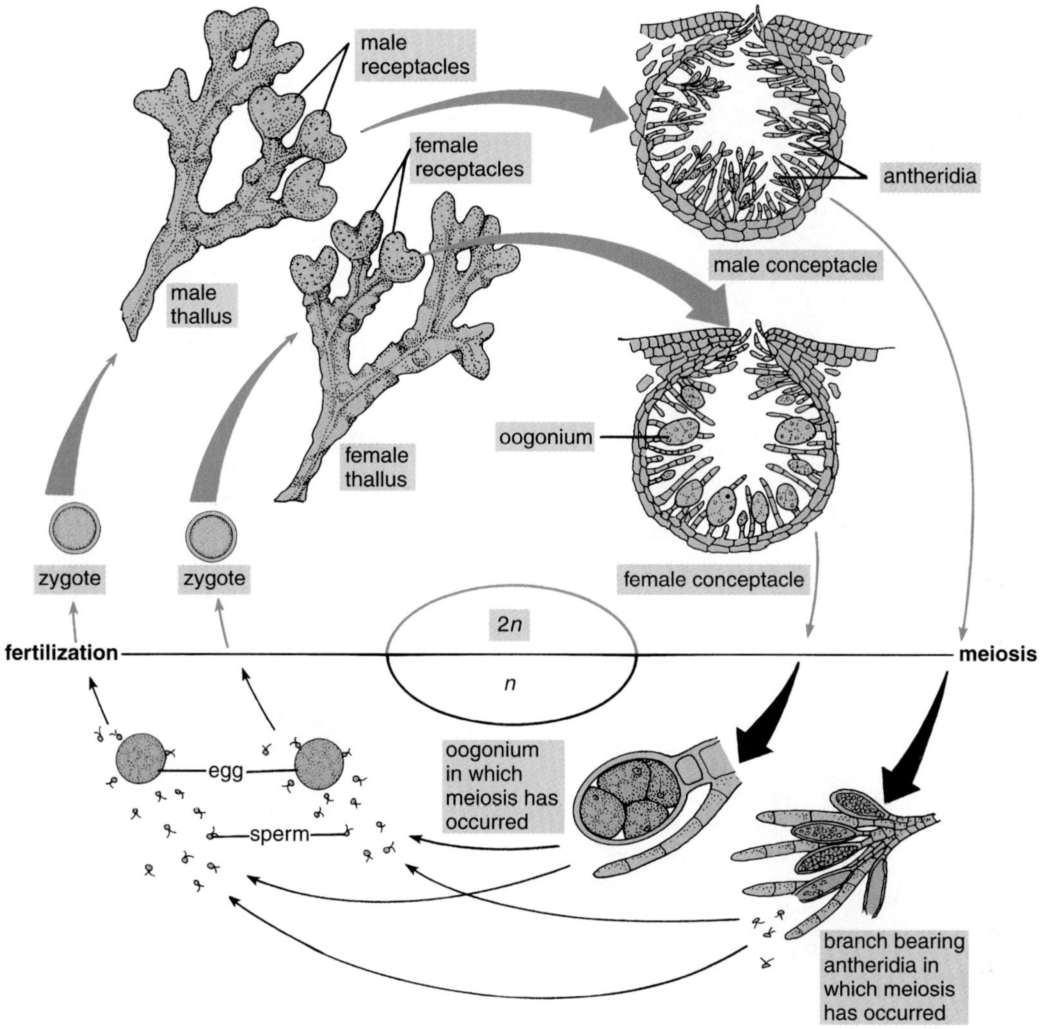

Figure 18.19 Life cycle of the common rockweed *Fucus.*

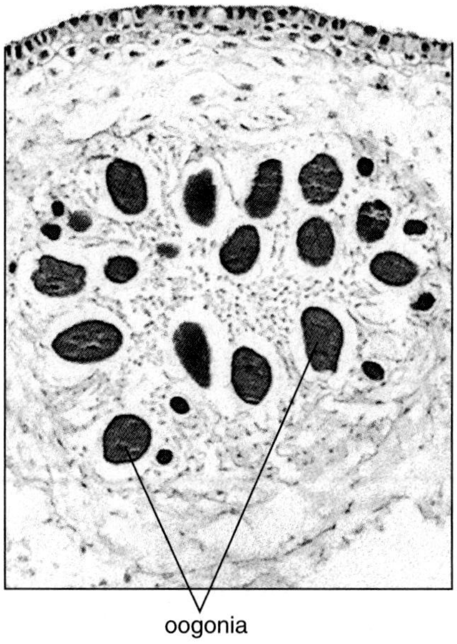

Figure 18.20 A sectioned female conceptacle of *Fucus,* ×100.

outwardly resemble one another. They are about 2 to 15 centimeters (1 to 6 inches) tall and are branched into many fine, threadlike segments. *Spermatangia,* the male sex structures, slightly resemble dense clusters of tiny grapes on slender branches of the male gametophyte thallus. Each spermatangium contains a single *spermatium* that functions as a nonmotile male gamete.

The female sex structures, called *carpogonia,* are produced on the female gametophyte thallus. Each carpogonium consists of a single cell that looks something like a microscopic bottle with a long neck called a *trichogyne.* A single nucleus at the base of the carpogonium functions as the female gamete, or *egg.* Since the spermatia have no flagella, they cannot move of their own accord, but currents may carry them considerable distances. If a spermatium should brush against a trichogyne, it may become attached. The walls between the spermatium and the trichogyne then break down, the nucleus of the spermatium migrates to the egg nucleus, and the two nuclei unite, forming a zygote.

Next, toward the base of a *pericarp* (an urn-shaped body, the outer part of which is formed by the female gametophyte thallus), the zygote begins to divide and eventually develops a

A.

B.

C.

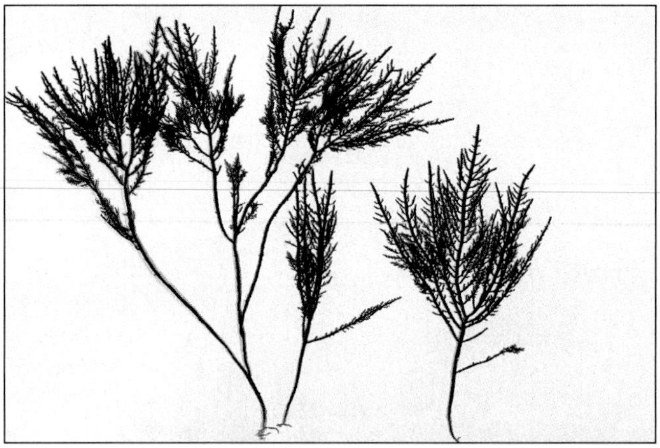

D.

Figure 18.21 Representative red algae. *A. Botryocladia. B. Stenogramme. C. Gigartina. D. Gelidium.*

cluster of clublike *carposporangia.* The pericarp and carposporangia combined constitute the *cystocarp.* Diploid asexual spores called *carpospores* are produced in the carposporangia and released, to be carried away by ocean currents.

When a carpospore lodges in a suitable location (e.g., a rock crevice or the hull of a ship), it usually germinates and grows into a *tetrasporophyte,* which closely resembles a gametophyte thallus. Tetrasporangia are formed along the branches of the tetrasporophytes. Each tetrasporangium undergoes meiosis, giving rise to four haploid *tetraspores.* When tetraspores germinate, they develop into male or female gametophytes, thereby completing the life cycle.

The red to purplish colors of most red algae are due to the presence of varying amounts of red and blue accessory

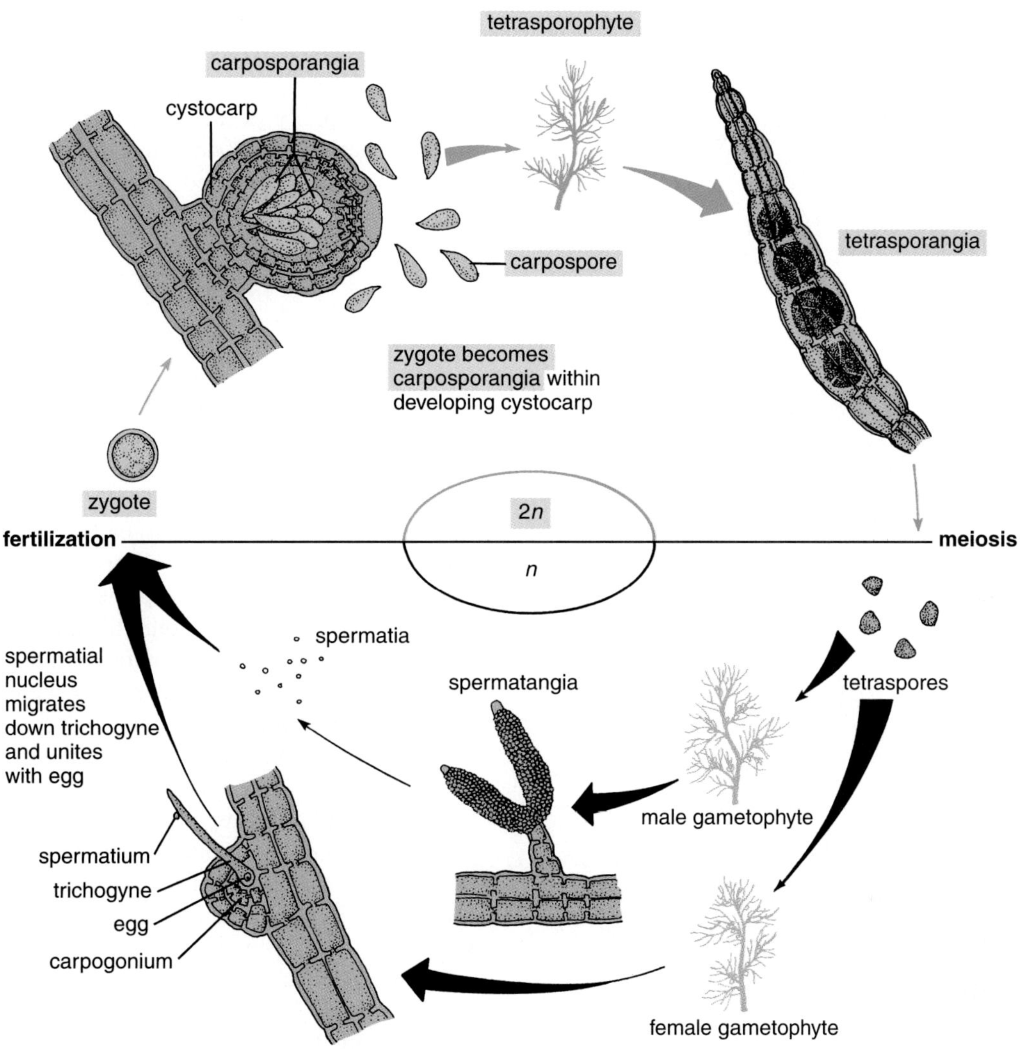

Figure 18.22 Life cycle of the red alga *Polysiphonia.*

pigments called *phycobilins,* similar to those found in the cyanobacteria. The similarity led to the belief of some *phycologists* (those who specialize in the study of algae) that the red algae may have been derived from the cyanobacteria. Several other pigments, including chlorophyll *a* and sometimes chlorophyll *d,* are also present in the chloroplasts. The principal reserve food is a carbohydrate called *floridean starch.* A number of red algae also produce **agar** and other important gelatinous substances discussed in "Agar" under the section entitled "Human and Ecological Relevance of the Algae," which begins on page 344.

PHYLUM EUGLENOPHYTA—THE EUGLENOIDS

Barnyard pools and sewage treatment ponds often develop a rich green bloom of algae. A superficial microscopic examination of water from such a pool usually reveals large numbers of active green cells, and a closer inspection may reveal one to several of the more than 750 species of euglenoids, of which *Phacus* and *Euglena* are common examples (Fig. 18.23A and B).

A *Euglena* cell is spindle shaped, has no rigid wall, and can be seen to change shape even as the organism moves along. Just beneath the plasma membrane are fine strips that spiral around the cell parallel to one another. The strips and the plasma membrane are devoid of cellulose and together are called a *pellicle.* A single flagellum, with numerous tiny hairs along one side, pulls the cell through the water. A second very short flagellum is present within a reservoir at the base of the long flagellum.

Other features of *Euglena* include the presence of a *gullet,* or groove, through which food can be ingested. The food of most of the 500 *Euglena* species is ingested, with only about a third having several to many mostly disc-shaped chloroplasts that permit photosynthesis to take place. A red **eyespot,** which along with the short flagellum is associated with light detection, is located in the cytoplasm near the base of the flagella. A carbohydrate food reserve called *paramylon* normally is present in the form of small whitish bodies of various shapes.

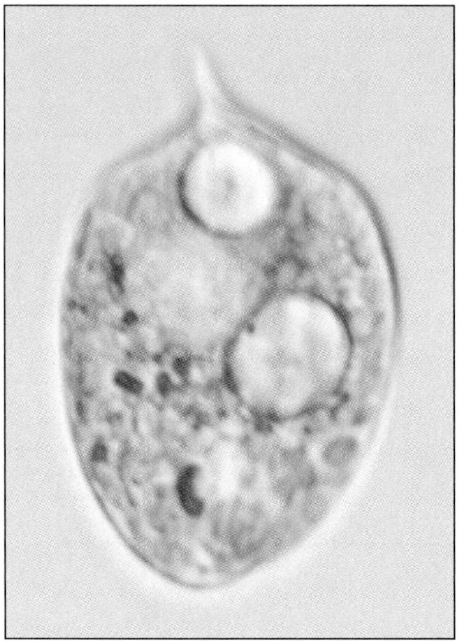

Figure 18.23A *Phacus,* a euglenoid often found in polluted waters. The single flagellum that pulls the cell through the water is not visible, ×2,000.

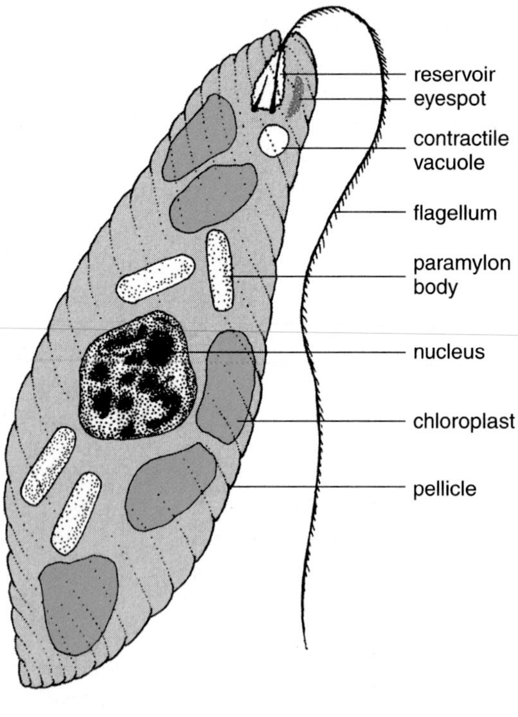

Figure 18.23B A single *Euglena.*

Reproduction is by cell division. The cell starts to divide at the flagellar end and eventually splits lengthwise, forming two complete cells. Sexual reproduction is suspected but has never been confirmed.

Some species of *Euglena* can live in the dark if appropriate food and vitamins are present. Others are known to reproduce faster than their chloroplasts under certain circumstances, so that some chloroplast-free cells are formed. As long as a suitable environment is provided, these cells also can survive indefinitely. In the past, when only two kingdoms were recognized, *Euglena's* capacity to satisfy its energy needs through either photosynthesis or ingestion of food resulted in its being treated as a plant in botany texts and as an animal in zoology texts.

PHYLUM DINOPHYTA—THE DINOFLAGELLATES

Occasionally, visitors to an ocean beach in midsummer may notice a distinctly reddish tint to the water, usually as a result of a phenomenon known as a **red tide.** Red tides are caused by the sudden and not fully understood multiplication of unicellular organisms called *dinoflagellates* (Fig. 18.24). There are over 3,000 species of dinoflagellates, with 300 of them known to be capable of producing red tides. When a red tide appears, some biologists dip a cup of sea water and save it for examination with a microscope. (The material can be preserved indefinitely with the addition of a few drops of formaldehyde, vinegar, or other weak acid.)

The cup of sea water just mentioned usually contains the *thecae* (shell-like remains) of hundreds of dinoflagellates, the best known representatives of Phylum Dinophyta. Some resemble armor-plated spaceships, while others may be smooth or have fine lengthwise ribs. The "armor plates" are located just inside the plasma membrane and are composed mostly of cellulose of varying thickness.

Dinoflagellates have two flagella that are distinctively arranged, usually being attached near each other in two adjacent and often intersecting grooves. One flagellum, which acts as a rudder, trails behind the cell. The other, which encircles the cell at right angles to the first groove, gives the cell a spinning motion as it undulates in its groove like a tiny snake.

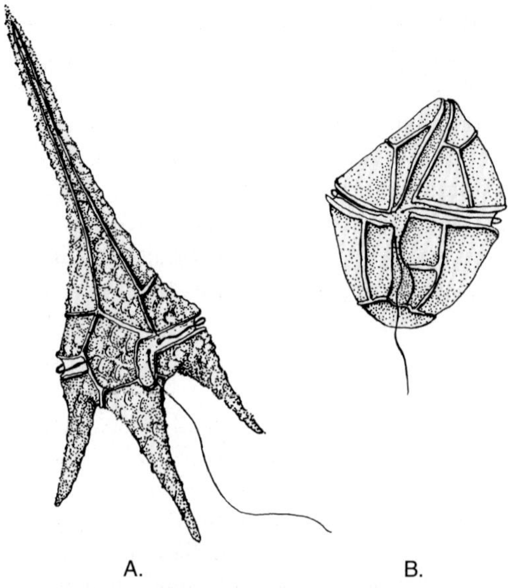

Figure 18.24 Dinoflagellates. *A. Ceratium. B. Gonyaulax.*

Most dinoflagellates have two or more disc-shaped chloroplasts, which contain distinctive brownish *xanthophyll* pigments in addition to various other pigments, including chlorophylls *a* and c_2. About 45% of the species are, however, nonphotosynthetic, and some ingest food particles, whether or not chlorophyll is present. Some have an *eyespot* (a pigmented organelle that is sensitive to light), and all have a unique nucleus in which the chromosomes remain condensed and clearly visible throughout the life of the cell. The chromosomes contain a disproportionately large amount of DNA—as much as 40 times that found in human cells. The food reserve is starch, which in dinoflagellates is stored outside the chloroplasts.

Many dinoflagellates have tiny projectiles that are fired out of the cells when they are irritated. The most common form of projectile is a *trichocyst,* which is rod-shaped and is fired within milliseconds of irritation. The cells move rapidly in the opposite direction of the discharge, leading to speculation that they aid in escape, but there is also some evidence that they spear other organisms.

Dinoflagellates occur in most types of fresh and salt water, but those that cause red tides have received the most publicity because about 40 of the species also produce powerful neurotoxins that accumulate in shellfish such as oysters, mussels, scallops, and clams. The poisons apparently don't harm the shellfish, but about 2,000 persons become ill each year—15% fatally—from eating contaminated shellfish. Although fish in open waters can swim away from affected waters, large numbers still die and wash ashore after a red tide, and the devastation to caged fish can be enormous. Even pelicans, dolphins, whales, and manatees have been poisoned by dinoflagellate toxins in their food chains.

Three major types of dinoflagellate toxins have been studied for possible use in chemical warfare. They are so potent that as little as half a milligram (barely enough to cover two of the periods printed on this page) can be fatal. Symptoms in humans begin with a tingling feeling in the fingertips and lips and eventually progress to paralysis of the diaphragm, followed by death through suffocation. Other symptoms include nausea, abdominal cramps, muscular paralysis, amnesia, hallucinations, and diarrhea. The toxin does harm the shellfish and causes massive death of clams in many cases.

The havoc to the fishing industry caused by major red tides is so great that several laboratories have conducted research with dinoflagellates and their marine habitats to try to determine the causes of their blooms and to find a way of preventing such destruction in the future.

Until 1970, blooms of toxin-producing dinoflagellates were known only from temperate waters of the Northern Hemisphere. By 1990, however, the blooms were also occurring with increasing frequency throughout both temperate and tropical waters of the Southern Hemisphere. As mentioned in the introduction to this chapter and discussed in Chapter 11, dinoflagellates play a major role in oceanic bioluminescence.

Reproduction is by cell division. Sexual reproduction appears to be rare.

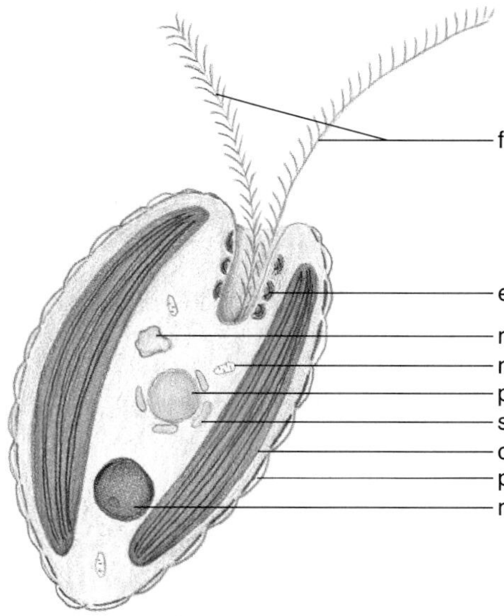

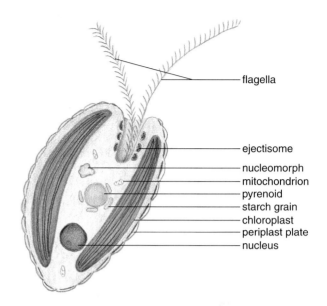

Figure 18.25 *Cryptomonas* sp., a cryptomonad, ×2000.

PHYLUM CRYPTOPHYTA—THE CRYPTOMONADS

The **cryptomonads** constitute a small group of asymmetrical, somewhat flattened, unicellular, marine and freshwater algae with two flagella, both flagella bearing short hairs throughout their length. Each cryptomonad cell has one or more plates on the inside of the plasma membrane, with a few also having plates outside of the plasma membrane. The plates may be green, blue, yellowish green, red, or brown. There is a single, two-lobed chloroplast with starch granules surrounding a central pyrenoid. In addition to the nucleus, there is a distinctive *nucleomorph* that possibly represents a vestigial nucleus of a primitive organism that may have, in ancestral forms, lived symbiotically within the larger cell. In most species, the *gullet,* which is analagous to that of a euglenoid, is lined with two or more rows of *ejectosomes.* Ejectosomes are structurally different from the stinging, harpoon-like projectiles (trichocysts) of protozoans and some dinoflagellates but unreel as long threads in similar fashion to trichocysts. Reproduction is by mitosis; sexual reproduction is unknown (Fig. 18.25).

PHYLUM PRYMNESIOPHYTA (HAPTOPHYTA)—THE HAPTOPHYTES

Most *haptophytes* are unicellular, with two smooth flagella of similar length inserted at the apex. Their pigments and carbohydrate reserves are similar to those of members of the Chromophyta, from which the majority differ in having a third flagellum—the *haptonema.* Unlike true flagella, however, the haptonema, which is located between the true flagella, does not function in propulsion; instead, it has a sticky tip that aids in capture of food. Haptophyte cells may be naked, but more often they are covered with scales (containing cellulose in some species) that are sloughed off into the water as they are replaced from beneath, where they are produced by dictyosomes. One group of haptophytes, the *coccolithophores,* has scales formed of calcite crystals. Two disc-shaped chloroplasts are present in each cell. Haptophytes, which occur in both salt and fresh water, are smaller than most diatoms and dinoflagellates but occur in huge numbers, especially in the tropics. They are a major component of marine plankton.

PHYLUM CHAROPHYTA—THE STONEWORTS

Stoneworts (Fig. 18.26) are primarily aquatic organisms of shallow, freshwater lakes and ponds; they often precipitate calcium salts on their surfaces. They consist of an axis with short lateral branches in whorls, making them loosely resemble small horsetail plants (discussed in Chapter 21). A number of plant and algal scientists place them in their own phylum (Charophyta), but others consider them a class of green algae because of their pigmentation and a few other green algal features. Their sexual reproduction is oogamous, and, unlike the antheridia of the algae discussed to this point, those of stoneworts are multicellular. Other features of both their vegetative growth and reproduction are more complex than those of any of the green algae. Some botanists have considered them more closely related to mosses than to algae.

Figure 18.26 *Chara,* a stonewort.

A comparative summary of the food reserves, chlorophylls, and flagella of the phyla of eukaryotic algae is given in Table 18.1.

HUMAN AND ECOLOGICAL RELEVANCE OF THE ALGAE

Diatoms

We have already noted that numbers of cyanobacteria are at the bottom of aquatic food chains, but members of all the protistan algal phyla play similar roles. Diatoms, for example, are consumed by fish that feed on plankton. Up to 40% of a diatom's mass consists of oils that are converted into cod and other liver oils that are rich sources of vitamins for humans. The oils also may in the past have contributed to petroleum oil deposits. The National Renewable Energy Laboratory in Golden, Colorado, is working on a project to convert oil from cultured diatoms into a "clean" diesel fuel substitute.

Diatoms also have other extensive and more direct industrial uses. In the past, they apparently have been at least as numerous as they are now. As billions upon billions of them have reproduced and died, their microscopic, glassy shells have accumulated on the ocean floor, forming deposits of *diatomaceous earth.* These deposits have accumulated to depths of hundreds of meters in some parts of the world and are quarried in several areas where past geological activity has raised them above sea level. At Lompoc, California, beds of diatomaceous earth are more than 200 meters (650 feet) deep, while in the Santa Maria oil fields of California, deposits reach a depth of 1,000 meters (3,280 feet).

Diatomaceous earth is light, porous, and powdery material that contains about 6 billion diatom shells per liter (1.057 quarts), yet a liter weighs only eight-tenths of a kilogram (1.76 pounds). It also has an exceptionally high melting point of 1,750°C (3,182°F) and is insoluble in most acids and other liquids. These properties make it ideal for a variety of industrial and domestic uses, including many types of filtration. The sugar industry uses diatomaceous earth in sugar refining, and its use for swimming-pool filters is widespread. It is also used in silver and other metal polishes, in toothpaste, and in the manufacture of paint that reflects light. Reflectorized paint is used on highway markers and signs and on the automobile license plates of some states. Diatomaceous earth is also packed as insulation around blast furnaces and boilers and is manufactured into the construction panels used in prefabricated housing.

Pacific Coast fish farmers have experienced losses of salmon and cod when dense concentrations of the diatom *Chaetoceros* have developed in the aquaculture pens. The diatoms have long hollow spines that break off and penetrate

TABLE 18.1

Comparison of the Phyla of Algae

PHYLUM	FOOD RESERVES	CHLOROPHYLLS[2]	FLAGELLA
Chlorophyta 7,500 species of green algae	Starch	Chlorophyll *b*	2, 4, or more at apex, or none
Chromophyta[1] 7,600 species of yellow-green algae, golden-brown algae, diatoms, and brown algae	Oils, fats, mannitol, laminarin	Chlorophyll c_1 Chlorophyll c_2	2, 1, or none, usually unequal, lateral, or at or near apex
Rhodophyta More than 5,000 species of red algae	Floridean starch, mannitol	Chlorophyll *d*	None
Euglenophyta 750 species of euglenoids	Paramylon	Chlorophyll *b*	2, 1, or 3, at or near apex
Dinophyta 3,000 species of dinoflagellates	Starch	Chlorophyll c_2	2; 1 trailing, 1 girdling; few with none
Cryptophyta 200 species of cryptomonads	Starch	Chlorophyll c_2	2; apical or lateral
Prymnesiophyta (Haptophyta)[2] 500 species of haptophytes	Laminarin	Chlorophyll c_1 Chlorophyll c_2	2; apical, equal
Charophyta 350 species of charophytes	Starch	Chlorophyll *b*	2; subapical

1. All phyla of algae have chlorophyll *a* and yellow to orange pigments called *carotenes* and *xanthophylls,* although the combinations and specific types of xanthophylls vary with the phyla. The *fucoxanthin* of the golden-brown and brown algae is a xanthophyll.

2. The name Haptophyta was changed to Prymnesiophyta because the international rules governing naming of organisms require that the name of a phylum be based on a generic name within the phylum, and the name Haptophyta was not based on a generic name.

the fish gills, disrupting gas exchange and causing bleeding. This damage, in turn, may permit secondary infections and excessive mucous production to occur.

Other Algae

A few green algae have occasionally been used for food, but members of this phylum have generally been used less by humans than algae of other phyla. With dwindling world food supplies, this could change. Sea lettuce has been used for food on a limited scale in Asian countries for some time, and several countries are experimenting with the suitability of plankton for human consumption. *Chlorella* contains most of the vitamins needed in human nutrition, except for vitamin C. Since *Chlorella* is so easy to culture, it may become an important protein source in many parts of the world. *Chlorella* has also been investigated as a potential oxygen source for atomic submarines, in addition to its possible use in space exploration.

Algin

Commercially produced ice cream, salad dressing, beer, jelly beans, latex paint, penicillin suspensions, paper, textiles, toothpaste, ceramics, and floor polish today all share a common ingredient, *algin,* produced by the giant kelps and other brown algae. It is now used in so many products that one might wonder how the world used to get along without it (Table 18.2).

Algin has the unique capacity to regulate water "behavior" in a wide variety of products. It can, for example, control the development of ice crystals in frozen foods, regulate the penetration of water into a porous surface, and generally stabilize any kind of suspension, such as an ordinary milkshake or other thick fluid containing water. It is produced by several kinds of seaweeds, but a major source is the giant kelp found in the cooler ocean waters of the world, usually just offshore where there are strong currents (see Fig. 18.16).

TABLE 18.2
Some Uses of Algin

Food

1. As a thickening agent in toppings, pastry fillings, meringues, potato salad, canned foods, gravies, dry mixes, bakery jellies, icings, dietetic foods, flavored syrups, candies, puddings

2. As an emulsifier and suspension agent in soft drinks and concentrates, salad dressings, barbecue sauces, frozen-food batters

3. As a stabilizer in chocolate drinks, eggnog, ice cream, sherbets, sour cream, coffee creamers, party dips, buttermilk, dairy toppings, milkshakes, marshmallows

Paper

1. Provides better ink and varnish holdout on paper surfaces; provides uniformity of ink acceptance, reduction in coating weight and improved holdout of oil, wax, and solvents in paperboard products. Makes improved coating for frozen food cartons

2. Used for coating greaseproof papers

Textiles

1. Thickens print paste and improves dye dispersal. Reduces weaving time and eliminates damage to printing rolls or screens

Pharmaceuticals and Cosmetics

1. As a thickening agent in weight-control products, cough syrups, suppositories, ointments, toothpastes, shampoos, eye makeup

2. As a smoothing agent in lotions, creams, lubricating jellies. Binder in manufacture of pills. Blood anticoagulant

3. As a suspension agent for liquid vitamins, mineral oil emulsions, antibiotics. Gelling agent for facial beauty masks, dental impression compounds

Industrial Uses

1. Used in manufacture of acidic cleaners, films, seed coverings, welding rod flux, ceramic glazes, boiler compounds (prevents minerals from precipitating on tubes), leather finishes, sizing, various rubber compounds (e.g., automobile tires, electric insulation, foam cushions, baby pants), cleaners, polishes, latex paints, adhesives, tapes, patching plaster, crack fillers, wall joint cement, fiberglass battery plates, insecticides, resins, tungsten filaments for light bulbs, digestible surgical gut (disappears by time incision is healed), oil well-drilling mud. Used in clarification of beet sugar. Mixed with alfalfa and grain meals in dairy and poultry feeds

Brewing

1. Helps create creamier beer foam with smaller, longer-lasting bubbles

This large seaweed is known to grow up to 92 meters (300 feet) or more long and sometimes grows at the rate of 3 to 6 decimeters (1 to 2 feet) per day. The basal meristems allow for blade regeneration and, accordingly, make the giant kelps valuable resources. Specially equipped oceangoing vessels (Fig. 18.27) harvest the kelp by mowing off the top meter (3 feet) of growth, taking the chopped material aboard, then transferring it to processing centers onshore where it is extracted and refined.

Minerals and Food

Brown algae also produce a number of other useful substances. Many seaweeds, but particularly kelps, build up concentrations of iodine to as much as 20,000 times that of the surrounding seawater. Although it is cheaper to obtain iodine from other sources in North America, dried kelp has been used in the treatment of goiter, which results from iodine deficiency, in other parts of the world. Kelps are relatively high in nitrogen and potassium and have been used as fertilizer for many years. Before such use, the seaweeds need to be rinsed to rid them of salt. They also have been used as livestock feed in northern Europe and elsewhere. In the Orient, many marine algae are used for food—in soups, confections, meat dishes, vegetable dishes, and beverages. In Japan, acetic acid is produced through fermentation of seaweeds.

During the Irish famine of 1845–1846, *dulse,* a red seaweed, became an important substitute for the potato crop that had been destroyed by blight. Dulse also occurs along both the Atlantic and Pacific coasts of North America. In Maine and eastern Canada, where it is a popular snack food, dulse is referred to as "Nova Scotia Popcorn." Another red seaweed, *purple laver* ("nori"), occurs in both American and Asian waters and is used extensively for food, particularly in the Orient. In Japan, it is cultured on nets or bamboo stakes set out in shallow marine bays (Fig. 18.28). It is harvested when the thin, crinkly, gelatinous blades are several centimeters (2 to 3 inches) in diameter and is used in meat and macaroni dishes; soups; and dry, spiced delicacies.

Figure 18.27 A specially equipped vessel harvesting kelp off the California coast. The ship moves backward through the kelp beds as the machinery at the stern mows the kelp and conveys it to the hold. *(Courtesy Kelco Company)*

Irish moss is another important edible red alga. It is also used in bulking laxatives, cosmetics, and pharmaceutical preparations. Blancmange is a dessert made from Irish moss and milk. *Carrageenan* is a mucilaginous substance extracted from Irish moss and used as a thickening agent in chocolate milk and other dairy products. *Funori,* obtained from yet another red alga, is used as a laundry starch, as an adhesive in hair dressings, and in some water-based paints.

Agar

One of the most important of all algal substances is **agar,** produced most abundantly by the red alga *Gelidium.* This substance, which has the consistency of gelatin, is used around the world in laboratories and medical institutions as a solidifier of nutrient culture media for the growth of bacteria. When other nutrients are added to it, it can also be used to solidify culture media for the growth of both plant and animal cells. Full-sized plants have been induced to develop from pollen grains sown on nutrient agar cultures. Orchid tissues are cultured commercially on such media and induced to grow into full-sized plants (as discussed in the section "Micropropagation" in Chapter 14), and its use in making the capsules containing drugs and vitamins is now worldwide. It is also used as an agent in bakery products to retain moistness, as a base for cosmetics, and as an agent in gelatin desserts to promote rapid setting.

Figure 18.28 Beds of purple laver ("nori") ready to harvest in shallow marine water at Sendai, Japan. *(Courtesy I. A. Abbott)*

Current research involving red algae and other seaweeds indicates they contain a number of substances of potential medicinal value. More than 20 seaweeds have been used in preparations designed for the expulsion of digestive-tract worms, control of diarrhea, and the treatment of cancer. Some have shown considerable potential as antibiotics and insecticides. Chemical relatives of DDT have been found to be produced by certain red algae. The sea hare and other marine animals feed on such algae, and it is possible that such animals may degrade (break down) the DDT-like compounds to simpler substances—a feat unknown among land animals.

OTHER MEMBERS OF KINGDOM PROTISTA

Protozoans (Phylum Protozoa) and *sponges* (Phylum Porifera) are included within Kingdom Protista; they have, however, traditionally been regarded as animals and are not covered in this book. The *slime molds,* which appear to be related to the protozoa, are discussed next.

PHYLUM MYXOMYCOTA—THE PLASMODIAL SLIME MOLDS

In the Ripley's "Believe It or Not" pavilion at the Chicago World's Fair in 1933, there was an exhibit of "hair growing on wood." The "hairs," while indeed superficially resembling short human hair, were actually the reproductive structures of a species of **slime mold** (Fig. 18.29). These curious organisms are totally without chlorophyll and are incapable of producing their own food. They are a bit of a puzzle to biologists because they are distinctly animal-like during much of their life cycle but also distinctly fungus-like when they reproduce.

The tiny, roundish spores of the more than 500 species of slime mold average only 10 to 12 micrometers in diameter and are individually invisible to the naked eye. Nevertheless, they are present nearly everywhere and are especially abundant in airborne dusts.

If you place almost any dead leaf or piece of bark in a dish and add a food source, such as a few dry oatmeal flakes to which a drop or two of water has been added, the odds are that slime mold spores will be present and will germinate. If the dish is covered, the spores will sometimes germinate in as little as 15 minutes, but germination usually takes several hours or longer. Within a few days, a glistening mass of active slime mold protoplasmic material, looking something like the netted veins of a leaf, may appear. This material, whose "veins" tend to merge into the shape of a fan at its leading edges, is a slime mold **plasmodium** (Fig. 18.30).

If the plasmodium is examined with a microscope, it becomes apparent that there are no cell walls present. The protoplasm in the veins, particularly toward the center, flows very rapidly and rhythmically. The protoplasmic movement may stop momentarily at regular intervals and then resume its flow, sometimes in the opposite direction.

Plasmodia are often white, but they also may be brilliantly colored in shades of yellow, orange, pink, blue, violet, or black. A few are colorless and more or less transparent. Plasmodia are found on damp forest debris, under logs, on old shelf or bracket fungi, sometimes on older mushrooms, and in other moist places where there is

Figure 18.29 Reproductive bodies (sporangia) of the plasmodial slime mold *Stemonitis,* ×20.

Ecological Review

The algae are the chief source of photosynthesis across the open ocean and the main basis for the food webs of the open ocean ecosystem, the most extensive on the planet. Photosynthetic algae are major contributors to the oxygen of our atmosphere and may make a similar contribution to the atmospheres of long-distance spacecraft. The Protista also include the heterotrophic slime molds, protozoans, and oomycetes, as well as the euglenoids, which are both photosynthetic and heterotrophic. Seaweeds are an important source of useful chemicals, ranging from antibiotics and insecticides to anti-cancer agents. In contrast to the human benefits derived from seaweeds, the water molds, another group of protists, cause diseases in many crop plants, such as the late blight of potatoes that created the Irish Potato Famine of 1846 in which over one million people starved to death and millions more were forced to emigrate.

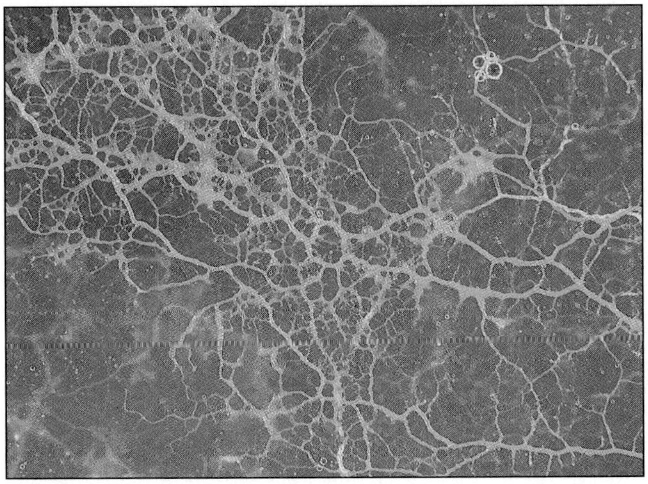

Figure 18.30 A plasmodium of the slime mold *Physarum,* ×10.

dead organic matter. Plasmodia tend to flow forward at a rate of up to 2.5 centimeters (1 inch) or more per hour, often against slow moisture seepage, feeding on bacteria and other organic particles as they go. They contain many diploid nuclei, all of which divide often and simultaneously as growth occurs. With an adequate food supply, a plasmodium may increase to 25 times its original size in just 1 week.

Dramatic events take place when significant changes in food supplies, available moisture, light, and other environmental features occur. The plasmodium usually is converted, often quite rapidly, into many separate small

A.

B.

Figure 18.31 Common slime mold sporangia. *A. Lamproderma,* ×25. *B. Lycogala,* ×10. *(B. courtesy L. L. Steimley)*

sporangia (Fig. 18.31), each containing thousands of minute, one-celled **spores.** The sporangia often are globe shaped, but in some species, they develop as long or wide stationary bodies that may resemble a jumbled network of tubes, or they may resemble erect hairs or end up as a shapeless blob. The sporangia may or may not have slender stalks, depending on the species. Others exhibit combinations of body forms. The spores are often distributed throughout a jumbled mass of threads called a *capillitium.*

When a spore is formed, a single nucleus and a little cytoplasm become surrounded by a wall. Meiosis takes place in the spore, and three of the four resulting nuclei degenerate. When the spore germinates, one or more amoebalike cells called *myxamoebae* emerge. Sometimes, these have flagella, in which case they are called *swarm cells.* Either form may become like the other through the development or loss of flagella. At first, myxamoebae or swarm cells feed on bacteria and other food particles. Sooner or later, however, they function as gametes, fusing in pairs and

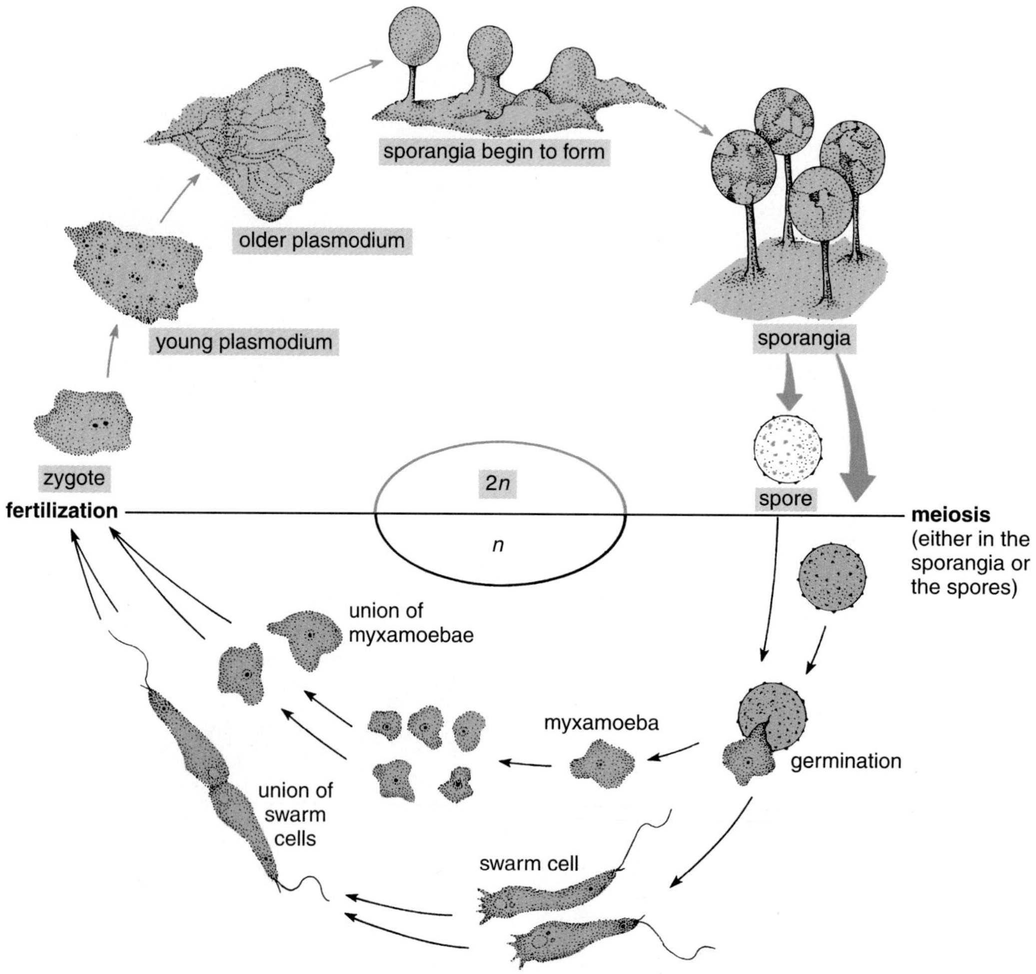

Figure 18.32 Life cycle of a plasmodial slime mold. *(After C.J. Alexopoulos, and C.W. Mims, 1979.* Introductory Mycology. *3d ed. © John Wiley & Sons, Inc. New York. Redrawn by permission of John Wiley & Sons, Inc.)*

forming zygotes. A new plasmodium usually develops from the zygote, although occasionally zygotes or small plasmodia may fuse and form larger plasmodia (Fig. 18.32).

PHYLUM DICTYOSTELIOMYCOTA—THE CELLULAR SLIME MOLDS

Two or more phyla of slime molds are recognized. Most of the species have typical plasmodia and follow the patterns just discussed. About two dozen species of *cellular slime molds* are, however, evidently not closely related to the other slime molds. Individual amoebalike cells of cellular slime molds feed independently, dividing and producing separate new cells from time to time. When the population reaches a certain size, they stop feeding and clump together, forming a mass called a *pseudoplasmodium.* The pseudoplasmodium resembles and crawls like a garden slug. It eventually becomes stationary and is transformed into a sporangium-like mass of spores.

Human and Ecological Relevance of the Slime Molds

The slime molds, like the bacteria, contribute to ecological balance in forests and woodlands by breaking down organic particles to simpler substances; they also reduce bacterial populations in their habitats. They can often be seen on bark mulch after a rain. One atypical species occasionally attacks cabbages and another causes powdery scab of potatoes. Yet another species causes a disease of watercress, but slime molds otherwise are of little economic significance.

PHYLUM OOMYCOTA—THE WATER MOLDS

Those who have kept tropical fish aquaria or have seen salmon at the end of a spawning run often have noticed cottony growths on cuts and bruises on a fish's body or have

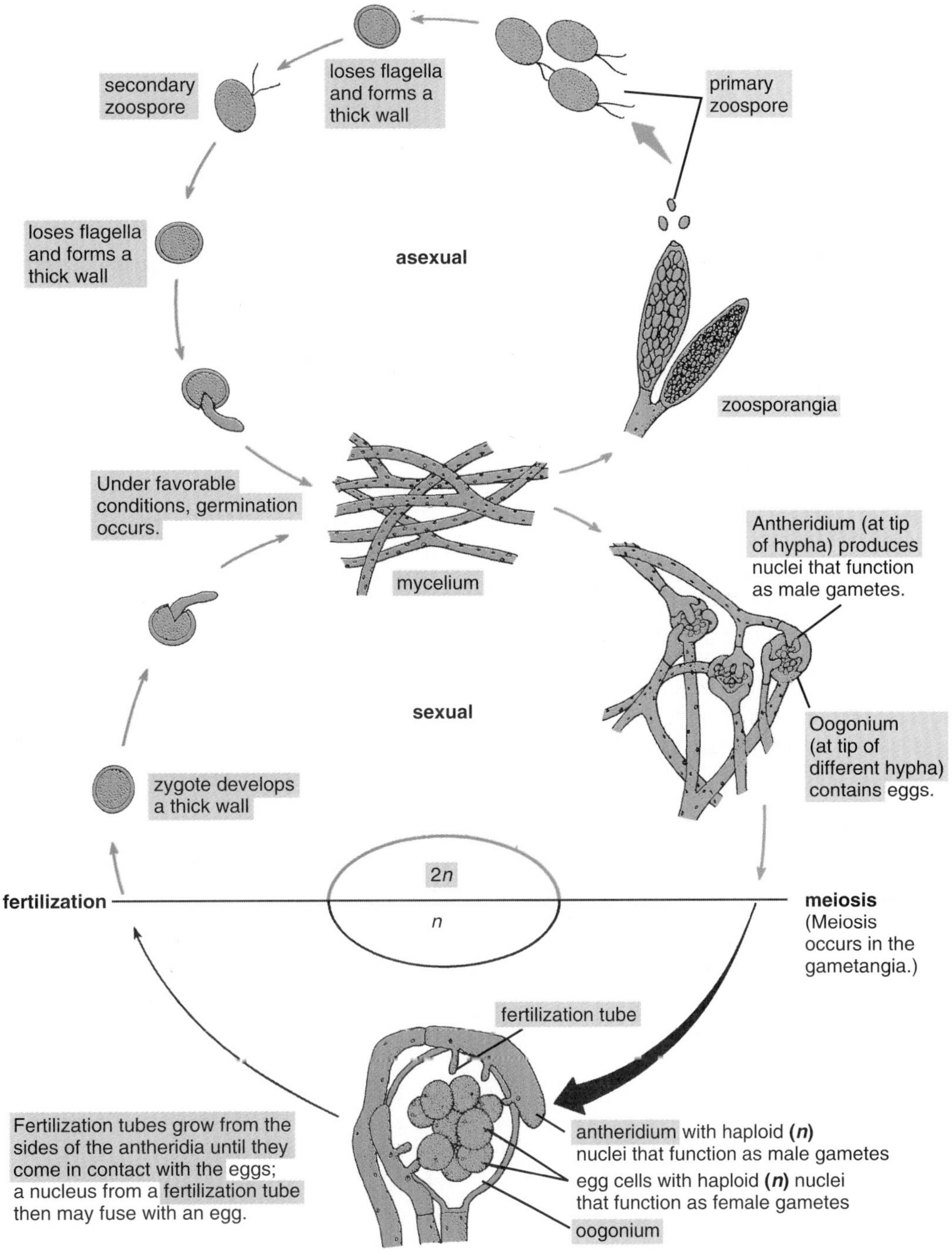
secondary zoospore
loses flagella and forms a thick wall
primary zoospore
loses flagella and forms a thick wall
asexual
zoosporangia
Under favorable conditions, germination occurs.
mycelium
Antheridium (at tip of hypha) produces nuclei that function as male gametes.
sexual
Oogonium (at tip of different hypha) contains eggs.
zygote develops a thick wall
2n
fertilization
n
meiosis (Meiosis occurs in the gametangia.)
fertilization tube
Fertilization tubes grow from the sides of the antheridia until they come in contact with the eggs; a nucleus from a fertilization tube then may fuse with an egg.
antheridium with haploid (n) nuclei that function as male gametes
egg cells with haploid (n) nuclei that function as female gametes
oogonium

Figure 18.33 Life cycle of the water mold *Saprolegnia.*

seen similar growths on the eyes of sick fish. These aquatic organisms, called **water molds,** or *oomycetes,* are also often found on dead insects such as houseflies and have been cultured on certain crushed flowering plant seeds placed in water. They range in form from single spherical cells to branching, threadlike, coenocytic hyphae. Coenocytic hyphae, which are not divided into individual cells, may form large masses of threads called *mycelia.* Motile cells with two flagella are produced at various stages of their life cycles (Fig. 18.33).

During asexual reproduction in water molds, crosswalls form just below the tips of certain hyphae. Numerous zoospores, each with two flagella, are produced in these special tip chambers. The zoospores, after emerging through a terminal pore, eventually give rise to new water mold mycelia. Sexual reproduction involves **oogonia** and **antheridia,** which arise on side branches under the influence of hormones produced within the organism. The mycelium is diploid, and meiosis takes place in the oogonia and antheridia. Zygotes formed in the oogonia eventually give rise to new mycelia.

Water molds share several features with brown algae, leading some phycologists to speculate they may have been derived from brown algae. Features common to water molds and brown algae include eggs (oogamy), cellulose in the cell walls, a predominantly diploid life cycle, and zoospores with two flagella.

Human and Ecological Relevance of the Water Molds

Two important members of Phylum Oomycota cause serious diseases of higher plants. Although dew or rain water is needed for their reproduction, neither grows under water. One, called *downy mildew of grapes,* completes its life cycle on grape leaves, usually killing the leaves and, if not controlled, the vine.

This disease seriously threatened the French wine industry toward the end of the 19th century, after it reached the vineyards on imported American cuttings. Within a few years, it was controlled when it was discovered that Bordeaux mixture, a combination of copper sulfate and lime (calcium oxide), inhibited the growth of downy mildew. This mixture, which makes the grapes look unappetizing, is the first substance known to have been used as a fungicide. It was originally sprayed on grape vines to discourage grape thieves.

Another important oomycete is called *late blight of potato.* In the past, this virulent organism plagued potato farmers so much that at times it altered the course of history. In the summer of 1846, the entire potato crop of Ireland was wiped out in one week, causing a famine during which over a million people starved to death and many more emigrated to the United States. The organism produces its mycelium inside the leaves, so once its spores land on the surface and send hyphae into the interior tissues, it is relatively immune to spraying. Spraying leaves with Bordeaux mixture or other fungicides before the spores have germinated, however, checks development of the oomycete.

Summary

1. There is evidence that all living organisms were confined to the ocean less than 1 billion years ago. It is believed that green algae began making the transition from water to land about 400 million years ago and gave rise to land plants.
2. Kingdom Protista includes many diverse organisms that all have eukaryotic cells. Members may be unicellular or multicellular and occur as either colonies or filaments. Modes of nutrition include photosynthesis, ingestion of food, a combination of both, or absorption of food in solution. Some members are nonmotile, but most are motile.
3. Green algae (Chlorophyta) have cells with the same pigments and reserve food (starch) as those of higher plants. The chloroplasts of *Chlamydomonas* have pyrenoids; the cells have two or more vacuoles and often a red eyespot. Asexual reproduction is by mitosis; sexual reproduction is isogamous.
4. *Ulothrix* is a filamentous green alga that may be attached to objects by means of a holdfast; each cell contains a curved plate chloroplast around the periphery. Asexual reproduction is by zoospores; sexual reproduction is isogamous.
5. *Spirogyra* is a floating, filamentous green alga with spiral, ribbonlike chloroplasts. Asexual reproduction is by fragmentation. Sexual reproduction is by conjugation.
6. *Oedogonium* is an epiphytic, filamentous green alga; it has cylindrical, netted chloroplasts. In asexual reproduction, zoospores are produced. Sexual reproduction is oogamous.
7. Other green algae include *Chlorella,* desmids, *Acetabularia, Volvox, Ulva,* and *Cladophora.*
8. The several classes of Chromophyta include the yellow-green algae, the true golden-brown algae, the diatoms, and the brown algae. Some members produce statospores.
9. Diatoms are very abundant; they have a glassy shell that consists of two "halves" that fit together like a box with a lid.
10. Fucoxanthin gives a golden-brown color to most chromophytes. Some diatoms move by contact of the cytoplasm with a surface as it protrudes through the pores.
11. Diatoms reproduce asexually by mitosis and sexually through the fusion of gametes that form an auxospore (zygote).
12. Many brown algae (Phaeophyceae) are large seaweeds; their thalli often may be differentiated into a stipe, flattened blades, and a holdfast.
13. The color of brown algae is largely due to fucoxanthin; the main carbohydrate food reserve is laminarin. Some brown algae produce algin, a useful gelatinous substance. The reproductive cells have lateral flagella. The common rockweed, *Fucus,* produces eggs and sperms that form zygotes in the water.
14. Most red algae (Rhodophyta), represented by *Polysiphonia,* are seaweeds with life cycles that involve three different types of thalli and nonmotile gametes.
15. Red and blue phycobilins are partially responsible for the colors of red algae; chlorophyll *d* may be present in the chloroplasts. The main carbohydrate food reserve is floridean starch. Some red algae produce agar, an economically important gelatinous substance.
16. Euglenoids (Euglenophyta) have no rigid cell wall, one functional flagellum, a gullet, and paramylon as a carbohydrate food reserve. Reproduction is by cell division. Sexual reproduction has not been confirmed.
17. Dinoflagellates (Dinophyta) are unicellular organisms with two flagella inserted at right angles to each other. Some cause red tides that can kill fish and poison

humans. Dinoflagellates in tropical waters exhibit bioluminescence (emission of light) when disturbed.

18. The cryptomonads (Cryptophyta) are biflagellated, unicellular algae with two chloroplasts and an extra vestigial nucleus (nucleomorph).
19. Haptophytes (Prymnesiophyta) have biflagellated cells with a flagellumlike haptonema that aids in the capture of food.
20. Stoneworts (Charophyta) are branched green organisms that superficially resemble horsetails.
21. Algae are ecologically and economically important. Diatomaceous earth is used for filtering, polishes, insulation, and reflectorized paint. *Chlorella* is a potential food and oxygen source. Algin is used as a stabilizer and thickening agent in hundreds of products.
22. Some brown algae are a source of fertilizer, iodine, and food. Red algae are a source of agar and food and have potential medicinal value.
23. The plasmodial slime molds, which are animal-like in their vegetative state, consist of a multinucleate mass of protoplasm called a plasmodium that flows over damp surfaces, ingesting food particles.
24. Slime molds form stationary sporangia that contain spores from which myxamoebae or swarm spores emerge upon germination. Myxamoebae or swarm spores function as gametes, with new plasmodia developing from the zygotes.
25. Cellular slime molds produce a pseudoplasmodium that crawls like a slug and can convert to a stationary, sporangium-like mass of spores.
26. Water molds have coenocytic mycelia and include organisms that cause diseases of fish and other aquatic organisms. Asexual reproduction involves zoospores; gametes are produced in oogonia and antheridia.

Review Questions

1. What forms of sexual and asexual reproduction occur in the green algae?
2. How would you distinguish *Chlamydomonas* from *Euglena*?
3. *Spirogyra, Ulothrix,* and *Oedogonium* all form filaments. How can you tell them apart?
4. In the green algae studied, where in the life cycles does the chromosome number change from *n* to 2*n,* and vice versa?
5. How do cells of diatoms differ from those of other organisms?
6. Which groups of algae produce the following important products: (1) agar, (2) algin, (3) nerve poisons, (4) abrasives for polishes?
7. Why are some green algae red and some red algae green?
8. Where and how is algin obtained?
9. Is there any difference in structure between the holdfasts of microscopic green algae and brown algae?
10. What is unique about the cells of cryptomonads and those of haptophytes?

Discussion Questions

1. Some algae are attached to solid objects or other organisms, while others are free-floating. What are the advantages and disadvantages of each type of growth?
2. The variety and sizes of algae found in the oceans are considerably greater than those of freshwater forms. Can you suggest reasons?
3. Seaweeds that grow in intertidal zones, where they may be exposed between tides, are often more gelatinous than their continually submerged counterparts. Explain.
4. Why do some algae grow on one side of a tree and not all around the trunk?
5. Should the bladders (floats) of some of the kelps give these algae any advantage over those algae that do not possess such structures?

Learning Online

Visit our web page at www.mhhe.com/botany for interesting case studies, practice quizzes, current articles, and animations within the Online Learning Center to help you understand the material in this chapter. You'll also find active links to these topics.

General References in Phycology
Diversity of Algae
Rhodophyta
Phaeophyceae
Kelp Forests
Chlorophyta
Chromophyta
Dinophyta
Economic and Ecological Importance of Algae
Pfisteria
Water Molds
Slime Molds

Additional Reading

Abbott, I. A. (Ed.). 1997. *Taxonomy of economic seaweeds with reference to some Pacific species,* vol. VI. La Jolla, CA: University of California Sea Grant College System.

Akatsuka, I. (Ed.). 1995. *Biology of economic algae.* Champaign, IL: Balogh Scientific Books.

Becker, E. W. 1994. *Microalgae: Biotechnology and microbiology.* New York: Cambridge University Press.

Bhattachyra, D. 1998. *Origins of algae and their plastids.* New York: Springer-Verlag.

Cox, E. J. 1996. *Identification of freshwater diatoms from live material.* Hingham, MA: Kluwer Academic Publications.

Fritsch, F. E. 1935, 1945. *Structure and reproduction of the algae,* 2 vols. Ann Arbor, MI: Books Demand.

Graham, L., and L. W. Wilcox. 1999. *Algae.* Old Tappan, NJ: Prentice-Hall.

Palmer, C. M. 1962. *Algae in water supplies.* U.S. Department of Health, Education and Welfare. Public Health Service. Washington, DC: Government Printing Office.

Scoging, A., and M. Bahl. 1997. Diarrhetic shellfish poisoning in the UK. *Lancet* 352 (9122).

Smith, G. M. 1950. *Freshwater algae of the United States,* 2d ed. New York: McGraw-Hill.

Sze, P. 1998. *A biology of the algae,* 3d ed. Dubuque, IA: WCB/McGraw-Hill.

Turner, N. J. 1995. *Food plants of coastal first peoples.* Seattle, WA: University of Washington Press.

Van Den Hoek, C., D. G. Mann, and H. M. Jahns. 1996. *Algae: An introduction to phycology.* New York: Cambridge University Press.

Fly agaric mushrooms (*Amanita muscaria*). (© Willard Clay Photography)

Kingdom Fungi

Chapter Outline

19

Overview

After a short introduction, the distinctions between Kingdoms Protista and Fungi are discussed. A summary of the features of Kingdom Fungi and a review of how the kingdom came to be recognized are given.

Kingdom Fungi includes five phyla of fungi. Life cycles and discussions of the economic importance of representative members of each phylum are presented. Nematode-trapping fungi, *Pilobolus*, truffles, morels, ergot, yeasts, stinkhorns, puffballs, bracket fungi, bird's-nest fungi, smuts, rusts, poisonous and hallucinogenic fungi, shiitake mushrooms, mushroom culture, antibiotics, industrial products obtained from fungi, and fungi in nature are discussed. The chapter concludes with an examination of various forms of lichens. Natural dyeing is briefly explored, and the economic importance of lichens is reviewed.

Some Learning Goals

1. Know the general features that distinguish Kingdom Fungi from the other kingdoms.
2. Distinguish the phyla and subphyla of fungi from one another on the basis of their cells or hyphae and their reproduction.
3. Learn the form and function of sporangium, conidium, coenocytic, mycelium, dikaryotic, zygospore, ascus, and basidium.
4. Assign each of the following to its phylum and/or subphylum of true fungi: athlete's foot, Dutch elm disease, *Pilobolus, Penicillium* mold, stinkhorn, yeast, ergot, chestnut blight fungus, puffball, smut.
5. Know five economically important fungi in each of the different major groups of true fungi.
6. Understand how lichens are classified and identified.
7. Learn the basic structure of a lichen.

When it comes to breaking down organic materials of all kinds and making them available for recycling, fungi and bacteria are the most important organisms known. The major oil spill that occurred in Alaska in 1989 intensified efforts to find better and faster ways of cleaning up such disasters than was possible in the past. Although there are bacteria that break down oil molecules, very few such organisms survive the salt water of the ocean for any length of time. Attention now, however, is being given to a fungus that breaks down many types of oils to harmless simpler molecules, such as carbon dioxide.

This fungus, known as *Corollospora maritima,* occurs naturally on almost all the beaches of the world and flourishes in ocean surf, which, as mentioned in Chapter 18, contains oils produced by diatoms. Ways of culturing the fungus inexpensively and in large quantities are being explored, and also strains that produce the oil-degrading enzymes in significantly larger amounts are being sought. This chapter examines the nature, reproduction, and other features of this and the thousands of other known true fungi found all over the world.

The word *fungus* may evoke images of mushrooms or some sort of powdery or spongy, creeping growth. Although mushrooms are indeed fungi, and while many fungi do appear to be creeping along the ground, the forms of fungi seem almost infinite. There are about 100,000 known species of mushrooms, rusts, smuts, mildews, molds, stinkhorns, puffballs, truffles, and other organisms in Kingdom Fungi, and over 1,000 new species are described each year. There may be as many as 1.5 million species yet undiscovered or undescribed, and our daily ravaging of rain forests and other fungal habitats undoubtedly is dooming many species to extinction before they can be found.

As they grow, most fungi produce an intertwined mass of delicate threads that tend to branch freely and often also fuse together. The individual, usually more or less tubular, threads are called **hyphae** (singular: **hypha**). Hyphae may or may not be partitioned into cylindrical cells (Fig. 19.1).

A mass of hyphae is collectively referred to as a **mycelium.** When appropriate food is available, fungi grow very rapidly. In fact, despite the microscopic size of individual hyphae, all those produced by a single fungus in one day when laid end to end could extend more than a kilometer (0.6 mile). Some fungi thrive in freezers if the temperature is not lower than –5°C (23°F), while others grow best at temperatures of 55°C (131°F) or higher.

Besides being vital to the natural recycling of dead organic material, fungi, along with bacteria, also cause huge economic losses through food spoilage and disease; these topics are explored in the sections on human and ecological relevance of fungi later in the chapter. Scientists who study fungi are known as *mycologists* (from the Greek word *myketos,* a fungus). Consumers of fungi are called *mycophagists.*

DISTINCTIONS BETWEEN KINGDOMS PROTISTA AND FUNGI

In the past, the true fungi, slime molds, and bacteria were all placed in a single division of the Plant Kingdom. Once the fundamental differences between prokaryotic and eukaryotic cells became known, however, the bacteria were placed in the prokaryotic Kingdom Monera. Then it became increasingly apparent that the metabolism, reproduction, and general lines

Figure 19.1A Typical fungal hyphae, ×10.

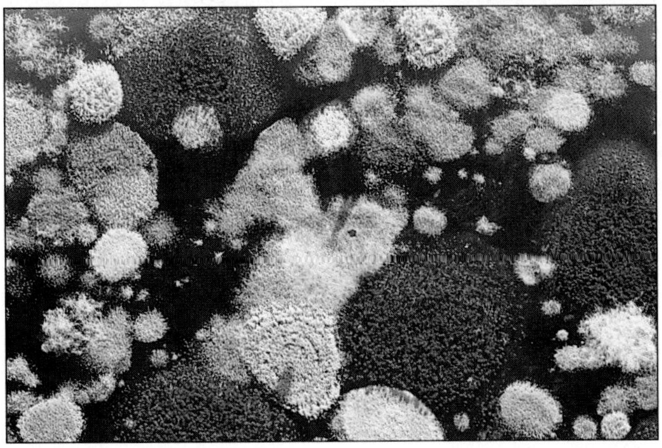

Figure 19.1B Colonies of fungi growing on the surface of stale brewed coffee.

of diversity of fungi were different from those of members of the Plant Kingdom, with the fungi evidently having been independently derived from ancestral unicellular organisms. Accordingly, fungi were placed in their own kingdom.

All true fungi are filamentous or unicellular heterotrophs, most of which absorb their food in solution through their cell walls. Some are *saprobes* (organisms that live on dead organic matter); others are parasitic decomposers; still others (mycorrhizal fungi) have a mutualistic relationship with plants. The slime molds engulf their food and were discussed in Chapter 18; they appear to be related to protozoa, which are single-celled protists included in Kingdom Protista. Chytrids and water molds are also funguslike, but their classification is controversial because, like most protists, their reproductive cells have flagella. However, chytrids, like other fungi, have chitin in their cell walls, and recent molecular research suggests the chytrids are probably more closely related to the true fungi than they are to members of Kingdom Protista; therefore, they are discussed here under Kingdom Fungi. The water molds, however, which have cellulose but no chitin in their cells walls, continue to be more appropriately included in Kingdom Protista.

The members of Kingdom Fungi are placed in five phyla. With the exception of some chytrids and all yeasts, they are filamentous. Most, but not all, fungi lack motile cells. Filamentous fungi produce hyphae that grow at their tips. Structures such as mushrooms are formed from hyphae tightly interwoven and packed together. The cell walls of true fungi consist primarily of *chitin,* a material also found in the shells of arthropods (e.g., insects, crabs). Fungi exhibit a variety of forms of sexual reproduction. The food substances that most fungi absorb through their cell walls are often broken down with the aid of enzymes secreted to the outside by the cells. Because of the great variety of form and reproduction throughout Kingdom Fungi, a neat pigeonholing of all the members into distinct groups is difficult. Broad groups can, however, be recognized.

KINGDOM FUNGI— THE TRUE FUNGI

Phylum Chytridiomycota— The Chytrids

If you were to immerse dead leaves or flowers, old onion bulb scales, dead beetle wing covers, or other organic material in water that has been mixed with a little soil, within a day or two, thousands of microscopic **chytrids** (pronounced kítt-ridds) probably would appear on the surfaces of the immersed objects (Fig. 19.2). These simple, mostly one-celled organisms include many parasites of protists, aquatic fungi, aquatic flowering plants, and algae. Some parasitize pollen grains, and many other species are *saprobic* (feed on nonliving organic material). Some of the most common chytrids consist of a spherical cell with colorless, branching threads called **rhizoids** at one end. The rhizoids anchor the organism to its food source.

Other chytrids may develop short hyphae or even a complete mycelium whose hyphae contain many nuclei. Such multinucleate mycelia without crosswalls are said to be **coenocytic** (pronounced seé-no-sitt-ik). The cell walls of a few chytrids have been reported to contain cellulose in addition to chitin.

Many chytrid species reproduce only asexually through the production of zoospores within a spherical cell. The zoospores, which each have a single flagellum, settle upon

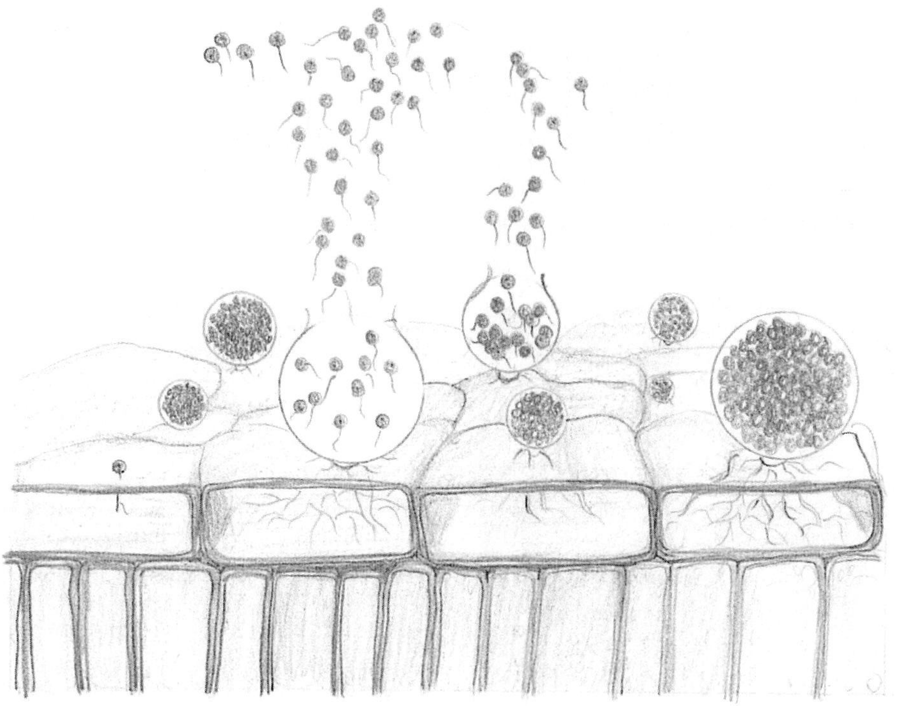

Figure 19.2 Chytrids on a dead leaf submerged in water.

release and grow into new chytrids. Some species undergo sexual reproduction by means of the fusion of two motile gametes with haploid nuclei or by the union of two non-motile cells whose diploid zygote nuclei undergo meiosis. The zygote commonly is converted into a resting spore. The origin of chytrids is unknown, but the presence of a flagellum on the motile cells has led some authorities to suggest they may have originated from the protozoa.

Phylum Zygomycota—The Coenocytic True Fungi

Although black bread molds are probably the best-known members of Phylum Zygomycota, they are not the only fungi that grow on bread. So many organisms can, in fact, contribute to bread spoilage that nearly all commercially baked goods have, in the past, had chemicals such as calcium propionate added to the dough to prevent or retard the growth of such organisms. There is now a trend toward eliminating preservatives from bread, pies, and other bakery items since the chemicals that have made the goods a less-suitable medium for the growth of fungi apparently are unhealthy for humans after prolonged use. Alternative ways of retarding fungal growth and spoilage are being sought.

Rhizopus (Fig. 19.3), a well-known representative black bread mold, has spores that are exceedingly common everywhere. They have been found in the air above the North Pole, over jungles, on the inside and outside of buildings, in soils, clothing, automobiles, and hundreds of kilometers out to sea, easily carried there by prevailing winds and breezes.

Asexual Reproduction

When a spore lands in a suitable growing area, it germinates and soon produces hyphae that may become an extensive mycelium. Like the mycelium of water molds, it is *coenocytic* (not partitioned into individual cells) and contains numerous haploid nuclei. After the mycelium has developed, certain hyphae called **sporangiophores** grow upright and produce globe-shaped **sporangia** at their tips (Fig. 19.4). Numerous black **spores** are formed within each sporangium. When these spores are released through the breakdown of the sporangium wall, they may blow away and repeat the cycle.

Sexual Reproduction

Black bread molds reproduce sexually by conjugation. Although there is no visible difference in form, black bread mold mycelia occur in two different mating strains. When a hypha of one strain encounters a hypha of the other, the chemicals they produce create an attraction, and swellings called *progametangia* develop opposite each other on the hyphae. The progametangia grow toward each other until they touch. A crosswall is formed a short distance behind each tip, and the two *gametangia* merge, becoming a single, large multinucleate *coenozygote* in which the nuclei of the two strains fuse in pairs. A thick, ornamented wall then develops around this coenozygote or zygospore with its numerous diploid nuclei. This structure, called a *zygosporangium,* is the characteristic sexual spore of members of this division. A zygosporangium may lie dormant for months, but eventually, it may crack open, and one or more sporangiophores with sporangia at their tips grow out. Meiosis apparently takes place just before

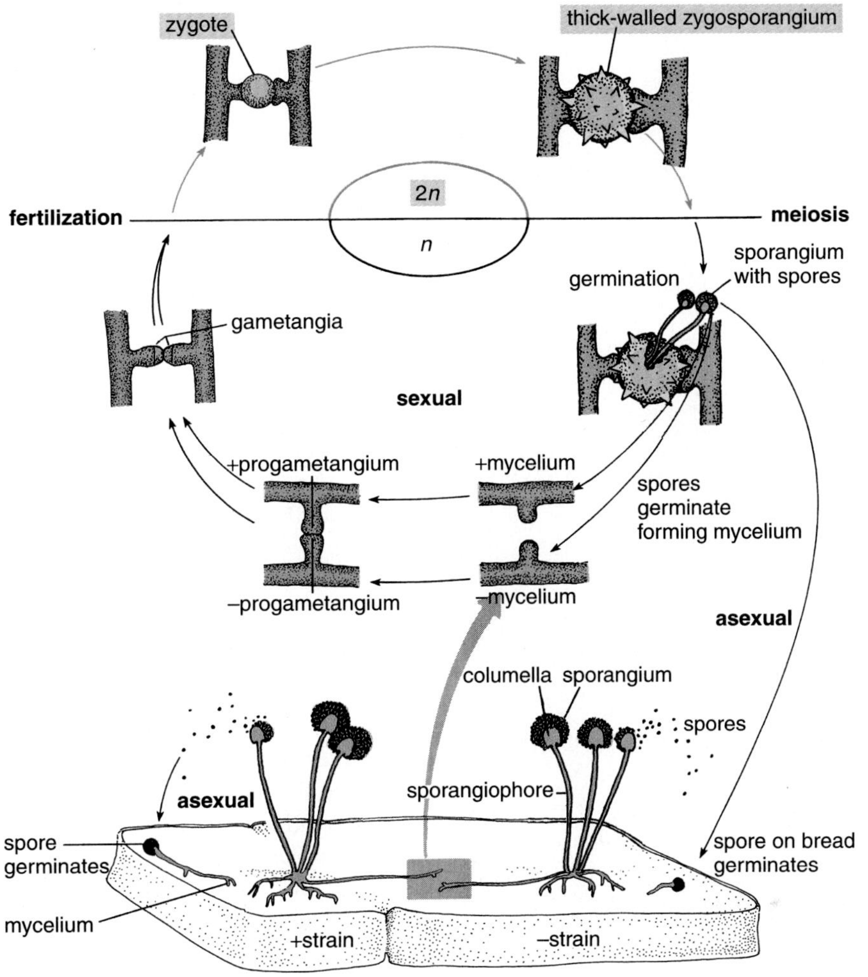

Figure 19.3 Life cycles of the black bread mold *Rhizopus.*

this occurs, and thousands of black spores are produced in the sporangia. In some species, the spores are produced externally on hyphae instead of being formed in sporangia.

One dung-inhabiting genus of fungi in this division has the scientific name of *Pilobolus,* derived from two Greek words meaning "cap thrower." The name is quite appropriate, as the mature sporangia are catapulted a distance of up to 8 meters (26 feet), where they adhere to grass or other vegetation (Fig. 19.5). When the vegetation is ingested by animals, the spores germinate in the digestive tract and are already growing in the dung when it is released.

The sporangia of *Pilobolus* fungi are forcibly released precisely in the direction of light, to which the sporangiophores are very sensitive. This action can be demonstrated by placing some horse dung (preferably at least two days old) in a glass dish that has a lid and then covering the dish with black paper. An opening of any shape is cut in the paper; the dish should then be set where it will receive adequate light for a few days. Any *Pilobolus* sporangia that have been produced and forcibly discharged will form a black pattern on the glass closely corresponding to the cut-out area.

Human and Ecological Relevance of the Coenocytic True Fungi

One species of bread mold is used in Indonesia and adjacent areas to produce a food called *tempeh.* Tempeh basically consists of boiled, skinless soybeans that have been inoculated with a bread mold and set aside for 24 hours. The mycelium that develops holds the soybeans together, produces enzymes that increase the content of several of the B vitamins, and renders the soybean protein more digestible. The tempeh is fried, roasted, or diced for soup and is prepared fresh daily. Other bread mold species are used with soybeans to make a Chinese cheese called *sufu.*

At least two species of *Rhizopus* (*R. arrhizus, R. nigricans*) are used commercially to carry out important steps in the manufacture of birth control pills and anesthetics, while others are used in the production of industrial alcohols and as a meat tenderizer. Yet another species produces a yellow pigment used for coloring margarine.

Figure 19.4 A zygomycete sporangium. Spores are visible in the split. Scanning electron micrograph, ×2000. *(Courtesy Kenneth D. Whitney)*

Phylum Ascomycota—The Ascomycetes (Sac Fungi)

Should you happen one summer to be touring or visiting in the south of France, you might be startled to see men and women with coils of rope slung around their shoulders pushing pigs in wheelbarrows. They happen to be wheeling the pigs into the woods to help them find *truffles.*

Truffles are gourmet "mushrooms," which grow mostly between 2.5 and 15 centimeters (1 and 6 inches) beneath the surface of the ground, usually near oak trees. They are somewhat prunelike in appearance and may be more than 10 centimeters (4 inches) in diameter, although most are smaller. They give off a tantalizing aroma that has been shown to contain pig sex *pheromones* (chemicals that produce specific responses). Pigs can detect truffles a meter (3 feet) below the surface and more than 15 meters (50 feet) away. The animals, in an attempt to get to them, strain at the ropes that are tied around their necks. The owners dig up the truffles, which were selling in the United States in 2001 for about $300 a pound, and reward the pigs with acorns or

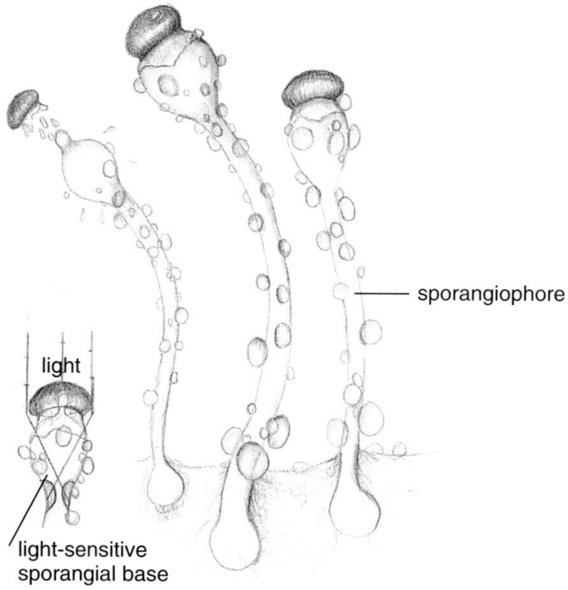

Figure 19.5 The dung fungus *Pilobolus.* The spores are released with force toward a light source.

other less interesting food. Trained dogs are now often substituted for pigs in truffle hunting, as they are less likely to eat the truffles.

Truffles are the reproductive bodies of representatives of a large and varied phylum of true fungi called **ascomycetes** (**sac fungi**). More than 30,000 species, including yeasts, powdery mildews, brown fruit rots, ergot, morels, microscopic parasites of insects, fungi associated with canned fruit, and many others, have thus far been described. Most produce mycelia superficially similar to those of the water molds and coenocytic true fungi, but the hyphae are partitioned into individual cylindrical cells. Each crosswall has one or more pores through which the single to many tiny nuclei can pass. Little bodies that serve as plugs are located near the pores. The "plugs" seal off individual cells if they become damaged.

Asexual Reproduction

Asexual reproduction is by means of **conidia** (singular: **conidium**). Conidia are spores that are produced externally—outside of a sporangium—either singly or in chains at the tips of hyphae called *conidiophores* (see Fig. 19.34). Asexual reproduction in yeasts is by **budding** (Fig. 19.6). As a yeast cell buds, the nucleus divides, and a small protuberance appears to balloon out slowly from the cell. One daughter nucleus moves into the bud, which becomes pinched off as it grows to full size.

Sexual Reproduction

Sexual reproduction involves the formation of tiny fingerlike sacs called **asci** (singular: **ascus**). When hyphae of two different "sexes" become closely associated in the more complex sac fungi, male *antheridia* may be formed on one

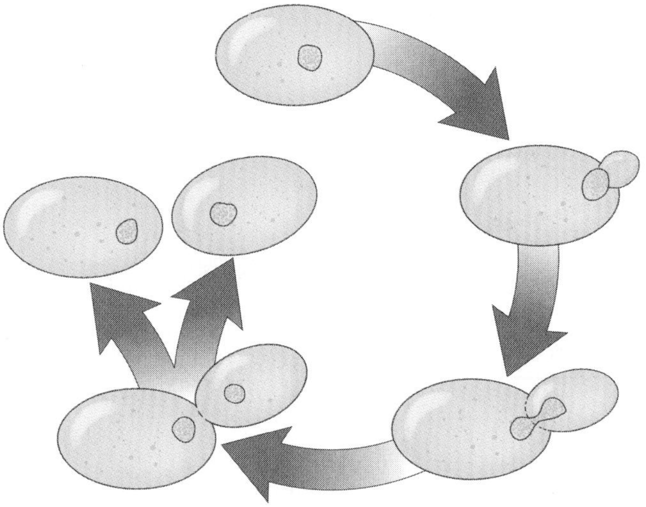

Figure 19.6 How a yeast cell reproduces asexually by *budding.*

and female *ascogonia* on the other, although in many species, antheridia and ascogonia may be produced on the same mycelium. Hyphae grow and connect an antheridium and an ascogonium to each other; male nuclei then migrate into the ascogonium. There, the male nuclei pair with the female nuclei present but do not unite. New hyphae (*ascogenous hyphae*), whose cells each contain one male and one female nucleus, grow from the ascogonium, the cells dividing in a unique way so that each cell has one of each kind of nucleus. This special cell division begins with the cell at the tip of an ascogenous hypha forming a hook called a *crozier.* The two nuclei in the crozier divide simultaneously, and crosswalls then form so that there are now three cells, the middle cell containing one male and one female nucleus. This middle cell becomes an ascus in which the two haploid (*n*) nuclei unite and become a diploid (2*n*) zygote nucleus. The fingerlike, tubular asci develop in a layer (referred to as the *hymenium*) at the surface of a structure called an *ascoma* (previously called an *ascocarp*). The zygote nucleus undergoes meiosis, and the resulting four haploid nuclei usually divide once more by mitosis so that there is a row of eight nuclei in each ascus. These nuclei become *ascospores* as they are walled off from one another with a little cytoplasm (Fig. 19.7).

Thousands of asci may be packed together in an ascoma, which often is cup shaped (*apothecium*—Fig. 19.8) but also may be completely enclosed (*cleistothecium*—Fig. 19.9) or flask shaped with a little opening at the top (*perithecium*). Truffles actually are enclosed ascomata (apothecia). Cup-shaped ascomata may be several centimeters (2 to 3 inches) in diameter and may be brilliantly colored on the inside. When ascospores are mature, they are often released with force from the asci. If an open ascoma is jarred at maturity, it may appear to belch fine puffs of smoke consisting of thousands of ascospores that are dispersed to new locations by air currents.

When an ascospore lands in a suitable area, it may germinate, producing a new mycelium, and then repeat the process. In many instances, however, a number of asexual generations involving conidia are produced between the sexual cycles.

Sexual reproduction in yeasts is somewhat streamlined in that individual yeast cells function as ascogonia and antheridia. When two haploid cells unite, they become a diploid fusion cell that may serve as an ascus in which meiosis takes place and ascospores are formed—all within the original cell wall. In some yeasts, however (including the common baking and brewing yeast *Saccharomyces cerevisiae*), the fusion cell may multiply asexually into a diploid colony, with all the cells eventually undergoing meiosis.

Human and Ecological Relevance of the Sac Fungi

Morels, which some people have called the world's most delicious mushrooms, and *truffles* have been prized as food for centuries. Most mushrooms are, however, included in the club fungi discussed in the next section "Phylum Basidiomycota—The Basidiomycetes (Club Fungi)." Wealthy Romans and Greeks used to insist on preparing morels personally according to various recipes they had concocted, and they are still prized as gourmet food today. Prior to 1982, numerous unsuccessful attempts were made to cultivate them under controlled conditions, but morels now can be mass-produced commercially and are presently being marketed. Morels (Fig. 19.10) are tan in color and have a spongelike, somewhat cone-shaped top on a stalk that resembles a miniature tree trunk. The numerous depressions between the ridges each contain thousands of asci. Although well-cooked morels are perfectly edible by themselves, some persons have become ill after consuming undercooked specimens or consuming them with alcohol. Caution in this regard is advised.

A related mushroom called a *false morel* or *beefsteak morel* is considered a delicacy by many but has caused death in others. False morels contain a volatile toxin (monomethyl hydrazine) that may render them poisonous to humans if cooking has failed to eliminate all of the toxin.

When rye, and to a lesser extent other grains, comes into flower in a field, it may become infected with ergot fungus (Fig. 19.11). This fungus seldom causes serious damage to the crop, but as it develops in the maturing grain, it produces several powerful drugs. If the infected grain is harvested and milled, a disease called **ergotism** may occur in those who eat the contaminated bread. The disease can affect the central nervous system, often causing hysteria, convulsions, and sometimes death. Another form of ergotism causes gangrene of the limbs, which can result in loss of the affected limb. It can also be a serious problem for cattle grazing in infected fields. They frequently abort their fetal calves and may succumb themselves.

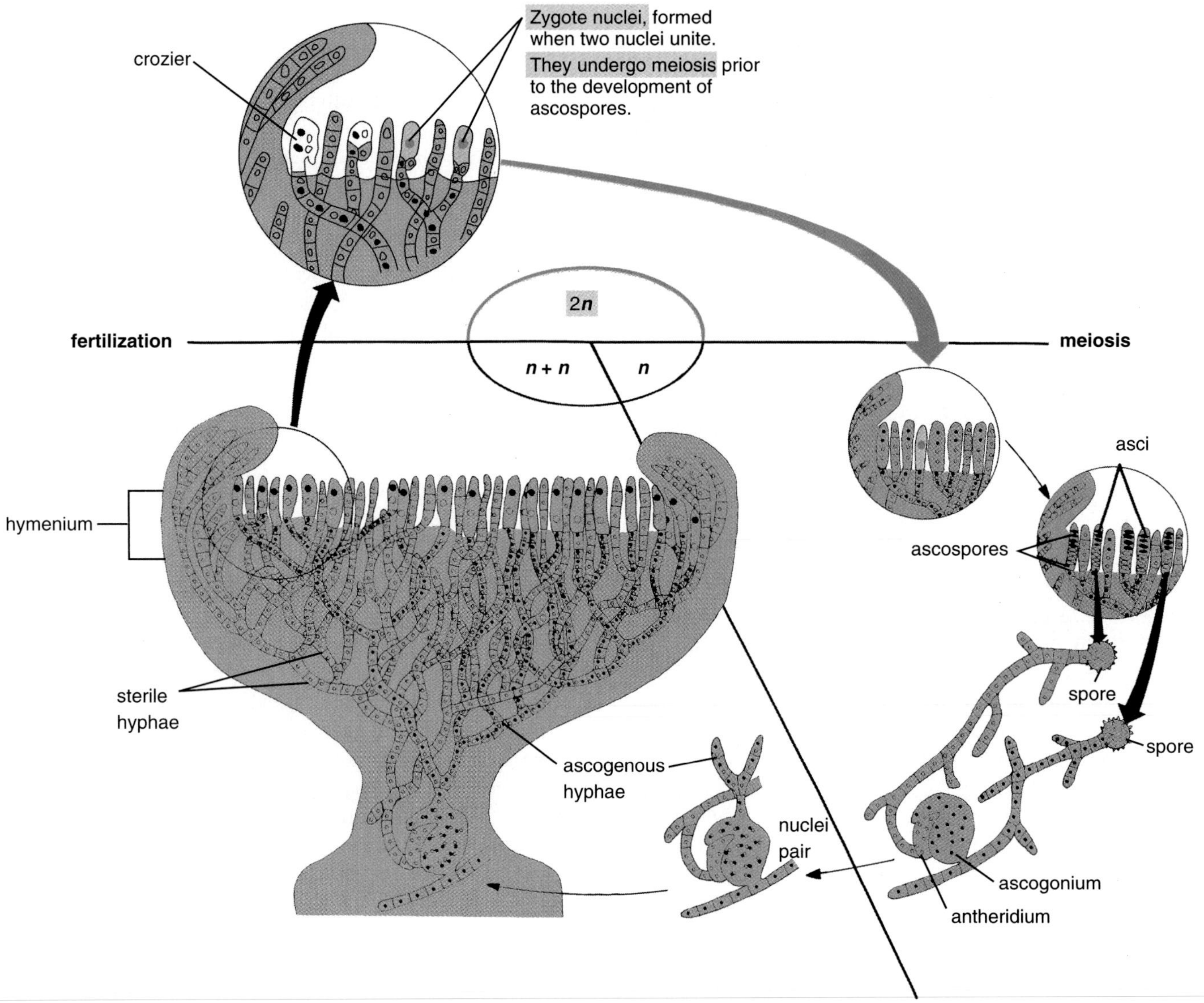

Figure 19.7 Life cycle of a sac fungus. When hyphae of two different strains of a sac fungus become closely associated, male antheridia may be formed on one and female ascogonia on the other. Male nuclei migrate into an ascogonium and pair but do not fuse with the female nuclei present. Then new hyphae (ascogenous hyphae), whose cells each contain a pair of nuclei, grow from the ascogonium. In a process involving fusion of the pairs of nuclei (followed by meiosis), fingerlike asci, each containing four or eight haploid nuclei, are formed in a layer called a *hymenium,* which lines an ascoma. The haploid nuclei become ascospores, which are discharged into the air. They are potentially capable of initiating new mycelia and repeating the process.

Ergotism was common in Europe in the Middle Ages. Known then as *St. Anthony's Fire,* it killed 40,000 people in an epidemic in A.D. 994. In 1722, the cavalry of Czar Peter the Great was felled by ergotism just as the czar was about to conquer Turkey, and the conquest never took place. Some historians have implicated ergotism in violent social upheavals such as the French Revolution and the Salem Witchcraft Trials. In 1951, an outbreak hospitalized 150 victims in a French village. Five persons died and 30 became temporarily insane, imagining they were being chased by snakes and demons.

In small, controlled doses, ergot drugs are medically useful. They stimulate contraction of the uterus to initiate childbirth and have been used in abortions and in the treatment of migraine headaches. Ergot is also an initial source for the manufacture of the hallucinogenic drug lysergic acid diethylamide, popularly known as LSD.

Fungi in this phylum play a basic role in the preparation of baked goods and alcohol. Enzymes produced by yeast aid in fermentation, producing ethyl alcohol and carbon dioxide in the process. The carbon dioxide produced in bread dough forms bubbles of gas, which are trapped, causing the dough to rise and giving bread its porous texture. Part of the flavor of individual wines, beers, ciders, sake, and other alcoholic beverages is imparted by the species, or strain, of yeast used to ferment the fruits or grains.

Yeasts are at least indirectly involved in the manufacture of a number of other important products. Ephedrine, a drug also produced by mormon tea, a leafless western desert

Figure 19.8 An ascoma of a sac fungus.

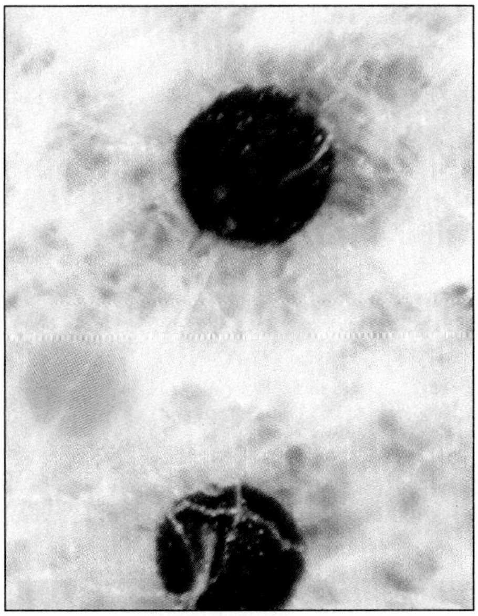

Figure 19.9 Closed ascomata (*cleistothecia*) of a powdery mildew on an oak leaf, ×200.

shrub, is obtained commercially from certain yeasts. The drug is used in nose drops and in the treatment of asthma. Yeasts are also a rich source of B vitamins and are used in the production of glycerol for explosives. Ethyl alcohol is used in industry as a solvent and in the manufacture of synthetic rubber, acetic acid, and vitamin D. Yeast contains about 50% protein and makes a nourishing feed for livestock. Its enzyme, invertase, catalyzes the conversion of

Figure 19.10 A morel. *(© Doug Sherman/Geofile)*

sucrose into glucose and fructose and is used to soften the centers of chocolate candies after the chocolate coating has been applied.

Several very important plant diseases are found in this phylum of fungi. *Dutch elm disease* (caused by *Ophiostoma ulmi* or *O. nova-ulmi*), a disease originally described in Holland, has devastated the once numerous and stately American elm trees in many towns and cities in the midwestern and eastern United States and has spread to the West and South. When Dutch elm disease was first discovered, limited control of the disease was achieved by spraying the trees with DDT to kill the elm bark beetles that spread the disease from tree to tree. DDT killed many useful organisms as well, however, and its use for such purposes was eventually banned.

Other sprays have since been used, again with limited success, and biological controls are now being sought. One such control was reported in 1980 by Gary Strobel of Montana State University. He injected *Pseudomonas* bacteria into 20 diseased trees and saved 7 of them when the bacteria multiplied and killed the Dutch elm disease fungus. In 1987, Dr. Strobel injected a genetically altered strain of *Pseudomonas* bacteria with more powerful toxins against

Figure 19.11 Ergotized rye. The prominent darker objects scattered throughout the rye are kernels infected with ergot fungus. *(Copyright © Loran Anderson)*

Figure 19.12 A peach leaf infected with peach leaf curl.

Dutch elm disease into 15 elm trees but then cut the trees down in protest to the red tape involved in not previously obtaining permission from the Environmental Protection Agency for his experiment. Whether or not this control technique will meet with widespread success remains to be seen, but it is the type of control that is much to be preferred to poisonous sprays that upset delicate ecological balances (see Appendix 2, which deals with biological controls).

Chestnut blight has virtually eliminated the once numerous American chestnut trees from the eastern deciduous forests. Attempts to control the disease have met with very limited success thus far. Much better success has been obtained in controlling *peach leaf curl,* a disease that attacks the leaves of some stone fruits, especially peach trees (Fig. 19.12). Sprays that contain copper or zinc salts apparently inhibit the germination of spores of peach leaf curl and have been effective in preventing serious damage when applied to trees before the dormant buds swell and open in the spring.

Phylum Basidiomycota—The Basidiomycetes (Club Fungi)

At the end of my first year in college, I did odd jobs during the summer, including various types of yard work. On one occasion while cleaning dead leaves and debris from a shaded garden area, I noticed a peculiar, unpleasant odor. On checking to see where the odor was coming from, I noticed two or three "growths," about the width of a pencil and the length of a finger, rising above the surface of the ground. On closer inspection, I saw that these "growths" had the consistency and appearance of a sponge and the tips were partially covered with a slimy and putrid-smelling substance. The "growths" turned out to be fungi called *stinkhorns* (Fig. 19.13), whose odor attracts flies; the flies disseminate the sticky spores that adhere to their bodies.

Stinkhorns are interesting but relatively unimportant representatives of another large phylum of true fungi, the **basidiomycetes** (**club fungi**). Other members of this phylum include mushrooms (Fig. 19.14A), or toadstools (the only distinctions between mushrooms and toadstools are based on folklore or tradition, with edible species being called mushrooms

Figure 19.13 A common stinkhorn. Note the slimy mass of spores toward the tip. *(Courtesy Leland Shanor)*

A.

B.

C.

Figure 19.14 *A.* A common *Russula* mushroom. *B.* Earth stars (*Geastrum* sp.) *C.* A shelf or bracket fungus (*Phacolus*). *(B. Courtesy Perry J. Reynolds; C. Courtesy Richard Critchfield)*

and poisonous species being called toadstools—mycologically, there is no difference), puffballs, earth stars (Fig. 19.14B), shelf or bracket fungi (Fig. 19.14C), rusts, smuts, jelly fungi, and bird's-nest fungi. They are called club fungi because in sexual reproduction, they produce their spores at the tips of swollen hyphae that often resemble small clubs. These swollen hyphal tips are called **basidia** (singular: **basidium**). The hyphae, like those of sac fungi, are divided into individual cells. These cells, however, have either a single nucleus or, in some stages, two nuclei. The crosswalls have a central pore (*dolipore*) that is surrounded by a swelling, and both the pore and swelling are covered by a cap. This cap, with some exceptions, blocks passage of nuclei between cells but allows cytoplasm and small organelles to pass through.

Asexual Reproduction

Asexual reproduction is much less frequent in club fungi than in the other phyla of fungi. When it does occur, it is mainly by means of conidia, although a few species produce buds similar to those of yeasts, and others have hyphae that fragment into individual cells, each functioning like a spore and forming a new mycelium after germination.

Sexual Reproduction

Sexual reproduction in many club fungi mushrooms begins in the same way as it does for members of the two fungal phyla previously discussed. When a spore lands in a suitable place—often an area with good organic material and humus in the soil—it germinates and produces a mycelium just beneath the surface. The hyphae of the mycelium are divided into cells that each contain a single haploid nucleus. Such a mycelium is said to be **monokaryotic.** Monokaryotic mycelia of club fungi often occur in four mating types, usually designated simply as types 1, 2, 3, and 4. Only types 1 and 3, or types 2 and 4, can mate with each other.

If the growth of the hyphae of compatible mating types happens to bring them close together, cells of each mycelium may unite, initiating a new mycelium in which each cell has two nuclei. Such a mycelium is said to be **dikaryotic.** Dikaryotic mycelia usually have little walled-off bypass loops called *clamp connections* between cells on the surface of the hyphae (Fig. 19.15). The clamp connections develop as a result of a unique type of mitosis that ensures each cell will have one nucleus of each original mating type within it.

After developing for a while, the dikaryotic mycelium may become very dense and form a compact, solid-looking mass called a *button.* This pushes above the surface and expands into a *basidioma* (formerly known as a *basidiocarp*), commonly called a **mushroom** (Fig. 19.16). Most mushrooms have an expanded umbrellalike cap (*pileus*) and a stalk (*stipe*). Some have a ring called an *annulus* on the

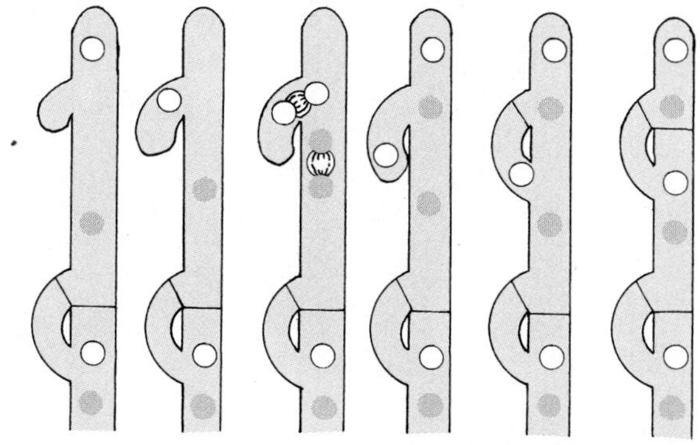

Figure 19.15 Development of a clamp connection. *Left* to *right.* A protrusion appears in the wall of a cell, and one nucleus migrates into the loop. Both nuclei undergo mitosis, and the loop carrying the daughter nucleus turns toward the cell wall. A clamp connection is established, and the daughter nuclei pair in two cells.

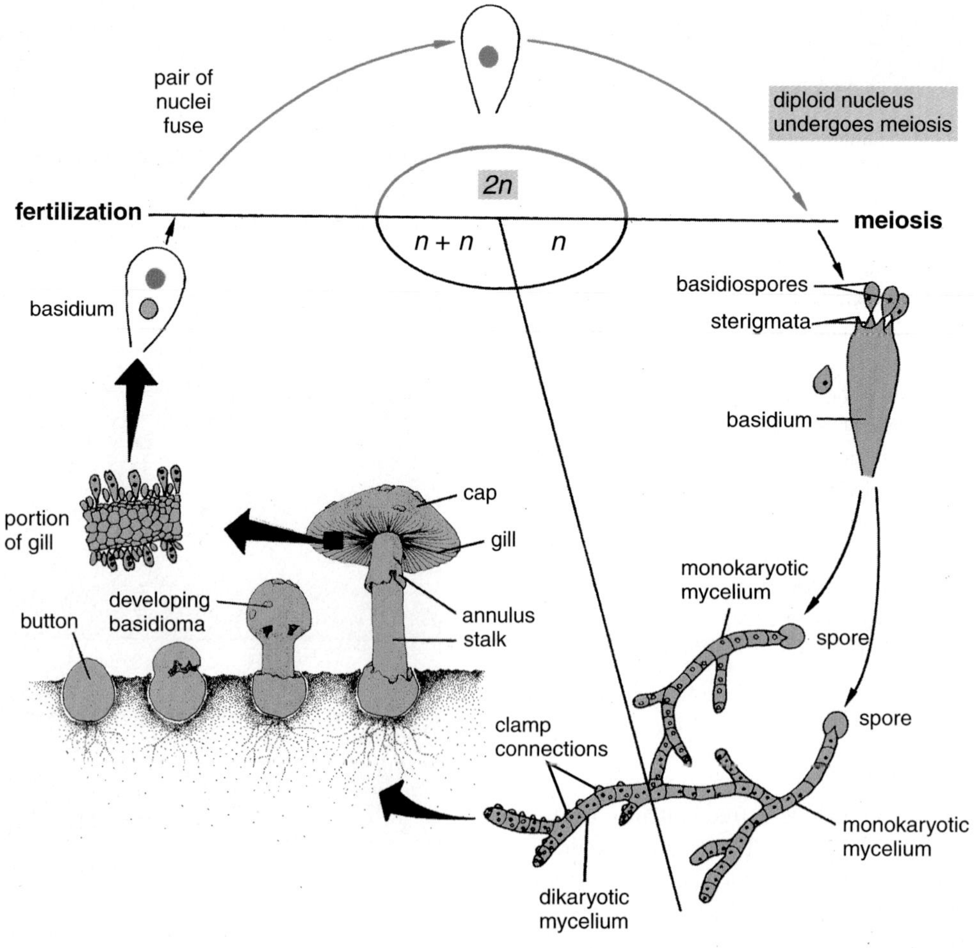

Figure 19.16 Life cycle of a typical mushroom.

Figure 19.17 A death angel mushroom (*Amanita*). Note the egglike volva at the base. *(Courtesy Dr. T. Duffy)*

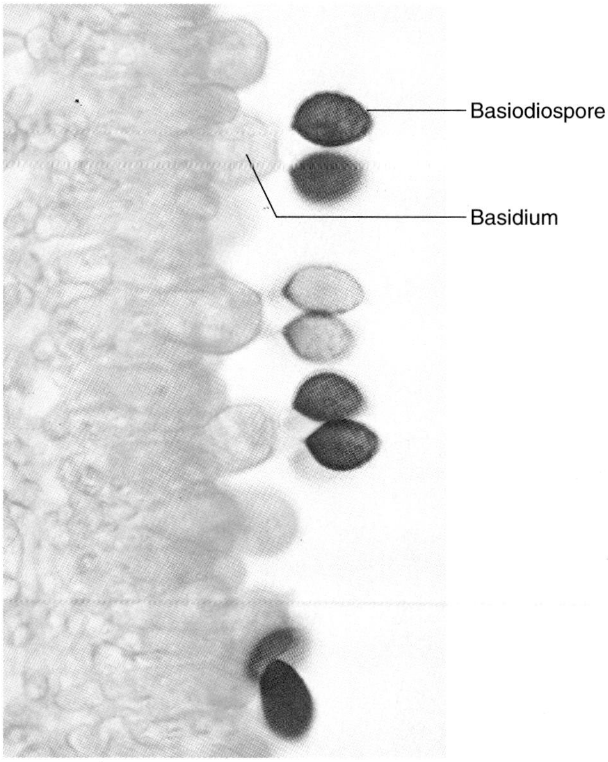

Figure 19.18 A small part of a mushroom gill showing basidiospores produced on the club-shaped basidia, ×500.

stalk. It is the remnant of a membrane that extended from the cap to the stalk and tore as the cap expanded. Some mushrooms, such as the notorious *death angel* and *destroying angel* mushrooms (*Amanita* spp.), also have a cup called a *volva* at the base (Fig. 19.17). Thin, fleshy-looking plates called **gills** radiate out from the stalk on the underside of the cap. Microscopic examination of a gill reveals it is composed of compacted hyphae, with large numbers of **basidia** oriented at right angles to the flat surfaces of the gill.

As each basidium matures, the two nuclei unite, and the diploid nucleus undergoes meiosis. The four nuclei that result from meiosis migrate through four (in a few species, two) tiny pegs at the tip of the basidium, walls forming around the nuclei in the process. The resultant cells are now **basidiospores.** The tiny pegs, called *sterigmata* (singular: *sterigma*), serve as stalks for the basidiospores (Fig. 19.18). One large mushroom may produce several billion basidiospores within a few days. These are forcibly discharged into the air between the gills. They then drift downward and blow away with the slightest breeze.

If you remove a mushroom stalk and place the cap gill-side down on a piece of paper, covering it with a dish to eliminate air currents, the spores will fall and adhere to the paper in a pattern perfectly reflecting the arrangement of the gills. The dish and cap can be removed a day later, and the *spore print* (Fig. 19.19) can be made more or less permanent with the application of a little clear varnish or shellac. Such spore prints can be used as an aid to identification, employing white paper for dark-colored spores and black paper for white or light-colored spores.

In nature, some of the basidiospores eventually repeat the reproductive cycle. Often a dikaryotic mycelium radiates out from its starting point, periodically producing basidiomata in

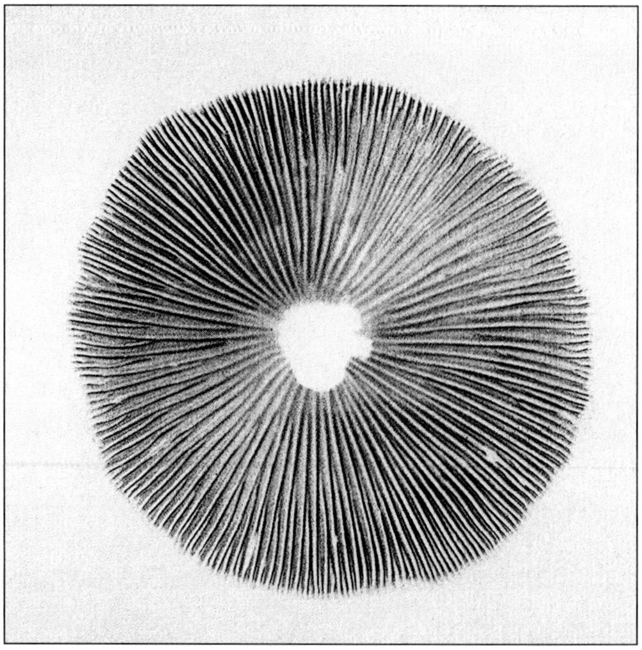

Figure 19.19 A spore print.

so-called *fairy rings* (Fig. 19.20). If conditions are favorable, the mycelium continues to grow at the edges for many years while dying in the center as food resources are depleted. Some mycelia have been known to grow in this fashion for over 500 years.

Some mushrooms such as *boletes* produce their spores on the surfaces of thousands of tiny pores instead of on gills (Fig. 19.21).

Shelf, or *bracket, fungi* (Fig. 19.22) grow out horizontally from the bark or dead wood from which they have grown, some adding a new layer of growth each year. Perennial species can become large enough and so securely attached that they can support the weight of a human adult.

Figure 19.20 A fairy ring of mushrooms (*Pholiota* sp.). The mushrooms appear periodically toward the edge of a dikaryotic mycelium that grows out from a point where the first dikaryotic cell was initiated.

Figure 19.21 A bolete (*Boletus* sp.) mushroom. The basidiospores are produced at the margins of pores instead of along gills.

Other members of this phylum produce spores within parchmentlike membranes, forming somewhat ball-like basidiomata called *puffballs.* Puffballs, which prior to developing their spores are generally edible, range in diameter from a few millimeters to 1.2 meters (0.125 inch to 4 feet) (Fig. 19.23). They have minimal or no stalks and rest in contact with the ground. Literally trillions of spores may be produced by a large puffball. These are released through a pore at the top or from random locations when the outer membrane breaks down. *Earth stars* (see Fig. 19.14B) are similar to puffballs but differ from them in having at the base an additional membrane that splits and resembles a set of woody flower petals.

Bird's-nest fungi (Fig. 19.24) grow on wood or manure and form nestlike cavities in which small egglike bodies containing basidiospores are produced. In some species, each "egg" has a sticky thread attached to it. When raindrops fall in the nests, the eggs may be splashed out, and as they fly through the air, the sticky threads catch on nearby vegetation, whipping the eggs around it (Fig. 19.25). When animals graze on the vegetation, the spores pass unharmed through the intestinal tract.

Smuts are parasitic club fungi that don't form basidiomata. They do considerable damage to corn, wheat, and other grain crops. In corn smut (Fig. 19.26), the mycelium grows between the cells of the host. The hyphae absorb nourishment from these cells and also secrete substances that stimulate them to divide and enlarge, forming tumors on

Figure 19.22 A shelf, or bracket, fungus growing out from the trunk of a tree.

Figure 19.23 A giant puffball. *(Courtesy John Hardy)*

Figure 19.24 Bird's nest fungi. Each "egg" in a nest contains basidiospores that are dispersed when raindrops splash them out.

the surfaces of the corn kernels. These eventually break open, revealing millions of sooty black spores, which are blown away by the wind. Some smuts affect only the flowering heads or grains, while others infect the whole plant.

Rusts are also parasites that don't form basidiomata. They attack a wide variety of plants. Some rusts grow and reproduce on only one species of flowering or cone-bearing plant. Others, however, require two or more different hosts to complete their life cycles. *Black stem rust of wheat,* which has reduced wheat yields by millions of bushels in a single year in the United States alone, has plagued farmers ever since wheat was first cultivated thousands of years ago. More than 300 races of black stem rust are now known. This rust requires both common barberry plants and wheat to complete its life cycle (Fig. 19.27).

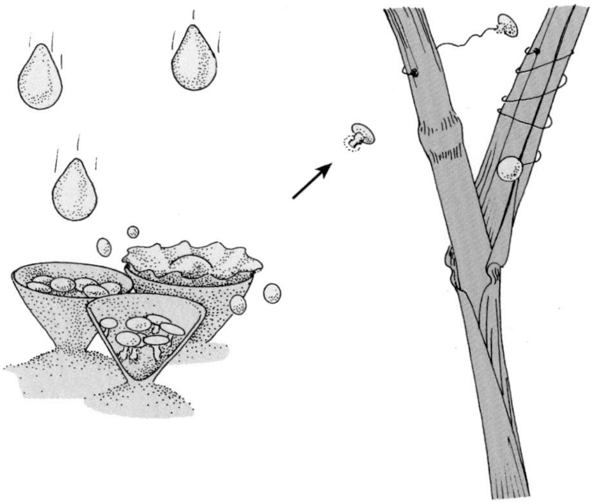

Figure 19.25 How the "eggs" in a bird's-nest fungus are dispersed. *After Harold J. Brodie. 1951. "The splash-cup dispersal mechanism in plants."* Canadian Journal of Botany *29:224–34. Reproduced by permission of the National Research Council of Canada from the* Canadian Journal of Botany, *29:224–34.*

Figure 19.26 Corn smut fungus on an ear of corn.

Since two hosts are necessary for black stem rust of wheat to complete its life cycle, it was believed that control of the disease could be accomplished through eradication of common barberry bushes. In an attempt to eradicate the disease, an estimated 600 million such plants were destroyed in the United

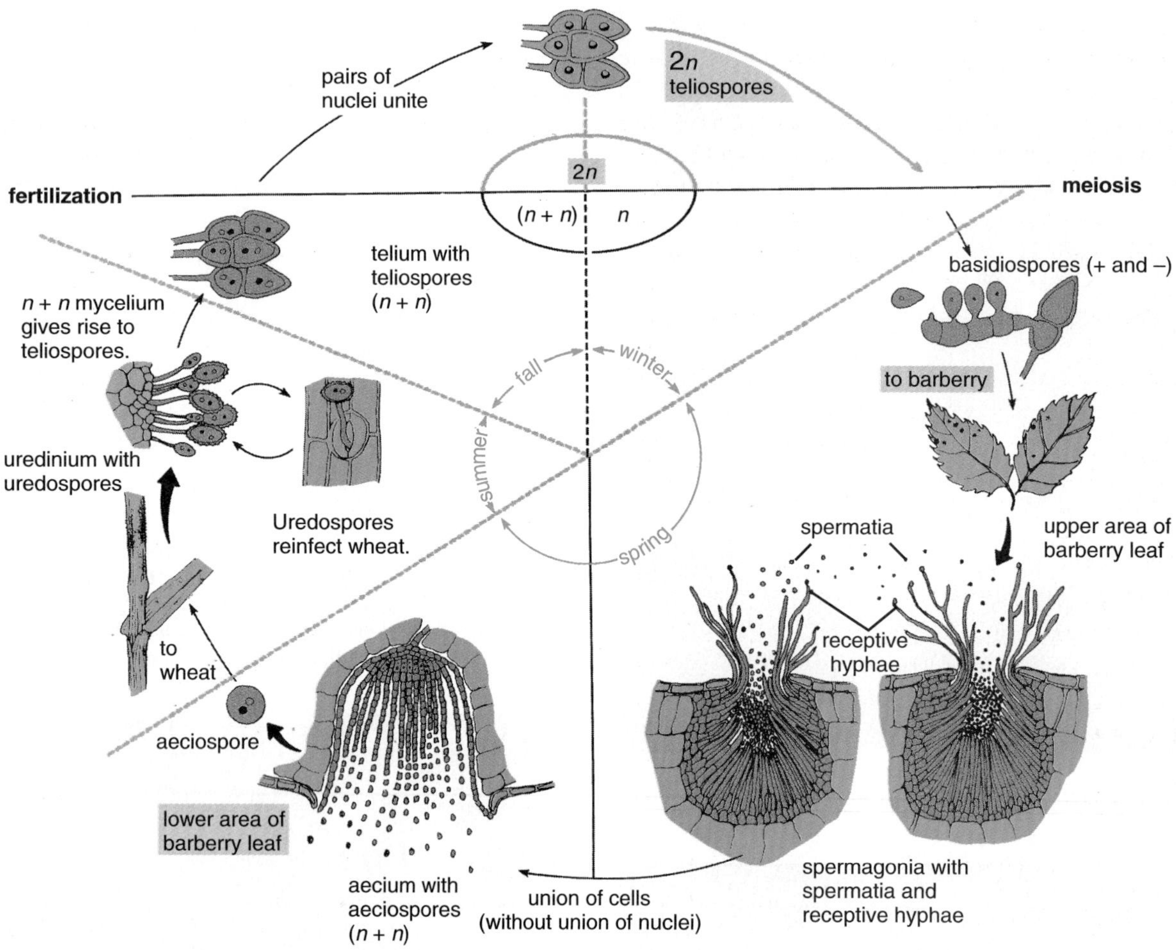

Figure 19.27 Life cycle of black stem rust of wheat.

States between 1918 and 1990, but it has proved impossible to eliminate the species altogether, partly because the *urediniospores* (binucleate spores produced in the summer) can overwinter in southern fields and reinfect wheat seedlings in the spring. Producing rust-resistant strains of wheat has helped, but even as new strains are developed, the rusts themselves hybridize or mutate, producing new races capable of attacking previously resistant varieties of cereals—a striking example of adaptation and natural selection in action.

Another serious rust with two hosts is the *white pine blister rust,* which has caused huge losses of valuable timber trees in both the eastern and western United States. Basidiospores infect the pine trees, and when the basidiospores germinate, other types of spores are produced. These different spores, in turn, infect currant and gooseberry bushes, and the spores formed on the currants and gooseberries eventually give rise to new basidiospores, completing the cycle. The U.S. Forest Service had a program of gooseberry bush eradication in operation for many years in an attempt to alleviate the problem, but the program was only partially successful. Spraying programs have more recently been implemented with some success, and rust-resistant trees also are being selected and bred as alternatives.

Other rusts with two hosts include *apple rust* (alternate host: cedar trees), *poplar leaf spot* (alternate host: larch or tamarack trees), and *corn rust* (alternate host: sorrel).

A relative of black stem rust was recently discovered in the Rocky Mountains of Colorado. This rust causes its rock cress host plants to produce fake flowers that look and smell so real many insects are fooled by them. When bees and butterflies visit the fake flowers, they find a sugary, nectarlike secretion. While gathering the sticky fluid, they inadvertently also pick up fungal sex cells, which are spread to other rock cress plants.

Human and Ecological Relevance of the Club Fungi

Of the approximately 25,000 described species of club fungi, fewer than 75 are known to be poisonous. Many of the latter are, however, common and not readily distinguishable by amateurs from edible species. Also, some edible forms, such as the inky cap mushrooms, cause no problems by themselves but may make one very ill if consumed with alcohol. Few of the poisonous forms normally are fatal, but unfortunately, some—such as *Amanita* spp., which cause 90% of the fatalities attributed to mushroom poisoning—are relatively common.

Poisoning from death angels and similar species is due to *alpha-amatin* (an alkaloid). The poison completely blocks RNA synthesis, and symptoms of poisoning usually take from 6 to 24 hours to appear. Treatment, which in the United States usually includes pumping of the stomach and ingestion of activated charcoal, is seldom successful by the time the intense stomachache, blurred vision, violent vomiting, and other symptoms occur.

In the late 1950s there were claims that administration of alpha lipoic acid (thioctic acid) in a few instances prevented a fatality, but subsequent research has failed to substantiate the claims. In Europe, injections of *silibinin* and *silymarin* (extracts of milk thistle—*Silybum marianum*) do appear to have been at least partially effective as antidotes. Oral forms of silymarin are available in North America, but they are much less effective than the injectable forms. Injectable silibinin and silymarin have not yet received FDA approval for treatment of mushroom poisoning in the United States. When a small, nonfatal ingestion has occurred, the mushroom poison usually leaves the patient with hypoglycemia (a blood sugar deficiency), which can be countered with the intravenous administration of glucose. Some wild-mushroom lovers have fed a part of their collection to a dog or cat, and when nothing happened to the animal after an hour or two, they have eaten the mushrooms themselves. Both they and the animals later succumbed. Others have mistakenly believed that the toxic substances are destroyed by cooking. Records show, however, that before the discovery of milk thistle antidotes, death ensued in 50% to 90% of those who had eaten just one or two of the deadly mushrooms, cooked or raw, and that even as little as 1 cubic centimeter (less than a 0.5 cubic inch) can be fatal.

There are many widespread but very unreliable beliefs about distinctions between edible and poisonous mushrooms. Some believe that a silver coin placed in the cooking pan will turn black while the mushrooms are cooking if any poison is present. However, there are both edible and poisonous species that will turn such a coin black, and others will not. Another superstition holds that edible species can be peeled while poisonous ones cannot. Again, this is a fallacy, for *Amanita* mushrooms peel quite easily. Still other erroneous beliefs maintain that poisonous mushrooms appear only in the fall or the early spring, or that all mushrooms eaten by snails and beetles are edible, or that all purplish-colored mushrooms are poisonous, or that all mushrooms growing in grassy areas are edible. Again, these notions simply are not supported by the facts. Wild mushrooms can be eaten with some confidence *only* if they have been identified by a knowledgeable authority. It is foolhardy for anyone to do otherwise.

Some poisonous mushrooms cause hallucinations in those who eat them. During the Mayan civilization in Central America, a number of *teonanacatl* ("God's flesh") sacred mushrooms (Fig. 19.28) were used in religious ceremonies. The consumption of these mushrooms has continued among native groups in Mexico and Central America to the present. Ingestion of small amounts results in sharply focused, vividly colored visions. Similar use of the striking fly agaric mushroom (Fig. 19.29; see also the chapter-opening photo) in Russia, and for a while in the Indus valley of India, dates back to many centuries B.C. Users appear to go into a state of intoxication. It is believed that the ancient Norwegian beserkers, who occasionally exhibited fits of exceptionally savage behavior, did so after consuming fly agaric mushrooms. Related species that occur in the United States have not produced the same effects but have, instead, caused the

Figure 19.28 Teonanacatl mushrooms. *(Courtesy Drug Enforcement Administration)*

Figure 19.29 Fly agaric mushrooms. *(Courtesy Perry J. Reynolds)*

user to become nauseated and to vomit. In Siberia, users have noted that the intoxicating principle is passed out in the urine, and some persons have adopted the practice of drinking the urine of persons who have consumed fly agaric mushrooms. Reindeer, incidentally, are reported to be obsessed with both fly agaric and human urine.

More than 90% of a mushroom is water, and mushrooms generally contain smaller quantities of nutritionally valuable substances than most foods. An apparent exception is the legendary *shiitake* mushroom, grown for centuries in China and Japan on oak logs and now cultured in the United States. It has more than double the protein of ordinary, commercially grown mushrooms and is very rich in calcium, phosphorus, and iron. It has excellent flavor and also contains significant amounts of B vitamins, vitamin D_2 (ergosterol), and vitamin C.

Ancient Chinese royalty believed that eating shiitake mushrooms would promote healthful vigor and retard the aging process, but no extensive research either supporting or refuting the claims has yet been conducted.

Lentinacin, an agent capable of lowering human cholesterol levels, has been obtained from shiitake mushrooms, and purified extracts from spores of the mushrooms have demonstrated antiviral activity against influenza and polio viruses in laboratory animals.

Since ancient times, many types of mushrooms have been cultivated for food. In the 2d century B.C., a Greek doctor by the name of Nicandros taught people how to grow mushrooms underneath fig trees on soil fertilized with manure. Andrea Cesalpino, a noted Italian botanist and physician of the 16th century, cultivated mushrooms by scattering pulverized poplar bark on very rich soil near poplar trees with which the mushrooms were associated. In more recent times, Italians have cultivated mushrooms on waste material from olives, on coffee grounds, on remnants of oak leaves after tannins have been extracted for leather tanning, and on laurel berries. Today, jelly fungi and various mushrooms are cultivated in the Orient on media composed of wood and manure.

In Geneva, Switzerland, there is a special market for wild mushrooms where more than 50 species are sold under the supervision of a state mycologist. Although wild mushrooms are consumed by many in North America, fewer such species are sold in markets. Only one species of mushroom (*Agaricus bisporus*—portabella mushroom) is extensively cultivated commercially, although several others such as shiitake and oyster are steadily gaining in popularity. Mushrooms have been grown in basements, caves, and abandoned mines, but contrary to widespread belief, light does not at all affect their growth.

Large-scale mushroom-growing operations generally use windowless warehouses with stacked rows of shallow planting beds because temperatures, humidity, and other climatic factors are easier to control in such buildings (Fig. 19.30). The mushrooms are grown on compost made from straw and horse or chicken manure. Prior to use, the compost is pasteurized for a week to destroy microorganisms, unwanted fungi, and insects and their eggs. Then it is inoculated with spawn, which is compact mycelium grown from germinated basidiospores sown on bran of wheat or other grains. After inoculation, the spawn is covered with a thin layer of soil. The moisture content of the compost is controlled with regular, light waterings, and a humidity of about 75% is maintained.

Figure 19.30 Mushroom beds in a commercial mushroom farm in Petaluma, California.

The mycelium grows in temperatures ranging from just above freezing to 33°C (91.4°F), but commercial growers try to keep the temperatures between 9°C and 13°C (48°F to 55°F) because the mushrooms are less subject to disease or insect attacks and are also firmer when grown under cool conditions. The mushroom buttons appear within a week to 10 days after spawn is planted and continue appearing for 6 months or more. About 1 kilogram (2.2 pounds) of mushrooms is obtained from each 10 square decimeters (slightly over 1 square foot) of growing area. An estimated 59,000 metric tons (65,000 tons) of mushrooms are produced annually in the United States, and lesser but significant amounts in Canada.

Fungi are constantly breaking down dead wood and debris and returning the components to the soil where they can be recycled. Sometimes, as we have seen, they can be very destructive from a human viewpoint, attacking everything from living plants and harvested or processed food to shoe leather, paper, cloth, construction timbers, paint, petroleum products, upholstery, and even glass, particularly in warmer humid climates.

Phylum Deuteromycota—The Deuteromycetes (Imperfect Fungi)

Any fungus for which a sexual stage has not been observed is classified as an **imperfect fungus.** Many otherwise unrelated fungi are grouped together in this artificial phylum, which includes several well-known disease organisms as well as fungi important in disease control and food processing. If a member of this group is studied further and a sexual stage is discovered, the fungus is reassigned to its appropriate division.

All imperfect fungi reproduce by means of conidia. One such group grows on dead leaves and debris at the bottom of streams with rapidly moving water. The spores often have four long extensions on them. These arms are arranged in such a way that three may catch like a tripod on a flat surface, keeping the fungus from being washed downstream.

One interesting group of imperfect fungi parasitizes protozoans and other small animals in various ways. Some develop an unbranched body inside their victim and slowly absorb nutrients until the host dies. They then produce a chain of spores that may stick to, or be eaten by, another victim. Others capture their prey on sticky hyphae to which passing amoebae adhere. One soil-dwelling group captures nematodes (eelworms) in hyphal rings or loops. In some, the rings are a little smaller than the circumference of a nematode, which tapers at both ends. When a nematode randomly happens to stick its head through such a loop, it frequently tries to struggle forward instead of backing out and becomes trapped (Fig. 19.31). The fungus then produces small rhizoidlike outgrowths called *haustoria* that grow into the worm's body and digest it.

More specialized species of nematode-trapping fungi have loops that constrict around the worm less than a tenth of a second after contact. The loops are spaced so that sometimes the tail of the worm gets caught in a second loop while the worm is thrashing around after being caught by the head. A number of these fungi can easily be grown on agar media, but they do not form loops unless a nematode is placed in the dish. The nematodes evidently produce a substance that promotes the development of the loops. Several species of imperfect fungi are found in ant and termite nests. The insects cultivate their fungus gardens by bringing in bits of leaves, other plant debris, caterpillar droppings, and their own feces, and in so doing, they form a rich growing medium. They also lick the hyphae and constantly probe them, using them for food as they grow.

Some wood-boring beetles have pouches in their bodies that function as fungus spore containers. The spores germinate in the wood tunnels, their mycelia forming a lining that produces yeastlike cells on which the beetles and their larvae feed. As discussed in Chapter 5, many higher plants have mycorrhizal fungi associated with their roots. These fungi greatly increase the absorptive surface area around the roots and may be far more important than root hairs in this regard, particularly in mature roots.

Human and Ecological Relevance of the Imperfect Fungi

Among the best known of the medically important fungi are the *Penicillium* molds (Fig. 19.32), which secrete *penicillin,* the well-known and widely used **antibiotic** (a substance produced by a living organism that interferes with the normal metabolism of another living organism). In 1929, Sir Alexander Fleming of England noticed that certain bacteria would not grow in the vicinity of the mycelium of a *Penicillium* mold and gave the name penicillin to the element in the mold that prevented the bacterial growth.

Fleming apparently did not grasp the significance of his findings, and the findings did not particularly excite the medical profession until the outbreak of World War II some 10 years later. At that time, a team of British and American scientists at Northern Regional Laboratories in Peoria,

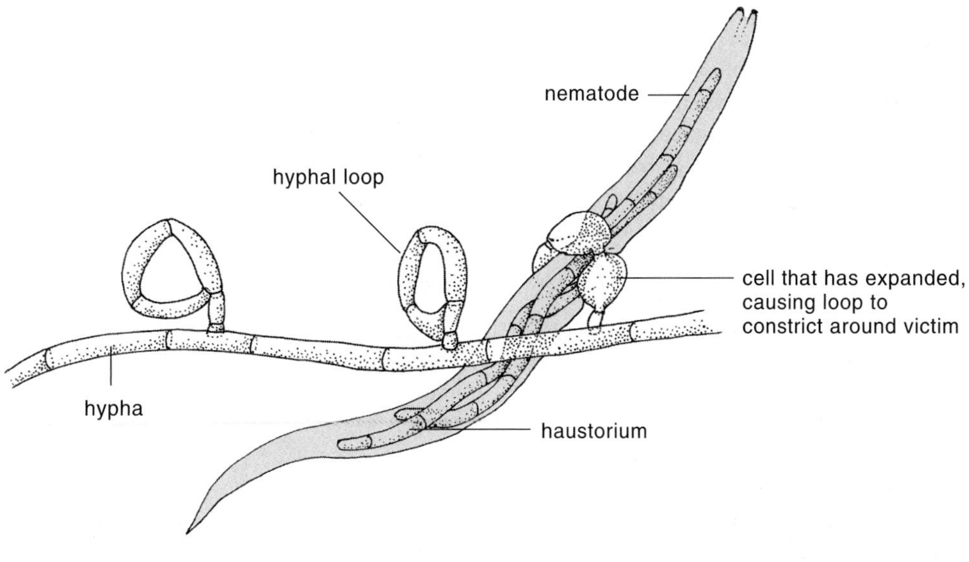

Figure 19.31 A nematode-trapping fungus with a victim. *After Drechsler, C. 1937. "Some hyphomycetes that prey on free-living terricolous nematodes."* Mycologia *29:447–552. Redrawn by permission.*

Figure 19.32 A *Penicillium* colony.

Illinois, set out to see if they could coax *Penicillium* molds into producing more of this antibacterial substance, primarily because war casualties created a need for greater quantities of more effective medicines that could keep wounds from becoming infected and to prevent the spread of venereal diseases.

The scientists began with cultures from the original mold observed by Fleming, but the amounts of penicillin it produced were so small that it was very expensive to obtain significant quantities for human use. They appealed to the general public, asking them to send in any material they found with a greenish or bluish mold on it. They received whole trainloads of moldy trash from all over the United States.

The breakthrough in the research came, however, when a different species of *Penicillium* mold—one that yielded 25 times the penicillin produced by the original culture—was found on a moldy cantaloupe from a local market. The scientists set to work germinating individual spores of this new mold on culture media, and by careful selection, they eventually were able to isolate a strain that produced more than 80 times the original quantity of penicillin. Later, when this strain was subjected to X-radiation, still other forms were produced that upped the penicillin output to 225 times that of Fleming's mold. Today, most of the penicillin produced around the world comes from descendants of that cantaloupe mold. Literally hundreds of other antibiotics effective in combating human and animal diseases have been discovered since the close of World War II, and the production of these drugs is a vast worldwide industry.

Penicillium molds are also used in other ways. Some are introduced into the milk of cows, sheep, and goats at stages in the production of gourmet cheeses, such as blue (Fig. 19.33), Camembert, Roquefort, Gorgonzola, and Stilton. The molds produce enzymes that break down proteins and fats in the milk, giving the cheeses their characteristic flavors.

Since the early 1980s, organ transplants have been aided by the discovery and production of a "wonder drug" from an imperfect fungus found in soil. Called *cyclosporine,* the drug suppresses immune reactions that cause rejection of transplanted organs without risking the development of leukemia and other undesirable side effects associated with other drugs.

Aspergillus is a genus of imperfect fungi whose species produce dark brown to blackish or yellow spores (Fig. 19.34). It is closely related to the *Penicillium* molds and is extensively used in industry. One or more species is used commercially for the production from sugar of citric acid, a substance for flavoring foods and for the manufacture of effervescent salts that were originally obtained from oranges. Citric acid is also used in the manufacture of inks and in medicines, and it is even used as a chicle substitute in some chewing gums.

Aspergillus fungi also produce gallic acid used in photographic developers, dyes, and indelible black ink. Other species are used in the production of artificial flavoring and perfume substances, chlorine, alcohols, and several acids. Further uses are in the manufacture of plastics, toothpaste, and soap and in the silvering of mirrors.

One species of *Aspergillus* is used in the Orient and elsewhere to make soy sauce, or *shoyu,* by fermenting soybeans with the fungus. A Japanese food called *miso* is made by fermenting soybeans, salt, and rice with the same fungus. More than one-half million tons of miso are consumed annually.

A number of diseases of both human and animals are caused by *Aspergillus* species. The diseases, called *aspergilloses,* attack the respiratory tract after the spores have been inhaled. One type thrives on and in human ears. Other diseases caused by different genera of imperfect fungi include those responsible for the widespread problem of athlete's foot and ringworm, for white piedra (a mild disease of beards and mustaches), and for tropical diseases of the hands and feet that cause the limbs to swell in grotesque fashion.

Figure 19.33 Blue cheese. The dark areas are conidia of a species of *Penicillium* that gives the cheese its unique flavor.

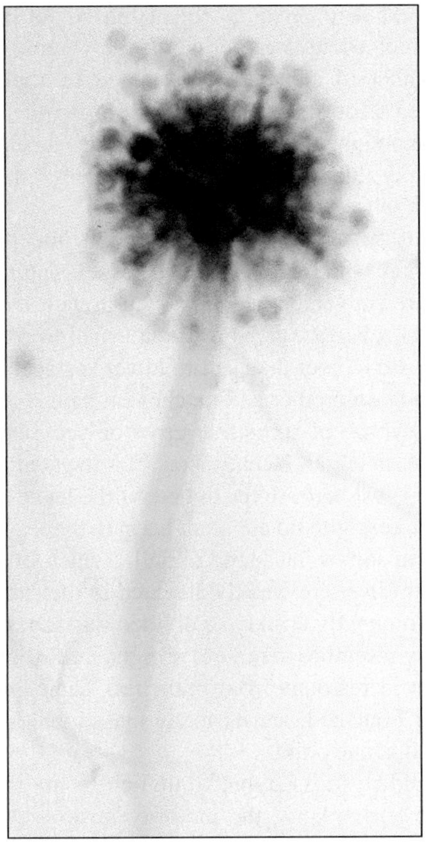

Figure 19.34 A conidiophore of *Aspergillus.* Rows of conidiospores (also called *conidia*) are radiating out from the conidiophore, ×500.

Aspergillus flavus, which grows on moist seeds, secretes *aflatoxin,* the most potent natural carcinogen known. The toxin causes liver cancer, and no more than 50 parts per billion is allowed in human food. In humid climates, such as those of the southeastern United States and adjacent Mexico, improperly stored grain can become moist enough to support the fungus so that foods such as peanuts, peanut butter, and peanut-based dairy feeds become carcinogenic. Dairy cattle feed is even more strictly controlled because concentrations of aflatoxin can accumulate in milk.

One serious disease called *valley fever* (*coccidiomycosis*), found primarily in the drier regions of the southwestern United States, usually starts with the inhalation of dust-borne spores of an imperfect fungus (*Coccidioides immitis*) that produces lesions in the upper respiratory tract and lungs. The disease may spread elsewhere in the body, with sometimes fatal results.

Two imperfect fungi show promise as biological controls of pest organisms. One has already been used with some success in controlling scale insects in Florida and other warm, humid regions. Another may be used to combat water hyacinths, which have caused serious clogging of waterways in areas of the world with mild to tropical climates.

LICHENS

A student who was interested in natural dyeing came to me a number of years ago and asked if she could experiment with the dye potential of local plants as a special project. In the course of her experimentation, she obtained beautiful shades of yellow, brown, and green from two dozen common local plants, using simple "recipes."[1] During the following summer, she extended the project to include lichens growing on the trees and rocks at a camp where she served as a counselor. The rich colors she obtained from the lichens were even more spectacular, which is perhaps not surprising since these organisms were, in the past, a major source of dyes and still are used in a minor way in commercial dyeing.[2]

Lichens traditionally have been referred to as prime examples of *symbiotic* relationships. Each consists of a fungus and an alga (or cyanobacterium) intimately associated in a spongy **thallus.** The thallus can range in diameter from less than 1 millimeter to more than 2 meters (0.04 inch to 6.5 feet). The photosynthetic component supplies the food for both organisms. The fungus protects the photosynthetic organisms from harmful light intensities, produces a substance that accelerates the rate of photosynthesis, and absorbs and retains water and minerals for both organisms. Supporting the belief that a lichen involves a symbiotic relationship is the fact that neither the fungus nor the photosynthetic organism grow independently where the lichen grows. The physiological evidence suggests, however, that it would probably be more correct to say that the fungus parasitizes the photosynthetic component in a controlled fashion, actually destroying chlorophyll-containing cells in some instances.

There are about 14,000 known species of lichens. The photosynthetic component is either a green alga or a cyanobacterium, a few lichens having two species of algae present. Three genera of green algae and one genus of cyanobacterium are involved in 90% of all lichen species, and one species of alga may be found in many different lichens. Each lichen, however, has its own unique species of fungus. With the exception of about 20 tropical species of lichens that have a club fungus and one species (associated with bald cypress trees) that has a bacterial component, lichens have members of the sac fungi for their fungal components. It is possible to isolate and culture the components separately. When this is done, however, the fungus takes on a very different, compact but indefinite shape, and the algae or cyanobacteria grow faster than they do when they are part of a lichen. The fungal component is very rarely found growing independently in nature, while the photosynthetic component is known to thrive independently of the lichen in

1. For further information on natural dyeing techniques, see Appendix 3.

2. Because the existence of many species of lichens is now threatened and they are also exceptionally slow growing, collecting of lichens for natural dyeing should no longer be encouraged.

some instances. Lichen species, therefore, are identified according to the fungus present.

Lichens grow very slowly, at a maximum rate of 1 centimeter (0.4 inch) and a minimum of as little as 0.1 millimeter (0.004 inch) per year. They are capable of living to an age of 4,500 or more years and are tolerant of environmental conditions that kill most other forms of life. They are found on bare rocks in the blazing sun or bitter cold in deserts, in both arctic and antarctic regions, on trees, and just below the permanent snow line of high mountains where nothing else will grow. One species grows completely submerged on ocean rocks. They even attach themselves to manufactured substances, such as glass, concrete, and asbestos.

Part of the reason for the wide range of tolerance of lichens is the presence in the lichens' thalli of gelatinous substances that enable them to withstand periods of rapid drying alternating with wet spells. While they are dry, their water content may drop to as low as 2% of their dry weight, and the upper part of the thallus becomes opaque enough to exclude much of the light that falls on them. In this state, most environmental extremes do not affect them at all as they temporarily become dormant and do not carry on photosynthesis.

The lichen thallus usually consists of three or four layers of cells or hyphae (Fig. 19.35). At the surface is a protective layer constituting the *upper cortex.* The hyphae are so compressed they resemble parenchyma cells, and it is here that the gelatinous substances aid in the layer's protective function. Below the upper cortex is an *algal layer* in which algal (or cyanobacterial) cells are scattered among strands of hyphae. Next is a *medulla* consisting of loosely packed hyphae, which occupies at least half the volume of the thallus. A number of substances produced by the lichen are stored here. A fourth layer, called the *lower cortex,* is frequently but not always present. It resembles the upper cortex but is usually thinner and is often covered with anchoring strands of hyphae called *rhizines.*

Lichens have been loosely grouped into three major growth forms, which have no basis in their natural relationships but are convenient as a first step in their identification (Fig. 19.36). *Crustose lichens* are attached to or embedded in their substrate over their entire lower surface. They often form brightly colored crusty patches on bare rocks and tree bark. The hyphae of some that grow on sedimentary rocks penetrate as much as 1 centimeter (0.4 inch) into the rock. Others grow just beneath the cuticle of the leaves of tropical hardwood trees, with no apparent harm to the leaves. *Foliose lichens* have somewhat leaflike thalli, which often overlap one another. They are weakly attached to the substrate. The edges are frequently crinkly or divided into lobes. *Fruticose lichens* may resemble miniature upright shrubs, or they may hang down in festoons from branches. Their thalli, which are usually branched, are basically cylindrical in form and are attached at one point.

It should be stressed that while lichens may be attached to trees or other plants, the majority in no way parasitize them. There are, however, a very small number of species that do produce parasitic rhizines that penetrate the cortical parenchyma cells of their hosts.

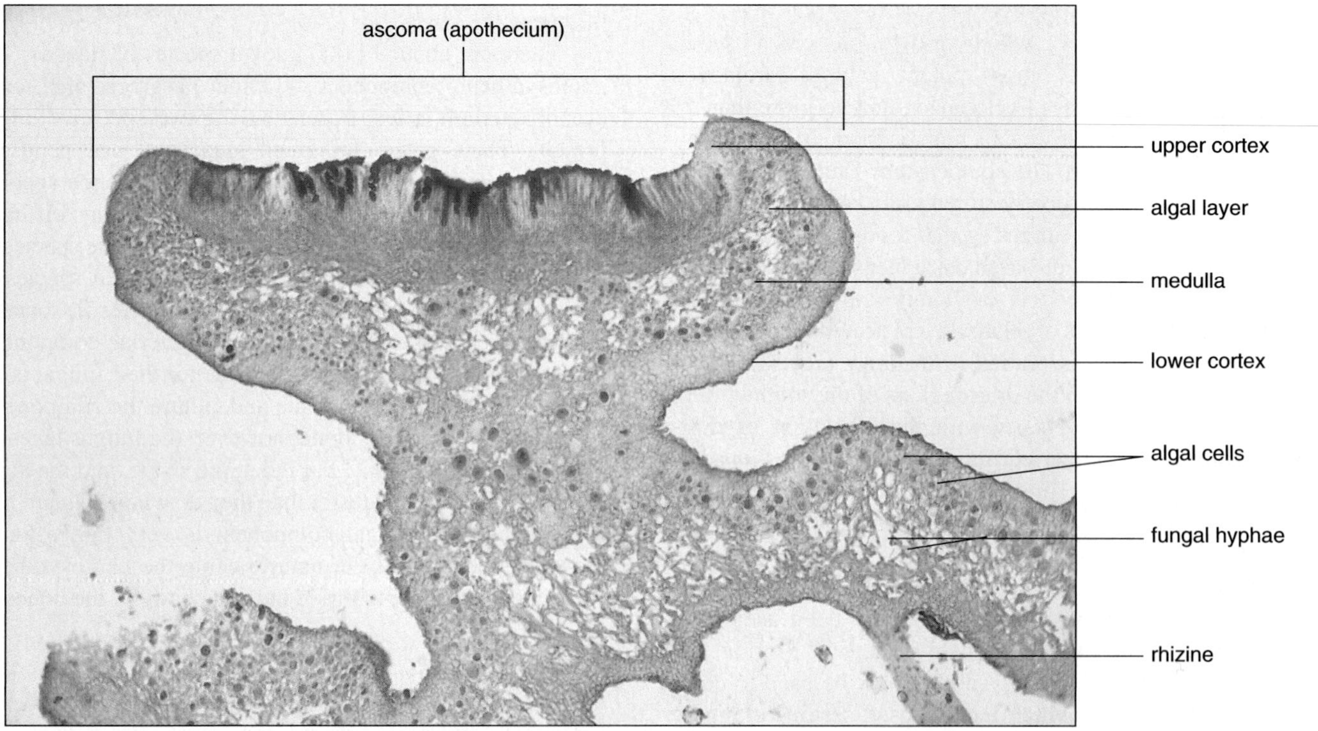

Figure 19.35 A section through a foliose lichen. *(Photomicrograph by G.S. Ellmore)*

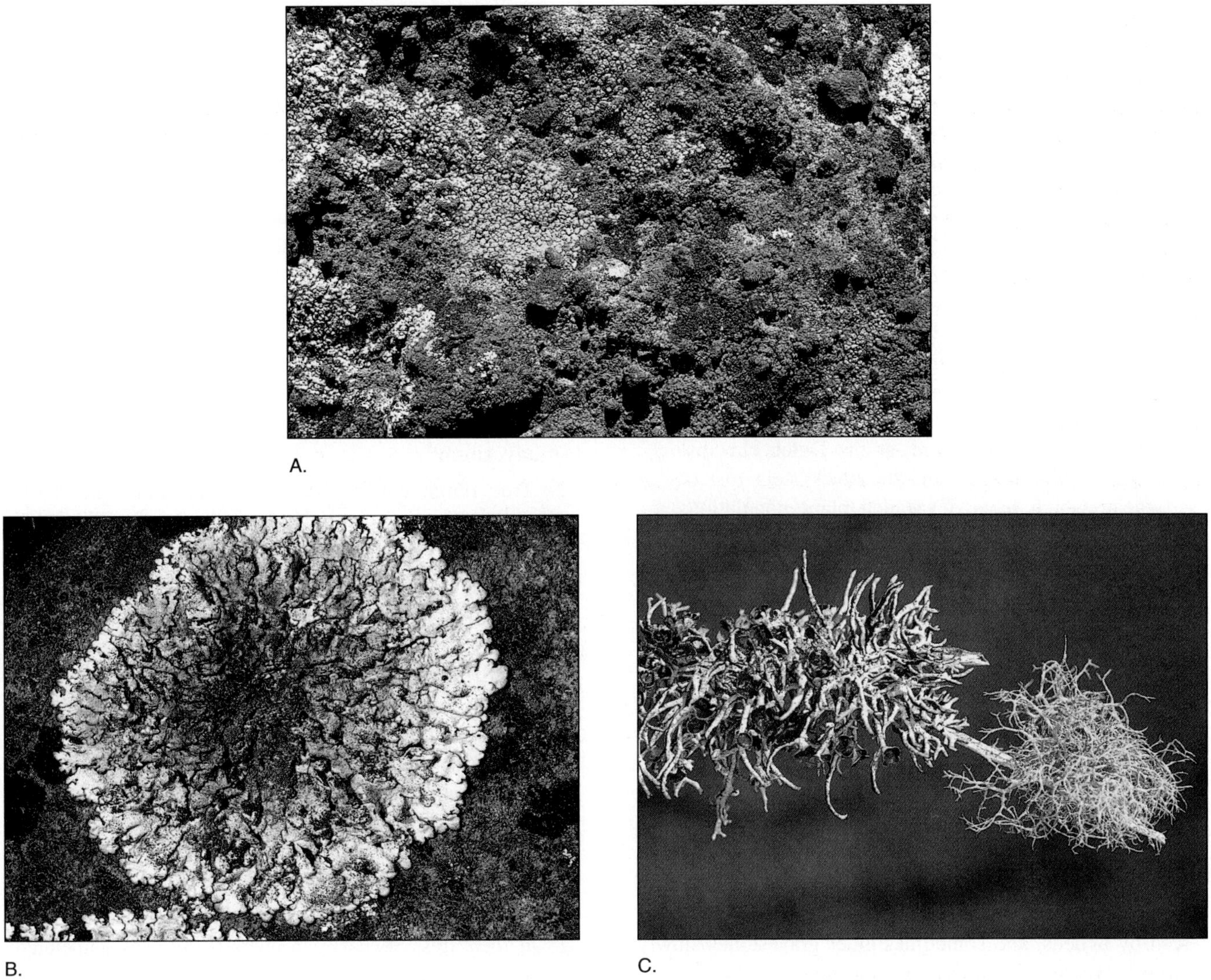

Figure 19.36 Three types of lichen thalli. *A.* Crustose lichens on the surface of a rock. *B.* A foliose lichen. *C.* Fruticose lichens.

Although the fungal component of a lichen usually reproduces sexually, lichens are naturally dispersed in nature primarily by asexual means. In about a third of the species, small powdery clusters of hyphae and algae called *soredia* (singular: *soredium*) are formed and cut off from the thallus in a set pattern as it grows. Rain, wind, running water, and animals act as agents of dispersal. In other lichens, specialized parts of the thallus, known as *isidia,* may break off or be separated by decay.

Sexual reproduction in lichens is similar to that of the sac fungi except that the ascomata produce spores continuously for many years. No one has yet observed the initiation of a new thallus in nature, but it is believed that they arise after ascospores carried by the wind come in contact with independently living algae, germinate, and parasitize them. Lichen algae reproduce by mitosis and simple cell division.

Human and Ecological Relevance of Lichens

One environmental factor from modern civilization has a marked effect on lichens. They are exceptionally sensitive to pollution, particularly that of sulfur dioxide. In fact, their sensitivity to this air pollutant is so great that it has been possible to calculate the amount of sulfur dioxide present in the air solely by mapping the occurrence or disappearance of certain lichens in a given area. Such studies have shown that some species of lichens have disappeared entirely from industrial areas and are now facing extinction due to air pollution. They are also very sensitive to nuclear radiation and were used on one occasion to monitor radioactive contamination when a satellite fell to earth in a remote area. After the nuclear meltdown in Chernobyl, reindeer moss lichens

Ecological Review

The sustainability of earth's ecosystems depends upon a balance between production of biomass by photosynthetic organisms and breakdown of that biomass and recycling of the nutrients it contains. In the process of recycling, the fungi play prominent roles in all major ecosystems. In addition, as food, as agents of disease, and as sources of antibiotics and other drugs long used by humans, the fungi have had important influences on the quality of the human environment and have even changed the course of history. Fungi are also a source of many plant diseases, such as the Dutch elm disease, chestnut blight, and the black stem rust of wheat, which have devastated plant populations across entire continents. Lichens, which can grow under some of the most extreme environmental conditions, are remarkably sensitive to air pollution and are useful indicators of air quality.

in Lapland absorbed radioactive elements and rendered the meat of reindeer that fed on the contaminated lichens unsafe to eat.

Lichens contribute to the degradation of historic ruins and other exposed rock or stone materials. Archaeological sites such as those of Troy (in modern Turkey) and Machu Picchu in Peru have suffered damage from the acids produced by lichens, and sometimes older gravestones slowly become covered by various lichen species.

Lichens provide food for many lower animals as well as large mammals. Reindeer and caribou can survive exclusively on fruticose lichens in Lapland when other food is unavailable, while North African sheep graze on a crustose lichen. By themselves, lichens do not make good food for human consumption, but they have been used as a food supplement (e.g., in soups) in parts of Europe. Most have acids that make them unpalatable, and some (e.g., rock tripe) have had harsh laxative effects on those who have tried them.

More than half of the lichens investigated have antibiotic properties. One lichen substance has been used in Europe in combination with another antibiotic in the treatment of tuberculosis. Europeans have also used lichen antibiotics to produce salves that have been effective in treating cuts and skin diseases.

Lichens were used for dyes by the Greeks and Romans, and a lichen dye industry persisted for many centuries in Europe. Native Americans and others also used lichens for dyes. Coal tar dyes now have largely replaced those of lichens, but lichens are still used in the manufacture of Scottish tweeds and East Indian cotton fabrics. Lichens are used in the preparation of the litmus paper used in elementary chemistry laboratories as an acid-alkaline indicator. The paper turns red under acid conditions and blue under alkaline ones.

Soaps are scented with extracts from lichens, and such extracts are still used in the manufacture of some European perfumes. Because of their resemblance to miniature trees and shrubs, some fruticose lichens are used by toy makers for the scenery of model railroads and car tracks. The importance of lichens in initiating soil formation is discussed in Chapter 25.

Summary

1. A mass of fungal threads collectively is called a mycelium.
2. True fungi, chytrids, water molds, and slime molds apparently were derived independently of each other.
3. Kingdom Fungi, whose members differ from those of other kingdoms in reproduction, metabolism, and lines of diversity, consists of five phyla of organisms.
4. True fungi are all filamentous or unicellular with chitinous walls; they are mostly nonmotile, and most produce spores. Sexual reproduction is varied. Fungi may thrive at temperatures below freezing or well above human comfort levels. They are major natural decomposers and cause huge economic losses through food spoilage and diseases.
5. Phylum Zygomycota (coenocytic fungi)—represented by *Rhizopus*—has coenocytic mycelia that produce sporangiophores with sporangia containing numerous spores at their tips. Sexual reproduction involves gametangia produced on hyphae of opposite strains that merge, creating a single large cell, which becomes a zygosporangium and in which nuclei fuse in pairs. Meiosis occurs prior to the zygosporangium cracking open.
6. Conidia are produced externally on hyphae in the Zygomycota, but conidia production is more common in other fungi. *Pilobolus* releases its sporangia with force toward a light source. Bread molds are used to make tempeh and sufu and are used for various industrial purposes.
7. The hyphae of true fungi are partitioned into cells with pores in their crosswalls.
8. Phylum Ascomycota (sac fungi) includes truffles, yeasts, powdery mildews, brown fruit rots, ergot, morels, and insect parasites. Asexual reproduction is by conidia or budding. Sexual reproduction involves the formation of asci following union of male and female structures.
9. Ergotism is a disease produced when rye infected with ergot is ingested. Yeasts produce carbon dioxide and alcohol used in baking and brewing, the drug ephedrine, and proteins and vitamins used for human and livestock

consumption. Dutch elm disease, chestnut blight, and peach leaf curl are caused by ascomycetes.

10. Phylum Basidiomycota (club fungi) includes mushrooms, bracket fungi, rusts, smuts, puffballs, stinkhorns, earth stars, jelly fungi, and bird's-nest fungi. The hyphae are partitioned into cells with either one nucleus (monokaryotic) or two nuclei (dikaryotic).
11. Asexual reproduction in basidiomycetes is usually by conidia. Sexual reproduction involves the fusion of cells of two mating types of monokaryotic hyphae. The union initiates the development of a dikaryotic mycelium that usually has clamp connections between each cell. The dikaryotic mycelium often develops basidiomata.
12. The cap of a mushroom is usually on a stalk, which may have an annulus around it and sometimes a volva at the base.
13. Basidia and basidiospores are produced on gills that radiate out from the stalk under the cap.
14. Puffballs and earth stars are stalkless club fungi that may each produce trillions of spores. Shelf fungi may add a new layer of growth each year.
15. Bird's-nest fungi produce their basidiospores in "eggs" that may become attached to vegetation ingested by animals. The eggs pass through the intestinal tract of the animals unharmed.
16. Smuts infest cereals. Rusts, which may have more than one host, infest cereals and other important plants.
17. Most club fungi are not poisonous. Death angel and destroying angel mushrooms cause about 90% of the fatalities attributed to mushroom poisoning. In Europe, milk thistle extracts appear to be at least partial antidotes to mushroom poisoning.
18. For safety, poisonous and edible mushrooms should be distinguished from each other only by authorities.
19. Some mushrooms have been used for intoxication and hallucinatory purposes. Most mushrooms are not high in nutritional value. *Agaricus bisporus* is the mushroom most commercially cultivated in the United States.
20. Fungi and bacteria play a major role in natural recycling processes. They are also destructive of many economically important substances.
21. Members of Phylum Deuteromycota (imperfect fungi) all lack known sexual reproductive phases but otherwise resemble members of the other phyla of true fungi.
22. Some imperfect soil fungi trap and digest nematodes. The *Penicillium* molds, which produce penicillin, and *Aspergillus,* some species of which are used in the production of shoyu, miso, and other commercial products, are primarily imperfect fungi. *Aspergillus* and other imperfect fungi also cause diseases, such as athlete's foot and ringworm.
23. Some imperfect fungi are cultivated by ants and termites in their nests and form mycorrhizal associations with the roots of higher plants.
24. The bodies (thalli) of lichens consist of an alga or cyanobacterium and a fungus in intimate symbiotic relationship. The photosynthetic component is usually a green alga or a cyanobacterium that may occur in many different species of lichens. Each lichen species has its own unique fungus (mostly sac fungus) component by which it is identified.
25. Lichens grow very slowly and may live to be thousands of years old. They are exceptionally sensitive to industrial pollution.
26. The lichen thallus usually consists of several layers of hyphae, one of which has algae or cyanobacteria between the threads.
27. Lichens occur in three growth forms: crustose, foliose, and fruticose.
28. Lichens reproduce sexually but are naturally dispersed primarily by soredia and other asexual means. Lichens provide food for animals, antibiotics, perfume, litmus paper, and natural dyes.

Review Questions

1. Why are fungi placed in a kingdom separate from protists and bacteria?
2. How do the cells of fungi differ from those of other organisms?
3. What does coenocytic mean? To which groups of organisms does it apply?
4. What means of asexual reproduction do the fungi exhibit?
5. If you were looking at hyphae of a bread mold, a sac fungus, and a club fungus through a microscope, could you distinguish among the three? How?
6. Which fungi produce zoospores?
7. Define conidia, haustoria, zygosporangium, sporangiophore, ascus, mycelium, ascoma, monokaryotic, dikaryotic, and clamp connection.
8. Is there any one best way to tell a poisonous mushroom from an edible one?
9. What is ergotism?
10. Discuss the human relevance of the sac fungi.
11. What is a spore print?
12. To which phylum of fungi do each of the following belong: truffles; puffballs; fungi involved with LSD; stinkhorns; peach leaf curl; late blight of potato; bird's-nest fungi; jelly fungi; downy mildew of grape; death angel; shelf, or bracket, fungi?
13. What is a fairy ring, and why does it form?

14. How do rusts and smuts attack their hosts?
15. Under what conditions do commercially produced mushrooms grow best?
16. Why are some fungi referred to as being "imperfect"?
17. Discuss the economic importance of imperfect fungi.
18. What is the relationship between an alga and the fungus in a lichen?
19. What makes up the basic structure of a typical lichen?
20. How is it possible for lichens to live to such great ages?

Discussion Questions

1. Since fungi produce so many trillions of spores, why is it we are not overrun by them?
2. Why are fungal diseases much more common now in the United States than they were at the time of the Declaration of Independence?
3. If all rusts had only one host, would they (theoretically, at least) be easier to control?
4. Since most lichens are not good for food and are not very important economically, should we be as concerned about some of them becoming extinct as we are about whales disappearing?

Learning Online

Visit our web page at www.mhhe.com/botany for interesting case studies, practice quizzes, current articles, and animations within the Online Learning Center to help you understand the material in this chapter. You'll also find active links to these topics:

General References in Mycology
Architectural Pattern of Fungi
Diversity of Fungi
Zygomycota
Basidiomycota
Ascomycota
Lichens
Economic and Ecological Importance of Fungi

Additional Reading

Ahmadjian, V. 1993. *The lichen symbiosis.* New York: John Wiley and Sons.

Alexopoulos, C. J., and C. W. Mims. 1995. *Introductory mycology,* 4th ed. New York: John Wiley and Sons.

Arora, D. K., and L. Ajello (Eds.). 1991. *Handbook of applied mycology,* vol. 2. New York: Marcel Dekker.

Batra, L. R. (Ed.). 1967. Insect-fungus symbiosis: Nutrition, mutualism and commensalism. *Scientific American* 217(5): 112–20.

Benjamin, D. R. 1995. *Mushrooms: Poisons and panaceas.* San Francisco: W. H. Freeman.

Bliss, A. 1994. *North American dye plants,* rev. ed. Loveland, CO: Interweave Press.

Cole, G. T., and H. C. Hoch (Eds.). 1991. *The fungal spore and disease initiation in plants and animals.* New York: Plenum.

Crowder, W. 1926. Marvels of the mycetozoa. *National Geographic Magazine* 49(4): 421–44.

Deacon, J. W. 1997. *Modern mycology,* 3d ed. Malden, MA: Blackwell Science.

Gow, N. A., et al. (Eds.). 1999. *The fungal colony.* New York: Cambridge University Press.

Hawksworth, D. L., et al. 1996. *Ainsworth & Bisby's dictionary of the fungi,* 8th ed. New York: Oxford University Press.

Lawrey, J. D., and M. E. Hale, Jr. 1984. *Biology of lichenized fungi.* Westport, CT: Praeger Publications.

Smith, A. H., and N. Weber. 1980. *The mushroom field hunter's guide,* enlarged ed. Ann Arbor, MI: University of Michigan Press.

Moss-covered rocks in a small stream.

Introduction to the Plant Kingdom: Bryophytes

Chapter Outline

20

Overview

This chapter opens with a discussion of the features of the Plant Kingdom and its members and then introduces bryophytes with a note about their past use as bandages. Liverworts are explored after the habitats and general life history of bryophytes are discussed. Next, the chapter examines hornworts and their life cycles and then goes over mosses in greater detail. The chapter concludes with observations on the human and ecological relevance of bryophytes.

Some Learning Goals

1. Know the features that distinguish the Plant Kingdom from other kingdoms.
2. Understand how bryophytes as a group differ from other plants.
3. Learn the basic differences between thalloid liverworts and "leafy" liverworts.
4. Explain how a liverwort thallus can be distinguished from that of a hornwort.
5. Know the structures involved in the life cycle of a moss and in which structures meiosis and fertilization occur.
6. Learn which features liverworts, hornworts, and mosses have in common and understand how their sporophytes differ.
7. Learn five uses of bryophytes by humans.

Older botany texts tended to apply the term **plants** to many of the simpler organisms mentioned in previous chapters. Today, however, the majority of botanists confine this term to bryophytes, ferns, cone-bearing plants, flowering plants, and various relatives of each of these groups. These groups are the subjects of the next several chapters.

Plants and green algae, discussed in Chapter 18, share several major pigments (e.g., chlorophyll *a,* chlorophyll *b,* carotenoids). Plants and green algae also have starch in common as the primary food reserve and cellulose in their cell walls. When plant and some algal cells divide, they develop *phragmoplasts* and *cell plates,* both of which are unknown in other organisms. These shared features suggest plants and green algae were derived from a common ancestor.

Fossils reveal that land plants first appeared about 400 million years ago, so any hypothetical ancestor probably progressed from an aquatic to a land habitat even earlier. By the time plants became established on land, they had developed several features that kept them from drying out. (1) A fatty cuticle that retards water loss developed on plant surfaces. (2) The **gametangia** (gamete-producing structures) and **sporangia** (spore-producing structures) of plants became multicellular and surrounded by a sterile cell jacket. (3) Plant zygotes developed into multicellular **embryos** (immature plantlets) within parental tissues that originally surrounded the egg.

Most members of the Plant Kingdom are more complex and varied in form than those of the kingdoms examined in the preceding chapters, and their tissues are correspondingly more specialized for photosynthesis, conduction, support, anchorage, and protection. The sporophyte phases of the life cycles are predominant in the more advanced members, and reproduction is primarily sexual.

The multicellular embryos of plants are unknown in the kingdoms discussed in the preceding chapters. This was recognized in past classifications when all organisms were placed in either the Plant Kingdom or the Animal Kingdom. The Plant Kingdom was then divided into two subkingdoms. The bacteria, protists, and fungi were put in the Subkingdom Thallophyta ("thallus plants"), and all other organisms except animals were put in the Subkingdom Embryophyta ("embryo plants").

In this text, three phyla of essentially nonvascular plants, the *bryophytes*, and several phyla of **vascular plants** (plants with xylem and phloem) are included in the Plant Kingdom. The bryophytes are discussed in this chapter, and the vascular plants are examined throughout the remaining chapters of the book.

INTRODUCTION TO THE BRYOPHYTES

In the midst of heavy battles in France during World War I, nurses, at what is now referred to as a M.A.S.H. unit, on one occasion ran out of bandages for the wounded soldiers. In desperation, they substituted some soft green plant material they found growing in the water at the edge of a nearby lake. To their surprise, the material turned out to be a great substitute for the bandages; there were fewer infections in the wounds with the plant bandages than in those with the cotton bandages.

The material the nurses used was a species of *Sphagnum* moss (bog or peat moss), which has since been experimentally demonstrated to have antiseptic properties. This moss has specialized water-absorbing "leaves" (see Fig. 20.10) and has been used as a packing material in the past. It is still widely used as a soil conditioner. The "bandage" *Sphagnum* is one of about 23,000 species of **bryophytes** that include *mosses, liverworts,* and *hornworts,* many of which frequently make a soft and cool-looking green covering on damp banks, trees, and logs that are shaded for at least a part of the day (Fig. 20.1).

Figure 20.1 Mosses and ferns growing on a tree trunk.

In contrast, some bryophytes can withstand long periods of desiccation and are also found on bare rocks in the scorching sun (Fig. 20.2), while others occur on frozen alpine slopes. The habitats of bryophytes range in elevation from sea level near ocean beaches up to 5,500 or more meters (18,000 or more feet) in mountains.

Some bryophytes are restricted to very specific habitats. For example, a few species are found only on the antlers and bones of dead reindeer. Others are confined to the dung of herbivorous animals, while still others grow only on the dung of carnivores. A few tropical bryophytes thrive only on large insect wing covers. The pygmy mosses, which appear annually on bare soil after rains, are only 1 to 2 millimeters (0.04 to 0.08 inch) tall and can complete their whole life cycle in a few weeks.

Bryophytes of all phyla often have mycorrhizal fungi associated with their rhizoids. In some instances, the fungi apparently are at least partially parasitic. One species of completely colorless liverwort that lives underneath mosses is totally dependent nutritionally on its fungal associate. The gametophytes of more advanced plants are completely dependent on their sporophytes for their nutrition.

The widespread peat mosses are ecologically very important in bogs and in the transformation of bogs to dry land. Peat mosses sometimes form floating mats over water and keep conditions acid enough to inhibit the growth of bacteria and fungi. Organisms that die in such waters or bogs are often preserved for hundreds or even thousands of years.

The luminous mosses are found in caves near the entrances and in other dark, damp places. They are called luminous because they glow an eerie golden-green in reflected light. The upper surfaces of their cells are slightly curved, and each cell functions as a tiny magnifying glass, concentrating the dim light on the chloroplasts at the base.

Figure 20.2 *Grimmia,* a rock moss that survives on bare rocks, often in scorching sun.

This allows photosynthesis to take place in light otherwise too faint for it to occur (Fig. 20.3).

None of the bryophytes have true xylem or phloem, and to be able to reproduce, all bryophytes must have external water, usually in the form of dew or rain. Many mosses do have special water-conducting cells called *hydroids* in the centers of their stems, and a few have food-conducting cells called *leptoids* surrounding the hydroids. Neither type of cell, however, conducts as efficiently as vessel elements or tracheids of xylem and sieve tube members of phloem, and most water is absorbed directly through the surface. The absence of xylem and phloem makes most bryophytes soft and pliable, and it is not surprising, therefore, that birds often use them to line their nests. In one study in the Appalachian Mountains of Virginia in 1975, for example, David Breil and Susan Moyle found birds native to the area used at least 65 species of bryophytes in the construction of their nests.

Alternation of Generations is more conspicuous in bryophytes and ferns than in most other organisms. In mosses, the "leafy" plant is a major part of the *gametophyte* generation that produces the *gametes*. The *sporophyte* generation, which grows from a "leafy" gametophyte, produces the *spores*. It usually resembles a tiny can with a rimmed lid at the tip of a slender upright stalk.

All bryophytes have similar life cycles, chromosomes, and habitats. However, based on their structure and reproduction, they are separated into three distinct phyla. None of the bryophytes appear closely related to other living plants, and fossils provide little evidence that members of each phylum are related to those of the other phyla. Botanists speculate that the three lines of bryophytes may have arisen independently from ancestral green algae.

Ecological Review

While most bryophytes are confined to moist, shaded areas, some can withstand long periods of desiccation on bare rocks in the scorching sun or survive in frozen alpine and arctic environments. Many bryophytes that are found in highly restricted environments play significant roles in the early stages of community succession. The peat mosses are very important in the successional processes that transform bogs into dry land, a transformation that stored vast amounts of atmospheric carbon in the peat bogs of northern environments. One reason that biomass accumulates in peat bogs rather than decomposing is that peat mosses, by establishing an acid pH, inhibit the growth of decomposer microbes. The antibiotic and absorbent properties of peat mosses have also been put to practical use in the past, when peat mosses have been used as bandages. Bryophytes are useful to other organisms in the community as well: Many birds use soft, pliable bryophytes for nesting materials, and many mycorrhizal fungi are associated with the rhizoids of bryophytes.

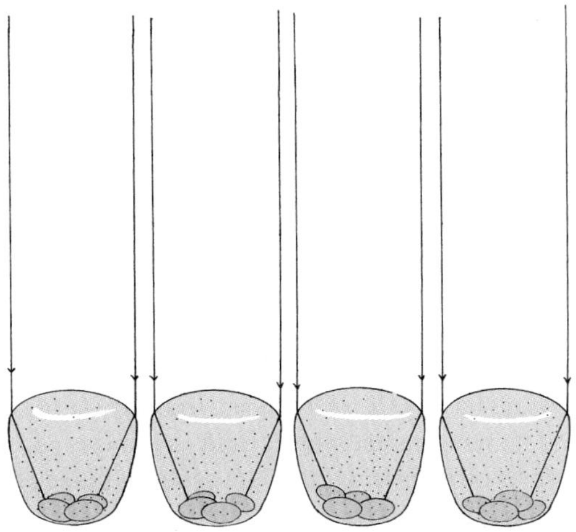

Figure 20.3 Lenslike cells of the protonema of the luminous moss. The cells concentrate light on the chloroplasts at their bases. *After D. von Denffer, W. Schumacher, K. Magdefrau, and F. Ehrendorfer. 1976. Strasburger's* Textbook of Botany, *30th German ed., translated by P. Bell and D. Coombe. 1976. Redrawn by permission of Gustav Fischer Verlag, Stuttgart, and Longman Group, London and New York.*

PHYLUM HEPATICOPHYTA—LIVERWORTS

The word *wort* simply means plant or herb. In medieval times, when the *Doctrine of Signatures*[1] held sway, the herbalists of the day thought some of the bryophytes—specifically those with flattened bodies and liver-shaped lobes (Fig. 20.4)—were useful for treating liver diseases. This belief proved to be without merit, but the name *liverwort* is still universally used today.

Structure and Form

There are about 8,000 known species of liverworts. The most common and widespread liverworts have flattened, lobed, somewhat leaflike bodies called **thalli** (singular: **thallus**). The *thalloid liverworts,* however, constitute only about 20% of the species. The other 80% are "leafy" and superficially resemble mosses.

Liverworts differ from mosses in several details and are considered less complex. Their thalli or "leafy" stages (gametophytes) develop from spores. When the spores germinate, they may produce a *protonema,* which is an immature gametophyte consisting of a short filament of cells. Typically, the protonema develops further into a more extensive gametophyte, or in some instances, the protonemal stage may be absent, the mature gametophyte then developing directly from the spore. Thalloid liverworts have smooth upper surfaces with various

1. The *Doctrine of Signatures* held that any plant part that resembled a human body part would be helpful in healing that body part if it was diseased or hurting.

A.

B.

Figure 20.4 *Marchantia* (a thalloid liverwort). *A.* Thalli with male gametophores. *B.* Thalli with female gametophores.

markings and pores, and the corners of cell walls of most liverworts are specially thickened. The lower surfaces have many one-celled *rhizoids*. Growth is prostrate instead of upright, and the rhizoids, which look like tiny roots, anchor the plants.

Thalloid Liverworts

The best-known species of thalloid liverworts are in the genus *Marchantia* (named in honor of French botanist N. Marchant) (see Fig. 20.4). The most widespread *Marchantia* species is often found on damp soil after a fire. The thallus, which is about 30 cells thick in the center and 10 cells thick at the margin, forks dichotomously as it grows. Each branch has a notch at the apex and a central groove that extends back lengthwise behind the notch. The thalli grow as meristematic cells at the notches divide. Older tissues at the rear decay as the new growth is added. The upper surface of the thallus is divided into diamond-shaped or polygonal segments, the segment lines marking the limits of chambers below. Each segment has a small bordered pore opening into the interior.

Seen through a microscope, a sectioned liverwort thallus looks like groups of short, erect, branching rows of cells with chloroplasts, sitting on a wall of colorless "bricks" (Fig. 20.5). The "brick wall," which may comprise most of the thallus, consists of parenchyma cells that have few, if any, chloroplasts. The tissue apparently stores substances produced in other cells. The bottom layer of cells is an epidermis from which rhizoids and scales arise. The individual groups of upright chlorenchyma cells are enclosed by vertical walls that are covered by a slightly dome-shaped layer of epidermal cells. A conspicuous pore is located in the center of each "roof" and remains open at all times. It resembles a tiny, short, suspended, open-ended barrel.

Marchantia—*Asexual Reproduction*

Marchantia reproduces asexually by means of **gemmae** (singular: **gemma**). Gemmae are tiny, lens-shaped pieces of tissue that become detached from the thallus. They are produced in small gemmae cups scattered over the upper surface of the liverwort gametophyte (Fig. 20.6). Raindrops may splash the gemmae as much as 1 meter (3 feet) away. While gemmae are in the cup, lunularic acid inhibits their further development, but each is capable of growing into a new thallus as soon as it leaves the cup. In addition, parts of an older thallus may die, isolating patches of active tissue, which may then continue to grow independently.

Marchantia—*Sexual Reproduction*

The gametangia of *Marchantia* are produced on separate male and female gametophytes and are more specialized than those of other liverworts. Both types of gametangia are formed on **gametophores** (umbrellalike structures borne on slender stalks rising from the central grooves of the thallus) (see Fig. 20.6). The top of the male gametophore, or **antheridiophore**, is disclike with a scalloped margin, while that of the female gametophore, or **archegoniophore,** looks like the hub and spokes of a wagon wheel.

Club-shaped male gametangia (**antheridia**) containing numerous *sperms* are produced in rows just beneath the upper surface of the antheridiophore. **Archegonia** are flasklike female gametangia, each containing a single *egg;* they are also produced in rows and hang neck downward beneath the spokes of the archegoniophore. Raindrops sometimes splash the released sperms, which have numerous flagella, more than 0.5 meter (1.5 feet) away. Fertilization may occur before the stalks of the archegoniophores have finished growing.

After fertilization, the zygote develops into a multicellular **embryo** (an immature *sporophyte*) that is totally dependent on the gametophyte for sustenance. A knoblike **foot** anchors the sporophyte (the diploid, spore-producing phase) in the tissues of the archegoniophore. The sporophyte hangs suspended by a short, thick stalk called the **seta.** The main part of the sporophyte, in which different types of tissues develop, is called a **capsule.** Liverwort sporophytes typically have no stomata.

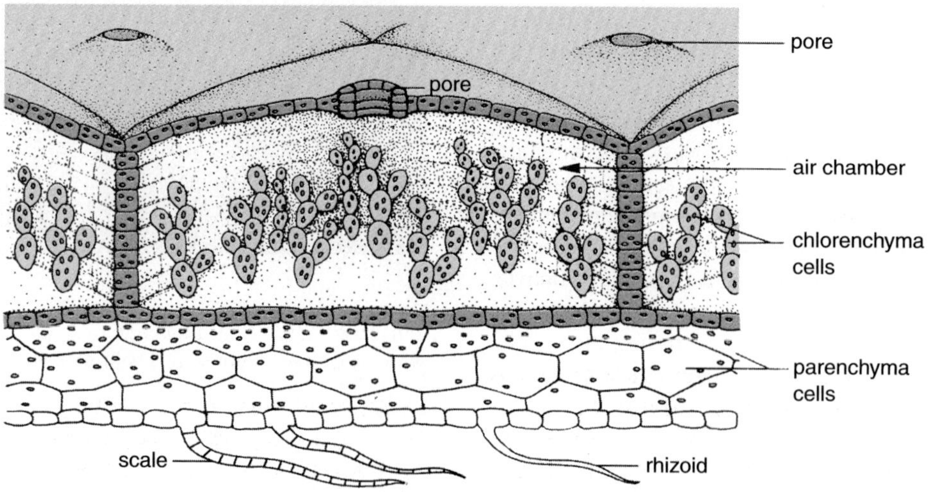

Figure 20.5 A section through a portion of a *Marchantia* thallus.

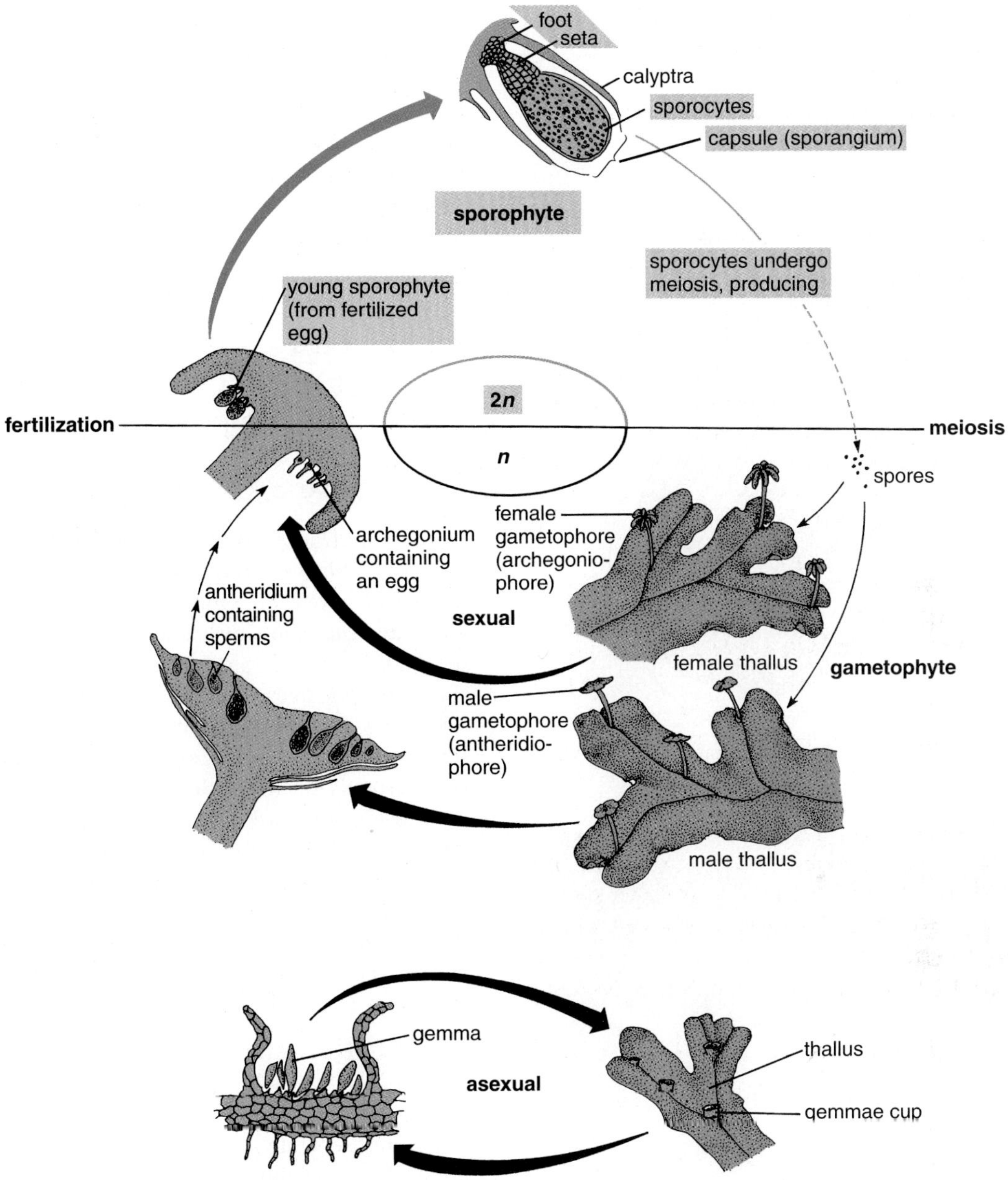

Figure 20.6 Life cycle of the thalloid liverwort *Marchantia*.

Sporocytes in the capsule undergo meiosis, producing haploid *spores*. Other capsule cells do not undergo meiosis but remain diploid and develop instead into long, pointed **elaters** with spiral thickenings. They are sensitive to changes in humidity (Fig. 20.7). Spore dispersal in *Marchantia* takes place as the elaters twist and untwist rapidly. In the sporophytes of other liverworts, the elaters may aid spore dispersal with a snapping action or by suddenly expanding.

Until the young sporophyte is mature, it is protected by the **calyptra**, a caplike tissue that grows out from the gametophyte, and by other membranes covering the capsule. The capsule splits at maturity, and air currents carry the spores away. Under favorable conditions, the spores germinate, producing new gametophytes.

Other thalloid forms, such as the floating or amphibious liverworts, do not produce gametophores. Instead, the archegonia and antheridia develop within the thallus beneath the central grooves, where the sporophytes also are formed. The spores are liberated from the submerged sporophytes as the thallus decays.

"Leafy" Liverworts

"Leafy" liverworts (Fig. 20.8) are often abundant in tropical forests and in fog belts. They always have two rows of partially overlapping "leaves" whose cells contain distinctive oil bodies. The "leaves" have no midribs, and unlike the "leaves" of mosses, they often have folds and lobes. In the

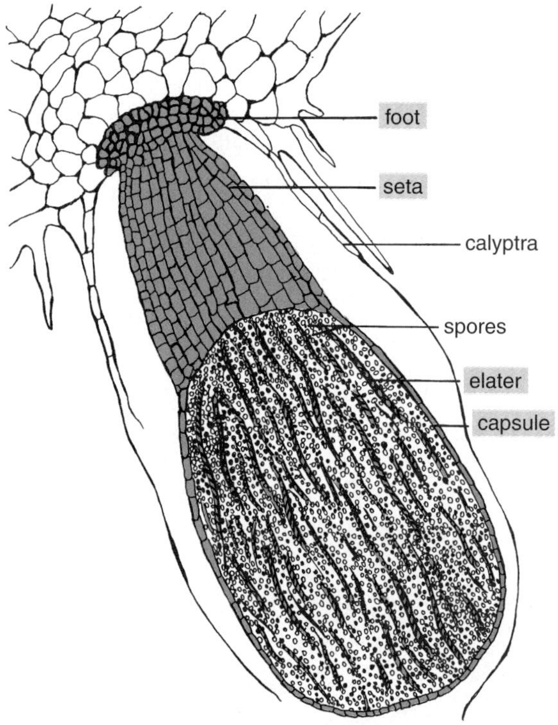

Figure 20.7 A longitudinal section through a sporophyte of *Marchantia*.

Figure 20.8 *Frullania*, a "leafy" liverwort.

tropics, the lobes form little water pockets in which tiny animals are nearly always present. It has been suggested these water pockets may function like the pitchers of pitcher plants (discussed in Chapter 7). A third row of "underleaves" is often present on the underside of "leafy" liverworts. The "underleaves" are smaller than the other "leaves" and not visible from the top. A few rhizoids that anchor the plants develop from the stemlike axis at the base of the "underleaves."

The archegonia and antheridia of the "leafy" liverworts are produced in cuplike structures composed of a few modified "leaves," either in the axils of "leaves" or on separate branches. At maturity, the sporophyte capsule may be pushed out from among the "leaves" as the seta elongates. When a spore germinates, it produces a **protonema** consisting of a short filament of photosynthetic cells. The protonema soon develops into a mature gametophyte plant.

PHYLUM ANTHOCEROPHYTA—HORNWORTS

Structure and Form

Hornworts, whose mature sporophytes look like miniature, greenish to blackish rods that may curve slightly, have gametophytes that resemble filmy versions of thalloid liverworts. They are usually less than 2 centimeters (0.8 inch) in diameter and thrive mostly on moist earth in shaded areas, although some occur on trees. There are only about 100 species worldwide; they are uncommon in arctic regions. They differ from liverworts and mosses in several respects and appear to be only distantly related to them. Hornworts usually have only one large chloroplast in each cell (a few species have up to eight). Each chloroplast has pyrenoids similar to those of green algae. The thalli have pores and cavities filled with mucilage, in contrast to the air-filled pores and cavities of thalloid liverworts. Nitrogen-fixing cyanobacteria often grow in the mucilage. Rhizoids anchor the plants.

Asexual Reproduction

Hornworts reproduce asexually primarily by fragmentation or as lobes separate from the main part of the thallus. A few hornworts form tiny tubers that are capable of becoming new gametophytes.

Sexual Reproduction

In sexual reproduction, archegonia and antheridia are produced in rows just beneath the upper surfaces of the gametophytes. Like both mosses and liverworts, some species of hornworts have *unisexual* plants, while other species are *bisexual*.

The distinctive sporophytes of hornworts have numerous stomata. They have no setae (stalks) and look like tiny green broom handles or horns rising through a basal sheath from a foot beneath the surface of the thallus (Fig. 20.9). A meristem above the foot continually increases the length of the sporophyte from the base when conditions are favorable. As growth occurs, sporocytes surrounding a central rodlike axis in the sporophyte undergo meiosis, producing spores. Diploid elaters similar to those of liverworts are intermingled

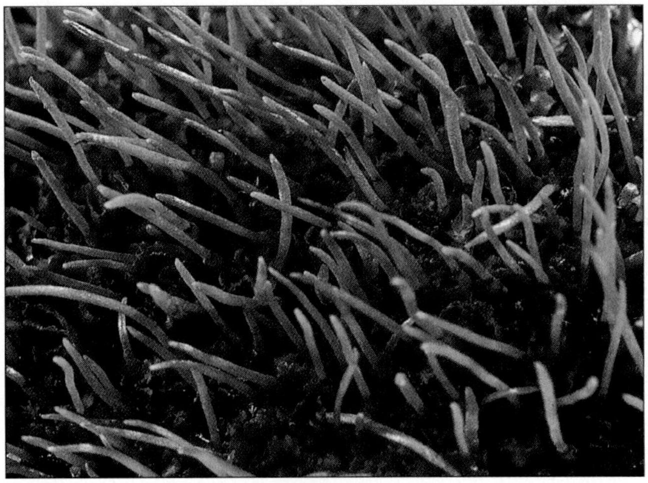

A.

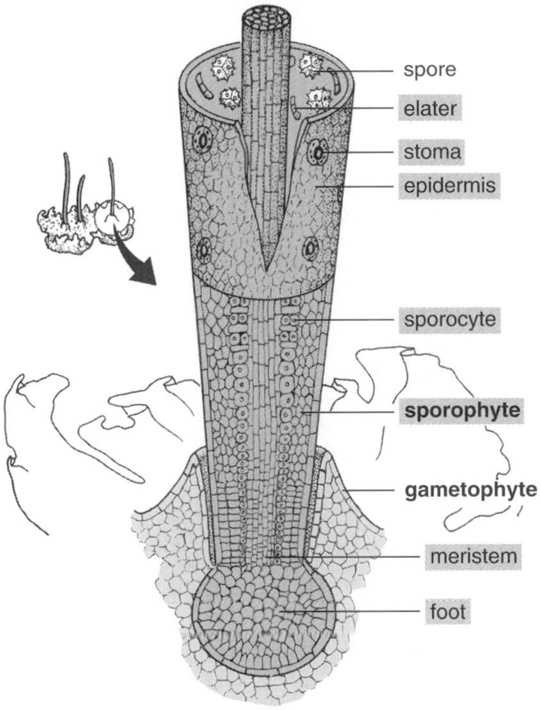

B.

Figure 20.9 *A.* Hornworts. *B.* Detail of a hornwort sporophyte.

with the spores. The tip of the sporophyte horn splits into two or three ribbonlike segments, releasing the spores, and the segments continue to peel back as long as the meristem is producing new tissue at the base.

PHYLUM BRYOPHYTA—MOSSES

Structure, Form, and Classes

Many different organisms have been called *mosses*. In fact, almost any greenish covering or growth on tree trunks and forest floors has probably been called "moss" at one time or another. Examples include lichens (e.g., reindeer moss), red algae (e.g., Irish moss), flowering plants (e.g., Spanish moss), and club mosses. Club mosses look somewhat like large true mosses but are vascular plants with xylem and phloem. About 15,000 species of mosses are currently known. These are divided into three different classes, commonly called *peat mosses, true mosses*, and *rock mosses*. Mosses are distinct, both in form and reproduction, and possibly in origin, from any other group of organisms.

The "leaves" of moss gametophytes have no mesophyll tissue, stomata, or veins such as those of the leaves of more complex plants. The blades are nearly always only one cell thick, except at the *midrib* that runs lengthwise down the middle, and they are never lobed or divided, nor do they have a *petiole* (leaf stalk). The midrib (absent in some genera) occasionally projects beyond the tip in the form of a hair or spine. The "leaf" cells usually contain numerous lens-shaped chloroplasts, except at the midrib. The "leaves" of peat mosses, however, have large transparent cells (without chloroplasts) that absorb and store water. Small green photosynthetic cells are sandwiched between the large cells (Fig. 20.10). The "leaves" are intially formed in three ranks and usually end up appearing to be arranged in a spiral or alternately on an axis that twists as it grows.

The axis is somewhat stemlike but has no xylem or phloem, although there is often a distinctive central strand of hydroids. At the base, there are rootlike rhizoids consisting of several rows of colorless cells that anchor the plant. Some water absorbed by rhizoids rises up the central strand, but most water used by the plant apparently travels up the outside of the plant by means of capillarity. The closely packed habit of many mosses and the fact that they rarely extend more than a few centimeters (2 to 3 inches) into the air favors such outside movements of water. Water is absorbed directly through the plant surfaces.

Sexual Reproduction

Sexual reproduction in mosses begins with the formation of multicellular gametangia, usually at the apices of the "leafy" shoots of gametophytes (see Fig. 20.13), although they frequently form on special separate branches. Both male and female gametangia are often produced on the same plant, but in some species, they occur on separate plants. The archegonia (female gametangia) are somewhat cylindrical and project upward from the base of the expanded gametophyte tip (Fig. 20.11). When certain cells break down in the swollen base of the archegonium (called the **venter**), a cavity develops in which a single *egg* cell is produced. The part of the archegonium above the venter is called the *neck*. The neck may taper toward the tip and contains a narrow *canal*. The canal is at first plugged with cells, but these break down as the archegonium matures, leaving an opening to the outside at the top. Several archegonia usually are produced at the same time, with sterile hairlike, multicellular filaments called *paraphyses* (singular: *paraphysis*) scattered among them.

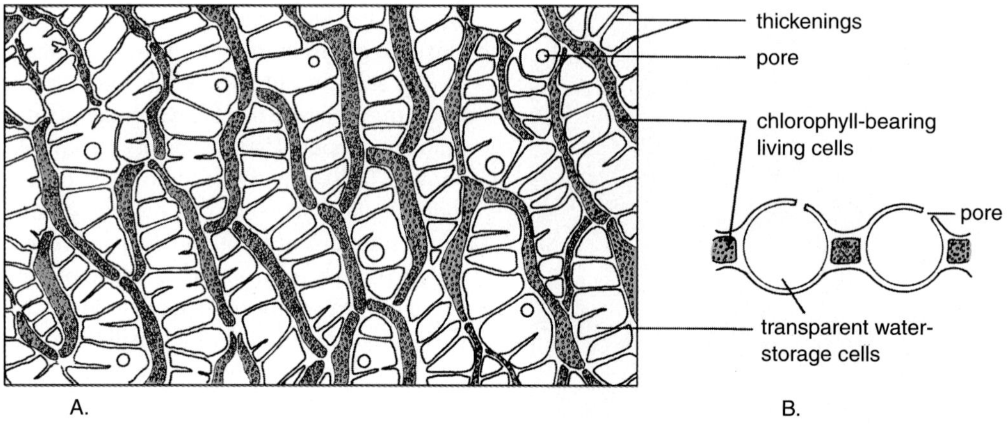

Figure 20.10 An enlargement of a portion of a peat moss (*Sphagnum*) leaf. *A.* Surface view. *B.* A further-enlarged cross section of living and dead cells.

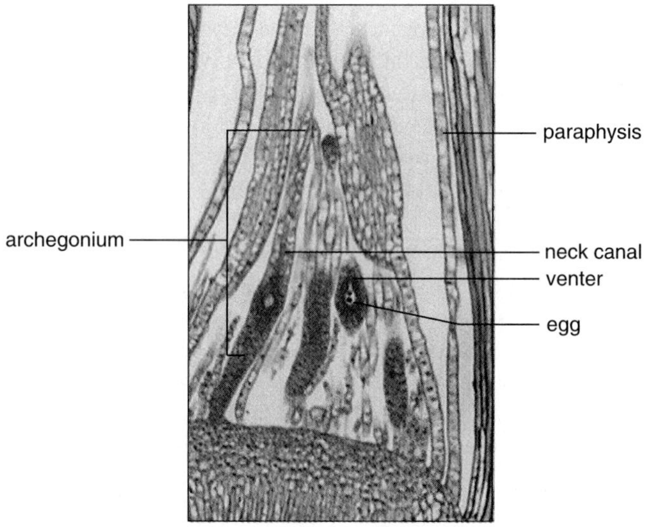

Fgure 20.11 A longitudinal section through the tip of a female gametophyte of the moss *Mnium*.

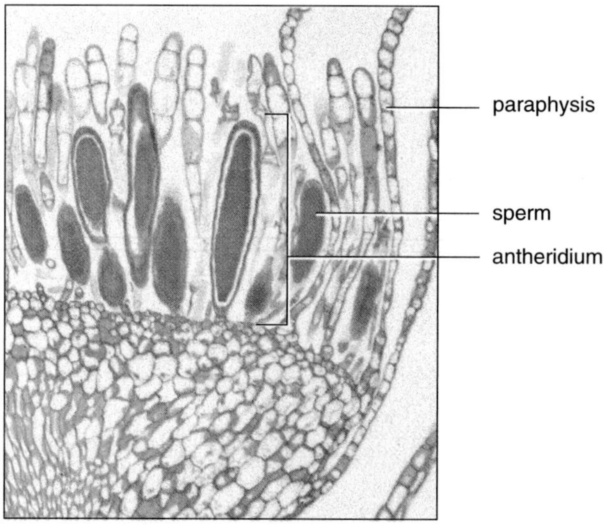

Figure 20.12 A longitudinal section through the tip of a male gametophyte of the moss *Mnium*.

Male gametangia also have paraphyses among them and are sausage shaped to roundish, with walls that are one cell thick. These *antheridia* (Fig. 20.12) are borne on short stalks. A mass of tissue inside each antheridium develops into numerous coiled or comma-shaped *sperm* cells. This mass of sperms is forced out of the top of the antheridium when it absorbs water and swells. After release, the sperm mass breaks up into individual cells, each with a pair of flagella. It is believed the breakup of the sperm mass is aided, in some cases, by fats produced by the moss, while in other instances, rain splash is responsible.

Archegonia release sugars, proteins, acids, or other substances that attract the sperm, and eventually a sperm, after swimming down the neck of an archegonium, unites with the egg, forming a diploid *zygote* (Fig. 20.13). The zygote usually grows rapidly into a spindle-shaped *embryo*. The embryo breaks down the cells at the base of the archegonium and becomes firmly established in the tissues of the stem by means of a swollen knob called a *foot*. As the embryo grows, cells around the venter divide, thereby accommodating its increasing size. The length of the embryo soon exceeds the length of the cavity in the venter. The top of the venter is split off and is left sitting like a pixie cap on top of the embryo. By this time, the embryo is a developing *sporophyte*. The pixie cap, called a *calyptra,* remains until the sporophyte is mature. In one genus with the common name of "extinguisher mosses," the calyptra looks just like a little candlesnuffer, and in the hairy cap mosses, it resembles a miniature, pointed, goatskin cap such as might be worn by a Shakespearean actor.

The cells of the sporophyte become photosynthetic as it develops, remaining so until maturity. The sporophyte, however, depends to varying degrees on the gametophyte for some of its carbohydrate needs as well as for at least a part of its water and minerals, which are absorbed through the foot.

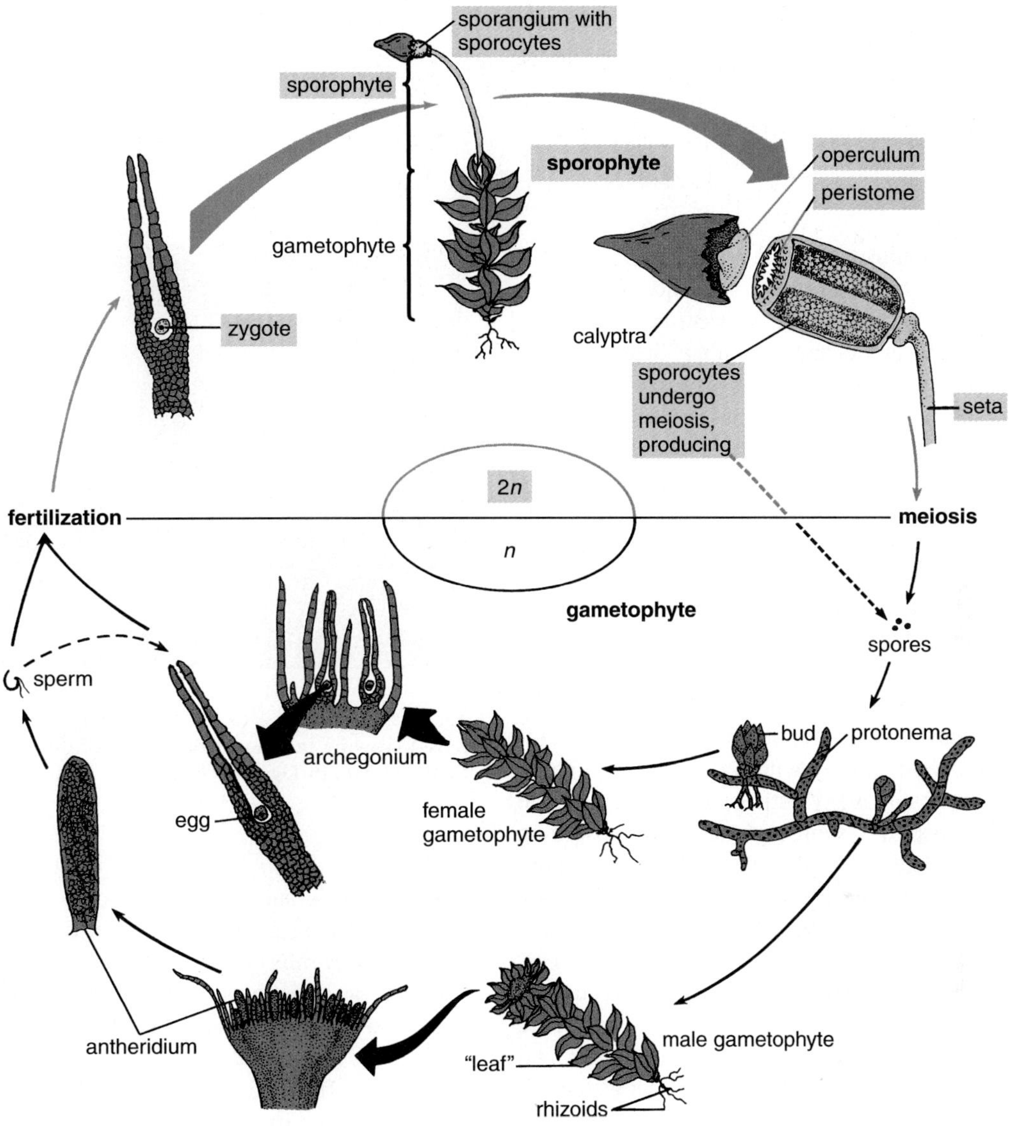

Figure 20.13 Life cycle of a moss.

The mature sporophyte is at first green and photosynthetic; it consists of a *capsule* located at the tip of a slender stalk called the *seta*. Depending on the species, the seta may be less than 1 millimeter (0.04 inch) long, or it may be up to 15 centimeters (6 inches) long. Most, however, are less than 5 centimeters (2 inches) long. The capsule may resemble a tiny apple, a pear, an urn, a box, or a wingtip fuel tank of an airplane and usually has from 3 or 4 to over 200 stomata at or near its surface. Unless extremely dry conditions prevail, the stomata normally remain open until the capsule begins to age, and then they close permanently. The free end of the capsule is usually protected by a little rimmed lid, the **operculum,** which falls off at maturity.

As the capsule matures, *sporocytes* inside it undergo meiosis, producing haploid *spores*. These spores, often numbering in the millions, are released from the capsule, usually through a structure called a **peristome,** after the operculum falls off. Most peristomes consist of a circular row or two of narrowly triangular and membranous teeth arranged around the rim of the capsule, each row having 16 teeth. The teeth are frequently colored orange or red and are often beautifully sculptured with bars and fringes. They open or close in response to changes in humidity. In a few species of mosses, the peristome is a cone-shaped structure with pores through which the spores are released.

Some rock mosses have neither a peristome nor an operculum. The spores in these mosses are released when the capsule splits lengthwise along four lines. In the dung mosses, a putrid odor is given off when the spores are ready for release. Some of the spores adhere to the legs and bodies of flies, which are attracted by the odor, and are disseminated as the insects clean themselves. Most moss spores are, however, simply blown away by the wind, and if they fall in a suitable damp location, they usually germinate relatively quickly.

Awareness

Hibernating Mosses

Mosses are the "amphibians" of the plant world, so-called because they are at home in either semiaquatic environments, such as moist stream banks, or drier habitats, such as rock surfaces or the arctic tundra. Unlike their towering cousins, the flowering plants, mosses are dependent on a watery landscape to reproduce. Because moss sperm are flagellated, a film of water is needed for the sperm to swim to the egg. But life on land is not easy, especially when water becomes limited or nonexistent for months. The genetic diversity of mosses is exceptional, and they inhabit some of the most inhospitable habitats on earth. Dry heaths, rock faces, tree trunks, and even deserts are home to these remarkable plants. Distinctive adaptations allow moss species this wide range of habitats, including the ability to tolerate drying out, a process known as **desiccation.**

Vascular plants have structural mechanisms that maintain an adequate water supply within plant tissues. These include an impermeable waxy cuticle on leaf surfaces, an internal water transport system (xylem), water-absorbing organs (roots), and leaf pores (stomata) that can close, conserving water. Parts of the life cycle, such as seeds, are especially tolerant of desiccation. Mosses do not possess these adaptations to living in a dry land environment. This means the internal water balance in mosses is in equilibrium with the atmosphere. When the air is dry, mosses are dry. When it rains, mosses quickly absorb water and become rehydrated. They are "opportunists" in this regard.

Some species of mosses are incapable of withstanding this desiccation-rehydration cycle, but those that have this ability can "return from the dead" with each cycle. It is as if they have been hibernating. There are several mechanisms that slow water loss from mosses that make the cycle less drastic. Many mosses form dense mats or tufts that create a moist microatmosphere within the tufts and over the plants. *Polytrichum commune* is a common moss that has moderate desiccation tolerance. It lives in habitats such as bogs or temperate, moist forests and can grow up to 40 centimeters tall. It grows luxuriantly during rainy periods but then twists into rusty red mats upon drying in the sun. These mosses have a thin, waxy cuticle covering their tiny "leaves" (8 to 10 millimeters in length) that retards water loss. They also have primitive water-conducting cells (hydroids) that move water through the plant when water is lost to the atmosphere by evaporation. Instead of root hairs, they have a rhizoid system that absorbs water from the soil. However, if dry conditions persist, all available water is lost to the atmosphere.

Additional mechanisms by which mosses are able to survive these desiccation-rehydration cycles are being discovered. *Tortula ruralis* is one of the most desiccation-tolerant mosses known, and research has centered on its ability to recover after prolonged or repeated desiccation events. Hydrated *T. ruralis* cells are similar to mesophyll cells in the leaves of vascular plants. There are chloroplasts with stacks of grana and a prominent nucleus. Mitochondria are numerous and similar in size and shape to those of higher plants, with internal membranes folded into cristae.

Box Figure 20.1 *Polytrichum commune* (hairy cap moss) growing in a deciduous forest of the northeastern United States. *(Photo by Daniel Scheirer)*

During desiccation, *T. ruralis* dries out, and its "leaves" fold up around the stem. Cells of desiccated plants are damaged. There is extensive plasmolysis as water is lost from the protoplast. Chloroplasts become smaller and more spherical, and starch is not present. Internal thylakoid membranes are collapsed and disorganized. Mitochondria in the hydrated cell are elongated with numerous cristae membranes. In the desiccated state, mitochondria are smaller and rounded, with few internal cristae membranes. The nucleoplasm of nuclei becomes dense, and chromatin is condensed. Compact-appearing nucleoli are prominent. The plasma membrane is damaged, and electrolytes leak out. In all desiccated mosses, photosynthesis stops, and respiration slows or ceases.

When the plants are rehydrated, dried "leaves" unfold and return to the normal hydrated state within 2 minutes. Most of the internal damage is repaired within a few more minutes. Activity of a drought-repair gene increases in the minutes after rehydration, and repair proteins are quickly manufactured and mobilized for action. Respiration resumes after a few minutes. Photosynthesis resumes up to 24 hours later.

Awareness Box Continued

Scientists at the USDA's Agricultural Research Service are attempting to locate the genes responsible for drought repair in *Tortula* species in the hope of transferring them to crop plants. With crops having more drought tolerance, rain shortage would be less of a problem than it is now. Arid lands currently unsuited for agriculture could blossom with crop plants genetically engineered with drought-tolerant genes. World population increases of 1.6% yearly means that an additional 78,000 metric tons of grain per day are required just to maintain current consumption levels. A lowly moss might someday be responsible for a revolution in food production.

D.C. Scheirer

In most mosses, fine, green tubular threads, consisting of single rows of cells with chloroplasts, first emerge from the spores. These soon branch and grow, forming an algalike *protonema*. The protonema can be distinguished from a filamentous green alga by the oblique crosswalls of its cells and by the lens-shaped chloroplasts. If light and other conditions are favorable, tiny "leafy" *buds* appear at intervals along the protonemal filaments after about 2 to 4 weeks of growth. These "leafy" buds develop rhizoids at the base and grow into new "leafy" gametophytes, completing the cycle.

Asexual Reproduction

Some reproduction in mosses does not depend on such a sexual cycle and its Alternation of Generations. It has been demonstrated under laboratory conditions that cells of archegonia and antheridia, paraphyses, "leaves," stems, and rhizoids all can develop protonemata. Two American biologists once collected bryophyte gametophyte fragments from a snowbed in the Canadian high arctic and found that 12% of their samples resumed growth in various ways when cultured in the laboratory. They calculated that for each cubic meter (1.3 cubic yards) of snow at their study site, there were over 4,000 bryophyte fragments capable of becoming new plants. They suggested that wind dispersal of fragments may be routine in arctic regions. Such dispersal and vegetative reproduction also occur widely in more temperate areas.

HUMAN AND ECOLOGICAL RELEVANCE OF BRYOPHYTES

Some bryophytes and lichens are pioneers on bare rock after volcanic eruptions or other geological upheavals and after the retreat of glaciers. They slowly accumulate mineral and organic matter that can then be inhabited or utilized by other organisms. This process, called *succession,* is discussed in Chapter 25. Mosses, in particular, retain moisture, slowly releasing it into the soil. They reduce flooding and erosion and contribute to humus formation. Some mosses grow only in soils that are rich in calcium; the presence of others indicates higher than usual soil salinity or acidity.

When certain mosses are present in a dry area, it is a good indication that running water occurs there some time during the year. A few mosses are occasionally a problem in water reservoirs, where they may plug entrances to pipes. A few bryophytes are reported to be grazed, along with lichens, by foraging mammals in arctic regions, but bryophytes are not generally edible. Some mosses have been used for packing dishes and stuffing furniture, and Native Americans are reported to have used mosses for diapers and under splints when setting broken limbs.

By far, the most important bryophytes to humans are the peat mosses. When allowed to absorb water, 1 kilogram (2.2 pounds) of dry peat moss will take up 25 kilograms (55 pounds) of water. Its extraordinary absorptive capacity has made it very useful as a soil conditioner in nurseries and as a component of potting mixtures. Live shellfish and other organisms are shipped in it. The natural acidity produced inhibits bacterial and fungal growth and gives it antiseptic properties. The absorbency, which is greater than that of cotton, combined with the antiseptic properties, has made it a useful poultice material for application to wounds. It was used for this purpose during the Crimean War of 1854 to 1856 and, as indicated in the chapter introduction, on an emergency basis during World War I. Extensive peat deposits have been formed from the remains of peat mosses that flourished in past eras. Peat, like the undecomposed peat mosses, is used around the world as a soil conditioner and as a fuel. In the manufacture of Scotch whiskey, sprouted barley is dried on a screen over a peat fire. The peat smoke permeates the barley and imparts a smoky flavor to the beverage. See Appendix 1 for the scientific names of all the bryophytes discussed.

Summary

1. Members of the Plant Kingdom have a cuticle and produce their gametes and spores in multicellular organs surrounded by a sterile jacket of protective cells. Their zygotes develop into embryos; and tissues specialized for photosynthesis, conduction, support, anchorage, protection, and reproduction are produced.
2. Cell plates and phragmoplasts appear when plant cells divide. Outside of the Plant Kingdom, these occur only

in certain green algae. The similarity in pigments, food reserve (starch), and occurrence of cell plates suggests a common ancestor for the green algae and plants. The Plant Kingdom includes three phyla of bryophytes and several phyla of vascular plants.

3. Bryophytes (liverworts, hornworts, mosses) occur in highly varied and also very specific habitats.
4. Water is essential to bryophyte reproduction. Most water is absorbed directly through the plant surfaces.
5. Liverwort gametophytes with flattened, dichotomously forking thalli are common, but about 80% of the liverwort species are "leafy." Liverworts have distinct upper and lower surfaces, with one-celled rhizoids that function in anchorage on the lower surface.
6. A *Marchantia* thallus has a central lengthwise groove along its upper surface and is chambered, each chamber containing chlorenchyma cells and having a surface pore. Rhizoids and scales arise from the thallus base.
7. *Marchantia* reproduces asexually by means of gemmae produced in surface cups and by thallus fragmentation.
8. *Marchantia* reproduces sexually by means of eggs and sperms produced in archegonia and antheridia on archegoniophores and antheridiophores that arise from the thallus.
9. The zygote develops into a sporophyte that is anchored to the archegoniophore by a foot, from which is suspended a capsule connected to the foot by a seta. Sporocytes in the capsule undergo meiosis, producing spores. Diploid elaters that aid in spore dispersal do not undergo meiosis.
10. The calyptra and other membranes protect the spores until they are released as the capsule splits; the spores may then develop into new gametophytes.
11. "Leafy" liverworts have two rows of overlapping "leaves" and frequently a third row of "underleaves" not visible from above. The "leaves" often have lobes that retain rain water.
12. Hornworts have one chloroplast with pyrenoids in each cell, and they resemble liverworts in their gametophytes. Their sporophytes are hornlike and have a meristem above the foot. Hornwort thalli have pores and cavities filled with mucilage, where cyanobacteria often grow.
13. Asexual reproduction in hornworts is by fragmentation. Sexual reproduction involves archegonia and antheridia produced in rows beneath the upper surface of a thallus. The tip of the hornlike sporophyte splits vertically, releasing the spores.
14. A moss gametophyte consists of an axis to which "leaves" are attached, with rhizoids at the base. The "leaves" are haploid and have no mesophyll, stomata, or veins. Water is absorbed primarily directly through the plant surfaces.
15. Multicellular archegonia and antheridia are produced at the tips of "leafy" shoots. Each archegonium has a cavity, the venter, containing a single egg and a neck through which a sperm gains access to the egg. Sperms are produced in antheridia.
16. After fertilization, the zygote grows into an embryo that is attached to the gametophyte by an embedded foot. The sporophyte developing from the embryo consists of a capsule and a seta. A calyptra derived from the gametophyte partially covers the capsule. Sporocytes in the capsule undergo meiosis, producing spores that are released through the teeth of the peristome, a structure at the tip of the capsule.
17. An operculum that falls off when the spores mature initially covers the peristome and protects sporocytes and spores. When moss spores germinate, protonemata with "leafy" buds develop. The buds grow into new gametophytes.
18. Mosses may be pioneers, along with lichens, on bare rocks. They are indicators of soil calcium, salinity, and acidity. Mosses are not generally edible, although a few are grazed in arctic regions. Some mosses are used for packing material, but the most significant use is that of peat mosses for soil conditioners. Peat mosses can absorb and retain large amounts of water, and their natural acidity gives them antiseptic properties. Peat deposits, from peat mosses that flourished in past eras, are used for fuel and also as a soil conditioner.

Review Questions

1. What basic features distinguish members of the Plant Kingdom from those of other kingdoms?
2. What features distinguish the bryophytes discussed in this chapter from other plants?
3. How could you tell a hornwort thallus from that of a thalloid liverwort?
4. Contrast the sporophytes of mosses, liverworts, and hornworts.
5. What is a protonema? Do all bryophytes have them? How would you tell a protonema from a green alga?
6. What adaptations do bryophytes have for their particular habitats?
7. Which parts of the life cycles of bryophytes have haploid (n) cells? Which parts have diploid ($2n$) cells?
8. Why is a bryophyte "leaf" technically not the same as a flowering plant leaf?
9. Define calyptra, operculum, capsule, peristome, paraphysis, foot, seta, archegoniophore, thallus, underleaves, and elaters.
10. If you were to single out one bryophyte as being the most important member of its phylum from a human viewpoint, which would you choose? Why?

Discussion Questions

1. Very few fossils of bryophytes have been found. Suggest reasons for this.
2. Do the multicellular sex structures of plants give them any advantages over other organisms with unicellular sex structures?
3. After reading about the characteristics and uses of peat mosses, can you suggest some possible new uses for these plants?
4. Some bryophytes produce unisexual gametophytes, while others produce bisexual gametophytes. Should one type have any survival or adaptive advantage over the other? Explain.

Learning Online

Visit our web page at www.mhhe.com/botany for interesting case studies, practice quizzes, current articles, and animations within the Online Learning Center to help you understand the material in this chapter. You'll also find active links to these topics:

Bryophytes
Economic and Ecological Importance of Lower Plants

Additional Reading

Bischler, H. 1998. *Systematics and evolution of the genera of the Marchantiales*. Champaign, IL: Balogh Scientific Books.

Chopra, R. N., and S. C. Bhatia (Eds.). 1990. *Bryophyte development: Physiology and biochemistry*. Boca Raton, FL: CRC Press.

Conard, H. S., and P. L. Redfearn, Jr. 1979. *How to know the mosses and liverworts,* 2d ed. Pictured Key Nature Series. Dubuque, IA: WCB/McGraw-Hill Publishers.

Crum, H., and L. E. Anderson. 1981. *Mosses of eastern North America*. New York: Columbia University Press.

Dyer, A. F., and J. G. Duckett (Eds.). 1984. *The experimental biology of bryophytes*. San Diego, CA: Academic Press.

Glime, J. M., and D. Saxzena. 1991. *Uses of bryophytes*. Houston, TX: Scholarly Publications.

Malcolm, B., and N. Malcolm. 2000. *Mosses and other bryophytes: An illustrated glossary*. Portland, OR: Timber Press.

Miller, N. G. 1988. *Bryophyte ultrastructure*. Champaign, IL: Balogh Scientific Books.

Schultze-Motel, W. (Ed.). 1984. *Advances in bryology,* vol. 2. Forestburgh, NY: Lubrecht & Cramer.

Schuster, R. M. 1966–1992. *The Hepaticae and Anthocerotae of North America*, 6 vols. New York: Columbia University Press.

Zinsmeister, H. D., and R. Mues. 1990. *Bryophytes: Their chemistry and chemical taxonomy*. New York: Oxford University Press.

'Ama'uma'u (*Sadleria cyatheoides*), a small Hawaiian tree fern.

The Seedless Vascular Plants: Ferns and Their Relatives

21

Chapter Outline

Overview

This chapter opens with a brief review of the features that distinguish the major groups of vascular plants without seeds from one another and from the bryophytes and then discusses representatives of each phylum. Included in the discussion are whisk ferns (*Psilotum*), club mosses (*Lycopodium, Selaginella*), quillworts (*Isoetes*), horsetails (*Equisetum*), and ferns. A digest of the human and ecological relevance of each group is given, and life cycles of representatives are illustrated. The chapter concludes with an examination of fossils, and each type is briefly described. A table showing the geologic time scale is provided.

Some Learning Goals

1. Describe the basic structural differences between bryophytes and vascular plants.
2. Distinguish from one another the four phyla of seedless vascular plants.
3. Contrast the differences in the life cycles of ground pines (*Lycopodium*) and spike mosses (*Selaginella*).
4. Summarize the structural features of horsetail (*Equisetum*) sporophytes.
5. Know how to recognize and explain the functions of all the structures involved in Alternation of Generations in a fern.
6. Identify and list ten important uses of vascular plants that do not produce seeds.
7. Explain what a fossil is and distinguish among the various types of fossils.

Thus far, our survey of all organisms traditionally regarded as plants has taken us from simple one-celled prokaryotic organisms, through more specialized eukaryotic protists and fungi, and on to the bryophytes. We now take up several phyla of plants whose members have internal conducting tissues but do not produce seeds.

In the introduction to the Plant Kingdom in Chapter 20, we mentioned that the ancestors of the bryophytes and the vascular plants probably were multicellular green algae that became established on the land over 400 million years ago. As green organisms moved ashore, they developed several new features, including sterile jackets of cells around gametangia, embryos within protective tissues, stomata for gas exchange, and a cuticle. Most of these features at least partly protected vital parts from drying out. Bryophytes have no significant internal tissues for conducting water, and any water used internally is absorbed directly through the aboveground parts. External water is required for fertilization.

During the early stages of vascular plant evolution, internal conducting tissues (xylem and phloem) began to develop, true leaves appeared, and roots that function in absorption as well as anchorage developed. At the same time, gametophytes became progressively smaller and more dependent on sporophytes that became progressively larger. Initially the sporophyte also afforded gametophyte protection.

Unlike conifers and flowering plants, the primitive vascular plants discussed in this chapter do not produce seeds. They include the ferns and a number of their relatives often referred to as "fern allies." Four phyla of seedless vascular plants are recognized:

1. **Phylum Psilotophyta** (*whisk ferns*). Features unique to members of this phylum of vascular plants include sporophytes that have neither true leaves nor roots, and stems and rhizomes that fork evenly.
2. **Phylum Lycophyta** (*club mosses* and *quillworts*). The stems of these plants are covered with **microphylls** (leaves with a single vein whose trace is not associated with a leaf gap—see Chapter 6). Most microphylls are photosynthetic.
3. **Phylum Equisetophyta** (*horsetails* and *scouring rushes*). The sporophytes of these plants have ribbed stems containing silica deposits and whorled scalelike microphylls that lack chlorophyll.
4. **Phylum Polypodiophyta** (*ferns*). The sporophytes of ferns have **megaphylls** (leaves with more than one vein and a leaf trace that is associated with a leaf gap—see Chapter 6) that are often large and much divided.

Fossils, discussed at the end of the chapter, give us clues to the ancestry of some of these plants and also to the origins of seed plants discussed in the chapters to follow.

PHYLUM PSILOTOPHYTA—THE WHISK FERNS

There is nothing very fernlike in the appearance of *whisk ferns,* commonly referred to by their scientific name of *Psilotum,* but they do loosely resemble small, green whisk brooms. Exactly where these plants fit in the Plant Kingdom is not clear. Traditionally, they have been associated with a number of extinct plants called *psilophytes* that flourished perhaps 400 million years ago, but there is not much fossil evidence to substantiate the link. Because whisk fern and true fern gametophytes share several features, some botanists classify them with the true ferns. They will be discussed first because, by most present criteria, they are among the simplest of all living seedless vascular plants.

Structure and Form

Psilotum sporophytes consist almost entirely of dichotomously forking (evenly forking) aerial stems and are unique among living vascular plants in having neither leaves nor roots. They usually grow up to 30 centimeters (1 foot) tall but occasionally grow as much as 1 meter (3 feet) or more tall (Fig. 21.1). The visible stems arise from short branching rhizomes just beneath the surface of the ground.

Enations, which are tiny, green, superficially leaflike, veinless, photosynthetic flaps of tissue, are spirally arranged along the stems. Photosynthesis takes place in the outer cells of the stem (epidermis and cortex). A central cylinder of xylem is surrounded by phloem. The xylem is star-shaped in cross section. Short roots, aided by mycorrhizal fungi, are scattered along the surfaces of the rhizomes. Some of the mycorrhizal fungi become established in cortical cells just beneath the epidermis.

Reproduction

Small *sporangia* that are fused together in threes and resemble miniature yellow pumpkins are produced at the tips of very short stubby branches in the upper parts of the angular stems. Spores released from the sporangia germinate slowly in the soil, in tree bark or tree fern "bark" crevices, or in other similar habitats. In Hawaii, they also grow in old lava flows.

The gametophytes, which lack pigmentation, develop from the spores beneath the soil surface and are easily overlooked. Sometimes they resemble tiny, transparent dog bones that are only about 2 millimeters (0.08 inch) wide and seldom more than 6 millimeters (0.25 inch) long. They are cylindrical and, like the aerial stems, may branch dichotomously. They have no chlorophyll and absorb their nutrients via one-celled, rootlike *rhizoids* aided by mycorrhizal fungi. Archegonia and antheridia develop randomly over the surface of the same gametophyte. After a sperm unites with an egg in an archegonium, the zygote develops a foot and a rhizome. As soon as the rhizome becomes established, upright stems are produced, and the rhizome separates from the foot (Fig. 21.2).

Whisk ferns are native to tropical and subtropical regions. In the United States, they are found in Florida, Louisiana, Texas, Arizona, and Hawaii. They are extensively cultivated in Japan and sometimes become a weed in greenhouses around the world. *Tmesipteris,* a close relative, has leaflike appendages; it is native to Australia and the South Pacific.

Fossil Whisk Fern Look-Alikes

Fossil plants that somewhat resemble whisk ferns have been found in Silurian geological formations (Table 21.1). These formations are estimated to be as much as 400 million years old. One group of these fossil plants, of which *Cooksonia* and *Rhynia* are examples (see Fig. 21.22), had naked stems and terminal sporangia. *Cooksonia* is the oldest plant known to have had xylem. A second group of fossils, represented by *Zosterophyllum,* had somewhat rounded sporangia produced along the upper parts of naked stems. *Zosterophyllum* and its relatives first appeared during the Devonian period (Table 21.1). They are thought to be ancestral to the club mosses, discussed in the next section, "Phylum Lycophyta—The Ground Pines, Spike Mosses, and Quillworts."

Whisk ferns are of little economic importance. Their spores have a slightly oily feel and were once used by Hawaiian men to reduce loincloth irritations of the skin. Hawaiians also made a laxative liquid by boiling whisk ferns in water.

Figure 21.1 *Psilotum* sporophytes. The small yellowish objects are three-lobed sporangia.

PHYLUM LYCOPHYTA—THE GROUND PINES, SPIKE MOSSES, AND QUILLWORTS

When I was in high school, my father was interviewed by a newspaper reporter in a hotel room of a large South American city. At the end of the interview, which took place

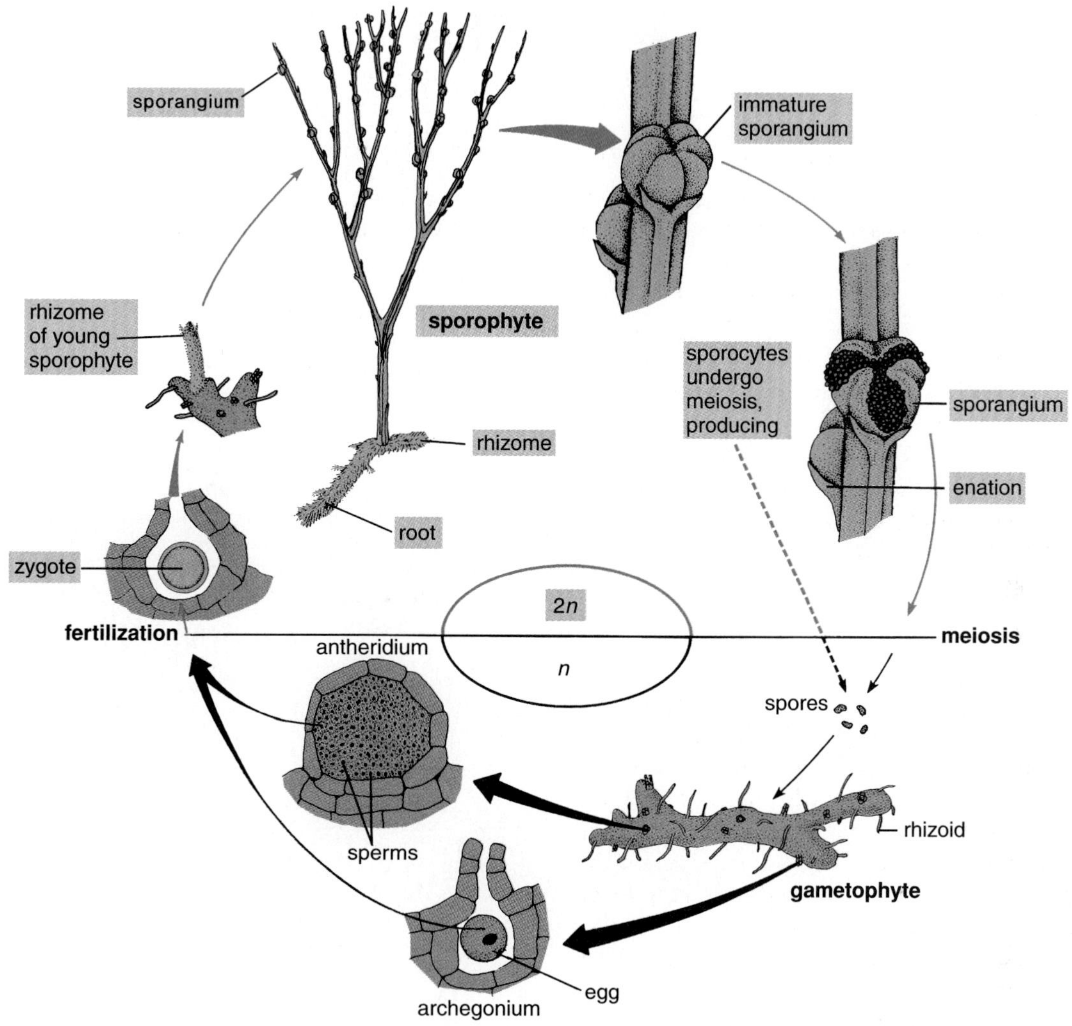

Figure 21.2 Life cycle of *Psilotum.*

with the rest of the family present, the reporter set up a press camera and said he wanted to take a picture of the family. After posing us, he took a vial from his pocket and poured a little powder along a metal bar attached to one end of a T-shaped device that otherwise looked like a flashlight. He then told us to smile, and a moment later, there was a flash of light followed by a large billow of smoke. I learned later that the *flash powder* (forerunner of flashbulbs and strobe lights) he had used contained millions of spores of primitive vascular plants collectively called **club mosses.**

The 950 or so known species of club mosses (referred to in this text as *ground pines* and *spike mosses*) mostly look enough like large true mosses that the Swedish naturalist Linnaeus lumped both together in a single class. Once details of the structure and the form of club mosses were known, however, it became obvious that they are quite unrelated to true mosses.

Today, there are living representatives of two major genera and two minor genera of club mosses. Several others became extinct about 270 million years ago. The sporophytes of all species have microphylls (leaves with a single unbranched vein that is not associated with a leaf gap—see Chapter 6) that are usually quite small; they also have true stems and true roots.

The two major genera of club mosses with living members are *Lycopodium,* with about 50 species, and *Selaginella,* with over 700 species. They are distributed throughout the world, but they are more abundant in the tropics and wetter temperate areas.

Lycopodium—Ground Pines

Structure and Form

Lycopodium plants often grow on forest floors. They are sometimes called *ground pines,* partly because they resemble little Christmas trees, complete with "cones" that are usually upright or, in a few species, hang down (Fig. 21.3). The stems of ground pine sporophytes are either simple or branched. The plants are mostly less than 30 centimeters (1 foot) tall, although some tropical species grow to heights of 1.5 meters (5 feet) or more. The upright, or sometimes pendent, stems

TABLE 21.1

Geologic Time Scale

ERA	PERIOD	AGE	EARLIEST EVIDENCE OF PLANTS	DURATION (MILLIONS OF YEARS)	MILLION YEARS BEFORE PRESENT
Cenozoic	Quaternary	Age of modern seed		63	
	Tertiary	plants			
					65
Mesozoic	Cretaceous			65	
					130
	Jurassic		Grasses Flowering plants	45	
					175
	Triassic	Age of ancient seed plants		45	
					220
Paleozoic	Permian		Ginkgoes	50	
			Conifers		
					270
	Carboniferous			80	
			Seed ferns		
					350
	Devonian	Age of spore-bearing plants	Lycopods Horsetails Bryophytes	50	
					400
	Silurian			40	
					440
	Ordovician	Age of bacteria and algae		60	
					500
	Cambrian		Marine algae	100	
					600
	Precambrian			1,700?	2,300?

develop from branching rhizomes. The leaves may be whorled or in a tight spiral and are rarely more than 1 centimeter (0.4 inch) long. Adventitious roots, whose epidermal cells often produce root hairs, develop along the rhizomes.

Reproduction

At maturity, some species of ground pines produce kidney-bean-shaped sporangia on short stalks in the axils of specialized leaves. Such sporangium-bearing leaves are called **sporophylls.** In other species, the sporophylls have no chlorophyll, are smaller than the other leaves, and are in terminal conelike clusters called **strobili** (singular: **strobilus**). In the sporangia, *sporocytes* undergo meiosis, producing spores that are released and carried away by air currents. The spores of some species germinate in a few days if they land in a suitable location, but spores of other species may not germinate for up to several years.

After germination, independent gametophytes develop from the spores. The gametophytes vary in shape, some

A.

B.

Figure 21.3 A. *Lycopodium cernum,* a club moss native to the tropics. *B. Lycopodium obscurum,* native to northern and eastern North America.

resembling tiny carrots. The gametophyte body usually develops in the ground in association with mycorrhizal fungi. Some gametophytes, however, develop primarily on the surface where the exposed parts turn green. All types produce both antheridia and archegonia on the same gametophyte that, in some species, may live for several years. Since the sperm are flagellated, water is essential for fertilization to occur.

Zygotes first become embryos with a foot, stem, and leaves and then develop into mature sporophytes (Fig. 21.4). If the gametophyte is underground, chlorophyll does not develop in the young sporophyte until it emerges into the light. Several sporophytes may be produced from a single gametophyte. A number of ground pines also reproduce asexually by means of small bulbils (bulbs produced in the axils of leaves), each of which is capable of developing into a new sporophyte.

Selaginella—Spike Mosses

Structure and Form

The sporophytes of *Selaginella,* the larger of the two major genera of living club mosses, are sometimes called *spike mosses* (Fig. 21.5). The approximately 700 species are widely scattered around the world in wetter areas, but they are especially abundant in the tropics. A few are common weeds in greenhouses. They tend to branch more freely than ground pines, from which they differ in several respects. The two most obvious differences are (1) their leaves each have a tiny extra appendage, or tongue, called a **ligule,** on the upper surface near the base and (2) they produce two different kinds of spores and gametophytes—an advanced feature referred to as **heterospory.** The seed-bearing coniferous and flowering plants discussed in chapters to follow are all heterosporous.

Reproduction

Sexual reproduction of spike mosses and ground pines both involve the production of sporangia, but spike moss sporangia develop on either *microsporophylls* or *megasporophylls* (Fig. 21.6). **Microsporophylls** bear *microsporangia* containing numerous **microsporocytes** that undergo meiosis, producing tiny **microspores.** The *megasporangia* of **megasporophylls** usually contain a **megasporocyte** that, after meiosis, becomes four comparatively large **megaspores.**

Each microspore may become a male gametophyte consisting simply of a somewhat spherical antheridium surrounded by a sterile jacket of cells within the microspore wall. Either 128 or 256 sperm cells with flagella are produced in each antheridium. A megaspore develops into a female gametophyte that is also relatively simple in structure. By the time this gametophyte is mature, however, it consists of many cells that have been produced inside the megaspore. As it increases in size, it eventually ruptures its thickened spore wall and produces

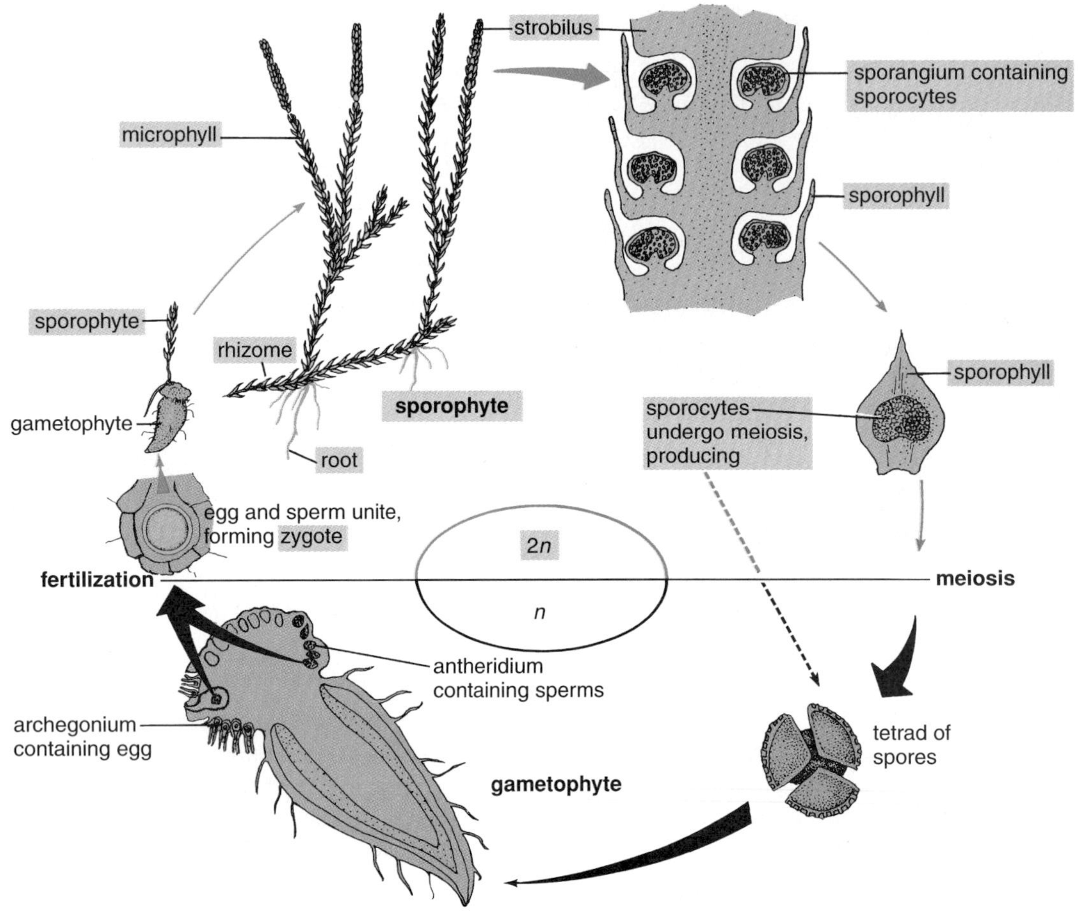

Figure 21.4 Life cycle of the ground pine *Lycopodium,* a homosporous lycopod.

Figure 21.5 *Selaginella,* a spike moss.

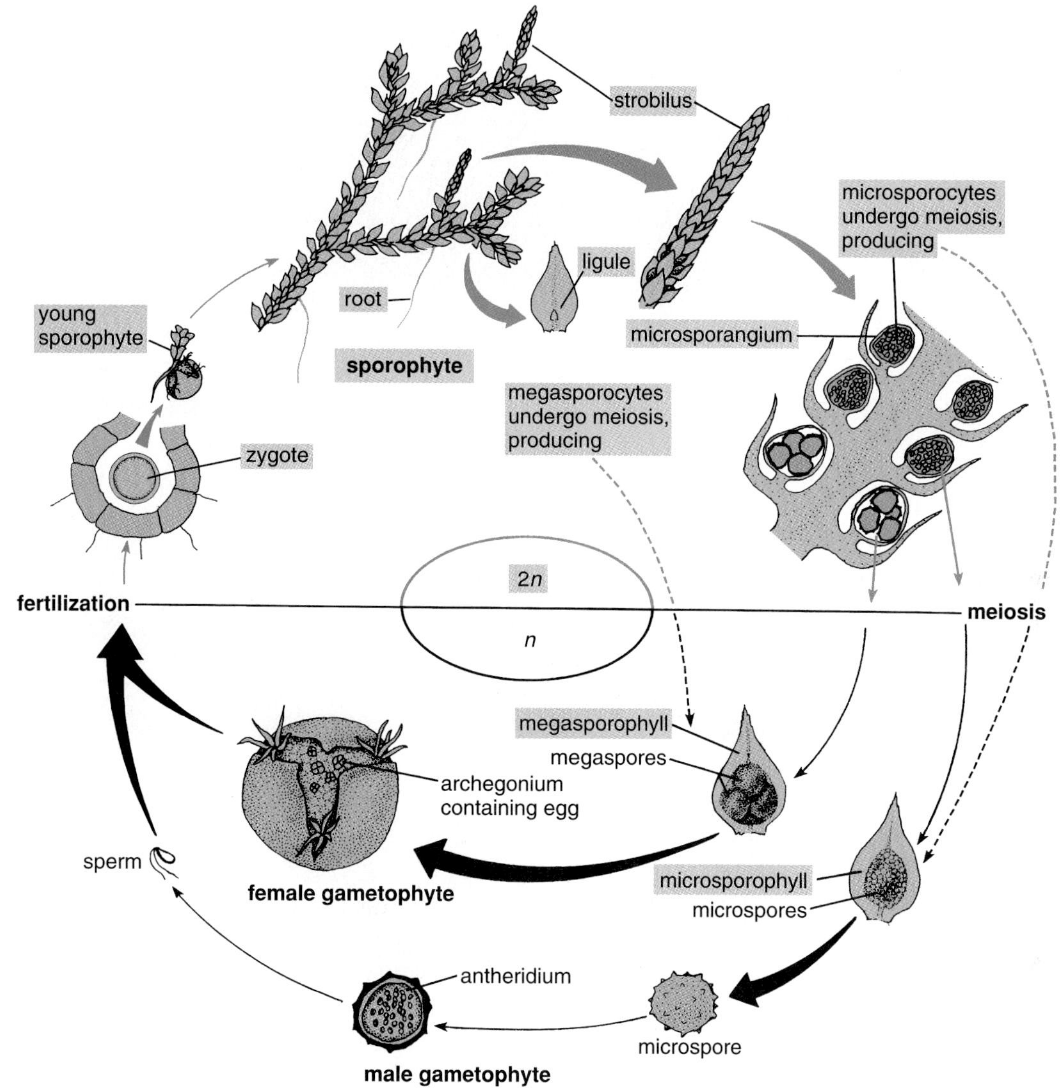

Figure 21.6 Life cycle of the spike moss *Selaginella,* a heterosporous lycopod.

several archegonia in the exposed seams. The development of both male and female gametophytes often begins before the spores are released from their sporangia. Fertilization and development of new sporophytes are similar to those of ground pines.

Isoetes—Quillworts

Structure and Form

There are about 60 species of *quillworts* (all in the genus *Isoetes*) (Fig. 21.7). Most are found in areas where they are at least partially submerged in water for part of the year. Their leaves (microphylls) are slightly spoon shaped at the base and look like green porcupine quills, although they are not stiff and rigid. They are arranged in a tight spiral on a stubby stem, which resembles the corm of a gladiola or a crocus. *Ligules* occur toward the leaf bases. The corms have a vascular cambium and may live for many years. Wading birds and muskrats often eat the corms. The plants are generally less than 10 centimeters (4 inches) tall, but the leaves of one species become 0.6 meter (2 feet) long.

Reproduction

Reproduction is similar to that of spike mosses, except that no strobili are formed. Both types of sporangia are produced at the bases of the leaves (Fig. 21.8). Up to 1 million microspores may occur in a single microsporangium.

Figure 21.7 Quillwort (*Isoetes*) sporophytes. *(Courtesy Robert A. Schlising)*

Ancient Relatives of Club Mosses and Quillworts

Some of the ancient and extinct relatives of the club mosses were large and treelike and grew up to 30 meters (100 feet) tall; their trunks were up to 1 meter (3 feet 3 inches) in diameter. They were dominant members of the forests and swamps of the Carboniferous period that reached its peak some 325 million years ago (see Table 21.1). Quillworts first appeared in the Cretaceous period, about 130 million years ago. Their fossils reveal that even the oldest were remarkably similar to their present-day relatives (Fig. 21.9). A knowledge of the life histories and ancestry of various members of this phylum enhances our understanding of the development and diversity of the Plant Kingdom as a whole.

Human and Ecological Relevance of Club Mosses and Quillworts

Like the whisk ferns, club mosses and quillworts are of little economic importance today. As mentioned earlier, large numbers of club moss spores produce a flash of light when ignited. This characteristic was exploited at one time in the manufacture of theatrical explosives and photographic flashlight powders. In the past, druggists mixed spore powder with pills and tablets to prevent them from sticking to one another. The spore powder itself has been used for centuries in folk medicine, particularly for the treatment of urinary disorders and stomach upsets. Some Native Americans used it as a talcum powder for babies, snuffed it to arrest nosebleeds, and applied it following childbirth to staunch hemorrhaging. It was also sometimes used to stop bleeding from wounds.

Club moss extracts have been used in several countries in the past to reduce fevers, but partly because of undesirable side effects, medicinal use of club mosses has now largely been abandoned. Native Americans of Washington, Oregon, and British Columbia became mildly intoxicated after chewing parts of one local species of club moss. It is reported that they became unconscious if they chewed too much.

Novelty stores sometimes sell a species of spike moss native to Mexico and to the southwestern United States. Known as the *resurrection plant,* it shrivels and rolls up in a ball when dry, appearing to be completely dead, but quickly unfolds and turns green when sprinkled with water. Other spike mosses have been placed on shelves indoors without water for nearly 3 years and then have resumed growth when given water. Ground pines and spike mosses have been used ornamentally indoors and outdoors as ground covers. Some ground pines are spray painted and used as Christmas ornaments or in floral wreaths. Several species of *Lycopodium* have been

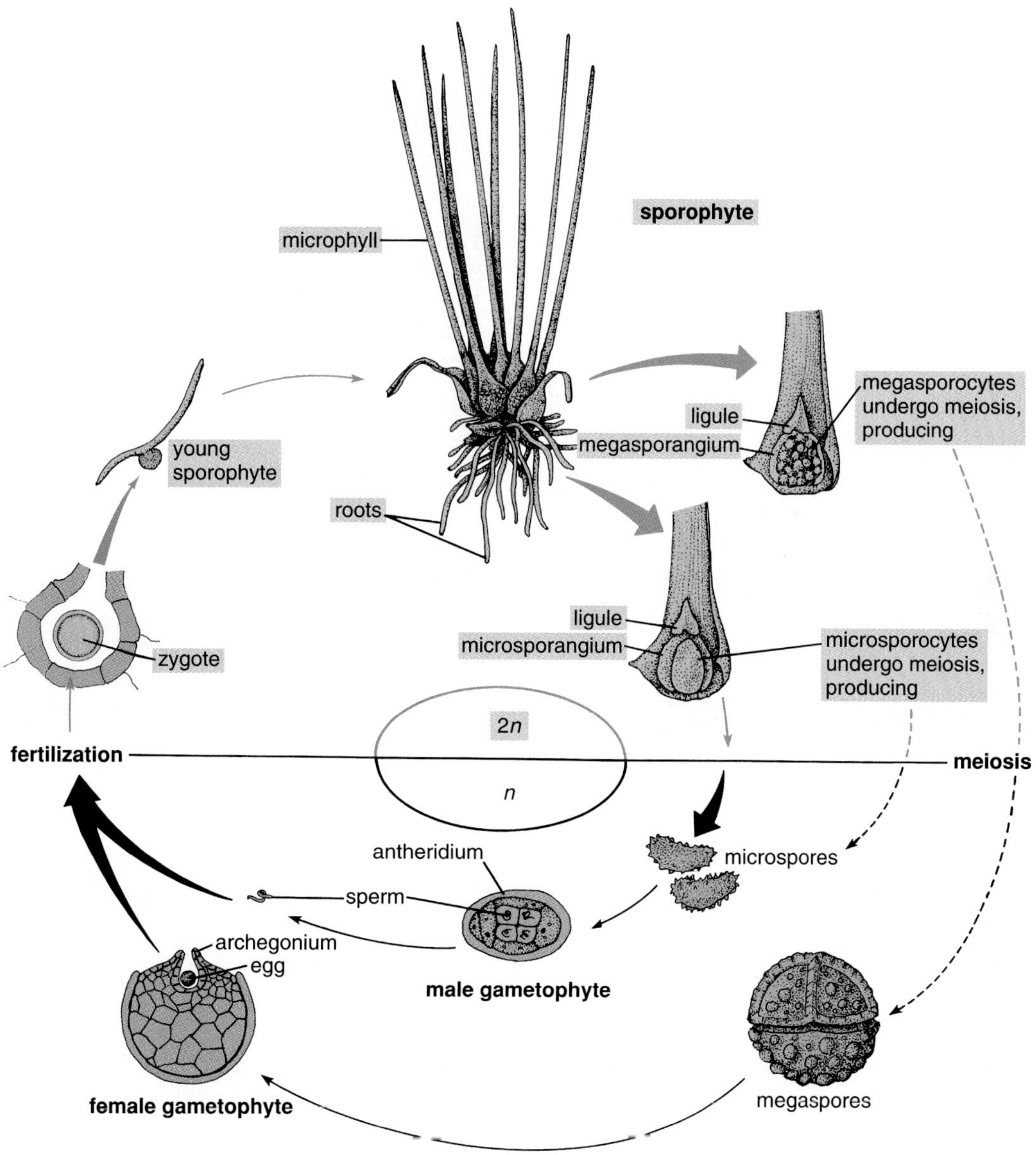

Figure 21.8 Life cycle of a quillwort (*Isoetes*), a heterosporous member of the phylum.

exploited to the extent that they are now on rare and endangered or threatened species lists; they should no longer be collected.

Quillwort corms have been eaten by domestic and wild animals, waterfowl, and humans.

PHYLUM EQUISETOPHYTA—THE HORSETAILS AND SCOURING RUSHES

Many backpackers and campers have become aware of a unique use of a relatively common and widespread genus of plants (*Equisetum*) known as *horsetails* or *scouring rushes.* Significant deposits of silica accumulate on the inner walls of the epidermal cells of the stems, which make excellent scouring material for dirty metal pots and pans. Native Americans were aware of this scouring property of the plants and used them extensively.

Structure and Form

About 25 species of horsetails (the name usually applied to branching forms that look a little like a horse's tail) and scouring rushes (unbranched forms) (Fig. 21.10) are scattered throughout all continents, including Australia, where they are weeds. They usually grow less than 1.3 meters (4 feet) tall, but some in the tropics and coastal redwood forests of California exceed 4.6 meters (15 feet) in height. Where branches occur,

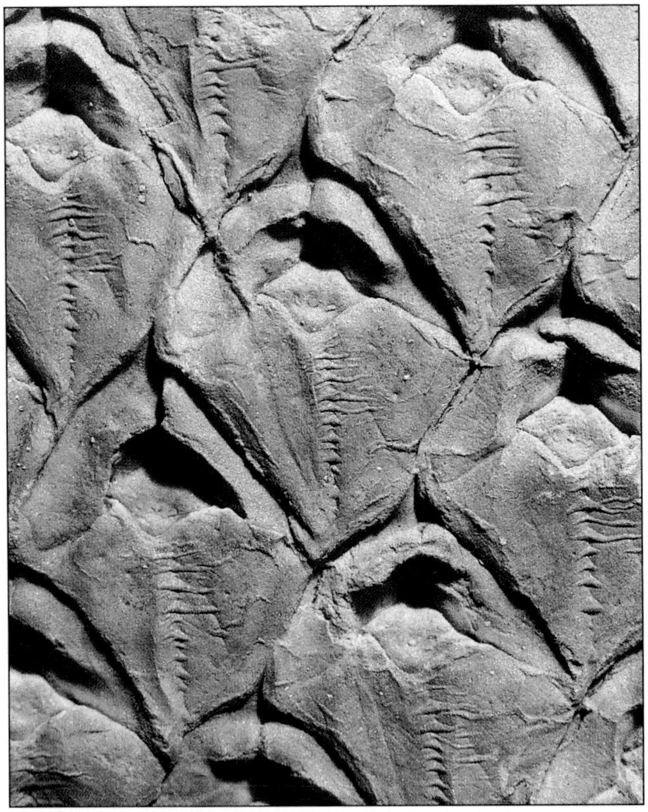

Figure 21.9 A small portion of the surface of the fossil lycopod, *Lepidodendron,* showing leaf (microphyll) bases similar to those seen in modern lycopods. *(Specimen courtesy University of Illinois Paleobotany Laboratory)*

A.

B.

Figure 21.10 Horsetails (*Equisetum*). *A*. An unbranched species. *B*. A branched species.

they are normally in whorls at regular intervals along the jointed stems. Both branched and unbranched species have tiny scalelike leaves (microphylls) in whorls at the nodes. These leaves are fused together at their bases, forming a collar. They are green when they first appear, but they soon wither and bleach, and virtually all photosynthesis occurs in the stems.

The stems are distinctly ribbed and have obvious nodes and internodes; there are numerous stomata in the grooves between the ribs. A cross section of a stem (Fig. 21.11) reveals that the pith breaks down at maturity, leaving a hollow central canal. There are two cylinders of smaller canals outside of the pith. The inner cylinder consists of water-conducting *carinal canals* that are aligned opposite the ribs of the stem. Each canal has a patch of xylem and phloem to the outside. The canals of the outer cylinder, called *vallecular canals,* contain air. They are larger than the carinal canals and are aligned opposite the "valleys" between the ribs.

The aerial stems develop from horizontal rhizomes, which also have regular nodes, internodes, and ribs. In some species, the rhizomes have adventitious roots and may form extensive branching systems as much as 2 meters (6.5 feet)

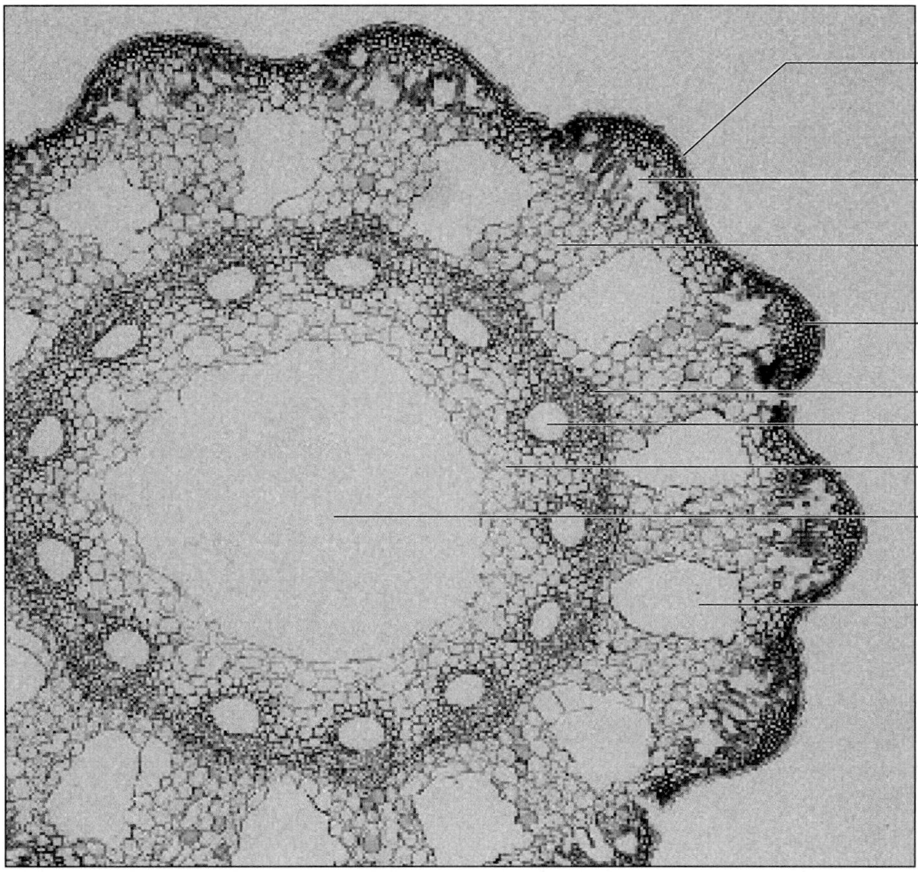

Figure 21.11 A horsetail (*Equisetum*) stem in cross section. *(Photomicrograph by G. S. Ellmore)*

below the surface. Both internal stem structure and external features help distinguish the various species of horsetails from one another.

Reproduction

If stems or rhizomes are broken up by a disturbance, such as a storm or foraging animals, the fragments can grow into new sporophyte plants. In most species, however, reproduction usually involves a sexual process. In the spring, some species produce special cream- to buff-colored non-photosynthetic stems from rhizomes (Fig. 21.12). Small conelike **strobili** develop at the tips of these special stems or, in other species, at the tips of regular photosynthetic stems.

The strobili are usually about 2 to 4 centimeters (0.75 to 1.5 inches) long. Hexagonal, dovetailing plates at the surface of the strobilus give it the appearance of an ellipsoidal honeycomb. Each hexagon marks the top of a *sporangiophore* that has 5 to 10 elongate sporangia connected to the rim. The stalks of the sporangiophores are attached to the central axis of the strobilus. The sporangia surround the sporangiophore stalks and point inward. These hidden sporangia are not visible until maturity when the sporangiophores separate slightly. The spores are then released.

When the sporocytes in the sporangia undergo meiosis, distinctive-appearing green spores are produced. At one pole, the spores have four ribbonlike appendages that are slightly expanded at the tips (Fig. 21.13). The appendages are called **elaters** (not structurally or otherwise related to the elaters of liverworts and hornworts). The elaters are very sensitive to changes in humidity and aid spores in their dispersal. While spores are being carried by an air current, the elaters are more or less extended like wings. If a spore enters a humid air pocket above a damp area below, the elaters coil, causing the spore to drop in an area that is more likely to support germination and growth.

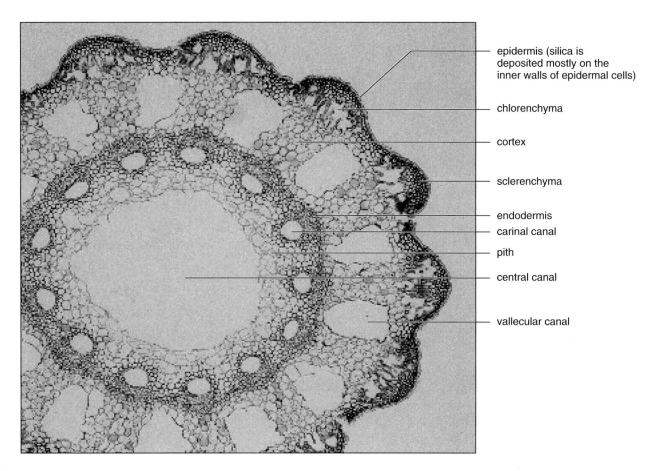

Germination of spores usually occurs within a week of their release. Lobed, cushionlike, green gametophytes develop and seldom grow to more than 8 millimeters (0.36 inch) in diameter. Rhizoids anchor them to the surface. At first, about half of the gametophytes are male with antheridia, and the other half are female with archegonia. After a month or two, however, the female gametophytes of most species become bisexual, producing only antheridia from then on. When water contacts mature antheridia, sudden changes in

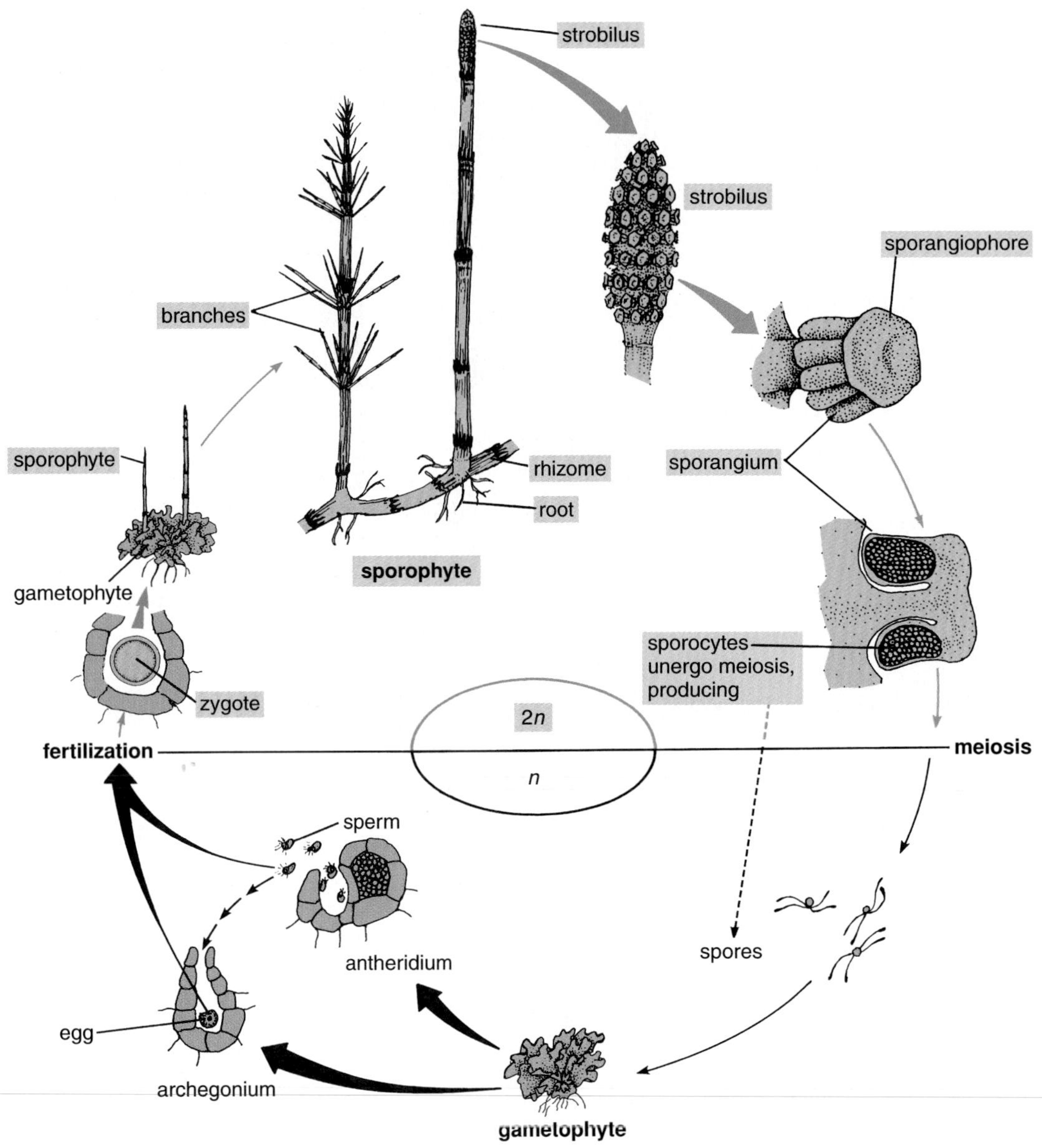

Figure 21.12 Life cycle of a horsetail (*Equisetum*) that has separate vegetative and reproductive shoots.

water pressure cause the sperms produced within to be explosively ejected. Several eggs on a female or bisexual gametophyte may be fertilized, and the development of more than one sporophyte is common.

Ancient Relatives of Horsetails

Horsetails and scouring rushes belong to the single remaining genus (*Equisetum*) of a phylum of several different orders that flourished in the Carboniferous period about 300 million years ago. At that time, some members of this phylum, like the ancient club mosses, were treelike (Fig. 21.14) and grew over 15 meters (50 feet) tall, while others were vinelike. Some were similar to present-day horsetails in having leaves in whorls (Fig. 21.15), jointed stems with internal canals, and sporangiophores with sporangia hanging down from the rims.

Human and Ecological Relevance of Horsetails and Scouring Rushes

The 7th-century Romans ate the boiled young strobili of field horsetails or fried them after mixing them with flour. Native Americans peeled off the tough epidermis of young stems and ate the inner parts either raw or boiled, while others cooked the rhizomes of the giant horsetail for food. Cows, goats, muskrats, bears, and geese also have been known to eat horsetails.

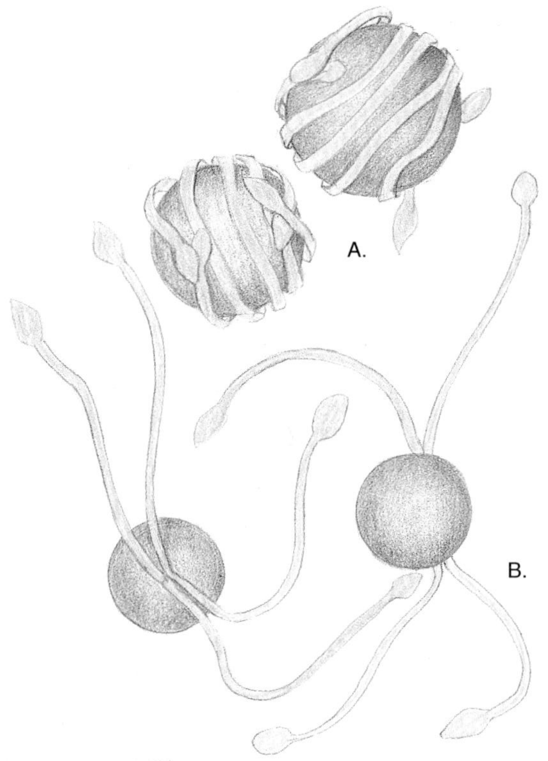

Figure 21.13 Horsetail spores. *A*. With elaters coiled. *B*. With elaters spread.

Hopi Indians ground dried horsetail stems to flour, which they mixed with cornmeal to make bread or mush. Horsetails have, however, occasionally been reported to be poisonous to horses, and they are not recommended for human consumption.

Native Americans and Asians had several other uses for these plants. One tribe drank the water from the carinal canals of the stem, and another thought the shoots were "good for the blood." It is not known if there was any basis for this latter idea, but there is an old unconfirmed report that field horsetail consumption "produces a decided increase in blood corpuscles." At least one or two species are known to have a mild diuretic effect (a *diuretic* is a substance that increases the flow of urine), and they have been used in the past in folk medicinal treatment of urinary and bladder disorders. Some have also been used as an antacid or an astringent (an *astringent* is a substance that arrests discharges, particularly of blood). One species was used in the treatment of gonorrhea, and others were used for tuberculosis.

At least two Native American tribes burned the stems and used the ashes to alleviate sore mouths or applied the ashes to severe burns. Members of another tribe ate the strobili of a widespread scouring rush to cure diarrhea, and still others boiled stems in water to make a hair wash for the control of lice, fleas, and mites.

During pioneer times when the covered wagons moved westward, the use of scouring rush stems for scouring and sharpening was widespread. They were used not only for

Figure 21.14 Reconstruction of a coal age (Carboniferous) forest. *(Photo by Field Museum of Natural History, Chicago)*

Figure 21.15 Reconstruction of the fossil giant horsetail, *Calamites.*

cleaning pots and pans but also for polishing brass, hardwood furniture and flooring, and for honing mussel shells to a fine edge. Scouring rushes are still in limited use for these purposes today. Some species of horsetails accumulate certain minerals in addition to silica. Veins of such minerals have been located beneath populations of horsetails by analyzing the plants' mineral contents. This process of analysis involves a chemical treatment of the tissues followed by the use of X-ray equipment.

In the geological past, the giant horsetails and club mosses were a significant part of the vegetation growing in vast swampy areas. In some instances, the swamps were stagnant and slowly sinking, permitting the gradual accumulation of plant remains, which, because of the lack of oxygen in the water, were not readily attacked by decay bacteria. Such circumstances, over aeons of time, were ideal for the formation of coal. Today, if you section a lump of coal thinly enough and examine it under the microscope, you can still see bits of tissue and spores of plants that were living hundreds of millions of years ago. One soft coal known as *cannel* consists primarily of the spores of giant horsetails and club mosses that, through the ages, were reduced to carbon.

For the scientific names of species discussed, see Appendix 1.

PHYLUM POLYPODIOPHYTA—THE FERNS

When I was a small child, my family had a large potted fern on each side of a fireplace. I became quite attached to the plants, and believing I was removing a "disease," I carefully scraped off the little brownish patches that appeared from time to time on the lower surfaces of the leaves. It wasn't until I got to college I learned that instead of controlling a disease, I had inadvertently been frustrating the sex life of my favorite plants!

If we could take a worldwide opinion poll about ornamental plants, ferns undoubtedly would rank high in popularity. In fact, in some parts of the world, it is difficult to find a household without at least one fern either inside or out in the garden. Their leaves are so infinitely varied in form and aesthetically pleasing that Thoreau was once moved to state, "God made ferns to show what He could do with leaves."

Structure and Form

The approximately 11,000 known species of ferns vary in size from tiny floating forms less than 1 centimeter (0.4 inch) in diameter to giant tropical tree ferns up to 25 meters (82 feet) tall. Fern leaves are *megaphylls* (leaves associated with leaf gaps and having branching veins—see Chapter 6) that are commonly referred to as **fronds.** They are typically divided into smaller segments and feathery in appearance (Fig. 21.16), but some are undivided, pleated, or tonguelike, and others resemble a four-leaf clover or grow in such a way as to form "nests." In the tropics, the "nest" ferns often accumulate enough humus to provide food and shelter for huge earthworms that are up to 0.6 meter (2 feet) long. Since ferns require external water for their sexual reproduction, they are most abundant in wetter tropical and temperate habitats, but a few are adapted to drier areas.

Reproduction

The sporophyte is the conspicuous phase of the life cycle in all the plants discussed in this chapter (Fig. 21.17). A fern sporophyte consists of the fronds, a stem in the form of a rhizome, and adventitious roots that arise along the rhizome. The fronds, regardless of their ultimate form, usually first appear tightly coiled at their tips. These *croziers* (unrelated to the croziers of sac fungi), or "fiddleheads" (Fig. 21.18), then unroll and expand, revealing the blades. At maturity, the blades are often divided into segments called **pinnae** (singular: **pinna**) that are attached to a midrib, or rachis. A stalk, or petiole, is usually present at the base.

When the fronds have expanded, small, often circular, rust-colored patches of powdery-looking material may appear on the lower surfaces of some or all of the blades. Because these patches appear similar to fungal rusts, some fern owners have thought their plants were diseased and have turned to sprays to deal with the "problem" or have even carefully scraped the patches off. The development of the patches is normal and healthy, however, and examination with a hand lens or dissecting microscope will reveal

Figure 21.16 Common ferns with typically dissected fronds. *A*. Cinnamon fern. The cinnamon-colored fertile fronds in the center are nonphotosynthetic and produce large numbers of sporangia. *B*. Holly fern. *C*. A maidenhair fern. *D*. 'Ama'uma'u, a small Hawaiian tree fern. (A, B *Courtesy Perry J. Reynolds*)

that they are actually clusters of *sporangia* (Fig. 21.19). The sporangia may be scattered evenly over the lower surfaces of the fronds, but they are often confined to the margins. The sporangia are mostly found in numerous discrete clusters called **sori** (singular: **sorus**). In many ferns, the sori, while they are developing, are protected by thin individual flaps of colorless tissue called **indusia** (singular: **indusium**). As the sporangia mature, the indusium, which often resembles a tiny semi-transparent umbrella attached by its base to the frond surface, shrivels and exposes the sporangia beneath.

Most of the sporangia are microscopic and stalked and look something like tiny transparent baby rattles with a conspicuous row of heavy-walled brownish cells along the edge. This row of cells, that looks like a tiny millipede, is called an **annulus**. It functions in catapulting spores out of the sporangium with a distinct snapping action influenced by moisture changes in the cells (Fig. 21.20).

Sporocytes undergo meiosis in the sporangia, usually producing either 48 or 64 spores per sporangium. Sporangia of some of the primitive adder's tongue ferns may have up to 15,000 spores, however, and the number of

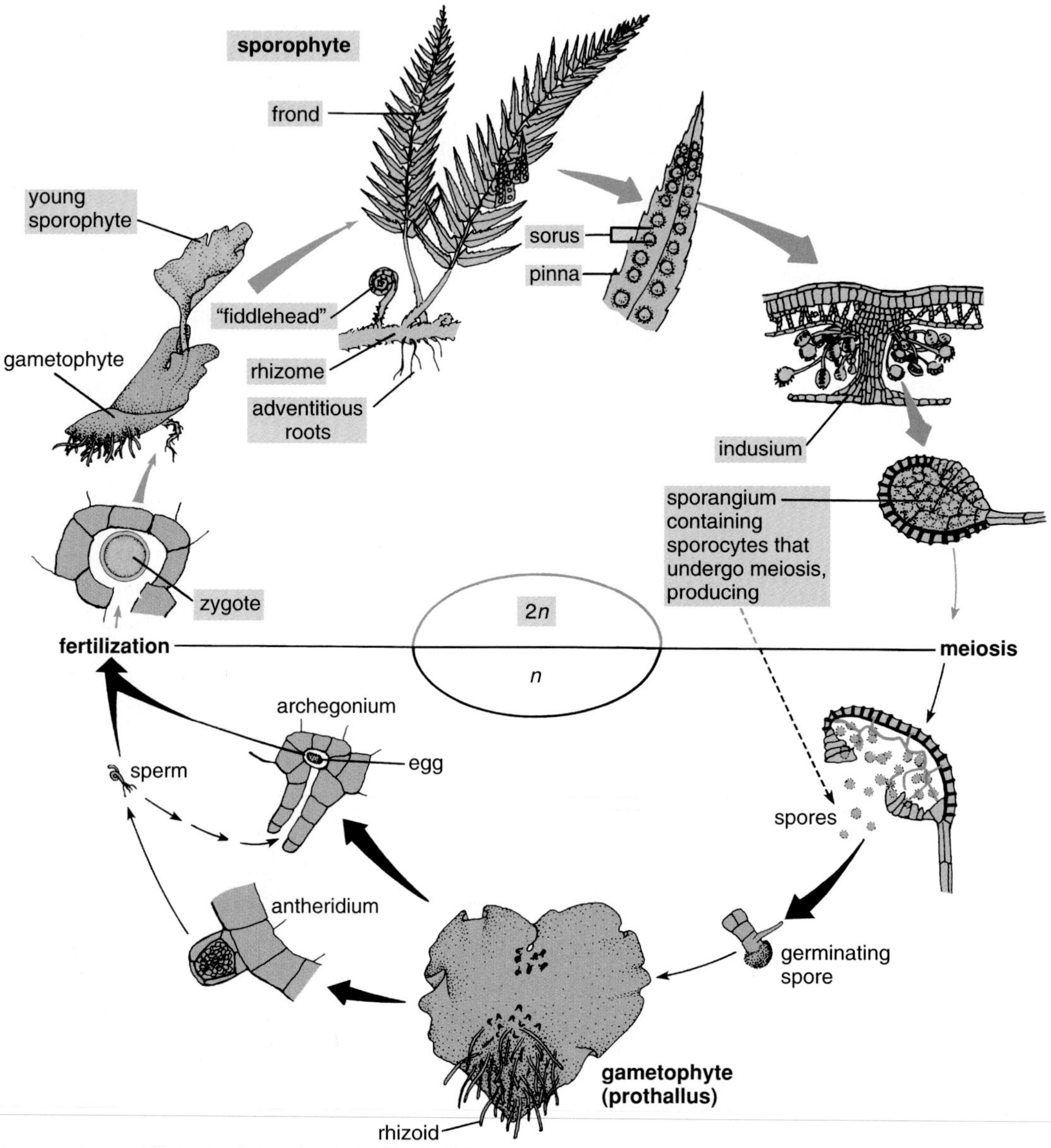

Figure 21.17 Life cycle of a fern.

Figure 21.18 A crozier (fiddlehead) of a tropical fern.

sporangia is often so great that it has been estimated that a single beech fern plant will produce a total of 50 million spores.

Two kinds of spores are produced in certain aquatic or amphibious ferns such as the clover fern (*Marsilea*) and *Salvinia*. Single large *megaspores* are produced in some sporangia, while numerous tiny *microspores* are produced in others. The vast majority of fern species, however, produce only one kind of spore.

After the spores have been flung out of their sporangia, they are dispersed by wind; relatively few end up in habitats suitable for their survival. Such habitats include shady, wet ledges and rock crevices or moist soil. The spores can also easily be germinated on damp clay flowerpots in the home or greenhouse. Those that germinate in favorable locations produce little "Irish valentines," or **prothalli** (sin-

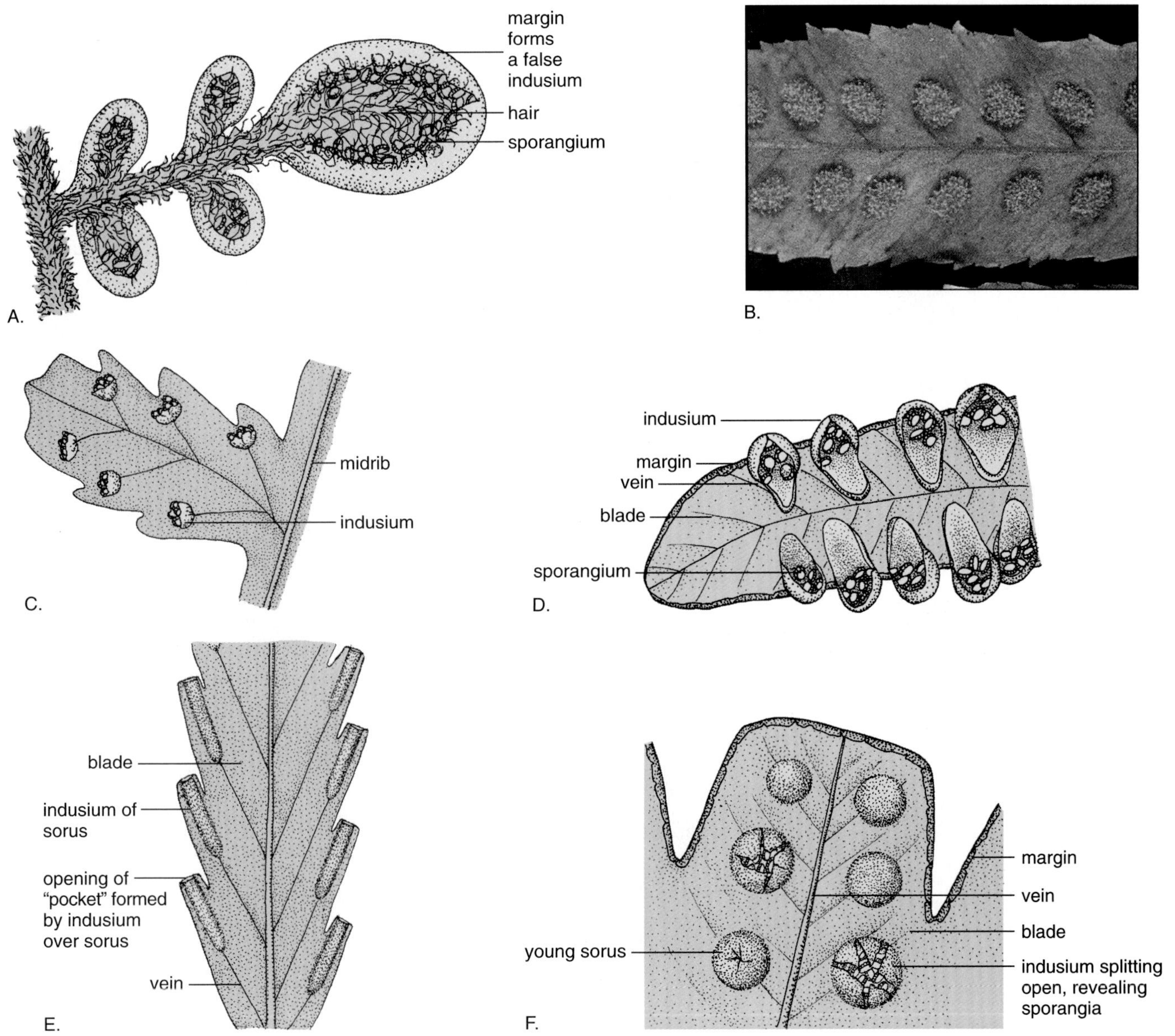

Figure 21.19 Pinnae of fern fronds, showing some types of arrangements of sporangia. *A. Cheilanthes. B. Polypodium. C. Cystopteris. D. Cibotium. E. Davallia. F. Cyathea.*

gular: **prothallus**), as the green heart-shaped gametophytes of these and other seedless vascular plants are more properly called (Fig. 21.21). These structures often curl slightly at their edges and may be 5 to 6 millimeters (0.25 inch) in diameter; they are visible without a microscope.

Prothalli are only one cell thick, except toward the middle where they are slightly thicker. Antheridia are interspersed among the rhizoids produced on the lower surface of the central area of most prothalli, with archegonia also being produced, usually closer to the notch of the heart-shaped gametophyte. The archegonia are somewhat flask shaped, with curving necks that protrude slightly above the surface, whereas the antheridia are more spherical and often elevated above the surface on short stalks. In a few species, antheridia and archegonia are produced on separate prothalli.

A single antheridium may produce from 32 to several hundred sperms, each with few to many flagella. Fertilization of an egg takes place within an archegonium and is chemically influenced so that it usually occurs with a sperm from a different prothallus. Only one zygote develops into a young sporophyte on any prothallus, regardless of the number of eggs that may be fertilized. This sporophyte usually has smaller, simpler fronds during its first season of growth, but typical full-sized fronds grow from the persisting rhizomes in succeeding years.

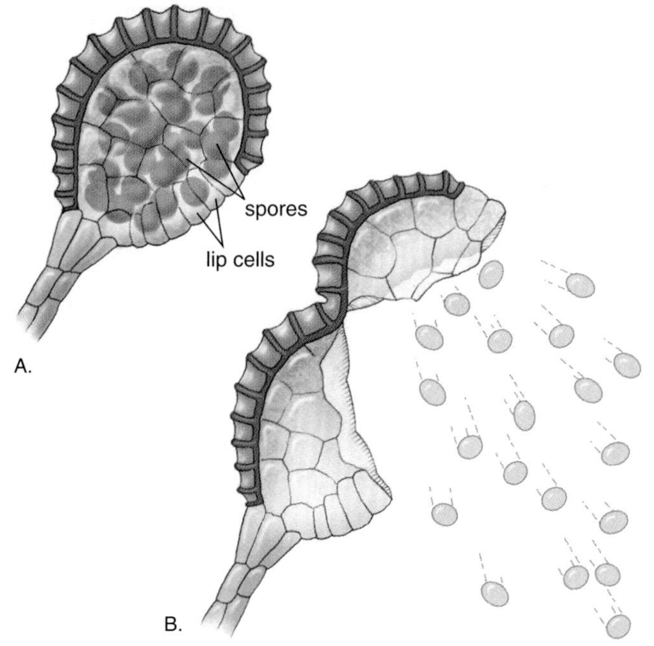

Figure 21.20 Release of spores from a fern sporangium. *A*. An intact sporangium. *B*. Spores being ejected as the sporangium splits; the annulus first draws back and then snaps forward.

Fossil Relatives of Ferns

Fossil remains of ferns and plants thought to be ancestors of ferns abound in ancient deposits. Possible ancestors of ferns (Fig. 21.22) are found in Devonian formations (see Table 21.1) estimated to be 375 million years old. Most resemble ferns in habit, but they have no broad fronds and otherwise are more like whisk ferns than true ferns. Well-preserved fossils of tree ferns related to large tropical ferns of the present day are found in geological strata dating back to between 250 and 320 million years ago (Fig. 21.23).

Ferns (especially tree ferns) became so abundant during the latter part of the Carboniferous period that this era in the past was referred to as the *Age of Ferns*. The discovery of seeds on some of them, however, raised questions about relationships, and the term *Age of Ferns* was dropped.

Human and Ecological Relevance of Ferns

In some parts of North America, it is unusual to find a house without at least a Boston fern growing within. Ferns make ideal house plants because many of them are adapted to

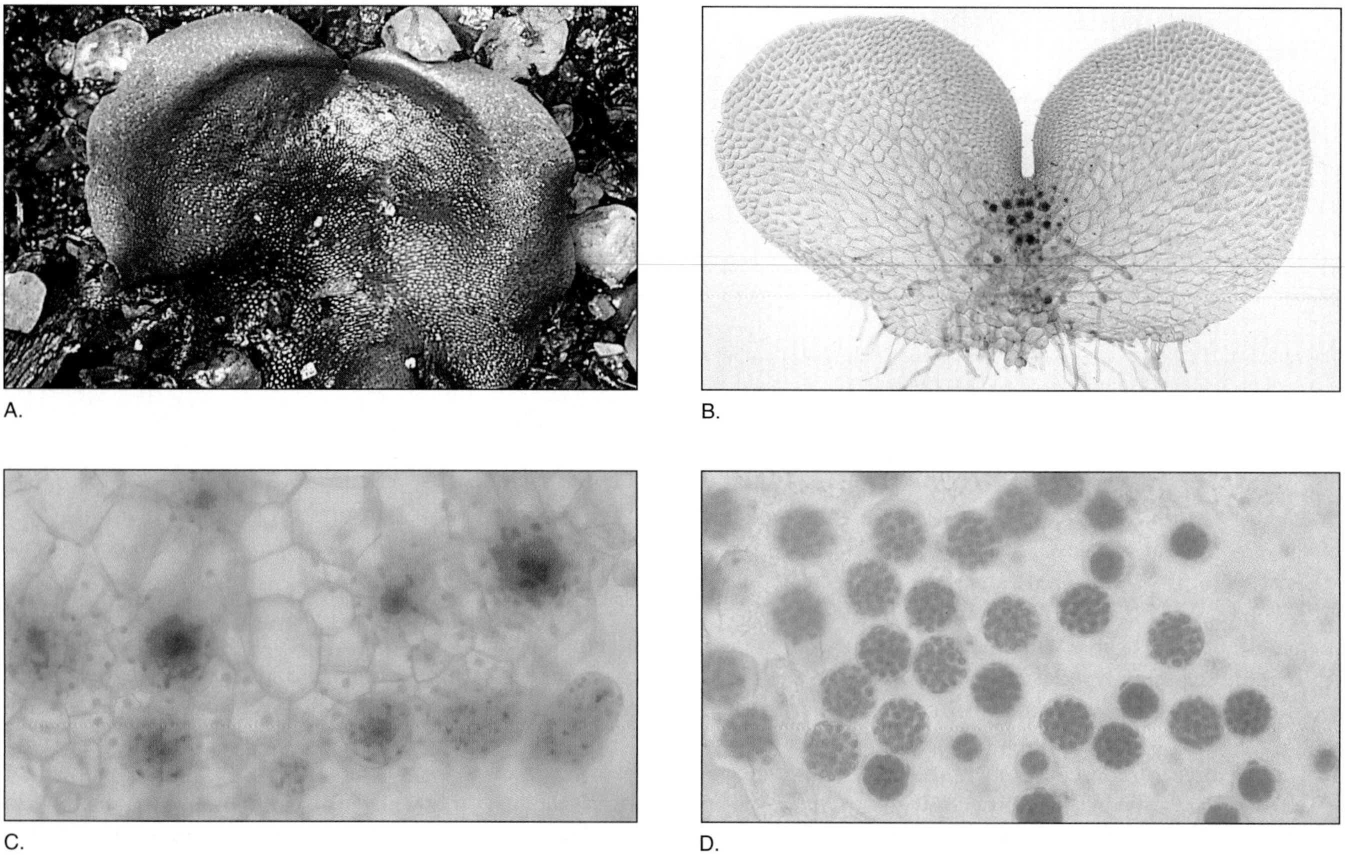

Figure 21.21 Fern prothalli. *A*. Surface view, ×10. *B*. A prothallus as seen with the aid of a microscope, ×20. *C*. Archegonia, ×100. *D*. Antheridia, ×100.

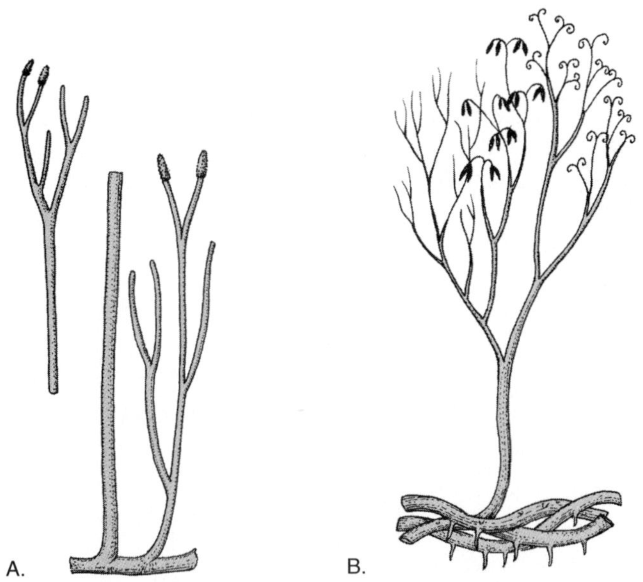

Figure 21.22 Possible ancestors of ancient ferns. /*A. Aglaophyton. B. Psilophyton* (?). The 19th-century origin of this drawing apparently was a composite of three distantly related genera.

growing in low light, and most are not as susceptible as other plants to aphids, mites, mealybugs, and similar pests. Outdoors, ferns are equally popular as ornamentals. Some eventually become subjects for an artist's brush or a natural history photographer's camera.

Apart from the pleasing aesthetic aspects of ferns, they function well as air filters. For example, during a single hour, one average-sized Boston fern can remove about 1,800 micrograms of formaldehyde (a common pollutant from carpets) from the air in a typical room measuring 3 meters by 3 meters (10 feet by 10 feet). According to the Environmental Protection Agency, if properly maintained, the fern can almost completely rid the air of pollutants on a daily basis.

Ferns also have other practical value. Commercial growers of the brilliantly colored anthuriums in Hawaii and elsewhere have found that native tree ferns provide the ideal amount of shade and other environmental conditions needed for bringing their flowers to perfection. Tree fern rhizome or root bark[1] and rhizome bark of certain other species, such as the royal fern (which produces osmunda bark), are a favorite medium of orchid, bromeliad, and staghorn fern growers. Its texture is well suited to the growth of the orchids' aerial roots, and as the bark slowly decomposes, rain water trickling over its surface picks up nutrients that are particularly appropriate for these plants. The demand for fern bark for orchids has exceeded the supply for a number of years, and it has become very expensive on the market (Fig. 21.24).

Figure 21.23 A fossil fern.

Figure 21.24 An orchid plant growing on bark.

As the young "fiddleheads" of many species of ferns unroll, a dense covering of hairs is visible on the petiole and rachis. In the past, the silky hairs of some of the larger tropical tree ferns (Fig. 21.25) were stripped and used for upholstery, pillow, and mattress stuffing. During the late 1800s, over 1,900 metric tons (2,094 tons) of this material were shipped from Hawaii to the mainland, and if it were not for

1. Ferns do not have a cork cambium and do not, therefore, produce bark that is typical of woody seed plants. The cells of the outer layers of fern roots and rhizomes, however, do become impregnated with suberin and consequently become bark-like in appearance and function.

A. B.

Figure 21.25 *A*. A Brazilian tree fern. *B*. The growing tip of a tree fern, showing the protective rust-colored hairs.

the eventual substitution of alternative materials, these magnificent plants might have been totally destroyed. Some tropical hummingbirds use these hairs along with scales of other ferns to line their nests, and at one time, Polynesians used them in a form of embalming for their dead. The trunks of tree ferns have been used in the construction of small houses in the tropics. Parts of one Hawaiian species of small tree fern and the fronds of an Asian fern yield red pigments used for dyeing cloth.

The bracken fern, which is distributed worldwide and is a weed in parts of Europe, has been used and even cultivated for human food for many years, particularly in Japan and New Zealand. It has, however, also long been known to be mildly poisonous to livestock. Research in both Europe and Japan has shown conclusively that bracken fronds fed to experimental animals produce intestinal tumors, and because of this, the consumption of these fronds for food is now actively discouraged.

Indigenous peoples of many areas where ferns occur have eaten the cooked rhizomes and young fronds of various ferns. Native Americans often baked the rhizomes of sword ferns, lady ferns, and others in stone-lined pits, removed the outer layers, and ate the starchy inner material. Similarly, native Hawaiians ate the starchy core of their tree ferns as emergency food. In Asia, the oriental water fern is still sometimes grown for food in rice paddies and used as a raw or cooked vegetable. In Malaysia, a relative of the lady fern is frequently used as a vegetable.

Uses of ferns in folk medicine abound. They have been used in the treatment of diarrhea, dysentery, rickets, diabetes, fevers, eye diseases, burns, wounds, eczema and other skin problems, leprosy, coughs, stings and insect bites, as a poison antidote, for labor pains, constipation, dandruff, and a host of other maladies. The male fern, which is more common in Europe than in North America, contains a drug that is effective in expelling intestinal worms (e.g., tapeworms). Its use for this purpose dates back to ancient times, and it is still occasionally used, although synthetic medicines have now largely replaced it. The licorice fern was used in the past by Native Americans of the Pacific Northwest for the treatment of sore throats and coughs and was also used as a flavoring agent and sugar substitute. Members of a tribe in

California chewed stalks of goldback ferns to quell toothaches and snuffed a liquid made from the fronds of the bird's-foot fern to arrest nosebleeds.

The fronds of bracken and other ferns have been used in the past for thatching houses. Anyone who has placed such fronds in compost piles knows they break down much more slowly than the leaves of other plants. Bracken fronds are still occasionally used as an overnight bedding base by fishermen and hunters. A substance extracted from these fronds has been used in the preparation of chamois leather, and the rhizome is used in northern Europe in the brewing of ale.

The chain fern has large fronds up to 2 meters (6.5 feet) long that have two flexible leathery strands in the petiole and rachis of each frond. Native Americans and others have gathered the fronds for many years to strip these strands for use in basketry and weaving. They do so by gently cracking the long axis with stones to expose the strands, which are then easily removed. The glossy black petioles of the five-finger fern have also been used in intricate basketry patterns by Native Americans. In Southeast Asia, the climbing fern has fronds with a rachis that may grow up to 12 meters (40 feet) long; it is still a favorite material (when available) for the weaving of baskets.

The mosquito fern, which is a floating water fern in the genus *Azolla*, forms tiny plants little bigger than duckweeds. It is found over wide areas where the climate is relatively mild. It sometimes forms such dense floating mats that it is believed to suffocate mosquito larvae that periodically need to reach the surface for air. This same fern frequently has cyanobacteria living symbiotically in cavities between cells. The cyanobacteria fix nitrogen, and it has been shown experimentally that plants without these organisms do not grow as well as those with them. In some parts of the world, mosquito ferns are thrown into rice paddies so that the cyanobacteria associated with them will fix nitrogen and reduce the need for added fertilizers.

Many ferns are now commercially propagated through tissue culture (discussed in Chapter 14). The scientific names of ferns mentioned in this chapter are given in Appendix 1.

Ecological Review

The ferns and their relatives were the first land plants with significant internal tissues for conducting water and roots for absorbing water and nutrients, morphological innovations that increased the environments in which these ancient plants could live. In the ancient geologic period known as the Carboniferous, giant horsetails and club mosses grew in great abundance in swampy areas. The biomass produced by these plants accumulated in huge deposits. The lack of oxygen inhibited bacteria-mediated decay of the plant biomass, which was eventually transformed into coal. These coal deposits represent large quantities of carbon dioxide removed from the ancient atmosphere by ferns and their relatives. The reintroduction of this carbon dioxide into the atmosphere in modern times, as we burn coal, may produce rapid climate change.

FOSSILS

Introduction

Several references to fossils have been made in this and preceding chapters, and other references occur in the chapters that follow. We have become increasingly aware of worldwide limitations of energy sources, and we now frequently hear references to fossil fuels as nonrenewable resources. Exactly what is a fossil, and how did it come into existence?

Originally, the word *fossil* was applied to anything unusual found in rocks, but its use is now more or less confined to any recognizable prehistoric organic object (or its impression) that has been preserved from past geological ages in the earth's crust (see Table 21.1). Such a definition includes teeth marks, borings, impressions of footprints and tracks, dung deposits, and deposits of chemicals that are evidence of the activities of algae and bacteria. Age itself does not make a fossil. Human remains thousands of years old that have been found in caves and tombs are not regarded as fossils, but it is conceivable that other remains of the same age could be preserved in rocks and regarded as fossils.

Fossils are formed in a number of different ways. The conditions for their formation almost always include the accumulation of sediments in an area where plants, animals, or other organisms are present or to which they have been transported. Such areas are found in swamps, oceans, lakes, or other bodies of water. Hard parts, such as wood or bones, are more likely to be preserved than soft parts. Other environmental factors, such as the presence of salt or antiseptic chemicals or stagnation resulting in low oxygen content of the water, all favor fossil formation.

Quick burial is another factor enhancing the chances of organic material becoming fossilized. Such burial might result if an organism was unable to avoid a rain of ash from a volcanic eruption or was trapped in a cave or sudden sandstorm. Other forms of quick burial occur when an organism falls into quicksand, a body of water, or asphalt. Descriptions of several common types of fossils follow.

Molds, Casts, Compressions, and Imprints

After silt or other sediment has buried an object and hardened into rock, the organic material may be slowly washed away by water percolating through the pores of the rock.

Figure 21.26 A compression fossil.

This leaves a space, which may then be filled in with silica deposits. If only a space is left, it is called a *mold;* if it is filled in with silica, it is called a *cast.* Artificial casts of plaster or wax compounds may be made from molds of the original objects. When objects such as leaves are buried by layers of sediment, the sheer weight of the overlying material may compress them to as little as 5% of their original thickness. When this occurs, all that is left is a thin film of organic material and an outline showing some surface details. Virtually no preservation of cells or other internal structure takes place. Such fossils are very common and are called *compressions* (Fig. 21.26). The image of a compression, like the details of a foot or a hand pressed into wet cement, is called an *imprint.* Coal is a special type of compression, often involving different plant parts that are thought to have been subjected to enormous pressures after the fallen plants slowly accumulated in a swamp.

Petrifactions

Petrifactions (Fig. 21.27) are uncompressed rocklike materials in which the original cell structure has been preserved. About 20 different mineral substances, including silica and the salts of several metals, are known to bring about petrifaction. At one time, it was believed the process occurred through the replacement of the plant parts by minerals in solution, one molecule at a time. It is now believed that chemicals in solution infiltrate the cells and cell walls, where they crystallize and harden, permanently preserving the original material.

Petrifactions can be studied by cutting thin sections with a diamond or carborundum saw and polishing the material with extremely fine grit powder until it is thin enough for light to pass through. Then it can be examined with a compound microscope. Another simpler method of studying petrifactions involves etching the cut surface with

Figure 21.27 Petrified wood. *(Courtesy Sharon Stern)*

a dilute acid and then applying a plastic or similar film. As soon as the film hardens, it can be peeled off. Such peels display lifelike microscopic details of the surface with which they were in contact. They are commonly used by paleobotanists (botanists who study fossil plant materials) in their research.

Coprolites

Coprolites are the fossilized dungs of prehistoric animals and humans. They may contain pollen grains and other plant and animal parts that provide clues to the food and feeding habits of past organisms and cultures.

Unaltered Fossils

Some plants or animals fell into bodies of oil or water that, because of substances present or a nearly total lack of oxygen, did not permit decay to occur. Some animals died in snowfields, and their bodies were permanently frozen. In such instances, preservation in the unaltered state occurred. Unaltered fossils, particularly frozen ones, are very rare.

Summary

1. Four phyla of seedless vascular plants are recognized: Psilotophyta (whisk ferns); Lycophyta (club mosses and quillworts); Equisetophyta (horsetails and scouring rushes); and Polypodiophyta (ferns).
2. Whisk ferns (*Psilotum*) are the simplest of all living vascular plants, consisting of evenly forking green stems that have small protuberances called enations, but no leaves; roots are also lacking. The stem contains a central cylinder of xylem and phloem.
3. Whisk fern spores germinate into tiny gametophytes, with antheridia and archegonia scattered over their surfaces. The zygote develops a foot and a rhizome. Upright stems are produced when the rhizome separates from the foot.
4. *Tmesipteris,* an Australian relative of whisk ferns, has leaflike appendages. Fossil plants resembling whisk ferns have been found in Silurian geological formations.
5. There are two genera of club mosses with living members. Ground pines (*Lycopodium*) develop sporangia in the axils of sporophylls. Several sporophytes may be produced from one gametophyte. Spike mosses (*Selaginella*) are heterosporous and have a ligule on each microphyll; microspores develop into male gametophytes with antheridia, and the megaspores develop into female gametophytes with archegonia.
6. Quillworts (*Isoetes*) are heterosporous and have quill-like microphylls that arise from a cormlike base. The corms have a cambium that remains active for many years.
7. Some fossil relatives of club mosses were large dominant members of the forests and swamps of the Carboniferous period.
8. Club moss spores have been used for flash powder, medicinal purposes, as talcum powder, and to staunch bleeding. The plants themselves have been used as ornamentals, novelty items, Christmas ornaments, and for intoxicating purposes.
9. Horsetails and scouring rushes (*Equisetum*) accumulate deposits of silica in their epidermal cells and have made good scouring material. They occur in both unbranched and branched forms. The stems are jointed and ribbed and have tiny scalelike leaves in whorls at each joint.
10. The stems of *Equisetum* are centrally hollow and contain cylindrically arranged carinal and vallecular canals. The stems arise from rhizomes that branch extensively below the surface of the ground.
11. Some species of *Equisetum* produce non-photosynthetic stems. Conelike strobili are produced in the spring in all species. *Equisetum* spores have ribbonlike elaters that are sensitive to changes in humidity.
12. Equal numbers of male and female gametophytes are produced, but female gametophytes may become bisexual. The development of more than one sporophyte from a gametophyte is common.
13. Ancient relatives of horsetails were the size of trees when they flourished in the Carboniferous period of 300 million years ago.
14. Horsetails have been used for food after the parts containing silica were removed, but they are not recommended for human consumption. They have also been used medicinally as a diuretic, as an astringent, and in the treatment of venereal disease and tuberculosis. Other uses include a hair wash, a mineral indicator, and a metal polish. Cannel coal consists primarily of spores of giant horsetails that were reduced to carbon.
15. Fern leaves (fronds) are typically divided and feathery in appearance but vary greatly in form. They usually first appear as croziers that unroll and expand.
16. Patches of sporangia appear on the lower surfaces of fern fronds. The sporangia commonly occur in sori, which may be protected by indusia. Each sporangium has an annulus that functions in catapulting mature spores out of the sporangium.
17. Fern gametophytes (prothalli) develop after spores germinate. Most prothalli contain both archegonia and antheridia; only one zygote develops into a sporophyte.
18. Possible ancestors of ferns are found in Devonian deposits estimated to be 375 million years old.
19. Ferns are used as ornamentals, air filters, a source of "bark" for growing orchids and other plants, a source of stuffing materials for bedding, in tropical construction, as food, and in numerous folk medicinal applications. Other uses include basketry and weaving material,

ingredient in brewing ale, and ingredient in the preparation of chamois leather. One floating fern forms dense mats and is believed to suffocate mosquito larvae.

20. Fossils are recognizable prehistoric organic objects that are formed in different ways. Molds, casts, compressions, and imprints are formed when material buried by silt or other sediment has hardened into rock and the organic material has slowly been washed away by water.
21. Petrifactions are uncompressed rocklike materials in which the original cell structure has been preserved. Coprolites are fossilized dungs that may contain pollen grains and other plant and animal parts. Unaltered fossils are those of plants or animals that may have fallen into bodies of oil or water or snowfields and were not subjected to decay.

Review Questions

1. What basic features of the ferns and their relatives distinguish them from any organisms studied thus far?
2. How does the gametophyte of a whisk fern differ from that of a true fern?
3. Which of the fern relatives have significantly functional leaves? In those without conspicuous leaves, how are the carbohydrate needs of the plants met?
4. How does one distinguish among ground pines, spike mosses, and quillworts?
5. How did the ancient ground pines differ from those of the present day?
6. How do the spores and the female gametophytes of horsetails differ from those of any other plants studied thus far?
7. What is the location and function of carinal canals?
8. In your opinion, which have the most human relevance today: club mosses and horsetails of the present or those of the geological past? Why?
9. Define crozier, rachis, pinna, indusium, and prothallus.
10. Diagram the life cycle of a typical fern.
11. Summarize present and past human uses of ferns.
12. How are fossils formed, and what different types are recognized?

Discussion Questions

1. Would you assume there is any significance to the fact that both the sporophytes and the gametophytes of whisk ferns branch in the same manner?
2. Do spike mosses, which produce two kinds of spores and gametophytes, have any advantage over ground pines, which produce only one kind of spore and gametophyte?
3. Some gametophytes of fern relatives develop underground, while others develop at the surface. If you were to be a gametophyte, which would you prefer? Why?
4. After looking at the internal structure of a horsetail stem, can you suggest a function for the silica in the ribs?
5. How would we be affected if all ferns were to become extinct in a few years?

Learning Online

Visit our web page at www.mhhe.com/botany for interesting case studies, practice quizzes, current articles, and animations within the Online Learning Center to help you understand the material in this chapter. You'll also find active links to these topics:

Ferns and Relatives
Economic and Ecological Importance of Lower Plants

Additional Reading

Bir, S. S. 1983. *Pteridophytes: Their morphology, cytology, taxonomy, and phylogeny*. Houston, TX: Scholarly Publications.

Cobb, B. 1975. *A field guide to the ferns and their related families*. Boston, MA: Houghton Mifflin.

Grillos, S. J. 1966. *Ferns and fern allies of California*. Berkeley, CA: University of California Press.

Lellinger, D. B. 1985. *A field manual of the ferns and fern allies of the United States and Canada*. Washington, DC: Smithsonian Institution Press.

Lloyd, R. M. 1964. Ethnobotanical uses of California pteridophytes by western American Indians. *American Fern Journal* 54: 76–82.

McHugh, A. 1992. *The cultivation of ferns*. North Pomfret, VT: Trafalgar.

Mickel, J. 1979. *How to know the ferns and fern allies*. Pictured Key Nature Series. Dubuque, IA: WCB/McGraw-Hill Publishers.

Montgomery, J. D., and D. E. Fairbrothers. 1992. *New Jersey ferns and fern allies*. New Brunswick, NJ: Rutgers University Press.

Raghaven, V. 1989. *Developmental biology of fern gametophytes*. New York: Cambridge University Press.

Seward, A. C. (Ed.). 1991. *Fossil plants for students of botany and geology*, 4 vols. Houston, TX: Scholarly Publications.

Stewart, W. N., and G. W Rothwell 1993. *Paleobotany and the evolution of plants*, 2d ed. New York: Cambridge University Press.

A female cycad cone (*Cycas* sp.). Cycads are slow-growing, superficially palmlike gymnosperms. Gymnosperm means naked-seeded, and the developing seeds can be seen here out in the open at the edges of the cone scales.

Introduction to Seed Plants: Gymnosperms

Chapter Outline

22

Overview

The chapter begins by exploring the differences between ferns and seed plants and then gives a brief geological history of gymnosperms. The leaves, roots, and stems of pine trees are discussed, and the life cycle of a typical gymnosperm (pine) is given. Other conifers, such as yews, podocarps, junipers, and redwoods, are mentioned, and a brief discussion of *Ginkgo,* cycads, *Ephedra, Gnetum,* and *Welwitschia* follows. The chapter concludes with a digest of the human relevance of gymnosperms, with particular emphasis on the conifers.

Some Learning Goals

1. Learn the features common to typical conifer pollen and seed strobili and how they differ.
2. Understand what distinguishes the phyla of living gymnosperms from one another.
3. Know the significance of seeds and their evolutionary importance.
4. Learn the pine leaf modifications that adapt them to a harsh environment.
5. Indicate where the following structures occur in the life cycle of a pine tree: archegonia, eggs, sperms with flagella, male and female gametophytes, the sporophyte, integument, vessels, sporocytes, embryo, and pollen grains.
6. Explain the function of each of the following: resin canals, mycorrhizal fungi, nucellus, generative cell, megaspore, microsporocyte.
7. Identify and learn a use for each of at least 10 different gymnosperms.

Some of the giant tree ferns have large, graceful leaves held above an unbranched trunk (see Fig. 21.25) and superficially resemble coconut palms. Anyone who has sat at the base of a coconut palm when a coconut has fallen, however, has been reminded of a basic difference between ferns and coconut palms. The palms are flowering plants that produce seeds such as those of pine trees and other cone-bearing plants, whereas the ferns produce no seeds at all.

The oldest known seeds were produced by plants that appeared late in the Devonian period, more than 350 million years ago. Seeds provided a significant adaptation for plants that had invaded the land. Unlike spores, seeds have a protective seed coat and a supply of food (usually endosperm) for the embryo. The embryo may be capable of lying dormant through long periods of freezing weather, drought, and may, in some instances, even survive fire. This survival value of seeds undoubtedly played a major role in seed plants becoming the dominant vegetation on earth today.

The first seed plants were so fernlike in appearance that they were originally classified as ferns. When fossils with obvious seeds on the fronds were discovered, however, these *pteridosperms* ("seed ferns") were reclassified as *gymnosperms.*

The name **gymnosperm** is derived from two Greek words: *gymnos,* meaning naked, and *sperma,* a seed. The name refers to the exposed nature of the seeds, which are produced on the surface of sporophylls or similar structures instead of being enclosed within a fruit as they generally are in the flowering plants (see the photograph of a cycad cone on page 421, and Fig. 22.1). The seed-bearing sporophylls of the sporophyte are often spirally arranged in strobili (seed cones) that develop at the same time as smaller pollen-bearing strobili (pollen cones). The pollen cones produce *pollen grains.*

The female gametophyte is produced inside an **ovule** that contains a fleshy, nutritive diploid tissue called the **nucellus** (see Fig. 22.7). The nucellus is itself enclosed within one or more outer layers of diploid tissue. These outer layers of tissue constitute an **integument** that becomes a **seed coat** after the fertilization and development of an embryo takes place (see Fig. 22.8).

The sporophytes of gymnosperms are mostly trees and shrubs, with a few species being vines. The gametophytes are proportionately even more reduced in extent than they

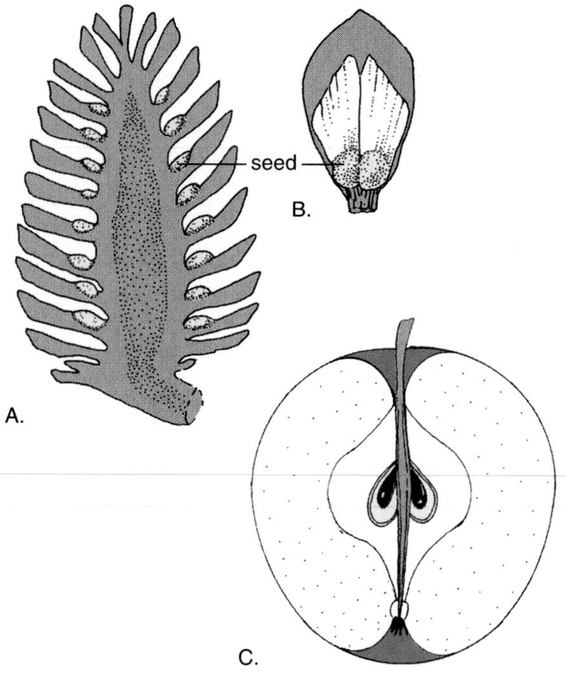

Figure 22.1 A comparison between exposed gymnosperm seeds and enclosed angiosperm seeds. *A.* Exposed seeds on a woody seed cone of a pine tree. *B.* A single seed cone scale with two seeds. *C.* A section through an apple (angiosperm fruit), showing the enclosed seeds.

are in ferns and their relatives. Unlike the gametophytes discussed so far, they don't grow independently but develop within sporophyte structures.

Four phyla of living gymnosperms are recognized. Phylum Pinophyta includes about 575 species of pines, firs, spruces, hemlocks, cedars, redwoods, and other coniferous woody plants. Fossils of some conifers extend back 290 million years to the late Carboniferous period. Phylum Ginkgophyta has a single living representative, *Ginkgo,* which has fan-shaped leaves and seeds enclosed in a fleshy covering. The superficially palmlike cycads are assigned to Phylum Cycadophyta. Phylum Gnetophyta includes three genera of gnetophytes that have wood with vessels—a structural element unknown in other gymnosperms.

PHYLUM PINOPHYTA—THE CONIFERS

Pines

The largest genus of conifers, *Pinus* (pines) has over 100 living species. They are the dominant trees in the vast coniferous forests of the Northern Hemisphere. They have also been planted extensively in the Southern Hemisphere, but only the Merkus pine occurs there naturally, and its distribution barely extends south of the equator. They include the world's oldest known living organisms, the bristlecone pines, native to the White Mountains of eastern central California and in the Snake range on the central Nevada-Utah border. Some trees still standing are about 4,600 years old, and one that was, unfortunately, cut down in 1964 was found to have been about 4,900 years old.

Structure and Form

Pine leaves are needlelike and are arranged in clusters or bundles of two to five leaves each (a handful of species have as many as eight or as few as one to a cluster). Regardless of the number of leaves, each cluster (*fascicle*) forms a cylindrical rod if the leaves are held together (Fig. 22.2). The fascicles are short shoots with restricted growth, a feature of some gymnosperms not found in flowering plants. Pines often live in areas where the topsoil is frozen for a part of the year, making it difficult for the roots to obtain water. In addition, the leaves may be exposed to high winds and bitterly cold temperatures. Accordingly, they have several modifications that enable them to survive in harsh environments.

Just beneath the epidermis there is a **hypodermis,** consisting of one to several layers of thick-walled cells. The epidermis itself is coated with a thick cuticle. The stomata, instead of being at the surface, are recessed or sunken in small cavities. The veins and their associated tissues are surrounded by an endodermis, and the mesophyll cells do not have the obvious air spaces typical of the spongy mesophyll of the leaves of flowering plants (see Figs. 7.6 and 7.11).

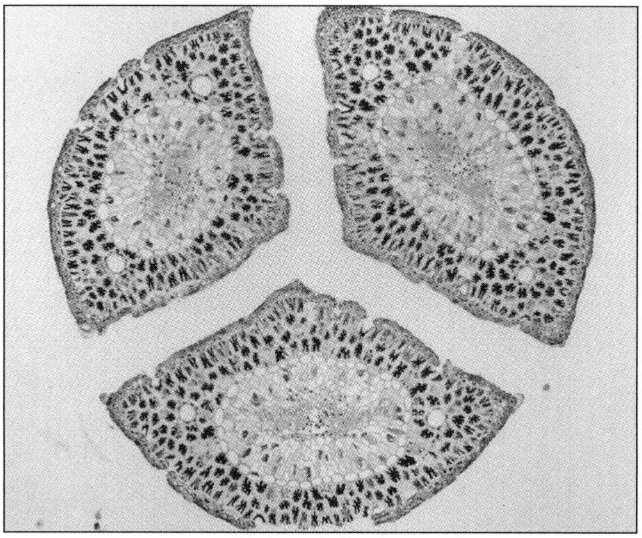

Figure 22.2 A cluster (fascicle) of pine needles in cross section, showing how the needles form a cylindrical rod.

Conspicuous **resin canals** develop in the mesophyll. These resin canals, which are found throughout other parts of the plant as well, consist of tubes lined with special cells that secrete *resin.* Resin is aromatic and antiseptic and prevents the development of fungi; it also deters insect attacks. Other conifers apparently produce resin canals in response to injury. Pine fascicles usually *absciss* (i.e., fall off—see Chapter 7) within 2 to 5 years of their maturing, but those of bristlecone pines persist for up to 30 years. The fascicles are

Ecological Review

The evolution of seeds by the seed plants, including the gymnosperms, facilitated their invasion of land. Encased in a seed, the embryo of seed plants can withstand the extremes of the terrestrial environment such as freezing weather, drought, and even fire. Other characteristics, such as a thick cuticle on the epidermis, sunken stomata, and a hypodermis, consisting of one to several layers of thick-walled cells, also enable gymnosperms to withstand harsh environmental conditions from the freezing winters of the far north to dry desert areas. Resins produced by gymnosperms provide defense against fungal and insect attacks, while mycorrhizal fungi increase availability of soil nutrients and water to gymnosperms. Mutualistic associations with squirrels and birds, such as nutcrackers, promote gymnosperm seed dispersal. Gymnosperms provide humans with a wealth of "nature's services," ranging from construction materials to cancer treatments.

Figure 22.3 A portion of a cross section of a pine stem, showing annual rings.

lost a few at a time so that some functional leaves are always present on a healthy tree.

Wood varies considerably in hardness. Most gymnosperm wood, including that of pines, consists primarily of tracheids and differs from the wood of dicots in having no vessel members or fibers. Conifer wood, because of the absence of thick-walled cells, is said to be *soft,* while the wood of broadleaf trees is described as *hard.*

In many conifers, the annual rings of the xylem are often fairly wide as a result of a comparatively rapid growth rate during the growing season. Resin canals are formed both vertically and horizontally throughout various tissues (Fig. 22.3). The bark includes the secondary phloem and may be relatively thick. It often becomes 7.5 centimeters (3 inches) or more wide, and in the giant redwood, it may become as much as 60 centimeters (2 feet) wide. Companion cells are absent from the phloem, but similar *albuminous cells* apparently perform the same function.

Mycorrhizal fungi are associated with the roots of most conifers. In fact, pine seedlings that germinate in sterilized soil do not grow well at all until the fungi are introduced or allowed to develop. The roots of adjacent pines often interweave. New England pioneers made use of this characteristic in eastern white pines when clearing land. After trees were felled, they would tip over the stumps with the roots still attached and use them for fences. The fences survived for many years.

Figure 22.4 A cluster of pine pollen cones.

Reproduction

Pines, like spike mosses, quillworts, and a few of the ferns, produce two kinds of spores (see Fig. 22.8). **Pollen cones** (male strobili) consist of papery or membranous scales arranged in a spiral or in whorls around an axis; they are usually produced in the spring. The pollen cones usually develop toward the tips of the lower branches in clusters of up to 50 or more (Fig. 22.4) and are mostly less than 4 centimeters (about 1.5 inches) long. **Microsporangia** develop in pairs toward the bases of the scales.

Microsporocytes in the microsporangia each undergo meiosis, producing four haploid **microspores.** These then develop into **pollen grains** that each consist of four cells and a pair of external air sacs. The air sacs look something like tiny water wings (Fig. 22.5) and give the pollen grains added buoyancy that may result in the grains being carried great distances by the wind.

Pines produce pollen grains in astronomical numbers. For example, it has been estimated that each of the 50 or more pollen cones commonly found in a single cluster may produce more than 1 million grains, and there may be hundreds of such clusters on one tree. The grains accumulate as a fine yellow dust on cars, shrubbery, or anything else in the vicinity, and they often form an obvious scum on pools and puddles. Within a few weeks after the pollen has been released, the now shriveled pollen cones fall from the trees.

Megaspores are produced in **megasporangia** located within ovules at the bases of the seed cone scales. The **seed cones** (female strobili) are much larger than the pollen cones, becoming as much as 60 centimeters (2 feet) long in sugar pines and weighing as much as 2.3 kilograms (5 pounds) in Coulter pines. When mature, they have woody scales, with inconspicuous bracts between them, arranged in a spiral around an axis. They are mostly produced on the upper branches of the same tree on which the pollen cones appear (Fig. 22.6).

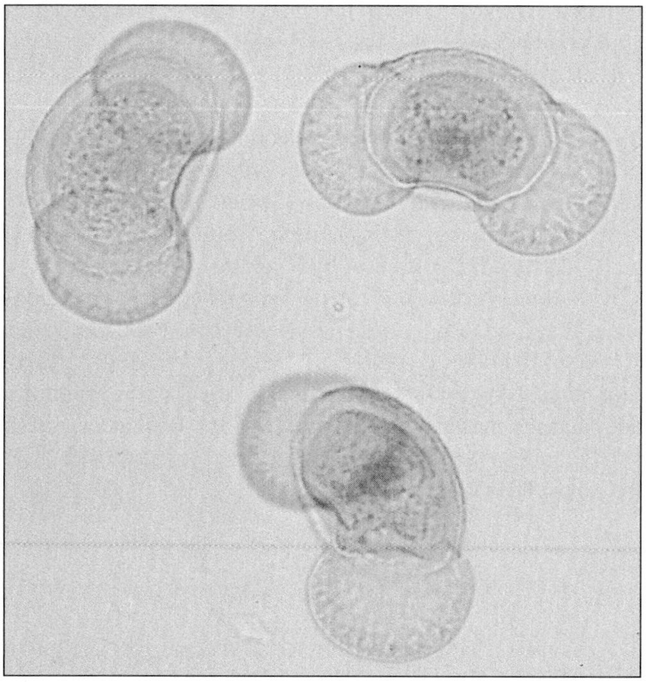

Figure 22.5 Pollen grains of a pine, as seen with the aid of a microscope. Each pollen grain has a pair of air sacs that provide added buoyancy, ×300.

A.

B.

Figure 22.6 Seed cones of pines. *A.* Immature cones shortly after being produced. *B.* A mature cone.

The ovules (Fig. 22.7) occur in pairs toward the base of each scale of the immature seed cones and are larger and more complex than the microsporangia of pollen cones. Each ovule has within it a *megasporangium* containing the **nucellus** and a single **megasporocyte.** This, in turn, is surrounded and enclosed by a thick, layered **integument.** The integument has a somewhat tubular channel or pore called a **micropyle** that is pointed toward the cone's central axis. One of the integument layers later becomes the **seed coat** of the seed.

A single megasporocyte within the megasporangium of each ovule undergoes meiosis, producing a row of four relatively large megaspores. Three of the megaspores soon degenerate. Over a period of months, the remaining one slowly develops into a *female gametophyte* that ultimately may consist of several thousand cells. The nucellus is used as the food source for the growing gametophyte

As gametophyte development nears completion, two to six *archegonia* differentiate at the end facing the micropyle. Each archegonium contains a single large egg. When a stained, thin, lengthwise section of a pine ovule is examined with a microscope, only one archegonium or egg may be seen, or the micropyle may appear to be missing. This is due to the section not having been sliced precisely through the middle of all the structures.

Seed cones are at first usually reddish or purplish in color and commonly take two seasons to mature into the green and finally the brownish woody structures with which we are all familiar (Fig. 22.8). During the first spring, the immature cone scales spread apart, and pollen grains carried by the wind sift down between the scales. There they catch in sticky drops of fluid (*pollen drops*) oozing out of the micropyles. As the fluid evaporates, the pollen is drawn down through the micropyle to the top of the nucellus.

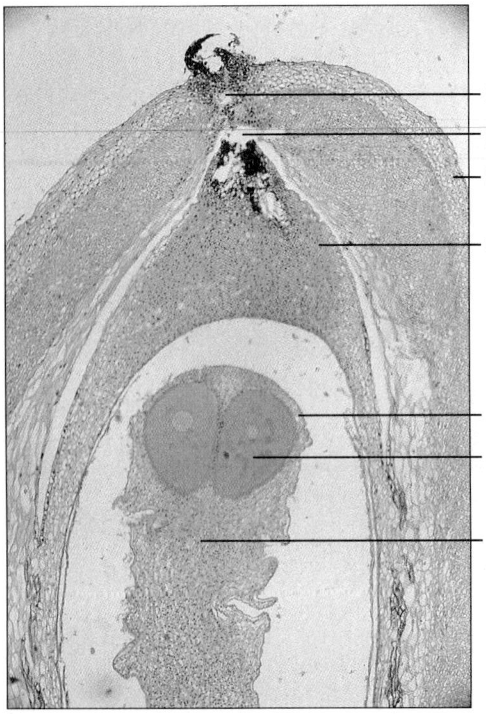

Figure 22.7 A longitudinal section through part of a pine ovule, ×100.

After pollination, the scales grow together and close, protecting the developing ovule. Meiosis and megaspore development don't occur until about a month after pollination. After a functional megaspore is produced, the female gametophyte and its archegonia don't mature until more than a year later.

Meanwhile, the pollen grain (immature *male gametophyte*) produces a **pollen tube** that slowly grows and digests its way through the nucellus to the area where the archegonia develop. While the pollen tube is growing, two of the original four cells in the pollen grain enter it. One of these, called the **generative cell,** divides and forms two more cells, called the **sterile cell** and the **spermatogenous cell.** The spermatogenous cell divides again, producing two male gametes, or **sperm.** The sperms have no flagella (unlike the sperms of other organisms discussed in the preceding chapters) and are confined to the pollen tube until just before fertilization occurs. The germinated pollen grain, with its pollen tube and two sperms, constitutes the mature *male gametophyte.* Notice that no antheridium has been formed.

About 15 months after pollination, the tip of the pollen tube arrives at an archegonium, unites with it, and discharges the contents. One sperm unites with the egg, forming a zygote. The other sperm and remaining cells of the pollen grain degenerate. The sperms of other pollen grains present may unite with the eggs of other archegonia, and each zygote begins to develop into an **embryo** that is nourished by the female gametophyte. This is similar to the development of fraternal twins or triplets in animals. At a later stage, an embryo may divide in such a way as to produce the equivalent of identical twins or quadruplets in animals. Normally, however, only one embryo completes development. While this development is occurring, one of the layers of the integument hardens, becoming a *seed coat.* A thin membranous layer of the cone scale becomes a "wing" on each seed. The wing may aid in the seed's dispersal. Squirrels and other animals may also help dispersal by breaking open the cones.

In other species, such as the lodgepole, jack, and knobcone pines, the cones remain on the tree with the scales closed until they are seared by fire or open with old age. Sometimes, these cones are slowly buried as the cambium adds tissues that increase the girth of the branch. Seeds of such engulfed cones have been reported to germinate after they have been dug out of the stem.

Other Conifers

Some conifers don't produce woody seed cones with conspicuous scales, nor do all produce both pollen and seed cones on the same plant. For example, yew (*Taxus*) and California nutmeg (*Torreya*) produce ovules singly at the tips of short axillary

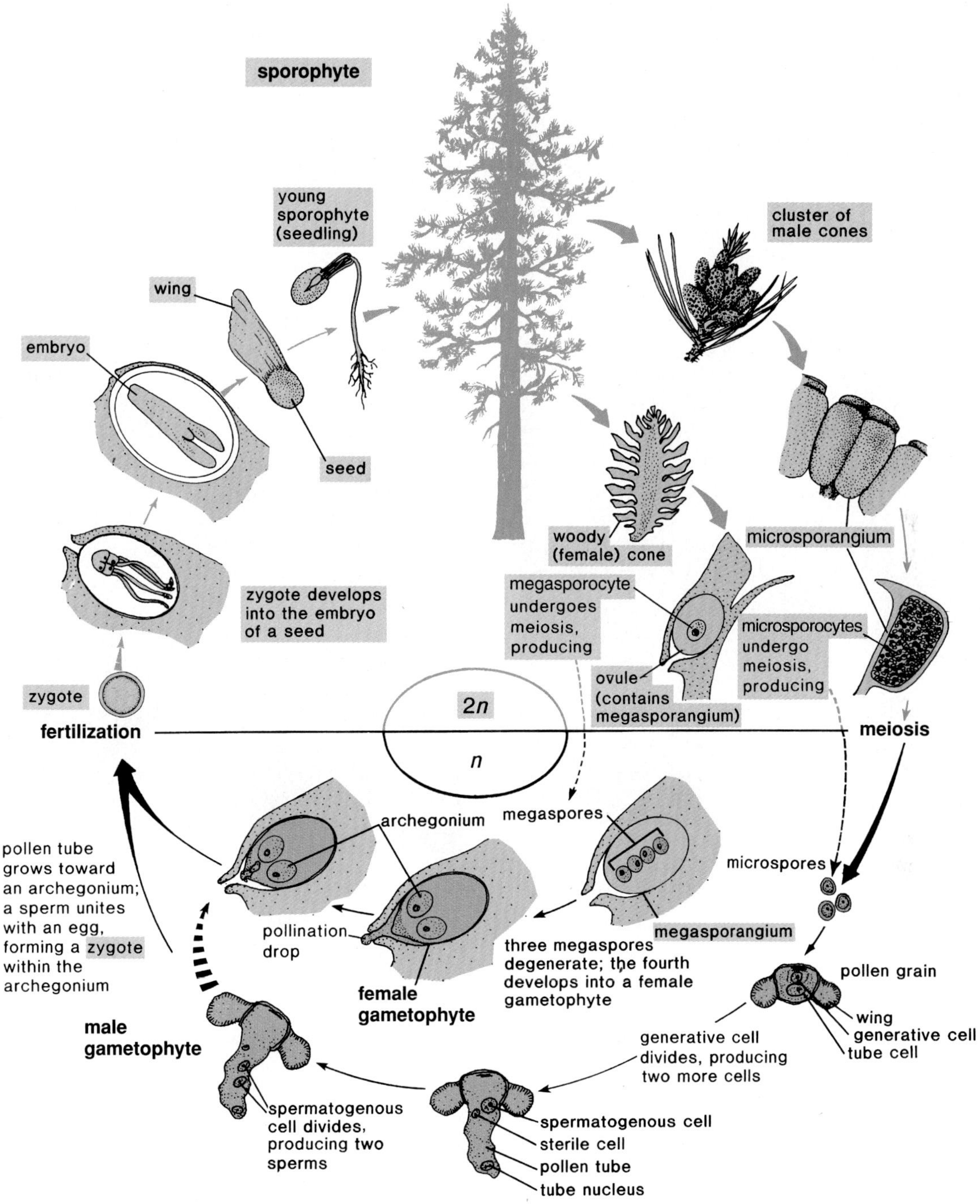

Figure 22.8 Life cycle of a pine.

shoots. Each ovule is at least partially surrounded by a fleshy, cuplike covering called an **aril.** In yews, this is bright red and open at one end, giving the fleshy seed the appearance of a small red hors d'oeuvre olive with its stuffing removed (Fig. 22.9). The seeds with their fleshy arils are produced only on female plants, while the pollen cones are produced only on male plants.

Podocarps are conifers of the Southern Hemisphere and are widely planted as ornamentals in regions with milder climates. Their fleshy-coated seeds are produced singly and are similar to those of yews. They are not, however, open at one end and have an additional larger appendage at the base (Fig. 22.10). The origin of these fleshy seeds is not clear and has led to speculation that yews and podocarps may have diverged from other conifers very early in the evolution of gymnosperms.

The scales of the seed cones of junipers tend to be fleshy at maturity, and they look more like berries than cones. Juniper pollen grains, as well as those of a number of other

Figure 22.9 The seeds of yew (*Taxus*) are not produced in cones and are surrounded at maturity by a red, fleshy, cuplike structure called an *aril.*

Figure 22.10 Fleshy seeds of a podocarp (*Podocarpus*). Note the large appendage at the base of each seed.

conifers, do not have air sacs. Cypress and redwood seed cone scales are flattened at the tips and narrow at the base, where they do not overlap one another as do pine cone scales.

The two California species of redwoods are both renowned for their size, height, and longevity. Coastal redwoods occasionally grow to a height of 90 meters (295 feet), and one tree in Humboldt County, California, is 111.6 meters (366.2 feet) tall; it is believed to be the tallest conifer in the world. The other species, usually referred to as *Big Tree* or *Giant Redwood,* is confined to the western slopes of California's Sierra Nevada range. It does not grow quite as tall as the coastal redwood but exceeds it in total mass. The General Sherman tree in Sequoia National Park, for example, is 31 meters (101.5 feet) in circumference at the base and over 24 meters (79 feet) in circumference 1.5 meters (5 feet) above the ground. It weighs an estimated 5,594 metric tons (6,167 tons). There are 600,000 board feet of timber in this single tree—enough to build more than 75 five-room houses (although the wood is generally not suitable for construction purposes) or to make 20 billion toothpicks. It is over 3,500 years old.

OTHER GYMNOSPERMS

Living representatives of other gymnosperms are not as numerous or as well known as the conifers and outwardly don't resemble them at all. In fact, some of them look more like leftover props from a science-fiction movie set. They include the **ginkgos,** the **cycads,** and the **gnetophytes** (pronounced née-toe-fytes).

Phylum Ginkgophyta—*Ginkgo*

There is only one living species of *Ginkgo* (Fig. 22.11), whose name is derived from Chinese words meaning "silver apricot." The fossil record indicates *Ginkgo* and other members of

A.

B.

C.

Figure 22.11 *Ginkgo. A.* A mature tree in the fall. *B.* Seeds and leaves. *C.* Male strobili.

its family (Ginkgoaceae) were once widely distributed, especially in the Northern Hemisphere. Despite isolated reports to the contrary, there are doubts that ginkgoes now exist anywhere they have not been cultivated, and the plant has often been called a living fossil.

Ginkgoes are often referred to as *maidenhair trees* because their notched, broad, fan-shaped leaves look like larger versions of the individual pinnae of maidenhair ferns. They are widely cultivated in the United States and are popular street trees in some areas. The leaves are mostly produced in a spiral on short, slow-growing spurs and have no midrib or prominent veins. Instead, hairlike veins branch *dichotomously* (fork evenly) and are relatively uniform in their width. They are deciduous and turn a bright golden yellow before abscission in the fall.

Ginkgo is dioecious, with a life cycle similar to that of pines. The sperms, however, have flagella, and being dioecious, the male and female reproductive structures are produced on separate trees. The mature seeds resemble small plums and are enclosed in fleshy seed coats. The flesh is, however, unrelated to that of true fruits. In North America, male trees (propagated from cuttings) are preferred for ornamental purposes because the seed flesh has a nauseating odor and is irritating to the skin of some individuals. In China and Korea, however, the seeds are considered a delicacy, and female trees predominate; a minimal number of males are propagated to ensure pollination.

Phylum Cycadophyta—The Cycads

Cycads have the appearance of a cross between a tree fern and a palm but are related to neither. These slow-growing plants of the tropics and subtropics have unbranched trunks that grow more than 15 meters (50 feet) tall in a few species and have a crown of large pinnately divided leaves. Several of the approximately 100 known living species of cycads are presently facing extinction. During the Mesozoic era, now extinct gymnosperms known as *cycadeoids* were abundant. Cycadeoids bore a superficial resemblance to cycads but had very different reproductive structures and are not related.

Cycad life cycles are similar to those of conifers, except that pollination of cycads is generally brought about by beetles instead of wind. In addition, each sperm of cycads has from 10,000 to 20,000 spirally arranged flagella. Cycads are dioecious, and both the pollen strobili and seed strobili of some species are huge (e.g., more than a meter [3 feet 3 inches] long with a weight of over 220 kilograms [100 pounds]) (Fig. 22.12). The scales of seed strobili of some species are covered with feltlike or woolly hairs.

A.

B.

Figure 22.12 *A.* A male cycad with a strobilus. *B.* A female cycad with a strobilus.

Phylum Gnetophyta—The Gnetophytes

The 70 known species of *gnetophytes* are distributed among three distinctive genera. They are unique among the gymnosperms in having vessels in the xylem. More than half of the gnetophytes are species of *joint firs* in the genus *Ephedra.* These shrubby plants inhabit drier regions of southwestern North America. Their tiny leaves are produced in twos and threes at a node and turn brown soon after they appear. The stems and branches, which are often whorled, are slightly ribbed; they are photosynthetic when they are young (Fig. 22.13).

Before pollination, the ovules of *Ephedra* produce a small tubular extension resembling the neck of a miniature bottle extending into the air. Sticky fluid oozes out of this extension, which constitutes the micropyle, and airborne pollen catches in the fluid. Male and female strobili may be produced on the same plant or on different ones, depending on the species.

Most of the remaining species in this division are in the genus *Gnetum* (Fig. 22.14), which has not been given an English common name. Its members occur in the tropics of South America, Africa, and Southeast Asia. Most are vine-like, with broad leaves similar to those of flowering plants. The best-known species of *Gnetum,* however, is a tree that grows up to 10 meters (33 feet) tall.

The third genus, *Welwitschia,* has only one species, which is confined to the temperate Namib and Mossamedes deserts of southwestern Africa. Here the average annual rainfall is only 2.5 centimeters (1 inch),

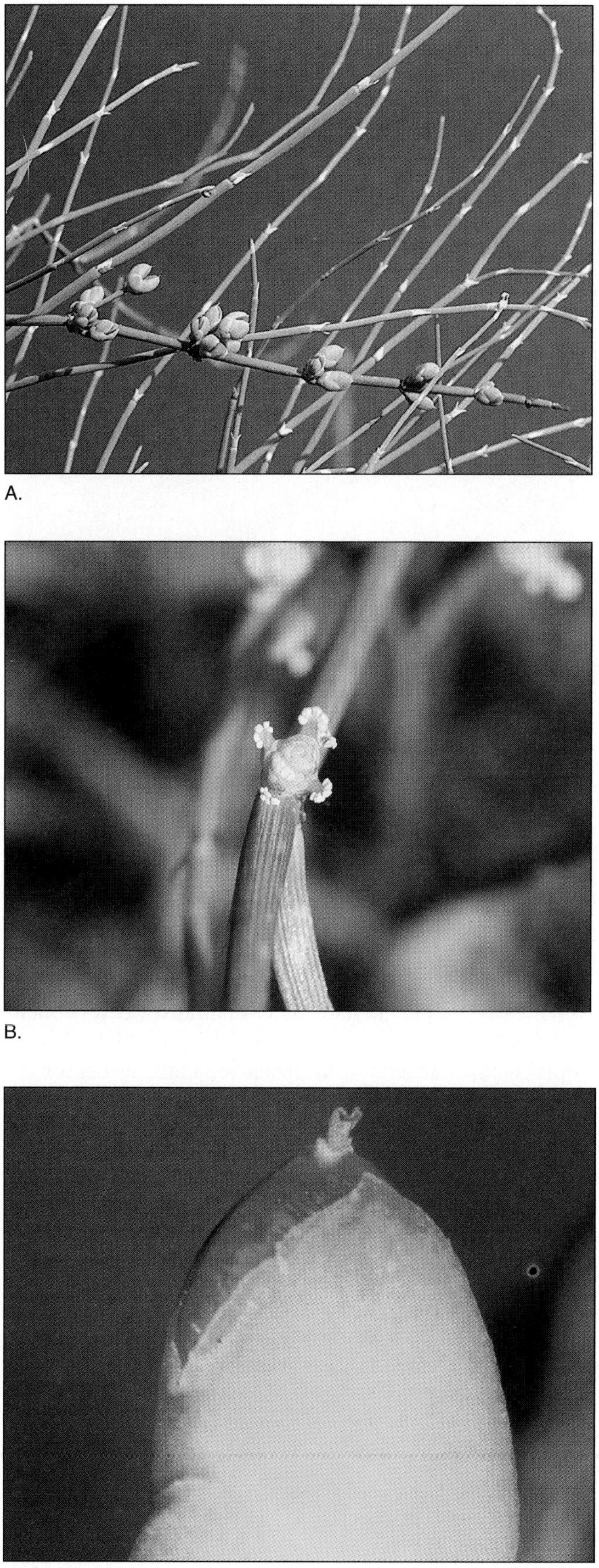

A.

B.

C.

Figure 22.13 Joint fir (*Ephedra*). *A.* Part of a single plant. *B.* Male strobili, ×20. *C.* A female strobilus, ×20.

Figure 22.14 A climbing *Gnetum.*

and in some years, it does not rain at all. The plants carry on CAM photosynthesis, and their stomata are open at night. The plants apparently survive much of the time on dew and condensate from fog that rolls in off the ocean at night.

Welwitschia plants are also truly extraordinary in appearance. The stem rises only a short distance above the ground's surface and is in the form of a large shallow cup that tapers at the base into a long taproot. The plants may live to be 100 years old and at maturity have a crusty, bark-like covering on the surface of the stem cup. The stems may be more than 1 meter (3 feet 3 inches) in diameter (Fig. 22.15).

Throughout their life span, *Welwitschia* plants usually produce only two leaves. The leaves are wide and straplike, each with a meristem at the base. The meristems constantly add to the length of the leaves, but as the leaves flap about in the wind, they become tattered and split, wearing off at the tips so that they are seldom more than 2 meters (6.5 feet) long. *Welwitschia* is dioecious, with both male and female strobili being produced on axes that emerge from the axils of the leaves. This makes the strobili appear to be growing around the rim of the stem cup.

Figure 22.15 A female *Welwitschia* plant in the Namibian Desert. *Photo by Margaret Marker, in the collection of the National Botanical Institute, Kirstenbosch. Courtesy Fiona Getliffe Norris.*

HUMAN RELEVANCE OF GYMNOSPERMS

What do we plant when we plant the tree?
We plant the ship, which will cross the sea.
We plant the mast to carry the sails;
We plant the planks to withstand the gales—
The keel, the keelson, the beam, the knee;
We plant the ship when we plant the tree.

What do we plant when we plant the tree?
We plant the houses for you and me.
We plant the rafters, the shingles, the floors,
We plant the studding, the lath, the doors,
The beams, the siding, all parts that be;
We plant the house when we plant the tree.

What do we plant when we plant the tree?
A thousand things that we daily see;
We plant the spire that out-towers the crag,
We plant the staff for our country's flag,
We plant the shade, from the hot sun free;
We plant all these when we plant the tree.

Henry Abbey

As a group, the gymnosperms are second only to the flowering plants in their impact on our daily lives. Space does not permit a detailed account of all they contribute, but the following sections provide an overview of some of their uses, past and present.

Conifers

In the early 1970s, the late author-naturalist Euell Gibbons filmed a series of television commercials in which he mentioned uses of several wild plants for food. In one of the commercials, he rhetorically inquired if his audience had ever eaten a pine tree and added, "many parts are edible."

The edibility of parts of many conifers was known to Native Americans long before Europeans set foot on the North American continent. In fact, early explorers found large numbers of pines stripped of their bark. For centuries, these inner parts (phloem, cambium) had been used for emergency food. The Adirondack Mountains of New York are believed to have received their name from a Mohawk Indian word meaning "tree eater," in reference to Native American use of the inner bark of eastern white pines. This material (specifically the phloem) contains sugars that make it taste sweet. Some tribes ate the material raw, some dried it and ground it to flour, and others boiled it or stored dried strips for winter food. Early settlers in New England candied strips of eastern white pine inner bark. To prevent scurvy, both they and local Native Americans drank a tea made of the needles, which are rich in vitamin C.

The seeds of nearly all pines are edible, but those of western North America include the larger and better-tasting species. The protein content of those analyzed generally ranges between 15% and 30%, with much of the remainder consisting of oils.

California Native Americans relished gray pine seeds in particular, but even the small seeds of ponderosa pine were eaten raw or made into a meal for soups and bread. Cones of pinyons were collected by tribes of the Southwest and thrown on a fire to loosen the seeds. These were then pounded and made into cakes or soup. The soup was often fed to infants. In Siberia, local residents crush the seeds of Siberian white pine to obtain a nutritious oil, but its use has declined since corn and cottonseed oils became available.

Italians and other Europeans cook "pignolias," the seeds of the stone pine, in stews and soups. The seeds are also used in cakes and cookies, and some are exported to the United States for this purpose. Many of the so-called nuts used by commercial American bakers in cakes and confectionery, however, are really seeds from the east Himalayan chilghoza pine. Other sources include the Mexican stone pine and a few pinyons.

Eastern white pines were often used as masts in sailing vessels. In colonial days, the royal surveyors marked certain trees for the use of the Crown, and severe penalties were imposed on colonists who ignored the ban on the use of any white pine not growing on private land. It was, however, legal for colonists to use white pines that had blown down, which gave rise to the term *windfall.* Eastern white pine wood contains less resin than that of other species and was extensively used for crates, boxes, matchsticks, furniture, flooring, and paneling. By the end of the 19th century, eastern white pines, which originally occurred over vast tracts of the northeastern United States and Canada, had been decimated by wholesale logging done with no thought to conservation. Bald cypress trees in the southeastern United States met a similar fate. White pine blister rust also took its toll. Although new growth is now being promoted, most white pine lumber used today comes from large stands of western white pine in the Pacific Northwest.

The trunks of lodgepole pines are used in both the United States and Canada for telephone poles; the straight-grained wood is also used for railroad ties, mine timbers, and pulp.

Smog has severely damaged ponderosa and other native pines in California. For a number of years, the U.S. Forest Service has experimented with Afghanistan pine, a smog- and drought-resistant pine from Russia and adjacent areas, as a replacement for native trees. Growth rates in tests have been very rapid. Rapid growth is a desirable commercial feature, since considerably more timber can be produced in the same time needed to obtain it from slower-growing species, but the wisdom of introducing non-native plants into natural communities is in question, since there are many examples of such activities thoroughly disrupting delicate ecological balances.

The resin produced in the resin canals of conifers is a combination of a liquid solvent called *turpentine* and a waxy substance called *rosin.* When a conifer tree is wounded or damaged by insects, resin usually covers the area, sometimes trapping the insects. Out in the air, the turpentine evaporates quickly, leaving a protective layer of rosin, which deters water loss and fungal attacks. Both turpentine and rosin are very useful products, and a large industry centered in the southern United States and in the south of France is devoted to their extraction and refinement. They are often referred to as *naval stores,* a term that originated when the British Royal Navy used large quantities for caulking and sealing their sailing ships. Today, most naval stores and a third or more of the lumber used in the United States come from a group of southern yellow pines, particularly slash pine.

Pitch pine, also a source of naval stores before slash and other yellow pines became more profitable, was used in the past for the waterwheels of mills. Pitch pine wood was also used as fuel for steam engines, as it produces considerable heat when it burns. Turpentine is considered a premier paint and varnish solvent; rosin is used by musicians on violin bows and by baseball pitchers to improve their grip on the ball. Batters apply pine tar to the handles of bats to minimize slippage.

In the past, pine pitch was used by Native Americans for patching canoes, and it has been suggested that Noah's ark was sealed with pitch from aleppo pine. Pine resin was used for purifying wine in the 1st century A.D., and today, Greeks still add it to certain wines. *Colofonia* is the Spanish word for a type of resin Monterey pines produced abundantly around the old Spanish capital of Monterey. The early California priest Padre Arroyo suggested during the first half of the 19th century that California received its name from this Spanish word. California was, however, the name of a mythical paradise in a Spanish novel published in the early 16th century, and no more than an interesting similarity between two Spanish words may be involved.

The huge kauri pines (*Agathis*) of New Zealand are in a family separate from that of true pines. They are the source of a mixture of resins called *dammar* that is used in high-quality, colorless varnishes. Dammar was also the resin originally used in the manufacture of linoleum.

In New Zealand, dammar, also called *amber,* is obtained primarily in fossil form from former or present kauri pine forest areas. Most amber, however, has come from extinct conifers that flourished 60 to 70 million years ago in the Baltic area of the former Soviet Union and from other extinct conifers in what is now the Dominican Republic. It occurs as lumps of translucent material with a deep orange-yellow tint. According to Greek mythology, amber was the congealed tears of Phaëthon's sisters who, while weeping over his death, were turned into trees. Some of the lumps weigh up to 45 kilograms (100 pounds). The supply was at its peak at the turn of the century but is now nearing exhaustion. Remarkably lifelike preservations of prehistoric insects millions of years old have occurred in amber (Fig. 22.16).

Other products refined from resin are used in the manufacture of menthol for cigarettes (menthol also occurs naturally in members of the Mint Family), floor waxes, printer's ink, paper coatings, varnishes, and perfumes.

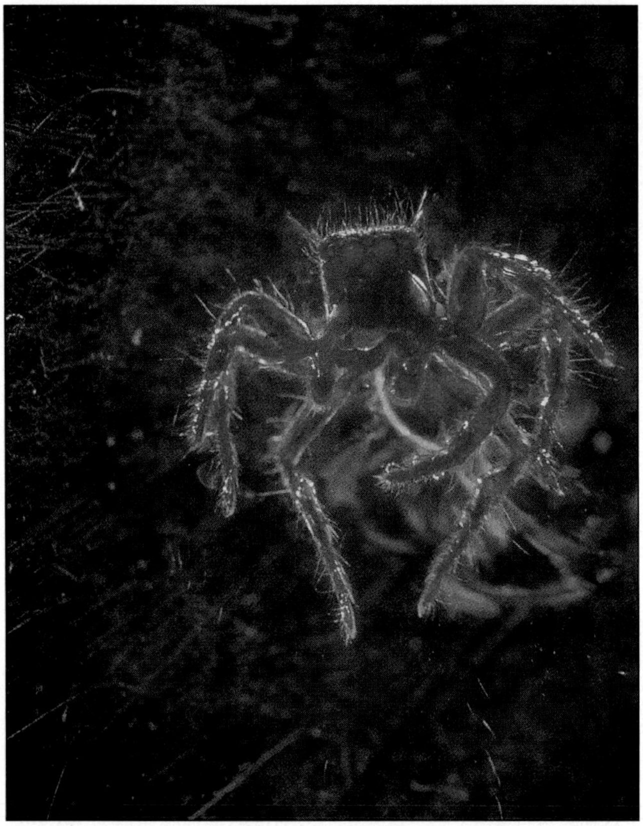

Figure 22.16 A prehistoric red wolf spider preserved in Dominican Republic amber. The amber is estimated to be between 20 and 30 million years old. *(Photograph by John Yellen)*

White spruce is the chief source of pulpwood for newsprint (Fig. 22.17) and other paper in North America. Enormous quantities of paper are used every day. A single midweek issue of a large metropolitan newspaper may use an entire year's growth of 50 hectares (123 acres) of these trees, and that amount may double for weekend editions. A large American publishing company, in an attempt to find ways of reducing paper consumption in the United States, tried trimming 2.5 centimeters (1 inch) from the width of all rolls of toilet paper in their building facilities. They found that the employees still used the same number of rolls per month as they had previously. From this, it was calculated that if all rolls of toilet paper were similarly trimmed in width, 1 million trees would be saved each year in the United States alone.

Split roots of the white spruce are quite pliable and were used by Native Americans for lashing canoes and for basketry. Spruce beer, brewed from young twigs and leaves with an added sugar source, such as honey or molasses, was once used as a remedy for scurvy. Resin of white, red, black, and Sitka spruces was used as a type of chewing gum by Native Americans, who sometimes hardened it slightly in cold water. Europeans who have tried it report that it has to be of the right consistency to be enjoyable, since it behaves like unhardened taffy if it is too soft and is bitter if it is too hard. In southeast Alaska, Sitka spruce buds are boiled in sugar water to make spruce bud syrup.

The tracheids of spruces have spiral thickenings on the inner walls. These apparently are responsible for giving the wood a resonance that makes it ideal for use as soundboards for musical instruments (Fig. 22.18). Sitka spruce of the Pacific Northwest produces a strong resilient wood that is a favorite material of manufacturers of light aircraft.

Larches, which along with the dawn redwood and bald cypress are exceptions to the rule that conifers are evergreens, have some of the most durable of all conifer woods. Fence posts of larch are known to last 20 years. In the southern and southwestern United States, posts of juniper wood and bald cypress last even longer, some remaining usable for 40 to 50 years or more. The resin of the western larch has been used in the manufacture of baking powder, and the European larch is the source of a special type of turpentine.

There are about 40 species of true firs that are widely used in the construction, plastic, and paper industries, as ornamentals, and as Christmas trees. The balsam fir produces on its bark blisters containing a clear resin. This resin, known as *Canada balsam,* was used in the past for cementing optical lenses and is still occasionally used for making permanent mounts on microscope slides. It has medicinal properties, too, and was used by New England colonists in sore throat medications.

Douglas fir, found in the mountains of the West, is not a true fir. In the Pacific Northwest, it grows into giant trees that are second only to the redwoods in size. It is probably the most desired timber tree in the world today. The wood is strong and relatively free of knots as a result of rapid growth, with less branching than most other conifers. It is heavily used in plywoods and is a major source of large beams. A useful wax is extracted from the bark of Douglas firs. Exploitation has nearly eliminated old-growth stands, but large numbers of new trees are being grown in managed forests.

Coastal redwoods are also prized for their wood, which contains substances that inhibit the growth of fungi and bacteria. The wood is light in weight, strong, and soft, but it splits easily. It is used for some types of construction, furniture, posts, greenhouse benches, and for many other purposes. California wineries use it extensively for wine barrels. The Giant Redwoods (Big Trees) are so huge that a double bowling alley was built on the log of the first specimen cut down. They are no longer logged and now serve almost exclusively as tourist attractions in California. They are, however, being planted as timber trees in Romania and Yugoslavia.

Wood of the bald cypress, found in southern swamps, is like that of the redwood in being very resistant to decay. In the past, it was used for railroad ties, coffins, general construction, guttering, and shingles. The trees are well known for their "knees," which rise above the water as tapering growths from the roots (Fig. 22.19). At one time, it was widely believed that these were a means of admitting oxygen

Figure 22.17 A modern newsprint factory. *(Courtesy International Paper Company)*

Figure 22.18 A violin with a soundboard made from red spruce.

to the roots, but this is now in doubt. They are favored for making knickknacks, such as wall ornaments and lamp bases. The leaves of the bald cypress yield a red dye.

A dull red dye can be obtained from the younger bark of the eastern hemlock, which is also a source of tannins for shoe leather. The tanning industry so depleted the native eastern hemlocks that it now has had to use tropical substitutes. The wood of these small trees contains exceptionally hard knots that can chip an axe blade. It sputters and throws out sparks freely when placed on a fire. Native Americans made a poultice for scrapes and cuts by pounding the inner bark. British Columbia Indians used to scrape out the cambium and phloem of both the western and mountain hemlocks for food, killing the trees in the process.

Northern white cedar, also known as *arborvitae,* is a favorite ornamental in temperate areas. The wood is pliable, and several Native American tribes used it for canoes. The Atlantic or southern white cedar was the first tree to be used for the construction of pipes for pipe organs in North America. During World War II, old logs of this species found in a swamp in New Jersey were milled and used in the construction of patrol torpedo boats.

The fleshy red aril surrounding the seeds of yews is sweet and edible, but the seeds themselves and other parts of the plants are very poisonous. The wood is tough and resilient and is favored for making bows.

Figure 22.19 Bald cypress trees in a southern swamp. Note the "knees" protruding above the surface.

English yew, the wood of long bows, changed history in 1415 and brought an end to the Middle Ages at the Battle of Agincourt, when the English long bow proved its superiority over heavily armored cavalry. A red or purple dye can be obtained from the bark and roots of yew. Podocarps, two species of which are valuable timber trees in New Zealand, have edible seeds.

In 1989, researchers at the Johns Hopkins Oncology Center reported that 12 women with advanced ovarian cancers that had not responded to traditional therapies (including, in some cases, surgery) had a decrease of 50% or more in the size of their tumors and one woman's tumor disappeared altogether after treatment with *taxol,* a drug obtained from the bark of the Pacific yew tree. Pacific yew trees are small and do not occur in extensive stands. They also grow so slowly it takes more than 70 years to attain their full size, and removal of their bark kills them. Despite the problems of the drug's availability, scientists are excited about the potential of the drug in the fight against ovarian cancer, and the trees are now being mass-produced in nurseries.

The wood of incense cedars has been used in the manufacture of venetian blinds and pencils. Red cedar wood is also used for pencils, as well as for cedar chests, closet lining, fence posts, and cigar boxes. It was used at one time in Germany for smoking hams. An aromatic oil used in floor-sweeping compounds is extracted from red cedar wood, and the "berries" of this and related junipers are widely eaten by birds. Many Native American tribes used the berries and inner bark as survival food during bleak winters. Some roasted the dried berries and brewed a beverage from them. Western red cedar was the most important single plant of Native Americans of the Northwest who used it for housing, clothing, nets, canoes, totem poles, medicines, and other purposes. Berries of the dwarf juniper are used to flavor gin. Some authorities indicate that the word *gin* may have been derived from *genievre,* the French word for juniper berry.

Other Gymnosperms

Despite the foul odor of the fleshy seed coat of seeds of *Ginkgo,* the starchy food reserves of the seeds themselves have a nutlike flavor punctuated with a hint of shrimp. In the Orient, *Ginkgo* seeds are widely used for food, either boiled

Awareness

A Living Fossil?

Just imagine a real-life Jurassic Park, where prehistoric dinosaurs long-thought extinct actually are alive. Well, something close to this happened in Australia in 1994. David Noble, while hiking on his Christmas holiday in the Wollemi National Park near Sydney, stepped into a stand of trees that, until that moment, were thought to have been extinct for over 150 million years. Known only as fossils, there they stood—about 40 pine trees growing in an isolated, rugged, rain-forest gully that had protected them all these years. This is one of the most significant botanical finds of the century—and the trees are among the rarest plants alive.

Now called the *Wollemi pine* (*Wollemia nobilis*), these conifer (cone-bearing) trees are so distinct they constitute a new genus. While most conifers have dark green leaves, Wollemi pine leaves are bright and lighter green, almost the color of green apples. The leaves are complex and unusual. The trees are bisexual, with bright green seed cones and brown cylindrical pollen cones on the tips of upper branches. Also distinctive is the corklike knobby bark, which is an unusual chocolate brown. The tallest tree is estimated to be 150 years old, towering 130 feet from the ground, with a trunk about 3 feet wide. The new trees belong to the plant family Araucariaceae, which, while found only in the Southern Hemisphere today (Norfolk Island pine is an example), had a worldwide distribution during the Jurassic and Cretaceous periods (208–66 million years ago). This suggests that the Wollemi pines' closest relatives lived when Australia, New Zealand, Africa, South America, and India were all parts of the supercontinent Gondwana. During this period, the eastern coast of Australia lay close to the South Pole, when worldwide climates were relatively warm to hot and wet.

Sydney's Royal Botanic Gardens and the New South Wales National Parks and Wildlife Service are jointly studying the Wollemi pine with methods ranging from scanning electron microscopy of the leaf and pollen, to DNA extraction and gene-sequencing studies. One priority is to study its propagation methods so the plant can be established in cultivation. Although the Wollemi site is within 200 kilometers of Sydney, Australia's largest city, the exact location remains a secret to protect the plant from seed collectors and poachers.

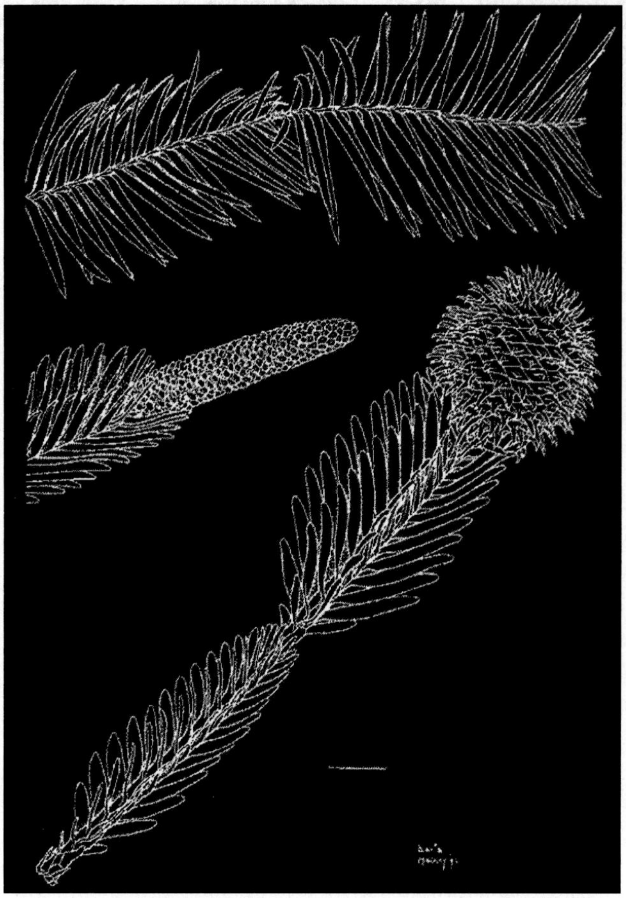

Box Figure 22.1 Drawings of Wollemi pine. *Top:* Leaves and branch; *Middle:* Pollen cone (male) at end of branch; *Bottom:* Seed cone (female). Scale bar = 2 cm. *Illustration by David MacKay, Royal Botanic Gardens, Sydney.*

The Wollemi pines, protected in their sheltered spot, not only escaped when their closest relatives died some 50 million years ago, but remained hidden until this exciting, remarkable find. *Wollemi* is an aboriginal word meaning "look around you." What an appropriate name for this real living fossil—it reminds all of us that there remains much to understand and explore in our world.

D. C. Scheirer

or roasted, and are found imported in canned form in Chinese food stores of large metropolitan areas in the United States and Canada. Extracts of *Ginkgo* plants have been clinically demonstrated to improve blood circulation in humans, and in 1998, more than 25 million Europeans were routinely taking *Ginkgo* extracts to counteract the effects of aging. Clinical studies have shown that *Ginkgo* leaf extracts do, indeed, improve blood supply to the brain and lungs, thereby improving both short-term memory and breathing.

Florida arrowroot starch was once obtained from the extensive cortex and pith of a species of cycad whose northernmost distribution occurs in Florida. Before it could be used for human consumption, however, a poisonous substance had to be leached out. Since cycads grow too slowly to make continued preparation of the arrowroot starch profitable, the practice has been abandoned. Today, cycads seen outside of their natural habitats are being grown primarily for ornamental or educational purposes. In Louisiana and elsewhere, however, the large compound leaves are used in Palm Sunday religious services. Some species are nearing extinction in the wild and may soon be known only in botanical gardens and conservatories.

In the southwestern United States, joint firs (*Ephedra*) are grazed by livestock, and the stems are still brewed into "Mormon tea." To offset a slight bitterness, a teaspoon of sugar, honey, or jam is added to each cup of tea. Native Americans and pioneers used a concentrated version of the tea in the treatment of venereal diseases. They also ground the seeds into flour from which a bitter bread was made. The drug *ephedrine,* widely used in the treatment of asthma and other respiratory problems, still is extracted from a Chinese species, but most of the drug now in use is synthetically produced. The Chinese used ephedrine medicinally more than 4,700 years ago. Continued medicinal use of ephedrine is now questionable because of potentially serious side effects.

One species of *Gnetum* is cultivated in Java for the shoots, which are eaten after being cooked in coconut milk. Fibers from the bark are made into a rope.

See Appendix 1 for scientific names of species discussed in this chapter.

Summary

1. The exposed seeds of gymnosperms are produced on sporophylls, forming a strobilus, or cone. The seed (female) cones and smaller pollen (male) cones are produced on the sporophytes.
2. A female gametophyte develops within a nucellus that is enclosed in an integument inside an ovule. The integument becomes the seed coat of a seed after fertilization, and the female gametophyte nurtures the development of the embryo.
3. Four phyla of living gymnosperms are recognized. Phylum Pinophyta includes the conifers. *Ginkgo* is the sole living representative of Phylum Ginkgophyta. The cycads are in Phylum Cycadophyta. Phylum Gnetophyta includes the gnetophytes *Ephedra, Gnetum,* and *Welwitschia.*
4. Pines are the most numerous conifers. The needlelike leaves are arranged in clusters of two to five and have modifications adapting them to harsh environments.
5. Resin canals, occurring throughout the plants, secrete resin that inhibits fungi and certain insect pests.
6. Pine xylem lacks vessel members and fibers and is relatively soft. The phloem lacks companion cells but has albuminous cells that apparently perform the same function. Pine roots are always associated with mycorrhizal fungi, which are essential to normal development of the plants.
7. Two kinds of spores are produced. Microspores are produced in papery pollen cones that, in turn, develop in clusters toward the tips of lower branches. The microsporangia develop in pairs toward the base of the pollen cone scales and give rise to four-celled pollen grains that occur in huge numbers.
8. Megaspores are formed in ovules at the bases of seed cone scales. The integument of the ovule has a pore called the micropyle. One megaspore develops into a female gametophyte. The mature female gametophyte contains archegonia.
9. Before the archegonia mature, pollen grains catch in sticky pollination drops between the cone scales. Each pollen grain produces a pollen tube that digests its way down to the developing archegonia, and two of the original four cells in the pollen grain migrate into the tube as it grows. The generative cell divides and produces a sterile cell and a spermatogenous cell that itself divides, producing two sperms.
10. After pollination, one sperm unites with the egg, forming a zygote. The zygote develops into an embryo of a seed that has a membranous wing formed from a layer of the cone scale.
11. Some conifers produce seeds enclosed in fleshy or berrylike coverings. Their evolutionary origin is not clear.
12. *Ginkgo* has small fan-shaped leaves with evenly forking veins. The life cycle of *Ginkgo* also is similar to that of cycads. The edible seeds are enclosed in a fleshy covering that has a rank odor at maturity.
13. Cycads superficially resemble palm trees with unbranched trunks and crowns of large, pinnately divided leaves. They have strobili and life cycles similar to those of conifers, but their sperms, unlike those of pines, have numerous flagella.
14. Gnetophytes all have vessels in their xylem. Half the species are in the genus *Ephedra,* whose members have jointed stems and leaves reduced to scales.

Gnetum species have broad leaves and occur in the tropics, primarily as vines. *Welwitschia* is confined to southwest African deserts. Its stem is in the form of a shallow cup with straplike leaves that extend from the rim; basal meristems on the leaves constantly add to their length.

15. The seeds and inner bark of pines are edible, and a tea has been made from the leaves. Eastern white pine stems were used as masts for sailing vessels and for crates, furniture, flooring, paneling, and matchsticks. Western white pine is the source of most such lumber today.
16. Resin from pines consists of turpentine and rosin. Turpentine is used as a solvent, and resin is used by musicians and by baseball players. Dammar from kauri pines is used in colorless varnishes. Amber is fossilized resin. Resin is also used in floor waxes, printer's ink, paper coatings, perfumes, and the manufacture of menthol.
17. White spruce is the chief source of newsprint. It was also used for basketry and canoe lashing by Native Americans, with molasses or honey for treating scurvy, and in brewing a beer. Spruce resin was used for a type of chewing gum. The wood is used as soundboards for musical instruments and in the construction of aircraft.
18. Larch and juniper woods are used for fence posts. Firs are used in the construction, paper, ornament, and Christmas-tree industries. Douglas fir is probably the most desired timber tree in the world today.
19. Coastal redwoods are also prized for their wood, which is resistant to fungi and insects. Bald cypress wood, used in the past for coffins and shingles, is also resistant to decay.
20. A dye and tannins are obtained from the eastern hemlock. Native Americans used parts of hemlocks for poultices and for food.
21. Eastern white cedar's wood was used for canoes, and that of the Atlantic cedar was used for construction of pipes for pipe organs. Yew wood is used for making bows, and an extract has potential for the treatment of human ovarian cancer.
22. Podocarps of New Zealand have edible seeds. Incense cedar wood is used for cedar chests, cigar boxes, pencils, and fence posts. Juniper berries are used to flavor gin and were used by Native Americans for food and a beverage.
23. *Ginkgo* seeds are edible, and *Ginkgo* plant extracts are used to improve blood circulation. Arrowroot starch was once obtained from a cycad. Mormon tea is brewed from the leaves and stems of joint firs (*Ephedra*), which, in the past, were also a source of the drug ephedrine and a venereal disease treatment. One *Gnetum* species is cultivated in Java for food.

Review Questions

1. What is a gymnosperm? How is it distinguished from any other kind of organism?
2. What is the difference between a seed and a spore?
3. How are the leaves of pines different from those of broadleaf flowering plants?
4. What is a resin canal, and what is its function? Where are resin canals found?
5. How do pines differ in their reproduction from ground pines and ferns?
6. How do pollen grains differ from spores or sperms?
7. Which conifers discussed in this chapter do not have woody seed cones?
8. If you had samples of leaves of a pine, *Ginkgo,* a cycad, a joint fir, and *Welwitschia* all together, indicate how you could tell them apart by constructing a key.
9. What parts of a pine are considered edible?
10. What is resin? Discuss some of its uses, past and present.

Discussion Questions

1. *Ginkgo* and cycads have broad leaves, whereas those of pines are needlelike. Can you suggest any significance of this in terms of the climates and habitats involved?
2. If no distinction were made at the level of kingdom between plants and animals, what would be the equivalent, if any, of sporophyte and gametophyte in humans?
3. Both bristlecone pines and redwoods can live to be thousands of years old. What do you suppose makes this possible?
4. Most of the old-growth stands of conifers in North America are now gone, and others will be gone soon. Much of what has been harvested is being replaced with new growth, sometimes with hybrids and non-native plants. Our forests are essential to our economy as we know it. If you had the power to change the way we manage and exploit our natural forest resources, what would you do differently, assuming you did not want to damage the economy?
5. If money were no object and you wished to landscape your yard primarily with gymnosperms, realistically, what would you include, taking into account your particular geographical area? Why?

Learning Online

Visit our web page at www.mhhe.com/botany for interesting case studies, practice quizzes, current articles, and animations within the Online Learning Center to help you understand the material in this chapter. You'll also find active links to these topics:

Diversity of Plants
Gymnosperms

Additional Reading

Beck, C. B. (Ed.). 1988. *Origin and evolution of gymnosperms.* New York: Columbia University Press.

Bever, D. N. 1981. *Northwest conifers: A photographic key.* Portland, OR: Binford and Mort.

Bold, H. C., C. S. Alexopoulos, and T. Delevoryas. 1987. *Morphology of plants and fungi,* 5th ed. New York: Harper & Row.

Farjon, A., and B. T. Styles. 1997. *Pinus (Pinaceae).* Brooklyn, NY: New York Botanical Garden.

Johri, B. M., and C. Biswas. 1997. *The gymnosperms.* New York: Springer-Verlag.

Krussman, G. 1985. *Manual of cultivated conifers.* Portland, OR: Timber Press.

Nimsch, H. 1995. *A reference guide to the gymnosperms of the world.* Champaign, IL: Balogh.

Smith, W. K., and T. M. Hinckley (Eds.). 1994. *Resource physiology of conifers: Acquisition, allocation, and utilization.* San Diego, CA: Academic Press.

Van Gelderen, D. M. 1996. *Conifers: The illustrated encyclopedia,* 2 vols. Portland, OR: Timber Press.

Van Pelt, R. 2001. *Forest giants of the Pacific Coast.* Seattle, WA: University of Washington Press.

Vidakovic, M. (Transl. by M. Soljan). 1991. *Conifers. Morphology and variation.* Zagreb, Croatia: Graficki Zavod Hrvatske. Distr. by University of Arizona Press.

Weiner, M. A., and J. A. Weiner. 1994. *Herbs that heal.* Mill Valley, CA: Quantum Books.

Flowers of the butterfly weed (*Asclepias curassavica*), a native of the tropics from Florida to South America. Parts of the plant are used medicinally in tropical areas. In Mexico, for example, root extracts have been used in the treatment of cancers of the intestinal tract, kidneys, and uterus.

Seed Plants: Angiosperms

Chapter Outline

23

Overview

This chapter begins with a probe of the differences between angiosperms and gymnosperms and then discusses the origin of flowering plants. The exceptional diversity of the flowering plants is reviewed next, and the chapter continues with a description of the parallel development of the gametophytes in the anthers and ovules. This leads up to pollination, followed by some details of fertilization and the development of a seed. A discussion of trends of specialization and classification in flowering plants is followed by a section on pollination ecology. Plant preservation, including simple herbarium techniques and practice, is next. The chapter closes with a brief survey on the uses of herbaria and a word of caution to plant collectors concerning unnecessary depletion of native floras.

Some Learning Goals

1. Understand the basic differences between angiosperms and gymnosperms.
2. Contrast two principal schools of thought concerning the origin of the flowering plants and the nature of the first flowers.
3. Diagram the life cycle of a flowering plant, indicating shifts from haploid to diploid cells, and vice versa.
4. Compare two types of female gametophyte development and learn how a male gametophyte develops.
5. Know the characteristics of flowers associated with specific types of pollinators.
6. Know major trends of specialization in the flowering plants.
7. Know the functions of a herbarium and the techniques of preparing herbarium specimens.

A flowering plant once saved my life in an unconventional way. While in Washington's Olympic National Park, I was conducting field research on a group of relatively small herbaceous plants known as *bleeding hearts* (Fig. 23.1). At one point, I found myself on a slope of loose shale above a steep cliff. Suddenly, the shale started to move, and before I could do anything about it, I was rapidly sliding on my back, feet first, directly toward the edge of the cliff. There were no bushes, trees, or anything stable I could cling to or grab that would at least slow my journey to doom in an avalanche of shale. Just a few feet before I was to hurtle over the cliff, however, I cannoned into a large clump of bleeding hearts that somehow had rooted securely enough beneath the shale surface to stop my speedy plunge. Obviously, I lived to tell the tale, and it won't surprise you to know that ever since the experience, I have had a special place in my garden for bleeding hearts!

Figure 23.1 Pacific bleeding hearts (*Dicentra formosa*).

The Pacific bleeding hearts that saved my life are only one of more than 250,000 known species of flowering plants that comprise, by far, the largest and most diverse of the phyla of the Plant Kingdom. These flowering plants are called **angiosperms.**

The term *angiosperm* is derived from two Greek words: *angeion,* meaning "vessel," and *sperma,* meaning "seed." The "vessel" is the **carpel,** which is like an inrolled leaf with seeds along its margins. A green pea pod, for example, is a carpel that resembles a leaf that has folded over and fused at the margins, enclosing the attached seeds. Many flowers have *pistils* (see Chapter 8) composed of either a single carpel or two more united carpels. A seed develops from an *ovule* within a carpel and is part of an *ovary* that becomes a *fruit.* Although the angiosperms generally have organs and tissues similar to those of the gymnosperms, the enclosed ovules and seeds of the angiosperms distinguish them from gymnosperms, which have exposed ovules and seeds.

All angiosperms are presently considered to be in the Phylum Magnoliophyta (previously known as the Anthophyta). Other classifications of flowering plants are discussed in Chapter 16. Phylum Magnoliophyta is divided into two large classes, the Magnoliopsida (dicots) and the Liliopsida (monocots). Table 8.1 on page 135 lists the distinguishing features of members of the two classes.

Since Darwin's *Origin of Species* appeared in 1859, there have been two major theories concerning the origin of angiosperms. The older and now more or less disregarded theory was held by the German botanist Adolph Engler and his followers. It suggested that flowering plants evolved from conifers and that primitive flowers are similar in structure to the strobili of conifers. Such flowers are inconspicuous and in clusters, such as those of the grasses, oaks, willows, and cattails.

ntemporary botanists, however, point to anatom- al, and fossil evidence, along with cladistic inter- hat indicate angiosperms evolved independently *teridosperms* (seed ferns—discussed in Chapter ilso hypothesize that a flower is really a modified ng modified leaves. The most primitive flower is thought to be one with a long receptacle and many spirally arranged flower parts that are separate and not differentiated into sepals and petals. In addition, the stamens and carpels are flattened and numerous. Such flowers are found among relatives of magnolias and buttercups, leading many modern botanists to postulate that all present-day flowering plants are derived from a primitive stock with such characteristics. A further discussion of primitive and advanced characteristics follows later in the chapter.

Figure 23.2 Snowplant flowers. *(Courtesy Robert A. Schlising)*

PHYLUM MAGNOLIOPHYTA—THE FLOWERING PLANTS

The plants of members of this phylum vary greatly in size, shape, texture, form, and longevity. The phylum includes, for example, the tiny duckweeds that may be less than 1 millimeter long, all the grasses and palms, many aquatic and epiphytic plants, and most shrubs and trees, including the huge *Eucalyptus regnans* trees of Tasmania that rival the redwoods in total volume.

A few flowering plants are parasitic. Dodders, for example, occasionally cause serious crop losses as they twine about their hosts and, by means of haustoria (shown in Fig. 5.15B), intercept food and water in the host xylem and phloem. Broomrapes also parasitize a variety of plants, as do mistletoes. Mistletoes produce some chlorophyll and depend only partially on their hosts for food. Still others, such as the beautiful snowplant (Fig. 23.2) and some of the orchids, are *saprophytes* (i.e., their nutrition comes mostly from the absorption in solution of dead organic matter). The vast majority of flowering plants, however, produce their food independently through photosynthesis.

Like the gymnosperms, the angiosperms are *heterosporous* (produce two kinds of spores), and the sporophytes are even more dominant than in the gymnosperms. The female gametophytes are wholly enclosed within sporophyte tissue and reduced to only a few cells. At maturity, the male gametophytes consist of a germinated pollen grain with three nuclei.

Development of Gametophytes

While the flower is developing in the bud, a diploid **megasporocyte** cell differentiates from all the other cells in the ovule (Fig. 23.3). This megasporocyte undergoes meiosis, producing four haploid *megaspores.* Soon after they are produced, three of these megaspores degenerate and disappear, but the nucleus of the fourth undergoes mitosis, and the cell enlarges. While the cell is growing larger, its two haploid nuclei divide once more. The resulting four nuclei then divide yet another time. Consequently, eight haploid nuclei in all are produced (without walls being formed between them). By the time these three successive mitotic divisions have been completed, the cell has grown to many times its original volume. At the same time, two outer layers of cells of the ovule differentiate. These layers, called **integuments,** later become the **seed coat** of the seed. As they develop, they leave a pore, or gap, called the **micropyle,** at one end.

At this stage, there are eight haploid nuclei in two groups, four nuclei toward each end of the large cell. One nucleus from each group then migrates toward the middle of the cell. These two **central cell nuclei** may become a binucleate cell, or they may fuse together, forming a single diploid nucleus. Cell walls also form around the remaining nuclei. In the group closest to the micropyle, one of the cells functions as the female gamete, or *egg.* The other two cells, called *synergids,* either are destroyed or degenerate during or after events that occur later. At the other end, the remaining three cells, called *antipodals,* have no apparent function, and later they also degenerate. The large sac, usually containing eight nuclei in seven cells, constitutes the *female gametophyte* (**megagametophyte**), formerly known as the *embryo sac* (Fig. 23.4).

Usually while the megagametophyte is developing, a parallel process that leads to the formation of male gametophytes takes place in the **anthers.** As an anther develops, four patches of tissue differentiate from the main mass of cells. These patches of tissue contain many diploid **microsporocyte** cells, each of which undergoes meiosis, producing a *quartet* (also referred to as a *tetrad*) of **microspores.** Four chambers or cavities (pollen sacs) lined with nutritive *tapetal* cells are visible in an anther cross section by the time the microspores have been produced. As the anther matures, the walls between adjacent pairs of chambers break down so that only two larger sacs remain.

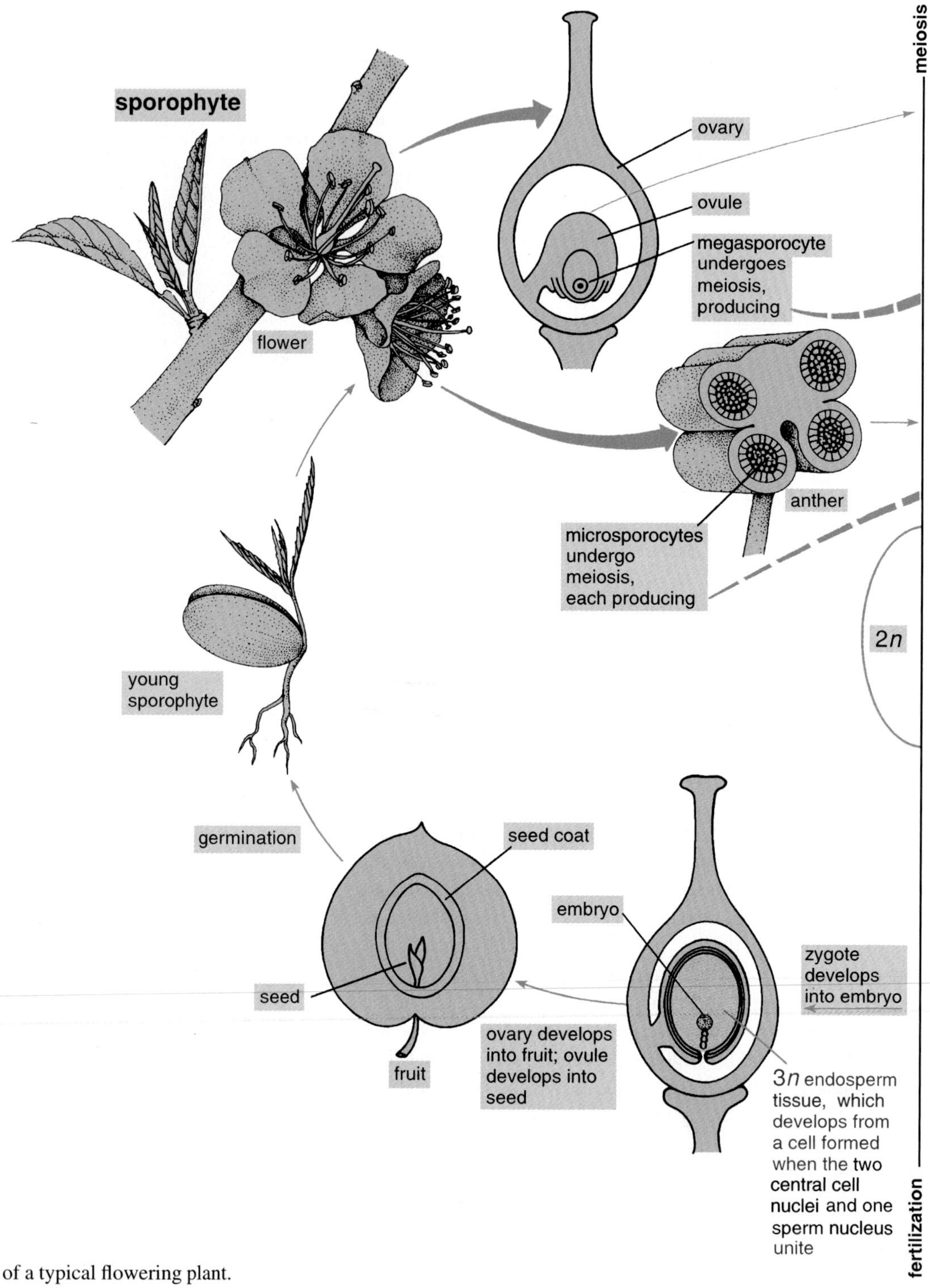

Figure 23.3 Life cycle of a typical flowering plant.

After meiosis, the haploid microspores in the pollen sacs undergo several changes more or less simultaneously, the processes usually taking from a day or two to a couple of weeks. The following three changes are the most important:

1. the nucleus in each microspore divides once by mitosis;
2. the members of each quartet of microspores separate from one another (in some species, the separation does not occur, but this is unusual);
3. a two-layered wall, whose outer layer is often finely sculptured, develops around each microspore.

When these events are complete, the microspores have become **pollen grains** (Fig. 23.5). The outer layer of the pollen grain wall, called the **exine,** which is often sculptured, contains chemicals that may later react with other chemicals in the stigma of a flower. As a result of these reactions, the pollen grain may germinate or its further develop-

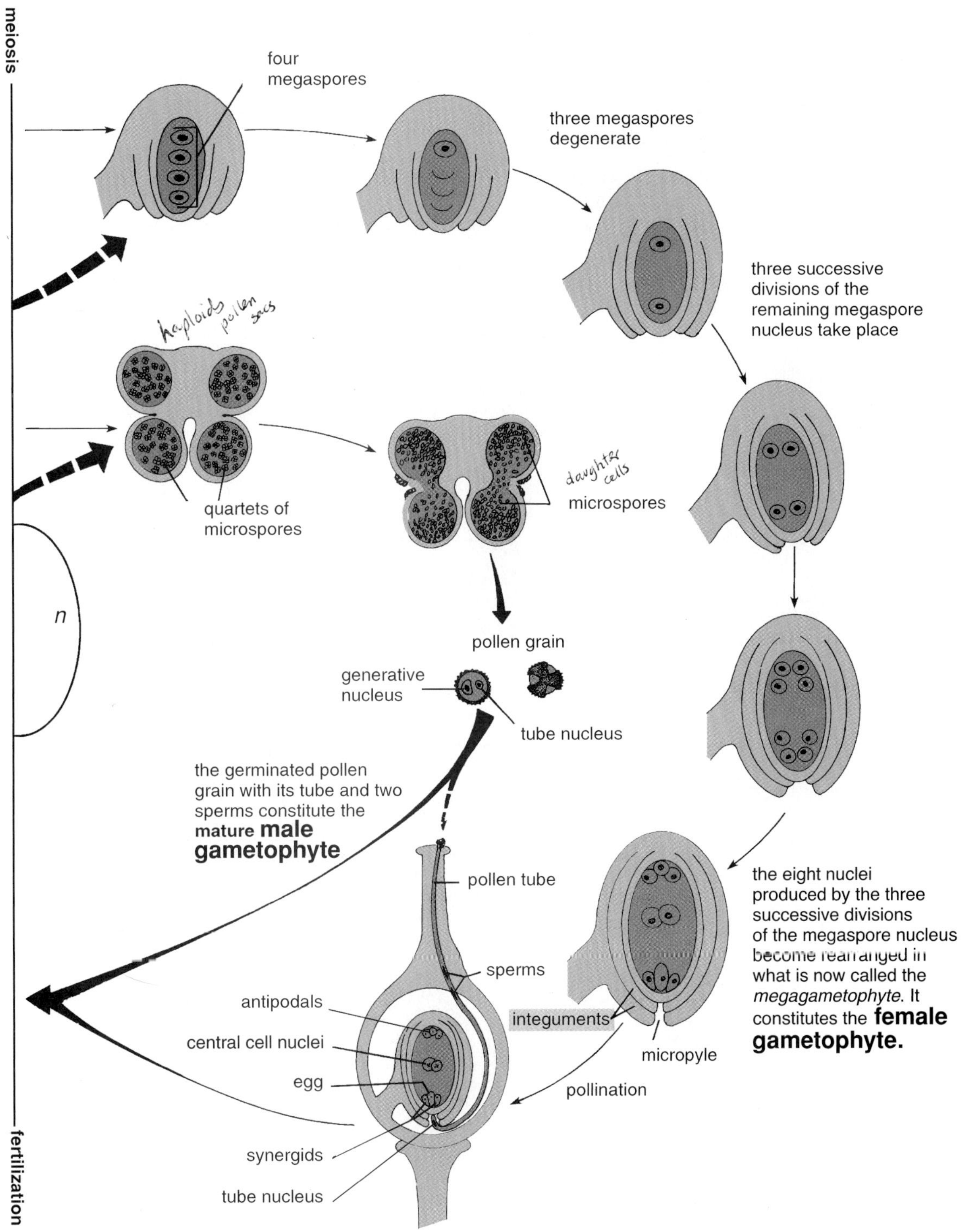
meiosis
four megaspores
three megaspores degenerate
three successive divisions of the remaining megaspore nucleus take place
haploids pollen sacs
quartets of microspores
daughter cells
microspores
n
pollen grain
generative nucleus
tube nucleus
the germinated pollen grain with its tube and two sperms constitute the mature male gametophyte
pollen tube
sperms
antipodals
central cell nuclei
egg
synergids
tube nucleus
integuments
micropyle
pollination
the eight nuclei produced by the three successive divisions of the megaspore nucleus become rearranged in what is now called the megagametophyte. It constitutes the female gametophyte.
fertilization

ment may be blocked, depending on whether or not it originated from the same plant, another plant of the same species, or a plant of a different species.

The cytoplasm of pollen grains is often rich in vitamins, and pollen is often collected and sold in health-food stores as a food supplement. The exine of many pollen grains is virtually indestructible, and pollen has been found preserved in deposits thousands of years old. This has proved invaluable in archaeological investigations. The sculpturing of pollen grain exine of various species is also distinctive enough to allow identification to family, genus, and even species in some instances. This feature has resulted in its use in forensic investigations of criminal cases.

Many pollen grains have three boat-shaped or porelike thin areas (*apertures*) in the wall, but the number of apertures may range from one to many. The development of the male gametophyte, discussed in the next section,

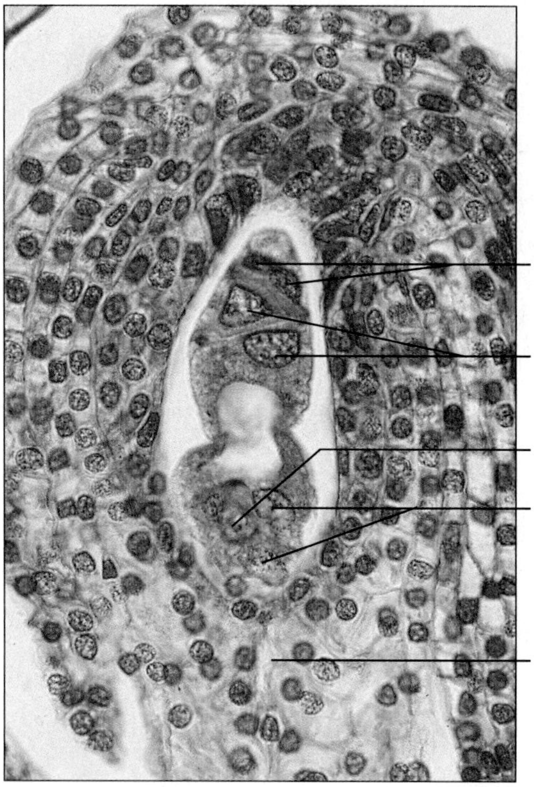

Figure 23.4 A mature megasporocyte of a lily (*Lilium* sp.).

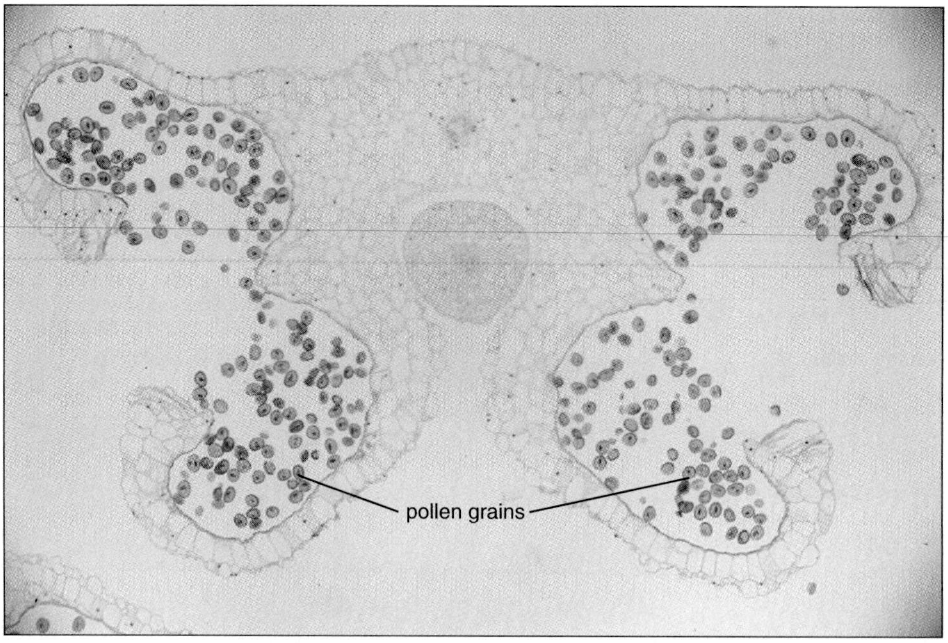

Figure 23.5 A cross section of a lily (*Lilium* sp.) anther.

"Fertilization and Development of the Seed," may involve any of the apertures. One of the pollen grain's two nuclei, the **generative nucleus,** will later divide, producing two nuclei that become surrounded by a plasma membrane and function as *sperm* cells. Unlike the sperm of many plants discussed in preceding chapters, angiosperm sperms have no flagella. The remaining **vegetative nucleus** (often referred to as the *tube nucleus*) is involved in events that take place after the pollen grain has left the anther (Fig. 23.6).

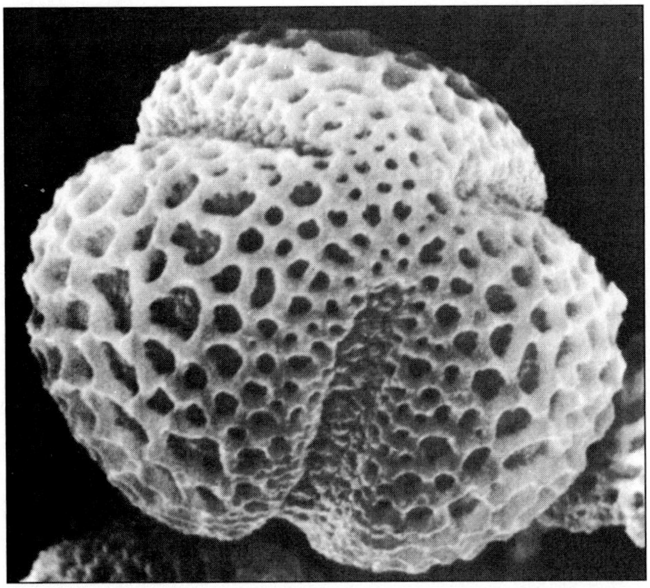

Figure 23.6 Scanning electron micrograph of a poppy pollen grain. The furrows are apertures through which a pollen tube may later emerge.

Ecological Review

Pollination occupies center stage in the life cycles of flowering plants, which depend on various aspects of their environment to pollinate their flowers. The most important vehicles of pollination are wind and insects, but pollinators also include birds, mammals, water, and gravity. Those flowering plants that are pollinated by animals have often evolved complex mutualistic relationships with their pollinators—flowers of particular colors, shapes, and fragrances attracting and rewarding them with appropriate amounts of nectar and/or pollen. However, pollinator behavior and pollination is manipulated by some plants through force, chemicals, or sexual enticement. Flowering plants, which are also attractive to people, are increasingly threatened by uncontrolled overcollecting by professional botanists, nursery personnel, and amateurs.

Pollination

One often reads in the press or popular magazines about flowers being *pollenized* (a coined word not found in dictionaries), and the implication is made that the dusting of the pollen of one flower upon another is the equivalent of fertilization. Not so! **Pollination** is simply the transfer of pollen grains from an anther to a stigma, nothing more. **Fertilization** involves the union of egg and sperm, and it may not occur until days or weeks or months after pollination has taken place, or it may not follow pollination at all.

Most pollination is brought about by insects or wind, but in many species, water, birds, bats, other mammals, and gravity act as agents or pollinators. The adaptations between the flower and its pollinators can be intricate and precise and may even in certain instances involve force, chemicals, or sexual enticement (as shown in Fig. 23.18). In some instances, self-pollination may occur, with the pollen grains germinating on the stigma of the same flower in which they were produced. The genetic implications of self-pollination and cross-pollination are discussed in Chapters 13 and 14. A discussion of pollination ecology is given later in this chapter.

Fertilization and Development of the Seed

After pollination, with the notable exception of a number of crop plants (e.g., peas, apricots), further development of the male gametophyte may not take place unless the pollen grain is (1) from a different plant of the same species or (2) from a variety different from that of the flower receiving it.

Under suitable conditions, the dense cytoplasm of the pollen grain absorbs fluids from the stigma and bulges out through one of the apertures in the form of a tube. This **pollen tube** then grows down between the cells of the stigma and style until it reaches the micropyle of the ovule. In corn, it may have to grow more than 50 centimeters (20 inches) before it arrives at its destination, but in most plants, the distance is considerably less.

The pollen tube's journey may last less than 24 hours and usually does not take more than 2 days, although there are a few plants in which growth takes over a year. As the tube grows, most of the contents of the pollen grain are discharged into it. The vegetative nucleus stays at the tip, while the generative nucleus (cell) lags behind and divides by mitosis, usually in the tube, producing two nuclei that become sperm cells; no flagella develop. Sometimes, the generative nucleus (cell) divides before the pollen tube has formed. The germinated pollen grain with its vegetative nucleus and two sperms constitutes the *mature male gametophyte* (*microgametophyte*) (Fig. 23.7).

When the pollen tube reaches the micropyle, it continues on to the *female gametophyte* (*megagametophyte*), which it enters, destroying a now degenerating synergid in the process; it then either bursts or forms a pore, discharging its contents. Next, an event unique to angiosperms (and a few gnetophytes), called **double fertilization** (or **double fusion**), takes place. One sperm migrates from the synergid to the egg, losing most of its protoplasm along the way. The sperm cell nucleus then unites with the egg nucleus, forming a **zygote.**

The other sperm cell also migrates from the synergid, and, upon reaching the central cell nuclei, unites with them, producing a $3n$ (triploid) *endosperm nucleus.* The endosperm

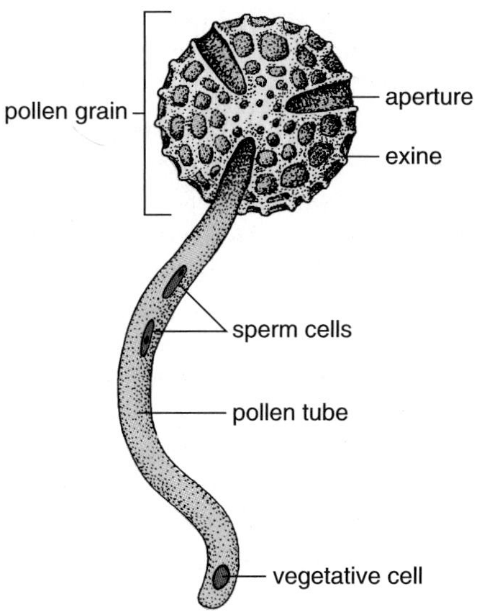

Figure 23.7 A mature male gametophyte of a flowering plant.

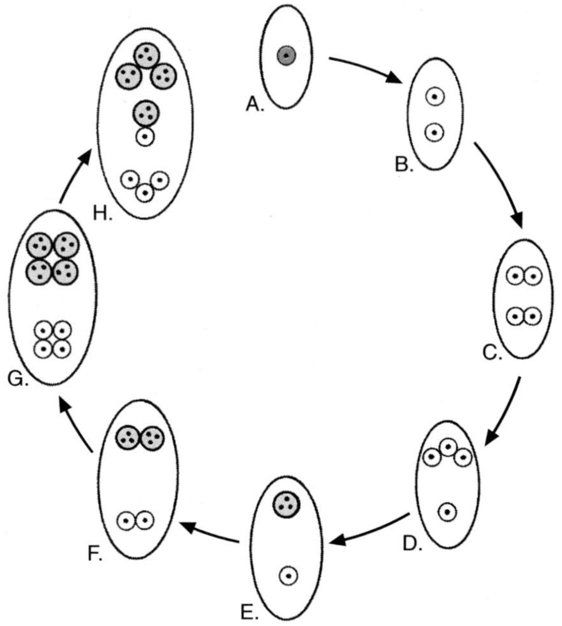

Figure 23.8 How the megasporocyte of a lily develops. *A.* The process begins with a diploid (2*n*) megasporocyte. *B.* and *C.* The megasporocyte undergoes meiosis, producing four haploid (*n*) megaspore nuclei. *D.* and *E.* The megaspore nuclei unite, forming a triploid (3*n*) nucleus; the other haploid nucleus remains separate. *F.* and *G.* Both the 3*n* and *n* nucleus undergo two consecutive mitotic divisions, resulting in four 3*n* and four *n* nuclei. *H.* Three of the 3*n* nuclei function as antipodals; the fourth 3*n* nucleus and one *n* nucleus function as central cell nuclei. The remaining haploid nuclei function as an egg and two synergids.

nucleus becomes exceptionally active and divides repeatedly by mitosis, with cell cycles being completed every few hours, usually without formation of cell walls. This nutritive 3*n* tissue, called **endosperm,** may eventually have hundreds of thousands of nuclei; it surrounds the embryo that develops from the zygote.

In some monocots, such as corn and other grasses, the endosperm tissue becomes an extensive part of the seed, but in most dicots, such as members of the Legume Family (Fabaceae; e.g., peas, beans), the endosperm provides nutriment for the embryo that develops from the zygote but is absorbed into the cotyledons by the time the seed is mature. The integuments harden, becoming a *seed coat,* and the remaining haploid nuclei or cells (antipodals, synergids, and tube nucleus) degenerate. At the conclusion of these various events, the ovule has become a seed, and at the same time, the ovary matures into a fruit. Seed dispersal and fruits are discussed in Chapter 8.

Other Types of (Female) Gametophyte Development

The process of female gametophyte development just described in the section "Fertilization and Development of the Seed" occurs in about 70% of the known flowering plants. The remaining 30% exhibit variations in which the female gametophyte has from 4 to 16 nuclei or cells at maturity, and the endosperm may be 5*n*, 9*n*, or even 15*n*.

One such variation occurs in lilies, a favorite of microscope-slide manufacturers for showing female gametophyte development. When a megasporocyte undergoes meiosis in a lily, all four of the haploid megaspore nuclei produced remain functional nuclei (not cells). Three of the megaspore nuclei unite, forming one 3*n* nucleus; the fourth megaspore nucleus remains *n*. Both the 3*n* nucleus and the *n* nucleus undergo mitosis twice, resulting in a large cell with four 3*n* nuclei plus four *n* nuclei. The large cell with eight nuclei becomes the female gametophyte; one of the central cell nuclei is 3*n*, and the other central cell nucleus is *n*. During fertilization when one sperm (*n*) unites with the two central cell nuclei, a 5*n* endosperm nucleus is produced (Fig. 23.8).

Apomixis and Parthenocarpy

Some embryos of seeds can develop *apomictically;* that is, without development or fusion of gametes (sex cells) but with the normal structures (e.g., ovaries) otherwise being involved. An embryo may develop, for example, from a 2*n* nutritive cell or other diploid cell of an ovule, instead of from a zygote. After germination, this makes the plant that develops from such a seed the equivalent of a vegetatively propagated plant. Fruits that develop from ovaries having unfertilized eggs are said to be **parthenocarpic.** Such fruits are seedless and are found in navel oranges, supermarket bananas, and certain varieties of figs and grapes. *Apomixis* was discussed in Chapter 15.

To further complicate matters, some seedless fruits aren't parthenocarpic. For example, when Thompson seedless grapes are fertilized, the ovules don't develop with the fruit. Also, applying dilute hormone sprays to flowers can bring about artificial parthenocarpy. Seedless tomatoes are often produced in this way. Crossing watermelon varieties with different chromosome numbers results in hybrid watermelons that are seedless because the chromosomes can't pair properly during meiosis, and fertilization and seed formation don't occur.

Trends of Specialization and Classification in Flowering Plants

Various classifications of plants have been proposed ever since the 4th century B.C. when Theophrastus grouped plants into trees, shrubs, and herbs. The first classifications were merely for convenience and did not necessarily reflect natural relationships. Even today, plants may be lumped together in unnatural groupings in order to make them easier to identify. For example, some wildflower books arrange together all white-flowered species or all yellow-flowered species. There is nothing wrong with such identification arrangements, but because such schemes do not reflect natural relationships (i.e., relationships based on heredity and evolution), it is often difficult to recognize family characteristics or lineage. We don't infer that all persons with red hair are more closely related to each other than they are to those with dark hair or that all long-haired dogs are more closely related to each other than they are to short-haired dogs. Modern botanists, therefore, try to group plants according to their natural relationships. These are based on evidence gleaned from breeding experiments, form and structure, chemical components, fossil records, and other features. There are, however, many interpretations of trends in the specialization of flowering and other plants that are based primarily on inference from evidence presently available.

Although the information is very incomplete, the fossil record suggests that the flowering plants first appeared about 160 million years ago during the late Jurassic period (see Table 21.1), although the oldest known fossil of a flowering plant (discovered in China in 1998) has been dated at 142 million years. The flowering plants then developed during the Cretaceous period and the ensuing periods of the Cenozoic era into the dominant elements of the flora they are today. Most botanists hypothesize that primitive flowering plants had their origins in the Jurassic period and had the following features: their leaves were simple; the flowers had numerous spirally arranged parts that were not fused to each other and were variable in number; the flowers were also *radially symmetrical,* or *regular* (i.e., the flowers could be divided into two equal halves along more than one lengthwise plane); and the flowers had both stamens and pistils.

Although today there still are many plants whose flowers have primitive features, various specializations and modifications have occurred since flowers first appeared. Flower parts have become fewer and definite in number. Some parts have fused, and spiral arrangements have been compressed to whorls (Fig. 23.9).

The first pistil is believed to have been formed from a leaflike structure with ovules along its margins. The edges of the blade, called a *carpel,* apparently rolled inward and fused together. In due course, the separate carpels of primitive flowers fused together, forming the common *compound pistil* consisting of several carpels, which is found in many of today's angiosperms (Fig. 23.10). Each segment of an orange, for example, represents a single carpel, and three carpels are easily distinguishable in a cross section of a cucumber.

In advanced flowers, the *receptacle* or other flower parts have fused to the ovary and grown up around it, so that the calyx and the corolla appear to be attached at the top of the ovary. When the ovary is embedded in the receptacle and other parts, it is said to be an **inferior ovary,** and the flower parts attached at the top of the ovary are said to be **epigynous.** A more primitive **superior ovary** is produced on top of the receptacle with the other flower parts attached around its base; such flower parts are said to be **hypogynous.** Some flowers have ovaries in an intermediate or *half-inferior position,* with **perigynous** flower parts, which are usually attached to a corolla tube of fused petals; the floral tube itself is generally not fused to the ovary (Fig. 23.11). Flowers have

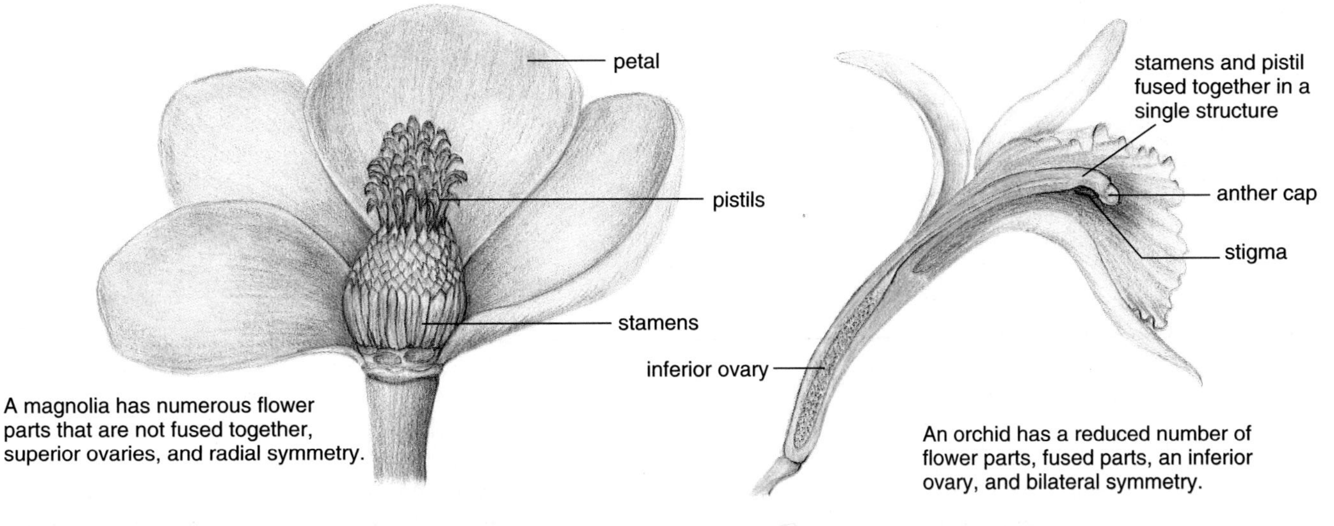
petal
pistils
stamens
A magnolia has numerous flower parts that are not fused together, superior ovaries, and radial symmetry.
stamens and pistil fused together in a single structure
anther cap
stigma
inferior ovary
An orchid has a reduced number of flower parts, fused parts, an inferior ovary, and bilateral symmetry.

Figure 23.9 Comparison between a primitive flower, magnolia (*left*) and an advanced flower, orchid (*right*).

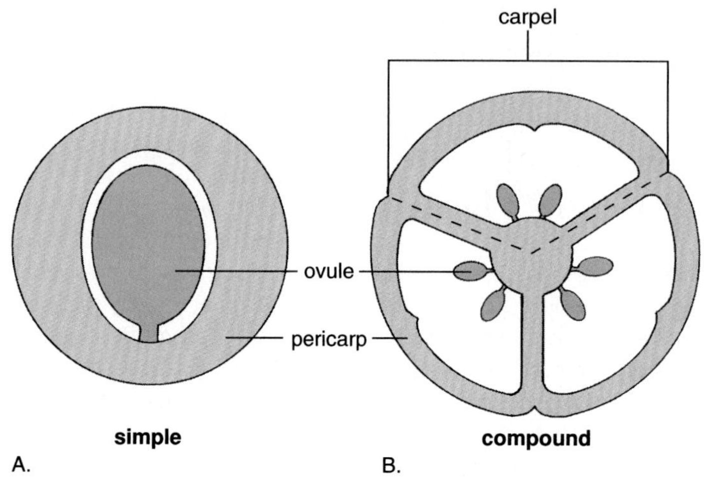

Figure 23.10 Ovaries in cross section. *A.* The ovary of a simple pistil as found in a peach. *B.* The ovary of a compound pistil as found in a lily.

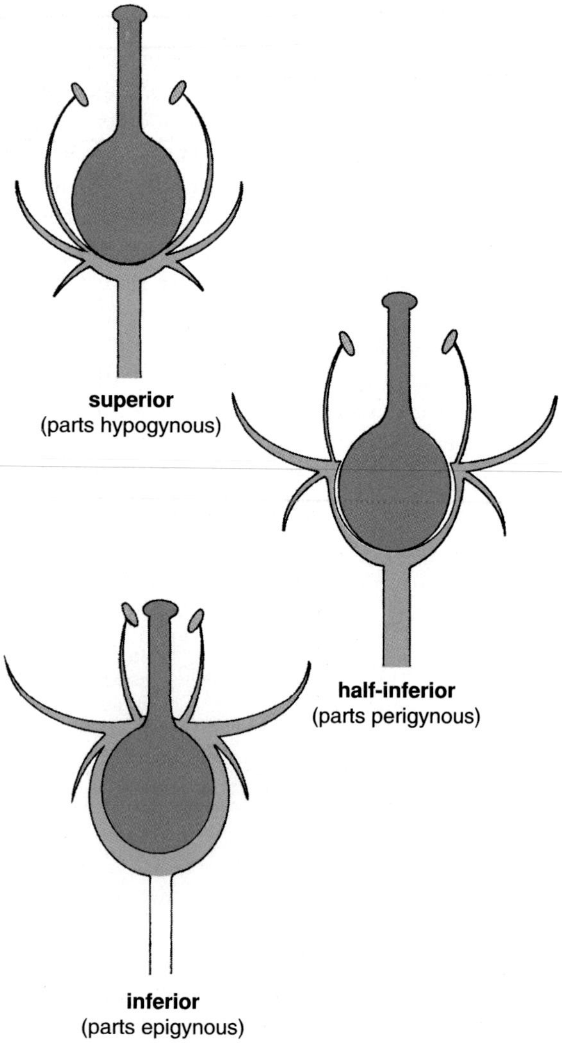

Figure 23.11 Ovary positions in flowers: *superior* (parts hypogynous; e.g. peach); *half-inferior* (parts perigynous; e.g., cherry); and *inferior* (parts epigynous; e.g., apple).

also tended to become *bilaterally symmetrical,* or *irregular* (i.e., capable of being divided into two symmetrical halves only by a single lengthwise plane passing through the axis), as in sweet peas and orchids.

If a flower has a calyx, corolla, stamens, and a pistil, it is *complete.* If, however, the corolla or other flower parts are missing, the flower is *incomplete.* If, as is the case with most flowers, both stamens and a pistil (or pistils) are present, it is said to be *perfect,* regardless of whether or not it also happens to be complete or incomplete. In some families, however, the flowers are not only incomplete but also have become *imperfect* (*unisexual*). Each imperfect flower has either stamens or a pistil but not both. The Pumpkin Family (Cucurbitaceae) includes pumpkins, squashes, watermelons, cantaloupes, cucumbers, and other species with imperfect flowers (Fig. 23.12). When both male and female imperfect flowers occur on the same plant (as is the case with many members of the Pumpkin Family), the species is **monoecious.** If a plant bears only male flowers and other plants of the same species bear only female flowers, the species is said to be **dioecious.**

In both the dicots and the monocots, evolutionary specialization has involved reduction and fusion of parts and a shifting of the ovary from a superior to an inferior position, as hypothetical progression is made from families with primitive characteristics through intermediates to families with advanced features.

Observations on the human and ecological relevance of the flowering plants occupy literally thousands of volumes. A brief overview of several aspects of this subject is given in Chapters 24 and 25.

POLLINATION ECOLOGY

When certain *consumers,* such as insects, forage for food among photosynthetic *producers,* such as plants, they often come in contact with flowers. Many insects and other ani-

A.

B.

Figure 23.12 Unisexual flowers of a squash. *A.* Male. *B.* Female. Note the inferior ovary beneath the corolla.

A.

B.

Figure 23.13 Flower markings on coneflowers (*Rudbeckia*). *A.* In ordinary light. *B.* In ultraviolet light.

mals become dusted with pollen, and as they feed or collect pollen and/or nectar, they unknowingly but effectively bring about pollination of the plants they visit. Throughout the evolutionary history of the flowering plants, the pollinators have evidently coevolved with plants. In some instances, the relationship between the two has become highly specialized.

Twenty thousand different species of bees are included among the pollinators of present-day flowering plants. By far, the best known of these are honey bees. Their chief source of food is nectar, but they also gather pollen for their larvae. The flowers that bees visit are generally brightly colored and mostly blue or yellow—rarely pure red. Pure red appears black to bees, and they generally overlook red flowers. Flowers often have lines or other distinctive markings, which may function as honey guides that lead the bees to the nectar. Bees can see ultraviolet light (a part of the spectrum not visible to humans), and some flower markings are visible only in ultraviolet light, making patterns seen by bees sometimes different from those seen by humans (Fig. 23.13).

Many bee-pollinated flowers are delicately sweet and fragrant. In contrast, flowers pollinated by beetles tend to have stronger, yeasty, spicy, or fruity odors. Beetles don't have keen visual senses, and flowers pollinated by them are usually white or dull in color. Some beetle-pollinated flowers don't secrete nectar but either furnish the insects with pollen or have food available on the petals in special storage cells, which the beetles consume.

Some flowers, including the *stapelias* of Africa (Fig. 23.14), smell like rotten meat. Short-tongued flies pollinate such flowers, which tend to be dull red or brown. These plants are related to our milkweeds, although superficially they don't resemble milkweeds at all. They are often called *carrion flowers* because of their foul odor and appearance. Flies with longer tongues may also pollinate bee-pollinated flowers.

Moth- and butterfly-pollinated flowers, like bee-pollinated flowers, often have sweet fragrances. Night-flying moths visit flowers that tend to be white or yellow—colors that stand out against dark backgrounds in starlight or moonlight.

Red flowers are sometimes pollinated by butterflies, some of which can detect red colors, but these insects are more often found visiting bright blue, yellow, or orange

Figure 23.14 A *Stapelia* (carrion flower) plant.

Figure 23.15 A hummingbird visiting a flower.

flowers. The nectaries of these flowers are at the bases of corolla tubes or spurs, where only moths and butterflies with longer tongues can reach. However, an enterprising bumblebee will occasionally bypass convention and chew through the base of a spur to get at the nectar.

Birds—particularly the hummingbirds of the Americas and the sunbirds of Africa—and the flowers they pollinate are also adapted to one another (Fig. 23.15). The birds do not have a keen sense of smell, but they have excellent vision. The flowers they visit are often bright red or yellow and usually have little if any odor. Bird-pollinated flowers also are typically large, are part of a large, sturdy inflorescence (flower cluster), or, in some cases, are produced on trunks of trees.

Birds are highly active pollinators and tend to burn energy rapidly. They must feed frequently to sustain themselves. Many of the flowers birds prefer produce copious amounts of nectar, thereby assuring repeated visits. The nectar is often produced in long floral tubes that keep most insects out. Some native California fuchsias and their relatives have long threads extending from each pollen grain (Fig. 23.16). When a hummingbird inserts its long bill into such a flower, the pollen grain threads catch on short, stiff hairs located toward the base of its bill, and in this way, the bird unknowingly transfers pollen from one flower to another.

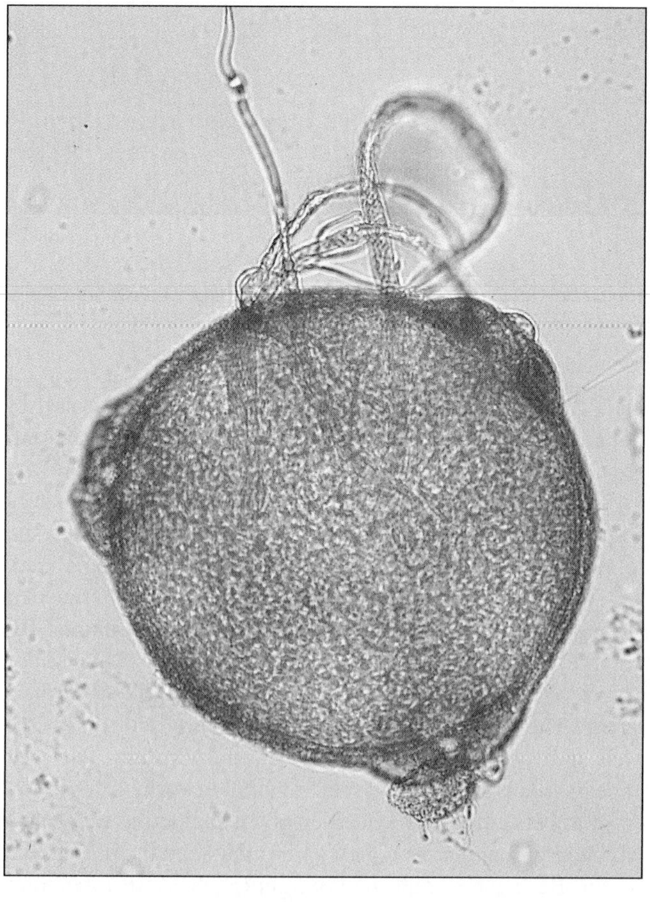

Figure 23.16 A pollen grain of a California fuchsia, showing the threads that catch in the short bristle hairs at the base of a hummingbird's bill, ×1,000.

Bat-pollinated flowers, found primarily in the tropics, tend to open only at night when the bats are foraging (Fig. 23.17). These flowers are dull in color, and like flowers pollinated by birds, they either are large enough for the animal to insert part of its head or consist of ball-like inflorescences containing large numbers of small flowers that dust the visitor with pollen.

The very large Orchid Family, with approximately 35,000 species, has pollinators among all the types mentioned. Some of the adaptations between orchid flowers and their pollinators are extraordinary (Fig. 23.18). The pollen grains of most orchids are produced in little sacs called **pollinia** (singular: **pollinium**) that typically have sticky pads at the bases. When a bee visits such a flower, the pollinia are usually deposited on its head. The "glue" of the sticky pads dries almost instantly, causing the pollinia to adhere tightly. In some orchids, the pollinia are forcibly slapped on the pollinator through a trigger mechanism within the flower.

Figure 23.17 A bat visiting a flower of an organ-pipe cactus. The bat's head is covered with pollen. *(Photo by Donna J. Howell)*

Members of *Ophrys,* a genus of orchids found in Europe and in North Africa, have a modified petal that resembles a female bumblebee or wasp. Male bees or wasps emerge from their pupal stage a week or two before the females and apparently mistake the flowers for potential mates. They try to copulate with the flowers, and while they are doing so, pollinia are deposited on their heads. When they visit other flowers, the pollinia catch in sticky stigma cavities. During a single visit by the pollinators, the pollinia removed from one flower are replaced with fresh pollinia from the next flower visited.

The pollinia of orchids pollinated by moths and butterflies are attached to their long tongues with sticky clamps instead of pads. The pollinia of some bog orchids become attached to the eyes of their pollinators, which happen to be female mosquitoes. After a few visits, the mosquitoes are blinded and unable to continue their normal activities.

Among the most bizarre of the orchid pollination mechanisms are those in which the pollinator is dunked in a pool of watery fluid secreted by the orchid itself; the pollinator escapes under water through a trapdoor. The route of the insect ensures contact with pollinia and stigma surfaces. In other orchids with powerful narcotic fragrances, pollinia are slowly attached to the drugged pollinator. When the transfer of pollinia has been completed, the fragrance abruptly disappears. The temporarily stupefied insect then recovers and flies away.

HERBARIA AND PLANT PRESERVATION

The botanical resources of many universities and other institutions include **herbaria** (singular: **herbarium**), which are often part of larger museums. Some of the larger herbaria contain literally millions of specimens. Unfortunately, today, many institutional herbaria are understaffed, underfunded, and underutilized.

Herbaria are essentially libraries of dried, pressed (and/or liquid-preserved) plants, algae, and fungi, arranged and labeled so that specific specimens can easily be retrieved. Properly prepared and maintained specimens should remain in excellent condition for 300 or more years. Formal training and experience are not needed to make one's own herbarium. The materials, some simple equipment (that may be homemade), and the ability to follow a few relatively elementary procedures are all that is necessary.

Methods

Fungi and bryophytes are usually dried and stored in small boxes or packets. Larger algae tend to have unique properties that lend themselves to preservation according to the

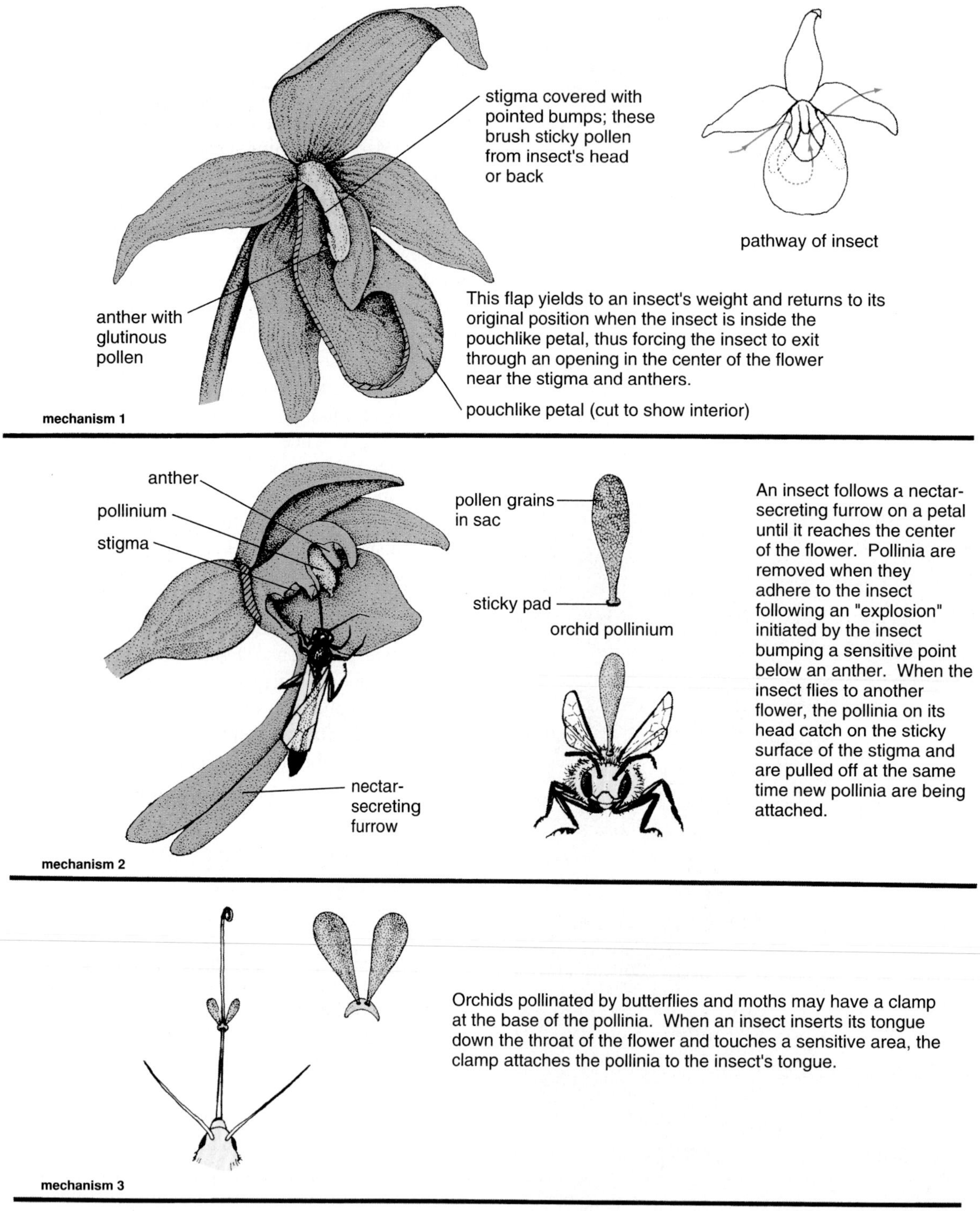

stigma covered with pointed bumps; these brush sticky pollen from insect's head or back
pathway of insect
anther with glutinous pollen
This flap yields to an insect's weight and returns to its original position when the insect is inside the pouchlike petal, thus forcing the insect to exit through an opening in the center of the flower near the stigma and anthers.
pouchlike petal (cut to show interior)
mechanism 1
anther
pollinium
stigma
pollen grains in sac
sticky pad
orchid pollinium
An insect follows a nectar-secreting furrow on a petal until it reaches the center of the flower. Pollinia are removed when they adhere to the insect following an "explosion" initiated by the insect bumping a sensitive point below an anther. When the insect flies to another flower, the pollinia on its head catch on the sticky surface of the stigma and are pulled off at the same time new pollinia are being attached.
nectar-secreting furrow
mechanism 2
Orchids pollinated by butterflies and moths may have a clamp at the base of the pollinia. When an insect inserts its tongue down the throat of the flower and touches a sensitive area, the clamp attaches the pollinia to the insect's tongue.
mechanism 3

Figure 23.18 Some pollinating mechanisms in orchids.

footnote on page 338. The specimens that comprise the bulk of the holdings of most herbaria, however, consist of vascular plants, and what follows pertains primarily to such plants.

The moisture content of flowers and other plant parts to be preserved should be reduced as quickly as possible, with a minimum of distortion. This is usually done with the aid of a *plant press* (Fig. 23.19). This simple device consists of two pieces of plywood (or other wood materials or thin metal plates) with dimensions of approximately 30 × 46 centimeters (12 × 18 inches) and a pair of webbing or leather straps to go around the boards. A number of *felts* (sheets of heavy blotting paper) of similar dimensions are placed between the boards. A folded page of newspaper is placed between each blotter, and a few sheets of stiff cardboard are interspersed between the blotters.

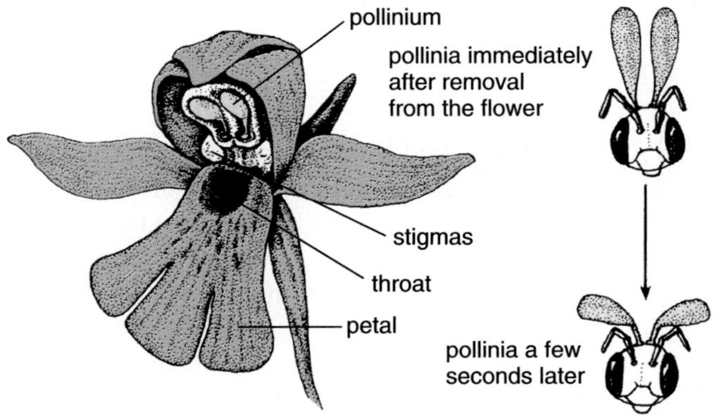

In the showy orchids, there are two separate stigma patches. When an insect visits a flower, the pollinia are attached to its head; then they twist outward and forward so that when the insect visits another flower, the pollinia are simultaneously deposited in the separate stigmas while new pollinia are attached to the insect's head.

mechanism 4

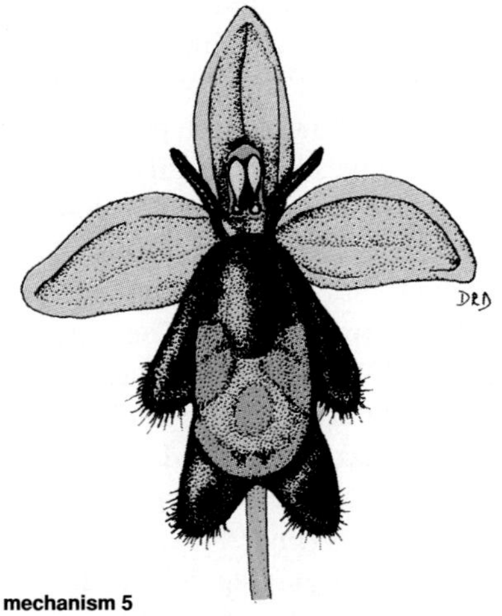

One genus of orchids native to Europe and North Africa has a petal modified to resemble a female wasp. Male wasps try to copulate with the "female" and, in doing so, bump a sensitive area containing pollinia. The pollinia are "glued" to a wasp's head in the process and carried to another flower. The wasps thus effectively bring about cross-pollination.

mechanism 5

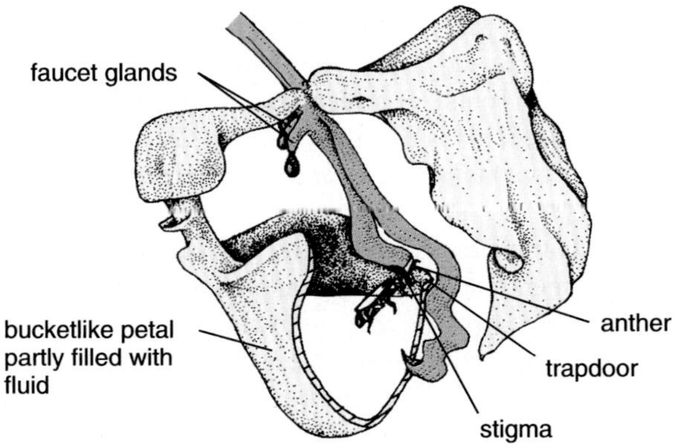

The bucket orchids of South America have faucet glands that partially fill a bucketlike petal with fluid. In visiting the flower, the insects may fall into the fluid. They escape by a trapdoor. The pathway of an insect assures contact with both the stigma and anthers containing pollinia, since they are located adjacent to the trapdoor.

mechanism 6

Any soil clinging to the roots of a specimen to be pressed is washed off, and the plant is laid out on one of the newspaper sheets. Leaves are carefully straightened out, as are petals and other plant parts, so that they are not folded during pressing. Notes on where, when, and by whom the specimen was collected are penciled on the newspaper, or a number corresponding to such notes in a separate field notebook is used. The newspaper is then folded over the specimen and placed between two blotters (felts).

Many specimens may be placed in a press at one time or, if there is space, between the same sheets of newspaper. Only one species should be placed in a single fold of newspaper, however, as mixing species invariably leads to some being pressed better than others, due mostly to the varying amounts of woody tissues within the stems. If the plant is small, press the entire plant, and try to have both upper and lower leaf or flower surfaces visible. If, however, the plant is large (e.g., a tree) press representative parts, especially flowers and/or fruits. After the newspaper and specimen are

Figure 23.19 A plant press.

returned to the press, the straps are tightened around it as much as possible, and it is left to dry in the sun or near a heater (but not close enough to scorch!) for 3 or 4 days. Unless the leaves were succulent or wet at the time they were placed in the press, they should be dry enough to mount on paper at this time. If a press is not available, plants may be pressed between newspapers and blotters by placing heavy weights on top of them.

If possible, 100% rag-content paper should be used for mounting the specimen, as papers made from wood pulp deteriorate with age. The paper also should be thick enough to support the now-brittle specimen. The pressed plant(s) may be attached to the paper in one of several ways. Some workers spread a good white library glue on a glass plate and then place the specimen on the wet glue so that one side is covered with it. They then transfer the specimen to herbarium paper that normally measures 29 × 42 centimeters (11.5 × 16.5 inches). Others place the specimen on the paper first and add glue or liquid plastic at strategic points. The bottom right-hand corner of the paper should be kept clear for a label indicating the scientific name of the plant, collection information, the collector's name, and the collection date (Fig. 23.20).

Professional botanists also number their individual collections, giving a collection number after the collector's name. The label should measure about 7.5 × 12.5 centimeters (3 × 5 inches), and its paper should also be of 100% rag content. Specimens are then placed in manila folders and stored in some systematic fashion so retrieval of individual specimens is easily accomplished. Major institutional herbaria usually arrange the plants by families in a presumed evolutionary sequence, and then the genera and species are alphabetically arranged within the families. Paradichlorobenzene, mothballs, or similar material should be added to the cabinet or storage area to prevent insect damage during storage.

Although other forms of flower preservation are not associated with herbaria or professional botanical activities, a few will be mentioned here simply for reference. For example, some people are interested in pressing flowers for use in dry arrangements on cards, place mats, or other decorative items. In such cases, the pressing is done as though they were herbarium specimens, but the material is then further manipulated in one of several ways. For wall art, the specimens can be mounted on a piece of smooth cardboard, covered with clear plastic or glass, matted, and framed (Fig. 23.21). Decorative notepaper can be made by placing pressed flowers on the paper and covering them with rice paper or facial tissue. A mixture of one part white glue to three parts water is then brushed over the tissue, causing it to become a permanent mount, with the specimen clearly visible through the thin paper film. Clear contact paper cut to the appropriate size makes a good mount for pressed flowers placed on cards or place mats. Pressed flowers can also be embedded in clear plastic poured into molds.

Flowers can also be dried without pressing. With a little patience, relatively lifelike three-dimensional preservations can be made, although flowers with petals that are not easily detached lend themselves more readily to three-dimensional drying than others. Most such drying is done in a shoe box, but almost any type of container can be used. The bottom of the box is covered with about 2 centimeters (0.8 inch) of sand, silica gel, or borax mixture. The fresh flower, with a little of the peduncle or stem still attached, is then gently laid on the surface. After this, more sand, silica gel, or borax mixture is slowly drizzled by hand into the box until the entire flower is buried, with care being taken not to create air pockets around any parts (Fig. 23.22).

Each of the three drying agents mentioned has certain advantages and drawbacks. Sand must be thoroughly washed several times to be certain it is perfectly clean before use. The sand also needs to be completely dry, uniformly fine, and if possible, have individual grains that are relatively rounded rather than angular. Such sand with rounded grains is found around the Great Salt Lake in Utah and is available in arts and crafts stores. It takes about 2 weeks to dry most flowers with sand. Silica gel is also available commercially and is the quickest drying agent of the three mentioned, usually completing the job in 4 to 5 days (if the box is placed in a microwave oven, the drying process may be completed in as little as 5 minutes). Its granules tend to be of different sizes, however, giving the flower surface a slightly irregular texture. It also tends to darken certain colors, and it is expensive. When borax is used, two parts of it should be mixed with one part sand or cornmeal. It dries flowers in about 3 weeks but is sometimes difficult to remove completely from the dried flower surfaces.

When drying has been completed, the container is tilted, and the sand or other drying material is slowly and gently poured out in an uninterrupted motion. If any petals have come loose, they may be glued in place later. Wire may be inserted in the stem to add rigidity, and after any clinging granules have been carefully removed, soft, powdered colored chalk may be dusted on to restore any fading of color. It is recommended that beginners use fairly large flowers whose petals are not easily detached (e.g., zinnias, rose buds) for their first attempts at three-dimensional flower drying.

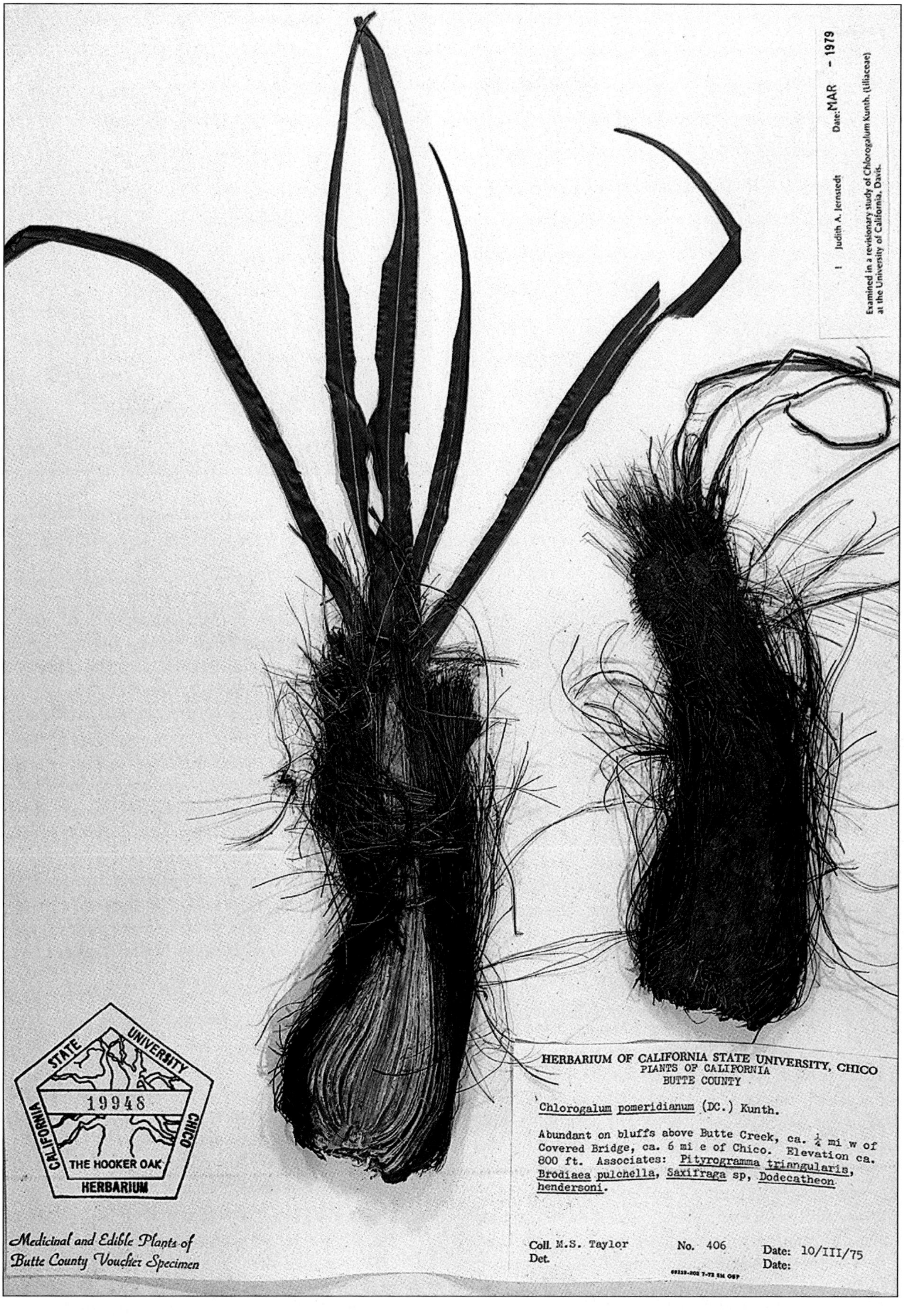

Figure 23.20 A mounted herbarium specimen.

Figure 23.21 Pressed flowers framed as wall art.

Figure 23.22 Three-dimensional drying. A flower placed in a box is embedded in sand, silica gel, or a borax mixture.

A Word of Caution to Collectors

Dried plant collections of herbaria, in particular, have proved invaluable in botanical research in the past and will continue to do so in the future. They have facilitated quick identifications of plants in emergency situations where children have eaten plants or plant parts suspected of being poisonous, or they have helped pin down specific plants that have caused allergic reactions. Herbaria also have been involved in archaeological research where the uses of plants by past cultures have been determined, and they have been used for teaching purposes at various educational levels. Herbarium specimens have been useful in criminal litigation for both the prosecution and the defense and have been the primary source of information on the distribution of plants with potential for new agricultural and horticultural crops or those with possible medicinal values. Most of the unraveling of problems pertaining to natural relationships of plants begins in a herbarium, and without these plant libraries, increasing our knowledge along many practical and theoretical lines would be severely restricted.

Literally hundreds of plants native to North America are now on rare and endangered species lists, and thousands more are in similar predicaments on other continents. *The day has come when both professional and amateur persons interested in plants must discipline themselves to exercise extreme caution in collecting native plants. Collectors should first know what they are collecting or otherwise refrain, and collecting for private collections without serious purpose should be strictly limited. Except for certain types of research, a good photograph of a plant may actually be preferable to a dried specimen and aesthetically more pleasing. It is sincerely hoped that as much as possible, each reader will confine a collection to photographs of native plants.* Contact the local plant conservation organizations (e.g., the Nature Conservancy) for information on threatened or endangered plants.

Summary

1. Flowering plants (angiosperms) have ovules and seeds completely enclosed within carpels; the carpels comprise ovaries that become fruits. There is one phylum of flowering plants (Magnoliophyta); it is divided into two classes (dicots and monocots).
2. Englerian botanists believed flowering plants evolved from conifers. Most contemporary botanists hypothesize they evolved separately from pteridosperms and that the first flowers had many separate, flattened parts spirally arranged on an elongate receptacle.

3. Flowering plants are heterosporous.
4. Flowering plant gametophytes develop in separate structures. The female gametophyte (megagametophyte) develops in the ovule. Integuments, which later become a seed coat, surround the megagametophyte. Pollen grains developed in anthers become male gametophytes.
5. Pollination is the transfer of pollen grains from an anther to a stigma; it is brought about by insects, wind, and other agents.
6. After pollination, a pollen tube may grow from a pollen grain to the female gametophyte; the tube nucleus remains at its tip, and the generative nucleus divides, producing two sperm nuclei.
7. Following the discharge of the contents of the pollen tube into the female gametophyte, a sperm nucleus unites with the egg, forming a zygote; the other sperm nucleus simultaneously unites with the two central cell nuclei, forming a $3n$ endosperm nucleus.
8. The endosperm nucleus becomes nutritive endosperm tissue that may become part of the seed or be absorbed by the seed's embryo.
9. Due to variations in how a female gametophyte develops, some flowering plants produce $5n$, $9n$, or $15n$ endosperm tissue.
10. Some artificial grouping of flowers to aid identification does not reflect natural relationships. Sources of evidence used to try to group plants naturally include fossils, which suggest the flowering plants first appeared about 160 million years ago.
11. Primitive flowering plants had simple leaves and numerous spirally arranged flower parts that were not fused to each other; they possessed both stamens and pistils and were radially symmetrical (regular).
12. Specializations include a reduction in the number of parts, fusion of parts, appearance of compound pistils composed of several individual carpels, inferior ovaries, bilateral symmetry (irregular flowers), and unisexual flowers.
13. Monoecious species have both male and female flowers on the same plants; dioecious species have male and female flowers on separate plants.
14. Bee-pollinated flowers are delicately sweet and fragrant and tend to be blue or yellow in color.
15. Beetle-pollinated flowers tend to have stronger odors and are usually white or dull in color.
16. Some fly-pollinated flowers emit foul odors.
17. Moth-pollinated flowers tend to be white or yellow.
18. Bird-pollinated flowers are usually bright red or yellow and have much nectar but little odor.
19. Most orchids produce pollen grains in pollinia that adhere or clamp onto parts of visiting insects. The flowers of some orchid species have developed bizarre pollination mechanisms.
20. Herbaria are essentially libraries of dried, pressed or otherwise preserved plants, fungi, and algae arranged so that specific specimens may be readily located. Properly preserved plants may last for hundreds of years.
21. A plant to be pressed is placed in a plant press between sheets of newspaper and absorbent material. Dry specimens are affixed to sheets of high-quality paper, with a label giving collection information.
22. Plant parts may be pressed for artwork, place mats, and so on. Flowers may be preserved three-dimensionally.
23. Because so many plants are now on rare and endangered species lists, collectors should try to confine their future collecting to photographs as much as possible.

Review Questions

1. What are the basic differences between gymnosperms and angiosperms?
2. What is the function of the integuments?
3. How are the pollen grains of flowering plants different from those of pine trees?
4. What is the functional equivalent of an archegonium in a flower?
5. What is the function of endosperm, and how does it originate?
6. What differences are there between the female gametophytes of peaches and lilies?
7. Distinguish between radial and bilateral symmetry.
8. What distinguishes flowers with primitive features from those that are advanced?
9. In general, how do flowers pollinated by bees differ from those pollinated by moths, flies, beetles, and birds?
10. What is an herbarium?

Discussion Questions

1. The world's largest flowers are on inconspicuous parasitic plants that resemble a fungus mycelium. If the flowers were not present, would it be possible to tell that the plant is not a fungus? How?
2. In most female gametophytes with eight nuclei, only the egg and the central cell nuclei do not degenerate. What would happen if the megaspore itself functioned as an egg and did not go on to produce other nuclei in a female gametophyte?
3. It takes only one pollen grain to initiate the development of an ovule into a seed, yet a single flower may produce many thousands of pollen grains. Do you suppose such huge numbers of pollen grains are really necessary? Why do you suppose wind-pollinated plants produce much more pollen than insect-pollinated plants?

4. Are such items as the name of the collector and the date significant on an herbarium specimen?
5. Is it really important that plants be classified? What would be the consequences if they were not?

Learning Online

Visit our web page at www.mhhe.com/botany for interesting case studies, practice quizzes, current articles, and animations within the Online Learning Center to help you understand the material in this chapter. You'll also find active links to these topics:

Diversity of Plants
Angiosperms
Economic and Ecological Importance of Angiosperms
Flowers and Fruits
Flowers
Fruits and Seeds

Additional Reading

Bold, H. C., C. S. Alexopoulos, and T. Delevoryas. 1987. *Morphology of plants and fungi,* 5th ed. New York: Harper & Row.

Crest, M., et al. 1999. *Fertilization in higher plants: Molecular and cytological aspects.* New York: Springer-Verlag.

Cronquist, A. 1992. *An integrated system of classification of flowering plants,* 2d ed. New York: New York Botanical Garden.

Endress, P. K., and E. M. Friis (Eds.). 1995. *Early evolution of flowers: Plant systematics and evolution—Supplement 8.* New York: Springer-Verlag.

Geesink, R. 1991. *Thommer's analytical key to the families of flowering plants.* New York: State Mutual Book and Periodical Service.

Greyson, R. I. 1994. *The development of flowers.* New York: Oxford University Press.

Hughes, N. F. 1994. *The enigma of angiosperm origins.* New York: Cambridge University Press.

Johri, B. M. (Ed.). 1984. *Embryology of angiosperms.* New York: Springer-Verlag.

Krassilov, V. A. 1997. *Angiosperm origins: Morphological and ecological aspects.* Bethesda, MD: International Scholars.

Kung, S. D. 1999. *Discoveries in plant biology,* 3 vols. Riveredge, NJ: World Scientific Publications.

Morris, D. L. 1997. *Flower drying handbook: Includes complete microwave drying instructions.* New York: Sterling.

Raghaven, V. 1999. *Developmental biology of flowering plants.* New York: Springer-Verlag.

Raven, P. H., R. F. Evert, and S. E. Eichhorn. 1999. *Biology of plants,* 6th ed. New York: Worth.

Stace, C. 1992. *Plant taxonomy and biosystematics,* 2d ed. New York: Cambridge University Press.

Takhtajan, A. 1996. *Diversity and classification of flowering plants.* New York: Columbia University Press.

Pods of a tropical cacao tree. (*Theobroma cacao*). The seeds are the source of chocolate.

Flowering Plants and Civilization

24

Chapter Outline

Overview

This chapter begins with comments on some of the problems involved in distinguishing between fact and fancy in reported past uses of plants. It continues with a brief discussion of Vavilov's centers of origin of cultivated plants and subsequent modifications of Vavilov's theories. A survey of 16 well-known flowering plant families is presented in phylogenetic sequence, and miscellaneous information is given, including brief comments on family characteristics. Some past, present, or possible future uses of family members are explored.

Some Learning Goals

1. Give reasons for basing scientific evaluation on more than a single sampling.
2. Learn major regions of distribution of cultivated plants and identify several plants from each region.
3. Know characteristics of 10 flowering plant families.
4. Identify which families have flowers with many separate parts and superior ovaries.
5. Know five useful plants in the Laurel, Rose, Legume, and Spurge Families.
6. Identify medicinal plants in the Poppy and Nightshade Families.
7. Construct a simple, original key to five flowering plant families.

Reminder: In this chapter, there are references to a very large number of plants. As in all chapters throughout the book, a few scientific names are given within the text, but the scientific names of all the plants mentioned are listed in Appendix 1.

Puccoons (*Lithospermum* spp.) are herbaceous plants that grow on dry plains and slopes throughout the western United States and British Columbia. One puccoon with greenish yellow flowers (*L. ruderale*) has seeds that are so hard, it was given the common name of *stoneseed* in Nevada and California.

Native American women of the Shoshoni tribe of Nevada reportedly drank a cold water infusion of the roots of stoneseed every day for 6 months to ensure permanent sterility. Biologist Clellan Ford became curious about these reports and gave extracts of stoneseed plants to mice. He found the extracts effectively eliminated the estrous cycle of the mice and decreased the weights of the ovaries, the thymus, and the pituitary glands.

Although this reported Native American use of a plant was demonstrated experimentally to have a basis in fact, distinguishing between fact and fantasy in recorded past uses of plants is often difficult, particularly if the plants have become rare or extinct. However, this type of research is essential today if we are going to save potential sources of medicinal drugs and other useful plant products before they are eliminated by clearing of land and other practices in the name of "progress."

Earlier generations, despite occasional misguided superstition and folklore, did cure some of the diseases they treated with plants, even though they did not know the scientific reasons for the results. As indicated earlier, botanists who have recognized this have teamed with anthropologists, medical doctors, and interpreters to interview tribal and other medical practitioners all over the world. They are gleaning and sifting through as much of this information as possible before it is too late. Their work is already leading to useful new medicinal and other discoveries.

ORIGIN OF CULTIVATED PLANTS

In the1880s, Alphonse de Candolle, a Swiss botanist, published a book entitled *Origin of Cultivated Plants,* based on data he had gathered from many sources. He deduced that cultivated plants probably originated in areas where their wild relatives grow.

In 1916, N. I. Vavilov, a Russian botanist and geneticist, began a follow-up of de Candolle's work. During the next 20 years, he expanded on de Candolle's work and modified his conclusions. Vavilov became persuaded, as a result of his research, that most cultivated plants differ appreciably from their wild relatives. He also concluded that dispersal centers of cultivated plants are characterized by the presence of dominant genes in plant populations, with recessive genes becoming apparent toward the margins of a plant's distribution.

Vavilov recognized eight centers of diversity of cultivated plants, with some plants originating in more than one center. The centers, some of which were subdivided, are shown in Figure 24.1.

Since the 1950s, a number of major world crops have been subjected to analysis at the molecular level. These studies have provided evidence that many of the cultivated plants did not originate in Vavilov's centers. A significant number of such studies were undertaken by agricultural geneticist Jack R. Harlan and his students at the Crop Evolution Laboratory of the University of Illinois. Harlan concluded that some crops do not have centers of origin, and he revised Vavilov's concept, preferring instead to associate

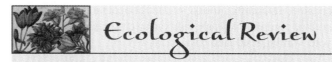

Ecological Review

Nothing is more important to the relationship of humans to the environment than cultivated plants that provide sources of food, fiber, animal forage, and medicines. Cultivated plants have been developed in nearly all climatic regions of the earth and reflect the wide diversity of environments occupied by humans. Cultivated plants appear to have originated in six major regions: Near Eastern (e.g., wheat, carrot, apple), Chinese (e.g., soybean, cucumber, peach), African (yam, cotton, coffee), South Asian and Pacific (e.g., rice, sugar cane, citrus fruits), North American (e.g., sunflower, tobacco), and South and Central American (e.g., white or Irish potato, squash, pineapple). Plants originating in these regions are now grown throughout the world. Today's most important cultivated plants are a tiny fraction of the thousands of species used by peoples around the world; preserving the knowledge of useful plants held by traditional societies is a major challenge for this generation of botanists.

crop origins with regions. These regions are illustrated in Figure 14.2. A sampling of cultivated plants that appear to have originated in six major regions follows.

Near-Eastern Region

The Near-Eastern region encompasses the Mediterranean area, northern Europe (including the former Soviet Union), Turkey, the Balkans, Pakistan, and Iran. Among the many cultivated plants that appear to have originated in this region are barley, wheat and other cereals, garden pea, lentil, chickpea, lupine, asparagus, beet, carrot, turnip, olive, stone fruits (cherry, plum, apricot), apple, pear, onion, garlic, cabbage, broccoli, lettuce, parsley, flax, clover, pistachio, several range grasses, digitalis, belladonna, opium poppy, and psyllium.

Chinese Region

Many of the cultivated plants believed to have orginated in this region apparently came from the temperate and southern parts of China. Examples of widely known plants from this part of the world include bamboo, peach, litchi, walnut, persimmon, ginger, ginseng, gourds, camphor, tea, tung, soybean, buckwheat, horseradish, Chinese cabbage, and cucumber. While China is known for its centuries-old cultivation and consumption of rice, this cereal probably originated elsewhere in Asia.

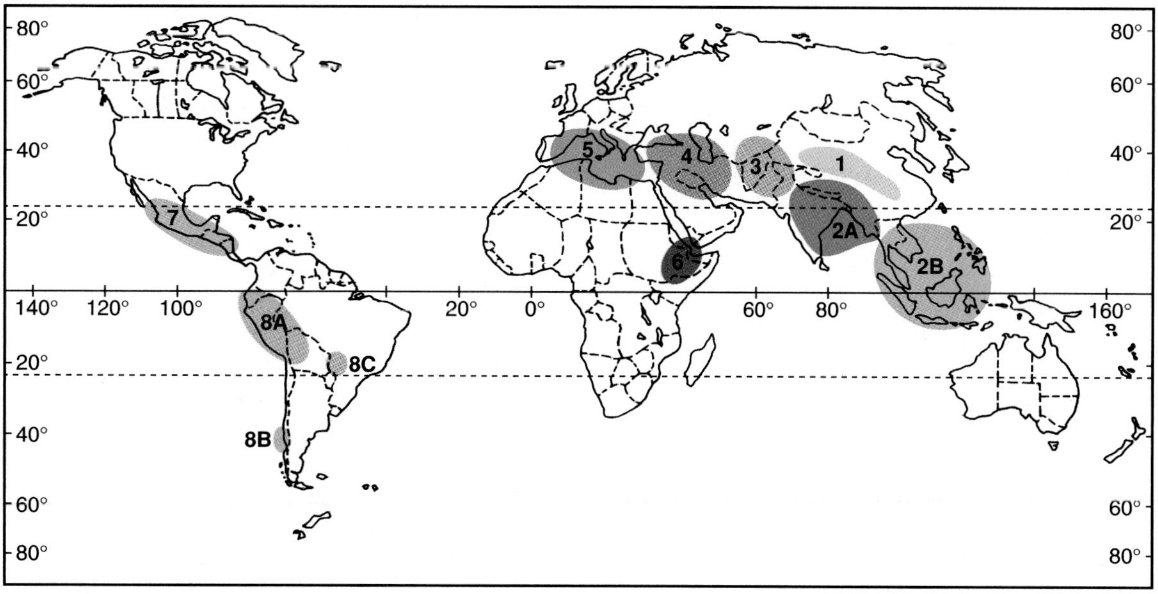

Figure 24.1 Major centers of diversity of cultivated plants according to N. I. Vavilov. Jack Harlan and his associates have abandoned these centers and have associated the plants with regions instead. See text. *After Vavilov, N. I. 1951.* The origin, variation, immunity and breeding of cultivated plants. *K. Starr Chester, Trans. New York: The Ronald Press Company, © 1951.*

African Continent

Although there is evidence that some cultivated plants originated in South and East Africa, most African plants appear to have come from Ethiopia, Eritrea, West Africa, and other areas between the Sahara and the Congo. Better-known plants include yam, sorghum, okra, oil palm, baobab, the sweet melons, coffee, some cotton, and castor bean.

South Asia and the Islands of the Pacific

This region encompasses South Asia, Malaysia, Indonesia, Melanesia, the Philippines, and thousands of islands extending across the southern Pacific Ocean. Among the plants from this region are sugar cane, eggplant, mango, banana, citrus fruits, safflower, nutmeg, clove, cardamon, turmeric, black pepper, jute, hemp fiber, coconut, taro, rice, mung bean, sesame, betel nut, onion, and jackfruit.

North America

Few cultivated plants are believed to have originated in North America and those that did came primarily from tropical and subtropical areas. Such plants include sunflower and tobacco.

South and Central America

There is no distinct line between South America and areas to the north with respect to some of the cultivated plants originating in the New World. The wild relatives of common garden beans, for example, appear to have been distributed throughout parts of both North and South America. Cultivated plants from the region include corn, peanut, white (Irish) potato, lima bean, common bean, cashew, pineapple, papaya, avocado, red pepper, tomato, cotton, cocaine, amaranth, quinoa, cacao (chocolate), guava, sweet potato, pumpkin, squash, rubber, prickly pear, peyote, chayote, vanilla, sisal, and cassava.

SELECTED FAMILIES OF FLOWERING PLANTS

The following is a survey of a few of the well-known plant families, indicating not only past uses of some of their members but also possible future uses in a few instances, along with biological notes and miscellaneous observations.

The reader is cautioned against assuming that past medicinal uses were always effective or without harmful side effects and is urged to refrain from experimenting with such plants.

More than 300 families of flowering plants are recognized today. Flowers and fruits are considerably more reliable and stable indicators of heredity than leaves or other vegetative plant parts, so flowering plant families are distinguished from one another mostly on the basis of flower and fruit parts and structure. Some families include a mere handful of species, while others are very large, their members numbering in the thousands. Space permits only brief discussion of a few of the larger, better-known families here, and the reader is referred to the additional readings at the end of the chapter for other sources of information.

The families are taken up in a more or less phylogenetic (evolutionary) sequence, beginning with those that have more generalized flower structure and progressing to those whose flowers are considered specialized. Generalized flowers tend to have numerous parts that are not fused together, and the ovaries are superior. In specialized flowers, the parts are reduced in number and often fused together; the ovary is inferior. (See Fig. 23.9 and the amplification of this subject that begins on page 449.)

The following is an abbreviated key to the families discussed. (See Chapter 16 for a discussion of keys and their uses.)

1. Flowers with parts in fours or fives or multiples thereof; seeds with two cotyledons (DICOTS)
    2. Petals separate from one another, or lacking
        3. Petals present
            4. Stamens more than twice as many as the petals
                5. Stamens, petals, and sepals attached to the rim of a cup surrounding the one to many pistils ....Rose Family (Rosaceae)
                5. Stamens, petals, and sepals not attached to the rim of a cup
                    6. Pistils several to many in each flower ....Buttercup Family (Ranunculaceae)
                    6. Pistil one
                        7. Ovary superior ....Poppy Family (Papaveraceae)
                        7. Ovary inferior..Cactus Family (Cactaceae)
            4. Stamens not more than twice as many as the petals
                        8. Herbaceous vines; fruit a pepo ....Pumpkin Family (Cucurbitaceae)
                        8. Primarily herbs, shrubs, and trees; fruit not a pepo

9. Fruit a legume . . . . .Legume Family (Fabaceae)
9. Fruit not a legume
10. Fruit a silique or silicle . . . . .Mustard Family (Brassicaceae)
10. Fruit not a silique or silicle
11. Ovary superior; stems square in cross section; leaves opposite; fruit of four nutlets . . . . .Mint Family (Lamiaceae)
11. Ovary inferior; stems rounded in cross section; leaves alternate; fruit a schizocarp . . . . .Carrot Family (Apiaceae)
3. Petals lacking; calyx sometimes petal-like
12. Ovary of three carpels and usually elevated on a gynophore; anthers splitting lengthwise . . . . .Spurge Family (Euphorbiaceae)
12. Ovary of one carpel; gynophore lacking; anthers splitting by raised flaps . . . . .Laurel Family (Lauraceae)
2. Petals fused together
13. Flowers not in a head; each flower with its own receptacle; ovary superior . . . . .Nightshade Family (Solanaceae)
13. Flowers in a head, several to many florets on a common receptacle, comprising a "flowerlike" inflorescence; ovary inferior . . . . .Sunflower Family (Asteraceae)
1. Flowers with parts in threes or multiples thereof; seed with one cotyledon (MONOCOTS)
14. Flowers inconspicuous; without petals or sepals . . . .Grass Family (Poaceae)
14. Flowers conspicuous; petals and sepals mostly similar in coloration
15 Ovary superior; petals all alike . . . . .Lily Family (Liliaceae)
15. Ovary inferior; one petal different in form from the other two . . . . .Orchid Family (Orchidaceae)

DICOTS

The Buttercup Family (Ranunculaceae)

Nearly all the 1,500 members of the Buttercup Family are herbaceous. The flowers, whose petals often vary in number, have numerous stamens and several to many pistils with superior ovaries (Fig. 24.2). Most have dissected leaves with no stipules and with petioles that are slightly expanded at the base. Well-known representatives include ornamental plants such as buttercup, columbine, larkspur, anemone, monkshood, and *Clematis* (Fig. 24.3). Most members of the Buttercup Family are concentrated in north temperate and arctic regions.

Columbine flowers have five spurred petals that resemble a circle of doves. The name comes from *columba,* the Latin word for dove. A blue and white species of columbine is the state flower of Colorado. Native Americans controlled diarrhea with a tea made from boiled columbine roots, and members of at least two tribes believed columbine seeds to have aphrodisiac properties. A man would pulverize the seeds, and after rubbing them in the palms of his hands, he would try to shake hands with the woman of his choice, believing the woman would then succumb to his advances. Others crushed and moistened the seeds and applied them to the scalp to repel lice.

Figure 24.2 A buttercup flower.

Most members of the family are at least slightly poisonous, but the cooked leaves of cowslips have been used for food, and the well-cooked roots of the European bulbous buttercup are considered edible. The European buttercup causes blistering on the skin of sensitive individuals. East Indian fakirs are reported to deliberately blister their skin with buttercup juice in order to appear more pitiful when begging. Native Americans of the West gathered buttercup

A. B. C. D.

Figure 24.3 Representatives of the Buttercup Family. *A.* Columbine. *B. Hepatica. C.* Monkshood. *D. Isopyrum. (C. Courtesy Donald E. Brink, Jr.)*

achenes, which they parched and ground into meal for bread. Others made a yellow dye from buttercup flower petals. Karok Indians made a blue stain for the shafts of their arrows from blue larkspurs and Oregon grape berries.

Goldenseal, sold in health-food stores, is a plant that was once abundant in the woods of temperate eastern North America. It has become virtually extinct in the wild because of relentless collecting by herb dealers. They sold the root for various medicinal uses, including remedies for inflamed throats, skin diseases, and sore eyes. At least one Native American tribe mixed the pounded root in animal fat and smeared it on the skin as an insect repellent.

Monkshood yields a drug complex called *aconite* that was once used in the treatment of rheumatism and neuralgia. Although popular as garden flowers, monkshoods are very poisonous. Death may follow within a few hours of ingestion of any part of the plant. Most species have purplish to bluish or greenish flowers, but one Asian monkshood, called *wolfsbane,* has yellow flowers. Wolf hunters in the past poisoned the animals with a juice obtained from wolfsbane roots.

The Laurel Family (Lauraceae)

The Laurel Family's primitive flowers have no petals but have six sepals that are sometimes petal-like. The stamens, which occur in three or four whorls of three each, are a curiosity because the anthers open by flaps that lift up. The ovary is superior. Most of the approximately 1,000 species in this family are tropical evergreen shrubs and trees, many with aromatic leaves. The family received its name from the famous laurel cultivated for centuries in Europe. Its foliage was used by the ancient Greeks to crown victors in athletic events and later was used in the conferring of academic honors.

Several important spices come from members of this family. Powdered cinnamon is the pulverized bark of a small tree native to India and Sri Lanka, although it is also grown commercially elsewhere. Cassia is very similar and is often sold interchangeably with cinnamon. Cinnamon oil is distilled from young leaves of the trees. Use of cinnamon and cassia dates back thousands of years. They were used in perfumes and anointing oils at the time of Moses, and other records reveal their use in Egypt at least 3,500 years ago.

Camphor has been used since ancient times. This evergreen tree, native to China, Japan, and Taiwan, is the main source of camphor essence still used in cold remedies and inhalants, insecticides, and perfumes. The essence is distilled from wood chips. Camphor trees are smog resistant, and some American cities and towns with milder climates have been using them as street trees.

Sassafras trees, native to the eastern United States and eastern Asia, also have spicy-aromatic wood. A flavoring widely used in toothpaste, chewing gum, mouthwashes, and soft drinks was obtained by distillation of wood chips and bark. Sassafras tea still is considered a refreshing beverage. Sassafras is also an ingredient of some homemade root beers, and in the southern states, an alcoholic beer has been made by adding molasses to boiled sassafras shoots and allowing the mixture to ferment. In Louisiana, powdered sassafras leaves (called filé) have been used as a thickening and flavoring agent for gumbo. In the past, it has been used by country physicians for treating hypertension and for inducing a sweat in those with respiratory infections. Reports indicate that in large doses, it has a narcotic-stimulant effect, and it is also reported to be carcinogenic. Most sassafras flavorings now in use are artificial.

The sweet bay, used as a flavoring agent in gravies, sauces, soups, and meat dishes, comes from the leaves of the laurel. Leaves of the related California bay (Fig. 24.4) are sometimes used as a substitute for sweet bay and for making Christmas wreaths. This tree, which is native to California, also occurs in southwestern Oregon, where it is known as myrtle (true myrtles, however, belong to a different family).

Figure 24.4 A fruit and leaves of a California bay tree (also known as Oregon myrtle).

California bay wood is hard and can be polished to a high luster; it is used for making a variety of bowls, ornaments, and other smaller wooden articles.

Early settlers in the West and Native Americans of the region used California bay for the relief of rheumatism by bathing in hot water to which a quantity of leaves had been added. The nutlike fruits (drupes) were roasted and used for winter food. A leaf was placed under a hat on the head to "cure" a headache (but even a small piece of leaf placed near the nostrils can produce an almost instant headache!). A few leaves placed on top of flour or grains in a canister will keep weevils away, and small branches have been used as chicken roosts to repel bird lice and fleas. A leaf or two rubbed on exposed skin functions as a mosquito repellent.

Avocados are also members of the Laurel Family. The fruits have more energy value by weight than red meats and are rich in vitamins and iron.

The Poppy Family (Papaveraceae)

Most members of the Poppy Family are herbs distributed throughout temperate and subtropical regions north of the equator, but several poppies occur in the Southern Hemisphere, and a number are widely planted as ornamentals. Poppies, like buttercups, tend to have numerous stamens, but most have a single pistil (Fig. 24.5). Most also have milky or colored sap, and their sepals usually fall off as the flowers open. All members produce alkaloidal drugs.

Bloodroot (*Sanguinaria canadensis*) is a pretty, early-flowering spring plant of eastern North American deciduous forests. A bright reddish sap that is produced in its rhizomes was used by some Native Americans as a facial dye, an insect

Figure 24.5 A prickly poppy flower.

Figure 24.6 Immature opium poppy capsules that were gashed with a razor blade. Note the opium-containing latex oozing from the gashes.

repellent, and a cure for ringworm; children today paint their nails with it. It has a very bitter taste, except when ingested in minute amounts. The bitterness made it effective in inducing vomiting, but members of one tribe treated sore throats with it after compensating for the bitterness by squeezing a few drops on a lump of maple sugar so that it could be held in the mouth.

Opium poppies have had a significant impact on societies of both the past and the present. Opium itself was described by Dioscorides in the 1st century A.D., and ancient Assyrian medical texts refer to both opium and opium poppies. Opium smoking, which does not extend back nearly as far as the use of the drug in other ways, became a major problem in China in the 1600s. Smoking opium has given way to other forms of use in recent years. The substance is obtained primarily by making small gashes in the green capsules of the poppies (Fig. 24.6). The crude opium appears as a thick, whitish fluid oozing out of the gashes and is scraped off. It contains two groups of drugs. One group contains the narcotic and addicting drugs morphine and codeine, best known for their widespread medicinal use as painkillers and cough suppressants. Members of the other group are neither narcotic nor addictive. They include papaverine, used in the treatment of circulatory diseases, and noscapine, used as a codeine substitute because it functions like codeine in suppressing coughs but does not have its side effects.

Heroin, a scourge of modern societies, is a derivative of morphine. It is from four to eight times more powerful than morphine as a painkiller, and less than 100 years ago, it was advertised and marketed in the United States as a cough suppressant. It is estimated that 75% of American drug addicts used heroin up until the early 1980s, but cocaine has now largely replaced it. The loss to society in terms of its economic, physical, and moral impact is enormous, since addicts frequently commit violent crimes to obtain the funds needed to support their habits that often cost well over $200 per day.

The seeds of opium poppies contain virtually no opium and are widely used in the baking industry as a garnish. However, there are reports that even the mere traces of opium in some poppy seeds have produced positive results in urine drug tests. They also contain up to 50% edible oils that are used in the manufacture of margarines and shortenings. Another type of oil obtained from the seeds after the edible oils have been extracted is used in soaps and paints.

The Mustard Family (Brassicaceae)

The original Latin name for the Mustard Family was *Cruciferae.* The name describes the four petals of the flowers that are arranged in the form of a cross. The flowers also have four sepals, usually four nectar glands, and six stamens, two of which are shorter than the other four. All members produce siliques or silicles that are unique to the family (see Fig. 8.17). All 2,500 species of the family produce a pungent, watery juice, and nearly all are herbs distributed primarily throughout the temperate and cooler regions of the Northern Hemisphere.

Among the widely cultivated edible plants of the Mustard Family are cabbage, Chinese cabbage, cauliflower, brussels sprouts, broccoli, radish, kohlrabi, turnip, horseradish, watercress, and rutabaga. The widely used condiment

Figure 24.7 Shepherd's purse.

mustard is a mixture of the ground, dried seeds of two species of *Brassica.* Some edible members are also widespread weeds. The leaves of shepherd's purse (Fig. 24.7), for example, can be cooked and eaten, and the seeds can be used for bread meal. Other wild edible members include several cresses, peppergrass, sea rocket, toothwort, and wild mustard. Wild mustards are often weeds in row crops. Their leaves are sometimes sold as vegetable greens in markets.

The seeds of wild mustard, shepherd's purse, and several other members of this family produce a sticky mucilage when wet. Biologists at the University of California at Riverside discovered a potential new use for these seeds. They fed pelleted alfalfa rabbit food to mosquito larvae in water tanks they were using for experiments on mosquito control. They noticed the larvae, which had to come to the surface at frequent intervals for air, often stuck to the pellets and suffocated. Curious, the workers examined the pellets under a microscope and found that they contained mustard seeds. Evidently, the field where the alfalfa had been harvested had also contained mustard plants. The scientists then tried heating the mustard seeds to kill them and found this did not affect production of mucilage by wet seeds. It was calculated that 0.45 kilogram (1 pound) of such seeds could kill about 25,000 mosquito larvae. A few mosquito abatement districts have used the seeds effectively, but experiments are needed to determine if there is a practical way to harvest many more seeds and control mosquitoes on a much larger scale by such non-polluting means.

Native Americans mixed the tiny seeds of several members of this family with other seeds and grains for bread meal and gruel. To prevent or reduce sunburn, Zuni Indians applied a water mixture of ground western wallflower plants to the skin. Watercress is widely known as a salad plant and has had many medicinal uses ascribed to it. During the 1st century A.D., for example, Pliny listed more than 40 medicinal uses. Native Americans of the west coast of the United States treated liver ailments with a diet consisting exclusively of large quantities of watercress for breakfast, abstinence from any further food until noon, and then resumption of an alcohol-free but otherwise normal diet for the remainder of the day. This was repeated until the disease, if curable, disappeared. Dyer's woad, a European plant that has become naturalized and established in parts of North America, is the source of a blue dye that was used for body markings by the ancient Anglo-Saxons. The seeds of other members of the family produce camelina and canola oils. Camelina oil has been used in soaps and was once used as an illuminant for lamps. Canola oil (obtained from rape seed) is a source of low LDL fats and is widely used in food preparation.

The Rose Family (Rosaceae)

The Rose Family includes more than 3,000 species of trees, shrubs, and herbs distributed throughout much of the world. The flowers characteristically have the basal parts fused into a cup, with petals, sepals, and numerous stamens attached to the cup's rim (Fig. 24.8). The family is divided into subfamilies on the basis of flower structure and fruits. The flowers of one group have inferior ovaries and produce pomes for

Figure 24.8 A Sitka rose.

fruits. Flowers of other groups have ovaries that are superior or partly inferior and produce follicles, achenes, drupes, or clusters of drupelets.

The economic impact of members of the Rose Family is enormous, with large tonnages of stone fruits (e.g., cherries, apricots, peaches, plums), pome fruits (e.g., apples, pears), and aggregate fruits (e.g., strawberries, blackberries, loganberries, raspberries) being grown annually in temperate regions of the world (Fig. 24.9).

Members of this family have been relevant to humans in many other ways in the past and still continue to be so. Roses themselves, for example, have for centuries been favorite garden ornamentals of countless numbers of gardeners, and the elegant fragrance of some roses delights many. In Bulgaria and neighboring countries, a major perfume industry has grown up around the production from damask roses of a perfume oil known as attar (or otto) of roses. In a valley near Sofia, more than 200,000 persons are involved in the industry, whose product brought more than $3,000 per kilogram ($1,360 per pound) during the 1990s. A considerable quantity of the oil is blended with less expensive substances in the perfume industry. Perfume workers are rarely reported to develop respiratory disorders, suggesting that the plant extracts may have medicinal properties.

The fruits of wild roses, called *hips* (Fig. 24.10), are exceptionally rich in vitamin C. In fact, they may contain as much as 60 times the vitamin C of a comparable quantity of citrus fruit. Native Americans from coast to coast included rose hips in their diet (except for members of a British Columbia tribe, who believed they gave one an "itchy seat"), and it is believed that this practice contributed to scurvy being unknown among Native Americans. During World War II when food supplies became scarce in some European countries, children in particular were kept healthy on diets that included wild rose hips. In addition to vitamin C, the hips contain significant amounts of iron, calcium, and phosphorus. Today many Europeans eat *Nyppon Sopa,* a sweet, thick pureé of rose hips, whenever they have a cold or influenza.

Figure 24.9 A raspberry.

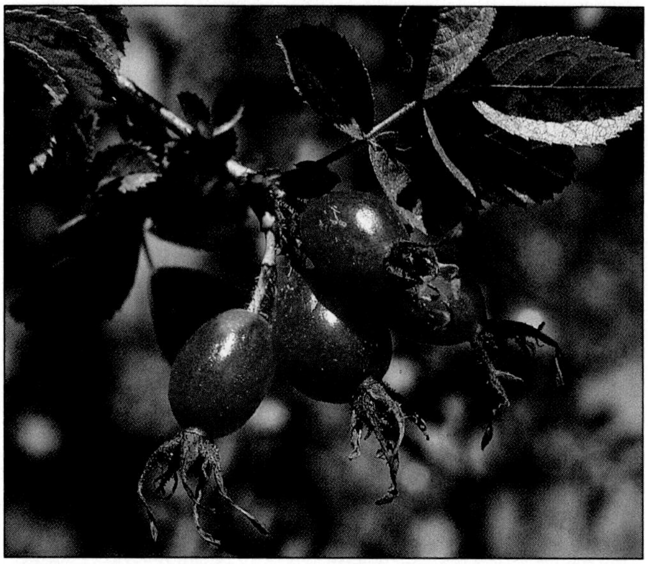

Figure 24.10 Mature rose hips.

After giving birth, the women of one western Native American tribe drank western black chokecherry juice to staunch the bleeding. Other tribes frequently made a tea from blackberry roots to control diarrhea. Five hundred Oneida Indians once cured themselves of dysentery with blackberry root tea, while many nearby white settlers, who refused to use "Indian cures," died from the disease. Men of certain tribes used older canes of roses for arrow shafts (presumably after removing the prickles!). Wild blackberries, raspberries, salmonberries, thimbleberries, dewberries, juneberries, and strawberries all provided food for Native Americans and early settlers, and they are still eaten today, either fresh or in pies, jams, and jellies. A spiced blackberry cordial is still a favorite for "summer complaints" in southern Louisiana. Wild strawberries are considered by many to be distinctly superior in flavor to cultivated varieties.

The Legume Family (Fabaceae)

The Legume Family, originally referred to as the *Leguminosae,* is the third largest of the approximately 300 families of flowering plants, with only the Sunflower and Orchid Families having more species. The 13,000 family members of the Fabaceae are cosmopolitan in distribution and include many important plants. The flowers range in symmetry from radial (regular) to bilateral (irregular). The irregular flowers have a boat-shaped *keel* composed of two fused petals that enclose the pistil, two *wing petals,* and a larger *banner* petal (Fig. 24.11). The stamens in such flowers are generally fused in the form of a tube around the ovary. The legume fruit is the common feature shared by all members of the family (see Fig. 8.16).

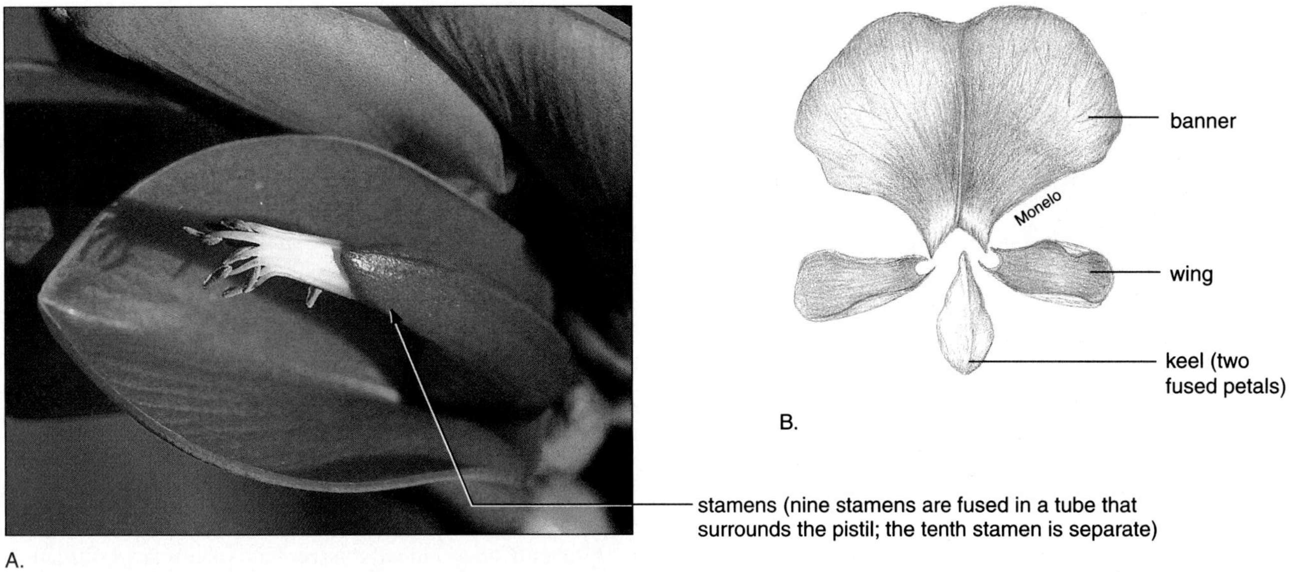

Figure 24.11 *A.* Irregular (bilaterally symmetrical) flowers of a coral tree. *B.* Parts of a sweet pea flower. *C.* An inflorescence of a powderpuff "flower" consisting of regular radially symmetrical legume flowers.

Important crop plants include peas, many kinds of beans (e.g., kidney, lima, garbanzo, broad, mung, tepary), soybeans, lentils, peanuts, alfalfa, sweet clover, jicama, licorice, and wattle. Wattle is an Australian tree that is grown commercially as a source of tannins for leather tanning. Carob, another member of this family, is widely used as a chocolate substitute. Several copals (hard resins used in varnishes and lacquers) are obtained from certain legume plants, as are gum arabic and gum tragacanth used in mucilages, pastes, paints, and cloth printing. The hard wood of several tropical leguminous trees (e.g., rosewood) is prized for furniture.

Important dyes, such as indigo, logwood (used in staining tissues for microscope slides and now scarce), and woadwaxen (a yellow dye), come from different legume plants. Locoweeds, belonging to *Astragalus*—a genus of about 1,600 species—have killed many horses, cattle, and sheep, particularly in the southwestern United States. The poisonous principle in those species affecting livestock seems to vary in concentration according to the soil type in which the plants are growing. Other poisonous legumes include lupines, jequirity beans, black locusts, and mescal beans.

About 90% of the members of the Legume Family exhibit leaf movements, but few are as rapid as those of the sensitive plant (*Mimosa pudica*), whose leaves fold within seconds in response to a disturbance. Sensitive plants grow as weeds in the tropics and the Deep South of the United States; they are discussed in Chapter 11 and are shown in Figure 11.14. Many of the movements of other legume plant leaves are correlated primarily with day length.

Clovers have been widely used in the past by gatherers of wild food plants. The leaves are difficult to digest in quantity, but the rhizomes were gathered and usually roasted or steamed in salt water, then dipped in grease before being eaten. The seeds of both clovers and vetches also were gathered and either ground for meal or cooked in a little water and eaten as a vegetable. Today, seeds of several legumes, including alfalfa and mung beans, are popular for their sprouts, which are widely used in salads and Oriental dishes. A tropical bean called *winged bean* (Fig. 24.12) has unusually high levels of protein, and all parts of the plant are edible. It is presently being grown in several widely scattered tropical and subtropical regions and also is being marketed on a limited scale in some temperate zones. It is believed to have great potential for improving the diet of undernourished peoples throughout the tropics.

Figure 24.12 A winged bean. Winged beans are highly nutritious, are easy to grow in the tropics, and hold promise for improving diets in developing countries.

The Spurge Family (Euphorbiaceae)

Although many of the members of the Spurge Family are tropical, they are widespread in temperate regions both north and south of the equator. The stamens and pistils are produced in separate flowers that often lack a corolla and are inconspicuous. In true spurges (*Euphorbia*), the female flower is elevated on a stalk called a *gynophore* and is surrounded by several male flowers that each consist of little more than an anther. Both the female and male flowers are inserted on a cup composed of fused bracts, the cup usually having distinctive glands on the rim. This type of inflorescence is called a *cyathium* (Fig. 24.13). Sometimes, the inconspicuous flowers are surrounded by brightly colored bracts (e.g., poinsettia, shown in Fig. 7.19) that give the inflorescence the appearance of a single, large flower. Most members of this large family produce a milky latex, and a number of species are poisonous.

Sooner or later, many gardeners experience "urges to purge spurges," as some members of the family (e.g., spotted spurge) are exceptionally aggressive weeds that reproduce very rapidly. Other large tropical spurges closely resemble columnar cacti.

Several economically important plants are cultivated, particularly in frost-free regions. For example, an estimated 90 million metric tons (100 million tons) of cassava are harvested annually from plants cultivated in South America, Africa, and eastern Asia. The roots develop thickened storage areas that resemble large sweet potatoes (shown in Fig. 5.8) and are a diet staple of the tropics, much as white potatoes and cereals are of temperate areas. Poisonous principles are removed by boiling, fermenting, or squeezing out the juice. In dried and powdered form, the cassava is known as farina. In Western countries, tapioca is prepared by forcing heated cassava pellets through a mesh while it is being agitated. Cassava starch is also used as a base for the production of alcohol, acetone, and other industrial chemicals.

Another cultivated spurge of the tropics is the Pará rubber tree, the source of the crude rubber from which most rubber products are made today. Although wild South American trees were the original commercial source of rubber, rubber trees have been widely planted in Indonesia, Africa, and adjacent areas. The trees vary in height from less than 5 to

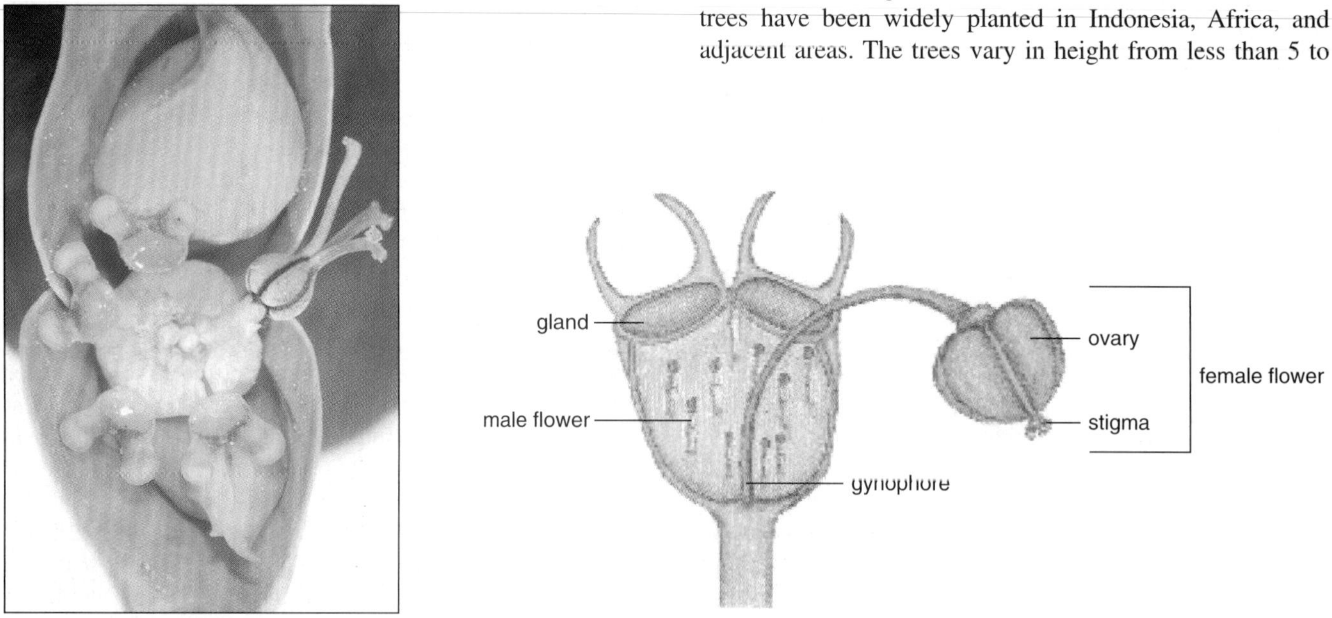

Figure 24.13 A spurge. *A.* Inflorescence. *B.* An individual cyathium.

over 50 meters (16 to 164 feet) and produce most of the latex in the inner bark. The laticifers in which latex is secreted spiral around inside the trunk at an angle of about 30 degrees. Accordingly, cuts are made at the same angle into the inner bark to obtain maximum yields of latex, which trickles down into collecting cups that are attached to the tree. After collection, the latex is coagulated by chemicals or smoke and then shipped in sheet or crumbled form to processing plants. Sometimes, an anticoagulant is mixed with it, and the liquid is transferred in tankers. Much of the world's rubber goes into the manufacture of automobile and aircraft tires, but other products made from rubber are legion.

The Pará rubber tree should not be confused with a broad-leaved ornamental known as the rubber plant. The rubber plant is popular as a house plant and also produces latex; it is, however, a member of the Fig Family (Moraceae).

The latex of other spurges may hold a key to future sources of fuel and lubricating oils. In 1976, Melvin Calvin, a University of California Nobel Prize winner, proposed the use of latex of gopher plants as a source of materials for oil. He estimated that such plants, which can grow in semidesert areas, would produce 10 to 50 barrels of oil per year on 0.4 hectare (1 acre) of land at a cost of $3 to $10 per barrel.[1]

A spurge called *candelilla* occurs in remote areas of Mexico. It produces a wax on its stems that is used in the making of candles and other wax products. Still another spurge produces seeds with a special oil used in plasticizers, and castor oil (from castor beans) is used in the manufacture of nylon, plastics, and soaps. Castor beans themselves are very poisonous—as few as one to three are sufficient to kill a child. The plants grow very rapidly and are popular as ornamentals despite their being one of the leading natural causes of poisoning among American children.

A Mexican jumping bean is the seed of a certain spurge in which a small moth has laid an egg. When the egg hatches, the grub periodically changes position with a jerk, causing the seed to jump. Crown-of-thorns is an ornamental plant with somewhat flexible twisting stems bearing vicious-looking spines. Some believe it to have been the plant from which the crown of thorns for the head of Christ was made. Poinsettias, or Christmas flowers, are favorite yuletide plants in various parts of the world. Tung oil, used in oil paints and varnishes, and Chinese vegetable tallow, a substance used in the manufacture of soap and candles, are two more commercially important products obtained from the seeds of cultivated members of the Spurge Family.

The Cactus Family (Cactaceae)

Cacti are native only to the Americas but include many highly regarded ornamentals that have been exported around the world. The flowers are usually showy (Fig. 24.14), with numerous stamens, petals, and sepals. The sepals are often colored like the petals, and the inferior ovary develops into a berry. There are possibly more than 1,500 species, most occurring in drier subtropical regions. The leaves of many are reduced in size or more often in the form of spines, with

A.

B.

Figure 24.14 Cactus flowers.

1. Jojoba, a member of the Box Family (Buxaceae), is a desert shrub with an acorn-sized capsule containing a large oily seed with about 50% liquid wax content. This high-grade wax and another found in the seeds of meadowfoams, which are members of the Meadowfoam Family (Limnanthaceae), are the only known natural substitutes for sperm whale oil, a vital ingredient in engine lubricants. The importing of sperm whale oil into the United States was banned in 1970 to protect the nearly extinct large ocean mammals, and an expensive synthetic substitute is being used. Experiments and research are in progress to find improved strains of jojoba and to test the feasibility of its being grown on a large scale. Two California counties have been using a mixture of petroleum oil and jojoba oil since 1981 in the transmissions of public transportation buses to see if the mixture will keep them from overheating. The results are promising and may reduce the need for transmission oil changes from once every 50,000 miles to once every 100,000 miles.

the fleshy, flattened or cylindrical, often fluted stems carrying on the photosynthesis of the plants (Fig. 24.15). Many cacti can tolerate high temperatures, and some can withstand up to several years without moisture. They vary in size from pinheadlike forms to the giant saguaro (see Fig. 26.7) that can attain heights of 15 meters (50 feet) and weigh more than 4.5 metric tons (5 tons). Cacti are generally exceptionally slow-growing and, because they need so little care, make good house plants for sunny windows.

In 1944, a marine pilot was forced to bail out of his aircraft over the desert near Yuma, Arizona. Until he was rescued 5 days later, he survived the intense heat and low humidity of the area by chewing the juicy pulp of barrel cacti in the vicinity. Since then, the use of cacti for emergency fluids and food has been recommended in most survival manuals.

Most cacti have edible fruits, and only three cacti (peyote, living rock, and hedgehog cactus) are known to be poisonous. Prickly pear fruits are occasionally sold in American

Figure 24.15 Cacti. *A*. Prickly pear cacti. *B*. Peyote. *C*. A barrel cactus. *D*. An organ-pipe cactus.

supermarkets and taste a little like pears. Prickly pear fruits also have seeds that Native Americans of the Southwest dried and ground for flour they mixed with water and used as an *atole* (thickened souplike) staple food. A good syrup is obtained from boiling the fruits of prickly pears and also those of the giant saguaros. In the past, cactus candy was made by partly drying strips of barrel cactus and boiling them in saguaro fruit syrup, but the cactus is now usually boiled in cane sugar syrup.

Native Americans of the Southwest used to scoop out barrel cacti, dry them, and use them for pots. They also mixed the sticky juice of prickly pear cacti in the mortar used in constructing their adobe huts. In Texas, a poultice of prickly pear stem was applied to spider bites. Hopi Indians chewed raw cholla cactus as a treatment for diarrhea, and the skeletons of these cacti were used for flower arrangements.

In the middle of the 19th century, Australians planted a few imported prickly pear cacti in the dry interior. These cacti found no natural enemies in their new environment and multiplied rapidly, infesting more than 24 million hectares (60 million acres) within 75 years. In 1925, in an effort to control them, Australia introduced an Argentine moth among the cacti. The moth's caterpillars feed on prickly pear cacti and gradually brought the plants under control, making the land usable again.

Another cactus parasite, the cochineal insect (related to the mealybug, a common house plant pest), feeds on prickly pear cacti in Mexico. At one time, the insects were collected for a crimson dye they produce; the dye was used in lipstick and rouge before aniline dyes derived from coal tar were introduced in the 1930s.

Peyote cacti are small button-like plants that have no spines, with roots resembling those of carrots. They contain several drugs, the best known of which is *mescaline,* a powerful hallucinogen. Dried slices of peyote have been used in native religious ceremonies in Mexico for centuries and more recently by at least 30 tribes of Native Americans.

The Mint Family (Lamiaceae)

The 3,000 members of the Mint Family are relatively easy to distinguish since they have a unique combination of angular stems that are square in cross section, opposite leaves, and bilaterally symmetrical (irregular) flowers (Fig. 24.16). Most also produce aromatic oils in the leaves and stems. The superior ovary is four-parted, with each of the four divisions developing into a nutlet. Included in the family are such well-known plants as rosemary, thyme, sage (not to be confused with sagebrush of the Sunflower Family), oregano, marjoram, basil, lavender, catnip, peppermint, and spearmint.

Mint oils can be distilled at home with ordinary canning equipment. Whole plants (or at least the leaves) are loosely packed to a depth of about 10 centimeters (4 inches) or more in the bottom of a large canning pot. Then a wire rack or other support is also put in the pot, and a bowl is placed in the middle on the rack. Enough water is added to cover the

Figure 24.16 Flowers of lamb's ear mint.

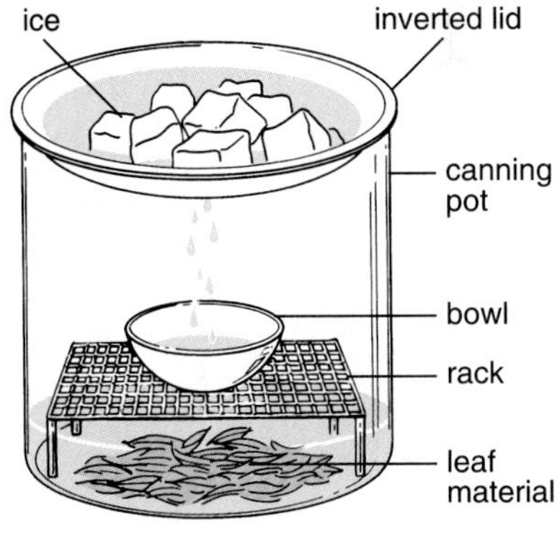

Figure 24.17 A simple apparatus for distilling mint oil at home. See text.

vegetation, the pot is placed on a range, and the lid is inverted over it. The water is brought to a boil, and as it does so, ice is placed on the inverted lid. The oils vaporize and condense when they contact the cold lid, dripping then from the low point into the bowl (Fig. 24.17). Of course, some moisture also condenses, but the oil, being lighter, floats on top. Peppermint oil is easy to collect this way and will keep for a year or two in a refrigerator.

Mint oils have been used medicinally and as an antiseptic in different parts of the world. Mohegan Indians used catnip tea for colds, and dairy farmers in parts of the midwestern

United States used local mint oils to wash their milking equipment before antibiotics became popular for this purpose. As a result, mastitis, a common disease of dairy cattle, was seldom encountered in their herds. Horehound, a common mint weed of Europe, has become naturalized on other continents and is cultivated in France. A leaf extract is still used in horehound candy and cough medicines. In England, it is a basic ingredient of horehound beer. Vinegar weed, also known as blue curls, is a common fall-flowering plant of western North America. Native Americans of the area used it in cold remedies, for the relief of toothaches, and in a bath for the treatment of smallpox. It was also used to stupefy fish.

Menthol, the most abundant ingredient of peppermint oil (Fig. 24.18), is widely used today in toothpaste, candies, chewing gum, liqueurs, and cigarettes. Most American mint is grown commercially in the Columbia River basin of Oregon and Washington. Geese are sometimes used in the mint fields to control both insects and weeds, since they do not interfere with the growth of the mint plants themselves.

Figure 24.18 A peppermint plant in flower. Oil from the leaves is a source of menthol.

Figure 24.19 A flowering head of chia.

Ornamental mints include salvias and the popular variegated-leaf *Coleus* plants, neither of which has typical mint oils in the foliage. *Chia* (Fig. 24.19), another relatively odorless mint, is confined to the drier areas of western North America. Native Americans parched chia seeds and used them in gruel. The seeds, which become mucilaginous when wet, were also ground into a paste that was placed in the eye to aid in the removal of dirt particles. The paste was also used as a poultice for gunshot wounds, and Spanish Californians made a refreshing drink from ground chia seeds, lemon juice, and sugar. Chia seeds reportedly contain an unidentified substance that has effects similar to those of caffeine. Before the turn of the century, one physician reported that a tablespoon of chia seeds was sufficient to sustain a man on a 24-hour endurance hike. Since that time, backpackers have experimented with the seeds, and results tend to support the earlier claim. A thorough scientific investigation of the matter is needed. Chia seeds are sold commercially for making into a paste that is spread on clay models and then watered. The seeds sprout and resemble green hair.

The Nightshade Family (Solanaceae)

The Nightshade Family is concentrated in the tropics of Central and South America. The flowers have fused petals, with the stamen filaments fused to the corolla so that they appear to be arising from it (Fig. 24.20). The superior ovary develops into a berry or a capsule. The more than 3,000 species of the family have alternate leaves and occur as herbs, shrubs, trees, or vines. Well-known representatives include tomato, white potato, eggplant, pepper, tobacco, and petunia.

Many nightshades produce poisonous drugs, some of which have medicinal uses. One of the best-known medicinal drug producers is the deadly nightshade of Europe. A drug complex called *belladonna* is extracted from its leaves. Belladonna was used in the "magic potions" of the past and also for dilating human pupils for cosmetic purposes. It is now the source of several widely used drugs, including atropine, scopolamine, and hyoscyamine. Atropine is used in shock treatment, for relief of pain, to dilate eyes, and to counteract muscle spasms. Scopolamine is used as a tranquilizer, and hyoscyamine has effects similar to those of atropine. Capsicum, obtained from a pepper, is used as a gastric stimulant and is also a principal ingredient of mace, which is used to repel human or animal assailants. Capsaicin, derived from peppers, is used in ointments for the relief of arthritic and neuropathic pain.

Jimson weed (see Fig. 8.6) is also a source of medicinal drugs that have been used in treatment of asthma and other ailments. The drugs can be fatal if ingested in sufficient quantities but were used in controlled amounts in Native American rituals of the past. Records indicate that users became temporarily insane but had no recollection of their activities when the effects of the drug wore off. The drug solanine is present in most, if not all, members of the family. Many arthritis sufferers apparently are sensitive to solanine, and a number of arthritics have reported partial relief through total abstinence of members of this family (including potatoes, tomatoes, peppers, and eggplant).

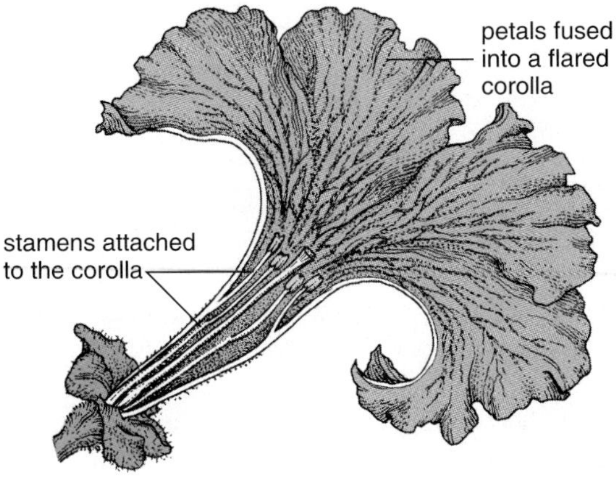

Figure 24.20 A sectioned petunia flower.

Tobacco cultivation occupies more than 800,000 hectares (2 million acres) of American farmland. In its dried form, tobacco contains 1% to 3% of the drug nicotine. Nicotine is used in certain insecticides, and it is used also for killing intestinal worms in farm livestock. It is, however, an addictive drug. The evidence of human tobacco use as a primary cause of heart and respiratory diseases including lung cancer and other cancers such as those of the mouth and throat mounts almost daily. The only "benefit" it may have to humans appears to be as a killer of leeches. It is said that leeches attaching themselves to heavy smokers will drop off dead within 5 minutes from nicotine poisoning—a very dubious justification for continued human use!

Tomatoes are among the most popular of all "vegetables." About 18 million metric tons (20 million tons) are grown annually around the world. The plants are day-neutral (day lengths are discussed in Chapter 11), and even though they require warm night temperatures (16°C or 60°F) to set fruit well, they are easily cultivated in greenhouses when natural conditions are unfavorable. Most commercially grown American tomatoes are processed into juice, tomato paste, and catsup. In Italy, a small amount of edible oil is extracted from the seeds after the pulp has been removed. Most American tomatoes are grown in California and Florida, where they are harvested with special machinery developed during the 1960s when inexpensive labor became unavailable (Fig. 24.21).

The white, or Irish, potato is one of the most important foods grown in temperate regions of the world, with annual production estimated at well over 270 million metric tons (300 million tons). The leading producers are China, Poland, the United States, and countries of the former Soviet Union, which account for about 30% of the total. It is believed that white potatoes originated on an island off the coast of Chile and were sent back to Europe by Spanish invaders of South America in the 16th century. In the 1840s, late blight infested and destroyed the potato crop of Ireland, causing severe famine. Irish settlers subsequently emigrated to the United States, Canada, Australia, and other parts of the world.

When potato tubers are exposed to the sun, they turn green at the surface. Poisonous drugs (e.g., solanine) are produced in the green areas. These have proved fatal to both animals and humans and should never be eaten.

The Carrot Family (Apiaceae)

Members of the Carrot Family are widely distrbuted in the Northern Hemisphere, and many have savory-aromatic herbage. The flowers tend to be small and numerous and are arranged in umbels. The ovary is inferior, and the stigma is two lobed. The leaves are generally dissected, and the bases of their petioles usually form sheaths around the stem. Included in the 2,000 members of the family are dill, celery, carrot, parsley, caraway, coriander, fennel, anise, and parsnip. Anise is one of the earliest aromatics mentioned in

Figure 24.21 Harvesting a field with a mechanical tomato harvester.

literature. It is used for flavoring cakes, curries, pastries, and candy. Pocket gophers apparently are attracted by its aroma, and some poison baits are enhanced with anise. A liqueur known as *anisette* is flavored with it.

Caraway seeds are used to flavor a Danish liqueur, called *kümmel*. The seeds are well known for their use in rye and pumpernickel breads.

Some members of the Carrot Family are poisonous. Water hemlock (Fig. 24.22) and poison hemlock are common weeds in ditches and along streams. All parts of the plants are deadly, often having been fatal to unwary wild-food lovers. Socrates is believed to have died as a result of ingesting poison hemlock, which should not be confused with cone-bearing hemlock trees.

Several members of the Carrot Family, such as cow parsnip, squawroot, and hog fennel, have edible roots and were used for food by Native Americans. The reader is advised, however, to be absolutely certain of the identity of such plants before experimenting with them.

The Pumpkin Family (Cucurbitaceae)

Although most species in the Pumpkin Family are tropical or subtropical, many occur in temperate areas of both the Northern and Southern Hemispheres. Plants are prostrate or climbing herbaceous vines with tendrils. The flowers have

Figure 24.22 A water hemlock in flower.

fused petals, and female flowers have an inferior ovary with three carpels. All are unisexual. Some species have both male and female flowers on the same plant, while others have only male or only female flowers on one plant. In male flowers, the stamens cohere to varying degrees, depending on the species. The family has about 700 members, several of which have many horticultural varieties.

This family includes many important edible plants, and some have been cultivated for so long that they are unknown in the wild state. Well-known members of the family include pumpkins, squashes, cucumbers, cantaloupes (Fig. 24.23B), and watermelons.

The vegetable sponge (Fig. 24.23C), when it is growing, resembles a large cucumber and has a highly netted fibrous skeleton that can serve as a bath sponge after the soft tissues have been removed.

Gourds found in Mexican caves and subjected to radiocarbon dating have proved to be many thousands of years old. Various types of gourds (Fig. 24.23A), serving many purposes, are still grown today. Some are scooped out and used for carrying liquids or for storing food, particularly grains. South Americans drink maté, a tea, from gourds, which are also used for several types of musical instruments. In parts of Africa, gourds are used to catch monkeys. A type of gourd with a narrow neck is scooped out and partly filled with corn or other grains. One end of a rope is then tied to the gourd and the other to a stake driven into the ground. When a monkey tries to grab a fistful of grain, it finds that the neck will not allow its bulging hand to be removed. Most do not realize that letting go of the grain would allow them to escape, and they stubbornly hang on until they are captured.

Melonette, a small cucumberlike vine of the southeastern United States, has seeds that can be *purgative* (drastically laxative). Other cucumberlike plants of the western states, manroots (see Fig. 5.9), produce huge water-storage roots, some weighing as much as 90 kilograms (200 pounds). These roots were crushed by Native Americans and thrown into dammed streams to stupefy fish. An oil from the seeds was applied to the scalp as a remedy for infections that caused hair loss.

The Sunflower Family (Asteraceae)

The Sunflower Family, with approximately 20,000 species, is the second largest of the flowering plant families in terms of number of species. The individual flowers are called **florets.** They are usually tiny and numerous but are arranged in a compact inflorescence so that they resemble a single flower. A sunflower or daisy, for example, consists of dozens if not hundreds of tiny flowers crowded together, with those around the margin having greatly developed corollas that extend out like straps, forming what appear to be the "petals" of the inflorescence (Fig. 24.24 gives details of a sunflower inflorescence). In dandelions, all the individual florets of the inflorescence have narrow straplike extensions.

A.

B.

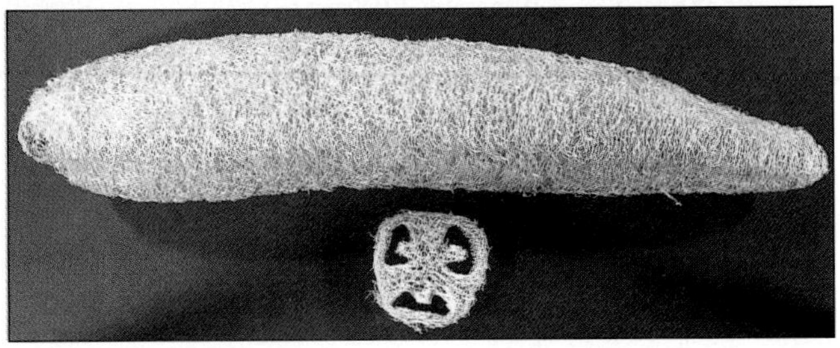

C.

Figure 24.23 Fruits and items associated with members of the Pumpkin Family (Cucurbitaceae). *A.* A Hawaiian ceremonial gourd with feathers attached and gourds used in South America for drinking maté. A metal straw that strains out the maté leaves is resting in one gourd. *B.* Cantaloupes. *C.* A luffa (vegetable sponge).

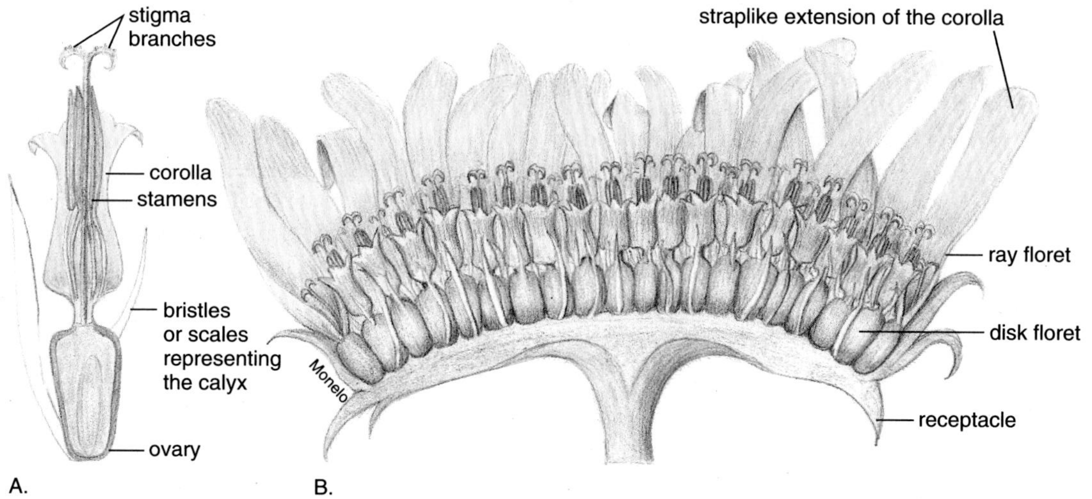

Figure 24.24 Parts of a sunflower. *A.* A section through a single floret. *B.* A section through an inflorescence.

Well-known members of this family include lettuce, endive, chicory, Jerusalem artichoke, globe artichoke, dahlia, chrysanthemum, marigold, sunflower, and thistle.

Santonin, obtained from flower buds of a relative of sagebrush that is native to the Middle East, is used as an intestinal worm remedy. Tarragon, used as a spice in meat dishes and pickles, comes from another relative of sagebrush. Pyrethrum is a natural insecticide obtained from certain chrysanthemum flowers. Fructose, a sugar, is obtained from the tubers of Jerusalem artichokes and dahlias. Dahlias are also renowned for their huge showy flowers, while Jerusalem artichokes are often eaten as a vegetable.

Marigolds are favorite plants of organic gardeners. Their roots are said to release a substance that repels nematodes, and the odor of the leaves repels white flies and other insects. Unfortunately, snails and slugs seem to be immune and voraciously consume the foliage.

Many members of this family were used widely by Native Americans. The dry fruits of balsamroot and mule ears were used for food. Balsamroot plants as a whole were eaten raw or cooked, and in the West, extracts of both mule ears and tarweeds were used to treat poison oak inflammations. Salsify and thistle roots were also used for food.

Young leaves and roots of dandelions have been eaten for centuries, and the flowers have been used to make wine. Roasted dandelion and chicory roots have been used as a coffee substitute. During World War II, chicory was grown as a crop specifically for use as a coffee adulterant.

The dried and boiled leaves of American yarrow are said to make a nourishing broth. European yarrow has become naturalized in North America. Its rhizomes contain an anaesthetic that, when chewed, numbs the tongue and gums, and it has been used to ease the discomfort of teething toddlers; the substance has also been used in suppression of menstruation. This plant is believed to have been used by Achilles in treating the wounds of his soldiers.

Sunflowers themselves were widely used by Native Americans for their dry fruits, which were ground into a meal for bread. They are grown commercially today primarily for the edible oil extracted from the seeds (after removal from the husks), but the seeds are increasingly being eaten by modern Americans.

MONOCOTS

The Grass Family (Poaceae)

Individual plants of the the grass family are more numerous and more widely distributed than plants of any other flowering plant family. There are, however, possibly only one-fourth as many species of grasses as there are members of the Sunflower Family. The flowers of grasses are highly specialized in structure and have a terminology all their own (Fig. 24.25). The calyx and corolla are represented by tiny, inconspicuous scales, and the flowers are protected by boat-shaped bracts. The stigmas, when they are exposed, are feathery, and the leaves sheathe the stem at their bases.

Nearly all of the cereals, including wheat, barley, rye, oats, rice, and corn, belong in this family, which includes nine of the ten most important crop plants in the world. Indeed, civilization as we know it would be vastly different without them. More than 900 million metric tons (1 billion tons) of cereals, feeding more than half of the world's population, are harvested each year, primarily in the Orient, North America, and Europe.

Sugar cane (Fig. 24.26), from which about 55 million metric tons (60 million tons) of sugar is extracted annually, is grown at lower elevations throughout humid tropical areas. It is a large grass, often growing to heights of 6 meters (20 feet). After the cane is harvested, raw liquid sugar is squeezed out by a milling process. The juice then is centrifuged, with the

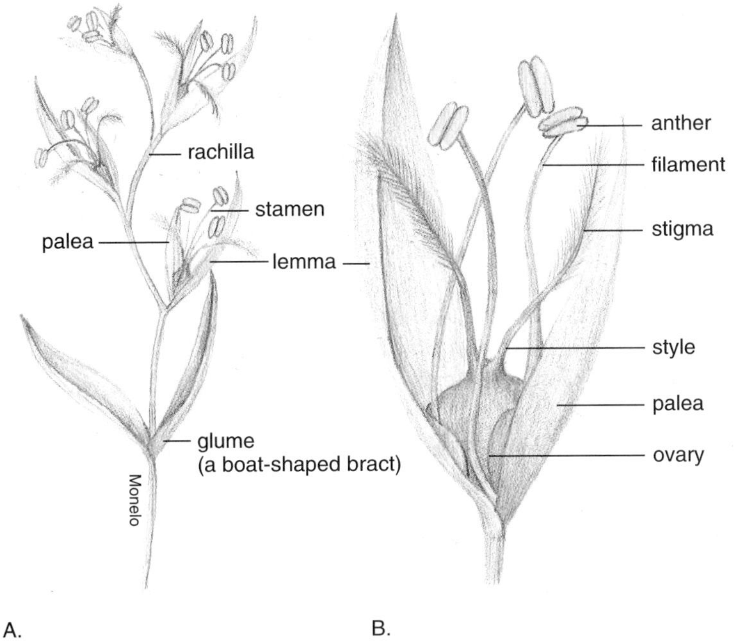

Figure 24.25 Grass flowers. *A.* An expanded grass spikelet. *B.* An enlargement of a single flower (floret).

Figure 24.26 Sugar cane plants.

solids being crystalized into table sugar. The dark remnant (molasses) may be used to produce rum or alcohol. The solid waste is sometimes made into paper or particle board, and also has been converted to gasoline in South Africa. In Hawaii, the waste is used in the production of electric power.

Grasses have been used widely by primitive peoples for making mats and baskets and for thatching huts. Some varieties of sorghum are grown for their fibers, which are made into brooms, although natural broom fiber now has been replaced to a considerable extent by synthetic materials. Some varieties are grown for the seeds, which are processed into cereal flours. Others are a source of silage and a carnauba-like wax. Citronella oil, once a common ingredient of mosquito repellents, is obtained from a grass grown in Indonesia. It is now used in cheaper soaps, cosmetics, and perfumes. Related grasses are the source of lemon grass oil, used for the same purposes as citronella oil.

Juice squeezed from fresh, young grass mowings has a high protein content and is being investigated as a source of protein for future human consumption.

The Lily Family (Liliaceae)

Although the 4,550 members of this family are particularly abundant in the tropics and subtropics, they occur in almost any area that supports vegetation. The flowers are often large, and their parts are all in multiples of three, with the sepals and petals frequently resembling each other in color and form (Fig. 24.27).

In addition to many types of lilies and daffodils used widely as ornamentals, the family includes asparagus, sarsaparilla, squill, meadow saffron, bowstring hemp, and *Aloe.* Sarsaparilla was at one time widely used for flavoring soft

Awareness

Coffee and Caffeine

Among all the foods, beverages, and products provided by flowering plants, coffee (*Coffea arabica*) is one that millions of people, particularly Americans, have contact with every day. We take our coffee break seriously. We drink some 350 million cups of coffee daily, which is enough coffee to turn the Metrodome into a deep caffeine pool. Here are some frequently asked questions concerning coffee and caffeine. So grab a cup of coffee, pull up your favorite chair, sit down, and relax while you read about one of the world's favorite beverages.

Where Does Coffee Grow?

Coffee is a small tree or shrub native to the mountains of Ethiopia but, because of its popularity as a drink, coffee was spread by traders to many tropical regions of the world. Today, Brazil and Colombia are the world's leading coffee producers. Coffee is derived from the seeds (beans) of *Coffea arabica* or *Coffea robusta,* two species of the 70 or more recognized varieties between them. The trees require tropical or subtropical climate and need about 60 to 100 inches of rain per year. They grow best at mountain elevations of 3,000 to 6,000 feet where the temperature is cool and relatively stable (68°F) because they are unable to tolerate frost. Trees start producing fruit around 3 to 5 years of age.

Botanically Speaking, What Is a Coffee "Bean"?

In Costa Rica, the crop season begins with the first "flowering rains" that come in March or April. Flowering buds that were produced at the end of the dry season are now conditioned to flower by the gentle rains. The white, delicate flowers have little aroma and bloom for 3 days before falling off the trees. Each pollinated flower will produce one fruit, which is called a cherry because of its red color. The coffee cherry contains two seeds that are tightly packed against each other, resulting in each seed having a flat side. The coffee "beans" are actually the two seeds of each fruit. Coffee fruits change color from dark green to deep red and are handpicked when ripe (red color). Each tree is usually harvested three or four times because of the uneven maturation of fruits. A single tree can yield up to 5 to 6 pounds of fruit per year.

How Much Caffeine Does a Cup of Coffee Have?

A 7-ounce cup of coffee prepared by the drip method contains 115 to 175 milligrams of caffeine, while instant coffee contains 65 to 100 milligrams. Decaffeinated instant coffee contains 2 to 3 milligrams of caffeine.

How Much Caffeine Is in Soft Drinks?

The following table compares the caffeine content of several brands of soft drinks.

Soft Drink (12-oz can)	Caffeine Content (mg)
Jolt	72.0
Mountain Dew	54.0
Coca-Cola	45.6
Dr. Pepper	39.6
Pepsi-Cola	38.4
RC Cola	36.0

Chemically Speaking, What Is Caffeine?

Caffeine is a plant alkaloid found not only in coffee beans but also in tea leaves and cola nuts. It works as a stimulant of the central nervous system and the respiratory system, as well as increasing heart rate and constricting blood vessels. As a stimulant, caffeine makes us more aware and alert and able to work without drowsiness. It is reported that coffee was discovered in the Near East (perhaps in Yemen) by goats, which became frisky and sleepless after eating the wild fruit. A shepherd is said to have located the shrub the goats had chewed and described it as "a shrub with leaves like laurel, flowers like jasmine, and little dark berries the chief object of the animals' nibbling."

What Are the Effects of Caffeine on the Body?

Caffeine stimulates or increases heartbeat rate, respiration, and basal metabolic rate and increases the production of stomach acid and urine. These effects translate into a stimulating "lift" that a person may feel. Generally, one feels less drowsy, less fatigued, and more alert and focused after consuming caffeine.

Where Does the Word "Coffee" Come From?

The word "coffee" is probably derived from the Arabic word "kahveh," which means "stimulating," and is thought to have been used to describe the effects of eating coffee seeds (beans).

What Does Caffeine Taste Like?

Caffeine is very bitter to the taste, as are other plant alkaloids. Caffeine is added as a flavoring agent to beverages such as root beer because it produces the sharp bitterness characteristic of the drink.

Awareness Box Continued

How Is Coffee Decaffeinated?

If green (unroasted) coffee beans are rinsed with hot water above 175°F, caffeine is extracted from the beans. However, in the process of removing caffeine, hot water alone is not used because it strips away too many of the essential flavors and aromatic elements contained in the beans. Along with water, a decaffeinating chemical such as methylene chloride or ethyl acetate is generally used. One common method involves placing the green beans in a rotating drum and softening them with steam for 30 minutes. They are then rinsed for approximately 10 hours with methylene chloride, which removes the caffeine from the beans. Methylene chloride containing the caffeine is drained off, and the beans are steamed a second time for 8 to 12 hours to remove any remaining methylene chloride solvent. Finally, the beans are air- or vacuum-dried to remove excess moisture before the roasting process. Most decaffeinated coffee contains less than 0.1 part per million (ppm) of residual methylene chloride, 100 times less than the maximum level of 10 ppm allowed by the U.S. Food and Drug Administration.

D. C. Scheirer

Figure 24.27 A tiger lily flower.

Figure 24.28 *Sansevieria* plants.

drinks and medicines and is obtained from the roots of a genus of woody vines (*Smilax*) whose stems are often covered with prickles. The bulbs of squills are the source of a rodent poison and also of a drug used as a heart stimulant.

Meadow saffron is the source of *colchicine,* a drug once used to treat rheumatism and gout but now much more widely used in experimental agriculture to interfere with spindle formation in cells so that the chromosome number of plants may be artificially increased. This increasing of the chromosome number can result in larger and more vigorous varieties of plants. (Meadow saffron should not be confused with true saffron, a member of the Iris Family and the source of the world's most expensive spice and a powerful yellow dye.)

Bowstring hemps are related to the familiar, seemingly indestructible house plants called *sansevierias* (Fig. 24.28), which have long, narrow, stiff, upright leaves. The plants are cultivated in tropical Africa for their long fibers, which are used for string, rope, bowstrings, mats, and cloth. New Zealand flax, a larger plant, is grown in South America and New Zealand for similar purposes but is also widely used in ornamental plantings.

Several *Aloe* species produce juices used in shampoos, cosmetics, sunburn lotions, and in the treatment of burns. African *Aloe* species are prized as ornamentals in areas with

Figure 24.29 An uprooted California soaproot plant.

Figure 24.30 Bamboo orchids growing wild in Hawaii.

milder climates. Their thick, fleshy leaves have short spines along the margins. The spines were once used for phonograph needles.

Many lily bulbs (e.g., onions and garlic) are edible, and wild lily bulbs were used for food extensively by Native Americans. Wild bulbs should no longer be eaten, however, as doing so may endanger the survival of native species. *(Caution: Lily bulbs should not be confused with those of daffodils and their relatives. Daffodil and related bulbs are highly poisonous).*

The California soaproot (Fig. 24.29) is confined to California and southern Oregon and was important for food to Native Americans of the region. It also had several other uses. The large bulbs are covered with coarse fibers that were removed and tied to sticks to make small brooms. The bulbs themselves produce a lather in water and were used for soap. Sometimes, numbers of bulbs were crushed and thrown in a small stream that had been dammed. Fish would be stupefied and float to the surface. The bulbs were generally eaten after being roasted in a stone-lined pit in which a fire had been made. While they were roasting, a sticky juice would ooze out. This was used for gluing feathers to arrow shafts.

A resin used in stains and varnishes exudes from the stem of dragon's blood plants. Grass trees of Australia yield resins used in sealing waxes and varnishes.

The Orchid Family (Orchidaceae)

This very large family has, according to some authorities, more than 35,000 species that are widely distributed. Like members of the Lily Family, they are especially abundant in the tropics. In many genera, the number of individual plants at any one location may be quite small—sometimes limited to a single plant.

The flowers are exceptionally varied in size and form, and the habitats of the plants are equally diverse. The flowers of one Venezuelan species have a diameter of less than 1 millimeter (1/25 of an inch), while those of a species native to Madagascar may be more than 45 centimeters (about 18 inches) long. One species of *Dendrobium* orchid from Java has flowers that are so delicate they wither within 5 or 6 minutes after opening. Many orchids are epiphytic on the bark of trees. During its 5-month flowering season, one epiphytic species of Malaysia and the Philippine Islands produces 10,000 flowers on plants that weigh more than 1 metric ton. Others are aquatic or terrestrial, and a saprophytic species (see Fig. 8.1) native to western Australia grows and flowers entirely underground.

Orchids have three sepals and three petals, with one of the petals (the *lip petal*) differing in form from the other two (Fig. 24.30). The stamens and pistil are united in a unique single structure, the *column* (see Fig. 23.9). The stigma usually consists of a sticky depression on the column. The anthers contain sacs of pollen called *pollinia* and are covered with a cap until they are removed by an insect or other pollinator. The specific adaptations between orchid flowers and their pollinators are extraordinary and sometimes bizarre (as illustrated in Fig. 23.18).

Orchids have minute seeds that are often produced in prodigious numbers (e.g., a single fruit of certain orchid species may contain up to 1 million seeds). Each seed consists of only a few cells, and in order for a seed to germinate, it must become associated with a specific mycorrhizal fungus that produces substances necessary for its development. Once a seed has germinated, it may take from 6 to 12 or more years before the first flower appears.

Contrary to popular belief, some orchids can be grown relatively easily on a windowsill that has bright light, but not direct sunlight (see Appendix 4). Because orchids are among the most beautiful and prized of flowers, a large industry has grown up around their culture and propagation, discussed in the section on mericloning and tissue culture in Chapter 14. One species, the vanilla orchid, is grown commercially in the tropics for its fruits, which are the source of true vanilla flavoring.

Summary

1. Distinguishing between fact and fantasy in reported past uses of plants is often difficult. Teams of specialists are interviewing tribal medicine men and women of the tropics to try to save potentially useful plants from extinction.
2. N. I. Vavilov proposed eight major centers of distribution of cultivated plants. More recent studies by Harlan and others indicate strict recognition of the centers is without merit.
3. Flowering plant families are surveyed in evolutionary sequence, discussing first those considered primitive.
4. The Buttercup Family (Ranunculaceae) includes buttercup, columbine, larkspur, anemone, monkshood, *Clematis,* goldenseal, and wolfsbane.
5. The Laurel Family (Lauraceae) includes cinnamon, camphor, sassafras, sweet bay, California bay (or myrtle), avocado, and laurel.
6. The Poppy Family (Papaveraceae) includes bloodroot and opium poppy (the source of medicinal drugs and heroin).
7. The Mustard Family (Brassicaceae) includes cabbage, cauliflower, brussels sprouts, broccoli, radish, kohlrabi, turnip, horseradish, watercress, rutabaga, shepherd's purse, western wallflower, dyer's woad, and camelina.
8. The Rose Family (Rosaceae) includes stone fruits (e.g., cherry, apricot, peach, plum), strawberry, raspberry, rose, and related wild species.
9. The Legume Family (Fabaceae) includes pea, bean, lentil, peanut, soybean, alfalfa, clover, licorice, wattle, indigo, logwood, locoweed, jicama, sensitive plant (*Mimosa pudica*), and winged bean.
10. The Spurge Family (Euphorbiaceae) includes poinsettia, spurge, cassava, Pará rubber, gopher plant, candelilla, castor bean, crown-of-thorns, Mexican jumping bean, and tung oil tree.
11. The Cactus Family (Cactaceae) includes all cacti (e.g., prickly pear, cholla, barrel cactus, peyote).
12. The Mint Family (Lamiaceae) includes rosemary, thyme, sage, oregano, marjoram, basil, lavender, catnip, peppermint, spearmint, horehound, salvia, *Coleus,* and chia.
13. The Nightshade Family (Solanaceae) includes tomato, white potato, eggplant, pepper, tobacco, belladonna, petunia, and jimson weed (a source of hallucinogenic drugs).
14. The Carrot Family (Apiaceae) includes dill, caraway, celery, carrot, parsley, coriander, fennel, anise, parsnip, water hemlock, poison hemlock, cow parsnip, squawroot, and hog fennel.
15. The Pumpkin Family (Cucurbitaceae) includes pumpkin, squash, cucumber, cantaloupe, watermelon, vegetable sponge, gourd, melonette, and manroot.
16. The Sunflower Family (Asteraceae) includes sunflower, dandelion, lettuce, endive, chicory, Jerusalem and globe artichokes, dahlia, chrysanthemum, marigold, thistle, sagebrush, pyrethrum, balsamroot, tarweed, and yarrow.
17. The Grass Family (Poaceae) includes nearly all cereals (e.g., wheat, barley, rye, oats, rice, corn), sugar cane, sorghum, citronella, and lemon grass.
18. The Lily Family (Liliaceae) includes lilies, asparagus, sarsaparilla, squill, meadow saffron, bowstring hemp, *Aloe,* and New Zealand flax.
19. The Orchid Family (Orchidaceae) has highly specialized flowers. It includes a species that is the source of vanilla flavoring.

Review Questions

1. To which flowering plant family does each of the following belong: poinsettia, lupine, columbine, peach, pear, cinnamon, sarsaparilla, belladonna, peyote, horehound, rubber, gourd, jimson weed, parsley, sorghum, asparagus, broccoli, lettuce, tomato, opium?
2. Make a list of the poisonous plants in the families discussed.
3. Which plants mentioned are or have been used for medicines?
4. Which plants mentioned have been used for tools or utensils?
5. Native Americans made extensive use of plants for a wide variety of purposes in the past. List such uses for as many plants as possible.

Discussion Questions

1. Scientific investigations often take a great deal of time and money. Is it worth the effort to check out scientifically the past uses of plants?
2. A return to exclusively herbal medicines is being advocated in some quarters. Is this a good idea? What are the pros and cons?
3. Would you expect drugs produced naturally by plants to be more effective or better than drugs produced synthetically? Why?
4. If you were asked to single out the three most important families of the 16 discussed, which would you choose? Why?
5. A number of wild edible plants were mentioned in this chapter. What would happen if a large portion of the population were to gather these wild plants as a major source of food?

Learning Online

Visit our web page at www.mhhe.com/botany for interesting case studies, practice quizzes, current articles, and animations within the Online Learning Center to help you understand the material in this chapter. You'll also find active links to these topics:

Origins of Agriculture
Temperate Fruits and Nuts
Tropical Fruits and Nuts
Cereal Grains and Forage Grasses
Legumes
Spices, Herbs and Perfumes
Vegetable Oils and Waxes
Hydrogels, Elastic Latexes, and Resins
Psychoactive Drugs and Poisons from Plants and Fungi
Stimulating Beverages Derived from Plants
Alcoholic Beverages from Plants and Fungi
Fibers, Dyes, and Tannins from Plants
Wood, Cork, and Bamboo
Ornamental Plants
Medicinal Plants

Additional Reading

Akerle, O., et al. (Eds.). 1991. *Conservation of medicinal plants.* New York: Cambridge University Press.

Brill, S., and E. Dean. 1994. *Identifying and harvesting edible and medicinal plants.* New York: Hearst Books.

Buchmann, S. L., and G. P. Nabham. 1997. *Forgotten pollinators.* Washington, DC: Island Press.

Gibbons, E. 1987. *Stalking the wild asparagus.* New York: David McKay Co.

Harlan, J. R. 1992. *Crops and Man,* 2d ed. Madison, WI: American Society of Agronomy.

Hornsok, L. (Ed.). 1993. *Cultivation and processing of medicinal plants.* New York: Wiley & Sons.

Kunkel, G. 1984. *Plants for human consumption: Annotated checklist of edible phanerogams and ferns.* Forestburgh, NY: Lubrecht & Cramer.

Lewis, W. H., and M. P. F. Elvin-Lewis. 1982. *Medical botany: Plants affecting man's health.* New York: John Wiley & Sons.

Mauseth, J. 1997. *Introduction to plant biology,* 2d ed. Sudbury, MA: Jones & Bartlett.

Simmons, F. J. 1998. *Plants of life, plants of death.* Madison, WI: University of Wisconsin Press.

Simpson, B. B., and M. Conner-Orgarzaly. 1994. *Economic botany: Plants in our world.* Hightstown, NJ: McGraw-Hill.

Sweet, M. 1976. *Common edible and useful plants of the West,* rev. ed. Healdsburg, CA: Naturegraph.

Van Allen Murphey, E. 1987. *Indian uses of native plants,* 3d ed. Glenwood, IL: Meyer Books.

Wijesekera, R. O. 1991. *The medicinal plant industry.* Boca Raton, FL: CRC Press.

Western flower and grass associations in the spring.

Ecology

25

Chapter Outline

Overview

This chapter explores some of the ecological topics not already discussed elsewhere in the text.

Regional issues, including acid deposition, water contamination, wetlands, and hazardous waste, are discussed first. Then global matters are introduced with a discussion of populations, communities, and ecosystems. This is followed by a brief look at producers, primary and secondary consumers, decomposers, and food chains. Then energy flow through an ecosystem is considered, and the nitrogen, carbon, and water cycles are explored. Discussions of succession and fire ecology follow. Human disruption of ecosystems, including ozone depletion, global warming, and loss of biodiversity, conclude the chapter.

Some Learning Goals

1. Know at least 10 ways in which humans have disrupted ecosystems.
2. Know the functions of producers, primary consumers, secondary consumers, and decomposers in an ecosystem.
3. Understand energy flow through an ecosystem.
4. Learn how nitrogen and carbon are cycled.
5. Define succession and know a succession that begins with bare rocks and one that begins with water.
6. Define ecotype, secondary succession, eutrophication, and climax vegetation.

Ecology is the broad discipline of study involving the relationships of organisms to each other and to their environment. The term *ecology,* which has become a household word, was first proposed by Ernst Haeckel in 1869, but the discipline traces its origins back to early civilizations when humans learned to modify their surroundings through the use of tools and fire.

Various aspects of ecology were touched on throughout earlier chapters. For example, soils and organisms associated with them were discussed in Chapter 5. Some of the adaptations and modifications of leaves associated with specific environments were discussed in Chapter 7. The effects of light and temperature on plants were covered in Chapters 10 and 11. Other important ecological topics comprise the bulk of this chapter, but the reader is referred to the Additional Reading section at the end of the chapter for a sampling of works covering a number of these themes in greater depth.

In recent years, scientists, and increasingly the general public, have become alarmed about the effects of human carelessness on our environment. It wasn't until the 1970s, however, that damage to forests and lakes caused by acid deposition (acid rain), the "greenhouse effect," contamination of ground water by nitrates and pesticides, reduction of the stratospheric ozone shield, major global climatic changes, loss of biodiversity in general, loss of tropical rain forests in particular, and other aspects of ecology gained widespread publicity.

Since it is possible to review only a few of the major aspects of ecological study and environmental considerations in a text of this nature, this chapter will briefly explore some of the issues at two broad levels: (1) *regional issues,* and (2) *global matters.*

REGIONAL ISSUES

Although funding for solving many types of ecological problems is considered by citizens interested in the environment to be inadequate, the allocations to regional issues usually gain better public support because they are highly visible, and progress in dealing with them can be followed.

In the past two decades alone, concern about the effects of pollution on lands, waters, and peoples has been widely reported in the media and in scientific journals. During this time, attempts to correct or stem environmental damage have resulted in requirements that many construction projects file environmental impact reports before proceeding. These reports provide information that helps various agencies evaluate the possible effects a proposed project may have on the flora, fauna, and physical environment and determine what *mitigations* (alleviations of the problem) may be necessary before the project can or should be approved. The process has sometimes resulted in a great deal of controversy and emotional debate between landowners who believe their rights are being usurped and those who are convinced that preservation of the environment should supersede material considerations. Brief discussions of some regional issues follow.

Acid Deposition

Acid deposition occurs after fossil fuels are burned. The burned fuels release sulfur and nitrogen compounds into the atmosphere. There, chemical reactions brought about by sunlight and rain convert the compounds to nitric acid (HNO_3). Burning may result in the conversion of sulfur dioxide (SO_2) to sulfuric acid (H_2SO_4). These may combine with moisture droplets and become a liquid acid deposition (*acid rain*), which has a pH of less than 5.5 as compared with a pH of about 5.7 for unpolluted rain water. Some of the acid deposition, however, may be in the form of solid particulates. The acid deposition, in turn, adversely affects or even kills many living organisms on which it falls and increases the concentration of other toxic elements. Mycorrhizal fungi are beneficially associated with the roots of many trees, and are susceptible to acidified soil, which may be linked to acid deposition. Higher acidity of soil water also reduces the capacity of plant roots to obtain needed mineral nutrients.

Trees have been dying prematurely in parts of the world's forests. Some have died because of a lack of water during successive years of low rainfall or from other causes, such as insect infestations. Salt scattered to melt ice and snow on roads has affected trees in the immediate vicinity. There seems to be little question that acid deposition has had a major role in stunting or destroying trees downwind from industrial sites. In recent years, there has been increasing international attention and cooperation in reducing industrial and automobile emissions that play a role in the production of acid deposition.

Acid deposition also affects nonliving materials. Dr. Merle Robertson, an authority on pre-Columbian art, reported that the natural weathering of ancient Mayan ruins in southern Mexico had been accelerated by acid deposition over the past decade or two. The acid deposition apparently originates when moisture-laden winds blow over major oil refineries and other industrial sites located upwind from the ruins.

Water Contamination

Water becomes contaminated in different ways. Much of the pollution in our lakes and streams comes from the dumping of toxic industrial wastes and from runoff over polluted land. Other sources include the spraying of pesticides, the exhaust and other emissions of aircraft and ships, and airborne pollutants originating with the combustion of fossil fuels. Ground-water supplies become polluted from pesticides and wastes trickling through the soil, from septic tanks, and from garden and farm fertilizers. Even deep wells, which seldom became polluted in the past, are increasingly becoming contaminated with unacceptably high levels of nitrates and other substances that harm the health of water-dependent humans and animals. Blue-baby syndrome is one example of the effects of high nitrate levels in drinking water. In response to the problem, many communities are improving their water treatment plants, and many families are installing water filtration systems in their homes.

Long-range goals for curbing water pollution include a tightening of restrictions on the dumping of wastes and greater improvements in municipal water purification plants and systems. Genetic engineering and bacteria also probably will play a major role in the future. For example, a bacterium that can remove more than 99% of 2,4,5-T (a major component of Agent Orange—discussed in Chapter 11) from a contaminated environment has already been bred, and several other bacteria are being genetically engineered to enhance their capacities for breaking down other toxic wastes.

Wetlands

Swamps, marshes, lagoons, river margins and estuaries, floodplains, and other wetland areas have historically been regarded as wastelands and have routinely been drained and converted to drier, more "productive" agricultural and industrial sites. For example, only 6% of California's original wetlands, 2% of Iowa's original wetlands, and less than 10% of the wetlands once along the shores of Lake Ontario remain today. Wetlands are, however, far from the wastelands they were once thought to be. Consider that the plants, algae, and other organisms in just 1 hectare (2.47 acres) of tidal wetland can perform the same recycling functions that about $150,000 of the latest wastewater treatment equipment is capable of—at no cost to taxpayers. The wetlands, at the same time, provide habitat, shelter, and breeding grounds for a wide variety of wildlife, many species of which are now threatened with extinction (Fig. 25.1).

Some attitudes toward wetlands have begun to change as the public has slowly become aware of the folly of eliminating them. City and regional planners in many areas now either ban new developments on wetlands or require a proposed development that would encroach on a wetland to provide mitigation measures such as the creation of new wet areas equal in extent and quality to those that may be lost. In addition, projects now in progress in many areas involve the restoration of damaged or lost wetlands.

Hazardous Waste

Members of earlier generations routinely drained the oil from their vehicle crankcases directly on to the ground or into storm drains or took the used oil and other hazardous wastes to the dump. Highly toxic industrial wastes were also disposed of both within and at the outskirts of cities and towns. In 1996, it was estimated that some 12 million children in the United States were living less than 4 miles from a hazardous waste site. Disasters have occurred when living organisms (including humans) have not been properly isolated from radioactive wastes produced by atomic energy plants. Even when hazardous wastes aren't unceremoniously dumped, serious accidents and spills take place, with the effects sometimes lingering indefinitely, as, for example, in and around the former Soviet Union's Chernobyl atomic meltdown site.

Figure 25.1 Butte Sink, California. A typical wetland, as seen from the air.

Today, there are concerted efforts to curb the disposal of hazardous wastes and to greatly reduce the probabilities of accidents and spills. At most solid-waste dumps, it is illegal to dispose of even empty latex paint cans, let alone more toxic materials, and heavy fines are levied on those found disposing of industrial wastes in an improper manner. Monies from a U.S. government "superfund" are being used, with some success, to clean up selected old hazardous waste sites. The process thus far is slow and inadequate, but the increased restrictions on disposal methods, and, as previously noted, the genetic engineering of bacteria that can dismantle and render harmless many types of wastes hold promise for the future.

GLOBAL MATTERS

There are many problems such as stratospheric ozone depletion, worldwide changes in climate, and loss of biodiversity, that are global in scope and long lasting in impact. Strategies for attacking the problems emphasize the need for multinational dialogue and cooperation between specialists in fields as diverse as biology, chemistry, economics, and political science.

To help comprehend the "big picture," we'll first examine some basic principles and then apply the principles to some of the global matters.

Populations, Communities, and Ecosystems

We recognize that plants, animals, and other organisms tend to be associated in various ways with one another and also with their physical environment. For example, forests consist of **populations** (groups of individuals of the same species) of trees or other plants that form a **plant community** (unit composed of all the populations of plants occurring in a given area). The lichen and moss flora on a rock also constitute a community, as do the various seaweeds in a tidepool (Fig. 25.2). However, these communities also invariably have animals and other living organisms associated with them. It is preferable, therefore, to refer to the unit composed of all the populations of living organisms in a given area as a **biotic community.** Considered together, the communities and their physical environments, which interact and are interconnected by physical, chemical, and biological processes, constitute **ecosystems.** Some populations, communities, and ecosystems may be microscopic in extent, while others can be much larger, or even global.

Populations

Populations may vary in numbers, in density, in genetic diversity, and in the total mass of individuals. Depending on circumstances, a field biologist may investigate a population

Figure 25.2 A tidepool community at low tide.

in various ways. If, for example, a conservation organization is concerned about the preservation of a rare or threatened species, the organization may simply count the *number* of individuals, although this may not always be feasible. If such a count is not feasible, the organization may estimate population *density* (number of individuals per unit volume—e.g., five blueberry bushes per square meter). If the individuals in a population vary greatly in size or are unevenly scattered, a better estimate of the population's importance to the ecosystem may be calculated by determining the **biomass** (total mass of the living individuals present).

Communities

Communities are composed of populations of one to many species of organisms living together in the same location. Similar communities occur under similar environmental conditions, although actual species composition can vary considerably from one location to another. A community is difficult to define precisely, because species of one community may also occur in other communities. Furthermore, species of one community may have specific genetic adaptations to that community; therefore, if individuals are transplanted to a second, different community where the same species occurs, the transplanted individuals may not necessarily be able to survive alongside their counterparts that are themselves adapted to this second community. Individuals adapted to specific communities within their overall distribution are called *ecotypes,* and areas transitional between communities are called *ecotones.*

Analysis and classification of communities are important in the preparation of maps that form the basis of activities such as land-use planning, forestry, natural resource management, and military maneuvers.

Ecosystems

Living organisms interacting with one another and with factors of the nonliving environment constitute an ecosystem. The nonliving factors of the environment (abiotic factors) include light, temperature, concentrations of oxygen and other gases, air circulation, fire, precipitation, rocks, and soil type. The distribution of a plant species in an ecosystem is controlled mostly by temperature, precipitation, soil type, and the effects of other living organisms (biotic factors). For example, in Mediterranean climates, such as those that occur in parts of California and Chile, nearly all precipitation occurs during the winter months, and the summers are dry. This type of climate favors spring annuals that complete their life cycles by summer and evergreen shrubs that can tolerate long periods of drought. Forests may occur in areas where melting heavy winter snowfall soaks deeply into the soil, compensating to a certain extent for the lack of summer moisture.

The distribution of plant species is influenced by the mineral content of soils. For example, serpentine soils occurring on the American west coast contain relatively high amounts of magnesium, iron, usually nickel, and chromium and low amounts of calcium. These soils often support species that are not found on nearby nonserpentine soils. Biotic interactions, such as competition for light and grazing by the animal members of the biotic community, and abiotic factors, such as mineral nutrients and available water, also influence the distribution of plant species.

The leaves and other parts of plant species that occur naturally in areas of low precipitation and high temperatures (*xerophytes*) generally are adapted to their particular environment through modifications that reduce transpiration. These modifications are discussed under "Specialized Leaves" in Chapter 7 and "Regulation of Transpiration" in Chapter 9. Plants of arid areas may also have specialized forms of photosynthesis, such as CAM photosynthesis. Similarly, plants that grow in water (*hydrophytes*) are modified for aquatic environments.

Ecosystems may sustain themselves entirely through photosynthetic activity, energy flow through **food chains** (discussed in the next paragraph), and the recycling of nutrients. Organisms, called **producers,** are capable of carrying on photosynthesis (e.g., plants, algae) and store energy that may be released by other organisms. Animals such as cows, caribou, caterpillars, and other organisms that feed directly on producers are called **primary consumers. Secondary consumers,** such as tigers, toads, and tsetse flies, feed on primary consumers. **Decomposers** break down organic materials to forms that can be reassimilated by the producers. The foremost decomposers in most ecosystems are bacteria and fungi. Note that there is not necessarily a sharp distinction between consumers and decomposers. Decomposers, for example, depend just as much as consumers on organic matter for their nutrition.

In any ecosystem, the producers and consumers interact, forming food chains or interlocking *food webs* that determine the flow of energy through the different levels (Fig. 25.3). In such chains or webs, each link feeds on the one below and is consumed by the one above, with photo-

Figure 25.3 A food web.

synthetic organisms at the bottom and the largest non-herbivorous animals at the top. Since most non-plant organisms have more than a single source of food and are themselves often consumed by a variety of consumers, there are considerable differences in the length and intricacy of food chains or webs, but there are rarely more than six links in any given chain.

Light energy itself, which enters at the producer level, can't be recycled in an ecosystem. Only about 1% of the light energy falling on a temperate zone community is involved in the production of organic material. As the organisms at each level respire, energy dissipates as heat into the atmosphere. Some parts of organisms may not be consumed at each level of the food chain. For example, leaves that are grazed by an herbivore are broken down and much of their energy is released within the animal. However, the energy stored within the uneaten leaves of the same plant isn't released until the leaves fall from the plant and decomposers (bacteria and fungi) break them down.

A significant proportion of farmland is used to raise animal feed instead of food crops for humans. It has been estimated that when cattle graze, however, only about 10% of the energy stored by the green plants they consume is stored in animal tissue, with most of the remaining energy dissipating as heat into the atmosphere. When we eat beef, only approximately 10% of its stored energy is used by our bodies to manufacture new blood cells and otherwise sustain life. The remaining energy is lost as heat. Obviously, then, in a long food chain, the final consumer gains only a tiny fraction of the energy originally captured by the producer at the bottom of the chain.

Conversely, there is proportionately much less loss of energy between levels in short food chains. Assume, for example, that for every 100 calories of light energy falling on corn plants each day, 10 calories are stored in corn tissue. Then suppose the corn is fed to cattle. Again, only 10%, or 1 calorie, of the original energy may be stored in animal tissue. If, as secondary consumers, we eat the beef, our bodies, in turn, may use 10% of the energy available in the beef, or only 0.1% of the original energy available. If, however, we eat the corn directly, we end up with the steer's percentage, which is 10 times more of the original solar energy than we would get if we ate the steer that ate the corn. (The actual amount of energy per gram of beef is considerably higher than that found in a gram of corn.)

From this, it follows that a vegetarian diet makes much more efficient use of solar energy than one that relies heavily on meats, and where food is scarce or humans are very abundant (as in India or Ethiopia), humans may be more or less forced to favor a vegetarian diet. It also follows that in terms of the numbers of individuals and the total mass, there is a sharp reduction at each level of the food chain. In a given portion of ocean, for example, there may be billions of microscopic algal producers supporting millions of tiny crustacean consumers, which, in turn, support thousands of small fish, which meet the food needs of scores of medium-sized fish, which are finally consumed by one or two large fish (Fig. 25.4). In other words, one large fish may very well depend on a billion tiny algae to meet its energy needs every day.

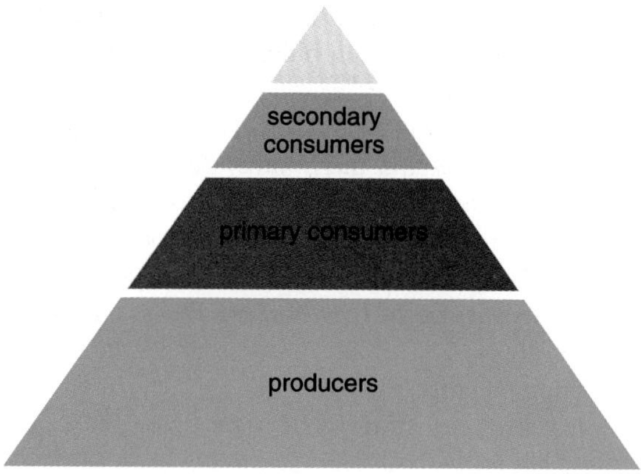

Figure 25.4 An energy pyramid of an ecosystem. There is much more energy at the bottom than at the top.

The interrelationships and interactions among the components of an ecosystem can be quite complex, but many function together in a somewhat regulatory fashion. An increase in food made available by producers can result in an increase in consumers, but the increased number of consumers and competition among them reduces the available food, which then inevitably leads to a reduction of consumers to earlier levels. While cyclical in nature, the net result is sustained self-maintenance of the ecosystem. This is the basis for the so-called *Balance of Nature.*

Interactions Among Plants, Herbivores, and Other Organisms

While it is easy to see the total mass of consumers is largely determined by the total mass of food made available by the producers, the interactions among different producers themselves, between predators and prey, and between the decomposers and the other members of the ecosystem are usually more subtle. Many flowering plants produce substances that either inhibit or promote the growth of other flowering plants. Black walnut trees, for example, produce a substance that wilts tomatoes and potatoes and injures apple trees that come in contact with black walnut roots. Many other plants produce phytoalexins (chemicals that kill or inhibit disease fungi or bacteria), making them resistant to various diseases (see discussion under "Use of Resistant Varieties" in Appendix 2).

Conversely, some bacteria and fungi limit higher plant growth by producing various inhibitory chemical compounds. Population size in other bacteria, fungi, and flowering plants may be limited by nutritional needs and availability. The degree to which this occurs varies considerably with the organisms involved. Some of the species of the Figwort Family (Scrophulariaceae), which includes snapdragons and similar plants, have no chlorophyll, and depend entirely on their flowering plant hosts for their energy and other nutritional needs. Other related species, as well as mistletoes, do have chlorophyll but apparently also require supplemental food from their hosts. Still other species often parasitize the roots of certain plants but are also capable of existing independently.

Mycorrhizal fungi are intimately associated with the roots of most woody and many other plants in such a way that both organisms derive benefit (such associations, called *mutualisms,* are a major part of life in general; they are briefly discussed in Chapter 5). The fungi greatly increase the absorptive surface of the root, usually playing a major role in the absorption of phosphorus and other nutrients, while obtaining energy from root cells.

Thomas Belt, a naturalist of more than 100 years ago, first called attention to an association between tropical ants and thorny, rapidly growing species of *Acacia* (Fig. 25.5). The Acacia has large, hollow thorns at the base of each leaf and is host to ants that feed on sugars, fats, and proteins produced by petiolar nectaries and special bodies at the tip of each leaflet. The ants live within the hollow thorns and vigorously attack any other organism, from insects to large animals, that come in contact with the plant. They also kill, by *girdling,* any plant that touches the *Acacia* (girdling is discussed under "bridge grafting" in Appendix 4). Experiments have shown that when ants are removed from these *Acacia* species, the plants grow very slowly and usually soon die from insect attacks or from shading by other plants.

Large herbivorous animals, such as deer and moose, feed on a wide variety of plants, each differing in nutritional value. Each plant species also produces different combinations, types, and amounts of chemical compounds in addition to proteins, fats, and carbohydrates. Many of the chemical compounds may be toxic, but some animals grazing on the plants that produce these chemicals do not display symptoms of poisoning because their digestive systems are capable of breaking down the compounds and eliminating or excreting them to a limited degree. The limitations imposed by such compounds result in the consumers varying their diet, seeking familiar foods and being wary of new ones.

Some plants are dependent for their existence on predators controlling populations of grazing animals. Populations of others such as oaks may become limited in extent when squirrels and acorn-consuming birds remove acorns that would potentially become trees. All plant species have some natural defense, such as chemical compounds or structural modifications (for example, spines). If this were not so, primary consumers of all kinds from insects to elephants could soon render a species extinct. In an ecosystem, the defenses that both producers and consumers have against each other have been developed through a process of coevolution resulting from natural selection and are more or less balanced.

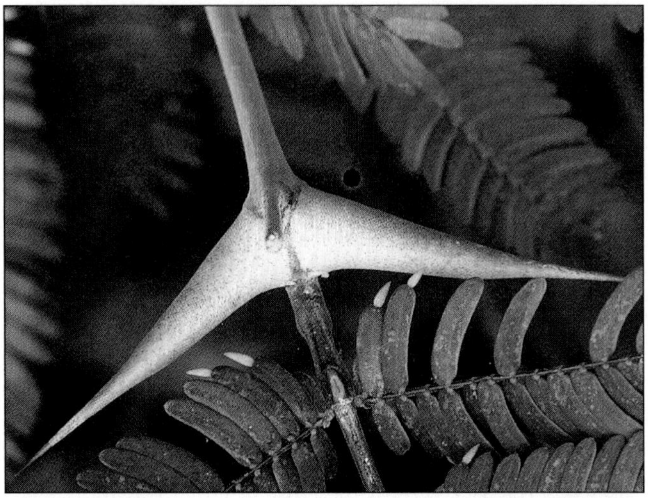

A.

B.

Figure 25.5 A species of *Acacia* that is host to ants that live in its hollow thorns. The ants attack any organisms, large or small, that come in contact with the plant. The plant provides food for the ants through nectaries at the bases of the thorns (*A*) and nitrogenous bodies at the tips of the leaflets (*B*).

NATURAL CYCLES

Water and elements such as carbon, nitrogen, and phosphorus are constantly cycling throughout nature. Such cycling involves transformation between organic and inorganic forms. We'll use the water, carbon, and nitrogen cycles as examples.

Nutrients constantly cycle and are recycled in the web of life. Carbon, nitrogen, water, phosphorus, and other molecules for eons have been passing through cycles. Some molecules that were a part of a primeval forest that became compressed and turned to coal may have become part of another plant after the coal burned. Then the new plant may have been eaten by an animal that, in turn, contributed molecules to a part of yet another living organism. Just think of where the molecules in your own body may have been in the past few billion years. They may well have been a part of some prehistoric seaweed, a saber-toothed tiger, a mighty dinosaur, or even all three!

The Water Cycle

Most of a living cell consists of water, and it is in this medium that the myriad of chemical reactions of living organisms takes place. In fact, life as we know it couldn't exist without water. The earth's water is constantly being recycled, and the total amount remains stable. Ninety-eight percent of the water is in oceans, rivers, lakes, and puddles that make up about two-thirds of the earth's surface. Most of the remaining water is in living organisms, glaciers, polar ice, water vapor, and the soil.

Some water that falls on land penetrates until it reaches an area of saturation known as the *water table.* Water in the water table may emerge from beneath the surface in the form of springs and artesian wells.

Water evaporates from bodies of water and is transpired from plants, which comprise more than 98% of the earth's biomass. Water also evaporates from the surfaces of animals and damp areas of the earth as the sun shines on them. The water vapor rises into the atmosphere, condenses, and falls back to earth in the form of rain, snow, and hail in a constant cycle—the *water cycle* (Fig. 25.6).

The Carbon Cycle

In the process of "feeding" (acquiring the energy they need to maintain themselves and grow, i.e., respire), bacteria process carbon (and nitrogen) in a way that allows these elements to be recycled (Fig. 25.7). Bacteria "eat" organic matter (carbohydrates), obtaining the energy present in the molecules. As a result of this activity (respiration), they convert carbohydrates to carbon dioxide. One of the two raw materials of photosynthesis, carbon dioxide constitutes 0.03% of our atmosphere. The combined plant life of the oceans and the land masses is estimated to use about 14.5 billion metric tons (16 billion

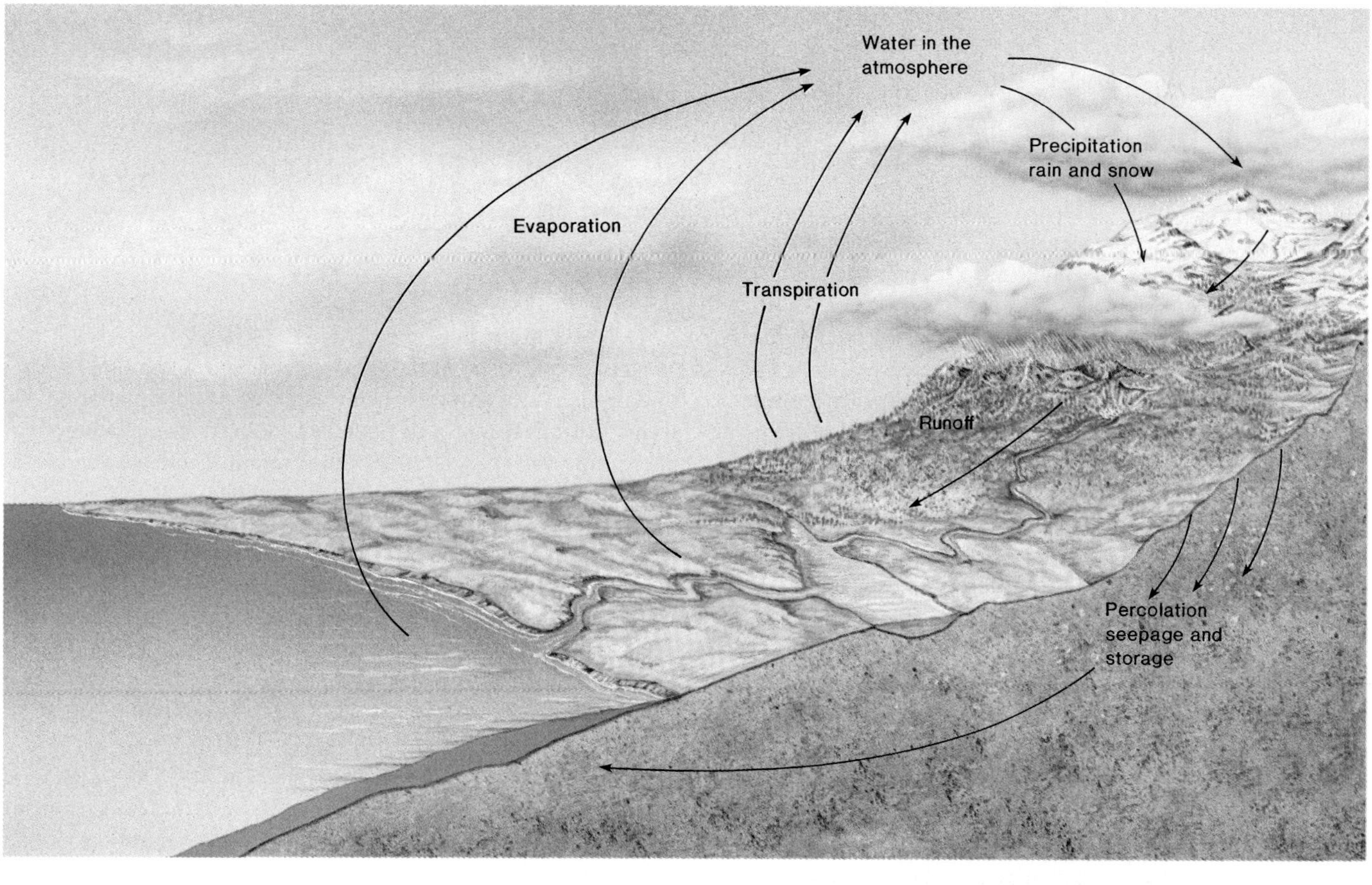

Figure 25.6 The water cycle.

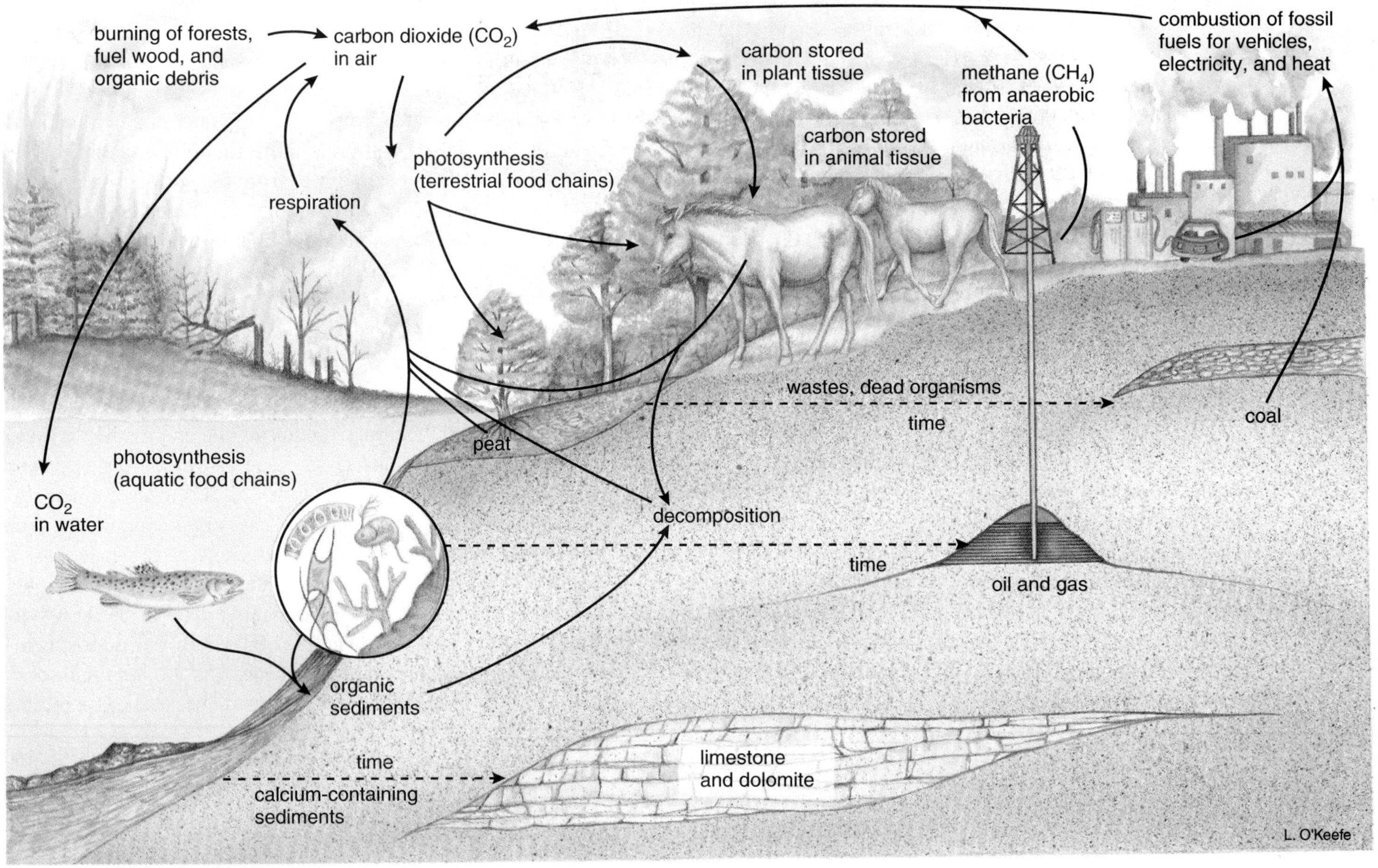

Figure 25.7 The carbon cycle.

tons) of carbon obtained from carbon dioxide every year. Respiration by all living organisms constantly replaces carbon dioxide, with perhaps as much as 90% or more of it being produced by the incredible numbers of decay bacteria and fungi as they decompose tissues.

The burning of fossil fuels by the internal combustion engines of industry and transportation releases lesser but significantly increasing amounts of carbon dioxide into the air, and a small amount originates with fires and volcanic activity. At the present rate of use by photosynthetic plants, it has been calculated that all of the carbon dioxide of our atmosphere would be used up in about 22 years if it were not constantly being recycled.

Scientists have found that plant growth initially can be accelerated by increasing levels of carbon dioxide in the air, and some commercial nurseries pump carbon dioxide along rows of bedding plants as a "fertilizer." However, the accelerated growth soon becomes limited by other factors in the environment and ultimately does not necessarily result in crop yield increases. Recent studies indicate that while C_3 plants may respond to increased carbon dioxide levels, C_4 plants may not, suggesting that in nature, the response of C_3 plants to higher levels of carbon dioxide may give them a competitive edge over C_4 plants.

Anaerobic bacteria produce large volumes of carbon-containing methane gas, which is discussed on page 504.

The Nitrogen Cycle

Most of the nitrogen in living organisms is in the protoplasmic proteins of their cells, with much of protoplasm being water. Nitrogen, the most abundant element in our atmosphere, constitutes about 18% of the protein. There are nearly 69,000 metric tons of nitrogen in the air over each hectare of land (35,000 tons per acre), but the total amount of nitrogen in the soil seldom exceeds 3.9 metric tons per hectare (2 tons per acre) and is usually considerably less. This discrepancy results from the nitrogen of the atmosphere being chemically inert, which is another way of saying that it will not combine readily with other molecules. It is, therefore, largely unavailable to plants and animals for their use in building proteins and other substances containing nitrogen.

Most of the nitrogen supply of plants (and indirectly, therefore, of animals) is derived from the soil in the form of inorganic compounds and ions taken in by the roots. These compounds and ions include those that contain nitrogen chemically combined with oxygen or hydrogen. Animals, through their digestive processes, and bacteria and fungi break down the more complex molecules of dead plant and animal tissues to simpler ones. Some nitrogen from the air is also *fixed,* that is, converted to ammonia or other nitrogenous compounds by various *nitrogen-fixing bacteria.* Some of these organisms gain access to various plants, particularly legumes (e.g., peas, beans, clover, alfalfa), through the root hairs, with the plant producing root nodules in which the bacteria multiply (root nodules are shown in Fig. 5.17). Others live free in the soil.

Figure 25.8 shows there is a constant flow of nitrogen from dead plant and animal tissues into the soil and from the soil back to the plants. Decay bacteria and fungi can break down enormous quantities of dead leaves and other tissues to tiny fractions of their original volumes within a few days to a few months. If they were abruptly to cease their activities, the available nitrogen compounds would be completely exhausted within a few decades and the carbon dioxide supply needed for photosynthesis seriously depleted. Forests and prairies would die as the accumulations of shed leaves, bodies, and debris buried the living plants and shielded their leaves from the light essential to photosynthetic activity. At present, even with the various bacteria involved in the nitrogen cycle functioning normally, the total amount of nitrogen in the soil is not being increased by their activities but is merely being recycled. Note, however, that addition of nitrogenous fertilizers will artificially increase soil nitrogen content.

Significant amounts of nitrogen are continually being lost as water leaches it out or carries it away through erosion of topsoil. More is lost with each harvest, the average crop removing about 25 kilograms per hectare (25 pounds per acre) per year. This nitrogen loss from the harvesting of crops can be sharply reduced if vegetable and animal wastes are recycled and returned to the soil each year. While bacteria are decomposing tissues, they use nitrogen, and little is available until they die and release their accumulations into the soil. Accordingly, until bacteria have completed their breakdown

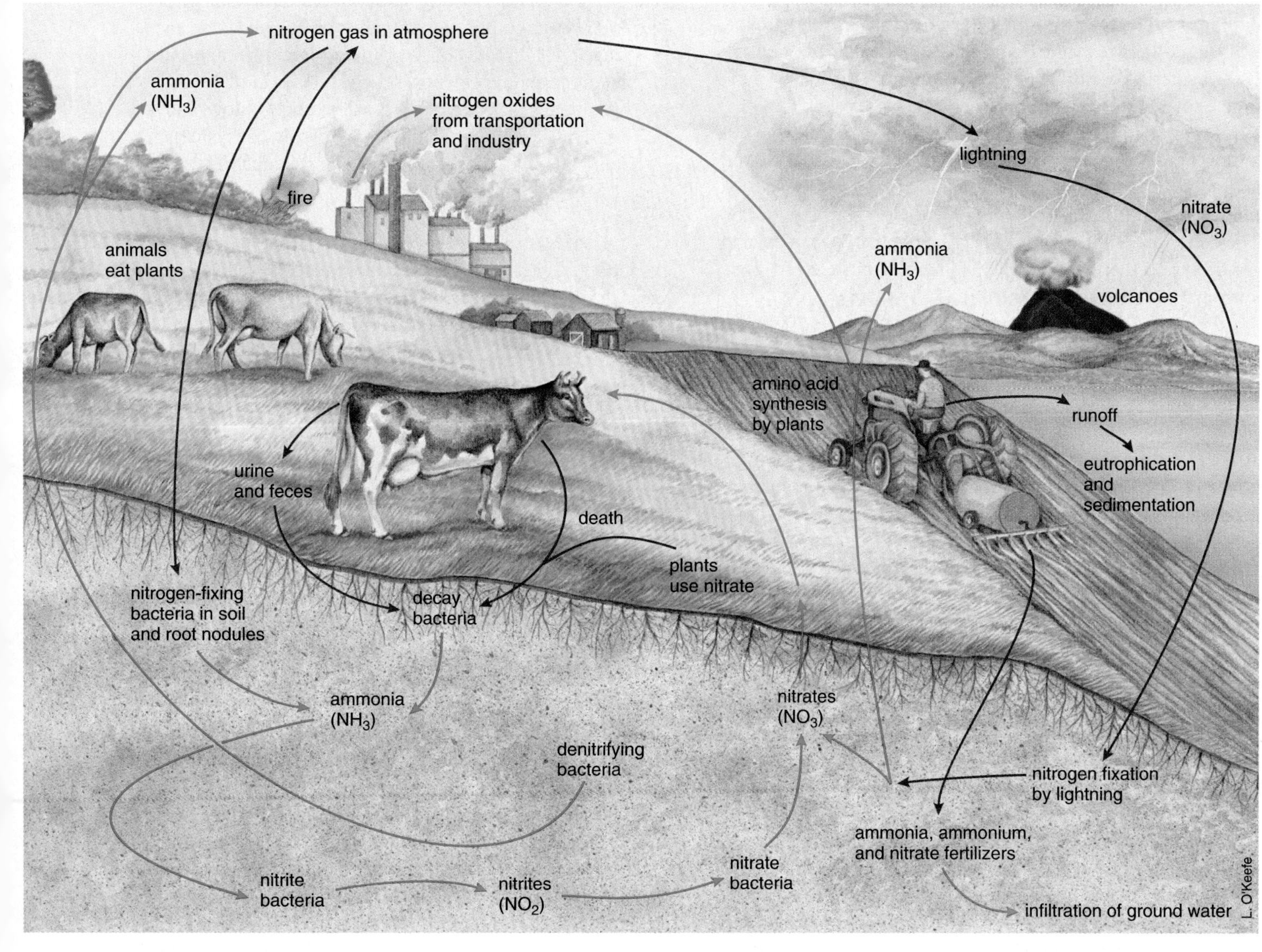

Figure 25.8 The nitrogen cycle.

of organic matter, crops should not be planted in soils to which only partially decomposed materials have been added. Likewise, when sawdust, straw, or other organic mulches are spread around plants in a garden to control weeds and conserve soil moisture, the soil nitrogen will be less available to the growing plants until the mulches have been decomposed.

Weeds and stubble are often controlled by burning. Fire, however, causes serious loss of nitrogen, which has to be replaced. It has been estimated that the annual combined loss of nitrogen from the soil in the United States from fire, harvesting, and other causes exceeds 21 million metric tons (23 million tons), and only 15.5 million metric tons (17 million tons) are replaced by natural means. To offset the net loss, some 32 million metric tons (35 million tons) of inorganic fertilizers are applied to the soils each year. If organic matter is not added at the same time, however, this application of inorganic fertilizers, combined with the annual burning of stubble, may eventually result in the creation of a *hardpan* soil.

Hardpan develops through the gradual accumulation of salt residues, which dissolve humus and disrupt the structure of the soil, causing the clay particles to clump and also producing colloids that are impervious to moisture. In hardpan soils and others low in oxygen (e.g., flooded areas), denitrifying bacteria use nitrates instead of oxygen in their respiration, depleting the remaining soil nitrogen.

Precipitation returns a little nitrogen to the soil from the atmosphere, where it has accumulated as a result of the action of light on industrial pollutants, fixation by flashes of lightning, and diffusion of ammonia released through decay. The activities of nitrogen-fixing bacteria and volcanoes also contribute to the natural replenishment of nitrogen by converting it to forms that can be utilized by plants.

Ecological Review

Unlike any other single species, explosive human population growth has disrupted ecosystems across the entire planet. These disruptions have included changes in atmospheric composition, over-enrichment of ecosystems with nutrients, and large-scale land clearing, which threatens to produce a massive extinction of species. Knowledge of the basic ecology of populations (groups of individuals of the same species), communities (all the populations occurring in a given area), and ecosystems (communties and the physical environments with which they interact) is necessary to understand the magnitude and significance of the ecological changes produced by the growing human population. Plant population ecology includes processes such as pollination, seed dispersal, germination, and seedling survival and establishment. Plant community classification and analysis is an important aspect of land-use planning, natural resource management, and conservation of biodiversity. Knowledge of succession, energy flow, and nutrient cycling are critical to assessing ecosystems and managing them for health and sustainability.

SUCCESSION

After a volcano spews lava over a landscape or after an earthquake or a landslide exposes rocks for the first time, there is initially no sign of life on the lava or rock surfaces. Within a few years or sometimes within a few months or even weeks, living organisms begin to appear, and a sequence of events known as **succession** takes place. During succession, the species of plants and other organisms that first appear gradually alter their environment as they carry on their normal activities, such as metabolism and reproduction. In time, the accumulation of wastes, dead organic material, and inorganic debris and other changes (such as of shade and water content in the habitat) favor different species that may replace the original ones. These, in turn, modify the environment further so that yet other species become established.

Succession occurs whenever and wherever there has been a disturbance of natural areas on land or in water. It proceeds at varying rates, depending on the climate, the soils, and the animals or other organisms in the vicinity. Ecologists recognize a number of variations of two basic types of succession. *Primary succession* involves the actual formation of soil in the beginning stages, while *secondary succession* takes place in areas that had previously been covered with *climax vegetation* (discussed in the next section, "Primary Succession") and had experienced a drastic environmental change such as a volcanic eruption, conversion to farmland, and so on.

Primary Succession

Xerosere

One of the most universal types of primary succession, called a **xerosere,** begins with bare rocks and lava that have been exposed through glacial or volcanic activity or through landslides. Initially, the rocks are sometimes subjected to alternate thawing and freezing, at least in temperate to colder areas. Tiny cracks or flaking may occur on the surface as a result. Lichens often become established on such surfaces (Fig. 25.9). They produce acids that very slowly etch the rocks, and as they die and contribute organic matter, they are replaced by other, larger lichens. Certain rock mosses adapted to long periods of desiccation also may become established, and a small amount of soil begins to build up. This is augmented by dust and debris blown in by the wind. Eventually, enough of a mat of lichen and moss material is present to permit some ferns or even seed plants to become established, and the pace of soil buildup and rock breakdown accelerates.

Figure 25.9 Early stages of succession on a Hawaiian road after an earthquake produced cracks and a hole. Ferns have become established in the cracks and hole less than 3 years after the disruption.

If deep cracks appear in the rocks, the larger seeds may widen them further as they germinate and the roots expand in girth. It has been calculated that germinating seedlings can exert a force of up to 31.635 kilograms per square centimeter (450 pounds per square inch). Indeed, instances are known of seedlings splitting rocks that weigh several tons (Fig. 25.10).

As soil buildup continues, larger plants take over, and eventually the vegetation reaches an equilibrium in which the associations of plants and other organisms remain the same until another disturbance takes place or climatic changes occur. Such relatively stable plant associations are referred to as **climax vegetation.**

The climax vegetation of deciduous forests in eastern North America is dominated by maples and beeches, oaks, hickories, and hemlocks or other combinations of trees. In desert regions, various cacti form a conspicuous part of the climax vegetation, while in the Pacific Northwest, large coniferous trees predominate. In parts of the Midwest, prairie grasses and other herbaceous plants form the climax vegetation, and in wet tropical regions, a complex association of trees and herbs constitutes the climax.

Occasionally, when a volcano produces ash instead of lava that buries existing landscape and associated vegetation, some of the successional stages involving lichens and mosses may be bypassed, with larger plants becoming the successional pioneers. This occurred following the series of ash eruptions, debris, and mudflows of Mount St. Helens in the state of Washington during the early 1980s.

Hydrosere

Succession takes place in wet habitats as well as in drier ones, and such succession is called a **hydrosere.** In the northern parts of midwestern states, such as Michigan, Wisconsin, and Minnesota, ponds and lakes of various sizes abound. Many

Figure 25.10 A blue oak that grew from an acorn lodged in a small crack of the rock. Over the years, the oak has split the rock apart.

were left behind by retreating glaciers and often have no streams draining them. The water that evaporates from them is replaced annually by precipitation runoff. They also grow a tiny bit smaller each year as a result of succession (Fig. 25.11).

This succession often begins with algae either carried in by the wind or transported on the muddy feet of waterfowl and wading birds. Although algae multiply throughout the upper sunlit levels of the entire pond, they tend to become concentrated in shallow water near the margin, and with each reproductive cycle, the dead parts sink to the bottom. Floating plants, such as duckweeds, may then appear, often forming a band around the body of water just offshore (Fig. 25.12). When nutrients, oxygen, pH, and temperatures are low, peat mosses encroach from the sides and become the dominant floating plants. There are presently about 4 million square kilometers (1.5 million square miles) of peat bogs throughout the world.

Water lilies and other rooted aquatic plants with floating leaves often become established, each group of plants contributing to the organic material on the bottom, which slowly turns to muck. Cattails and other flowering plants that produce their inflorescences above the water often take root in the muck around the edges, and the accumulation of organic material accelerates.

Meanwhile, the algae, duckweeds, peat mosses, and other plants move farther out, and the surface area of exposed water gradually diminishes. Grasslike sedges become established along the damp margins and sometimes form floating mats as their roots interweave with one another. Dead organic material accumulates and fills in the area under the sedge mats, and herbaceous and shrubby plants then move in. As the margins become less marshy, coniferous trees whose roots can tolerate considerable moisture (e.g., tamaracks or eastern white cedars) gain a foothold, eventually growing across the entire site as the pond or lake disappears. The trees continue to aid in the formation of true soil, and in due course, the climax vegetation takes over. No visible trace of the pond or lake now remains, and the only evidence of its having been there lies beneath the surface, where fossil pollen grains, bits of wood, fossilized fish skeletons, and other material reveal the past history. Such succession may take thousands of years and has never been witnessed from start to finish. The evidence that it does occur, however, is extensive and compelling.

Under natural conditions, some stream-fed lakes and ponds eventually become filled with silt and debris, although this, too, may take thousands of years to occur. The streams that feed these lakes bring in silt, and the nutrient content of

Figure 25.11 Succession in a pond in northern Wisconsin. *(Courtesy Robert A. Schlising)*

Figure 25.12 Duckweeds floating on a pond in early stages of succession.

the water rises as dissolved organic and inorganic materials (particularly nitrogen and phosphorus) are brought in. This gradual to relatively rapid enrichment, called **eutrophication,** facilitates the growth of algae and other organisms that add their debris to the bottom of the lake. When sewage and other pollutants enter the lake, the process of eutrophication may be greatly accelerated through stimulation of the growth of aquatic organisms. Eutrophication may also accelerate when trees are cleared from land surrounding lakes prior to the construction of summer homes and resorts. The cleared land erodes more readily, with precipitation runoff carrying soil into the water. Regardless of size, all bodies of water, including rivers, are subjected to these processes.

Secondary Succession

Secondary succession, which occurs more rapidly than primary succession, may take place if soil is already present and there are surviving species in the vicinity. In fact, survivors strongly condition subsequent succession. Many secondary successions follow human disturbances (e.g., land that was cleared when timber was harvested or land converted to farmland). Other secondary successions follow fires. Grasses and other herbaceous plants become established on burned or logged land. These usually are followed by trees and shrubs that have widely dispersed seeds (e.g., aspen and sumac in the Midwest and East, and chaparral plants, such as chamise and gooseberries, in the West). After going through fewer stages than are typical of primary succession, the climax vegetation takes over, often within 100 years.

Fire Ecology

Natural fires, started primarily by lightning and the activities of prehistoric humans, have occurred for thousands of years in North America and other continents. Trees, such as the giant redwoods and ponderosa and lodgepole pines, although scarred by certain types of fires, often survive; the dates of fires can be determined by the proximity of the scars to specific annual rings (see Fig. 6.7). In the West, growth rings of ponderosa pines show that in the past, forests of such pines used to burn on an average of every 6 to 7 years. These fires and the climate, topography, and soil combined to have a profound effect on various biomes.

As humans cultures developed, major and largely successful efforts to control fires were made, and this, in turn, significantly altered vegetational patterns. As knowledge of the role of fires in the maintenance of ecosystems has accumulated, it has become apparent that trying to eliminate fires, at least in certain areas, disrupts natural habitats more in the long run than allowing them to occur, and agencies such as the U.S. National Park Service may allow fires at higher elevations to run their natural course.

Fires also play a role in the composition of forests. In the mountains of east-central California, gooseberry and deerbrush appear in abundance after a fire, but their numbers stabilize within 15 to 30 years when larger trees return to the area. Ponderosa, jack and southern longleaf pines, and Douglas firs (which do not tolerate shade) are among the species that repeatedly replace themselves after fires; seeds of some species rarely germinate until they have been exposed to fire. The majority of chaparral species, both woody and herbaceous, are so adapted to fire that their seeds also will not germinate, as a rule, until fires remove accumulated litter and toxic wastes produced by the plants during growth.

Fires actually benefit grasslands, chaparral, and forests by converting accumulated dead organic material to mineral-rich ash, whose nutrients are recycled within the ecosystem. If the soil has been subjected to fire, some of its nutrients and organic matter may have been lost, and the composition of microorganisms originally present is likely to have been altered. Losses are offset, however, by the fact that soil bacteria, including cyanobacteria, which are capable of fixing nitrogen from the air, increase in numbers after a fire, and there is a decrease in fungi that cause plant diseases.

In some areas, such as the prairie states of the Midwest, grasses are better adapted to fire than woody plants, producing seeds within a year or two after germination. Perennial grass buds that are at the tips of rhizomes close to or beneath the surface, where they escape the most intense heat of fires, usually survive, producing new growth the first season after a fire. Accordingly, a fire destroys only one season's growth of grass, often after growth has been completed. Shrubs, however, have much of their living tissue above ground, and a fire may destroy several years' growth. In addition, woody plants often

Figure 25.13 Secondary succession on a burned area. California bay trees are resprouting the first season after a burn.

do not produce seeds until several years after a seed germinates. Many shrubs do sprout from burned stumps, particularly in chaparral areas (Fig. 25.13), but repeated burning keeps them small, thereby favoring grasses. There is evidence that at least some of the North American grasslands originated and were maintained because of fire. Since grassland fires have largely been controlled, many of the areas have now been invaded by shrubs that were once confined to watercourses.

GLOBAL WARMING

A global rise in temperature due to the accumulation in the atmosphere of gases that permit radiation from the sun to reach the earth's surface but prevent the heat from escaping back into space is often referred to as the *greenhouse effect.* Carbon dioxide and methane, the two most common gases involved in potential global warming, have been part of our atmosphere for millions of years, but others, such as chlorofluorocarbons, are relatively recently produced by-products of the manufacture of refrigerants, plastics, and aerosol cans.

Carbon Dioxide

D. L. Lindstrom of the University of Illinois at Chicago and D. R. MacAyeal of the University of Chicago examined records of cores of ancient ice from Siberia, Scandinavia, and the Arctic Ocean. Using computers to simulate the status of ice and atmosphere going back 30,000 years, they found the earth's temperature had increased as the levels of carbon dioxide had increased. At the end of the most recent ice age, the rise in temperature was sufficient to melt the ice. Their findings suggest that cycles of ice ages followed by shorter warm periods may have been caused solely by rising and falling levels of carbon dioxide.

In 1986, worldwide carbon dioxide emissions from transportation and industrial sources totaled somewhat less than 5 billion tons, but in 1987, the total rose to more than 5.5 billion tons and has continued to rise through the early 2000s. The burning of fossil fuels and deforestation has two significant effects on carbon dioxide content of the air: (1) it eliminates the photosynthesizing organisms that remove carbon dioxide from the atmosphere, and (2) it results in carbon stored in wood and other biological molecules being released into the air as carbon dioxide. Burning of fossil fuels and deforestation have been largely responsible for a 25% increase of carbon dioxide in the atmosphere since 1850. In the last 25 years alone, the earth's atmosphere has become 0.4°C (0.7°F) warmer, and between 1983 and 1990, the surface temperature of the ocean rose about 0.8°C (1.5°F).

Those may not seem to be significant amounts, but during the last ice age in North America when ice covered the northern United States and Canada, the average temperature of the earth at sea level was only 4°C (7°F) colder than it is now. In 1989, Mostafa Tolba, head of the United Nations Environment Program, estimated that if current levels of gas released into the atmosphere continue, the earth's temperature is likely to rise between 1.4°C and 4.3°C (2.5°F and 8°F) by 2039.

Higher temperatures melt ice at the poles and after melting cause water to expand. Higher ocean levels cause inundation of low-lying, often densely populated, coastal areas. The Environmental Protection Agency estimates that for each 30 centimeters (11.8 inches) the sea rises, it encroaches 30 meters (100 feet) inland, and already during the past century, worldwide ocean levels have risen 12.7 centimeters (5 inches). It is estimated that 7,000 square miles of land in the United States alone will be flooded if the temperatures rise as predicted. Major shifts in population to higher latitudes could follow, with the grain belts of the midwestern United States shifting into Ontario, and the fertile crescents of Asia shifting north into Russia. Higher temperatures can also have major effects on winds, currents, and weather patterns, causing droughts and creating deserts in some areas, while bringing about heavy rainfall in others.

Awareness

John Muir, Father of America's National Park System

Today, America's national parks are overcrowded as some 270 million people enjoy their beauty each year, relaxing in the great outdoors, hiking in fields and mountains, or navigating their rivers. Whether it is Yosemite, the Grand Canyon, Acadia, or one of the many others, each park has a unique and special appeal. These magnificent parks are the result of many people's vision and foresight; however, perhaps none more so than John Muir who lived from 1838 to 1914. Often called the "Father of the National Park System," he influenced presidents, members of Congress, and "just plain folks" by his love of nature and his writings about it. Always traveling and exploring various ecosystems, he visited Alaska five times, walked 1,000 miles from Indianapolis to the Gulf of Mexico, as well as walking across California's San Joaquin Valley to explore the Sierra Nevada. Later he wrote of viewing the Sierra Nevada range for the first time, "Then it seemed to me the Sierra should be called not the Nevada, or 'Snowy Range,' but the Range of Light, the most divinely beautiful of all the mountain chains I have ever seen."

Wherever he roamed in the wilderness, he noticed increasing waste and destruction from various sources, whether overgrazing by cattle and sheep or overcutting by loggers. It was his love for the American forest that drove him to work tirelessly for its preservation. He wrote some 300 articles and 10 books that told of his travels and contained his naturalist philosophy. He called to everyone to "climb the mountains and get their good tidings." He wrote that the wilderness should be preserved, viewed as a treasure rather than a resource to be exploited. Through his writing, he drew attention to the destruction forests were facing and worked to find remedies.

One solution was the formation of national parks. Largely through Muir's efforts, Congress, in 1890, created Yosemite National Park. He was also personally involved in the creation of Sequoia, Grand Canyon, Mount Rainer, and Petrified Forest national parks. He went one step further by founding the Sierra Club in 1892 and served as its first president until his death in 1914. He and his friends conceived the Sierra Club with the mission to "do something for wilderness and make the mountains glad." His idea was to form an association that would work to protect Yosemite National Park.

This brought him to the attention of President Theodore Roosevelt, who visited Muir in Yosemite in 1903. Together with Muir, under the trees of the park, Roosevelt's innovative conservation programs began to take form.

John Muir's words and life inspire environmental activism today; they also teach the importance of enjoying nature and protecting it. Muir wrote a warning that is just as timely today as it was 100 years ago.

Box Figure 25.1 Shafts of light penetrate the coastal redwoods of Muir Woods.

> The axe and saw are insanely busy, chips are flying thick as snowflakes and every summer thousands of acres of priceless forests, with their underbrush, soil, springs, climate, scenery and religion, are vanishing away in clouds of smoke, while, except in the national parks, not one forest guard is employed. Any fool can destroy trees. They cannot run away; and if they could, they would still be destroyed—chased and hunted down as long as fun or a dollar could be got out of their bark hides, branching horns or magnificent bole backbones. Few that fell trees plant them; or would planting avail much towards getting back anything like the noble primeval forest.[1]

A living memorial to John Muir is located 12 miles north of San Francisco, Muir's home in his later years. Muir Woods is a 560-acre national monument consisting of

1. *Atlantic Monthly, No. 80,* 1897.

Awareness Box Continued

a grove of majestic coastal redwoods along a canyon floor. The towering redwoods and lush canyon ferns growing along Redwood Creek make this a forest of tranquility. Only a 30-minute drive from San Francisco, Muir Woods is visited by over 1.5 million people each year. He declared that this grove of coastal redwoods was the "best tree-lovers monument that could possibly be found in all the forests of the world."

John Muir—lover of America's forests, roamer of the wilderness, inspired writer, and fighter for conservation.

D. C. Scheirer

Methane

Swamps and wetlands have long been known to be sources of methane produced by anaerobic bacteria. Many animals produce methane during digestive processes, and large amounts of this gas are produced by the wood-digesting organisms within the guts of termites. Anaerobic bacteria in rice paddies produce significant quantities of methane. The total annual production of methane in the atmosphere has been slowly increasing in recent years. A small part of this increase may stem from the prodigious mushrooming in numbers of termites in cleared tropical rain forest areas, which, in 1998, were being destroyed at the rate of more than 100 acres per minute, 24 hours a day.

Ozone Depletion

Methane gas and chlorofluorocarbons (CFCs), the inert chemicals used for refrigeration and other industrial purposes, are broken down into active compounds by sunlight at high altitudes. The breakdown products destroy ozone, a form of oxygen that in the stratosphere provides a natural shield for living organisms against ultraviolet radiation. Increased ultraviolet radiation correspondingly increases skin cancers. There is evidence the ozone shield has gone through lengthy cycles of expansion and contraction over millions of years, but there appears to be little question that ozone breakdown has been accelerated by industrial pollutants. The weakening of the ozone shield has been recognized as a serious global problem by North American countries and the European Economic Community. In 1987, the United States proposed a 50% reduction on production and uses of chlorofluorocarbons by the year 2000, and in 1989, the European Economic Community proposed a total ban on uses of chlorofluorocarbons, also by the year 2000. In 1998, however, developing nations, such as India, China, and Brazil, still had plans to expand the production of chlorofluorocarbons and contended that a ban would place them at an economic disadvantage. Since global cooperation is urgently needed, the major industrial nations have been seeking ways to allay the economic concerns of third-world countries. An international meeting, with mixed results, was convened in 1992 to try to foster global cooperation on this issue, but by the early 2000s, concrete evidence of such cooperation was still not in evidence.

Chlorofluorocarbons apparently are not the only force actively destroying the protective ozone layer. Bromine-based compounds called *halons,* which are commonly found in electronic equipment, such as computer protection systems, and in portable fire extinguishers, are reported to be as much as 3 to 10 times more destructive of ozone than chlorofluorocarbons. Halon concentrations in the atmosphere increased about 20% a year between 1980 and 1986, according to the Environmental Protection Agency. Some scientists believe the actual concentrations to be as much as 50% higher than that, and as a result of their strong recommendations, powders and other inert gases are being substituted for halons in fire extinguishers now being made.

LOSS OF BIODIVERSITY

Most plants and animals are adapted, sometimes in subtle ways, to the habitats in which they occur. When their natural habitats are destroyed, a few species may be able to adapt to new habitats, but most are not capable of doing so and ultimately perish. Ever since living organisms first appeared on this planet, various species have become extinct as climates and other environmental factors changed. However, the rates of extinction have accelerated enormously in just the last half-century as many types of habitats, from tropical rain forests and swamps to deserts, have been damaged or destroyed. Literally thousands of species are now facing extinction, many within the next 10 years.

If food and fiber plants are not particularly endangered by loss of natural habitats, why should we be concerned if seemingly unimportant wild species disappear? One of the many consequences of such losses pertains to the origins of our crop plants. Virtually all food, fiber, medicinal, and other useful plants have been developed from wild species, and keeping crops from succumbing to diseases often depends on our ability to breed new disease-resistant strains by tapping the gene pools of the plants' wild cousins. The gene pools usually have developed over thousands or even millions of years, and if they disappear, the most skillful plant breeders may be no match for rapidly mutating fungal and other diseases that attack our crops.

Summary

1. Ecology is the study of the relationships of organisms and their environment. It is a vast field with many facets.
2. Regional ecological issues include acid deposition, water contamination, wetlands, and hazardous waste.
3. Global ecological subjects include populations, communities, ecosystems, interactions between organisms, loss of biodiversity, and the nitrogen and carbon cycles.
4. Acid deposition damages or kills living organisms; it is produced when sulphur and nitrogen compounds released by the burning of fossil fuels are converted to nitric acid and sulphur dioxide by sunlight and rain.
5. Water contamination occurs when toxic wastes, pesticides, septic tanks, and fertilizers wash or leach into surface and ground water. Bacteria may in the future be used to break down various contaminants.
6. Wetlands were once considered wastelands. They have been drained but now increasingly are being protected.
7. Hazardous waste sites are gradually being detoxified, but progress is slow.
8. Populations vary in numbers, density, and in the total mass of individuals. Populations of one community may also occur in another community, and ecotypes may be specifically adapted to a single community.
9. Precipitation, temperature, soils, and biotic factors play roles in determining the distributions of plant species in an ecosystem.
10. Xerophytes and hydrophytes have modifications of leaves and other organs.
11. Species distribution is influenced by soil mineral content and biotic factors, such as competition for light, nutrients, and water.
12. Ecosystems sustain themselves through photosynthetic activity, energy flow through food chains, and the recycling of nutrients. Producers carry on photosynthesis; consumers feed on producers. Primary consumers feed directly on producers, and secondary consumers feed on primary consumers.
13. Decomposers break down organic materials to forms that can be reassimilated by the producers. In any ecosystem, the producers and consumers comprise food chains that determine the flow of energy through the different levels. Food chains vary in length and intricacy and, because of their interconnections, form food webs.
14. Energy itself is not recycled in an ecosystem; it escapes in the form of heat as it passes from one level to another. In a long food chain, the final consumer gains only a tiny fraction of the energy originally captured by the producer at the bottom of the chain.
15. When producers increase the amount of food available, consumers may increase correspondingly and compete for the food; this reduces what is available, resulting in a self-maintaining ecosystem. The composition of an ecosystem may be influenced by its living components through the secretion of growth-inhibiting and growth-promoting substances.
16. Some nitrogen from the air is fixed by the nitrogen-fixing bacteria found in legumes and other plants. In the nitrogen cycle, there is a constant flow of nitrogen from dead plant and animal tissues into the soil and from the soil back to the plants.
17. Water leaches nitrogen from the soil and carries it away when erosion occurs. Other nitrogen is lost from harvesting crops, but the loss can be reduced if wastes are decomposed and annually returned to the soil. Fire also causes nitrogen loss.
18. Replacement of nitrogen loss by the application of chemical fertilizers may eventually create hardpan by altering the soil structure.
19. Bacteria also recycle carbon and other substances, such as water and phosphorus.
20. Succession occurs whenever there has been a disturbance of natural areas on land or in water.
21. Primary succession involves the formation of soil in the beginning stages, while secondary succession takes place in areas previously covered with vegetation. A xerosere is a primary succession in which bare rock or lava is converted to soil through the activities of lichens, plants, and physical forces in an orderly progression of events over a period of time.
22. Climax vegetation becomes established at the conclusion of succession and remains until or unless a disturbance disrupts it.
23. A hydrosere is initiated in a wet habitat and culminates in a climax vegetation. As a lake or other body of water is filled in with silt and debris, eutrophication facilitating the growth of algae and other organisms occurs.
24. Secondary succession, which proceeds more rapidly than primary succession, may take place if soil is present. It may occur after fires.
25. A global rise in temperature due to carbon dioxide, methane, and other gases preventing the sun's radiant heat from escaping back into space is referred to as the greenhouse effect.
26. The carbon dioxide and methane levels in the earth's atmosphere and the earth's temperature have been rising. It is predicted that polar ice will, as a result, melt, and flooding of low-lying coastal areas will occur.
27. In the stratosphere, sunlight converts methane, chlorofluorocarbons, and halons into active compounds that destroy the ozone shield that protects us from intense ultraviolet radiation.

28. Loss of biodiversity has serious consequences, including the potential loss of means with which to combat crop diseases.

Review Questions

1. How does acid deposition occur?
2. What are the principal threats to global drinking-water sources?
3. Distinguish among a plant community, a biotic community, an ecosystem, and a biome.
4. What is the difference between a primary consumer and a secondary consumer?
5. How are short food chains more efficient in the use of solar energy than long ones?
6. Why is the diet of larger herbivorous animals varied?
7. What types of checks and balances exist in an ecosystem?
8. How are nitrogen and carbon recycled in nature?
9. What is primary succession as opposed to secondary succession?
10. What are general characteristics of climax vegetation as compared with nonclimax vegetation?
11. What is the greenhouse effect?
12. What is the ozone shield, and what significant role does it play? What threats are there to its existence?

Discussion Questions

1. Humans have disrupted ecosystems almost everywhere they have established themselves, at least in industrialized countries. Do you believe that humans could also improve an ecosystem?
2. If a vegetarian diet makes more efficient use of solar energy, should we all strive to become vegetarians for this reason?
3. Besides the hypothetical example given (in which it was observed that eating corn directly was a more efficient use of solar energy than feeding it to cattle and then eating the beef), can you think of other ways in which food chains might be shortened?
4. Peanut butter has many of the nutrients needed in human nutrition. On the basis of what you have learned about the diet of animals in an ecosystem, do you think it would be a good idea to live on peanut butter and water as a means of saving money?
5. Could succession take place in an abandoned swimming pool?

Learning Online

Visit our web page at www.mhhe.com/botany for interesting case studies, practice quizzes, current articles, and animations within the Online Learning Center to help you understand the material in this chapter. You'll also find active links to these topics:

General Ecology Sites
Acid Deposition
Estuaries
Food Webs
Nutrient Cycling
Succession and Stability
Global Ecology
Global Warming
Estuaries
Beaches and Mud Flats
Shallow Subtidal Communities
Solid, Toxic, and Hazardous Waste
Biodiversity
Endangered Species
Conservation and Management of Habitats and Species

Additional Reading

Barbour, M., et al. 1998. *Terrestrial plant ecology,* 3d ed. Menlo Park, CA: Benjamin/Cummings.

Barbour, M. G., and W. D. Billings. 1998. *North American terrestrial vegetation,* 2d ed. New York: Cambridge University Press.

Colegate, S. M., and P. R. Dorling (Eds.). 1994. *Plant associated toxins: Agricultural, phytochemical and ecological aspects.* New York: CAB International.

Cousens, R., and M. Mortimer. 1995. *Dynamics of weed populations.* New York: Cambridge University Press.

Crawley, M. J. (Ed.). 1997. *Plant ecology,* 2d ed. Malden, MA: Blackwell Scientific Publications.

Cronk, Q. C. B., and J. L. Fuller. 2001. *Plant invaders: The threat to natural ecosystems.* Herndon, VA: Stylus Publications.

Jeppesen, E., et al. 1997. *The structuring role of submerged macrophytes in lakes.* New York: Springer-Verlag.

Jolivet, P. 1998. *Interrelationships between insects and plants.* Boca Raton, FL: St. Lucie Press.

Koch, G. W., and H. A. Mooney (Eds.). 1995. *Carbon dioxide and terrestrial ecosystems.* San Diego, CA: Academic Press.

Krochmal, C., and A. Krochmal. 1995. *Acid rain: The controversy.* Kettering, OH: PPI.

Lerner, H. R. 1999. *Plant responses to environmental stress: From phytohormones to genome reorganization.* New York: Dekker Marcel.

Mooney, H. A., and G. W. Koch. 1995. *Carbon dioxide and terrestrial ecosystems.* San Diego, CA: Academic Press.

Narwal, S. S., and P. Tauro. 1999. *Allelopathy in agriculture and forestry,* 2 vols. Bridgehampton, NY: State Mutual Book & Periodical Service.

Pugnaire, F. I., and F. Valladares (Eds.). 1999. *Handbook of functional plant ecology.* New York: Dekker Marcel.

Silvertown, J. 1993. *Introduction to plant population ecology,* 3d ed. New York: Halsted Press.

Wilkinson, R. E. (Ed.). 1994. *Plant environment interactions.* New York: Dekker Marcel.

Wilson, E. O. 1992. *The diversity of life.* New York: W. W. Norton & Co.

Wooward, F. I., and D. J. Beerling. 2001. *Vegetation and the terrestrial carbon cycle: Modeling the first 400 million years.* New York: Cambridge University Press.

Fynbos (pronounced "fane-boss"), the name given to the Mediterranean scrub biome in South Africa. Chaparral, discussed in Chapter 25, is the North American representation of this biome. The flowering plant in the foreground is a species of *Cotyledon,* a member of the Stonecrop Family (Crassulaceae).

Biomes

26

Chapter Outline

Overview

This chapter examines and discusses the major biomes of North America. Characteristic climatic and other factors associated with each biome are mentioned. Descriptions and listings of principal plant and animal life are included. A discussion of the future of the tropical rain forest, which is presently being destroyed at an accelerated pace, concludes the chapter.

Learning Goal

Learn the major biomes of North America and describe the characteristics and principal living members of each.

Biotic communities considered on a global or at least on a continental scale are referred to as *biomes.* Several important biomes occupy the North American continent (Fig. 26.1), and most are also represented on other continents.

MAJOR BIOMES OF NORTH AMERICA

Tundra

Tundra occupies vast areas of the earth's land surfaces (about 20% in all), primarily above the Arctic Circle (Fig. 26.2). It has the appearance of being treeless, although miniature willow and birch only 2.5 to 50 centimeters (1 inch to 2 feet) tall do survive in some areas. Another characteristic of arctic tundra is the presence of *permafrost* (permanently frozen soil) at a depth of from 10 to 20 centimeters (from 4 to 8 inches) to about 1 meter (3 feet 3 inches) below the surface. The level of the permafrost determines the depth to which plant roots can penetrate. Precipitation averages less than 25 centimeters (10 inches) a year, but because of the permafrost, it is largely held at or near the surface of the land. There is a very short growing season of only 2 to 3 months, with frost possible on any day of the year. However, temperatures can soar to 27°C (81°F) or higher during a long midsummer day. The vegetation of arctic tundra is dominated by dwarf shrubs, sedges, grasses, lichens, and mosses. During the brief growing season, tiny perennial

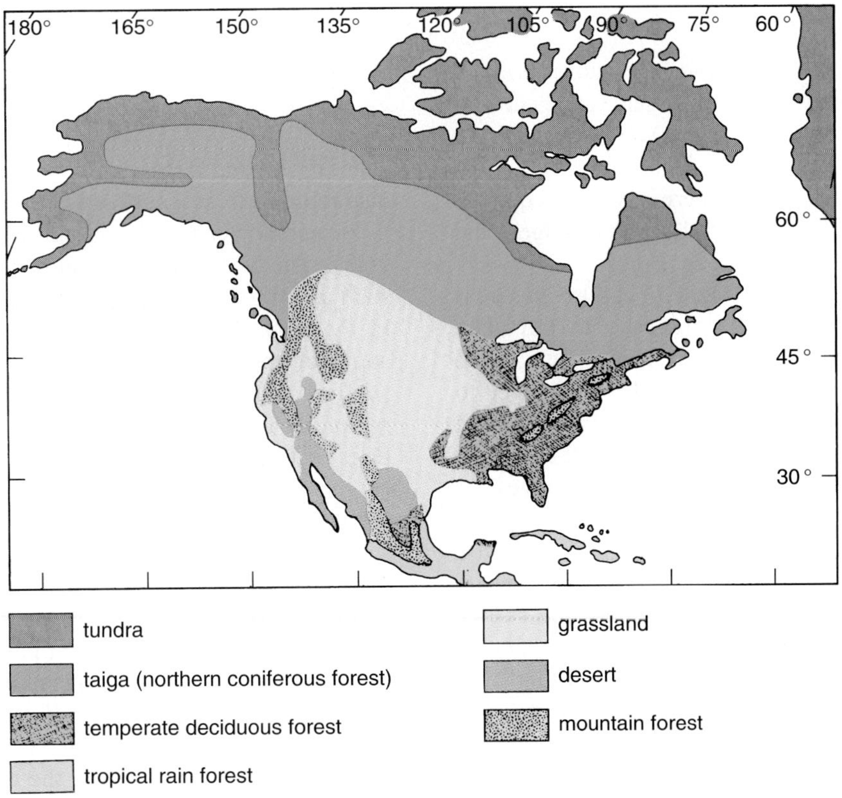

Figure 26.1 Major biomes of North America.

Figure 26.2 Alaskan tundra in the summer.

plants produce brightly colored flowers and form brilliant mats over the topsoil, which is largely organic and generally only 5.0 to 7.5 centimeters (2 to 3 inches) deep.

Patches of alpine tundra occur above timberline in mountains below the Arctic Circle. Permafrost is frequently absent from alpine tundra areas, and grasses, sedges, and herbs tend to predominate.

The biome is exceptionally fragile. A car driven across the tundra compresses the soil sufficiently to kill plant roots, and the tracks are evident many years later. Occasionally, sheep grazing tundra have been observed to pull up patches of the matted vegetation, leaving exposed edges. High winds then can catch the exposed edges and rip away larger segments of mat, leaving barren patches called *blowouts*.

Animals of the tundra include lemmings, the little rodents that reproduce prodigiously every 3 or 4 years and then decline in numbers. The arctic fox and snowy owl also fluctuate in numbers, the fluctuations being correlated with the size of the populations of lemmings that constitute their principal food. Other animals of the tundra include caribou, polar bears (near the coast), shrews, marmots, ptarmigans, loons, plovers, jaegers, and arctic terns.

Taiga

The *taiga* (from a Russian word meaning "wet forest") is also referred to as a northern coniferous or boreal forest. It is located adjacent to and south of the arctic tundra (Fig. 26.3). The vegetation is dominated by evergreen trees, such as spruce, fir, and pine. Tamarack (larch) also occurs in the taiga. Birch, aspen, and willow may be found in some of the wetter areas. Many perennials and a few shrubs occur, but there are few annuals. Snow blankets the region during the long and severe winters, with temperatures that drop to –50°C (–58°F) or lower during the coldest months. In summer, the temperatures often reach 20°C to 30°C (68°F to 86°F). Most precipitation occurs in the summer, ranging from about 25 centimeters (10 inches) to more than 100 centimeters (39 inches) in parts of western North America. A variety of birds, including jays, warblers, and nuthatches, inhabit the region, which is dotted with many lakes, ponds, and marshes. Rodents, such as shrews and jumping mice, and larger mammals, notably moose and deer, are found in the taiga. Ermine and wolverines also make their home in the taiga, and caribou overwinter in this biome.

Figure 26.3 Taiga in Alaska.

Figure 26.4 A scene within the eastern deciduous forest. *(Courtesy Robert A. Schlising)*

Temperate Deciduous Forest

Most deciduous trees are broad-leaved species that shed their leaves annually during the fall and remain dormant during the shorter days and colder temperatures of winter. Most *temperate deciduous forests* (Fig. 26.4) occur, like the taiga, on large continental masses in the Northern Hemisphere. In North America, an example of this type of forest occurs from the Great Lakes region south to the Gulf of Mexico and roughly extends from the Mississippi River to the eastern seaboard, although some do not consider the eastern coastal plain to be part of the temperate deciduous forest biome.

Temperatures within the area vary a great deal but normally fall below 4°C (39°F) in midwinter and rise above 20°C (68°F) in summer. The climax trees of the forest are well adapted to subfreezing temperatures as long as the cold is accompanied by precipitation or snow cover. Most of the annual precipitation, which totals between 50 and 165 centimeters (20 to 65 inches), occurs in the summer.

Some of the most beautiful of all the broad-leaved trees are found in a variety of associations in this biome. In the upper Midwest, sugar maple and American basswood predominate. Sugar maple, birch, and oak are also found to the Northeast, where they tend to be associated with the stately American beech. In the west-central part of the forest, oak and hickory abound. Oaks also are abundant along the eastern slopes of the Appalachian Mountains, where the American chestnut was once a conspicuous part of the flora. The chestnut has now virtually disappeared, having succumbed to chestnut blight disease that was introduced and began killing the trees during the early 1900s. Oak and hickory extend into the southeastern United States, where they become associated with pine and other tree species, such as the bald cypress.

Before the arrival of European immigrants, a mixture of large deciduous trees that included maple, ash, basswood, beech, buckeye, hickory, oak, tulip tree, and magnolia was found on the eastern slopes and valleys of the Appalachian Mountains. Some of the trees grew over 30 meters (100 feet) tall and had trunks up to 3 meters (10 feet) in diameter. Except for a few protected pockets in the Great Smoky Mountains National Park, the largest trees of this rich forest have been all but eliminated through logging. A smaller-treed extension of this forest is found in an area northeast of Baton Rouge, Louisiana, western Tennessee and Kentucky, and in southern Illinois. The American elm, also once a part of the forest, is rapidly disappearing as Dutch elm disease fells both wild trees and those planted along city streets. One small midwestern town that had some 600 elm trees planted along its streets found itself left with only six live trees within a year or two after Dutch elm disease struck.

A mixture of deciduous trees and evergreens occurs on the northern and southeastern borders of the forest. Hemlock and eastern white pine are found from New England west to Minnesota and south to the Appalachians. The once vast stands of eastern white pine are now largely gone, their valuable lumber having been used for construction and other purposes. Some have been lost to still another tree disease, white pine blister rust, but scattered trees remain, particularly where they have been protected. Various pines dominate the eastern coastal plain from New Jersey to Florida and westward to east Texas, with pitch pine being common in New Jersey. Some of the southern pines (e.g., long-leaf, slash, loblolly) are now cultivated in the southeastern United States for wood pulp, turpentine, lumber, and other commercially valuable products.

During the summer, the trees of the deciduous forests form a relatively solid canopy that keeps most direct sunlight from reaching the floor. Many of the showiest spring flowers of the region (such as bloodroot, hepatica, trillium, and violet) flower before the trees have leafed out fully and complete most of their growth within a few weeks. Other plants that can tolerate more shade, principally members of the Sunflower Family (e.g., aster and goldenrod), flower in succession in forest openings from midsummer through fall.

Animal life in the eastern deciduous forest includes red and gray foxes; raccoons; opossums; many rodents, such as gray squirrels; snakes, such as copperheads and black rat snakes; salamanders; and a wide variety of birds, including hawks, flickers, and mourning doves.

Grassland

Naturally occurring *grasslands* are found toward the interiors of continental masses (Fig. 26.5). They tend to intergrade with forests, woodlands, or deserts at their margins, depending on precipitation patterns and amounts. A grassland may receive as little as 25 centimeters (10 inches) of rainfall or as much as 100 centimeters (39 inches) annually. Temperatures can range from 45°C (113°F) in midsummer to –45°C (–50°F) in midwinter. North American grasslands known as *prairies* grew on fertile soils and provided grazing for huge herds of buffalo. The buffalo disappeared, however, as the settlers cultivated more and more of the land and hunters increasingly slaughtered the large animals. By 1889, there were only 551 left, but breeding and conservation subsequently increased their numbers. Large areas are now used for growing cereal crops (particularly corn and wheat) and for grazing cattle.

Before they were destroyed, American prairies were a remarkable sight. In Illinois and Iowa, the grasses grew over 2 meters (6 feet 6 inches) tall during an average season and another meter taller during a wet one. A dazzling display of wildflowers began before the young perennial grasses emerged in the spring and continued throughout the growing season. I have counted over 50 species of flowering plants in bloom at the same time on 1 hectare (2.47 acres) of protected grassland in the middle of spring.

Areas with a Mediterranean climate (e.g., the Great Central Valley of California), where most of the precipitation takes place in the winters, generally support grasses that are shorter than those of the midwestern prairies. Such grasslands usually include *vernal pools,* some of which may be more than 50,000 years old. The vernal pools are temporary accumulations of rain water that evaporate after the rains come to an end. Their unique floras include an orderly sequence of flowering plants, some appearing initially at the pool margins, with each species forming a distinct zone or band until the water is gone (Fig. 26.6). Some species flower only in the damp soil and drying mud that remains. The seeds of many species germinate under water.

Grassland animals include cottontails, jackrabbits, gophers, mice, and pronghorns. As indicated earlier, buffalo were once abundant. More than 20,000 of these large animals now are protected in national parks, game preserves,

Figure 26.5 Undisturbed western grassland in the spring.

Figure 26.6 Zonation in a vernal pool in northern California. *(Courtesy Robert A. Schlising)*

and private ranches. Various sparrows (e.g., vesper and savannah sparrows) and other birds still find homes in uncultivated tracts of this biome.

Desert

If you ask the average person to describe a *desert,* the response will probably include words such as *sand, heat, mirage, oasis,* and *camels* (Fig. 26.7). These are indeed found in some of the world's large deserts that are located in the vicinity of 30° latitude both north and south of the equator, but deserts occur wherever precipitation is consistently low or the soil is too porous to retain water. Most receive less than 12.5 centimeters (5 inches) of rain per year. The low humidity results in wide daily temperature ranges. On a summer day when the temperature has reached over 35°C (95°F), it will generally fall below 15°C (59°F) the same night. The light intensities reach higher peaks in the dry air than they do in areas where atmospheric water vapor filters out some of the sun's rays. Some desert plants have become adapted to these higher light intensities through the evolution of crassulacean acid metabolism (CAM) photosynthesis (discussed in Chapter 10), a process by which certain organic acids accumulate in the chlorophyll-containing parts of the plants during the night and are converted to carbon dioxide during the day. This permits much more photosynthesis to take place than would otherwise be possible, since most of the carbon dioxide of the atmosphere is excluded from the plants during the day by the stomata, which remain closed (thereby also retarding water loss).

Other adaptations of desert plants include thick cuticles, water-storage tissues in stems and leaves, leaves with a leathery texture and/or reduced in size, or even a total absence of leaves. In cacti and other succulents without functional leaves, the stems take over the photosynthetic activities of the plants. Desert perennials are adapted to the biome in various ways. Cacti and similar succulents have widespread shallow root systems that can rapidly absorb water from the infrequent rains. The water can then be stored for long periods in the interior of the stems, which are modified for such storage. Other perennials grow from bulbs that are dormant for much of the year. Annuals provide a spectacular display of color and variety, particularly during an occasional season when above-average precipitation has occurred. The seeds of the annuals often germinate after a fall or winter rain, and the plants then grow slowly or remain in a basal circular cluster of leaves (rosette) for several months before producing flowers in the spring. Literally hundreds of different species of desert annuals may occur within a few square kilometers (one or two square miles) of typical desert in the southwestern United States.

Desert animals are adapted, in many instances, to foraging at night when it is cooler. These include various mice and kangaroo rats, snakes (notably rattlesnakes and king snakes), chuckwallas, and lizards (e.g., gila monster). Various thrashers, doves, and flycatchers are included in the bird life.

Figure 26.7 A desert community in the southwestern United States.

Mountain and Coastal Forest

In the geologic past, deciduous forests extended to western North America. As the climate changed and summer rainfall was reduced, conifers largely replaced the deciduous trees, although some (e.g., maple, birch, aspen, oak) still remain, particularly at the lower elevations. Today, coniferous forests occupy vast areas of the Pacific Northwest and extend south along the Rocky Mountains and the Sierra Nevada and California coast ranges. Isolated pockets of the *mountain and coastal* biome also occur in other parts of the West, particularly toward the southern limits of the mountains.

The trees tend to be very large, particularly in and to the west of the Cascade Mountains of Oregon and Washington and on the western slopes of the Sierra Nevada. Part of the reason for the huge size of trees such as Douglas fir is the high annual rainfall, which exceeds 250 centimeters (100 inches) in some areas. The world's tallest conifers, the coastal redwoods of California (Fig. 26.8), are at low elevations between the Pacific Ocean and the California outer coast ranges. Here their size and longevity apparently depend more on fog, which reduces transpiration rates, than on large amounts of rain.

One of the characteristics of mountain forests is a fairly conspicuous altitudinal zonation of species. In other words, one encounters different associations of plants as one proceeds from sea level up the mountain sides. At lower elevations in both the Rocky Mountains and the Sierra Nevada, the predominant conifer is ponderosa pine. At lower elevations in the northern part of the Cascades, Douglas fir, western red cedar, and western hemlock are more common. At intermediate elevations in the Sierra Nevada, sugar pine, white fir, and Jeffrey pine take over, while at higher elevations, other species of pine, fir, and hemlock predominate.

Most of the mountain forest biome has comparatively dry summers. This has led to frequent forest fires, even before human carelessness became a factor. Several tree species are well adapted to fires. The Douglas fir, for example, has a very thick protective bark that can be charred without transmitting sufficient heat to the interior to damage more delicate tissues, and its seedlings thrive in open areas after a fire. When the bark of the giant redwoods of the Sierra Nevada is burned, the trees are rarely killed. This undoubtedly has contributed to the great age and size of many of the trees. The cones of some of the pine trees (e.g., knobcone pine) remain closed and do not release their seeds until a fire opens them, while seeds of several other species germinate best after they have been exposed to fire. These attributes of members of the mountain forest biome, as mentioned in the section on fire ecology in Chapter 25, have led to the practice in some of our national parks of occasionally allowing fires at higher elevations to run their natural course. The higher-than-normal incidence of fires occurring since humans came in large numbers to the forest has made it necessary, however, to control fires in most instances, even though in doing so, we may be interfering with natural cycles that would otherwise occur.

Figure 26.8 Coastal redwoods in northern California.

Animal life in the mountain forests includes many different rodents (especially chipmunks and voles), bears, mountain lions, bobcats, mountain beavers (not related to true beavers), mule deer, and elk. Large birds, such as the golden eagle, and many small birds, including mountain chickadees, warblers, and juncos, are an integral part of this biome.

Tropical Rain Forest

About 5% of the earth's surface, representing nearly half of the forested areas of the earth, are included in the *tropical rain forest* biome (Fig. 26.9). The biome is distributed throughout those areas of the tropics where annual rainfall amounts normally range between 200 and 400 centimeters (79 and 157 inches) and where temperatures range between 25°C and 35°C (77°F and 95°F). Although monthly rainfall amounts vary, there is no dry season, and some precipitation

Figure 26.9 A tropical rain forest scene. *(Courtesy Mary Lane Powell)*

occurs throughout all 12 months of the year, frequently in the form of afternoon cloudbursts. The humidity seldom drops below 80%.

Such climatic conditions support a diversity of flora and fauna so great that the number of species exceeds those of all the other biomes combined. The forests are dominated by broad-leaved evergreen trees, whose trunks are often unbranched for as much as 40 or more meters (160 feet), with luxuriant crowns that form a beautiful dark green and several-layered canopy. The root systems are shallow and the tree trunks often buttressed (buttress roots are shown in Fig. 5.14). There are literally hundreds of species of such trees, each usually represented by widely scattered specimens.

Most of the plants of the rain forests are woody, although not all of them are evergreen. Several of the deciduous tree species shed their leaves from some branches, retain the leaves on others, and flower on yet other branches all at the same time, while branches of adjacent trees of the same species are losing their leaves or flowering at different times. Many hanging woody vines and even more numerous *epiphytes*—especially orchids and bromeliads—can be seen on or attached to tree branches. The epiphyte roots are not parasitic and have no contact with the ground. The plants are sustained entirely by rain water that accumulates in their leaf bases and by their own photosynthesis. Traces of minerals, also necessary to the growth of the epiphytes, accumulate in the rain water as it trickles over decaying bark and dust.

The multilayered canopy is so dense that very little light penetrates to the floor, and the few herbaceous plants that survive are generally confined to openings in the forest. Despite the lush growth, there is little accumulation of litter or humus, and the soil is relatively poor. Decomposers rapidly break down any leaves or other organic material on the forest floor, and the nutrients released by decomposition are quickly recycled or leached by the heavy rains.

A few larger animals, with adaptations for moving through the mesh of branches in the rain forest, are found on the forest floor. Such animals include peccaries, tapirs, and anteaters. Most of the great numbers of animals, however, live out their lives in the canopies. Tree frogs, with adhesive pads on their toes, are common in tropical rain forests, as are various monkeys, bats (especially fruit bats), sloths, opossums, tree snakes, and lizards. Ants abound, and many of the extraordinary variety of other insects are adapted to their environment through excellent camouflages. Large flocks of parrots and other birds feed on the abundance of insects as well as the available fruits.

Ecological Review

Biomes, which are biotic communities viewed on global or continental scales, are associated with particular climates determined mainly by latitude, altitude, and distance from coastal areas. While these large-scale communities include all groups of organisms, biomes are distinguished primarily by their predominant plants and plant growth forms. The biomes of North America and their characteristic plant growth forms include tundra (few trees, lichens, mosses and grasses), taiga or boreal forest (coniferous trees, deciduous trees in wetter areas), temperate deciduous forest (deciduous trees, diverse understory herbaceous plants, conifers at geographical margins), grassland (perennial and annual grasses), deserts (drought-adapted plants), mountain forests (coniferous trees), and tropical rain forests (high diversity of trees and vines). The biomes, which encompass the earth's terrestrial biodiversity, are an excellent focus for conservation efforts. Saving large tracts of biomes, such as the tropical rain forests, is a practical approach to preserving biodiversity for future generations.

The Future of the Tropical Rain Forest Biome

In the 1960s, major plans were developed to convert the Amazon rain forest into large farms, hydroelectric plants, and mines. During the 1990s and early 2000s, gold-mining activities were filling the rivers with silt, and, as indicated in the discussion of the greenhouse effect in Chapter 25, the tropical rain forests were being destroyed or damaged for commercial purposes at the rate of more than 100 acres per minute. In one study of Amazon rain-forest birds conducted in 1989, it was found that a population of low-flying species in a 10-hectare (25-acre) section of the forest fell by 75% within just 6 weeks after adjacent land had been cleared, and 10 of the 48 species apparently disappeared completely.

This situation has been repeated on numerous occasions and confirmed by a project cosponsored by the Smithsonian Institution and Brazil's Institute for Research in the Amazon.

During 10 years of study, this project demonstrated that when large expanses of forest are cut into smaller pieces separated by as little as 10 meters (33 feet) of cleared land, it can have disastrous effects on the ecology of the entire forest, with most of the bird and other animal species originally present disappearing permanently. Very little of this biome that is the home of more than 50% of all living species of organisms is presently protected from commercial development. Many of the organisms are doomed to extinction, often before they have been described for the first time, and the biome will essentially have vanished within 20 years if governments and individuals do not take definitive action to halt the large-scale destruction.

Summary

1. Biomes are biotic communities considered on a global or at least a continental scale. Major biomes of North America include tundra, taiga, temperate deciduous forests, grasslands, deserts, mountain forests, and tropical rain forests.
2. Tundra is found primarily above the Arctic Circle. It includes few trees and many lichens and grasses and is characterized by the presence of permafrost below the surface. Alpine tundra occurs above timberline in mountains below the Arctic Circle.
3. Taiga is dominated by coniferous trees, with birch, aspen, and willow in the wetter areas. Many perennials and few annuals occur in taiga.
4. Temperate deciduous forests are dominated by trees such as sugar maple, American basswood, beech, oak, and hickory, with evergreens such as hemlock and eastern white pine toward the northern and southeastern borders.
5. Grasslands are found toward the interiors of continental masses. Those located in Mediterranean climatic zones usually include vernal pools with unique annual floras. Many grasslands have been converted to agricultural use.
6. Deserts are characterized by low annual precipitation and wide daily temperature ranges, with plants adapted both in form and metabolism to the environment.
7. Mountain and coastal forests occupy much of the Pacific Northwest and extend south along the Rocky Mountains and California mountain ranges. The primarily coniferous tree species tend to be in zones determined by altitude. The forests have mostly dry summers, and some of the species associated with it have thick bark that protects the trees from fires that occur frequently in this type of climate. Other species (e.g., knobcone pine) depend on fires for normal distribution and germination of seeds.
8. The tropical rain forests constitute nearly half of all forest land and contain more species of plants and animals than all the other biomes combined. Numerous woody plants and vines form multilayered canopies that permit very little light to reach the floor. The soils are poor, with nutrients released during decomposition being rapidly recycled. The biome is being destroyed so rapidly that it will disappear within 20 years if the destruction is not halted.

Review Questions

1. Characterize desert and tundra biomes.
2. What is the function of fire in grassland and mountain forest biomes?
3. What kinds of plants predominate in tropical rain forests?
4. Why does the loss of the tropical forest biome have very serious implications for the future?

Discussion Questions

1. Fire has been a natural phenomenon in several biomes for thousands of years, and most plant species are adapted to it in various ways in the ecosystem where it occurs regularly. Should we then not extinguish forest and grassland fires when they occur?
2. Much of the tropical rain forest biome is located in third-world countries, which have precarious economies. Bearing that in mind, what could be done to halt the destruction of this priceless and vital part of our ecosystem?

Learning Online

Visit our web page at http://www.mhhe.com/botany for interesting case studies, practice quizzes, current articles, and animations within the Online Learning Center to help you understand the material in this chapter. You'll also find active links to these topics:

Biomes and Environmental Habitats
Tundra
Taiga
Temperate Forests
Savannah
Grasslands
Deserts
Tropical Rain Forests
Tropical Rain Forests and Land Use Issues
Tropical Forests and Extinction
North American Habitats and Extinction Issues

Additional Reading

Baskin, J. M. et al. (Eds.). 1999. *Savannas, barrens, and rock outcrop plant communities of North America.* New York: Cambridge University Press.

Chapin, F. S. III, and C. K. Orner (Eds.). 1995. *Arctic and alpine biodiversity: Patterns, causes, and ecosystem consequences.* New York: Springer-Verlag.

Dallman, P. R. 1998. *Plant life in Mediterranean climates: California, Chile, South Africa, Australia, and the Mediterranean basin.* Santa Barbara, CA: University of California, Santa Barbara, Art Museum.

Dawes, C. J. 1998. *Marine botany,* 2d ed. New York: John Wiley & Sons.

Furley, P. A., et al. 1993. *The nature and dynamics of forest-savanna borders.* New York: Chapman and Hall.

Mueller-Dombois, D., and F. R. Fosberg. 1997. *Vegetation of the tropical Pacific islands,* vol. 132. New York: Springer-Verlag.

Packham, J. R., et al. 1992. *Functional ecology of woodlands.* New York: Chapman and Hall.

Whitmore, T. C. 1990. *An introduction to tropical rain forests.* Fair Lawn, NY: Oxford University Press.

Appendix 1

Scientific Names of Organisms Mentioned in the Text

This is an alphabetical list of the organisms whose scientific names may not be mentioned in the text. The common names are listed alphabetically along with the scientific names. Common and scientific names of organisms mentioned in Appendices 2 through 4 are provided within the respective appendices.

Common Names and Scientific Names of Organisms

COMMON NAME	SCIENTIFIC NAME
Aardvark	*Orycteropus* spp.
Abrasives, horsetail source of	*Equisetum* spp.
Absinthe liqueur, source of ingredients	*Pimpinella anisum, Artemisia absinthium,* and others
Acacia	*Acacia* spp.
Aconite, source of	*Aconitum* spp.
Actinomycetes	*Actinomyces* spp. and others
Adder's tongue fern, reticulate	*Ophioglossum reticulatum* (has highest known diploid chromosome number—1,260)
Adder's tongue ferns	*Ophioglossum* spp.
Afghanistan pine	*Pinus eldarica*
Aflatotoxin, source of	*Aspergillus flavus*
African sausage tree	*Kigelia pinnata*
Agar, source of	*Chondrus crispus, Eucheuma* spp., *Gelidium* spp., *Gracilaria* spp., and other red algae
Agave	*Agave angustifolia, A. palmeri, A. tequilana,* and other *Agave* spp.
Air plant—*see also* Bromeliad	*Kalanchoë* spp.
Alder	*Alnus* spp.
Alfalfa	*Medicago sativa*
Alfalfa caterpillar	*Colias philodice*
Algae	members of Kingdom Protista—all phyla
Algae, agar-producing	*Acanthopeltis* spp., *Ahnfeltia* spp., *Gelidium* spp. (principal source), *Gracilaria* spp., *Pterocladia* spp., and others
Algae, alginate-producing	*Ascophyllum* spp., *Durvillea* spp., *Ecklonia* spp., *Laminaria* spp., *Macrocystis* spp., and others
Algae, bark	*Pleurococcus* spp. and others (see footnote under *Bark, green algae that inhabit*)
Algae, brown	members of Phylum Chromophyta, Kingdom Protista
Algae, carrageenan-producing	*Chondrus crispus, Eucheuma* spp., and others
Algae, coralline	*Bossiella* spp., *Corallina* spp., *Lithothamnion* spp., and others
Algae/cyanobacteria, edible[1]	*Chlorella,* Irish moss (*Chondrus crispus*), kelp (*Laminaria* spp.), laver or nori (*Porphyra* spp.), spirulina (*Spirulina* spp.), wakame (*Undaria* spp.), and others

1. More than 150 species of algae and cyanobacteria are known to be edible, but most of the approximately half million tons of dried algae consumed annually consists of species of *Laminaria, Porphyra,* and *Undaria.* Most commercially grown nori consists of the fronds of *Porphyra tenera,* but other species of *Porphyra* are edible. During their life cycles, *Porphyra* spp. alternate between the familiar frond (bladed) form and a relatively inconspicuous filamentous form that was discovered after a British phycologist germinated spores of *Porphyra umbilicaulis* in a culture dish in her laboratory. The filamentous form previously had been considered a distinct species that had been named *Conchocelis rosea.* Species of *Ascophyllum, Fucus, Laminaria,* and *Macrocystis* are harvested for animal and poultry feeds. *Dunaliella bardawil* is cultured commercially as a source of beta-carotene and glycerol. *Spirulina* spp. (cyanobacteria) have a protein content of up to 70%; they are commercially cultivated for human consumption, particularly in Mexico and Israel, and have a been a staple food of natives of the Lake Chad region in Africa for centuries.

Common Names and Scientific Names of Organisms

COMMON NAME	SCIENTIFIC NAME
Algae/cyanobacteria, toxic	*Anabaena* sp., *Caulerpa* sp., *Chlorella* sp., *Chondria armata, Gambierdiscus toxicus, Hizikia* sp., *Lyngbya majusculis, Oscillatoria nigroviridis, Protogonyaulax (Gonyaulax)* sp., *Prototheca* sp., *Prymnesium parvum, Ptychodiscus (Gymnodinium brevis), Schizothrix calcicola,* and others
Algae/cyanobacteria used as fertilizers or soil conditioners	*Anabaena azollae, Chlamydomonas mexicana,* kelps, and others
Algae, flatworm	*Platymonas* spp.
Algae, golden brown	members of Phylum Chromophyta, Kingdom Protista
Algae, green	members of Phylum Chlorophyta, Kingdom Protista
Algae, green colonial	*Chaetopeltis* spp., *Eudorina* spp., *Pandorina* spp., *Pediastrum* spp., *Scenedesmus* spp., *Volvox* spp., and others
Algae, green filamentous	*Oedogonium* spp., *Spirogyra* spp., *Ulothrix* spp., *Zygnema* spp., and others
Algae, medicinal	*Laminaria* spp., *Digenia* spp., and many others
Algae, metal-removing	*Chlamydomonas reinhardtii*
Algae, red	members of Phylum Rhodophyta, Kingdom Protista
Algae, snowbank	*Chlamydomonas nivale* and others
Algae, sponge	*Chlorella* spp., *Zoochlorella* spp.
Algae, yellow-green	members of Phylum Chromophyta, Kingdom Protista
Almond	*Prunus amygdalus*
Aloe juice, source of	*Aloe barbadensis, A. ferox, A. vera,* and others
Amaranth	*Amaranthus* spp.
Amaryllis	*Amaryllis* spp.
Ama'uma'u	*Sadleria cyatheoides*
American chestnut	*Castanea dentata*
American elm	*Ulmus americana*
Amoeba	*Amoeba proteus* and others
Amoeba, fungal internal parasites of	*Cochlonema verrucosum* and others
Amoeba, fungal trappers of	*Dactylella* spp. and others
Anabaena	*Anabaena* spp. (including nitrogen-fixing spp. such as *A. azollae*)
Anemone	*Anemone* spp.
Angelica	*Angelica archangelica*
Anise	*Pimpinella anisum*
Anise swallowtail butterfly	*Papilio zelicaon*
Annatto	*Bixa orellana*
Ant	*Formica* spp. and many others
Anteater	*Myrmecophaga jubata*
Ants, bullhorn *Acacia*	*Pseudomyrmex ferruginea*
Aphid	*Anuraphis* spp., *Aphis* spp., and others
Aphid, root (pest of grape vines)	*Phylloxera* spp.
Apple[2]	*Malus domestica* (= *Malus pumila*)[2]
Apple brown rot, causal agent	*Monolinia fructigena*
Apple scab, causal agent	*Venturia inaequalis*
Apricot	*Prunus armeniaca*
Apricot brown rot	*Sclerotinia fructicola*
Arabidopsis (Mouse-ear cress)	*Arabidopsis thaliana*
Arborvitae (American/Northern)	*Thuja occidentalis*
Archaebacteria	members of Phylum Archaebacteria, Kingdom Archaea
***Archaefructus,* extinct plant believed to be the earliest flowering plant**	
Arrowroot	*Maranta arundinacea, Tacca leontopetaloides*
Arrowroot, Florida, source of	*Zamia floridana*
Artichoke, Chinese (Crosne)	*Stachys affinis*
Artichoke, globe	*Cynaria scolymus*
Artichoke, Jerusalem	*Helianthus tuberosus*
Arum Lily (Arum) Family	Araceae
Ascomycete	member of Phylum Ascomycota, Kingdom Fungi
Ash, blue	*Fraxinus quadrangulata*
Ash, Oregon	*Fraxinus latifolia*
Ash, white	*Fraxinus americana*

2. There are more than 1,000 varieties of apples, mostly of hybrid origin. The principal ancestors of *Malus pumila* probably include *M. sylvestris, M. dasyphylla,* and *M. praecox.* Some authorities include *Malus* within the genus *Pyrus* and refer to most cultivated apples as *Pyrus malus.* Others distinguish between the two genera on the basis of leaf pubescence and stone cells within the fruit, referring those cultivars with leaf pubescence and sclereids to *Malus* and those without these features to *Pyrus.*

Common Names and Scientific Names of Organisms

COMMON NAME	SCIENTIFIC NAME
Asparagus	*Asparagus officinalis*
Aspen, quaking	*Populus tremuloides*
Aspergillosis, causal agent(s) of	*Aspergillus fumigatus* and other *Aspergillus* spp.
Aster	*Aster* spp.
Astringent, horsetail source of	*Equisetum arvense, E. debile,* and others
Athlete's foot, fungal causal agent of	*Trichophyton* spp.
Autograph tree (Fig. 8.18*C*)	*Clusia rosea*
Avocado	*Persea americana* and others
Azalea	*Rhododendron* spp.
Baby blue eyes	*Nemophila menziesii*
Baby powder, ground pine source of	*Lycopodium clavatum*
Bacteria, acetone-producing	*Clostridium acetobutylicum* and others
Bacteria, acidophilus	*Lactobacillus acidophilus*
Bacteria, ammonifying	*Clostridium* spp., *Micrococcus* spp., *Proteus* spp., *Pseudomonas* spp., and others
Bacteria, anthrax	*Bacillus anthracis*
Bacteria, blue-green—*see* Cyanobacteria	
Bacteria, botulism	*Clostridium botulinum*
Bacteria, brucellosis	*Brucella abortus, B. suis, B. melitensis*
Bacteria, Bt	*Bacillus thuringiensis*
Bacteria, bubonic plague	*Yersinia pestis*
Bacteria, buttermilk	*Streptococcus lactis, S. cremoris, Leuconostoc citrovorum,* and others
Bacteria, butyl alcohol	*Clostridium acetobutylicum* and others
Bacteria, cholera	*Vibrio cholerae*
Bacteria, decay/decomposer	*Clostridium* spp., *Micrococcus* spp., *Proteus* spp., *Pseudomonas* spp., and others
Bacteria, denitrifying	*Micrococcus denitrificans, Thiobacillus denitrificans,* and others
Bacteria, dextran	*Leuconostoc mesenteroides*
Bacteria, diphtheria	*Corynebacterium diphtheriae*
Bacteria, ensilage	*Lactobacillus delbrueckii, L. plantarum,* and others
Bacteria, ethanol-producing	*Bacillus stearothermophilus* (mutant form)
Bacteria, frost-damage preventing	*Pseudomonas syringiae*
Bacteria, gas gangrene	*Clostridium novyi, C. perfringens, C. septicum*
Bacteria, giant	*Epulopiscium fishelsonii*
Bacteria, glutamic acid-producing	*Arthrobacter* spp., *Brevibacterium* spp., *Micrococcus* spp.
Bacteria, gonorrhea	*Neisseria gonorrhoeae*
Bacteria, grease- and oil-dissolving	*Pseudomonas aeruginosa*
Bacteria, green sulfur	*Chlorobium* spp., *Chloropseudomonas* spp., *Prosthecochloris* spp., and others
Bacteria, hot water (Sulfolobus)	*Pyrodictium* spp.
Bacteria, human ulcer-causing	*Heliobacter pylori*
Bacteria, hydrogen	*Hydrogenomonas* spp.
Bacteria, ice-minus	*Pseudomonas syringiae*
Bacteria, iron	*Gallionella* spp., *Sphaerotilus* spp.
Bacteria, kefir	*Lactobacillus bulgaricus, Streptococcus lactis*
Bacteria, lactic acid	*Lactobacillus delbrueckii* and others
Bacteria, Legionnaire's disease	*Legionella pneumophilia*
Bacteria, luminescent	*Achromobacter* spp., *Flavobacterium* spp., *Photobacterium* spp., *Pseudomonas* spp., *Vibrio* spp., and others
Bacteria, meningitis	*Neisseria meningitidis* and others
Bacteria, methane	*Methanobacterium* spp., *Methanococcus* spp., *Methanosarcina* spp., and others
Bacteria, milky spore disease	*Bacillus popilliae*
Bacteria, mosquito-killing	*Bacillus thuringiensis* var. *israelensis*
Bacteria, nitrate (nitrifying)	*Nitrobacter* spp.
Bacteria, nitrite (nitrosifying)	*Nitrosomonas* spp.
Bacteria, nitrogen-fixing	*Azorhizobium* spp., *Azotobacter* spp., *Brachyrhizobium* spp., *Clostridium pasteurinum, Rhizobium* spp., *Sinorhizobium* spp., and others.
Bacteria, paratyphoid fever	*Salmonella paratyphi*
Bacteria, pneumonia (some forms of pneumonia are viral)	*Streptococcus pneumoniae* and others
Bacteria, PPLO (mycoplasmas)	*Mycoplasma pneumoniae*
Bacteria, pseudomonad	*Pseudomonas* spp.

Common Names and Scientific Names of Organisms

COMMON NAME	SCIENTIFIC NAME
Bacteria, purple nonsulfur	*Rhodomicrobium* spp., *Rhodopseudomonas* spp., *Rhodospirillum* spp.
Bacteria, purple sulfur	*Amoebobacter* spp., *Lamprocystis* spp., *Rhodothece* spp., and others
Bacteria, salmonella (food-poisoning bacteria)	*Salmonella* spp.
Bacteria, salt	*Halococcus* spp., *Halobacterium* spp.
Bacteria, sauerkraut	*Leuconostoc* spp. and others
Bacteria, sorbose	*Acetobacter suboxydans*
Bacteria, spotted fever	*Rickettsia rickettsii*
Bacteria, strep throat	*Streptococcus* spp.
Bacteria, sulfolobus	*Sulfolobus* spp., *Thermoplasma* spp., *Thermoproteus* spp.
Bacteria, sulfur	*Desulfovibrio* spp., *Thiobacillus* spp., and others
Bacteria, syphilis	*Treponema pallidum*
Bacteria, tetanus	*Clostridium tetani*
Bacteria, tularemia	*Francisella tularensis*
Bacteria, typhoid fever	*Salmonella typhi*
Bacteria, typhus fever	*Rickettsia prowazeki* and others
Bacteria, vinegar	*Acetobacter* spp.
Bacteria, whooping cough	*Bordetella pertussis*
Bacteria, yogurt	*Streptococcus thermophilus*
Bald cypress	*Taxodium distichum*
Balsa	*Ochroma lagopus*
Balsam fir	*Abies balsamea*
Bamboo	*Bambusa* spp., *Phyllosytachys* spp.
Banana	*Musa paradisiaca* and others[3]
Banana Family	*Musaceae*
Banyan tree	*Ficus* spp.
Baobab, African	*Adansonia digitata*
Baobab, Australian	*Adansonia gregorii*
Barbasco	*Lonchocarpus nicou* var. *utilis*, *Derris elliptica*, and others
Barberry	*Berberis verruculosa* and other *Berberis* spp.
Barberry, Common/European	*Berberis vulgaris*

COMMON NAME	SCIENTIFIC NAME
Bark, green algae that inhabit	*Protococcus* spp.[4]
Barley	*Hordeum vulgare*
Barn swallow	*Hirundo rustica erythrogaster*
Barrel cactus	*Ferocactus* spp., *Mammillaria* spp., and others
Barrel cactus, Coville's	*Ferrocactus covillei*
Basil	*Ocimum basilicum*
Basswood	*Tilia* spp.
Basswood, American	*Tilia americana*
Bat	*Eidolon* spp., *Epomophorus* spp., and others
Bat (Fig. 23.15)	*Leptonycteris sanbornii*
Bay, California (also known as Oregon myrtle)	*Umbellularia californica*
Bay laurel	*Lauris nobilis*
Bay, sweet	*Laurus nobilis*
Beach strawberry	*Fragaria chinensis*
Bean, broad	*Vicia faba*
Bean, castor	*Ricinus communis*
Bean Family	Fabaceae (formerly Leguminosae)
Bean, garbanzo	*Cicer arietinum*
Bean, garden	*Phaseolus vulgaris*
Bean, green	*Phaseolus vulgaris*
Bean, jequirity	*Abrus precatorius*
Bean, kidney	*Phaseolus vulgaris*
Bean, lima	*Phaseolus lunatus*
Bean, mescal—*see* Mescal button	
Bean, Mexican jumping	*Sebastiana* spp. and others
Bean, mung	*Phaseolus aureus* (= *Vigna radiata*)
Bean, navy	*Phaseolus vulgaris*
Bean, pinto	*Phaseolus vulgaris*
Bean, scarlet runner	*Phaseolus coccineus*
Bean, tepary	*Phaseolus acutifolius* var. *latifolius*
Bean, winged	*Psophocarpus tetragonolobus*
Bear	*Ursus* spp. and others
Bear, polar	*Thalarctos maritimus*
Bearberry (Kinnikinick)	*Arctostaphylos uva-ursi*
Beaver, mountain	*Aplodontia rufa*

3. The domestic banana was developed from hybrids between *Musa acuminata* and *M. balbisiana,* and its genetic history is complex. N. W. Simmonds, a recognized authority on the genus *Musa,* believes that only cultivars and not species of domestic banana should be recognized; others prefer to retain Linnaeus's species name of *Musa paradisiaca.*

4. These algae are known under several names (*Desmococcus, Phytoconis, Pleurococcus, Protococcus*), and uncertainty exists as to which name has priority. The green algal component of certain lichens, *Trebouxia,* also occurs independently on bark.

Common Names and Scientific Names of Organisms

COMMON NAME	SCIENTIFIC NAME
Bedstraw	*Galium* spp.
Bee, honey	*Apis mellifera*
Beech, American	*Fagus grandifolia*
Beefsteak morel	*Helvella* sp.
Beet, garden	*Beta vulgaris*
Beet, sugar	*Beta vulgaris* (horticulturally selected strains)
Beetle	member of Order Coleoptera, Class Insecta, Phylum Arthropoda, Kingdom Animalia
Beetle, scarab	member of Family Scarabaeidae—see *Beetle*
Begonia	*Begonia* spp.
Belladonna, source of	*Atropa belladonna*
Bermuda grass	*Cynodon dactylon*
Betel nut	*Areca catechu*
Betony, wood	*Pedicularis canadensis*
Bigleaf maple (Fig. 8.20)	*Acer macrophyllum*
Big tree	*Sequoiadendron giganteum*
Birch	*Betula papyrifera* and others
Bird's-nest fungus (Fig. 19.23)	*Crucibulum levis*
Birth control pills, fungi used in manufacture of	*Rhizopus nigricans, R. arrhizus*
Bison	*Bison bison*
Bittersweet	*Celastrus scandens*
Blackberry	*Rubus argutus, R. laciniatus, R. procerus, R. ursinus,* and others
Blackbird	*Euphagus* spp. and others
Black bread mold	*Rhizopus stolonifer* and others
Black locust	*Robinia pseudo-acacia*
Black stem rust of wheat	*Puccinia graminis*
Bladderwort	*Utricularia minor* and other *Utricularia* spp.
Blazing star	*Liatris ligulistylis*
Bleeding, ground pine used to arrest	*Lycopodium clavatum*
Bleeding heart	*Dicentra* spp.
Bleeding heart, eastern	*Dicentra eximia*
Bleeding heart, Pacific	*Dicentra formosa*
Bloodroot	*Sanguinaria canadensis, S. isabellinus*
Blueberry	*Vaccinium* spp.
Blue curls	*Trichostema* spp.
Blue-green algae—*see* Cyanobacteria	
Blue-green bacteria—*see* Cyanobacteria	
Blue jay	*Cyanocitta cristata*
Bobcat	*Felis rufus*
Bolete	*Boletus* spp. and others
Bollworm	*Pectinophora gossypiella*
Bowstring fibers, source of	*Sansevieria metalaea*
Bowstring hemp, source of	*Sansevieria* spp.
Box elder	*Acer negundo*
Boysenberry	*Rubus* hybrids, with *R. ursinus* as one parent
Bracken fern	*Pteridium aquilinum*
Brazil nut	*Bertholettia excelsa*
Breadfruit	*Artocarpus altilis*
Bridalwreath	*Spiraea vanhouttei* hybrids and others
Broccoli	*Brassica oleracea* var. *botrytis*
Bromeliad ("Air plant")	Member of the Bromeliad Family (Bromeliaceae)[5]
Broomrape	*Orobanche* spp.
Brown algae	Member of Phylum Chromophyta, Kingdom Protista. Representative genera include *Ascophyllum, Durvillea, Ecklonia, Ectocarpus, Hizikia, Laminaria, Undaria,* and others
Brussels sprouts	*Brassica oleracea* var. *gemmifera*
Bryophyte (*see also* **individual listings)**	member of Phyla Anthocerophyta, Hepaticophyta, or Bryophyta, Kingdom Plantae
Bryopsid	member of Phylum Chlorophyta, Kingdom Protista
Bt	*Bacillus thuringiensis*
Buckeye	*Aesculus* spp.
Buckwheat	*Fagopyrum esculentum*
Buffalo	*Bison bison*
Bullhorn acacia	*Acacia cornigera*
Bunchberry	*Cornus canadensis*
Burn treatment, horsetail source of ashes for	*Equisetum hyemale* and others

5. There are more than 2,000 species of bromeliads, which include pineapple (*Ananus comosus*), Spanish moss (*Tillandsia usneoides*), and many popular house plants in genera such as *Aechmea, Bilbergia, Cryptanthus* (not to be confused with *Cryptantha,* which is in the Boraginaceae), *Neoregelia, Nidularium, Quesnelia,* and *Vriesia*.

Common Names and Scientific Names of Organisms

COMMON NAME	SCIENTIFIC NAME
Butcher's broom	*Ruscus aculeata*
Buttercup	*Ranunculus* spp.
Buttercup, European bulbous	*Ranunculus bulbosa*
Buttercup Family	Ranunculaceae
Butterfly	member of Superfamily Papilionoidea, Order Lepidoptera, Phylum Arthropoda, Kingdom Animalia
Butterwort	*Pinguicula grandiflora, P. vulgaris,* and other *Pinguicula* spp.
Button snakeroot	*Eryngium* spp.
Cabbage (green or red)	*Brassica oleracea* var. *capitata*
Cabbage, Chinese	*Brassica chinensis*
Cabbage Family	Brassicaceae (formerly Cruciferae)
Cabbage looper	*Trichoplusia ni*
Cabbage worm	*Pieris rapae*
Cacao	*Theobroma cacao*
Cactus (Fig. 24.14*A*)	*Hamatocactus setispinus*
Cactus, barrel	*Mamillaria* spp., *Ferocactus* spp., and others
Cactus, cholla	*Opuntia* spp. (cylindrical forms)
Cactus family	Cactaceae
Cactus, giant saguaro	*Carnegia gigantea*
Cactus, hedgehog	*Echinocereus* spp. and others
Cactus, living rock	*Ariocarpus fissuratus* and others
Cactus, organ-pipe	*Lemaireocereus* spp.
Cactus, prickly pear	*Opuntia* spp.
Cajuput, source of	*Melaleuca cajuputi*
Calabash	*Lagenaria siceraria*
Calabazilla	*Cucurbita foetidissima*
California bay (also known as Oregon myrtle)	*Umbellularia californica*
California poppy	*Eschscholzia californica*[6]
Camel	*Camelus* spp.
Camelina	*Camelina sativa*
Camellia[7]	*Camellia* spp.[7]
Camphor, source of	*Cinnamomum camphora*
Candelilla	*Euphorbia antisyphilitica*
Candlenut	*Aleurites moluccana*
Cankerworm	*Alsophila pometaria* and others
Canna	*Canna edulis* and other *Canna* spp. and hybrids
Cantaloupe	*Cucumis melo*
Caraway	*Carum carvi*
Cardamon/Cardamom	*Elettaria cardamomum*
Caribou	*Rangifer tarandus*
Carnation	*Dianthus caryophyllus*
Carnaubalike wax, source of	*Stipa tenacissima*
Carnauba wax, source of	*Copernicia cerifera*
Carob	*Ceratonia siliqua*
Carpetweed Family	Molluginaceae
Carrot	*Daucus carota*
Carrot Family	Apiaceae (formerly Umbelliferae)
Cashew	*Anacardium occidentale*
Cassava	*Manihot esculenta*
Cassia[8]	*Cinnamomum cassia*[8]
Catalpa	*Catalpa* spp.
Caterpillar	larval stage of member of Order Lepidoptera, Phylum Arthropoda, Kingdom Animalia
Catnip	*Nepeta cataria*
Cattail	*Typha* spp.
Cattle—*see* Cow	
Cauliflower	*Brassica oleracea* var. *botrytis* (= *B. oleracea* var. *cauliflora*)[9]
Caussu wax, source of	*Calathea lutea*
Cedar, Atlantic white	*Chamaecyparis thyoides*
Cedar, eastern red	*Juniperus virginiana*
Cedar, northern white	*Thuja occidentalis*
Cedar, incense	*Calocedrus decurrens*
Cedar, southern white	*Chamaecyparis thyoides*
Cedar, western red	*Thuja plicata*
Celery, Celeriac	*Apium graveolens*
Cell-from-hell (dinoflagellate)	*Pfiestera piscicida*
Cellular slime mold	member of Phylum Acrasiomycota, Kingdom Protista

6. Although the generic name was given in honor of Johann Friedrich Eschscholtz, an early 19th century German naturalist and surgeon, the name was first published as *Eschscholzia,* making the spelling *Eschscholtzia* an orthographic variant.
7. More than 80 species of *Camellia* and 2,000 horticultural varieties are recognized, with most of the ornamental varieties having been derived from *C. japonica* and *C. sasanqua.* The late George Petersen of Chico, California, produced 700 of the horticultural varieties. Other important members of the genus include *C. sinensis* (tea), and *C. oleifera,* whose seeds yield tea tree oil.
8. This should not be confused with the genus *Cassia,* the source of senna in the Legume Family, or cassie, a perfume oil whose source is *Acacia farnesiana,* another member of the Legume Family.
9. Broccoli and cauliflower are two different forms of the same variety.

Common Names and Scientific Names of Organisms

COMMON NAME	SCIENTIFIC NAME
Century plant	*Agave americana* and others
Chamise	*Adenostoma fasciculatum*
Chara	*Chara* spp.
Chard	*Beta vulgaris* var. *cicla*
Cheese bacteria—*see* Bacteria, buttermilk	
Cheese fungi	*Penicillium camembertii* (for Camembert cheese), *P. roqueforti* (for blue, Gorgonzola, Roquefort, and Stilton cheeses)
Cherry, sour	*Prunus cerasus*
Cherry, sweet	*Prunus avium*
Chestnut, American	*Castanea dentata*
Chia	*Salvia columbariae*
Chickadee, mountain	*Parus gambeli*
Chickpea	*Cicer arietinum*
Chickweed (Himalayan)	*Stellaria decumbens*
Chicle, source of	*Manilkara zapota*
Chicory	*Cichorium intybus*
Chimpanzee	*Pan troglodytes* and others
China grass	*Boehmeria nivea*
Chinese vegetable tallow	*Sapium sebiferum*
Chipmunk	*Eutamias* spp., *Tamias* spp., and others
Chlamydomonas	*Chlamydomonas* spp.
Chloroxybacteria	member of Chloroxybacteriae, Phylum Eubacteria, Kingdom Bacteria
Chocolate, source of	*Theobroma cacao*
Chokecherry	*Prunus virginiana* var. *melanocarpa*
Cholla (cactus)	*Opuntia* spp. (cylindrical forms)
Christmas flower	*Euphorbia pulcherrima*
Chrysanthemum	About 160 spp.; many garden cultivars are hybrids of *Chrysanthemum frutescens* and *C. morifolium*
Chuckwalla	*Sauromalus obesus*
Chufa	*Cyperus esculentus*
Chytrid	*Allomyces arbusculus* and many other members of Phylum Chytridiomycota, Kingdom Fungi
Cilantro	*Coriandrum* sp.
Cinnamon, cassia	*Cinnamomum cassia, C. burmannii, C. loureii*
Cinnamon, true	*Cinnamomum zeylanicum*
Citric acid, fungal producers of	*Aspergillus niger* and others
Citronella oil, source of	*Cymbopogon nardus*

COMMON NAME	SCIENTIFIC NAME
Citrus	*Citrus* spp.
Citrus Family	Rutaceae
Cladophora	*Cladophora* spp.
Clematis	*Clematis* spp.
Clover	*Trifolium* spp.
Clover, bur	*Medicago polymorpha*
Cloves	*Syzygium aromaticum* (formerly *Eugenia caryophyllus*)
Club fungus	member of Phylum Basidiomycota, Kingdom Fungi
Club moss	member of Phylum Lycophyta, Kingdom Plantae
Club moss (Fig. 21.3*A*)	*Lycopodium cernuum*
Club moss (Fig. 21.3*B*)	*Lycopodium obscurum*
Coastal redwood	*Sequoia sempervirens*
Cobra plant	*Darlingtonia californica*
Coca/Cocaine, source of	*Erythroxylum* (often misspelled *Erythroxylon*) *coca. E. novogravatense* is a lesser source.
Cochineal insect	*Dactylopius coccus*
Cocklebur	*Xanthium strumarium*
Cockroach	*Blatta orientalis, Blatella germanica,* and others
Cockroach plant	*Haplophyton cimicidum*
Cockscomb	*Celosia* spp.
Coffee, Arabian	*Coffea arabica*
Coffee, Liberian	*Coffea liberica*
Coffee, robusta	*Coffea canephora*
Coffee Family (= Madder Family)	Rubiaceae
Coleus	*Coleus blumei, C. x hybrida,* and others
Columbine	*Aquilegia* spp.
Columbine (Fig. 24.3A)	*Aquilegia formosa*
Compass plant (Fig. 7.13)	*Lactuca serriola*; (*Silphium laciniatum* is also known as Compass plant)
Coneflower	*Rudbeckia* sp.
Coneflower, Asian	*Strobilanthes* spp.
Copal, sources of	*Agathis alba, Copaifera demeussei, Hymenea coubaril, Trachylobium verrucosum,* and others
Copperhead	*Ancistrodon contortrix*
Coral tree	*Erythrina crista-galli*
Cordage fibers, source of	*Agave sisalina, A. heterocantha, A. lophantha, Phormium tenax,* and others

Common Names and Scientific Names of Organisms

COMMON NAME	SCIENTIFIC NAME
Coriander	*Coriandrum sativum*
Corn (Maize)	*Zea mays*
Corn borer, European	*Pyrausta nubialis*
Cotton	*Gossypium arboreum, G. barbadense, G. herbaceum, G. hirsutum, G. raimondii*
Cottonwood	*Populus deltoides, P. fremontii,* and others
Cow	*Bos* sp.
Cow parsnip	*Heracleum lanatum*
Cowslip	*Caltha palustris*
Crabapple	*Crataegus* spp., *Malus cortonaria* (= *Malus sylvestris?)*
Crab grass	*Digitaria sanguinalis*
Cranberry, American	*Vaccinium macrocarpon*
Cress, garden	*Lepidium sativum, Barbarea verna,* and others
Cress, rock	*Arabis* spp.
Crocus, autumn/fall	*Colchicum autumnale* and other *Crocus* spp.
Crown of thorns	*Euphorbia milii* var. *splendens* and others
Crozier, tropical tree fern (Chapter 11 opener)	*Sadleria cyatheoides*
Crustacean	member of Class Crustacea, Phylum Arthropoda, Kingdom Animalia
Cryptomonad	member of Phylum Cryptophyta, Kingdom Protista
Cucumber	*Cucumis sativus*
Cucumber, squirting	*Ecballium elaterium*
Cyanobacteria	member of Kingdom Bacteria; common genera include *Anabaena, Lyngbya, Oscillatoria, Phormidium, Schizothrix,* and *Spirulina*
Cyanobacteria, Lake Chad edible	*Spirulina* sp.
Cyanobacteria, Red Sea	*Trichodesmium erythraeum*
Cyanobacteria, thermal	*Bacillosiphon induratus, Synechococcus* spp., and others
Cycad (Chapter 22 opener)	*Cycas* sp.
Cycad (Fig. 22.12*A*)	*Dioon edule*
Cycad (Fig. 22.12*B*)	*Encephalartos altensteinii*
Cycadeoid (extinct gymnosperm with palmlike leaves)	*Cycadeoidea* and other genera
Cyclamen	*Cyclamen* spp.
Cypress	*Cupressus* spp.
Cypress, bald	*Taxodium distichum*
Daffodil	*Narcissus* spp. (see note under *Narcissus*)
Dahlia	*Dahlia* spp.
Daisy	*Dimorphotheca* spp., *Layia* spp., and others
Daisy fleabane	*Erigeron* spp.
Dandelion	*Taraxacum officinale* (Scandinavia only), elsewhere, *Taraxacum* sp. aff.
Dandruff, fern(s) used in treatment of	*Adiantum capillus-veneris, Polystichum munitum*
Date	*Phoenix dactylifera*
DDT-like compound, algal producers of	*Laurencia* spp. and others
Death angel (Death cap)	*Amanita phalloides*
Deer	*Odocoileus* spp. and others
Deer, mule	*Odocoileus hemionus*
Dendrobium (orchid)	*Dendrobium* spp. and hybrids
Desmids	*Closterium* spp., *Cosmarium* spp., and others
Destroying angel	*Amanita virosa*
Dewberry	*Rubus* hybrids with *R. ursinus* as one parent
Diatom	*Biddulphia* spp., *Cymbella* spp., *Navicula* spp., *Cymatopleura solea* (Fig. 18.14); *Thalassiosira elsayedii, Delphineis karstenii, Pseudonitzchia australis,* and many others
Dicot	member of Class Magnoliopsida, Phylum Magnoliophyta, Kingdom Plantae
Digitalis, source of	*Digitalis purpurea, D. lanata*
Dill	*Anethum graveolens*
Dinoflagellate	member of Phylum Dinophyta, Kingdom Protista. Representative genera include *Gambierdiscus, Gonyaulax,* and *Gymnodinium*
Dinoflagellate, midnight-bioluminescent	*Gonyaulax polyedra*
Dischidia	*Dischidia rafflesiana*
Divi-divi	*Caesalpinia coriaria*
Dodder	*Cuscuta* spp.
Dogbane	*Apocynum* spp.
Dogwood	*Cornus* spp.
Douglas fir	*Pseudotsuga menziesii*
Dove	member of Family Columbidae, Class Aves, Phylum Vertebrata, Kingdom Animalia

Common Names and Scientific Names of Organisms

COMMON NAME	SCIENTIFIC NAME
Dove, mourning	*Zenaidura macroura*
Downy mildew of grape	*Plasmopora viticola*
Dragon's blood	*Dracaena* spp., *Daemonorops* spp.
Drimys	*Drimys winteri* and other *Drimys* spp.
Duckweed	*Lemna* spp., *Wolffia* spp., and others
Dulse	*Rhodymenia* spp.
Dung mosses (on dung of carnivores)	*Tayloria* spp.
Dung mosses (on dung of herbivores)	*Splachnum* spp.
Dutch elm disease, causal agent of	*Ophiostoma ulmi/O. nova-ulmi*
Dutchman's breeches	*Dicentra cucullaria*
Dyer's woad	*Isatis tinctoria*
Dyes, sources of—*see listing in* Appendix 3	
Eagle, golden	*Aguila chrysautos*
Earth star	*Geaster* spp. and others
Earthworm	*Lumbricus* spp. and others
Ebony	*Diospyros ebenum*
Eelworm (nematode)	member of Class Nematoda, Phylum Aschelminthes, Kingdom Animalia
Eelworm (nematode) fungi, those that trap with constricting rings	*Dactylaria* spp., *Arthrobotrys actyloides*
Eelworm (nematode) fungi, those that trap with passive rings	*Dactylella* spp.
Eggplant	*Solanum melongena*
Elephant	*Elephas* spp., *Loxodonta* spp.
Elephant ears	*Colocasia* spp.
Elk	*Cervus canadensis*
Elm, American	*Ulmus americana*
Elm bark beetle	*Hylurgopinus rufipes, Scolytus multistriatus*
Endive	*Cichorium endivia* ssp. *divaricatum*
Endive, Belgian	*Cichorium intybus*
Ergot	*Claviceps purpurea*
Ermine	*Mustela erminea*
Eucalyptus, source of bark/wood for tannins	*Eucalyptus wandoo*
Eucalyptus, Tasmanian giant	*Eucalyptus regnans*
Eucalyptus oil, source of	*Eucalyptus* spp.; there are more than 250 spp. of *Eucalyptus*
Euglenoid	member of Phylum Euglenophyta, Kingdom Protista
Fennel	*Foeniculum vulgare*
Fenugreek	*Trigonella foenum-graecum*
Fern(s), adder's tongue	*Ophioglossum* spp.
Fern(s), amphibious	*Marsilea* spp. and others
Fern(s), aquatic (floating)	*Azolla* spp., *Salvinia* spp.
Fern(s), source of astringent	*Actiniopteris radiata, Drynaria quercifolia, Pteridium aquilinum,* and others
Fern, bird's foot	*Pellaea mucronata*
Fern, bird's nest	*Asplenium nidus*
Fern, Boston	*Nephrolepis exaltata*
Fern, bracken	*Pteridium aquilinum*
Fern, Brazilian tree (Fig. 21.25)	*Cyathea* sp.
Fern used in treating burns	*Polystichum munitum*
Fern, chain	*Woodwardia fimbriata*
Fern, cinnamon	*Osmunda cinnamomea*
Fern, climbing (Asian)	*Lygodium salicifolium*
Fern(s) used in treating coughs	*Adiantum aethiopicum, A. lunulatum, Polypodium glycyrrhiza*
Fern(s) used in treating dandruff	*Adiantum capillus-veneris, Polystichum munitum*
Fern used in treating diabetes	*Adiantum caudatum*
Fern(s) used in treating diarrhea	*Botrychium lunaria, B. ternatum, Pteridium aquilinum,* and others
Fern(s) used as diuretic	*Adiantum venustum, Lygodium japonicum*
Fern(s) source of dyes	*Sadleria cyatheoides* (trunk), *Sphenomeris chusana* (fronds)
Fern(s) used in treating dysentery	*Botrychium lunaria, B. ternatum, Pteridium aquilinum,* and others
Fern used in treating eczema	*Lygodium flexuosum*
Fern used in treating eye diseases	*Asplenium adiantum-nigrum*
Fern used to reduce fevers	*Marsilea quadrifolia*
Fern, five-finger	*Adiantum pedatum*
Fern(s) used as food	*Athyrium filix-femina, Dryopteris austriaca, D. filix-mas, Polystichum munitum,* and others
Fern(s), fossil	*Psaronius* spp., *Thamnopteris* spp., and others
Fern, goldback	*Pentagramma triangularis*
Fern, holly	*Polystichum lonchitis*
Fern(s) used by hummingbirds	*Cyathea arborea, Lophosoria quadripinnata, Nephelea mexicana*

Common Names and Scientific Names of Organisms

COMMON NAME	SCIENTIFIC NAME
Fern used for treating insect stings and bites	*Adiantum capillus-veneris*
Fern used for easing labor pains	*Athyrium filix-femina*
Fern, lady	*Athyrium filix-femina*
Fern(s) used as laxative	*Asplenium trichomanes, Polypodium vulgare*
Fern used in treating leprosy	*Marsilea quadrifolia*
Fern, licorice	*Polypodium glycyrrhiza*
Fern(s) poisonous to livestock	*Onoclea sensibilis, Pteridium aquilinum*
Fern, edible Malaysian (relative of Lady fern)	*Athyrium esculentum*
Fern, male	*Dryopteris filix-mas*
Fern, mosquito	*Azolla caroliniana*
Fern, nest	*Asplenium nidus*
Fern used to arrest nosebleeds	*Pellaea mucronata*
Fern(s) used for orchid bark	*Cibotium* spp., *Osmunda* spp.
Fern, Oriental water	*Ceratopteris thalictroides*
Fern used as poison antidote	*Polystichum squarrosum*
Fern(s) used in treating rickets	*Asplenium ruta-muraria, Osmunda regalis*
Fern(s) used for stuffing mattresses, pillows, upholstery	*Cibotium* spp., *Sadleria* spp.
Fern, sword	*Polystichum munitum*
Fern used in treating toothache	*Pentagramma triangularis*
Fern(s), Hawaiian tree	*Cibotium* spp., *Sadleria* spp.
Fern, tree	*Cyathea* spp., *Ctenitis* spp., *Dicksonia* spp., *Marattia* spp., *Sphaeropteris* spp., and others
Fern, tropical (Fig. 21.18)	*Dicranopteris linearis*
Fern, tropical tree (Fig. 21.25)	*Cibotium* sp.
Fern used for expelling worms	*Dryopteris filix-mas*
Fern(s) used for treating wounds	*Lygodium circinatum, Ophioglossum vulgatum*
Fevers, fern used to reduce	*Marsilea quadrifolia*
Fevers, ground pine used to reduce	*Lycopodium clavatum*
Fig, common	*Ficus carica*
Fig, tropical	*Ficus* spp.
Fig, tropical (Fig. 5.14)	*Ficus macrophyllus*
Figwort Family	Scrophulariaceae
Filaree	*Erodium* spp.
Fir, balsam	*Abies balsamea*

COMMON NAME	SCIENTIFIC NAME
Fir, Douglas	*Pseudotsuga menziesii*
Fir, white	*Abies concolor*
Fireweed	*Epilobium angustifolium*
Fish	member of Class Pisces, Phylum Vertebrata, Kingdom Animalia
Fish, flashlight	*Anomalops katoptron, Photoblepharon palpebratus*
Fish molds	*Saprolegnia* spp. and others
Five-finger fern	*Adiantum pedatum*
Flashlight powder, ground pine source of	*Lycopodium* spp.
Flatworm	*Convoluta roscoffensis*
Flax	*Linum* spp.
Flax, New Zealand	*Phormium tenax*
Flea	member of Order Siphonaptera, Phylum Arthropoda, Kingdom Animalia
Flicker	*Colaptes* spp.
Florida arrowroot	*Zamia integrifolia*
Flour, Hopi Indian horsetail source of	*Equisetum laevigatum*
Flowerpot leaf plant	*Dischidia rafflesiana*
Fly	member of Order Diptera, Phylum Arthropoda, Kingdom Animalia
Fly agaric	*Amanita muscaria*
Flycatcher	*Empidonax* spp., *Myiarchus* spp., and others
Fly, tsetse	*Gossinia morsitans, G. palpalis*
Fly, white	*Aleurocanthus woglumi* and others
Foliose lichen (Fig. 19.34)	*Physcia* sp.
Fossil, compression (Fig. 21.26)	*Annularia radiata*
Fossil, ground pine (*Lycopodium*) (Fig. 21.9)	*Lepidodendron*
Four-o'clock Family	*Nyctaginaceae*
Fox, arctic	*Alopex lagopus*
Fox, gray	*Urocyon cinereoargentus*
Fox, red	*Vulpes fulva*
Foxglove	*Digitalis purpurea*
Frangipanni	*Plumeria rubra* and other *Plumeria* spp.
Frog	*Rana* spp., and others
Fruit fly, common	*Drosophila melanogaster* (there are many other species of fruit fly)
Fuchsia, California	*Epilobium canum*

Common Names and Scientific Names of Organisms

COMMON NAME	SCIENTIFIC NAME
Fumitory, Himalayan	*Corydalis gerdae*
Fungi that produce antibiotics	*Penicillium* spp., *Cephalosporium* spp., and others
Fungi that cause aspergilloses	*Aspergillus fumigatus, Candida albicans, Coccidiodes immitis,* and others
Fungi that cause athlete's foot	*Trichophyton* spp.
Fungi used by beetles for food	*Ambrosiella* spp., *Monilia* spp.
Fungi, bird's-nest	*Nidularia* spp., *Crucibulum levis*
Fungi used in manufacturing birth control pills	*Rhizopus nigricans, R. arrhizus*
Fungi, cap-thrower	*Pilobolus* spp.
Fungi, cheese	*Penicillium camembertii* (for Camembert cheese), *P. roquefortii* (for blue, Gorgonzola, Roquefort, and Stilton cheeses)
Fungi, flavor-producing	*Aspergillus* spp.
Fungi, hallucinogenic	*Amanita muscaria, Conocybe* spp., *Panaeolus* spp., *Psilocybe* spp., and others
Fungi, horse dung	*Pilobolus* spp.
Fungi, industrial alcohol-producing	*Aspergillus* spp.
Fungi, insect-parasitizing	members of Order Laboulbeniales, Phylum Ascomycota, Kingdom Fungi, and others
Fungi, meat-tenderizing	*Thamnidium* spp.
Fungi, ringworm	*Epidermophyton* spp., *Microsporium* spp., *Trichophyton* spp.
Fungi, shelf—*see* Fungi, bracket	
Fungi, shoyu	*Aspergillus oryzae, A. soyae*
Fungi used in silvering of mirrors	*Aspergillus* spp.
Fungi used in manufacturing soap	*Penicillium* spp.
Fungi, soil	*Fusarium* spp., and others
Fungi, soy sauce	*Aspergillus oryzae, A. soyae*
Fungi, sufu	*Actinomucor elegans, Mucor* spp.
Fungi, teonanacatl (sacred)	*Conocybe* spp., *Panaeolus* spp., *Psilocybe* spp., and others
Fungus, bolete (Fig. 19.20)	*Serillus pungens*
Fungus, bracket (Fig. 19.13*C*)	*Phacolus* sp.
Fungus, bracket (Fig. 19.21)	*Fomes* sp.
Fungus, bracket/shelf	*Grifola sulphurea*
Fungus, downy mildew of grape	*Plasmopora viticola*
Fungus, "foolish seedling" (of rice)	*Gibberella fujikuroi*
Fungus used in Beadle & Tatum genetic experiments	*Neurosopora crassa*
Fungus, jelly	*Auricularia* spp., *Exidia* spp., *Tremella* spp., and others
Fungus, kidney bean leaf (production of fungal inhibitors stimulator)	*Colletotrichum lindemuthianum*
Fungus, miso	*Aspergillus oryzae*
Fungus, causal agent of Panama disease (of bananas)	*Fusarium oxysporum*
Fungus used in producing plastics	*Aspergillus terreus*
Fungus, sac (Fig. 19.7)	*Caloscypha fulgens*
Fungus, tempeh	*Rhizopus oligosporus*
Fungus used in *r* manufacturing toothpaste	*Aspergillus nige*
Fungus, white piedra	*Trichosporon beigeli*
Fungus used in manufacturing yellow food-coloring agent	*Blakeslea trispora*
Funori, source of	*Gloiopeltis* spp.
Fur, green algae that inhabit animal	*Trentepohlia* spp.
Gentian, source of	*Gentiana* spp.
Geranium	*Geranium* spp., *Pelargonium* spp.
Geranium Family	Geraniaceae
Gila monster	*Heloderma suspectum*
Ginger	*Zingiber officinale* and others
Ginseng, source of	*Panax quinquefolium* and others
Giraffe	*Giraffa camelopardalis*
Gladiolia/Gladiolus	*Gladiolus* spp.
Gloeocapsa	*Gloeocapsa* spp.
Goat	*Capra* spp.
Goldback fern	*Pentagramma triangularis*
Golden brown algae	members of Phylum Chromophyta, Kingdom Protista
Golden chain tree	*Laburnum anagyroides*
Goldenrod	*Solidago* spp.
Goldenseal	*Hydrastis canadensis*

Common Names and Scientific Names of Organisms

COMMON NAME	SCIENTIFIC NAME
Goldenweed	*Haplopappus gracilis*[10]
Goose	*Branta* spp. and others
Gooseberry	*Ribes* spp.
Goosefoot Family	Chenopodiaceae
Gopher plant	*Euphorbia lathyrus* (*E. lathyris* = *E. lathyrus*)
Gopher, pocket	*Geomys bursarius, Thomomys* spp., and others
Gourd	*Lagenaria siceraria* and others
Grape	*Vitis* spp.
Grapefruit	*Citrus paradisi*
Grape, wine/table	*Vitis vinifera*
Grass (including lawn grasses)	*Bromus* spp. and others[11]
Grass, Bermuda	*Cynodon dactylon*
Grass, crested wheat	*Agropyron cristatum*
Grass Family	Poaceae (formerly Gramineae)
Grass, Indian	*Sorghastrum nutans*
Grass, pampas (Fig. 7.5)	*Cortaderia selloana*
Grass tree (Australian)	*Xanthorrhea* spp.
Gray pine[12]	*Pinus sabiniana*
Green algae	member of Phylum Chlorophyta, Kingdom Protista; representative genera include *Caulerpa, Chlorella, Codium, Dunaliella, Enteromorpha, Hydrodictyon, Microcystis, Pandorina, Pithophora, Scenedesmus, Spirogyra, Tetraselmis;* there are more than 200 genera and about 7,500 species
Greenbrier	*Smilax* spp.
Ground pine	*Lycopodium* spp.
Ground pine, fossil relatives of	*Baragwanathia* spp., *Drephanophycus* spp., *Protolepidodendron* spp., and others
Ground pine used for baby powder	*Lycopodium clavatum*
Ground pine used to arrest bleeding	*Lycopodium clavatum*
Ground pine used as intoxicant	*Lycopodium selago*
Ground pine used for ornaments	*Lycopodium clavatum, L. complanatum, L. obscurum,* and other *Lycopodium* spp.
Ground pine used to reduce fevers	*Lycopodium clavatum*
Guava	*Psidium guajava*
Gum arabic, source of	*Acacia senegal*
Gum tragacanth, source of	*Astragalus echidenaeformis, A. gossypinus, A. gummifer,* and others
Guppy	*Lebistes reticulatus*
Hairy cap moss (p. 392)	*Polytrichum commune*
Haptophyte	member of Phylum Prymnesiophyta, Kingdom Protista
Hawk	*Buteo* spp., *Falco* spp., and others
Hazelnut	*Corylus* spp.
Hazelnut, European	*Corylus avellana*
Heath	*Erica* spp. and others
Heath Family	Ericaceae
Hemlock, eastern	*Tsuga canadensis*
Hemlock, mountain	*Tsuga mertensiana*
Hemlock, poison	*Conium maculatum*
Hemlock, water	*Cicuta* spp.
Hemlock, western	*Tsuga heterophylla*
Hemp	*Cannabis sativa*
Hemp, Manila	*Musa textilis*
Hemp, Mauritius	*Furcraea gigantea*
Henbit	*Lamium amplexicaule*
Henna	*Lawsonia inermis*
Hepatica	*Hepatica* spp.
Hepatica (Fig. 24.3*C*)	*Hepatica americana*
Hickory	*Carya* spp.
Hog	*Sus scrofa,* and others
Hog fennel	*Lomatium* spp.
Holly, American	*Ilex opaca*
Honey bee	*Apis mellifera*
Hop hornbeam	*Ostrya virginiana*
Hops	*Humulus lupulus*
Horehound	*Marrubium vulgare*
Hornwort	*Anthoceros* spp.
Horse	*Equus caballus*
Horse chestnut	*Aesculus hippocastanum*
Horsetail	*Equisetum* spp.
Horsetail (Fig. 21.10*A*)	*Equisetum hyemale*

10. This species has a diploid number of $2n = 4$; i.e., each body cell has four chromosomes.
11. The Grass Family (Poaceae) comprises about 4,500 species of grasses. Some plants with *grass* in their name are in other families and are not true grasses, e.g., Grass of parnassus (*Parnassia californica*); Grass pink (*Petrorhagia dubia*).
12. Gray pines were formerly known as Digger pines. The common name was changed in deference to Native Americans who consider *digger* a derogatory term.

Common Names and Scientific Names of Organisms

COMMON NAME	SCIENTIFIC NAME
Horsetail (Fig. 21.10*B*)	*Equisetum telmateia*
Horsetail used as abrasive	*Equisetum* (all spp.)
Horsetail used as astringent	*Equisetum arvense, E. debile,* and other *Equisetum* spp.
Horsetail used for treating burns	*Equisetum hyemale* and others
Horsetail used for treating diarrhea	*Equisetum hyemale*
Horsetail used as diuretic	*Equisetum arvense, E. debile,* and others
Horsetail used for treating dysentery	*Equisetum hyemale*
Horsetail, field	*Equisetum arvense*
Horsetail, fossil	*Equisetites* spp., *Hyenia* spp., *Sphenophyllum* spp., and others
Horsetail, giant	*Equisetum telmateia*
Horsetail, Hopi Indian flour source	*Equisetum laevigatum*
Horsetail, treelike fossil	*Calamites* spp.
Horsetail used as hair wash	*Equisetum hyemale*
Horsetail used as water source	*Equisetum telmateia*
Hot springs, blue-green bacteria (cyanobacteria) of	*Bacillosiphon induratus, Synechococcus* spp., and others
"Human hair" slime mold	*Stemonitis* spp.
Hummingbird	*Archilocus* spp. and others
Hummingbird, Oasis (Fig. 23.15)	*Rhodopis vesper*
Hummingbirds, ferns used by (for nest material)	*Cyathea arborea, Lophosoria quadripinnata, Nephelea mexicana*
Hummingbirds, tropical	*Chlorostilbon maugaeus* and others
Hyacinth	*Hyacinthus orientalis* and other *Hyacinthus* spp.
Hyacinth, grape	*Muscari* spp.
Hyacinth, water	*Eichhornia crassipes*
Hyssop	*Hyssopus officinalis*
Ice plant	*Carpobrotus* spp. (esp. *C. edulis*), *Mesembryanthemum crystallinum,* and others
India, toxic blue-green bacteria (cyanobacteria) of	*Lyngbya majuscula*
Indian pipe	*Monotropa uniflora*
Indian warrior	*Pedicularis densiflora*
Indigo	*Indigofera tinctoria*
Insects—*see individual entries*	
Insects, fern used for treating stings and bites of	*Adiantum capillus-veneris*
Ipecac, source of	*Cephaelis ipecacuanha*
Iris	*Iris* spp.
Iris, butterfly	*Moraea* spp.
Iris Family	Iridaceae
Ironwood, South American	*Krugiodendron ferreum*
Isopyrum	*Isopyrum occidentale*
Ivy, Algerian	*Hedera canariensis*
Ivy, Boston	*Parthenocissus tricuspidata*
Ivy, English	*Hedera helix*
Ivy, poison	*Toxicodendron radicans* (formerly *Rhus toxicodendron*)
Jacaranda	*Jacaranda* spp.
Jaeger	*Stercorarius* spp.
Japanese yew (Fig. 22.9)	*Taxus cuspidata*
Jicama	*Pachyrhizus erosus*
Jimson weed (Fig. 8.6)	*Datura stramonium*
Jojoba	*Simmondsia californica, S. chinensis*
Joshua tree	*Yucca brevifolia*
Jumping mouse	*Zapus hudsonius, Napaeozapus insignis*
Junco	*Junco* spp.
Junco, slate-colored	*Junco hyemalis*
Juneberry	*Amelanchier* spp.
Juniper	*Juniperus* spp.
Juniper, dwarf	*Juniperus communis* and others
Jute	*Corchorus capsularis* and others
Kauri pine/resin	*Agathis australis, A. robusta*
Kelp	*Alaria* spp., *Dictyoneurum* spp., *Ectocarpus* spp., *Egregia* spp., *Laminaria* spp., *Lessoniopsis* spp., *Nereocystis* spp., and others
Kelp, giant	*Macrocystis pyrifera*
Knotweed	*Polygonum aviculare, P. arenastrum*
Kohlrabi	*Brassica oleracea* var. *caulorapa* (= *B. oleracea* var. *gongyloides*)
Koonwarra angiosperm (extinct angiosperm whose fossil was discovered in Australia) (appears to be similar to members of the pepper family—Piperaceae)	
Kudzu	*Pueraria thunbergiana* (= *P. lobata*)
Kumquat	*Fortunella japonica*
Lamb's ears	*Stachys byzantina*
Larch, eastern	*Larix laricina*
Larch, European	*Larix decidua*
Larch, western	*Larix occidentalis*
Larkspur, blue	*Delphinium* spp.

Common Names and Scientific Names of Organisms

COMMON NAME	SCIENTIFIC NAME
Larkspur, red	*Delphinium nudicaule*
Late blight of potato	*Phytophthora infestans*
Laurel	*Laurus nobilis*
Laurel Family	Lauraceae
Lavender	*Lavandula officinalis, L. angustifolia*
Leaf hopper	member of Order Homoptera, Phylum Arthropoda, Kingdom Animalia
Leaf miner	*Agromyza* spp. and others
Leaf roller	*Archips argyrospila* and others
Leafy liverwort—*see* Liverwort, leafy	
Legume Family	Fabaceae (formerly Leguminosae)
Lemming	*Lemmus* spp., *Dicrostonyx groenlandicus*
Lemon	*Citrus limon*
Lemongrass oil, source of	*Cymbopogon citratus, C. flexuosus*
Lentil	*Lens esculenta* (= *Lens culinaris*)
Lettuce	*Lactuca sativa*
Lichen (symbiotic association of an alga and a fungus)	member of Phylum Ascomycota, Kingdom Fungi[13]
Lichen, foliose (Fig. 19.34)	*Physcia* sp.
Lichen, foliose (Fig. 19.35*B*)	*Parmelia* sp.
Lichen, fruticose (Fig. 19.39*C*)	*Usnea* sp.
Lichen, grazed by North African sheep	*Lecanora* spp.
Lichen, litmus	*Rocella* spp.
Lichen, natural dye	*Parmelia* spp., *Usnea* spp., and others
Lichen, perfume stabilizer	*Evernia* spp.
Lichen, reindeer (reindeer moss)	*Cladonia* spp., *Cetraria islandica*
Lichens, crustose (Fig. 19.35*A*)	
black	*Rinodina* sp.
chartreuse	*Acarospora citrina*
gray	*Psora* sp.
orange-red	*Caloplaca elegans*
yellow	*Candelariella vitellina*
Lichens used as miniature trees and shrubs	*Cladonia* spp. and others
Licorice, source of	*Glycyrrhiza glabra*
Lignum vitae	*Guaiacum officinale*
Lilac, common	*Syringa vulgaris*
Lily	*Lilium* spp. and others
Lily	*Lilium regale, L. auratum, L. martagon*
Lily, kaffir	*Clivia* sp.
Lily, tiger	*Lilium pardalinum*
Lily, wood	*Lilium superbum*
Lily Family	Liliaceae
Lime	*Citrus aurantifolia*
Litchi	*Litchi sinensis*
Litmus indicator dye, source of	*Rocella* spp.
Live oak	*Quercus chrysolepis, Q. virginiana, Q. wislizenii,* and others
Liverwort	member of Phylum Hepaticophyta, Kingdom Plantae
Liverwort, leafy (Fig. 20.8)	*Porella* sp.
Liverworts, leafy	*Calopogeia* sp., *Bazzania trilobata, Frullania* spp., *Jungermannia* spp., *Porella* spp., and others[14]
Liverworts, thalloid	*Conocephalum* spp., *Lunularia* spp., *Marchantia* spp., and others
Lizard	*Sceloporus* spp. and others
Lobeline sulfate, source of (used in formulas to assist in stopping smoking)	*Lobelia inflata*
Locoweed	*Astragalus mollisimus* and other *Astragalus* spp.
Locust, black	*Robinia pseudo-acacia*
Loganberry	*Rubus* hybrids, with *R. ursinus* as one parent; *Rubus vitifolius*
Logwood	*Haematoxylon campechianum*
Loon	*Gavia* spp.
Louse	Orders Mallophaga and Anaplura, Class Insecta, Phylum Arthropoda, Kingdom Animalia
Love-lies-bleeding	*Amaranthus caudatus*
Lucerne—*see* Alfalfa	
Luffa	*Luffa cylindrica, L. acutangula*
Lupine	*Lupinus* spp.

13. The lichens are arbitrarily treated under Phylum Ascomycota within Kingdom Fungi because (1) the vast majority of fungal components of each species are ascomycetes, and (2) the fungal component of each species of lichen is unique to the species, while the algal component is often common to more than one species of lichen.
14. There are thousands of species of leafy liverworts assigned to about 200 genera.

Common Names and Scientific Names of Organisms

COMMON NAME	SCIENTIFIC NAME
Lupine, tree with seed valves	*Lupinus arboreus*
Madder Family	Rubiaceae
Magnolia	*Magnolia* spp.
Magnolia Family	Magnoliaceae
Mallow	*Malva* spp.
Mallow Family	Malvaceae
Mango	*Mangifera indica*
Mangrove	*Rhizophora mangle, R. candelaria,* and others
Mangrove, black	*Avicennia germinans, A. nitida* (Fig. 5.10)
Manila hemp	*Musa textilis*
Manioc—*see* Cassava	
Manroot	*Marah* spp.
Maple	*Acer* spp.
Maple, bigleaf (Fig. 8.20)	*Acer macrophyllum*
Maple, hard	*Acer saccharum*
Maple, red	*Acer rubrum*
Maple, silver	*Acer saccharinum*
Maple, sugar	*Acer saccharum*
Marigold	*Tagetes erecta* and other *Tagetes* spp.
Marijuana	*Cannabis sativa*
Marjoram	*Majorana hortensis* (= *Origanum majorana),* pot marjorum = *Origanum onites*
Maté	*Ilex paraguariensis*
Meadow foam	*Limnanthes* spp.
Mealy bugs	*Pseudococcus* spp.
Melon	*Cucumis melo*
Melon, honeydew	*Cucumis melo* (variety)
Melonette	*Melothria pendula*
Mermaid's wineglass	*Acetabularia* spp.
Mescal bean	*Sophora secundiflora*
Mesquite	*Prosopis glandulosa*
Milkweed	*Asclepias syriaca* and other *Asclepias* spp.
Milkweed, swamp	*Asclepias incarnata*
Millet[15]	
Millipede	member of Class Diplopoda, Phylum Arthropoda, Kingdom Animalia
Mint—*see* Peppermint, Spearmint, etc.	
Mint Family	Lamiaceae (formerly Labiatae)
Mistletoe	*Phoradendron* spp.
Mistletoe, dwarf	*Arceuthobium* spp.
Mite	member of Order Acarina, Phylum Arthropoda, Kingdom Animalia
Mock orange	*Philadelphus x virginalis* and other *Philadelphus* spp. and hybrids
Mollusc	member of Phylum Mollusca, Kingdom Animalia
Monkey	*Ateles dariensis* and many others
Monkey flower	*Mimulus* spp.
Monkshood	*Aconitum columbianum*
Monocot	member of Class Liliopsida, Phylum Magnoliophyta, Kingdom Plantae
Moose	*Alces americana, A. alces*
Morel	*Morchella esculenta* and other *Morchella* spp.
Morel, false	*Helvella* sp.
Morning glory	*Ipomoea violacea* and others
Mosquito	*Anopheles* spp., *Culex* spp., and others
Moss	member of Phylum Bryophyta, Kingdom Plantae
Moss, annual (bare soil)	*Acaulon* spp., *Ephemerum* spp., and others
Moss, antler and bone	*Tetraplodon* spp.
Moss used by Indians to treat burns	*Bryum* spp., *Mnium* spp.
Moss, copper-rich substrate-inhabiting	*Mielichhoferia* spp., *Scopelophila* spp.
Moss, carnivore dung-inhabiting	*Tayloria* sp.
Moss, calcium absence indicator	*Andreaea* sp., *Rhacomitrium lanuginosum*
Moss, calcium presence indicator	*Didymodon* spp., *Desmatotodon* spp., and others
Moss, exceptionally desiccation-resistant	*Tortura ruralis*
Moss, European roof-waterproofing	*Dicranoweisia* sp.
Moss, extinguisher	*Encalypta* spp.
Moss, hair(y) cap	*Polytrichum* spp.
Moss , herbivore dung-inhabiting	*Splachnum* spp.

15. Several species of grain are called *millet,* but the most extensively cultivated taxa are *Pennisetum glaucum* (pearl millet) and *Eleusine coracana* (finger millet). Other millets, some of which are used as pasturage, include *Panicum milliaceum* (broomcorn millet); *P. maximum; P. obtusum; P. purpurascens; P. ramosum; P. texanum; P. virgatum; Echinochloa colona; E. crus-galli; Paspalum* sp.; *Eragrostis* sp.; *Setaria italica;* and others.

Common Names and Scientific Names of Organisms

COMMON NAME	SCIENTIFIC NAME
Moss, luminous	*Schisostega pennata, Mittenia* sp.
Moss, mammal dung-inhabiting	*Splachnum luteum*
Moss, peat	*Sphagnum* spp.
Moss, pollution-sensitive	*Hypnum* spp. and others
Moss, pygmy—*see* Moss, annual	
Moss, rock	*Andreaea* spp., *Grimmia* spp., and others
Moss rose	*Portulaca grandiflora*
Moss, saline (salty) soil indicator	*Pottia* spp.
Moss, seasonal running-water indicator	*Fontinalis* spp.
Moss, sphagnum	*Sphagnum* spp.
Moth	member of Order Lepidoptera, Class Insecta, Phylum Arthropoda, Kingdom Animalia
Moth, Argentine, used to control cactus in Australia	*Cactoblastus cactorum*
Moth, codling	*Carpocapsa pomonella*
Moth, gypsy	*Porthetria dispar*
Moth, Mexican jumping bean	*Carpocaps asaltitans*
Moth, Yucca	*Pronuba* spp., *Tegeticula* spp.
Moth mullein	*Verbascum blattaria*
Mountain beaver	*Aplodontia rufa*
Mouse	*Mus musculus, Peromyscus* spp., and others
Mouse, jumping	*Zapus hudsonius, Napaeozapus insignis*
Mulberry	*Morus* spp.
Mulberry, red	*Morus rubra*
Mulberry, white	*Morus alba*
Mule ears	*Wyethia* spp.
Mullein	*Verbascum thapsus*
Mullein, moth	*Verbascum blattaria*
Mushroom[16]	*Agaricus* spp. and others
Mushroom, common red (Fig 19.13*A*)	*Russula* sp.
Mushroom, fairy ring (Fig. 19.19)	*Pholiota* sp.
Mushroom, common cultivated edible	*Agaricus bisporus*
Mushroom, inky cap	*Coprinus* spp.
Mushroom, oyster	*Pleurotus ostreatus*
Mushroom, pore (Fig. 19.20)	*Serillus pungens*
Mushroom, portabella	*Agaricus bisporus* (variety)
Mushroom, shaggy mane	*Coprinus comatus*
Mushroom, shiitake	*Lentinus edodes*
Muskrat	*Ondatra zibesthicus*
Mustard	*Brassica campestris, B. nigra,* and others
Mustard, cultivated	*Brassica alba, B. juncea,* and others
Mustard Family	Brassicaceae (formerly Cruciferae)
Myrrh, source of	*Commiphora abyssinica*
Myrtle[17]	*Umbellularia californica*[17]
Narcissus	*Narcissus* spp. and hybrids[18]
Nasturtium (garden)	*Tropaeolum majus*
Nectarine	*Prunus persica*
Neem tree	*Azadirachta indica*
Nematode	member of Class Nematoda, Phylum Aschelminthes, Kingdom Animalia
Nettle	*Urtica* spp.
Nicotine relative (nornicotine), source of	*Duboisia hopwoodii, Nicotiana tabacum*
Nightshade, deadly	*Atropa belladonna*
Nightshade Family	Solanaceae
Nori—*see* Purple laver	
Nostoc	*Nostoc* spp.
Nutmeg	*Myristica fragrans*
Nutmeg, California	*Torreya californica*
Nutmeg Family	Myristicaceae
Oak	*Quercus* spp.
Oak, black	*Quercus velutina*
Oak, blue	*Quercus douglasii*
Oak, cork	*Quercus suber*
Oak, Hooker	*Quercus lobata*
Oak, live (Fig. 9.7)	*Quercus wislizenii* (other live oaks include *Quercus chrysolepis* and *Q. virginiana*)

16. *Mushroom* is a term generally applied to the fruiting bodies with stalked, caplike structures produced by members of Phylum Basidiomycota, Kingdom Fungi. The term is also loosely applied to some of the fruiting bodies of members of other classes of true fungi. There are thousands of known species.
17. This plant, also known as the California bay, is in the Laurel Family (Lauraceae). True myrtles are in the Myrtle Family (Myrtaceae).
18. The 27 known species of *Narcissus* (native to Europe and the Mediterranean regions) have been extensively cultivated and hybridized. There is a botanical classification for wild forms and a horticultural classification based primarily on the extent of the corona, flower color, and fragrance. The horticultural forms are known by common names such as narcissi, daffodils, jonquils, pheasant's eye, angel's tears, etc.

Common Names and Scientific Names of Organisms

COMMON NAME	SCIENTIFIC NAME
Oak, poison	*Toxicodendron diversilobum*
Oak, red	*Quercus borealis*
Oak, white	*Quercus alba*
Olibanum tree	*Boswellia* spp.
Olive	*Olea europaea*
Onion	*Allium cepa*
Oomycete	member of Phylum Oomycota, Kingdom Fungi
Opuntia—*see* Prickly pear	
Orchid	*Cattleya* spp., and many others[19]
Orchid, bamboo	*Arundina graminifolia*
Orchid, Bletilla (Fig. 8.18*B*)	*Bletilla* sp.
Orchid, bucket	*Coryanthes* spp.
Orchid "bark," fern sources of	*Cibotium* spp., *Osmunda* spp.
Orchid with cladophylls	*Epidendrum* spp.
Orchid, showy	*Orchis* spp.
Orchid, underground-flowering	*Rhizanthella gardneri*
Orchid, vanilla	*Vanilla planifolia* and others
Orchid Family	Orchidaceae
Oregano	*Origanum vulgare* and others
Oregon grape	*Berberis aquifolium* (= *Mahonia aquifolium*) and other *Mahonia* spp.
Organpipe cactus	*Lemaireocereus* spp.
Osage orange	*Maclura pomifera*
Oscillatoria	*Oscillatoria* spp.
Our Lord's Candle	*Yucca whipplei*
Owl, snowy	*Nyctea scandiaca*
Painted lady	*Echeveria derenbergii*
Palm, coconut	*Cocos nucifera*
Palm, date	*Phoenix dactylifera*
Palm, oil	*Elaeis guineensis*
Palm, panama hat	*Carludovica palmata*
Palm, Seychelles Island	*Lodoicea maldivica*
Palm, carnauba wax	*Copernicia cerifera*
Palm Family	Arecaceae (formerly Palmae)
Pansy	*Viola tricolor*
Papaya	*Carica papaya*
Pará rubber tree	*Hevea brasiliensis*
Parsley	*Petroselinum crispum*
Parsley Family	Apiaceae (formerly Umbelliferae)
Parsnip	*Pastinaca sativa*
Passion fruit	*Passiflora edulis, P. mollissima,* and other *Passiflora* spp.
Patchouli oil, source of	*Pogostemon cablin* and others
Pea (garden)	*Pisum sativum*
Pea, sweet	*Lathyrus odoratus*
Peach	*Prunus persica*
Peach leaf curl	*Taphrina deformans*
Peanut	*Arachis hypogaea*
Pear	*Pyrus communis*
Peat moss	*Sphagnum* spp.
Pecan	*Carya illinoensis*
Peccary	*Pecari angulatus, Tayassus pecari*
Penicillin mold[20]	*Penicillium* spp.[20]
Pennyroyal	*Hedeoma pulegioides*
Peony	*Paeonia* spp.
Peperomia	*Peperomia* spp.
Pepper	*Capsicum annuum, C. frutescens*[21]
Pepper, black	*Piper nigrum*
Pepper, red	*Capsicum annuum, C. baccatum, C. chinense, C. frutescens, C. pubescens*
Peppergrass	*Lepidium* spp.
Peppermint	*Mentha piperita*
Persimmon	*Diospyros* spp.
Petitgrain oil, source of	*Citrus aurantium* var. *amara*
Petunia	*Petunia* spp. and hybrids
Peyote	*Lophophora williamsii*
Phoebe	*Sayornis phoebe*

19. Depending on which authorities are followed, the number of known orchid species (all in the family Orchidaceae) may exceed 30,000. Popularly cultivated orchids include species of *Cattleya, Cymbidium, Dendrobium, Odontoglossum, Oncidium, Paphiopedilum, Phalaenopsis, Vanda,* and both interspecific and intergeneric hybrids.

20. The original producer of penicillin discovered by Sir Alexander Fleming was *Penicillium notatum;* current commercially used producers of penicillin are strains of *Penicillium chrysogenum.* Other commercially cultivated *Penicillium* species include *P. roquefortii* (used to make roquefort cheese), *P. camembertii* (used to make blue cheese), and *P. griseofulvum* (used for the production of a ringworm and athlete's foot antibiotic known as *griseofulvin*).

21. The drug *capsicum,* whose active ingredient is the oleoresin *capsaicin,* is derived from these species, and garden peppers include these and other species of *Capsicum.*

Common Names and Scientific Names of Organisms

COMMON NAME	SCIENTIFIC NAME
Pigweed[22]	*Amaranthus* spp., *Chenopodium* spp.
Pigweed Family[22]	Amaranthaceae, Chenopodiaceae
Pillbug	*Cylisticus convexus* and others
Pine	*Pinus* spp.
Pine, Afghanistan	*Pinus eldarica*
Pine, Aleppo	*Pinus halepensis*
Pine, bristlecone	*Pinus longaeva*
Pine, Chilghoza	*Pinus gerardiana*
Pine, Colorado bristlecone	*Pinus aristata*
Pine, Coulter	*Pinus coulteri*
Pine, eastern white	*Pinus strobus*
Pine, European stone	*Pinus pinea*
Pine, gray (formerly Pine, digger)	*Pinus sabiniana*
Pine, jack	*Pinus banksiana*
Pine, jeffrey	*Pinus jeffreyi*
Pine, kauri	*Agathis australis, A. robusta*
Pine, knobcone	*Pinus attenuata*
Pine, loblolly	*Pinus taeda*
Pine, lodgepole	*Pinus contorta*
Pine, longleaf	*Pinus palustris*
Pine, Merkus	*Pinus merkusii*
Pine, Mexican pinyon	*Pinus cembroides*
Pine, Mexican stone	*Pinus cembroides*
Pine, Monterey	*Pinus radiata*
Pine, pinyon	*Pinus edulis, P. monophylla, P. quadrifolia*
Pine, pitch	*Pinus rigida*
Pine, ponderosa	*Pinus ponderosa*
Pine, red	*Pinus resinosa*
Pine, Scotch	*Pinus sylvestris*
Pine, shortleaf	*Pinus echinata*
Pine, Siberian white	*Pinus sibirica*
Pine, slash	*Pinus caribaea, P. elliottii*
Pine, southern yellow—*see* Pine, loblolly, Pine, longleaf, Pine, shortleaf, and Pine, slash	
Pine, stone—*see* Pine, European stone, and Pine, Mexican stone	
Pine, sugar	*Pinus lambertiana*
Pine, western white	*Pinus monticola*
Pine, western yellow	*Pinus ponderosa*
Pine, Wollemi	*Wollemia nobilis*
Pineapple	*Ananas comosus*
Pinedrops	*Pterospora* spp.
Pistachio	*Pistacia vera*
Pitcher plant	*Sarracenia* spp. and others
Pitcher plant, Asian	*Nepenthes* spp. and others
Plantain	*Plantago* spp. (cooking bananas, also called plantains, are mostly *Musa x paradisiaca*)
Plastic, fungus used in production of	*Aspergillus terreus*
Plasticizers, source of oil for	*Euphorbia agascae*
Plover	*Charadrius* spp. and others
Plum, European[23]	*Prunus domestica*[23]
Podocarps, New Zealand timber	*Podocarpus dacrydoides, P. totara*
Podocarps, ornamental	*Podocarpus macrophylla, P. nagi,* and others
Poinsettia	*Euphorbia pulcherrima*
Poison ivy	*Toxicodendron radicans*
Poison oak	*Toxicodendron diversilobum*
Poison sumac	*Toxicodendron vernix*
Polyanthus	*Primula polyanthus* and hybrids
Pomegranate	*Punica granatum*
Poor man's pepper	*Lepidium virginicum*
Popcorn	*Zea mays* (horticultural variety)
Poplar	*Populus* spp.
Poppy (*see also* California poppy)	*Papaver* spp. and others
Poppy Family	Papaveraceae
Poppy, bush	*Dendromecon rigida*
Poppy, Mexican	*Hunnemannia* spp.
Poppy, opium	*Papaver somniferum*
Poppy, Oriental	*Papaver orientale*
Poppy, prickly (Fig. 24.5)	*Argemone glauca*
Porcupine	*Erethizon* spp., *Hystrix* spp.[24]

22. Species of *Amaranthus* in the Amaranth family (Amaranthaceae) and *Chenopodium* in the Goosefoot family (Chenopodiaceae) have been called *pigweeds,* and both families have also been referred to as the Pigweed family. Because of this many botanists prefer to refer to the Amaranthaceae as the Amaranth family and the Chenopodiaceae as the Goosefoot family. Most *Amaranthus* spp. are called amaranths and most *Chenopodium* spp. are called *goosefoot,* although the widespread weed *Chenopodium album* is popularly called lamb's quarters. Pig's weed (*Oryza* sp.) is completely unrelated to either of the two families just mentioned; it is a grass related to rice.
23. Although the European plum was involved in the development of the majority of older plum varieties more than 600 varieties of plum have been developed from American plums such as *Prunus subcordata* and *P. americana,* along with various hybrids involving at least one American parent.
24. *Hystrix* is also a name for a genus of grasses.

Common Names and Scientific Names of Organisms

COMMON NAME	SCIENTIFIC NAME
Portulaca Family	Portulacaceae
Potato, Irish	*S. tuberosum* ssp. *tuberosum*
Potato, sweet	*Ipomea batatas*
Potato vine	*Solanum jasminoides*
Powderpuff flower (Fig. 24.11*C*)	*Calliandra inaequilatera*
Powdery mildew	*Erysiphe* spp. and others
Prayer plant	*Maranta* spp.
Preferns	*Cladoxylon* spp., *Protopteridium* spp. and others
Prickly pear	*Opuntia* spp.
Primrose	*Primula vulgaris* and about 400 other *Primula* spp.
Pronghorn	*Antilocarpa americana*
Psyllium	*Plantago ovata*
Ptarmigan	*Lagopus* spp.
Pteridosperms	*Lyginopteris* spp., *Medullosa* spp., and others
Puffball	*Calvatia cyathiformis* and other *Calvatia* spp., *Lycoperdon* spp.
Pulque, source of	*Agave* spp.
Pumpkin	*Cucurbita pepo*
Pumpkin Family	Cucurbitaceae
Puncture vine	*Tribulus terrestris*
Purple laver (Fig. 18.28)	*Porphyra tenera*
Puya (rare)	*Puya raimondii*
Pyrethrum	*Chrysanthemum cinerariifolium, C. coccineum, C. marschallii*
Quillwort	*Isoetes* spp.
Quillwort, fossil relatives of	*Cinchona ledgeriana* and other *Cinchona* spp.
Quince	*Cydonia oblonga*
Quinine, source of	*Cinchona officinalis, C. ledgeriana*
Quinoa	*Chenopodium quinoa*
Rabbit	*Oryctolagus cuniculus*
Rabbit, cottontail	*Sylvilagus* spp.
Rabbit, jack	*Lepus* spp.
Raccoon	*Procyon lotor*
Radish	*Raphanus sativus*
Rafflesia (Fig. 8.2)	*Rafflesia micropylora*

COMMON NAME	SCIENTIFIC NAME
Ragweed	*Ambrosia* spp.
Rape/rapeseed	*Brassica napus*
Raspberry, red	*Rubus idaeus, R. strigosus,* and their hybrids
Rat	*Rattus norvegicus, R. rattus,* and others
Rat, kangaroo	*Dipodomys* spp.
Rat snake, black	*Elaphe obsoleta*
Rattlesnake	*Crotalus* spp.
Red algae	Member of Phylum Rhodophyta, Kingdom Protista; representative genera include *Chondrus, Eucheuma, Gelidium, Gigartina, Gracilaria, Polysiphonia, Porphyra,* and *Pterocladia;* there are about 3,900 spp. of red algae
Redbud, eastern	*Cercis canadensis*
Redbud, western	*Cercis occidentalis*
Redwood, coastal	*Sequoia sempervirens*
Redwood, dawn	*Metasequoia glyptostroboides*
Redwood, giant	*Sequoiadendron giganteum*
Reindeer	*Rangifer* spp.
Reserpine, source of	*Rauvolfia serpentina*
Resurrection plant	*Selaginella lepidophylla*
Rhododendron	*Rhododendron* spp.
Rhubarb	*Rheum rhaponticum*
Rice[25]	*Oryza sativa*[25]
Rice-paper plant	*Tetrapanax papyriferus* (some authors refer to *Fatsia japonica [F. papyrifera]* as rice-paper plant
Rice, wild	*Zizania aquatica*
Robin	*Turdus migratorius*
Rock cress	*Arabis* sp.
Rock-rose, European	*Helianthemum vulgare*
Rock tripe	*Umbilicaria* spp.
Rockweeds	*Fucus* spp., *Pelvetia* spp., and others
Rose	*Rosa odorata* and other *Rosa* spp. (there are more than 25,000 rose cultvars)
Rose, damask	*Rosa damascena*
Rose Family	Rosaceae
Rose, Sitka (Fig. 24.8)	*Rosa rugosa*

25. At the beginning of the year 2000, the International Rice Research Institute in the Philippines had in storage the seeds of 20 species and more than 81,000 different varieties of rice. The staple food of nearly 2 billion people, rice cultivation presently occupies 11% of agricultural land. Rice has been cultivated in Asian countries for more than 7,000 years. Although the great majority of rice cultivated is *Oryza sativa,* some forms of *Oryza glaberrima* are also cultivated.

Common Names and Scientific Names of Organisms

COMMON NAME	SCIENTIFIC NAME
Rosemary	*Rosmarinus officinalis*
Rotenone, source of	*Derris elliptica, Lonchocarpus nicou*
Rotenone, relative of	*Tephrosia vogelii*
Rubber, Pará	*Hevea brasiliensis*
Rubber plant	*Ficus elastica*
Ruellia	*Ruellia portellae* and others
Rust, apple	*Gymnosporangium juniperi-virginianum*
Rust, black stem of wheat	*Puccinia graminis*
Rust, corn	*Puccinia sorghi*
Rust, poplar leaf spot	*Melampsora medusae*
Rust, rock cress	*Puccinia monoica*
Rust, white pine blister	*Cronartium ribicola*
Rutabaga	*Brassica campestris* var. *napo-brassica* (= *Brassica napus*)
Rye	*Secale cereale*
Ryegrass	*Lolium* spp.
Safflower	*Carthamus tinctorius*
Saffron (true)	*Crocus sativus*
Saffron, meadow	*Colchicum autumnale*
Sage[26]	*Salvia officinalis*[26]
Sagebrush	*Artemisia tridentata*
Sage, Jerusalem	*Phlomis fruticosa*
Saguaro	*Carnegia gigantea*
Salmon	*Oncorhynchus* spp., *Salmo salar,* and others
Salmonberry	*Rubus spectabilis*
Salsify	*Tragopogon* spp.
Saltbush/Saltscale	*Atriplex* spp.
Salvia	*Salvia* spp.
Sansevieria	*Sansevieria trifasciata* and other *Sansevieria* spp.
Santonin, source of	*Artemisia cina*
Sargassum (Fig. 18.18)	*Sargassum* sp.
Sarsaparilla, source of	*Smilax* spp.
Sassafras	*Sassafras albidum*
Sausage tree, African	*Kigelia pinnata*
Savory	*Satureia hortensis*
Saxifrage	*Saxifraga* spp.
Screw pine	*Pandanus veitchii* and other *Pandanus* spp.
Sea anemone	*Stephanauge* spp. and others

COMMON NAME	SCIENTIFIC NAME
Sea hare	*Aplysia californica*
Sea lettuce	*Ulva* spp.
Sea palm	*Postelsia palmaeformis*
Sea rocket	*Cakile edentula*
Sedge	*Carex* spp. and others
Seed ferns (Pteridosperms)	*Lyginopteris* spp., *Medullosa* spp., and others
Senna	*Cassia senna* and others
Sensitive plant	*Mimosa pudica*
Sesame	*Sesamum indicum*
Shallot[27]	*Allium cepa*[27]
Sheep	*Ovis* spp.
Shepherd's purse	*Capsella bursa-pastoris*
Shrimp	*Crago* spp. and others
Sisal	*Agave sisalina*
Skunk	*Mephitis* spp.
Slime mold	member of Phyla Myxomycota and Acrasiomycota, Subkingdom Myxobionta, Kingdom Protista
Slime mold (Fig. 18.31*A*)	*Lamproderma* sp.
Slime mold (Fig. 18.31*B*)	*Lycogala epidendrum*
Slime mold, cellular	member of Phylum Acrasiomycota, Kingdom Protista
Slime mold, human-hair	*Stemonitis* spp.
Sloth	*Bradypus* spp., *Choleopus* spp.
Smut	*Ustilago* spp. and others
Smut, corn	*Ustilago maydis*
Snail	*Haplotrema concava* and others
Snapdragon	*Antirrhinum majus*
Snowplant	*Sarcodes sanguinea*
Snowy owl	*Nyctea scandiaca*
Soaproot, California	*Chlorogalum pomeridianum*
Sorghum	*Sorghum bicolor* and other *Sorghum* spp.
Sorrel	*Oxalis* spp.
Southern yellow pine—*see* Pine, loblolly, Pine, longleaf, Pine, shortleaf, and Pine, slash	
Soybean	*Glycine max*
Spanish moss	*Tillandsia usneoides*
Sparrow, savannah	*Passerculus sandwichensis*

26. This sage, which is in the Mint Family (Lamiaceae), should not be confused with *sagebrush,* which is in the Sunflower Family (Asteraceae).

27. Linnaeus applied the name *Allium ascalonicum* to what was probably an *Allium* cultigen believed to have originated in Asia Minor, and the name *Allium ascalonium* since has generally been applied to shallots. Shallots seldom set seed, however, and seeds sold under the name *Allium ascalonicum* have sometimes proved to be those of other *Allium* spp. The current practice of considering shallots to be a form of *Allium cepa* should lessen the confusion.

Common Names and Scientific Names of Organisms

COMMON NAME	SCIENTIFIC NAME
Sparrow, song	*Melospiza melodia*
Sparrow, vesper	*Pooecetes gramineus*
Spearmint	*Mentha spicata*
Spiderwort	*Tradescantia virginiana* and other *Tradescantia* spp.
Spiderwort, European	*Tradescantia paludosa*
Spike moss	*Selaginella* spp.
Spike moss, fossil relatives of	*Lepidodendron* spp., *Sigillaria* spp., and others
Spinach	*Spinacia oleracea*
Spirogyra	*Spirogyra* spp.
Sponge	*Spongilla* spp. and others
Sponge, vegetable	*Luffa cylindrica*
Spring beauty	*Claytonia virginica*
Spruce, black	*Picea mariana*
Spruce, Norway	*Picea abies*
Spruce, red	*Picea rubens*
Spruce, Sitka	*Picea sitchensis*
Spruce, white	*Picea glauca*
Spurge	*Euphorbia* spp.
Spurge (Fig. 24.13)	*Euphorbia peplus*
Spurge Family	Euphorbiaceae
Squash	*Cucurbita maxima, C. mixta, C. moschata, C. pepo,* and others
Squawroot	*Perideridia* spp.
Squill	*Scilla* spp.
Squills	*Urginea maritima*
Squirrel	*Citellus* spp., *Sciuris* spp., and others
Squirrel corn	*Dicentra canadensis*
Squirrel, gray	*Sciurus carolinensis*
Squirting cucumber	*Ecballium elaterium*
Stapelia (Fig. 23.14)	*Stapelia similis*
Stinkhorn	*Mutinus* spp., *Phallus impudicus,* and others
Stinkhorn, common (Fig. 19.12)	*Mutinus caninus*
Stonecrop	*Sedum* spp., *Crassula* spp., and others
Stoneseed	*Lithospermum ruderale*
Stonewort	*Chara* spp., *Nitella* spp.
Strawberry	*Fragaria ananassa* and other *Fragaria* spp. and hybrids
String-of-pearls	*Senecio rowellianus*
Strychnine, source of	*Strychnos nox-vomica* and other *Strychnos* spp.
Sugar cane	*Saccharum officinarum*
Sumac	*Rhus* spp.
Sunbird	*Anthodiaeta* spp., *Notiocinnyris* spp., and others
Sundew	*Drosera* spp.
Sundew relative used for flypaper	*Drosophyllum lusitanicum*
Sunflower	*Helianthus annuus, H. debilis*
Sunflower Family	Asteraceae (formerly Compositae)
Sweet pea	*Lathyrus odoratus*
Sword fern	*Polystichum munitum*
Sycamore	*Platanus occidentalis* and others
Tamarack	*Larix* spp.
Tamarisk	*Tamarix* spp.
Tangerine	*Citrus reticulata*
Tapir	*Tapirus* spp.
Taro	*Colocasia esculenta*
Tarragon	*Artemisia dracunculus*
Tarweed	*Grindelia* spp.
Tarweed, western (Fig. 4.13*A*)	*Calycadenia* sp.
Tea	*Camellia sinensis*
Tent caterpillar	*Malacosoma americanum* and others
Teosinte (annual)	*Zea mexicana*
Teosinte (perennial)	*Zea diploperennis*
Tequila, source of	*Agave* spp.
Termite	*Odontotermes* spp., *Reticulitermes* spp., and others
Thalloid liverworts	*Marchantia* spp., *Conocephalum* spp., and others
Thimbleberry	*Rubus parviflorus*
Thistle	*Cirsium* spp. and others
Thistle, Canada	*Cirsium arvense*
Thrasher	*Toxostoma* spp.
Thyme	*Thymus vulgaris* and others
Ti (Ki) plant	*Cordyline fruticosa*
Tiger	*Panthera tigris*
Toad	*Bufo americanus*
Tobacco	*Nicotiana tabacum, N. rustica*
Tomato[28]	*Solanum esculentum*
Tomato fruitworm	*Heliothis armigera*

28. DNA evidence indicates the tomato, long known as *Lycopersicum esculentum,* belongs in the genus *Solanum* and should be transferred to that genus, making the correct name *Solanum esculentum.*

Common Names and Scientific Names of Organisms

COMMON NAME	SCIENTIFIC NAME
Tomato, Galápagos	*Solanum esculentum* var. *minor, S. pimpinellifolium, S. cheesmanii* (salt tolerant sp.)
Tomato hornworm	*Protoparce quinquemaculata*
Toothwort	*Dentaria* spp.
Tortoise, giant Galápagos	*Testudo elephantopus porteri*
Touch-me-not	*Impatiens glandulifera* and others
Tree fern (Fig. 21.25*B*)	*Cibotium glaucum*
Tree fern, small Hawaiian (Fig. 21.16*D*)	*Sadleria cyatheoides*
Tree-of-heaven	*Ailanthus altissima*
Trillium	*Trillium* spp.
Truffles	*Tuber* spp.
Tulip	*Tulipa* spp.
Tulip tree	*Liriodendron tulipifera*
Tumbleweeds	*Amaranthus albus, Salsola pestifera,* and others
Tung oil, source of	*Aleurites fordii*
Turmeric, source of	*Curcuma longa, C. domestica*
Turnip	*Brassica rapa*
Turtle	*Chelydra* spp., *Chrysemys* spp., and others
Twinflower	*Linnaea borealis*
Ulothrix	*Ulothrix* spp.
Ultraviolet light, flowers seen in (Fig. 23.13)	*Rudbeckia* sp.
Unicorn plant	*Proboscidea* spp.
Venus's flytrap	*Dionaea muscipula*
Vetch	*Vicia* spp.
Vetchling, yellow	*Lathyrus aphaca*
Vinegar weed	*Trichostema* spp.
Violet	*Viola odorata* and other *Viola* spp.
Violet, African	*Saintpaulia ionantha* and other *Saintpaulia* spp.
Violet, gold	*Viola douglasii*
Virginia creeper	*Parthenocissus quinquefolia*
Virus[29]	
Vole	*Microtus* spp. and others
Wahoo	*Euonymus alata* and others
Wake-robin	*Trillium* spp.
Wallflower, western	*Erysimum capitatum*
Walnut	*Juglans* spp.
Walnut, black	*Juglans nigra*
Warbler	*Dendroica* spp. and others
Watercress	*Nasturtium officinale*
Water fern, oriental	*Ceratopteris thalictroides*
Watermelon	*Citrullus lanatus*
Water mold	member of Phylum Oomycota, Subkingdom Mastigobionta, Kingdom Protista
Water net	*Hydrodictyon* spp.
Watersilk	*Spirogyra* spp.
Water weed	*Elodea* spp.
Water weed, yellow	*Ludwigia repens*
Wattle	*Acacia decurrens, A. mearnsii,* and others
Weaver birds	*Anaplectes* spp., *Hyphantoris* spp., and others
Webworm, fall	*Hyphantria cunea*
Welwitschia	*Welwitschia mirabilis*
Whale, sperm	*Physeter catodon*
Wheat[30]	
Wheel tree	*Trochodendron aralioides*
Whisk fern	*Psilotum* spp.
Whisk fern, fossil relatives of	*Asteroxylon* spp., *Psilophyton* spp., *Rhynia* spp., and others
Whisk fern, living relatives of	*Tmesipteris* spp.
White pine blister rust	*Cronartium ribicola*
Willow	*Salix* spp.
Willow Family	Salicaceae
Window leaves, plants with	*Fenestraria* spp. and others
Wintergreen oil, sources of	*Gaultheria procumbens* and others

29. Depending on the classification used, viruses may not have a scientific name. Many are named after the disease they cause; e.g., tobacco mosaic virus causes tobacco mosaic disease. One classification attempts to give them at least a Latin prefix, so that the virus for warts is *Papavovirus;* for smallpox, *Poxvirus;* for polio, *Picornavirus;* for measles and mumps, *Paramyxovirus.*

30. More than 20,000 varieties of cultivated bread wheat, which has a history dating back thousands of years, are presently recognized. The ancestry and cytology are complex and still not fully understood. The principal ancestors appear to have been *Triticum monococcum* (which, after mutant forms were incorporated, became known as *einkorn* wheat) and species of *Aegilops,* especially *A. speltoides,* with several other mutations and natural hybridizations having occurred throughout the past several thousand years. Emmer wheat has been recognized as *Triticum dicoccom* or *T. turgidum* var. *dicoccum*; durum wheat as *T. durum* or *T. turgidum* var. *durum;* Polish wheat (also known as Jerusalem rye) as *T. polonicum*; and common bread wheat as *T. aestivum* (which is believed to be have been derived from *T. turgidum* and a genome from *Aegilops tauschii*). Other taxa believed to have played a role in the development of cultivated wheat include *T. longissima* and *T. searsii.* Uncertainty as to the precise evolutionary history of wheat persists, however, and awaits further investigation.

Common Names and Scientific Names of Organisms

COMMON NAME	SCIENTIFIC NAME
Wisteria	*Wisteria sinensis* and other *Wisteria* spp.
Witch hazel	*Hamamelis virginiana*
Woad, dyer's	*Isatis tinctoria*
Woadwaxen	*Genista tinctoria*
Wolfsbane	*Aconitum vulparia*
Wolverine	*Gulo luscus, G. gulo*
Woodpecker	*Dendrocopus* spp. and others
Wormwood	*Artemisia annua, A. absinthium*
Yam	*Dioscorea alata, D. cayensis, D. composita, D. esculenta, D. floribunda, D. rotundata, D. trifida*
Yareta	*Azorella yareta*
Yarrow, American	*Achillea lanulosum*
Yarrow, European	*Achillea millefolium*
Yeast, baking/brewing	*Saccharomyces cerevisiae*
Yellow-green algae	member of Phylum Chromophyta, Kingdom Protista
Yew	*Taxus* spp.
Yew, Japanese (Fig. 22.9)	*Taxus cuspidata*
Zebra	*Equus zebra* and others
Zinnia	*Zinnia elegans* and others

Appendix 2

Biological Controls

If you were to ask the average farmer or backyard gardener how to control a particular insect or plant pest, you might be given the name of some poisonous spray or bait that has proved "effective" in the past. Evidence that spraying with such substances yields only temporary results, however, has been mounting for many years, and the spraying is frequently followed by even larger invasions of pests. Also, the residues of poisonous sprays often accumulate in the soil and disrupt the microscopic living flora and fauna essential to the soil's health. The problem is compounded and the ecology further upset when large amounts of inorganic fertilizers are added. As increasing numbers of people become aware of the devastating effects on the environment of pesticides and herbicides, they have been turning to **biological controls** as an alternative to the use of poisonous sprays. To the surprise of some, such controls are often more effective than traditional controls.

Poisonous sprays often promote pest invasions because the sprays usually kill beneficial insects along with the undesirable ones. In addition, the pests, through mutations, often become resistant to the sprays. In undisturbed natural areas, weeds are never a problem, and even though pests may be present, they seldom destroy the community. Why is this so? You may recall from your reading that all members of a community are in ecological balance with one another. The plants produce a variety of substances that may either repel or attract insects, inhibit or promote the growth of other plants, and generally contribute to the health of the community as a whole.

Virtually all insects have their own pests and diseases, as do most other living organisms. Each pest ensures, at least indirectly, that the various species of a community are perpetuated. This principle of nature can be applied, to a certain extent, to farming and gardening. The following are some general and specific biological controls either now in widespread use or are in various tests showing promise for the future.

GENERAL CONTROLS

Establishment of Beneficial Organisms

Ladybugs (Family Coccinellidae)

The small and often colorful beetles called *ladybugs,* and particularly their larval stages, consume large numbers of aphids, thrips, insect eggs, weevils, and other pests. They are obtainable from various commercial sources (e.g., Planet Natural, P.O. Box 3146, Bozeman, MT 59772 [1-800-289-6656 or http://www.planetnatural.com/]; Nature's Way Organics, P.O. Box 228, Rimrock, AZ 86335) but, if given a chance, they probably will establish themselves without being imported. When obtained from outside of the local area, they should be placed in groups at the bases of plants on which pests are present, preferably in the early evening after watering.

Lacewings (Families Chrysopidae and Hemerobiidae)

Lacewings are slow-flying, delicate-winged insects that consume large numbers of aphids, mealybugs, and other pests. They lay their eggs on the undersides of leaves, each egg being borne at the tip of a slender stalk. The larvae consume the immature stages of leafhoppers, bollworms, caterpillar eggs, mites, scale insects, thrips, aphids, and other destructive pests. Commercial sources include Nature's Way Organics, P.O. Box 228, Rimrock, AZ 86335 (1-800-493-1885); All Pest Control, 6030 Grenville Lane, Lansing, MI 48910.

Praying Mantis (Family Mantidae)

About 20 species of *praying mantis* are now established in the United States. These are voracious feeders that prey somewhat indiscriminately on flying insects and sometimes even on other mantises. They can be established by tying their egg cases to tree branches or at other locations above the ground. The egg cases, which form compact masses about 2.5 to 5.0 centimeters (1 to 2 inches) long, are obtainable from various commercial sources, including Peaceful Valley Farm Supply, P.O. Box 2209, Grass Valley, CA 95945; Planet Natural, P.O Box 3146, Bozeman, MT 59772 (1-800-289-6656 or http://www.planetnatural.com/).

Trichogramma Wasps (Family Trichogrammatidae)

Trichogramma wasps are minute insects, mostly less than 1 millimeter (1/25 inch) long; they parasitize insect eggs and are known to have significantly reduced populations of well over 100 different insect pests, including alfalfa caterpillars, armyworms, cabbage loopers, cutworms, hornworms, tent caterpillars, and the larvae of many species of moths. As with other insects used as biological controls, trichogramma wasps should not be released unless there are pest eggs in the vicinity, as the wasps may otherwise parasitize eggs of beneficial butterflies and other useful insects. They are available from commercial sources such as Unique Insect Control, P.O. Box 15376, Sacramento, CA 95851; New Earth, 4422 East Hwy. 44, Shepherdsville, KY 40165.

Ichneumon Wasps (Family Ichneumonidae)

The *ichneumon wasps* belong to a very large family of wasps that are mostly stingless. These tiny wasps tend to be slender and have long ovipositors that are sometimes longer than the body. Most insects are parasitized by at least one species of ichneumon; many species parasitize the larval stages of insects, consuming the host internally after hatching from eggs deposited on the body; alternatively, they may complete development in a later stage. Ichneumons will usually appear naturally in a backyard or farm population of pests if toxic sprays and other unnatural conditions have not interfered with their normal activities.

Tachinid Flies (Family Tachinidae)

Many members of the large family of *tachinid flies* resemble houseflies or bumblebees. All parasitize other insects, including a large variety of caterpillars, Japanese beetles, European earwigs, grasshoppers, gypsy moths, tomato worms, sawflies, and various beetles. Contact Mad Planter Beneficials at 1-800-548-8199 for further information.

Mites (Phytoseiidae and others)

Species of several genera of mites prey on red spider mites and have been used successfully to control other mites and scale insects.

Mosquito Fish (Gambusia affinis)

Mosquito fish have been added to bodies of fresh water all over the world to control mosquitoes. The fish feed on mosquito larvae, particularly as the larvae rise to the surface for air. One mosquito fish can consume thousands of larvae per day.

Use of Pathogenic Bacteria

Bacillus thuringiensis (Bt) is one of several pathogenic bacteria registered for use on edible plants in the United States. It reproduces only in the digestive tracts of caterpillars and is harmless to humans and all other wildlife, including earthworms, birds, and mammals. It is exceptionally effective against a wide range of caterpillars, such as tomato hornworms and fruitworms, cabbage worms and loopers, grape leaf rollers, corn borers, cutworms, fall webworms, and tent caterpillars. It is mass-produced and sold in a powdered spore form at nurseries and garden supply stores under the trade names of *Dipel, Biotrol,* and *Thuricide.* The powder is mixed with water and applied as a spray. Beneficial bacteria are available from Solutions Unlimited, Sharon Springs, NY 13459.

Establishment of Toads and Frogs

It has been estimated that a single adult toad will consume about 10,000 insects and slugs in one growing season. Toads and frogs feed at night when snails, slugs, sowbugs, earwigs, and other common pests are active.

Use of Beneficial Nematodes

Several species of these abundant microscopic roundworms are notorious for damaging economically important crops when they invade plant roots and other underground organs. Most species, however, are either harmless or beneficial to plants. They have been used successfully in parasitizing cabbage worm caterpillars, codling moth larvae, Japanese beetle grubs, and tobacco budworms and have shown considerable potential against other pests. One species that has been particularly effective in controlling ants, beetles, bugs, flies, wasps, and many other insects is the caterpillar nematode (*Neoaplectana carpocapsae*). It carries a symbiotic bacterium (*Xenorhabdus nematophilus*), which multiplies rapidly in the host, killing most insects within 24 hours after initial contact. It may be obtained from Nematode Farm, Inc., 2617 San Pablo Avenue, Berkeley, CA 94702.

Use of Limonoid Sprays

Limonoids are bitter substances found in the rinds, seeds, and juice of citrus fruits (especially grapefruit). If the rinds and seeds of two or three fruits are ground up, soaked overnight in a pint of water, and the solid material is strained out, the liquid may then be sprayed on plants. The bitter principle apparently stops or reduces the feeding of larvae on the foliage. In experiments, limonoid sprays have proved effective against corn earworm, fall armyworm, tobacco budworm, and pink bollworm, but undoubtedly will deter many other pests as well.

Use of Liquefied Pest Sprays

Jeff Cox, an editor of Rodale's *Organic Gardening* magazine, called attention to this method of pest control in the magazine in October 1976, and again in May 1977. Insect pests or slugs are gathered in small quantities and liquefied with a little water in a blender. The material is then further diluted with water and sprayed throughout the infested area. It is not known why spraying with "bug juice" is effective against pests. It is known, however, that virtually all organisms harbor viruses. It has been theorized that even the inactive viruses carried by healthy insects and slugs may somehow be activated in the process of liquefaction. The viruses would be spread throughout an entire yard or farm if all parts of the area were sprayed. Most viruses are highly specific, generally attacking a single species of organism. M. Sipe, a Florida entomologist who recommended the "bug juice" technique, also suggested that the odor of the liquefied insects possibly attracts their predators and parasites or that the insects' distress **pheromones** (naturally produced insect chemicals that influence sexual or other behavior) are released by the blender, with the pheromones acting as an insect repellent. Possibly the observed effects of spraying "bug juice" are the result of a combination of viruses, predator attraction, and repellent pheromones. Sipe warns that if one tries this method of pest control, care should be taken to use only pest species and only those that are doing significant damage. Failure to heed this warning could disrupt the activities of natural predators and other natural controls present. This approach needs extended testing and investigation of its safety for use by humans, but test results over the past 20 years in various areas of North America have yielded impressive results with no evidence of harm to humans or beneficial organisms.

Use of Resistant Varieties

Many plants may kill or inhibit disease fungi or bacteria with chemicals known as *phytoalexins.* Phytoalexins are synthesized at the point of attack or invasion by the pathogen and are toxic to the fungus or bacterium. In selecting for improved fruit quality, vigor of growth, or other desirable characteristics, horticulturists in the past have sometimes unknowingly bred out a plant's capacity to produce certain phytoalexins, although general vigor is usually accompanied by disease resistance. Now that this aspect of a plant's defense mechanisms is known, breeders are concentrating on developing varieties capable of producing phytoalexins against various fungi, bacteria, and even nematodes. Several tomato varieties, for example, are listed as being *VFN.* The letters *V* and *F* indicate a resistance to *Verticillium* and *Fusarium* (common pathogenic fungi), while the letter *N* denotes a resistance to *root-knot nematodes.*

Other aspects of plant disease resistance include thick cuticles; the secretion of gums, resins, and other metabolic products that may interfere with fungal and bacterial spore germination; and the presence within all the cells of the plant of chemical compounds toxic to pathogens.

Interplanting with Plants That Produce Natural Insecticides or Substances Offensive to Pests

Many plant species produce substances that repel a significant number of pests, but none produce anything that repels all pests. Among the best-known plant producers of insect repellents are marigolds; garlic; and members of the Mint Family, such as pennyroyal, peppermint, basil, and lavender. An expanded discussion of this subject is given in Appendix 3.

SPECIFIC CONTROLS

Weeds

In 1974, the Weed Science Society of America published a special committee report (*Weed Science* 22: 490–95) on the biological control of weeds, summarizing the status of projects on the biological control of weeds with insects and plant pathogens in the United States and Canada. Table A2.1 is condensed from that report and supplemented with additional information. Many other biological controls for these and other weeds are currently under investigation.

Insects

The maintenance of ecological balance in nature includes a vast array of predator-prey relationships between animals, birds, insects, and other organisms. Specific biological controls for several types of insect pests, in addition to the general controls previously discussed, are given in Table A2.2.

COMPANION PLANTING

The "Additional Reading" list reveals that the literature on the chemical interactions among plants and also among plants and their consumers is already extensive. Despite the scientific evidence on the subject to date, however, a significant amount of the "backyard biological control" that is practiced today is based primarily on empirical information. Such information has been obtained from thousands of gardeners and farmers who have tried various techniques with their plantings and pest controls. As a result, they have come to conclusions that certain things work, while others do not, but they have not deliberately set up controlled experiments, nor have they necessarily understood the scientific basis for what

TABLE A2.1
Specific Weeds and Agents Involved in Their Biological Control

WEED	AGENT(S) OF BIOLOGICAL CONTROL
Alligator weed (*Alternanthera philoxeroides*)	Flea beetles (*Agasicles hygrophila*)
Bladder campion (*Silene cucubalus*)	Tortoise beetle (*Cassida hemisphaerica*)
Brazil peppertree (*Schinus terebinthifolius*)	Weevil (*Bruchus atronotatus*) and others
Brushweed (*Cassia surattensis*)	Imperfect fungus (*Cephalosporium* sp.)
Curly dock (*Rumex crispus*)	Rust (*Uromyces rumicis*)
Curse (*Clidemia hirta*)	Thrip (*Liothrips urichi*) and others
Cypress spurge (*Euphorbia cyparissias*)	Sphinx moth (*Hyles euphorbiae*)
Dalmatian toadflax (*Linaria dalmatica*)	Leaf miner (*Stagmatophora serratella*) and others
Emex (*Emex australis*)	Seed weevils (*Apion antiquum*) and others
Gorse (*Ulex europaeus*)	Seed weevils (*Apion ulicis*) and others
Halogeton (*Halogeton glomeratus*)	Casebearer (*Coleophora parthenica*) and others
Hawaiian blackberry (*Rubus penetrans*)	Sawflies (*Pamphilius sitkensis, Priophorus morio*) and others
Jamaica feverplant (puncture vine) (*Tribulus terrestris*)	Weevils (*Microlarinus* spp.)
Joint vetch (*Aeschynomene virginica*)	Imperfect fungus (*Colletotrichum gloeosporioides*)
Klamath weed (*Hypericum perforatum*)	Leaf beetles (*Chrysolina* spp.), buprestid beetle (*Agrilus hyperici*)
Lantana (*Lantana camara*)	Seed weevil (*Apion* sp.), ghost moth (*Hepialus* sp.), plume moth (*Platyptilia pusillidactyla*), hairstreaks (*Strymon* spp.), and others
Leafy spurge (*Euphorbia esula*)	Wood-boring beetle (*Oberea* sp.) and others
Mediterranean sage (*Salvia aethiopis*)	Snout beetles (*Phrydiuchus spp.*)
Milkweed vine (*Morrenia odorata*)	Oomycete fungus (*Phytophthora citrophthora*), rust (*Aecidium asclepiadinum*)
Prickly pear (*Opuntia* spp.)	Moth (*Cactoblastis cactorum*), cochineal insects (*Dactylopius* spp.), and others
Puncture vine (*see* Jamaica feverplant)	Weevils (*Microlarinus* spp.)
Scotch broom (*Cytisus scoparius*)	Seed weevil (*Apion fuscirostre*) and others
Skeleton weed (*Chondrilla juncea*)	Gall mite (*Aceria chondrillae*), root moth (*Bradyrrhoa gilveolella*), rust (*Puccinia chondrillina*), powdery mildews (*Erysiphe cichoracearum, Leveillula taurica*)
Spiny emex (*Emex spinosa*)	Seed weevil (*Apion antiquum*)
Tansy ragwort (*Senecio jacobaea*)	Seed fly (*Hylemya seneciella*), cinnabar moth (*Tyria jacobaeae*), leaf beetle (*Longitarsus jacobaeae*)
Thistles:	
Bull thistle (*Cirsium vulgare*)	Weevil (*Ceuthorrhynchidius horridus*), tortoise beetle (*Cassida rubiginosa*)
Canada thistle (*Cirsium arvense*)	Weevil (*Ceutorhynchus litura*), flea beetle (*Altica carduorum*), stem gall fly (*Urophora cardui*)
Diffuse knapweed (*Centaurea diffusa*)	Seed fly (*Urophora affinis*)
Italian thistle (*Carduus pycnocephalus*)	Flea beetles (*Rhinocyllus conicus, Psylliodes chalcomera*), weevil (*Ceutorhynchus trimaculatus*)
Milk thistle (*Silybum marianum*)	Flea beetle (*Rhinocyllus conicus*)
Musk thistle (*Carduus nutans*)	Weevils (*Ceutorhynchus trimaculatus, Ceuthorrhynchidius horridus, Rhinocyllus conicus*), flea beetle (*Psylliodes chalcomera*)
Perennial sowthistle (*Sonchus arvensis*)	Peacock fly (*Tephritis dilacerata*)
Plumeless thistle (*Carduus acanthoides*)	Tortoise beetle (*Cassida rubiginosa*), seed weevil (*Rhinocyllus conicus*), weevil (*Ceuthorrhynchidius horridus*)
Russian thistle (*Salsola kali* var. *tenuifolia*)	Casebearer (*Coleophora parthenica*) and others
Slenderflower thistle (*Carduus tenuiflorus*)	Weevil (*Ceutorhynchus trimaculatus*)
Spotted knapweed (*Centaurea maculosa*)	Seed fly (*Urophora affinis*) and others
Star thistle (*Centaurea nigrescens*)	Weevil (*Ceuthorrhynchidius horridus*)
Yellow star thistle (*Centaurea solstitialis*)	Seed fly (*Urophora siruna-seva*)
Water hyacinth (*Eichhornia crassipes*)	Weevils (*Neochetina bruchi, N. eichhorniae*), moth (*Sameodes albiguttalis*)
Water purslane (*Ludwigia palustris*)	Snout beetle (*Nanophyes* sp.)

TABLE A2.2
Specific Biological Controls for Several Types of Insect Pests

INSECT	CONTROL
Ants (about 8,000 spp. within the Superfamily Formicoidea)	Ants that carry aphids into trees and consume ripening fruits can be prevented from getting farther than the trunk by applying a band of sticky material around the trunk. A commercial preparation sold under the trade name of *Tanglefoot* is particularly effective. A water suspension of ground hot peppers (*Capsicum* spp.) used as a spray can act as an ant deterrent. *Caution:* Many ants are beneficial to a balanced ecology; they should not be decimated indiscriminately.
Grasshoppers (there are several families of grasshoppers, but the insects that usually constitute the most serious pests are species of *Melanoplus,* Family Acrididae)	In 1980, the Environmental Protection Agency permitted private companies to begin the mass culture of a protozoan, *Nosema locustae,* for use in controlling rangeland grasshoppers. Tests have shown that properly timed applications of spores mixed with wheat bran can reduce grasshopper populations by up to 50%.
Gypsy moths (*Porthetria dispar*)	Parasitic wasps (*Apanteles flavicoxis, A. indiensis*) imported from India lay their eggs in gypsy moth caterpillars and kill large numbers.
Japanese beetles (*Popillia japonica*)	The pathogenic bacterium *Bacillus popillae,* which is sold commercially, is specific for Japanese beetle larvae. It causes what is known as "milky spore disease" in the grubs while they are still in the soil, and it is very destructive. It is available from St. Gabriel Laboratories, 14540 John Marshall Hwy., Gainesville, VA 20155 (1-800-801-0061 or http://www.milkyspore.com)
Mealybugs (*Pseudococcus* spp.)	The small brown beetles called *crypts* (*Cryptolaemus montrouzieri*) effectively control mealybugs in greenhouses and also outdoors on apple, pear, peach, and citrus trees. Order from Rincon-Vitova Insectaries, Inc., P.O. Box 95, Oak View, CA 93022.
Mosquitoes (*Culex* spp., *Anopheles* spp., and others)	The bacterium *Bacillus thuringiensis* var. *israelensis* has proved to be very effective in destroying mosquito larvae. A fungus (*Lagenidium giganteum*) has also proved highly effective against mosquito larvae if the temperature is above 20°C (68°F). The bacterium is available from several sources, including Abbott Laboratories, Dept. 95-M, 1400 Sheridan Rd., N. Chicago, IL 60064; Sandoz, Inc., 480 Camino del Rio S., San Diego, CA 92108.
Red spider mites (*Tetranychus telarius*)	Predatory mites (*Phytolesius persimilis,* which works best when weather is not hot, and *Amblyseius californicus,* which is more effective in hot weather) effectively control populations of red spider mites.
White flies (*Trialeurodes vaporariorum*)	A minute wasp, *Encarsia formosa,* parasitizes white flies exclusively. The wasps have been known to be very effective in greenhouses. They are obtainable from White Fly Control Co., Box 986, Milpitas, CA 95035; Rincon-Vitova Insectaries, Inc., P.O. Box 95, Oak View, CA 93022. White flies are attracted to the color yellow. Large numbers of white flies are trapped when a yellow board is sprayed or painted with any sticky substance and placed in the vicinity of the pests.

they have observed. This does not mean that their observations are not useful or that they are invalid. In fact, such empirical observations have often been the inspiration for investigations and experiments by scientists. The scientific investigations have sometimes revealed that the empirical observations were biased or not carefully made or that erroneous conclusions had been drawn, but frequently, sound scientific bases for these observations have been uncovered.

Further insights into how plants inhibit or enhance the growth of others and into the nature of their resistance to disease or insect-repelling mechanisms continue to be discovered. Observations of such phenomena in the past have led organic gardeners and others to the practice of *companion planting,* which involves the interplanting of various crops and certain other plants in such a way that each species derives some benefit from the arrangement. The following companion planting list, based primarily on empirical information, appeared in the February 1977 issue of *Organic Gardening and Farming* magazine. It is included here with the permission of Rodale Press, Inc.

Table A2.3 is a list of combinations of vegetables, herbs, flowers, and weeds that are mutually beneficial, according to current reports of organic gardeners and to companion-planting traditions.

TABLE A2.3
Companion Plants

PLANT	COMPANIONS AND EFFECTS
Asparagus	Tomatoes, parsley, basil
Basil	Tomatoes (improves growth and flavor); said to dislike rue; repels flies and mosquitoes
Beans	Potatoes, carrots, cucumbers, cauliflower, cabbage, summer savory, most other vegetables and herbs; around house plants when set outside
Beans (bush)	Sunflowers (beans like partial shade, sunflowers attract birds and bees), cucumbers (combination of heavy and light feeders), potatoes, corn, celery, summer savory
Beets	Onions, kohlrabi
Borage	Tomatoes (attracts bees, deters tomato worm, improves growth and flavor), squash, strawberries
Cabbage Family	Potatoes, celery, dill, chamomile, sage, thyme, mint, pennyroyal, rosemary, lavender, beets, onions. Aromatic plants deter cabbage worms
Carrots	Peas, lettuce, chives, onions, leeks, rosemary, sage, tomatoes
Catnip	Plant in borders; protects against flea beetles
Celery	Leeks, tomatoes, bush beans, cauliflower, cabbage
Chamomile	Cabbage, onions
Chervil	Radishes (improves growth and flavor)
Chives	Carrots; plant around base of fruit trees to discourage insects from climbing trunk
Corn	Potatoes, peas, beans, cucumbers, pumpkin, squash
Cucumbers	Beans, corn, peas, radishes, sunflowers
Dill	Cabbage (improves growth and health), carrots
Eggplant	Beans
Fennel	Most plants are supposed to dislike it
Flax	Carrots, potatoes
Garlic	Roses and raspberries (deters Japanese beetle); with herbs to enhance their production of essential oils; plant liberally throughout garden to deter pests
Horseradish	Potatoes (deters potato beetles); around plum trees to discourage curculios
Lamb's quarters	Nutritious edible weed; allow to grow in modest amounts in the corn
Leek	Onions, celery, carrots
Lettuce	Carrots and radishes (lettuce, carrots, and radishes make a strong companion team), strawberries, cucumbers
Lovage	Plant here and there in garden
Marigolds	The workhorse of pest deterrents; keeps soil free of nematodes; discourages many insects; plant freely throughout garden
Marjoram	Here and there in garden
Mint	Cabbage family; tomatoes; deters cabbage moth
Mole plant	Deters moles and mice if planted here and there throughout the garden
Nasturtium	Tomatoes, radishes, cabbage, cucumbers; plant under fruit trees. Deters aphids and pests of cucurbits
Onion	Beets, strawberries, tomato, lettuce (protects against slugs), beans (protects against ants), summer savory
Parsley	Tomato, asparagus
Peas	Squash (when squash follows peas up trellis), plus grows well with almost any vegetable; adds nitrogen to the soil
Petunia	Protects beans; beneficial throughout garden
Pigweed	Brings nutrients to topsoil; beneficial growing with potatoes, onions, and corn; keep well thinned
Potato	Horseradish, beans, corn, cabbage, marigold, limas, eggplant (as trap crop for potato beetle)
Pot marigold	Helps tomato, but plant throughout garden as deterrent to asparagus beetle, tomato worm, and many other garden pests
Pumpkin	Corn
Radish	Peas, nasturtium, lettuce, cucumbers; a general aid in repelling insects
Rosemary	Carrots, beans, cabbage, sage; deters cabbage moth, bean beetles, and carrot fly
Rue	Roses and raspberries; deters Japanese beetle; keep it away from basil
Sage	Rosemary, carrots, cabbage, peas, beans; deters some insects
Southernwood	Cabbage; plant here and there in garden

TABLE A2.3

Companion Plants

PLANT	COMPANIONS AND EFFECTS
Soybeans	Grows with anything, helps everything
Spinach	Strawberries
Squash	Nasturtium, corn
Strawberries	Bush beans, spinach, borage, lettuce (as a border)
Summer savory	Beans, onions; deters bean beetles
Sunflower	Cucumbers
Tansy	Plant under fruit trees; deters pests of roses and raspberries; deters flying insects; also Japanese beetles, striped cucumber beetles, squash bugs, deters ants
Tarragon	Good throughout garden
Thyme	Here and there in garden; deters cabbage worm
Tomato	Chives, onion, parsley, asparagus, marigold, nasturtium, carrot, limas
Turnip	Peas
Valerian	Good anywhere in garden
Wormwood	As a border, keeps animals from the garden
Yarrow	Plant along borders, near paths, near aromatic herbs; enhances essential oil production of herbs

SOME SOURCES OF HERB PLANTS AND SEEDS

China Herb Co., 428 Soledad, Salinas, CA 93901
Cottage Herbs, P.O. Box 100, Troy, ID 83871
De Giorgi Co., 6011 N St., Omaha, NE 68117
Fragrant Fields, Dongola, IL 62926
Herbs-Liscious, 1702 S. Sixth St., Marshalltown, IA 50158
Hsu's Ginseng Enterprises, P.O. Box 509, Wausau, WI 54402
Jude Herbs, Box 56360, Huntington Station, NY 11746
Otto Richter and Sons, Box 260, Goodwood, Ontario, LOC 1A0
PG Nursery, R18, Box 470, Bedford, IN 47421
Putney Nursery, Putney, VT 05346
Rawlinson Garden Seed, 269 College Rd., Truro, Nova Scotia B2N 2P6
Sanctuary Seeds, 2388 West Fourth Avenue, Vancouver, British Columbia V6K 1P1
Sea Island Savory Herbs, 5920 Chisolm, John's Island, SC 29455
Shoestring Seeds, P.O. Box 2261, Martinsville, VA 24113
Story House Herb Farm, Route 7, Box 246, Murray, KY 42071
Sunnybrook Farms Nursery, Box 6, Chesterland, OH 44026
Sunshine Herbs and Flowers, Rt. 1, Box 234, Comer, GA 30629
Thompson and Morgan, Inc., P.O. Box 1308, Jackson, NJ 08527
The Thyme Garden, 20546-0 Alsea Hwy., Alsea, OR 97324
Wildwood Herbal, P.O. Box 746, Albemarle, NC 28002
Willhite Seed Company, Box 23, Poolville, TX 76076

ADDITIONAL READING

Ananthakrishnan, T. N. 1998. *Technology in biological control.* Enfield, NH: Science Pubs.

Barboso, P. 1998. *Conservation biological control.* San Diego, CA: Academic Press.

Boyland and Kuykendall (Eds.). 1998. *Plant-microbe interactions and biological control.* New York: Dekker, Marcel Press.

Carson, R. 1999. *Silent spring.* Boston: Houghton Mifflin.

Cook, R. J., and K. F. Baker. 1983. *Nature and practice of biological control of plant pathogens.* St. Paul, MN: American Phytopathological Society.

Goeden, R. D., et al. 1981. Natural and applied control of insects by protozoa. *Annual Review of Entomology* 26: 49–73.

Hawkins, B. A., and H. V. Cornell (Eds.). 1999. *Theoretical approaches to biological control.* New York: Cambridge University Press.

Hoy, M., and G. L. Cunningham. 1983. *Biological control of pests by mites: Proceedings of a conference.* Oakland, CA: Agricultural and Natural Resources, University of California.

Huffaker, C. B., and A. P. Gutierrez (Eds.). 1998. *Ecological entomology,* 2d ed. New York: John Wiley and Sons.

Jutsum, A. R., and R. F. S. Gordon (Eds.). 1989. *Insect pheromones in plant protection.* New York: John Wiley and Sons.

Pickett, C. H. 1998. *Enhancing biological control: Habitat management to promote natural enemies of agricultural pests.* Berkeley, CA: University of California Press.

Rechcigl, J. E., and N. A. Rechcigl. 1999. *Biological and biotechnological control of insect pests.* Los Angeles, CA: Lewis.

Rice, E. L. 1995. *Biological control of weeds and plant diseases: Advances in applied allelopathy.* Norman, OK: University of Oklahoma Press.

Appendix 3

Useful and Poisonous Plants, Fungi, and Algae

WILD EDIBLE PLANTS, FUNGI, AND ALGAE

Words of Caution

At least some parts of literally thousands of native and naturalized plants, fungi, and algae have been used for food and other purposes by Native Americans. Many were also used by the immigrants who came later from other areas of the world. A representative compilation of wild edible plants and fungi is shown in Table A3.1.

The list of plants and fungi in Table A3.1 has been compiled from a variety of sources; the author has had opportunities to sample only a fraction of these himself and cannot confirm the edibility of all of the organisms listed. *The reader is cautioned to be certain of the identity of a plant or fungus before consuming any part of it.* Cow parsnip (*Heracleum lanatum*) and water hemlocks (*Cicuta* spp.), for example, resemble each other in general appearance, but although cooked roots of cow parsnip have been used for food for perhaps many centuries, those of water hemlocks are very poisonous and have caused many human fatalities.

As indicated in Chapter 21, many species of organisms are now on rare and endangered species lists, and a number of them will become extinct within the next few years. Although the wild edible plants and fungi discussed here may not presently be included in such lists, it might not take much indiscriminate gathering to endanger their existence as well. Because of this, one should exercise the following rule of thumb: *Never reduce a population of wild plants by more than 10% when collecting them for any purpose! If the population consists of less than 10 plants, do not disturb it.*

TABLE A3.1

Wild Edible Plants and Fungi

ORGANISM	SCIENTIFIC NAME	USES
Amaranth	*Amaranthus* spp.	Young leaves used like spinach; seeds ground with others for flour
Arrow grass	*Triglochin maritima*	Seeds parched or roasted (***Caution: All other plant parts are poisonous***)
Arrowhead	*Sagittaria latifolia*	Tubers used like potatoes
Balsamroot	*Balsamorhiza* spp.	Whole plant edible, especially when young, either raw or cooked
Basswood	*Tilia* spp.	Fruits and flowers ground together to make a paste that can serve as a chocolate substitute; winter buds edible raw; dried flowers used for tea
Bearberry (Kinnikinik)	*Arctostaphylos uva-ursi*	Berries are edible but much more palatable when cooked
Bedstraw (Cleavers)	*Galium aparine*	Roasted and ground seeds make good coffee substitute
Beechnuts	*Fagus grandifolia*	Seeds used as nuts; oil extracted from seeds for table use
Biscuit root	*Lomatium* spp.	Roots eaten raw or dried and ground into flour; seeds edible raw or roasted
Bitterroot	*Lewisia rediviva*	Outer coat of the bulbs should be removed to eliminate the bitter principle; bulbs are then boiled or roasted
Blackberry (wild)	*Rubus* spp.	Fruits edible raw, in pies, jams, and jellies

TABLE A3.1 Wild Edible Plants and Fungi

ORGANISM	SCIENTIFIC NAME	USES
Black walnut	*Juglans nigra*	Nut meats edible
Bladder campion	*Silene cucubalus*	Young shoots (less than 5 cm tall) cooked as a vegetable
Blueberry	*Vaccinium* spp.	Fruits edible raw, frozen, and in pies, jams, and jellies
Bracken fern	*Pteridium aquilinum*	Young uncoiling leaves ("fiddleheads") cooked like asparagus; rhizomes also edible but usually tough (***Caution: Evidence indicates that frequent consumption of bracken fern can cause cancer of the intestinal tract***)
Broomrape	*Orobanche* spp.	Entire plant eaten raw or roasted
Bulrush (Tule)	*Scirpus* spp.	Roots and young shoot tips edible raw or cooked; pollen and seeds also edible
Butternut	*Juglans cinerea*	Nut meats edible
Camas	*Camassia quamash*	Roasted bulbs considered a delicacy
Caraway	*Carum carvi*	Young leaves in salads; seeds for flavoring baked goods and cheeses
Cattail	*Typha* spp.	Copious pollen produced by flowers in early summer is rich in vitamins and can be gathered and mixed with flour for baking; rhizomes can be cooked and eaten like potatoes
Chicory	*Cichorium intybus*	Leaves eaten raw or cooked; dried, ground roots (roasted) make good coffee substitute
Chokecherry	*Prunus virginiana*	Fruits make excellent jelly or can be cooked with sugar for pies and cobblers
Clover	*Trifolium* spp.	Roots edible
"Coffee" (wild)	*Triosteum* spp.	Berries dried and roasted make good coffee substitute
Common chickweed	*Stellaria media*	Plants cooked as a vegetable
Corn lily	*Clintonia borealis*	Youngest leaves can be used as a cooked vegetable
Cow parsnip	*Heracleum lanatum*	Roots and young stems cooked (***Caution: Be certain of identity; some other members of the family that are similar in appearance to cow parsnip are highly toxic***)
Cowpea	*Vigna sinensis*	"Peas" and young pods cooked as a vegetable (plant naturalized in southern U.S.)
Crab apple	*Pyrus* spp.	Jelly made from fruits
Cranberry (wild, bog)	*Vaccinium* spp.	Berries edible cooked, preserved, or in drinks; adding a small amount of salt while cooking significantly reduces amount of sugar needed to counteract acidity
Crowberry	*Empetrum nigrum*	Fruits should first be frozen, then cooked with sugar
Dandelion	*Taraxacum* sp. aff.	Leaves rich in vitamin A; dried roots make good coffee substitute; wine made from young flowers
Dock	*Rumex* spp.	Leaves cooked like spinach; tartness of leaves varies from species to species and sometimes from plant to plant—tart forms should be cooked in two or three changes of water
Douglas fir	*Pseudotsuga menziesii*	Cambium and young phloem edible; tea made from fresh leaves
Elderberry	*Sambucus* spp.	Fresh flowers used to flavor batters; fruits used in pies, jellies, wine (***Caution: Other parts of the plant are poisonous***)
Evening primrose	*Oenothera hookeri, O. biennis,* and others	Young roots cooked
Fairy bells	*Disporum trachycarpum*	Berries can be eaten raw
Fennel	*Foeniculum vulgare*	Leaf petioles eaten raw or cooked
Ferns	Most (but not all) spp.	Young coiled fronds (fiddleheads) may be cooked as a vegetable
Fireweed	*Epilobium angustifolium*	Young shoots and leaves boiled as a vegetable
Fritillary	*Fritillaria* spp.	Cooked bulbs are edible

TABLE A3.1
Wild Edible Plants and Fungi

ORGANISM	SCIENTIFIC NAME	USES
Ginger (wild)	*Asarum* spp.	Rhizomes can be used as substitute for true ginger
Gooseberry	*Ribes* spp.	Berries eaten cooked, dried, or raw; make excellent jelly
Grape (wild)	*Vitis* spp.	Berries usually tart but can be eaten raw; make good jams and jellies
Grass	Many genera and species	Seeds of most can be made into flour; rhizomes of many perennial species can be dried and ground for flour
Greenbrier	*Smilax* spp.	Roots dried and ground; refreshing drink made with ground roots, sugar, and water
Groundnut	*Apios americana*	Tubers cooked like potatoes
Hawthorn	*Crataegus* spp.	Fruits edible raw and in jams and jellies
Hazelnut	*Corylus* spp.	Nuts eaten raw or roasted
Hickory	*Carya* spp.	Nuts edible
Highbush cranberry	*Viburnum trilobum*	Fruits make excellent jellies and jams
Huckleberry	*Vaccinium* spp.	Berries eaten raw or in jams and jellies
Indian paintbrush	*Castilleja* spp.	Flowers of many species edible (***Caution: On certain soils, plants absorb toxic quantities of selenium***)
Indian pipe	*Monotropa* spp.	Whole plant edible raw or cooked
June berries	*Amelanchier* spp.	Fruit edible fresh, dried, or preserved
Juniper	*Juniperus* spp.	"Berries" dried, ground, and made into cakes
Labrador tea	*Ledum* spp.	Tea made from young leaves
Lamb's quarters	*Chenopodium album*	Leaves and young stems used as cooked vegetable
Licorice	*Glycyrrhiza lepidota, G. glabra*	Roots edible raw or cooked
Mallow	*Malva* spp.	Leaves and young stems used as vegetable (use only small amounts at one time)
Manzanita	*Arctostaphylos* spp.	Berries eaten raw, in jellies or pies, or made into "cider" (***Caution: Raw berries can be somewhat indigestible***)
Maple	*Acer* spp.	Sugar maples (*Acer saccharum*) well known for the sugar content of the early spring sap; other species (e.g., box elder—*A. negundo,* bigleaf maple—*A. macrophyllum*) also contain usable sugars in their early spring sap
Mariposa lily	*Calochortus* spp.	Bulbs edible raw or cooked
Mayapple	*Podophyllum peltatum*	Fruit good raw or cooked (***Caution: Other parts of the plant are poisonous***)
Maypops	*Passiflora incarnata*	Fruits edible raw or cooked
Miner's lettuce	*Claytonia perfoliata*	Leaves eaten raw as a salad green
Mint	*Mentha arvensis* and others	Leaves of several mints used for teas
Mormon tea	*Ephedra* spp.	Tea from fresh or dried leaves (add sugar to offset bitterness); seeds for bitter meal
Mulberry	*Morus* spp.	Fruits of the red mulberry (*M. rubra*) are used raw and in pies and jellies; fruits of white mulberry (*M. alba*) edible but insipid
Mushrooms	Many genera and species	***Utmost caution should be exercised in identifying mushrooms before consuming them;*** although poisonous species are in the minority, they are common enough; edible forms that are relatively easy to identify include morels (*Morchella esculenta*), most puffballs (*Lycoperdon* spp.), and inky cap mushrooms (*Coprinus* spp.)
Mustard	*Brassica* spp.	Leaves used as vegetable; condiment made from ground seeds
Nettles	*Urtica* spp.	Leaves and young stems cooked like spinach
New Jersey tea	*Ceanothus americanus*	Tea from leaves

TABLE A3.1
Wild Edible Plants and Fungi

ORGANISM	SCIENTIFIC NAME	USES
Nutgrass	*Cyperus esculentus* and others	Tubers can be eaten raw
Oak	*Quercus* spp.	Acorns were ground for flour and widely used by native North Americans; all contain bitter tannins that must be leached out before use
Onion (wild)	*Allium* spp.	Bulbs edible raw or cooked
Orach	*Atriplex patula* and others	Leaves and young stems cooked as a vegetable
Oregon grape	*Berberis aquifolia, B. nervosa*	Berries edible raw or preserved
Ostrich fern	*Matteuccia struthiopteris*	Young coiled fronds cooked as a vegetable
Pawpaw	*Asimina triloba*	Fruit edible raw or cooked
Pennycress	*Thlaspi arvense*	Young leaves are edible raw
Peppergrass	*Lepidium* spp.	Immature fruits add zest to salads; seeds spice up meat dressings
Persimmon	*Diospyros virginiana*	Fully ripened fruits can be eaten raw or cooked
Pickerel weed	*Pontederia cordata*	Fruits edible raw or dried
Pigweed (*see* Amaranth)		
Pines	*Pinus* spp.	Cambium, young phloem, and seeds edible; tea from fresh needles rich in vitamin C
Pipsissewa	*Chimaphila umbellata*	Drink made from boiled roots and leaves (cool after boiling)
Plantain	*Plantago* spp.	Young leaves eaten in salads or as cooked vegetable
Poke	*Phytolacca americana*	Fresh young shoots boiled like asparagus (***Caution: Older parts of plant poisonous***)
Prairie turnip	*Psoralea esculenta*	Turniplike roots cooked like potatoes
Prickly pear	*Opuntia* spp.	Fruits and young stems peeled and eaten raw or cooked
Psyllium	*Plantago ovata*	Seed husks widely used as a bulking laxative
Purple avens	*Geum rivale*	Liquid from boiled root has chocolate-like flavor
Purslane	*Portulaca oleracea*	Leaves and stems cooked like spinach
Quackgrass	*Elytrigia repens*	Noxius weed whose rhizomes can be used as emergency food
Raspberry (wild)	*Rubus* spp.	Fruits edible raw or in pies, jams, and jellies
Redbud	*Cercis* spp.	Flowers used in salads; cooked young pods edible
River-beauty	*Epilobium latifolium*	Young shoots and fleshy leaves can be cooked as a vegetable
Rose (wild)	*Rosa* spp.	Fruits (hips) exceptionally rich in vitamin C; hips can be eaten raw, pureed, or candied
Salal	*Gaultheria procumbens, G. shallon*	Ripe berries edible raw, dried, or preserved
Salmonberry	*Rubus spectabilis*	Fruits edible raw, dried, or cooked
Salsify	*Tragopogon* spp.	Roots edible raw or cooked
Saltbush	*Atriplex* spp.	Seeds nutritious (***Caution: On certain soils, plants can absorb toxic amounts of selenium***)
Sassafras	*Sassafras albidum*	Tea from roots (***Caution: Large quantities have narcotic effect***); leaves and pith used for Louisiana filé
Serviceberry	*Amelanchier* spp.	All fruits edible (mostly bland)
Sheep sorrel	*Rumex acetosella*	Raw leaves have a pleasant sour taste; leaves can be used as seasoning in other dishes

TABLE A3.1

Wild Edible Plants and Fungi

ORGANISM	SCIENTIFIC NAME	USES
Shepherd's purse	*Capsella bursa-pastoris*	Leaves cooked as vegetable; seeds eaten parched or ground for flour
Showy milkweed	*Asclepias speciosa*	Flowers eaten raw or cooked; young shoots cooked
Silverweed	*Potentilla anserina*	Cooked roots edible
Soap plant	*Chlorogalum pomeridianum*	Bulbs slow-baked and eaten like potatoes after fibrous outer coats are removed
Solomon's seal	*Polygonatum* spp.	Rootstocks dried and ground for bread flour
Sorrel	*Oxalis* spp.	Leaves mixed in salads
Spatterdock	*Nuphar polysepalum*	Seeds placed on hot stove burst like popcorn and are edible as such; peeled tubers eaten boiled or roasted
Speedwell	*Veronica americana* and others	Leaves and stems used in salads
Spring beauty	*Claytonia* spp.	Bulbs edible raw or roasted
Strawberry (wild)	*Fragaria* spp.	Fruits superior in flavor to cultivated varieties
Sunflower	*Helianthus annuus*	Seeds eaten raw or roasted; seeds yield cooking oil
Sweet cicely	*Osmorhiza* spp.	Roots have aniselike flavor
Sweet flag	*Acorus calamus*	Young shoots used in salads; roots candied
Thimbleberry	*Rubus parviflorus*	Fruits edible raw, cooked, dried, or preserved; dried leaves used for tea
Thistle	*Cirsium* spp.	Peeled stems edible; roots edible raw or roasted
Vetch	*Vicia* spp.	Tender green pods edible baked or boiled
Watercress	*Nasturtium officinale*	Leaves edible raw in salads or cooked as a vegetable
Waterleaf	*Hydrophyllum* spp.	Young shoots raw in salads; shoots and roots cooked as vegetable
Water plantain	*Alisma* spp.	The bulblike base of the plant is dried and then cooked
Water shield	*Brasenia schreberi*	Tuberlike roots are peeled and then dried to be ground for flour or boiled
Winter cress	*Barbarea* spp.	Leaves and young stem edible as cooked vegetable
Yarrow	*Achillea lanulosa*	Plant dried and made into broth (***Caution: The closely related and widespread European yarrow—****A. millefolium****—is somewhat poisonous***)
Yellow pond lily (*see* Spatterdock)		
Yew	*Taxus* spp.	Bright red pulpy part of berries edible (***Caution: Seeds and leaves are poisonous***)

POISONOUS PLANTS AND FUNGI

Literally thousands of plants and fungi contain varying amounts of poisonous substances. In many instances, the poisons are not present in sufficient quantities to cause adverse effects in humans when only moderate contact or consumption is involved and cooking may destroy or dissipate the substance. Some plants and fungi have substances that produce toxic effects in some organisms but not in others. Ordinary onions (*Allium cepa*), for example, occasionally poison horses or cattle yet are widely used for human food, and poison ivy (*Toxicodendron radicans*) or poison oak (*Toxicodendron diversilobum*) produce dermatitis in some individuals but not in others. Table A3.2 and Table A3.3 include plants and fungi that are native to, or cultivated in, the United States and Canada.

TABLE A3.2

Plants and Fungi Known to Have Caused Human Fatalities

ORGANISM	SCIENTIFIC NAME	POISONOUS PARTS
Angel's trumpet	*Datura suaveolens*	All parts, especially seeds and leaves
Azalea	*Rhododendron* spp.	Leaves and flowers (however, poisoning is rare)
Baneberry	*Actaea* spp.	Berries and roots
Belladonna	*Atropa belladonna*	All parts, especially fruits and roots
Black cherry	*Prunus serotina*	Bark, seeds, leaves (***Caution: Seeds of most cherries, plums, and peaches contain a poisonous principle***)
Black locust	*Robinia pseudo-acacia*	Seeds, leaves, inner bark
Black snakeroot	*Zigadenus* spp.	Bulbs
Buckeye	*Aesculus* spp.	Seeds, shoots, flowers, leaves, roots; even the honey that bees make from buckeye flowers is poisonous
Caladium	*Caladium* spp.	All parts
Carolina jessamine	*Gelsemium sempervirens*	All parts; even visiting honey bees can be poisoned
Castor bean	*Ricinus communis*	Seeds
Chinaberry	*Melia azedarach*	Fruits and leaves
Daphne	*Daphne mezereum*	All parts
Death angel (Fly agaric)	*Amantia muscaria*	All parts (as little as one bite can be fatal)
Death camas (*see* Black snakeroot)		
Destroying angel	*Amanita verna*	All parts (as little as one bite can be fatal)
Dieffenbachia (Dumb cane)	*Dieffenbachia* spp.	All parts
Duranta	*Duranta repens*	Berries
Dutchman's breeches	*Dicentra cucullaria*	All parts
English ivy	*Hedera helix*	Berries and leaves
False hellebore	*Veratrum* spp.	All parts
Foxglove	*Digitalis purpurea*	All parts
Gloriosa lily	*Gloriosa superba* and other *Gloriosa* spp.	All parts (especially tubers)
Golden chain	*Laburnum anagyroides*	Seeds and flowers
Jequirity bean	*Abrus precatorius*	Seeds
Jimson weed	*Datura stramonium* and other *Datura* spp.	All parts, especially seeds
Lantana	*Lantana camara*	Unripe fruits
Lily of the valley	*Convallaria majalis*	All parts
Lobelia	*Lobelia* spp.	All parts
Mistletoe	*Phoradendron* spp.	Berries
Monkshood	*Aconitum* spp.	All parts
Moonseed	*Menispermum canadense*	Fruits
Mountain laurel	*Kalmia latifolia*	Leaves, shoots, flowers
Mushrooms	Many genera and species, especially *Amanita* spp.	All parts
Nightshade	*Solanum* spp.	Unripened fruits (***Caution: A poisonous principle is produced in common potato [Solanum tuberosum] tubers exposed to light long enough for the skins to turn green or greenish***)
Oleander	*Nerium oleander*	All parts
Poison hemlock	*Conium maculatum*	All parts
Poke	*Phytolacca americana*	Roots and mature stems
Rhododendron (*see* Azalea)		

TABLE A3.2
Plants and Fungi Known to Have Caused Human Fatalities

ORGANISM	SCIENTIFIC NAME	POISONOUS PARTS
Rhubarb	*Rheum rhaponticum*	Leaf blades (***Caution: Although young petioles are widely eaten, dangerous accumulations of a poisonous substance can occur in leaf blades***)
Rubber vine	*Cryptostegia grandiflora*	All parts
Sandbox tree	*Hura crepitans*	Milky sap and seeds
Tansy	*Tanacetum vulgare*	Leaves, flowers
Tung tree	*Aleurites fordii*	All parts, especially seeds
Water hemlock	*Cicuta* spp.	Roots
White snakeroot	*Eupatorium rugosum*	All parts
Yellow oleander	*Thevetia peruviana*	All parts, especially fruits
Yew	*Taxus* spp.	All parts except "berry" pulp

TABLE A3.3
Other Organisms Producing Significant Quantities of Poisonous Substances

ORGANISM	SCIENTIFIC NAME	POISONOUS PARTS
Amaryllis	*Amaryllis* spp.	Bulbs
Autumn crocus	*Colchicum autumnale*	All parts
Bittersweet	*Celastrus scandens*	Seeds
Bleeding hearts	*Dicentra* spp.	All parts
Bloodroot	*Sanguinaria canadensis*	All parts
Blue cohosh	*Caulophyllum thalictroides*	Fruits, leaves
Boxwood	*Buxus sempervirens*	Leaves
Buckthorn	*Rhamnus* spp.	Fruits
Bushman's poison	*Acokanthera* spp.	All parts
Buttercup	*Ranunculus* spp.	All parts; toxicity varies from species to species; mostly cause blistering
Buttonbush	*Cephalanthus occidentalis*	Leaves
Caladium	*Caladium* spp.	All parts
Carolina jessamine	*Gelsemium sempervirens*	All parts (even visiting honey bees can be poisoned)
Chincherinchee	*Ornithogalum thyrsoides*	All parts
Crown of thorns	*Euphorbia milii*	Milky latex
Culver's root	*Veronicastrum virginicum*	Root
Daffodil	*Narcissus* spp.	Bulbs
Desert marigold	*Baileya radiata*	All parts
Fly poison	*Amianthemum muscaetoxicum*	Leaves, roots
Four-o'clock	*Mirabilis jalapa*	Seeds, roots
Goldenseal	*Hydrastis canadensis*	Rhizomes, leaves
Holly	*Ilex aquifolium*	Berries

TABLE A3.3

Other Organisms Producing Significant Quantities of Poisonous Substances

ORGANISM	SCIENTIFIC NAME	POISONOUS PARTS
Horse chestnut	*Aesculus hippocastanum*	Seeds, flowers, leaves
Hyacinth	*Hyacinthus* spp.	Bulbs
Hydrangea	*Hydrangea* spp.	Buds, leaves
Jack-in-the-pulpit	*Arisaema triphyllum*	Roots, leaves
Jessamine	*Cestrum* spp.	Leaves, young stems
Jonquil (*see* Daffodil)		
Karaka nut	*Corynocarpus laevigata*	Seeds
Kentucky coffee tree	*Gymnocladus dioica*	Fruits
Larkspur	*Delphinium* spp.	Young plants, seeds
Lignum vitae	*Guaiacum officinale*	Fruits
Locoweed	*Astragalus* spp.	Location of poisonous principles varies from species to species; plants more of a problem for livestock than for humans
Lupine	*Lupinus* spp.	Location of poisonous principles varies from species to species; primarily in pods and seeds
Marijuana	*Cannabis sativa*	Resins secreted by glandular hairs among flowers
Mayapple	*Podophyllum peltatum*	All parts except ripe fruits
Mescal bean	*Sophora secundiflora*	Seeds
Narcissus (*see* Daffodil)		
Ngaio	*Myoporum laetum*	Leaves
Opium poppy	*Papaver somniferum*	Unripe fruits
Philodendron	*Philodendron* spp.	Leaves, stems
Pittosporum	*Pittosporum* spp.	Fruits, leaves, stems
Poinsettia	*Euphorbia pulcherrima*	Milky latex
Poison hemlock	*Conium maculatum*	All parts
Poison ivy	*Toxicodendron radicans*	Leaves
Poison oak	*Toxicodendron diversilobum*	Leaves
Poison sumac	*Toxicodendron vernix*	Leaves
Poke	*Phytolacca americana*	Roots, leaves, stems (uncooked fruits may be slightly poisonous)
Prickly poppy	*Argemone* spp.	Seeds, leaves
Privet	*Ligustrum vulgare*	Fruits
Rhododendron	*Rhododendron* spp.	All parts
Sneezeweed	*Helenium* spp.	All parts
Snow-on-the-mountain	*Euphorbia marginata*	Milky latex
Squirrel corn	*Dicentra canadensis*	All parts
Star-of-Bethlehem	*Ornithogalum umbellatum*	All parts
Sweet pea	*Lathyrus* spp.	Seeds
Tobacco	*Nicotiana tabacum*	Leaves (when eaten)
Water hemlock	*Cicuta* spp.	All parts

MEDICINAL PLANTS, FUNGI, AND ALGAE

Many modern medicines prescribed by Western doctors are synthetic or include synthetic substances, but a significant number still contain drugs naturally produced by plants, fungi, and algae. As recently as 50 years ago, the vast majority of medicines used in the treatment of human diseases and ailments were plant or fungus produced, and both Chinese and Indian Ayurvedic medicine, practiced for thousands of years, are still today largely plant based. Plants, fungi, and algae rarely, if ever, produce medicinal drugs in pure form, but the mixtures of drugs, vitamins, and minerals found in them tend to work together synergistically. This often results in the combinations being more effective than they would be if the substances were each used medicinally in isolated form. Oriental medicinal practices carry the synergistic aspects further by combining up to several different organisms in prescriptions for given ailments. It should be noted that the amounts and potencies of medicinal drugs produced by plants, fungi, or algae can vary considerably from population to population or even from organism to organism. Accordingly, many reputable purveyors of plant, fungal, or algal medicines routinely have batches tested to ensure standardization and quality for the user. Table A3.4 includes a sampling of plants, fungi, and algae associated with past and some present medicinal uses. Some of the drugs concerned are prescribed for specific ailments by modern medical practitioners, while others are a part of folk medicine still practiced in rural areas. ***Caution:** Do not use any of the plants, fungi, or algae listed here for medicinal purposes without consulting a reputable, qualified health practictioner.*

TABLE A3.4
Plants, Fungi, and Algae Associated with Medicinal Uses

ORGANISM	SCIENTIFIC NAME	USES
Agrimony	*Agrimonia* spp.	High silica content makes aerial parts of plant useful as an astringent to stop bleeding
Alfalfa	*Medicago sativa*	Leaf concentrates shown to promote desirable balance between LDL and HDL cholesterol levels
Aloe	*Aloe* spp. (esp. *Aloe vera*)	Juice from leaves contains chrysophanic acid, which promotes healing of burns; used for relief of constipation
American mountain ash	*Pyrus americana*	Liquid made from steeping inner bark in water used as an astringent; tea of berries used as a wash for piles; berries eaten to prevent or cure scurvy
***Anemarrhena asphodelioides* (*see* Chinese lily)**		
Anemone	*Anemone canadensis*	Pounded boiled root applied to wounds as an antiseptic
Angelica	*Angelica archangelica*	Fruits used in treating colds and fevers; leaves used to stimulate appetite
***Angelica dahurica* (*see* Dahurian angelica)**		
***Angelica polymorpha* (*see* Chinese angelica)**		
Anise	*Pimpinella anisum*	Seed oil used to relieve indigestion, colds, and respiratory problems such as sinusitis
***Apocynum androsaemifolium* (*see* Dogbane, bitter)**		
***Apocynum venetium* (*see* Dogbane, venetian)**		
Apple	*Malus domestica*	Source of polyphenols and enzyme inhibitors that exhibit antioxidant and bactericidal activity (e.g., against gingivitis bacteria), especially when working in concert with bioflavonoids
Apricot	*Prunus armeniaca*	Seed extract said to function as a bronchodilator
Arnica	*Arnica* spp.	Plants applied as a poultice to bruises and sprains
Ashwagandha	*Withania somnifera*	Reported to have multiple immune system-boosting effects as well as alleviating menopausal symptoms
Asian epimedium	*Epimedium grandiflorum*	Plant extracts are said to have a stimulatory hormone-like effect on the prostate gland and testes
Asian skullcap	*Scutellaria baicalensis*	Plant extracts contain flavonoids and antioxidants; some components function together as a bronchodilator and bactericide; they also reduce blood pressure and LDL cholesterol levels

TABLE A3.4

Plants, Fungi, and Algae Associated with Medicinal Uses

ORGANISM	SCIENTIFIC NAME	USES
Asparagus, wild	*Asparagus racemosus*	Root flavonoid extracts used to relieve menopausal problems
Astragalus	*Astragalus* spp.	Root extracts used to boost the immune system; said to be good for colds, flu, and immune-deficiency disorders; also lowers blood pressure. (***Caution: Some Astragalus spp. sequester toxic amounts of selenium; should not be taken if a fever is present***)
***Astragalus membranaceus* (*see* Membranous milk-vetch)**		
***Atractylodes lancea* (*see* Southern tsangshu)**		
***Atractylodes macrocephala* (*see* Southern tsangshu)**		
Balm of Gilead	*Populus* x *gileadensis*	Buds used as an ingredient in cough syrups
Balsam poplar	*Populus balsamifera*	Buds made into ointment, which Native Americans placed in nostrils for relief of congestion
Barberry	*Berberis vulgaris*	Slows heartbeat rate
Beefsteak plant (*see* Perilla)		
Bifidobacteria	*Bifidobacterium bifidum (Helicobacter pylori, B. breve, B. infantis, B. longum, and others)*	Bifidobacteria destroy the bacteria that cause ulcers in humans
Bilberry/Blueberry	*Vaccinium* spp.	Evidence that regular consumption of fruit, which contains more than a dozen anthocyanosides, increases oxygen flow to eyes, slowing progression of cataracts, glaucoma, and macular degeneration; helps to balance insulin levels
Bitter melon	*Momordica charantia*	Plant extracts promote increased insulin production and are believed to reduce sugar damage to pancreatic cells
Bittersweet nightshade	*Solanum dulcamara*	Plant extracts used to treat skin problems such as acne, eczema, and boils. (***Caution: The fruits and other plant parts are poisonous***)
Blackberry	*Rubus* spp.	Tea of roots used by northern California Native Americans to cure dysentery
Black cohosh	*Cimicifuga racemosa*	Dried rhizomes used in cough medicines and for rheumatism; counters effects of declining estrogen levels in women (e.g., hot flashes, sleep disturbances); alleviates urinary tract problems
Black currant	*Ribes nigrum*	Oil from seeds used to improve suppleness of skin and to reduce skin dryness
Black haw	*Viburnum prunifolium*	Bark used in treatment of asthma and for relieving menstrual irregularities
Bloodroot	*Sanguinaria canadensis*	Native Americans used rhizome for ringworm, as an insect repellent, and for sore throat
Blue cohosh	*Caulophyllum thalictroides*	Tea of root drunk by Native Americans and early settlers a week or two before giving birth to promote rapid parturition
Boneset	*Eupatorium perfoliatum*	Water infusion of dried plant tops widely used to treat fevers and colds
Borage	*Borago officinalis*	Oil from seeds contains gamma linoleic acid (GLA) and other oils beneficial in human nutrition
Boswellia	*Boswellia serrata*	Extract of resin from this East Indian tree inhibits substances that cause joint swelling
Broom snakeweed	*Gutierrezia sarothrae*	Navajo Indians applied chewed plant to insect stings and bites of all kinds
Buckthorn	*Rhamnus catharticus*	Fruits used as a laxative
***Bupleurm chinense* (*see* Chinese thoroughwax)**		
Burdock	*Arctium lappa*	Used as an insulin substitute in folklore; root extract used in 17th century for venereal diseases

TABLE A3.4
Plants, Fungi, and Algae Associated with Medicinal Uses

ORGANISM	SCIENTIFIC NAME	USES
Butcher's broom	*Ruscus aculeatus*	Plant extracts shown in clinical trials to strengthen capillaries and to relieve hemorrhoid symptoms; improve flow of blood to the hands and feet.
Button snakeroot	*Eryngium* spp.	Natchez Indians inserted chewed stem in nostrils to arrest nosebleed
Cajuput	*Melaleuca cajuputi*	Oil obtained from leaves and twigs is used in treatment of muscular pain and as an antiseptic
California bay	*Umbellularia californica*	Yuki Indians put leaves in bath of hot water and bathed for relief of rheumatism; leaves used as an insect repellent
Camphor	*Cinnamomum camphora*	Oil from leaves and wood used in cold remedies and liniments
Camptotheca	*Camptotheca acuminata*	Extracts from flowers and immature fruits yield *camptothecin,* which has given evidence of being effective against certain forms of cancer
Cardamon	*Elettaria cardamomum*	Seed oil has antibiotic properties and is used in treatment of colds, coughs, and other respiratory problems
Cascara	*Rhamnus purshiana*	Bark extract widely used as a laxative
Catnip	*Nepeta cataria*	Leaf tea used for treatment of colds and to relieve infant colic
Cat's claw	*Uncaria tomentosa*	Root bark extracts used in treatment of intestinal problems, including diverticulosis and Crohn's disease; extracts also shown to have anti-inflammatory properties; *hirsutin* component lowers blood pressure; *rhynchophylline* (alkaloid) inhibits clumping of blood platelets and has been shown in animals to increase brain serotonin levels
Cayenne pepper	*Capsicum frutescens*	Used to reduce mucous drainage (recent evidence, however, suggests it may be carcinogenic); capsaicin extracts used in ointments to relieve pains of arthritis and neuropathy; aids digestion
Celery	*Apium graveolens*	Seed contains an essential oil that acts like an antioxidant that fights free radicals that attack joints; oil believed to have sedative properties
Chamomile	*Chamaemelum nobile*	Tea used as a digestive aid
Chaste tree	*Vitex agnus-castus*	Extract of berries reduces symptoms of premenstrual syndrome
Chaulmoogra	*Hydnocarpus* spp.	Seed oil used in the treatment of skin diseases such as eczema, psoriasis, and leprosy
Cherry (wild)	*Prunus serotina*	Tea brewed from bark used for coughs and colds
Chia	*Salvia columbariae*	Mucilaginous seeds used by Spanish Californians to make a refreshing drink; seeds contain a caffeine-like principle that enabled Native Americans to perform unusual feats of endurance; seed paste used in eye irritated by foreign matter
Chinese angelica	*Angelica polymorpha*	Root extracts used to suppress or relieve asthma
Chinese cinnamon	*Cinnamomum cassia*	Pulverized bark ingested to improve urinary flow and reduce more than normal frequencies of urination
Chinese club moss	*Lycopodium serratum*	Source of an alkaloid (huperzine A) that inhibits destruction of acetyl-choline involved in nerve transmissions, and thereby enhances memory
Chinese lily	*Anemarrhena asphodelioides*	Plant extracts apparently can aid in controlling blood glucose levels, hayfever, dermatitis, and other allergic symptoms; rhizome extracts used to quench thirst caused by fevers
Chinese magnolia	*Magnolia officinalis, Magnolia quinquepeta*	Bark extracts used to reduce nasal stuffiness and discharge, to drain sinuses, and to alleviate asthma and sinus headaches
Chinese rubber tree	*Eucommia ulmoides*	Bark extract improves circulation to the hands and feet; reduces high blood pressure; alleviates frequent urination problems
Chinese thoroughwax	*Bupleurum chinense*	Root extracts found to have general calming effect and to promote sound sleep; usually used in combination with other herbal extracts

TABLE A3.4

Plants, Fungi, and Algae Associated with Medicinal Uses

ORGANISM	SCIENTIFIC NAME	USES
Chlorella	*Chlorella* spp.	Green algae that boost the immune system, relieve constipation, and can remove heavy metals from food
Chocolate	*Theobroma cacao*	Seed extracts are good source of L-arginine and magnesium and are believed (when combined with other chocolate constituents) to elevate serotonin levels; chocolate also contains theobromin (somewhat similar to caffeine in action) and phenylethylene, which are believed to produce sustained elevation of mood (*Note:* These attributes pertain primarily to chocolate that does not have milk, sugar, or other products added to it)
Chrysanthemum indicum **(*see* Indian chrysanthemum)**		
Cinchona	*Cinchona* spp.	Bark yields quinine drugs used in treating malaria
Cinquefoil (Eurasian)	*Potentilla erecta*	Dried rhizome used to control diarrhea
Club moss	*Lycopodium clavatum*	Spores dusted on wounds or inhaled by Native Americans to arrest nosebleeds
Coca	*Erythroxylon coca*	Cocaine from leaves used as a local anesthetic; South American laborers use it as a stimulant
Cola	*Cola nitida, C. acuminata*	Seeds contain up to 3.5% caffeine and 1% theobromine, which may lessen fatigue
Coleus	*Coleus forskolii*	Plant extracts used in treatment of hypertension, allergies, glaucoma, and psoriasis
Cordyceps	*Cordyceps sinensis*	This fungus, which parasitizes caterpillars, apparently alleviates respiratory problems and elevates phagocyte action
Cornsilk	*Zea mays*	Cornsilk extracts used for centuries as a diuretic
Corydalis	*Corydalis turtschaninovii*	One isolate of several produced by the plant has multiple anti-inflammatory and calming effects; it has been used to relieve joint pain (*Note:* This plant also produces poisonous alkaloids)
Cotton	*Gossypium* spp.	Cotton root "bark" used by black slaves and Native Americans to induce abortions
Cranberry	*Vaccinium oxycoccum*	Fruit juice drunk to treat female yeast infections
Creosote bush	*Larrea divaricata*	Decoction from leaves used as a cure-all by Native Americans but especially for respiratory problems
Cubebs	*Piper cubeba*	Dried fruit best known as a condiment but is also used in treatment of asthma
Cynanchum	*Cynanchum* spp.	Plant extracts of these relatives of milkweeds are said to reduce mucous congestion of the lungs
Dahurian angelica	*Angelica dahurica*	Plant extracts used to treat allergy symptoms
Damiana	*Turnera diffusa*	Dried leaves used for minor pain, as a laxative, a flavoring for a liqueur, and as a sexual stimulant; also said to improve blood circulation
Dandelion	*Taraxacum* sp. aff.	Root extracts said to stimulate the liver and facilitate its natural functioning in detoxification
Deadly nightshade	*Atropa belladonna*	Belladonna, a drug complex extracted from leaves, contains the drugs atropine, hyoscyamine, and scopolamine; these are used as an opium antidote, for shock treatments, and for dilation of pupils; scopolamine is also used as a tranquilizer and for "twilight sleep" in childbirth
Devil's claw	*Harpagophytum procumbens*	Tuber extracts reported to have anti-inflammatory and pain-relieving properties
Di-huang (*see* Rehmannia)		
Dogbane, bitter	*Apocynum androsaemifolium*	Roots boiled in water and resulting liquid used as a heart medication (contains a drug similar in action to digitalis)
Dogbane, venetian	*Apocynum venetium*	Leaves smoked for bronchitis relief; extracts lower blood pressure

TABLE A3.4

Plants, Fungi, and Algae Associated with Medicinal Uses

ORGANISM	SCIENTIFIC NAME	USES
Dogwood	*Cornus* spp.	Inner bark boiled in water and resulting liquid drunk to reduce fevers
Dong quai	*Angelica sinensis*	Root extracts (which contain flavonoids) used in the alleviation of hot flashes and other menopausal symptoms; also used to treat premenstrual syndrome
Echinacea	*Echinacea purpurea*	Leaves and roots have antiviral and anti-inflammatory properties; used to boost the immune system
Elderberry	*Sambucus* spp.	Source of *Sambucol,* which is reported to have antiviral properties, especially in controlling those viruses involved in the common cold
Ephedra	*Ephedra* spp.	Drug ephedrine, widely used to relieve nasal congestion and low blood pressure; obtained from inner bark, berries, flowers, leaves (most ephedrine now in use is synthetic); also known as *ma huang* **(*Caution: Stems contain toxic amounts of cyanide*)**
***Epimedium grandiflorum* (*see* Asian epimedium)**		
Ergot	Source: *Claviceps purpurea* on cereal grains	Used to treat migraine headaches and to control bleeding after childbirth
Eucalyptus	*Eucalyptus* spp.	Oil extracted from leaves used to alleviate bronchitis and coughs
***Eucommia ulmoides* (*see* Chinese rubber tree)**		
European birch	*Betula pendula*	Oil distilled from bark and leaves used in treatment of kidney stones and urinary tract infections
Evening primrose	*Oenothera* spp.	Seeds are source of GLA oils beneficial in human nutrition
Eyebright	*Euphrasia officinalis*	Plant extracts used as an eyewash and in the relief of allergic itching of the eyes
Fennel	*Foeniculum vulgare*	Extracts of roots, stems, and fruits used as an appetite suppressant and as an eyewash
Fenugreek	*Trigonella foenum-graecum*	Seeds used in bulking laxatives; reduces mucus resulting from asthma and sinus problems; reduces skin inflammation
Feverfew	*Chrysanthemum parthenium*	Dried flowers used in treatment of migraine headaches, to induce abortion and menstruation, and as an insecticide; dilutes bronchial mucus; keeps body from producing histamines
Field mint	*Mentha arvensis*	Oil distilled from aerial parts of plants used to alleviate symptoms of colds, fevers, bronchitis; also used as an antiseptic
Flax	*Linum usitatissimum*	Cold-processed seed oils are rich source of gamma linoleic acid (GLA), beneficial in human nutrition, and in suppressing or reversing atherosclerosis; crushed seeds used as a laxative and for treating bronchial problems
Flowering ash	*Chionanthus virginicus*	Bark used as a laxative and in treatment of liver ailments
Forsythia	*Forsythia suspensa*	Plant extract has anti-inflammatory properties; is used to relieve sinus congestion and headaches
Foxglove	*Digitalis purpurea*	Drug *digitalis,* widely used as a heart stimulant, obtained from leaves
Frankincense	*Boswellia serrata*	Used to reduce joint pain and stiffness
Gambir	*Uncaria rhyncophylla*	Experimentally shown to relax blood vessels
Garlic	*Allium sativum*	Evidence that *allicin* and other sulfur-containing compounds extracted from bulbs inhibit common cold and other viruses; decreases artery-plugging fibrin levels, and regular consumption appears to enhance general cardiovascular health, including the lowering of LDL cholesterol levels and the inhibition of blood platelet clumping; the risk of stomach cancer also appears to be lowered
Gastrodia orchid	*Gastrodia elata*	Used in treating epilepsy and blood circulation problems; glucosides lower blood pressure

TABLE A3.4

Plants, Fungi, and Algae Associated with Medicinal Uses

ORGANISM	SCIENTIFIC NAME	USES
Gentian	*Gentiana catesbaei*	Catawba Indians applied hot-water extract of roots to sore backs; liquid drunk as a remedy for stomach aches; aids digestion and boosts circulation
Geranium (wild)	*Geranium maculatum*	Dried roots used for dysentry, diarrhea, and hemorrhoids
Ginger	*Zingiber officinale*	Powerful antioxidant; aids digestion and reduces nausea (including that of motion sickness); helps promote normal bladder control
Ginger (wild)	*Asarum* spp.	Extract of rhizome used as a broad-spectrum antibiotic
Ginkgo	*Ginkgo biloba*	Concentrated leaf extract improves oxygen-carrying capacity of capillaries, especially those of the brain, and improves memory; used for treating vertigo and tinnitus
Ginseng (*see also* **Siberian ginseng)**	*Panax* spp.	Considered a general panacea, especially in the Orient; strengthens the reproductive and adrenal glands; said to increase stamina
Globe artichoke	*Cynara scolymus*	Leaf and root extract used to inhibit gallstone formation and to alleviate digestion problems
Goldenrod	*Solidago canadensis*	Inflorescences used in the treatment of kidney stones and urinary tract infections
Goldenseal	*Hydrastis canadensis*	Rhizome source of alkaloidal drugs used in treatment of inflamed mucous membranes; also used as a tonic (*Note:* Pregnant women should not use goldenseal)
Goldthread	*Coptis groenlandica*	Native Americans boiled plant and gargled the liquid for sore or ulcerated mouths
Gotu kola	*Centella asiatica*	Shown in clinical trials to reduce swelling of ankles, generally improve blood circulation, and accelerate healing of wounds
Grape	*Vitis vinifera*	Seed extract source of powerful antioxidants (including quercetin) that also improve blood flow to the retina, thereby retarding macular degeneration; red grapes in particular produce significant amounts of *reservatrol,* which has been demonstrated to enhance enzyme activity associated with the regeneration and stimulation of nerve cells
Grapefruit	*Citrus* x *paradisi*	Seed extract used to combat bacterial or fungal infections
Green hellebore	*Helleborus viridis*	Plant extract used to treat hypertension (drug now synthesized); Thompson Indians used plant in small amounts to treat syphilis
Green tea	*Camellia sinensis*	Unfermented leaves source of polyphenols, which appear to reduce incidence of cancers in regular users through neutralization of free radicals; epigallocatechin gallate (EGCG) ingredient of green tea demonstrated by the Mayo Clinic to be particularly effective in control of prostate cancer
Gum plant	*Grindelia camporum*	Liquid from freshly and briefly boiled plants effective in treating poison oak and poison ivy rashes; used in treatment of coughs
Gymnema	*Gymnema sylvestre*	Extracts used to stabilize insulin levels in diabetics
Gynostemma	*Gynostemma pentaphylla*	Chinese plant related to melons; extracts believed to stimulate the immune system and aid in metabolism of fats that contribute to strokes
Hawthorn	*Crataegus oxycantha*	Anti-inflammatory that lowers LDL cholesterol levels; dilates coronary blood vessels
Hemlock	*Tsuga* spp.	Native Americans made tea of inner bark to treat colds and fevers (*Note:* Do not confuse this tree with poison hemlock [*Cicuta* spp.])
Hops	*Humulus lupulus*	Extracts are sedative and used in treating insomnia and nervous tension
Horehound	*Marrubium vulgare*	Extract from dried tops of plants used in lozenges for relief of sore throats and colds; dilutes mucus in bronchial tubes

TABLE A3.4
Plants, Fungi, and Algae Associated with Medicinal Uses

ORGANISM	SCIENTIFIC NAME	USES
Horse chestnut	*Aesculus hippocastanum*	Seed and leaf extracts used to improve blood flow; night cramps of legs; reduce varicose veins and leg swelling (***Caution: Plant is poisonous and only standardized extracts of demonstrated therapeutic value should be used; a coumarin component of horse chestnut leaves can interact adversely with aspirin and other anticoagulants***)
Horseradish	*Armoracia rusticana*	Roots used to treat infections of the urinary tract
Horsetail	*Equisetum* spp.	Plants boiled in water; liquid used as a delousing hairwash or as a gargle for mouth ulcers
Huperzia serrata = Lycopodium serratum **(*see* Chinese club moss)**		
Hydrangea	*Hydrangea paniculata*	Essential oil from roots acts as diuretic (***Caution: Leaves contain toxic amounts of cyanide***)
Hypericum perforatum **(*see* St. John's wort)**		
Indian chrysanthemum	*Chrysanthemum indicum*	Glucoside extract said to lower blood pressure
Indigo (wild)	*Baptisia tinctoria*	Native Americans boiled plant and used liquid as an antiseptic for skin sores
Ipecac	*Cephaelis ipecacuana*	Drug from roots and rhizome used to treat amoebic dysentery; also used as an emetic
Java plum	*Syzygium cumini*	Powdered seeds used to counter excessive thirst and excretion of sugar in the urine, characteristic of diabetics
Jimson weed	*Datura* spp.	Drugs atropine, hyoscyamine, and scopolamine obtained from seeds, flowers, and leaves; drug stramonium used for knockout drops and in treatment of asthma (***Caution: Jimson weeds are highly poisonous [see*** **Deadly nightshade])**
Joe-pye weed	*Eupatorium purpureum*	Dried root said to prevent formation of gallstones
Joshua tree	*Yucca brevifolia*	Cortisone and estrogenic hormones made from sapogenins produced in the roots
Jujube	*Ziziphus jujuba*	Fruit extracts shown to promote restful sleep and to aid in balancing irregular heartbeat
Juniper	*Juniperus* spp.	Tea of "berries" drunk by Zuni Indian women to relax muscles following childbirth (***Caution: Internal consumption can interfere with absorption of iron and other minerals***)
Kansas snakeroot	*Echinacea angustifolia*	Dried roots used as antiseptic in treatment of sores and boils, periodontal disease, and sinus drainage problems
Kava kava	*Piper methysticum*	Leaf tea used as a sedative, a muscle relaxant, and as a pain reliever
Kirilow's cucumber	*Trichosanthes kirilowii*	Used to inhibit mucous production in the lungs
Lactobacillus	*Lactobacillus acidophilus, L. rhamnosus, L. salivarius,* and others	Lactobacilli normally populate the gastrointestinal tract, where they function in various ways to boost the immune system and destroy pathogenic bacteria; antibiotics destroy these useful bacteria, and repopulation is facilitated by ingestion of lactobacilli capsules or products such as yogurt, which contain the useful bacteria
Lemon balm	*Melissa officinalis*	Leaf extracts and oils used for colds
Licorice	*Glycyrrhiza glabra*	Rhizomes source of licorice used in cough drops and for the soothing of inflamed mucous membranes; stimulates interferon production; relieves allergic symptoms; boosts immune system; deglycyrrhizinated licorice increases protection of upper digestive tracts by augmenting the mucous coating (***Caution: Undeglycyrrhizinated licorice can elevate blood pressure***)
Ligusticum	*Ligusticum wallichii*	Extract has been demonstrated to relax blood vessels
Lily of the valley	*Convallaria majalis*	All parts of plant contain a heart stimulant similar to digitalis; used to control irregular heartbeat (***Caution: Plants are poisonous***)

TABLE A3.4

Plants, Fungi, and Algae Associated with Medicinal Uses

ORGANISM	SCIENTIFIC NAME	USES
Lilyturf plant	*Ophiopogon japonicus*	Root extracts said to aid in diluting thick mucous secretions in the lungs
Lobelia	*Lobelia inflata*	Drug lobeline sulfate obtained from dried leaves; drug used in preparations to aid in cessation of smoking and in treatment of respiratory disorders (***Caution: Has effects similar to those of nicotine; more than 10 grams (one-third ounce) of dried plant can produce a coma***)
Lycopodium serratum **(*see* Chinese club moss)**		
Maca	*Lepidium meyenii*	Root extracts said to elevate testosterone levels and improve sexual performance in men
Madagascar periwinkle	*Catharanthus roseus*	A semisynthetic extract (*vinpocetine*) derived from *vincamine* produced by this plant said to be a significant memory enhancer
Magnolia vine	*Schisandra chinensis*	Plant extracts contain a powerful antioxidant that appears to protect healthy tissues (liver in particular) from damage caused by higher than normal blood sugar levels. Synergistic effect when combined with ginseng
Ma huang (*see* Ephedra)		
Maitake mushrooms	*Grifola frondosa*	A substance (beta-glucan) produced by these mushrooms evidently stimulates the production of cells that aid in the inhibition of cancer cells
Malabar kino	*Pterocarpus marsupium*	Leaf extracts contain *epicatechin,* which promotes oxygen uptake and better processing of sugar by body tissues
Mandrake	*Mandragora officinarum*	Extracts of plant used in folk medicine as a painkiller (drugs hyoscyamine, podophyllin, and mandragorin have been isolated; podophyllin used experimentally in treatment of paralysis)
Mangosteen	*Garcinia mangostana*	Fruit acid is believed to aid in weight reduction
Manroot	*Marah* spp.	Native Americans used oil from seeds to treat scalp problems and the crushed roots for relief from saddle sores
Marginal fern	*Dryopteris marginalis*	Rhizomes contain oleoresin used in expulsion of tapeworms from the intestinal tract
Marijuana	*Cannabis sativa*	Tetrahydrocannabinol obtained from resinous hairs in inflorescences; ancient medicinal drug of China
Mayapple	*Podophyllum peltatum*	Podophyllin obtained from rhizomes used experimentally in treatment of paralysis; dried rhizome powder used on warts (***Caution: Plant is poisonous, and extracts are very irritating to the skin***)
Maypop	*Passiflora incarnata*	Dried leaves used as sedative; Native Americans used juice as treatment for sore eyes
Melia	*Melia toosendan*	Used in traditional Chinese medicine to relieve joint pain
Membranous milk vetch	*Astragalus membranaceus*	Extracts strengthen the immune system, especially that of the upper respiratory tract; promote interferon production and repair of damaged bronchial tubes; there is evidence it can counter bone loss (osteoporosis) resulting from extended use of corticosteroids (***Caution: Some other*** *Astragalus spp.* ***also known as milk vetch are toxic***)
Mesquite	*Prosopis glandulosa*	Native Americans mixed dried leaf powder with water and used liquid to treat sore eyes
Mexican yam	*Dioscorea floribunda*	Tuberous roots produce up to 10% *diosgenin,* a precursor of progesterone and cortisone, and are a source of DHEA (dihydroepiandrosterone), a complex hormone naturally produced by humans; DHEA levels decline with aging; there is some evidence that controlled DHEA supplementation in older persons retards some aspects of aging
Milk thistle	*Silybum marianum*	Silymarin extracted from plants has antioxidant properties that appear to be especially beneficial to the liver
Milk vetch (*see* Astragalus)		

TABLE A3.4

Plants, Fungi, and Algae Associated with Medicinal Uses

ORGANISM	SCIENTIFIC NAME	USES
Milkweed	*Asclepias syriaca*	Quebec Indians promoted temporary sterility by drinking infusion of pounded roots
Mistletoe	*Phoradenron flavescens*	Native Americans reported to use small amounts as a contraceptive and sedative (***Caution: Plants are toxic and should not be taken internally***)
Monkshood	*Aconitum napellus*	Source of aconite once used in treatment of rheumatism and neuralgia (***Caution: Plant is highly toxic***)
Mormon tea (*see* Ephedra)		
Muira puama	*Ptychopetalum olacoides*	Leaf extracts said to stimulate hormone production and improve circulation; some consider it to be an aphrodisiac
Mukul myrrh	*Commiphora mukul*	Resin from tree stabilizes cholesterol levels and reduces osteoarthritis pain
Mulberry (red)	*Morus rubra*	Rappahannock Indians applied milky latex of leaf petioles to scalp to control ringworm
Mulberry (white)	*Morus alba*	Bark extract said to function as a bronchodilator
Mullein	*Verbascum thapsus*	Native Americans smoked leaves for respiratory ailments and asthma; flowers once widely used in cough medicines
Nettle (*see* Stinging nettle)		
Noni	*Morinda citrifolia*	Used in treatment of diabetes, high blood pressure, kidney disorders, and other ailments
Oats	*Avena sativa*	Extract from green oat seeds said to enhance both physical and sexual health
Olive	*Olea europaea*	Leaf extract contains *oleuropein* (calcium elenolate), which is a wide-spectrum bactericide and a virucide; it evidently enhances the production of phagocytes, thereby strengthening the immune system
Onion	*Allium* spp.	Cheyenne Indians applied bulbs in poultice to boils; juice and olive oil used to cure earaches; can lower blood pressure and help dissolve blood clots; the active principle, *allicin,* is also produced by garlic
***Ophiopogon japonicus* (*see* Lilyturf plant)**		
Opium poppy	*Papaver somniferum*	Morphine and codeine obtained from latex of immature fruits
Oregon grape	*Berberis aquifolium*	Bark tea drunk by Native Americans to settle upset stomach; used in strong doses for treatment of venereal diseases
Pacific yew	*Taxus brevifolia*	*Taxol,* a promising anticancer agent, is extracted from bark
Panax ginseng	*Panax pseudoginseng, Panax ginseng*	Root extract strengthens respiratory immune system, resulting in reduction of respiratory infections; strong antioxidant
Pansy (wild)	*Viola* spp.	Plants ground up and applied to skin sores or inflammations
Papaya	*Carica papaya*	Exudate of scarified unripe fruit is source of *papain* (a protein that is used to digest ruptured back discs and to facilitate digestion of food, as a meat tenderizer, for termite control, and for reduction of cloudiness in beer); *papain,* which is also believed to have antibiotic properties, may be used with *bromelain* from pineapples and trypsin to facilitate breakdown of cardiovascular plaque
Parsley	*Petroselinum crispum*	Richer in vitamin C than citrus fruits; inhibits proliferation of tumor cells; suppresses halitosis; general organ tonic
Pau d'arco	*Tabebuia heptaphylla*	General immune system booster
Peanut	*Arachis hypogoea*	*Reservatrol* extracted from peanut and mulberry plants said to be effective in inhibiting several types of cancers
Pennyroyal	*Mentha pulegium*	Native Americans used leaf tea in small amounts for relief of headaches and flatulence and to repel chiggers (***Caution: Pennyroyal is toxic in larger amounts***)

TABLE A3.4

Plants, Fungi, and Algae Associated with Medicinal Uses

ORGANISM	SCIENTIFIC NAME	USES
Peppermint	*Mentha piperita*	Peppermint oil is used to alleviate symptoms of respiratory infections and inflammation
Perilla	*Perilla frutescens*	Seeds are the source of perilla oil, which is exceptionally rich in omega-3 fatty acids essential to cardiac health
Persimmon	*Diospyros virginiana*	Liquid from boiled fruit used as an astringent; fruits with high beta-carotene content; leaves have high vitamin C content
Peruvian balsam	*Myroxylon balsamum*	Resin obtained from scorched or incised tree trunks is used as an antiseptic on burns, wounds, and hemorrhoids
Peyote	*Lophophora williamsii*	Alcoholic extract of plant used as an antibiotic
Pine	*Pinus* spp.	Pycnogenols extracted from bark have powerful antioxidant properties
Pineapple	*Ananas comosus*	*Bromelain* extracted from pineapple decreases clumping of blood platelets and fibrin, thereby improving circulation; *bromelain* also accelerates healing and can relieve pain, all without the side effects of aspirin, which is widely used for the same purposes; repeatedly chewing or holding fresh pineapple in the mouth may cure mouth ulcers
Pinkroot	*Spigelia marilandica*	Powdered root very effective in expulsion of roundworms from intestinal tract
Pipssisewa	*Chimaphila umbellata*	Native Americans steeped plant in water and used liquid to draw out blisters
Pitcher plant	*Sarracenia purpurea*	Native Americans used root widely as smallpox cure (records indicate it was effective)
Plantain	*Plantago ovata* and other *Plantago* spp.	(Not related to banana-like plantains.) Seed husks (known as *psyllium*) absorb water and are widely used in bulking laxatives; said to lower LDL cholesterol levels;
Pleurisy root	*Asclepias tuberosa*	Liquid from roots boiled in water used in treatment of respiratory problems
Polypore fungus	*Grifola umbellata*	All parts enhance kidney and bladder function; also believed to have anticancer immune system-boosting properties
Prickly ash	*Zanthoxylum americanum*	Bark and berries widely used by Native Americans for toothache (pieces inserted in cavities); liquid infusion drunk for venereal diseases
Psoralea	*Psoralea corylifolia*	Flavonoids used in Chinese medicine to facilitate relief of urinary tract problems
Psyllium (*see* Plantain)		
Pumpkin	*Cucurbita pepo*	Seed oil used to promote prostate health
Puncture vine	*Tribulus terrestris*	Plant extracts believed to elevate testosterone levels and promote muscle gain in men and to elevate estrogen levels in women
Purple coneflower	*Echinacea purpurea*	Plant extracts used to boost the immune system
Pygeum	*Pygeum africanum*	Bark extracts used to promote shrinkage of benign swelling of the prostate gland in men; there is evidence that a combination of pygeum and stinging nettle (*Urtica dioica*) can significantly reduce urgency for night urination
Psyllium	*Plantago ovata*	Ground seed husks absorb water and function as a bulking laxative; lowers LDL cholesterol levels
Quassia	*Picraea excelsa*, *Quassia amara*	Wood extracts used as pinworm remedy and as insecticides
Rauvolfia[1]	*Rauvolfia serpentina*	*Reserpine* obtained from roots; drug used in treatment of mental illness and in counteracting effects of LSD
***Rauvolfia*[1] *yunnanensis* (*see* Yunnan rauvolfia)**		

1. Frequently misspelled *Rauwolfia.*

TABLE A3.4
Plants, Fungi, and Algae Associated with Medicinal Uses

ORGANISM	SCIENTIFIC NAME	USES
Red-rooted sage	*Salvia miltiorhiza*	Plant extracts elevate blood oxygen content and are used to enhance blood circulation, particularly in the lungs; inhibits blood platelet clumping
Red yeast	*Monascus purpureus*	This yeast is cultured on rice; the combination improves circulation and balances cholesterol levels
Rehmannia	*Rehmannia glutinosa*	Experiments with animals indicates efficacy in strengthening kidney function and in lowering blood pressure
Rice (brown)	*Oryza sativa*	Inositol hexaphosphate (IP6), a B vitamin that is produced by rice, has been shown to control growth of cancer cells
Rye	*Secale cereale*	An Australian patented ryegrass extract known as *Oralmat* is proving to be effective in treating asthma, allergies, and other disorders, without the side effects of steroids
Saffron (meadow)	*Colchicum autumnale*	Drug *colchicine* from corms used in past for treatment of gout and back disc problems but now mostly used experimentally to induce doubling of chromosome numbers in plants
St. John's wort	*Hypericum perforatum*	Extracts used in the treatment of depression; boosts serotonin production in the brain; the serotonin suppresses cravings for carbohydrates and tends to promote normal sleep patterns (*Note:* St. John's wort can interfere with the normal metabolic activities of some prescription medications, and its use should be supervised by informed personnel)
***Salvia miltiorhiza* (*see* Red-rooted sage)**		
Sarsaparilla	*Aralia nudicaulis*	Cough medicines made from roots
Sassafras	*Sassafras albidum*	Tea of root bark used to induce sweating; used externally as a liniment
Saw palmetto	*Serenoa repens*	Berry extracts clinically demonstrated to aid in shrinkage of benign swelling of prostate gland, increase urine flow, and normalize frequency of urination in men
***Schisandra chinensis* (*see* Magnolia vine)**		
***Scutellaria baicalensis* (*see* Asian skullcap)**		
Self-heal	*Prunella vulgaris*	Native Americans applied plants in poultices to boils; plant glucosides said to tone blood vessels
Seneca snakeroot	*Polygala senega*	Liquid from bark boiled in water applied to snakebites; taken internally as an abortifacient; used in a cough remedy
Senna	*Cassia senna* and other spp.	Leaf extract used as a laxative or purgative
Siberian ginseng	*Eleutherococcus senticosus*	Liquid extract of rhizome and roots used as an immune system and stamina booster; in some individuals, it appears to counteract chronic fatigue syndrome
Sicklepod	*Cassia obtusifolia*	Plant extracts said to lower both blood pressure and LDL cholesterol levels
Skeleton weed	*Lygodesmia juncea*	Widely used by Native American women to increase milk flow
Skullcap	*Scutellaria laterifolia*	Dried plant used as an anticonvulsive in treatment of epilepsy and as a sedative (*see also* **Asian skullcap**)
Slippery elm	*Ulmus fulva*	Dried inner bark, which contains an aspirin-like substance, used to soothe inflamed membranes
Southern tsangshu	*Atractylodes lancea*, *Atractylodes macrocephala*	Plant extracts used as a diuretic; also used to balance blood sugar levels and promote spleen health

TABLE A3.4

Plants, Fungi, and Algae Associated with Medicinal Uses

ORGANISM	SCIENTIFIC NAME	USES
Soy	*Glycine max*	Isoflavones from plants (especially fruits) have medicinal value; ipriflavone appears to inhibit development of osteoporosis and increase bone density; genistein apparently diminishes production of cellular stress protein and inhibits the growth of human prostate cancer cells; reduces symptoms of menopause
Spicebush	*Lindera benzoin*	Berries, buds, and bark brewed for tea used to reduce fevers
Spruce	*Picea* spp.	Cree Indians ate small, immature female cones for treatment of sore throat; spruce leaf oil and spruce shoots used in Europe to alleviate cold, bronchitis, and fever symptoms
Squills	*Urginea maritima*	Bulbs of red variety are source of a heart stimulant; bulbs of white variety are widely used as a rodent killer
Stevia	*Stevia rebaudiana*	Stevia extracts are 30–100 times sweeter than sugar; used as a sweetener by diabetics and others needing to reduce their sugar intake
Stinging nettle	*Urtica dioica*	Used in treatment of allergic disorders and inflammatory conditions of the lungs; the roots are rich in vitamin C, which apparently inhibits breakdown of testosterone, consequently increasing testosterone levels in the body (*see also* **Pygeum**)
Stoneseed	*Lithospermum ruderale*	Shoshoni women reported to have drunk cold-water infusion of roots daily for 6 months to ensure permanent sterility (experiments with mice suggest substance to the reports)
Strophanthus	*Strophanthus* spp.	Seeds are major source of cortisone and also source of a heart stimulant
Strychnine plant	*Strychnos nox-vomica*	Strychnine extracted from seeds widely used as an insect and animal poison and is the principal ingredient in blowgun darts used by South American aborigines; minute amounts stimulate the central nervous system and relieve paralysis
Sumac	*Rhus* spp. (especially *R. glabra*)	Native Americans applied leaf decoction as a remedy for frostbite; fruits and liquid made from leaves applied to poison ivy rash and gonorrhea sores; root chewed for treament of mouth ulcers
Sweet flag	*Acorus calamus*	Boiled root applied to burns; root chewed for relief of colds and toothache
Sweet gum	*Liquidambar styraciflua*	Bud balsam used to treat chigger bites; balsam also used in insect fumigating powders
Sword fern	*Polystichum munitum*	Boiled rhizome used by Native Americans to treat dandruff; sporangia and spores applied to burns
Tamarind	*Tamarindus indica*	Fruit pulp used as laxative
Tinospora	*Tinospora cordifolia*	Reported to boost the immune system
Tomato	*Solanum esculentum*	Rich source of lycopene, which is involved with health of the prostate gland and eyes
Tree peony	*Paeonia suffruticosa*	Bark extract said to reduce production of blood platelets
Trichosanthes kirilowii (*see* **Kirilow's cucumber**)		
Turmeric	*Curcuma longa*	Rhizome extracts appear to lower LDL cholesterol levels, prevent blood clots, suppress cancer proliferation, and reduce joint pain
Uncaria	*Uncaria rhynchophilla*	Glucoside extracts used in China to lower blood pressure
Uzara	*Xysmalobium undulatum*	Root extracts used to control diarrhea (***Caution: Uzara glycosides may react with other glycosides; drug should be taken only at recommended doses and not with other medications***)

TABLE A3.4

Plants, Fungi, and Algae Associated with Medicinal Uses

ORGANISM	SCIENTIFIC NAME	USES
Valerian	*Valeriana septentrionalis*	Pulverized plant applied to wounds; extracts taken internally have sedative effect and are used to treat insomnia
Velvet bean	*Mucuna* spp.	Seeds contain L-dopa used in treatment of Parkinson's disease
Virginia snakeroot	*Aristolochia serpentaria*	Native Americans used tea of plant for reducing high fevers
Wahoo	*Euonymus atropurpurea*	Bark steeped in water; liquid has digitalis-like effect on heart
Walnut (English)	*Juglans regia*	Extracts have been demonstrated to kill or inhibit a wide range of pathogenic microorganisms and to reduce LDL cholesterol levels
Watercress	*Rorippa nasturtium-aquaticum*	Some evidence that daily consumption retards development of lung cancer in smokers
Water plantain	*Alisma plantago-aquatica*	Rhizome extracts have diuretic properties; believed to improve bladder and kidney function
Western wallflower	*Erysimum capitatum*	Zuni Indians ground plant with water and applied it to skin to prevent sunburn
White mulberry	*Morus alba*	Fruit believed to improve kidney and liver function; also alleviates respiratory problems including asthma and mucous production
Willow	*Salix* spp.	Chickasaw Indians snuffed infusion of roots as a remedy for nosebleed; Pomo Indians boiled bark in water and applied liquid for relief of skin itches; fresh inner bark contains *salicin,* an aspirin-like compound used to reduce fevers
Wintergreen	*Gaultheria procumbens*	Oil from leaves used as a folk remedy for body aches and pains
Witch hazel	*Hamamelis virginiana*	Oil distilled from twigs and leaves used primarily as an external astringent that staunches bleeding
Wormseed	*Chenopodium ambrosioides*	Oil from seeds used to expel intestinal worms
Wormwood	*Artemisia* spp.	Yokia Indians made tea from leaves to treat bronchitis; other Native Americans used tea as a cold remedy
Yarrow	*Achillea millefolium*	Native Americans used plant infusion for treating wounds, earaches, and burns; infusion drunk to relieve upper respiratory tightness; fresh leaf inserted in nostril to staunch nosebleed
Yellow lady's slipper	*Cypripedium calceolus*	Dried root used for relief of insomnia and as sedative
Yellow nut grass	*Cyperus esculentus*	Paiute Indians pounded tubers with tobacco leaves and applied mass in wet dressing for treatment of athlete's foot
Yerba santa	*Eriodictyon californicum*	Native Americans smoked leaves or drank leaf tea for treatment of colds or asthma
Yucca	*Yucca* spp.	Plants produce saponins used in birth control preparations and to treat inflammation and other conditions that might otherwise be treated with steroids such as cortisone
Yunnan rauvolfia	*Rauvolfia yunnanensis*	Plant extracts said to lower blood pressure
Zhi-mu (*see* Chinese lily)		

HALLUCINOGENIC PLANTS

Although a few hallucinogenic substances produced by animals have been isolated and some have been synthesized, the majority of known hallucinogens are produced by plants. Table A3.5 is not a complete list, but it includes the better-known sources. The reader is referred to the "Additional Reading" for further information.

TABLE A3.5

Hallucinogenic Substances Produced by Plants and Fungi

ORGANISM	SCIENTIFIC NAME	PART USED	PRINCIPAL ACTIVE SUBSTANCE
Ajuca	*Mimosa hostilis*	Roots	Nigerine
Belladonna	*Atropa belladonna*	Leaves	Hyoscyamine, scopolamine
Caapi	*Banisteriopsis caapi*	Wood	Harmine
Canary broom (*Genista* of florists, but true *Genista* is a broom genus that differs from *Cytisus* in that *Cytisus* spp. have a seed appendage whereas those of *Genista* do not)	*Cytisus canariensis*	Seeds	Cytisine
Catnip	*Nepeta cataria*	Leaves	Unknown
Cohoba	*Piptadenia peregrina, P. macrocarpa*	Seeds, pods (snuff)	Tryptamines
Coral bean	*Erythrina* spp.	Seeds	Unknown
Cubbra borrachera	*Brugmansia* spp.	Leaves	Scopolamine
Ergot fungus	*Claviceps purpurea*	Rhizomorph	Ergine (LSD)
Fly agaric	*Amanita muscaria*	Mushroom cap	Ibotenic acid, muscimol
Henbane	*Hyoscyamus* spp.	Leaves	Hyoscyamine, scopolamine
Iboga	*Tabernanthe iboga*	Root bark	Ibogaine
Jimson weed	*Datura* spp.	All parts	Scopolamine
Kava kava	*Piper methysticum*	Root (large amounts of beverage produce hallucinations)	Myristicin-like compound
Mace	*Myristica fragrans*	Aril of seed	Myristicin
Mescal bean	*Sophora secundiflora*	Seeds	Cytisine
Morning glory	*Ipomoea violacea*	Seeds	Ergine
Nutmeg	*Myristica fragrans*	Seeds	Myristicin
Ololiuqui	*Rivea corymbosa*	Seeds	Turbicoryn
Peyote	*Lophophora williamsii*	Stems	Mescaline
Psilocybe and related mushrooms	*Psilocybe* spp., *Conocybe* spp., *Panaeolus* spp., and others	All parts	Psilocybin, psilocin
Rape dos Indios	*Maquira sclerophylla*	Dried plant (snuff)	Unknown
San Pedro	*Trichocereus pachanoi*	Stems	Mescaline
Sassafras	*Sassafras albidum*	Root bark (large amounts of tea)	Safrole
Sweet flag	*Acorus calamus*	Dried root	Asarone, β-asarone
Syrian rue	*Peganum harmala*	Seeds	Harmine
Vygie	*Mesembryanthemum expansum*	All parts	Mesebrine
Wood rose	*Argyreia nervosa*	Seeds	Ergoline alkaloids
Yakee (Parica)	*Virola* spp.	Resin from inner surface of freshly removed bark (snuff)	Tryptamine
Yohimbe	*Corynanthe* spp., *Pausinystalia johimbe*	Bark	Yohimbine

SPICE PLANTS

The word *spice* describes any aromatic plant or part of a plant used to flavor or season food; spices are also used to add scent or flavor to manufactured products (Table A3.6). Although spices have no nutritional value, they add a pleasurable zest to meals, and before food preservation was possible, they helped make palatable food that was still edible but unappealing.

The value placed on spice plants was responsible for changing the course of Western civilization as a principal motive behind the voyages of discovery.

TABLE A3.6

Plants Used to Season or Flavor

SPICE	SCIENTIFIC NAME OF PLANT	PARTS USED; REMARKS	PRINCIPAL SOURCE
Allspice	*Pimento officinalis*	Powdered dried fruit	Jamaica
Almond	*Prunus amygdalus*	Oil from seed used for flavoring baked goods	Mediterranean; U.S.
Angelica	*Angelica archangelica*	Stems candied; oil from seeds and roots used in liqueurs	Europe; Asia
Anise	*Pimpinella anisum*	Oil distilled from fruits used for flavoring	Widely cultivated
Arrowroot	*Maranta arundinacea*	Powdered root used in milk puddings, baked goods	South America
Asafoetida	*Ferula asafoetida*	Powdered gum from stems and roots used in minute quantities with fish	Middle East
Balm (Melissa)	*Melissa officinalis*	Oil from leaves used in beverages; leaves used as food flavoring	U.S.; Mediterranean
Basil	*Ocimum basilicum*	Leaves used in meat dishes, soups, sauces	Mediterranean
Bay	*Laurus nobilis*	Leaves used in soups, sauces	Europe
Bell pepper	*Capsicum frutescens*	Dried, diced fruit used in chip dips, salad dressings	Widely cultivated
Bergamot	*Monarda didyma*	Leaves used with pork (*Note:* A perfume oil obtained from a variety of orange—*Citrus aurantium* var. *bergamia*—is also called bergamot)	North America (*Monarda*); Italy (*Citrus*)
Black pepper	*Piper nigrum*	Dried fruits used as a condiment	India; Indonesia
Borage	*Borago officinalis*	Leaves used as a beverage flavoring	England
Burnet	*Sanguisorba minor*	Used in soups and casseroles	Eurasia
Calamus	*Acorus calamus*	Powdered rhizome used for flavoring	Europe; Asia; North America
Capers	*Capparis spinosa*	Flower buds used for flavoring relishes, pickles, sauces	Mediterranean
Caraway	*Carum carvi*	Seeds used in breads, cheeses; seed oil used in the liqueur kümmel	North America; Europe
Cardamon	*Elletaria cardamomum*	Dried fruit and seeds used for flavoring baked goods (*Note:* Several false cardamons—*Amomum* spp.—are sold commercially)	India; Sri Lanka; Central America
Cassia	*Cinnamomum cassia*	Powdered bark used as cinnamon substitute	Southeast Asia
Cayenne pepper	*Capsicum* spp.	Powdered dried fruits used in chili powder, Tabasco sauce	American tropics
Celery	*Apium graveolens*	Seeds used in celery salt, soups	Europe; U.S.
Chervil	*Anthriscus cerefolium*	Used as a parsley substitute	Europe; Near East
Chicory	*Cichorium intybus*	Ground, dried root added to coffee	Mediterranean
Chives	*Allium schoenoprasum*	Leaves, bulbs used with sour cream, butter	Widely cultivated
Chocolate	*Theobroma cacao*	Ground seeds used for flavoring	Africa; South America
Cilantro	*Coriandrum sativum*	Leaves used in avocado dip and with poultry	Europe
Cinnamon	*Cinnamomum zeylanicum*	Ground bark used for flavoring baked goods; oil from leaves used as flavoring, clearing agent	Seychelles; Sri Lanka

TABLE A3.6

Plants Used to Season or Flavor

SPICE	SCIENTIFIC NAME OF PLANT	PARTS USED; REMARKS	PRINCIPAL SOURCE
Citrus	*Citrus* spp.	Fruits, especially rinds, source of flavoring oil	Mediterranean; South Africa; U.S.
Cloves	*Syzygium aromaticum*	Dried flower buds used to flavor cooked fruits, toothpaste, candy	Moluccas (Spice Islands)
Coffee	*Coffea arabica*	Roasted seeds source of mocha-coffee flavoring	Tropics
Coriander	*Coriandrum sativum*	Ground seed used in German frankfurters, curry powders	Mediterranean
Costmary	*Chrysanthemum balsamita*	The leaves used sparingly in salads add a mint flavor	Europe, Asia
Cubebs	*Piper cubeba*	Dried fruits used as seasoning	East Indies
Cumin	*Cuminum cyminum*	Ground seed used with meats, pickles, cheeses, curry	Mediterranean
Curry	A blend of parts of plants of several different spp.	A spicy condiment containing several ingredients, such as turmeric, cumin, fenugreek, and zedoary	India
Dill	*Anethum graveolens*	Seeds used in pickling brines; leaves used for seasoning meat loaves, sauces	Europe; Asia
Dittany	*Origanum dictamnus*	Leaves used as seasoning for poultry, meats	Crete
Eucalyptus	*Eucalyptus* spp.	Oil from leaves used in toothpastes, flavoring agents	Australia
Fennel	*Foeniculum vulgare*	Seeds used in baked goods	Europe
Fenugreek	*Trigonella foenum-graecum*	Oil distilled from seeds used in pickle, chutney, curry powders, imitation maple flavoring	Widely cultivated
Filé (*see* Sassafras)			
Fruit-scented sage	*Salvia dorisiana*	Plant used with beef and fish; it adds a grapefruit-pineapple flavor to the meat	Honduras
Garlic	*Allium sativum*	Fresh or dry bulbs used for meat seasonings	Widely cultivated
Ginger	*Zingiber officinale* and others	Dried rhizomes used for flavoring many foods and drinks	India; Taiwan
Grains of paradise	*Afromomum melegueta*	Seeds used to flavor beverages and medicines	West Africa
Hops	*Humulus Iupulus*	Dried inflorescences of female plants used in brewing beer	Europe; North America
Horseradish	*Rorippa armoracia*	Grated fresh root used as a condiment	Europe; North America
Juniper	*Juniperus* spp.	"Berries" used to season beef roasts, poultry, sauces	North America
Lemon balm	*Melissa officinalis*	Leaves give a lemon-mint flavor to stews and desserts	Southern Europe
Licorice	*Glycyrrhiza glabra*	Dried rhizome and root used to flavor pontefract cakes, candies	Middle East
Lovage	*Ligusticum scoticum*	Stems candied; seeds used in pickling sauces; celery substitute	Europe
Mace	*Myristica fragrans*	Aril of seed used for flavoring beverages, foods	Grenada; Indonesia; Sri Lanka
Marigold	*Tagetes* spp.	Petals substituted for saffron in rice dishes, stews	Widely cultivated
Marjoram	*Origanum hortensis*	Leaves used in stews, dressings, sauces	Mediterranean
Mugwort	*Artemisia douglasiana*	Fatty meat flavored with leaves	West Coast of North America
Mustard	*Brassica* spp.	Ground seeds used in meat condiment	Europe; China
Nasturtium	*Tropaeolum majus*	Flowers, seeds, leaves used in salads	Widely cultivated
Nutmeg	*Myristica fragrans*	Seeds used for flavoring foods, beverages	Grenada; Indonesia; Sri Lanka

TABLE A3.6
Plants Used to Season or Flavor

SPICE	SCIENTIFIC NAME OF PLANT	PARTS USED; REMARKS	PRINCIPAL SOURCE
Oregano	*Origanum vulgare* and others	Leaves used as seasoning with poultry, meats	Europe
Paprika (*see* Cayenne pepper)			
Parsley	*Petroselinum crispum*	Leaves used as meat garnish and flavoring in sauces	Widely cultivated
Peppermint	*Mentha piperita*	Oil from leaves used for food, drink, dentifrice flavoring (much commercial menthol is derived from *Mentha arvensis* grown in Japan)	U.S.; Russia
Pimiento	*Capsicum* spp.	Bright red fruits of a cultivated variety of pepper used in stuffing olives and in cold meats, cheeses	Central and South America
Poppy	*Papaver somniferum*	Seeds used in baking	Widely cultivated
Rosemary	*Rosmarinus officinalis*	Oil from leaves used in perfumes, soaps	Mediterranean
Rue	*Ruta graveolens*	Flavoring for fruit cups, salads	Europe
Saffron	*Crocus sativus*	Dried stigmas used to flavor oriental-style dishes	Spain; India
Sage	*Salvia officinalis*	Leaves used in poultry and meat dressings	Yugoslavia
Salad burnet	*Poterium sanguisorba*	Leaves impart a cucumber-like flavor to salads	Europe; W. Asia
Sarsaparilla	*Smilax* spp.	Roots are source of flavoring for beverages, medicines	American tropics
Sassafras	*Sassafras albidum*	Bark and wood yield flavoring for beverages, toothpaste, gumbo	U.S.
Savory (summer)	*Satureia hortensis*	Leaves used in green bean and bean salads, lentil soup, with fish	Mediterranean
Savory (winter)	*Satureia montana*	Leaves used as seasoning in stuffings, meat loaf, stews	Europe
Scallion	*Allium fistulosum*	Leaves used in wine cookery, soups	Widely cultivated
Sesame	*Sesamum indicum*	Seeds used in baking	Asia
Shallot	*Allium ascalonicum*	Bulbs, leaves used in Colbert butter, wine cookery	Widely cultivated
Southernwood	*Artemisia abrotanum*	Leaves used to flavor cakes	Europe
Star anise	*Illicium verum*	Fruits used in candy and cough drops	China
Stonecrop	*Sedum acre*	Dried leaves (ground) used as pepper substitute	Europe
Sweet Cicely	*Osmorhiza* spp.	Leaves have sweet, slight anise flavor; used to flavor dishes and baked goods that incorporate cooked fruits	North America; E. Asia
Sweet woodruff	*Galium odoratum*	Plants used to flavor fruit punches and strawberries	Europe; N. Africa; Asia
Tansy	*Tanacetum vulgare*	Leaves used in small amounts to flavor baked goods, pancakes, and puddings	Europe; Asia
Tarragon	*Artemisia dracunculus*	Leaves and flowering tops used in pickling sauces	Europe
Thyme	*Thymus vulgaris*	Leaves used in meat and poultry dishes, soups, sauces	Widely cultivated
Tonka bean	*Dipteryx* spp.	Seeds source of flavoring for tobacco; vanilla substitute (now largely synthesized)	American tropics
Turmeric	*Curcuma longa*	Rhizomes powdered and used in curry powders, meat flavoring	India; China
Vanilla	*Vanilla planifolia*	Flavoring extracted from fruits; used in foods, drinks	Malagasay Republic
Wintergreen	*Gaultheria procumbens*	Oil from leaves, bark used as flavoring for confections, toothpaste	U.S.
Zedoary	*Curcuma zedoaria*	Dried rhizome used in liqueurs, curry powders	India

DYE PLANTS

In the recent and the ancient past, dyes from many different plants were used to color cotton, linen, and other fabrics. Since the middle of the 19th century, however, natural dyes have been almost completely replaced in industry by synthetic dyes, and today, the use of natural dyes is largely confined to individual hobbyists.

Any reader interested in experimenting with natural dyes is encouraged to choose not only those plant materials included in Table A3.7, but to try any local plants available. The experimenter will soon find that quite unexpected colors may be derived from plants, as the colors of fresh flowers, bark, or leaves often bear little relationship to the colors of the dyes. For methods of dyeing, see the footnote given under the heading of "Lichens" in Chapter 19 and references in the "Additional Reading" at the end of Appendix 3.

TABLE A3.7

Plant and Fungal Sources of Natural Dyes

ORGANISM OR DYE	SCIENTIFIC NAME OF SOURCE	REMARKS
Acacia	*Acacia* spp.	Brown dyes from bark and fruits
Alder	*Alnus* spp.	Brownish dyes from bark
Alkanet	*Alkanna tinctoria*	Red dye from roots
Annatto	*Bixa orellana*	Yellow or red dye from pulp surrounding seeds
Bamboo	*Bambusa* spp.	Light green dye from leaves
Barberry	*Berberis vulgaris*	Grayish dye from leaves
Barwood	*Baphia nitida*	Purplish dyes from wood
Bearberry	*Arctostaphylos uva-ursi*	Yellowish dye from leaves
Bedstraw	*Galium* spp.	Light reddish brown dyes from roots
Birch	*Betula* spp.	Light brown to black dyes from bark
Black cherry	*Prunus serotina*	Red dye from bark; gray to green dyes from leaves
Black walnut	*Juglans nigra*	Rich brown dye from bark; brown dye from walnut hulls
Bloodroot	*Sanguinaria canadensis*	Red dye from rhizomes
Blueberry	*Vaccinium* spp.	Blue to gray dye from mature fruits (tends to fade)
Bougainvillea	*Bougainvillea* spp.	Light brownish dyes from floral bracts
Brazilwood	*Caesalpinia* spp.	Reddish dyes from wood
Buckthorn	*Rhamnus* spp.	Green dyes from fruits
Buckwheat	*Fagopyrum esculentum*	Blue dye from stems
Buckwheat (wild)	*Eriogonum* spp.	Dark gold, pale yellow, and beige dyes from stems and flowers
Buffaloberry	*Shepherdia argentea*	Red dye from fruit
Butternut	*Juglans cinerea*	Yellow to grayish brown dyes from fruit hulls
Cocklebur	*Xanthium strumarium*	Dark green dye from stems and leaves
Coffee	*Coffea arabica*	Light brown dye from ground, roasted seeds
Cudbear (Archil)	*Rocella* spp. (lichen)	Red dye obtained by fermentation of thallus
Cutch	*Acacia* spp.; *Uncaria gambir*	Brown to drab green dyes from stem gums
Dock	*Rumex* spp.	Light brown dyes from stems and leaves
Dogwood	*Cornus florida*	Red dye from bark; purplish dye from root
Doveweed	*Eremocarpus setigerus*	Light to olive green dye from entire plant
Dyer's rocket	*Reseda luteola*	Orangish dye from all parts
Elderberry	*Sambucus* spp.	Blackish dye from bark; purple, blue, or dark brown dyes from fruits
Eucalyptus	*Eucalyptus* spp.	Beige dyes from bark
Fennel	*Foeniculum vulgare*	Yellow dyes from shoots
Fig	*Ficus carica*	Green dyes from leaves and fruits

TABLE A3.7

Plant and Fungal Sources of Natural Dyes

ORGANISM OR DYE	SCIENTIFIC NAME OF SOURCE	REMARKS
Fustic	*Chlorophora tinctoria*	Yellow, bright orange, and greenish dyes from heartwood
Gamboge	*Garcinia* spp.	Yellow dye from resins that ooze from cuts made on stems
Giant reed	*Arundo donax*	Pale yellow dye from leaves
Grape	*Vitis* spp.	Bright yellow to olive green dyes from leaves
Hawthorn	*Crataegus* spp.	Pink dye from ripe fruits
Hemlock	*Tsuga* spp.	Reddish brown dye from bark
Henna	*Lawsonia inermis*	Orange dye from shoots and leaves
Hickory	*Carya tomentosa*	Yellow dye from bark
Hollyhock	*Althaea rosea*	Purplish black dye from flower petals
Horsetail	*Equisetum* spp.	Tan dyes from all green parts
Indigo	*Indigofera tinctoria*	Bright blue dyes from leaves
Kendall green (*see* Woadwaxen)		
Larkspur	*Delphinium* spp.	Blue dyes from petals
Lichens	Many genera and species	Many lichens yield (with various mordants) brilliant shades of yellows, golds, and browns
Litmus	*Rocella tinctoria*	Widely used pink-to-blue pH indicator dye from thallus
Logwood	*Haematoxylon campechianum*	Dark blue-purple dye from heartwood (widely used for staining tissues in microscope slides)
Lokao	*Rhamnus* spp.	Green dye from wood
Lupine	*Lupinus* spp.	Greenish dyes from flowers
Madder	*Rubia tinctorium*	Bright red dye from roots
Madrone	*Arbutus menziesii*	Brown dye from bark
Manzanita	*Arctostaphylos* spp.	Beige to dull yellow dyes from dried fruits
Maple	*Acer* spp.	Pink dye from bark
Marsh marigold	*Caltha palustris*	Yellow dye from petals
Milkweed	*Asclepias speciosa*	Pale yellow dyes from leaves
Morning glory	*Ipomoea violacea*	Gray green dye from blue flowers
Mullein	*Verbascum thapsus*	Gold dyes from leaves
Oak	*Quercus* spp.	Yellow dye from bark
Onion	*Allium cepa*	Reddish brown dyes from dry outer bulb scales of red onions; yellow dyes from similar parts of yellow onions
Oregon grape	*Berberis aquifolium*	Yellow dyes from roots
Osage orange	*Maclura pomifera*	Yellow, gray, and green dyes from fruits; yellow-orange dye from wood
Peach	*Prunus persica*	Green dyes from leaves
Poke	*Phytolacca americana*	Red dyes from mature fruits
Pomegranate	*Punica granatum*	Dark gold dye from fruit rinds
Prickly lettuce	*Lactuca serriola*	Green dye from leaves
Privet	*Ligustrum vulgare*	Yellow-green dye from leaves; deep gray dye from berries
Quercitron	*Quercus velutina*	Bright yellow dye from bark
Rhododendron	*Rhododendron* spp.	Tan dyes from leaves
Safflower	*Carthamnus tinctorius*	Reddish dye from flower heads
Saffron	*Crocus sativus*	Powerful yellow dye from stigmas
Sage	*Salvia officinalis*	Yellow dye from shoots
St. John's wort	*Hypericum* spp.	Light brownish dyes from leaves

TABLE A3.7

Plant and Fungal Sources of Natural Dyes

ORGANISM OR DYE	SCIENTIFIC NAME OF SOURCE	REMARKS
Sandalwood	*Pterocarpus santalinus*	Red dye from wood
Sappanwood	*Caesalpinia sappan*	Red dye from heartwood
Sassafras	*Sassafras albidum*	Orange brown dye from bark
Scotch broom	*Cytisus scoparius*	Yellow dye from all parts of plant
Smoke tree	*Cotinus coggyria*	Orange-yellow dye from wood (dye sometimes called "young fustic")
Smooth sumac	*Rhus glabra*	Grayish brown dye from bark
Tansy	*Tanacetum* spp.	Yellow and green dyes from leaves
Toyon	*Heteromeles arbutifolia*	Reddish brown dyes from leaves
Turmeric	*Curcuma longa*	Orangish dye from rhizome
Woad	*Isatis tinctoria*	Blue dye from leaves
Woadwaxen	*Genista tinctoria*	Yellow dye from all parts
Yerba santa	*Eriodictyon californicum*	Rich dark brown dyes from leaves

ADDITIONAL READING

Berry, A. 1995. *Know your spices: An alphabetical guide to your spice rack.* Franklin, TN: Runaway Press.

Bliss, A. 1994. *North American dye plants,* rev. ed. New York: Loveland, CO: Interweave Press.

Buchanan, R. 1995. *A dyer's garden: From plant to pot. Growing dyes for natural fibers.* Loveland, CO: Interweave Press.

Elias, J. 1999. *The A to Z guide to healing herbal remedies.* New York: Random House Value Publishing.

Furst, P. E. 1992. *Mushrooms: Psychedelic fungi,* rev. ed. Edgemont, PA: Chelsea House Publications.

Graedon, J., and T. Graedon. 2001. *The people's pharmacy guide to home and herbal remedies.* New York: St. Martin's Press.

Heatherly, A. N. 1998. *Healing Plants: A medicinal guide to native North American plants and herbs.* New York: Lyons Press.

Murray, M., and J. Pizzorno. 1998. *Encyclopedia of natural medicine,* rev. 2d ed. Rocklin, CA: Prima.

Null, G. 1998. *The complete encyclopedia of natural healing.* New York: Kensington.

Sanecki, K. N. 1998. *Growing and using herbs.* New York: Sterling.

Schneider, E. 1998. *Uncommon fruits and vegetables.* New York: William Morrow & Co, Inc.

Schultes, R. E., and A. Hoffman. 1980. *The botany and chemistry of hallucinogens,* 2d ed. Springfield, IL: Charles C. Thomas.

Vogel, V. J. 1990. *American Indian medicine.* Norman, OK: University of Oklahoma Press.

See also the Additional Reading entries in Chapter 24.

For a list of edible tropical and uncommon fruits, along with pertinent comments about them, check the following web site: www.mhhe.com/botany

Appendix 4

House Plants and Home Gardening

GROWING HOUSE PLANTS

If sales volume is an indication, house plants have never been more popular in the United States than they are now. Many are easy to grow and will brighten windowsills, planters, and other indoor spots for years if a few simple steps are followed to ensure their health and vigor.

Water

House plants are commonly overwatered, resulting in the unnecessary development of rots and diseases (see Table A4.2). As a rule, the surface of the potting medium should be dry to the touch before watering, but the medium should not be allowed to dry out completely unless the plant is dormant. Care should be taken, particularly during the winter, that the water is at room temperature. Rain water, if available, is preferred over tap water, particularly if the water has a high mineral content or is chlorinated. Broad-leaved plants should periodically have house dust removed with a damp sponge (never use detergents to clean surfaces—they remove protective waxes). Many plants benefit from a daily misting with water, particularly in heated rooms.

Containers

In time, plants may develop too extensive a root system for the pots in which they are growing (commonly called *becoming root bound*). Nutrients in the potting medium may become exhausted, salts and other residues from fertilizers and water may build up to the point of inhibiting growth, or the plants themselves may produce substances that accumulate until they interfere with the plant's growth. To resolve these problems, the plants should be periodically repotted and, if necessary, divided at the time of repotting.

Temperatures

Most house plants don't thrive where the temperatures are either too high or too low (see Table A4.1). In general, they tend to do best with minimum temperatures of about 13°C (55°F) and maximum temperatures of about 29°C (84°F). Many house plants that prefer warmer temperatures while actively growing also benefit from a rest period at lower temperatures after flowering.

Light

Next to overwatering, insufficient light most commonly contributes to the decline or death of house plants (see Table A4.2). This does not mean that house plants do better in direct sunlight—such light frequently damages them—but filtered sunlight (as, for example, through a muslin curtain) is usually better for the plant than the light available in the middle of the room. Plants can also thrive in artificial light of appropriate quality. Ordinary incandescent bulbs have too little light of blue wavelengths, and ordinary fluorescent tubes emit too little red light. A combination of the two, however, works very well. Generally, the wattage of the incandescent bulbs should be only one-fourth that of the fluorescent tubes in such a combination. Several types of fluorescent tubes specially balanced to imitate sunlight are also available.

Humidity

Dry air is hard on most house plants. The level of humidity around the plants can be raised by standing the pots in dishes containing gravel or crushed rock to which water has been added. The humidity level can also be raised through the use of humidifiers, which come in a variety of sizes and capacities. Daily misting, as mentioned previously, can also help.

COMMON HOUSE PLANTS

Explanation of symbols given with each plant

Water

= needs little water (applies primarily to cacti and succulents; these plants store water in such a way that the soil can be completely dry for a week or two without their being adversely affected)

= water regularly but not excessively; wait until the potting medium surface is dry to the touch before watering

= immerse pot in water for a few minutes each week and water frequently, never allowing the potting medium to become dry; do not, however, leave the base of the pot standing in water

= little to regular

= regular to frequent

Minerals in hard water are taxing on house plants, and commercial water softeners do not improve water for the plants. Use rain water or filtered water if at all possible; otherwise repot more often.

Temperature

= cool. Maximum 13°C–16°C (55°F–61°F); minimum 5°C–7°C (41°F–45°F)

= cool to medium. Maximum 18°C–21°C (65°F–70°F); minimum 10°C–13°C (50°F–55°F)

= medium. Maximum 22°C–29°C (72°F–84°F); minimum 14°C–17°C (57°F–63°F)

= medium to warm. Maximum 30°C–32°C (86°F–89°F); minimum 18°C–20°C (64°F–68°F)

= warm. Maximum 33°C–37°C (91°F–99°F); minimum 21°C–24°C (69°F–75°F)

Many house plants are native to the tropics, where they thrive under year-around warm temperatures, while cacti and succulents thrive in warm summers and cool winters, with wide daily temperature fluctuations. The closer one is able to imitate a plant's natural environment, the better the plant will grow (see Table A4.1).

Light

= needs shading or indirect daylight

= prefers bright light but needs to be screened from direct sunlight

= prefers direct sunlight

= shading to bright

= bright to direct light

As mentioned previously, improper lighting is second only to overwatering as a cause of problems for house plants. They are frequently given too little light or occasionally too much. A south-facing windowsill may be ideal for certain plants in midwinter but excessively bright in midsummer; conversely, a north-facing windowsill may have enough light for certain plants in midsummer but not in midwinter. Adjustable screens permit manipulation of daylight to suit the plants involved.

Humidity

= will tolerate dry air

= dry to regular

= will tolerate the air in most houses provided it is mist-sprayed occasionally

= regular to humid

= needs high humidity; use a humidifier if possible

Virtually all plants with or symbols benefit from having a pan of gravel with water beneath the pot.

Potting Medium

= requires a porous, slightly acid medium that drains immediately

= requires a loam that is slightly alkaline (e.g., a mixture of sand and standard commercial potting medium)

= requires a peaty potting mixture and acid fertilizer

TABLE A4.1

Environments Suitable for Common House Plants

PLANT	SCIENTIFIC NAME	ENVIRONMENTAL REQUIREMENTS	REMARKS
Abutilon	*Abutilon* spp. and hybrids		Avoid rangy growth by pinching out branch tip buds
Aechmea	*Aechmea fasciata*		*See* **Bromeliad;** produces side shoots that should be propagated as main plant dies after flowering
African lily	*Agapanthus* spp.		Do not repot until pot is full; keep cool in winter
African violet	*Saintpaulia* spp.		Let rest under cooler conditions after flowering; dislikes cold water
Agave	*Agave* spp.		Keep cool and dry in winter
Algerian ivy	*Hedera canariensis*		Resembles a larger-leaved English ivy; leaves often variegated
Aloe	*Aloe* spp.		Keep cool and dry in winter
Aluminum plant	*Pilea cadierei*		Plants do not usually survive long in houses
Amaryllis	*Amaryllis* spp.		Let leaves die back in fall; put bulb in cool, dark place until early spring; then repot, water, and fertilize weekly
Amazon lily	*Eucharis grandiflora*		
Anthurium	*Anthurium* spp.		If it grows but doesn't flower, try putting it in a cooler location for a few weeks
Aphelandra	*Aphelandra squarrosa*		Mist-spray frequently; fertilize regularly
Arabian violet	*Exacum affine*		Needs well-fertilized soil with humus content
Aralia (*see* **Fatsia**)			
Asparagus fern	*Asparagus plumosus*		Not a true fern; repot annually; fertilize weekly
Aspidistra	*Aspidistra* spp.		Sometimes called "cast iron plant" because it can stand neglect
Aucuba	*Aucuba japonica*		Must be kept cool in winter
Avocado	*Persea* spp.		Easily propagated from seed; provides good greenery but will not produce fruit indoors

TABLE A4.1

Environments Suitable for Common House Plants

PLANT	SCIENTIFIC NAME	ENVIRONMENTAL REQUIREMENTS	REMARKS
Azalea	*Rhododendron* spp.		Needs acid fertilizer; avoid warm locations
Bamboo (dwarf)	*Bambusa angulata*		Needs good air circulation, bright light
Basket grass	*Oplismenus hirtellus*		Hanging-basket cultivated variety sometimes sold as *Panicum variegatum*
Begonia	*Begonia* spp.		Easily propagated from leaves; repot regularly
Bilbergia	*Bilbergia* spp.		*See* **Bromeliad;** tough plant that can tolerate some neglect
Birdcatcher plant	*Pisonia umbellifera*		Sticky exudate on fruits attracts birds in the plant's native habitat of New Zealand; strictly a foliage plant in houses
Bird of paradise plant	*Strelitzia reginae*		Can be grown outdoors in milder climates
Bird's nest fern	*Asplenium nidus*		Produces a spongelike mass of roots at base; requires much water and regular fertilizing
Black-eyed Susan	*Thunbergia alata*		Annual climbing vine; grow from seed
Bloodleaf	*Iresine herbstii*		Easily propagated
Boston fern	*Nephrolepis exaltata*		Needs regular watering and fertilizing
Brake	*Pteris cretica*		*See* **Ferns** for propagation notes
Bromeliads	Many species		These plants absorb virtually all their water and nutrients through their leaves; they should not be placed in regular potting soil (regular potting soil usually kills them) nor should they be watered with high-calcium-content water; they produce offshoots that should be propagated as main plant dies after flowering

TABLE A4.1

Environments Suitable for Common House Plants

PLANT	SCIENTIFIC NAME	ENVIRONMENTAL REQUIREMENTS	REMARKS
Cacti	Many species		Contrary to popular belief, these slow-growing plants should not be grown in pure sand; add some humus to potting mixture and withhold water in winter; keep cool in winter
Caladium	*Caladium* spp.		Must have high humidity; keep root at no less than 18°C (65°F) in pot during winter
Calceolaria	*Calceolaria crenatiflora*		Discard after flowering
Calla lily	*Zantedeschia aethiopica*		After flowering, allow plant to dry up; repot in fall and start over
Cape jasmine	*Gardenia jasminoides*		Night temperatures below 22°C (72°F) needed to initiate flowering; needs cool temperatures in winter
Carpet plant	*Episcia lilacina*		
Carrion flower	*Stapelia* spp.		Cactuslike plants with foul-smelling flowers
Century plant (*see* **Agave**)			
Chinese evergreen	*Aglaonema costatum*		Needs warm temperatures and much water all year
Chrysanthemum	*Chrysanthemum* spp.		Plants may be artificially dwarfed through use of chemicals; flowering initiated by short days
Cineraria	*Senecio* x *hybridus* (other plants called cinerarias include *Chrysanthemum ptarmiciflorum, Senecio vira-vira,* and *Senecio cineraria*)		Needs cool temperatures; discard after flowering
Cliff brake	*Pellaea rotundifolia*		Hanging-basket fern; needs minimum temperature above 10°C (50°F)
Coffee	*Coffea arabica*		Handsome foliage plant that will produce fruit if self-pollinating variety is obtained
Coleus	*Coleus blumei, C. hybridus,* and others		To control size, restart plants from cuttings annually
Columnea	*Columnea* spp.		See **African violet** for culture conditions

TABLE A4.1

Environments Suitable for Common House Plants

PLANT	SCIENTIFIC NAME	ENVIRONMENTAL REQUIREMENTS	REMARKS
Copperleaf	*Acalypha wilkesiana*		Seldom survives average house environment for long
Corn plant	*Dracaena massangeana*		Uses much water when large; easy to grow
Croton	*Codiaeum* spp.		Needs constant high humidity and bright light
Crown of thorns	*Euphorbia milii* and *E. splendens*		Deviation from watering routine may result in loss of leaves, but plant generally recovers
Cyclamen	*Cyclamen* spp.		Fertilize weekly; keep cool; withhold water after flowering for few weeks, then start over
Donkey tail	*Sedum morganianum*		Keep cool and dry in winter
Dracaena	*Dracaena* spp.		Many kinds—all easy to grow and tolerant of some neglect
Dumbcane	*Dieffenbachia* spp.		Needs regular fertilizing; keep away from small children (poisonous)
Dwarf banana	*Musa cavendishii*		Will produce small, edible bananas if given enough light, water, and humidity
Dwarf cocos palm	*Microcoleum weddelianum*		Keep temperature above 18°C (65°F) at all times
Echeveria	*Echeveria* spp.		Keep cool in winter
English ivy	*Hedera helix*		Needs cool temperatures to grow at its best but will usually survive higher temperatures
False aralia	*Dizygotheca elegantissima*		Benefits from frequent mist-spraying
Fatshedera (= hybrid between *Fatsia* and *Hedera*)	*Fatshedera lizei*		Climbing shrub
Fatsia	*Fatsia japonica*		Also called *Aralia*
Ferns	Many species		Water regularly; propagate from spores or rhizomes
Figs:			
Climbing fig	*Ficus pumila*		Damp-sponge leaves regularly
Fiddleleaf fig	*Ficus lyrata*		Damp-sponge leaves regularly
Weeping fig	*Ficus benjamina*		Damp-sponge leaves regularly

TABLE A4.1
Environments Suitable for Common House Plants

PLANT	SCIENTIFIC NAME	ENVIRONMENTAL REQUIREMENTS	REMARKS
Fingernail plant	*Neoregelia* spp.		*See* **Bromeliad;** name from red tips of leaves
Fittonia	*Fittonia* spp.		Strictly terrarium plants—humidity too low elsewhere
Flame violet	*Episcia cupreata*		Add charcoal and peat to potting medium
Flowering maple	*Abutilon pictum*		Needs bright light to flower
Fuchsia	*Fuchsia* spp.		Soil must be alkaline
Gardenia (*see* Cape jasmine)			
Geranium	*Pelargonium* spp.		Make cuttings annually and discard parent plants each fall; keep cool through winter; available with various scents, including orange, rose, or coconut
Gloxinia	*Sinningia speciosa*		Fertilize heavily and water frequently; after flowering, withhold water and keep bulb cold for few weeks
Goldfish plant	*Columnea* spp.		Pot in mixture of leaf mold, fern bark, peat moss, and charcoal; use only rain water or filtered water
Goosefoot plant (not related to true goosefoots in the Goosefoot family—Chenopodiaceae)	*Nephthytis* spp.		Soil should have high organic content and be kept evenly moist; humidity must be maintained
Grape ivy	*Rhoicissus rhomboidea*		Tolerates low light better than most plants
Gynura (*see* Velvetleaf)			
Hart's tongue fern	*Phyllitis scolopendrium*		Doesn't thrive if water is of poor quality
Haworthia	*Haworthia* spp.		Aloe-like plants that need minimum temperatures above 10°C (50°F) in winter
Hen and chickens	*Sempervivum tectorum* and hybrids		Keep cool in winter
Hibiscus	*Hibiscus rosa-sinensis*		Fertilize weekly; does better outdoors in mild climates
Hippeastrum (*see* Amaryllis)			
Holly fern	*Cyrtomium falcatum*		Relatively tough fern; keep cool in winter

TABLE A4.1

Environments Suitable for Common House Plants

PLANT	SCIENTIFIC NAME	ENVIRONMENTAL REQUIREMENTS	REMARKS
Houseleek	*Sempervivum* spp.		Keep cool in winter
Hydrangea	*Hydrangea* spp.		Prune after flowering; keep cool in winter; blue-flowering plant can be converted to pink by changing the soil to acid, and vice versa (***Caution: Leaves and buds are poisonous***)
Hypocyrta	*Nematanthus* spp.		Plant several in a hanging basket
Hypoestes (*see* Pink polka dot plant)			
Impatiens	*Impatiens* spp.		Exceptionally easy to propagate from cuttings
Ivy-arum	*Epipremnum aureum*		Can tolerate some neglect
Jade plant	*Crassula argenta* and others		Keep cool in winter
Jamaican pansy	*Achimenes* spp.		Requires treatment similar to that for **Gloxinia**
Kaffir lily	*Clivia miniata*		Save the plant's energy by removing flowers as they wither
Kalanchoë	*Kalanchoë* spp.		Withhold water and fertilizer for a few weeks after flowering, then repot and start over
Lantana	*Lantana camara*		Fertilize twice a month; can be espaliered
Madagascar jasmine (*see* Stephanotis)			
Maidenhair fern	*Adiantum* spp.		Mist-spray regularly
Meyer fern (not a true fern)	*Asparagus densiflora* var. *meyeri*		Fertilize regularly; repot annually
Moneywort	*Lysimachia nummularia*		Hanging-pot plant; needs bright light
Moonstones	*Pachyphytum* spp.		Keep cool in winter
Moses in the cradle	*Rhoeo* spp.		Can tolerate some neglect
Mother-in-law's tongue (*see* Sansevieria)			
Mother of thousands	*Saxifraga* spp.		Also called **Saxifrage;** hanging plant; plantlets formed on runners can be removed and grown separately

TABLE A4.1

Environments Suitable for Common House Plants

PLANT	SCIENTIFIC NAME	ENVIRONMENTAL REQUIREMENTS	REMARKS
Neanthe palm	*Chamaedorea elegans*		Sometimes also called *Parlor palm;* stays less than 1 meter tall
Nerine	*Nerine* spp.		Water freely until flowering completed, then rest bulbs for a few months
Norfolk Island pine	*Araucaria* spp.		Needs cold temperatures 2°C–3°C (36°F–38°F) in winter
Octopus tree	*Schefflera octophylla*		Does best under cool conditions
Oleander	*Nerium oleander*		Keep pot cool in winter for better flowering; keep away from small children (very poisonous)
Orchid	Thousands of species	No single set of environmental conditions applies	Contrary to popular belief, the common *Cattleya* and related orchids do not need high temperatures and humidity; most can get along with a minimum temperature of 13°C (56°F) at night and a minimum humidity of 40%; most need bright light; they should never be placed in soil; pot them in sterilized pots with chips of fir bark or shreds of tree fern bark (*see* "Additional Reading" at the end of this appendix for culture references)
Oxtongue	*Gasteria* spp.		Can tolerate some neglect; needs a cool and relatively dry winter to flower
Palms	Many species		Use deep pots; fertilize regularly
Parlor palm	*Chamaedorea elegans* (*see also* Neanthe palm)		Palm is relatively easy to grow
Peperomia	*Peperomia* spp.		Many varieties; keep warm, humid; fertilize regularly
Persian violet	*Exacum affine*		Needs good air circulation
Philodendron	*Philodendron* spp.		Relatively tough plants; repot each spring
Piggyback plant	*Tolmiea menziesii*		Plantlets formed on leaves can be separated and propagated
Pilea (*see* Aluminum plant)			

TABLE A4.1
Environments Suitable for Common House Plants

PLANT	SCIENTIFIC NAME	ENVIRONMENTAL REQUIREMENTS	REMARKS
Pineapple	*Ananas comosus*		*See* **Bromeliad;** easily grown from the top of a pineapple; if plant has not flowered after one year, enclose it in a plastic bag with a ripe apple for a few days (ethylene from apple should initiate flowering); no temperature below 15°C (59°F)
Pink polka dot plant	*Hypoestes sanguinolenta*		Susceptible to diseases and pests; usually does not last long as a house plant
Pittosporum	*Pittosporum* spp.		Put several cuttings in one pot for bushy appearance
Pleomele (*see* Dracaena)			
Poinsettia	*Euphorbia pulcherrima*		After flowering, let plant dry under cool conditions until it loses its leaves, place in total darkness for a month, then restart
Prayer plant	*Maranta leuconeura*		Name derived from fact that leaves fold together in evening
Primrose	*Primula* spp.		Needs much water; does well outside in cool weather
Pteris (*see* Brake)			
Purple tiger	*Calathea amabilis*		Use pots that are broader than deep
Rosary plant	*Ceropegia woodii*		Hanging-pot plant whose potting medium must drain well or plant will not survive
Rubber plant	*Ficus elastica*		Do not overwater!
Sago palm	*Cycas revoluta*		Very slow growing (not a palm but a gymnosperm); never allow to dry out, but be certain water drains
Sansevieria	*Sansevieria* spp.		Perhaps the toughest of all houseplants—nearly indestructible
Satin pothos	*Scindapsus pictus*		Basket or pot plant
Schefflera	*Schefflera* spp.		Plants can get quite large if growth is not controlled
Screw pine	*Pandanus* spp.		If given space, can become large; mist-spray often
Selaginella (*see* Spike moss)			

TABLE A4.1

Environments Suitable for Common House Plants

PLANT	SCIENTIFIC NAME	ENVIRONMENTAL REQUIREMENTS	REMARKS
Sensitive plant	*Mimosa pudica*		Leaves fold when touched; does not usually last more than a few months in most houses
Sentry palm	*Howea fosterana*		Palm that is easy to grow
Setcreasea	*Setcreasea pallida*		Don't overwater
Shrimp plant	*Beloperone guttata*		Winter temperatures should be above 15°C (59°F)
Spathe flower	*Spathiphyllum* spp.		Prefers warm winters and even warmer summers; potting soil should have high organic content
Spider plant	*Chlorophytum comosum*		Plantlets formed at tips of stems can be propagated separately
Spiderwort	*Tradescantia* spp.		Easy to grow; do not overwater
Spike moss	*Selaginella* spp.		Can become a weed in greenhouses
Splitleaf philodendron	*Monstera deliciosa*		Plant adapts to various indoor locations quite well
Sprenger fern (not a true fern)	*Asparagus densiflora* var. *sprengeri*		Fertilize weekly; repot annually
Staghorn fern	*Platycerium* spp.		Tough plant; immerse in water weekly
Stephanotis	*Stephanotis floribunda*		Use very little fertilizer; keep cool in winter but water sparingly
Stonecrop	*Sedum* spp., *Crassula* spp.		Keep cool in winter
Stove fern	*Pteris cretica*		Water and fertilize regularly
String-of-pearls	*Senecio rowleyanus*		Keep cool in winter
Stromanthe	*Stromanthe* spp.		Needs warm shaded environment
Sundew	*Drosera* spp.		Sterilize pots; grow only on sphagnum moss (plants die soon in soil)
Syngonium	*Syngonium podophyllum*		Repot annually in spring

TABLE A4.1

Environments Suitable for Common House Plants

PLANT	SCIENTIFIC NAME	ENVIRONMENTAL REQUIREMENTS	REMARKS
Tillandsia	*Tillandsia* spp.		*See* **Bromeliad;** best-known species is called *Spanish moss,* which does best suspended from wires or other plants
Ti plant	*Cordyline terminalis*		Needs high humidity; seems to do better with other plants in pot
Treebine	*Cissus antarctica*		Plant does not do well in acid potting medium
Umbrella plant	*Cyperus* spp.		One of very few plants that need to stand in water
Velvetleaf	*Gynura sarmentosa*		Gets "stringy" but is easily restarted from cuttings; flowers have unpleasant odor
Venus's flytrap	*Dionaea muscipula*		Sterilize pots; grow only in sphagnum moss; repot annually
Vriesia	*Vriesia* spp.		*See* **Bromeliad**
Wandering Jew	*Zebrina pendula*		Easy to grow; do not overwater
Wax plant	*Hoya* spp.		Climber; leave pot in one place—does not like to be moved
Zebrina (*see* Wandering Jew)			

GROWING VEGETABLES

General Tips on Vegetable Growing

Seed Germination

Many gardeners germinate larger seeds (e.g., squash, pumpkin) in damp newspaper. A few sheets of newspaper are soaked in water for a minute and then hung over a support for about 15 minutes or until the water stops dripping. The seeds are then lined up in a row on the newspaper, wrapped, and the damp mass is placed in a plastic bag. The bag is tied off or sealed and placed in a warm (not hot!), shaded location (the floor beneath a running refrigerator is an example). Depending on the species, germination should occur within 2 to several days.

Tiny seeds (e.g., carrots, lettuce) may be mixed with clean sand before sowing to bring about a more even distribution of the seed in the rows.

Transplanting

Roots should be disturbed as little as possible when seedlings or larger plants are transplanted. Even a few seconds' exposure to air will kill root hairs and smaller roots. They should be shaded (e.g., with newspaper) from the sun and transplanted late in the day or on a cool, cloudy day if at all possible. To minimize the effects of transplanting, immediately water the seedlings or plants in their new location with a dilute solution of vitamin B/hormone preparation (e.g., Superthrive).

Bulb or Other Plants with Food-Storage Organs

All plants with food-storage organs (e.g., beets, carrots, onions) develop much better in soil that is free of lumps and rocks. If possible, the areas where these plants are to be grown should be dug to a minimum depth of 30 centimeters (12

TABLE A4.2

Common Ailments of House Plants

PROBLEM SYMPTOMS	POSSIBLE CAUSES
Wilting or collapse of whole plant	Lack of water; too much heat; too much water or poor drainage resulting in root rot
Yellowish or pale leaves	Insufficient light; too much light; microscopic pests (especially spider mites); too much or too little fertilizer
Brown, dry leaves	Humidity too low; too much heat; poor air circulation; lack of water
Tips and margins of leaves brown	Mineral content of water; drafts; too much sun or heat; too much or too little water
Ringed spots on leaves	Water too cold
Leaves falling off	Improper watering or water too cold; excessive use of fertilizer or wrong fertilizer; too much sun or, if lower leaf drop only, too little light
Stringy growth	Needs more light; too much fertilizer
Base of plant soft or rotting	Overwatering
No flowers or flower buds drop	Too much or too little light; night temperatures too high
Water does not drain	Drain hole plugged; potting mixture has too much clay
Mildew present	Fungi present—arrest with sulfur dust

COMMON PESTS	CONTROLS
Aphids	Wash off under faucet or spray with soapy (not detergent) water; pyrethrum or rotenone sprays also effective
Mealybugs	Remove with cotton swabs dipped in alcohol; spray with Volck oil
Scale insects	Remove by hand; spray with Volck oil
Spider mites	Use sprays containing small amounts of xylene (act as soon as possible—spider mites multiply very rapidly)
Thrips	Spray with pyrethrum/rotenone or Volck oil sprays
White flies	Spray with soapy water or pyrethrum/rotenone sprays every 4 days for 2 weeks (only the adults are susceptible to the sprays)

For additional controls, see Appendix 2, "Biological Controls"

inches) and the soil sifted through a 0.7 centimeter (approximately 0.25 inch) mesh before planting. Obviously such a procedure is not always practical, but it can yield dramatic results.

Cutworms

Cutworms forage at or just beneath the surface of the soil. Their damage to young seedlings can be minimized by a collar placed around each plant. Tuna cans with both ends removed make effective collars when pressed into the ground to a depth of about 1 to 2 centimeters (0.4 to 0.8 inch).

Protection Against Cold

Some seedlings can be given an earlier start outdoors if plastic-topped coffee cans with the bottoms removed are placed over them. The plastic lids can be taken off during the day and replaced at night during cool weather. Conical paper frost caps can serve the same purpose.

Watering

Proper watering promotes healthy growth. It is much better to water an area thoroughly (e.g., for 20 to 30 minutes) every few days than to wet the surface for a minute or two daily. Shallow daily watering promotes root development near the surface, where midsummer heat can damage the root system. Conversely, too much watering can leach minerals out of the topsoil.

Fertilizers

Manures, bone and blood meals, and other organic fertilizers, which release the nutrients slowly and do not "burn" young plants, are preferred. Plants will utilize minerals from any available source, but in the long run, the plants will be healthier and subject to fewer problems if they are not given sudden boosts with liquid chemical fertilizers.

Pests and Diseases

Preferred means of dealing with pests are listed in Appendix 2 "Biological Controls." If sprays are necessary, biodegradable substances such as rotenone and pyrethrum should be used.

COMMON VEGETABLES AND THEIR NUTRITIONAL VALUES

Note: The nutritional values (NV) given are per 100 grams (3.5 ounces), edible portion, as determined by the United States Department of Agriculture.

Asparagus

NV (spears cooked in water): 20 calories; protein, 2.2 gm; fats, 0.2 gm; vit. A, 900 I.U.; vit. B1, 0.16 mg; vit. B2, 0.18 mg; niacin, 1.5 mg; vit. C, 26 mg; fiber, 0.7 gm; calcium, 21 mg; phosphorus, 50 mg; iron, 0.6 mg; sodium, 1.0 mg; potassium, 208 mg

Asparagus can be started from seed, but time until the first harvest can be reduced by a year or two if planting begins with 1-year-old root clusters of healthy, disease-resistant varieties (e.g., "Mary Washington"). Asparagus requires little care if appropriate preparations are made before planting in the permanent location. Seeds should be sown sparsely, and the seedlings thinned to about 7.5 centimeters (3 inches) apart. Before transplantation the following spring, dig a trench 30 to 60 centimeters (12 to 24 inches) deep and about 50 centimeters (20 inches) wide in an area that receives full sun, usually along one edge of the garden. If the soil is heavy, place crushed rock or gravel on the bottom of the trench to provide good drainage. Add a layer of steer manure about 10 centimeters (4 inches) thick, followed by about 7.5 centimeters (3 inches) of rich soil that has been prepared by thorough mixing with generous quantities of steer manure and bone meal. Place root clusters about 45 centimeters (18 inches) apart in the trench and cover with about 15 centimeters (6 inches) of prepared soil (be sure not to allow root clusters to dry out). As the plants grow, gradually fill in the trench. If 1-year-old roots are planted, wait for 2 years before harvesting tips; if 2-year-old roots are planted, some asparagus may be harvested the following year. In all cases, no harvesting should be done after June so that the plants may build up reserves for the following season. Cut shoots below the surface but well above the crown before the buds begin to expand. Cut all stems to the ground after they have turned yellow later in the season.

Beans

String, or Snap, Beans

NV (young pods cooked in water): 25 calories; protein, 1.6 gm; fats, 0.2 gm; vit. A, 540 I.U.; vit. B1, 0.07 mg; vit. B2, 0.09 mg; niacin, 0.5 mg; vit. C, 12 mg; fiber, 1.0 gm; calcium, 50 mg; phosphorus, 37 mg; iron, 0.6 mg; sodium, 4 mg; potassium, 151 mg

String, or snap, beans are warm-weather plants, although they can be grown almost anywhere in the United States. Wait until all danger of frost has passed and the soil is warm. Prepare the soil, preferably the previous winter, by digging to a depth of 30 centimeters (12 inches) and mixing in aged manure and bone meal. Pulverize the soil just before sowing; if soil has a low pH, add lime. Plant seeds thinly in rows about 40 to 50 centimeters (16 to 20 inches) apart; thin plants to 10 centimeters (4 inches) apart when the first true leaves have developed. Beans respond unfavorably to a very wet soil—do not overwater! In areas with hot summers, beans also prefer some light shade, particularly in midafternoon. As the bean plants grow, nitrogen-fixing bacteria invade the roots and supplement the nitrogen supply. Early vigorous growth can be enhanced by inoculating the seeds with such bacteria, which are available commercially in a powdered form. To maintain a continuous supply of green beans, plant a new row every 2 to 3 weeks during the growing season until 2 months before the first predicted frost. Cultivate regularly to control weeds, taking care not to damage root systems. Do not harvest or work with beans while the plants are wet, as this may invite disease problems.

Pole Beans

NV similar to those of string beans

Soil preparation and cultivation are the same as for string beans. Plant beans in hills around poles that are not less than 5 centimeters (2 inches) in diameter and at least 2 meters (6.5 feet) tall. As beans twine around their supports, it helps to tie them to the support with plastic tape as they grow. If harvested before the pods are mature, pole beans will produce over a longer period of time than bush varieties.

Lima Beans

NV (immature seeds cooked in water): 111 calories; protein, 7.6 gm; fats, 0.5 gm; vit. A, 280 I.U.; vit. B1, 0.18 mg; vit. B2, 0.10 mg; niacin, 1.3 mg; vit. C, 17 mg; fiber, 1.8 gm; calcium, 47 mg; phosphorus, 121 mg; iron, 2.5 mg; sodium, 1.0 mg; potassium, 422 mg

Lima beans take longer to mature than other beans and are more sensitive to wet or cool weather. They definitely need warm weather to do well. Prepare soil and cultivate as for string beans.

Soybeans

NV (dry, mature seeds): 403 calories; protein, 34.1 gm; fats, 17.7 gm; vit. A, 80 I.U.; vit. B1, 1.10 mg; vit. B2, 0.31 mg; niacin, 2.2 mg; vit. C[1]; fiber, 4.9 gm; calcium, 226 mg; phosphorus, 554 mg; iron, 8.4 mg; sodium, 5.0 mg; potassium, 1,677 mg

Prepare soil and cultivate as for string beans.

---

1. Values not available.

Broad (Fava) Beans

NV (dry, mature seeds): 338 calories; protein, 25.1 gm; fats, 1.7 gm; vit. A, 70 I.U.; vit. B1, 0.5 mg; vit. B2, 0.3 mg; niacin, 2.5 mg; vit. C[1]; fiber, 6.7 gm; calcium, 47 mg; phosphorus, 121 mg; iron, 2.5 mg; sodium, 1.0 mg; potassium, 422 mg

Unlike other beans, broad beans need cool weather for their development. Sow as early as possible (in mild climates, they may be sown in the fall, as they can withstand light frosts). Since the plants occupy a little more space than bush beans, plant in rows about 0.9 to 1 meter (3 feet or more) apart and thin to about 20 centimeters (8 inches) apart in the rows. After the first pods mature, pinch out the tips to promote bushier development. To most palates, broad beans do not taste as good as other types of beans.

Beets

NV (cooked in water): 32 calories; protein, 1.1 gm; fats, 0.1 gm; vit. A, 20 I.U.; vit. B1, 0.03 mg; vit. B2, 0.04 mg; vit. C, 6 mg; fiber, 0.8 gm; calcium, 14 mg; phosphorus, 23 mg; iron, 0.5 mg; sodium, 43 mg; potassium, 208 mg

Beets will grow in a variety of climates but do best in cooler weather. They can tolerate light frosts and can be grown on a variety of soil types, although they prefer a sandy loam supplemented with well-aged organic matter. As with any bulb or root crop, they develop best in soil that is free of rocks and lumps.

Beet "seeds" are really fruits containing several tiny seeds. Plant them in rows 40 to 60 centimeters (16 to 24 inches) or more apart and thin to about 10 centimeters (4 inches) apart in the rows after germination. After harvesting, the beets will keep in cold storage for up to several months. The leaves, if used when first picked, make a good substitute for spinach.

Broccoli

NV (spears cooked in water): 26 calories; protein, 3.1 gm; fats, 0.3 gm; vit. A, 2,500 I.U.; vit. B1, 0.09 mg; vit. B2, 0.2 mg; niacin, 0.8 mg; fiber, 1.5 gm; calcium, 88 mg; phosphorus, 62 mg; iron, 0.8 mg; sodium, 10 mg; potassium, 267 mg

Broccoli is a cool-weather plant that will thrive in any well-prepared soil, providing that it has not been heavily fertilized just prior to planting (fresh fertilizer promotes rank growth). The plants can stand light frosts and are planted in both the spring and fall in areas with mild climate. Although broccoli may continue to produce during the summer, most growers prefer not to keep the plants going during warm seasons because of the large numbers of pest insects they may attract.

Sow seeds indoors and transplant outdoors after danger of killing frosts has passed. Place plants about 0.9 to 1 meter (3 feet or more) apart and keep well watered. Keep area weeded and pests under control. Harvest heads (bundles of spears) while they are still compact. Smaller heads will develop very shortly after the first harvest; if these are removed regularly, the plants will continue to produce for some time, although the heads become smaller as the plants age.

Cabbage

NV (raw): 24 calories; protein, 1.3 gm; fats, 0.2 gm; vit. A, 130 I.U.; vit. B1, 0.05 mg; vit. B2, 0.05 mg; niacin, 0.3 mg; vit. C, 47 mg; fiber, 1.0 gm; calcium, 49 mg; phosphorus, 29 mg; iron, 0.4 mg; sodium, 20 mg; potassium, 233 mg

Growth requirements of cabbage are similar to those of broccoli.

Carrots

NV (raw): 42 calories; protein, 1.1 gm; fats, 0.2 gm; vit. A, 11,000 I.U.; vit. B1, 0.6 mg; vit. B2, 0.5 mg; niacin, 0.6 mg; vit. C, 8 mg; fiber, 1.0 gm; calcium, 37 mg; phosphorus, 36 mg; iron, 0.7 mg; sodium, 47 mg; potassium, 341 mg

Carrots are hardy plants that can tolerate a wide range of climate and soils, but the soil must be well prepared, free of rocks and lumps, and preferably not too acidic. The seeds are slow to germinate. Plant in rows a little more than 30 centimeters (12 inches) apart and thin seedlings to about 5 centimeters (2 inches) apart in the rows. Weed the rows regularly until harvest. Carrots keep well in below-ground storage containers when freezing weather arrives.

Cauliflower

NV (cooked in water): 22 calories; protein, 2.3 gm; fats, 0.2 gm; vit. A, 60 I.U.; vit. B1, 0.09 mg; vit. B2, 0.08 mg; niacin, 0.6 mg; vit. C, 55 mg; fiber, 1.0 gm; calcium, 21 mg; phosphorus, 42 mg; iron, 0.7 mg; sodium, 9 mg; potassium, 206 mg

Growth requirements of cauliflower are similar to those of broccoli except that heavier fertilizing is required. As cauliflower heads develop, protect them from the sun by tying the larger leaves over the tender heads. Harvest while the heads are still solid.

Corn

NV (cooked sweet corn kernels): 83 calories; protein, 3.2 gm; fats, 1.0 gm; vit. A, 400 I.U. (yellow varieties; white varieties have negligible vit. A content); vit. B1, 0.11 mg; vit. B2, 0.10 mg; niacin, 1.3 mg; vit. C, 7 mg; fiber, 0.7 gm; calcium, 3 mg; phosphorus, 89 mg; iron, 0.6 mg; sodium, trace; potassium, 165 mg

---

1. Values not available.

There are several types of corn (e.g., popcorn, flint corn, dent corn), but sweet corn is the only type grown to any extent by home gardeners. It can be grown in any location where there is at least a 10-week growing season and warm summer weather.

Corn does best in a fertile soil, which should be prepared by mixing with compost and liberal amounts of chicken manure or fish meal. Since corn is wind-pollinated, it can be helpful to grow the plants in several short rows at right angles to the prevailing winds rather than in a single long row. For best results, use only fresh seeds and plant in rows 60 centimeters (24 inches) apart for dwarf varieties and 90 centimeters (36 inches) apart for standard varieties. Thin to 20 to 30 centimeters (8 to 12 inches) apart in the rows after the plants have produced three to four leaves. Cultivate frequently to control weeds. The corn is ready to harvest when the silks begin to wither.

Cucumber

NV (raw, with skin): 15 calories; protein, 0.9 gm; fats, 0.1 gm; vit. A, 250 I.U.; vit. B1, 0.03 mg; vit. B2, 0.04 mg; niacin, 0.2 mg; vit. C, 11 mg; fiber, 0.6 mg; calcium, 25 mg; phosphorus, 27 mg; iron, 1.1 mg; sodium, 6 mg; potassium, 160 mg

Until drought- and disease-resistant varieties were developed in recent years, cucumbers were considered rather temperamental plants to grow. The newer varieties are no more difficult to raise than those of most other common vegetables.

The soil should be a light loam—neither too heavy nor too sandy. It should be mixed with well-aged manure and compost and heaped into small mounds about 2 meters (6.5 feet) apart. Five to six seeds should be planted in each mound about 2.5 centimeters (1 inch) below the surface in the middle of the spring. When the plants are about 1 decimeter (4 inches) tall, thin to three plants per mound. Cultivate regularly, and to promote continued production, pick all cucumbers as soon as they attain eating size.

Eggplant

NV (cooked in water): 19 calories; protein, 1.0 gm; fats, 0.2 gm; vit. A, 10 I.U.; vit. B1, 0.05 mg; vit. B2, 0.04 mg; niacin, 0.5 mg; vit. C, 3 mg; fiber, 0.9 gm; calcium, 11 mg; phosphorus, 21 mg; iron, 0.6 mg; sodium, 1 mg; potassium, 150 mg

Eggplant is strictly a hot-weather plant that is sensitive to cold weather or dry periods and needs heavy fertilizing. Since seedling development is initially slow, plant the seeds indoors about 2 months before the plants will be set out, which should be about 5 to 6 weeks after the last average date of frost.

Eggplants do best in enriched sandy soils that are supplemented with additional fertilizer once a month. Never permit them to dry out. Place the seedlings about 70 to 80 centimeters (28 to 32 inches) apart in rows 0.9 to 1 meter (3 feet or more) apart. Some staking of the plants may be desirable. The fruits are ready to harvest when they have a high gloss. They are still edible after greenish streaks appear and the gloss diminishes, but they are not as tender at this stage.

Lettuce

NV (crisp, cabbage-head varieties): 13 calories; protein, 0.9 gm; fats, 0.1 gm; vit. A, 330 I.U.; vit. B1, 0.06 mg; vit. B2, 0.06 mg; niacin, 0.3 mg; vit. C, 6 mg; fiber, 0.5 gm; calcium, 20 mg; phosphorus, 22 mg; iron, 0.5 mg; sodium, 9 mg; potassium, 175 mg

NV (leaf varieties): 18 calories; protein, 1.3 gm; fats, 0.3 gm; vit. A, 1,900 I.U.; vit. B1, 0.05 mg; vit. B2, 0.08 mg; niacin, 0.4 mg; vit. C, 18 mg; calcium, 68 mg; phosphorus, 25 mg; iron, 1.4 mg; sodium, 9 mg; potassium, 264 mg

This favorite salad plant comes in a wide variety of types and forms, all of which do better in cooler weather, although a few of the leaf types (e.g., oak leaf) can tolerate some hot periods. As long as the individual plants are given room to develop and the soil is not too acid, most varieties can be grown on a wide range of soil types.

Since lettuce can stand some frost, sow the seeds outdoors as early in the spring as the ground can be cultivated. Do not cover the seeds with more than a millimeter or two of soil—they need light to germinate. Mix the soil with a well-aged manure and a general-purpose fertilizer a week or two before sowing. Plant seedlings about 30 centimeters (12 inches) apart in rows 30 to 40 centimeters (12 to 16 inches) apart. For best results, do not allow the soil to dry out, and plant only varieties suited to local conditions. The most common crisp, cabbage-head varieties found in produce markets will not form heads in hot weather, and many others will bolt (begin to flower) during hot weather and longer days. Cultivate weekly between rows to promote rapid growth and to control weeds.

Onion

NV (raw): 38 calories; protein, 1.5 gm; fats, 0.1 gm; vit. A, 40 I.U. (yellow varieties only); vit. B1, 0.03 mg; vit. B2, 0.04 mg; niacin, 0.2 mg; vit. C, 10 mg; fiber, 0.6 gm; calcium, 27 mg; phosphorus, 36 mg; iron, 0.5 mg; sodium, 10 mg; potassium, 157 mg

These easy-to-grow vegetables do best in fertile soils that are free of rocks and lumps, are well drained, and are not too acid or sandy.

Onions may take several months to mature from seed. The viability of the seed decreases rapidly after the first year. Bulb formation is determined by day length rather than by the total number of hours in the ground. Because of these characteristics of onions, most gardeners prefer to purchase *sets* (young plants that already have a small bulb) from commercial growers, although green or bunching onions are still easily grown from seed.

Plant the sets upright 6 to 7.5 centimeters (2.5 to 3 inches) apart in rows and firm in place. Except for weeding, watering, and occasional shallow cultivation, they will need little care until harvest about 14 weeks later. The onions are mature when the tops fall over. After they are pulled from the ground, allow them to dry in the shade for 2 days. Then remove the tops 2 to 3 centimeters (about 1 inch) above the bulbs, and spread out the bulbs to continue curing for 2 to 3 more weeks. After this, store them in sacks or other containers that permit air circulation until needed.

Peas

NV (green, cooked in water): 71 calories; protein, 5.4 gm; fats, 0.4 gm; vit. A, 540 I.U.; vit. B1, 0.28 mg; vit. B2, 0.11 mg; niacin, 2.3 mg; vit. C, 20 mg; fiber, 2.0 gm; calcium, 23 mg; phosphorus, 99 mg; iron, 1.8 mg; sodium, 1 mg; potassium 196 mg

Peas are strictly cool-weather plants that generally produce poorly when the soil becomes too warm. Seeds should be planted in the fall or very early spring. As is the case with beans, peas receive a better start if the seeds are inoculated with nitrogen-fixing bacteria (see discussion of beans) at planting time. Prepare the soil by mixing thoroughly with liberal amounts of aged manure and bone meal. Plant the seeds about 2.5 centimeters (1 inch) deep in heavy soil or 5 centimeters (2 inches) deep in light, sandy soil, about 2.5 centimeters (1 inch) apart in single rows for dwarf bush varieties or in double files 15 centimeters (6 inches) apart for standard varieties, with intervals of 80 to 90 centimeters (32 to 36 inches) between the rows. After germination, thin the plants to 10 centimeters apart. Place support wires, strings, or chicken wire between the rows at the time of planting; peas will not do well without such support.

Green peas should be picked while still young and then cooked or frozen immediately, as the sugars that make them sweet are converted to starch within 2 to 3 hours after harvest.

Peppers

NV (raw sweet or bell peppers): 22 calories; protein, 1.2 gm; fats, 0.2 gm; vit. A, 420 I.U.; vit. B1, 0.08 mg; vit. B2, 0.08 mg; niacin, 0.5 mg; vit. C, 128 mg; fiber, 1.4 gm; calcium, 9 mg; phosphorus, 22 mg; iron, 0.7 mg; sodium, 13 mg; potassium, 213 mg

Peppers, like eggplants, are strictly hot-weather plants for most of their growing season, but unlike eggplants, they actually do better toward the end of their season if temperatures have moderated somewhat. Sweet or bell peppers are closely related and have similar cultural requirements.

Plant seeds indoors 8 to 10 weeks before the outdoor planting date, which is generally after the soil has become thoroughly warm. They will grow in almost any sunny location in a wide variety of soils, but to obtain the large fruits seen in produce markets, the plants need to be fertilized heavily and watered regularly. Plant seedlings 50 to 60 centimeters (20 to 24 inches) apart in rows that are 60 to 90 centimeters (24 to 36 inches) apart. Sweet peppers can be harvested at almost any stage and are still perfectly edible after they have turned red.

Potatoes

NV (baked in skin): 93 calories; protein, 2.6 gm; fats, 0.1 gm; vit. A, trace; vit. B1, 0.10 mg; vit. B2, 0.04 mg; niacin, 1.7 mg; vit. C, 20 mg; fiber, 0.6 gm; calcium, 9 mg; phosphorus, 65 mg; iron, 0.7 mg; sodium, 4 mg; potassium, 503 mg

Potatoes grow best in a rich, somewhat acid, well-drained soil that has had compost or well-aged manure added to it. They are subject to several diseases, and it is advisable to use disease-free seed potatoes purchased from a reputable dealer. Two to 3 weeks before the average date of the last spring frost, plant the seed potatoes whole or cut into several pieces, making sure that each piece has at least one eye. Place the potato pieces about 30 centimeters (12 inches) or more apart at a depth of about 12 to 15 centimeters (5 to 6 inches) in rows 0.9 to 1 meter (about 3 feet) apart. Later plantings are feasible. Spread a thick mulch (e.g., straw) over the area after planting to keep soil temperatures down and to retain soil moisture.

Potatoes are ready for harvest when the tops start turning yellow, but they may be left in the ground for several weeks after that if the soil is not too wet. After harvest, wash the potatoes immediately and place them in a dry, cool, dark place until needed. If left exposed to light, the outer parts of the potato turn green; poisonous substances are produced in these tissues, and such potatoes should be discarded.

Spinach

NV (raw): 26 calories; protein, 3.2 gm; fats, 0.3 gm; vit. A, 8,100 I.U.; vit. B1, 0.10 mg; vit. B2, 0.20 mg; niacin, 0.6 mg; vit. C, 51 mg; fiber, 0.6 gm; calcium, 93 mg; phosphorus, 51 mg; iron, 3.1 mg; sodium, 71 mg; potassium, 470 mg

Spinach is a cool-season crop that goes to seed as soon as the days become long and warm. It should be planted in the fall or early spring. If protected by straw or other mulches, it will overwinter in the ground in most areas and be ready for use early in the spring. Spinach has a high nitrogen requirement and reacts negatively to acid soils. It is otherwise easy to grow. Mix the soil thoroughly with aged manure and bone meal before planting. Plant seedlings 3 to 5 centimeters (1 to 2 inches) apart in rows 40 to 50 centimeters (16 to 20 inches) apart. Keep the plants supplied with adequate moisture and their growing area free of weeds. Harvest the whole plant when a healthy crown of leaves develops.

Squash

NV (cooked zucchini): 12 calories; protein, 1.0 gm; fats, 0.1 gm; vit. A, 300 I.U.; vit. B1, 0.05 mg; vit. B2, 0.08 mg; niacin, 0.8 mg; vit. C, 9 mg; fiber, 0.6 gm; calcium, 25 mg; phosphorus, 25 mg; iron, 0.4 mg; sodium, 1 mg; potassium, 141 mg

All varieties of squash are warm-weather plants, and all are targets of a variety of pests. Thorough preparation of the soil before planting pays dividends in production and in the health of the plants. Mix compost and aged manure with the soil and heap the soil in small hills about 1.2 meters (4 feet) apart from one another. Plant four to five seeds in each hill and thin the seedlings to three after they are about 10 centimeters (4 inches) tall. Summer squashes (e.g., zucchini) mature in about 2 months, while winter squashes (e.g., acorn) can take twice as long to mature. Summer squashes should be harvested while very young—they can balloon, seemingly overnight, into huge fruits. Winter squashes should be harvested before the first frost; only clean, undamaged fruits will store well. Keep such squashes laid out in a cool, dry place until use and not piled on top of one another. Check them occasionally for the development of surface fungi.

Tomatoes

NV (raw, ripe): 22 calories; protein, 1.1 gm; fats, 0.2 gm; vit. A, 900 I.U.; vit. B1, 0.06 mg; vit. B2, 0.04 mg; niacin, 0.7 mg; vit. C, 23 mg; fiber, 0.5 gm; calcium, 13 mg; phosphorus, 27 mg; iron, 0.5 mg; sodium, 3 mg; potassium, 244 mg

These almost universally used fruits are easy to grow, providing one understands a few basic aspects of their cultural requirements:

(1) Many tomato plants normally will not initiate fruit development from their flowers when night temperatures drop below 14°C (57°F) or day temperatures climb above 40°C (104°F). For the earliest yields, seeds may be germinated indoors several weeks before the plants are to be placed outside, but little is accomplished by transplanting before the night temperatures begin to remain above 14°C (57°F); in addition, some growers insist that plants given an early start indoors do not always do as well later as those germinated outdoors.

(2) Tomatoes require considerably more phosphorus than nitrogen from any fertilizers added to the soil where they are to be grown. Many inexperienced gardeners make the mistake of giving the plants lawn or general-purpose fertilizers that are proportionately high in nitrogen. As a result, the plants may grow vigorously but produce very few tomatoes. Give tomatoes bone meal, tomato food (Magamp is an excellent commercial slow-release preparation), or steer manure mixed with superphosphate.

(3) Tomato plants seem more susceptible than most to soil fungi and to root-knot nematodes. The damage caused by these organisms may not become evident until the plants begin to bear. Then the lower leaves begin to wither, and yellowing progresses up the plant or there seems to be a general loss of vigor and productivity. Using disease- and nematode-resistant varieties (usually indicated by the letters V, F, and N on seed packets) is by far the simplest method of controlling these problems. Another effective control involves dipping the seedling roots in an emulsion of 0.25% corn oil in water when transplanting; experiments have shown that the corn oil greatly reduces root-knot nematode infestation.

(4) Many garden varieties of tomatoes need to be staked to keep fruits off the ground where snails and other organisms can gain easy access to them. When using wooden stakes, be sure they have a diameter of 5 centimeters (2 inches) or more, and tie the plants securely to the stakes with plastic tape. Thinner stakes are likely to break or collapse when the plants grow to a height of 2 meters (6.5 feet) or more. Some growers prefer to use heavy wire tomato towers instead of stakes.

(5) Hornworms and tomato worms almost invariably appear on tomato plants during the growing season. They can virtually strip a plant and ruin the fruits if not controlled. Fortunately, control is simple and highly effective with the use of *Bacillus thuringiensis,* which is discussed in Appendix 2.

(6) The eating season for garden tomatoes can be extended for about 2 months past the first frost if all the green tomatoes are picked before frost occurs. Place the tomatoes on sheets of newspaper on a flat surface in a cool, dry place, where they will ripen slowly a few at a time. Generally, the taste of tomatoes ripened this way is superior to that of hothouse tomatoes sold in produce markets. Be sure when picking the green tomatoes to handle them very gently, as they bruise very easily and molds quickly develop in the bruised areas.

For additional illustrated information on pruning fruit trees, grapevines, roses, and berry plants, as well as an introduction to bonsai, grafting, and lawn care, check the following web site: www.mhhe.com/botany

ADDITIONAL READING

Adams, C. F. 1981. *Nutritional value of American foods in common units.* Washington, DC: Government Printing Office.

Brickell, C. 1999. *American Horticultural Society: Pruning and training.* New York: DK Publishing.

Brickell, C., and J. D. Zuk (Eds.). 1996. *American Horticultural Society A to Z encyclopedia of garden plants.* New York: DK Publishing.

Bubel, N. 1998. *The new seed-starters handbook.* Emmaus, PA: Rodale Press.

Hartmann, H. T., D. E. Kester, and F. T. Davies, Jr. 1997. *Plant propagation: Principles and practices,* 6th ed. Upper Saddle River, NJ: Prentice-Hall.

Hessayon, D. G. 1998. *The house plant expert.* New York: Sterling.

Hill, L. 1992. *Fruits and berries for the home garden,* rev. ed. Pownal, VT: Garden Way Publishing.

Jeavons, J. 1995. *How to grow more vegetables,* 5th ed. Berkeley, CA: Ten Speed Press.

Kramer, J. 1992. *Know your houseplants.* New York: Lyons Press.

Norman, K. 1998. *Essential bonsai.* New York: Anness Publishing.

Sunset Editors. 1999. *Basic gardening illustrated.* Menlo Park, CA: Lane Publishing.

Sunset Editors. 2001. *Northeast garden book.* Menlo Park, CA: Lane Publishing.

Sunset Editors. 2001. *Western garden book.* Menlo Park, CA: Lane Publishing.

Appendix 5

Metric Equivalents and Conversion Tables

Metric System of Measurement

APPLICATION	INTERNATIONAL SYSTEM OF UNITS	ENGLISH SYSTEM EQUIVALENTS
Length	Kilometer	0.62137 miles
	Meter	39.37 inches
	Centimeter	0.3937 inch
	Millimeter	0.03937 inch
	Micrometer	0.00003937 inch
	Nanometer	0.00000003937 inch
	Angstrom	0.0000000003937 inch
Mass (Weight)	Metric ton	2,200 pounds
	Kilogram	2.2 pounds
	Gram	0.03527 ounce
	Milligram	0.00003527 ounce
Volume	Liter	1.06 quart
	Milliliter	0.00106 quart
	Cubic meter	35.314 cubic feet
	Cubic centimeter	0.061 cubic inch
Area	Hectare	2.471 acres
Temperature	To convert Celsius to Fahrenheit, multiply the Celsius figure by 9, divide the total by 5, and add 32.	

Conversion Table

TO CONVERT	TO	MULTIPLY BY
Millimeters	Inches	0.039
Centimeters	Inches	0.39
Inches	Centimeters	2.54
Feet	Centimeters	30.48
Ounces	Grams	28.35
Pounds	Grams	453.6
Pounds	Kilograms	0.4536
Grams	Ounces	0.035
Kilograms	Pounds	2.205
Fluid ounces	Milliliters	29.57
Quarts	Liters	0.9463
Milliliters	Fluid ounces	0.03
Liters	Quarts	1.057

Temperature Conversion Scale

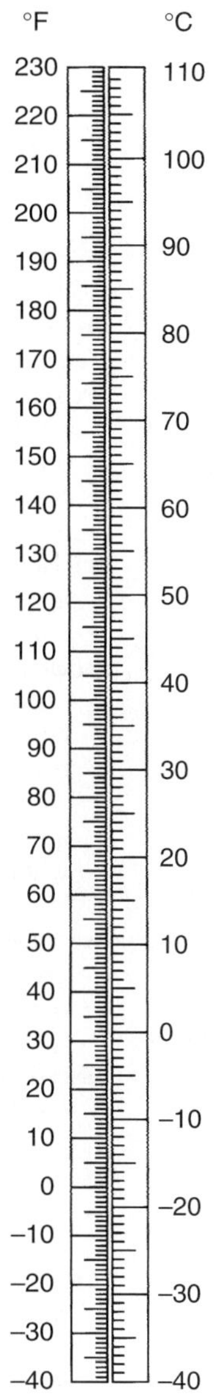

To convert Fahrenheit to Celsius, use the following formula:

$C = \frac{5}{9}(°F - 32)$

To convert Celsius to Fahrenheit, use the following formula:

$F = (\frac{9}{5}°C) + 32$

Glossary

a

abscisic acid (ab-siz′ik as′id) (ABA) a growth-inhibiting hormone of plants; it is involved with other hormones in dormancy (p. 203)
abscission (ab-sizh′un) the separation of leaves, flowers, and fruits from plants after the formation of an abscission zone at the base of their petioles, peduncles, and pedicels (p. 125)
achene (uh-keen′) a single-seeded fruit in which the seed is attached to the pericarp only at its base (p. 140)
acid (as′id) a substance that dissociates in water, releasing hydrogen ions (p. 19)
active transport (ak′tiv trans′port) the expenditure of energy by a cell in moving a substance across a plasma membrane against a diffusion gradient (p. 159)
adventitious (ad-ven-tish′uss) said of buds developing in internodes or on roots, or of roots developing along stems or on leaves (p. 266)
aerobic respiration (air-oh′bik res-puh-ray′shun) respiration that requires free oxygen (p. 186)
agar (ah′gur) a gelatinous substance produced by certain red algae and also a few brown algae; it is often used as a culture medium, particularly for bacteria (p. 199, 341, 347)
aggregate fruit (ag′gruh-git froot) a fruit derived from a single flower having several to many pistils (p. 142)
air layering (air lay′urr-ing) an asexual plant propagation technique whereby aerial stems are induced to form roots. The rooted portion of the stem is then cut and planted (p. 267)
algin (al′jin) a gelatinous substance produced by certain brown algae; it is used in a wide variety of food substances and in pharmaceutical, industrial, and household products (p. 337)
allele (uh-leel′) one of at least two alternative forms of a gene (p. 242)
Alternation of Generations (ol-tur-nay′shun uv jen-ur-ay′shunz) alternation between a haploid gametophyte phase and a diploid sporophyte phase in the life cycle of sexually reproducing organisms (p. 227, 384)
amino acid (ah-mee′noh as′id) one of the organic, nitrogen-containing units from which proteins are synthesized; there are about 20 in all proteins (p. 23)
anaerobic respiration (an-air-oh′bik res-puh-ray′shun) respiration in which the hydrogen removed from the glucose during glycolysis is combined with an organic ion (instead of oxygen) (p. 186)
aneuploid (an′you-ploid) an aberration in normal chromosome number in which one or more extra chromosomes are present or one or more chromsomes are missing (p. 239)
angiosperm (an′jee-oh-spurm) a plant whose seeds develop within ovaries that mature into fruits (p. 442)
annual (an′you-ul) a plant that completes its entire life cycle in a single growing season (p. 91, 132)
annual ring (an′you-ul ring) a single season's production of xylem (wood) by the vascular cambium (p. 93)
annulus (an′yu-luss) a specialized layer of cells around a fern sporangium; it aids in spore dispersal through a springlike action; also a membranous ring around the stipe of a mushroom (p. 411)
anther (an′thur) the pollen-bearing part of a stamen (p. 133, 443)
antheridiophore (an-thur-id′ee-oh-for) a stalk that bears an antheridium (p. 386)
antheridium (pl. **antheridia**) (an-thur-id′ee-um; pl. an-thur-id′ee-ah) the male gametangium of certain algae, fungi, bryophytes, and vascular plants other than gymnosperms and angiosperms (p. 332, 351, 386)
anthocyanin (an-thoh-sy′ah-nin) a water-soluble pigment found in cell sap; anthocyanins vary in color from red to blue (p. 43)
antibiotic (an-tee-by-ot′ik) a substance produced by a living organism that interferes with the normal metabolism of another living organism (p. 373)
anticodon (an-tee-koh′don) the three-nucleotide sequence in a tRNA molecule that base pairs with the complementary mRNA codon for the amino acid carried by that specific tRNA (p. 237)
apical dominance (ay′pi-kul dom′i-nunts) suppression of growth of lateral buds by hormones (p. 205)
apical meristem (ay′pi-kul mair′i-stem) a meristem at the tip of a shoot or root (p. 54)
apomixis (ap-uh-mik′sis) reproduction without fusion of gametes or meiosis in otherwise normal sexual structures (p. 282)
archegoniophore (ahr-kuh-goh′nee-oh-for) a stalk bearing an archegonium (p. 386)
archegonium (pl. **archegonia**) (ahr-kuh-goh′nee-um; pl. ahr-kuh-goh′nee-ah) the multicellular female gametangium of bryophytes and most vascular plants other than angiosperms (p. 386)
aril (air′il) an often brightly colored appendage surrounding the seed of certain plants (e.g., yew) (p. 427)
ascus (pl. **asci**) (as′kus; pl. as′eye) one of often numerous, frequently fingerlike hollow structures in which the fusion of two haploid nuclei is followed by meiosis; a row of ascospores (usually eight) is ultimately produced in each ascus on or within the sexually initiated reproductive bodies of cup (sac) fungi (p. 360)
asexual reproduction (ay-seksh′yule ree-proh-duk′shun) any form of reproduction not involving the union of gametes (p. 222)
assimilation (uh-sim-i-lay′shun) cellular conversion of raw materials into protoplasm and cell walls (p. 14, 192)
atom (at′um) the smallest individual unit of an element that retains the properties of the element (p. 15)
ATP (ay-tee-pee) adenosine triphosphate, a molecule with three phosphate groups found in all living cells; it is the principal vehicle for energy storage and exchange in cell metabolism (p.173, 176)
autotrophic (aw-toh-troh′fik) descriptive of an organism capable of sustaining itself through conversion of inorganic substances to organic material (p. 304)
auxin (awk′sin) a growth-regulating substance produced either naturally by plants or synthetically (p. 200, 266)
axil (ak′sil) the angle formed between a twig and the petiole of a leaf; normally the site of an *axillary bud* (also called *lateral bud*) (p. 87)

b

backcross (bak′kross) a cross involving a hybrid and one of its parents (p. 245)
bacteriophage (bak-teer′ee-oh-fayj) a virus whose host is a bacterium (p. 319)
bark (bahrk) tissues of a woody stem between the vascular cambium and the exterior (p. 95)

base (bayss) a substance that dissociates in water, releasing hydroxyl (OH^-) ions (p. 19, 232)

basidiospore (buh-sidd′ee-oh-spor) a spore produced on a basidium (p. 367)

basidium (pl. **basidia**) (buh-sid′ee-um; pl. buh-sid′ee-ah) one of usually numerous, frequently club-shaped hollow structures in which the fusion of two haploid nuclei is followed by meiosis, the four resulting nuclei becoming externally borne basidiospores; basidia are produced on or within sexually initiated reproductive bodies of the club fungi (e.g., mushrooms, puffballs) (p. 365, 367)

berry (bair′ee) a thin-skinned fruit that usually develops from a compound ovary and commonly contains more than one seed (p. 137)

biennial (by-en′ee-ul) a plant that normally requires two seasons to complete its life cycle, the first season's growth being strictly vegetative (p. 132)

biological controls (by-oh-loj′i-kull kun-trohlz′) the use of natural enemies and inhibitors in combating insect pests and other destructive organisms (p. 542)

biomass (by-oh-mass) the total mass of living organisms present (p. 491)

biome (by′ohm) similar biotic communities considered on a worldwide scale (e.g., desert biome, grassland biome) (p. 509)

biotechnology (by-oh-tek-nol′-oh-jee) the manipulation of organisms, tissues, cells, or molecules for specific applications primarily intended for human benefit (p. 259–264)

biotic community (by-ot′ik kuh-myu′nit-ee) an association of plants, animals, and other organisms (e.g., woodland) (p. 490)

blade (blayd) the conspicuous, flattened part of a leaf (also called *lamina*) or seaweed (p. 87, 110, 337)

bond (bond) a force that holds atoms together (p. 18)

bonsai (bon-sy′) container-grown plants (usually trees) that have been dwarfed artificially through skillful pruning and manipulation of the growing medium (p. 594)

botany (bot′an-ee) science involving the study of plants (p. 7)

botulism (bot′yu-lizm) poisoning from consumption of food infected by botulism bacteria (p. 306)

bract (brakt) a structure that is usually leaflike and modified in size, shape, or color (p. 121)

bryophyte (bry′oh-fyt) a photosynthetic, terrestrial, aquatic, or epiphytic, embryo-producing plant without xylem and phloem (e.g., mosses, liverworts, hornworts) (p. 382)

budding a form of asexual reproduction in which a new cell develops to full size from a protuberance arising from a mature cell, as in yeasts (p. 360)

bulb (buhlb) an underground food-storage organ that is essentially a modified bud consisting of fleshy leaves that surround and are attached to a small stem (p. 100, 268)

bundle scar (bun′dul skahr) a small scar left by a vascular bundle within a leaf scar when the leaf separates from its stem through abscission (p. 88)

bundle sheath (bun′dul sheeth) the parenchyma and/or sclerenchyma cells surrounding a vascular bundle (p. 115)

C

callose (kal′ohs) a complex carbohydrate that develops in sieve tubes following an injury; it is commonly associated with the sieve areas of sieve tube members (p. 14)

callus (kal′uss) undifferentiated tissue that develops around injured areas of stems and roots; also the undifferentiated tissue that develops during tissue culture (p. 14)

Calvin cycle (kal′vin sy′kuhl) see *light-independent reactions*

calyptra (kuh-lip′truh) tissue from the enlarged archegonial wall of many mosses that forms a partial or complete cap over the capsule (p. 387)

calyx (kay′liks) collective term for the sepals of a flower (p. 132)

cambium (kam′bee-um) a meristem producing secondary tissues; see *vascular cambium, cork cambium* (p. 54)

capillary water (kap′i-lair-ee waw′tur) water held in the soil against the force of gravity; capillary water is available to plants (p. 81)

capsule (kapp′sool) a dry fruit that splits in various ways at maturity, often along or between carpel margins; also the main part of a sporophyte in which different types of tissues develop (p. 140, 386)

carbohydrate (kahr-boh-hy′drayt) an organic compound containing carbon, hydrogen, and oxygen, with twice as many hydrogen as oxygen atoms per molecule (p. 21)

carpel (kahr′pul) an ovule-bearing unit that is a part of a pistil (p. 133, 136, 442)

caryopsis (kare-ee-op′siss) a dry fruit in which the pericarp is tightly fused to the seed; it does not split at maturity (p. 141)

Casparian strip (kass-pair′ee-un strip) a band of suberin around the radial and transverse walls of an endodermal cell (p. 70)

cell (sel) the basic structural and functional unit of living organisms; in plants, it consists of protoplasm surrounded by a cell wall (p. 29)

cell biology (sel by-ol′uh-jee) the biological discipline involving the study of cells and their functions (p. 9)

cell cycle (sel sy′kul) a sequence of events involved in the division of a cell (p. 44)

cell division see *cytokinesis*

cell membrane (sel mem′brayn) see *plasma membrane*

cell plate (sel playt) the precursor of the middle lamella; it forms at the equator during telophase (p. 50)

cell sap (sel sap) the liquid contents of a vacuole (p. 43)

cell wall (sel wawl) the relatively rigid boundary of cells of plants and certain other organisms (p. 32)

central cell nuclei (sen-truhl sell new′klee-eye) nuclei, frequently two in number, that unite with a sperm in an embryo sac, forming a primary endosperm nucleus (p. 443)

centromere (sen′truh-meer) the dense, constricted portion of a chromosome to which a spindle fiber is attached (p. 45, 223)

chemiosmosis (kem-ee-oz-moh′siss) a theory that energy is provided for phosphorylation by protons being "pumped" across inner mitochondrial and thylakoid membranes (p. 190)

chiasma (pl. **chiasmata**) (kyaz′mah; pl. ky-az′mah-tah) the X-shaped configuration formed by two chromatids of homologous chromosomes as they remain attached to each other during prophase I of meiosis (p. 223)

chlorenchyma (klor-en′kuh-mah) tissue composed of parenchyma cells that contain chloroplasts (p. 55)

chlorophyll (klor′uh-fil) green pigments essential to photosynthesis (p. 41, 175)

chloroplast (klor′uh-plast) an organelle containing chlorophyll, found in cells of most photosynthetic organisms (p. 40)

chromatid (kroh′muh-tid) one of the two strands of a chromosome; they are united by a centromere (p. 45, 223)

chromatin (kroh′muh-tin) a readily staining complex of DNA and proteins found in chromosomes (p. 38)

chromoplast (kroh′muh-plast) a plastid containing pigments other than chlorophyll; the pigments are usually yellow to orange (p. 41)

chromosome (kroh′muh-sohm) a body consisting of a linear sequence of genes and composed of DNA and proteins; chromosomes are found in cell nuclei and appear in contracted form during mitosis and meiosis (p. 38)

cilium (pl. **cilia**) (sil′ee-um; pl. sil′ee-uh) a short hairlike structure usually found on the cells of unicellular aquatic organisms, normally in large numbers and arranged in rows; the most common function of cilia is propulsion of the cell (p. 213)

circadian rhythm (sur-kay′dee-an rith′um) a mostly daily rhythm of growth and activity found in living organisms (p. 211)

citric acid cycle (sit′rik-ass-id sy′kul) a complex series of reactions following glycolysis in aerobic respiration that involves ATP, mitochondria, and enzymes and that results in the combining of free oxygen with protons and electrons from pyruvic acid to make water (p. 188)

cladistics (kluh-dis′tiks) analysis of shared features (p. 296)

cladophyll (klad′uh-fil) a flattened stem that resembles a leaf; also called *phylloclade* (p. 100)

class (klas) a category of classification between a division and an order (p. 289, 290)

climax vegetation (kly′maks vej-uh-tay′shun) vegetational association that perpetuates itself indefinitely at the culmination of ecological succession (p. 499)

cloning vector (kloh′ning vek′torr) a DNA molecule that can replicate and transfer DNA between cells (p. 260)

codon (koh′donn) the sequence of three nucleotides in an mRNA molecule that constitutes the code for a specific amino acid or a stop signal in protein synthesis; it is complementary to an anticodon (p. 233)

coenocytic (see′-no-sitt-ik) multinucleate, the nuclei not individually separated from one another by crosswalls, as in the hyphae of water molds (p. 357)

cohesion-tension theory (koh-hee′zhun ten′shun thee′uh-ree) theory that explains the rise of water in plants through a combination of cohesion of water molecules in vessels and tracheids and tension on the water columns brought about by transpiration (p. 161)

coleoptile (koh-lee-op′tul) a protective sheath surrounding the emerging shoot of seedlings of the Grass Family (Poaceae) (e.g., corn, wheat) (p. 148)

coleorhiza (koh-lee-uh-ry′zuh) a protective sheath surrounding the emerging radicle (immature root) of members of the Grass Family (Poaceae) (e.g., corn, wheat) (p. 148)

collenchyma (kuh-len′kuh-muh) tissue composed of cells with unevenly thickened walls (p. 55)

colloid (kol′oyd) a substance consisting of a medium in which fine particles are permanently dispersed (p. 81)

community (kuh-myu′nit-ee) a collective term for all the living organisms sharing a common environment and interacting with one another (p. 491)

companion cell (kum-pan′yun sel) a specialized cell derived from the same parent cell as the closely associated sieve tube member immediately adjacent to it (in angiosperm phloem) (p. 60)

compost (kom′post) a mixture of decomposed organic matter, particularly decomposed plant materials (p. 305)

compound (kom′pownd) a substance whose molecules are composed of two or more elements (p. 17)

compound leaf (kom′pownd leef) a leaf whose blade is divided into distinct leaflets (p. 110)

conidium (pl. **conidia**) (kuh-nid′ee-um; pl. kuh-nid′ee-uh) an asexually produced fungal spore formed outside of a sporangium (p. 360)

conifer (kon′i-fur) a cone-bearing tree or shrub (p. 423)

conjugation (kon-juh-gay′shun) a process leading to the fusion of isogametes in algae, fungi, and protozoa; also the means by which certain bacteria exchange DNA (p. 331)

conjugation tube (kon-juh-gay′shun t(y)oob) a tube permitting transfer of a gamete or gametes between adjacent cells, as in *Spirogyra* or desmids (p. 331)

consumer (kon-soo′muhr) organisms that feed on producers (p. 492)

cork (kork) tissue composed of cells whose walls are impregnated with suberin at maturity; the outer layer of tissue of an older woody stem; produced by the cork cambium (p. 62)

cork cambium (kork kam′bee-um) a narrow cylindrical sheath of cells between the exterior of a woody root or stem and the central vascular tissue; it produces *cork* to its exterior and *phelloderm* to its interior; it is also called *phellogen* (p. 55, 90)

corm (korm) a vertically oriented, thickened food-storage stem that is usually enveloped by a few papery nonfunctional leaves (p. 100)

corolla (kuh-rahl′uh) collective term for the petals of a flower (p. 132)

cortex (kor′teks) a primary tissue composed mainly of parenchyma; the tissue usually extends between the epidermis and the vascular tissue (p. 70, 88)

cotyledon (kot-uh-lee′dun) an embryo leaf ("seed leaf") that usually either stores or absorbs food (p. 91, 147)

covalent bond (koh-vay′luhnt bond) a force provided by pairs of electrons that travel between two or more atomic nuclei; holding atoms together and keeping them at a stable distance from each other (p. 18)

crossing-over (kross′ing oh′vur) the exchange of corresponding segments of chromatids between homologous chromosomes during prophase I of meiosis (p. 223)

crown division (krown duh-vizh′unn) the asexual production of multiple plants by division of the base of a stem (crown) (p. 266)

cuticle (kyut′i-kul) a waxy or fatty layer of varying thickness on the outer walls of epidermal cells (p. 60, 112)

cutin (kyu′tin) the waxy or fatty substance of which a cuticle is composed (p. 60, 112)

cutting (kutt′ing) any vegetative plant part used for asexual propagation (p. 266)

cyclosis (sy-kloh′sis) the flowing or streaming of cytoplasm within a cell (p. 44)

cytochrome (sy′toh-krohm) iron-containing protein involved in molecule transfer in an electron transport system (p. 180)

cytogenetics (sy′toh-juh-net-iks) the study of the genetic effects of chromosome structure and behavior (p. 239)

cytokinesis (sy-toh-kuh-nee′sis) division of a cell, usually following mitosis (p. 45)

cytokinin (syt-uh-ky′nin) a growth hormone involved in cell division and several other metabolic activities of cells (p. 203)

cytology (sy-tol′uh-jee) see *cell biology*

cytoplasm (sy′tuh-plazm) the protoplasm of a cell exclusive of the nucleus (p. 32)

cytoplasmic streaming (sy-tuh-plaz′mik streem′ing) see *cyclosis*

cytoskeleton (sy-toh-skel′uh-ton) a network of microtubules and microfilaments involved in movement within a cell (p. 44)

cytosol (sy′toh-sol) fluid, living part of a cell; organelles are distributed within it (p. 32)

d

dark reactions (dahrk ree-ak′shunz) see *light-independent reactions*

day-neutral plant (day new′trul plant) a plant that is not dependent on specific day lengths for the initiation of flowering (p. 215)

deciduous (duh-sij′yu-wuss) shedding leaves annually (p. 88, 110, 125)

decomposer (dee-kuhm-poh′zur) organism (e.g., bacterium, fungus) that breaks down organic material to forms capable of being recycled (p. 492)

dedifferentiate (dee-diff-urr-en′shee-ayt) to become less specialized (usually pertains to cells) (p. 266)

development (duh-vel′up-ment) changes in the form of a plant resulting from growth and differentiation of its cells into tissues and organs (p. 198)

dicotyledon (dy-kot-uh-lee′dun) a class of angiosperms whose seeds commonly have two cotyledons; frequently abbreviated to *dicot* (p. 91)

dictyosome (dik′tee-oh-sohm) an organelle consisting of disc-shaped, often branching hollow tubules that function in accumulating and packaging substances used in the synthesis of materials by the cell (p. 40)

differentially permeable membrane (dif-uh-rensh′uh-lee pur′mee-uh-bul mem′brayn) a membrane through which different substances diffuse at different rates (p. 156)

differentiation (dif-uh-ren-shee-ay′shun) the change of a relatively unspecialized cell to a more specialized one (e.g., the change of a cell just produced by a meristem to a vessel member or fiber) (p. 198)

diffusion (dif-fyu′zhin) the random movement of molecules or particles from a region of higher concentration to a region of lower concentration, ultimately resulting in uniform distribution (p. 156)

digestion (duh-jes′jin) an enzyme-controlled conversion of complex, usually insoluble substances to simpler, usually soluble substances (p. 14, 192)

dihybrid cross (dy-hy′brid kross) a cross involving two different pairs of genes and heterozygous parents (p. 245)

dikaryotic (dy-kair-ee-ot′ik) having a pair of nuclei in each cell or a type of the mycelium in club fungi (p. 366)

dioecious (dy-ee′shuss) having unisexual flowers or cones, with the male flowers or cones confined to certain plants and the female flowers or cones of the same species confined to other different plants (p. 450)

diploid (dip′loyd) having two sets of chromosomes in each cell; the 2*n* chromosome number characteristic of the sporophyte generation (p. 226)

disinfest (diss′in-fest) the removing of surface contaminants from a plant surface (p. 269)

diuretic (dy-yu-ret′ik) a substance tending to increase the flow of urine (p. 409)

division (duh-vizh′un) the largest undivided category of classification of organisms within a kingdom; considered synonymous with *phylum* (p. 290)

DNA (dee-en-ay) standard abbreviation of deoxyribonucleic acid, the carrier of genetic information in cells and viruses (p. 26)

dominance (dom′uh-nints) a condition in which one allele of a gene (dominant allele) masks the phenotypic expression of another allele (recessive allele) (p. 242)

dormancy (dor′man-see) a period of growth inactivity in seeds, buds, bulbs, and other plant organs even when environmental conditions normally required for growth are met (p. 148, 217)

double fusion (dub′ul fu′shun) the more or less simultaneous union of one sperm and egg (forming a zygote) and union of another sperm and central cell nuclei (forming a primary endosperm nucleus) that occur in the megagametophyte of flowering plants (p. 447)

drupe (droop) a simple fleshy fruit whose single seed is enclosed within a hard endocarp (p. 137)

e

ecology (ee-kol′uh-jee) the biological discipline involving the study of the relationships of organisms to each other and to their environment (p. 488)

ecosystem (ee′koh-sis-tim) a system involving interactions of living organisms with one another and with their nonliving environment (p. 490)

egg (eg) a nonmotile female gamete (p. 222)

elater (el′uh-tur) a straplike appendage (usually occurring in pairs) attached to a horsetail (*Equisetum*) spore (p. 407); also, a somewhat spindle-shaped sterile cell occurring in large numbers in liverwort sporangia (p. 387); both types of elaters facilitate spore dispersal

electron (ee-lek′tron) a negatively charged particle of an atom (p. 15)

element (el′uh-mint) one of more than 100 types of matter, most existing naturally but some human-made, each of which is composed of one kind of atom (p. 15)

embryo (em′bree-oh) immature sporophyte that develops from a zygote within an ovule or archegonium after fertilization (p. 66, 382, 386, 426)

enation (ee-nay′shun) one of the tiny, green leaflike outgrowths on the stems of whisk ferns (*Psilotum*) (p. 398)

endocarp (en′doh-kahrp) the innermost layer of a fruit wall (p. 136)

endodermis (en-doh-dur′mis) a single layer of cells surrounding the vascular tissue (stele) in roots and some stems; the cells have Casparian strips (p. 70)

endoplasmic reticulum (en-doh-plaz′mik ruh-tik′yu-lum) a complex system of interlinked double-membrane channels subdividing the cytoplasm of a cell into compartments; parts of it are lined with ribosomes (p. 38)

endosperm (en′doh-spurm) a food-storage tissue that develops through divisions of the primary endosperm nucleus; it is digested by the sporophyte after germination in some species (e.g., corn) or before maturation of the seed in other species (e.g., beans) (p. 448)

endosymbiont hypothesis (en–doh-sim′bee-ont hy-poth′uh-sis) the theory that mitochondria and chloroplasts were free-living bacteria that became incorporated in cells. (p. 248)

energy (en′ur-jee) the capacity to do work; some forms of energy are heat, light, and kinetic (p. 20)

enzyme (en′zym) one of numerous complex proteins that speeds up a chemical reaction in living cells without being used up in the reaction (i.e., it catalyzes the reaction) (p. 24, 174, 233)

epicotyl (ep′uh-kaht-ul) the part of an embryo or seedling above the attachment point of the cotyledon(s) (p. 148)

epidermis (ep-uh-dur′mis) the exterior tissue, usually one cell thick, of leaves, young stems and roots, and other parts of plants (p. 60)

epigynous (ee-pidj′uh-nuss) having flower parts attached above the ovary (p. 449)

epiphyte (ep′uh-fyt) an organism that is attached to and grows on another organism without parasitizing it (p. 326)

ergotism (ur′got-izm) a disease resulting from consumption of goods made with flour containing ergot fungus (p. 361)

essential element (eh-sen′shul el′uh-mint) one of 18 elements generally considered essential to the normal growth, development, and reproduction of most plants (p. 166)

ethylene (eth′uh-leen) a simple, naturally produced, gaseous hormone that inhibits plant growth and promotes the ripening of fruit (p. 204)

etiolation (ee-tee-oh-lay′shun) a condition characterized by long internodes, poor leaf development, and pale, weak appearance due to a plant's having been deprived of light (p. 215)

eukaryotic (yu-kair-ee-ot′ik) pertaining to cells having distinct membrane-bound organelles, including a nucleus with chromosomes (p. 32)

eutrophication (yu-troh-fuh-kay′shun) the gradual enrichment of a body of water through the accumulation of nutrients, resulting in a corresponding increase in algae and other organisms (p. 501)

evolution (ev-oh-loo′shun) the accumulation of genetic changes in populations of living organisms through many generations (p. 274)

exine (ek′syne) the outer layer of the wall of a pollen grain or spore (p. 444)
exocarp (ek′soh-kahrp) the outermost layer of a fruit wall (p. 135)
explant (eks′plant) an excised portion of leaf or stem tissue used for tissue culture (p. 269)
extranuclear DNA (ex-truh-nyu′klee-ahr dee-en′ay) DNA found outside the nucleus, typically in plastids and mitochondria (p. 248)
eyespot (eye′spot) a small, often reddish structure within a motile unicellular organism; it appears to be sensitive to light (also called *stigma*) (p. 341)

f

F_1 (first filial generation) (eff wun) the offspring of a cross between two parent plants (p. 242)
F_2 (second filial generation) (eff too) the offspring of the F_1 generation (p. 242)
FAD (eff-ay-dee) flavin adenine dinucleotide, a hydrogen acceptor molecule involved in the Krebs cycle of respiration and in photosynthesis (p. 180)
family (famm′uh-lee) a classification category between genus and order (p. 290)
fat (fat) an organic compound containing carbon, hydrogen, and oxygen but with proportionately much less oxygen than is present in a carbohydrate molecule (p. 22)
fermentation (fur-men-tay′shun) respiration in which the hydrogen removed from the glucose during glycolysis is transferred back to pyruvic acid, creating substances such as ethyl alcohol or lactic acid (p. 186)
fertilization (fur-til-i-zay′shun) formation of a zygote through the fusion of two gametes (p. 228, 447)
fiber (fy′bur) a long thick-walled cell whose protoplasm often is dead at maturity (p. 55)
filament (fil′uh-mint) threadlike body of certain bacteria, algae, and fungi (p. 301); also the stalk portion of a stamen (p. 133)
fission (fish′un) the division of cells of bacteria and related organisms into two new cells (p. 301, 302)
flagellum (pl. **flagella**) (fluh-jel′um; pl. fluh-jel′uh) a fine threadlike structure protruding from a motile unicellular organism or the motile cells produced by multicellular organisms; functions primarily in locomotion (p. 213, 301)
floret (flor′et) a small flower that is a part of the inflorescence of members of the Sunflower Family (Asteraceae) and the Grass Family (Poaceae) (p. 479)
florigen (flor′uh-jen) one or more hormones once thought from circumstantial evidence to initiate flowering but which have never been isolated or proved to exist (p. 216)
follicle (foll′uh-kuhl) a dry fruit that splits along one side only (p. 138)
food chain (food chayn) a natural chain of organisms of a community wherein each member of the chain feeds on members below it and is consumed by members above it, with autotrophic organisms (*producers*) being at the bottom; interconnected food chains are referred to as *food webs* (p. 492)
foot (foot) the basal part of the embryo of bryophytes and other plants; it is attached to and absorbs food from the gametophyte (p. 386)
fossil (fos′ul) the remains or impressions of any natural object that has been preserved in the earth's crust (p. 417)
frond (frond) a fern leaf; term occasionally also applied to palm leaves (p. 410)
fruit (froot) a mature ovary usually containing seeds; term also somewhat loosely applied to the reproductive structures of groups of plants other than angiosperms (p. 134)
fucoxanthin (fyu-koh-zan′thin) a brownish pigment occurring in brown and other algae (p. 334, 337)

g

gametangium (pl. **gametangia**) (gam-uh-tan′jee-um; pl. gam-uh-tan′jee-ah) any cell or structure in which gametes are produced (p. 338, 382)
gamete (gam′eet) a sex cell; one of two cells that unite, forming a *zygote* (p. 222)
gametophore (guh-me′toh-for) a stalk on which a gametangium is borne (p. 386)
gametophyte (guh-me′toh-fyte) the haploid (*n*) gamete-producing phase of the life cycle of an organism that exhibits Alternation of Generations (p. 227)
gemma (pl. **gemmae**) (jem′uh; pl. jem′ee) a small outgrowth of tissue that becomes detached from the parent body and is capable of developing into a complete new plant or other organism; gemmae are produced in cuplike structures on liverwort thalli and are also produced by certain fungi (p. 386)
gene (jeen) a unit of heredity; part of a linear sequence of such units occurring in the DNA of chromosomes (p. 26, 233)
gene bank (jeen bank) a collection of plants or seeds maintained for their germ plasm (p. 259)
generative cell (jen′uh-ray-tiv sel) the cell of the male gametophyte of angiosperms that divides, producing two sperms; also, the cell of the male gametophyte of gymnosperms that divides, producing a *sterile cell* and a *spermatogenous cell* (p. 426)
gene synthesizer (jeen sin′thuh-size-urr) a machine that creates specific DNA sequences (p. 260)
genetic drift (juh-net′ik drift) a change in the genetic makeup of a population that may take place by chance alone (p. 278)
genetic engineering (juh′net′ik en-juh-neer′ing) the introduction, by artificial means, of genes from one form of DNA into another form of DNA (p. 259)
genetics (juh-net′iks) the biological discipline involving the study of heredity (p. 9)
genome (jee′nohm) the sum total of DNA in an organism's chromosomes (p. 233)
genotype (jeen′oh-typ) the genetic constitution of an organism; may or may not be visibly expressed, as contrasted with phenotype (p. 242)
genus (pl. **genera**) (jee′nus; pl. jen′er-ah) a category of classification between a family and a species (p. 288)
germination (jur′min-ay-shun) the beginning or resumption of growth of a seed or spore (p. 148)
germ-line mutation (jurm′-lyn mew-tay′shun) a mutation in a cell from which gametes are derived; the mutation can be passed on to offspring (p. 238)
germ plasm (jurm plaz′im) the sum total of all the genes of a species or group of organisms (p. 258)
gibberellin (jib-uh-rel′in) one of a group of plant hormones that have a variety of effects on growth; they are particularly known for promoting elongation of stems (p. 202)
gill (gil) one of the flattened plates of compact mycelium that radiate out from the stalk on the underside of the caps of most mushrooms (p. 367)
girdling (gurd′ling) the removal of a band of tissues extending inward to the vascular cambium on the stem of a woody plant (p. 267)
gland (gland) a small body of variable shape and size that may secrete certain substances but that also may be functionless (p. 61, 113)
glycolysis (gly-kol′uh-sis) the initial phase of all types of respiration in which glucose is converted to pyruvic acid without involving free oxygen (p. 186)
graft (graft) the union of a segment of a plant, the *scion,* with a rooted portion, the *stock* (p. 267)

grain (grayn) see *caryopsis*

granum (pl. **grana**) (gra′num; pl. gra′nuh) a series of stacked thylakoids within a chloroplast (p. 40)

gravitational water (grav-uh-tay′shun-ul waw′tur) water that drains out of the pore spaces of a soil after a rain (p. 81)

gravitropism (grav-uh-troh′pism) growth response to gravity (p. 208)

ground meristem (grownd mair′i-stem) meristem that produces all the primary tissues other than the epidermis and stele (e.g., cortex, pith) (p. 54, 68, 88)

growth (grohth) progressive increase in size and volume through natural development (p. 13)

guard cell (gahrd sel) one of a pair of specialized cells surrounding a stoma (p. 61, 114)

guttation (guh-tay′shun) the exudation from leaves of water in liquid form due to root pressure (p. 110, 164)

gymnosperm (jim′noh-spurm) a plant whose seeds are not enclosed within an ovary during their development (e.g., pine tree) (p. 422)

h

half-life (haf-lyf) the amount of time it takes for a radioactive element to lose half of its radioactivity (p. 276)

haploid (hap′loyd) having one set of chromosomes per cell, as in gametophytes; also referred to as having n chromosomes (as contrasted with $2n$ chromosomes in the diploid cells of sporophytes) (p. 226)

haustorium (pl. **haustoria**) (haw-stor′ee-um; pl. haw-stor′ee-uh) a protuberance of a fungal hypha or plant organ such as a root that functions as a penetrating and absorbing structure (p. 75)

heartwood (hahrt′wood) nonliving, usually darker-colored wood whose cells have ceased to function in water conduction (p. 94)

heirloom variety (air′loom vuh-rye′it-ee) a previously popular plant variety that is currently being maintained because of certain desirable qualities (p. 258)

herbaceous (hur-bay′shuss or ur-bay′shuss) referring to nonwoody plants (p. 91)

herbarium (pl. **herbaria**) (hur-bair′ee-um or ur-bair′ee-um; pl. hur-bair′ee-uh) a collection of dried pressed specimens, usually mounted on paper and provided with a label that gives collection information and an identification (p. 453)

heterocyst (het′uh-roh-sist) a transparent, thick-walled, slightly enlarged cell occurring in the filaments of certain cyanobacteria (p. 313)

heterosis (hett-urr-oh′sis) hybrid vigor; superior qualities of heterozygous offspring as compared with those of their homozygous parents (p. 258)

heterospory (het-uh-ross′por-ee) the production of both microspores and megaspores (p. 401)

heterotrophic (het-ur-oh-troh′fick) incapable of synthesizing food and therefore dependent on other organisms for it (p. 304)

heterozygous (het-uh-roh-zy′guss) having two different alleles at the same locus on homologous chromosomes (p. 243, 257)

holdfast (hold′fast) attachment organ or cell at the base of the thallus or filament of certain algae (p. 329, 337)

homologous chromosomes (hoh-mol′uh-guss kroh′muh-sohmz) pairs of chromosomes that associate together in prophase I of meiosis; each member of a pair is derived from a different parent (p. 223)

homozygous (hoh-moh-zy′guss) having two identical alleles at the same locus on a pair of homologous chromosomes (p. 243, 256)

hormone (hor′mohn) an organic substance generally produced in minute amounts in one part of an organism and transported to another part of the organism where it controls or affects growth and development (p. 199)

hybrid (hy′brid) heterozygous offspring of two parents that differ in one or more inheritable characteristics (p. 244, 258, 281)

hydathode (hy′duh-thohde) structure at the tip of a leaf vein through which water is forced by root pressures (p. 164)

hydrolysis (hy-drol′uh-sis) the breakdown of complex molecules to simpler ones as a result of the union of water with the compound; the process is usually controlled by enzymes (p. 192)

hydrosere (hy′droh-sear) a primary succession that is initiated in a wet habitat (p. 499)

hygroscopic water (hy-gruh-skop′ik waw′tur) water that is chemically bound to soil particles and therefore unavailable to plants (p. 81)

hypha (pl. **hyphae**) (hy′fuh; pl. hy′fee) a single, usually tubular, threadlike filament of a fungus; *mycelium* is a collective term for hyphae (p. 356)

hypocotyl (hy-poh-kot′ul) the portion of an embryo or seedling between the radicle and the cotyledon(s) (p. 148)

hypodermis (hy-poh-dur′mis) a layer of cells immediately beneath the epidermis and distinct from the parenchyma cells of the cortex in certain plants (p. 118, 422)

hypogynous (hi-podj′un-nuss) having flower parts attached below the ovary (p. 449)

hypothesis (hy-poth′uh-sis) a postulated explanation for some observed facts that must be tested experimentally before it can be accepted as valid or discarded if it proves to be incorrect (p. 7)

i

imbibition (im-buh-bish′un) adsorption of water and subsequent swelling of organic materials because of the adhesion of the water molecules to the internal surfaces (p. 157)

inbreeding (in′breed-ing) mating between individuals with a common ancestry (p. 256)

inbreeding depression (in′breed-ing dee-presh′unn) poor performance and low fertility of inbred individuals (p. 258)

incomplete dominance (in′kom-pleet dom′in-uns) a condition in which the heterozygous phenotype is intermediate to the two homozygous phenotypes as a result of one allele only partly masking another allele (p. 247)

indusium (pl. **indusia**) (in-dew′zee-um; pl. in-dew′zee-uh) the small, membranous, sometimes umbrellalike covering of a developing fern sorus (p. 411)

inferior ovary (in-feer′ee-or oh′vuh-ree) an ovary to which parts of the calyx, corolla, and stamens have become more or less united so they appear to be attached at the top of it (p. 133, 449)

inflorescence (in-fluh-res′ints) a collective term for a group of flowers attached to a common axis in a specific arrangement (p. 133)

inorganic (in-or-gan′ik) descriptive of compounds having no carbon atoms (p. 21)

integument (in-teg′yu-mint) the outermost layer of an ovule; usually develops into a seed coat; a gymnosperm ovule usually has a single integument, and an angiosperm ovule usually has two integuments (p. 422, 426, 443)

intermediate-day plant (in-tur-me′dee-ut day plant) a plant that has two critical photoperiods; it will not flower if the days are either too short or too long (p. 214)

internode (in′tur-nohd) a stem region between nodes (p. 87, 110)

inversion (chromosomal) (in-verzh′un) a chromosome rearrangement as a result of a segment having been removed, rotated 180°, and then reinserted (p. 239)

in vitro (in vee′troh) "in glass"; growing or being maintained on artificial media, usually in glass test tubes or flasks (p. 269)

ion (eye′on) a molecule or atom that has become electrically charged through the loss or gain of one or more electrons (p. 18)

isogamy (eye-sog′uh-me) sexual reproduction in certain algae and fungi having gametes that are alike in size (p. 330)

isotope (eye′suh-tohp) one of two or more forms of an element that have the same chemical properties but differ in the number of neutrons in the nuclei of their atoms (p. 16)

k

kinetochore (kuh-net′uh-kor) specialized protein complexes that develop on the vertical faces of a centromere during late prophase; spindle fibers are attached to them (p. 45)

kingdom (king′dum) the highest category of classification (e.g., Plant Kingdom, Animal Kingdom) (p. 290)

knot (not) a portion of the base of a branch enclosed within wood (p. 104)

l

lamina (lam′uh-nuh) see *blade*

lateral bud (lat′uh-rul bud) see *axil*

laticifer (luh-tis′uh-fur) specialized cells or ducts resembling vessels; they form branched networks of latex-secreting cells in the phloem and other parts of plants (p. 97)

leaf (leef) a flattened, usually photosynthetic structure arranged in various ways on a stem (p. 110)

leaf gap (leef gap) a parenchyma-filled interruption in a stem's cylinder of vascular tissue immediately above the point at which a branch of vascular tissue (*leaf trace*) leading to a leaf occurs (p. 90)

leaflet (leef′lit) one of the subdivisions of a compound leaf (p. 110)

leaf scar (leef skahr) the suberin-covered scar left on a twig when a leaf separates from it through abscission (p. 88)

leaf trace (leef trays) see *leaf gap*

legume (leg′yoom) a dry fruit that splits along two "seams," the seeds being attached along the edges (p. 138)

lenticel (lent′uh-sel) one of usually numerous, slightly raised, somewhat spongy groups of cells in the bark of woody plants; lenticels permit gas exchange between the interior of a plant and the external atmosphere (p. 62, 90)

leucoplast (loo′kuh-plast) a colorless plastid commonly associated with starch accumulation (p. 41)

light-dependent reactions (lyt-dee-pendent ree-ak′shunz) a series of chemical and physical reactions through which light energy is converted to chemical energy with the aid of chlorophyll molecules; in the process, water molecules are split, with hydrogen ions and electrons being produced and oxygen gas being released; ATP and NADPH also are created (p. 176)

light-independent reactions (lyt in-dee-pen-dent ree-ak′shunz) a cyclical series of chemical reactions that utilizes carbon dioxide and energy generated during the light-dependent reactions of photosynthesis, producing sugars, some of which are stored as insoluble carbohydrates, while others are recycled; the reactions are independent of light and occur in the stroma of chloroplasts (p. 176)

lignin (lig′nin) a polymer with which certain cell walls (e.g., those of wood) become impregnated (p. 55)

ligule (lig′yool) the tiny tongue-like appendage at the base of a spike moss (*Selaginella*) or quillwort (*Isoetes*) leaf (p. 401)

linked genes (linked jeens) genes located on the same chromosome (p. 245, 248)

lipid (lip′id) a general term for fats, fatty substances, and oils (p. 22)

locule (lok′yool) a cavity within an ovary or a sporangium (p. 140)

locus (loh′kuss) the position of a gene on a chromosome (p. 242, 249)

long-day plant (long-day plant) a plant in which flowering is not initiated unless exposure to more than a critical day length occurs (p. 214)

m

map unit (map you′nit) a unit of measure equivalent to 1% recombination (p. 249)

mass-flow hypothesis (mass flo hy-poth′uh-sus) see *pressure-flow hypothesis*

mass selection (mass suh-lek′shun) a plant breeding technique in which seeds of plants in a population are used to create each generation (p. 257)

maternal inheritance (muh-terr′nal in-hair′it-ans) inheritance in which the female gamete contributes extranuclear genes to the offspring (p. 248)

megagametophyte (meg-uh-ga-mee′toh-fyt) the female gametophyte of angiosperms, which, in approximately 70% of the species investigated, contains eight nuclei (p. 443)

megaphyll (meg′uh-fill) a leaf having branching veins; it is associated with a *leaf gap* (p. 410)

megasporangium (meg-uh-spor′an-jee-um) a sporangium in which megaspores are formed (p. 425)

megaspore (meg′uh-spor) a spore that develops into a female gametophyte (megagametophyte) (p. 401, 425, 443)

megasporocyte (meg uh-spor′oh-syt) a diploid cell that produces megaspores upon undergoing meiosis (p. 401, 426, 443)

meiocyte (my′oh-syt) see *sporocyte*

meiosis (my-oh′sis) the process of two successive nuclear divisions through which segregation of genes occurs and a single diploid ($2n$) cell becomes four haploid (n) cells (p. 222)

mericloning (mair′i-kloh-ning) see *micropropagation*

meristem (mair′i-stem) a region of undifferentiated cells in which new cells arise (p. 54)

mesocarp (mez′uh-karp) the middle region of the fruit wall that lies between the exocarp and the endocarp (p. 136)

mesophyll (mez′uh-fil) parenchyma (chlorenchyma) tissue between the upper and lower epidermis of a leaf (p. 115, 397)

metabolism (muh-tab′uh-lizm) the sum of all the interrelated chemical processes occurring in a living organism (p. 171)

microfilament (my′kroh-fil′uh-mint) a protein filament involved with cytoplasmic streaming and with contraction and movement in eukaryotic cells (p. 44)

microphyll (my′kroh-fil) a leaf having a single unbranched vein not associated with a *leaf gap* (p. 397)

micropropagation (my-kroh-prop-uh-gay′shun) propagation of plants *in vitro* (p. 269)

micropyle (my′kroh-pyl) a pore or opening in the integuments of an ovule through which a pollen tube gains access to an embryo sac or archegonium of a seed plant (p. 426, 443)

microshoot (my′kroh-shoot) one of several to many shoots produced by a plant growing *in vitro* (p. 270)

microsporangium (my-kro-spor-anj′ee-um) a sporangium in which microspores are formed (p. 425)

microspore (my′kroh-spor) a spore that develops into a male gametophyte (microsporocyte) (p. 401, 443)

microsporocyte (my′kroh-spor′oh-syt) a diploid cell that produces microspores upon undergoing meiosis (p. 401, 425, 443)

microsporophyll (my-kroh-spor′uh-fil) a leaf, usually reduced in size, on or within which microspores are produced (p. 401)

microtubule (my′kroh-t(y)oo-byul) an unbranched tubelike proteinaceous structure commonly found inside the plasma membrane where it apparently regulates the addition of cellulose to the cell wall (p. 44)

middle lamella (mid′ul luh-mel′uh) a layer of material, rich in pectin, that cements two adjacent cell walls together (p. 35)
midrib (mid′rib) the central (main) vein of a pinnately veined leaf or leaflet (p. 112)
mitochondrion (pl. **mitochondria**) (my-toh-kon′dree-un; pl. my-toh-kon′dree-uh) an organelle containing enzymes that function in the citric acid cycle and the electron transport chain of aerobic respiration (p. 42)
mitosis (my-toh′sis) nuclear division, usually accompanied by cytokinesis, during which the chromatids of the chromosomes separate and two genetically identical daughter nuclei are produced (p. 44)
molecule (mol′uh-kyul) the smallest unit of an element or compound retaining its own identity; consists of two or more atoms (p. 14, 17)
monocotyledon (mon-oh-kot-uh-lee′dun) a class of angiosperms whose seeds have a single cotyledon; commonly abbreviated to *monocot* (p. 91)
monoecious (moh-nee′shuss) having unisexual male flowers or cones and unisexual female flowers or cones both on the same plant (p. 450)
monohybrid cross (mon-oh-hy′brid kross) a cross involving a single pair of genes and heterozygous parents (p. 244)
monokaryotic (mon-oh-kair-ee-ot′ik) having a single nucleus in each cell or unit of the mycelium in club fungi (p. 366)
monomer (mon′oh-mur) a simple individual molecular unit of a polymer (p. 21)
motile (moh′tul) capable of independent movement (p. 301)
multiple fruit (mul′tuh-pul froot) a fruit derived from several to many individual flowers in a single inflorescence (p. 142)
mushroom (mush′room) a sexually initiated phase in the life cycle of a club fungus, usually consisting of an expanded *cap* and a *stalk (stipe)* (p. 366)
mutagen (mew′tuh-jenn) an agent that causes a mutation to occur (p. 238)
mutation (myu-tay′shun) a heritable change in a gene or chromosome (p. 238)
mycelium (my-see′lee-um) a mass of fungal hyphae (p. 356)
mycorrhiza (pl. **mycorrhizae**) (my-kuh-ry′zuh; pl. my-kuh-ry′zee) a symbiotic association between fungal hyphae and a plant root (p. 75)

n

n (en) having one set of chromosomes per cell (*haploid*) (p. 226)
NAD (en-ay-dee) nicotinamide adenine dinucleotide, a molecule that during respiration temporarily accepts electrons whose negative charges are balanced by also accepting protons and thereby hydrogen atoms (p. 188)
NADP (en-ay-dee-pee) nicotinamide adenine dinucleotide phosphate, a high-energy storage molecule that temporarily accepts electrons from photosystem I in the light reactions of photosynthesis (p. 176, 180)
nastic movement (nass′tik moov′mint) a nondirected movement of a flat organ (e.g., petal, leaf) in which the organ alternately bends up and down (p. 207)
neutron (new′tron) an uncharged particle in the nucleus of an atom (p. 15)
node (nohd) region of a stem where one or more leaves are attached (p. 55, 87)
noncoding DNA (non-koh′ding dee-en-ay) DNA that does not code for a gene (p. 237)
nucellus (new-sel′us) ovule tissue within which an embryo sac develops (p. 422, 426)
nuclear envelope (new′klee-ur en′vuh-lohp) a porous double membrane enclosing a nucleus (p. 37)
nucleic acid (new-klay′ik as′id) see *DNA, RNA*
nucleolus (pl. **nucleoli**) (new-klee′oh-luss; pl. new-klee′oh-ly) a somewhat spherical body within a nucleus; contains primarily RNA and protein; there may be more than one nucleolus per nucleus (p. 38)
nucleotide (new′klee-oh-tyd) the structural unit of DNA and RNA (p. 26, 232)
nucleus (new′klee-uss) the organelle of a living cell that contains chromosomes and is essential to the regulation and control of all the cell's functions; also, the core of an atom (p. 15, 37)
nut (nutt) one-seeded dry fruit with a hard, thick pericarp; a nut develops with a cup or cluster of bracts at the base (p. 140)
nutrient (noo′tree-uhnt) a substance that furnishes the elements and energy for the organic molecules that are the building blocks from which an organism develops (p. 198)

o

oil (oyl) a fat in a liquid state (p. 22)
oogamy (oh-og′uh-mee) sexual reproduction in which the female gamete, or egg, is nonmotile and larger than the male gamete, or sperm, which is motile (p. 332)
oogonium (pl. **oogonia**) (oh-oh-goh′nee-um; pl. oh-oh-goh′nee-ah) a female sex organ of certain algae and fungi; it consists of a single cell that contains one to several eggs (p. 332, 351)
operculum (oh-per′kyu-lum) the lid or cap that protects the peristome of a moss sporangium (p. 391)
orbital (or′buh-till) a volume of space in which a given electron occurs 90% of the time (p. 15)
order (or′dur) a category of classification between a class and a family (p. 290)
organelle (or-guh-nel′) a membrane-bound body in the cytoplasm of a cell; there are several kinds, each with a specific function (e.g., mitochondrion, chloroplast)[1] (p. 32)
organic (or-gan′ik) pertaining to or derived from living organisms and pertaining to the chemistry of carbon-containing compounds (p. 21)
osmosis (oz-moh′sis) the diffusion of water or other solvents through a differentially permeable membrane from a region of higher concentration to a region of lower concentration (p. 156)
osmotic potential (oz-mot′ik puh-ten′shil) potential pressure that can be developed by a solution separated from pure water by a differentially permeable membrane (the pressure required to prevent osmosis from taking place) (p. 157)
osmotic pressure (oz-mot′ik presh′ur) see *osmotic potential*
outcrossing (out′kross-ing) cross-pollination between individuals of the same species (p. 258)
ovary (oh′vuh-ree) the enlarged basal portion of a pistil that contains an ovule or ovules and usually develops into a fruit (p. 133)
ovule (oh′vyool) a structure of seed plants that contains a female gametophyte and has the potential to develop into a seed (p. 133, 422)
oxidation-reduction reactions (ok-suh-day′shun ree-duk′shun) chemical reactions involving gain or loss of electrons to or from a compound (p. 172)

p

palisade mesophyll (pal-uh-sayd′ mez′uh-fil) mesophyll having one or more relatively uniform rows of tightly packed, elongate, columnar parenchyma (chlorenchyma) cells beneath the upper epidermis of a leaf (p. 115)
palmately compound; palmately veined (pahl′mayt-lee kom′pownd; pahl′mayt-lee vaynd) having leaflets or principal veins radiating out from a common point (p. 110, 112)

---

1. Ribosomes, which are considered organelles, are an exception in that they are not bounded by a membrane.

papilla (pl. **papillae**) (puh-pil′uh; pl. puh-pill′ay) a small, usually rounded or conical protuberance (p. 331)
parenchyma (puh-ren′kuh-muh) thin-walled cells varying in size, shape, and function; the most common type of plant cell (p. 55)
parental type (pah-renn′tuhl typ) an offspring with the same combination of alleles as one of its parents (p. 249)
parthenocarpic (par-thuh-noh-kar′pik) developing fruits from unfertilized ovaries; the resulting fruit is, therefore, usually seedless (p. 448)
particle gun (pahrt′ik-kuhl gunn) a machine capable of changing the genetic makeup of plant tissue by shooting DNA-coated particles into it (p. 260)
passage cell (pas′ij sel) a thin-walled cell of an endodermis (p. 70)
pectin (pek′tin) a water-soluble organic compound occurring primarily in the middle lamella; when combined with organic acids and sugar, it becomes a jelly (p. 35)
pedicel (ped′i-sel) the individual stalk of a flower that is part of an inflorescence (p. 132)
peduncle (pee′dun-kul) the stalk of a solitary flower or the main stalk of an inflorescence (p. 132)
peptide bond (pep′tyd bond) the type of chemical bond formed when two amino acids link together in the synthesis of proteins (p. 24)
perennial (puh-ren′ee-ul) a plant that continues to live indefinitely after flowering (p. 132)
perianth (pari′ee-anth) the calyx and corolla of a flower (p. 132)
pericarp (per′uh-karp) collective term for all the layers of a fruit wall (p. 136)
pericycle (per′uh-sy-kul) tissue sandwiched between the endodermis and phloem of a root; often only one or two cells wide in transverse section; the site of origin of lateral roots (p. 71)
periderm (pair′uh-durm) outer bark; composed primarily of cork cells (p. 62)
perigynous (purr-idj′uh-nuss) having flower parts attached around the ovary; the flower parts are usually attached to a cup (p. 449)
peristome (per′uh-stohm) one or two series of flattened, often ornamented structures (*teeth*) arranged around the margin of the open end of a moss sporangium; the teeth are sensitive to changes in humidity and facilitate the release of spores (p. 391)
petal (pet′ul) a unit of a corolla; it is usually both flattened and colored (p. 132)
petiole (pet′ee-ohl) the stalk of a leaf (p. 87, 110)
$P_{far\text{-}red}$, or P_{fr} (pee-far-red or pee-ef-ahr) a form of phytochrome (*which see*) (p. 215)
pH (pee-aitch) a symbol of hydrogen ion concentration indicating the degree of acidity or alkalinity (p. 82)
phage (fayj) see *bacteriophage*
phellogen (fel′uh-jun) see *cork cambium*
phenotype (fee′noh-typ) the physical appearance of an organism (p. 242)
pheromone (fer′uh-mohn) something produced by an organism that facilitates chemical communication with another organism (p. 544)
phloem (flohm) the food-conducting tissue of a vascular plant (p. 59)
photon (foh′ton) a unit of light energy (p. 177)
photoperiodism (foh-toh-pir′ee-ud-izm) the initiation of flowering and certain vegetative activities of plants in response to relative lengths of day and night (p. 214)
photosynthesis (foh-toh-sin′thuh-sis) the conversion of light energy to chemical energy; water, carbon dioxide, and chlorophyll are all essential to the process, which ultimately produces carbohydrate, with oxygen being released as a by-product (p. 14, 171)
photosynthetic unit (foh-toh-sin-thet′ik yew′nit) one of two groups of about 250 to 400 pigment molecules each that function together in chloroplasts in the light reactions of photosynthesis; the units are exceedingly numerous in each chloroplast (p. 176)
photosystem (foh′toh-sis-tum) collective term for a specific functional aggregation of photosynthetic units (p. 178)
phragmoplast (frag′mo-plast) a complex of microtubules and endoplasmic reticulum that develops during telophase of mitosis (p. 50)
phytochrome (fy′tuh-krohm) protein pigment associated with the absorption of light; it is found in the cytoplasm of cells of green plants and occurs in interconvertible active and inactive forms ($P_{far\ red}$ and P_{red}); it facilitates a plant's capacity to detect the presence (or absence) and duration of light (p. 215)
pilus (pl. **pili**) (py′lis; pl. py′lee) the equivalent of a conjugation tube in bacteria (p. 302)
pinna (pl. **pinnae**) (pin′uh; pl. pin′ee) a primary subdivision of a fern frond; the term is also applied to a leaflet of a compound leaf (p. 410)
pinnately compound; pinnately veined (pin′ayt-lee kom′pownd; pin′ayt-lee vaynd) having leaflets or veins on both sides of a common axis (e.g., rachis, midrib) to which they are attached (p. 110, 112)
pistil (pis′tul) a female reproductive structure of a flower, composed of one or more carpels and consisting of an ovary, style, and stigma (p. 133)
pit (pit) a more or less round or elliptical thin area in a cell wall; pits occur in pairs opposite each other, with or without shallow, domelike borders (p. 59)
pith (pith) central tissue of a dicot stem and certain roots; it usually consists of parenchyma cells that become proportionately less of the volume of woody plants as cambial activity increases the organ's girth (p. 88)
plankton (plank′ton) free-floating aquatic organisms that are mostly microscopic (p. 327)
plant anatomy (plant uh-nat′uh-mee) the botanical discipline that pertains to the internal structure of plants (p. 8)
plant community (plant kuh-myu′nuh-tee) an association of plants inhabiting a common environment and interacting with one another (p. 490)
plant ecology (plant ee-koll′uh-jee) the science that deals with the relationships and interactions between plants and their environment (p. 9)
plant geography (plant jee-og′ruh-fee) the botanical discipline that pertains to the broader aspects of the space relations of plants and their distribution over the surface of the earth (p. 9)
plant morphology (plant mor-fol′uh-jee) the botanical discipline that pertains to plant form and development (p. 9)
plant physiology (plant fiz-ee-ol′uh-jee) the botanical discipline that pertains to the metabolic activities and processes of plants (p. 8)
plant taxonomy (plant tak-son′uh-mee) the botanical discipline that pertains to the classification, naming, and identification of plants (p. 9)
plasma membrane (plaz′muh mem′brayn) the outer boundary of the protoplasm of a cell; also called *cell membrane,* particularly in animal cells (p. 36)
plasmid (plaz′mid) one of up to 30 or 40 small, circular DNA molecules usually present in a bacterial cell (p. 260)
plasmodesma (pl. **plasmodesmata**) (plaz-muh-dez′muh; pl. plaz-muh-dez′muh-tah) minute strands of cytoplasm that extend between adjacent cells through pores in the walls (p. 35)
plasmodium (pl. **plasmodia**) (plaz-moh′dee-um; pl. plaz-moh′dee-ah) the multinucleate, semiviscous liquid, active form of slime mold; it moves in a "crawling-flowing" motion (p. 348)
plasmolysis (plaz-mol′uh-sis) the shrinking in volume of the protoplasm of a

cell and the separation of the protoplasm from the cell wall due to loss of water via osmosis (p. 157)

plastid (plas′tid) an organelle associated primarily with the storage or manufacture of carbohydrates (e.g., leucoplast, chloroplast) (p. 40)

plumule (ploo′myool) the terminal bud of the embryo of a seed plant (p. 147)

pneumatophore (noo-mat′oh-for) spongy root extending above the surface of the water, produced by a plant growing in water; pneumatophores facilitate oxygen absorption (p. 73)

pole (pohl) an invisible focal point toward each end of a cell from which spindle fibers extend in arcs during mitosis or meiosis (p. 224)

pollen grain (pahl′un grayn) a structure derived from the microspore of seed plants that develops into a male gametophyte (p. 133, 425, 444)

pollen tube (pahl′un t(y)oob) a tube that develops from a pollen grain and conveys the sperms to the female gametophyte (p. 426, 447)

pollination (pahl-uh-nay′shun) the transfer of pollen from an anther to a stigma (p. 447)

pollinium (pl. **pollinia**) (pah-lin′ee-um; pl. pah-lin′ee-ah) a cohesive mass of pollen grains commonly found in members of the Orchid Family (Orchidaceae) and the Milkweed Family (Asclepiadaceae) (p. 453)

polymer (pahl′i-mur) a large molecule composed of many monomers (p. 21)

polymerase (poh-limm′err-ace) an enzyme that creates a polymer (e.g., DNA polymerase synthesizes DNA) (p. 234)

polypeptide (pahl-ee-pep′tide) a chain of amino acids (p. 23)

polyploidy (pahl′i-ploy-dee) having more than two complete sets of chromosomes per cell (p. 282)

pome (pohm) a simple fleshy fruit whose flesh is derived primarily from the receptacle (p. 138)

population (pop-yew-lay′shun) a group of organisms, usually of the same species, occupying a given area at the same time (p. 490)

P_{red} or P_r (pee-red or pee-ahr) a form of phytochrome (which see) (p. 215)

pressure-flow hypothesis (presh′ur floh hy-poth′uh-sis) the theory that food substances in solution in plants flow along concentration gradients between the sources of the food and *sinks* (places where the food is utilized) (p. 165)

prickle (prik′uhl) a pointed outgrowth from an epidermis or cortex beneath the epidermis (p. 119)

primary consumer (pry′mer-ree kon-soo′mur) organism that feeds directly on producers (p. 492)

primary tissue (pry′mer-ee tish′yu) a tissue produced by an apical meristem (e.g., epidermis, cortex, primary xylem and phloem, pith) (p. 54)

primordium (pry-mord′ee-um) an organ or structure (e.g., leaf, bud) at its earliest stage of development (p. 88)

procambium (proh-kam′bee-um) a tissue produced by the primary meristem that differentiates into primary xylem and phloem (p. 54, 68, 88)

producer (pruh-dew′sur) an organism that manufactures food through the process of photosynthesis (p. 492)

prokaryotic (proh-kair-ee-ot′ik) having a cell or cells that lack a distinct nucleus and other membrane-bound organelles (e.g., bacteria) (p. 32)

promoter region (proh-moh′turr ree′jin) the DNA sequence to which RNA polymerase binds to initiate transcription (p. 237)

proplastid (proh-plas′tid) a tiny, undifferentiated organelle that can duplicate itself and that may develop into a chloroplast, leucoplast, or other type of plastid (p. 42)

protein (proh′tee-in or proh′teen) a polymer composed of many amino acids linked together by peptide bonds (p. 23)

protein sequencer (proh′tee-in or proh′teen see′kwens-urr) a machine that reveals the sequence of amino acids in a protein (p. 260)

prothallus (pl. **prothalli**) (proh-thal′us; pl. proh-thal′eye) the gametophyte of ferns and their relatives; also called *prothallium* (p. 412, 413)

protoderm (proh′tuh-durm) the primary meristem that gives rise to the epidermis (p. 54, 68, 88)

proton (proh′ton) a positively charged particle in the nucleus of an atom (p. 15)

protonema (proh-tuh-nee′muh) a green, usually branched, threadlike or sometimes platelike growth from a bryophyte spore; it gives rise to "leafy" gametophytes (p. 388)

protoplast (proh′toh-plast) the unit of protoplasm within a plant cell wall (p. 259)

protoplast fusion (proh′toh-plast few′szhinn) the process of combining *in vitro* two protoplasts in one cell (p. 259)

pruning (proon′ing) removal of portions of plants for aesthetic purposes, for improving quality and size of fruits or flowers, or for elimination of diseased tissues (p. 594)

pyrenoid (py′ruh-noyd) a small body found on the chloroplasts of certain green algae and hornworts; pyrenoids are associated with starch accumulation; they may occur singly on a chloroplast, or they may be numerous (p. 328)

pyruvic acid (py-roo′vik as′id) the organic compound that is the end product of the glycolysis phase of respiration (p. 186)

q

quantitative trait (kwan′tuh-tay-tiv trait) a trait controlled by several genes and influenced by the environment; it is usually measured on a continuous scale (p. 248)

quiescence (kwy′ess-ens) a state in which a seed or other plant part will not germinate or grow unless environmental conditions normally required for growth are present (p. 217)

r

rachis (ray′kiss) the axis of a pinnately compound leaf or frond extending between the lowermost leaflets or pinnae and the terminal leaflet or pinna (corresponds with the midrib of a simple leaf) (p. 110)

radicle (rad′i-kuhl) the part of an embryo in a seed that develops into a root (p. 66, 148)

ray (ray) radially oriented tiers of parenchyma cells that conduct food, water, and other materials laterally in the stems and roots of woody plants; they are generally continuous across the vascular cambium between the xylem and the phloem; the portion within the wood is called a *xylem ray,* while the extension of the same ray in the phloem is called a *phloem ray* (p. 59)

receptacle (ree-sep′tuh-kuhl) the commonly expanded tip of a peduncle or pedicel to which the various parts of a flower (e.g., calyx, corolla) are attached (p. 132)

recessive (ree-ses′iv) a condition in which the phenotypic expression of one allele of a gene is masked by the phenotypic expression of another (dominant) allele (p. 242)

recombinant DNA (ree-komm′bin-int dee-en-ay) a molecule created *in vitro* containing DNA from at least two organisms (p. 259)

recombinant type (ree-komm′bin-int typ) an individual offspring that due to recombination has a combination of alleles different from either of its parents (p. 249)

red tide (red tyd) the marine phenomenon that results in the water becoming temporarily tinged with red due to the sudden proliferation of certain dinoflagellates that produce substances poisonous to animal life and humans (p. 342)
reproduction (ree-proh-duk′shun) the development of new individual organisms through either sexual or asexual means (p. 13)
resin canal (rez′in kuh-nal′) a tubular duct of many conifers and some angiosperms that is lined with resin-secreting cells (p. 422)
respiration (res-puh-ray′shun) the cellular breakdown of sugar and other foods, accompanied by release of energy; in aerobic respiration, oxygen is utilized (p. 14, 171, 186, 189)
restriction enzyme (ruh-strikt′shunn en′zym) an enzyme capable of severing a DNA molecule at a specific site (p. 260)
rhizoid (ry′zoyd) a delicate root- or root-hair-like structure of algae, fungi, the gametophytes of bryophytes, and certain structures of a few vascular plants; functions in anchorage and absorption but have no xylem or phloem (p. 357)
rhizome (ry′zohm) an underground stem, usually horizontally oriented, that may be superficially rootlike in appearance but that has definite nodes and internodes (p. 100)
ribosome (ry′boh-sohm) a granular particle composed of two subunits consisting of RNA and proteins; ribosomes lack membranes, are the sites of protein synthesis, and are very numerous in living cells (p. 40, 237)
RNA (ar-en-ay) the standard abbreviation for *ribonucleic acid,* an important cellular molecule that occurs in three forms, all involved in communication between the nucleus and the cytoplasm and in the synthesis of proteins (p. 26)
root (root) a plant organ that functions in anchorage and absorption; most roots are produced below ground (p. 54)
root cap (root kap) a thimble-shaped mass of cells at the tip of a growing root; functions primarily in protection (p. 67)
root hair (root hair) a delicate protuberance that is part of an epidermal cell of a root; root hairs occur in a zone behind the growing tip (p. 68)
root nodule (root nodd′yewl) a small swelling associated with nitrogen-fixing bacteria that invade the roots of leguminous plants and alders (p. 78)
runner (run′ur) a stem that grows horizontally along the surface of the ground; typically has long internodes; see also *stolon* (p. 100)

S

salt (salt) a substance produced by the bonding of ions that remain after hydrogen and hydroxyl ions of an acid and a base combine to form water (p. 19)
samara (sah-mair′uh) a dry fruit whose pericarp extends around the seed in the form of a wing (p. 142)
saprobe (sap′rohb) an organism that obtains its food directly from nonliving organic matter (p. 304)
sapwood (sap′wood) outer layers of wood that transport water and minerals in a tree trunk; sapwood is usually lighter in color than heartwood (p. 94)
schizocarp (skit′soh-karp) a twin fruit unique to the Parsley Family (Apiaceae) (p. 142)
science (sy′ints) a branch of study involved with the systematic observation, recording, organization, and classification of facts from which natural laws are derived and used predictively (p. 7)
scion (sy′un) a segment of plant that is grafted onto a *stock* (p. 268)
sclereid (sklair′id) a sclerenchyma cell that usually has one axis not conspicuously longer than the other; it may vary in shape and is heavily lignified (p. 55)
sclerenchyma (skluh-ren′kuh-muh) tissue composed of lignified cells with thick walls; the tissue functions primarily in strengthening and support (p. 55)
secondary consumer (sek′on-dair-ee kon-soo′mer) an organism that feeds on other consumers (p. 492)
secondary tissue (sek′un-der-ee tish′yu) a tissue produced by the vascular cambium or the cork cambium (e.g., virtually all the xylem and phloem in a tree trunk) (p. 68)
secretory cell, tissue (see′kruh-tor-ee sel, tish′yu) cell or tissue producing a substance or substances that are moved outside the cells (p. 63)
seed (seed) a mature ovule containing an embryo and bound by a protective *seed coat* (p. 133, 222)
seed coat (seed′ koht) the outer boundary layer of a seed; it is developed from the *integument(s)* (p. 422, 426, 443)
semiconservative replication (semm′ee-kon-surv-uh-tiv repp-lee-kay′shun) DNA replication mechanism that ensures each daughter molecule has one parental strand and one new strand (p. 234)
semipermeable membrane (sem-ee-pur′-me-uh-bil mem-brayn) see *differentially permeable membrane*
senescence (suh-ness′ints) the breakdown of cell components and membranes that leads to the death of the cell (p. 205)
sepal (see′puhl) a unit of the calyx that frequently resembles a reduced leaf; sepals often function in protecting the unopened flower bud (p. 132)
sessile (sess′uhl) without petiole or pedicel; attached directly by the base (p. 110)
seta (see′tuh) the stalk of a bryophyte sporophyte (p. 386)
sexual reproduction (seksh′yule ree-proh-duk′shun) reproduction involving the union of gametes (p. 222)
short-day plant (short-day plant) a plant in which flowering is initiated when the days are shorter than its critical photoperiod (p. 214)
sieve plate (siv playt) an area of the wall of a sieve tube member that contains several to many perforations that permit cytoplasmic connections between similar adjacent cells, the cytoplasmic strands being larger than plasmodesmata (p. 60)
sieve tube (siv t(y)oob) a column of sieve-tube members arranged end to end; food is conducted from cell to cell through *sieve plates* (p. 60)
sieve tube member (siv t(y)oob mem′bur) a single cell of a sieve tube (p. 60)
silique (suh-leek′) a dry fruit that splits along two “seams,” with the seeds borne on a central partition (p. 138)
simple fruit (sim′pul froot) a fruit that develops from a single pistil (p. 136)
simple leaf (sim′pul leef) a leaf with the blade undivided into leaflets (p. 110)
slime mold (slym mold) a simple organism that moves like an amoeba but resembles a fungus when reproducing (p. 348)
solvent (sol′vent) a substance (usually liquid) capable of dissolving another substance (p. 156)
somatic hybrid (soh-matt′ik hy′brid) a plant produced by protoplast fusion (p. 259)
somatic mutation (soh-matt′ik mew-tay′shun) a mutation in a somatic (body) cell; such a mutation is not passed on to offspring (p. 238)
sorus (pl. **sori**) (sor′uss; pl. sor′eye) a cluster of sporangia; the term is most frequently applied to clusters of fern sporangia (p. 411)
speciation (spee-see-ay′shun) the origin of new species through evolution (p. 239)
species (spee′seez; *species* is spelled and pronounced the same way in either singular or plural form; there is no such thing as a *specie*) the basic unit

of classification; a population of individuals capable of interbreeding freely with one another but because of geographic, reproductive, or other barriers, do not in nature interbreed with members of other species (p. 288)

sperm (spurm) a male gamete; except for those of red algae and angiosperms, sperms are frequently motile and are usually smaller than the corresponding female gametes (p. 222, 332, 426)

spice (spyss) an aromatic organic plant product used to season or flavor food or drink (p. 571)

spindle (spin′duhl) an aggregation of fiber-like threads (*microtubules*) that appears in cells during mitosis and meiosis; some threads are attached to the centromeres of chromosomes, whereas other threads extend directly or in arcs between two invisible points designated as *poles* (p. 48, 49)

spine (spyn) a relatively strong, sharp-pointed, woody structure usually located on a stem; it is usually a modified leaf or stipule (p. 119)

spongy mesophyll (spun′jee mez′uh-fil) mesophyll having loosely arranged cells and numerous air spaces; it is generally confined to the lower part of the interior of a leaf just above the lower epidermis (p. 115)

sporangiophore (spuh-ran′jee-uh-for) the stalk on which a sporangium is produced (p. 358)

sporangium (pl. **sporangia**) (spuh-ran′jee-um; pl. spuh-ran′jee-uh) a structure in which spores are produced; it may be either unicellular or multicellular (p. 349, 358, 382)

spore (spor) a reproductive cell or aggregation of cells capable of developing directly into a gametophyte or other body without uniting with another cell (*Note:* a bacterial spore is not a reproductive cell but is an inactive phase that enables the cell to survive under adverse conditions); sexual spores formed as a result of meiosis are often called *meiospores;* spores produced by mitosis may be referred to as *vegetative spores* (p. 227, 349, 358)

sporocyte (spor′oh-site) a diploid cell that becomes four haploid spores or nuclei as a result of undergoing meiosis (p. 227, 387)

sporophyll (spor′uh-fil) a modified leaf that bears a sporangium or sporangia (p. 400)

sporophyte (spor′uh-fyt) the diploid (2*n*) spore-producing phase of the life cycle of an organism exhibiting Alternation of Generations (p. 227)

stamen (stay′min) a pollen-producing structure of a flower; it consists of an anther and usually also a filament (p. 133)

stele (steel) the central cylinder of tissues in a stem or root; usually consists primarily of xylem and phloem (p. 90)

stem (stem) a plant axis with leaves or enations (p. 54)

stigma (stig′muh) the pollen receptive area of a pistil (p. 133)

stipe (styp) the supporting stalk of seaweeds, mushrooms, and certain other stationary organisms (p. 337)

stipule (stip′yool) one of a pair of appendages of varying size, shape, and texture present at the base of the leaves of some plants (p. 87)

stock (stok) the rooted portion of a plant to which a scion is grafted (p. 268)

stolon (stoh′lun) a stem that grows vertically below the surface of the ground; it typically has relatively long internodes; see also *runner* (p. 100)

stoma (pl. **stomata**) (stoh′muh; pl. stoh′mah-tuh) a minute pore or opening in the epidermis of leaves, herbaceous stems, and the sporophytes of hornworts (*Anthoceros*); it is flanked by two guard cells that regulate its opening and closing and thus regulate gas exchange and transpiration (p. 61, 113)

strobilus (pl. **strobili**) (stroh′buh-luss; pl. stroh′buh-leye) an aggregation of sporophylls on a common axis; it usually resembles a cone or is somewhat conelike in appearance (p. 400, 407)

stroma (stroh′muh) a region constituting the bulk of the volume of a chloroplast or other plastid; it contains enzymes that in chloroplasts play a key role in carbon fixation, carbohydrate synthesis, and other photosynthetic reactions (p. 40)

style (styl) the structure that connects a stigma and an ovary (p. 133)

subculture (subb′kull-choor) the transfer of tissue culture plantlets or plant parts to a new medium, usually as a form of propagation (p. 270)

suberin (soo′buh-rin) a fatty substance found primarily in the cell walls of cork and the Casparian strips of endodermal cells (p. 62, 90)

succession (suk-sesh′un) an orderly progression of changes in the composition of a community from the initial development of vegetation to the establishment of a climax community (p. 498)

sucrose (soo′krohs) a disaccharide composed of glucose and fructose; the primary form in which sugar produced by photosynthesis is transported throughout a plant (p. 21)

superior ovary (soo-peer′ee-or oh′vuh-ree) an ovary that is free from the calyx, corolla, and other floral parts, so the sepals and petals appear to be attached at its base (p. 133, 449)

symbiosis (sim-by-oh′siss) an intimate association between two dissimilar organisms that benefits both of them (mutualism) or is harmful to one of them (parasitism) (p. 300)

syngamy (sin′gam-mee) a union of gametes; fertilization (p. 228)

t

2*n* (too-en) having two sets of chromosomes; diploid (p. 227)

3*n* (three-en) having three sets of chromosomes; triploid (p. 227, 447)

tendril (ten′dril) a slender structure that coils on contact with a support of suitable diameter; it usually is a modified leaf or leaflet and aids the plant in climbing (p. 103, 118)

thallus (pl. **thalli**) (thal′uss; pl. thal′eye) a multicellular plant body that is usually flattened and not organized into roots, stems, or leaves (p. 337, 375, 385)

thorn (thorn) a pointed specialized stem (p. 119)

thylakoid (thy′luh-koyd) coin-shaped membranes whose contents include chlorophyll; they are arranged in stacks that form the grana of chloroplasts (p. 41)

tip layering (tipp lay′urr-ing) asexual propagation involving the burying of the tip of a flexible stem in soil to induce the formation of adventitious roots; the rooted portion is then cut from the parent plant and grown separately (p. 267)

tissue (tish′yu) an aggregation of cells having a common function (p. 54)

tissue culture (tish′yu kul′chur) the culture of isolated living tissue on an artificial medium (p. 269)

totipotency (toh-tuh-poh′ten-see) the potential of a cell to develop into a complete plant (p. 269)

tracheid (tray′kee-id) a xylem cell that is tapered at the ends and has thick walls containing pits (p. 59)

transcript (tran′skriptt) the RNA molecule formed by transcription (p. 237)

transcription (trans-krip′shun) the copying of a sequence of DNA nucleotides into an RNA sequence (p. 236)

transformation (trans-forr-may′shun) the transfer of DNA from one organism to another (p. 260)

transgenic plant (trans-jeen′ik plant) a plant containing recombinant DNA (p. 259)

translation (trans-lay′shun) the process of decoding RNA into protein (p. 237)

translocation (trans-loh-kay′shun) a chromosomal rearrangement resulting from a segment of one chromosome being moved to another chromosome (p. 240)

transpiration (trans-puh-ray′shun) loss of water in vapor form; most transpiration takes place through the stomata (p. 110, 266)

transposable genetic element (trans-poh′suh-bil juh-nett′ik el′uh-mint) a DNA sequence (*transposon*) capable of being moved from one chromosomal location to another (p. 231)

transposition (trans-poh-zish′unn) the movement of a transposable genetic element (p. 231)

tropism (troh′pizm) response of a plant organ or part to an external stimulus, usually in the direction of the stimulus (p. 207)

tuber (t(y)oo′bur) a swollen, fleshy underground stem (e.g., white potato) (p. 100)

turgid (tur′jid) firm or swollen because of internal water pressures resulting from osmosis (p. 157)

turgor movement (turr′gor moov′mint) the movement that results from changes in internal water pressures in a plant part (p. 210)

turgor pressure (tur′gur presh′ur) pressure within a cell resulting from the uptake of water (p. 157)

u

unisexual (yu-nih-seksh′yu-ul) a term usually applied to a flower lacking either stamens or a pistil (p. 450)

v

vacuolar membrane (vak-yu-oh′lur mem′brayn) the delimiting membrane of a cell vacuole; also called *tonoplast* (p. 43)

vacuole (vak′yu-ohl) a pocket of fluid that is separated from the cytoplasm of a cell by a membrane; it may occupy more than 99% of a cell's volume in plants; also, food-storage or contractile pockets within the cytoplasm of unicellular organisms (p. 43)

vascular bundle (vas′kyu-lur bun′dul) a strand of tissue composed mostly of xylem and phloem and usually enveloped by a bundle sheath (p. 91)

vascular cambium (vas′kyu-lur kam′bee-um) a narrow cylindrical sheath of cells that produces secondary xylem and phloem in stems and roots (p. 54, 90, 268)

vascular plant (vas′kyu-lur plant) a plant having xylem and phloem (p. 382)

vein (vayn) a term applied to any of the vascular bundles that form a branching network within leaves (p. 115)

velamen root (vel′uh-min root) an aerial root with a multilayered epidermis believed to function in retarding moisture loss (p. 60)

venter (ven′tur) the site of the egg in the enlarged basal portion of an archegonium (p. 389)

vessel (ves′uhl) one of usually very numerous cylindrical "tubes" whose cells have lost their cytoplasm; occur in the xylem of most angiosperms and a few other vascular plants; each vessel is composed of vessel members laid end to end; the perforated or open-ended walls of the vessel members permit water to pass through freely (p. 59)

vessel element (ves′uhl el′uh-ment) a single cell of a vessel (p. 59)

viability (vy-uh-bill′it-ee) capacity of a seed or spore to germinate (p. 265)

virus (vy′riss) a minute particle consisting of a core of nucleic acid, usually surrounded by a protein coat; it is incapable of growth alone and can reproduce only within, and at the expense of, a living cell (p. 317, 318)

vitamin (vyt′uh-min) a complex organic compound produced primarily by photosynthetic organisms; various vitamins are essential in minute amounts to facilitate enzyme reactions in living cells (p. 198)

w

water-splitting (photolysis) (waw′tuhr split-ing foh-tohl′uh-siss) a process in photosystem II of photosynthesis whereby water molecules are split with the release of oxygen (p. 180)

whorled (wirld) having three or more leaves or other structures at a node (p. 110)

x

xerosere (zer′roh-sear) a primary succession that initiates with bare rock (p. 498)

xylem (zy′lim) the tissue through which most of the water and dissolved minerals utilized by a plant are conducted; it consists of several types of cells (p. 59)

z

zoospore (zoh′uh-spor) a motile spore occurring in algae and fungi (p. 329)

zygote (zy′goht) the product of the union of two gametes (p. 222, 226, 447)

Index

Italic Type Indicates Ilustrations.

a

b

C

d

e

f

m

t